ANNUAL REVIEW OF MICROBIOLOGY

ANNUAL REVIEW OF MICROBIOLOGY

VOLUME 52, 1998

L. NICHOLAS ORNSTON, *Editor*
Yale University

ALBERT BALOWS, *Associate Editor*
Centers for Disease Control, Atlanta

E. PETER GREENBERG, *Associate Editor*
University of Iowa, Iowa City

http://annurev.org   science@annurev.org   650-493-4400

ANNUAL REVIEWS   4139 EL CAMINO WAY,   P.O. BOX 10139   PALO ALTO, CALIFORNIA 94303-0139

ANNUAL REVIEWS
Palo Alto, California, USA

International Standard Serial Number: 0066-4227
International Standard Book Number: 0-8243-1152-3
Library of Congress Catalog Card Number: 49-432

⊗ The paper used in this publication meets the minimum requirements of American
National Standard for Information Sciences—Permanence of Paper for Printed Library
Materials, ANSI Z39.48-1992.

TYPESET BY TECHBOOKS, FAIRFAX, VA
PRINTED AND BOUND IN THE UNITED STATES OF AMERICA

PREFACE

It is ironic that the art of classification underlies the most rigorous of scientific disciplines. Whatever we do, we must begin by placing the subject of our attention within a category, and the design of categories is a subjective activity. Difficulties arise because scientists are most comfortable presenting their efforts as objective, and sharply defined categories can provide effective shields against fresh inquiry. Problems begin early in science education. There is a disease of youth known as hardening of the categories, and it is exacerbated by the frequent obligation to apply soft pencil to a stream of boxes supplied in objective tests during which questioning impedes progress. Swift categorization can cancel thought at many scientific levels. We may read that structure of an enzyme has been determined. Which structure? An enzyme, being a biological entity, moves, and flexibility has much to do with its function. In the words attributed to Galileo, "it moves". We take comfort in knowledge that the genome of a bacterium has been sequenced. Which bacterium? A representative of a bacterial species? Can the concept of a species, perhaps successful with some subsets of genes, be applied to bacteria at the level of organisms? Even if this were so, evolutionary trees, increasingly rootless, have come to look more and more like evolutionary mycelia. Categories crumble as knowledge grows.

It would be an exaggeration to say that the charge of the editorial board is to crumble categories, but established categories of knowledge do not serve as constraints during the meetings that shape the *Annual Review of Microbiology*. The quest for creative contradiction is always present, and the highest value is placed upon topics and authors that will allow us to perceive biology from a fresh perspective. The board welcomed new members Joan Bennett and Linc Sonenshein. Joining the editorial board in the cheerful quest leading to this volume were guests Mary Lidstrom and Donald Krogstad. This volume of *Annual Review of Microbiology* is particularly notable because it was made possible by flawless passing of the baton by three production editors, Naomi Lubick, Bob Johnson, and Nancy Donham. We are grateful to all of them.

NICK ORNSTON
EDITOR

Annual Review of Microbiology
Volume 52 (1998)

CONTENTS

Edward A. Adelberg

Annu. Rev. Microbiol. 1998. 52:1–40

THE RIGHT PLACE
AT THE RIGHT TIME

Edward A. Adelberg
Department of Genetics, Yale University School of Medicine, New Haven CT 06520;
e-mail: eadelber@aol.com

KEY WORDS: amino acid biosynthesis, bacterial genetics, genetic regulation, membrane
              transport, somatic cell genetics

CONTENTS

INTRODUCTION

People of extraordinary talent can launch and develop their careers wherever they may find themselves, within a broad range of possible times. For the rest of us, however it helps to be in the right place at the right time—for me, the right place was Berkeley and the right time was the 1950s.

Microbiology was blossoming rapidly during the 50s: Microorganisms were becoming the organisms of choice for developing the fundamental principles of both genetics and biochemistry (which would gradually merge to become the modern field of molecular biology); virology was making rapid strides; new antibiotics and their mechanisms of action were being discovered; and

1

0066-4227/98/1001-0001$08.00

the fields of bacterial taxonomy, evolution and ecology were being developed, particularly by CB van Niel and his students.

For someone interested in these subjects, there was no better place to be at that time than the University of California at Berkeley. Plant, animal, and bacterial viruses were being studied in Wendell Stanley's Virus Laboratory; the mechanism of carbon-dioxide fixation in algal photosynthesis was being worked out by Melvin Calvin's group; the aquatic fungi were under investigation in Ralph Emerson's laboratory; the biochemistry of bacterial fermentations was being elucidated by Horace Barker; new pathways of carbohydrate metabolism in bacteria were being discovered by Michael Doudoroff; Roger Stanier was isolating new groups of microorganisms from nature and unraveling their nutrition and biochemistry; and microbiology was flourishing in many other laboratories.

Roger Stanier and Michael Doudoroff had joined the Department of Bacteriology in the 1940s—a very small department, in which the other faculty members were medical microbiologists and immunologists. The GI Bill, which paid the tuition costs for every veteran of World War II who wished to attend an accredited college, swelled the enrollment at Berkeley to over 30,000 students, and another faculty member was needed to teach the 300 or more students who were required (and sometimes even wanted) to take basic bacteriology each year. How I came to fill that position is described in the following section.

This memoir will have two themes: One describes the research that I and my associates carried out between 1946, when I began my graduate studies, and 1991, the date of publication of my last experimental work. The other, with which I shall begin, describes the scientific and academic settings in which this work was done.

THE SETTINGS

The Road to Berkeley

I was born in 1920 in Cedarhurst, NY. There were no scientists in the family tree on my father's side, but my maternal grandfather, David Ehrlich, was Paul Ehrlich's first cousin, a kinship which I modestly reveal at every opportunity. For genetic or other reasons I seemed to have a natural bent for science, which was greatly encouraged by the faculty of the secondary school I attended, Woodmere Academy. I was particularly influenced by our science teacher, David Harrower, who managed to earn a PhD in ornithology while a full-time high-school teacher, and who encouraged my interest in birds.

I entered Yale University in 1938, with the vague idea of majoring in chemical engineering or chemistry, but was influenced by an older student and birding companion to prepare for a career in forestry instead. Yale had one of the few graduate schools of forestry in the country, and the preeminent one at that, and I thought I might specialize there in wildlife management.

At that time Yale allowed undergraduates with the appropriate course background to combine the first year of a two-year forestry master's degree program with the senior year in college. I took that route, majoring in Plant Science, and was in my first year of Forestry School (and in my fourth undergraduate year) when the Japanese attacked Pearl Harbor and the United States went to war. From then on, it was just a question of which branch of the military to join, and when.

Early in 1942 I and some of my forestry classmates learned of an Army Air Force program for the training of meteorological officers. David Smith, Paul Burns, and I applied and were admitted in March of 1942. Yale was good enough to grant me my BS degree a few months early, and I went to NYU to take an intensive graduate-level training course in meteorology. I received my commission in November of that year (having been married the day before to Marion Sanders, now my wife of 55 years) and took my first post in what was to be four years in the Army Air Force weather service.

The last two years or so of that service were spent overseas; at the end I held the rank of major and was in command of a weather squadron of 400 officers and men, scattered over the islands of the western Pacific. I returned to Yale in March of 1946, with the intention of completing my master's degree in forestry, but two conversations with Yale faculty members led me to change career goals. The first, with the Dean of the Forestry School, convinced me that a job with a paper company in the deep South (the most likely prospect for a Yale forestry graduate at that time) was not for me. The second, with Professor Paul Burkholder of the Botany Department, convinced me that a career in microbiological research would be rewarding.

Microbiology was then a brand-new field at Yale, represented by Burkholder (who had recently switched from plant physiology to microbial physiology as a research field) and by two new faculty members—Norman Giles and Edward Tatum—all three in the Department of Botany. Tatum had been recruited to Yale in 1945 as a full professor on the occasion of the break-up of George Beadle's group at Stanford, Beadle moving to Caltech to head up Biology.

As a new graduate student, I was exposed almost immediately to the pioneering work of Beadle and Tatum on the biochemical genetics of *Neurospora*, the work for which they would receive the Nobel Prize some years later. It is almost impossible to convey the drama and excitement surrounding their work at that time: I think it is no exaggeration to say that their "one gene–one enzyme" theory, which postulated that genes function by determining the structure of corresponding proteins, is one of the great paradigms of biological science, and one of the cornerstones of modern molecular biology [for an account of their work, see Joshua Lederberg's biography of Edward Tatum (41)]. I lost no time in asking Ed Tatum if I could spend my first summer in his laboratory.

One of the major projects in Ed's laboratory was the isolation of *Neurospora* strains with mutational blocks in biosynthetic pathways. Beadle and Tatum had shown that analyzing such mutants—e.g. characterizing the intermediates that they accumulated and the compounds that could substitute for the missing endproducts in stimulating their growth—would permit the reconstruction of those pathways. They had had some success with this approach in their work on the biosynthesis of the eye-pigments of *Drosophila*, but had turned to *Neurospora crassa* as a much more expedient experimental system.

It was known that some microorganisms, including *Neurospora*, could grow in a medium containing only inorganic salts and a carbon source such as a simple sugar, whereas other microorganisms required numerous growth factors such as amino acids and vitamins. Beadle and Tatum reasoned that the former group possessed a full complement of enzymes for the biosynthesis of these compounds, while the latter group lacked one or more enzymes in each of the biosynthetic pathways (a crucial insight). From their work on *Drosophila*, and from the work of others on the biosynthesis of certain flower pigments, they hypothesized that these pathways were under genetic control.

This reasoning led to the following experimental design: *N. crassa* conidia were exposed to mutagenic treatment (usually X-rays at that time), and the surviving conidia were applied to cultures of the opposite mating type where nuclear fusions and subsequent meioses ensued. The individual ascospores arising from these meioses were then manually dissected under the microscope and inoculated onto complete medium to produce new, haploid isolates. (This procedure was employed as a way of producing cultures deriving from single conidial nuclei. *Neurospora* conidia are multinucleate, whereas each ascospore arises from a single meiotic nucleus.)

Because the ascospores were cultivated on complete medium (in which yeast extract and peptone provided a wide array of amino acids, vitamins, and other growth factors) many types of mutants blocked in biosynthetic pathways were expected to survive and grow. Each isolate was then used to inoculate a tube of minimal medium, containing inorganic salts, sucrose as carbon source, and biotin (the one growth factor required by *N. crassa*). Those isolates able to grow on complete medium but not on minimal medium were by definition nutritional mutants (requiring growth factors not required by the parental, or wildtype culture), and were set aside for biochemical and genetic analysis. About one-half of one percent of the ascospore cultures were of this type.

In June of 1946, Ed Tatum handed me a *Neurospora* culture that had emerged from the above process, and suggested that as a summer project I determine its nutritional requirement and do the standard genetic crosses to determine if it was a single-gene mutant. I did those things, finding that the strain he had given me was an apparent single-gene mutant requiring both isoleucine and valine for growth. I found this work exciting, and by the end of the summer it was clear to me that I wanted to do my PhD dissertation in Ed's laboratory. Ed accepted me: As I recall, I was his second graduate student, Ray Barratt having started with Ed a few months earlier.

There was no problem deciding on a dissertation project: David Bonner, who was then a research associate of Ed's, had earlier isolated an isoleucine-valine mutant, designated 16117, and they were keenly interested in understanding how a single gene mutation could lead to a block in the biosynthesis of two different amino acids. In the fall of 1946 I set to work on that problem; my results, which formed the basis of my PhD dissertation, are included in the second part of this memoir.

Life in the Tatum laboratory was very good indeed. Ed Tatum was unassuming, friendly, good-natured—an ideal person to have as one's mentor. He worked at the bench every day, and taught his students (by example) the frugality which had been a necessity throughout his earlier career. We learned glassblowing, carpentry, and machine shop work (at all of which Ed was expert) so that we could make all of our own equipment to the extent possible, including such things as simple thermostats, distillation glassware, pipette washers, laboratory furniture, even room partitions. We made testtube racks out of hardware cloth. Ed himself had one pipette of each size for nonsterile work; he rinsed each one out after each use and put it back on his home-made rack.

Needless to say, we washed our own glassware. When we needed a large supply of asparagine (which was limiting for *Neurospora*'s growth) we stopped work for a week and isolated it from etiolated plant stems under the guidance of friends at the Connecticut Agricultural Station. All of this frugality was born of necessity more than of virtue. There was no NIH or NSF in those days, and what little research money Ed had (I have no idea where he got it) was needed for essentials.

As I said at the beginning, I joined Ed's laboratory when he was at Yale, in 1946. In 1948 he returned to Stanford to accept a professorship, and several of his graduate students (I among them) went with him. In my case, I was granted a leave from Yale's PhD program for the purpose, receiving my PhD from Yale a year later.

Ed maintained close relations with the microbiologists at Berkeley. During my last year in Ed's Stanford laboratory, Roger Stanier paid a visit and mentioned to Ed that the Bacteriology Department at Berkeley had an open faculty position. After some soul-searching during which I reconsidered my original intent to accept a job that I had been offered in the pharmaceutical industry, I accepted the position of Instructor in July of 1949. (In those days, a two-year term as Instructor preceded the Assistant Professorship.) I was at Berkeley!

The Berkeley Years

After I accepted the position at Berkeley but before I moved from Stanford, Roger visited our laboratory again and advised me to apply for a research grant. It seems that the Office of Naval Research had begun a grant program in the basic sciences, including biochemistry.

I agreed to do so, and suggested that I ask for $500 per year—the amount that I knew Ed Tatum spent on my supplies. Roger said no, I should think big, and persuaded me to ask for $5,000 per year, to me an unimaginable sum. The ONR grant came through, and I was launched at Berkeley on an independent career. I was assigned space in a 500-square-foot room, which not only had to accommodate me and my first graduate student (Irving Miller), but which we had to share with the technician who prepared all the microbiological materials for three big undergraduate laboratory courses. Irv Miller and I shared a desk, a hood, and a total of seven feet of laboratory bench.

A year later, however, a suite of laboratories on the fifth floor of the Life Sciences Building was renovated for the shared use of Roger Stanier, Mike Doudoroff, and myself, an arrangement which continued happily until I left Berkeley in 1961. Each of us had a small office, but all other rooms were shared: Our post-docs, graduate students, and technicians were mingled in every room, and we had a common set of support facilities: chemicals, supplies, reprints, dishwashing, sterilizing, etc. We hired a manager for the joint operation (Bill Rood, a graduate student in political science) and empowered him to pool our grant funds in order to buy supplies and equipment and hire personnel. (Such an arrangement today would lead to swift and severe censure and retribution by the federal government!)

Even though we each had a small research grant when we moved into our shared quarters, money was so tight that when I needed bookshelves for my new office, I bought the lumber and built them myself over the weekend. Roger and I had adjoining offices, but could afford only one telephone for the two us. So I came in on another Sunday and cut a pass-through in the partition between our offices, with a little sliding door. We put a table on each side of the partition, and moved our shared phone back and forth through the pass-through. It never occurred to us to complain.

The mutually supportive atmosphere of the fifth-floor laboratories was such that when John Clark arrived to do post-doctoral work with Roger, he became interested in the bacterial genetics that my group was doing, and ended up working on bacterial conjugation with me as well as on streptomycin-resistance with Roger. Conversely, when Irving Miller came to Berkeley to do a PhD with me (on isoleucine-valine biosynthesis) he became more interested in the work of Stanier's group on kynurenine metabolism, on which he proceeded to do his thesis work.

The labs were laid out so that to get from any part to any other part one had to pass through a conference area with a perpetual coffee pot, an arrangement that greatly stimulated informal conversations between members of the three groups. Mike Doudoroff particularly enjoyed that aspect of the lab, and I got some of my best ideas from Mike in front of the blackboard in the coffee room. Although I had enormous (by today's standards) teaching responsibilities, I found time to work at the bench every day, plus many evenings and weekends, and I continued my work on the biosynthetic pathways of isoleucine and valine.

Writing this in 1997, it occurs to me that some of my readers may not be aware of what a privilege it was to share laboratories with Roger Stanier and Mike Doudoroff. Considering that my own development as a microbiologist was inseparable from my association with these two giants, I want to say a bit more about their influence—not only on me, but on the field of microbiology in general. For full accounts of their careers, I would recommend Roger's memoir that appeared in this journal in 1980 (60), and the biography of Mike written by Horace Barker (13).

Mike Doudoroff, the son of a White Russian admiral who was the naval attaché to the Russian embassy in Japan, came to California with his parents at the age of twelve, in 1923. He received his undergraduate degree from Stanford and his PhD as well, earning the latter in the laboratory of CB van Niel at Stanford's Hopkins Marine Station. Mike joined the Bacteriology Department at Berkeley in 1940; during his last year at the Marine Station, where he assisted in van Niel's famous microbiology course, Roger Stanier arrived to begin his own PhD research with van Niel, and the two became close friends. In 1947, Mike prevailed upon his colleagues in the Bacteriology Department to recruit Roger to the faculty, and an association began that was to continue until 1971, when Roger moved to the Pasteur Institute in Paris.

In his obituary of Roger Stanier, SE Luria pointed out that Roger was "one of the brilliant scions of the Delft school of microbiology" (43). By "the Delft school" he meant the succession running from Beijerinck through Kluyver to van Niel and ultimately to van Niel's most gifted students, Doudoroff and Stanier. During the 1950s, when I was in their company, Mike and Roger were reaching the peaks of their careers, and I would like to mention just a few of their projects that were going on in our shared laboratories.

Mike was continuing in his pioneering work on the bacterial metabolism of carbohydrates. Much of his work was done with *Pseudomona saccharophila*, which he had originally isolated as a hydrogen-oxidizing bacterium. During the years that we shared laboratories, Mike and his students elucidated the metabolic pathways and enzymatic mechanisms involved in the bacterial oxidation of sucrose, maltose, glucose, gluconic acid, arabinose, and fructose; he studied oxidative assimilations and, with Roger, demonstrated the role of poly-β-hydroxybutyric acid in the assimilation of organic compounds by bacteria and the role of organic substrates in bacterial photosynthesis.

During this same period, Roger continued his groundbreaking studies on the bacterial oxidation of aromatic compounds. He and his students and collaborators characterized a number of bacterial groups, including *Leucothrix* and *Thiothrix*, *Caulobacter*, pseudomonads, and cyanobacteria. With Howard Schachman and Arthur Pardee he discovered bacterial ribosomes and chromatophores and, with his graduate student, Charles Spotts, he deduced that streptomycin acts by binding to the bacterial ribosome. He began his brilliant series of studies on bacterial photosynthesis with Germaine Cohen-Bazire, who was later to become his wife. During that same period he continued to influence the thinking of microbiologists about the place of bacteria in the microbial world, and brought about the renaming of "blue-green algae" as cyanobacteria, in view of their procaryotic nature.

In those years, the opportunities at Berkeley for interaction with other microbiologists, geneticists, and biochemists were for all practical purposes unlimited. Although the Bacteriology Department had its own PhD program, Roger, Mike, and I were also on the faculty of two other university-wide graduate programs: one in Biochemistry, and one in Microbiology. The latter was run jointly with colleagues at the UC Davis campus, including Mort Starr, Monty Reynolds, Herman Phaff, and Bob Hungate. The Berkeley and Davis groups visited back and forth regularly, and we got together at least once a year with van Niel and his group at the Hopkins Marine Station, where Barbara Bachmann was a graduate student. On the Berkeley campus itself we had close connections with the Virus Laboratory group, one of whom—Gunther Stent—became an active joint member of our own department.

As I mentioned earlier, I had heavy teaching duties from the day I arrived. Mike and Roger taught the courses for the students majoring in bacteriology, and I was assigned "Bact. 2," the introductory course for the nonmajors. There were about 300 of these each year, varying from a minority possessing both a suitable background and a real interest in the subject, to a majority lacking both. (The latter were required to take bacteriology to give a veneer of scientific respectability to their majors, such as Home Economics and Optometry.) There were three lectures a week and two three-hour laboratories, with six teaching assistants under my daily supervision. My first year was made especially traumatic by a subgroup of about 70 students who only had to take one three-hour laboratory session per week. I was the sole laboratory supervisor for this group, and the exercise consisted of demonstrating to them a variety of bacteria and other microorganisms, most of which I had never seen before myself.

My teaching improved in subsequent years as a result of my daily association with Mike and Roger and attending their courses. I developed my own lecture syllabus, on the basis of which I was invited by Ernest Jawetz (a faculty member at UC San Francisco and a good friend) to write the introductory chapters for a highly condensed review-style textbook of medical microbiology to be published by the Lange publishing company. Jack Lange was a faculty colleague of Ernest's who had organized and published the very successful *Physician's Handbook*; he conceived of a series of reviews for medical students, of which ours was to be the second, and the series eventually grew to cover over thirty subjects. Joseph Melnick, at Baylor, volunteered to write the chapters on viruses, and *Review of Medical Microbiology* was born (38). Following Lange's practice of biennial revision, the book is now in its twentieth edition; at some point along the way we celebrated its millionth copy, including translations into a dozen or more languages.

Roger and Mike had themselves thought about writing a general microbiology textbook for some time, to replace the one written by our friend in Seattle, Erling Ordal, which was the best of its time though becoming out of date. One day in 1954 a field representative for Prentice-Hall came to see me and asked me if I'd like to do a book for them. I demurred, but suggested that he go after Roger and Mike; he did, and with some pressuring from me they decided the time was ripe and agreed to sign up with Prentice-Hall, inviting me to join them as the most junior author.

We began meeting evenings, to prepare chapter outlines and assign the writing of various chapters to one or another of us. The book was to have three parts, the first of which—dealing with the major groups of microorganisms—Roger proposed be called The Microbial World. We hadn't given much thought to the title for the book itself, but it probably would have been rather conventional, such as Introduction to Microbiology or something of the sort. In the midst of our writing, however, Prentice Hall called to say that a Science Book Club had come into being, and that our book might have a shot at being adopted by it if it had a jazzier title. And so the whole book became *The Microbial World*, a phrase that caught on and is still popular today. (It was not adopted by the book club, by the way.)

With a detailed outline in hand, we each set about drafting our assigned chapters and preparing three typed sets. By June of 1955 these were ready, and we agreed to spend the summer editing them. This we did by meeting every day from early morning until late afternoon on the patio of Roger's hillside house, stopping only for brown-bag lunches. Each of us had a copy of the chapter to be edited in front of us, and the author of that chapter read it aloud while the other two interrupted with questions and comments.

In the case of Mike's chapters and mine, the interruptions came at the end of almost every sentence, when Roger would say "Hold it! What you really want to say here is ....", and he'd dictate a much improved sentence. Often, we would argue about the need for substantive changes and/or major changes in the sequence of the concepts or facts being presented. These would be hammered out on the spot, if possible; otherwise the author of that chapter would be sent home that evening to do a major rewrite. Roger's chapters needed little, if any, change, but Mike and I had to rewrite much of our material to bring it up to Roger's standards. In the end, the style of the writing bore Roger's imprint throughout the entire book. Needless to say, this word-by-word editing process took up the entire summer, which proved to be one of the greatest learning experiences (and one of the most enjoyable) of my life. The book finally appeared late in 1957 (61), and we collaborated on several later editions before others took it over.

Another book grew out of a collaboration with Gunther Stent in the teaching of bacterial genetics. Some years earlier, Joshua Lederberg had published an annotated reprint collection called *Papers in Microbial Genetics: Bacteria and Bacterial Viruses*. It was the sort of thing that we thought could be useful for our students, but Josh's collection had become sadly outdated as bacterial genetics gathered steam in the late 1950s. Gunther and I decided to put out a new version, which ultimately took the shape of two volumes: mine, called *Papers on Bacterial Genetics* (6), and Gunther's, called *Papers on Bacterial Viruses* (62). Each had a lengthy introduction that served as a minitext on the subject; writing my own volume's introduction forced me to organize my thinking, and I acquired an overview of bacterial genetics that was to serve me well in later years. (I also learned a lot about writing, thanks to Gunther's thorough editing of my text.)

I remained at Berkeley until 1961, becoming chairman of the department in 1958. (I was only an Associate Professor, but the more senior faculty were divided into factions that would not support each other for the chair.) The 12 years at Berkeley included some of the happiest years of my professional and personal life, and it was with considerable ambivalence that, in the fall of 1960, I read a letter asking me if I would consider taking the chairmanship of the Microbiology Department at Yale. After some months of agonizing, my wife and I decided to move to Yale, a homecoming of sorts: We had met in New Haven when I was an undergraduate at Yale and she was attending the Women's College of New Haven (later transferring to Mount Holyoke), and our first child was born while I was a graduate student in 1947. We were also influenced in our decision by the fact that both our families were on the East Coast, by the excellence of the schools that would be available in the New Haven area for our three sons, Michael, David, and Arthur (then ages 14, 12, and 10), and by New Haven's relative proximity to Woods Hole. And so we left Berkeley, which only a call from Yale could have induced us to do.

I should say a word here about Woods Hole. Although I lived in the northeast until I was almost 30 years old, I had never heard of the place until I reached Berkeley, where I found that several colleagues were commuting to Woods Hole from the West Coast to spend their summers at the Marine Biological Laboratory there. One of them, Dan Mazia, convinced me that leaving Berkeley for Yale would be considerably less painful if I could arrange to spend my own summers there.

Accordingly, I called Bill McElroy, who I had learned was in charge of the Physiology Course at the MBL, and asked if he could use my services in any way. Bill responded most graciously, inviting me to assist Phil Hartman in teaching the bacterial genetics section of the course that summer, and to succeed Phil when his term in the course ended the following year. That began a string of summers at Woods Hole that is unbroken to this day: After my own term of teaching in the course expired, I spent the next 25 or so summers in the MBL Library, writing papers, reports, and grant applications, preparing my lectures for the coming year, and catching up on my scientific reading. I chaired MBL's Library Committee for many years, and served a term as a Trustee. We acquired a small house near the laboratory, and even after my retirement in 1991 we have continued to summer there, albeit with a much lighter work load.

Return to Yale

I assumed the chair of the Microbiology Department at the Yale School of Medicine in July of 1961, and set up my laboratory with the help of several students and assistants who moved with me, particularly Ann Templin. It was a small department, made even smaller by the departure of David Bonner for La Jolla and by the imminent retirement of Phillip Cowles.[1] When I came to Yale, my dowry included the renovation (according to my own design) of an entire floor of the Brady Laboratory above the department's existing floor, and a full professorship postion to fill. I decided to go after an immunologist, and asked my friend Elvin Kabat, who he would say (after himself) was the best immunologist in the country. "After me," he said, "Byron Waksman," whom I duly recruited from Massachusetts General Hospital and with whom I proceeded to share the new floor. Byron quickly built a prestigious group, and over the next two years we added several junior faculty working in bacterial physiology and genetics to round out the department.

---

[1] Phil had been in the department for his entire professional life. I'll never forget the following incident, which should impress historians of microbiology. One day in the early 1960s, Sydney Brenner was visiting and asked me to introduce him to Professor Cowles. I did so, wondering what the two could have in common. After the usual pleasantries, Sydney asked Phil if he could help him locate d'Herelle's phage stocks. "Sure," said Phil, opening his refrigerator and handing Sydney a rack of tubes. They were indeed a set of d'Herelle's original phage stocks, which d'Herelle had left with Phil when he worked briefly at Yale in the 1920s.

Like most medical microbiology departments of that era, we were a diverse group, including specialists in immunology, virology, parasitology, microbial physiology, and bacterial genetics, with joint members representing medical mycology and laboratory medicine. Although it was a congenial group, we found the diversity to be a problem: The different groups had little in common, tending not to appear at each other's seminars or to interact scientifically. We were grappling with this problem, exploring some reorganizational options, when—in the late 1960s—Leon Rosenberg approached me with the idea of starting up a university-wide Department of Genetics.

At that time, there were geneticists scattered among several departments of the medical school, as well as in the biology department across town. Leon and I initiated discussions with all of these colleagues and found considerable enthusiasm for the idea among the medical school geneticists. Those in the biology department, however, expressed a preference for staying where they were, wishing us well in establishing a genetics department in the medical school.

To make a long story short, we succeeded in doing so in (as I remember) 1972.[2] Among those who wanted to transfer into the new department as primary appointees were no fewer than five members of the Microbiology Department— roughly half its faculty. This, of course, would devastate the department, and I was personally unwilling to proceed without assurance from the Dean that he would rebuild it. At that time Lewis Thomas had just assumed the deanship; Byron Waksman and I met with him, and he agreed that he would add the necessary positions, reshaping the department to emphasize research and teaching in the mechanisms, diagnosis, and treatment of infectious disease.

This was just what I thought the school needed, and I happily joined my other genetics colleagues in organizing the new department. But six months after Lew Thomas became Dean, he left Yale to head the Sloan-Kettering Institute in New York. His successor was Bob Berliner,[3] and Bob decided not to invest the considerable resources (postions, space, budget) that it would take to rebuild Microbiology. Instead, he dispersed the nongeneticists among several existing

---

[2]We were not allowed to call ourselves the Department of Genetics because the Biology Department objected on the grounds that it would convey to the outside world the impression that genetics was not a significant activity of their own department. So we named ourselves the Department of Human Genetics, although most of those who joined the new department were molecular and cellular geneticists with little background in human genetics. This immediately became a problem in recruiting graduate students, and remained so until the department was allowed to change its name in 1992 to the Department of Genetics.

[3]Five years my senior, he was the oldest and I the youngest member of our high-school bird club, organized by the teacher I mentioned earlier in this memoir, David Harrower. At the time of this writing I learned that David Kessler, Yale's new Medical School dean, is also a graduate of that tiny high school, Woodmere Academy.

departments, including Pathology and Epidemiology/Public Health. Although I had been a leader of the enterprise that led to the demise of the Microbiology Department at Yale, I keep reminding myself that I acted only when I had every reason to believe that the department would not only survive but would prosper. As of this writing in 1997, however, some 25 years later, plans to restore the department exist only on paper, and I have never shaken off a feeling of guilt.

The establishment of the new Department of Human Genetics coincided with my own major switch in research interests (to mammalian cell genetics and membrane physiology), as described in the second part of this memoir. My strategy was to employ the methods of microbial genetics to isolate mutants of cultured mammalian cells affected in membrane transport functions. In this work I collaborated with my colleague, Carolyn Slayman, who had also transferred to the new Human Genetics Department, and who was to teach me all I would ever come to know about membrane transport kinetics. Although we were reasonably successful in this endeavor, I found the progress to be painfully slow: To isolate and characterize in a preliminary way just one mammalian cell membrane transport mutant, for example, took many months, in contrast with my experience in bacterial genetics where an idea for a selection on Monday could produce a basketful of partially characterized mutants by Friday.

I was thus more than ready when I was asked, in 1983, to join the Office of the Provost of the university, as Deputy Provost for the Biomedical Sciences. Unlike most scientists, I actually enjoyed administration, and felt I had some talent for it. I started out half-time, but within a few years I found it to be a full-time responsibility, and discontinued active research. Carolyn took over our grant-supported projects, and over the next few years my participation in planning and supervising experiments was phased out. The last experimental paper with my name on it appeared in 1991.

At Yale, the Provost is essentially the chief operating officer of the University, overseeing an operating budget of a billion dollars a year. The Provost is responsible for the allocation of faculty positions, space, and budget to a dozen graduate and professional schools, an even larger numer of Arts and Science departments, and a host of ancillary organizations ranging from athletics to campus police. He or she sets policies ranging from faculty conflicts of interest to compliance with governmental regulation. At any given time, there are from four to six deputy and associate provosts who act as the first line of contact between the Provost and Yale's extraordinary collection of administrative units.

During the eight years I spent as Deputy Provost, I was responsible for the School of Medicine, the School of Forestry and Environmental Studies, and the School of Nursing; the Departments of Biology, Molecular Biophysics and Biochemistry, Psychology, and Anthropology (in the Faculty of Arts and Sciences); the Peabody Museum of Natural History; the Office of University

Safety; and a variety of related smaller units and activities. I chaired countless committees, whose charges ranged from advising the President on deer-hunting on Yale's forest land (prompted by a protest by the local animal-rights organization) to reviewing the basis of Yale's indirect cost rate on federal grants.

Many, if not most, faculty see the administration as a bureaucratic burden at best, and the enemy at worst. My goal as Deputy Provost, on the other hand, was to give my colleagues in the biomedical sciences all the support they needed— or at least all I could manage to get for them. Of my many responsibilities, I felt the most important to be assisting the departments in recruiting (or holding onto) the very best scientists in their fields. The stakes were high: to meet the competition from the likes of Harvard, Stanford, and Caltech, we had to come up with painfully high amounts of money for salaries, laboratory renovations, major equipment, and computing resources, as well as (in many cases) helping to find employment for spouses and housing. In all of this we had a very high success rate, a result in which I took considerable pride and gratification. I am not ashamed to say that I enjoyed my role as a university administrator as much as I did my roles as researcher and teacher.

The Escherichia Coli Genetic Stock Center

By the time I had restarted my laboratory at Yale, my collection of genetic strains of E. coli K12 had grown to the point where my technicians were finding it difficult to keep up with the duties of preserving them, cataloguing them, and sending cultures of them to colleagues around the world. In 1971 I persuaded the National Science Foundation to fund the collection as a separate entity, making it (I believe) the first Genetic Stock Center with NSF support. The grant included funds for a full-time curator (and ultimately Director), which position was accepted by Barbara Bachmann.

Barbara had earned her PhD with van Niel during the days when the Stanford, Berkeley, and Davis microbiology groups formed an extended family, and was thus a long-standing colleague and friend. She had run the microbiology teaching laboratories for us at Berkeley, and had done so at Yale when she and her husband moved east. She brought to the Stock Center an inexhaustible supply of energy and talent, both scientific and organizational, and under her leadership the Center expanded to take in a large number of E. coli K12 collections from other laboratories. Within a few years the collection numbered over 7,000 strains, and Barbara's staff was sending out over 4,000 samples each year to scientists in academic, governmental, and industrial laboratories around the world.

Almost every sample sent out was chosen on the basis of one or more telephone conversations with Barbara, who nevertheless found the time to work out the pedigree of every strain in the collection, resolving hundreds of disagreements over locus and allele designations in the process, and to publish the complete genetic map of *E. coli* K12 with periodic revisions. (The trick, I discovered, is to work over sixty hours a week with no vacations for twenty or more years.) A major innovation was her ability to persuade K12 geneticists all over the world to accept assignments by her of blocks of allele numbers, so that each published allele would be uniquely identified. Without Barbara's efforts, K12 genetics would be chaotic.

My role in all of this, once I had turned my collection over to Barbara, was to give the Center whatever support I could—including raising funds for expanded space and working with the Biology Department in establishing a position in the Center for Mary Berlyn, who succeeded Barbara as Director in 1993. I consider my launching of the Stock Center and my recruitment of Barbara to run it to be one of my most useful contributions to the field of microbiology.

Retirement

I retired in 1991. I was not one of those with a set of hobbies or writing plans just waiting for my retirement to get the attention they deserved. My wife and I enjoy travel, and during my first retirement years we visited the Galapagos Islands, Ecuador, Russia, Norway, Denmark, Holland, Hawaii, and some less exotic domestic spots. Nevertheless, I had a lot of available time and energy, and was delighted whenever the university called me back for temporary duty.

The first such time was about a year after I retired. As Deputy Provost some years earlier I had organized Yale's Office of University Safety, which is responsible for biological, chemical, radiation, and physical safety in the laboratories and service units on the campus. I was so successful that the person I recruited for Director, Larry Gibbs, was stolen by Stanford, and Yale needed someone to run the Office until a successor could be found. I did so for about six months, while chairing the search committee for the new Director.

A year or two later the University once again found itself shorthanded, when the Director of the Peabody Museum, Alison Richard, was made Provost. Yale's new President, Rick Levin, asked me to be Acting Director until a permanent replacement could be found, and I accepted with pleasure. One of my most agreeable duties as Deputy Provost had been overseeing the Museum with Alison reporting to me, and now I had the pleasure of reporting to her. For about nine months I happily dealt with problems ranging from the design of new spreadsheets for the museum's budget to the organization of exhibits and the appointment of new curators, not to mention the early planning stages for a new building to house a large fraction of the 11,000,000 or so items in the museum's collections. Once again, my appetite for administration and organization was satisfied.

But even that pleasant experience was only temporary, and I found myself once again out of work. In an attempt to find some useful occupation, I called on Mary Berlyn.

In the eight years since Mary had come to the *E. coli* Genetic Stock Center, she had developed an elegant online electronic database incorporating not only all the data on the genetic stocks, but also on all *E. coli* genes and gene products (including the genetic map), with full bibliographic information. Keeping up with the current literature was proving to be an increasing problem, and I volunteered to take over that activity. We worked out a process whereby I would search Medline's monthly update for all references on "coli," select those dealing with *E. coli*'s own genes and gene products, import them into a bibliographical database program, and edit the references for ultimate export into Mary's database.

There is, of course, another major database dealing with *E. coli* genes and gene products, and that is the Ecocyc database run by Monica Riley at Woods Hole and her collaborators. It turned out that Monica, too, could use help in keeping up with the current literature, and—with the help of Jan Glover at Yale's medical library—we found a way to export my monthly updates into Monica's bibliography as well.

I have been doing this for two years now, and find it absorbing. If my memory didn't have such a short half-life, I should know more about *E. coli* than anyone else in the world (with the exception of Mary and Monica). In any case, I am gratified for the opportunity to stay active, even in a small way, as a microbiologist.

THE SCIENCE

My independent research career began in 1949 with my appointment to the Berkeley faculty and continued unbroken until 1991, the date of my last scientific publication—a period of 42 years. In the following pages, I will organize my narrative not strictly chronologically, but rather according to the special areas of research in which my laboratory became involved. These included the following:

1. The biosynthesis of isoleucine and valine in *Neurospora* and *Escherichia coli*;

2. Genetic regulation of amino acid biosynthesis in *E. coli*;

3. Mechanisms of mutation in bacteria;

4. The bacterial chromosome, sex factors, and conjugation;

5. Membrane transport in cultured mammalian cells.

The Biosynthesis of Isoleucine and Valine in Neurospora and E. coli

As I mentioned at the beginning of this memoir, David Bonner had isolated a *Neurospora* isoleucine-valine mutant, designated 16117, and he and Ed Tatum were keenly interested in understanding how a single gene mutation could lead to a block in the biosynthesis of two different amino acids. In the fall of 1946 I set to work on that problem as my doctoral dissertation project.

Bonner had published evidence that 16117 accumulated an isoleucine precursor that he believed to be the corresponding keto-acid (which we informally called "keto-isoleucine"). I isolated an organic acid fraction from filtrates of submerged cultures of 16117, but could detect no keto-acids. The fraction did, however, show strong activity in supporting the growth of an isoleucine-dependent *E. coli* mutant, which presented an ideal assay system for the isolation of the active compound from filtrates of the *Neurospora* strain. We were off and running!

I was able to produce a small amount of the crystalline p-bromphenacyl ester of the active material, enough to obtain an elemental analysis. The analysis showed relative percentages of carbon, hydrogen, and bromine that agreed with those expected for the ester of the keto-acid corresponding to isoleucine—but with one extra molecule of water. (Bonner interpreted that to mean a water of crystallization, but as we shall see, he was wrong.)

My big break was the appearance on the scene of paper chromatography: I adapted the technique to the separation of organic acids and scaled it up to produce sufficient quantities for chemical and biological tests. Paper chromatograms revealed that 16117 accumulated roughly equal amounts of two organic acids, one of which supported growth of the isoleucine-dependent bacterial mutant. I isolated this compound by preparation-scale paper chromatography, and characterized it as α,β-dihydroxy, β-methylvaleric acid: i.e. a dihydroxy acid with the carbon skeleton of isoleucine. This structure is indeed equivalent to the keto acid plus one molecule of water.

Similar analyses of the second organic acid seen on paper chromatograms of 16117 filtrates showed it to be the dihydroxy acid corresponding to valine (α,β-dihydroxy-isovaleric acid). I will refer to these compounds as dihydroxy-isoleucine (DHI) and dihydroxy-valine (DHV) for convenience.

In 1948, while I was in the middle of this work, a paper appeared showing that threonine and isoleucine appeared to share a common four-carbon precursor. I tested a number of *E. coli* and *Neurospora* isoleucine-dependent mutants and found that they fell into three groups: Group A responded only to isoleucine; Group B to isoleucine and DHI; and Group C to those two compounds as well as to C_4 compounds, including α-ketobutyrate, α-aminobutyrate, and threonine. On the basis of this data, the following crude scheme for the biosynthesis of isoleucine was constructed:

$$\rightarrow C_4 \rightarrow DHI \rightarrow Isoleucine$$

The work described above formed the basis of my PhD dissertation, which was submitted to Yale in 1949, three years after I began graduate work. The reader will note that, aside from some minor genetic crosses and some nutritional testing, the bulk of the work was pure organic chemistry, including synthesis of the isoleucine dihydroxy acid precursor in the course of proving its structure.[4] This dissertation work led to my first four publications (2, 7, 63, 64).

Not long after my arrival in Berkeley I learned that Ed Umbarger, a graduate student in Mueller's laboratory at Harvard, was also working on isoleucine and valine biosynthesis. We began corresponding and exchanging data as well as preprints, and soon we were actively collaborating by mail—even though it was to be many years before we were to meet face to face. (There was no money for travel in those days, or even for long-distance telephone calls.)

The first work we did together concerned the role of Bonner's keto-isoleucine. The biosynthetic scheme published by Bonner, Tatum and myself made no mention of keto-isoleucine, despite the fact that it would have been plausible to postulate the dehydration of the dihydroxy acid to the keto acid, followed by transamination to isoleucine (the pathway we later showed to be correct). Bonner, Tatum, and I had been led (or rather misled) to discard this scheme, however, by finding that certain isoleucine-dependent bacterial mutants could use dihydroxy-isoleucine (DHI) but not Bonner's keto-isoleucine (KI) for growth, while others could use either one. This would suggest that KI preceded DHI in the biosynthetic pathway to isoleucine, which would make little biochemical sense.

The confusion surrounding the role of so-called keto-isoleucine was compounded by the fact that Bonner had obtained differing results, depending on whether he used keto-isoleucine obtained by the enzymatic oxidation of isoleucine, or keto-isoleucine that he synthesized. The most serious discrepancies involved work with the latter material. To investigate this problem, I sent Ed Umbarger a sample of Bonner's synthetic compound, which we compared chromatographically and chemically with keto-acids synthesized in Umbarger's laboratory. As Ed found [and as we published in our first joint article (11)],

---

[4] A few years later, Karl Folkers and his team at Merck used a different route to synthesize and resolve the optical isomers of both dihydroxy acids; Ed Tatum and I tested the L-isomer of the synthetic isoleucine precursor and found it to be biologically active (58).

Bonner's synthetic material was actually α-ketobutyric acid, an unfortunate by-product of the synthesis he had bungled. ("Unfortunate" because α-ketobutyrate turned out to itself be a precursor of isoleucine, much earlier in the pathway, and thus active for certain isoleucine-dependent mutants.)

With this clarification, Ed and I were able to establish that the correct sequence was indeed

C_4 → dihydroxy-isoleucine → keto-isoleucine → isoleucine,

where the C_4 precursor could be supplied as α-ketobutyrate among other C_4 compounds.

Throughout this early period, the double requirement (for both isoleucine and valine) in various microbial mutants was generally believed, as first proposed by Bonner, to reflect a block in one of the pathways, such that the accumulated intermediate inhibited an analogous enzymatic step in the other pathway. Later, however, Umbarger and I found that both the glutamic acid-valine and glutamic acid-isoleucine transaminase activities were missing in the *E. coli* mutant that accumulated the keto acids (11). Now, my graduate student Jack Myers and I made crude enzymatic preparations from *N.crassa* and *E. coli* and demonstrated that extracts of wild-type cells were capable of enzymatic dehydration of both dihydroxy acids to their corresponding keto acids, and that both activities were missing in each of the mutants that accumulated the dihydroxy acids (47).

We concluded that the isoleucine and valine biosynthetic pathways constitute two parallel series of reactions catalyzed by a single set of dually active enzymes. This generalization was proved to be correct when other laboratories extended these observations to the earlier steps in the pathways as well.

In 1954, I decided to make a direct determination of the contributions of threonine carbons to the carbon skeleton of isoleucine in *Neurospora*. I did this by biologically labeling DHI with C^{14}-1,2-acetate (which would label all DHI carbons) in the presence and absence of unlabeled L-threonine, and quantitating threonine's competition for each carbon in DHI. The result were striking: The distribution of label could only be explained by a condensation of a C_4 compound derived from threonine with a C_2 compound, followed by an intramolecular rearrangement (3, 4).

I presented this finding at a symposium, at which Murray Strassman, Sidney Weinhouse, and their colleagues presented data leading to the same conclusion. From their presentation, I learned (to my chagrin) that I had missed an earlier paper of theirs in which they had shown an analogous intramolecular rearrangement in the biosynthesis of valine. Thus, they—and not I—get credit for the first demonstration of such a biochemical reaction. (Much later, other laboratories showed that the rearrangement in both pathways is brought about by an enzyme called reductoisomerase, which converts α-acetolactate to DHV and α-aceto-β-hydroxybutyrate to DHI.)

The Regulation of Amino Acid Biosynthesis

While the studies on the isoleucine-valine biosynthetic pathways were going on, I began working with E. coli strain 11A16, a strain that Umbarger had shown accumulated the keto-acids corresponding to isoleucine and valine. In testing different ratios of isoleucine and valine in the medium to optimize keto-acid accumulation, I found to my surprise that the more valine in the medium, the less "keto-valine" accumulated. I hypothesized that high valine concentrations selected for variants in the bacterial population that accumulated less keto-acid. I was at the blackboard with one of my students designing an experiment to test this idea when Mike Doudoroff drifted in, coffee cup in hand. He listened for a few minutes, and then suggested that when the endproduct (valine) is in the medium, the cells save energy by shutting down unnecessary endogenous synthesis.

"Nonsense," I said, "How could cells be that smart?" or words to that effect. We immediately designed an experiment to disprove his seemingly preposterous theory: A heavy suspension of cells grown in isoleucine and valine was washed and resuspended in minimal medium without amino acids, and their accumulation of keto-valine was followed as a function of added valine. (Without isoleucine, the cells could not grow, and variants could not be selected).

Doudoroff proved to be right: In nongrowing cells, valine shut off its own biosynthesis! This was, I believe, the first demonstration of the regulation of a biosynthetic pathway by feedback inhibition. (Actually, two similar phenomena had been reported in the literature, but the authors of those papers appeared not to recognize the significance of "feedback inhibition" as a regulatory process.)

I wrote to Ed Umbarger about these results, and he informed me about some related results of his own: In growing cells, valine inhibited the formation (rather than activity) of one of the enzymes in its own biosynthetic pathway, alanine-α-ketoisovalerate transaminase. We decided to publish a second paper together that reported both sets of data (11); one of the statements in the summary was: "These results show . . . that an end-product, in this case valine, can regulate the rate of its own biosynthesis." I believe that this was the first paper to recognize endproduct regulation of biosynthesis, both by feedback inhibition of enzyme activity and feedback repression of enzyme synthesis.

I was to return to the regulation of isoleucine-valine biosynthesis ten years later. Meanwhile, I had a different encounter with the subject of biosynthetic regulation as a consequence of my first sabbatical leave, which I spent at the Pasteur Institute in Paris in 1956–57. My choice of the Pasteur followed from the correspondence that I had begun with Georges Cohen, who had also been publishing papers in the field of amino-acid biosynthesis and regulation. I wrote to Georges and asked if I could spend my sabbatical in his laboratory and he kindly agreed.

When I arrived in Georges' laboratory, he was in the midst of a new project on the incorporation of amino-acid analogues into proteins, which he and Munier had shown caused a shift in *E. coli* from exponential to linear growth. Their conclusion was that some key protein components of the protein-synthesizing machinery became nonfunctional as a consequence of incorporating the analogues: New proteins continued to be made, but only at the rate determined by the preexisting, functional components. Georges suggested that we isolate mutants resistant to amino acid analogues, in the hope that at least some of these would be altered in the structure of amino acid activating enzymes (later to be named amino acyl-tRNA synthetases)—the existence of which had recently been inferred.

We readily isolated a mutant resistant to p-fluorophenylalanine (FPA), an analogue of tyrosine. Our experiments with the mutant showed, however, that it owed its resistance not to alteration of an activating enzyme but rather to its overproduction of tyrosine, which competed with FPA effectively (16). It was later found that a mutant resistant to thienylalanine similarly overproduced phenylalanine. It occurred to me: "Since the wild-type strain used in our studies does not excrete detectable amounts of phenylalanine or tyrosine, some mechanism of control must operate in the biosynthesis of these amino acids. This mechanism, whatever it may be, has presumably been interfered with in the mutants, which have thus become overproducers of the amino acids. The study of such mutants promises to provide information on mechanisms of control of biosynthesis and on gene-enzyme relationships, as well as on drug-resistance" [quoted from (5)]. I therefore decided to conduct a wide screen for mutants of *E. coli* that excrete metabolites conferring resistance to antimetabolites, and found over a dozen different types. I showed, for example, that the mutant resistant to ethionine excreted methionine (5), but did not carry the project further.[5]

The genetic regulation of amino acid biosynthesis became a hot issue a few years later, following Jacob and Monod's discovery of operons and their control

[5]This work had one unexpected consequence for my career. Shortly after I had completed the experiments described above, I received a visit from a microbiologist from the Eli Lilly Research Laboratories in Indianapolis. He was out on a fishing expedition (looking for ideas that might be commercialized at Lilly), and I told him that I thought the sort of overproducing mutants I was working on might be used to produce amino acids and other natural metabolites on an industrial scale. The result was an invitation to give a seminar at Lilly, which led to my appointment there as a consultant—a relationship that lasted for eight years and that I enjoyed immensely. (Lilly never did follow up on the overproducing mutants, by the way.)

by operators and repressors. I decided to screen for regulatory mutants in the isoleucine-valine (*ilv*) biosynthetic pathway, by testing valine-resistant mutants for derepression of *ilv* enzymes (guessing, from my previous experience with antimetabolites, that one route to valine resistance would be derepression of isoleucine biosynthesis). Table 1 lists the set of *E. coli* mutants that we had in our laboratory at that time.

We also had an F-prime plasmid carrying the entire set of the above *ilv* genes, which we found to be tightly clustered. Using a strain carrying this plasmid, T Ramakrishnan and I selected for valine-resistance and found several in which the resistance determinant was transferred by the plasmid in close linkage with the *ilv* genes. Several of the mutants were found to be coordinately derepressed for the enzyme products of *ilvE*, *ilvD*, and *ilvA*, but not of *ilvB* or *ilvC*. These mutants excreted (i.e. overproduced) isoleucine but not valine. We ultimately showed that *ilvE, D*, and *A* indeed constituted an operon, and that *ilvB* and *ilvC* were each independently regulated (51, 52).

Mechanisms of Mutation in Bacteria

My training with Ed Tatum instilled in me a deep interest in all aspects of genetics—not simply in the use of genetic mutants for the analysis of biosynthetic pathways and their regulation. During the 1950s and 1960s, following Watson and Crick's discovery in 1953 of the structure of DNA, a great many laboratories plunged into work on mechanisms of mutation; bacteria and bacteriophage were favorite experimental materials. A 1954 paper from Seymour Cohen's laboratory, showing that thymine-requiring mutants of *E. coli* died when deprived of thymine ("thymine-less death") piqued my interest, and led me to ask whether thymine starvation would induce mutations among the surviving cells.

Carol Coughlin and I found that thymine deprivation is indeed highly mutagenic, as measured by the number of induced mutants among the survivors (17). We published this paper back-to-back with one from Arthur Pardee's laboratory, which demonstrated that 5-bromouracil is mutagenic for bacteriophage,

Table 1   *E. coli* mutants in the Adelberg laboratory, 1958–1959

Locus	Enzyme of the *ilv* pathway
ilvA	Threonine deaminase
ilvB	Acetohydroxy acid synthase ("condensing enzyme")
ilvC	Isomeroreductase
ilvD	Dihydroxy acid dehydratase
ilvE	Branched chain amino acid aminotransferase ("transaminase B")

commenting that both papers showed that mutation can be a consequence of disrupted DNA metabolism. This may seem obvious today, but it should be remembered that these papers were published only four years after the discovery of the structure of DNA.

It was to be ten years before another opportunity presented itself in my laboratory for an investigation of mutagenesis. Acridines such as proflavin had been shown by Brenner and others to induce frameshift mutations in bacteriophages replicating within their bacterial hosts; acridines, however, had been found not to be mutagenic for the bacterial genome itself. Lerman found that proflavin interacted with DNA by intercalating between adjacent nucleotide pairs; he suggested that insertions and deletions of base-pairs are produced as a result of recombination between paired DNA molecules having short regions of nonhomologous pairing, which would be promoted by the intercalation of an acridine molecule in one or the other of the paired molecules. Thus, acridines would be mutagenic for replicating bacteriophage, in which genetic recombination takes place at a high rate, but not in the bacterial genome where recombination does not normally occur.

Suzanne Sesnowitz and I decided to test Lerman's hypothesis by treating conjugating bacteria with proflavin, and testing certain classes of recombinants for mutation in the *ara* operon. We were able to show that under these conditions (but not in nonconjugating, haploid cells) mutations were indeed induced by proflavin, but only when the *ara* region itself was transitorily diploid. Furthermore, we showed that the majority of the proflavin-induced mutants were of the frame-shift type, as predicted by Lerman's hypothesis (56, 57). Mutations were not induced in nondiploid regions of the chromosome, even in conjugating cells.

From the work on phage genetics it was known that the phenotypic effect of a primary mutation could be reversed by a secondary mutation at a different position in the genome. Such secondary mutations were called "suppressors," and during the course of our work on mutation we carried out two studies on the suppression phenomenon. In one, Gudmundur Eggertsson and I carried out the first systematic selection for suppressors in the *E. coli* genome, identifying and mapping six widely scattered suppressor loci (27); in the other, Manny Murgola and I showed that streptomycin could be used to suppress conditionally expressed lethal mutations, such as the loss of aminoacyl-tRNA synthetase activity (44).

My last (and least) contribution to the mutation field itself came when Mort Mandel, Grace Chen, and I published a paper on the optimal conditions for mutagenizing bacteria with N-methyl-N^1-nitro-N-nitrosoguanidine (9). This became a *Citation Index* "citation classic," because nitrosoguanidine proved to be the most powerful chemical mutagen known and was widely used for many years: Hundreds of papers cited it. It was a very simple exercise, however, carried out as a brief summer project as part of the Physiology Course at the Marine Biology Laboratory at Woods Hole, where I was then an Instructor. The editor of *Citation Index* asked me to write a 500-word article about it; I couldn't think of more than 300 words to say, so the article was rejected.

The Bacterial Chromosome, Sex Factors, and Conjugation

F-PRIMES   As recounted above, I was to spend a sabbatical year in 1956–57 at the Pasteur Institute in Paris, working with Georges Cohen. Shortly before I was to leave for Paris, however, François Jacob gave a seminar at Berkeley describing the work that he and Élie Wollman were doing on bacterial conjugation. As he reported there, they had discovered that F^+ populations of *E. coli* K12 (see below) owe their fertility to the presence of rare "Hfr" mutants, in which the F plasmid has become attached to the chromosome. Their results were compatible with the hypothesis that an Hfr cell injects its chromosome linearly into a genetic recipient cell; each different Hfr mutant starts the injection at a different place on the chromosome, with sequential orders of gene transfer compatible with the F^+ chromosome being a circular structure that is broken at a different place in each Hfr strain.

This was exciting stuff! As a graduate student, I had shared a laboratory bench with another of Ed Tatum's students, Joshua Lederberg, whose dissertation announced the discovery of genetic recombination in bacteria (more precisely, in *E. coli* K12). Josh had gone on to show that two cells of strain K12 would mate only if one of them possessed a particular extrachromosomal genetic element (for which he later coined the generic term "plasmid"). The plasmid in question he designated F (for "fertility"), and cells carrying it were designated F^+. F, he showed, was transferred autonomously during the act of conjugation, along with one or another chromosomal gene. For this and his later discovery of phage-mediated transduction of bacterial genes, he shared the Nobel Prize with Beadle and Tatum.

So I had been present at the birth of modern bacterial genetics, and had followed Josh's subsequent work on bacterial conjugation with keen interest. His later work, however, had become mired in a set of puzzling genetic linkage results, and all of this was now becoming clarified by Jacob and Wollman's electrifying discoveries. Indeed, I was so excited by Jacob's seminar that I immediately asked him if I could work not only with Georges Cohen but also in his laboratory during my coming stay at the Pasteur. It turned out that this would work out well, since Georges was planning to be on leave in the United States during the second half of my year there; it was arranged that I would work with Georges during the first half, and with Jacob and Wollman during the second half.

I have already described the work I did with Georges. My time in François's and Élie's laboratory was spent mainly in learning their techniques, and using a special modification of the penicillin selection process that Jack Myers and I had developed (46) to isolate a set of auxotrophic mutants for them in one of their K12 strains. When the time came for me to return to Berkeley I was hooked on bacterial genetics, and determined to continue work in that field. The first thing I wanted to do was to test Jacob and Wollman's hypothesis that, during bacterial conjugation, the donor chromosome undergoes breakage at random times, such that different recipient cells receive different lengths of DNA. An alternative hypothesis was that all cells receive the full chromosome, and that breakage and integration of random lengths of DNA take place post-conjugationally in the recipient cells. François and Élie had evidence for the former view, but I felt that it was not conclusive and had designed additional experiments that I believed would be decisive.

I therefore took with me, when I left Paris, one of their well-studied Hfr strains called P4x, and a number of their recipient (so-called F$^-$) strains. When I started work the following fall semester in Berkeley, I began the experiments I had planned, using P4x as the model donor strain, in collaboration with my assistant Sarah Burns.

To describe what happened next, I must first explain (or remind the reader) that Hfr cells, unlike F$^+$ cells, do not transfer F as an autonomous plasmid. Rather, as François and Élie had shown, F appears to be attached to the chromosome in Hfr cells, and recombinants emerging from Hfr X F$^-$ matings are usually themselves F$^-$. (Remember that in F$^+$ X F$^-$ matings the recombinants are always themselves F$^+$, as a consequence of the simultaneous transfer of the autonomous F plasmid.) The exception occurs in those rare cases where the recipient cell receives from the Hfr cell an entire chromosome: the recombinants in such cases are themselves Hfr.

This had led François and Élie to propose that in conjugating Hfr cells the chromosome break is adjacent to the F attachment site, such that the attached F is at the distal end of the transferred chromosome, and is transferred only in those rare matings in which the entire chromosome enters the recipient. With this as background, I can now return to my own experiments.

In Paris, the crosses I had carried out with P4x had always given the standard result for an Hfr X F$^-$ cross: the recombinant progeny were F$^-$, unless the entire chromosome had been transferred. In Berkeley, however, such crosses yielded Hfr progeny in every case: i.e. F was being transferred in every mating pair!

At first glance this would seem to be the simple result of a reversion from the Hfr state to the F$^+$ state—but if that had been the correct explanation, P4x should now show the very low rate of recombination and random orientation of transfer characteristic of F$^+$ populations. (Remember that in F$^+$ populations the only true donor cells are the rare Hfr mutants in which F has become attached to the chromosome. The rate of recombination in F^+ X F^- crosses is typically five orders of magnitude lower than in Hfr X F^- crosses.) This, however, was not the case: my Berkeley Hfr transferred its chromosome at a moderately high frequency, as well as transferring F to every recipient.

The progeny of such crosses behaved exactly like their Hfr parent (which I now designated P4x-1): they transferred chromosomal genes at a moderately high frequency, like an Hfr, but transferred F with 100% efficiency, like an F^+. Furthermore, they transferred chromosomal loci in the same sequence as did their Hfr parent.

These observations we interpreted to mean that in one or more P4x cells, F had detached from the chromosome, and that the detached F we were observing was a genetic variant: it differed heritably from the original, wild-type F in having a high affinity for a specific site on the chromosome. I described these findings to Élie Wollman, who was in Berkeley that year working with Gunther Stent. I suggested that the variant F (which I designated F^2) had arisen by mutation, but Élie quickly pointed out a much more attractive hypothesis: namely, that F^2 had arisen by a genetic exchange between the attached F and the chromosome, in a manner analogous to the acquisition of the *gal* locus by lambda phage as had recently been demonstrated by Alan Campbell. This would suggest that F^2 had acquired a segment of chromosomal DNA adjacent to its original attachment site, and would explain the specific affinity of F^2 for that site. In our published paper (8), we credited Élie with proposing this hypothesis—which was later to be fully confirmed.

The hypothesis of genetic exchange suggested that the chromosome from which F^2 had detached might have retained a portion of F. We confirmed this by "curing" P4x-1 of its sex factor, thereby converting it to an F^- strain. When these F^- cells were infected with F^1 (the wild-type F, which normally confers low-frequency donorship on its host cells) they became high frequency donors, showing that they had indeed acquired a chromosomal locus having high affinity for the sex factor. We called such a locus *sfa* (for "sex factor affinity") and were able to demonstrate linkage of *sfa* with a nearby marker.

I had also told my findings to Joshua Lederberg, who was now at Stanford University. Josh's student, Yukinori Hirota, immediately screened Josh's collection of Hfr cultures for variant Fs similar to F^2. He found many: in each case, the variant F exhibited high affinity for the chromosomal site at which it had previously been attached. Josh subsequently proposed the generic term "F-prime" for plasmids of this type, a term that became standard in the literature.

We submitted our paper to the *Journal of Bacteriology* in July of 1959, and it appeared in the March, 1960, issue (8). Meanwhile, however, news of our discovery reached François Jacob in Paris, who immediately saw in it the potential for making partially diploid cells of *E. coli*: exactly what he and Jacques

Monod needed to test their new ideas concerning operator loci and the regulation of operons. What they wanted was an F-prime carrying the *lac* locus, so that they could construct strains carrying one *lac* allele on the chromosome and a different *lac* allele on the F-prime in the same cell.

François mated a *lac*$^+$ Hfr (of the P4x type, in which *lac* is the last marker transferred, and is thus closely linked to the attached F) with a *lac*$^-$ F$^-$. In such a cross, *lac*$^+$ is transferred as a chromosomal marker more than 90 minutes after conjugation begins. On the basis of our findings, François reasoned that, as a rare event, F might detach from the chromosome bearing the *lac*$^+$ locus itself. If so, he should detect it in a *lac*$^+$ recombinant appearing within the first ten minutes after the commencement of mating (since it takes less than ten minutes for an F or a short F-prime to be transferred). Indeed, such "early" *lac*$^+$ recombinants did appear in his crosses in low numbers; on isolation, they proved to be cells harboring the F-*lac* plasmid that he sought.[6]

François isolated F-*lac* so quickly after hearing about our discovery of F-primes that he was ready to publish his results even before our paper was ready to go to press. Furthermore, he was going to publish in *Comptes Rendus de Séances de l'Académie des Sciences*, which meant his paper would appear many months before ours. To avoid the appearance of scooping us, he sent me a copy of his manuscript and asked that he be allowed to put my name on as co-author. I wrote back to say that I could not accept such an honor, having had nothing to do with his development of a method for isolating defined F-primes; he wrote back saying that I was holding up his publication and insisting on my co-authorship. I reluctantly agreed, and the paper by Jacob and Adelberg duly appeared (37). It is certainly one of the most important papers bearing my name, and I am embarrassed to this day about my gratuitous authorship.

F-primes carrying known regions of the chromosome came to be known as F-genotes. Jacob's laboratory showed that variant cells could be selected in which a temperature sensitive F-genote became temperature resistant, and that this resistance was due to the stable integration of the F-genote into the chromosome to form an Hfr. Peter Bergquist and I decided to analyze this phenomenon, and produced our own Hfr (which we called PB15) in which a temperature-sensitive F-*gal* had been stably integrated into the chromosome. We showed that stabilization was due to the inversion of one of the two *gal* genes (6a, 13a).

THE SEX FACTOR (F)   The attachment of F to the chromosome in Hfr cells had been postulated by Jacob and Wollman to explain the fact that, in Hfr X F$^-$

---

[6]Jacob and Monod went on to use F-lac to construct [lac$^+$/lac$^-$] partial diploids, which they then used to demonstrate the existence of operator loci, repressors, and operons. This confluence of Jacob's bacterial genetics and Monod's bacterial physiology led to their winning the Nobel Prize a few years later (sharing it with Jacob's mentor, André Lwoff).

crosses, only recombinants receiving the most distal part of the transferred chromosome were males. Since F is the genetic determinant of maleness, they had in essence shown a linkage of F (or at least part of it) to the most distal marker of any Hfr. In confirmation of this finding, Suzanne Dewitt and I showed that we could use a bacteriophage to co-transduce F with the most distal marker of a particular Hfr, but not with any of its other markers. The transductants receiving the attached F behaved as Hfr males with the same transfer origin as that of the Hfr on which the transducing phage had been grown (25).

The fact that F could attach to the bacterial chromosome, and could undergo genetic variation [as described in (8), above] suggested strongly that it consists of DNA—but at the time of our experiments (the early 1960s) this had not been demonstrated. Patricia Driskell (later Driskell-Zamenhof) and I reasoned that if F were indeed a DNA element, it should be inactivated by the decay of ^{32}P atoms incorporated during replication in the presence of ^{32}P-phosphate, as Stent and his colleagues had shown for bacteriophages.

We thus designed the following experimental protocol (26): Cells carrying F^2 (see above) were allowed to incorporate ^{32}P, and allowed to transfer their sex factors to non-radioactive female (F^-) cells. The latter were stored in the frozen state for various lengths of time, plated out, and the resulting colonies tested for maleness (retention of viable F^2 particles) by Lederberg's technique of "replica plating." If F^2 had indeed incorporated ^{32}P, the cells harboring it should lose maleness with time in storage, as a result of ^{32}P decay. Our results were positive, and from the rate of decay of maleness we were able to estimate the number of P atoms in F^2 DNA as lying between 1.9 and 5.4×10^5. We ruled out RNA as the chemical structure by showing that we could inhibit the incorporation of ^{32}P into F^2 with concentrations of mitomycin C that inhibited DNA synthesis, but not RNA synthesis.

We were able to do these experiments because cells harboring F^2 are high-frequency donors, so that colonies arising from them can be scored for maleness using replica plating. This approach would not normally work for F^1 (wild-type F), however, since colonies of F^1 cells contain too few donors. We were able to get around this obstacle by transferring ^{32}P-containing F^1 particles to non-radioactive F^- cells carrying an *sfa* locus (see above). Again, the results were positive; the rate of decay indicated that F^1 contains between 0.85 and 2.5×10^5 P atoms. These experiments provided strong chemical evidence for the DNA nature of F.

THE BACTERIAL CHROMOSOME   Most of our work on the bacterial chromosome was concerned with Jacob and Wollman's inference that the linkage map formed a closed circle, and that the chromosome was therefore itself a physically closed, circular structure. Our first foray into this field followed upon Larry Taylor's isolation of three new Hfr strains, which proved to transfer distal markers with exceptionally high frequency. It must be remembered that the Hfr chromosome undergoes spontaneous, random breakage during transfer, so that there is a gradient of marker transfer frequencies from the proximal to the distal end of the chromosome. In ordinary Hfrs this gradient is very steep: The recombination frequency for the earliest transferred marker exceeds 10 percent, while for the latest markers it may be only 0.001 to 0.01 percent, or more than 1000-fold lower. In our three new donor strains, however, the most distal markers were inherited with frequencies of one to two percent, only 5 to 10-fold lower than the most proximal markers. Larry and I called these donors Vhf, for "very high frequency."

We did not discover the basis for Vhf behavior—Vhf males might form unusually stable conjugation bridges, or the chromosome itself might be less prone to rupture—nor did we discover the genetic basis for this characteristic. The Vhf males, however, allowed for the first time a true demonstration of marker linkage around the entire chromosome, and a confirmation of the circular nature of the F^- chromosome (65).

Jacob and Wollman had already inferred the circular nature of the F^+ as well as of the F^- chromosome based on linkage analysis with a variety of Hfr donors. The Hfr chromosome, on the other hand, behaves as an open (linear) structure during transfer: whether it is circular or linear in non-conjugating cells was unknown, although it was generally believed to be linear. We found a way to test this, by crossing (with each other) two Hfr strains having the same transfer origin (point of breakage) but different genetic markers. In these experiments one of the two Hfrs had been pregrown under conditions that converted it to the physiologically recipient state. In such crosses, the most proximal and most distal markers of the physiological donor showed strong linkage, indicating that the chromosome of the recipient Hfr was a closed structure (66). We had thus shown that in all cells of *E. coli* K12, whether F^-, F^+, or Hfr, the chromosome is circular. In another series of papers, Jim Pittard, John Loutit and I showed that F-genotes (and thus, by inference, F itself) also appear to be circular, and capable of crossing over with the chromosome (48–50).

CONJUGATION: RESTRICTION AND MODIFICATION   In 1963, Herbert Boyer joined my laboratory as a post-doctoral fellow, having done his PhD research in Ellis Englesberg's laboratory. We agreed that he would continue working on a project involving the arabinose operon that he had initiated in Ellis's laboratory, but there was a hitch: The strains he had brought with him had been constructed in *E. coli* B, and my laboratory was set up to work exclusively with *E. coli* K12. These two strains, independently isolated from nature, were known to have a number of different genetic properties.

This did not seem to be a major problem: We would simply transfer the desired mutant alleles from B to K12 by conjugation. First, however, we had to confirm that K12 and B would indeed conjugate with each other (it had not been previously tried, to our knowledge). We found that we could indeed obtain some gene transfer from a K12 Hfr donor to a B cell recipient—but the data were anomolous. For example, a series of markers on the donor chromosome did not appear to enter the B recipient in an ordered progression, but rather to enter simultaneously after about 30 minutes. Also, the frequencies of recombination were much lower than in K12 × K12 crosses, and linkages were significantly reduced. When some (but not all) of the recombinants were backcrossed with the parental K12 Hfr, however, the results resembled those of a K12 × K12 cross in every way.

We then made crosses in the opposite direction, making B into a genetic donor by infecting it with F-*lac*; again, low frequencies of recombination were observed, some of the hybrids behaving like B strains in further matings with K12 donors, and some behaving like K12 strains.

All of this was inexplicable until Herb recalled the recent work of Arber and his colleagues on restriction and modification, and suggested that restriction between K12 and B strains might underlie the anomolous results of our crosses. He went on to verify this, and showed that in both K12 and B the genetic determinants of restriction and modification were linked to each other and to the *thr* locus: By transferring these loci between K12 and B in one or the other direction, one could make these strains fully compatible in conjugational recombination.

Herb published two papers on this phenomenon while in my laboratory (14, 15). Since the crucial ideas were entirely his own, I did not put my name on these papers, but I enjoy remembering that Herb's work on restriction and modification began in my laboratory, and led ultimately to the founding of Genentech and to the Boyer-Cohen method of gene-splicing.

Bacterial Genetic Nomenclature

I cannot leave the subject of my group's work on bacterial genetics without mentioning a nonexperimental paper that I believe had a significant impact on the advance of the field, namely, the paper entitled "A Proposal for a Uniform Nomenclature in Bacterial Genetics" (24). As with so many of the successful ideas that were developed in my laboratory, this one grew out of discussions with John Clark, who had worked with me in Berkeley and, during my second year at Yale, in New Haven.

John and I were concerned over the then-growing confusion in bacterial genetic nomenclature; it was John who pointed out that the solution would be to use symbols and abbreviations that clearly distinquish between genotype and phenotype. The term "lac⁻" ("lac-minus"), for example, was used sometimes to mean the phenotypic inability to use lactose as carbon-source, and sometimes to mean a genotypic mutation in a locus coding for beta-galactosidase.

Demerec and Hartman had already introduced the usage of three-letter symbols (such as "lac"); John and I drafted a proposal according to which genetic loci would have italicized three-letter lower-case symbols, followed by a capital letter when necessary to distinquish between two mutant loci conferring the same general phenotype (e.g. *lacZ* and *lacY* to denote the loci coding for beta-galactosidase and galactoside permease, respectively). Different mutant alleles would be given different numbers, so that—for example—*lacZ1* and *lacZ2* would represent two different mutant alleles of the *lacZ* locus. The wild-type locus would, in all cases, be designated with a superscript plus sign (e.g. *lacZ*$^+$). Symbols for phenotypes, on the other hand, would be as different as possible; at the least, they would start with a capital letter if a three-letter symbol were to be used, and would not be italicized. Thus, Lac$^-$ would refer to the inability to use lactose; it might be caused by mutations in any of several different loci, including *lacZ* and *lacY*.

John and I sent our draft to Demerec and Hartman, who readily agreed to sponsor it. The paper was duly published with Demerec as senior author in honor not only of his pioneering work in the field of bacterial genetics nomenclature, but of his pioneering work in the field of bacterial genetics in general. The principles laid out in that paper were eventually adopted by all *E. coli* geneticists.

Membrane Transport in Mammalian Cells

In about the summer of 1969 I participated in a symposium in Aspen, Colorado, where I presented the results of our experiments on conjugal chromosome transfer in *E. coli*. One of the other participants was Henry Harris, from Oxford University; I was tremendously impressed by his account of mammalian cell fusion, and his use of that system to establish the dominance of certain nuclear functions in fusion heterocaryons. At that time I was casting about for a new direction in which to take my own research, having found that bacterial genetics was becoming more and more a subject for molecular biologists, with whom I was poorly prepared to compete. By the end of the week in Aspen I had arranged to spend my next academic leave working in Henry's laboratory, with the intention of switching to mammalian cell genetics on my return.

Yale provides its faculty with the opportunity to take a one-semester leave every three years, and I spent the period from January to June of 1970 at Oxford.[7]

---

[7] I accomplished little at Oxford, unless one counts brass rubbing, because I could not get my cells to grow and form clones at low cell density. After four months of failure, which I blamed on my own incompetency, it turned out the CO_2 incubator that I shared with other members of Henry's group was being charged with industrial-grade CO_2, which is contaminated with enough sulfite to kill isolated cells—although allowing the passage of cultures at high cell density. At that time I was the only one trying to clone cells, so no one else had noticed the problem.

During that period my senior associate, Dr. Joan (Jody) Stadler was a guest in Frank Ruddle's laboratory at Yale, with the result that both of us became familiar with the culture and fusion of mammalian cells.

On my return to Yale, Jody and I carried out a number of experiments on the cell fusion process (59) and I collaborated with Irv Miller, Jerry Eisenstadt, and Jerry Degnen on gene transfer between lines of cultured mammalian cells (22, 23). While this work was going on, I was following with interest a small but growing literature on membrane transport systems in established lines of cultured mammalian cells. There were several such systems, with complex, overlapping specificities, and it seemed to me that the isolation of genetically defective transport mutants would be of great help in defining the number and characteristics of the different systems.

AMINO ACID TRANSPORT   I was struck by a paper showing that cells of the mouse lymphocytic cell line, L5178Y (our favorite line at the time) could be killed (over time, during storage in liquid nitrogen) by internal radiation from accumulated tritium-labeled amino acids—a process we called "tritium suicide." (Actually, I don't remember whether we or someone else coined that name.) I recalled that ^{32}P-suicide had been used to select bacterial mutants defective in phosphate transport, and putting these findings together I proposed that we use tritium suicide to select for mutants of L5178Y defective in amino acid transport. Cells would be allowed to take up a tritium-labeled amino acid and stored under liquid nitrogen for a period long enough to allow the cells to be killed by internal radiation. Transport mutants, defective in amino acid uptake, would survive and be selected.

I was joined in this project by my colleague Carolyn Slayman, with whom I collaborated on all future work on membrane transport systems in mammalian cells, and by our graduate student, Morris Finkelstein. We decided to start this line of work by selecting for mutants of L5178Y that were defective for the uptake of α-aminobutyric acid (AIB). Christensen and his colleagues had shown AIB to be a model substrate for a transport system in Ehrlich ascites cells that they designated the "A system," a system with preference for small, neutral amino acids. Before embarking on the actual tritium-suicide experiments, however, it was necessary for us to characterize the uptake of AIB in L5178Y mouse lymphocytes. This we did, showing that AIB was taken up by a sodium-dependent, A-like system specific for small, neutral amino acids (29).

Armed with this knowledge, we used tritium-labeled AIB to select two A-system mutants in L5178Y; both were found to have normal K_m values for AIB and other small, neutral amino acids, but greatly reduced V_{max} values. Competition experiments verified the A-like nature of the defective system (30).

As part of his doctoral dissertation, Morris Finkelstein had shown that L-glutamic acid was taken up by both a low-affinity and a second, high-affinity system in L5178Y. Ann Dantzig had now joined our group, and together we made a detailed analysis of glutamate uptake by our parental line and the two A-system mutants. We found that the low-affinity system was in fact our A system: the V_{max} (but not the K_m) for glutamate was reduced in the mutants to the same extent as for the small, neutral amino acids. The data for the high-affinity uptake of glutamate indicated a small reduction in one of the mutants but not in the other (18, 19).

Over the next few years our laboratory used a number of different selection systems (tritium suicide, replica plating, killing with visible light after the uptake of bromodeoxyuridine) to isolate additional amino acid transport mutants, including mutants defective in proline, glycine, and leucine uptake. For these experiments, we switched to a line of Chinese hamster ovary (CHO) cells, which are susceptible to replica plating. These studies resulted in the isolation of what appeared to be a mutant in the structural gene coding for the A system (20); a mutant in a previously unrecognized system specific for glycine and sarcosine (28); and a mutant in the L system (specific for neutral amino acids with bulky, hydrophobic side chains, such as leucine), which we showed was resistant to the anti-cancer drug, melphalan (21). In many of these studies we were joined by our post-doctoral fellow, Jim Mullin, and our assistant, Margaret Fairgrieve.

In the course of our work on amino acid transport, I had the unusual pleasure of collaborating with my own son, David, who did his MD thesis in our laboratory. With David, Ann Dantzig and I showed that the increase in AIB uptake that had been reported by Ted Puck for CHO cells treated with "reverse transformation agents" (agents such as dibutyryl cyclic AMP plus testololactone that restored normal phenotypic properties to transformed cells) was due to an increase in the V_{max} of the A system—but that this effect was unrelated to the effect on transformation properties, since the agents brought about the same effect in non-transformed cells (1).

POTASSIUM TRANSPORT   Stimulated by the research activities of many of my colleagues at Yale and elsewhere, I became interested in the ouabain-sensitive Na^+, K^+ ATPase, the pump that allows mammalian cells to maintain high intracellular concentrations of K^+ and low intracellular concentrations of Na^+ against their transmembrane gradients. My approach to the understanding of this system was, as usual, genetic: Following François Jacob's dictum (which I believe he borrowed from someone else, possibly Sal Luria) "whatever biological system you're working on, you're better off if you have a mutant for it than if you don't," I began thinking about ways to isolate mutants that were altered in the activity of the Na/K pump.

It finally struck me that, if the pump were essential for maintaining high intracellular concentrations of K, and if those high concentrations were necessary for cell survival, then we should be able to find a threshold concentration of external K below which the pump would function inadequately and the cells would not survive. If we then selected for mutants able to grow at below-threshold K concentrations, we might find some with altered pumps, e.g. pumps with an increased affinity for potassium.

The selection was carried out by Jay Gargus, an MD/PhD student working both in our laboratory and that of Joe Hoffman, using 0.2M K as the threshold below which we found that cells of the mouse fibroblastic line LMTK$^-$ could not grow. A number of clones were isolated that could grow at this potassium concentration, and two of them were singled out for further study: LTK-1 and LTK-5. To our surprise (it seems we were always being surprised, which is a good thing for making new discoveries) neither mutant showed any alteration in the pump. Rather, one of the mutants (LTK-1) showed a pronounced decrease in the activity of a specific component of K efflux, while the other (LTK-5) appeared to have an increase in a furosemide-sensitive influx system (32). After leaving Yale Jay went on to show that LTK-1 lacked a specific K channel (31).

We returned to studies on LTK-5 when David Jayme joined our group as a post-doctoral fellow. We decided that LTK-5's ability to grow at 0.2M extracellular K could not be due to an increase in furosemide-sensitive K influx, for two reasons: (a) we found that influx via the furosemide-sensitive system was barely detectable at 0.2M K, and (b) high concentrations of furosemide failed to inhibit the growth of LTK-5 at 0.2M K. We therefore analyzed K efflux in LTK-5, and found that it had a severely reduced rate of efflux via a system sensitive to two diuretics: furosemide, and the much more potent inhibitor, bumetanide. If the loss of this system were responsible for the ability of LTK-5 to grow at 0.2M K, then inhibiting the system with diuretics should confer a similar ability on the parent strain, LMTK$^-$. We showed that this was indeed the case: thus, the reduction of K efflux either by mutation or by diuretics inhibition permitted growth in K$^+$ deficient medium (39). In a later paper, we showed that the diuretic-sensitive system requires both Na$^+$ and Cl$^-$, and can be described as a K$^+$ Na$^+$Cl$^-$ cotransport system (40).

In 1982 I spent the spring semester in the laboratory of Enrique ("Henry") Rozengurt, at the Imperial Cancer Research Fund in London. Henry's laboratory was a leading center for work on the cell cycle in cultured mammalian cells, and particularly for work on cation fluxes and concentrations in relation to growth regulation. In his earlier studies, Henry had implicated K$^+$ influx in the mechanism of mitogenesis by a combination of three growth factors: insulin, epidermal growth factor, and vasopressin (IEG). We agreed that I would team up with a visiting post-doctoral scientist from Spain, Abelardo Lopez-Rivas, in evaluating the precise relationship between intracellular levels of K^+ and the IEG-stimulated initiation of DNA synthesis in quiescent (serum-starved) 3T3 mouse fibroblasts. .

We devised a procedure for artificially maintaining intracellular K^+ at different levels over extended periods and found that IEG-stimulated DNA synthesis in quiescent cells commenced only when K^+ exceeded a certain threshold level. We showed that a sigmoidal dependence of DNA synthesis on K^+ was generated in early G_1, and that it reflected the control of protein synthesis by K^+. In serum-free medium we found the K^+ content of the cells to be just below the threshold concentration, so that small changes in intracellular K^+ have a critical effect on the ability of the cells to initiate DNA synthesis (42).

THE SODIUM/PROTON ANTIPORTER   While in Rozengurt's laboratory, I followed closely the work that he and his colleagues were doing on the Na/H antiporter, which they had also implicated in the regulation of the cell cycle. When I returned to Yale, I decided to look for genetic mutants affecting the antiporter as a way of investigating this presumed relationship, and drew up a series of possible selection schemes. All of them were based on the assumption that we could get 3T3 cells to grow in dialyzed serum supplemented with IEG, which Rozengurt had found would stimulate passage of serum-starved cells through the G_1 and S phases of the cell cycle. To my disappointment, we were unable to obtain sustained cell cycling by this treatment: For reasons we still do not understand, these growth factors could stimulate only one turn of the cycle.

A way to select for mutants affecting the Na/H antiporter was devised, however, by Jacques Pouysségur in France. Called "proton suicide," it involved preloading cells in neutral medium with high intracellular concentrations of Na^+, then transferring them to an acidic medium. Under these conditions, the cells use the antiporter to exchange intracellular Na for H, lowering their internal pH to a lethal level. Cell populations surviving this treatment are enriched for mutants with decreased antiporter activity; after several cycles of selection and re-growth of the survivors, antiporter-deficient mutants are readily isolated.

We decided to use Pouysségur's method to select antiporter mutants in LLC-PK$_1$, an established line of pig kidney epithelial cells, which Kurt Amsler had brought from John Cook's laboratory when he joined us at Yale for post-doctoral work. We had originally chosen this strain because we had reason to believe it might survive six or more hours in a cell-sorter, which we had hoped to use to select mutants using dyes sensitive to internal pH. (This was before we knew of Pouysségur's method.) Another reason, however, was that LLC-PK$_1$ undergoes differentiation when plated in culre dishes: As the isolated, relatively undifferentiated cells reach confluence, they form a polarized epithelial sheet with tight junctions, capable of the vectorial transport of water and solutes and exhibiting well-differentiated apical and basolateral surfaces. We hoped that we could use antiporter mutants of LLC-PK$_1$ to investigate the role of this transport system not only in growth but in differentiation as well.

First, however, we had to demonstrate the presence of the antiporter in LLC-PK$_1$ cells (which had earlier been claimed by other workers to lack it). With John Haggerty, a new post-doctoral fellow, we found that if we preloaded cells with protons, we could easily identify amiloride-sensitive Na$^+$ uptake (amiloride sensitivity was characteristic of all previously studied antiporters). The uptake of sodium ions was totally dependent on exchangable internal protons. The amiloride derivative 5-N-ethyl-N-isopropyl amiloride (EIPA) was found to be almost 200 times more potent an inhibitor of the antiporter than amiloride itself: the half-maximal inhibitory concentration of EIPA was 0.02 μm (36).

Like fibroblasts, our pig kidney epithelial cells could be brought into a classical quiescent state by depriving them of serum for 6 days, and stimulated to reenter the cell cycle at a point early in G$_1$ by the restoration of serum to the medium. We found the antiporter to be relatively inactive during quiescence, but to increase 2- to 3-fold within 4 hours after serum addition. Pouysségur's group had shown that blockade of the antiporter prevented hormone-induced initiation of DNA synthesis in quiescent fibroblasts; similarly, we found that EIPA blocked the initiation of DNA synthesis in quiescent LLC-PK$_1$ cells when serum was added to cells in bicarbonate-free medium, but not when bicarbonate was present. We concluded that alkalinization of the cytoplasm is essential for hormonal stimulation, and that a bicarbonate-dependent transport system can substitute for the antiporter in bringing this about (33). In this work we were joined not only by John Haggerty, but also by postdoctoral fellow Neeraj Agarwal and by Kurt Amsler.

Armed with this information, we proceeded to use Pouysségur's method to isolate mutants of LLC-PK$_1$ with altered antiporter activities. The first such mutant, designated PKE5, had greatly reduced antiporter activity. The mutant grew normally at pH 7.0, but required bicarbonate for growth at pH 6.5; thus, the antiporter is essential for growth at acidic pH in the absence of bicarbonate (12), in agreement with the results obtained with EIPA discussed above.

The second type of mutant was one with an increase in antiporter activity. Again following Pouysségur's lead (this was becoming a bit tiresome) we selected for mutants that could recover from an acid load in the presence of 100 μm EIPA. One such mutant, designated PKE20, we studied in detail: The V$_{max}$ of its antiporter was 2 1/2 times higher than that of the parental type, and the antiporter was 100 times less sensitive to EIPA (34).

The confluent monolayer formed by cultured LLC-PK$_1$ cells has many characteristics of the proximal tubule in the kidney. There, the antiporter is restricted to the apical cell surface, but there were contradictory reports of its location in LLC-PK$_1$ cells: One group found it to be exclusively apical, while another found that, in most cells examined, it resided on the basolateral surface.

By growing our LLC-PK$_1$ cells on Nuclepore filters, which permit access to the apical and basolateral surfaces independently, we were able to resolve this question. We were able to demonstrate antiporter activity on both surfaces in almost equal amounts. The antiporters on the two surfaces showed clear differences, however: the apical form exhibited an IC$_{50}$ for EIPA of 13 μM, and the basolateral form, an IC$_{50}$ of 44 nM. Furthermore, the PKE20 mutant was found to overexpress only the apical form at confluence. Taken together, these results suggested the existence of two distinct forms of the antiporter, under separate genetic control (35). Later, under Bob Reilly's leadership, we characterized the antiporter in apical membrane vesicles prepared from wild-type and PKE20 cells and found that the latter contained the amiloride-sensitive form in increased amounts (54) as predicted from the whole-cell experiments.

One of our long-range goals had been to clone and sequence the antiporter gene, but Pouysségur accomplished this (for a human cell line) long before we had completed the necessary preliminary experiments. Bob Reilly (who had gone on to a faculty position in the Department of Medicine) sequenced the basolateral antiporter of LLC-PK$_1$ cells using the PCR procedure with primers based on Pouysségur's published human sequence, and found the pig and human sequences to be 95% identical (55).

At this point I retired from the laboratory and devoted full time to my administrative duties in the Provost's Office (I had been half-time in that office since 1983; see above). I kept in touch with Carolyn's and Bob Reilly's laboratories, where the work was continuing, and I am happy to report that Bob was ultimately able to clone and sequence the apical antiporter of LLC-PK$_1$ (RF Reilly, personal communication).

EPILOGUE

Am I—or rather, was I—a microbiologist? That seems to be the general impression, but the question has intrigued me all of my professional life.

What is a microbiologist? The simple answer would be "one who studies microorganisms." Begging the question (not all that simple) of what is a microorganism, however, I would make a fairly sharp distinction between the *study* of a microorganism (or group of microorganisms) for its own sake, and the *use* of a microorganism for the study of a universally applicable biological phenomenon. The former practice I would call microbiology: an attempt to learn something that uniquely applies to the microorganism(s) under study. The latter I would call biochemistry, or genetics, or molecular biology, or whatever most closely describes the real interest of the investigator.

What does that make me? For the first ten years of my professional life I used *N. crassa* to study the biosynthesis of isoleucine and valine. For the next fifteen years or so I used *E. coli* to study such phenomena as the genetic regulation of amino acid biosynthesis, mechanisms of genetic mutation and suppression, and bacterial conjugation. For the last twenty years or so I worked exclusively with a variety of cultured mammalian cell lines, studying membrane transport systems by a genetic approach.

By my definition, only a small part of that research activity could be classified as microbiology. If I am to be remembered as a microbiologist, however, it will be because of the time I spent with Roger Stanier and Michael Doudoroff at Berkeley (the right place) in the 1950s (the right time)—immersed daily in discussing, learning, teaching, and writing about microorganisms.

Visit the *Annual Reviews home page* at
http://www.AnnualReviews.org.

Literature Cited

1. Adelberg DE, Dantzig AH, Adelberg EA. 1980. The effect of reverse transformation agents on α-aminoisobutyric acid uptake in transformed and non-transformed cells. *Biochem. Biophys. Res. Commun.* 97:642–48

2. Adelberg EA. 1951. Studies on the isoleucine precursor α,β-dihydroxy-β-ethylbutyric acid. *J. Bacteriol.* 61:365–73

3. Adelberg EA. 1954. Isoleucine biosynthesis from threonine. *J. Am. Chem. Soc.* 76:4241–42

4. Adelberg EA. 1955. The biosynthesis of isoleucine and valine. III: tracer experiments with L-threonine. *J. Biol. Chem.* 216:431–37

5. Adelberg EA. 1958. Selection of bacterial mutants which excrete antagonists of antimetabolites. *J. Bacteriol.* 76:326

6. Adelberg EA. 1961. *Papers on Bacterial Genetics.* Boston: Little-Brown

6a. Adelberg EA, Bergquist P. 1972. The stabilization of episomal integration by genetic inversion: a general hypothesis. *Proc. Natl. Acad. Sci. USA* 69:2061–65

7. Adelberg EA, Bonner DM, Tatum EL. 1951. A precursor of isoleucine obtained from a mutant strain of *Neurospora crassa. J. Biol. Chem.* 190:837–41

8. Adelberg EA, Burns SN. 1960. Genetic variation in the sex factor of *Escherichia coli. J. Bacteriol.* 79:321–30

9. Adelberg EA, Mandel M, Chen GC. 1965. Optimal conditions for mutagenesis by N-methyl-N'-nitro-N-nitrosoguanidine in *Escherichia coli* K12. *Biochem. Biophys. Res. Commun.* 18:788–95

10. Adelberg EA, Umbarger HE. 1951. The role of α-keto-β-ethylbutric acid in the biosynthesis of isoleuncine. *J. Biol. Chem.* 192:883–89

11. Adelberg EA, Umbarger HE. 1953. Isoleucine and valine metabolism in *Escherichia coli.* V: α-ketoisovaleric acid accumulation. *J. Biol. Chem.* 205:475–82

12. Agarwal N, Haggerty JG, Adelberg EA, Slayman CW. 1986. Isolation and characterization of a Na-H antiporter-deficient mutant of LLC-PK$_1$ cells. *Am. J. Physiol.* 251:C825–30

13. Barker HA. 1993. Michael Doudoroff. In *Biographical Memoirs of the National Academy of Sciences* 62:119–41. Washington, DC: Natl. Acad. Press

13a. Bergquist P, Adelberg EA. 1972. Abnormal excision and transfer of chromosomal segments by a strain of *Escherichia coli* K-12. *J. Bacteriol.* 111:119–98

14. Boyer HJ. 1964. Genetic control of restriction and modification in *Escherichia coli. J. Bacteriol.* 88:1652–60

15. Boyer HJ. 1966. Conjugation in *Escherichia coli. J. Bacteriol.* 91:1762–72

16. Cohen GN, Adelberg EA. 1958. Kinetics of incorporation of p-fluorophenylalanine by a mutant of *Escherichia coli* resistant to this analogue. *J. Bacteriol.* 76:328–30

THE RIGHT PLACE AT THE RIGHT TIME   39

17. Coughlin CA, Adelberg EA. 1956. Bacterial mutation induced by thymine starvation. *Nature* 178:5–10
18. Dantzig AH, Finkelstein MC, Adelberg EA, Slayman CW. 1978. The uptake of L-glutamic acid in a normal mouse lymphocyte cell line and in a transport mutant. *J. Biol. Chem.* 253:5813–19
19. Dantzig AH, Adelberg EA, Slayman CW. 1979. Properties of two mouse lymphocyte cell lines genetically defective in amino acid transport. *J. Biol. Chem.* 254:8988–93
20. Dantzig AH, Slayman CW, Adelberg EA. 1982. Isolation of a spontaneous CHO amino acid transport mutant by a combination of tritium suicide and replica plating. *Somat. Cell Genet.* 8:509–20
21. Dantzig AH, Fairgrieve M, Slayman CW, Adelberg EA. 1984. Isolation and characterization of a CHO amino acid transport mutant resistant to melphalan (L-phenylalanine mustard). *Somat. Cell Genet.* 10:113–21
22. Degnen GE, Miller IL, Eisenstadt JM, Adelberg EA. 1976. Chromosome-mediated gene transfer between closely-related strains of cultured mouse cells. *Proc. Natl. Acad. Sci. USA* 73:2838–42
23. Degnen GE, Miller IL, Adelberg EA, Eisenstadt JM. 1977. Overexpression of an unstably inherited gene in cultured mouse cells. *Proc. Natl. Acad. Sci. USA* 74:3956–59
24. Demerec M, Adelberg EA, Clark AJ, Hartman PE. 1966. A proposal for a uniform nomenclature in bacterial genetics. *Genetics* 54:51–76
25. Dewitt SK, Adelberg EA. 1961. Transduction of the attached sex factor of *Escherichia coli*. *J. Bacteriol.* 83:673–78
26. Driskell-Zamenhof PJ, Adelberg EA. 1963. Studies on the chemical nature and size of sex factors of *Escherichia coli* K12. *J. Mol. Biol.* 6:483–97
27. Eggertsson G, Adelberg EA. 1965. Map positions and specificities of suppressor mutations in *Escherichia coli* K12. *Genetics* 52:319–40
28. Fairgrieve M, Mullin JM, Dantzig AH, Slayman CW, Adelberg EA. 1987. Isolation and characterization of a glycine transport mutant in an established mammalian cell line, CHO(PEOT/1). *Somat. Cell Mol. Genet.* 13:505–12
29. Finkelstein MC, Adelberg EA. 1977. Neutral amino acid transport in an established mouse lymphocytic cell line. *J. Biol. Chem.* 252:7107–8
30. Finkelstein MC, Slayman CW, Adelberg EA. 1977. Tritium suicide selection of mammalian cell mutants defective in the transport of neutral amino acids. *Proc. Natl. Acad. Sci. USA* 74:4549–51
31. Gargus JJ, Coronado R. 1985. A selectable mutation alters the conductance of a mammalian K+ channel. *Fed. Proc.* 44:1901 (Abstr.)
32. Gargus J, Miller IL, Slayman CW, Adelberg EA. 1978. Genetic alterations in potassium transport in L cells. *Proc. Natl. Acad. Sci. USA* 75:5589–93
33. Haggerty JG, Agarwal N, Amsler K, Slayman CW, Adelberg EA. 1987. Stimulation by serum of the Na+/H+ antiporter in quiescent pig kidney (LLC-PK1) cells, and the role of the antiporter in the reinitiation of DNA synthesis. *J. Cell Physiol.* 132:173–77
34. Haggerty JG, Agarwal N, Cragoe EJ Jr, Adelberg EA, Slayman CW. 1988. LLC-PK1 mutant with increased Na+/H+ exchange and decreased sensitivity to amiloride. *Am. J. Physiol.* 255:C495-501
35. Haggerty JC, Agarwal N, Reilly RF, Adelberg EA, Slayman CW. 1988. Pharmacologically different Na/H antiporters on the apical and basolateral surfaces of cultured kidney cells (LLC-pK1). *Proc. Natl. Acad. Sci. USA* 85:6797–6801
36. Haggerty JG, Cragoe EJ Jr, Slayman CW, Adelberg EA. 1985. Na+/H+ exchanger activity in the pig kidney epithelial cell line, LLC-pK1: inhibition by amiloride and its derivatives. *Biochem. Biophys. Res. Commun.* 127:759–67
37. Jacob F, Adelberg EA. 1959. Transfert de caractères génétiques par incorporation au facteur sexuel d' *Escherichia coli*. *C. R. Acad. Sci.* 249:189–191
38. Jawetz E, Melnick J, Adelberg EA. 1954. *Review of Medical Microbiology*. Los Altos, CA: Lange Med.
39. Jayme DW, Adelberg EA, Slayman CW. 1981. Reduction of K+ efflux in cultured mouse fibroblasts, by mutation or by diuretics, permits growth in potassium-deficient medium. *Proc. Natl. Acad. Sci. USA* 78:1057–61
40. Jayme DW, Slayman CW, Adelberg EA. 1984. Furosemide-sensitive potassium efflux in cultured mouse fibroblasts. *J. Cell Physiol.* 120:41–48
41. Lederberg J. 1990. Edward Lawrie Tatum. In *Biographical Memoirs of the National Academy of Sciences*. 59:357–86. Washington, DC: Natl. Acad. Press
42. Lopez-Rivas A, Adelberg EA, Rozengurt E. 1982. Intracellular K+ and the mitogenic response of 3T3 cells to peptide factors in serum-free medium. *Proc. Natl. Acad. Sci. USA* 79:6275–79

43. Luria SE. 1983. Obituary: Roger Y Stanier. *ASM News* 49:62–63

44. Murgola EJ, Adelberg EA. 1970. Streptomycin-suppressible lethal mutations in *Escherichia coli*. *J. Bacteriol.* 103:20–26

45. Murgola EJ, Adelberg EA. 1970. Mutants of *Escherichia coli* K-12 with an altered glutamyl-transfer ribonucleic acid synthetase. *J. Bacteriol.* 103:178–83

46. Myers JW, Adelberg EA. 1953. Modification of the penicillin technique for the selection of auxotrophic bacteria. *J. Bacteriol.* 65:348–53

47. Myers JW, Adelberg EA. 1954. The biosynthesis of isoleucine and valine. I: Enzymatic transformation of the dihydroxy acid precursors to the keto-acid precursors. *Proc. Natl. Acad. Sci. USA* 40:493–99

48. Pittard J, Adelberg EA. 1963. Gene transfer by F' strains of *Escherichia coli* K12. II: Interaction between F-merogenote and chromosome during transfer. *J. Bacteriol.* 85:1402–8

49. Pittard J, Adelberg EA. 1964. Gene transfer by F' strains of *Escherichia coli* K12. III: An analysis of the recombination events occurring in the F' male and in the zygotes. *Genetics* 49:995–1007

50. Pittard J, Loutit JS, Adelberg EA. 1963. Gene transfer by F' strains of *Escherichia coli* K12. I: Delay in initiation of chromosome transfer. *J. Bacteriol.* 85:1394–1401

51. Ramakrishnan T, Adelberg EA. 1964. Regulatory mechanisms in the biosynthesis of isoleucine and valine. I: Genetic derepression of enzyme formation. *J. Bacteriol.* 87:566–73

52. Ramakrishnan T, Adelberg EA. 1965. Regulatory mechanisms in the biosynthesis of isoleucine and valine. II: Identification of two operator genes. *J. Bacteriol.* 89:654–60

53. Ramakrishnan T, Adelberg EA. 1965. Regulatory mechanisms in the biosynthesis of isoleucine and valine. III: Map order of the structural genes and operator genes. *J. Bacteriol.* 89:661–79

54. Reilly RF, Haggerty JG, Aronson PS, Adelberg EA, Slayman CW. 1991. Increased Na^+-H^+ antiporter activity in apical membrane vesicles from mutant LLC-K_1 cells. *Am. J. Physiol.* 260 (Cell Physiol. 29): C738–44

55. Reilly RF, Hidelbrandt F, Biemesderfer D, Sardet C, Pouysségur J, et al. 1992. cDNA cloning and immunolocalization of a Na(+)-H+ exchanger in LLC-PK_1 renal epithelial cells. *Am. J. Physiol.* 261:F1088–94

56. Sesnowitz-Horn S, Adelberg EA. 1969. Proflavin-induced mutations in the L-arabinose operon of *Escherichia coli*. I: Production and genetic analyses of the mutations. *J. Mol. Biol.* 46:1–15

57. Sesnowitz-Horn S, Adelberg EA. 1969. Proflavin-induced mutations in the L-arabinose operon of *Escherichia coli*. II: Enzyme analyses of the mutants. *J. Mol. Biol.* 46:17–23

58. Sjolander J, Folkers K, Adelberg EA, Tatum EL. 1954. α, β-dihydroxyisovaleric acid and α, β-dihydroxy-o-methylvaleric acid, precursors of valine and isoleucine. *J. Am. Chem. Soc.* 76:1085–87

59. Stadler JK, Adelberg EA. 1972. Cell cycle changes and the ability of cells to undergo virus-induced fusion. *Proc. Natl. Acad. Sci. USA* 69:1929–33

60. Stanier RY. 1980. The journey, not the arrival, matters. *Annu. Rev. Microbiol.* 34:1–48

61. Stanier RY, Doudoroff M, Adelberg EA. 1957. *The Microbial World.* Englewood Cliffs, NJ: Prentice-Hall

62. Stent GS. 1961. *Papers on Bacterial Viruses.* Boston: Little-Brown

63. Tatum EL, Adelberg EA. 1950. Characterization of a valine analogue accumulated by a mutant strain of *Neurospora crassa*. *Arch. Biochem.* 29:235–36

64. Tatum EL, Adelberg EA. 1951. Origin of the carbon skeletons of isoleucine and valine. *J. Biol. Chem.* 190:843–52

65. Taylor AL, Adelberg EA. 1960. Linkage analysis with very high frequency males of *Escherichia coli*. *Genetics* 45:1233–43

66. Taylor AL, Adelberg EA. 1961. Evidence for a closed linkage group in Hfr males of *Escherichia coli* K-12. *Biochem. Biophys. Res. Commun.* 5:400–4

Annu. Rev. Microbiol. 1998. 52:41–79

LANTIBIOTICS: Biosynthesis and Biological Activities of Uniquely Modified Peptides from Gram-Positive Bacteria

Hans-Georg Sahl and Gabriele Bierbaum

Institut für Medizinische Mikrobiologie und Immunologie, Universität Bonn, D-53105 Bonn, Germany; e-mail: sahl@mibi03.meb.uni-bonn.de, gabi@mibi03.meb.uni-bonn.de

KEY WORDS: antibiotic peptides, nisin, mersacidin, lantibiotic biosynthesis, peptide modifying enzymes

---

ABSTRACT

A plethora of novel gene-encoded antimicrobial peptides from animals, plants and bacteria has been described during the last decade. Many of the bacterial peptides possess modified building blocks such as thioethers and thiazoles or unsaturated and stereoinverted amino acids, which are unique among ribosomally made peptides. Genetic and biochemical studies of many of these peptides, mostly the so-called lantibiotics, have revealed the degree to which cells are capable of transforming peptides by posttranslational modification. The biosynthesis follows a general scheme: Precursor peptides are first modified and then proteolytically activated; the latter may occur prior to, concomitantly with or after export from the cell. The genes for the biosynthetic machinery are organized in clusters and include information for the antibiotic prepeptide, the modification enzymes and accessory functions such as dedicated proteases and ABC transporters as well as immunity factors and regulatory proteins. These fundamental aspects are discussed along with the biotechnological potential of the peptides and of the biosynthesis enzymes, which could be used for construction of novel, peptide-based biomedical effector molecules.

CONTENTS

41

INTRODUCTION

In recent years antibiotic peptides came into the focus of both basic and applied research disciplines. Intensive research has revealed that production of ribosomally made antimicrobial peptides is a well-conserved feature in evolution; apparently such peptides serve multiple functions in innate immunity of eukaryotes and in microbial interactions. Virtually any group of organisms studied was found to produce entire sets of defense peptides either constitutively or in reponse to a challenge with live bacteria or bacterial cell-wall components. To name but the best studied examples, there are cecropins from insects and tachyplesins of crabs, the magainins from amphibia and the well-defined families of α- and β-defensins and protegrins in all mammals (for a review see 14); finally, various groups of plant peptides were found, e.g. the thionins in monocotyledonous plants (21).

Among the prokaryotes, production of substances that can antagonize competitors is widespread, and antibiotic peptides are one of the major tools in such an ecological context. The variety of peptide structures in this area is immense because particularly Gram-positive bacteria can overcome restrictions of the genetic code by using nonribosomal synthesis mechanisms and rather exotic amino acids as building blocks (77, 168). However, ribosomally made antibiotic peptides and proteins are also frequently produced and usually designated as bacteriocins. While generally bacteriocins of Gram-negative bacteria are large domain–structured protein toxins (e.g. colicins) with receptor-mediated narrow-spectrum activity, the bacteriocins of Gram-positive bacteria are mostly peptides (63), which share a great deal of similarities with eukaryotic defense peptides. They are of similar size (approximately 20–40 amino acids), mostly cationic and amphiphilic, and derive from prepeptides that contain negatively charged, N-terminal extensions that keep the prepeptides inactive and may serve various functions in biosynthesis and targeting. The majority of the peptide bacteriocins are plain peptides, which may or may not contain cysteines and disulfide bonds (63). However, a significant number of the peptide bacteriocins are subject to unique posttranslational modifications resulting in unusual structural features. The best-studied examples of such modified bacteriocins are microcin B17 (92), produced by *Escherichia coli* strains, and almost 30 peptides from various Gram-positive genera that contain, among other modified residues, the thioether amino acids lanthionine and methyllanthionine, and which, therefore, are collectively designated *lantibiotics* from "*lan*thionine-containing an*tibiotic* peptides." This review summarizes the principal features of lantibiotics and focuses on recent findings in biosynthesis, biological activities, and engineering of lantibiotics. More detailed coverage of the entire area or of certain aspects such as the chemistry or applications may be found in other reviews (56, 133), in a review collection (79), or dedicated books (62, 67).

STRUCTURAL ASPECTS

Unusual Residues

Lantibiotics for which sufficient structural information is available consist of a minimum of 19 and a maximum of 38 amino acids. The percentage of residues involved in modifications ranges from 24% in lactocin S to 47% in the cinnamycin group of lantibiotics. This unprecedented degree of posttranslational peptide modification is mostly based on just three amino acids, the hydroxyl amino acids Ser and Thr and the sulfhydryl analog Cys. In only a few cases are Lys, Asp, and Ile residues also found in a modified form. The key reaction is the selective dehydration of Ser and Thr in the propeptide segment of the lantibiotic prepeptide. Although the enzymes that catalyze this reaction have been identified (see below), the biochemistry of the water elimination remains to be elucidated. The resulting α,β-unsaturated amino acids didehydroalanine (Dha, from Ser) and didehydrobutyrine (Dhb, from Thr) are then targets for nucleophilic additions; however, the number of suitable reaction partners is usually smaller than the number of didehydro residues and, as a result, almost all mature lantibiotics contain one or more didehydroamino acids.

The most common addition reaction to the double bonds of Dha and Dhb is that of SH groups of suitably positioned Cys residues; this reaction results in the formation of acid stable thioethers and of the characteristic amino acids lanthionine (Lan) and 3-methyllanthionine (MeLan). In several lantibiotics Cys residues are located at the C-terminus which may be oxidized and decarboxylated before the addition to Dha and Dhb, resulting in 2-aminovinyl-D-cysteine (AviCys) in epidermin (89) and cypemycin (101) or 2-aminovinyl-3-methyl-D-cysteine (AviMeCys) in mersacidin (78), respectively. ϵ-amino groups of Lys may also add to Dha as found in cinnamycin and related lantibiotics; here, a lysinoalanine (LysN-Ala) bridge connects the C-terminus with residue 6 and contributes to the globular structure of these lantibiotics. In one particular case, lactocin S (147), the double bond in Dha is modified via addition of H_2 to form D-Ala; this provides an elegant mechanism for transformation of L-Ser into D-Ala.

Dha and Dhb, when in position 1 of the mature peptide, i.e. when N-terminally exposed after proteolytic removal of the leader peptide segment, are not stable. After addition of a water molecule they spontaneously deaminate to 2-oxo-propionyl and 2-oxobutyryl residues, respectively, as found in lactocin S and Pep5 (70, 147). In epilancin K7, and presumably in epicidin 280 (52), the oxopropionyl is reduced to the corresponding 2-hydroxypropionyl residue (153). It is obvious that these residues block Edman degradation; cypemycin was also reported to be blocked, although in this case the N-terminus appears to be twofold methylated (101). Another unusual modification occurs in the cinnamycin subgroup at the Asp15 residue which is hydroxylated to *erythro*-3-hydroxy aspartic acid. Finally, *allo*isoleucine (aIle) was found in cypemycin; currently, there are no clues as to how these rare modifications are introduced.

Gram-negative bacteria have not been found to produce lantibiotics; however, the heavily modified microcin B17 may well be regarded as an analog. Again, the Ser and Cys residues undergo modifications, although in this case it is a peptide backbone rather than a side-chain modification; the sulfhydryl and hydroxyl groups add to the backbone carbonyl of adjacent Gly residues, which results in the formation of oxazole and thiazole rings (3, 92).

A Survey of Peptides and Structures

Generally, structure elucidation of lantibiotics poses serious problems and it cannot be excluded that further modified amino acids could be detected. However, such residues escape routine amino acid analysis and peptide sequencing and cannot be deduced from gene sequences that can be obtained by standard procedures. Also, N-terminal modifications and didehydro residues create analytical problems by blocking Edman degradation. With this respect, the method of Meyer et al (100) is particularly useful; it allows Edman sequencing after subjecting the lantibiotics to a series of chemical modifications. Although this procedure does not reveal the exact nature of a modified residue, it gives information on its position and usually enables the construction of a gene probe. Both the partial peptide sequence and the gene sequence then allow many structural conclusions and may leave open only the bridging pattern. In this respect it is helpful when related peptides are known and simple proteolytic fragmentation may give hints as to which residues are bridged. Thus, Table 1,

Table 1 Lantibiotics characterized as of November 1997

Lantibiotic	Producing species	Molecular mass (Da)	% Modified residues[a]	Total rings	Lan	MeLan	Dha	Dhb	Others
Type-A Lantibiotics									
Nisin A	*Lactococcus lactis*	3353	38	5	1	4	2	1	0
Nisin Z	*Lactococcus lactis*	3330	38	5	1	4	2	1	0
Subtilin	*Bacillus subtilis*	3317	40	5	1	4	2	1	0
Epidermin	*Staphylococcus epidermidis*	2164	41	4	2	1	0	1	AviCys
Gallidermin	*Staphylococcus gallinarum*	2164	41	4	2	1	0	1	AviCys
[Val1, Leu6]-epidermin	*Staphylococcus epidermidis*	2151	41	4	2	1	0	1	AviCys
Mutacin B-Ny266	*Streptococcus mutans*	2270	41	4	2	1	1	1	AviCys
Pep5	*Staphylococcus epidermidis*	3488	26	3	2	1	0	2	Oxobutyryl
Epicidin 280	*Staphylococcus epidermidis*	3133	27	3	1	2	0	1	Oxo/Hydroxypropionyl
Epilancin K7	*Staphylococcus epidermidis*	3032	32	3	2	1	2	2	Hydroxypropionyl
Lactocin S	*Lactobacillus sake*	3764	24	2	2	0	0	1	3 D-Ala, Oxopropionyl
SA-FF22	*Streptococcus pyogenes*	2795	27	3	1	2	0	1	0
Lacticin 481	*Lactococcus lactis*	2901	26	3	2	1	0	1	0
Salivaricin A	*Streptococcus salivarius*	2315	27	3	1	2	0	0	0
[Lys2, Phe7]-Salivaricin A	*Streptococcus salivarius*	2321	27	3	1	2	0	0	0
Variacin	*Micrococcus varians*	2658	28	3	2	1	0	1	0
Cypemycin	*Streptomyces* ssp	2094	41	1	0	0	0	4	2 alle, AviCys, Me_2N-Ala
Type-B Lantibiotics									
Cinnamycin	*Streptomyces cinnamoneus*	2042	47	4	1	2	0	0	HyAsp, LysN-Ala
Duramycin	*Streptomyces cinnamoneus*	2014	47	4	1	2	0	0	HyAsp, LysN-Ala
Duramycin B	*Streptoverticillium* ssp	1951	47	4	1	2	0	0	HyAsp, LysN-Ala
Duramycin C	*Streptomyces griseoluteus*	2008	47	4	1	2	0	0	HyAsp, LysN-Ala
Ancovenin	*Streptomyces* ssp	1959	37	3	1	2	1	0	0
Mersacidin	*Bacillus* ssp	1825	42	4	0	3	1	0	AviMeCys
Actagardine	*Actinoplanes* ssp	1890	45	4	1	2	0	0	MeLan-Sulfoxide
Structure incomplete									
Cytolysin L1	*Enterococcus faecalis*	4164	NK	NK	Present	Present	Present	Present	?
Cytolysin L2	*Enterococcus faecalis*	2631	NK	NK	Present	Present	Present	Present	?
Mutacin T8	*Streptococcus mutans*	3245	NK	NK	Present	Present	Present	Present	?
Carnocin UI49	*Carnobacterium piscicola*	4635	NK	NK	Present	Present	Present	Present	?

NK: not known; Present: conclusive evidence for the presence of thioether amino acids and didehydro residues available; [a]Bridge forming amino acids (Lan, MeLan, AviCys, MeAviCys and LysAla) are counted as two amino acids. For relevant literature see text (when published later than 1995) or (62, 133) when published earlier.

which contains a list of all currently known lantibiotics and a summary of their structural features, lists [Val1, Leu6]-epidermin (60), mutacin B-Ny266 (105), epicidin 280 (52), salivaricin A (128), [Lys2, Phe7]-salivaricin A (144), and variacin (117) with their related peptides, although experimental proof for the structures is missing.

According to a proposal by Jung (66), the lantibiotics are grouped into type-A and type-B peptides. These categories are based on structural and functional aspects and take into account that some lantibiotics are elongated, flexible amphiphiles that form pores in bacterial membranes, while others are globular, conformationally defined peptides that inhibit enzyme functions. Table 1 maintains the type-A and -B scheme, although with the characterization of a significant number of new lantibiotics with intermediate features, categorization becomes more and more difficult. However, groups of related peptides can be clearly distinguished.

The prototype lantibiotic and best studied type-A peptide is certainly nisin, which was discovered already in the 1940s. Despite an early report on the presence of lanthionine in this peptide (7) and numerous subsequent studies by Hurst and associates (reviewed in 56), it was not before 1970 that Gross and coworkers published the complete structure of nisin (48; Figure 1) and that of the related peptide subtilin (47). Both share an identical ring pattern and about 60% sequence similarity; nisin Z, [Asn27]-nisin, is a frequently occurring natural variant (106). Although epidermin and its variants [Leu6]-epidermin (gallidermin), [Val1, Leu6]-epidermin and [Phe1, Lys2, Trp4, Dha5, Phe6]-epidermin (mutacin B-Ny266) are much shorter and completely different at the C-terminus, their 12 N-terminal residues are highly related to nisin and subtilin. In this entire subgroup, the setup of the first two rings has been completely conserved and the sequence similarity of residues 1–7 is higher even between the epidermin-variant mutacin B-Ny266 and subtilin than between the mutacin and epidermin itself.

Another group of related lantibiotics is formed by Pep5, epicidin 280, and epilancin K7. These peptides are strongly charged (between six and eight positively charged residues), contain only three rings and are N-terminally blocked by oxo-(or hydroxy-)propionyl or -butyryl groups. The recently discovered epicidin 280 (52; C Heidrich, U Pag, M Josten, J Metzger, RW Jack, G Bierbaum,

---→

Figure 1 Selected structures of representative lantibiotics. Nisin (*A*), epidermin (*B*), Pep5 (*C*), and lactocin S (*D*) are typical elongated flexible peptides. Lacticin 481 (*E*) represents a group with a crossbridged C-terminus and an unbridged N-terminal part. The type-B peptides mersacidin (*F*), actagardine (*G*), and cinnamycin (*H*) are conformationally well-defined, globular peptides (see text for details).

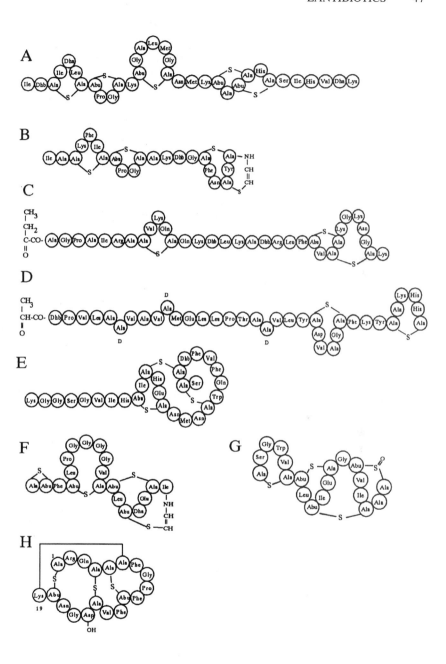

G Jung, HG Sahl, submitted for publication) appears like a natural variant of Pep5 in which several amino acid exchanges and sequence inversions occur, as well as a four-residue deletion in the central flexible segment. The C-terminal double ring system of epilancin K7 is similar to that of nisin and subtilin, with one ring in epilancin and subtilin even being identical. Like this group of lantibiotics, lactocin S has comparatively few thioethers and an N-terminal oxopropionyl; however, it has a net charge of -1 at pH 7, and there is no sequence similarity to the Pep5-like peptides; as a unique feature, lactocin S contains D-Ala.

A structurally homogenous subgroup comprises streptococcin A-FF22 (SA-FF22), lacticin 481 (also described as lactococcin DR; 124), salivaricin A and its [Lys2, Phe7]-variant, as well as variacin. This group apparently shares the same bridging pattern, although this has only been proven for lacticin 481 (156) and SA-FF22 (RW Jack, personal communication). Like lactocin S, they have little or no net charge at neutral pH, but contain His residues that could become protonated at slightly acidic pH. Their grouping as type-A peptides is open to discussion as the three intertwined rings cover the C-terminal two-thirds of the molecules and leave only some 6-8 N-terminal residues unbridged; this certainly induces a rather globular conformation that is not seen with the typical elongated cationic type-A peptides of the nisin-like lantibiotics.

Cypemycin is a unique peptide (101) that contains only one Cys, and hence only one thioether; the Cys residue is at the C-terminus and oxidatively decarboxylated as in epidermin and mersacidin. Thus, four Dhb residues in the molecule do not have a reaction partner, leaving most of the peptide in an unbridged and most likely flexible structure. Otherwise, the peptide is in the same size range as the cinnamycin group of peptides and is also produced by a *Streptomyces* strain that was so far found to produce only type-B lantibiotics.

The duramycins, cinnamycin and ancovenin, form a group of natural variants (i.e. they have an identical bridging pattern and few conservative amino acid exchanges) of which cinnamycin was isolated as early as 1958 (5). Difficulties in structure determination and several reisolations under different names (for details see 65, 66) may cause some confusion when reading literature published before 1990. However, Fredenhagen et al (40) introduced structural revisions for some peptides and proposed the consistent nomenclature used in Table 1. The cinnamycin group is characterized by a three thioether-based bridging pattern and all peptides, except ancovenin, have an additional LysN-Ala bridge connecting the C-terminus with Dha in position 6. This extraordinary degree of crossbridging in a head-to-tail direction has severe implications on the spatial structure of these peptides and seems also responsible for the pronounced protease stability of the type-B peptides. In contrast to the type-A lantibiotics, bridge formation also occurs from N-terminally positioned cysteines to C-terminally located didehydro residues.

Jung (66) did not include mersacidin and actagardine in the A and B categories because they are neither cationic, elongated pore forming lantibiotics, nor do they have the cinnamycin-like head-to-tail bridging. However, their defined spatial structure and the overall similarity of the mersacidin and cinnamycin prepeptides (e.g. the unusual size of the leader segment) clearly show their relationship. Actagardine was also described as gardimycin (66); its NMR-derived primary structure as originally published by Kettenring et al (71) required substantial revisions, (167) after which the significant similarity to mersacidin (78) became obvious. Although the overall bridging pattern is quite different, one of the four rings is almost entirely conserved in both peptides (Figure 1). Mersacidin contains an unusually small ring that just connects Cys1 and Dhb2, as well as an epidermin-analogous C-terminus with the oxidized and decarboxylated Cys adding to, in this case, a Dhb residue to form an AviMeCys bridge.

For several lantibiotics, the available information on structures is incomplete and related peptides that could give clues for a bridging pattern are not known. This group includes the cytolysins from enterococci, although their structural genes and the entire biosynthesis gene cluster have been sequenced and well studied. The cytolysins are remarkable in many ways. They are the only lantibiotics with cytotoxic activity; very early, they were identified as hemolysin, a major pathogenicity factor of enterococci that also can kill bacteria; only recently they were found to contain Lan/MeLan (15). So far, they represent the only two-component lantibiotic in which two structurally different peptides have to interact in order to kill bacteria. Such two-peptide systems have been already described for unmodified bacteriocins, e.g. lactococcin G (108), and it may well be that such systems are also more widespread among the lantibiotics: After complete purification, staphylococcin C55 turned out to be a small peptide bacteriocin that contains Lan/MeLan, and seems to consist of two synergistically active peptides (D Navaratna and JR Tagg, personal communication).

Another so-far-unique feature of the cytolysins is that their maturation requires two processing steps. The positions of the thioethers (at least 2 in CylA1 and 1 in CylA2) and of several didehydro residues remain unknown. Both prepeptides share significant sequence similarity and CylA2 differs mainly through a substantial deletion in the propeptide part and an N-terminal extension (15). For carnocin UI49 (152) and the mutacin of *S. mutans* T8 (109) only the N-terminal sequences up to the first didehydro residue, the molecular masses and the identification of several thioether amino acids have been published.

Conformations

Meanwhile, the spatial structures of several lantibiotics have been studied. Generally, it proved difficult to obtain crystals and X-ray structures of these peptides.

Thus, all published conformational studies are based on 2D-NMR techniques and molecular dynamics simulation with additional information coming from circular dichroism spectroscopy. In aqueous solution, type-A lantibiotics were found to be more flexible than previously anticipated; defined structural elements were only identified in small rings, particularly in intertwined double ring systems (93, 154). In membrane-mimicking solvents such as trifluoroethanol or DMSO the peptides adopted rod-shaped amphiphilic structures (41, 154) with charged residues and hydrophobic side chains aligning on opposite sides of the molecule; such a conformation is considered a prerequisite for the membrane-depolarizing activity. Structure determinations of nisin in the presence of SDS or DOPC micelles strongly supported this view and gave detailed insight into the interaction of this lantibiotic with model membranes (155, 157).

Type-B lantibiotics are strong amphiphiles as well, but in contrast to the type-A peptides, they have much less conformational freedom as shown for duramycin B and C (165), mersacidin (116), and actagardine (166). Apparently, in the cinnamycin group the head-to-tail bridging scaffold, supported by the additional LysN-Ala bridge, imposes strong conformational constraints. A similar degree of rigidity is found in mersacidin and actagardine, although both have only one double ring system; however, unlike in type-A peptides, virtually all residues are members of at least one ring system. The conserved ring present in both peptides forms a U-shaped cleft in the molecule, which could be important for function.

HOW TO MAKE LANTIBIOTICS

Prepeptides and Modification Reactions

Lantibiotics are synthesized from ribosomally made precursor peptides or "prepeptides," which are encoded by the structural genes *lanA* (the locus symbol *lan* refers collectively to homologous genes of different lantibiotic gene clusters). In comparison to the mature lantibiotics, the LanA prepeptides carry an N-terminal extension, which is called leader peptide (Figure 2). The C-terminal segment of the prepeptide, in which the Ser, Thr, and Cys residues are modified to the rare amino acids, is designated the propeptide part (28). The Lan and MeLan residues are introduced into the prepeptides by a two-step posttranslational modification procedure that was first proposed by Ingram (58, 59). In the first step, the hydroxyl amino acids Ser and Thr are dehydrated to yield Dha and Dhb (163). The thioethers are formed in a subsequent step by an intramolecular Michael addition that involves the thiol groups of neighboring Cys residues and the double bonds of the didehydroamino acids (Figure 2).

In the producer strains of nisin-like lantibiotics (see also Table 2), the lantibiotic modification enzyme LanB is presumably involved in the dehydration

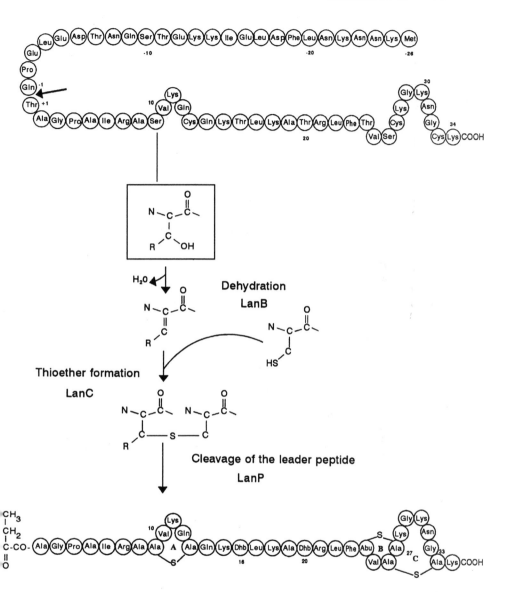

Figure 2 Schematic representation of the posttranslational modification of Pep5 (see text). *Top*, the unmodified Pep5 prepeptide. *Arrow*, the protease cleavage site. *Middle*, the dehydration of a hydroxyl amino acid (R=H for Ser or R=CH₃ for Thr) and the formation of the thioether (R=H for Lan or R=CH₃ for MeLan). *Bottom*, cleavage of the leader peptide and deamination of the N-terminus then result in mature Pep5.

Table 2   Lantibiotics and their biosynthesis machinery

Lantibiotic	Overall charge	Leader peptide		Modification enzymes		Processing and export	
		FNLD-Type	GG-Type	LanB, LanC	LanM	LanP, LanT	LanT(P)[1]
Nisin	+3	+		+		+	
Subtilin	+2	+		+		+	
Pep5	+7	+		+		+	
Epicidin 280	+4	+		+		$+^2$	
Epilancin K7	+5	+				(+)	
Epidermin	+3	+		+		$+^3$	
Lactocin S	−1				+	+	
Lacticin 481	0		+		+		+
Cytolysins	0		$+^4$		+		+
Mersacidin	−1		?		+		(+)

[1] LanT(P): ABC transporter with an additional cysteine protease domain.

[2] LanT is missing in the epicidin gene cluster (52).

[3] EpiT is inactive; a functional LanT is in the gallidermin cluster (114).

[4] The cytolysins are exported by a LanT(P) protein and cleaved at a GG-site, but further processed outside the cells (15).

(+) Incomplete but conclusive sequence information available (153; K Altena & A Guder, unpublished). The lactocin S leader peptide does not align with either type.

reaction and a second enzyme, LanC, catalyzes the ring formation (99). In contrast, in all other producer strains, both reactions are probably performed by a single modification enzyme, LanM.

With those lantibiotics that carry an C-terminal AviCys or an AviMeCys residue, e.g. epidermin or mersacidin, the Cys residue is oxidatively decarboxylated by the LanD enzyme before formation of the thioether (88, 89, 91). In the case of lactocin S, three D-Ala residues are introduced into the prepeptide, presumably by a stereospecific hydration of three Dha residues by a so-far unidentified enzyme (147). The leader peptide is removed proteolytically from the prepeptide after the modification reactions have been completed, thus releasing the mature peptide. This proteolytic processing can take place before, during, or after export. For example, those lantibiotics that carry an oxobutyryl, oxopropionyl, or hydroxypropionyl residue in position +1, such as lactocin S, epilancin K7, Pep5, and epicidin 280, are processed within the cell (52, 99, 145, 153). These unusual residues derive from didehydroamino acids that become N-terminally exposed after removal of the leader peptide. The hydroxypropionyl residue of epilancin K7 is formed by a reduction of an oxopropionyl group (153) and an enzyme (EciO) that could catalyze a similar reaction is apparently encoded in the epicidin 280 biosynthetic gene cluster (52).

In other type-A lantibiotic systems, the protease LanP is located extracellularly, as e.g. in the case of nisin (158) and epidermin (42), and activates the lantibiotic only after export by the ABC-(ATP binding cassette-) transporter LanT. All nisin-like lantibiotics that are processed by a LanP enzyme are distinguished by a conserved cleavage site between leader and propeptide part (66) that contains a positively charged or polar amino acid in position -1, a Pro residue in position -2 (with the exception of epicidin 280), a negatively charged or polar amino acid in position -3, and a hydrophobic amino acid in position -4 (Figure 3). Other lantibiotics, such as the cytolysins and the peptides that are related to streptococcin A-FF22, are processed concomitantly with export, and this reaction is performed by a chimeric protease-transporter. These peptides also possess a characteristic cleavage site with Gly in position -2 and Ala, Ser, or Gly in position -1, which is called the "double glycine" cleavage site (Figure 3; 51).

Prepeptides seem to be elusive products that are quickly processed. Thus, only the prepeptides of the lantibiotic Pep5 have been isolated from the cytoplasm of the producer strain; they were found to be in intermediate stages of modification, containing various numbers of didehydroamino acids but no thioethers in their propeptide region (163). Other prepeptides are now available as fusion proteins (90) or were purified from cells that did not produce an active processing protease (158).

The lantibiotic leader peptides do not show any sequence similarity to signal sequences that direct the peptides to *sec*-dependent transport systems. Generally they vary between 23 (nisin) and 59 (cinnamycin) amino acids in length, and seem to adopt a helical conformation in structure-inducing hydrophobic solvents (4). Type-A leader sequences are generally hydrophilic with a high proportion of charged amino acids and a net negative charge. Three possibilities have been discussed with regard to the function of leader peptides. First, results obtained by site-directed mutagenesis of nisin and Pep5 indicate that the conserved FNLD-motif (Figure 3), which is located between position (-20) and (-15) of the nisin-like lantibiotics, might be involved in interaction with the modifying enzymes and/or transport system. The removal of the Phe residue abolished the excretion of nisin (159) and reduced the yields of Pep5 (107). Second, for those lantibiotics that are processed after export from the cell, e.g. nisin, it was demonstrated that the presence of the leader peptide keeps the mature lantibiotic inactive (158). Third, there is some evidence that the leader peptides of Pep5 and gallidermin interact with their propeptide parts (4), indicating that this interaction could stabilize a hairpin conformation of the prepeptide that may be essential for its interaction with the modification enzymes. Nisin-subtilin chimeras fused in the leader peptide (125), at the protease cleavage site (83) or within the propeptide part (22) were tested for correct modification in nisin or subtilin expression systems; the results indicate that in different modification systems the prepeptides have to meet different specificities.

FNLD-Type:

NisinA	- - - - M S T K D F N L D L V S V S K K - - D S G A S P R
NisinZ	- - - - M S T K D F N L D L V S V S K K - - D S G A S P R
Subtilin	- - - M S K F D D F D L D V V K V S K Q - D S K I T P Q
Epidermin	M E A V K E K N D L F N L D V V K V N A K E S N D S G A E P R
Gallidermin	M E A V K E K N N L F D L E I K K E T S Q - N T D E L E P Q
Pep5	- - - - M E K N N K D L F D L E I K K D N M E - N N E L E A Q
Epicidin 280	- - - - M E N K K D L F D L E N K G V E T Q - K S D L E P Q
Epilancin K7	- - - - M N N S L F D L E I K K D N M E - N N E L E A Q

GG-Type:

Variacin	- - M T N A - - - - - - - - - F Q A L D E V T D A E L D A I L G G
Lacticin	- M K E Q N S - - - - - - - - F N L L Q E V T E S E E L D L V A G G
Salivaricin	M N A M K N S K D - - - - - - - I L N N A I Q E V V S E K E L M E V A G G
Salivaricin A1	M S F M K E N S K D - - - - - - - I L T N A I Q E V S L E K E L D I Q A G A
SA-FF22	- - M E K N L S - - - - - - - - - V I N S L Q E E L S V E E E M E A I - G A
CytolysinL1	V L N K E N Q E N Y Y S N K L E L V V V P S F E E L S V E E E M E A I Q G S
CytolysinL2	V L N K E N Q E N Y Y S N K L E L V V G P S F E E L S L E E E M E A I Q G S

Cinnamycin	M A S I L Q A S V V D A D F R A A L L E N P A A F G A S A A A L P T P V E A Q D Q A S D F W T K D I A A T E A F A
Mersacidin	M S Q E A I R S W K D P F S R E N S T Q N P A G N - - - - - - - P F S E L K E A Q M D K L V G A G D M E A A - -

Lactocin S	M K T E K K V L D E L S L H A S A K M G A R D V E S S M N A D

Figure 3 Comparison of different types of lantibiotic leader peptides. The FNLD-type, as first described for nisin (159), is typical for the type-A peptides, which are modified by LanB/LanC and processed at a conserved site with Pro in position −2. The GG-("double glycine")-type is associated with LanM proteins and hybrid transporter proteases; these leader segments are also characterized by an excess of Glu and Asp residues. The cinnamycin and mersacidin leader peptides are unusually long and do not significantly align with other leader segments, although elements of the GG-type may be detectable. The lactocin S peptide does not compare with any other known peptide (see also Table 2).

Molecular Genetics of Biosynthesis

BIOSYNTHETIC GENE CLUSTERS The structural genes of the LanA prepeptides have been sequenced for most type-A lantibiotics (2, 20, 46, 52, 57, 69,
106, 115, 117, 128, 136, 137, 147, 153), and for the type-B lantibiotics cinnamycin (68) and mersacidin (8). The genes that are necessary for modification
i.e. *lanB/lanC*, or *lanM* and *lanD*, for proteolytic processing (*lanP*), transport
(*lanT*), producer self-protection, and regulation are found in close proximity
to the structural genes, forming biosynthetic gene clusters (for recent detailed
reviews see 29, 62, 133, 142). Some representative biosynthetic gene clusters
are shown in Figure 4.

The self-protection ("immunity") factors include proteins or peptides (LanI),
that are associated with the membrane, and/or dedicated ABC transporters
(LanEFG). Regulation is usually achieved through two-component regulatory
systems (LanR, LanK) or, in the case of epidermin, through a single protein
(EpiQ) (111). Many biosynthetic gene clusters consist of several transcription
units; the structural gene is often separated from the genes of the biosynthetic
proteins by a weak terminator structure, which allows moderate readthrough,
thus ensuring a high level of transcription of the prepeptide mRNA in comparison to the mRNA encoding the biosynthetic enzymes. Concerning their
composition, the gene clusters fall into two groups (see also Table 2), for
example all nisin-like lantibiotics are processed by separate LanB and LanC
enzymes, whereas this function is performed by a single LanM enzyme in other
lantibiotic subclasses. Also, nisin-like lantibiotics never possess a chimeric
transporter with an associated protease activity that again is found in the other
groups. A protease-associated transporter and a LanM protein are also present
in the mersacidin gene cluster, the only type-B lantibiotic for which sequence
information is available (K Altena, A Guder, unpublished results).

The gene clusters are found on the bacterial chromosome or large plasmids.
Sometimes, transposase genes are present in close proximity (52, 115). The
nisin biosynthetic gene cluster is located on three classes of 70 kb transposons
that also confer sucrose utilization, and most of these transposons can be propagated by conjugation (54, 121). The structural gene of salivaricin, *salA*, has also
been detected by PCR with high frequency in nonproducing strains of *Streptococcus salivarius* and *Streptococcus pyogenes* (144), and similar findings have
been reported for nisin (104).

REGULATION OF BIOSYNTHESIS Many lantibiotics are only expressed during
late exponential or early stationary phase and regulatory genes are present in
several gene clusters. For example, the gene clusters of subtilin (75), nisin (39),
and mersacidin (K Altena & G Bierbaum, unpublished results) contain a pair
of genes encoding two-component regulatory systems. The first component,

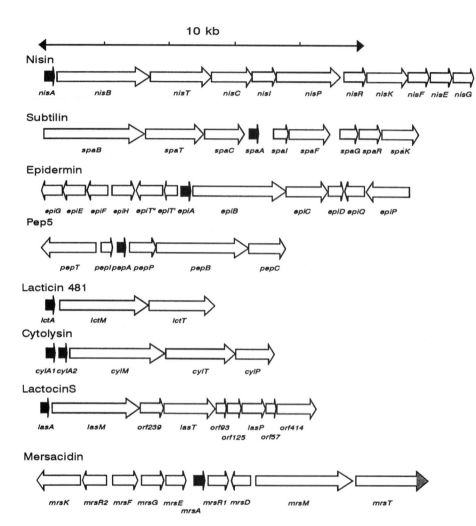

Figure 4   Representative biosynthetic gene clusters of lantibiotics. The structural genes are marked in *black*. The accession numbers of the gene clusters in the databanks are listed in Jack et al (62) and Siezen et al (142). The sequence of the C-terminal part of *mrsT*, which is marked by *shading*, needs to be confirmed (A Guder, personal communication). To avoid confusion we used the systematic gene nomenclature as proposed by de Vos et al (28) and Siezen (142) throughout the review, although it is not consistent with some of the original publications.

a membrane-bound sensor kinase (LanK) of 380–480 residues is thought to autophosphorylate a His residue in its intracellular domain in response to an extracellular signal. This phosphate residue is then transferred to a conserved Asp of an intracellular response regulator protein (LanR) and provokes a conformational change that enables the response regulator to activate transcription. Only in the case of nisin is the signal molecule that triggers transcription of the biosynthetic gene cluster known; here nisin itself autoregulates its own biosynthesis (33, 80). With the exception of *nisRK*, which is transcribed independently of the presence of nisin A, only a very low transcription of the nisin A biosynthetic gene cluster takes place during exponential growth (26). This transcription leads to a slow accumulation of nisin A in the culture supernatant, which after reaching a critical concentration level, triggers transcription of the nisin biosynthetic genes and immunity system by activating the promoters of *nisABTCIP* and *nisFEG* (26). This regulation system of nisin biosynthesis has already been adapted for biotechnological applications. For controlled overexpression of proteins in *Lactococcus*, a system has been developed that depends on the presence of *nisRK*, the *nisA* promoter fused to the gene to be expressed, and low concentrations of nisin as inducer (27). The signals that activate the LanK/LanR systems of subtilin and mersacidin are not yet known. The production of epidermin is regulated by EpiQ, which possesses some similarity to response regulators, mainly in its C-terminal part, but is missing the highly conserved phosphate acceptor Asp residue (135). Nevertheless, it has been conclusively demonstrated that EpiQ binds to an inverted repeat immediately upstream of the −35 region of the structural gene *epiA* and thereby activates transcription (111). Similar inverted repeats (ANAATTACN$_6$GTAATTNT) are found in nearly all promoters of the epidermin biosynthetic gene cluster and the transcription of these genes is increased in the presence of EpiQ (112, 114).

Modification Enzymes

DEHYDRATING AND THIOETHER FORMING ENZYMES   The gene clusters of the nisin-like lantibiotics contain two genes (*lanB, lanC*) encoding proteins that are involved in biosynthesis as shown by inactivation experiments. However, their function could not be deduced from comparison with protein databases. It was therefore concluded that these enzymes most probably catalyze the unprecedented modification reactions, the dehydration of hydroxyl amino acids, and formation of thioethers (1, 38, 76, 150). After inactivation of *pepC* in the Pep5 biosynthetic gene cluster, truncated but dehydrated peptides that still contained Cys residues instead of thioethers were isolated from the culture supernatant, indicating that most probably PepC is the thioether forming enzyme and that, in conclusion, PepB functions in dehydration (99). The LanB proteins possess about 1000 residues and seem to be associated with the membrane, whereas the LanC proteins consist of about 400–450 amino acids in alternating hydrophobic and hydrophilic segments (38, 49, 99, 135). Because LanB and LanC are novel enzymes with interesting biotechnological applications, attempts have been made to study their activity in vitro. EpiB has been expressed in *S. carnosus* and was enriched to 50% of the total protein (113) and EpiC was purified to homogeneity after expression in *E. coli* (85), but in vitro activity tests employing the EpiA prepeptide failed so far with both enzymes. For the nisin and subtilin systems, evidence has been obtained that the LanB and LanC enzymes may form a biosynthetic membrane-associated complex with the transporter LanT (72, 141), which may explain the failure of the in vitro tests with the isolated enzymes.

A single modification enzyme, LanM, of 900 to 1000 amino acids is encoded in the gene clusters of lantibiotics that are not closely related to nisin and inactivation or expression experiments have shown that these proteins are necessary for the biosynthesis of e.g. lactocin S (145, 146), cytolysin (46) or lacticin 481 (124). The C-terminus of the LanM proteins is characterized by sequence similarity to the LanC enzymes. In contrast, the N-terminus does not display any similarity to the LanB proteins, which excludes the possibility that *lanM* originates from a gene fusion of *lanB* and *lanC* (142). Because of the similarity of LanM and LanC it has been proposed that the C-terminus of LanM fulfills the function of LanC and that the N-terminus of the LanM enzymes could be involved in the dehydration reaction, but this hypothesis remains to be proven experimentally.

OXIDATIVE DECARBOXYLASES   The gene clusters of those lantibiotics that possess a C-terminal AviCys or AviMeCys residue (epidermin and mersacidin) are characterized by a further modification enzyme, LanD. The LanD proteins are small enzymes, containing 181 (EpiD) or 194 amino acids (MrsD) (135; K Altena, unpublished results). EpiD was purified to homogeneity (91) and was extensively studied in vitro. The experiments showed that EpiD is involved in the oxidation and decarboxylation of the C-terminal Cys residue of preepidermin (88). During this reaction FMN is reduced and a double bond is formed in the Cys residue. The peptide is then decarboxylated, yielding a (Z)-enethiol compound that most probably converts to an enethiolate anion (86). This residue subsequently reacts with the Dha residue in position 19 to form the C-terminal AviCys residue of epidermin. The substrate specificity of EpiD was tested with heptapeptide libraries: It is located in the last three residues of the propeptide region and the presence of a leader peptide is not necessary (89). The activity of EpiD has also been demonstrated in vivo by expression of EpiD and an affinity-tag-labeled EpiA in *E. coli* (87).

Transport and Protease Functions

The removal of the leader peptide from the modified prepeptide leads to activation of the lantibiotic. For the nisin-like lantibiotics and lactocin S, this reaction is catalyzed by subtilisinlike serine proteases LanP; this event can take place before or after export from the cell by the ABC-transporter LanT. Lantibiotics, which are characterized by the double-Gly cleavage site in their prepeptides (Figure 3), possess LanT transporters with an N-terminal protease domain and are processed and activated concomitantly with export. The only exception to this rule is cytolysin, because here the processing by the chimeric transporter does not lead to activation; after secretion, the truncated prepeptides have to be cleaved a second time, through the action of the specialized serine protease CylP (15).

The lanP genes vary greatly in size, encoding proteins that range from 266 (LasP) (145) to 682 amino acids (NisP) (158). These differences in size are caused by the presence or absence of preprosequences, which direct the enzyme to the *sec*-system and serve as intramolecular chaperones. In addition, NisP has a unique C-terminal extension, which consists of a spacer (100 amino acids) and a C-terminal transmembrane sequence with the Gram-positive cell wall anchor consensus motif (LPXTGX); thus, NisP is most probably coupled to the peptidoglycan (42, 158). EpiP (42) and CylP (15) are active in the culture supernatant, whereas LasP (145), PepP (99) and ElkP (153) are intracellular enzymes as judged from the absence of a *sec*-dependent secretion signal and a prosequence. Because the three-dimensional structure of several subtilisinlike serine proteases has been elucidated, it was possible to characterize the catalytic sites of NisP, EpiP, and CylP through homology modeling. Here, the dominant interaction between substrate and active site is mediated by the residue in position -1 of the prepeptide (15, 143). These results have been confirmed by activity assays with site-directed mutants of prenisin and preepidermin as substrate; the exchanges in position -1 and -4 inhibited processing (42, 159).

The first group of *lanT* genes usually encodes transport proteins of 500 to 600 amino acids that belong to the group-A ABC-transporters. The N-terminal domain of the proteins consists of six membrane spanning helices, whereas the intracellular C-terminal domain contains the ATP-binding site, which is characterized by two highly conserved amino acid sequences (Walker motifs); ATP hydrolysis provides the energy for the export process. The function of several LanT genes has been tested by inactivation experiments. NisT and LasT are essential for export (119, 145). EpiT does not constitute an active protein, since the *epiT* gene contains one frameshift and two deletions. However, introduction of *gdmT*, the transporter gene of the natural variant gallidermin, into the epidermin producer strain increased production yields. Besides, an additional hydrophobic protein, LanH, seems to be necessary for optimal activity of the

GdmT transporter (114). SpaT was not essential for export of a fusion protein, in which the leader of subtilin had been fused to alkaline phosphatase, in *B. subtilis* 168 (61). Similarly, Pep5 was excreted in the absence of PepT, albeit at a reduced level (107) and a *lanT* gene was not found in the epicidin 280 biosynthetic gene cluster (52).

The second group of LanT transport proteins, comprising LctT (124), CylT (45, 51), and MrsT (K Altena, A Guder, unpublished data) contains an N-terminal extension of 100 to 200 amino acids that is characterized by two conserved motifs. Similar proteins are present in many gene clusters of nonlantibiotic bacteriocins (51). Experiments with the N-terminal extension of LagD, the exporter of the nonlantibiotic bacteriocin lactococcin G, have shown that this domain possesses protease activity and cleaves the leader peptide concomitantly with export at a conserved double-Gly cleavage site. Most probably the proteolytic domain is located on the inside of the cytoplasmatic membrane and belongs to the family of cysteine proteases (51). A similar double-Gly cleavage site is present in the SalA and VarA prepeptides and it is expected that similar transporters will be found in their gene clusters (117, 128).

Producer Self Protection

Lantibiotics are bacteriocins, i.e. they are active against strains that are closely related to the producer strain, and therefore, potentially harmful substances for the producer. Protection is mediated by the so-called "immunity" peptides or proteins, LanI, and, in some systems, specialized ABC-transport proteins, LanEFG, which are encoded in two or three separate open reading frames.

Immunity peptides of about 57–69 amino acids are found in the gene clusters of Pep5 (122), epicidin 280 (52), and lactocin S (145), interestingly, those lantibiotics that appear to be proteolytically activated inside the cells. These peptides display an N-terminal hydrophobic and C-terminal hydrophilic segment and, as shown for PepI, are most probably attached to the outside of the cytoplasmic membrane (122). Immunity proteins of 165 (SpaI) and 245 (NisI) amino acids are found in the gene clusters of subtilin (74) and nisin (81). Both proteins possess an N-terminal lipoprotein signal sequence, and NisI was shown to be coupled to the cell membrane (118). Although nisin and subtilin share about 60% of sequence similarity with an identical arrangement of the thioether rings, NisI and SpaI do not show any homologies.

The nisin, subtilin, epidermin and mersacidin gene clusters also contain LanEFG transporters, which consist of two (SpaF and SpaG) (74) or three separately encoded subunits, NisFEG (140), EpiFEG (112) and MrsFGE (K Altena & G Bierbaum, unpublished data). LanF always constitutes the intracellular ATP-binding domain, whereas LanG and LanE represent the membrane-spanning subunits. The only exception is SpaF, where the N-terminus contains the ATP-binding motifs and the C-terminus is a membrane spanning domain. A nonproducing *Lactococcus lactis* strain that carries *nisRKFEG* displays immunity to nisin (37), which confirms results of inactivation experiments (74, 140) and of heterologous expression of EpiFEG (112). The mechanism by which the LanEFG transporters function is not yet clear, but the proteins could act by transporting those molecules that have inserted into the membrane back to the culture supernatant.

BIOLOGICAL ACTIVITIES

Pore Formation

The first clues regarding the mode of action of nisin were obtained by Ramseier (120), who treated *Clostridia* cells and observed leakage of intracellular compounds. Linnett and Strominger (95) reported that nisin interferes with cell wall biosynthesis in in vitro systems and Reisinger et al (123) demonstrated that the lantibiotic forms a complex with the lipid-bound peptidoglycan precursor. The conclusion, however, that nisin kills bacteria through inhibition of cell wall synthesis seems not valid in view of subsequent experiments with intact cells. Nisin and other type-A peptides clearly promote rapid efflux of ions, solutes and small metabolites such as amino acids and nucleotides which rapidly depolarizes the cytoplasmic membrane and leads to an instant stop of all biosynthetic processes (129, 131, 132). Also, killing of cells sets in immediately after addition of the peptides, while inhibitors of cell wall biosynthesis (e.g. mersacidin, see below) usually induce slow lysis, and killing does not occur before the completion of at least one cell cycle. These results strongly suggest that nisin and related peptides kill by disruption of the cytoplasmic membrane and that cell wall biosynthesis inhibition, although prominent in in vitro systems, may not primarily contribute to killing in vivo.

Subsequent experiments were designed to study the molecular mechanisms of the membrane disruption. Results obtained with various physiological and artificial membrane systems (intact bacterial cells and cytoplasmic membrane vesicles, red blood cells, liposomes, micelles, and planar lipid bilayers) strongly indicated that the bactericidal action is based on the formation of short-lived transmembrane pores that are nonselective, oligodynamic, and require energy for formation and opening (for reviews see 6, 130). Generally, the electrical transmembrane potential, as generated by metabolizing bacterial cells, represents the major driving force for activity; however, nisin is also active at a sufficiently high ΔpH (102). To describe such an activity, the intensively studied peptide alamethicin, which forms voltage-dependent pores in a barrel-stave setup (13), was considered the most suitable model system (130, 134). Essentially, such a model for the activity of type-A lantibiotics is still valid, although a number of recent studies revealed important details contributing to an improved model as proposed by Driessen et al (36). The elegant studies of van den Hooven et al (155, 157), who determined the conformation of nisin in the presence of membrane-mimicking micelles, demonstrated that, due to its amphiphilic nature, a peptide molecule not only interacts with the phospholipid head groups via ionic forces, but also inserts with its hydrophobic side into the outer leaflet of a bilayer. Several peptide molecules could then form a pore by moving through the membrane in response to the electrical potential, while remaining surface-bound and carrying the lipids across (103; Figure 5). A similar wedge or "wormhole" model was recently proposed for the frog defensin peptide magainin (98). It is obvious that, in addition to amphiphilicity, a considerable flexibility of the peptides in the central segment is another prerequisite for such an activity; indeed, it was shown through mutational analysis of lantibiotics (e.g. 82) and with synthetic peptides (e.g. 162) that reduction of flexibility in this part of the peptides is detrimental to activity.

A number of recent studies, mainly conducted with wild-type nisin and site-directed mutants, fully agree with such a model and revealed interesting details on the behavior of nisin in various model membrane systems (31, 43); e.g. it became clear that the C-terminal part of the molecule (16) and the overall negative surface charge of the membrane (31, 44, 164) are important for binding and activity.

Although pore formation is generally accepted as the primary mode of killing, and in spite of the rather clear picture that is emerging from the model membrane studies, there are still a number of in vivo effects that remain enigmatic. One such mystery is the striking difference in sensitivity that can be observed even among different strains of just one bacterial species. Again, the membrane composition seems to play a crucial role (161), but there may be more to explain in this respect. Structure-function studies with intact bacterial cells demonstrate that the inactive nisin1-12 fragment is able to antagonize the activity of the lantibiotic (24); it was concluded that nisin may bind specifically to a particular site, which could be blocked by the fragment. In this context it is interesting to come back to the early observation that nisin binds to the cell wall precursors (123); indeed, we have recently found that the lipid-bound peptidoglycan precursors may be helpful, though not essential, in the formation of pores (H Brötz, M Josten, I Wiedemann, U Schneider, F Götz, G Bierbaum, HG Sahl, manuscript in preparation). It is well conceivable that in vivo the interaction of lantibiotics with these and other yet unidentified molecules may be important for membrane disruption and killing.

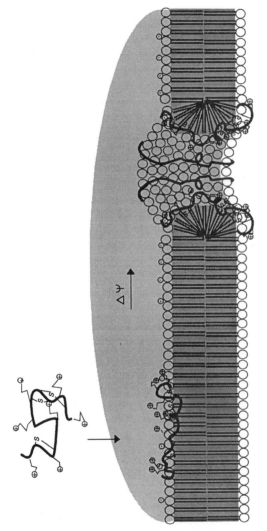

Figure 5   Model for the formation of pores by cationic type-A lantibiotics (36, 103): Peptides are unstructured in aqeous solution but adopt an amphiphilic conformation when binding to membranes. This enables insertion into the outer leaflet of the membrane in a surface-bound fashion. When membranes are sufficiently energized, the peptides move through the membrane, remaining surface-bound and carrying the phospholipids across. Accumulation of several peptides at the same site leads to the formation of transient pores.

Inhibition of Phospholipase and of Peptidoglycan Biosynthesis

Inhibitory activities of cinnamycin-like type-B lantibiotics have been identified in a number of biological test systems, not only in screenings for antimicrobials; such activities include e.g. the inhibition of the angiotensin-converting enzyme by cinnamycin and ancovenin (139) or the immunomodulating effects through inhibition of phospholipase A2, and hence interference with prostaglandin and leucotriene biosynthesis (40). Antibacterial activities were observed with *Bacillus* strains, and effects on membrane functions, on ATP-dependent proton translocation and calcium uptake or on ATPases were described (for reviews see 62, 133). Also, the formation of defined pores in phosphatidylethanolamine-containing planar membranes was reported (138). These effects can be explained on the molecular level by the specific binding of the lantibiotics to phosphatidylethanolamine (25, 40, 55). However, in contrast to type-A peptides, no efflux of solutes from cinnamycin-treated staphylococci was observed (H-G Sahl, unpublished results), so that pore formation may not play a role in vivo with intact bacterial cells.

Apparently, the cinnamycin-like peptides exert their activity through interfering with enzyme activities by blocking the respective substrate. This is also true for the type-B peptides mersacidin and actagardine. Both were shown to inhibit biosynthesis of peptidoglycan (18, 148) and transglycosylation was identified as the target reaction (19). The substrate of this reaction is the lipid-bound cell wall precursor undecaprenyl-pyrophosphoryl-MurNAc(pentapeptide)-GlcNAc, the so-called lipid II, which was recently found to form a high affinity complex with mersacidin (17). In contrast to the glycopeptide vancomycin, which binds to lipid II via the C-terminal D-Ala-D-Ala of the pentapeptide side-chain, complex formation of the lantibiotic probably involves the disaccharide-PP moiety (Figure 6), a new target binding site not used by any current antibacterial drug.

Secondary Effects

Electron microscopic investigations of staphylococcal cells treated with nisin and Pep5 revealed massive cell wall degradation, particularly in the septum area, and it was noticed that incubation of cells with these lantibiotics liberated autolytic enzymes (10). Detailed kinetic studies with isolated cell walls, purified enzymes, and peptides demonstrated that the cationic peptides replace the enzymes from their cell wall (i.e. substrate) intrinsic inhibitors, the teichoic and teichuronic acids, which results in an apparent activation of cell-wall hydrolases and an induction of autolysis (11). Although the antibiotic effect of such an activity certainly sets in more slowly than pore formation, it may well contribute to the overall bactericidal activity of amphiphilic peptides by reducing

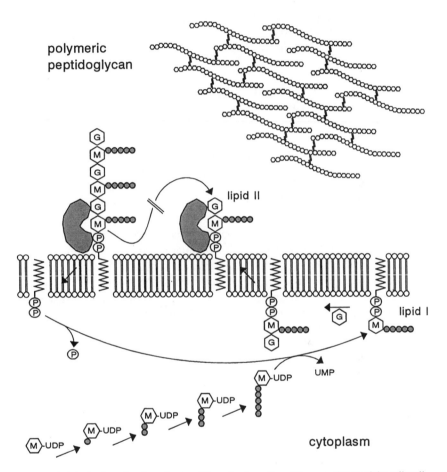

Figure 6 Mechanism of action of mersacidin and actagardine. After completion of the cell wall precursor synthesis, the lipid-bound intermediate (lipid II) is transferred to the outside of the membrane and becomes accessible for the lantibiotics. The peptides (marked by *shading*) tightly bind to lipid II and prevent its incorporation into polymeric peptidoglycan. Binding, which could also occur to a growing peptidoglycan strand, involves at least the disaccharide-PP moiety of lipid II, but does not include interactions with the peptide side chain (17).

the number of survivors that are always observed in killing assays with such peptides.

Another well-documented antibiotic effect is the inhibition of spore outgrowth by subtilin and nisin (23, 97). In this case, the activity can clearly be attributed to the didehydroalanine residue in position 5 of both peptides; it is assumed that the double bond provides a reactive group for an interaction with a spore-associated factor that is essential for outgrowth.

Quorum Sensing

The recent finding by Kuipers et al (80) that nisin biosynthesis is an autoregulated process (see also paragraph on regulation of biosynthesis) provides another answer to the question as to why bacteria may produce antibiotics. Apparently, in connection with a two-component signal transduction system, lantibiotics can contribute to cell-cell communication and serve as signal molecules for cell density, a phenomenon called "quorum sensing." Through low-level production by uninduced cells the lantibiotic accumulates to a threshold concentration (equivalent to a certain cell density); this serves as an input signal that, in the case of nisin, results in strong upregulation of its own biosynthesis through the action of a nisin-binding sensor kinase (NisK) and a corresponding transcription activator (NisR). In principle, a similar pheromone-like function can be envisaged for those lantibiotics that contain a two-component regulatory system in their gene cluster. Similar roles in cell-density–dependent gene expression have been demonstrated for unmodified bacteriocins, the plantaricin bacteriocins of *L. plantarum* C11 (32), and for nonantibiotic peptides such as the pneumococcal competence factor (50).

APPLICATIONS

Peptides and Peptide Engineering

So far, only nisin is applied commercially; it is used as a food additive in over 50 countries, including the EU (food additive E234) and the USA (for reviews see 30, 56). As a preservative in processed cheese, nisin is employed at a concentration of 5–20 μg g^{-1} in combination with heat treatment (85–106°C for 6–10 min), which kills Gram-negative cells, and functions mainly by preventing the outgrowth of clostridial spores. Nisin was also shown to be beneficial in canned vegetables, salad dressings, products containing liquid pasteurized eggs, and natural cheeses. Recent experiments have demonstrated that nisin could have an application in the therapy of peptic ulcer caused by *Helicobacter pylori* and of bovine mastitis (30).

Another lantibiotic with a potential medical application is gallidermin, which is very active against *Propionibacterium acnes* and could be effective in a topical formulation against acne (65). The type-B lantibiotic mersacidin shows good in vivo activity in experimental infections of mice with methicillin-resistant *Staphylococcus aureus* (MRSA), which pose an ever growing challenge to antimicrobial chemotherapy (94). Effective treatment of infections with MRSA is currently only possible with the glycopeptide antibiotics vancomycin and teicoplanin; however, low-level resistant strains seem to be emerging and therapeutic failure of vancomycin has been reported in 1997 (53). In this context, mersacidin and the closely related lantibiotic actagardine with their novel mode of action could become the lead structures for a new class of antibiotic substances.

Lantibiotics possess structural genes and are principally amenable to the alteration of their structure by site-directed mutagenesis. Because of the post-translational modifications that are introduced after ribosomal biosynthesis, the mutated genes have to be expressed in the presence of the modification apparatus, i.e. in most cases in the wild-type producer or a closely related strain. Different expression systems have been developed for nisin in *Lactococcus lactis* subspec. *lactis* (34, 35, 81, 84), for epidermin in *S. carnosus* (110), for subtilin in *Bacillus subtilis* 168 (96), and for Pep5 in *S. epidermidis* 5 Pep5$^-$ (9). Employing these systems it was even possible to introduce modified residues, as for example didehydroamino acids into nisin (84), gallidermin (110), and Pep5, or to construct a novel, fourth ring structure in Pep5 (12).

On the other hand, the limitations of these systems were discovered. Some mutated peptides, especially those that affect the formation of the ring structures in epidermin (S19A epidermin) and nisin (T23A nisin) are not produced (82, 110), and mutations in the neighboring position of a hydroxyl amino acid may prevent the dehydration reaction (160).

Site-directed mutagenesis has been used extensively in basic research to investigate the structure-function relationship of lantibiotics and to identify the functional elements necessary for activity. The role of the modified residues was probed by removing single modified amino acids from the peptides. From these studies it has become clear that the ring structures stabilize the antibacterial conformation of lantibiotics and, at least in the case of Pep5, protect them against the action of host proteases by imposing conformational restraints upon the peptides (12, 82). In contrast, no general function could be attributed to the presence of the didehydro residues. The Dha in position 5 in nisin and subtilin seems to mediate the sporicidal action of the peptides; upon exchange of this residue for Ala any sporicidal effect of subtilin and nisin is lost (23, 97). On the other hand, Dha residues are intrinsically instable and addition of a water molecule to the double bond may lead eventually to cleavage of the peptide chain. Especially in subtilin, the degradation of the Dha residue in position 5, which is catalyzed by the neighboring Glu4 residue, is the reason for the rapid loss of activity at room temperature (96). Similarly, the degradation products of nisin include [2-hydroxy-Ala5]nisin, [Ile4-amide, pyruvyl-Leu6]des-Dha5-nisin and nisin (1-32)-peptide amide, which result from water addition to the Dha residues in position 5 and 33 (93, 127). In contrast, the Dhb residues in the central part of Pep5 seem to stabilize the three-dimensional conformation of this peptide, and an exchange for Ala leads to a rather drastic loss of antibacterial activity (12), whereas Dha5 and Dha33 in nisin can be exchanged for Ala nearly without effect on the antibacterial (pore-forming) activity (34). The flexible hinge region in the central part of the nisin-like lantibiotics is also necessary for activity as shown for nisin, epidermin/gallidermin and Pep5 (12, 82, 110). A few mutant peptides have been employed in detailed mode of action studies (43).

For biotechnological or medical applications, the optimization of lantibiotics is the most interesting field of peptide engineering. It was possible to increase the stability of subtilin by exchanging Glu4 for Ile (96), to engineer nisin peptides with improved stability (Dha5Dhb nisin Z) or solubility (N27K nisin Z, H31K nisin Z) (126) or to enhance the stability of gallidermin against trypsin (Dhb14P gallidermin, A12L gallidermin) (110), and of Pep5 against chymotrypsin (A19C Pep5) (12). T2S nisin Z, L6V gallidermin, and M17Q/G18T nisin Z show an enhanced antibacterial activity with at least some indicator strains (82, 84, 110). The engineering of peptides with increased activity has proven to be the most difficult part of site-directed mutagenesis and, so far, no peptide has been created that is more active against all indicator strains tested. The reason for this phenomenon may be in the rather undefined events taking place during pore-formation by nisin-like lantibiotics: e.g. the cytoplasmic membrane varies from species to species with regard to phospholipid composition, and especially the amount of negatively charged lipids in the membrane influences the action of nisin decisively (16). In addition, as already discussed above, pore formation is a relatively poorly characterized process that consists of several steps, which comprise binding of the peptides to the membrane, possible aggregation of monomers, passage through the membrane and opening of the pore. Therefore, an increased ability to disrupt artificial membranes may be balanced by the decreased binding kinetics of an engineered peptide. In K12L nisin A, for example, these two properties result in a net antibacterial activity that is comparable to that of wild-type nisin (34, 43). Additional properties of the indicator strain, such as the formation of slime capsules, of an active autolytic system (11), the production of yet another lantibiotic (151) or the presence of resistance genes (37) also modify the antibacterial activity of a given peptide. Such factors complicate rational optimization of the peptides by site-directed mutagenesis, and a random mutagenesis, system by PCR in the presence of dITP was developed for nondirected improvement of nisin (149).

Biosynthetic Enzymes

The above experiments have shown that the introduction of novel modified residues into lantibiotics by site-directed mutagenesis is possible, indicating that the modifying systems are not restricted to the biosynthesis of wild-type thioether bridges but that their substrate specificity is more relaxed. To introduce didehydroamino acids and thioether bridges into a great variety of peptides or proteins, it has been proposed to chemically synthesize the substrates for the lantibiotic modification machinery and then to transform these peptides in vitro with the isolated enzymes. This method would also solve problems associated with the construction of expression systems (35, 96), or with insufficient producer self protection against peptides with increased activity (82) and exclude the destructive action of host proteases that degrade peptides that are not stabilized by ring structures (12). The immense potential of such an in vitro approach has already been demonstrated by experiments that were performed using EpiD and peptide libraries which resulted in the introduction of an enethiolate group into a great variety of peptides (89). In contrast to these promising results with EpiD, the in vitro experiments that employed the EpiB and EpiC enzymes have been disappointing so far (85, 113). However, the overwhelming majority of lantibiotic modification enzymes has not yet been tested in vitro, and especially in the case of the LanM enzymes, which presumably catalyze both modification reactions, dehydration and thioether formation, further research is warranted and necessary. A possible alternative approach that represents a system positioned halfway between in vitro processing of chemically synthesized peptides and site-directed mutagenesis, comprises the coordinated expression in *E. coli* of the peptide to be modified and the enzymes needed for the modification reaction. The experiment was successfully conducted with affinity-tag-labeled EpiA and EpiD (87) and could be useful for further research on and applications of lantibiotic modification enzymes.

RELATIONSHIPS: LANTIBIOTICS, UNMODIFIED BACTERIOCINS, PEPTIDE PHEROMONES

The type-A and type-B categories introduced by Jung (66) are based on structural and functional aspects of the peptides and do not necessarily reflect evolutionary relationships. In Table 1, subgroups of peptides are compiled within the type-A and -B peptide groups that are based on (besides general features and ring patterns) primary sequence similarities in both the propeptide and—as far as information is available—leader peptide segments. These subgroups are certainly related in an evolutionary sense. However, it is interesting to note that some structural elements are common and occur in several subgroups, such as the oxidized C-terminal Cys residue and the resulting Avi(Me)Cys rings. In epilancin K7, the C-terminal double ring system is almost identical to the subtilin and nisin rings, whereas the N-terminus of epilancin is completely different. It remains to be studied whether this observation indicates that during evolution certain building blocks have been horizontally moved and rearranged to yield new lantibiotics (similar to the construction of novel enzymes through mixing and matching of functional protein domains), or whether it is merely a consequence of substrate specificities of the biosynthetic enzymes that may only handle certain amino acids in a structurally defined and restricted environment with respect to the choice of neighboring amino acids.

Interesting evolutionary relationships between lantibiotics are revealed when the entire biosynthetic gene clusters are taken into consideration. Two groups can be clearly distinguished (Table 2) that only partially overlap Jung's type-A and type-B categories (66). The first group comprises the strongly cationic peptides of the nisin- and Pep5-like lantibiotics that are modified by LanB/LanC enzymes, processed by LanP, and exported by LanT proteins. The second group, for which unfortunately much less information is available, includes peptides with little or no net charge such as the lacticin 481-like peptides, mersacidin, and the enterococcal cytolysins; these lantibiotics are modified by LanM proteins and exported and processed by one hybrid transporter with an additional cysteine protease domain. Obviously, it is very relevant in this context (and may answer the question as to what the primary function of the leader peptide is) that the different biosynthetic machineries coincide with different conserved leader peptide types (Figure 3), the FNLD-type and the GG-type ("double Gly") leader peptides. Lactocin S seems to be in an intermediate position: it is modified by a LanM protein but has regular *lanP* and *lanT* genes in its gene cluster; its leader peptide segment does not align with both the FNLD-and GG-type leader peptides.

Generally, the principal setup of the biosynthetic gene clusters of lantibiotics is very similar to that of unmodified bacteriocins (63). Particularly, the GG-leader-containing lantibiotics share with a number of unmodified bacteriocins not only the same type of leader peptide, but also the hybrid protease transporter (51). Indeed, the striking similarity leads us to assume that during evolution the capability of producing a lanthionine-containing (rather than an unmodified) bacteriocin was gained (or lost) by receiving (or losing) a *lanM* gene.

Another aspect for possible evolutionary relationships emerges from the signalling function of nisin. There are peptide pheromones such as the pneumococcal competence factor that are produced as prepeptides with a GG-type leader peptide; these peptides are also transported and processed by a hybrid protease-transporter (for review see 73), although other routes of activation for pheromones involved in cell density monitoring exist, e.g. for the staphylococcal AgrD-derived modified octapeptide (64).

Apparently, intimate associations between pheromone activities and antibiotic functions occur frequently among bacteria, although the degree of association may vary considerably: While nisin carries both functions, in the plantaricin C11 regulon one of the amphiphilic peptides serves only as the signal for the upregulation of those genes that code for the antibiotic peptides (32); in the staphylococcal Agr-system, the pheromone has only an indirect antagonizing activity in that it binds in a nonfunctional way to the sensor kinase of other strains, thus blocking their signaling pathways (64); finally, the competence factor only serves in cell-signaling, and no antibacterial effect has been detected so far. It is the striking similarity in the organization of the biosynthesis and the functional interrelationship that suggests evolutionary links for these groups of peptides. It seems justified to consider these peptides as ancestral forms of both eukaryotic peptide hormones and innate immunity systems, of which so many have been discovered during the last decade (14). The advantages for bacteria of combining signaling and antimicrobial functions in one molecule or one pathway are obvious; future research will show what benefits can be obtained from such peptides and their biosynthesis machinery regarding the development of novel antimicrobials for biomedical and agro-food applications.

ACKNOWLEDGMENTS

We greatly appreciate continuous financial support by the Deutsche Forschungsgemeinschaft (several grants to both of us), by the Bundesministerium für Forschung und Technologie (grants 01KI9404 and 01KI9705/8), and by the BONFOR programme of the Medizinische Einrichtungen, University of Bonn. We would like to thank all members of our research group for excellent contributions, strong teamwork, and a stimulating research atmosphere; H Brötz and E Molitor are acknowledged for providing Figures 3 and 6.

> Visit the *Annual Reviews home page* at
> http://www.AnnualReviews.org.

Literature Cited

1. Augustin J, Rosenstein R, Wieland B, Schneider U, Schnell N, et al. 1992. Genetic analysis of epidermin biosynthetic genes and epidermin-negative mutants of *Staphylococcus epidermidis. Eur. J. Biochem.* 204:1149–54
2. Banerjee S, Hansen JN. 1988. Structure and expression of a gene encoding the precursor of subtilin, a small protein antibiotic. *J. Biol. Chem.* 263:9508–14
3. Bayer A, Freund S, Jung G. 1995. Post-translational heterocyclic backbone modifications in the 43-peptide antibiotic microcin B17: structure elucidation and NMR study of a 13C, 15N-labelled gyrase inhibitor. *Eur. J. Biochem.* 234:414–26
4. Beck-Sickinger AG, Jung G. 1991. Synthesis and conformational analysis of

lantibiotic leader-, pro- and pre-peptides. In *Nisin and Novel Lantibiotics*, ed. G Jung, H-G Sahl, pp. 218–30. Leiden: Escom. 490 pp.

5. Benedict RG, Dvonch W, Shotwell OL, Pridham TG, Lindenfelser LA. 1952. Cinnamycin, an antibiotic from *Streptomyces cinnamoneous nov. sp. Antibiot. Chemother.* 2:591–94

6. Benz R, Jung G, Sahl H-G. 1991. Mechanism of channel formation by lantibiotics in black lipid membranes. In *Nisin and Novel Lantibiotics*, ed. G Jung, H-G Sahl, pp. 359–72. Leiden: Escom. 490 pp.

7. Berridge NJ, Newton GGF, Abraham EP. 1952. Purification and nature of the antibiotic nisin. *Biochem. J.* 52:529–35

8. Bierbaum G, Brötz H, Koller K-P, Sahl H-G. 1995. Cloning, sequencing and production of the lantibiotic mersacidin. *FEMS Microbiol. Lett.* 127:121–26

9. Bierbaum G, Reis M, Szekat C, Sahl H-G. 1994. Construction of an expression system for engineering of the lantibiotic Pep5. *Appl. Environ. Microbiol.* 60:4332–38

10. Bierbaum G, Sahl H-G. 1985. Induction of autolysis of staphylococci by the basic peptide antibiotics Pep5 and nisin and their influence on the activity of autolytic enzymes. *Arch. Microbiol.* 141:249–54

11. Bierbaum G, Sahl H-G. 1987. Autolytic system of *Staphylococcus simulans*: influence of cationic peptides on activity of N-Acetylmuramoyl-L-alanine amidase. *J. Bacteriol.* 169:5452–58

12. Bierbaum G, Szekat C, Josten M, Heidrich C, Kempter C, et al. 1996. Engineering of a novel thioether bridge and role of modified residues in the lantibiotic Pep5. *Appl. Environ. Microbiol.* 62:385–92

13. Boheim G, Hanke W, Jung G. 1983. Alamethicin pore formation: voltage-dependent flip-flop of α-helix dipoles. *Biophys. Struct. Mech.* 9:181–91

14. Boman HG. 1995. Peptide antibiotics and their role in innate immunity. *Annu. Rev. Immunol.* 13:61–92

15. Booth MC, Bogie CP, Sahl H-G, Siezen RJ, Hatter KL, Gilmore MS. 1996. Structural analysis and proteolytic activation of *Enterococcus faecalis* cytolysin, a novel lantibiotic. *Mol. Microbiol.* 21:1175–84

16. Breukink E, van Kraaij C, Demel RA, Siezen RJ, Kuipers OP, de Kruijff B. 1997. The C-terminal region of nisin is responsible for the initial interaction of nisin with the target membrane. *Biochemistry* 36:6968–76

17. Brötz H, Bierbaum G, Leopold K, Reynolds PE, Sahl H-G. 1998. The lantibiotic mersacidin inhibits peptidoglycan synthesis by targeting lipid II. *Antimicrob. Agents Chemother.* 42:154–60

18. Brötz H, Bierbaum G, Markus A, Molitor E, Sahl H-G. 1995. Mode of action of the lantibiotic mersacidin—inhibition of peptidoglycan synthesis via a novel mechanism? *Antimicrob. Agents Chemother.* 39:714–19

19. Brötz H, Bierbaum G, Reynolds PE, Sahl H-G. 1997. The lantibiotic mersacidin inhibits peptidoglycan biosynthesis at the level of transglycosylation. *Eur. J. Biochem.* 246:193–99

20. Buchmann GW, Banerjee S, Hansen JN. 1988. Structure, expression, and evolution of a gene encoding the precursor of nisin, a small protein antibiotic. *J. Biol. Chem.* 263:16260–66

21. Cammue BPA, de Bolle MFC, Schoofs HME, Terras FRG, Thevissen K, et al. 1994. Gene-encoded antimicrobial peptides from plants. In *Antimicrobial Peptides*, ed. HG Boman, J Marsh, JA Goode, Ciba Found. Symp. 186:91–101. Chichester: Wiley

22. Chakicherla A, Hansen JN. 1995. Role of the leader and structural regions of prelantibiotic peptides as assessed by expressing nisin-subtilin chimeras in *Bacillus subtilis* 168, and characterization of their physical, chemical, and antimicrobial properties. *J. Biol. Chem.* 270:23533–39

23. Chan WC, Dodd HM, Horn N, Maclean K, Lian L-Y, et al. 1996. Structure-activity relationships in the peptide antibiotic nisin: role of dehydroalanine 5. *Appl. Environ. Microbiol.* 62:2966–69

24. Chan WC, Leyland M, Clark J, Dodd HM, Lian L-Y, et al. 1996. Structure-activity relationships in the peptide antibiotic nisin: antibacterial activity of fragments of nisin. *FEBS Lett.* 390:129–32

25. Choung S-Y, Kobayashi T, Takemoto K, Ishitsuka H, Inoue K. 1988. Interaction of a cyclic peptide, Ro 09–0198, with phosphatidylethanolamine in liposomal membranes. *Biochim. Biophys. Acta* 940:180–87

26. De Ruyter PGGA, Kuipers OP, Beerthuyzen MM, van Alen-Boerrigter I, de Vos WM. 1996. Functional analysis of promoters in the nisin gene cluster of *Lactococcus lactis. J. Bacteriol.* 178:3434–39

27. De Ruyter PGGA, Kuipers OP, de Vos WM. 1996. Controlled gene expression systems for *Lactococcus lactis* with the food-grade inducer nisin. *Appl. Environ. Microbiol.* 62:3662–67

28. De Vos WM, Jung G, Sahl H-G. 1991.

Appendix: definitions and nomenclature of lantibiotics. In *Nisin and Novel Lantibiotics*, ed. G Jung, H-G Sahl, pp. 457–63. Leiden: Escom. 490 pp.

29. De Vos WM, Kuipers OP, van der Meer JR, Siezen RJ. 1995. Maturation pathway of nisin and other lantibiotics: post-translationally modified antimicrobial peptides exported by gram-positive bacteria. *Mol. Microbiol.* 17:427–37

30. Delves-Broughton J, Blackburn P, Evans RJ, Hugenholtz J. 1996. Applications of the bacteriocin, nisin. Anton van Leeuwenhoek *Int. J. Gen. Microbiol.* 69: 193–202

31. Demel RA, Peelen T, Siezen RJ, de Kruijff B, Kuipers OP. 1996. Nisin Z, mutant nisin Z and lacticin 481 interactions with anionic lipids correlate with antimicrobial activity—a monolayer study. *Eur. J. Biochem.* 235:267–74

32. Diep DB, Håvarstein LS, Nes IF. 1996. Characterization of the locus responsible for the bacteriocin production in *Lactobacillus plantarum* C11. *J. Bacteriol.* 178:4472–83

33. Dodd HM, Horn N, Chan WC, Giffard CJ, Bycroft BW, et al. 1996. Molecular analysis of the regulation of nisin immunity. *Microbiology* 142:2385–92

34. Dodd HM, Horn N, Giffard CJ, Gasson MJ. 1996. A gene replacement strategy for engineering nisin. *Microbiology* 142:47–55

35. Dodd HM, Horn N, Hao Z, Gasson MJ. 1992. A lactococcal expression system for engineered nisins. *Appl. Environ. Microbiol.* 58:3683–93

36. Driessen AJM, van den Hooven HW, Kuiper W, van de Kamp M, Sahl H-G, et al. 1995. Mechanistic studies of lantibiotic-induced permeabilization of phospholipid vesicles. *Biochemistry* 34:1606–14

37. Duan K, Harvey ML, Liu C-Q, Dunn NW. 1996. Identification and characterization of a mobilizing plasmid, pND300, in *Lactococcus lactis* M189 and its encoded nisin resistance determinant. *J. Appl. Bacteriol.* 81:493–500

38. Engelke G, Gutowski-Eckel Z, Hammelmann M, Entian K-D. 1992. Biosynthesis of the lantibiotic nisin: genomic organization and membrane localization of the NisB protein. *Appl. Environ. Microbiol.* 58:3730–43

39. Engelke G, Gutowski-Eckel Z, Kiesau P, Siegers K, Hammelmann M, Entian K-D. 1994. Regulation of nisin biosynthesis and immunity in *Lactococcus lactis* 6F3. *Appl. Environ. Microbiol.* 60:814–25

40. Fredenhagen A, Märki F, Fendrich G, Märki W, Gruner J, et al. 1991. Duramycin B and C, two new lanthionine-containing antibiotics as inhibitors of phospholipase A2 and structural revision of duramycin and cinnamycin. In *Nisin and Novel Lantibiotics*, ed. G Jung, H-G Sahl, pp. 131–40. Leiden: Escom. 490 pp.

41. Freund S, Jung G, Gutbrod O, Folkers G, Gibbons WA, et al. 1991. The solution structure of the lantibiotic gallidermin. *Biopolymers* 31:803–11

42. Geissler S, Götz F, Kupke T. 1996. Serine protease EpiP from *Staphylococcus epidermidis* catalyzes the processing of the epidermin precursor peptide. *J. Bacteriol.* 178:284–88

43. Giffard CJ, Dodd HM, Horn N, Ladha S, Mackie AR, et al. 1997. Structure-function relations of variant and fragment nisins studied with model membrane systems. *Biochemistry* 36:3802–10

44. Giffard CJ, Ladha S, Mackie AR, Clark DC, Sanders D. 1996. Interaction of nisin with planar lipid bilayers monitored by fluorescence recovery after photobleaching. *J. Membrane Biol.* 151:293–300

45. Gilmore MS, Segarra RA, Booth MC. 1990. An HlyB-type function is required for expression of the *Enterococcus faecalis* hemolysin/bacteriocin. *Infect. Immun.* 58:3914–23

46. Gilmore MS, Segarra RA, Booth MC, Bogie CP, Hall LR, Clewell DB. 1994. Genetic structure of the *Enterococcus faecalis* plasmid pAD1-encoded cytolytic toxin system and its relationship to lantibiotic determinants. *J. Bacteriol.* 176: 7335–44

47. Gross E, Kiltz HH, Nebelin E. 1973. Subtilin. VI: Die Struktur des Subtilins. *H-Z. Z. Physiol. Chem.* 354:810–12

48. Gross E, Morell JL. 1971. The structure of nisin. *J. Am. Chem. Soc.* 93:4634–35

49. Gutowski-Eckel Z, Klein C, Siegers K, Bohm K, Hammelmann M, Entian K-D. 1994. Growth-phase dependent regulation and membrane localization of SpaB, a protein involved in biosynthesis of the lantibiotic subtilin. *Appl. Environ. Microbiol.* 60:1–11

50. Håvarstein LS, Coomaraswamy G, Morrison DA. 1995. An unmodified heptadecapeptide pheromone induces competence for genetic transformation in *Streptococcus pneumoniae*. *Proc. Natl. Acad. Sci. USA* 92:11140–44

51. Håvarstein LS, Diep DB, Nes IF. 1995. A family of bacteriocin ABC transporters carry out proteolytic processing of their

substrates concomitant with export. *Mol. Microbiol.* 16:229–40

52. Heidrich C. 1997. Sequenz und Funktionsanalyse von Lantibiotika-Biosynthesegenen aus *Staphylococcus epidermidis.* PhD thesis. Rheinische Friedrich-Wilhelms-Universität, Bonn Germany. 142 pp.

53. Hiramatsu K, Hanaki H, Ino T, Yabuta K, Oguri T, Tenover FC. 1997. Methicillin-resistant *Staphylococcus aureus* clinical strain with reduced vancomycin susceptibility. *J. Antimicrob. Chemother.* 40:135–36

54. Horn N, Swindell S, Dodd H, Gasson M. 1991. Nisin biosynthesis genes are encoded by a novel conjugative transposon. *Mol. Gen. Genet.* 228:129–35

55. Hosoda K, Ohya M, Kohno T, Maeda T, Endo S, Wakamatsu K. 1996. Structure determination of an immunopotentiator peptide, cinnamycin, complexed with lysophosphatidylethanolamine by 1H-NMR. *J. Biochem.* 119:226–30

56. Hurst A. 1981. Nisin. *Adv. Appl. Microbiol.* 27:85–123

57. Hynes WL, Ferretti JJ, Tagg JR. 1993. Cloning of the gene encoding streptococcin A-FF22, a novel lantibiotic produced by *Streptococcus pyogenes*, and determination of its nucleotide sequence. *Appl. Environ. Microbiol.* 59:1969–71

58. Ingram LC. 1969. Synthesis of the antibiotic nisin: formation of lanthionine and β-methyl-lanthionine. *Biochim. Biophys. Acta* 184:216–19

59. Ingram LC. 1970. A ribosomal mechanism of synthesis for peptides related to nisin. *Biochim. Biophys. Acta* 224:263–65

60. Israil AM, Jack RW, Jung G, Sahl H-G. 1996. Isolation of a new epidermin variant from two strains of *Staphylococcus epidermidis*—frequency of lantibiotic production in coagulase-negative staphylococci. *Zbl. Bakteriol.* 284:285–96

61. Izaguirre G, Hansen JN. 1997. Use of alkaline phosphatase as a reporter polypeptide to study the role of the subtilin leader segment and the SpaT transporter in the posttranslational modifications and secretion of subtilin in *Bacillus subtilis* 168. *Appl. Environ. Microbiol.* 63:3965–71

62. Jack RW, Bierbaum G, Sahl H-G. 1998. *Lantibiotics and Related Peptides.* Austin: Landes. In press

63. Jack RW, Tagg JR, Ray B. 1995. Bacteriocins of gram-positive bacteria. *Microbiol. Rev.* 59:171–200

64. Ji G, Beavis R, Novick RP. 1997. Bacterial interference caused by autoinducing peptide variants. *Science* 276:2027–30

65. Jung G. 1991. Lantibiotics—ribosomally synthesized biologically active polypeptides containing sulfide bridges and α,β-didehydroamino acids. *Angew. Chem. Int. Ed. Engl.* 30:1051–68

66. Jung G. 1991. Lantibiotics: a survey. In *Nisin and Novel Lantibiotics*, ed. G Jung, H-G Sahl, pp. 1–34. Leiden: Escom. 490 pp.

67. Jung G, Sahl H-G, eds. 1991. *Nisin and Novel Lantibiotics.* Leiden: Escom. 490 pp.

68. Kaletta C, Entian K-D, Jung G. 1991. Prepeptide sequence of cinnamycin (Ro 09–0198): the first structural gene of a duramycin-type lantibiotic. *Eur. J. Biochem.* 199:411–15

69. Kaletta C, Entian K-D, Kellner R, Jung G, Reis M, Sahl H-G. 1989. Pep5, a new lantibiotic: structural gene isolation and prepeptide sequence. *Arch. Microbiol.* 152:16–19

70. Kellner R, Jung G, Sahl H-G. 1991. Structure elucidation of the tricyclic lantibiotic Pep5 containing eight positively charged amino acids. In *Nisin and Novel Lantibiotics*, ed. G Jung, H-G Sahl, pp. 141–58. Leiden: Escom. 490 pp.

71. Kettenring JK, Malabarba A, Vékey K, Cavalleri B. 1990. Sequence determination of actagardine, a novel lantibiotic, by homonuclear 2D NMR spectroscopy. *J. Antibiot.* 43:1082–88

72. Kiesau P, Eikmanns U, Gutowski-Eckel Z, Weber S, Hammelmann M, Entian K-D. 1997. Evidence for a multimeric subtilin synthetase complex. *J. Bacteriol.* 179:1475–81

73. Kleerebezem M, Quadri LEN, Kuipers OP, de Vos WM. 1997. Quorum sensing by peptide pheromones and two-component signal-transduction systems in gram-positive bacteria. *Mol. Microbiol.* 24:895–904

74. Klein C, Entian K-D. 1994. Genes involved in self-protection against the lantibiotic subtilin produced by *Bacillus subtilis* ATCC 6633. *Appl. Environ. Microbiol.* 60:2793–801

75. Klein C, Kaletta C, Entian K-D. 1993. Biosynthesis of the lantibiotic subtilin is regulated by a histidine kinase/response regulator system. *Appl. Environ. Microbiol.* 59:296–303

76. Klein C, Kaletta C, Schnell N, Entian K-D. 1992. Analysis of genes involved in biosynthesis of the lantibiotic subtilin. *Appl. Environ. Microbiol.* 58:132–42

77. Kleinkauf H, von Döhren H. 1996. A non-

ribosomal system of peptide biosynthesis. *Eur. J. Biochem.* 236:335–51

78. Kogler H, Bauch M, Fehlhaber H-W, Griesinger C, Schubert W, Teetz V. 1991. NMR-spectroscopic investigations on mersacidin. In *Nisin and Novel Lantibiotics*, ed. G Jung, H-G Sahl, pp. 159–70. Leiden: Escom. 490 pp.

79. Konings RNH, Hilbers CW, eds. 1996. Lantibiotics: a unique group of antibiotic peptides. Anton Leeuwenhoek *Int. J. Gen. Microbiol.* 69:87–202

80. Kuipers OP, Beerthuyzen MM, de Ruyter PGGA, Luesink EJ, de Vos WM. 1995. Autoregulation of nisin biosynthesis in *Lactococcus lactis* by signal transduction. *J. Biol. Chem.* 270:27299–304

81. Kuipers OP, Beerthuyzen MM, Siezen RJ, de Vos WM. 1993. Characterization of the nisin gene cluster nisABTCIPR of *Lactococcus lactis*: requirement of expression of the nisA and nisI genes for producer immunity. *Eur. J. Biochem.* 216:281–92

82. Kuipers OP, Bierbaum G, Ottenwälder B, Dodd HM, Horn N, et al. 1996. Protein engineering of lantibiotics. Anton Leeuwenhoek *Int. J. Gen. Microbiol.* 69:161–69

83. Kuipers OP, Rollema HS, de Vos WM, Siezen RJ. 1993. Biosynthesis and secretion of a precursor of nisin Z by *Lactococcus lactis*, directed by the leader peptide of the homologous lantibiotic subtilin from *Bacillus subtilis*. *FEBS Lett.* 330:23–27

84. Kuipers OP, Rollema HS, Yap WMGJ, Boot HJ, Siezen RJ, de Vos WM. 1992. Engineering dehydrated amino acid residues in the antimicrobial peptide nisin. *J. Biol. Chem.* 267:24340–46

85. Kupke T, Götz F. 1996. Expression, purification, and characterization of EpiC, an enzyme involved in the biosynthesis of the lantibiotic epidermin, and sequence analysis of *Staphylococcus epidermidis* epiC mutants. *J. Bacteriol.* 178:1335–40

86. Kupke T, Götz F. 1997. The enethiolate anion reaction products of EpiD: pK$_a$ value of the enethiol side chain is lower than that of the thiol side chain of peptides. *J. Biol. Chem.* 272:4759–62

87. Kupke T, Götz F. 1997. In vivo reaction of affinity-tag-labelled epidermin precursor peptide with flavoenzyme EpiD. *FEMS Microbiol. Lett.* 153:25–32

88. Kupke T, Kempter C, Gnau V, Jung G, Götz F. 1994. Mass spectroscopic analysis of a novel enzymatic reaction: oxidative decarboxylation of the lantibiotic precursor peptide EpiA catalyzed by the flavo-protein EpiD. *J. Biol. Chem.* 269:5653–59

89. Kupke T, Kempter C, Jung G, Götz F. 1995. Oxidative decarboxylation of peptides catalyzed by flavoprotein EpiD. Determination of substrate specificity using peptide libraries and neutral loss mass spectrometry. *J. Biol. Chem.* 270:11282–89

90. Kupke T, Stevanović S, Ottenwälder B, Metzger JW, Jung G, Götz F. 1993. Purification and characterization of EpiA, the peptide substrate for post-translational modifications involved in epidermin biosynthesis. *FEMS Microbiol. Lett.* 112: 43–48

91. Kupke T, Stevanović S, Sahl H-G, Götz F. 1992. Purification and characterization of EpiD, a flavoprotein involved in the biosynthesis of the lantibiotic epidermin. *J. Bacteriol.* 174:5354–61

92. Li Y-M, Milne JC, Madison LL, Kolter R, Walsh CT. 1996. From peptide precursors to oxazole and thiazole-containing peptide antibiotics: microcin B17 synthase. *Science* 274:1188–93

93. Lian L-Y, Chan WC, Morley SD, Roberts GKC, Bycroft BW, Jackson D. 1991. Solution structures of nisin and its two major degradation products determined by NMR. *Biochem. J.* 283:413–20

94. Limbert M, Isert D, Klesel N, Markus A, Seibert G, et al. 1991. Chemotherapeutic properties of mersacidin in vitro and in vivo. In *Nisin and Novel Lantibiotics*, ed. G Jung, H-G Sahl, pp. 448–56. Leiden: Escom. 490 pp.

95. Linnett PE, Strominger JL. 1973. Additional antibiotic inhibitors of peptidoglycan synthesis. *Antimicrob. Agents Chemother.* 4:231–36

96. Liu W, Hansen JN. 1992. Enhancement of the chemical and antimicrobial properties of subtilin by site-directed mutagenesis. *J. Biol. Chem.* 267:25078–85

97. Liu W, Hansen JN. 1993. The antimicrobial effect of a structural variant of subtilin against outgrowing *Bacillus cereus* T spores and vegetative cells occurs by different mechanisms. *Appl. Environ. Microbiol.* 59:648–51

98. Ludtke SJ, He K, Heller WT, Harroun TA, Yang L, Huang HW. 1996. Membrane pores induced magainin. *Biochemistry* 35:13723–28

99. Meyer C, Bierbaum G, Heidrich C, Reis M, Süling J, et al. 1995. Nucleotide sequence of the lantibiotic Pep5 biosynthetic gene cluster and functional analysis of PepP and PepC: evidence for a role of PepC in thioether formation. *Eur.*

J. Biochem. 232:478–89

100. Meyer HE, Heber M, Eisermann B, Korte H, Metzger JW, Jung G. 1994. Sequence analysis of lantibiotics: chemical derivatization procedures allow a fast access to complete Edman degradation. *Anal. Biochem.* 223:185–190

101. Minami Y, Yoshida K-I, Azuma R, Urakawa A, Kawauchi T, et al. 1994. Structure of cypemycin, a new peptide antibiotic. *Tetrahedron Lett.* 35:8001–4

102. Moll GN, Clark J, Chan WC, Bycroft BW, Roberts GCK, et al. 1997. Role of transmembrane pH gradient and membrane binding in nisin pore formation. *J. Bacteriol.* 179:135–40

103. Moll GN, Roberts GCK, Konings WN, Driessen AJM. 1996. Mechanism of lantibiotic-induced pore formation. Anton Leeuwenhoek *Int. J. Gen. Microbiol.* 69:185–91

104. Moschetti G, Villani F, Blaiotta G, Baldinelli A, Coppola S. 1996. Presence of non-functional nisin genes in *Lactococcus lactis* subsp. *lactis* isolated from natural starters. *FEMS Microbiol. Lett.* 145:27–32

105. Mota-Meira M, Lacroix C, LaPointe G, Lavoie MC. 1997. Purification and structure of mutacin B-Ny266: a new lantibiotic produced by Streptococcus mutans. *FEBS Lett.* 410:275–79

106. Mulders JWM, Boerrigter IJ, Rollema HS, Siezen RJ, de Vos WM. 1991. Identification and characterization of the lantibiotic nisin Z, a structural nisin variant. *Eur. J. Biochem.* 201:581–84

107. Neis S, Bierbaum G, Josten M, Pag U, Kempter C, et al. 1997. Effect of leader peptide mutations on biosynthesis of the lantibiotic Pep5. *FEMS Microbiol. Lett.* 149:249–55

108. Nissen-Meyer J, Holo H, Håvarstein LV, Sletten K, Nes IF. 1992. A novel lactococcal bacteriocin whose activity depends on the complementary action of two peptides. *J. Bacteriol.* 174:5686–92

109. Novák J, Caufield PW, Miller EJ. 1994. Isolation and biochemical characterization of a novel lantibiotic mutacin from *Streptococcus mutans. J. Bacteriol.* 176:4316–20

110. Ottenwälder B, Kupke T, Brecht S, Gnau V, Metzger J, et al. 1995. Isolation and characterization of genetically engineered gallidermin and epidermin analogs. *Appl. Environ. Microbiol.* 61:3894–903

111. Peschel A, Augustin J, Kupke T, Stevanović S, Götz F. 1993. Regulation of epidermin biosynthetic genes by EpiQ.

Mol. Microbiol. 9:31–39

112. Peschel A, Götz F. 1996. Analysis of the *Staphylococcus epidermidis* genes epiF, -E, and -G involved in epidermin immunity. *J. Bacteriol.* 178:531–36

113. Peschel A, Ottenwälder B, Götz F. 1996. Inducible production and cellular location of the epidermin biosynthetic enzyme EpiB using an improved staphylococcal expression system. *FEMS Microbiol. Lett.* 137:279–84

114. Peschel A, Schnell N, Hille M, Entian K-D, Götz F. 1997. Secretion of the lantibiotics epidermin and gallidermin: sequence analysis of the genes gdmT and gdmH, their influence on epidermin production and their regulation by EpiQ. *Mol. Gen. Genet.* 254:312–18

115. Piard J-C, Kuipers OP, Rollema HS, Desmazeaud MJ, de Vos WM. 1993. Structure, organization, and expression of the lct gene for lacticin 481, a novel lantibiotic produced by *Lactococcus lactis. J. Biol. Chem.* 268:16361–68

116. Prasch T, Naumann T, Markert RLM, Sattler M, Schubert W, et al. 1997. Constitution and solution conformation of the antibiotic mersacidin determined by NMR and molecular dynamics. *Eur. J. Biochem.* 244:501–12

117. Pridmore D, Rekhif N, Pittet A-C, Suri B, Mollet B. 1996. Variacin, a new lanthionine-containing bacteriocin produced by *Micrococcus varians*: comparison to lacticin 481 of Lactococcus lactis. *Appl. Environ. Microbiol.* 62:1799–802

118. Qiao M, Immonen T, Koponen O, Saris PEJ. 1995. The cellular location and effect on nisin immunity of the NisI protein from *Lactococcus lactis* N8 expressed in *Escherichia coli* and *L. lactis. FEMS Microbiol. Lett.* 131:75–80

119. Qiao M, Saris PEJ. 1996. Evidence for a role of NisT in transport of the lantibiotic nisin produced by *Lactococcus lactis* N8. *FEMS Microbiol. Lett.* 144:89–93

120. Ramseier HR. 1960. Die Wirkung von Nisin auf *Clostridium butyricum* Prazm. *Arch. Mikrobiol.* 37:57–94

121. Rauch PJG, Beerthuyzen MM, de Vos WM. 1994. Distribution and evolution of nisin-sucrose elements in *Lactococcus lactis. Appl. Environ. Microbiol.* 60:1798–804

122. Reis M, Eschbach-Bludau M, Iglesias-Wind MI, Kupke T, Sahl H-G. 1994. Producer immunity towards the lantibiotic Pep5: identification of the immunity gene pepI and localization and functional analysis of its gene product. *Appl. Environ. Microbiol.* 60:2876–83

123. Reisinger P, Seidel H, Tschesche H, Hammes WP. 1980. The effect of nisin on murein synthesis. *Arch. Microbiol.* 127: 187–93

124. Rince A, Dufour A, Le Pogam S, Thuault D, Bourgeois CM, Le Pennec JP. 1994. Cloning, expression, and nucleotide sequence of genes involved in production of lactococcin DR, a bacteriocin from *Lactococcus lactis*, subsp. *lactis. Appl. Environ. Microbiol.* 60:1652–57

125. Rintala H, Graeffe T, Paulin L, Kalkkinen N, Saris PEJ. 1993. Biosynthesis of nisin in the subtilin producer *Bacillus subtilis* ATCC6633. *Biotechnol. Lett.* 15:991–96

126. Rollema HS, Kuipers OP, Both P, de Vos WM, Siezen RJ. 1995. Improvement of solubility and stability of the antimicrobial peptide nisin by protein. engineering. *Appl. Environ. Microbiol.* 61:2873–78

127. Rollema HS, Metzger JW, Both P, Kuipers OP, Siezen RJ. 1996. Structure and biological activity of chemically modified nisin A species. *Eur. J. Biochem.* 241: 716–22

128. Ross KF, Ronson CW, Tagg JR. 1993. Isolation and characterization of the lantibiotic salivaricin A and its structural gene salA from *Streptococcus salivarius* 20P3. *Appl. Environ. Microbiol.* 59:2014–21

129. Ruhr E, Sahl H-G. 1985. Mode of action of the peptide antibiotic nisin and influence on the membrane potential of whole cells and on artificial membrane vesicles. *Antimicrob. Agents Chemother.* 27:841–45

130. Sahl H-G. 1991. Pore formation in bacterial membranes by cationic lantibiotics. In *Nisin and Novel Lantibiotics*, ed. G Jung, H-G Sahl, pp. 347–58. Leiden: Escom. 490 pp.

131. Sahl H-G, Brandis H. 1982. Mode of action of the staphylococcin-like peptide Pep5 and culture conditions affecting its activity. *Zbl. Bakt. Hyg., I. Abtl. Orig. A* 252:166–75

132. Sahl H-G, Brandis H. 1983. Efflux of low Mr substances from the cytoplasm of sensitive cells caused by the staphylococcin-like agent Pep5. *FEMS Microbiol. Lett.* 16:75–79

133. Sahl H-G, Jack RW, Bierbaum G. 1995. Biosynthesis and biological activities of lantibiotics with unique post-translational modifications. *Eur. J. Biochem.* 230:827–53

134. Sahl H-G, Kordel M, Benz R. 1987. Voltage-dependent depolarization of bacterial membranes and artificial lipid bilayers by the peptide antibiotic nisin. *Arch. Microbiol.* 149:120–24

135. Schnell N, Engelke G, Augustin J, Rosenstein R, Ungermann V, et al. 1992. Analysis of genes involved in the biosynthesis of lantibiotic epidermin. *Eur. J. Biochem.* 204:57–68

136. Schnell N, Entian K-D, Götz F, Hörner T, Kellner R, Jung G. 1989. Structural gene isolation and prepeptide sequence of gallidermin, a new lanthionine containing antibiotic. *FEMS Microbiol. Lett.* 58:263–68

137. Schnell N, Entian K-D, Schneider U, Götz F, Zähner H, et al. 1988. Prepeptide sequence of epidermin, a ribosomally synthesized antibiotic with four sulphiderings. *Nature* 333:276–78

138. Sheth TR, Henderson RM, Hladky SB, Cuthbert AW. 1992. Ion-channel. formation by duramycin. *Biochim. Biophys. Acta* 1107:179–85

139. Shiba T, Wakamiya T, Fukase K, Ueki Y, Teshima T, Nishikawa M. 1991. Structure of the lanthionine peptides nisin, ancovenin and lanthiopeptin. In *Nisin and Novel Lantibiotics*, ed. G Jung, H-G Sahl, pp. 113–22. Leiden: Escom. 490 pp.

140. Siegers K, Entian K-D. 1995. Genes involved in immunity to the lantibiotic nisin produced by *Lactococcus lactis* 6F3. *Appl. Environ. Microbiol.* 61:1082–89

141. Siegers K, Heinzmann S, Entian K-D. 1996. Biosynthesis of lantibiotic nisin. Posttranslational modification of its prepeptide occurs at a multimeric membrane-associated lanthionine synthetase complex. *J. Biol. Chem.* 271:12294–301

142. Siezen RJ, Kuipers OP, de Vos WM. 1996. Comparison of lantibiotic gene clusters and encoded proteins. Anton Leeuwenhoek *Int. J. Gen. Microbiol.* 69:171–84

143. Siezen RJ, Rollema HS, Kuipers OP, de Vos WM. 1995. Homology modelling of the *Lactococcus lactis* leader peptidase NisP and its interaction with the precursor of the lantibiotic nisin. *Protein Eng.* 8:117–25

144. Simpson WJ, Ragland NL, Ronson CW, Tagg JR. 1995. A lantibiotic gene family widely distributed in *Streptococcus salivarius* and *Streptococcus pyogenes. Dev. Biol. Stand.* 85:639–43

145. Skaugen M, Abildgaard CIM, Nes IF. 1997. Organization and expression of a gene cluster involved in the biosynthesis of the lantibiotic lactocin S. *Mol. Gen. Genet.* 253:674–86

146. Skaugen M, Nes IF. 1994. Transposition in *Lactobacillus sake* and its abolition of lactocin S production by insertion of IS *1163*, a new member of the IS*3* family. *Appl. Environ. Microbiol.* 60:2818–25

147. Skaugen M, Nissen-Meyer J, Jung G, Stevanović S, Sletten K, et al. 1994. In vivo conversion of L-serine to D-alanine in a ribosomally synthesized polypeptide. *J. Biol. Chem.* 269:27183–85

148. Somma S, Merati W, Parenti F. 1971. Gardimycin, a new antibiotic inhibiting peptidoglycan synthesis. *Antimicrob. Agents Chemother.* 11:396–401

149. Spee JH, de Vos WM, Kuipers OP. 1993. Efficient random mutagenesis method with adjustable mutation frequency by use of PCR and dITP. *Nucl. Acids Res.* 21:777–78

150. Steen MT, Chung YJ, Hansen JN. 1991. Characterization of the nisin gene as part of a polycistronic operon in the chromosome of *Lactococcus lactis* ATCC 11454. *Appl. Environ. Microbiol.* 57:1181–88

151. Stoffels G, Guðamundsdóttir Á, Abee T. 1994. Membrane-associated proteins encoded by the nisin gene cluster may function as a receptor for the lantibiotic carnocin UI49. *Microbiology* 140:1443–50

152. Stoffels G, Nissen-Meyer J, Guðamundsdóttir Á, Sletten K, Holo H, Nes IF. 1992. Purification and characterization of a new bacteriocin isolated from a *Carnobacterium* spp. *Appl. Environ. Microbiol.* 58:1417–22

153. Van de Kamp M, van den Hooven HW, Konings RNH, Hilbers CW, van de Ven FJM, et al. 1995. Elucidation of the primary structure of the lantibiotic epilancin K7 from *Staphylococcus epidermidis* K7: cloning and characterisation of the epilancin-K7–encoding gene and NMR analysis of mature epilancin K7. *Eur. J. Biochem.* 230:587–600

154. Van de Ven FJM, van den Hooven HW, Konings RNH, Hilbers CW. 1991. NMR-studies of lantibiotics: the structure of nisin in aqueous solution. *Eur. J. Biochem.* 202:1181–88

155. Van den Hooven HW, Doeland CCM, van de Kamp M, Konings RNH, Hilbers CW, van de Ven FJM. 1996. Three-dimensional structure of the lantibiotic nisin in the presence of membrane-mimetic micelles of dodecylphosphocholine and of sodium dodecylsulphate. *Eur. J. Biochem.* 235:382–93

156. Van den Hooven HW, Lagerwerf FM, Heerma W, Haverkamp J, Piard J-C, et al.

1996. The structure of the lantibiotic lacticin 481 produced by *Lactococcus lactis*: location of the thioether bridges. *FEBS Lett.* 391:317–22

157. Van den Hooven HW, Spronk CAEM, van de Kamp M, Konings RNH, Hilbers CW, van de Ven FJM. 1996. Surface location and orientation of the lantibiotic nisin bound to membrane-mimicking micelles of dodecylphosphocholine and of sodium dodecylsulphate. *Eur. J. Biochem.* 235:394–403

158. Van der Meer JR, Polman J, Beerthuyzen MM, Siezen RJ, Kuipers OP, de Vos WM. 1993. Characterization of the *Lactococcus lactis* nisin A operon genes nisP, encoding a subtilisin-like serine protease involved in precursor processing, and nisR, encoding a regulatory protein involved in nisin biosynthesis. *J. Bacteriol.* 175:2578–88

159. Van der Meer JR, Rollema HS, Siezen RJ, Beerthuyzen MM, Kuipers OP, de Vos WM. 1994. Influence of amino acid substitutions in the nisin leader peptide on biosynthesis and secretion of nisin by *Lactococcus lactis*. *J. Biol. Chem.* 269:3555–62

160. Van Kraaij C, Breukink E, Rollema HS, Siezen RJ, Demel RA, et al. 1997. Influence of charge differences in the C-terminal part of nisin on antimicrobial activity and signaling capacity. *Eur. J. Biochem.* 247:114–20

161. Verheul A, Russel NJ, van't Hof R, Rombouts FM, Abee T. 1997. Modifications of membrane phospholipid composition in nisin-resistant *Listeria monocytogenes* Scott A. *Appl. Environ. Microbiol.* 63:3451–57

162. Wade D, Andreu D, Mitchell SA, Silveira AMV, Boman A, et al. 1992. Antibacterial peptides designed as analogs or hybrids of cecropins and melittin. *Int. J. Peptide Protein Res.* 40:429–36

163. Weil H-P, Beck-Sickinger AG, Metzger J, Stevanović S, Jung G, et al. 1990. Biosynthesis of the lantibiotic Pep5: isolation and characterization of a prepeptide containing dehydroamino acids. *Eur. J. Biochem.* 194:217–23

164. Winkowski K, Ludescher RD, Montville TJ. 1996. Physicochemical characterization of the nisin-membrane interaction with liposomes derived from *Listeria monocytogenes*. *Appl. Environ. Microbiol.* 62:323–27

165. Zimmermann N, Freund S, Fredenhagen A, Jung, G. 1993. Solution structures of the lantibiotics duramycin B and C. *Eur. J. Biochem.* 216:419–28

166. Zimmermann N, Jung G. 1997. The three-dimensional solution structure of the lantibiotic murein-biosynthesis-inhibitor actagardine determined by NMR. *Eur. J. Biochem.* 246:809–19

167. Zimmermann N, Metzger JW, Jung G. 1995. The tetracyclic lantibiotic actagardine: 1H-NMR and 13C-NMR assign-ments and revised primary structure. *Eur. J. Biochem.* 228:786–97

168. Zuber P, Nakano MM, Marahiel MA. 1993. Peptide antibiotics. In *Bacillus Subtilis and Other Gram-Positive Bacteria*, ed. AL Sonenshein, JA Hoch, R Losick, pp. 897–916. Washington: Am. Soc. Microbiol.

Annu. Rev. Microbiol. 1998. 52:81–104
Copyright © 1998 by Annual Reviews. All rights reserved

THINKING ABOUT BACTERIAL POPULATIONS AS MULTICELLULAR ORGANISMS

James A. Shapiro

Department of Biochemistry and Molecular Biology, University of Chicago, 920 East 58 Street, Chicago, Illinois 60637; e-mail: jsha@midway.uchicago.edu

KEY WORDS: intercellular communication, coordinated multicellular behavior, metabolic coordination, cellular differentiation, adaptive benefits of multicellularity

ABSTRACT

It has been a decade since multicellularity was proposed as a general bacterial trait. Intercellular communication and multicellular coordination are now known to be widespread among prokaryotes and to affect multiple phenotypes. Many different classes of signaling molecules have been identified in both Gram-positive and Gram-negative species. Bacteria have sophisticated signal transduction networks for integrating intercellular signals with other information to make decisions about gene expression and cellular differentiation. Coordinated multicellular behavior can be observed in a variety of situations, including development of *E. coli* and *B. subtilis* colonies, swarming by *Proteus* and *Serratia*, and spatially organized interspecific metabolic cooperation in anaerobic bioreactor granules. Bacteria benefit from multicellular cooperation by using cellular division of labor, accessing resources that cannot effectively be utilized by single cells, collectively defending against antagonists, and optimizing population survival by differentiating into distinct cell types.

CONTENTS

81

0066-4227/98/1001-0081$08.00

INTRODUCTION: A DECADE OF CHANGES

Ten years ago, *Scientific American* published an article entitled "Bacteria as Multicellular Organisms" (141). It was based largely on the observation of pattern and organized cellular differentiation in the colonies of many bacterial species, including *E. coli.* While stimulating interest in the idea that bacteria may be more interactive than generally realized, that 1988 article did not convince most microbiologists that multicellularity should be considered a basic tenet in our thinking about how bacteria operate. Multicellularity was still widely considered a specialized adaptive strategy of particular groups, such as the *Myxobacteria* or *Actinomycetes.* The historical tradition of single-cell, pure-culture microbiology derived from Koch's postulates and medical bacteriology still held sway, despite an old and well-established alternative interactive, multicellular tradition based on environmental microbiology (42).

Today, the intellectual landscape is dramatically different. Intercellular communication and concerted multicellular activities are now generally accepted to be common among bacteria (62, 81, 184). The change in thinking is due mainly to the discovery that "quorum-sensing" signal molecules, in particular the *N*-acyl homoserine lactones (AHLs) in Gram-negative species, are used throughout the eubacterial kingdom to regulate the expression of a wide variety of phenotypes (40, 54, 55, 90, 129). The past decade has also witnessed the discovery of new phenomena, such as autoaggregation of chemotactic bacteria (20, 24, 25, 185) and coordinated behaviors in complex colony morphogenesis (12, 14, 53, 104, 109). The molecular basis of intercellular coordination is being clarified in multicellular taxa such as the *Myxobacteria* (152) and *Streptomycetes* (33), and homologies have been discovered between intercellular and unicellular regulatory circuits (177). Our knowledge of biochemical and thermodynamic coordination within microbial consortia has grown, particularly during anaerobic biotransformations (e.g. 47, 130). And new optical and molecular technologies reveal pattern and physiologically significant spatial organization among bacterial populations (e.g. 52).

This review emphasizes the conceptual basis for thinking about bacterial populations as multicellular organisms. I present my view of a few core ideas underlying an integrated view of bacterial multicellularity and then provide selected examples to illustrate them. A more detailed discussion of particular aspects can be found in the recent book dedicated to this theme (148).

CORE CONCEPTS OF BACTERIAL MULTICELLULARITY

Briefly stated, the core concepts of bacterial multicellularity may be summarized as follows:

1. Bacterial cells have communication and decision-making capabilities that enable them to coordinate growth, movement, and biochemical activities.

2. Examples of communication and coordinated behaviors are widespread (possibly ubiquitous) among bacterial taxa and are not limited to a few groups with a specialized multicellular vocation.

3. Bacterial populations derive adaptive benefits from multicellular cooperation and their ability to integrate the diverse activities of different cells. These benefits include (but are not limited to):

   (*a*) More efficient proliferation resulting from a cellular division of labor;

   (*b*) Access to resources and niches that cannot be utilized by isolated cells;

   (*c*) Collective defense against antagonists that eliminate isolated cells; and

   (*d*) Optimization of population survival by differentiation into distinct cell types.

INTERCELLULAR COMMUNICATION

Diverse Classes of Signal Molecules

Intercellular communication is the basis of coordinated multicellular function. Because the AHL field has been so well reviewed (40, 54, 55, 62, 62a, 129, 168), it is not discussed in detail here. But it is worthwhile pointing out several basic characteristics of AHL signaling. Because short-chain AHLs cross bacterial membranes freely, their concentration reflects total production and, consequently, total density of the producing population (82). Thus, a major function of at least some AHL molecules is to provide a quorum-sensing function (54). The canonical quorum-sensing function is assumed to assure an individual cell of a critical population density before undertaking expression of specialized functions. AHL production is determined by genetic loci homologous to *Vibrio fischeri luxI* , and AHL concentration is sensed by proteins homologous to *V. fischeri* luxR. Significantly, modules of luxI–luxR homologues operate in many different bacterial control circuits governing a wide range of phenotypes, and the same AHL molecule can regulate distinct functions in different bacterial species (40, 54, 55, 62, 62a, 129, 168). Thus, *AHL/luxI/luxR* signaling systems are standard molecular routines that can be incorporated into regulatory circuits whenever multicellular control is adaptively useful. Some AHLs stimulate their own production and can establish positive autoinducer feedback loops, in effect serving as more than simple quorum sensors (40).

AHLs are far from unique as intercellular signals between bacteria. Density- and growth-phase phenomena are common, and their study has led to the identification of many additional classes of signaling molecules (Table 1).

The γ-butyrolactones of *Streptomyces* species may be considered analogues of the homoserine lactones (16), but other signaling molecules are quite distinct, including some ordinarily classed as toxins (antibiotics and bacteriocins). Overdose of a particular signal can be lethal, as with mammalian hormones like insulin. Interestingly, a *Rhizobium leguminosarum* bacteriocin has recently been found to be an AHL (133). In Gram-positive bacteria, antimicrobial peptides are signaling molecules as well, often stimulating their own production (Table 1; 90).

The oligopeptides are a widely utilized class of intercellular signals in Gram-positive bacteria (Table 1; 90), just as they are in eukaryotes. Oligopeptide pheromones can be considered the first class of prokaryotic communication molecule to be detected (in the 1960s) as competence factors in *S. pneumoniae* and *B. subtilis* (32, 46, 171, 172), but their structures were only recently determined (71, 102, 113, 156, 176). The tremendous diversity of oligopeptides makes them especially suitable when a high degree of discrimination is required, as in the use of distinct oligopeptide mating pheromones and inhibitors involved in regulating the transfer of each of many conjugative plasmids in gram-positive cocci (35, 124).

There are important differences between the properties of oligopeptides and AHLs, especially the fact that oligopeptides do not transfer freely across bacterial membranes, as do short-chain AHLs (82). In general, Gram-positive oligopeptide signals are synthesized from larger precursor proteins, transported outside the cell by ATP-binding-cassette (ABC) transporters, and detected by surface receptors that belong to two-component protein kinase regulatory systems (90). Because some receptors in Gram-positive bacteria show homology

Table 1 Signaling molecules other than N-acyl homoserine lactones

Bacterial species	Signal	Molecular class	Phenotype affected	Citation
Myxococcus xanthus	A factor	Amino acid (mixture)	Fruiting body development	83
Escherichia coli	Aspartate, glutamate	Amino acid	Chemotactic autoaggregation	24, 25
Streptomyces griseus	A-factor	γ-butyrolactone	Sporulation, streptomycin production	16, 75
Streptomyces virginiae	VB-A	γ-butyrolactone	Virginiamycin production	75
Streptomyces sp. *FR1-5*	IM-2	γ-butyrolactone	Staphylomycin production	16, 75
Streptomyces sp. Y-86,36923	Butalactin	γ-butyrolactone	Antibiotic production	16
Streptomyces viridochromogenes	Factor I	γ-butyrolactone	Anthracycline production	16
Streptomyces alboniger	Pamamycin-C	Macrolide	Aerial mycelium formation	75
Streptoverticillium sp.	Carbazomycinal	β-lactam derivatives	Aerial mycelium formation	75
Nocardia sp.	B-factor	Adenosine derivative	Rifamycin production	75
Cylindrospermum licheniforme	(Unnamed)	Fused lactam, thioketon rings	Akinete formation	75
Stigmatella aurantica	Lipid pheromone	2,5,8-trimethyl-8-hydroxy-nonan-4-one	Fruiting body formation	77, 161
Myxococcus xanthus	E-factor	Branched-chain fatty acids	Fruiting body formation	169, 184
Staphylococcus aureus	Rap	Octapeptide	Toxic exoprotein, virulence factor secretion	78
Bacillus subtilis	ComX	Decapeptide, modified tryptophan	Transformation competence	102
Bacillus subtilis	CSF	Pentapeptide	Transformation competence, sporulation	156
Bacillus subtilis	Sporulation factor	Oligopeptide	Sporulation	176
Streptococcus pneumoniae	ComC	Heptadecapeptide	Transformation competence	71
Lactococcus lactis	Nisin	Oligopeptide	Nisin (lantibiotic) production	90
Lactobacillus plantarum	Bacteriocin inducing factor	26 amino acid oligopeptide	Plantaricin (class II antimicrobial)	90
Lactobacillus sake	Bacteriocin inducing factor	Oligopeptide	Sakacin (class II antimicrobial)	90
Carnobacterium piscicola	Bacteriocin inducing factors	24 and 49 amino acid oligopeptides	Carnobacteriocin (class II antimicrobial)	90
Enterococcus faecalis	Sex pheromones	Oligopeptides, plasmid-specific	Agglutination, plasmid transfer	35, 124
Enterococcus faecalis	Sex pheromone inhibitors	Oligopeptides, plasmid-specific	Inhibit sex pheromone binding	35, 124
Myxococcus xanthus	C factor	Protein	Fruiting body formation	88, 152

to oligopeptide uptake systems from enteric bacteria, it will not be surprising to discover oligopeptide signals in the Gram-negatives as well.

Not all the small diffusible molecules used for interbacterial communication belong to special classes of dedicated signals. Amino acids also serve as communication molecules in the initiation of fruiting body formation in *Myxococcus xanthus* (83, 88, 152) and during autoaggregation in chemotactic *E. coli* (24, 25). In *M. xanthus*, the extracellular A signal is a mixture of several amino acids generated by surface protease activity (83). When the mixture reaches a density threshold of 10 μM at $\geq 3 \times 10^8$ cells per ml, it activates a cellular signal transduction system leading to expression of early functions needed for fruiting body development. By serving as a signal to ensure that development only initiates above a critical cell density, A signal fills the canonical quorum-sensing function. The role of amino acids in chemotactic autoaggregation is different. Cells of chemotactic *E. coli* can be observed to form punctate aggregates in fluid medium, often generating striking spatial patterns of spots, lines, and circles (24, 25). Autoaggregation can be blocked by analogues or mutations blocking Tar chemotaxis receptor (24), and excretion of aspartate and glutamate, powerful chemoattractants, occurs in stressed cultures (25). Both amino acids thus signal *E. coli* cells to change their density under conditions when it may be advantageous to do so.

Our knowledge of intercellular signaling by larger proteins is still limited to a few examples (Table 1). The best characterized is the C factor of *M. xanthus* (88, 152). Although we do have a great deal of information about pili and type IV transport systems in interbacterial DNA exchange (36), the role of complex surface organelles such as pili, macromolecular transport complexes, and extracellular fibrils in bacteria-bacteria signaling remains largely to be explored in all but a few bacteria. Pili are required for social motility, fruiting body development, and spore formation in *M. xanthus* (80, 186). Similarly, *M. xanthus* mutants defective in the formation of extracellular fibrils are blocked in cohesion in liquid suspension and are defective in motility, predation, and all stages of fruiting body formation (6, 9, 10). These deficits can be corrected by addition of purified fibrils to a fibril-less mutant culture (30). Piliation is also an important determinant of *N. gonorrhoeae* aggregation and colony morphology (167). In scanning electron micrographs, networks of pili can be seen to connect the gonococcal cells (170). There has long been known to be a correlation between *N. gonorrhoeae* colony morphology and virulence (87). It is ironic that our understanding of the strictly prokaryotic roles of complex surface organelles in cell-cell communication lags behind our understanding of their roles in communication between prokaryotic and eukaryotic cells during symbiosis and pathogenesis (7, 48). Such elaborate systems must surely also be utilized to coordinate the activities of distinct bacterial cells.

Given our growing knowledge of the sophisticated ways that nodulating bacteria use specific exopolysaccharides to alter plant cell behavior (7, 39, 96), we can also expect to find these highly diversified chemical structures used for interbacterial communication. Exopolymers are certainly an important component of many multicellular populations, such as *P. putida* and *E. coli* colonies, where they are readily visualized by scanning electron microscopy (SEM) (139, 140), and they play a critical role in collective motility.

Signal Response Systems: Interpreting Chemical Messages in an Informationally Rich Environment

From the perspective of the core multicellularity concepts outlined above, it is to be expected that intracellular systems responding to intercellular signals will be molecular computing networks allowing each cell to make appropriate decisions and adjust its activity to coordinate with other cells in the group. This expectation is very much in line with current thinking about the decision-making capabilities of cells in higher organisms (5, 22). Beppu (16) has pointed out parallels between the use of protein kinases by *Streptomyces griseus* in the γ-butyrolactone response and in eukaryotic signal transduction systems. The *Myxobacteria* and *Bacillus* communities have documented two good examples of complex signal-processing networks that respond to intercellular cues.

Fruiting body development and sporulation in *M. xanthus* involve a cascade of signaling events (152). Commitment to fruiting body formation must occur under suitable conditions of cell density and spatial organization (88, 125, 126). Transfer of C-factor signal requires cell motility to achieve the proper alignment of the signaling bacteria (127) and reinforces this alignment by modulating the activity of the intracellular Frz network, which controls the reversal of *M. xanthus* cell movement (153, 177). The C-factor–Frz interaction appears to generate the periodic rippling that precedes formation of fruiting bodies (127) and is thought to accomplish two important goals: building up cell density and aligning cells for subsequent morphogenetic movements. Genetic studies show that the C-factor affects other signal transduction components as well as the Frz system and also positively stimulates its own production (154).

Both *B. subtilis* competence and sporulation depend on population density and extracellular factors (63, 64, 155, 176). There are two extracellular factors stimulating competence: the ComX pheromone, a modified decapeptide (102), and competence stimulating factor (CSF), a pentapeptide (156). There are at least two extracellular sporulation factors as well (64), and one of these is CSF, which serves both as a competence and sporulation factor (156). Commitment to competence and sporulation involve major cellular changes at the end of exponential growth in response to external, internal, and intercellular conditions (63, 157). They represent mutually exclusive cellular differentiation outcomes. Besides extracellular signals, factors determining the competence/sporulation decision include nutritional deprivation, glucose, TCA cycle activity, and status of the genome with respect to cell-cycle and DNA damage. The key competence regulator is the ComK transcription factor, and the key sporulation regulator is phosphorylated SpoOA transcription factor. Common elements lead to activation of both factors, such as SpoOK permease, the CSF receptor. The ability of an individual cell to integrate all the signals and decide between competence and sporulation requires the operation of a highly interconnected regulatory network including negative feedback between the competence and sporulation pathways, a four-step phosphorelay leading to Spo0A phosphorylation, and positive feedback loops on the final expression of ComK and Spo0A~P (63). This elaborate molecular network is the kind of distributed computing system described by Bray (22).

A strong prediction of the multicellular view is that the complexities of the *M. xanthus* and *B. subtilis* signal response systems will prove to be typical rather than exceptional among bacteria. This prediction is being realized with the AHL molecules. Further sophistication in the cellular responses to AHLs has become evident with the discovery of the use of multiple AHLs to influence a particular phenotype, such as bioluminescence (40) or virulence (183), and of sequential cascades of AHL signaling linked to signal transduction functions such as RpoS in *P. aeruginosa* (92, 112, 134). AHL signaling must involve more than quorum-sensing in cases where exogenous AHLs are not sufficient to stimulate exoprotein production in low-density cultures of *E. carotovora* (107) and *P. aeruginosa* (31).

COORDINATED MULTICELLULAR BEHAVIORS—A GENERAL BACTERIAL TRAIT

We often forget that the well-agitated suspension culture is largely a laboratory construct. Many microbiologists no longer remember that most bacteria proliferate and survive attached to surfaces (182). When we examine surface cultures, we find that bacteria differentiate biochemically and morphologically, and they interact in ways that produce spatially organized populations. It is worth pointing out that colony development and collective motility phenomena in bacteria (e.g. 41, 72, 151) hold valuable lessons for understanding the formation and development of biofilms, perhaps the most widespread multicellular prokaryotic structures in nature (38, 115). It has recently been reported that an AHL plays an essential role in the spatial organization of *P. aeruginosa* biofilms in the laboratory (38a).

E. coli Colony Development

Cell-cell interactions in an *E. coli* microcolony begin after the first cell division. The two daughters elongate alongside each other unless one of the sibling bacteria is attracted by a third nearby bacterium (150). Attractions and fusions of microcolonies are invariably observed, even when the chemotactic sensory system is absent (144, 150). The standard rule for *E. coli* microcolonies is to maximize cell-to-cell contact, i.e. population density, rather than individual cell access to substrate. Nonetheless, cells on agar divide just as rapidly as cells in comparable well-aerated liquid medium (142). Thus, *E. coli* has evolved to reproduce efficiently in a multicellular context.

Multicellular, density-dependent aspects of initial colony spreading can be observed most readily with colonies inoculated as small spots from a micropipette (140). The cells at the edge of a spot inoculum are initially disordered but align themselves during the first two hours of growth after inoculation (150). Before the inoculated spot expands, it fills in and cells around the periphery pile up to create a multilayered mound (140). Once this mound has formed, active expansion over the substrate begins (146). This structure includes exopolymers visible in micrographs (140, 146) and develops into a distinct peripheral zone demarcated by a deep groove, which expands with the colony (140). The groove appears to define the region of active spreading over the substrate, consistent with geometrical interpretations of colony growth dynamics (117). Using small fragments of glass wool as probes, it can be confirmed that colony spreading involves the development of multicellular structure rather than autonomous cell divisions at the periphery. When a spreading colony encounters such a fiber, motile cells at the periphery swim into the liquid around the fiber, coat it, and form a younger population. With short fibers (<100 microns), the coating population does not have time to develop the structure needed for spreading and stays in a small zone along the fiber, which is engulfed by the advancing colony (146).

After 24 hours or more of development, *E. coli* colonies display considerable spatial organization. Organized cellular differentiation was first detected as concentric patterns of beta-galactosidase staining (136, 137). Some of these patterns resulted from expression of stable *lacZ* fusions, including those to *polA* (143, 146), but other patterns reflected the formation of concentric zones of derepression and replication of Mu*dlac* elements under control of the Mu*cts62* repressor (145, 149). Concentric patterning was also visible in surface contours of colonies lacking genetic modifications (143–145). SEM examination revealed zones within colonies characterized by cells of distinct sizes, shapes, and patterns of multicellular arrangement (140). Vertical sections through colonies revealed stratification into layers of cells with different protein contents, many of which appeared to be nonviable (144). Such zones containing dead bacteria are analogous to the stalks of Myxobacterial fruiting bodies created by cell lysis (152). Programmed cell death clearly plays a role in colony and fruiting body morphogenesis, and bacterial examples of this phenomenon are rapidly accumulating (4, 29, 51, 187). All these data indicate an unanticipated capacity for cellular differentiation and creation of discrete zones of differentiated cells ("tissues") in *E. coli* K12 colonies. The *E. coli* results parallel the earliest systematic observations of colony development showing spatially organized cellular differentiation (95).

B. subtilis Colony Development

Morphogenetic studies of *B. subtilis* colonies were pioneered by physicists interested in pattern formation (12, 14, 53, 104). By altering nutrient levels and agar concentrations, they observed transitions between distinct colony morphologies corresponding to forms described in inorganic material. At lowest nutrient levels and highest agar concentrations, *B. subtilis* colonies assumed a fractal shape typical of structures generated by diffusion-limited aggregation (DLA) (12, 14, 53, 104, 110). DLA shapes occur when the physical processes of nutrient diffusion govern bacterial growth, i.e. when the bacteria are starved, incapable of active motility, and have lost capabilities to control morphogenesis. As nutrient concentrations increase (but still at high agar concentrations), colonies assume the so-called Eden configuration and expand by creating arrays of highly elongated cells encased in exopolymer around the periphery (12, 104, 175). At intermediate nutrient and agar concentrations, dendritic patterns develop that increase colony surface area to permit more efficient uptake of nutrients. Inside the dendrites, groups of moderately elongated, highly motile cells are found encased in an envelope of exopolymer (15). The dendrites advance as the cell groups push on the exopolymer envelope. At certain agar and nutrient concentrations, transitions occur from highly branching to more compact colonies. These transitions have been modeled as resulting from activation of long-range negative chemotaxis functions (13). The physics-inspired approach thus implicates both group motility, with an important role for exopolymers, and intercellular chemotactic signaling.

Besides fascinating colony shapes, patterns of differential *lacZ* expression from synthetic gene fusions analogous to those in *E. coli* colonies have recently been observed in *B. subtilis* (108, 109, 128). When colonies carrying the fusion constructs are placed on agar capable of inducing different colony morphologies, the beta-galactosidase patterns are observed to be "nested" inside the overall colony shape, suggesting a connection between the control of colony expansion and *lacZ* fusion expression (108).

Proteus and Serratia Swarming

B. subtilis motility in colonies is related to a collective process observed in both Gram-positive and Gram-negative species, called swarming (72). Swarming involves rapid migration over a surface by groups of elongated, hyperflagellated "swarmer" cells encased in exopolymers (11, 166, 179). Length and hyperflagellation distinguish swarmer cells from "swimmer" cells, which resemble motile *E. coli* and are capable only of swimming in fluid medium, not of migrating over agar surfaces (74). Starting with the book Hauser published over a century ago (70), swarming has been studied most intensively in *Proteus mirabilis* and *Proteus vulgaris*. Recent studies have also focused genetic and molecular techniques on *Serratia marcescens* (106, 111) and *Serratia liquefaciens* (43, 44, 57, 58).

The key aspect of swarming motility is its collective nature. Isolated swarmer cells do not migrate over agar, only groups or "rafts" of aligned swarmers do (165). Raft size is proportional to the hardness of the agar because swarming is more difficult at higher agar concentrations (119, 163, 164). Swarmers are also encased in exopolysaccharide as they migrate (159). The acidic capsular polysaccharide produced by *Proteus mirabilis* plays a key role in swarming mobility; mutants that lack it are inhibited in their migration (67). The inherent multicellularity of *Proteus* swarm colony morphogenesis provides part of the explanation for the formation of symmetrical colonies with periodically spaced terraces (18). Terracing results from cycles of alternating swarming and consolidation phases (11). Swarming periodicity is not based on cycles of nutrient exhaustion and chemotactic migration (119). Instead, it is possible to explain the clock-like behavior of swarming *Proteus* colonies by variation in a multicellular parameter, the age-weighted swarmer cell population density (45).

Genetic studies with *Serratia liquefaciens* have revealed dual genetic control of swarming in that species (57, 58). Swarmer cell differentiation is triggered by ectopic expression of the FlhDC motility/chemotaxis regulators (43). A second level of swarming control in *Serratia* involves AHL signaling. Mutation of the *swrI* homologue of *luxI* leads to a swarm-defective phenotype (44). The AHL does not play any role in swarmer cell differentiation but instead stimulates production of a surfactant essential for *Serratia* swarming (58). *Serratia* swarm colonies produce cyclic peptide surfactants critical to motility (105), and the *swrI* mutant phenotype can be reversed by the addition of detergents (PW Lindum, U Anthoni, G Christoffersen, L Eberl, S Molin, M Givskov, manuscript in preparation).

Granule Development in Anaerobic Bioreactors

In nature and industry, most biotransformations are carried out by microbial consortia, not by monocultures (100, 130). Such consortia have definite physical organizations, most commonly biofilms or granular aggregates. A paradigm of spatially organized consortia demonstrating self-organization is the class of biotransforming granules that form in upflow anaerobic sludge blanket (UASB) reactors (132). UASB reactors are the most widely used form of high-rate reactor for anaerobic biological wastewater treatment to achieve the biodegradation of organic substances to CH_4 and CO_2. Ancillary objectives include the breakdown of toxic pollutants, such as halogenated hydrocarbons (21).

In the UASB granules, several different groups of bacteria carry out sequential metabolic processes (132, 160): (*a*) conversion of xenobiotics to biodegradable molecules, as in dehalogenation; (*b*) hydrolysis of polymers to small molecules;

(c) fermentation of small molecules to H_2, CO_2, acetate, and short-chain volatile fatty acids (mainly propionate and butyrate); (d) oxidation of volatile fatty acids to acetate and H_2/CO_2; and (e) conversion of acetate, H_2, and CO_2 to methane by aceticlastic methanogens.

The organisms that carry out these diverse processes are organized in granules of about 0.1–5 mm in diameter. The size and composition of the granules depend on the wastewater composition and conditions such as temperature (3, 99, 160). Much of the physical integrity of the granules is due to large amounts of exopolymers, particularly proteins and polysaccharides (65).

The microbial flora of these granules is rich in the aceticlastic methanogens, including *Methanosarcina* spp. (66, 132). *Methanosarcina* can transition between a multicellular aggregate and a disaggregated single-celled form (37). In addition, many propionate and butyrate-fermenting *Syntrophobacter* spp. and *Syntrophomonas* spp. have been identified in microcolonies in intimate association with the methanogens (3, 132, 160). The spatial organization of the bacteria is critical for thermodynamic reasons. The partial pressure of hydrogen must be kept low to ensure efficient fermentation of the volatile fatty acids (130, 131). Many interspecific syntrophic reactions are only energetically beneficial if hydrogen transfers occur over distances of a few microns or less (160). Because of the need for such close proximity, random cell–cell associations would lower metabolic efficiency. On this basis, signaling mechanisms to organize the syntrophic species can be predicted. Larger-scale organization is observed in the distribution of distinct species (99) and of distinct metabolic processes (3) within the UASB granules. By introducing new species, the metabolic capabilities and substrate range of the granules can be extended. Dechlorination ability was acquired by a bioreactor seeded with *Desulfomonile tiedjei*, and the *Desulfomonile* cells were seen by immunofluorescence to be incorporated into the granules (2).

ADAPTIVE BENEFITS FROM MULTICELLULAR COOPERATION

More Efficient Proliferation from Cellular Division of Labor

In a 1988 *Scientific American* article, nitrogen-fixing heterocyst formation in filamentous photosynthetic cyanobacteria was cited to illustrate how two cell types could cooperate in a monospecific multicellular population (1, 141). Likewise, the ability of *Proteus* to spread rapidly over an agar surface depends on swarmer cell differentiation, and the velocity of spreading is directly related to the rate of biomass production (119). Monocultures also display larger-scale functional differentiation of various cell groups, such as the stalks and sporangia in Myxobacterial fruiting bodies (152) and substrate and aerial mycelia in *Streptomyces* (33).

Microbiologists know myriad cases where biochemical cycles involve the participation of multiple different species, and the example of *Methanobacillus omelianskii*, supposedly a single organism that is actually a symbiotic association of two different species (23), is a classic reminder of the limitations of the pure culture approach (42). Laboratory studies document increased biodegradation by mixed cultures (e.g. 47). Syntrophic processes mediated by microbial guilds or consortia underlie the mineral cycles and thermodynamics of biogeochemistry (100). As with the UASB granules, we need to find out how capabilities for communication and spatial organization are used interspecifically. Interspecific plasmid transfer and communication of symbiotic and pathogenic organisms with their eukaryotic hosts demonstrate that many bacteria can coordinate their behaviors with other species (7, 48). In some cases, such as the type IV transport systems related to plasmid transfer mechanisms, the parallel is explicit between prokaryote-eukaryote and interspecific communication among bacteria (7, 181). An interesting variant on interspecific signaling is the excretion of chemoattractants by predatory *M. xanthus* to entrap prey *E. coli* (151).

Access to Resources and Niches That Require a Critical Mass and Cannot Effectively Be Utilized by Isolated Cells

In common laboratory media, bacteria are supplied with simple growth substrates readily utilized by individual cells. In nature, many bacteria break down complex organic polymers, requiring the concerted action of many cells. The predatory *Mxyobacteria* utilize a "wolf pack" strategy to attack and lyse their prey organisms by liberating digestive extracellular enzymes and absorbing the cell contents (122). Using *M. xanthus* and casein as a model substrate, substrate utilization was found to be dependent on population density (121). Interestingly, groups of *M. xanthus* cells can migrate chemotactically but individual cells cannot (151). *M. xanthus* grazes on cyanobacteria in ponds (27). However, the aqueous environment can dilute both the lytic exoenzymes and liberated nutrients. Thus, the predators construct spherical colonies and trap prey organisms in pockets where lysis and feeding can occur efficiently (27), showing that multicellular behavior to permit resource utilization includes the capacity for morphogenesis of organized macroscopic structures.

Other bacteria illustrate the role of population density in resource utlization. The phytopathogen *Erwinia carotovora* synthesizes exoenzymes for plant cell wall degradation, but does so under control of an AHL quorum-sensing system (79, 116). The circuit apparently ensures that the bacteria will only invest their cellular capital in digestive exoenzyme production once there are sufficient cells for effective attack on the plant structure. A similar strategy apparently accounts for the AHL control of synthesis of elastase, toxins, and other virulence factors by the opportunistic pathogen *Pseudomonas aeruginosa* (92, 112). An extra rationale for quorum-sensing by pathogens is to restrain synthesis of virulence factors that, if synthesized constitutively, could be detected by host defense systems at subeffective concentrations and lead to the destruction of low-density populations (62a, 162).

Collective Defense Against Antagonists That Eliminate Isolated Cells

Many agents can effectively kill isolated bacterial cells in suspension but are ineffective against dense or organized populations of the same bacteria. One example is catalase protection against oxidative damage. There is no difference in the survival to hydrogen peroxide challenge of *cat* (catalase-defective) and *cat*+ (catalase-positive) *E. coli* strains when tested in dilute suspension, but in thick suspensions or in microcolonies on an agar surface, the *cat*+ strain shows much greater resistance (98). Appropriately, catalase expression is RpoS-dependent (91). Another example is a penicillin-resistant biofilm on a pacemaker composed of penicillin-sensitive *S. aureus* bacteria; the biofilm fed a recurring septicemia that could be cleared but never eliminated by antibiotic therapy (103). It appears to be a general rule that organization in biofilms and colonies provides enhanced resistance to a wide range of antibacterials (38, 89, 180).

Multicellular defense also has an aggressive aspect. Many bacteria produce antibiotics, generally under the control of intercellular communication and quorum-sensing systems. In the *Actinomycetes*, antibiotic synthesis is frequently regulated by γ-butyrolactone signaling molecules and is coupled with morphological differentiation (Table 1). Similarly, the synthesis of carbapenem by *Erwinia carotovora* to eliminate competitors for nutrients liberated by degraded plants is regulated by a two-stage AHL hierarchy (107). In the Gram-positive bacteria, a number of antimicrobial peptides are also subject to autoinduction (Table 1) (90).

Optimization of Population Survival by Differentiation into Distinct Cell Types

Inevitably, bacterial populations encounter new circumstances in which they will have to survive diverse physical, chemical, nutritional, and biological challenges. From an ecological point of view, the population is the key biological entity. Its survival depends upon having the right cells when confronted by phage attack, antibiotics, desiccation, or the need to utilize novel growth substrates. Populations have several distinct mechanisms for creating new cell types.

SPORULATION AND FORMATION OF DORMANT CELLS   The most obvious diversification is formation of spores or other resistant, dormant forms. Sporulation is subject to intercellular signaling and multicellular regulation in all cases that have been closely investigated, including *B. subtilis* (63), *M. xanthus* (152), and *Streptomyces coelicolor* (33). Moreover, spore formation is often connected with elaborate processes of multicellular morphogenesis. Thus, spore formation can be considered a function of the entire interactive population. Among nonsporulating bacteria, dormant forms can survive for long periods under difficult conditions. In *E. coli*, survival in stationary phase involves several factors, including the RpoS sigma factor (91), which is subject to quorum-sensing control in *P. aeruginosa* (92) and possibly also in *E. coli* (76). An interesting corollary to multicellular control of dormancy is the suggestion that exit from dormancy involves intercellular signaling and "wake-up" pheromones, both in isolated cells (84, 85) and in biofilm populations (8). This may explain why many dormant cells are difficult to culture from dilute suspension but not from denser inocula. An extracellular protein needed for resuscitation of dormant *Micrococcus luteus* cells has been reported (86).

EXCHANGE OF GENETIC INFORMATION   Besides sporulation, *Bacillus subtilis* populations can develop subpopulations competent for DNA uptake with potential to incorporate new genetic information and thereby novel proliferation and survival abilities (155). Development of competence is a multicellular process involving intercellular signaling and elaborate signal processing. In *B. subtilis*, competent cells can take up DNA from any source (155). A different kind of competence occurs in *Neisseria gonorrhoeae*. Competence is constitutive (19) and is restricted to the uptake of DNA fragments carrying an *N. gonorrohoea*–specific sequence tag (60). *Neisseria* populations use DNA exchange for the purpose of stimulating recombination (gene conversion) events that alter the primary structures of their surface proteins (including pili) (167). Such phase and antigenic variations in surface protein structure allow *Neisseria* populations to modulate aggregation and virulence properties and to evade immune surveillance.

DNA exchange mechanisms are ecologically significant. Plasmids, phages, transposons, and other mobile genetic elements play major roles in the evolution of antibiotic resistance (17, 26, 178), pathogenicity determinants (34, 48), and new catabolic pathways (174). Taking our knowledge of genetic exchange to its logical conclusion, Sonea & Panisset (158) developed the radically multicellular concept of the distributed prokaryotic genome. They argued that there is one large bacterial genome encoding more functions than can be accommodated in any individual cell, but that every bacterial species has access to the whole through mobile elements. Thus, bacteria can be tailor-made for new ecological niches (cf. 93).

MUTATION   DNA restructuring frequently involves the same mobile elements that participate in intercellular genetic exchange (17, 26, 138, 147). Even purely intracellular processes of genetic change serve to diversify the parental population. For survival of the individual bacterium, loss of surface receptors is almost always not beneficial, but the reliable appearance of receptorless mutants means the population as a whole will survive phage attack. Similar logic applies to bacteria with impaired ribosome efficiency plus streptomycin resistance (61). Many natural bacterial populations, such as pathogens, have a mutator phenotype, suggesting advantages to elevated rates of spontaneous mutation in highly variable environments (94).

The phenomenon of adaptive mutation indicates that bacteria can increase their mutational activity in response to starvation and other stress conditions (49, 123, 147). Enhanced mutagenesis under selection or starvation is almost always studied in dense, postlogarithmic populations wherein intercellular signaling is most intense: lawns on selective plates (28, 135), papillae forming on aging colonies (68, 69), and saturated liquid cultures (101). For example, in the *lac33* frameshift reversion system, there is a role for F'*lacpro* transfer functions that are activated by the conditions prevailing in dense, highly aerobic surface populations (56, 114, 118).

Several suggested mechanisms for adaptive mutation can reasonably be considered in a multicellular context. One is the proposal that some cells in stressed populations undergo extensive DNA changes, do not survive, but donate fragments of their rearranged genomes to other cells, which then go on to proliferate as the experimentally detected adaptive "mutants" on selective media (73, 120). The view that signal transduction networks stimulate the action of natural genetic engineering systems in response to starvation or other stresses (147) has been strengthened by the observation that the growth phase regulators, RpoS and HNS, have opposite effects on the *araB-lacZ* fusion system (59). The RpoS requirement for fusion formation suggests potential quorum-sensing involvement (76), even though an *E.coli* AHL signal has not been detected (P Williams, GSAB Stewart, personal communication).

The recent observation that many revertants of the *lac33* frameshift contain additional unselected mutations provides experimental support for a hypermutable state under selective conditions (50, 173). Hall's original proposal for a hypermutable state (in which individual cells undergoing elevated random mutagenesis perish unless they produce the right mutation for the current selective conditions) has a multicellular logic (69). A single hypermutable cell has little probability of survival, but a population containing many hypermutable bacteria can multiply the chances of success by orders of magnitude and thus has a clear advantage, when in trouble, to sacrifice a subpopulation to the hypermutable state.

Critical tests of a possible multicellular dimension to intracellular processes of genetic change must come from experiments on the regulatory mechanisms underlying adaptive mutation. A key focus will be the roles, if any, of intercellular communication molecules.

CONCLUDING COMMENTS: IS THERE A CONFLICT BETWEEN CELLULAR AND MULTICELLULAR VIEWS OF BACTERIAL POPULATIONS?

Through studies of collective behaviors and intercellular signaling molecules, we are beginning to appreciate the extensive capacities for communication and coordination that enhance bacterial power to operate in the biosphere. Collectively and coordinately, bacteria act far more efficiently than they could as autonomous agents. Any effort to provide a formal definition of bacterial multicellularity or draw boundaries between single-celled and multi-celled perspectives would be counterproductive. The key to bacterial multicellularity resides in the ability of each individual cell to receive, interpret, and respond to information from its neighbors. In other words, recognizing bacterial multicellularity deepens our appreciation of the information-processing capabilities of individual bacterial cells. Meaningful information transfer between components and the system as a whole is integral to the notion of organism. I predict the concept of organism will increasingly be seen as a fundamental idea throughout science, which in all fields is moving away from Cartesian reductionism toward a more connectionist, interactive view of natural phenomena. Exploring this more organic view of nature will require detailed investigation of self-organizing complex systems. Bacteria provide some of the best experimental material available (145, 148). Thus, thinking about bacterial populations as multicellular organisms may help put microbiology at the top of the scientific agenda in the 21st Century.

ACKNOWLEDGMENTS

I thank my long-time collaborator Martin Dworkin, my colleagues Alexander Zehnder, Paul Williams, Gordon Stewart, Douglas Kell, Harley McAdams, Adam Arkin, and Alan Grossman, and my collaborators in the CASE (Complex Adaptive Systems Ecology) consortium, Søren Molin, Michael Givskov, Birgit Ahring, and Julian Wimpenny, for brainstorming, for invaluable comments on the manuscript, and for pointing out many multicellular phenomena previously unknown to me. My research on bacterial multicellularity and adaptive mutation has been supported by the National Science Foundation.

Visit the *Annual Reviews home page* at
http://www.AnnualReviews.org.

Literature Cited

1. Adams DG. 1997. Cyanobacteria. See Ref. 148, pp. 109–48
2. Ahring BK, Christiansen N, Mathrani I, Hendriksen HV, Macario AJ, Conway de Macario E. 1992. Introduction of a de novo bioremediation ability, aryl sulfate dechlorination, into anaerobic granular sludge by inoculation of sludge with *Desulfomonile tiedjei. Appl. Environ. Microbiol.* 58:3677–82
3. Ahring BK, Schmidt JE, Winther-Nielsen M, Macario AJ, de Macario EC. 1993. Effect of medium composition and sludge removal on the production, composition, and architecture of thermophilic (55 degrees C) acetate-utilizing granules from an upflow anaerobic sludge blanket reactor. *Appl. Environ. Microbiol.* 59:2538–45
4. Aizenman E, Engelberkulka H, Glaser G. 1996. An *Escherichia coli* chromosomal addiction module regulated by $3',5'$-bispyrophosphate—a model for programmed cell death. *Proc. Natl. Acad. Sci. USA* 93:6059–63
5. Alberts B, Bray D, Lewis J, Raff M, Roberts K, Watson JD. 1994. *Molecular Biology of the Cell.* New York: Garland. 1294 pp. 3rd ed.
6. Arnold JW, Shimkets LJ. 1988. Cell surface properties correlated with cohesion in *Myxococcus xanthus. J. Bacteriol.* 170:5771–77
7. Baker B, Zambryski P, Staskawicz B, Dinesh-Kumar SP. 1997. Signalling in plant-microbe interactions. *Science* 276:726–33
8. Batchelor SE, Cooper M, Chhabra SR, Glover LA, Stewart GSAB, et al. 1997. Cell-density regulated recovery of starved biofilm populations of ammonia oxidising bacteria. *Appl. Environ. Microbiol.* 63:2281–86
9. Behmlander RM, Dworkin M. 1991. Extracellular fibrils and contact mediated cell interactions in *Myxococcus xanthus. J. Bacteriol.* 173:7810–21
10. Behmlander RM, Dworkin M. 1994. Biochemical and structural analysis of the ex-
tracellular matrix fibrils of *Myxococcus xanthus. J Bacteriol.* 176:6295–303
11. Belas R. 1997. *Proteus mirabilis* and other swarming bacteria. See Ref. 148, pp. 183–219
12. Ben-Jacob E, Cohen I. 1997. Cooperative formation of bacterial patterns. See Ref. 148, pp. 394–416
13. Ben-Jacob E, Schochet O, Tenenbaum A, Cohen I, Czirók A, Vicsek T. 1994. Generic modelling of cooperative growth patterns in bacterial colonies. *Nature* 368:46–48
14. Ben-Jacob E, Shmueli H, Schochet O, Tenenbaum A. 1992. Adaptive self-organization during growth of bacterial colonies. *Physica A* 187:378–424
15. Ben-Jacob E, Tenenbaum A, Schochet O, Avidan O. 1994. Holotransformations of bacerial colonies and genome cybernetics. *Physica A* 202:1–47
16. Beppu T. 1995. Signal transduction and secondary metabolism: prospects for controlling productivity. *Trends Biotechnol.* 13:264–69
17. Berg DE, Howe MM, eds. 1989. *Mobile DNA.* Washington, DC: ASM. 972 pp.
18. Bisset KA. 1973. The zonation phenomenon and structure of the swarm colony in *Proteus mirabilis. J. Med. Microbiol.* 6:429–33
19. Biswas GD, Thompson SA, Sparling PF. 1989. Gene transfer in *Neisseria gonorrhoeae. Clin. Microbiol. Rev.* 2(Suppl.): S24–28
20. Blat Y, Eisenbach M. 1995. Tar-dependent and -independent pattern formation by *Salmonella typhimurium. J. Bacteriol.* 177:1683–91
21. Bouwer EJ, Zehnder AJB. 1993. Bioremediation of organic compounds—putting microbial metabolism to work. *Trends Biotechnol.* 11:360–67
22. Bray D. 1990. Intracellular signalling as a parallel distributed process. *J. Theoret. Biol.* 143:215–31
23. Bryant MP, Wolin EA, Wolin MJ, Wolfe RS. 1967. *Methanobacillus omelianskii,*

a symbiotic association of two species of bacteria. *Arch. Microbiol.* 59:20–31

24. Budrene EO, Berg HC. 1991. Complex patterns formed by motile cells of *Escherichia coli. Nature* 349:630–33

25. Budrene EO, Berg HC. 1995. Dynamics of formation of symmetrical patterns by chemotactic bacteria. *Nature* 376:49–53

26. Bukhari AI, Shapiro JA, Adhya SL, eds. 1977. *DNA Insertion Elements. Plasmids and Episomes.* Cold Spring Harbor Lab. 782 pp.

27. Burnham JC, Collart SA, Highison BW. 1981. Entrapment and lysis of the cyanobacterium *Phormidium luridum* by aqueous colonies of *Myxococcus xanthus* PCO2. *Arch. Microbiol.* 129:285–94

28. Cairns J, Foster PL. 1991. Adaptive reversion of a frameshift mutation in *Escherichia coli. Genetics* 128:695–701

29. Chaloupka J, Vinter V. 1996. Programmed cell death in bacteria. *Folia Microbiol.* 41:451–64

30. Chang B-Y, Dworkin M. 1994. Isolated fibrils rescue cohesion and development in the Dsp mutant of *Myxococcus xanthus. J. Bacteriol.* 176:7190–96

31. Chapon-Herve V, Akrim M, Latifi A, Williams P, Ladzunski A, Bally M. 1997. Regulation of xcp secretion pathway by multiple quorum-sensing modulations in *Pseudomonas aeruginosa. Mol. Microbiol.* 24:1169–78

32. Charpak M, Dedonder R. 1965. Production d'un "facteur de competence" soluble par *Bacillus subtilis* Marburg *ind*–168. *CR Acad. Sci. Paris* 260:5638–41

33. Chater KF, Losick R. 1997. The mycelial life-style of *Streptomyces coelicolor* A3(2) and its relatives. See Ref 148, pp. 149–82

34. Cheetham BF, Katz ME. 1995. A role for bacteriophages in the evolution and transfer of bacterial virulence determinants. *Mol. Microbiol.* 18:201–8

35. Clewell DB. 1993. Bacterial sex pheromone-induced plasmid transfer. *Cell* 73: 9–12

36. Clewell DB. 1993. *Bacterial Conjugation.* New York: Plenum. 413 pp.

37. Conway de Macario E, Macario AJ, Mok T, Beveridge TJ. 1993. Immunochemistry and localization of the enzyme disaggregatase in *Methanosarcina mazei. J. Bacteriol.* 175:3115–20

38. Costerton JW, Lewandowski Z, Caldwell DE, Korber DR, Lappin-Scott HM. 1995. Microbial biofilms. *Annu. Rev. Microbiol.* 49:711–45

38a. Davies DG, Parsekk MR, Pearson JP, Iglewski BH, Costerton JW, Greenberg EP. 1998. The involvement of cell-to-cell signals in the development of a bacterial biofilm. *Science* 179: In press

39. Denarie J, Debelle F, Rosenberg C. 1992. Signaling and host range variation in nodulation. *Annu. Rev. Microbiol.* 46:497–531

40. Dunlap PV. 1997. *N*-acyl-L-homoserine lactone autoinducers in bacteria: unity and diversity. See Ref. 148, pp. 69–108

41. Dworkin M. 1993. Tactic behavior of *Myxococcus xanthus. J. Bacteriol.* 154: 452–59

42. Dworkin M. 1997. Multiculturism versus the single microbe. See Ref. 148, pp. 3–13

43. Eberl L, Christiansen G, Molin S, Givskov M. 1996. Differentiation of *Serratia liquefaciens* into swarm cells is controlled by the expression of the *flhD* master operon. *J. Bacteriol.* 178:554–59

44. Eberl L, Winson MK, Sternberg C, Stewart GSAB, Christiansen G, et al. 1996. Involvement of *N*-acyl-L-homoserine lactone autoinducers in controlling the multicellular behavior of *Serratia liquefaciens. Mol. Microbiol.* 20:127–36

45. Esipov S, Shapiro JA. 1998. Kinetic model of *Proteus mirabilis* swarm colony development. *J. Math. Biol.* 36:249–68.

46. Felkner IC, Wyss O. 1964. A substance produced by competent *Bacillus cereus* 569 cells that affects transformability. *Biochem. Biophys. Res. Commun.* 16:94–99

47. Field JA, Stams AJ, Kato M, Schraa G. 1995. Enhanced biodegradation of aromatic pollutants in cocultures of anaerobic and aerobic bacterial consortia. *Ant. v. Leeuwenhoek J. Microbiol. Serol.* 67:47–77

48. Finlay BB, Falkow S. 1997. Common themes in microbial pathogenicity. *Microbiol. Mol. Biol. Rev.* 61:136–69

49. Foster PL. 1993. Adaptive mutation: the uses of adversity. *Annu. Rev. Microbiol.* 47:467–504

50. Foster PL. 1997. Nonadaptive mutations occur on the F' episome during adaptive mutation conditions in *Escherichia coli. J. Bacteriol.* 179:1550–54

51. Franch T, Gerdes K. 1996. Programmed cell death in bacteria—translational repression by messenger-RNA end-pairing. *Mol. Microbiol.* 21:1049–60

52. Fry NK, Raskin L, Sharp R, Alm EW, Mobarry BK, Stahl DA. 1997. In situ analyses of microbial populations with molecular probes: the phylogenetic dimension. See Ref. 148, pp. 292–338

53. Fujikawa H, Matsushita M. 1989. Fractal growth of *Bacillus subtilis* on agar plates. *J. Phys. Soc. Jpn.* 58:3875–78
54. Fuqua WC, Winans SC, Greenberg EP. 1994. Quorum sensing in bacteria: the LuxR-LuxI family of cell density-responsive transcriptional regulators. *J. Bacteriol.* 176:269–75
55. Fuqua C, Winans SC, Greenberg EP. 1996. Census and consensus in bacterial ecosystems: the LuxR-LuxI family of quorum-sensing transcriptional regulators. *Annu. Rev. Microbiol.* 50:727–51
56. Galitski T, Roth JR. 1995. Evidence that F plasmid transfer replication underlies apparent adaptive mutation. *Science* 268:421–23
57. Givskov M, Eberl L, Molin S. 1997. Control of exoenzyme production, motility and cell differentiation in *Serratia liquefaciens*. *FEMS Microbiol. Lett.* 148:115–22
58. Givskov M, Östling J, Ebert L, Lindum PW, Christensen AB, et al. 1998. The participation of two separate regulatory systems in controling swarming motility of *Serratia liquefaciens*. *J. Bacteriol.* 180: In press
59. Gómez-Gómez JM, Blázquez J, Baquero F, Martinez JL. 1997. H-NS and RpoS regulate emergence of LacAra+ mutants of *Escherichia coli* MCS2. *J. Bacteriol.* 179:4620–22
60. Goodman SD, Scocca JJ. 1988. Identification and arrangement of the DNA sequence recognized in specific transformation of *Neisseria gonorrhoeae*. *Proc. Natl. Acad. Sci. USA* 85:6982–86
61. Gorini L, Davies J. 1968. The effect of streptomycin on ribosomal function. *Curr. Topics Microbiol. Immunol.* 44:100–22
62. Gray KM. 1997. Intercellular communication and group behavior in bacteria. *Trends Microbiol.* 5:184–88
62a. Greenberg EP. 1997. Quorum sensing in gram-negative bacteria. *ASM News* 63:371–77
63. Grossman AD. 1995. Genetic networks controlling the initiation of sporulation and the development of genetic competence in *Bacillus subtilis*. *Annu. Rev. Genet.* 29:477–508
64. Grossman A, Losick R. 1988. Extracellular control of spore formation in *Bacillus subtilis*. *Proc. Natl. Acad. Sci. USA* 85:4369–73
65. Grotenhuis JTC, Smit M, van Lammeren AAM, Stams AJM, Zehnder AJB. 1991. Localization and quantification of extracellular polymers in methanogenic granular sludge. *Appl. Microbiol. Biotechnol.* 36:115–19
66. Grotenhuis JTC, Smit M, Plugge CM, Yuansheng X, van Lammeren AAM, et al. 1991. Bacteriological composition and structure of granular sludge adapted to different substrates. *Appl. Environ. Microbiol.* 57:1942–49
67. Gygi D, Rahman MM, Lai H-C, Carlson R, Guard-Petter J, Hughes C. 1995. A cell surface polysaccharide that facilitates rapid population migration by differentiated swarm cells of *Proteus mirabilis*. *Mol. Microbiol.* 17:1167–75
68. Hall BG. 1988. Adaptive evolution that requires multiple spontaneous mutations. I: Mutations involving an insertion sequence. *Genetics* 120:887–97
69. Hall BG. 1990. Spontaneous point mutations that occur more often when advantageous than when neutral. *Genetics* 126:5–16
70. Hauser G. 1885. *Über Fäulnissbacterien und deren Beziehungen zur Septicämie*. Leipzig: Vogel. 94 pp.
71. Havarstein LS, Coomaraswamy G, Morrison DA. 1995. An unmodified heptadecapeptide pheromone induces competence for genetic transformation in *Streptococcus pneumoniae*. *Proc. Natl. Acad. Sci. USA* 92:11140–44
72. Henrichsen J. 1972. Bacterial surface translocation: a survey and a classification. *Bacteriol. Rev.* 36:478–503
73. Higgins NP. 1992. Death and transfiguration among bacteria. *Trends Biochem. Sci.* 17:207–11
74. Hoeniger JF. 1964. Cellular changes accompanying the swarming of *Proteus mirabilis*. I: Observations on living cultures. *Can. J. Microbiol.* 10:1–9
75. Horinouchi S, Beppu T. 1992. Autoregulatory factors and communication in *Actinomycetes*. *Annu. Rev. Microbiol.* 46:377–98
76. Huisman GW, Kolter R. 1994. Sensing starvation: a homoserine lactone-dependent signaling pathway in *Escherichia coli*. *Science.* 265:537–39
77. Hull WE, Berkessel A, Stamm I, Plaga W. 1997. *Intercellular signalling in Stigmatella aurantiaca: proof, purification and structure of a myxobacterial pheromone*. Presented at Annu. Meet. Biol. Myxobacteria, 24th, New Braunfels, Texas
78. Ji GY, Beavis RC, Novick RP. 1995. Cell density control of staphylococcal virulence mediated by an octapeptide pheromone. *Proc. Natl. Acad. Sci. USA* 92:12055–59
79. Jones S, Yu B, Bainton NJ, Birdsall M,

Bycroft BW, et al. 1993. The lux autoinducer regulates the production of exoenzyme virulence determinants in *Erwinia carotovora* and *Pseudomonas aeruginosa. EMBO J.* 12:2477–82

80. Kaiser D. 1979. Social gliding is correlated with the presence of pili in *Myxococcus xanthus. Proc. Natl. Acad. Sci. USA* 76:5952–56

81. Kaiser D, Losick R. 1993. How and why bacteria talk to each other. *Cell* 73:873–85

82. Kaplan HB, Greenberg EP. 1985. Diffusion of autoinducer is involved in regulation of the *Vibrio fischeri* luminescence system. *J. Bacteriol.* 163:1210–14

83. Kaplan HB, Plamann L. 1996. A *Myxococcus xanthus* cell density-sensing system required for multicellular development. *FEMS Microbiol. Lett.* 139:89–95

84. Kaprelyants AS, Kell DB. 1996. Do bacteria need to communicate with each other for growth? *Trends Microbiol.* 4:237–42

85. Kell DB, Kaprelyants AS, Grafen A. 1995. Pheromones, social behaviour and the functions of secondary metabolism in bacteria. *Trends Ecol. Evol.* 10:126–29

86. Kell DB, Kaprelyants AS, Mukamolova GV, Davey HM, Young M. 1997. *Dormancy and social resuscitation in nonsporulating bacteria—the identification of a molecular wake-up call.* Presented at Eur. Congr. Biotechnol., 8th, Budapest

87. Kellog D, Peacock WL Jr, Deacon WE, Brown L, Pirkle CI. 1963. *Neisseria gonnorhoeae.* I: Virulence genetically linked to clonal variation. *J. Bacteriol.* 85:1274–79

88. Kim SK, Kaiser D, Kuspa A. 1992. Control of cell density and pattern by intercellular signalling in *Myxococcus* development. *Annu. Rev. Microbiol.* 46:117–39

89. Kinniment S, Wimpenny JWT. 1990. Biofilms and biocides. *Int. Biodeterior. Bull.* 26:181–94

90. Kleerebezem M, Quadri LE, Kuipers de vos OP. 1997. Quorum sensing by peptide pheromones and two component signal-transduction systems in Gram-positive bacteria. *Mol. Microbiol.* 24:895–904

91. Kolter R, Siegele DA, Tormo A. 1993. The stationary phase of the bacterial life cycle. *Annu. Rev. Microbiol.* 47:855–74

92. Latifi A, Foglino M, Tanaka K, Williams P, Lazdunski A. 1996. A hierarchical quorum-sensing cascade in *Pseudomonas aeruginosa* links the transcriptional activators LasR and RhlR (VsmR) to expression of the stationary-phase sigma factor RpoS. *Mol. Microbiol.* 21:1137–46

93. Lawrence JG. 1977. Selfish operons and speciation by gene transfer. *Trends Microbiol.* 5:355–59

94. LeClerc JE, Li B, Payne WL, Cebula TA. 1996. High mutation frequencies among *Escherichia coli* and *Salmonella* pathogens. *Science* 274:1208–11

95. Legroux R, Magrou J. 1920. État organisé des colonies bactériennes. *Ann. Inst. Pasteur* 34:417–31

96. Leigh JA, Coplin DL. 1992. Exopolysaccharides in plant-bacterial interactions. *Annu. Rev. Microbiol.* 46:307–46

97. Deleted in proof

98. Ma M, Eaton JW. 1992. Multicellular oxidant defense in unicellular organisms. *Proc. Natl. Acad. Sci. USA* 89:7924–28

99. Macario AJL, Visser FA, van Lier JB, Conway de Macario E. 1991. Topography of methanogenic subpopulations in a microbial consortium adapting to thermophilic conditions. *J. Gen. Microbiol.* 137:2179–89.

100. Madigan MT, Martinko JM, Parker J. 1997. *Brock Biology of Microorganisms.* Upper Saddle River, NJ: Prentice-Hall. 986 pp. 8th ed.

101. Maenhaut-Michel G, Shapiro JA. 1994. The roles of starvation and selective substrates in the emergence of *araB-lacZ* fusion clones. *EMBO J.* 13:5229–39

102. Magnuson R, Solomon J, Grossman AD. 1994. Biochemical and genetic characterization of a competence pheromone from *B. subtilis. Cell* 77:207–16

103. Marrie TJ, Costerton JW. 1982. A scanning and transmission electron microscopic study of an infected endocardial pacemaker lead. *Circulation* 66:1339–43

104. Matsushita M. 1997. The formation of colony patterns by a bacterial cell population. See Ref. 148, pp. 366–93

105. Matsuyama T, Kaneda K, Nakagawa Y, Isa K, Hara-Hotta H, Isuya Y. 1992. A novel extracellular cyclic lipopeptide which promotes flagellum-dependent and -independent spreading growth of *Serratia marcescens. J. Bacteriol.* 174:1769–76

106. Matsuyama T, Bhasin A, Harshey RM. 1995. Mutational analysis of flagellum-independent surface spreading of *Serratia marcescens* 274 on low-agar medium. *J. Bacteriol.* 177:987–91

107. McGowan S, Sebaihia M, Jones S, Yu B, Bainton N, et al. 1995. Carbapenem antibiotic production in *Erwinia carotovora* is regulated by CarR, a homologue of the LuxR transcriptional activator. *Microbiology* 141:541–50

108. Mendelson N, Salhi B. 1996. Patterns of

reporter gene expression in phase diagram of *Bacillus subtilis* colony forms. *J. Bacteriol.* 178:1980–89

109. Mendelson NH, Salhi B, Li C. Physical and genetic consequences of multicellularity in *Bacillus subtilis*. See Ref. 148, pp. 339–65

110. Ohgiwari M, Matsushita M, Matsuyama T. 1992. Morphological changes in growth phenomena of bacterial colony patterns. *J. Phys. Soc. Jpn.* 61:816–22

111. O'Rear JL, Alberti L, Harshey RM. 1992. Mutations that impair swarming motility in *Serratia marcescens* 274 include but are not limited to those affecting chemotaxis or flagellar function. *J. Bacteriol.* 174:6125–37

112. Pesci EC, Iglewski BH. 1997. The chain of command in *Pseudomonas* quorum sensing. *Trends Microbiol.* 5:132–34

113. Pestova EV, Havarstein LS, Morrison DA. 1996. Regulation of competence for genetic transformation in *Streptococcus pneumoniae* by an auto-induced peptide pheromone and a two-component regulatory system. *Mol. Microbiol.* 21:853–62

114. Peters JE, Benson SA. 1995. Redundant transfer of F′ plasmids occurs between *Escherichia coli* cells during nonlethal selection. *J. Bacteriol.* 177:847–50

115. Peyton BM, Characklis WG. 1995. Microbial biofilms and biofilm reactors. *Bioprocess Technol.* 20:187–231

116. Pirhonen M, Flego D, Heikinheimo R, Palva ET. 1993. A small diffusible signal molecule is responsible for the global control of virulence and exoenzyme production in the plant pathogen *Erwinia carotovora*. *EMBO J.* 12:2467–76

117. Pirt SJ. 1967. A kinetic study of the mode of growth of surface colonies of bacteria and fungi. *J. Gen. Microbiol.* 47:181–97

118. Radicella JP, Park PU, Fox MS. 1995. Adaptive mutation in *Escherichia coli*: a role for conjugation. *Science* 268:418–20

119. Rauprich O, Matsushita M, Weijer K, Siegert F, Esipov S, Shapiro JA. 1996. Periodic phenomena in *Proteus mirabilis* swarm colony development. *J. Bacteriol.* 178:6525–38

120. Redfield RJ. 1988. Evolution of bacterial transformation: Is sex with dead cells ever better than no sex at all? *Genetics* 119:213–21

121. Rosenberg E, Keller KH, Dworkin M. 1977. Cell density-dependent growth of *Myxococcus xanthus* on casein. *J. Bacteriol.* 129:770–77

122. Rosenberg E, Varon M. 1984. Antibiotics and lytic enzymes. In *Myxobacteria: Development and Cell Interactions*, ed. M Rosenberg, pp. 109–25. New York: Springer

123. Roseneberg SM, Harris RS, Torkelson J. 1995. Molecular handles on adaptive mutation. *Mol. Microbiol.* 18:185–89

124. Ruhfel RE, Leonard BAB, Dunny GM. 1997. Pheromone-inducible conjugation in *Enterococcus faecalis*: mating interactions mediated by chemical signals and direct contact. See Ref. 148, pp. 53–68

125. Sager B, Kaiser D. 1993. Spatial restriction of cellular differentiation. *Genes Dev.* 7:1645–53

126. Sager B, Kaiser D. 1993. Two cell-density domains within the *Myxococcus xanthus* fruiting body. *Proc. Natl. Acad. Sci. USA* 90:3690–94

127. Sager B, Kaiser D. 1994. Intercellular C-signaling and the traveling waves of *Myxococcus*. *Genes Dev.* 8:2793–804

128. Salhi B, Mendelson NH. 1993. Patterns of gene expression in *Bacillus subtilis* colonies. *J. Bacteriol.* 175:5000–8

129. Salmond GPC, Bycroft BW, Stewart GSAB, Williams P. 1995. The bacterial "enigma": cracking the code of cell-cell communication. *Mol. Microbiol.* 16:615–24

130. Schink B. 1997. Energetics of syntrophic cooperation in methanogenic degradation. *Microbiol. Mol. Biol. Rev.* 61:262–80

131. Schmidt JE, Ahring BK. 1993. Effects of hydrogen and formate on the degradation of propionate and butyrate in thermophilic granules from an upflow anaerobic sludge blanket reactor. *Appl. Environ. Microbiol.* 59:2546–51

132. Schmidt J, Ahring BK. 1996. Granular sludge formation in upflow anaerobic sludge blanket (UASB) reactors. *Biotechnol. Bioeng.* 49:229–46

133. Schripsema J, de Rudder KE, van Vliet TB, Lankhorst PP, de Vroom E, et al. 1996. Bacteriocin small of *Rhizobium leguminosarum* belongs to the class of *N*-acyl-L-homoserine lactone molecules, known as autoinducers and as quorum sensing co-transcription factors. *J. Bacteriol.* 178:366–71

134. Seed PC, Passador L, Iglewski BH. 1995. Activation of the *Pseudomonas aeruginosa lasI* gene by LasR and the *Pseudomonas* autoinducer PAI: an autoinduction regulatory hierarchy. *J. Bacteriol.* 177:654–59

135. Shapiro JA. 1984. Observations on the formation of clones containing *araB-lacZ* cistron fusions. *Mol. Gen. Genet.* 194:79–90

136. Shapiro JA. 1984. Transposable elements, genome reorganization and cellular differentiation in Gram-negative bacteria. *Symp. Soc. Gen. Microbiol.* 36:169–93
137. Shapiro JA. 1984. The use of Mu*dlac* transposons as tools for vital staining to visualize clonal and non-clonal patterns of organization in bacterial growth on agar surfaces. *J. Gen. Microbiol.* 130:1169–81
138. Shapiro JA. 1985. Intercellular communication and genetic change in bacteria. In *Engineered Organisms in the Environment: Scientific Issues*, ed. HO Halvorson, D Pramer, M Rogul, pp. 63–69. Washington, DC: ASM. 239 pp.
139. Shapiro JA. 1985. Scanning electron microscope study of *Pseudomonas putida* colonies. *J. Bacteriol.* 164:1171–81
140. Shapiro JA. 1987. Organization of developing *E. coli* colonies viewed by scanning electron microscopy. *J. Bacteriol.* 197:142–56
141. Shapiro JA. 1988. Bacteria as multicellular organisms. *Sci. Am.* 256:82–89
142. Shapiro JA. 1992. Concentric rings in *Escherichia coli* colonies. In *Oscillations and Morphogenesis*, ed. L Rensing, pp. 297–310. New York: Marcell Dekker. 501 pp.
143. Shapiro JA. 1992. Differential action and differential expression of *E. coli* DNA polymerase I during colony development. *J. Bacteriol.* 174:7262–72
144. Shapiro JA. 1994. Pattern and control in bacterial colony development. *Sci. Prog.* 76:399–424
145. Shapiro JA. 1995. The significances of bacterial colony patterns. *BioEssays* 17:597–607
146. Shapiro JA. 1997. Multicellularity: the rule not the exception. Lessons from *Escherichia coli* colonies. See Ref. 148, pp. 14–49
147. Shapiro JA. 1997. Genome organization, natural genetic engineering, and adaptive mutation. *Trends Genet.* 13:98–104
148. Shapiro JA, Dworkin M, eds. 1997. *Bacteria As Multicellular Organisms*. New York: Oxford Univ. Press. 466 pp.
149. Shapiro JA, Higgins NP. 1989. Differential activity of a transposable element in *E. coli* colonies. *J. Bacteriol.* 171:5975–86
150. Shapiro JA, Hsu C. 1989. *E. coli* K-12 cell-cell interactions by time-lapse video. *J. Bacteriol.* 171:5963–74
151. Shi W, Kohler T, Zusman DR. 1993. Chemotaxis plays a role in the social behaviour of *Myxococcus xanthus*. *Mol. Microbiol.* 9:601–11
152. Shimkets LJ, Dworkin M. 1997. Myxo-bacterial multicellularity. See Ref. 148, pp. 220–44
153. Søgaard-Andersen L, Kaiser D. 1996. C factor, a cell-surface–associated intercellular signaling protein, stimulates the cytoplasmic Frz signal transduction system in *Myxococcus xanthus*. *Proc. Natl. Acad. Sci. USA* 93:2675–79
154. Søgaard-Andersen L, Slack FJ, Kimsey H, Kaiser D. 1996. Intercellular C-signaling in *Myxococcus xanthus* involves a branched signal transduction pathway. *Genes Dev.* 10:740–54
155. Solomon JM, Grossman AD. 1996. Who's competent and when: regulation of natural genetic competence in bacteria. *Trends Genet.* 12:150–55
156. Solomon JM, Lazazzera BA, Grossman AD. 1996. Purification and characterization of an extracellular peptide factor that affects two different developmental pathways in *Bacillus subtilis*. *Genes Dev.* 10:2014–24
157. Solomon JM, Magnuson R, Srivastava A, Grossman AD. 1995. Convergent sensing pathways mediate response to two extracellular competence factors in *Bacillus subtilis*. *Genes Dev.* 9:547–58
158. Sonea S, Panisset M. 1983. *A New Bacteriology*. Boston: Jones & Bartlett. 140 pp.
159. Stahl SJ, Stewart KR, Williams FD. 1983. Extracellular slime associated with *Proteus mirabilis* during swarming. *J. Bacteriol.* 154:930–37
160. Stams AJM, Grotenhuis JTC, Zehnder AJB. 1989. Structure-function relationship in granular sludge. In *Recent Advances in Microbial Ecology*, ed. T Hattori, Y Ishida, Y Maruyama, RY Morita, A Uchida, pp. 440–45. Tokyo: Jpn. Sci. Soc.
161. Stephens K, Hegeman GD, White D. 1982. Pheromone produced by the myxo-bacterium *Stigmatella aurantiaca*. *J. Bacteriol.* 149:739–47
162. Stewart GSAB. 1997. *Molecular languages for bacterial communication and their role in pathogenesis*. Presented at Eur. Congr. Biotechnol., 8th, Budapest
163. Sturdza SA. 1973. Développement des cultures de *Proteus* sur gélose nutritive après être mise en contact avec un milieu neuf. *Arch. Roum. Pathol. Exp. Microbiol.* 32:179–83
164. Sturdza SA. 1973. Expansion immédiate des *Proteus* sur milieux gélosés. *Arch. Roum. Pathol. Exp. Microbiol.* 32:543–62
165. Sturdza SA. 1973. La reaction d'immobilisation des filaments de *Proteus* sur les milieux geloses. *Arch. Roum. Pathol. Exp. Microbiol.* 32:575–80

166. Sturdza SA. 1978. Recent notes on the mechanism of the *Proteus* swarming phenomenon. A review. *Arch. Roum. Pathol. Exp. Microbiol.* 37:97–111
167. Swanson J, Koomey JM. 1989. Mechanisms for variation of pili and outer membrane protein II in *Neisseria gonorrhoeae.* See Ref. 17, pp. 743–61
168. Swift S, Bainton NJ, Winson MK. 1994. Gram-negative bacterial communication by *N*-acyl homoserine lactones: a universal language? *Trends Microbiol.* 2:193–98
169. Toal DR, Clifton SW, Roe BA, Downard J. 1995. The *esg* locus of *Myxococcus xanthus* encodes the E1 alpha and E1 beta subunits of a branched-chain keto acid dehydrogenase. *Mol. Microbiol.* 16:177–89
170. Todd WJ, Wray GP, Hitchcock PJ. 1984. Arrangement of pili in colonies of *Neisseria gonorrhoeae. J. Bacteriol.* 159:312–20
171. Tomasz A. 1965. Control of the competent state in *Pneumococcus* by a hormone-like cell product: an example of a new type of regulatory mechanism in bacteria. *Nature* 208:155–59
172. Tomasz A, Hotchkiss RD. 1964. Regulation of the transformability of pneumococcal cultures by macromolecular cell products. *Proc. Natl. Acad. Sci. USA* 51:480–87
173. Torkelson J, Harris RS, Lombardo M-J, Nagendran J, Thulin C, Rosenberg SM. 1997. Genome-wide hypermutation in a subpopulation of stationary-phase cells underlies recombination-dependent adaptive mutation. *EMBO J.* 16:3303–11
174. van der Meer JR. 1997. Evolution of novel metabolic pathways for the degradation of chloroaromatic compounds. *Ant. v. Leeuwenhoek J. Microbiol. Serol.* 71:159–78
175. Wakita J-I, Itoh H, Matsuyama T, Matsushita M. 1997. Self-affinity for the growing interface of bacterial colonies. *J. Phys. Soc. Jpn.* 66:67–72
176. Waldburger C, Gonzalez D, Chambliss GH. 1993. Characterization of a new

sporulation factor in *Bacillus subtilis. J. Bacteriol.* 175:6321–27
177. Ward MJ, Zusman DR. 1997. Regulation of directed motility in *Myxococcus xanthus. Mol. Microbiol.* 24:885–93
178. Watanabe T. 1963. Infective heredity of multiple drug resistance in bacteria. *Bacteriol. Rev.* 27:87–115
179. Williams FD, Schwarzhoff RH. 1978. Nature of the swarming phenomenon in *Proteus. Annu. Rev. Microbiol.* 32:101–22
180. Wimpenny JWT, Kinniment SL, Scourfield MA. 1993. The physiology and biochemistry of biofilm. In *Microbial Biofilms: Formation and Control,* ed. SP Denyer, SP Gorman, M Sussman. Soc. Appl. Bacteriol. Tech. Ser. 30:51–94. Oxford: Blackwell Sci. 336 pp.
181. Winans SC, Burns DL, Christie PJ. 1996. Adaptation of a conjugal transfer system for the export of pathogenic macromolecules. *Trends Microbiol.* 4:64–68
182. Winogradsky S. 1949. *Microbiologie du Sol: Problèmes et Méthodes.* Paris: Masson. 861 pp.
183. Winson MK, Camara M, Latifi A, Foglino M, Chhabra SR, et al. 1995. Multiple *N*-acyl-L-homoserine lactone signal molecules regulate production of virulence determinants and secondary metabolites in *Pseudomonas aeruginosa. Proc. Natl. Acad. Sci. USA* 92:9427–31
184. Wirth R, Muscholl A, Wanner G. 1996. The role of pheromones in bacterial interactions. *Trends Microbiol.* 4:96–103
185. Woodward DE, Tyson R, Myerscough MR, Murray JD, Budrene EO, Berg HC. 1995. Spatio-temporal patterns generated by *Salmonella typhimurium. Biophys. J.* 68:2181–89
186. Wu SS, Kaiser D. 1995. Genetic and functional evidence that Type IV pili are required for social gliding motility in *Myxococcus xanthus. Mol. Microbiol.* 18:547–58
187. Yarmolinsky M. 1995 Programmed cell death in bacterial populations. *Science* 267:836–37

Annu. Rev. Microbiol. 1998. 52:105–26

BACTERIA AS MODULAR ORGANISMS

John H. Andrews
Department of Plant Pathology, University of Wisconsin, Madison, Wisconsin 53706;
e-mail: jha@plantpath.wisc.edu

KEY WORDS: ecology, indeterminate, morphogenesis, clone, developmental biology

ABSTRACT

The body plan of modular organisms is based on an indeterminate structure composed of iterated units or modules arrayed at various levels of complexity (such as leaves, twigs, and branches). Examples of modular organisms include plants and many sessile benthic invertebrates. In contrast, the body of unitary organisms is a determinate structure consisting usually of a strictly defined number of parts (such as legs or wings) established only during embryogenesis. Mobile animals are examples. Unlike that of unitary creatures, the form of a modular organism derives from a characteristic pattern of branching or budding of modules, which may remain attached or become separated to live physiologically independent lives as parts of a clone. Modular organisms tend to be sessile or passively mobile and, as genetic individuals, have the capacity for exponential increase in size. They do not necessarily undergo systemic senescence, and do not segregate somatic from germ line cells. It is argued here that bacteria are essentially modular organisms where the bacterial cell, microcolony, and macrocolony are modules of different levels of complexity analogous to modules of macroorganisms. This interpretation provides a broad conceptual basis for understanding the natural history of bacteria, and may illuminate the evolutionary origins and developmental biology of modular creatures.

CONTENTS

105

INTRODUCTION

"The most important feature of the modern synthetic theory of evolution is its foundation upon a great variety of biological disciplines."

GL Stebbins (92)

Modular organisms are those in which the product of the zygote (or its functional equivalent) is an indeterminate structure composed of parts that are iterated throughout the life cycle. The most obvious examples include plants and many of the sessile benthic invertebrates such as hydroids, corals, bryozoans, and colonial ascidians. Unitary organisms, by contrast, are those in which the zygote develops into a determinate structure wherein the number of body parts generally is fixed within strict limits established during embryogenesis at the start of the life cycle. Mobile animals in general and vertebrates in particular are the classic examples of the unitary category of organism.

Although the distinction between the modular and unitary types of general body plan has major evolutionary and ecological implications, its significance was largely unexplored by plant and animal ecologists until the 1970s (38). And, with few exceptions (3–7, 94), no attempt has been made to consider the prokaryotes (eubacteria and archaebacteria) or other microorganisms (the fungi and most protists) in the modular/unitary division of the living world. This article discusses (*a*) the characteristics and evolutionary consequences of modularity as shown by modular macroorganisms, (*b*) the case that bacteria are modular, and (*c*) implications of the modular[1] concept for understanding the natural history of bacteria.

How does viewing bacteria from a modular perspective give us novel insights into their biology or advance the understanding of modularity? As will be developed here, interpreting bacteria as modular creatures allows microbiologists to go beyond the conventional, clonal (asexual) concept of bacteria to a broader perspective that also includes growth form and the arrangement of modules at successive levels of complexity. More importantly, for biologists

[1]Note that the term "module" as used by ecologists differs from the module of developmental biologists, which is taken to be a discrete subunit with respect to gene expression (see e.g. 75, pp. 326–334).

Table 1 Major attributes of unitary and modular organisms[a]

Attribute	Unitary organisms	Modular organisms
Growth pattern	Noniterative; determinate	Iterative; indeterminate
Development[b]	Typically preformistic	Typically somatic embryogenesis
Germ plasm	Segregated from soma	Not segregated
Branching	Generally nonbranched	Generally branched
Mobility	Mobile; active	Sedentary[c]; passive
Internal age structure	Absent	Present
Reproductive value	Increases with age, then decreases; generalized senescence	Increases; senescence delayed or absent; directed at module
Role of environment in development	Relatively minor	Relatively major, especially among sessile forms
Examples	Mobile animals generally, especially the vertebrates	Many of the sessile invertebrates such as bryozoans, hydroids, corals, colonial ascidians; also plants, fungi, bacteria

[a]Modified from Andrews (5). These are generalizations. There are exceptions (see text for examples).
[b]Pertains to degree to which embryonic cells are irreversibly determined. Preformistic = all cell lineages so determined in early ontogeny; somatic embryogenesis = organisms capable of regenerating a new individual from some cells at any life-cycle stage (cells totipotent or pluripotent). See Buss (18).
[c]Juvenile or dispersal phases mobile. Many bacteria and protists are potentially somewhat mobile.

in general, this view interprets modularity as one of the earliest and most fundamental evolutionary trends—on a par, for example, with the ancient origin of central metabolic pathways. Further, it provides insight into the relationship between prokaryotes and eukaryotes, and the commonality in architecture of multicellular microorganisms and modular macroorganisms.

WHAT IT MEANS TO BE MODULAR

The main differences between unitary and modular organisms are summarized in Table 1 (see 3, 5 for details, caveats, and exceptions). Arguably most significant among these is the nature of and repetition of the subunit or module through time and space. Modules have been defined variously (95, 96) but, as conceived by macroecologists, were an iterated *multicellular* unit of growth and organization (40). This is visualized most easily in the case of plants as being a leaf, together with its axillary bud and internode. Higher-order levels of modules are shoots and then branches (95).

Frequently, modules remain attached to the organism, which derives much of its bulk and shape from the living and accumulated dead modules (corals and trees). Alternatively, they may slough away, in which case the creature literally falls to pieces as it grows (the floating aquatic plants *Lemna*, *Salvinia*, and *Azolla*). Modules that can function as separate physiological units have been termed ramets by botanists (38, 95); examples include the floating aquatics noted above or the individual plantlets, termed rosettes, of a strawberry (Figure 1).

Because modules can be iterated indefinitely, growth is characteristically indeterminate in modular organisms. This implies lack of a genetically fixed upper limit on size and the ability to change size drastically in response to environmental factors such as weather, nutrition, competition, or predation (80). However, the converse is not true: Indeterminately growing organisms are not necessarily modular—the echinoderms and mollusks, for instance, grow indeterminately but are unitary organisms.

Closely related to modular and indeterminate growth is clonal growth, the production of genetically similar or identical units (see *Immortality*, below), which may be aggregated (zooids of a bryozoan) or disaggregated (the floating aquatics noted above). Colonies result in cases where modules are aggregated. Strictly speaking, a clone whose origin can be traced to a genetic exchange event, i.e. arising as the product of a zygote, is termed a genetic individual or genet (2, 38, 50), though frequently (e.g. 3, 41, 42, 44), as is the case here, genet and clone are used interchangeably. The term genet was coined to emphasize that two levels of population structure exist in plant communities: One arises from the original zygotes present; the other from the genetically identical modules produced from each zygote, especially where they are capable of independent existence (ramets; 38). Thus, together, the ramets of strawberry, or aspen, or a floating aquatic plant, constitute one genet, even though they may be physically disconnected. In aggregate they constitute "the individual" of evolutionary biologists.

Modular organisms may or may not be clonal and clonality does not necessarily imply modularity. For instance, annelid worms, some starfish, and parthenogenetic fishes can reproduce asexually, thus are clonal (42), but they are fundamentally unitary in construction. Clonal growth is possible in modular creatures because embryonic cells are not irreversibly determined early in ontogeny, enabling a new physiological "individual" to be regenerated from certain cells at any life cycle stage (somatic embryogenesis) (Table 1) (18–20).

----→

Figure 1   Some modular macroorganisms and microorganisms. (*a*) A clonal terrestrial plant (strawberry); (*b*) a clonal, floating aquatic plant (*Salvinia* sp.); (*c*) a sea fan coral (*Gorgonia* sp.); (*d*) a colonial bryozoan (*Membranipora* sp.); (*e*) mycelium of a fungus; (*f*) microcolony of bacteria. [From Andrews (7) with permission from the American Society for Microbiology.]

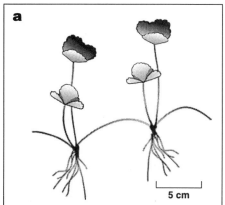

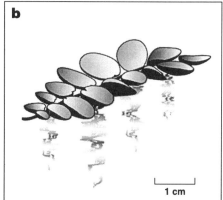

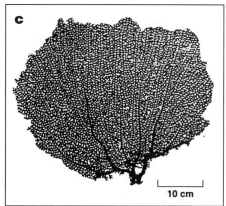

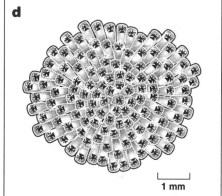

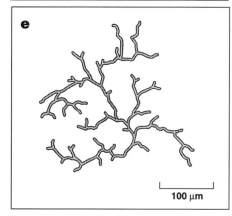

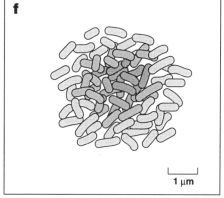

Modular organisms generally branch or have a characteristic budding pattern (14). Growth may be concentrated vertically (trees and many corals) or horizontally (crustose lichens or bryozoans). The shape of the organism follows typically from this pattern of branching or budding (along with fusion or fission in colonial forms; 43), which, in turn, depends on birth and death rates of the modules and how they are put together.

Is the modular/unitary distinction relevant to microorganisms? Even a casual reading of the features discussed above suggests, on one hand, that it is. The obvious analogies between modular macroorganisms and bacteria include (*a*) growth by iteration of a basic structural unit; (*b*) growth that is clonal (and, typically, colonial), where the clone consists of innumerable functionally independent ramets (colonies or microcolonies); (*c*) indefinite life span of the clone or genetic individual and absence of generalized senescence; and (*d*) marked influence of the environment on growth rate, life-cycle phases, and appearance of ramets. On the other hand, at first glance, bacteria do not appear to be modular for the following reasons: (*a*) Their unit of iteration is an individual cell, not a multicellular entity, and thus is not a conventional module; (*b*) many bacteria are at least potentially motile and most are unbranched; and (*c*) their haploid nature and irregular means of sexual reproduction, including a proclivity to acquire genes by horizontal transfer mechanisms, implies that a strictly defined origin and endpoint to a genetic individual (i.e. clonal lineage) is conceptual at best. I take up these caveats in the next section, and use as examples what most people think of as typical bacteria, i.e. eubacteria and archaebacteria conventionally depicted as unicellular rods and spheres. The multicellular, filamentous cyanobacteria, the myxobacteria, and the actinomycetes are more obviously modular and thus less controversial. For most practical purposes in this context they can be considered with the filamentous fungi, which have been discussed elsewhere (4, 6).

ARE BACTERIA MODULAR ORGANISMS?

Cells versus Modules

The individual cell in unicellular organisms is analogous to the multicellular module of macroorganisms because both are a defined unit of structure iterated indefinitely. A unicellular interpretation of module cannot, however, be applied to macroorganisms for two reasons. First, since all organisms are composed of cells, by definition all would be modular. Second, the cell of unicellular creatures is iterated indefinitely, potentially without constraint, whereas, barring disease states such as cancer, cell division in multicellular organisms is subordinated to fairly tight overall regulation. In fact, multicellular assemblages (colonies and frequently biofilms) of bacteria are the common situation in nature (22), although in general they are undifferentiated and lack the degree of coordination and integration of cells of the filamentous fungi, macroalgae, plants, and animals. The scope of interactions among bacterial cells within and even between colonies in nature is just beginning to be appreciated (47, 83; see related article by Shapiro, this volume). So, clumps of bacterial cells exist and are moved about; they are in effect a higher-order level of module, just as clumps of clonal modular *Lemna* or *Salvinia* are pushed across the surface of a lake by the wind.

Although most bacteria do not branch in the sense that a tree does, frequently they do have an organized pattern of division within colonies (17, 83), not unlike the multiserial (forming sheets rather than chains of cells) budding pattern of encrusting organisms such as certain lichens or marine invertebrates (23). For instance, the multiserial, multilaminate (multilayers of cells) colony forms of certain cheilostome bryozoans are very similar in organization and overall appearance to colonies of bacteria or yeasts; the uniserial, unilaminar ("runner") forms resemble the actinomycetes and filamentous fungi.

Motility and Mobility

Motile organisms *actively* search for food and mates, and can move to escape being eaten or injured. Thus they see quite a different world than do sessile creatures which, as relatively fixed objects, must endure the vicissitudes of their environments, including physical disturbances, local exhaustion of nutrients, and parasitism or predation. Directional growth may "move" them into new habitat (a clone of aspen moves up a hillside; rhizomorphs extend a fungal clone into new terrain), and a dispersal phase in the life cycle acts to move gametes or progeny to distant sites. As noted above, unitary organisms generally are motile, whereas modular organisms typically are sessile. Are bacteria sufficiently motile in nature to be disqualified as modular organisms on this basis alone?

Clearly, many bacteria are potentially motile by virtue of possessing axial filaments, various gliding mechanisms, or flagella. Moreover, all bacteria are mobile, if not actively motile, by *passive* dispersal processes at least at some point in their life cycle. However, bacterial mobility is no different in principle from the passive transport of certain life-cycle phases of modular organisms, while the prevalence of motility is subject to several qualifications. Motility is likely to be of minor significance in many common situations. For instance, bacteria tend to grow at surfaces, which implies adhesion, whether or not conditions favor the formation of extensive biofilms. Somewhat freer than they would be in biofilms, but still constrained, are bacteria in highly viscous environments such as the intestine. Although little is known about the viscosity of intestinal contents, *E. coli* probably faces very viscous surroundings much of the time. Koch (55) has discussed how viscosity limits the diffusion rate and, in turn, bacterial growth rate. Cells cannot offset these boundaries by elevating transport capability beyond a certain point. Another situation concerns dry environments such as the aerial surface of plants (especially hydrophobic leaves; 59). Here, bacterial movement would be limited to periods of free water provided by rain or dew. Even in habitats where water is abundant, potentially motile cells may rarely move actively. Fewer than 10% of the bacteria were motile in coastal waters off Australia (66). Apparently this is due to energy limitations, because >80% of the cells became motile 30 hours after the addition of nutrients. Motility is energetically expensive: in *E.coli*, 40 genes are required "for flagellar assembly, structure, and function in addition to the genes involved in sensory reception and transduction" (60). Thus, Mitchell et al (66) suggest that in some environments it may be advantageous for cells to cease movement and then resume it as conditions improve.

To summarize, while bacteria are not strictly sessile, frequently they more or less approximate this condition. Consequently, bacteria affect and are in turn affected by their environment much as are modular organisms. This includes the tendency to exhaust nutrients locally, forming resource depletion zones (39) not unlike the nutrient concentration gradients under bacterial colonies modeled by Pirt (73).

Immortality

Modular creatures tend to be long-lived, and this characteristic is maximized in those that are clonal (40, 44, 93, 96). Because asexual reproduction, generation time, and clonality are dominant themes in bacterial population biology (see below), it is useful to begin by placing these biological characteristics in a broader context.

Clonal organisms (e.g. aspens, corals, water hyacinths) are very large as genetic individuals, often amassing weights on the order of 10^6–10^9 g or areas of occupation on the order of dozens of km (25, 38). Even more interesting is that, according to dogma, they are potentially immortal (3, 40). As examples of what would be considered minimum ages, genetic individuals of huckleberry (*Gaylussacia brachycerium*) have been estimated at 13,000+ years (99); the alpine sedge *Carex* at 2000 years (93); certain clonal fish (unisexual female *Poeciliopsis*; 74) at 150,000 years; certain fungal clones (89) at 1500+ years; and salamanders at 5,000,000 years (90). This prompts the question, "How old are bacterial clones?"

The prospect of immortality arises because senescence evolves very slowly, if at all, where reproductive value ($\approx$ fecundity) increases with size (age) as it does for a clonal lineage (for caveats see 31, 32, 64). Of course, somatic mutations will tend to accumulate through time and the consequences of these on the evolutionary history of clonal organisms is debated by ecologists (19, 20, 31, 64, 87, 88, 100). To what extent has a 10,000-year-old clone changed over its life span and how genetically different are its component modules? To what extent is an aspen clone, or for that matter the different shoots on a maple tree, a genetic mosaic, and are these mutated genes passed sexually to progeny?

That such variation does not overrun individuals, especially clonal individuals, is probably because of natural selection acting on small differences in division rate among cells (intraorganism selection; 19, 69). Selection at the cellular level should be especially effective in haploid organisms and in those with what Otto & Orive (69) called "fairly unstructured development," i.e. what are referred to here as modular organisms. This phenomenon amounts to what has long been known as periodic selection in bacteria, i.e. the recurrent replacement of an existing clone (or clones) in a population by a superior clone (9, 54, 57). The parallels to the population biology of bacteria seem obvious. The only real issue is whether, given their haploid condition, together with mutation and varying degrees of recombination (12, 62, 63), a bacterial clone can persist long enough to be at all comparable with, say, the genetic individual of a coral or aspen.

Clonality generally implies that all members of a clone have descended entirely from a specific common ancestor (by mitosis or asexual reproduction without recombination), not genetic or phenotypic identity (65, 101). Clones can be nested based on their immediacy to a particular ancestor as reference point (65). In other words, differences among descendants occur because of point mutations and other genetic events such as deletions and transpositions (101), and these accumulate through time giving a collection of progressively diverging lines that have been referred to as a meroclone (65), clonal lineage (2), or clonal complex (101).

Operationally, a bacterial "clone" is routinely taken to be represented by an identical protein-banding pattern (electrophoretic type, ET) at about 20 loci as revealed by multilocus enzyme electrophoresis (61, 101). The shortcomings of this assumption have been discussed (62) and relate generally to the fact that neither the entire chromosome nor every cell in a population can be analyzed (61). A common ET may occur simply from random reassortment (62) and, further, it implies nothing about identity at other loci. Whittam & Slobodchickoff (101) concluded from genetic variation detected by RAPD analysis of enteropathogenic *E. coli* clones that nucleotide diversity *within* a clone could be about 0.5% variation (see also 49) and *between* clones approximately 3–4%. Based on such assessment the general conclusion is that, while no bacterial species is invariably clonal (35), some—such as *E. coli* and *Salmonella*—are highly clonal, while others—*Neisseria* and *Bacillus*—undergo relatively frequent recombination (63).

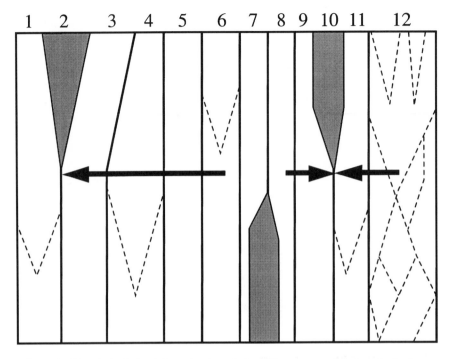

Figure 2 Hypothesized clonal dynamics through time. Clones are numbered on the horizontal axis. The vertical axis is evolutionary time with most clones originating below the bottom of the box. Clones 2 and 10 arise by genetic events (gene transfer and recombination, respectively). One clone (at position 7.5) goes extinct. *Dotted lines*, rounds of periodic selection. This is the "niche-adapted" clone model, from Reeves (76). (Reprinted from *FEMS Microbiology Letters* 100 (1992) p. 514, with kind permission of Elsevier Science-NL, Sara Burgerhartstraat 25, 1055 KV Amsterdam, The Netherlands.)

In overview, for those bacteria that are predominately clonal, the size and longevity of the clones are indefinite, potentially very great but unknown, and estimates ultimately subjective (depending on the number of nucleotide changes and their perceived phenotypic impact). As noted, periodic selection in the effectively nonrecombining species will cause favorable mutations to be selected, carrying with them unselected alleles as hitchhikers. What limits the otherwise absolute dominion by "favored" clones? In nature bacterial clones are probably maintained, as well as restricted, by the distribution of the microhabitats to which they are best adapted ("niche-specific selection"; 56, 76), just as environmental constraints probably restrict the spread of clonal macroorganisms (Figures 2, 3) (102). The existing information comes largely from epidemiological studies and is discussed later.

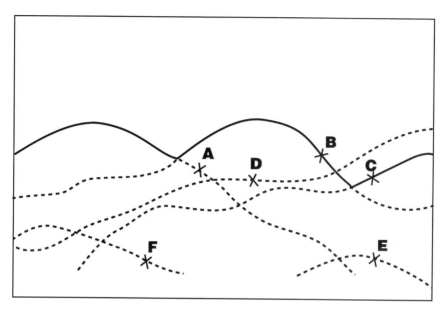

Figure 3   Hypothesized clonal dynamics through space. The fitness of six clones (*A* to *F*) is shown as a function of habitat gradient in cross-section. *Solid lines,* areas actually occupied; *dotted lines,* potential occupancy in absence of competition; *X*'s, sites of arrival. *A, B, C,* survivors; *D, E, F,* did not colonize or died following colonization. Note that though the fitness of *B* > *A* in the zone of *A*'s arrival, presumably *A* got there first and spread to the left, where it remains competitive, before *B* arrived. *E* and *F* have low general fitness; *D* has reasonable fitness overall but is less fit than *B* at its site of arrival. Figure and interpretation based on Williams (102) (Williams GC, *Sex and Evolution*, Figure 3, copyright 1975 by Princeton University Press. Reprinted by permission of Princeton University Press.)

SO WHAT?

The Antiquity of Morphogenes

Are the modular organisms that we see about us today descended from modular bacteria such as the myxobacteria? Are the genes for branching and programmed development of modules conserved over vast evolutionary time?

With growing interest in understanding how animal (86) and plant (48) development evolved, the formerly isolated fields of evolution and developmental biology recently have fused, leading to new insights (75). One general consensus, both for microorganisms and macroorganisms, is that there are apparently few true morphogenes per se (36, 37). (It is also true in any case that, strictly speaking, genes do not "control" development nor do genomes "program" development; 68.) Rather, the genetic program tends to act indirectly and from afar, typically through pleiotropic effects and the interaction of multiple gene products. Even in bacteria, multiple gene expression drives morphogenesis and is, in turn, affected by the morphogenetic state (84). The simple paradigm genes → proteins → form is not informative because it ignores the involvement of epigenetic effects and is not relatable to how a particular body plan results (33, 68)—as Harold (36, p. 387) says, "the interesting events in morphogenesis take place on a higher plane than that of the genes." Thus there is the danger that searches for mechanistic explanations of form may be too reductionist.

So what role do individual genes, or a coordinated genetic program, play in morphogenesis? Carroll & colleagues (71, 86) have shown that many of the same or very similar genes (*Hox; Distal-less*) are involved in appendage formation across a wide range of animal phyla. (Conversely, similar genes may function in producing different body plans in different organisms; 86.) They speculate that the similar genetic regulatory system can be traced to a common ancestor and that the unique plan of any given taxon may result from tinkering with the original program established in primitive metazoans. Without suggesting that appendage formation in mobile (unitary) animals is related to branching patterns in modular creatures, the questions I raise here are: How far back does an ancestral branching gene go? Is there a branching program unique to modular micro- and macroorganisms?

Present genomes include an assemblage of genes descended through time from an original common ancestor, more or less modified by duplication, divergence, and acquisition of parts acquired by horizontal transfer from the same or different species (52, 58). Genes influencing modularity, in particular the branching habit of modular organisms, may have arisen by conservation (from the original function in a common prokaryotic ancestor) or convergence (independent co-option from some other function in various lineages). The extent to which various "shape" genes may have molded modular organisms is debatable, and the idea that one or more such genes [or more likely, a common *genetic circuitry* (cf. 8, 26, 71, 86)] may be shared exclusively by these creatures is even more controversial. Perhaps more probable is that modularity has evolved several times in various lineages and what is shared is the overall result. To date the entire genomes only of *Saccharomyces* and certain eubacteria and archaea have been sequenced. To identify common genes, gene functions, and evolutionary relationships it would be interesting to compare the genomes of putatively modular creatures representing the filamentous fungi, yeasts, bacteria, plants, and sessile benthic invertebrates.

Evolutionary Origins of a Modular-Based Architecture

In the prokaryotes, we see not only the beginnings of multicellularity and differentiation but also of a branching growth form. That bacteria in nature function as loosely multicellular, communicative complexes (83) was alluded to earlier. The case for true multicellularity, together with the origins of differentiation and branching, is strongest for the streptomycetes and myxobacteria (see e.g. 78). Here I continue the theme of the previous section, taking up issues pertaining to the evolutionary relationships between prokaryotes and eukaryotes and the evolution of multicellularity as they relate to the commonality of form among modular organisms.

Whether the universal ancestor resembled a prokaryote, a eukaryote, or something unlike either is unknown and probably never will be known with any certainty. The dogma is that eukaryotes evolved from prokaryotes. Cavalier-Smith (21), for example, roots the universal tree in the eubacteria and argues that the big step in evolution of both the eukaryotes and the archaebacteria may have been loss of the cell wall. Generally, universal trees are unrooted (53, 69a, 70), implying an unknown but approximately similar time of origin of the three primary kingdoms—archaebacteria, eubacteria, and eukaryotes. While confirming that mitochondria and chloroplasts are derived from eubacteria, recent evidence from molecular phylogenies suggests that the eukaryotic nuclear line of descent did not originate with either the eubacteria or the archaebacteria (53, 69a, 70).

There are 17 multicellular protist taxa traceable to the unicellular Protista; exactly where these multicellular groups originated is not clear, but generally they are taken to have independent origins (28). Probably the myxobacteria predate the multicellular eukaryotes (46). Kaiser (46) speculates that multicellularity arose sometime after the myxobacteria and their closest relatives (sulfate reducers and bdellovibrios) diverged. This has been estimated variously at between about 650 million to 2 billion years ago (46, 85), possibly when the atmosphere became aerobic. Given the relatively complex social and developmental patterns of myxobacteria, it is probably more than coincidental that the myxobacterial genome, at about 9000 kbp, is some fourfold larger than that of their relatives (85). The origin of the developmental genes remains unknown.

The overall similarity in life cycle and fruiting structures between the myxobacteria and slime molds is striking (15, 20, 46) (Figure 4). The general evolutionary trend characterized by a phase of the life cycle dedicated to dispersal, and the differentiated, occasionally complex structures specialized to accomplish this, are evident throughout the fungi and in many protozoa. Indeed, it is a recurring theme for sessile organisms in general, including plants and benthic invertebrates. Whether this represents merely convergent evolution, as is popularly believed, or tinkering with a master set of "growth form" genes in a common ancestor, as may have occurred for appendages in animals (see above and 86), remains to be seen.

Whether early experiments with a modular type growth form originated with the bacteria, the prokaryotes were unable to exploit three-dimensional designs

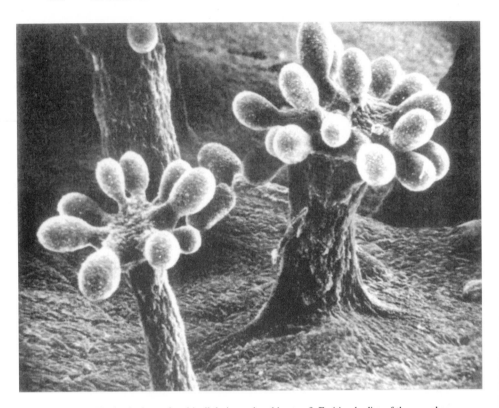

Figure 4   The beginnings of multicellularity and architecture? Fruiting bodies of the myxobacterium *Chondromyces crocatus*. The fruiting bodies of myxobacteria, although much shorter than those of the cellular slime molds, can contain roughly the same number of cells (10^5–10^6 for *Myxococcus* and *Dictyostelium*). The forms embrace similar geometries, including various combinations of cylinders, stalks, and spheres (46). Illustration courtesy of Pat Grilione.

larger or more complex than mounds or simple trees exhibited by the myxobacteria and actinomycetes. The three groups—fungi, animals, and plants—in which cell differentiation and growth form are well developed are, of course, all eukaryotic. The differences between prokaryotic and eukaryotic cells as building blocks are numerous; among these are (*a*) genome size, organization, and packaging; (*b*) functional organization of the cytoplasm based on an endomembrane system and cytoskeleton; (*c*) nature and mode of synthesis of the cell envelope; and (*d*) degree of intercellular communication (septal pores, plasmodesmata, and gap junctions, respectively, in the fungi, plants, and animals).

   Cavalier-Smith (21) argues that neither genome size nor organization is responsible for the limits on bacterial architectural design, but rather presence of a rigid peptidoglycan wall. Lacking a wall, animal cells are plastic. Plants and fungi possess walls but these are expansive, being engineered of interwoven polysaccharide microfibrils embedded within an amorphous matrix. This architecture allows for different shapes as ecological circumstances dictate. It accommodates the seemingly incompatible demands of plasticity for morphogenesis and growth, yet rigidity for containment of turgor pressure and penetration of substrata (48). Construction of this type of wall is a dynamic process enabled by the interplay of (*a*) an exocytotic system of wall-destined vesicles, which carry the necessary enzymes and wall precursors; (*b*) a cytoskeleton, which guides the vesicles to the appropriate area of the wall, thereby contributing to spatial regulation of wall synthesis; and (*c*) deployment of enzyme complexes at the appropriate site for synthesis (11, 21, 98). This cytoskeleton-cytosis–based process evidently is the fundamental requirement for building architecturally complex cells and modular (or unitary) organisms (21).

Bacteria Bigger Than Dinosaurs: The Benefits of Being Large

Modularity together with clonality have immense implications for the natural history of bacteria. Perhaps most important among these is size of the genetic individual and, consequently, the ability to sample many environments. Size thus influences competitive ability, and the potential to harvest patchy resources (45), such as hosts if you are a microbe or water if you are a tree. As noted above, clones may remain intact or break into pieces of physiologically independent parts (ramets) and there are tradeoffs with either strategy (24). The former situation can be illustrated by aspen (*Populus tremuloides*). A clone originates from buds along the spreading root system and in a particular case (estimated to comprise 47,000 trees and be of aggregate age >10,000 years) has been observed to occupy 43 ha (51). As a single, massive, cohesive unit it has sampled, up and down countless hillsides, the local environment over all those centuries. Individual ramets were born, grew, and occasionally died, but the genetic individual persisted. Clearly, there is adaptive benefit to integration.

Consider now the clonal organism in which the ramets are independent. At the cellular level this condition is not met perfectly by bacteria because usually the fission products are not independent. But it is approximated by groups of bacterial cells, all clonal members that, as a result of transport processes, happen to colonize at widely separated sites. The fate of one microcolony in this situation does not influence another. Thus the likelihood that an entire genotype would die is the product of the probabilities of the extinction of the independent parts. With each round of asexual division and fragmentation (doubling), the probability declines further by up to one half; i.e. as size of the clone increases, risk of death decreases. Hence, by dispersing rather than aggregating the products of asexual reproduction, the organism spreads the risk of extinction among the units (24). So, there are also advantages to existing as separate parts.

Separation of physiological individuals is accomplished extremely effectively by microorganisms. Clones of fungi (34) and bacteria (30) are commonly distributed across continents, if not globally, which is resounding testimony to the power of short generation times and effective dispersal mechanisms (and anthropogenic influence). In aggregate, the size of such patchy but cosmopolitan clones would be vast, and can only be imagined; any estimates would be subject to major sources of error related to environmental and host reservoirs. The existing data instead relate to distribution and have been gleaned from decades of research on the epidemiology of bacterial pathogens of humans. For instance, major epidemics and pandemics of meningitis in this century are attributable to several clones of serogroup A *Neisseria meningitidis* (61, 82, 91) (Figure 5). In the USA, much attention has focused in the past five years on *E. coli* O157:H7. This clone (101) causes severe and potentially fatal hemorrhagic colitis, a euphemism for bloody diarrhea. The bacterium typically is acquired by eating undercooked hamburger. Tracing the source of infection from patients back through fast-food restaurants, wholesalers, meat-packing companies, slaughter houses, and carcasses has been a challenging public exercise in scientific sleuthing with major political and economic ramifications. Yet another clone, Sonnei (known colloquially as *Shigella sonnei*, but in fact a clone of *E. coli*), is the causal agent of bacillary dysentery, and has been tracked for decades over several continents (49). A final example concerns the fascinating life history of *Legionella pneumophila*, causal agent of legionnaire's disease or legionellosis. The normal host for this organism apparently is aquatic protozoa, but humans can be infected by inhalation of contaminated aerosols (10, 103). What is striking in our context is not only that *L. pneumophila* is distributed worldwide but, as genetic individuals, the pathogenic clones can sample environments as divergent as the human body and natural bodies of water (81).

In overview, the ability to expose oneself to diverse environments is of adaptive value and most apparent in modular clonal organisms, especially (as in bacteria) where parts of the genetic individual are functionally independent and disperse freely. Even if much of the clone becomes extinct in parts of its range, portions multiply elsewhere to carry on the lineage. The practical implications in medical microbiology relate directly to vaccination strategies; in plant pathology they pertain to the development, deployment, and durability of resistant crop varieties.

Bacteria at the Organismal Level!

Our perception of bacteria has come largely from laboratory culture and is typically a reductionist one of genes, transcriptional regulation, and metabolic pathways. If we consider, instead, bacteria as organisms in their natural habitats,

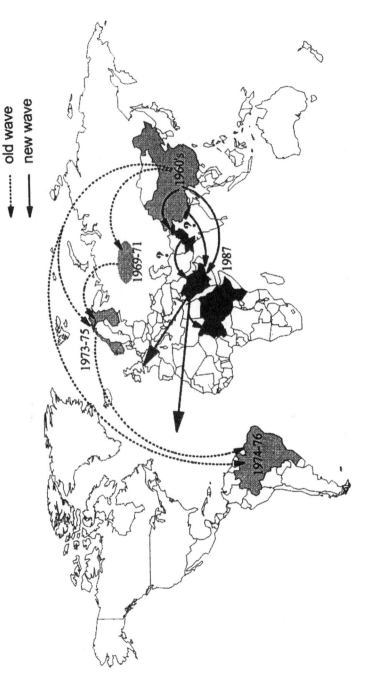

Figure 5   Old and new wave pandemics in the global spread of serogroup A meningococci of subgroup III *Neisseria meningitidis*. Both waves originated in China. Pulsed field gel electrophoresis shows that strains from the old wave are distinct from, though closely related to, the new wave strains. Most of the new wave strains are identical, confirming the wide distribution of this clone. (From Maiden and Feavers (61), with permission from Cambridge University Press.)

modularity underpins many of the characteristics that we think of as fundamentally bacterial. For example, bacteria are perhaps the classic example of organisms that grow by iteration. One ecological implication of iterative potential related to size (exposure of the same genetic individual to natural selection in many environments concurrently) was discussed above. Another also pertains to growth, but on a local scale. Like other modular organisms, bacteria in nature are essentially sedentary, tending to grow most actively at surfaces and often as biofilms [this is so even in bloodstreams (22) and aquatic systems (1)]. In the ensuing cooperative and competitive interactions at close quarters, and between the bacterial cell and its substratum, much of what we know about bacterial physiology assumes practical significance (47, 83). Examples include antibiotic or enzyme production (97), autoinduction (quorum sensing) (29), or signals related to host recognition and invasion (13). This means of interaction, with parts of one genetic individual influencing *parts* of another, is typical of how modular organisms in general behave and is fundamentally unlike competition among unitary creatures.

The phenotypic plasticity of bacteria can be interpreted, along with that of other modular organisms, in the context of the decisive role of the environment on development of modular creatures (cf. Table 1). First, in a strictly morphological sense, it is well known that bacterial cells increase in size with growth rate (67). Although this variation can be substantial, it is not as striking as the morphological plasticity of plants or fungi (3, 4, 6, 16). What *is* striking is the extreme morphological plasticity of myxobacteria which, like the cellular slime molds, essentially alternate between being a disaggregated organism and an aggregated one (46, 77). Second, at least as striking but not as visible, is plasticity at the metabolic level for which bacteria are famous and without peer (3, 27, 72). Versatility is possible largely by the ability to alter entire metabolic pathways to an extent and speed of response that eukaryotes cannot match (3). This is obviously of adaptive value to bacteria in coping with a rapidly changing environment. The bacterial genetic individual is also extremely versatile in that cells in different parts of the clone, at different habitat sites separated by perhaps a few dozens of μm to thousands of km, each respond uniquely to the ambient conditions of the particular location in which they find themselves. In contrast, the cells of multicellular organisms (especially the unitary metazoans), interconnected and lying within a soma, are relatively buffered from the external environment.

CONCLUSION

Until recently, general models in theoretical population biology tended to be based on assumptions limiting their broad applicability. These premises included (*a*) sexual (rather than clonal or asexual) reproduction, (*b*) demarcated germ and somatic cell lines within an individual, (c) senescence of the genetic individual and, associated with this, (d) an early rise, followed by a sharp decline, in reproductive value (Table 1) (40, 79). Whether imposed consciously or unconsciously, these limits circumscribed a fairly small set of the world's biota, i.e. mainly the mobile animals or, broadly speaking, what have come to be known as unitary organisms. Meanwhile, the models developed with microorganisms did little more than explain either very general phenomena— such as logistic population growth in a limited environment—or, conversely, phenomena unique to bacteria.

Adopting a modular perspective allows us to transcend the artificial limits of "microorganism versus macroorganism" and reinterpret the biology of bacteria in a broader perspective. The element of bacterial ecology missing from conventional descriptions and clonal theory relates to growth form. Whether or not a direct link between prokaryotes and eukaryotes by way of ancestral morphogenes is eventually shown, we see in the bacterial world the beginnings of several important evolutionary paths in addition to clonality: multicellularity and aggregation of cells into distinctive, three-dimensional, architectural arrays; differentiation and hence division of labor; and dispersal mechanisms together with a dispersal phase as part of the life cycle. These attributes are typical modular features and need to be recognized by microbiologists if the evolutionary ecology of bacteria is to be properly interpreted. And, for their part, ecologists and evolutionary biologists may find it enlightening to look to the microbial world in searching for the origins of modularity.

ACKNOWLEDGMENTS

I thank Grace Panganiban and Jeremy Ahouse for helpful discussions, and Mark Bailey, Keith Clay, Patricia McManus, and Bernhard Schmid for comments on a draft of the manuscript.

Visit the *Annual Reviews home page* at
http://www.AnnualReviews.org.

Literature Cited

1. Alldredge AL, Cohen Y. 1987. Can microscale chemical patches persist in the sea? Microelectrode study of marine snow, fecal pellets. *Science* 235:689–91
2. Anderson JB, Kohn LM. 1995. Clonality in soilborne, plant pathogenic fungi. *Annu. Rev. Phytopathol.* 33:369–91
3. Andrews JH. 1991. *Comparative Ecology of Microorganisms and Macroorganisms.* New York: Springer-Verlag
4. Andrews JH. 1992. Fungal life-history strategies. In *The Fungal Community: Its Organization and Role in the Ecosystem,* ed. GC Carroll, DT Wicklow, pp. 119–45. New York: Marcel Dekker. 2nd ed.
5. Andrews JH. 1994. All creatures unitary and modular. In *Ecology of Plant Pathogens,* ed. P Blakeman, B Williamson, pp. 3–16. Wallingford, UK: CAB Int.
6. Andrews JH. 1995. Fungi and the evolution of growth form. *Can. J. Bot.* 73: S1206–12 (Suppl. 1)

7. Andrews JH. 1995. What if bacteria are modular organisms? *ASM News* 61(12): 627–32

8. Arnone MI, Davidson EH. 1997. The hardwiring of development: organization and function of genomic regulatory systems. *Development* 124:1851–64

9. Atwood KC, Schneider LK, Ryan FJ. 1951. Selective mechanisms in bacteria. *Cold Spring Harbor Sympos. Quant. Biol.* 16:345–55

10. Barbaree JM, Breiman RF, Dufour AP, eds. 1993. *Legionella: Current Status and Emerging Perspectives.* Washington, DC: ASM Press

11. Bartnicki-Garcia S. 1987. The cell wall: a crucial structure in fungal evolution. In *Evolutionary Biology of the Fungi*, ed. ADM Rayner, CM Brasier, D Moore, pp. 389–403. Cambridge: Cambridge Univ. Press

12. Baumberg S, Young JPW, Wellington EMH, Saunders JR, eds. 1995. *Population Genetics of Bacteria*. Cambridge: Cambridge Univ. Press

13. Beachey EH, ed. 1980. *Receptors and Recognition.* New York: Chapman Hall

14. Bell AD. 1986. The simulation of branching patterns in modular organisms. *Philos. Trans. R. Soc. London Ser. B* 313:143–59

15. Bonner JT. 1982. Evolutionary strategies and developmental constraints in the cellular slime molds. *Am. Nat.* 119:530–52

16. Bradshaw AD. 1965. Evolutionary significance of phenotypic plasticity in plants. *Adv. Genet.* 13:115–55

17. Budrene EO, Berg HC. 1995. Dynamics of formation of symmetrical patterns by chemotactic bacteria. *Nature* 376:49–53

18. Buss LW. 1983. Evolution, development, and the units of selection. *Proc. Natl. Acad. Sci. USA* 80:1387–91

19. Buss LW. 1985. The uniqueness of the individual revisited. See Ref. 44, pp. 467–505

20. Buss LW. 1987. *The Evolution of Individuality.* Princeton: Princeton Univ. Press

21. Cavalier-Smith T. 1988. Eukaryote cell evolution. *Proc. Int. Bot. Congr., 14th, Berlin, 1987*, ed. W Greuter, B Zimmer, pp. 203–23. Königstein: Koeltz

22. Cheng K-J, Irvin RT, Costerton JW. 1981. Autochthonous and pathogenic colonization of animal tissues by bacteria. *Can. J. Microbiol.* 27:461–90

23. Coates AG, Jackson JBC. 1985. Morphological themes in the evolution of clonal and aclonal marine invertebrates. See Ref. 44, pp. 67–106

24. Cook RE. 1979. Asexual reproduction: a further consideration. *Am. Nat.* 113:769–72

25. Cook RE. 1983. Clonal plant populations. *Am. Sci.* 71:244–53

26. Davidson EH, Peterson KS, Cameron RA. 1995. Origin of bilateral body plans: evolution of developmental regulatory mechanisms. *Science* 270:1319–25

27. Dijkhuizen L. 1996. Evolution of metabolic pathways. In *Evolution of Microbial Life*, ed. DM Roberts, P Sharp, G Alderson, MA Collins, pp. 243–65. Cambridge: Cambridge Univ. Press

28. Dobzhansky T, Ayala FJ, Stebbins GL, Valentine JW. 1977. *Evolution.* San Francisco: Freeman

29. Fuqua C, Winans SC, Greenberg EP. 1996. Census and consensus in bacterial ecosystems: the LuxR-LuxI family of quorum-sensing transcriptional regulators. *Annu. Rev. Microbiol.* 50:727–51

30. Gabriel DW, Hunter JE, Kingsley MT, Miller JW, Lazo GR. 1988. Clonal population structure of *Xanthomonas campestris* and genetic diversity among citrus canker strains. *Mol. Plant-Microbe Interact.* 1:59–65

31. Gabriel W, Lynch M, Bürger R. 1993. Muller's ratchet and mutational meltdowns. *Evolution* 47:1744–57

32. Gardner SN, Mangel M. 1997. When can a clonal organism escape senescence? *Am. Nat.* 150:462–90

33. Goodwin BC. 1986. What are the causes of morphogenesis? *BioEssays* 3:32–36

34. Goodwin SB, Cohen BA, Fry WE. 1994. Panglobal distribution of a single clonal lineage of the Irish potato famine fungus. *Proc. Natl. Acad. Sci. USA* 91:11591–95

35. Guttman DS, Dykhuizen DE. 1994. Clonal divergence in *Escherichia coli* as a result of recombination, not mutation. *Science* 266:1380–83

36. Harold FM. 1990. To shape a cell: an inquiry into the causes of morphogenesis of microorganisms. *Microbiol. Rev.* 54:381–431

37. Harold FM. 1995. From morphogenes to morphogenesis. *Microbiology* 141:2765–78

38. Harper JL. 1977. *Population Biology of Plants.* London: Academic

39. Harper JL. 1985. Modules, branches, and the capture of resources. See Ref. 44, pp. 1–33

40. Harper JL, Rosen BR, White J. 1986. The growth and form of modular organisms. *Philos. Trans. R. Soc. London Ser. B* 313:3–5

41. Hughes RN. 1989. *A Functional Biology*

of Clonal Animals. London: Chapman & Hall

42. Hughes RN, Cancino JM. 1985. An ecological overview of cloning in metazoa. See Ref. 44, pp. 153–86

43. Hughes TP, Jackson JBC. 1980. Do corals lie about their age? Some demographic consequences of partial mortality, fission, and fusion. *Science* 209:713–15

44. Jackson JBC, Buss LW, Cook RE. 1985. *Population Biology and Evolution of Clonal Organisms.* New Haven: Yale Univ. Press

45. Janzen DH. 1977. What are dandelions and aphids? *Am. Nat.* 111:586–89

46. Kaiser D. 1986. Control of multicellular development: *Dictyostelium* and *Myxococcus. Annu. Rev. Genet.* 20:539–66

47. Kaiser D, Losick R. 1993. How and why bacteria talk to each other. *Cell* 73:873–85

48. Kaplan DR, Hagemann W. 1991. The relationship of cell and organism in vascular plants. *BioScience* 41:693–703

49. Karaolis DKR, Lan R, Reeves PR. 1994. Sequence variation in *Shigella sonnei* (Sonnei), a pathogenic clone of *Escherichia coli,* over four continents and 41 years. *J. Clin. Microbiol.* 32:796–802

50. Kays S, Harper JL. 1974. The regulation of plant and tiller density in a grass sward. *J. Ecol.* 62:97–105

51. Kemperman JA, Barnes BV. 1976. Clone size in American aspens. *Can. J. Bot.* 54:2603–7

52. Kimura M, Ohta T. 1974. On some principles governing molecular evolution. *Proc. Natl. Acad. Sci. USA* 71:2848–52

53. Knoll AH. 1992. The early evolution of eukaryotes: a geological perspective. *Science* 256:622–27

54. Koch AL. 1974. The pertinence of the periodic selection phenomenon to prokaryote evolution. *Genetics* 77:127–42

55. Koch AL. 1976. How bacteria face depression, recession, and depression. *Perspect. Biol. Med.* 20:44–63

56. Koch AL. 1987. Evolution from the point of view of *Escherichia coli.* In *Evolutionary Physiological Ecology,* ed. P Calow, pp. 85–103. Cambridge: Cambridge Univ. Press

57. Levin BR. 1981. Periodic selection, infectious gene exchange and the genetic structure of *E. coli* populations. *Genetics* 99:1–23

58. Li W-H, Graur D. 1991. *Fundamentals of Molecular Evolution.* Sunderland: Sinauer

59. Lindow SE. 1991. Determinants of epiphytic fitness in bacteria. In *Microbial Ecology of Leaves,* ed. JH Andrews, SS

Hirano, pp. 295–314. New York: Springer-Verlag

60. Macnab RM. 1996. Flagella and motility. In *Escherichia coli and Salmonella: Cellular and Molecular Biology,* ed. FC Neidhardt, 1:123–45. Washington, DC: ASM Press. 2nd ed.

61. Maiden MCJ, Feavers IM. 1995. Population genetics and global epidemiology of the human pathogen *Neisseria meningitidis.* See Ref. 12, pp. 269–93

62. Maynard Smith J. 1995. Do bacteria have population genetics? See Ref. 12, pp. 1–12

63. Maynard Smith J, Smith NH, O'Rourke M, Spratt BG. 1993. How clonal are bacteria? *Proc. Natl. Acad. Sci. USA* 90:4384–88

64. Melzer AL, Koeslag JH. 1991. Mutations do not accumulate in asexual isolates capable of growth and extinction—Muller's ratchet re-examined. *Evolution* 45:649–55

65. Milkman R, McKane M. 1995. DNA sequence variation and recombination in *E. coli.* See Ref. 12, pp. 127–42

66. Mitchell JG, Pearson L, Bonazinga A, Dillon S, Khouritt H, et al. 1995. Long lag times and high velocities in the motility of natural assemblages of marine bacteria. *Appl. Environ. Microbiol.* 61:877–82

67. Neidhardt FC, Ingraham JL, Schaechter M. 1990. *Physiology of the Bacterial Cell: A Molecular Approach.* Sunderland: Sinauer

68. Nijhout HF. 1990. Metaphors and the role of genes in development. *BioEssays* 12:441–46

69. Otto SP, Orive ME. 1995. Evolutionary consequences of mutation and selection within an individual. *Genetics* 141:1173–87

69a. Pace NR. 1997. A molecular view of microbial diversity of the biosphere. *Science* 276:734–40

70. Pace NR, Olsen GJ, Woese CR. 1986. Ribosomal RNA phylogeny and the primary lines of evolutionary descent. *Cell* 45:325–26

71. Panganiban G, Irvine SM, Lowe C, Roehl H, Corley LS, et al. 1997. The origin and evolution of animal appendages. *Proc. Natl. Acad. Sci. USA* 94:5162–66

72. Pardee AB. 1961. Response of enzyme synthesis and activity to environment. In *Microbial Reaction to Environment,* ed. GC Meynell, H Gooder, pp. 19–40. Cambridge: Cambridge Univ. Press

73. Pirt SJ. 1975. *Principles of Microbe and Cell Cultivation.* Oxford: Blackwell Sci.

74. Quattro JM, Avise JC, Vrijenhoek RC.

1992. An ancient clonal lineage in the fish genus *Poeciliopsis* (Atheriniformes: Poeciliidae). *Proc. Natl. Acad. Sci. USA* 89:348–52

75. Raff RA. 1996. *The Shape of Life: Genes, Development and the Evolution of Animal Form.* Chicago: Univ. Chicago Press

76. Reeves PR. 1992. Variation in O-antigens, niche-specific selection and bacterial populations. *FEMS Microbiol. Lett.* 100:509–16

77. Reichenbach H. 1984. Myxobacteria: a most peculiar group of social prokaryotes. In *Myxobacteria: Development and Cell Interactions,* ed. E Rosenberg, pp. 1–50. New York: Springer-Verlag

78. Russo VEA, Brody S, Cove D, Ottolenghi S. 1992. *Development: The Molecular Genetic Approach.* New York: Springer-Verlag

79. Schmid B. 1990. Some ecological and evolutionary consequences of modular organization and clonal growth in plants. *Evol. Trends Plants* 4:25–34

80. Sebens KP. 1987. The ecology of indeterminate growth in animals. *Annu. Rev. Ecol. System.* 18:371–407

81. Selander RK, McKinney RM, Whittam TS, Bibb WF, Brenner DJ, et al. 1985. Genetic structure of populations of *Legionella pneumophila. J. Bacteriol.* 163:1021–37

82. Selander RK, Musser JM. 1990. Population genetics of bacterial pathogenesis. In *The Bacteria: A Treatise on Structure and Function. Vol. XI: Molecular Basis of Bacterial Pathogenesis,* ed. BH Iglewski, VL Clark, pp. 11–36. New York: Academic

83. Shapiro JA, Dworkin M, eds. 1997. *Bacteria As Multicellular Organisms.* Oxford: Oxford Univ. Press

84. Shapiro L, Losick R. 1997. Protein localization and cell fate in bacteria. *Science* 276:712–18

85. Shimkets LJ. 1993. The myxobacterial genome. In *Myxobacteria II,* ed. M Dworkin, D Kaiser, pp. 85–107. Washington DC: ASM

86. Shubin N, Tabin C, Carroll S. 1997. Fossils, genes, and the evolution of animal limbs. *Nature* 388:639–48

87. Silander JA. 1985. Microevolution in clonal plants. See Ref. 44, pp. 107–52

88. Slatkin M. 1985. Somatic mutations as an evolutionary force. In *Evolution: Essays in Honour of John Maynard Smith,* ed. PJ Greenwood, PH Harvey, M Slatkin, pp. 19–30. Cambridge: Cambridge Univ. Press

89. Smith ML, Bruhn JN, Anderson JB. 1992. The fungus *Armillaria bulbosa* is among the largest and oldest living organisms. *Nature* 356:428–31

90. Spolsky CM, Phillips CA, Uzzell T. 1992. Antiquity of clonal salamander lineages revealed by mitochondrial DNA. *Nature* 356:706–8

91. Spratt BG, Smith NH, Zhou J, O'Rourke M, Feil E. 1995. The population genetics of the pathogenic *Neisseria.* See Ref. 12, pp. 143–60

92. Stebbins GL. 1968. Integration of development and evolutionary progress. In *Population Biology and Evolution,* ed. RC Lewontin, pp. 17–36. Syracuse: Syracuse Univ. Press

93. Steinger T, Kötner C, Schmid B. 1996. Long-term persistence in a changing climate: DNA analysis suggests very old ages of alpine *Carex curvula. Oecologia* 105:94–99

94. Trinci APJ, Cutter EG. 1986. Growth and form in lower plants and the occurrence of meristems. *Philos. Trans. R. Soc. London Ser. B* 313:95–113

95. Tuomi J, Vuorisalo T. 1989. Hierarchical selection in modular organisms. *Trends Ecol. Evol.* 4:209–13

96. Watkinson AR, White J. 1985. Some life-history consequences of modular construction in plants. *Philos. Trans. R. Soc. London Ser. B* 313:31–51

97. Weller DM, Thomashow LS. 1990. Antibiotics: evidence for their production and sites where they are produced. In *New Directions in Biological Control,* ed. RR Baker, PE Dunn, pp. 703–11. New York: Liss

98. Wessels JGH. 1993. Wall growth, protein excretion and morphogenesis in fungi. *New Phytol.* 123:397–413

99. Wherry ET. 1972. Box-huckleberry as the oldest living protoplasm. *Castanea* 37: 94–95

100. Whitham TC, Slobodchickoff CN. 1981. Evolution by individuals, plant-herbivore interactions, and mosaics of genetic variability: the adaptive significance of somatic mutations in plants. *Oecologia* 49: 287–92

101. Whittam TS. 1995. Genetic population structure and pathogenicity in enteric bacteria. See Ref. 12, pp. 217–45

102. Williams GC. 1975. *Sex and Evolution.* Princeton: Princeton Univ. Press

103. Winn WC Jr. 1995. *Legionella.* In *Manual of Clinical Microbiology,* ed. PR Murray, pp. 533–44. Washington, DC: ASM Press. 6th ed.

Annu. Rev. Microbiol. 1998. 52:127–64

REGULATION OF ACETATE METABOLISM BY PROTEIN PHOSPHORYLATION IN ENTERIC BACTERIA

Alain J. Cozzone
Institut de Biologie et Chimie des Protéines, Centre National de la Recherche Scientifique, Université de Lyon, 7 passage du Vercors, 69007 Lyon, France

KEY WORDS: carbon flux, glyoxylate bypass, bacterial enzyme phosphorylation

ABSTRACT

Growth of enteric bacteria on acetate as the sole source of carbon and energy requires operation of a particular anaplerotic pathway known as the glyoxylate bypass. In this pathway, two specific enzymes, isocitrate lyase and malate synthase, are activated to divert isocitrate from the tricarboxylic acid cycle and prevent the quantitative loss of acetate carbons as carbon dioxide. Bacteria are thus supplied with the metabolic intermediates they need for synthesizing their cellular components. The channeling of isocitrate through the glyoxylate bypass is regulated via the phosphorylation/dephosphorylation of isocitrate dehydrogenase, the enzyme of the tricarboxylic acid cycle which competes for a common substrate with isocitrate lyase. When bacteria are grown on acetate, isocitrate dehydrogenase is phosphorylated and, concomitantly, its activity declines drastically. Conversely, when cells are cultured on a preferred carbon source, such as glucose, the enzyme is dephosphorylated and recovers full activity. Such reversible phosphorylation is mediated by an unusual bifunctional enzyme, isocitrate dehydrogenase kinase/phosphatase, which contains both modifying and demodifying activities on the same polypeptide. The genes coding for malate synthase, isocitrate lyase, and isocitrate dehydrogenase kinase/phosphatase are located in the same operon. Their expression is controlled by a complex dual mechanism that involves several transcriptional repressors and activators. Recent developments have brought new insights into the nature and mode of action of these different regulators. Also, significant advances have been made lately in our understanding of the control of enzyme activity by reversible phosphorylation. In general,

127

0066-4227/98/1001-0127$08.00

analyzing the physiological behavior of bacteria on acetate provides a valuable approach for deciphering at the molecular level the mechanisms of cell adaptation to the environment.

CONTENTS

INTRODUCTION

"... et le rêve de toute cellule: devenir deux cellules."

—Jacques Monod

Bacteria are able to live in environmental conditions that change frequently and, most often, rapidly. For doing so, they have evolved a series of adaptive mechanisms through which they constantly monitor their surroundings and adjust their physiology accordingly. In particular, they possess an array of regulatory devices that facilitate the conversion of available nutrients into components of central metabolic pathways and, thereby, the production of precursors of new cell material as well as energy for substrate uptake, biosynthesis, and growth. One of these pathways is the tricarboxylic acid cycle, or Krebs cycle, which plays two essential roles. First, it is responsible for the total oxidation to CO_2

of acetyl coenzyme A, which is derived mainly from the pyruvate produced by glycolysis. Second, it provides intermediates that are required for the biosynthesis of amino acids, and therefore of cellular macromolecules, as well as heme biosynthesis. Bacteria must continuously replenish the pools of these intermediates in order to maintain them at sufficient levels for active metabolism. For this purpose, they utilize different anaplerotic reaction pathways according to the chemical nature of the carbon source being used. For example, when aerobic heterotrophs such as *Escherichia coli* or *Salmonella typhimurium* are required to grow on acetate as the sole source of carbon and energy, they activate a specific anaplerotic pathway, termed the glyoxylate bypass. The net effect of this bypass is the formation of 4-carbon dicarboxylic acids from acetyl coenzyme A molecules, in the ratio of 1:2, such that bacteria are supplied with the precursors they need to synthesize carbohydrates, proteins, and nucleic acids.

Since the pioneering work of Kornberg in the late 50s (99–101), an appreciable amount of data have been published on bacterial growth in the presence of acetate and operation of the glyoxylate shunt, concerning the enzymes involved, the energy required, and the biochemical reactions that direct the fluxes of carbon molecules. These different features have been presented in detail in a number of reviews (25, 35, 137, 140). A considerable advance in understanding the mechanisms that control the differential functioning of the Krebs cycle and the glyoxylate bypass was made, in the late 1970s, by demonstrating the influence of phosphorylation on the activity of the pivotal enzyme, NADP-linked isocitrate dehydrogenase (59, 60). In recent years, special attention has been given to the regulation of 2-carbon compound metabolism at the genetic level.

The aim of this review is to summarize the decisive findings pertaining to these various aspects of acetate metabolism. First, an overview is presented of acetate utilization by enteric bacteria, then the present status of knowledge of the regulation by enzyme phosphorylation is assessed, and finally, the genetic facets of this regulation are analyzed.

GROWTH OF BACTERIA ON ACETATE

When acetate is the sole source of carbon and energy, it must first be activated to acetyl coenzyme A (acetyl CoA) to become usable for replenishing intermediates of the tricarboxylic acid cycle via the glyoxylate bypass. This activation is also required when acetate is used for oxidation, for lipid synthesis, or for amino acid synthesis (25).

Conversion of Acetate to Acetyl CoA

Two enzymes successively catalyze first the conversion of acetate to acetyl phosphate with cleavage of ATP to ADP (reaction 1), then the transfer of the acetyl moiety from acetyl phosphate to CoA, with liberation of inorganic phosphate (reaction 2):

(Reaction 1)   acetate + ATP → acetyl phosphate + ADP,

(Reaction 2)   acetyl phosphate + CoA → acetyl CoA + Pi.

The first reaction is catalyzed by acetate kinase (ATP acetate phosphotransferase) encoded by the *ackA* gene, and the second reaction is mediated by phosphotransacetylase (acetyl CoA [CoA]: orthophosphate acetyl transferase) encoded by the *pta* gene. The levels of these two enzymes in *E. coli* and *S. typhimurium* vary little with different carbon sources; expression of the *ackA* and *pta* genes is thus neither induced by acetate nor catabolite repressed by glucose (13, 105). In addition, these enzymes are both equally active in aerobic and anaerobic conditions and they are operative not only during aerobic growth on acetate as carbon source, but also during anaerobic growth on sugars when acetate is a major fermentation product (13, 105).

The finding that *ackA* and *pta* mutants, under aerobic conditions, are severely impaired in the utilization of acetate as sole carbon source, while they can incorporate labeled acetate when grown on glycerol, indicates the presence of a second acetate uptake system (13). Cells of *E. coli* thus contain an inducible acetyl CoA synthetase (acetate:CoA ligase AMP forming) that catalyzes the following reaction:

acetate + ATP + CoA → acetyl CoA + AMP + PPi.

No acetyl CoA synthetase activity is measured in cells grown on glucose, which suggests that this enzyme is regulated by catabolite repression, contrary to acetate kinase and phosphotransacetylase.

In connection with the role of protein phosphorylation in the control of the glyoxylate bypass (see below), it is interesting to note that acetyl phosphate, the product of reaction 1 in the conversion of acetate to acetyl CoA, may also be a global signal of energy metabolism acting through two-component phosphorylation-dependent signal transduction switches. This role has been proposed from the observation that acetyl phosphate, as well as carbamyl phosphate and phosphoramidate, can act as a phosphate donor to the conserved aspartate residue of most response regulator proteins (120). The presence of acetyl phosphate would thus bypass the need for the cognate sensor histidine kinases (95, 148).

Catabolism of Fatty Acids

The degradation of fatty acids by β-oxidation and thiolytic cleavage results in their complete conversion to acetyl CoA (96, 145, 184). Consequently, bacteria can grow on fatty acids as sole carbon source by operating the same pathways as those required for growth on acetate (7). Cells of *E. coli* and *S. typhimurium* can grow only when the fatty acids are 12 or more carbons in length (long-chain fatty acids) and then only after a lag period allowing induction of the *fad* regulon. Fatty acids of only six to ten carbon atoms (medium-chain fatty acids) are less readily metabolized, and carboxylic acids shorter than six carbons (short-chain fatty acids) cannot be used by bacteria solely via the *fad* system (149, 169, 184). Upon degradation, one molecule of acetyl CoA is generated from each two-carbon segment of a fatty-acid molecule. This reaction results in the concomitant production of one equivalent each of $FADH_2$ and NADH, which may be used in ATP synthesis (25). In contrast, during acetate degradation this ATP is not formed, and more ATP is consumed to activate this substrate. Hence, growth on fatty acids is energetically more favorable, per carbon atom, than growth on acetate (25). Intriguingly, *E. coli* strains grow more rapidly on oleate than on acetate, whereas *S. typhimurium* strains have the opposite behavior (41).

THE GLYOXYLATE BYPASS

The glyoxylate bypass is the only pathway allowing growth on acetyl CoA. This pathway is observed in all organisms that can employ acetate or fatty acids as their sole carbon source. It represents an example par excellence of an anaplerotic sequence in operation.

Nature and Role

When the tricarboxylic acid cycle is in operation, acetyl CoA condenses with oxaloacetate to form citrate, and subsequent reactions effect the stepwise oxidation of citrate to regenerate the oxaloacetate. In each turn of this cycle, there is a concomitant loss of two molecular units of carbon dioxide for each molecular unit of acetate entering the cycle (103, 104). When acetyl CoA is the only carbon source available, no net assimilation of carbon can occur by this means. Therefore, growth on acetate or fatty acids requires the operation of a separate pathway to provide the cell with the 4-carbon, and thus 3-carbon, intermediates necessary for the biosynthesis of cellular components. This is accomplished by an epicycle of the tricarboxylic acid cycle, the glyoxylate bypass, also called the Krebs-Kornberg cycle (Figure 1), which diverts part of the flux of carbon molecules at isocitrate (99–101).

It is generally considered that only microorganisms (including bacteria, protozoa, yeasts, and molds) and higher plants, but not animal cells, are able to use the glyoxylate shunt. In that sense, the so-called lower organisms possess a savoir-faire that higher organisms do not, except for a few such as chicken, toad,

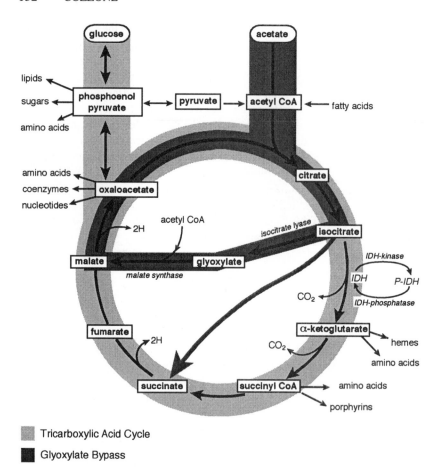

Figure 1 The glyoxylate bypass and related reactions.

and rat (37a, 65a, 152a). In plants the glyoxylate bypass occurs in glyoxysomes that are special peroxisomes present in germinating oil-storage seeds, e.g. castor beans, peanuts, cotton, and corn. There, it serves to convert the fatty acids of seed lipids into succinate and thus into the sugars needed by the growing seedlings. Because the glyoxylate bypass involves two oxidative steps, linked to electron transport, it also provides some energy. But most of the cell energy is derived from the simultaneous oxidation of other acetate molecules via the tricarboxylic acid cycle. Whenever a more easily metabolizable carbon source, such as glucose or succinate, becomes available in the growth medium, the glyoxylate bypass is turned off because of catabolite repression (165).

Enzymes and Regulation

The glyoxylate cycle consists of six of the eight reactions of the tricarboxylic acid cycle (Figure 1) but differs from it in bypassing the two oxidative steps in which carbon dioxide is evolved (101). Two unique enzymes of the glyoxylate cycle, isocitrate lyase and malate synthase, are induced upon growth of bacteria on acetate, as well as phosphoenolpyruvate carboxykinase to generate phosphoenolpyruvate. These enzymes have been studied in both *E. coli* and *S. typhimurium*, even though their expression is considerably lower in the latter species (187). Isocitrate lyase, a tetrameric protein encoded by the *aceA* gene, first catalyzes the aldol cleavage of isocitrate to succinate (an intermediate of the tricarboxylic acid cycle) and glyoxylate (1, 46, 47). This reaction requires a rather large intracellular pool of isocitrate because isocitrate lyase has a low affinity for this substrate (79). Then malate synthase, encoded by the *aceB* gene, catalyzes the condensation of acetyl CoA with glyoxylate to yield malate (another intermediate of the tricarboxylic acid cycle) (144). The overall reaction of the glyoxylate bypass is summarized in Figure 2. In each turn of the cycle, one mole of malate is generated from two moles of acetyl CoA. Malate is converted to oxaloacetate, which yields phosphoenolpyruvate by decarboxylation and phosphorylation reactions and, subsequently glucose-6-phosphate. Four moles of acetyl CoA and, hence, two moles of phosphoenolpyruvate are thus required to give one mole of glucose.

$$\text{isocitrate} \xrightarrow{\text{isocitrate lyase}} \text{succinate} + \text{glyoxylate}$$

$$\text{glyoxylate} + \text{acetyl CoA} \xrightarrow{\text{malate synthase}} \text{malate}$$

$$\text{succinate} + \text{acetyl CoA} \xrightarrow[\text{(Krebs cycle)}]{-2 \times (2H)} \text{isocitrate}$$

Net :   2 acetyl CoA - 2 x (2H) $\longrightarrow$ malate

Figure 2   The overall reaction of the glyoxylate bypass.

The concept that the tricarboxylic acid cycle and the glyoxylate bypass operate together, although concurrently, to allow catabolism of acetate is supported by genetic data. For example, mutants of *E. coli* deficient in citrate synthase are unable to grow on acetate (63, 72) indicating that at least this step of the tricarboxylic acid cycle is essential to acetate degradation. Mutants deficient in isocitrate lyase likewise are unable to grow on acetate, although they grow normally on other carbon and energy sources such as glucose (4, 102).

The possibility that regulation of the glyoxylate bypass could be exerted, at least in part, through inhibition of activity and/or repression of synthesis of the glyoxylate enzymes has been examined. Isocitrate lyase has been reported to be allosterically inhibited by phosphoenolpyruvate and this effect has been considered for some time to be physiologically significant in the regulation of flux through the glyoxylate bypass (4). However, detailed kinetic studies have shown that phosphoenolpyruvate rather behaves like an analog of succinate, and that allosteric effects are not involved. These results, originally obtained with the enzyme from *Pseudomonas indigofera* (186), have been confirmed in the case of *E. coli* (53, 116, 124), and consideration of the intracellular concentrations of isocitrate and phosphoenolpyruvate suggests that the inhibition of isocitrate lyase by phosphoenolpyruvate has no physiological significance. Similarly, no correlation seems to exist between the intracellular levels of oxaloacetate, pyruvate, citrate, or isocitrate, and the activity of the glyoxylate enzymes (106). Therefore the only mechanism that has, so far, been demonstrated to control the glyoxylate bypass concerns the reversible phosphorylation of isocitrate dehydrogenase, as discussed below.

Alternate Pathways

The glyoxylate bypass cannot operate under anaerobic conditions (90, 91). However, the cofactor ferredoxin permits anaerobes and certain photosynthetic bacteria to form pyruvate, and in some cases α-ketoglutarate, via a reductive carboxylation (2, 14). This activity can occur because of the low redox potential of ferredoxin. The irreversibility of the oxidative decarboxylation of pyruvate to acetyl CoA and CO_2, and of α-ketoglutarate to succinyl CoA and CO_2, observed in most microorganisms, is circumvented. Thus, the uptake of acetate can occur in those organisms that produce ferredoxin.

When glyoxylate is the only source of carbon available, it can be utilized as is for bacterial growth by activating a particular malate synthase, termed malate synthase G ("G" for induction by glyoxylate or glycolate) (123). These two 2-carbon compounds, which are products of plant and algal metabolism, can serve as carbon energy sources to a variety of bacterial species including *E. coli* and *S. typhimurium*. The first step of the pathway is the oxidation of glycolate to glyoxylate by the glycolate oxidase. Then glyoxylate is metabolized by either of two divergent condensation reactions, one leading to the glycolytic pathway and the other to the tricarboxylic acid cycle (144, 179). In the latter case, glyoxylate reacts with acetyl CoA to form malate by using the malate synthase G. In fact this reaction is basically identical to that of the glyoxylate cycle except that the condensation enzyme is malate synthase G instead of malate synthase A ("A" for induction by acetate). The two malate synthases are readily distinguished by differing stabilities and inhibitor patterns (144). They are regulated by distinct mechanisms. Malate synthase G is strongly induced, over 1000-fold, by glycolate (123), whereas malate synthase A is induced 20-fold by acetate or fatty acids (179).

Although it does not concern enteric bacteria, a novel alternate anaplerotic pathway to the glyoxylate cycle has been recently described (71). Cells of *Streptomyces collinus* can grow on acetate alone even though they do not harbor any isocitrate lyase activity. Two particular enzymes appear to play a central role in this pathway: the crotonyl-CoA reductase and an enzyme that resembles methylmalonyl-CoA mutase. Interestingly, a similar situation is encountered in the methylotroph *Methylobacterium extorquens* (22).

ISOCITRATE DEHYDROGENASE

Isocitrate dehydrogenase (IDH) is an enzyme of the tricarboxylic acid cycle that catalyzes the conversion of isocitrate to α-ketoglutarate. When bacteria are grown on acetate, IDH competes with isocitrate lyase for their common substrate, isocitrate. The flux of carbon is thus divided between the tricarboxylic acid cycle and the glyoxylate bypass (Figure 1). The partition of flux is regulated by the reversible phosphorylation of IDH, which represents the first example of control of a key metabolic enzyme by phosphorylation in prokaryotes.

Nature and Properties

In contrast with eukaryotic systems where IDHs are NAD-linked (152), most bacteria possess an NADP-linked enzyme. Those bacteria harboring NAD-dependent IDHs are incapable of growth on acetate and lack either a respiratory chain or a complete tricarboxylic acid cycle (20). Bacteria capable of growth on acetate use NADP-dependent enzymes to generate 90% of the NADPH necessary for biosynthesis (182, 183), and there appears to be no other major source of NADPH. Thus, the flux through NADP-dependent malic enzymes provides approximately 10% of the NADPH (182, 183), whereas transhydrogenase, a membrane-bound enzyme that uses energy in the form of the protonmotive force to reduce NADP, has less than 5% of the activity of IDH (115, 189). Hence, the NADP dependence of bacterial IDHs appears to be an adaptation to growth on acetate. A recent evolutionary analysis suggests that eubacterial NADP-dependent enzymes evolved from an NAD-dependent precursor about 3.5 billion years ago (38). Selection in favor of utilizing NADP would be a result of niche expansion during growth on acetate.

Several groups have reported the purification of IDH from various strains of *E. coli* (135). Since the original procedure described by Reeves et al (156), other protocols have been developed and, in particular, dye-ligand chromatography has proven to permit rapid purification of a homogenous enzyme with high specific activity (11, 15, 181). *E. coli* cells possess only one NADP-linked IDH consisting of two identical subunits (Table 1). The complete nucleotide sequence of the corresponding gene, *icd*, has been determined (178). The enzyme has no significant sequence homology with other IDHs or nucleotide binding proteins. The highest homology is with 3-isopropylmalate dehydrogenase from *Thermus aquaticus* (178). The two forms of IDH, phosphorylated and dephosphorylated, can be separated by ion-exchange chromatography or non-denaturing gel electrophoresis (9, 60). Analysis of the intracellular localization of the enzyme by transmission electron microscopy and immunochemical detection shows that it is dispersed in the cytoplasmic region (177).

Unlike many NAD-linked IDHs (152), the NADP-linked enzyme from *E. coli* is not regulated by allosteric effectors such as AMP or ADP. The enzyme is strongly inhibited by glyoxylate plus oxaloacetate in combination and it has been admitted for some time that this "concerted inhibition" could be of physiological significance (135). A detailed kinetic analysis in vitro of the inhibition mechanism has shown that it is, in fact, a rather complex phenomenon (136). The results depend on the order of addition of the assay components. When IDH or NADP is added last, the rate slowly decreases until a new, inhibited, steady state is obtained. When isocitrate is added last, the initial rate is almost zero, but it increases slowly until the same steady-state value is obtained. Glyoxylate and oxaloacetate give competitive inhibition against isocitrate and uncompetitive inhibition against NADP. IDH obeys a compulsory-order mechanism, with coenzyme binding first. Glyoxylate and oxaloacetate bind to and dissociate from the isocitrate-binding site of the enzyme slowly. The condensation of glyoxylate and oxaloacetate produces a very potent inhibitor of IDH, oxalomalate, which is, however, unstable and decomposes spontaneously to a less potent inhibitor, 4-hydroxy-2-oxoglutarate. Oxalomalate inhibits IDH competitively with respect to isocitrate and is a poor substrate for the enzyme. Therefore, both the instability of oxalomalate and the relatively high intracellular concentration of isocitrate during growth on acetate indicate that the inhibition of IDH by glyoxylate and oxaloacetate is not likely to be physiologically significant (136, 137). There do not appear to be any well-established non covalent regulators of *E. coli* IDH.

Table 1 Genes and proteins involved in the glyoxylate bypass and its regulation in *E. coli*

	Gene			Protein			
Name	Length (bp)	Map location (min)	Accession #	Name	Molecular weight of monomer[c]	Amino acids per monomer	Reference
icd	1248	25.6	J02799[a]	Isocitrate dehydrogenase (homodimer)	45,756	416	(178)
aceB	1599	90.85	X12431[a] P08997[b]	Malate synthase A	60,205	533	(16)
aceA	1302	90.85	X07543[a] P05313[b]	Isocitrate lyase (homotetramer)	47, 700	434	(161a)
aceK	1731	90.85	M18974[a] P11071[b]	Isocitrate dehydrogenase kinase/phosphatase (homodimer)	66, 528	577	(27)
iclR	822[d]	90.95	M34937[a] P16528[a] M31761[b] P76782[b]	Transcriptional repressor (homodimer)	29,741	274	(132, 175)
fadR	714	26.54	X08087[a] P09371[b]	Transcriptional repressor (homodimer)	26,837	238	(42)
fruR	1002	2.2	X55457[a] P21168[b]	Transcriptional activator (homotetramer)	37,999	334	(94)
himA	297	38.6	X04864[a] P06984[b]	Integration host factor (α subunit)	11,354	99	(54)
himD	282	20.4	X04864[a] P08756[b]	Integration host factor (β subunit)	10,651	94	(54)

[a]Accession number in GenBank/EMBL data bank.
[b]Accession number in Swiss-Prot data bank.
[c]Deduced from the nucleotide sequence of the relevant gene.
[d]The *iclR* gene of *S. typhimurium* extends over 822 bp encoding a protein of 29,517 mol. weight (57).

Reversible Phosphorylation

The original observations indicating that IDH can be reversibly inactivated in vivo were made in the course of a study of the levels of tricarboxylic acid cycle enzymes in *E. coli* during growth on different carbon sources (6, 80). After cessation of growth on limiting glycerol, the specific activities of malate dehydrogenase, α-ketoglutarate dehydrogenase, and IDH remain constant for several hours. However, after cessation of growth on limiting glucose, the specific activities of malate dehydrogenase and α-ketoglutarate dehydrogenase remain stable, whereas that of IDH is reduced fivefold over two hours, then rises again to 75% of its original value over the next two hours (80). The reason for this behavior is that *E. coli* cells excrete acetate during growth on glucose but not on glycerol (12). After cessation of growth on glucose, the enzymes of the glyoxylate bypass are induced, and acetate is oxidized. The specific activity of IDH decreases during the period of adaptation to acetate and during its utilization, but rises again after exhaustion of acetate. The latter increase does not simply result from induction of IDH expression, since it occurs even in the presence of protein synthesis inhibitors. Similar conclusions have been reached in another study using a quite different approach (116). It has been found that addition of glucose to a culture growing on acetate produces a metabolic crossover at IDH; the cellular level of isocitrate declines while that of α-ketoglutarate goes up. This metabolic crossover has suggested that IDH is activated during this transition (116).

The first demonstration that the activity of IDH is controlled by phosphorylation/dephosphorylation was made by Garnak & Reeves (59, 60). Addition of acetate to a stationary phase culture of *E. coli* in glycerol medium containing ^{32}P-labeled orthophosphate results in rapid loss of IDH activity and concomitant incorporation of radioactivity into the enzyme (59). The phosphorylated enzyme can be purified to electrophoretic homogeneity using a procedure involving binding to and elution from a dye-ligand matrix (60). The dephosphorylated active form of IDH can be resolved from the phosphorylated inactive form by non-denaturing gel electrophoresis (10). The active form has a slightly lower electrophoretic mobility than the inactive form, and only the former can be stained for enzyme activity using the tetrazolium salt method (10, 137). Unlike many phosphorylated enzymes, the phosphorylated form of IDH is completely inactive (10). The proportions of active and inactive forms in cell extracts can be determined by a specific method based on immunological cross-reaction (49). Thus, when bacteria are grown on acetate, about 75% of IDH is inactivated by phosphorylation.

The inactivation of IDH by phosphorylation forces isocitrate through the glyoxylate cycle and diverts some of the carbon flux from the tricarboxylic acid cycle. A precise balance between the two competing cycles is essential for growth; because isocitrate lyase has a much lower affinity for isocitrate ($K_m = 604 \ \mu M$) than does IDH ($K_m = 8 \ \mu M$), the flux through IDH must be restricted to allow the intracellular concentration of isocitrate to rise to a level sufficient to sustain flux through isocitrate lyase (53, 78). Mathematical analyses demonstrate that saturation of IDH with isocitrate, which occurs during growth on acetate, makes it impossible to directly regulate isocitrate lyase activity. Regulation can only be achieved indirectly through control of the enzyme activity (112). Mutant strains that cannot phosphorylate IDH fail to grow on acetate, demonstrating that the phosphorylation of the enzyme is required for use of the glyoxylate bypass (111).

Regulation by phosphorylation occurs also during transitions between carbon sources. For instance, addition of a preferred carbon source, such as glucose or pyruvate, to a culture growing on acetate makes the glyoxylate bypass unnecessary, and bacteria then shut this pathway down by dephosphorylating IDH (10,51). The inhibition of the glyoxylate bypass is due to the fact that the activation of the enzyme draws isocitrate through the tricarboxylic acid cycle. Isocitrate concentration therefore declines drastically, which decreases the velocity of isocitrate lyase accordingly (112). It must be noted that, in general, growth on different carbon sources is achieved without any obvious enzyme acting as a regulator of metabolic flux, as shown by analyzing the effects of 11 different single carbon sources (77). The only exception concerns IDH during growth on acetate. Of interest are the results obtained from mathematical modeling, which reveal that IDH is not "rate limiting" during growth on acetate and that flux through isocitrate lyase is essential to sustain a high intracellular level of isocitrate (51). According to this study, above a certain threshold concentration of isocitrate lyase, the tricarboxylic acid cycle and the glyoxylate bypass work in concert and the partition of carbon flux between IDH and isocitrate lyase is no longer a problem.

Mechanism of Inactivation

IDH from *E. coli* is phosphorylated at one amino acid residue, serine 113, per protein subunit (10, 178). The mechanism of enzyme inhibition is unusual because, unlike other phosphorylations, it leads to complete, rather than partial, loss of activity (10, 60, 110). To understand the precise mechanism of inactivation, the binding properties of the substrates and products to the native and phosphorylated enzymes have been analyzed. It has been observed that NADP can protect active IDH against attack by several proteases (58). The dephosphorylated inactive form is much less susceptible to proteolysis than the active form, and it is not protected by NADP. This finding suggests that binding of NADP to, or phosphorylation of, active IDH induces similar conformational changes. In addition, it has been proposed from fluorescence titration experiments that NADPH binds to active but not to inactive IDH, which would indicate that the phosphorylation of the enzyme occurs close to or at the coenzyme-binding site (58, 121). However these observations have been questioned by equilibrium binding studies that demonstrate that NADPH binds with an almost equal affinity to the active and inactive (phosphorylated) enzymes, and that it is isocitrate binding, not NADPH, that is the target substrate affected by the phosphate (40). Indeed, phosphorylation of IDH blocks isocitrate binding by disrupting the hydrogen bond between serine 113 and isocitrate, and by introducing electrostatic repulsion between two negatively charged groups: the phosphate of phosphoserine and the γ-carboxylate of isocitrate (39, 40, 82). This is the main structural basis for the inhibition of IDH. Introduction of a negatively charged aspartate or glutamate for the serine completely inactivates the enzyme. Replacement of the serine with other amino acids (lysine, cysteine, alanine, asparagine, or glutamine) produces active enzymes that bind both isocitrate and NADPH. Neither threonine nor tyrosine, when substituted for serine at the phosphorylation site, is detectably phosphorylated (39, 178).

The structure of the phosphorylated form of IDH from $E.$ $coli$ has been solved and refined at 2.5 Å resolution. Comparison with the structure of the dephosphorylated enzyme confirms that there are no large-scale conformational changes and that small conformational changes are highly localized around the site of phosphorylation at serine 113 (83). In the three-dimensional structure of the dephosphorylated enzyme, serine 113 is found in a pocket containing a number of positively charged and other polar residues that constitute the active site (83, 84). Asparagine 115 is postulated to be the second key determinant of specificity in IDH. Like serine 113, asparagine 115 interacts with the γ-carboxylate of bound isocitrate that, in turn, forms a salt bridge to the nicotinamide ring of the coenzyme (8, 172). Aspargine 115 also interacts with the phosphoserine of inactivated enzyme and with the carboxylate of an inactive Ser 113 Glu mutant in complex with isocitrate (82, 83). It has recently been demonstrated that replacing asparagine 115 by leucine reactivates the Ser 113 Glu mutant, as well as a Ser 113 Asp mutant, by completely suppressing the electrostatic repulsion according to a second-site mutation process (19).

A different mechanism for IDH inhibition appears to occur in the particular case of anaerobic cell growth. Expression of the glyoxylate bypass operon, like tricarboxylic acid cycle gene expression, responds to aerobic respiration but is repressed under anaerobic conditions. The principal elements of this regulation are protein ArcA and protein ArcB (Arc for "aerobic respiration control"), which are members of a two-component phosphorylation system (90–92). When $E.$ $coli$ cells are cultured anaerobically, the synthesis of IDH is considerably reduced owing to the inhibitory action of ArcA, which functions as a classical repressor by binding at a site overlapping the icd promoter (18).

Comparison of Different Species

The most extensively studied bacterial IDH is that of $E.coli$, though regulation by phosphorylation/dephosphorylation has been also reported for enzymes of other enteric bacteria such as $S.$ $typhimunium$, $Aerobacter$ $aerogenes$, and $Serratia$ $marcescens$ (6, 137). IDHs have been purified and characterized from a variety of species (20). Most bacteria possess only an NADP-dependent enzyme consisting of two identical subunits with molecular weights ranging from 40,000 to 57,000 (20). There are other prokaryotic NADP-linked IDHs that are monomeric enzymes with molecular weights of 80,000 to 100,000, as in *Corynebacterium glutamicum* (50) or in *Rhodomicrobium vannielii* (114). In some exceptional cases, both types of enzymes coexist in the same organism, as in *Vibrio* sp. strain ABE-1, which contains one homodimeric IDH I with a thermostability comparable to that of the enzymes from mesophiles, and one monomeric protein II with extreme lability above 20°C (88, 142, 176). The genes encoding these two structurally different isoenzymes have been cloned and sequenced (89). The amount of enzyme I increases when bacteria are grown on acetate minimal medium, whereas the level of enzyme II remains unchanged. During adaptation of the strict aerobe *Acinetobacter calcoaceticus* to growth on acetate, the specific activity of NADP-dependent IDH increases (158, 159). This response is unique, as compared to all other bacterial species, where the enzyme activity generally decreases under the same culture conditions. Moreover, in this bacterium, which contains two distinct isoenzymes, adaptation to acetate is accompanied by an increase in the proportion of the larger, allosteric, isoenzyme with a concomitant reduction of the level of the smaller, non-allosteric isoenzyme (158, 159). In the unicellular cyanobacterium *Synechocystis* sp. strain PCC6803, the highest NADP-dependent IDH activity and protein accumulation are obtained under nitrogen starvation, while maximum activity of the corresponding enzyme in the filamentous *Anabaena* sp. strain PCC7120 is observed under dinitrogen-fixing conditions (126, 127). Several genes encoding IDH have been sequenced from prokaryotic and eukaryotic sources [reviewed in (127)]. Comparison of the deduced amino acid sequences reveals conserved regions among dimeric and among monomeric enzymes, but no similarity can be detected between those two groups (50, 127).

ISOCITRATE DEHYDROGENASE KINASE/PHOSPHATASE

Isocitrate dehydrogenase kinase/phosphatase (IDH kinase/phosphatase) is the bifunctional enzyme that specifically catalyzes both the phosphorylation and dephosphorylation of IDH. The differential activity of this unusual enzyme determines the extent of phosphorylation of IDH, which in turn controls the balance of the carbon flux between the tricarboxylic acid cycle and the glyoxylate bypass. IDH represents the raison d'être of IDH kinase/phosphatase because it is the only substrate that the phosphorylating/dephosphorylating enzyme is able to recognize.

Nature and Properties

The first suggestion that IDH kinase/phosphatase from *E. coli* is bifunctional became apparent during the development of a purification protocol (109). The two activities co-elute from a variety of chromatographic media, including ion exchange, gel filtration, and affinity columns. Furthermore, the purified protein, which yields a single band following dodecylsulfate-gel electrophoresis, efficiently catalyzes both phosphorylation and dephosphorylation. The bifunctional nature of IDH kinase/phosphatase has been further elucidated following cloning of the encoding gene (108). Identification of this clone was achieved by complementation of an *aceK* mutation, restoring the ability of the host strain to grow on acetate. Physical and functional mapping of the clone has shown that both kinase and phosphatase activities are encoded by a single gene (Table 1). These two opposing activities are physically associated with the same polypeptide, which is present in duplicate in the native form of the enzyme (108).

In addition to the kinase and phosphatase activities, the enzyme from *E. coli* also exhibits a specific ATPase activity (173). Mutant strains devoid of IDH phosphatase activity retain both the kinase and the ATPase activities, indicating that ATP hydrolysis does not result from the cyclic phosphorylation of IDH. However, the kinase and ATPase activities of these mutants differ significantly from those of the wild-type IDH kinase/phosphatase expressed from the parental allele (173). This shows that the kinase and phosphatase do not reside on structurally independent domains. In contrast to many enzymes that catalyze kinetically unfavorable side reactions, the maximum velocity of the ATPase substantially exceeds those of the kinase and the phosphatase. ATP hydrolysis is only partially inhibited by phosphorylated and dephosphorylated IDH, with saturating levels of the phosphorylated form decreasing the role of ATP hydrolysis by a factor of five. Even in the presence of near-saturating concentrations of phospho-IDH, the rate of ATP hydrolysis is fourfold greater than the rate of the cyclic phosphorylation of IDH.

Active Sites

Kinetic analyses of different *E. coli* proteins produced by mutant alleles of the *aceK* gene encoding IDH kinase/phosphatase have provided evidence for a single active site on the enzyme (86). The products of one class of alleles retain the kinase activity but suffer reductions in phosphatase by factors of 200 to 400. The kinase, phosphatase, and intrinsic ATPase of these proteins all exhibit striking reductions in their affinity for phosphorylated IDH even though they retain affinity for the dephosphorylated form. This observation suggests that there is a single binding site for phospho-IDH, which is common to the kinase and phosphatase activities (107).

An unusual feature of IDH phosphatase is its absolute requirement for ATP or ADP for activity, with a corresponding lack of activity in the presence of non-hydrolyzable ATP analogs (109). Since both IDH kinase and IDH phosphatase require adenine nucleotides, identification of the number and location of the nucleotide binding sites can provide substantial insights into the organization of the active sites. Kinetic analysis of the mutants mentioned above (86) has shown that mutations at the ATP binding site eliminate both kinase and phosphatase activities, which confirms that the reactions catalyzed by these activities occur at the same binding site. Also, site-directed mutagenesis of the *aceK* gene has proven that mutations affecting the consensus amino acid sequence of IDH kinase/phosphatase for ATP binding eliminate simultaneously both the kinase and phosphatase activities without destabilizing the protein (174). Similarly, ultraviolet irradiation of IDH kinase/phosphatase in the presence of the photoaffinity labeling reagent 8-azido ATP results in parallel losses of its kinase and phosphatase activities, and in covalent attachment of the reagent to the protein at a single site (180).

The structural organization of IDH kinase/phosphatase has been investigated in detail by taking advantage of the fluorescence of the tryptophan residues of the protein (162). The quenching of intrinsic fluorescence measured upon binding of various nucleotide analogs shows that each monomer contains a single binding site and that the affinity for ATP and ADP is tenfold higher than that for GTP and GDP. Moreover, using limited proteolysis of IDH kinase/phosphatase by endoproteinase Lys-C, the nucleotide binding site has been localized within residues 315–340. Furthermore, a recent comparative analysis of the kinetic properties of IDH kinase, IDH phosphatase, and ATPase activities in wild-type and two mutant strains of *E. coli* supports the concept that all three activities share the same active site (122). In a recent study, we have found that the ATPase activity of *E. coli* IDH kinase/phosphatase is severely inhibited by the ATP-analog 5′-*p*-fluorosulfonyl-benzoyl-adenosine (FSBA). This inhibition is prevented by ATP and can be completely reversed by reducing agents, suggesting the involvement of cysteine residues, namely cysteine 356 and cysteine 523, in the mechanism of inhibition by FSBA (C Oudot, JC Cortay, AJ Cozzone, unpublished data).

Differential Activity and Regulation

The main role of IDH phosphorylation/dephosphorylation is to regulate the branch point between the tricarboxylic acid cycle and the glyoxylate bypass during steady-state growth on acetate, and also to control the glyoxylate bypass during transition between carbon sources. Besides responding to changes in the external environment, the reversible phosphorylation of IDH allows adaptation to different intracellular conditions. For example, the cellular level of IDH can vary substantially between different strains of *E. coli*. However, the IDH phosphorylation cycle responds to these differences by altering the fractional phosphorylation of IDH so that a constant level of IDH activity is maintained. When the level of IDH is increased by 15-fold using a clone of the *icd* gene that encodes this enzyme, the phosphorylation system is able to efficiently compensate for this enzyme by converting all of the excess IDH into the inactive phosphorylated form (111).

How to identify the factors that make the kinase prevail over the phosphatase and, conversely, to determine the conditions under which the phosphatase predominates? No clear answer has been given so far. As described above, evidence has accumulated that the kinase and phosphatase are present on the same polypeptide and share the same active site. Nevertheless, some reports indicate that certain mutations in the *aceK* gene produced by random mutagenesis can inhibit selectively the phosphatase activity. These mutations correspond to a 113-amino acid region of this 577-residue protein, which might act as a regulatory domain that would control a conformational equilibrium between the kinase and phosphatase forms of the protein (85, 86). A variety of metabolites are known to affect IDH kinase/phosphatase in vitro. In general, they inhibit the kinase and activate the phosphatase as follows: ADP, AMP, isocitrate, oxaloacetate, α-ketoglutarate, phosphoenolpyruvate, 3-phosphoglycerate, and pyruvate (53, 110, 138). Exceptions include ATP, which is a co-substrate for the kinase and an activator of the phosphatase, and NADPH, which inhibits both activities. The inhibition of the kinase by isocitrate is sigmoid, but the other effects are hyperbolic. Each kinase inhibitor alone is capable of giving complete inhibition, while the phosphatase activators give two- to threefold activation at saturation. Further, citrate, fructose-6-phosphate, glyoxylate, and NADP inhibit the kinase but have little effect on the phosphatase.

There is indirect evidence that the effects of these various metabolites are mediated by binding to the kinase/phosphatase rather than to IDH itself (138). The observation that NADP inhibits the kinase might be due, however, to the binding of the coenzyme to the dehydrogenase. During growth on acetate, both isocitrate and 3-phosphoglycerate participate in the control of IDH phosphorylation. These effectors may act as general indicators of the levels of metabolic intermediates and thus of the need for isocitrate to be directed to the glyoxylate bypass (35, 107). It can be envisaged, for instance, that when the pools of these metabolites are depleted, the phosphorylation of IDH increases and, consequently, more isocitrate is directed through the glyoxylate bypass, even though phosphorylation of IDH can also occur in mutants devoid of a functional glyoxylate bypass (157).

It is worth noting that in any case the minimum level of IDH kinase/phosphatase required for growth on acetate is very low. Indeed, when *E. coli* cells are mutated in the consensus ATP binding site of the enzyme, the kinase and the phosphatase activities are reduced by a factor of at least 100. But still, the protein retains sufficient kinase activity to support growth on acetate (174). This means that in wild-type cells IDH kinase/phosphatase is present in a large excess over the level required for steady-state phosphorylation of IDH. The precise function of this excess IDH kinase/phosphatase remains unclear.

Mechanisms of Sensitivity Amplification

Because IDH kinase/phosphatase is obviously capable of monitoring a large array of metabolites, one can expect that the reversible phosphorylation of IDH will be particularly responsive to the general metabolic state of the cell. A number of theoretical analyses and experiments performed in vitro indicate indeed that several mechanisms appear to amplify the sensitivity with which the glyoxylate bypass is controlled. These mechanisms combine to produce a system in which discrete changes in metabolic signals induce striking modifications in the flux through the glyoxylate bypass. They constitute a particularly effective feedback control machinery.

First, the fact that several metabolites affect the activities of the kinase and phosphatase in opposite directions makes the response of the system very sensitive to small changes in the intracellular concentrations of these effectors. This type of sensitivity enhancement is termed "multistep effect" because, in each case, the same effector acts at more than one step in the pathway (65). It has a theoretical limit based on the number of steps that the effector controls but it can approach close to that limit. Thus, a single effector acting at two steps of a cascade will approach the ultrasensitivity of an allosteric dimer if all the parameters are in the optimal range (65).

Second, the IDH phosphorylation cycle can also be subject to "zero-order ultrasensitivity" (65, 110). If either or both of the converter enzymes in a covalent modification system are close to saturation with their protein substrate, i.e. in the zero-order range, the system can respond more sensitively than if both of the converter enzymes are far from saturation, i.e. in the first-order range. It has been shown that such ultrasensitivity can occur in the IDH system in vitro (110). An analysis of the interplay between the zero-order effect and the multistep effect indicates that these two effects can act synergistically (64, 65). The interplay tends to enhance the sensitivity gained for either effect alone, and the zero-order ultrasensitivity tends to be the stronger of the two effects.

Third, another mechanism of sensitivity amplification, termed a "branch point effect", results from the striking difference in the affinities of IDH and isocitrate lyase for isocitrate. These correspond to Michaelis-Menten constants that differ by nearly two orders of magnitude (112, 182, 183). The consequence of this effect is that the flux through the glyoxylate bypass is particularly sensitive to the phosphorylation state of IDH. This flux, which is very active during growth on acetate, decreases by a factor of approximately 150 upon addition of glucose (112). This inhibition is brought about by two relatively modest events: a fourfold increase in the maximum velocity of IDH and a 5.5-fold decrease in the rate of isocitrate production.

THE ACETATE OPERON

Major advances have been made in our understanding of acetate metabolism over the last decade by analyzing the control of the glyoxylate bypass at the genetic level. The structure of the acetate operon has been determined and the dual regulation of its expression by negative and positive effectors has been demonstrated.

Structural Genes

The genes coding for malate synthase (*aceB*), isocitrate lyase (*aceA*), and IDH kinase/phosphatase (*aceK*), respectively, are present in that order in the same acetate (*aceBAK*) operon (23, 52, 119) located at 90.85 min on the *E. coli* K-12 linkage map (5) (Figure 3). The complete nucleotide sequence of these three genes and the corresponding intergenic regions have been determined (Table 1). The sequence of *aceK* was first reported by our group (27), and similar results have since been reported by other authors (97). We have also determined the sequence of the *aceA* gene (161a) and, at the same time, the sequence of *aceB* has been described (16). Each of the three structural genes of the *aceBAK* operon is preceded by a ribosome-binding site (168): AGAGG for *aceB*, GGAG for *aceA*, and GAGG for *aceK*.

The intergenic region between *aceB* and *aceA* consists of 32 base pairs (bp) with no particular structural pattern (Figure 3). In contrast, the 184-bp region between *aceA* and *aceK* contains two consecutive long dyad symmetries almost identical in sequence that are particularly prone to yielding stable stem-loop units (27). Each of these symmetries exhibits high sequence homology with elements of the REP (repetitive extragenic palindrome) family found in the intercistronic or postcistronic regions of various operons in *E. coli* and *S. typhimurium* (75, 76). These elements are considered to be involved in the transcriptional and/or translational regulation of gene expression, especially by stabilizing upstream messenger RNA through the inhibition of the processive action of $3' \rightarrow 5'$ exonucleases (134, 171). Another possible function would be to interact with gyrase to maintain the DNA supercoiling (188). To characterize the regulatory sequences located upstream from *aceB* gene that control the expression of the *aceBAK* operon, the sequence of the *metA* gene that precedes *aceB* has been determined, as well as that of the intergenic region between these

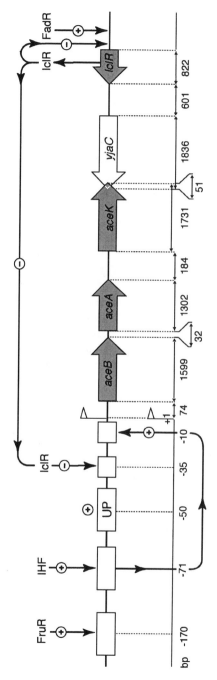

Figure 3 Schematic organization of the *aceBAK* operon and regulatory elements. The location and length of structural genes, intergenic regions, and regulatory elements are given in base pairs (not drawn at scale). *Arrows* in genes *aceBAK* and *iclR*, direction of transcription; +1, initiation site of transcription; *positive numbers*, region downstream from the initiation site; *negative numbers*, region upstream from the initiation site. Regulators exert either positive (+) or negative (−) control over the operon expression. FadR represses the *aceBAK* operon expression by stimulating (+) the expression of *iclR*. Note that in *S. typhimurium* the intergenic region between *aceK* and *iclR* is shorter than that in *E. coli* because it encompasses only 420 bp and encodes no open reading frame (56).

two genes (48). The *metA* gene contains 927 bp encoding homoserine *trans*-succinylase, an enzyme of 35,673 molecular weight involved in the synthesis of methionine. It is separated from *aceB* by 271 bp.

The *aceK* mutations generally render the cells unable to grow on acetate because of an increase in IDH activity, which severely reduces flux through the glyoxylate bypass. However, the effects of such mutations seem to depend on the genetic background of the mutated strains (111). In some strains, the *aceK* mutations retard but do not abolish growth on acetate, apparently because the level of IDH activity in these strains is unusually low. This observation indicates that phosphorylation of IDH is not required for growth on acetate if the IDH activity is already low. Another interesting finding is that some mutants that lack IDH kinase/phosphatase are capable of spontaneously reverting to growth on acetate by second-site mutations (111). One of these is particularly intringuing, since it has undergone second-site mutations that do not map to the genes for any of the enzymes involved in the glyoxylate bypass, which suggests the existence of a separate pathway under distinct control. One possible explanation of this has been proposed (78). The glyoxylate generated by isocitrate lyase may be oxidized in a cyclic pathway, termed the glyoxylate-oxidizing cycle, involving malate synthase, malic enzyme, and pyruvate dehydrogenase. The only requirement for flux through IDH would then be for the generation of α-ketoglutarate for biosynthesis. This glyoxylate-oxidizing cycle might have preceded the tricarboxylic acid cycle during evolution (78).

Differential Expression

Despite the fact that the three structural genes of the *aceBAK* operon are under control of the same promoter, located upstream from *aceB*, their enzyme products are expressed in quite different proportions. Indeed, the relative cellular levels of malate synthase, isocitrate lyase, and IDH kinase/phosphatase are approximately 0.3:1:0.003 (23, 26, 111). This estimation has been made from direct measurement of enzyme activities in extracts and/or from the amount of protein produced by relevant plasmids. Also, the expression of the operon has been examined in vitro using an original transcription-translation coupled system (17). In this system, plasmid-directed protein synthesis is limited to the formation of the specific N-terminal dipeptide of the gene product instead of the completed protein. Such control is obtained by restricting the supply of aminoacyl-tRNAs to those corresponding selectively to the first two triplets in the messenger RNA coding sequence. The assessments made from these different techniques concur in showing that there occurs a small upshift in expression between *aceB* and *aceA*, then a drastic downshift in *aceK* expression. This differential expression may reflect the fact that the enzymes produced play fundamentally different roles in the functioning of the glyoxylate bypass:

isocitrate lyase and malate synthase are metabolic enzymes, whereas IDH kinase/phosphatase is a regulatory enzyme whose intracellular concentration, like that of most regulatory proteins, need not be high. However, the significance of the slight increase in expression between *aceB* and *aceA* remains unclear.

The polarity of expression of the structural genes may be connected with their codon usage. A relationship exists between codon usage and gene expressivity, mainly because of the degree of codon usage adaptation to the cellular frequencies of tRNAs (68, 87). A dynamic modeling of the translation process allows quantitative measurement of this adaptation in terms of the mean number of tRNA discriminations that are necessary to translate each codon of the gene or the corresponding messenger RNA (67). Based on this method, the expressivity of *aceK* appears to be much lower than that of *aceB* and *aceA* (26, 27), suggesting that the low level of IDH kinase/phosphatase is due, at least in part, to a stringent regulation of polypeptide elongation. Another possible explanation for the profound downshift observed in *aceK* expression refers to the repetitive extragenic palindromic (REP) elements found in the intergenic region between *aceA* and *aceK* (see above). Because each element of this type is likely to lead to a decrease in gene expression (75, 134), the presence of two such elements in succession immediately upstream of *aceK* can be expected to result in strong downregulation (26). To evaluate the contribution of transcriptional or translational effects to the differential expression of the *aceBAK* genes, a series of gene and operon fusions between these genes and *lacZ* have been analyzed (24). It appears that the upshift between *aceB* and *aceA* results from differences in translational efficiency. The striking downshift between *aceA* and *aceK* appears to result from two effects: inefficient translation and premature transcriptional termination occurring in an apparently coupled process. The sequences responsible for inefficient translation of *aceK* lie in its ribosome binding site.

Negative Control by IclR

The *aceBAK* operon of *E. coli* is negatively controlled at transcription by the IclR repressor, a dimeric protein encoded by the *iclR* gene located between genes *aceK* and *metH* in the 90.95-min region of the chromosome (132, 175) (Table 1). The *iclR* gene is transcribed in the opposite direction to that of *aceBAK*. The corresponding gene of *S. typhimurium* has also been cloned and sequenced (57). The deduced amino acid sequences of these two IclR proteins exhibit 89% identity. In addition, the IclR protein of *E. coli*, which contains a helix-turn-helix motif characteristic of DNA-binding proteins, has a score of similarity of over 43% with GylR, a transcriptional regulator of the glycerol operon of *Streptomyces coelicolor* (132). In *E. coli*, *iclR* is separated from *aceK* by a 2386-bp region (Figure 3), which contains an open reading frame (ORF) of

1836 bp in the same orientation as *iclR*, but opposite to that of *aceK*, and which overlaps the last 51 bp of the coding region of *aceK*. This ORF encodes a protein of 69,355 molecular weight whose identity and function are still unknown (56). Designated YjaC, this protein contains several repeats similar to ankyrin motifs in eukaryotic proteins (133). Because these repeats have recently been shown to interfere with acetate metabolism by inhibiting both isocitrate lyase and malate synthase, a possible role for protein YjaC in the regulation of the glyoxylate pathway has been proposed (45). In contrast with *E. coli*, the intergenic region between *aceK* and *iclR* in *S. typhimurium* is only 420 bp long, and it does not contain an open reading frame, which might be of physiological significance (132).

The IclR repressor of *E. coli* has been overproduced and purified to homogeneity in a one-step procedure by cation exchange chromatography (28). The analysis of the interaction between IclR and the operator/promoter region of the *aceBAK* operon shows that the repressor binds to a 35-nucleotide sequence that largely overlaps the −35 binding site of RNA polymerase (28). Such topological coincidence between the operator and promoter regions indicates that the IclR repressor and the transcription enzyme will compete for the same binding site on DNA. This reinforces the concept that repression is a matter of blocking access to the promoter.

The sites of interaction between IclR and its operator have been examined (131). The number and nature of nucleotides essential to repressor binding have been determined by scanning populations of DNA previously modified by various chemicals acting selectively on purines or pyrimidines. A total of 46 nucleotides, distributed almost equally between the two strands of the operator region, are functionally important. These are clustered in two successive domains that expand from nucleotide −54 to nucleotide −27 and can organize in a palindrome-like structure containing an exceptionally large proportion of adenine and thymine residues. Similar results have been obtained subsequently using an in vitro oligonucleotide selection technique to determine the consensus recognition sequence for IclR (146).

The *iclR* gene and *aceBAK* operon seem to compete for a common repressor protein in view of the fact that IclR binds to its own promoter and thus regulates its own expression (70). The autoregulation of *iclR* is relatively insensitive to the carbon source, which is rather favorable to the cell, for the following reason: For growth on acetate, bacteria must induce *aceBAK* and concomitantly reduce the activity of IclR. Because IclR inhibits its own production, adaptation to acetate might be expected to lower such autorepression and lead to an increase of the level of IclR. If it were so, the induction of *aceBAK* would be thwarted, which is not the case (70). The reason IclR controls *aceBAK* differently from *iclR* deserves further investigation. Another interesting feature of *iclR* expression is the fact that it is activated by the regulatory protein FadR (69) (see below).

Negative Control by FadR

In *E. coli*, the FadR protein (Table 1) is a transcriptional factor that plays two main roles in fatty acid metabolism. First, it functions as a repressor of the β-oxidation pathway and, second, as an activator of unsaturated fatty acid biosynthesis (7, 73, 139). In fatty acid synthesis, FadR decreases the activity of the synthetic pathway when exogenous fatty acids are available for membrane lipid synthesis (74), whereas in the β-oxidation process it behaves like a classical transcriptional repressor (96, 153). The different actions of FadR on transcription can be explained by the positions of the operator sites to which the protein binds (34, 74). When FadR acts as a repressor, the binding sites are within the promoter region, either overlapping the -10 or -35 region, or lying just downstream of the -10 sequence (7, 74). When FadR acts as an activator, the binding sites are immediately upstream of the -35 region of the promoter (43, 44, 74). DNA binding by FadR is preferentially antagonized by CoA esters of long-chain fatty acids (34, 73).

The first evidence that FadR plays a role in acetate metabolism came from the observation that acetate is incorporated into *fadR* strains at a much greater rate than in isogenic *fadR*[+] strains (118). Biochemical studies demonstrate that the increased rate of acetate incorporation by *fadR* mutants is not due to an increased rate of macromolecular synthesis or degradation, to differences in the activities of enzymes required for acetate transport or oxidation, or to differences in the acetyl CoA pool size (140). Still, this increased rate of acetate incorporation by *fadR* mutants requires a functional glyoxylate bypass. Enzyme analysis has further shown that the levels of glyoxylate bypass enzymes are elevated in *fadR* mutants under noninducing growth conditions, e.g. glucose (117, 140). The activity of these enzymes is therefore negatively regulated by the *fadR* gene, considering that mutations in this gene result in increased expression of *aceBAK* on repressing media. However, the effects of these mutations are much smaller than those observed for *iclR* (119).

A significant advance in our understanding of the inhibitory action of FadR on *aceBAK* expression has been made by the observation that FadR activates the expression of the *iclR* gene (69). This activation is connected with the binding of FadR to a site on the DNA located just upstream of the *iclR* promoter(Figure 3). Therefore, the effects of *fadR* mutations on *aceBAK* expression can be explained by a decreased production of IclR resulting from the loss of FadR. Consistent with this interpretation, mutation of the FadR binding site of *iclR* induces an increase in *aceBAK* expression similar to that produced by disruption of *fadR* (69).

Positive Control by FruR

The repressor of the fructose regulon, FruR, is a pleiotropic transcriptional regulator that controls the expression of several major pathways involved in carbon and energy metabolism in enteric bacteria, namely *E. coli* and *S. typhimurium* (21, 61, 62). These pathways include glycolysis, the tricarboxylic acid cycle, the glyoxylate bypass, gluconeogenesis, and electron transport, as well as the Entner-Doudoroff and pentose phosphate pathways [reviewed in (165)]. The expression of several genes is thus controlled by FruR, either negatively or positively. For example, FruR represses the synthesis of the fructose and mannitol catabolic enzymes of phosphofructokinase, and of the energy coupling proteins of the phosphotransferase system (PTS). In contrast, it activates the production of the cytochrome *d* complex, an electron carrier and a terminal oxidase, and of three gluconeogenic enzymes: phosphoenolpyruvate synthase, phosphoenolpyruvate carboxykinase, and fructose-1,6-diphosphatase.

Concerning acetate metabolism (Figure 3), FruR stimulates the synthesis of isocitrate lyase, malate synthase, and isocitrate dehydrogenase, as shown for instance by the fact that a null mutation in the *fruR* gene of *S. typhimirium* blocks the expression of *aceBAK* (21, 154). It therefore appears that this protein serves as a switch, determining whether fermentative or oxidative and gluconeogenic conditions will prevail (165).

The *fruR* gene is located at 2.2 min on the *E. coli* map (Table 1), whereas the *fru* operon is located at approximately 47 min, which suggests per se that FruR does not function exclusively in *fru* operon regulation. This gene has been cloned in the pT7-5 expression vector and overexpressed, and the FruR protein has been purified to homogeneity using a one-step chromatographic procedure (29). In solution, FruR exists as a tetramer of four identical subunits (29). On the basis of sequence similarity with other bacterial regulatory proteins, FruR has been classified in the GalR-LacI family (113, 185). Genetic and biochemical studies show that each member of this family contains two functional domains: an N-terminal domain that exhibits a structural helix-turn-helix motif responsible for the binding of the regulator to its operator, and a C-terminal portion that displays inducer-binding properties and subunit interaction (93, 143, 166)

Using the purified protein together with labeled DNA fragments encompassing the regulatory regions of several operons, experiments have been conducted in vitro to ascertain the nature of FruR DNA-binding sequences (155). FruR has been shown to bind to two operators within the regulatory region preceding the structural genes of the fructose operon. These two operators, O_1 and O_2, comprise nearly identical palindromes of 12 bp with a half-site of TGAAAC. The two operators are located between the single putative promoter of the fructose operon and the translational initiation site of the *fruB* gene. Other regulated operons interact with FruR at a single site upstream of the first structural gene, as follows: (*a*) *ppsA* encoding phosphoenolpyruvate synthase (positive regulation); (*b*) *icd* encoding isocitrate dehydrogenase (positive regulation); (*c*) *aceBAK* encoding the glyoxylate bypass enzymes (positive regulation); and (*d*) *pts* encoding protein Hpr and enzyme I of the PTS system (negative regulation).

FruR can bind both upstream of certain promoters to activate their expression and to sites overlapping other promoters to block transcription. Experiments performed with a series of operons have led to the definition of a precise consensus DNA-binding site for the FruR protein of *E. coli* (129). For binding, FruR requires an 8-bp left half-site motif and a 3-bp conserved right half-site with the sequence 5′-GNNGAATC/GNT-3′, which means that it interacts asymmetrically with the two half-sites of its operator. Such asymmetry is also observed in the binding of FruR to the operator of *aceBAK*, which involves a 16-bp DNA sequence located 170 bp upstream from the transcriptional start point of the operon (29). In the course of experiments on the transcription of the *ppsA* gene encoding phosphoenolpyruvate synthase, we have recently observed that, upon binding to its specific site, FruR induces a sharp bend of 120° in the DNA helix, which may play an important role in the mechanism of promoter activation (D Nègre, AJ Cozzone, unpublished results). The mode of FruR action would then resemble that of CRP activation, in which the binding of the protein bends the DNA and facilitates its direct contact with RNA polymerase (98, 151).

The N-terminal domain of *E. coli* FruR, containing the first 60 amino acids, has been overproduced and purified to homogeneity after cloning the corresponding DNA fragment in frame with the fused gene of glutathione *S*-transferase (166). Its three-dimensional structure has been solved by ^{1}H and ^{15}N nuclear magnetic resonance (150). This DNA-binding domain exhibits a typical helix-turn-helix motif stabilized by a third helix. The topology of these structural elements is similar to that of other regulatory proteins such as LacI or PurR (167, 170), which confirms the universality of these features in the LacI protein family. However, the structure of the connecting fragment between helix 2 and helix 3 in FruR is different from that of LacI and PurR. This fragment contains three additional residues in FruR, without equivalent in the other repressors, which are stabilized by several hydrogen bonds and aromatic ring stacking. The shape of this structural element may be responsible for the specificity of FruR binding to DNA (150).

Positive Control by Other Effectors

Another positive regulator of *aceBAK* expression is the integration host factor (IHF). This factor is a small, heterodimeric protein encoded by genes *himA* and *himD* (128) (Table 1). It belongs to a family of DNA-bending proteins that share amino acid similarity with the bacterial HU protein (141). IHF binds to DNA in a sequence-specific manner that induces a severe bend in the DNA helix and thus facilitates the interaction of other components in a nucleoprotein array (66, 161). Although first discovered as a host factor for bacteriophage λ integration, IHF assists in many processes that involve higher-order protein-DNA complexes, e.g. replication, transcriptional regulation, and a variety of site-specific recombination systems (55, 147). The first indication that IHF may affect *aceBAK* expression has come from the identification of two regions upstream of the promoter that match the consensus sequence for IHF binding sites (160). These sites are centered at position -161 (IHFa) and position -74 (IHFb) relative to the initiation point of transcription. It has also been observed that mutation in *himA* produces a sixfold reduction in isocitrate lyase activity compared with the isogenic wild-type strain growing on acetate. The stimulatory effect of IHFb on *aceBAK* expression is much higher than that of IHFa (Figure 3), as shown by mutational analysis of the two binding sites. Furthermore, in the absence of IHFb, the IclR repressor retains the ability to inhibit the *aceBAK* promoter even under inducing conditions. IHF may provide a mechanism that allows differential control of *aceBAK* and *iclR* by IclR. The autorepression of *iclR* by IclR is relatively insensitive to carbon sources, while the regulation of *aceBAK* by IclR is sensitive (70). The finding that IclR regulates these two promoters differently may be due to the fact that *aceBAK* possesses an IHF site, whereas *iclR* does not (160). We have recently determined the number and nature of the nucleotides that constitute the major binding site of IHF by using the different chemical interference techniques previously used for analyzing the binding site of IclR (28, 131). We have also found that IHF alone, in the absence of any other activator including FruR, can stimulate *aceBAK* expression. This stimulatory effect is connected with a bend of 170° of the DNA helix (JF Prost, AJ Cozzone, unpublished data).

Stimulation of *aceBAK* expression is also induced by the interaction of RNA polymerase with a particular DNA region termed the upstream (UP) module. It has been shown that, in addition to its two major determinants (the -10 and -35 hexamers), the promoter strength of the *E. coli* RNA polymerase can be greatly increased by a third *cis*-acting recognition element, the UP module, which spans an $(A+T)$-rich region interacting with the α subunit of the enzyme (125, 164). The promoter region of *aceBAK* contains an UP element centered around base -50 (Figure 3), which can form a stable and specific complex with either the RNA polymerase holoenzyme or with its α subunit alone (130). The removal of certain bases between positions -32 to -50 of the *ace* promoter interferes with the binding of the α subunit, while the elimination of single bases from either strand of the UP element results, in contrast, in an enhanced binding affinity of RNA polymerase. It therefore appears that disruption of the DNA helix in this particular region promotes a local DNA flexibility that stabilizes the RNA polymerase-promoter complex and, consequently, increases the *aceBAK* expression (130). More generally, it is interesting to note that the different factors (UP, IHF, FruR) that stimulate *aceBAK* are all inducing conformational changes in DNA, which indicates that the functioning of this operon strongly depends on the structural organization of the DNA matrix.

Effects of Metabolites

The nature of the primary effector(s) of *aceBAK* expression is still to be determined. A number of metabolites have been assayed for their capacity of affecting, directly or indirectly, the functioning of this operon, but no definite conclusion can be drawn as yet. Acetate itself can be ruled out because when bacteria are cultured on fatty acids—a carbon source whose utilization does not proceed through acetate—the glyoxylate bypass is nevertheless induced (139). Acetyl CoA also can be ruled out, considering that when bacteria are grown simultaneously on acetate and on a preferred carbon source such as glucose, the glyoxylate bypass is not operative even though the ability of cells to convert acetate to acetyl CoA remains nearly the same as during growth on acetate alone. Furthermore, formation of the complex between IclR and the operator/promoter region of *aceBAK* is insensitive to both acetate and acetyl CoA (28). Similarly, acetyl phosphate, pyruvate, and oxaloacetate have no effect either. In contrast, the formation of this complex is severely impaired by phosphoenolpyruvate, which thus might be a good candidate as an inducer of the operon transcription (28). This possibility is not supported, however, by the early observation that phosphoenolpyruvate, or some metabolite readily derived therefrom, seems to inhibit isocitrate lyase synthesis (101). But, as discussed above, other reports suggest that the inhibition of isocitrate lyase by phosphoenolpyruvate has no physiological significance (53, 116, 124).

Another efficient metabolite could be fructose-1-phosphate or fructose-1,6-biphosphate. Indeed, in the different assays performed to evaluate the binding capacity of FruR to various operons of *E. coli* and *S. typhimurium*, these two molecules have been shown to always displace the protein from the DNA (154, 155), even at low concentrations. This would mean that in the presence of fructose-1-phosphate or fructose-1,6-biphosphate, operons under negative FruR control are derepressed, whereas operons under positive FruR control, such as *aceBAK*, are not activated. However, to date no evidence of such an effect has been obtained in vivo.

CONCLUDING REMARKS

The analysis of acetate metabolism is of particular interest in many respects. First, it refers to a biological system that is especially appropriate for understanding at the molecular level the adaptation mechanisms of bacteria to their nutritional environment, both extracellular and intracellular. Also, it provides rewarding insights into one of the most fundamental processes occurring within the cell: the regulation of the fluxes of carbon molecules into different metabolic pathways. This includes regulation of enzyme activities as well as gene expression in an operon whose functioning is exquisitely sophisticated. In addition, it raises certain basic questions in enzymology such as the mode of action and the biological advantage, if any, of bifunctional enzymes. Moreover, it deals with a variety of biochemical reactions that are specific to microorganisms and plants, and therefore can be interpreted in terms of evolution. And, last but not least, it demonstrates the key role of protein phosphorylation in the control of bacterial physiology. In this regard, the importance of the reversible modification of isocitrate dehydrogenase is obvious. But equally interesting is the phosphorylation of other enzymes of carbon metabolism, which has not yet been investigated in detail, but might play an important regulatory role as well. These phosphorylatable enzymes are (a) isocitrate lyase in the glyoxylate shunt (81, 163), (b) enolase in the first step of gluconeogenesis (36, 37), and (c) citrate lyase ligase during anaerobic citrate metabolism (3). Protein phosphorylation is considered to be a universal phenomenon among bacteria in view of the fact that it has been observed in nearly 100 different species (32, 33, 95). In a bacterium such as E. coli, 130 different phosphoproteins have been detected (30) but only a dozen have, so far, been identified (31, 33). Therefore, it would be worth determining the identity of the still-unknown proteins and enzymes, and checking whether some of them participate in acetate metabolism and, more generally, in carbon metabolic pathways. Looking for such phosphorylated molecules and searching metabolites that affect the expression of the acetate operon are two major future prospects that should lead to significant progress in the field.

ACKNOWLEDGMENTS

I thank David Hulmes for his careful reading of the manuscript, Emmanuelle Duglas for her excellent secretarial help, and Christian Van Herrewege for his efficient assistance in iconography. Research in my laboratory has been supported by grants from the Centre National de la Recherche Scientifique (UPR 412), the Université de Lyon, and the Institut Universitaire de France.

> Visit the *Annual Reviews home page* at
> http://www.AnnualReviews.org.

Literature Cited

1. Abeysinghe SIB, Baker PJ, Rice DW, Rodgers HF, Stillman RJ, et al. 1991. Use of chemical modification in the crystallization of isocitrate lyase from *Escherichia coli. J. Mol. Biol.* 220:13–16
2. Adman ET, Sieker LC, Jensen LH. 1973. Structure of a bacterial ferredoxin. *J. Biol. Chem.* 248:3987–96
3. Antranikian G, Gottschalk G. 1989.

Phosphorylation of citrate lyase ligase in *Clostridium sphenoides* and regulation of anaerobic citrate metabolism in other bacteria. *Biochimie* 71:1029–37
4. Ashworth JM, Kornberg HL. 1963. Fine control of the glyoxylate cycle by allosteric inhibition of isocitrate lyase. *Biochim. Biophys. Acta* 73:519–22
5. Bachmann BJ. 1990. Linkage map of

Escherichia coli K-12, edition 8. *Microbiol. Rev.* 54:130–97

6. Bennet PM, Holms WH. 1975. Reversible inactivation of the isocitrate dehydrogenase of *Escherichia coli* ML308 during growth on acetate. *J. Gen. Microbiol.* 87:37–51

7. Black PN, Di Russo CC. 1994. Molecular and biochemical analyses of fatty acid transport, metabolism, and gene regulation in *Escherichia coli. Biochim. Biophys. Acta* 1210:123–45

8. Bolduc JM, Dyer DH, Scott WG, Singer P, Sweet RM, et al. 1995. Mutagenesis and Laue structures of enzyme intermediates: isocitrate dehydrogenase. *Science* 268:1312–18

9. Borthwick AC, Holms WH, Nimmo HG. 1984. Isolation of active and inactive forms of isocitrate dehydrogenase from *Escherichia coli* ML308. *Eur. J. Biochem.* 141:393–400

10. Borthwick AC, Holms WH, Nimmo HG. 1984. The phosphorylation of *Escherichia coli* isocitrate dehydrogenase in intact cells. *Biochem. J.* 222:797–804

11. Borthwick AC, Holms WH, Nimmo HG. 1984. Amino acid sequence round the site of phosphorylation in isocitrate dehydrogenase from *Escherichia coli. FEBS Lett.* 174:112–15

12. Britten RJ. 1954. Extracellular metabolic products of *Escherichia coli* during rapid growth. *Science* 118:578

13. Brown TDK, Jones-Mortimer MC, Kornberg HL. 1977. The enzymatic interconversion of acetate and acetyl coenzyme A in *Escherichia coli. J. Gen. Microbiol.* 102:327–36

14. Buchanan BB, Arnon DI. 1970. Ferredoxins: chemistry and function in photosynthesis, nitrogen fixation, and fermentative metabolism. *Adv. Enzymol. Relat. Areas Mol. Biol.* 33:119–76

15. Burke WF, Johanson RA, Reeves HC. 1974. NADP-specific isocitrate dehydrogenase of *Escherichia coli*. II: subunit structure. *Biochim. Biophys. Acta* 351:333–40

16. Byrne C, Stokes HW, Ward KA. 1988. Nucleotide sequence of the *aceB* gene encoding malate synthase A in *Escherichia coli. Nucleic Acids Res.* 16:9342

17. Cenatiempo Y, Robakis N, Meza-Basso L, Brot N, Weissbach H, Reid BR. 1982. Use of different tRNASer isoacceptor species in vitro to discriminate between the expression of plasmid genes. *Proc. Natl. Acad. Sci. USA* 79:1466–68

18. Chao G, Shen J, Tseng CP, Park SJ,

Gunsalus RP. 1997. Aerobic regulation of isocitrate dehydrogenase gene (*icd*) expression in *Escherichia coli* by the *arcA* and *fnr* gene products. *J. Bacteriol.* 179:4299–304

19. Chen R, Grofler JA, Hurley JH, Dean AM. 1996. Second-site suppression of regulatory phosphorylation in *Escherichia coli* isocitrate dehydrogenase. *Prot. Sci.* 5:287–95

20. Chen RD, Gadal P. 1990. Structure, function and regulation of NAD and NADP dependent isocitrate dehydrogenase in higher plants and in other organisms. *Plant Physiol. Biochem.* 28:411–27

21. Chin AM, Feldheim DA, Saier MH Jr. 1989. Altered transcriptional patterns affecting several metabolic pathways in strains of *Salmonella typhimurium* which overexpress the fructose region. *J. Bacteriol.* 171:2424–34

22. Chistoserdova LV, Lidstrom ME. 1996. Molecular characterization of a chromosomal region involved in the oxidation of acetyl CoA to glyoxylate in the isocitrate-lyase-negative methylotroph *Methylobacterium extorquens* AM1. *Microbiology* 142:1459–68

23. Chung T, Klumpp DJ, LaPorte DC. 1988. Glyoxylate bypass operon of *Escherichia coli*: cloning and determination of the functional map. *J. Bacteriol.* 170:386–92

24. Chung T, Resnik E, Stueland C, LaPorte DC. 1993. Relative expression of the products of glyoxylate bypass operon: contributions of transcription and translation. *J. Bacteriol.* 175:4572–75

25. Clark DP, Cronan JE Jr. 1996. Two-carbon compounds and fatty acids as carbon sources. In *Escherichia coli and Salmonella typhimurium: Cellular and Molecular Biology*, ed. FC Neidhardt, 1:343–57. Washington, DC: ASM. 2nd ed.

26. Cortay JC, Bleicher F, Duclos B, Cenatiempo Y, Gautier C, et al. 1989. Utilization of acetate in *Escherichia coli*: structural organization and differential expression of the *ace* operon. *Biochimie* 71:1043–49

27. Cortay JC, Bleicher F, Rieul C, Reeves HC, Cozzone AJ. 1988. Nucleotide sequence and expression of the *aceK* gene coding for isocitrate dehydrogenase kinase/phosphatase in *Escherichia coli. J. Bacteriol.* 170:89–97

28. Cortay JC, Nègre D, Galinier A, Duclos B, Perrière G, Cozzone AJ. 1991. Regulation of the acetate operon in

Escherichia coli: purification and functional characterization of the IclR repressor. *EMBO J.* 10:675–79

29. Cortay JC, Nègre D, Scarabel M, Ramseier TM, Vartak NB, et al. 1994. In vitro asymmetric binding of the pleiotropic regulatory protein, FruR, to the *ace* operator controlling glyoxylate shunt enzyme synthesis. *J. Biol. Chem.* 269:14885–91

30. Cortay JC, Rieul C, Duclos B, Cozzone AJ. 1986. Characterization of the phosphoproteins of *Escherichia coli* cells by electrophoretic analysis. *Eur. J. Biochem.* 159:227–37

31. Cozzone AJ. 1988. Protein phosphorylation in prokaryotes. *Annu. Rev. Microbiol.* 42:97–125

32. Cozzone AJ. 1993. ATP-dependent protein kinases in bacteria. *J. Cell. Biochem.* 51:7–13

33. Cozzone AJ. 1997. Diversity and specificity of protein-phosphorylating systems in bacteria. *Folia Microbiol.* 42:165–70

34. Cronan JE Jr. 1997. In vivo evidence that acyl coenzyme A regulates DNA binding by the *Escherichia coli* FadR global transcription factor. *J. Bacteriol.* 179:1819–23

35. Cronan JE Jr, LaPorte D. 1996. Tricarboxylic acid cycle and glyoxylate bypass. In *Escherichia coli and Salmonella typhimurium: Cellular and Molecular Biology,* ed. FC Neidhardt, 1:206–16. Washington, DC: ASM. 2nd ed.

36. Dannelly HK, Cortay JC, Cozzone AJ, Reeves HC. 1989. Identification of phosphoserine in in vivo-labeled enolase from *Escherichia coli. Curr. Microbiol.* 19:237–40

37. Dannelly HK, Duclos B, Cozzone AJ, Reeves HC. 1989. Phosphorylation of *Escherichia coli* enolase. *Biochimie* 71:1095–100

37a. Davis WL, Jones RG, Farmer GR, Dickerson T, Cortinas E, et al. 1990. Identification of glyoxylate cycle enzymes in chick liver: the effect of vitamin D3. cytochemistry, and biochemistry. *Anat. Rec.* 227:271–84

38. Dean AM, Golding GB. 1997. Protein engineering reveals ancient adaptive replacements in isocitrate dehydrogenase. *Proc. Natl. Acad. Sci. USA* 94:3104–9

39. Dean AM, Koshland DEJ. 1990. Electrostatic and steric contributions to regulation at the active site of isocitrate dehydrogenase. *Science* 249:1044–46

40. Dean AM, Lee MH, Koshland DEJ. 1989. Phosphorylation inactivates *Es-*

cherichia coli isocitrate dehydrogenase by preventing isocitrate binding. *J. Biol. Chem.* 264:20482–86

41. DeVeaux LC, Cronan JE, Smith TL. 1989. Genetic and biochemical characterization of a mutation (*fatA*) that allows *trans*-unsaturated fatty acids to replace the essential *cis*-unsaturated fatty acids of *Escherichia coli. J. Bacteriol.* 171:1562–68

42. DiRusso CC. 1988. Nucleotide sequence of the *fadR* gene, a multifunctional regulator of fatty acid metabolism in *Escherichia coli. Nucleic Acids Res.* 16:7995–8009

43. DiRusso CC, Heimert TL, Metzger AK. 1992. Characterization of FadR, a global transcriptional regulator of fatty acid metabolism in *Escherichia coli*: interaction with the *fadB* promoter is prevented by long-chain fatty acyl CoAs. *J. Biol. Chem.* 267:8685–91

44. DiRusso CC, Metzger AK, Heimert TL. 1993. Regulation of transcription of genes required for fatty acid transport and unsaturated fatty acid biosynthesis in *Escherichia coli* by FadR. *Mol. Microbiol.* 7:311–22

45. Diaz-Guerra M, Esteban M, Martinez JL. 1997. Growth of *Escherichia coli* in acetate as a sole carbon source is inhibited by ankyrin-like repeats present in the 2',5'-linked oligoadenylate-dependent human RNase L enzyme. *FEMS Microbiol. Lett.* 149:107–13

46. Diehl P, McFadden BA. 1993. Site-directed mutagenesis of lysine 193 in *Escherichia coli* isocitrate lyase by use of unique restriction enzyme site elimination. *J. Bacteriol.* 175:2263–70

47. Diehl P, McFadden BA. 1994. The importance of four histidine residues in isocitrate lyase from *Escherichia coli. J. Bacteriol.* 176:927–31

48. Duclos B, Cortay JC, Bleicher F, Ron EZ, Richaud C, et al. 1989. Nucleotide sequence of the *metA* gene encoding homoserine *trans*-succinylase in *Escherichia coli. Nucleic Acids Res.* 17:2856

49. Edlin JD, Sundaram TK. 1989. Regulation of isocitrate dehydrogenase by phosphorylation in *Escherichia coli* K-12 and a simple method for determining the amount of inactive phosphoenzyme. *J. Bacteriol.* 171:2634–38

50. Eikmanns BJ, Rittmann D, Sahm H. 1995. Cloning, sequence analysis, expression, and inactivation of the *Corynebacterium glutamicum icd* gene encoding isocitrate dehydrogenase and biochemical characterization of the

enzyme. *J. Bacteriol.* 177:774–82

51. El-Mansi EMT, Dawson GC, Bryce CF. 1994. Steady-state modelling of metabolic flux between the tricarboxylic acid cycle and the glyoxylate bypass in *Escherichia coli. Comput. Appl. Biosci.* 10:295–99

52. El-Mansi EMT, MacKintosh C, Duncan K, Holms WH, Nimmo HG. 1987. Molecular cloning and over-expression of the glyoxylate bypass operon from *Escherichia coli* ML308. *Biochem. J.* 242:661–65

53. El-Mansi EMT, Nimmo HG, Holms WH. 1985. The role of isocitrate in control of the phosphorylation of isocitrate dehydrogenase in *Escherichia coli* ML308. *FEBS Lett.* 183:251–55

54. Flamm EL, Weisberg RA. 1985. Primary structure of the *hip* gene of *Escherichia coli* and of its product, the beta subunit of integration host factor. *J. Mol. Biol.* 183:117–28

55. Friedman DI. 1988. Integration host factor: a protein for all reasons. *Cell* 55:545–54

56. Galinier A, Bleicher F, Nègre D, Perrière G, Duclos B, et al. 1991. Primary structure of the intergenic region between *aceK* and *iclR* in the *Escherichia coli* chromosome. *Gene* 97:149–50

57. Galinier A, Nègre D, Cortay JC, Marcandier S, Maloy SR, Cozzone A. 1990. Sequence analysis of the *iclR* gene encoding the repressor of the acetate operon in *Salmonella typhimurium. Nucleic Acids Res.* 18:3656

58. Garland D, Nimmo HG. 1984. A comparison of the phosphorylated and unphosphorylated forms of isocitrate dehydrogenase from *Escherichia coli* ML308. *FEBS Lett.* 165:259–64

59. Garnak M, Reeves HC. 1979. Phosphorylation of isocitrate dehydrogenase of *Escherichia coli. Science* 203:1111–12

60. Garnak M, Reeves HC. 1979. Purification and properties of phosphorylated isocitrate dehydrogenase of *Escherichia coli. J. Biol. Chem.* 254:7915–20

61. Geerse RH, Izzo F, Postma PW. 1989. The PEP: fructose phosphotransferase system in *Salmonella typhimurium*: FPr combines enzyme III^Fru^ and pseudo-HPr activities. *Mol. Gen. Genet.* 216:517–25

62. Geerse RH, Van der Pluijm J, Postma PW. 1989. The repressor of the PEP: fructose phosphotransferase system is required for the transcription of the *pps* gene of *Escherichia coli. Mol. Gen. Genet.* 218:348–52

63. Gilvarg C, Davis BD. 1956. The role of the tricarboxylic acid cycle in acetate oxidation in *Escherichia coli. J. Biol. Chem. Biophys.* 143:461–70

64. Goldbeter A, Koshland DE Jr. 1981. An amplified sensitivity arising from covalent modification in biological systems. *Proc. Natl. Acad. Sci. USA* 78:6840–44

65. Goldbeter A, Koshland DE Jr. 1984. Ultrasensitivity in biochemical systems controlled by covalent modification: interplay between zero-order and multistep effects. *J. Biol. Chem.* 259:14441–47

65a. Goodman DB, Davis WL, Jones RG. 1980. Glyoxylate cycle in toad urinary bladder: possible stimulation by aldosterone. *Proc. Natl. Acad. Sci. USA* 77:1521–25

66. Goosen N, van de Putte P. 1995. The regulation of transcription initiation by integration host factor. *Mol. Microbiol.* 16:1–7

67. Gouy M, Gautier C. 1982. Codon usage in bacteria: correlation with gene expressivity. *Nucleic Acids Res.* 10:7055–74

68. Grantham R, Gautier C, Gouy M, Jacobzone M, Mercier R. 1981. Codon catalog usage is a genome strategy modulated for gene expressivity. *Nucleic Acids Res.* 9:R43–74

69. Gui L, Sunnarborg A, LaPorte DC. 1996. Regulated expression of a repressor protein: FadR activates *iclR. J. Bacteriol.* 178:4704–9

70. Gui L, Sunnarborg A, Pan B, LaPorte DC. 1996. Autoregulation of *iclR*, the gene encoding the repressor of the glyoxylate bypass operon. *J. Bacteriol.* 178:321–24

71. Han L, Reynolds KA. 1997. A novel alternate anaplerotic pathway to the glyoxylate cycle in streptomycetes. *J. Bacteriol.* 179:5157–64

72. Helling RB. 1995. *icdB* mutants of *Escherichia coli. J. Bacteriol.* 177:2592–93

73. Henry MF, Cronan JE Jr. 1991. An *Escherichia coli* transcription factor that both activates fatty acid synthesis and represses fatty acid degradation. *J. Mol. Biol.* 222:843–49

74. Henry MF, Cronan JE Jr. 1992. A new mechanism of transcriptional regulation release of an activator triggered by small molecule binding. *Cell* 70:671–79

75. Higgins CF, Ferro-Luzzi Ames G, Barnes WM, Clement JM, Hofnung M. 1982. A novel intercistronic regulatory

element of prokaryotic operons. *Nature* 298:760–62

76. Higgins CF, McLaren RS, Newbury SF. 1988. Repetitive extragenic palindromic sequences, mRNA stability and gene expression: evolution by gene conversion? A review. *Gene* 72:3–14

77. Holms H. 1996. Flux analysis and control of the central metabolic pathways in *Escherichia coli. FEMS Microbiol. Rev.* 19:85–116

78. Holms WH. 1986. Evolution of the glyoxylate bypass in *Escherichia coli*—an hypothesis which suggests an alternative to the Krebs cycle. *FEMS Microbiol. Lett.* 34:123–27

79. Holms WH. 1987. Control of flux through the citric acid cycle and the glyoxylate bypass in *Escherichia coli. Biochem. Soc. Symp.* 54:17–31

80. Holms WH, Bennett PM. 1971. Regulation of isocitrate dehydrogenase activity in *Escherichia coli* on adaptation to acetate. *J. Gen. Microbiol.* 65:57–68

81. Hoyt JC, Reeves HC. 1992. Phosphorylation of *Acinetobacter* isocitrate lyase. *Biochem. Biophys. Res. Commun.* 182: 367–71

82. Hurley JH, Dean AM, Sohl JL, Koshland DE Jr, Stroud RM. 1990. Regulation of an enzyme by phosphorylation at the active site. *Science* 249:1012–16

83. Hurley JH, Dean AM, Thorsness PE, Koshland DE Jr, Stroud RM. 1990. Regulation of isocitrate dehydrogenase by phosphorylation involves no long-range conformational change in the free enzyme. *J. Biol. Chem.* 265:3599–602

84. Hurley JH, Thorsness PE, Ramalingam V, Helmers NH, Koshland DE Jr, Stroud RM. 1989. Structure of a bacterial enzyme regulated by phosphorylation, isocitrate dehydrogenase. *Proc. Natl. Acad. Sci. USA* 86:8635–39

85. Ikeda TP, Houtz E, LaPorte DC. 1992. Isocitrate dehydrogenase kinase/phosphatase: identification of mutations which selectively inhibit phosphatase activity. *J. Bacteriol.* 174:1414–16

86. Ikeda TP, LaPorte DC. 1991. Isocitrate dehydrogenase kinase/phosphatase: *aceK* alleles that express kinase but not phosphatase activity. *J. Bacteriol.* 173: 1801–6

87. Ikemura T. 1981. Correlation between the abundance of *Escherichia coli* transfer RNAs and the occurrence of the respective codons in its protein genes: a proposal for a synonymous codon choice that is optimal for *E. coli* translational system. *J. Mol. Biol.* 151:389–409

88. Ishii A, Imagawa S, Fukunaga N, Sasaki S, Minowa O, et al. 1987. Isozymes of isocitrate dehydrogenase from an obligately psychrophilic bacterium, *Vibrio* sp. strain ABE-1: purification and modulation of activities by growth conditions. *J. Biochem.* 102:1489–98

89. Ishii A, Suzuki M, Sahara T, Takada Y, Sasaki S, Fukunaga N. 1993. Genes encoding two isocitrate dehydrogenase isozymes of a psychrophilic bacterium, *Vibrio* sp. strain ABE-1. *J. Bacteriol.* 175:6873–80

90. Iuchi S, Cameron DC, Lin ECC. 1989. A second global regulator gene (*arcB*) mediating repression of enzymes in aerobic pathways of *Escherichia coli. J. Bacteriol.* 171:868–73

91. Iuchi S, Lin ECC. 1988. *arcA* (*dye*), a global regulatory gene in *Escherichia coli* mediating repression of enzymes in aerobic pathways. *Proc. Natl. Acad. Sci. USA* 85:1888–92

92. Iuchi S, Matsuda Z, Fujiwara T, Lin ECC. 1990. The *arcB* gene of *Escherichia coli* encodes a sensor-regulator protein for anaerobic repression of the *arc* modulon. *Mol. Microbiol.* 4:715–27

93. Jahreis K, Lengeler JW. 1993. Molecular analysis of two ScrR repressors and of a ScrR-FruR hybrid repressor for sucrose and d-fructose specific regulons from enteric bacteria. *Mol. Microbiol.* 9:195–209

94. Jahreis K, Postma PW, Lengeler JW. 1991. Nucleotide sequence of the *ilvH-fruR* gene region of *Escherichia coli* K12 and *Salmonella typhimurium* LT2. *Mol. Gen. Genet.* 226:332–36

95. Kennelly PJ, Potts M. 1996. Fancy meeting you here! A fresh look at "prokaryotic" protein phosphorylation. *J. Bacteriol.* 178:4759–64

96. Klein K, Steinberg R, Fiethen B, Overath P. 1971. Fatty acid degradation in *Escherichia coli*: an inducible system for the uptake of fatty acids and further characterization of *old* mutants. *Eur. J. Biochem.* 19:442–50

97. Klumpp DJ, Plank DW, Bowdin LJ, Stueland CS, Chung T, LaPorte DC. 1988. Nucleotide sequence of *aceK*, the gene encoding isocitrate dehydrogenase kinase/phosphatase. *J. Bacteriol.* 170: 2763–69

98. Kolb A, Busby S, Buc H, Garges S, Adhya S. 1993. Transcriptional regulation by cAMP and its receptor protein. *Annu. Rev. Biochem.* 62:749–95

99. Kornberg HL. 1959. Aspects of terminal

respiration in microorganisms. *Annu. Rev. Microbiol.* 13:49–78

100. Kornberg HL. 1966. Anapletoric sequences and their role in metabolism. *Essays Biochem.* 2:1–31

101. Kornberg HL. 1966. The role and control of the glyoxylate cycle in *Escherichia coli. Biochem. J.* 99:1–11

102. Kornberg HL, Smith J. 1966. Temperature-sensitive synthesis of isocitrate lyase in *Escherichia coli. Biochim. Biophys. Acta* 123:654–57

103. Krebs HA, Johnson WA. 1937. The role of citric acid in intermediate metabolism in animal tissues. *Enzymologia* 4:148–56

104. Krebs HA, Lowenstein JM. 1960. The tricarboxylic acid cycle. In *Metabolic Pathways*, ed. DM Greenberg, 1:129–203. New York: Academic. 2nd ed.

105. Kwan HS, Chui HW, Wong KK. 1988. *ack*::Mu d1–8(Apr^{ac}) operon fusions of *Salmonella typhimurium* LT2. *Mol. Gen. Genet.* 211:183–85

106. Lakshmi TM, Helling RB. 1978. Acetate metabolism in *Escherichia coli. Can. J. Microbiol.* 24:149–53

107. LaPorte DC. 1993. The isocitrate dehydrogenase phosphorylation cycle: regulation and enzymology. *J. Cell. Biochem.* 51:14–18

108. LaPorte DC, Chung T. 1985. A single gene codes for the kinase and phosphatase which regulate isocitrate dehydrogenase. *J. Biol. Chem.* 260:15291–97

109. LaPorte DC, Koshland DE Jr. 1982. A protein with kinase and phosphatase activities involved in regulation of tricarboxylic acid cycle. *Nature* 300:458–60

110. LaPorte DC, Koshland DE Jr. 1983. Phosphorylation of isocitrate dehydrogenase as a demonstration of enhanced sensitivity in covalent regulation. *Nature* 305:286–90

111. LaPorte DC, Thorsness PE, Koshland DE. 1985. Compensatory phosphorylation of isocitrate dehydrogenase: a mechanism for adaptation to the intracellular environment. *J. Biol. Chem.* 260: 10563–68

112. LaPorte DC, Walsh K, Koshland DE Jr. 1984. The branch point effect: ultrasensitivity and subsensitivity to metabolic control. *J. Biol. Chem.* 259:14068–75

113. Leclerc G, Noel G, Drapeau GR. 1990. Molecular cloning, nucleotide sequence, and expression of *shl*, a new gene in the 2-minute region of the genetic map of *Escherichia coli. J. Bacteriol.* 172: 4696–700

114. Leyland ML, Kelly DJ. 1991. Purification and characterization of a monomeric isocitrate dehydrogenase with dual coenzyme from the photosynthetic bacterium *Rhodomicrobium vannielii. Eur. J. Biochem.* 202:85–93

115. Liang A, Houghton RL. 1981. Coregulation of oxidized nicotinamide adenine dinucleotide (phosphate) transhydrogenase and glutamate dehydrogenase activities in enteric bacteria during nitrogen limitation. *J. Bacteriol.* 146:997–1002

116. Lowry OH, Carter J, Ward JB, Glaser L. 1971. The effect of carbon and nitrogen sources on the level of metabolic intermediates in *Escherichia coli. J. Biol. Chem.* 246:6511–21

117. Maloy SR, Bohlander M, Nunn WD. 1980. Elevated levels of glyoxylate shunt enzymes in *Escherichia coli* strains constitutive for fatty acid degradation. *J. Bacteriol.* 143:720–25

118. Maloy SR, Nunn WD. 1981. Role of gene *fadR* in *Escherichia coli* acetate metabolism. *J. Bacteriol.* 148:83–90

119. Maloy SR, Nunn WD. 1982. Genetic regulation of the glyoxylate shunt in *Escherichia coli* K-12. *J. Bacteriol.* 149: 173–80

120. McCleary WR, Stock JB, Ninfa AJ. 1993. Is acetyl phosphate a global signal in *Escherichia coli? J. Bacteriol.* 175: 2793–98

121. McKee JS, Hlodan R, Nimmo HG. 1989. Studies of the phosphorylation of *Escherichia coli* isocitrate dehydrogenase: recognition of the enzyme by isocitrate dehydrogenase kinase/phosphatase and effects of phosphorylation on its structure and properties. *Biochimie* 71:1059–64

122. Miller SP, Karschnia EJ, Ikeda TP, LaPorte DC. 1996. Isocitrate dehydrogenase kinase/phosphatase: kinetic characteristics of the wild-type and two mutant proteins. *J. Biol. Chem.* 271: 19124–28

123. Molina I, Pellicer MT, Badia J, Aguilar J, Baldoma L. 1994. Molecular characterization of *Escherichia coli* malate synthase G: differentiation with the malate synthase A isoenzyme. *Eur. J. Biochem.* 224:541–48

124. Morikawa M, Izui K, Taguchi M, Katsuki H. 1980. Regulation of *Escherichia coli* phosphoenolpyruvate carboxylase by multiple effectors in vivo. *J. Biochem.* (Tokyo) 87:441–49

125. Murakami K, Kimura M, Owens JT, Meares CF, Ishihama A. 1997. The two α

subunits of *Escherichia coli* RNA polymerase are asymmetrically arranged and contact different halves of the DNA upstream element. *Proc. Natl. Acad. Sci. USA* 94:1709–14

126. Muro-Pastor MI, Florencio FJ. 1994. NADP$^+$-isocitrate dehydrogenase from the cyanobacterium *Anabaena* sp. strain PCC 7120: purification and characterization of the enzyme and cloning, sequencing, and disruption of the *icd* gene. *J. Bacteriol.* 176:2718–26

127. Muro-Pastor MI, Reyes JC, Florencio FJ. 1996. The NADP$^+$-isocitrate dehydrogenase gene (*icd*) is nitrogen regulated in cyanobacteria. *J. Bacteriol.* 178:4070–76

128. Nash H, Robertson C. 1981. Purification and properties of the *E. coli* protein factor required for λ integrative recombination. *J. Biol. Chem.* 256:9246–53

129. Nègre D, Bonod-Bidaud C, Geourjon C, Deléage G, Cozzone AJ, Cortay JC. 1996. Definition of a consensus DNA-binding site for the *Escherichia coli* pleiotropic regulatory protein, FruR. *Mol. Microbiol.* 21:257–66

130. Nègre D, Bonod-Bidaud C, Oudot C, Prost JF, Kolb A, et al. 1997. DNA flexibility of the UP element is a major determinant for transcriptional activation at the *Escherichia coli* acetate promoter. *Nucleic Acids Res.* 25:713–18

131. Nègre D, Cortay JC, Galinier A, Sauve P, Cozzone A. 1992. Specific interactions between the IclR repressor of the acetate operon of *Escherichi coli* and its operator. *J. Mol. Biol.* 228:23–29

132. Nègre D, Cortay JC, Old IG, Galinier A, Richaud C, et al. 1991. Overproduction and characterization of the *iclR* gene product of *Escherichia coli* K-12 and comparison with that of *Salmonella typhimurium* LT2. *Gene* 97:29–37

133. Neuwald AF, Green P. 1994. Detecting patterns in protein sequences. *J. Mol. Biol.* 239:698–712.

134. Newbury SF, Smith NH, Robinson EC, Hiles ID, Higgins CF. 1987. Stabilization of translationally active mRNA by prokaryotic REP sequences. *Cell* 48: 297–310

135. Nimmo HG. 1984. The control of bacterial isocitrate dehydrogenase by phosphorylation. In *Enzyme Regulation by Reversible Phosphorylation*, ed. P Cohen, pp. 123–41. Amsterdam, The Netherlands: Elsevier

136. Nimmo HG. 1986. Kinetic mechanism of *Escherichia coli* isocitrate dehydrogenase and its inhibition by glyoxylate

and oxaloacetate. *Biochem. J.* 234:317–23

137. Nimmo HG. 1987. The tricarboxylic acid cycle and anaplerotic reactions. In *Escherichia coli and Salmonella typhimurium: Cellular and Molecular Biology*, ed. FC Neidhardt, 1:156–69. Washington, DC: ASM. 1st ed.

138. Nimmo GA, Nimmo HG. 1984. The regulatory properties of isocitrate dehydrogenase kinase and isocitrate dehydrogenase phosphatase from *Escherichia coli* ML308 and the roles of these activities in the control of isocitrate dehydrogenase. *Eur. J. Biochem.* 141:409–14

139. Nunn WD. 1986. A molecular view of fatty acid catabolism in *Escherichia coli*. *Microbiol. Rev.* 50:179–92

140. Nunn WD. 1987. Two-carbon compounds and fatty acids as carbon sources. In *Escherichia coli and Salmonella typhimurium: Cellular and Molecular Biology*, ed. FC Neidhardt, 1:285–301. Washington, DC: ASM. 1st ed.

141. Oberto J, Drlica K, Rouvière-Yaniv J. 1994. Histones, HMG, HU, IHF: même combat. *Biochimie* 76:901–8

142. Ochiai T, Fukunaga N, Sasaki S. 1984. Two structurally different NADP-specific isocitrate dehydrogenases in an obligately psychrophilic bacterium, *Vibrio* sp. strain ABE-1. *J. Gen. Appl. Microbiol.* 30:479–87

143. Ogata RT, Gilbert W. 1978. An amino-terminal fragment of the *lac* repressor binds specifically to *lac* operator. *Proc. Natl. Acad. Sci. USA* 75:5851–54

144. Ornston LN, Ornston MK. 1969. Regulation of glyoxylate metabolism in *Escherichia coli* K-12. *J. Bacteriol.* 98: 1098–108

145. Overath P, Pauli G, Schairer HU. 1969. Fatty acid degradation in *Escherichia coli*: an inducible acyl CoA synthetase, the mapping of *old*-mutations, and the isolation of regulatory mutants. *Eur. J. Biochem.* 7:559–74

146. Pan B, Unnikrishnan I, LaPorte DC. 1996. The binding site of the IclR repressor protein overlaps the promoter of *aceBAK*. *J. Bacteriol.* 178:3982–84

147. Parekh BS, Sheridan SD, Hatfield GW. 1996. Effect of integration host factor and DNA supercoiling on transcription from the *ilvP*$_G$ promoter of *Escherichia coli*. *J. Biol. Chem.* 271:20258–64

148. Parkinson JS. 1993. Signal transduction schemes of bacteria. *Cell* 73:857–71

149. Pauli G, Overath P. 1972. *ato* operon: a highly inducible system for acetoacetate

and butyrate degradation in *Escherichia coli*. *Eur. J. Biochem.* 29:553–62

150. Penin F, Geourjon C, Montserret R, Böckmann A, Lesage A, et al. 1997. Three-dimensional structure of the DNA-binding domain of the fructose repressor from *Escherichia coli* by ^{1}H and ^{15}N NMR. *J. Mol. Biol.* 270:496–510

151. Pérez-Martín J, de Lorenzo V. 1997. Clues and consequences of DNA bending in transcription. *Annu. Rev. Microbiol.* 51:593–628

152. Plaut GWE. 1970. DPN-linked isocitrate dehydrogenase of animal tissues. *Curr. Top. Cell. Regul.* 2:1–27

152a. Popov VN, Igamberdiev AU, Schnarrenberger C, Volvenkin SV. 1996. Induction of glyoxylate cycle enzymes in rat liver upon food starvation. *FEBS Lett.* 390:258–60

153. Raman N, Di Russo CC. 1995. Analysis of acyl coenzyme A binding to the transcription factor FadR and identification of amino acid residues in the carboxyl terminus required for ligand binding. *J. Biol. Chem.* 270:1092–97

154. Ramseier TM, Bledig S, Michotey V, Feghali R, Saier MH Jr. 1995. The global regulatory protein, FruR, modulates the direction of carbon flow in *Escherichia coli*. *Mol. Microbiol.* 16:1157–69

155. Ramseier TM, Nègre D, Cortay JC, Scarabel M, Cozzone AJ, Saier MH Jr. 1993. In vitro binding of the pleiotropic transcriptional regulatory protein, FruR, to the *fru*, *pps*, *ace*, *pts* and *icd* operons of *Escherichia coli* and *Salmonella typhimurium*. *J. Mol. Biol.* 234:28–44

156. Reeves HC, Danmy GO, Chen CL, Houston M. 1972. NADP-specific isocitrate dehydrogenase of *Escherichia coli*. Purification and characterization. *Biochim. Biophys. Acta* 258:27–39

157. Reeves HC, Malloy PJ. 1983. Phosphorylation of isocitrate dehydrogenase in *Escherichia coli* mutants with a nonfunctional glyoxylate cycle. *FEBS Lett.* 158:239–42

158. Reeves HC, O'Neil S, Weitzman PDJ. 1983. Modulation of isocitrate dehydrogenase activity in *Acinetobacter calcoaceticus* by acetate. *FEBS Lett.* 163:265–68

159. Reeves HC, O'Neil S, Weitzman PDJ. 1986. Changes in NADP-isocitrate dehydrogenase isoenzyme levels in *Acinetobacter calcoaceticus* in response to acetate. *FEMS Microbiol. Lett.* 35:229–32

160. Resnik E, Pan B, Ramani N, Freundlich M, LaPorte DC. 1996. Integration host

factor amplifies the induction of the *aceBAK* operon of *Escherichia coli* by relieving IclR repression. *J. Bacteriol.* 178:2715–17

161. Rice PA, Yang S, Mizuuchi K, Nash HA. 1996. Crystal structure of an IHF-DNA complex: a protein-induced DNA U-turn. *Cell* 87:1295–306

161a. Rieul C, Bleicher F, Duclos B, Cortay JC, Cozzone AJ. 1988. Nucleotide sequence of the *aceA* gene coding for isocitrate lyase in *Escherichia coli*. *Nucleic Acids Res.* 16:5689

162. Rittinger K, Nègre D, Divita G, Scarabel M, Bonod-Bidaud C, et al. 1996. *Escherichia coli* isocitrate dehydrogenase kinase/phosphatase: overproduction and kinetics of interaction with its substrates by using intrinsic fluorescence and fluorescent nucleotide analogues. *Eur. J. Biochem.* 237:247–54

163. Robertson EF, Reeves HC. 1989. Phosphorylation of isocitrate lyase in *Escherichia coli*. *Biochimie* 71:1065–70

164. Ross W, Gosink KK, Salomon J, Igarashi K, Zou C, et al. 1993. A third recognition element in bacterial promoters: DNA binding by the alpha subunit of RNA polymerase. *Science* 262:1407–13

165. Saier MH, Ramseier TM, Reizer J. 1996. Regulation of carbon utilization. In *Escherichia coli and Salmonella typhimurium: Cellular and Molecular Biology*, ed. FC Neidhardt, 1:1325–43. Washington, DC: ASM. 2nd ed.

166. Scarabel M, Penin F, Bonod-Bidaud C, Nègre D, Cozzone AJ, Cortay JC. 1995. Overproduction, purification and structural characterization of the functional N-terminal DNA-binding domain of the *fru* repressor from *Escherichia coli* K-12. *Gene* 153:9–15

167. Schumacher MA, Choi KY, Zalkin H, Brennan RG. 1994. Crystal structure of LacI member, PurR, bound to DNA: minor groove binding by α helices. *Science* 266:763–70

168. Shine J, Dalgarno L. 1974. The 3'-terminal sequence of *Escherichia coli* 16S ribosomal RNA: complementarity to nonsense triplets and ribosome binding sites. *Proc. Natl. Acad. Sci. USA* 71:1342–46

169. Simons RW, Egan PA, Chute HT, Nunn WD. 1980. Regulation of fatty acid degradation in *Escherichia coli*: isolation and characterization of strains bearing insertion and temperature-sensitive mutations in gene *fadR*. *J. Bacteriol.* 142:621–32

170. Slijper AM, Bonvin JJ, Boelens R,

Kaptein R. 1996. Refined structure of *lac* repressor head-piece (1–56) determined by relaxation matrix calculation from 2D and 3D NOE data: change of tertiary structure upon binding to the *lac* operator. *J. Mol. Biol.* 259:761–73

171. Stern MJ, Ferro-Luzzi Ames G, Smith NH, Robinson EC, Higgins CF. 1984. Repetitive extragenic palindromic sequences: a major component of the bacterial genome. *Cell* 37:1015–26

172. Stoddard BL, Dean AM, Koshland DE Jr. 1993. The 2.5 è structure of isocitrate dehydrogenase with isocitrate, NADP and calcium: a pseudo-Michaelis ternary complex. *Biochemistry* 32:9310–16

173. Stueland CS, Eck KR, Stieglbauer KT, LaPorte DC. 1987. Isocitrate dehydrogenase kinase/phosphatase exhibits an intrinsic adenosine triphosphatase activity. *J. Biol. Chem.* 262:16095–99

174. Stueland CS, Ikeda TP, LaPorte DC. 1989. Mutation of the predicted ATP binding site inactivates both activities of isocitrate dehydrogenase kinase/phosphatase. *J. Biol. Chem.* 264:13775–79

175. Sunnarborg A, Klumpp D, Chung T, LaPorte DC. 1990. Regulation of the glyoxylate bypass operon: cloning and characterization of *iclR*. *J. Bacteriol.* 172: 2642–49

176. Suzuki M, Sahara T, Tsuruha JI, Takada Y, Fukunaga N. 1995. Differential expression in *Escherichia coli* of the *Vibrio* sp. strain ABE-1 *icdI* and *icdII* genes encoding structurally different isocitrate dehydrogenase isozymes. *J. Bacteriol.* 177:2138–42

177. Swafford JR, Malloy PJ, Reeves HC. 1983. Immunochemical localization of NADP-specific isocitrate dehydrogenase in *Escherichia coli*. *Science* 221: 295–96

178. Thorsness PE, Koshland DE Jr. 1987. Inactivation of isocitrate dehydrogenase by phosphorylation is mediated by the negative charge of the phosphate. *J. Biol. Chem.* 262:10422–25

179. Vanderwinkel E, De Vlieghere M. 1968.

Physiologie et génétique de l'isocitritase et des malate synthases chez *Escherichia coli*. *Eur. J. Biochem.* 5:81–90

180. Varela I, Nimmo HG. 1988. Photoaffinity labelling shows that *Escherichia coli* isocitrate dehydrogenase kinase/phosphatase contains a single ATP-binding site. *FEBS Lett.* 231:361–65

181. Vasquez B, Reeves HC. 1979. NADP-specific isocitrate dehydrogenase of *Escherichia coli*. IV: purification by chromatography on Affi-gel Blue. *Biochim. Biophys. Acta* 578:31–40

182. Walsh K, Koshland DE Jr. 1984. Determination of flux through the branch point of two metabolic cycles: the tricarboxylic acid cycle and the glyoxylate shunt. *J. Biol. Chem.* 259:9646–54

183. Walsh K, Koshland DE Jr. 1985. Branch point control by the phosphorylation state of isocitrate dehydrogenase: a quantitative examination of fluxes during a regulatory transition. *J. Biol. Chem.* 260:8430–37

184. Weeks G, Shapiro M, Burns RO, Wakil SJ. 1969. Control of fatty acid metabolism. I: Induction of the enzymes of fatty acid oxidation in *Escherichia coli*. *J. Bacteriol.* 97:827–36

185. Weickert MJ, Adhya S. 1992. A family of bacterial regulators homologous to Gal and Lac repressors. *J. Biol. Chem.* 267:15869–74

186. Williams JO, Roche TE, McFadden BA. 1971. Mechanism of action of isocitrate lyase from *Pseudomonas indigofera*. *Biochemistry* 10:384–90

187. Wilson RB, Maloy SR. 1987. Isolation and characterization of *Salmonella typhimurium* glyoxylate shunt mutants. *J. Bacteriol.* 169:3029–34

188. Yang Y, Ames GF. 1988. DNA gyrase binds to the family of prokaryotic repetitive extragenic palindromic sequences. *Proc. Natl. Acad. Sci. USA* 85:8850–54

189. Zahl KJ, Rose C, Hanson RL. 1978. Isolation and partial characterization of a mutant of *Escherichia coli* lacking pyridine nucleotide transhydrogenase. *Arch. Biochem. Biophys.* 190:598–602

Annu. Rev. Microbiol. 1998. 52:165–90

ANAEROBIC GROWTH OF A "STRICT AEROBE" (*BACILLUS SUBTILIS*)

Michiko M. Nakano and Peter Zuber

Department of Biochemistry and Molecular Biology, Louisiana State University
Medical Center, Shreveport, Louisiana 71130-3932;
e-mail: mnakan@lsumc.edu; pzuber@lsumc.edu

KEY WORDS: anaerobiosis, nitrate respiration, fermentation, gene regulation

ABSTRACT

There was a long-held belief that the gram-positive soil bacterium *Bacillus subtilis* is a strict aerobe. But recent studies have shown that *B. subtilis* will grow anaerobically, either by using nitrate or nitrite as a terminal electron acceptor, or by fermentation. How *B. subtilis* alters its metabolic activity according to the availability of oxygen and alternative electron acceptors is but one focus of study. A two-component signal transduction system composed of a sensor kinase, ResE, and a response regulator, ResD, occupies an early stage in the regulatory pathway governing anaerobic respiration. One of the essential roles of ResD and ResE in anaerobic gene regulation is induction of *fnr* transcription upon oxygen limitation. FNR is a transcriptional activator for anaerobically induced genes, including those for respiratory nitrate reductase, *narGHJI*. *B. subtilis* has two distinct nitrate reductases, one for the assimilation of nitrate nitrogen and the other for nitrate respiration. In contrast, one nitrite reductase functions both in nitrite nitrogen assimilation and nitrite respiration. Unlike many anaerobes, which use pyruvate formate lyase, *B. subtilis* can carry out fermentation in the absence of external electron acceptors wherein pyruvate dehydrogenase is utilized to metabolize pyruvate.

CONTENTS

165

INTRODUCTION

Bacteria often encounter drastic changes in their environment, including fluctuations in the levels of external oxygen. Unlike strict aerobes or strict anaerobes, which can survive only either in the presence or absence of oxygen,
facultative anaerobes can cope with changes in environmental oxygen levels
by sensing oxygen concentration and shifting cellular metabolism accordingly
(28, 36, 37, 45, 76). The changes in metabolism in response to changes in oxygen availability include adjustments to the rate and route of carbon source
utilization, to the pathways of electron flow to maintain an oxidation-reduction
balance, and to the mechanisms of energy production and of certain biosynthetic reactions. These changes are achieved by modulating protein activity, by
regulating the expression of the appropriate genes, or both. As one approach
to studying these differences in the pattern of gene expression, proteins whose
concentrations fluctuate in response to changes in oxygen levels have been
identified by two-dimensional gel electrophoresis of extracts obtained from
aerobically and anaerobically grown cultures of *Escherichia coli* (65, 71, 72)
and *Salmonella typhimurium* (1). Another approach, using random gene fusions to a promoterless *lacZ*, also resulted in the identification of *E. coli* genes
induced by oxygen limitation (8, 9, 87). Aerobically induced proteins include
the enzymes of the pyruvate dehydrogenase complex, several tricarboxylic acid
cycle enzymes, and superoxide dismutase (72). Anaerobically induced proteins
include some glycolytic enzymes and pyruvate formate lyase (71), required for

the anaerobic disposal of electrons in the form of formic acid. Most but not all of the aerobically induced genes were also induced under anaerobic conditions in the presence of nitrate, and the majority of anaerobically induced genes were repressed in the presence of alternative electron acceptors such as nitrate, and to a lesser extent by trimethylamine N-oxide (TMAO) and dimethylsulfoxide (DMSO). The repressive effect of the electron acceptors on some anaerobically induced genes reflects two anaerobic growth modes of *E. coli*, respiration and fermentation. In the absence of electron acceptors, *E. coli* carries out mixed-acid fermentation. In fermentative growth where NADH is not reoxidized by the electron transfer pathway that is coupled to ATP generation, NADH is reoxidized by conversion of pyruvate to various end products during which ATP is produced by substrate-level phosphorylation. Genes whose products function in this less energy-generating fermentation pathway are repressed when electron acceptors such as oxygen and nitrate are available to drive electron transfer.

Microbiology textbooks have described the gram-positive bacterium *Bacillus subtilis* as a strict aerobe and this, until recently, has been the common understanding among most investigators studying *B. subtilis*. That *B. subtilis* could grow under anaerobic conditions was stated for the first time, to our knowledge, by Priest in a review that appeared in 1993 (58), although membrane-bound nitrate reductase was isolated three decades ago from *B. subtilis* cultured under low aeration (7, 47a) (as described later, the enzyme is very likely respiratory nitrate reductase). The description by Priest prompted a fair number of investigators to start to examine anaerobiosis in *B. subtilis*. The *Bacillus* genome sequencing project has contributed significantly to this study by identifying genes that show similarity to those known to function in anaerobic metabolism in *E. coli*.

The natural habitat of *E. coli* is the large intestine of mammals, and a shift from an aerobic to an anaerobic environment is the usual result upon its entry into the host. *B. subtilis* is abundant in soil and is probably transferred from soil to other associated environments. Therefore, in the case of *B. subtilis*, fluctuations in the availability of oxygen are not necessarily the result of changes in habitat, but instead are derived easily by changes in the soil's water content. The conditions under which *E. coli* and *B. subtilis* must grow and survive in nature may account for the differences in the way the two organisms control anaerobiosis. In this review, we summarize recent findings in this new area of study.

NITRATE RESPIRATION

Nitrate is the preferred terminal electron acceptor when oxygen is absent because of its high midpoint redox potential ($E'^0 = +430$ mV). Nitrate respiration and nitrite respiration (described in the next section) are the only anaerobic

Table 1 Genes involved in nitrate/nitrite respiration of *B. subtilis* and their predicted products

Gene	Probable function of product	Homolog in *Escherichia coli*
resD	Response regulator; aerobic/anaerobic respiration	
resE	Sensor kinase; aerobic/anaerobic respiration	
fnr	Anaerobic regulator	*fnr*
narK	Nitrite extrusion	*narK*
narG	Subunit of nitrate reductase; molybdoprotein	*narG*; *narZ*
narH	Subunit of nitrate reductase; iron-sulfur protein	*narH*; *narY*
narJ	Assembly of nitrate reductase complex?	*narJ*; *narW*
narI	Subunit of nitrate reductase; cytochrome *b*	*narI*; *narV*
moaA	Molybdopterin synthesis; nitrate reductase cofactor	*moaA*
nasD	Subunit of NADH-dependent nitrite reductase	*nirB*
nasE	Subunit of NADH-dependent nitrite reductase	*nirD*
nasF	Siroheme synthesis; nitrite reductase cofactor	*cysG*
nirC	Integral membrane protein; nitrite transporter?	*nirC*
hmp	Flavohemoglobin	*hmp*

forms of respiration known thus far in *B. subtilis*. The genes involved in nitrate and nitrite respiration are listed in Table 1, and possible regulatory pathways are summarized in Figure 1.

The utilization of nitrate (and nitrite) as an electron acceptor is reasonable, considering that nitrate and nitrite normally are present in sufficiently high concentrations in soil, being products of nitrification. *B. subtilis* has no fumarate respiration (49, 66), and the existence of other electron acceptors during anaerobic growth has not been reported. Anaerobic nitrate reduction was shown to be coupled to energy generation, because *B. subtilis* grows anaerobically in minimal medium containing glycerol, a nonfermentable carbon source, if nitrate is present (49).

Nitrate Reductase

Unlike *E. coli*, which is unable to assimilate nitrate or nitrite during aerobiosis, *B. subtilis* can grow aerobically on nitrate or nitrite as its sole nitrogen source by converting nitrate or nitrite to ammonium. The reduction of nitrate to nitrite and nitrite to ammonium is catalyzed by nitrate and nitrite reductase, respectively. These enzymes are named assimilatory nitrate and nitrite reductases for the purpose of discriminating them from respiratory enzymes involved in anaerobic respiration. Genes encoding the subunits of assimilatory nitrate (*nasB* and *nasC*) and nitrite (*nasD* and *nasE*) reductases constitute an operon together with the *nasF* gene. The operon is transcribed divergently from the *nasA* gene (52).

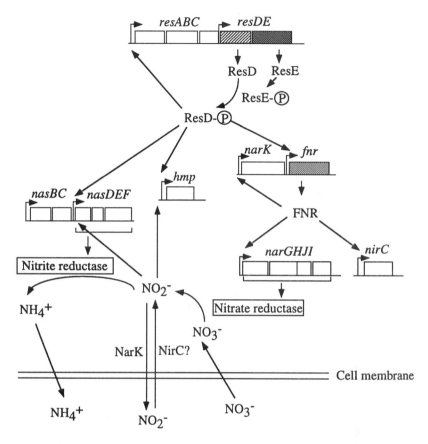

Figure 1   Regulatory pathways of nitrate and nitrite respiration in *B. subtilis. Boxes,* genes; *shaded boxes,* genes involved in anaerobic regulation; *arrows,* regulatory flows; *arrows above boxes,* start sites and direction of transcription. References and experimental details for this figure are described in the text.

Mutations in *nasA, nasB,* or *nasC* abolished growth on medium containing nitrate as the sole nitrogen source but had no effect on growth when nitrite alone was present. In contrast, mutations in *nasD, nasE,* or *nasF* led to a growth defect in glucose minimal medium containing either nitrate or nitrite (52). These results, taken together with data from amino acid sequence similarity searches, indicate that *nasB* and *nasC* encode the assimilatory nitrate reductase and *nasD* and *nasE* encode the assimilatory nitrite reductase, while *nasF* codes for an enzyme involved in the synthesis of siroheme, a cofactor of nitrite reductase. The *nasA* gene product is probably a nitrate transporter based on primary structure similarities (52). Nitrate reductase also requires a molybdenum cofactor for its activity. A gene (*moaA*) likely to encode an enzyme that functions in the biosynthesis of molybdopterin, the pterin component of the molybdenum cofactor (60), was isolated at a different locus on the *B. subtilis* chromosome than the *nas* genes (25). The statement by Priest (58) that *B. subtilis* grows anaerobically if nitrate is supplied prompted us to examine whether anaerobic growth of *B. subtilis* in the presence of nitrate, if it actually occurs, is affected by mutations in the genes required for nitrate reductase activity. Our results showed that *nasB* and *nasC* mutants grow anaerobically on nitrate as well as do wild-type; in contrast, *moaA* mutants were unable to grow by nitrate respiration. In keeping with this finding, the *nasB* and *nasC* mutants have lost assimilatory nitrate reductase activity but still retain nitrate reductase activity induced under anaerobic conditions, while neither nitrate reductase activities were detected in the *moaA* mutant (25). These results suggest that assimilatory and respiratory nitrate reductases in *B. subtilis* are genetically and biochemically distinct.

Shortly after these findings were made, genes (*narGHJI*) encoding respiratory nitrate reductase were isolated during three independent investigations involving (*a*) the *Bacillus* genome sequencing project (14), (*b*) the use of oligonucleotides deduced from conserved amino-acid sequences of *E. coli* respiratory nitrate reductase (34), and (*c*) a search of mutants unable to grow anaerobically by nitrate respiration (43). The respiratory nitrate reductase thus identified is very likely the membrane-bound enzyme isolated 30 years ago from semi-aerobic cultures (7, 47a) . The nitrate reductase was produced in high amounts in mutants blocked at stage zero of sporulation (Spo0), and it was postulated that membrane alteration of Spo0 mutants caused the hyperproduction (7). That *spo0* mutations affect expression of *narGHJI* is an alternative, but not mutually exclusive, possibility.

Comparison of amino acid sequences of proteins encoded by *narGHJI* with those of *E. coli* nitrate reductases (5, 6) indicated that NarG, H, and I are subunits of nitrate reductase and NarJ is required for the assembly of the enzyme (34). NarG is likely a catalytic subunit and shows homology with various molybdoproteins. NarH possesses an arrangement of cysteine residues typical of iron-sulfur centers and NarI is likely to be the apoprotein of a *b*-type cytochrome (34). It is thought that NarI receives electrons from the quinones and transfers them first to the iron-sulfur centers of the NarH subunit, and then to the molybdenum cofactor carried by the NarG protein, where nitrate is reduced to nitrite. The essential role of the nitrate reductase encoded by the *nar* genes in nitrate respiration of *B. subtilis* was confirmed by the observed defect of *nar* mutants in anaerobic growth in the presence of nitrate (34, 43). DNA sequence analysis indicated that the *narGHJI* genes probably constitute an operon bearing a putative FNR-binding sequence within its upstream, regulatory region. As discussed later, the expression of *narGHJI* is strongly induced by anaerobiosis and the induction is dependent on FNR.

NarK

The *narK* gene resides upstream of the *narGHJI* operon and encodes a protein similar to *E. coli* NarK (14). *E. coli* NarK is involved in extrusion of nitrite produced during nitrate respiration (17, 61). The *B. subtilis narK* mutant, like the *E. coli* mutant, was shown not to excrete nitrite when grown anaerobically with nitrate. The *B. subtilis narK* gene is able to complement an *E. coli narK* mutant. The *narK* gene constitutes an operon along with a downstream gene *fnr* encoding a protein homologous to an *E. coli* global anaerobic regulator FNR. Transcription of *narK* was shown to be highly induced by oxygen limitation, and the induction requires FNR, as expected by the presence of a putative FNR-binding consensus sequence within the *narK* regulatory region (14).

FNR

The second gene of the *narK* operon encodes a protein homologous to *E. coli* FNR, a member of the catabolite gene activator protein (CAP) family of transcriptional regulators (69). It possesses a DNA-binding domain with the characteristic helix-turn-helix motif (75, 77). Disruption of *fnr* in *B. subtilis* abolishes nitrate-stimulated anaerobic growth, and respiratory nitrate reductase activity is not detected, indicating that FNR is indispensable for nitrate respiration in *B. subtilis* (14), as in *E. coli*.

The differences in the *E. coli* CAP and the FNR binding sites are correlated with differences in the amino-acid sequence of the respective DNA binding motifs in the CAP and FNR proteins. Substitution of the specific base pairs within the CAP and FNR binding sites results in recognition by the other regulator (4, 64, 91). In CAP, Arg[180] contacts a 5' GC base pair of the conserved TGTGA motif of the core binding sequence. The residue at the corresponding position in *E. coli* FNR is valine (Val[208]), and instead of a GC base pair, a TA pair is found in the FNR half-site (5' TA base pair of the TTGAT motif). Evidence for the involvement of Val[208] of FNR in interaction with the TA base pair is ambiguous, and instead it was proposed that Ser[212] of FNR is crucial for DNA target specificity (75). In *Bacillus* FNR, the residue corresponding to Ser[212] of *E. coli* FNR is asparagine and the position corresponding to Arg[180] of *E. coli* CAP is also an arginine residue. Accordingly, a GC base pair is found at the predicted FNR binding site of *B. subtilis*, as in the *E. coli* CAP binding site. In fact, expression of *lacZ* driven from the *B. subtilis narK* promoter was activated 50-fold by CAP-cAMP complex in *E. coli*, indicating that CAP can recognize the *B. subtilis* FNR binding site and activates transcription at the *narK* promoter (14). Cruz Ramos and coauthors (14) suggested that the regulator DNA binding domain and its targets are co-evolved on the chromosome and that the absence of CAP in *B. subtilis* allows FNR to recognize a target sequence identical to a CAP site.

The mechanism of FNR-dependent transcriptional regulation in response to oxygen level is not fully understood in spite of intensive studies in *E. coli*, in which FNR functions under anaerobic conditions both as a repressor and an activator. Expression of *E. coli fnr* is not induced but is repressed by about 50% when a culture undergoes a shift from aerobic metabolism to anaerobiosis (74). Therefore, FNR activity should be modulated by the presence or absence of oxygen. Four of the five cysteine residues in the FNR primary structure are clustered in an amino-terminal domain, but no such cysteine cluster is found in CAP.

Based on this observation, a model was proposed that the redox state of iron bound to the cysteine residues modulates FNR activity (26, 39, 83). A change in oligomeric state of FNR in response to oxygen was also proposed as a mechanism to regulate FNR activity (44). The *B. subtilis* FNR also contains a similar cysteine cluster at its carboxyl terminus, suggesting that a similar mechanism involving iron activates FNR as a transcriptional regulator during anaerobiosis (14).

In contrast to *E. coli fnr*, *B. subtilis fnr* was shown to be positively regulated at the transcriptional level in response to low oxygen concentration (14). Thus, oxygen-dependent control of FNR could involve two different mechanisms, i.e. control of transcription initiation and modulation of protein activity, as was observed in studies of FixK, an FNR homologue of *Rhizobium meliloti* (3). *fnr* transcriptional regulation is exerted in *B. subtilis* at two sites, one at the FNR-dependent *narK* operon promoter, and the other at an intergenic *fnr*-specific promoter independent of FNR (14, 51). More details follow.

ResD-ResE Two-Component Signal Transduction System

The FNR-independent anaerobic induction of *fnr* expression described above indicated that *B. subtilis* has another regulatory pathway or other regulatory pathways that is or that are involved in anaerobic gene expression. *resD* and *resE*, identified first in the *Bacillus* genome sequencing project (73), were later shown to be required for FNR-independent *fnr* expression (51). Amino-acid sequence analysis suggested that ResE is a histidine kinase and ResD is a response regulator, both being members of a two-component signal transduction system. *resD* and *resE* reside within an operon along with upstream genes *resA*, *resB*, and *resC*, which encode proteins similar to those that function in cytochrome *c* biogenesis. ResD and ResE were shown to play a role as global regulators in aerobic and anaerobic respiration because of their essential role in expression of genes whose products function in respiration such as *ctaA* (required for heme A synthesis) and the *petCBD* operon (encoding subunits of the cytochrome *bf* complex), as well as *resABC* (80). Furthermore, *resD* mutants (and *resE* mutants to a lesser extent) exhibited pleiotropic mutant phenotypes related to respiratory functions, including streptomycin resistance, lack of heme A-containing aa_3 and caa_3 terminal oxidases, and a requirement of 6-carbon sugars for normal growth. The mutational defects are not restricted to aerobic respiration, considering that *resDE* mutants are unable to grow anaerobically in the presence of nitrate (80). Essential roles of the ResD-ResE two-component signal transduction system in anaerobic respiration is discussed in detail below.

NITRITE RESPIRATION

Most bacilli, including *Bacillus licheniformis*, are assumed to be denitrifying bacteria, generating nitrogen from nitrate. However, some bacilli such as *Bacillus macerans* produce ammonia from nitrate during nitrate respiration (66). A recent study showed that *B. subtilis* converts nitrate to ammonia with no evidence of denitrification products detected during anaerobic growth (33). The study further led to the finding that *B. subtilis* wild-type strain and a nitrate reductase mutant grew anaerobically in the presence of nitrite. Mutation in *resDE* but not *fnr* abolished nitrite-stimulated anaerobic growth and respiratory nitrite reductase activity (33). Although it remains to be determined if nitrite reduction in *B. subtilis* is coupled to energy production through proton motive force, this process can be referred to as nitrite respiration according to the simple definition of a process utilizing an inorganic electron acceptor.

Nitrite Reductase

A search of the recently completed *B. subtilis* genome nucleotide sequence (42) uncovered only one nitrite reductase, the one encoded by the *nasD* and *nasE* genes described above. This prompted us to determine if "assimilatory nitrite reductase" also has a role in anaerobic nitrite respiration (MM Nakano, T Hoffmann, D Jahn, unpublished data). As mentioned before, mutations in the *nas* genes have no effect on nitrate respiration. However, nitrite respiration was impaired by mutations in the *nasD*, -*E*, or -*F* genes. *nasB* and *nasC* mutants grow well anaerobically in the presence of nitrite. These results indicate that nitrite reductase encoded by *nasD* and *nasE* is required for nitrite respiration.

E. *coli* has two similar membrane-bound nitrate reductase genes, [*narGHJI* (5) and *narZYWV* (6)] (11), the products of which catalyze reduction of nitrate, a reaction that is coupled to ATP synthesis. There is also a periplasmic nitrate reductase encoded by the *napFDAGHBC* operon (27). In addition, E. *coli* possesses two nitrite reductases, one that is soluble and NADH-dependent encoded by *nirBDCcysG* (30, 56), and a formate-dependent nitrite reductase encoded by *nrfABCDEFG* (35). The NADH-dependent enzyme functions together with the respiratory nitrate reductase encoded by *narGHJI* when nitrate is abundant and its major role is to detoxify nitrite by converting it to ammonium ion. The formate-dependent nitrite reductase functions in ATP generation when nitrite is present and nitrate is limited (53). *B. subtilis,* in which the function of nitrate/nitrite reduction is not only respiratory but also assimilatory, possesses nitrate and nitrite reductases of different composition than those of *E. coli.* *B. subtilis* has two distinct nitrate reductases, one functioning in nitrogen assimilation and the other in respiration. In contrast, one nitrite reductase appears to function both in nitrogen assimilation and nitrite respiration. The subunits of nitrite reductase, NasD and NasE, show high amino acid sequence similarity to NirB and NirD, the subunits of *E. coli* Nir nitrite reductase (52). Interestingly, a *B. subtilis* homologue of *nirC*, a gene in the *nirBDC* operon of *E. coli*, is located not in the *nas* operon but at a different locus of the *Bacillus* genome. *nirC*, the gene product of which is an integral membrane protein, is not required for nitrite reduction or siroheme synthesis in *E. coli* (30). NirC is thought to be a nitrite transporter, although there is no direct evidence that *nirC* encodes a nitrite efflux or uptake system (11). Assignment of the function of *nirC* might provide a clue to why the *B. subtilis nirC* gene is not linked to the genes for nitrite reductase, as is the case for its counterpart in *E. coli.*

Our previous studies showed that the *nasB* promoter is under global nitrogen regulation, i.e. the expression is high when cells are grown in the presence of poor nitrogen sources such as nitrate, proline, or glutamate and very low when ammonium or glutamine is provided as the sole nitrogen source. Mutational analysis indicated that a dyad symmetry sequence ($TGTNAN_7TNACA$) located between 41 and 59 bp upstream of the *nasB* transcription start site is essential for the activation of *nasB* during growth under nitrogen-limiting conditions (50). A transcription factor required for global nitrogen regulation, TnrA, was shown to be required for the activation of *nasB* during nitrogen-limiting conditions (89). Since transcription from the *nasB* promoter is not induced by anaerobiosis, *nasDEF* should be transcribed from an internal promoter, at least under anaerobic conditions. The same dyad symmetry required for the nitrogen-regulation of the *nasB* promoter is found in the intergenic region located upstream of the *nasD* gene. A DNA fragment containing the dyad symmetry element exhibits promoter activity, indicating that nitrite reductase genes *nasD* to *nasF* are likely transcribed from the *nasD* promoter as well as from the *nasB* promoter. Expression of the *nasDEF* genes from the *nasD* promoter is derepressed during aerobic nitrogen-limited growth, as observed for transcription initiation from the *nasB* promoter. But unlike the *nasB* promoter, the *nasD* promoter is activated by oxygen limitation even in the presence of excess nitrogen. The anaerobic induction is partly dependent on FNR when cells are grown in the presence of nitrate, while induction of transcription is FNR-independent when cells encounter exogenous nitrite under anaerobic conditions. In contrast, *resDE* mutations impaired anaerobic induction during growth either by nitrate or nitrite respiration (MM Nakano, unpublished data). These results imply that nitrite stimulates anaerobic induction of *nasDEF* and that the induction is dependent on ResD/E. Requirement of FNR for *nasD* expression during anaerobic growth with nitrate but not with nitrite reflects the requirement of respiratory nitrate reductase (production of which is FNR-dependent) for nitrite production. This demonstrates that the presence of the intergenic *nasD* promoter ensures expression of nitrite reductase genes under anaerobic conditions. Are anaerobic and nitrogen control of *nasD* coupled? If so, is the anaerobic regulation mediated partly through the TnrA-target site in the *nasD* promoter region? If not, how is *nasDEF* induced by oxygen limitation? How is ResD involved in the regulation, and how might a nitrite-derived signal be transduced to the ResDE regulatory system? These are intriguing questions that remain to be answered.

REGULATORY PATHWAYS IN NITRATE/NITRITE RESPIRATION

Probably the most interesting issue in the study of anaerobiosis is how external oxygen limitation is sensed by cells and how the cells then adjust their cellular metabolism to promote growth in an anaerobic environment. Does oxygen itself or some other molecule (or molecules) function as a signal? How is the signal transferred to "sensor" molecules and then processed to eventually affect gene expression? A current model has been drawn from studies by our laboratory and by others (Figure 1).

ResD and ResE Are Required for Expression of Anaerobically Induced Genes

As discussed above, the ResD and ResE two-component regulatory proteins play important roles in both aerobic and anaerobic respiration. At present, two genes and one operon involved in nitrate/nitrite respiration are known to be controlled by ResDE: *fnr*, *hmp*, and *nasDEF*. Because regulation of *hmp* and *nasDEF* by ResDE is described elsewhere in this review, we describe only ResDE-dependent *fnr* expression in this section.

A *lacZ* fusion to the *fnr*-specific promoter (*fnr-lacZ*) was constructed and inserted into an SPβ prophage locus to examine regulation of FNR-independent anaerobic expression of *fnr* (51; see previous section). While *fnr* expression from the *narK* operon promoter was barely detectable in aerobic cultures and dramatically induced by a shift to anaerobiosis (14, 51), *fnr* expression, dependent on transcription from the *fnr* promoter (*fnr-lacZ*), was low but easily detected during aerobiosis and further induced by oxygen limitation. The anaerobic induction of *fnr-lacZ* was completely dependent on *resD* and *resE*, indicating transcription from the *fnr*-specific promoter was dependent on phosphorylated ResD produced mainly from a reaction catalyzed by the ResE kinase. Considering that the ResD-ResE signal transduction pathway is not activated exclusively by anaerobiosis as described earlier, how might an oxygen-limitation signal enter the ResDE-*fnr* regulatory pathway? This is one of the important questions on which future studies should be focused. In order to determine if ResDE alone is required for *fnr* expression during nitrate respiration, *fnr* was placed under the control of the IPTG (isopropyl-β-D-thiogalactopyranoside)-inducible promoter, P*spac*. A *resDE* mutant carrying P*spac-fnr* cannot grow anaerobically in the presence of IPTG, indicating that *resDE* has an additional role to play in nitrate respiration (51).

Analysis of the fnr Promoter Region Required for ResDE-Dependent Activation

Although ResD is probably a transcription factor, there is no direct evidence that ResD protein, either phosphorylated or unphosphorylated, interacts directly with the regulatory regions of ResD-controlled genes (*resABCDE*, *petCBD*, *ctaA*, *fnr*, *hmp*, and *nasDEF*). Nor has it been determined whether ResD is the direct activator for the transcription of these genes or is indirectly engaged by activating other regulators that play subordinate roles in aerobic/anaerobic gene control. There is no apparent consensus sequence within the regulatory regions of the ResD-controlled promoters. To unravel ResDE-dependent control, the *fnr* promoter was subject to mutational analysis to localize the region required for anaerobic *fnr* induction (15). Deletion of the *fnr* promoter up to −61 relative to the transcription start site had no significant effect on anaerobic, ResDE-dependent induction. However, deletion to position −58 resulted in a severe reduction in anaerobic *fnr* expression. The −58 position lies in a region of dyad symmetry (TNACAAN$_2$TTGTNA) located between −59 and −46 of the *fnr* regulatory region. A change of the second adenine in the left arm of the palindrome to guanine has no effect on the *fnr* expression. In contrast to, other changes in the left arm, i.e. the leftmost thymine to guanine, the first adenine to cytosine and a change of the third thymine in the right arm to guanine led to reduction in anaerobic expression. The mutant promoters that were affected in anaerobic induction were shown not to be regulated by ResDE (MM Nakano, unpublished data). These results indicate that the dyad symmetry element has an essential role for ResDE-dependent regulation of *fnr*. However, the identical dyad symmetry sequence is not found in regulatory regions of the other ResDE-controlled genes, including *hmp* and *nasDEF*. Three possibilities could explain this observation. First, phosphorylated ResD binds to the dyad symmetry element in the *fnr* promoter, and the other genes are controlled indirectly via the ResDE signal transduction system. The second possibility is that another regulatory protein binds to the sequence and cooperative interaction with ResD is necessary for the *fnr* induction under anaerobic conditions. Finally, ResDE is essential for the expression of a gene encoding an unidentified anaerobic regulator that binds to the dyad symmetry element. Analysis of *cis*-regulatory regions of other ResDE-controlled genes and DNase I footprinting using purified, phosphorylated ResD protein should provide some insight into the mechanism of the ResDE-dependent control.

Isolation and Analysis of a resE Suppressor Mutant

In the hope of understanding the activation mechanism of the ResDE signal transduction pathway, we isolated a spontaneous suppressor mutant that can grow anaerobically in the absence of ResE (MM Nakano, K Haga, H Yoshikawa, AL Sonenshein, P Zuber, unpublished data). The suppressor mutation (*sre-1*) can bypass the requirement of ResE but not of ResD for nitrate respiration, including anaerobic induction of *fnr*. These results suggest that the suppression is due to ResE-independent ResD phosphorylation, either by loss of phosphatase activity directed against ResD-phosphate, activation of another kinase to phosphorylate ResD, or kinase-independent ResD phosphorylation. The *sre-1* mutation was also shown to bypass the requirement of ResE for the transcription of *resA*, another ResDE-controlled gene. *resA* expression is higher under anaerobic conditions than aerobic conditions. This high level of the *resA* expression was also observed under aerobic conditions only when *resE* and *sre-1* are mutated. Thus, ResE appears to exert a negative effect on *resA-lacZ* expression in *sre-1* cells under aerobic conditions. These results imply that *sre-1* allows ResE-independent phosphorylation of ResD and that ResE is both a kinase for ResD and a phosphatase for phosphorylated ResD. If so, the higher anaerobic *resA* expression could be due to inhibition of the phosphatase activity of ResE under anaerobic conditions. This also raises the possibility that *fnr* induction by oxygen limitation is due, at least in part, to the higher concentration of phosphorylated ResD caused by reduced ResE phosphatase activity under anaerobic conditions.

The wild-type and mutant alleles of *sre-1* have been isolated and sequenced in an attempt to understand the role the locus plays in anaerobic growth as well as on *resA* and *fnr* expression in *resE* mutants (MM Nakano, K Haga, H Yoshikawa, AL Sonenshein, P Zuber, unpublished data). The sequencing results indicated that *sre-1* is *pgk* (phosphoglycerate kinase) and the *sre-1* mutation is caused by an insertion of an adenine base in a stretch of seven adenine bases immediately after the translation start codon, resulting in a truncated product. When the wild-type *pgk* gene was supplied in *trans*, the phenotypes of the *resE sre-1* mutant were restored to those of the *resE* strain with respect to anaerobic growth and expression of *resA* and *fnr*. This indicates that the *sre-1* mutation is recessive and likely a loss-of-function mutation. Why does the lack of phosphoglycerate kinase lead to ResE-independent phosphorylation of ResD? One possibility is that accumulation of 1,3-diphosphoglycerate, a substrate of phosphoglycerate kinase, leads to phosphorylation of ResD independent of the ResE kinase. The cognate kinase-independent phosphorylation of a response regulator by a small phosphorylated molecule has been reported (i.e. in the phosphorylation of PhoB by acetyl phosphate) (86). 1,3-diphosphoglycerate, like acetyl phosphate, is a high-energy phosphate compound. In attempts to test this hypothesis, a mutation in the *gap* (glyceraldehyde 3-phosphate dehydrogenase) gene that resides just upstream of the *pgk* genes was introduced into the *resE sre-1* mutant. Given the fact that the enzyme catalyzes a reaction to generate 1,3-diphosphoglycerate from glyceraldehyde-3-phosphate, the mutation in *gap* should inhibit the accumulation of 1,3-diphosphoglycerate in the *resE sre-1* mutant. Therefore, the *resE sre-1 gap* mutant would be expected to have the same phenotype as the *resE* mutant if accumulation of 1,3-diphosphoglycerate causes the observed suppression. The result showed that the introduction of the *gap* mutation into the *resE sre-1* mutant had no effect on *resA* expression. However, two more probable *gap* genes were identified in the recently published sequence of the *B. subtilis* genome (42), stressing the need for a multiple *gap* null mutant to determine the role of 1,3-diphosphoglycerate in ResD phosphorylation. The role of low molecular-weight phospho-donors in response to regulator phosphorylation in vivo is unclear and sometimes is influenced by other factors. It was shown that activation of PhoB by acetyl phosphate, which is independent of the cognate kinase PhoR, requires another sensor kinase EnvZ under certain conditions (40). Even if 1,3-diphosphoglycerate is indeed responsible for phosphorylation of ResD, future studies are necessary to answer some important questions. First, does ResE-independent phosphorylation require another kinase or is it kinase-independent? Second, is 1,3-diphosphoglycerate the donor of phosphate for ResD or is it the signal for activation of ResD by another kinase? Third, if the effect of the *sre-1* mutation is not caused by accumulation of 1,3-diphosphoglycerate, how might Pgk protein function in *resA* regulation? The answers to these questions could shed light on the mechanism of ResD-ResE signal transduction.

What Is the Specific Stimulus Affecting ResE Activity?

Two-component signal transduction systems are known to be widespread in bacteria, and a considerable body of information is now available that has provided insights into their mechanisms of action (54, 79). It is apparent that each system has a specific stimulus, either environmental or intracellular, to which a sensor kinase responds, leading to autophosphorylation. However, these signals are often difficult to identify. How sensor-signal interaction results in heightened autokinase activity is also poorly understood.

Hydrophobicity profiles of the ResE protein primary structure revealed two transmembrane regions near the amino terminus that are separated by a "periplasmic" region of around 142 amino acids. The periplasmic domain of many sensor kinases has been shown to be essential for signal response. However, the presence of transmembrane domains in sensor kinases does not always mean that the kinase senses an extracellular signal. For example, the membrane-bound hemoprotein kinase FixL responds to oxygen limitation. FixL and the cognate response regulator FixJ are required for activation of genes involved in nitrogen fixation. A truncated FixL of *Rhizobium meliloti* that lacks transmembrane domains still retains oxygen-regulated activity (46). ResE also contains an elongated cytoplasmic region that lies between the second transmembrane domain and the histidine kinase domain. The extended cytoplasmic domain was proposed to be an internal sensing domain in *E. coli* PhoR for phosphate regulation (67). At present, it is not known whether ResE senses an extracellular or an intracellular signal. Involvement of the ResD-ResE two-component signal transduction system in both aerobic and anaerobic respiration in *B. subtilis* is different from the control of respiration within *E. coli*, wherein genes required for aerobic and anaerobic respiration are often regulated by the opposing activities of FNR and the ArcA-ArcB two-component regulatory proteins, but sometimes by both regulators (28, 36, 76). This implies that the stimulus sensed by ResE is not oxygen per se but a signal derived from the cell's need to activate a respiratory pathway. The decision of which respiratory pathway to utilize, aerobic or anaerobic, would be made later, depending on which electron acceptor (oxygen, nitrate, or nitrite) is available. This scenario would favor the existence of an intracellular signal. Although no conclusive results are available that provide insights regarding the nature of the signal sensed by ResE, a description of our preliminary results addressing this important question appears below.

As already mentioned, aerobic and anaerobic *resA* transcription requires ResD and ResE. We found that aerobic *resA* expression also depends on the expression of TCA cycle enzyme genes, whereas anaerobic expression is independent of the TCA cycle (MM Nakano, P Zuber, AL Sonenshein, unpublished data). Our original hypothesis explaining the effect of TCA cycle activity on *resA* expression was the following: Because NADH is mainly generated by glycolysis and the TCA cycle, NADH or $NADH/NAD^+$ ratio may be the signal to activate the ResDE system. Under anaerobic conditions where TCA cycle enzyme activities are reduced (see below), NADH may be produced by enhanced glycolytic activity, and hence, the requirement of the TCA cycle for *resA* expression could be bypassed. If the NADH/NAD$^+$ ratio is the signal, mutations in genes that encode terminal oxidases would result in derepression of *resA*, because NADH is not reoxidized as a result of the block downstream in the electron transfer pathway. Three such mutations, *qoxABCD*, *ctaCD*, and *cydABCD*, were introduced. The *qox* operon encodes aa_3-600 quinol oxidase that is predominant during vegetative growth and is necessary for regenerating oxidized menaquinone, a major electron carrier utilized in *B. subtilis* electron transfer (62). The cytochrome caa_3-605 oxidase is encoded by the *cta* genes (63, 84). The *cydABCD* operon encodes cytochrome *d* oxidase. *ctaCD* and *cydABCD* mutations had no effect on aerobic *resA* expression; in contrast, *qoxABCD* mutation resulted in higher constitutive *resA* expression either in the presence or absence of TCA cycle gene mutations (MM Nakano, P Zuber, AL Sonenshein, unpublished data). This derepression is still dependent on *resE*. To our surprise, however, the patterns of the NADH/NAD$^+$ ratio in the *qox* and *qox* TCA cycle mutants were similar to those in the TCA cycle mutants in which NADH is not accumulated in stationary growth. Further examination has shown that expression of the TCA cycle enzyme genes, especially the aconitase gene, *citB*, is repressed in the *qox* mutant, resulting in the observed defect in NADH production. These results suggest an alternative hypothesis, that reduced menaquinone but not NADH may be important for *resA* expression and for repression of the TCA cycle enzymes. Cells, when detecting an abundance of reduced menaquinone, turn off the expression of the TCA cycle enzymes in order to stop generating more NADH and oxidize menaquinone by activating terminal oxidases via the ResD-ResE regulatory pathway. Because *B. subtilis* contains only menaquinone and not ubiquinone (12), both aerobic and anaerobic respiration may be activated by ResDE through a mechanism mediated by reduction of menaquinone. This hypothesis is now being tested experimentally.

FNR Is an Anaerobic Regulator in Nitrate Respiration

fnr transcription is highly induced by anaerobiosis via the ResD-ResE signal transduction pathway, as described. In addition, anaerobiosis appears to be necessary to maintain the activity of FNR by virtue of iron-cysteine cluster. FNR activates transcription of the *narK* operon, and thus expression of *fnr* located downstream of the *narK* gene, probably by direct binding to a putative FNR consensus sequence in the *narK* promoter region (14). The FNR binding sites (TGTGAN$_6$TCACA) were found in the regulatory regions of five anaerobically induced genes/operons in *B. subtilis*. These are the *narK* and *narG* operons, *nirC*, *ywiC*, and *ywiD*. *ywiC* and *ywiD* are located between the *narK* and *narG* operons and are divergently transcribed. The two genes are likely to share an FNR-binding site, but their functions in anaerobiosis are unknown at present.

The expression of *narK*, *narG*, *nirC*, and *ywiD* is strongly activated by anaero-biosis, while *ywiC* expression is only weakly induced (15). The FNR-binding sequence is centered at position −41.5 relative to the transcription start sites of these genes except for *ywiC*, which has the FNR box at position −61.5 (14, 15). Direct binding of FNR to the putative FNR-binding sequence has not yet been demonstrated.

Nitrite-Dependent Induction of hmp Expression

Another type of anaerobic regulation has been demonstrated for the flavohe-moglobin gene *hmp* (43). The *hmp* gene was isolated as an anaerobically indu-ced gene among members of an SPβ phage-borne library containing *B. subtilis* chromosomal DNA fused to a promoterless *lacZ* gene. The *hmp* gene product is a member of a family of flavohemoglobins that contain heme and flavin binding domains (2). Flavohemoglobin genes have been isolated from many organisms including *E. coli* (85), *Alcaligenes eutrophus* (13), *Erwinia chrysanthemi* (21), and *Saccharomyces cerevisiae* (92). The function of this ubiquitous protein is uncertain. Although *B. subtilis hmp* transcription is dramatically induced by oxygen limitation, *hmp* mutations have no significant effect on anaerobic growth (43).

Anaerobic *hmp-lacZ* expression was reduced 15-fold in *resDE* mutants as compared to wild-type and was severely repressed in *fnr* and *narGHJI* mu-tants. The *cis*-regulatory region required for anaerobic induction of *hmp* does not contain a consensus FNR-binding site, suggesting that the requirement of *fnr* for *hmp* transcription is probably indirect. In fact, addition of nitrite can bypass the requirements of FNR and NarGHJI for *hmp* expression; however, the expression still depends on ResDE. This result was similar to that obtained when the involvement of FNR in anaerobic *nasD* transcription was examined, demonstrating that the function of FNR and nitrate reductase in anaerobic in-duction of *hmp* is to produce nitrite (FNR is necessary to activate *narGHJI* transcription), which may act as a transcriptional co-inducer (43).

Deletion analysis of the *hmp* promoter region was performed to localize the *cis*-acting site required for anaerobic induction of *hmp* (MM Nakano, M LaCelle, P Zuber, unpublished information). Deletion to −67 relative to the transcription start site had no significant effect on anaerobic *hmp* induction (200-fold induction), but deletion to −60 caused a significant reduction in the anaerobic induction (40-fold induction). Induction was further reduced (8-fold induction) by deletion to −49. This indicates that the region between −67 and −49 is mainly important for anaerobic induction of *hmp*. No character-istic sequence features such as dyad symmetries or tandem repeats are found in this region, and this region shows no apparent similarity to the regulatory regions of the *fnr* and *nasDEF* promoters that are also anaerobically induced by ResDE. In *E. coli*, dual response regulators (NarL and NarP) and dual sensor kinases (NarX and NarQ) control nitrate- and nitrite-regulated gene expression (78). Anaerobic expression of the *E. coli hmp* gene was shown to be stimulated 6-fold by nitrate and 25-fold by nitrite. Interestingly, the induction of *E. coli hmp* by nitrate/nitrite was unaffected by *narL* and/or *narP* mutations, indicating an alternative pathway of anaerobic nitrate/nitrite regulation (57). Future investigation is required to identify the *cis*-element and *trans*-factors involved in anaerobic induction in response to nitrite.

FERMENTATION

In the absence of external electron acceptors, some bacteria can grow by fermentation, wherein energy is generated by substrate-level phosphorylation and pyruvate is used to reoxidize NADH, thereby allowing glycolysis to continue. Because a respiratory electron transfer chain linked to terminal electron acceptors is absent in fermentation, recycling of NADH is accomplished by conversion of pyruvate to fermentation products. The fermentation pathways in other bacteria are described fully in several reviews (10).

B. subtilis lacks or has a very inefficient glucose fermentation pathway (49, 66, 68); however, it grows anaerobically by fermentation either when both glucose and pyruvate are provided or when glucose and mixtures of amino acids are present (49). In contrast, *E. coli* can ferment either glucose or pyruvate. One of the main reasons *B. subtilis* needs both glucose and pyruvate to grow by fermentation might be the apparent lack of pyruvate formate lyase (PFL). In *E. coli*, synthesis of pyruvate dehydrogenase (PDH) complex, which catalyzes conversion of pyruvate to acetyl-CoA during aerobic growth, is repressed during fermentative growth (70, 90) and residual PDH activity is inhibited by NADH (29). Instead, formation of acetyl-CoA (and formate) from pyruvate is catalyzed by PFL. No NADH is generated in this reaction, thus a redox balance is maintained. On the other hand, three lines of evidence show that PDH functions in the conversion of pyruvate during fermentation in *B. subtilis* (49). First, an *ace* mutant that lacks PDH activity (24, 32) cannot grow by fermentation. Second, in vivo NMR scans of fermentative cultures indicate that *B. subtilis* does not produce formate, one of the products of the PFL-catalyzed reaction, as one of its fermentation products. Finally, the *ace* mutant produces little if any fermentation products, while the wild-type *B. subtilis* produces ethanol, acetate, lactate, small amounts of acetoin, and 2,3-butanediol. [While we have been preparing this review, the completion of *B. subtilis* genome sequence was reported (42) and no *pfl* gene was found in the *B. subtilis* genome.] These results indicate that *B. subtilis* carries out a mixed acid/butanediol fermentation pathway in which PDH plays a key role (49).

Most of the genes involved in nitrate respiration were shown to be dispensable for fermentation (49). These include *fnr* and those required for nitrate reductase activity, suggesting that genes for enzymes involved in fermentation are induced independently of FNR. In fact, transcription of *lctE*, which encodes L-lactate dehydrogenase, was activated by anaerobiosis in an FNR-independent manner and this induction was abolished in the presence of nitrate (15). *resD* mutations showed moderate effects on fermentation. A search for genes essential for fermentation in *B. subtilis* uncovered the *ftsH* gene. *ftsH* mutations also affect nitrate respiration (49). The *ftsH* gene encodes an integral membrane protein with a putative ATP-binding domain and an amino-acid sequence similar to an active-site motif of zinc metalloproteases (81, 82). *ftsH* is an essential gene in *E. coli*, while the *B. subtilis ftsH* gene is not essential for survival under aerobic conditions but is required for entry into sporulation, for competence development, for survival after heat and osmotic shock, and for secretion of exoproteins (18, 19, 47). How *ftsH* mutations affect these diverse cellular functions, including anaerobiosis, remains to be determined.

METABOLISM DURING ANAEROBIC GROWTH

The changes in metabolism accompanying a shift from an aerobic environment to one of oxygen limitation have yet to be characterized in any detail. A few studies relevant to this issue are summarized below.

Pyruvate Metabolism

Studies in *E. coli* showed that the overall rate of glycolysis is expected to be higher under anaerobic than aerobic conditions, in view of the fact that some glycolytic enzymes are induced by anaerobiosis (especially under conditions that favor fermentation) (72). The most striking difference between glycolytic pathways of anaerobically and aerobically grown cells is the fate of pyruvate. In *E. coli*, either PFL or PDH can be used to catabolize pyruvate in nitrate respiration (38), while in fermentation, as described in the previous section, pyruvate is metabolized to formate and acetyl-CoA by PFL. In contrast, *B. subtilis* apparently uses PDH for conversion of pyruvate during anaerobic growth under fermentation as well as during nitrate respiration (49).

Krebs Cycle

Studies of facultative anaerobes have demonstrated that other major metabolic changes in response to oxygen level are changes in the direction and the activity of the Krebs cycle. The rationale for these changes relates to the three major metabolic functions of the Krebs cycle: (*a*) supplying biosynthetic precursors (α-ketoglutarate, succinyl CoA, and oxaloacetate), (*b*) generating energy, and (*c*) generating reducing power. Considering that carbon and electron flow—as well as energy generation—are altered by shifts between aerobiosis and anaerobiosis, the production of the Krebs cycle enzymes is also regulated in response to oxygen availability. When *E. coli* grows under anaerobic conditions, especially during fermentative growth in which NADH cannot be reoxidized by the respiratory chain, the α-ketoglutarate dehydrogenase complex is strongly repressed and succinate dehydrogenase is replaced by fumarate reductase. The net effect of these changes is that the Krebs cycle is transformed into two opposingly oriented half-cycles. Additionally, Krebs-cycle enzyme activities are reduced to the minimal level necessary to supply biosynthetic precursors (28, 36, 76).

Citrate synthase and aconitase activities in anaerobically grown *B. subtilis* in the presence of nitrate are reduced 10-fold and 30-fold, respectively, from levels in aerobic cultures (MM Nakano, P Zuber, AL Sonenshein, unpublished data). But isocitrate dehydrogenase activity showed only a twofold reduction as compared to the observed activity in aerobic growth. This reduction of the Krebs cycle enzyme activities is mainly the result of reduced transcription of the Krebs cycle enzyme genes, regulation that does not involve ResDE and FNR. Previous studies identified a dyad symmetry sequence centered at position -66 relative to the transcription start site of *citB* (aconitase gene) as a target of carbon catabolite repression (22, 23). A deletion of the upstream arm of the dyad symmetry element abolished anaerobic repression of *citB*, indicating that catabolite and anaerobic repression of *citB* are regulated by a single mechanism that does not involve FNR and ResDE (MM Nakano, P Zuber, AL Sonenshein, unpublished data).

Components in the Electron Transfer Pathway

In *E. coli*, ubiquinone is the intermediate electron carrier during respiration using oxygen or nitrate, whereas menaquinone is used during anaerobic respiration with other terminal electron acceptors. As mentioned above, *B. subtilis* possesses only menaquinone, which may explain the unique regulation of respiration in *B. subtilis*. Molecular genetic and biochemical studies of cytochromes during anaerobic respiration in *B. subtilis* need to be studied.

Phospholipids

The phospholipids of *B. subtilis* consist largely of phosphatidylethanolamine and phosphatidylglycerol. The remaining species are mainly the lysine ester of phosphatidylglycerol and a small amount of cardiolipin (diphosphatidylglycerol) (16). Cardiolipin synthesized from phosphatidylglycerol has been shown to accumulate under anaerobiosis mainly because its degradation is inhibited (59). The significance of this feature for anaerobiosis is unknown.

OTHER FACTORS

It is known that some environmental signals induce changes in chromosomal DNA supercoiling, which in turn influences the expression of certain classes of genes. Conditions that affect DNA supercoiling include anaerobiosis, high osmolarity, temperature, nutrient shifts, and growth stage (20, 31). Anaerobiosis results in an increase of the plasmid linking number in *B. subtilis*, whereas in *E. coli* it leads to a decrease in linking number (41). Involvement of DNA supercoiling in anaerobic gene expression in *B. subtilis* has not been reported at the time of this writing.

Aerotaxis, the migratory response toward or away from oxygen, was reported in *B. subtilis* (48, 88) but is not covered in this review.

CONCLUDING REMARKS

This review has summarized recent studies on anaerobiosis in *B. subtilis* and emphasized some of its unique characteristics. These include the Res system required for both aerobic and anaerobic respiration, a nitrite reductase that appears to function in both nitrogen assimilation and respiration, and the central role of PDH complex in fermentation. Many genes have been identified that are involved in anaerobiosis of *B. subtilis*. Some genes that are constitutively expressed or induced by oxygen limitation encode regulatory proteins that are necessary for the appropriate expression of anaerobiosis-specific loci. Another class of genes that are highly induced under oxygen-limiting conditions are required more directly for anaerobic metabolism or energy generation. The natural habitat of *B. subtilis* is soil, in which several critical factors influence oxygen availability. The overall aeration of soil is not as important as the oxygen concentration within individual microenvironments (55). The number of anaerobic bacteria in the upper few centimeters of soil can be 10 times higher than that at greater depth as a result of consumption of oxygen by growing aerobes. Soil water is another important factor in determining whether a microenvironment promotes aerobic or anaerobic metabolism. A water-saturated soil aggregate larger than 3 mm in radius has no oxygen within its interior (55).

An increase in water content in soil may also cause other changes, such as osmolarity, pH, and the nature and amount of soluble nutrition. This suggests that some anaerobically regulated genes are likely interwoven in regulatory cascades governing the expression of other stress-induced genes. At present, no studies have been reported to show coordinate regulation of anaerobically induced genes with those induced upon stress, with the exception of our finding that *ftsH* mutations cause pleiotropic phenotypes, including a defect in anaerobiosis. The widely studied processes associated with stationary phase phenomena such as sporulation, competence development, and antibiotic and degradative enzyme production have not been studied under conditions promoting anaerobic growth (anaerobic sporulation occurs at highly reduced frequency under conditions thus far examined). Reexamination of such well-characterized pathways in anaerobically grown *B. subtilis* will contribute to the understanding of anaerobiosis as well as that of stationary phase events.

Addition in Proof

Our preliminary results of DNase I protection analysis showed that the phosphorylated form of ResD protected sequences within the promoter regions of *nasD* and *hmp* from DNase I cleavage. No discrete regions of protection in the *fnr* promoter were observed (MM Nakano, X Zhang, FM Hulett, unpublished information). This result suggests that ResD directly binds to putative consensus sequences in the *nasD* and *hmp* regulatory regions to activate transcription. In contrast, *fnr* transcription might be activated by another regulator, the expression of which is dependent upon ResD. Alternatively, binding of the regulator to the dyad symmetry sequence described above recruits ResD to the promoter region.

ACKNOWLEDGMENTS

We thank Joan McDermott and Mitsuo Ogura for critical reading of the manuscript. Research at Louisiana State University is supported by the National Institutes of Health grant GM45898 and the National Science Foundation grant MCB9722885.

Visit the *Annual Reviews home page* at
http://www.AnnualReviews.org.

Literature Cited

1. Alibadi Z, Park YK, Slonczewski JL, Foster JW. 1988. Novel regulatory loci controlling oxygen- and pH-regulated gene expression in *Salmonella typhimurium. J. Bacteriol.* 170:842–51
2. Andrews SC, Shipley D, Keen JN, Findlay JBC, Harrison PM, Guest JR. 1992. The haemoglobin-like protein (HMP) of *Escherichia coli* has ferrisiderophore reductase activity and its C-terminal domain shares homology with ferredoxin NADP$^+$ reductases. *FEBS Lett.* 302:247–52
3. Anthamatten D, Scherb B, Hennecke H. 1992. Characterization of a *fixLJ*-regulated *Bradyrhizobium japonicum* gene sharing similarity with the *Escherichia coli fnr* and *Rhizobium meliloti fixK*

genes. *J. Bacteriol.* 174:2111–20
4. Bell AI, Gaston KL, Cole JA, Busby SJW. 1989. Cloning of binding sequences for the *Escherichia coli* transcription activators, FNR and CRP: location of bases involved in discrimination between FNR and CRP. *Nucl. Acids Res.* 17:3865–74
5. Blasco F, Iobbi C, Giordano G, Chippaux M, Bonnefoy V. 1989. Nitrate reductase of *Escherichia coli*: completion of the nucleotide sequence of the *nar* operon and reassessment of the role of the α and β subunits in iron binding and electron transfer. *Mol. Gen. Genet.* 218:249–56
6. Blasco F, Iobbi C, Ratouchniak J, Bonnefoy V, Chippaux M. 1990. Nitrate reductases of *Escherichia coli*: sequence of

the second nitrate reductase and comparison with that encoded by the *narGHJI* operon. *Mol. Gen. Genet.* 222:104–11

7. Bohin J-P, Bohin A, Schaeffer P. 1976. Increased nitrate reductase A activity as a sign of membrane alteration in early blocked asporogenous mutants of *Bacillus subtilis*. *Biochimie* 58:99–108

8. Choe M, Reznikoff WS. 1991. Anaerobically expressed *Escherichia coli* genes identified by operon fusion techniques. *J. Bacteriol.* 173:6139–46

9. Clark DP. 1984. The number of anaerobically regulated genes in *Escherichia coli*. *FEMS Microbiol. Lett.* 24:251–54

10. Clark DP. 1989. The fermentation pathways of *Escherichia coli*. *FEMS Microbiol. Rev.* 63:223–34

11. Cole J. 1996. Nitrate reduction to ammonia by enteric bacteria: redundancy, or a strategy for survival during oxygen starvation? *FEMS Microbiol. Lett.* 136:1–11

12. Collins MD, Jones D. 1981. Distribution of isoprenoid quinone structural types in bacteria and their taxonomic implication. *Microbiol. Rev.* 45:316–54

13. Cramm R, Siddiqui RA, Friedrich B. 1994. Primary sequence and evidence for a physiological function of the flavohemoprotein of *Alcaligenes eutrophus*. *J. Biol. Chem.* 269:7349–54

14. Cruz Ramos H, Boursier L, Moszer I, Kunst F, Danchin A, Glaser P. 1995. Anaerobic transcription activation in *Bacillus subtilis*: identification of distinct FNR-dependent and -independent regulatory mechanisms. *EMBO J.* 14:5984–94

15. Cruz Ramos H, Nakano MM, Danchin A, Glaser P. 1997. *Three different mechanisms for anaerobic transcription activation in* Bacillus subtilis. Presented at Int. Conf. Bacilli, 9th, Lausanne

16. de Mendoza D, Grau R, Cronan JE Jr. 1993. Biosynthesis and function of membrane lipids. In *Bacillus Subtilis and Other Gram-Positive Bacteria: Biochemistry, Physiology, and Molecular Genetics*, ed. AL Sonenshein, JA Hoch, R Losick, pp. 411–21. Washington, DC: Am. Soc. Microbiol.

17. DeMoss JA, Hsu P-Y. 1991. NarK enhances nitrate uptake and nitrite excretion in *Escherichia coli*. *J. Bacteriol.* 173:3303–10

18. Deuerling E, Mogk A, Richter C, Purucker M, Schumann W. 1997. The *ftsH* gene of *Bacillus subtilis* is involved in major cellular processes such as sporulation, stress adaptation and secretion. *Mol. Microbiol.* 23:921–33

19. Deuerling E, Paeslack B, Schumann W. 1995. The *ftsH* gene of *Bacillus subtilis* is transiently induced after osmotic and temperature upshift. *J. Bacteriol.* 177:4105–12

20. Drlica K. 1992. Control of bacterial DNA supercoiling. *Mol. Microbiol.* 6:425–33

21. Favey S, Labesse G, Vouille V, Boccara M. 1995. Flavohaemoglobin HmpX: a new pathogenicity determinant in *Erwinia chrysanthemi* strain 3937. *Microbiology* 141:863–71

22. Fouet A, Jin S, Raffel G, Sonenshein AL. 1990. Multiple regulatory sites in the *Bacillus subtilis citB* promoter region. *J. Bacteriol.* 172:5408–15

23. Fouet A, Sonenshein AL. 1990. A target for carbon source-dependent negative regulation of the *citB* promoter of *Bacillus subtilis*. *J. Bacteriol.* 172:835–44

24. Freese E, Fortnagel U. 1969. Growth and sporulation of *Bacillus subtilis* mutants blocked in the pyruvate dehydrogenase complex. *J. Bacteriol.* 99:745–56

25. Glaser P, Danchin A, Kunst F, Zuber P, Nakano MM. 1995. Identification and isolation of a gene required for nitrate assimilation and anaerobic growth of *Bacillus subtilis*. *J. Bacteriol.* 177:1112–15

26. Green J, Sharrocks AD, Green B, Geisow M, Guest JR. 1993. Properties of FNR proteins substituted at each of the five cysteine residues. *Mol. Microbiol.* 8:61–68

27. Grove J, Tanapongpipat S, Thomas G, Griffiths L, Crooke H, Cole J. 1995. *Escherichia coli* K-12 genes essential for the synthesis of *c*-type cytochromes and a third nitrate reductase located in the periplasm. *Mol. Microbiol.* 19:467–81

28. Gunsalus RP, Park S-J. 1994. Aerobic-anaerobic gene regulation in *Escherichia coli*: control by the ArcAB and Fnr regulons. *Res. Microbiol.* 145:437–50

29. Hansen RG, Henning U. 1966. Regulation of pyruvate dehydrogenase activity in *Escherichia coli* K12. *Biochim. Biophys. Acta* 122:355–58

30. Harborne NR, Griffiths L, Busby SJW, Cole JA. 1992. Transcriptional control, translation and function of the products of the five open reading frames of the *Escherichia coli nir* operon. *Mol. Microbiol.* 6:2805–13

31. Higgins CF, Dorman CJ, Ni Bhriain N. 1989. Environmental influences on DNA supercoiling: a novel mechanism for the regulation of gene expression. In *The Bacterial Chromosome*, ed. K Drlica, M Riley, pp. 421–32. Washington, DC: Am. Soc. Microbiol.

32. Hodgson JA, Lowe PN, Perham RN.

1983. Wild-type and mutant forms of the pyruvate dehydrogenase multienzyme complex from *Bacillus subtilis*. *Biochem. J.* 211:463–72

33. Hoffmann T, Frankenberg N, Marino M, Jahn D. 1998. Ammonification in *Bacillus subtilis* utilizing dissimilatory nitrite reductase is dependent on *resDE*. *J. Bacteriol.* 180:186–89

34. Hoffmann T, Troup B, Szabo A, Hungerer C, Jahn D. 1995. The anaerobic life of *Bacillus subtilis*: cloning of the genes encoding the respiratory nitrate reductase system. *FEMS Microbiol. Lett.* 131:219–25

35. Hussain H, Grove J, Griffiths L, Busby S, Cole J. 1994. A seven-gene operon essential for formate-dependent nitrite reduction to ammonia by enteric bacteria. *Mol. Microbiol.* 12:153–63

36. Iuchi S, Lin ECC. 1993. Adaptation of *Escherichia coli* to redox environments by gene expression. *Mol. Microbiol.* 9:9–15

37. Iuchi S, Weiner L. 1996. Cellular and molecular physiology of *Escherichia coli* in the adaptation to aerobic environments. *J. Biochem.* 120:1055–63

38. Kaiser M, Sawers G. 1994. Pyruvate formate-lyase is not essential for nitrate respiration by *Escherichia coli*. *FEMS Microbiol. Lett.* 117:163–68

39. Khoroshilova N, Beinert H, Kiley PJ. 1995. Association of a polynuclear iron-sulfur center with a mutant FNR protein enhances DNA binding. *Proc. Natl. Acad. Sci. USA* 92:2499–503

40. Kim SK, Wilmes-Riesenberg MR, Wanner BL. 1996. Involvement of the sensor kinase EnvZ in the in vivo activation of the response regulator PhoB by acetyl phosphate. *Mol. Microbiol.* 22:135–47

41. Krispin O, Allmansberger R. 1995. Changes in DNA supertwist as a response of *Bacillus subtilis* towards different kinds of stress. *FEMS Microbiol. Lett.* 134:129–35

42. Kunst F, Ogasawara N, Moszer I, Albertini AM, Alloni G, et al. 1997. The complete genome sequence of the Gram-positive bacterium *Bacillus subtilis*. *Nature* 390:249–56

43. LaCelle M, Kumano M, Kurita K, Yamane K, Zuber P, Nakano MM. 1996. Oxygen-controlled regulation of flavohemoglobin gene in *Bacillus subtilis*. *J. Bacteriol.* 178:3803–8

44. Lazazzera BA, Bates DM, Kiley PJ. 1993. The activity of the *Escherichia coli* transcription factor FNR is regulated by a change in oligomeric state. *Genes Dev.* 7:1993–2005

45. Lin ECC, Iuchi S. 1991. Regulation of gene expression in fermentative and respiratory systems in *Escherichia coli* and related bacteria. *Annu. Rev. Genet.* 25:361–87

46. Lois AF, Weinstein M, Ditta GS, Helinski DR. 1993. Autophosphorylation and phosphatase activities of the oxygen-sensing protein FixL of *Rhizobium meliloti* are coordinately regulated by oxygen. *J. Biol. Chem.* 268:4370–75

47. Lysenko E, Ogura T, Cutting SM. 1997. Characterization of the *ftsH* gene of *Bacillus subtilis*. *Microbiology* 143:971–78

47a. Michel JF, Cami B, Schaeffer P. 1968. Sélection de mutants de *Bacillus subtilis* bloqués au début de la sporulation: I. Mutants asporogènes pléotropes sélectionnés par croissance en milieu au nitrate. *Ann. Inst. Pasteur* 114:11–20

48. Miller JB, Koshland DE Jr. 1977. Sensory electrophysiology of bacteria: relationship of the membrane potential to motility and chemotaxis in *Bacillus subtilis*. *Proc. Natl. Acad. Sci. USA* 74:4752–56

49. Nakano MM, Dailly YP, Zuber P, Clark DP. 1997. Characterization of anaerobic fermentative growth in *Bacillus subtilis*: identification of fermentation end products and genes required for the growth. *J. Bacteriol.* 179:6749–55

50. Nakano MM, Yang F, Hardin P, Zuber P. 1995. Nitrogen regulation of *nasA* and the *nasB* operon, which encode genes required for nitrate assimilation in *Bacillus subtilis*. *J. Bacteriol.* 177:573–79

51. Nakano MM, Zuber P, Glaser P, Danchin A, Hulett FM. 1996: Two-component regulatory proteins ResD-ResE are required for transcriptional activation of *fnr* upon oxygen limitation in *Bacillus subtilis*. *J. Bacteriol.* 178:3796–802

52. Ogawa K, Akagawa E, Yamane K, Sun Z-W, LaCelle M, et al. 1995. The *nasB* operon and *nasA* gene are required for nitrate/nitrite assimilation in *Bacillus subtilis*. *J. Bacteriol.* 177:1409–13

53. Page L, Griffiths L, Cole JA. 1990. Different physiological roles of two independent pathways for nitrite reduction to ammonia by enteric bacteria. *Arch. Microbiol.* 154:349–54

54. Parkinson JS. 1995. Genetic approaches for signaling pathways and proteins. In *Two-Component Signal Transduction*, ed. JA Hoch, TJ Silhavy, pp. 9–23. Washington, DC: Am. Soc. Microbiol.

55. Paul EA, Clark FE. 1989. *Soil Microbiology and Biochemistry*. San Diego: Academic

56. Peakman T, Crouzet J, Mayaux JF, Busby S, Mohan S, et al. 1990. Nucleotide sequence, organisation and structural analysis of the products of genes in the *nirB-cysG* region of the *Escherichia coli* K-12 chromosome. *Eur. J. Biochem.* 191:315–23

57. Poole RK, Anjum MF, Membrillo-Hernández J, Kim SO, Hughes MN, Stewart V. 1996. Nitric oxide, nitrite, and Fnr regulation of *hmp* (flavohemoglobin) gene expression in *Escherichia coli* K-12. *J. Bacteriol.* 178:5487–92

58. Priest FG. 1993. Systematics and ecology of *Bacillus.* See 16, pp. 3–16

59. Rigomier D, Lacombe C, Lubochinsky B. 1978. Cardiolipin metabolism in growing and sporulating *Bacillus subtilis. FEBS Lett.* 89:131–35

60. Rivers SL, McNairn E, Blasco F, Giordano G, Boxer DH. 1993. Molecular genetic analysis of the *moa* operon of *Escherichia coli* K-12 required for molybdenum cofactor biosynthesis. *Mol. Microbiol.* 8:1071–81

61. Rowe JJ, Ubbink-Kok T, Molenaar D, Konings WN, Driessen AJM. 1994. NarK is a nitrite-extrusion system involved in anaerobic nitrate respiration by *Escherichia coli. Mol. Microbiol.* 12:579–86

62. Santana M, Kunst F, Hullo MF, Rapoport G, Danchin A, Glaser P. 1992. Molecular cloning, sequencing, and physiological characterization of the *qox* operon from *Bacillus subtilis* encoding the aa_3-600 quinol oxidase. *J. Biol. Chem.* 267:10225–31

63. Saraste M, Metso T, Nakari T, Jalli T, Lauraeus M, van der Oost J. 1991. The *Bacillus subtilis* cytochrome *c* oxidase: variations on a conserved protein theme. *Eur. J. Biochem.* 195:517–25

64. Sawers G, Kaiser M, Sirko A, Freundlich M. 1997. Transcriptional activation by FNR and CRP: reciprocity of binding-site recognition. *Mol. Microbiol.* 23:835–45

65. Sawers RG, Zehelein E, Böck A. 1988. Two-dimensional gel electrophoretic analysis of *Escherichia coli* proteins: influence of various anaerobic growth conditions and the *fnr* gene product on cellular protein composition. *Arch. Microbiol.* 149:240–44

66. Schirawski J, Unden G. 1995. Anaerobic respiration of *Bacillus macerans* with fumarate, TMAO, nitrate and nitrite and regulation of the pathways by oxygen and nitrate. *Arch. Microbiol.* 163:148–54

67. Scholten M, Tommassen J. 1993. Topology of the PhoR protein of *Escherichia coli* and functional analysis of internal deletion mutants. *Mol. Microbiol.* 8:269–75

68. Shariati P, Mitchell WJ, Boyd A, Priest FG. 1995. Anaerobic metabolism in *Bacillus licheniformis* NCIB 6346. *Microbiology* 141:1117–24

69. Shaw DJ, Rice DW, Guest JR. 1983. Homology between CAP and Fnr, a regulator of anaerobic respiration in *Escherichia coli. J. Mol. Biol.* 166:241–47

70. Smith MW, Neidhardt FC. 1983. 2-Oxoacid dehydrogenase complexes of *Escherichia coli*: cellular amounts and patterns of synthesis. *J. Bacteriol.* 156:81–88

71. Smith MW, Neidhardt FC. 1983. Proteins induced by anaerobiosis in *Escherichia coli. J. Bacteriol.* 154:336–43

72. Smith MW, Neidhardt FC. 1983. Proteins induced by aerobiosis in *Escherichia coli. J. Bacteriol.* 154:344–50

73. Sorokin A, Zumstein E, Azevedo V, Ehrlich SD, Serror P. 1993. The organization of the *Bacillus subtilis* 168 chromosome region between the *spoVA* and *serA* genetic loci, based on sequence data. *Mol. Microbiol.* 10:385–95

74. Spiro S, Guest JR. 1987. Regulation and over-expression of the *fnr* gene of *Escherichia coli. J. Gen. Microbiol.* 133:3279–88

75. Spiro S, Guest JR. 1990. FNR and its role in oxygen-regulated gene expression in *Escherichia coli. FEMS Microbiol. Rev.* 75:399–428

76. Spiro S, Guest JR. 1991. Adaptive responses to oxygen limitation in *Escherichia coli. Trends Biochem.* 16:310–14

77. Spiro S, Roberts RE, Guest JR. 1989. FNR-dependent repression of the *ndh* gene of *Escherichia coli* and metal ion requirement for FNR-regulated gene expression. *Mol. Microbiol.* 3:601–8

78. Stewart V. 1993. Nitrate regulation of anaerobic respiratory gene expression in *Escherichia coli. Mol. Microbiol.* 9:425–34

79. Stock JB, Surette MG, Levit M, Park P. 1995. Two-component signal transduction systems: structure-function relationships and mechanisms of catalysis. See 54, pp. 25–51

80. Sun G, Sharkova E, Chesnut R, Birkey S, Duggan MF, et al. 1996. Regulators of aerobic and anaerobic respiration in *Bacillus subtilis. J. Bacteriol.* 178:1374–85

81. Tomoyasu T, Yamanaka K, Murata K, Suzaki T, Bouloc P, et al. 1993. Topology and subcellular localization of FtsH protein in *Escherichia coli. J. Bacteriol.* 175:1352–57

82. Tomoyasu T, Yuki T, Morimura S, Mori H, Yamanaka K, et al. 1993. The *Escherichia coli* FtsH protein is a prokaryotic member of a protein family of putative ATPases involved in membrane functions, cell cycle control, and gene expression. *J. Bacteriol.* 175:1344–51

83. Trageser M, Unden G. 1989. Role of cysteine residues and of metal ions in the regulatory functioning of FNR, the transcriptional regulator of anaerobic respiration in *Escherichia coli. Mol. Microbiol.* 3:593–99

84. van der Oost J, von Wachenfeld C, Hederstedt L, Saraste M. 1991. *Bacillus subtilis* cytochrome oxidase mutants: biochemical analysis and genetic evidence for two aa_3-type oxidases. *Mol. Microbiol.* 5:2063–72

85. Vasudevan SG, Armarego WLF, Shaw DC, Lilley PE, Dixon NE, Poole RK. 1991. Isolation and nucleotide sequence of the *hmp* gene that encodes a haemoglobin-like protein in *Escherichia coli* K-12. *Mol. Gen. Genet.* 226:49–58

86. Wanner BL, Wilmes-Riesenberg MR. 1992. Involvement of phosphotransacetylase, acetate kinase, and acetyl phosphate synthesis in control of the phosphate regulation in *Escherichia coli. J. Bacteriol.* 174:2124–30

87. Winkelman JW, Clark DP. 1996. Anaerobically induced genes of *Escherichia coli. J. Bacteriol.* 167:362–67

88. Wong LS, Johnson MS, Zhulin IB, Taylor BL. 1995. Role of methylation in aerotaxis in *Bacillus subtilis. J. Bacteriol.* 177:3985–91

89. Wray LV, Ferson AE, Rohrer K, Fisher SH. 1996. TnrA, a transcription factor required for global nitrogen regulation in *Bacillus subtilis. Proc. Natl. Acad. Sci. USA* 93:8841–45

90. Yamamoto I, Ishimoto M. 1975. Effect of nitrate reduction on the enzyme levels in carbon metabolism in *Escherichia coli. J. Biochem.* 78:307–15

91. Zhang X, Ebright RH. 1990. Substitution of 2 base pairs (1 base pair per DNA half-site) within the *Escherichia coli lac* promoter DNA site for catabolite gene activator protein places the *lac* promoter in the FNR regulon. *J. Biol. Chem.* 265:12400–3

92. Zhu H, Riggs AF. 1992. Yeast flavohemoglobin is an ancient protein related to globins and a reductase family. *Proc. Natl. Acad. Sci. USA* 89:5015–19

Annu. Rev. Microbiol. 1998. 52:191–230

SOMETHING FROM ALMOST NOTHING: Carbon Dioxide Fixation in Chemoautotrophs

Jessup M. Shively,[1] Geertje van Keulen,[2] and Wim G. Meijer[3]

[1]Department of Biological Sciences, Clemson University, Clemson, South Carolina 29634; e-mail: sjessup@clemson.edu; [2]Department of Microbiology, University of Groningen, Biomolecular Sciences and Biotechnology Institute, 9750 AA Haren, The Netherlands; e-mail: G.van.Keulen@biol.rug.nl; and [3]Department of Industrial Microbiology, University College Dublin, Belfield, Dublin-4, Ireland; e-mail: wim.meijer@ucd.ie

KEY WORDS: bacteria, Calvin cycle, symbiosis, carboxysomes, gene location/organization, regulation

---

ABSTRACT

The last decade has seen significant advances in our understanding of the physiology, ecology, and molecular biology of chemoautotrophic bacteria. Many ecosystems are dependent on CO_2 fixation by either free-living or symbiotic chemoautotrophs. CO_2 fixation in the chemoautotroph occurs via the Calvin-Benson-Bassham cycle. The cycle is characterized by three unique enzymatic activities: ribulose bisphosphate carboxylase/oxygenase, phosphoribulokinase, and sedoheptulose bisphosphatase. Ribulose bisphosphate carboxylase/oxygenase is commonly found in the cytoplasm, but a number of bacteria package much of the enzyme into polyhedral organelles, the carboxysomes. The carboxysome genes are located adjacent to *cbb* genes, which are often, but not always, clustered in large operons. The availability of carbon and reduced substrates control the expression of *cbb* genes in concert with the LysR-type transcriptional regulator, CbbR. Additional regulatory proteins may also be involved. All of these, as well as related topics, are discussed in detail in this review.

CONTENTS

191

0066-4227/98/1001-0191$08.00

INTRODUCTION

Since the previous comprehensive reviews (79, 159) were published, there has been an explosion of information regarding bacterial CO_2 fixation. The vast majority of the articles referenced in these two reviews were published prior to 1990. Thus, the preparation of a new review seemed timely. In view of the mass of information, space limitations, and our own particular interest, the scope of this review has been limited to CO_2 fixation in chemoautotrophs. Other autotrophs are compared/contrasted only where deemed appropriate.

Chemoautotrophic bacteria were some of the first bacteria studied. Now, over a century later, they still retain a great deal of scientific interest. In recent years, major advances have been made in understanding CO_2 fixation in these organisms—including the physiology, biochemistry, and molecular biology of the fixation process; the regulation of, and interaction between, the Calvin cycle and central metabolic pathways; and the involvement of the specialized organelle unique among prokaryotes, the carboxysome—and a clearer understanding of the various ecological niches occupied by these organisms has been gained.

THE CALVIN CYCLE

Occurrence in Bacteria

FACULTATIVE AND OBLIGATE AUTOTROPHS   Chemoautotrophic bacteria are subdivided into two major groups: obligate chemoautotrophic bacteria, which are completely dependent on CO_2 fixation, and facultative chemoautotrophs, which assimilate CO_2 via the Calvin cycle and in addition have the ability to use a wide range of other growth substrates.

Obligate chemoautotrophs are specialists. Their metabolism is optimized for the utilization of a small number of reduced substrates (6, 75). In general these bacteria have a high affinity for reduced-growth substrates and display high growth rates compared with nonspecialists. In addition, obligate chemoautotrophs growing under nutrient limiting conditions are able to oxidize 300–400% more reduced substrate than is available in the medium. The high growth rate and respiratory overcapacity allow the specialist to rapidly respond to sudden changes in the availability of reduced substrates because the metabolic pathways required for their utilization are already present and need not be induced.

Facultatively chemoautotrophic bacteria display an enormous metabolic versatility, which allows them to grow on a wide range of substrates. In contrast to obligate autotrophs, catabolic and anabolic pathways of facultatively autotrophic bacteria are inducible. A characteristic of these bacteria is the ability to grow mixotrophically, i.e. substrates supporting autotrophic and heterotrophic growth are used simultaneously (e.g. 25).

Both specialist and facultative strategies can be advantageous, depending on the growth conditions (6). As a consequence of their metabolic characteristics, obligate and facultative chemoautotrophs have different environmental niches. The former are encountered where there is a continuous or fluctuating supply of reduced inorganic compounds with a low turnover of organic compounds, whereas the latter thrives when both inorganic reduced substrates and organic compounds are present.

AUTOTROPHIC GROWTH SUBSTRATES   Molecular hydrogen and reduced nitrogen (e.g. NH_4^+, NO_2^-), sulfur (e.g. $S_2O_3^{2-}$, H_2S), metals (e.g. Fe^{2+}, Mn^{2+}), and carbon (e.g. CO, CH_4, CH_3OH) compounds serve as electron donors for chemoautotrophic bacteria. A variety of anthropogenic, biological, and geological processes are responsible for the introduction of reduced compounds in the biosphere. These include industry and agriculture, anaerobic metabolism in sediments and animal guts, and volcanic activity (75).

Phylogenetic studies based on 16S rRNA sequences of autotrophic bacteria showed that nitrifying bacteria and sulfur and iron oxidizers are not limited to one phylogenetic group. Ammonia oxidizers belong to the β- and γ-subdivision of the proteobacteria; nitrite oxidizers are found in the α-, β-, γ-, and δ-subdivisions (163). Representatives of iron and sulfur oxidizers are present in all subdivisions of the proteobacteria and in the gram-positive division (53, 84, 108).

FREE-LIVING AUTOTROPHIC BACTERIA   Free-living autotrophic bacteria require a sufficient supply of reduced organic and inorganic compounds to serve as electron donors as well as a supply of substrates, which may serve as electron acceptors. Because many reduced substrates are produced in vastly different,

sometimes harsh environments, chemoautotrophic bacteria display a stagger-
ing physiological diversity that includes acidophiles, thermophiles, and psy-
chrophiles. For example, the optimum growth temperature and pH of sulfur
reducing bacteria varies between $25°$–$72°C$ and pH 2–8.

Many reduced substrates are produced anaerobically, in sediments for example, which subsequently accumulate and diffuse to aerobic layers. Substrates
such as sulfide and ferrous iron are rapidly and spontaneously oxidized under aerobic conditions (16, 152). Chemoautotrophic bacteria generally require
oxygen as an electron acceptor and are therefore limited to the oxic layer,
where they have to compete with the chemical oxidation of reduced inorganic
compounds. Thus, it is not surprising that chemoautotrophic bacteria depending on oxygen-labile compounds as electron donors position themselves at the
oxic/anoxic interface of sediments or stratified lakes (33, 107). Because this
interface is usually highly dynamic, autotrophic bacteria are generally chemotactic toward the chemocline. A number of H_2S-oxidizing bacteria, such as
Beggiatoa and *Thiovulum* spp., form microbial mats or veils, which are surrounded by a thin, unstirred layer. This provides the bacteria with a stable
microenvironment, which offers protection against sudden changes in H_2S and
O_2 concentrations and therefore represents an adaptation to life at the interface
between H_2S and O_2 (66). Some chemoautotrophic bacteria are able to replace oxygen as terminal electron acceptors with, for example, nitrate or ferric
iron. *Thiobacillus ferrooxidans* is able to use ferric iron as an electron acceptor
during growth on elemental sulfur, although the growth yield is about twofold
lower than during aerobic growth (122). Interestingly, these alternative electron
acceptors are products of aerobic chemolithoautotrophic oxidation.

A remarkable adaptation required for the use of reduced iron is the ability
to grow at a pH between 0.5 and 4. Iron in the biosphere is usually present as
ferric iron in the form of insoluble complexes. The concentration of ferrous iron
at neutral pH is generally low because at this pH ferric iron is poorly soluble
and ferrous iron is rapidly oxidized. The solubility of ferric iron increases
and the rate of ferrous iron oxidation strongly decreases below pH 4 (152),
which explains why many iron-oxidizing bacteria are acidophiles. Iron oxidizers
growing at neutral pH, such as *Gallionella ferruginea*, are usually encountered
at the oxic/anoxic interface because of the rapid oxidation of ferrous iron in the
presence of oxygen.

SYMBIOTIC AUTOTROPHIC BACTERIA   In 1977, hydrothermal vents with an
abundance of associated fauna were discovered 2500 m below the surface of
the sea (22). Because these communities thrive in complete darkness, it was
obvious that the primary production was not dependent on photosynthesis. This
led to the discovery of symbiosis between sulfide-oxidizing chemoautotrophic

bacteria and invertebrates, in which CO_2 fixation by the Calvin cycle provides the invertebrate host with a source of organic carbon. Since then, it has been shown that this type of symbiosis is commonplace in habitats where oxygen-labile sulfide and oxygen are both present. To date, symbiotic associations involving over 200 bacterial species and hosts belonging to six different phyla have been discovered (12).

The fact that symbiosis is widespread indicates that there are distinct advantages to a symbiotic relationship. As discussed above, sulfide-oxidizing chemoautotrophic bacteria grow at the oxic/anoxic interface because of the extreme oxygen-labile nature of sulfide. The association with an invertebrate host helps sulfide-oxidizing bacteria overcome these limitations. The host will—by means of its size, movement, or burrowing—bridge the aerobic-anaerobic interface that gives the sulfide oxidizer access to both oxygen and sulfide (12). The host benefits from the presence of chemoautotrophic bacteria because they provide a source of organic substrates. In a number of cases, the symbiotic host lacks a digestive system and is completely dependent on the autotrophic activity of its symbiotic partner.

It is unclear how these symbiotic relationships originated. Phylogenetic studies showed that the symbiotic autotrophs within the bivalve families Vesicomyidae and Lucinacea are monophyletic, i.e. have a common ancestor. In addition, the available fossil record of the host and the 16S rRNA sequence data of the bacterial symbiont indicated that cospeciation of bacterial symbiont and host had occurred (27). This supports the hypothesis that symbiosis was established relatively early, with a single symbiotic ancestor from which the present symbiotic associations evolved. However, this may not be true for all symbiotic associations. Closely related chemoautotrophs were discovered in host species belonging to distantly related phyla, indicating that these bacterial symbionts have the ability to associate with more than one host (29). Conversely, symbionts of clams belonging to the bivalve genus *Solemya* are polyphyletic, indicating that symbiosis in *Solemya* originated from independent events (74). These data and the observation that symbiotic chemoautotrophic bacteria have different phylogenetic origins in the γ- or ε-subdivision of the proteobacteria indicate that many chemoautotrophic bacteria had the ability to enter a symbiotic relationship (29, 32, 74, 116, 117). Methanotrophic bacteria are also able to form a symbiotic association with marine invertebrates and, in some cases, coexist in the same cell with symbiotic chemoautotrophic bacteria (13, 28, 39). These observations indicate that current diversity in symbiotic associations results from numerous independent symbiotic events that occurred throughout evolution.

COMMUNITIES DEPENDING ON CHEMOAUTOTROPHY   Primary production in the terrestrial environment is largely dependent on photosynthesis by cyanobacteria, plants, and algae. Marine organisms at depths greater than 300 m live in constant darkness and are therefore dependent on input of organic material from surface waters. The deep sea is therefore similar to a desert environment. The discovery of hydrothermal vents and associated invertebrate communities 2500 m below the surface showed that oases exist in the ocean-desert (12, 52). Because these communities thrive in total darkness, it is unlikely that they are dependent on the influx of organic matter from surface waters. The fluid excreted by hydrothermal vents contains H_2S and smaller amounts of H_2, NH_4^+, and CH_4; the fluid is anoxic, highly reduced, with temperatures ranging from 5°–25°C for "warm vents" to 350°C for "black smokers" (63). The chemical and physical characteristics of hydrothermal fluid immediately suggested that chemoautotrophic metabolism may be the basis of the food chain in these deep-sea communities (184). Large numbers of bacteria are present in the milky-bluish waters flowing from warm vents, in microbial mats that form on virtually all areas exposed to the warm-vent plumes, and in symbiotic association with invertebrates (discussed above).

Given the variety in growth conditions of hydrothermal vents, it is hardly surprising that microbial communities associated with hydrothermal vents display a large physiological and phylogenetic diversity (110, 111, 136). However, the majority of chemoautotrophs are sulfur oxidizers because reduced-sulfur compounds are the predominant constituents of hydrothermal fluid usable as electron donors in chemosynthesis. Heterotrophic, chemoheterotrophic, and obligately chemoautotrophic sulfur-oxidizing bacteria were isolated from deep-sea hydrothermal vents of the Galapagos Rift. The latter were characterized as *Thiomicrospira* spp. and the former two as *Pseudomonas*-like and *Thiobacillus*-like bacteria, respectively. In addition, microbial mats resembling those formed by sulfur-oxidizing *Thiothrix* and *Beggiatoa* spp. were observed (136). A phylogenetic analysis, based on 16S rRNA sequences isolated from microbial mats at a hydrothermal vent system in Hawaii, showed that this community was dominated (60% of isolates) by bacteria affiliated with *Thiovulum* spp. that are in the ε-subdivision of the proteobacteria. These bacteria produce veils, which is an adaptation to the oxic/anoxic interface, as discussed earlier. Bacteria belonging to the γ- and δ-subdivisions of the proteobacteria formed the second and third largest groups (110). In contrast to the bacterial diversity of these hydrothermal vents, a mid-Atlantic Ridge hydrothermal vent community is dominated by a single bacterial species belonging to the ε-subdivision (116).

The only terrestrial ecosystem completely dependent on chemoautotrophy was recently discovered in a Romanian cave. As in the hydrothermal vents, the chemoautotrophic bacteria in this system oxidize H_2S, which is present in a stream feeding the cave system (138). The bacteria in microbial mats that form on the surface of the H_2S-rich water are predominately *Thiobacillus thioparus* (176).

Overview of the Calvin Cycle (97, 159)

Carbon dioxide fixation by the Calvin cycle is dependent on 13 enzymatic reactions (Figure 1). The enzyme responsible for the actual fixation of CO_2, ribulose-1,5-bisphosphate carboxylase/oxygenase (RuBisCO), catalyzes the carboxylation of ribulose-1,5-bisphosphate (RuBP) to form two molecules of 3-phosphoglycerate. The other enzymes of the Calvin cycle are dedicated to the regeneration of RuBP. Three glycolytic enzymes—phosphoglycerate kinase, glyceraldehyde-3-phosphate dehydrogenase, and triosephosphate isomerase—convert two molecules of 3-phosphoglycerate to glyceraldehyde-3-phosphate and dihydroxyacetone phosphate at the expense of two molecules of ATP and two of NADH. This is followed by a series of rearrangement reactions that result in the production of ribulose-5-phosphate.

The rearrangement reactions are divided into two similar metabolic units, made up of an aldolase, a phosphatase, and a transketolase (APT). The first APT unit converts glyceraldehyde-3-phosphate and dihydroxyacetone phosphate into xylulose-5-phosphate and erythrose-4-phosphate, which is followed by the formation of xylulose-5-phosphate and ribulose-5-phosphate from dihydroxyacetonephosphate and erythrose-4-phosphate by the second unit. The phosphatase used in the second APT unit [sedoheptulose bisphosphatase (SBPase)] catalyzes the dephosphorylation of SBP, which is, in contrast to the gluconeogenic fructose bisphosphatase (FBPase), an activity unique to the Calvin cycle. In principle, the pentose phosphate cycle would provide an alternative for these rearrangement reactions. In this scenario, the aldolase and phosphatase of the second APT unit are replaced by transaldolase. The main difference is that the use of a phosphatase in the aldolase/SBPase variant renders the second APT unit irreversible, whereas it is reversible in the transaldolase variant. All autotrophic organisms studied to date employ the aldolase/SBPase variant of the Calvin cycle.

The sequential reactions of the two APT units result in the formation of xylulose-5-phosphate and ribose-5-phosphate. These are converted to ribulose-5-phosphate by pentose phosphate epimerase and pentose phosphate isomerase, respectively. The final step in the regeneration of RuBP is catalyzed by another unique Calvin cycle enzyme, phosphoribulokinase (PRK), which phosphorylates ribulose-5-phosphate at the expense of an ATP. The net result of the activity of the Calvin cycle is the formation of one molecule of triose phosphate from three molecules of CO_2 at the expense of nine molecules of ATP and six of NADH.

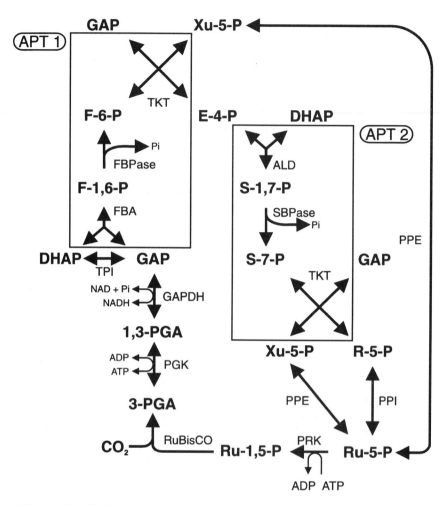

Figure 1 The Calvin cycle. Metabolites: DHAP, dihydroxyacetonephosphate; E-4-P, erythrose-4-phosphate; F-6-P, fructose-6-phosphate; F-1,6-P, fructose-1,6-bisphosphate; GAP, glyceraldehyde-3-phosphate; 3-PGA, 3-phosphoglycerate; 1,3-PGA, 1,3-diphosphoglycerate; R-5-P, ribose-5-phosphate; Ru-5-P, ribulose-5-phosphate; Ru-1,5-P, ribulose-1,5-bisphosphate; S-7-P, sedoheptulose-7-phosphate; S-1,7-P, sedoheptulose-1,7-bisphosphate; Xu-5-P, xylulose-5-phosphate; Pi, inorganic phosphate. Enzymes: ALD, aldolase; FBA, fructose-1,6-bisphosphate aldolase; FBPase, fructose-1,6-bisphosphatase; GAPDH, glyceraldehyde-3-phosphate dehydrogenase; PGK, phosphoglycerate kinase; PPE, pentose-5-phosphate epimerase; PPI, pentose-5-phosphate isomerase; PRK, phosphoribulokinase; RuBisCO, ribulose-1,5-bisphosphate carboxylase/oxygenase; SBPase, sedoheptulose bisphosphatase; TKT, transketolase; TPI, triosephosphate isomerase.

In addition to the carboxylation of RuBP, RuBisCO catalyzes the oxygenation of this substrate, resulting in the formation of phosphoglycolate and 3-phosphoglycerate. The incorporation of O_2 instead of CO_2 is an intrinsic characteristic of RuBisCO, the extent of which depends on the properties of the particular enzyme studied (65). The production of phosphoglycolate during aerobic autotrophic growth is therefore unavoidable. Phosphoglycolate itself is of little use to the organism and may even be harmful, as it is a known inhibitor of triosephosphate isomerase, one of the enzymes of the Calvin cycle (89). Chemoautotrophic bacteria dispose of phosphoglycolate by using phosphoglycolate phosphatase, which catalyzes the formation of glycolate, which is subsequently either metabolized or excreted, depending on the organism (see e.g. 21).

Enzymes (55, 153)

UNIQUE CALVIN CYCLE ENZYMES   Three enzymatic activities are unique to the Calvin cycle: RuBisCO, PRK, and SBPase. The latter are discussed in the section on Isoenzymes.

We only briefly discuss RuBisCO because recent reviews provide excellent overviews of its biochemistry, engineering, and structure-function relationships. Two types of RuBisCO are encountered in chemoautotrophic bacteria. The dominant form I enzyme, found in most chemoautotrophs, is similar to the enzyme present in plants, algae, and cyanobacteria. It is composed of large (L) (50–55 kDa) and small (S) (12–18 kDa) subunits in a hexadecameric (L_8S_8) structure. The form II enzyme, which consists of large subunits only (L_2, L_4, or L_8), was initially discovered in phototrophic purple non-sulfur bacteria, but it has since been found in certain symbiotic and free-living chemoautotrophic bacteria and, recently, in eukaryotic dinoflagellates (14, 19, 36, 60, 109, 124, 135, 155; JM Shively, unpublished data). *Thiobacillus denitrificans, Thiobacillus intermedius, Thiobacillus neapolitanus*, and *Hydrogenovibrio marinus* possess both a form I and form II RuBisCO, which, with the exception of *Rhodospirillum rubrum*, is also the case for purple non-sulfur bacteria. The bacterial form I enzymes are similar to enzymes encountered in higher plants with respect to their structure and biochemical properties. In contrast, even though the residues involved in catalysis and activation are conserved, the amino acid sequence similarity between the form I and form II large subunits is only 35%.

The efficiency of RuBisCO is commonly evaluated by determining its specificity factor, tau (τ), i.e. the enzyme's ability to discriminate between carboxylation and oxygenation (65). The form II enzymes have a characteristically low τ value between 10 and 20 (128, 160). The form I enzymes of higher plants generally have τ values above 80 (128, 168). Interestingly, very high τ values have been recently reported in several eukaryotic algae (128, 167, 168). The form I enzymes of bacteria, including chemo- and photoautotrohic bacteria and cyanobacteria, as well as green algae, are intermediate, commonly falling between 30 and 60 (56, 128, 160). A number of factors have been shown to alter specificity, including mutations resulting in alterations of both the large and small subunits, the creation of hybrid RuBisCO, and the interaction of the enzyme with other proteins (18, 126, 127, 153). The τ values for the form I and form II enzymes makes it understandable why *H. marinus* predominantly synthesizes form II RuBisCO when grown in an atmosphere containing 10% CO_2, whereas form I RuBisCO is the dominant enzyme during growth in the presence of 2% CO_2 (60). A similar differential regulation of form I and form II RuBisCO was observed earlier in the phototroph *Rhodobacter sphaeroides* (68).

The second unique enzyme of the Calvin cycle, PRK, has received considerably less attention. PRK from *Ralstonia eutropha* and the photosynthetic bacteria, *Rhodopseudomonas acidophila* and *R. sphaeroides*, are octameric proteins with subunits of 32–36 kDa (47, 133, 150, 151, 158). In contrast to RuBisCO, bacterial PRK is not related to the dimeric enzyme (45-kDa subunits) from higher plants (73). The activity of PRK in bacteria is tightly regulated by the concentration of intracellular metabolites. With the exception of the enzyme from *T. neapolitanus* (93), activity of phosphoribulokinases from chemoautotrophic bacteria is dependent on activation by NADH (1, 70, 92, 151), which, with few exceptions, is also true for the PRK from phototrophic purple bacteria (132, 133, 158). Phosphoenolpyruvate and AMP are common inhibitors of PRK (1, 4, 46, 91, 93). The regulation of PRK activity by NADH and AMP can be rationalized as a response of the enzyme to the redox and energy state of the cell. The concentration of phosphoenolpyruvate may reflect the amount of carbon available. In addition to allosteric regulation, the activity of PRK of *R. eutropha* is apparently regulated by an additional mechanism that may involve covalent modification of the enzyme. Addition of pyruvate to a culture growing autotrophically resulted in the rapid inactivation of PRK, which was paralleled by a decrease in CO_2 fixation by whole cells. Because the activity of RuBisCO remained high, it is likely that flux control of the Calvin cycle is exerted at the level of PRK (86, 87).

ISOENZYMES   The most obvious change in the metabolism of facultatively autotrophic bacteria following the transition from heterotrophic to autotrophic growth is the induction of RuBisCO and PRK, which are unique to the Calvin cycle. In addition, the activity of Calvin cycle enzymes other than RuBisCO and PRK increase dramatically following the transition to autotrophic growth (102). Mutants with strongly reduced activities in phosphoglycerate kinase and triosephosphate isomerase have been isolated that only had marginally reduced growth rates while growing on heterotrophic growth substrates such as succinate and gluconate. In contrast, neither mutant was able to grow autotrophically, despite the fact that the other Calvin cycle enzymes were fully induced (99, 103). The sharp increase in activity of enzymes other than RuBisCO and PRK has therefore been interpreted as a reflection of the increased carbon flux required to assimilate sufficient CO_2 to allow autotrophic growth to proceed.

In addition to the significant increase in activity of Calvin cycle enzymes other than RuBisCO and PRK, biochemical and genetic analysis have identified isoenzyme forms of some of these enzymes. The first to be discovered was FBPase. Johnson & MacElroy (64) noted that in contrast to the gluconeogenic enzyme from heterotrophic organisms, the FBPase of *T. neapolitanus* was not stimulated by acetyl coenzyme A but was inhibited by phosphoenolpyruvate. It was later shown that *Thiobacillus versutus*, *Nocardia opaca*, and *Xanthobacter flavus* contained different FBPase activities following heterotrophic and autotrophic growth that were catalyzed by two different proteins (2, 105, 172, 185). The subsequent characterization of the FBPase proteins from *N. opaca* and *X. flavus* showed that the enzyme induced during autotrophic growth was twice as active with sedoheptulose bisphosphate as the substrate in comparison to fructose bisphosphate. The constitutive FBPase of *N. opaca* and *X. flavus* dephosphorylated both substrates in a 1:4.5 and 1:1 ratio, respectively. In addition to the differences in substrate specificity, the two forms of FBPase also displayed differences in the regulation of their activity. ATP caused a twofold increase in activity of the inducible enzyme of *X. flavus*; the constitutive enzyme was unaffected. The autotrophic enzyme of *N. opaca* was inhibited by ATP and RuBP, whereas AMP and phosphoenolpyruvate had no effect. In contrast, the constitutive form was inhibited by the latter two metabolites.

Two unrelated and mechanistically distinct types of aldolase are encountered in Bacteria, Archaea, and Eukarya. Class I aldolase forms a Schiff base between the substrate and the ε-amino group of a lysine residue during catalysis, whereas class II enzymes depend on a divalent cation as electrophile in the catalytic cycle (94). Following heterotrophic growth of *X. flavus*, aldolase activity was independent of the addition of Fe^{2+}, whereas aldolase activity was stimulated 14-fold by Fe^{2+} following autotrophic growth (170). These data strongly suggest that *X. flavus* employs a constitutive class I aldolase and induces a class II aldolase during autotrophic growth.

The function of most of the isoenzymes is unknown because specific gene disruptions to study their role in metabolism have not been constructed. It is possible that these isoenzymes are merely needed to increase activity of the enzymes of the Calvin cycle, which is necessary to allow for the high rate of CO_2 fixation required to sustain autotrophic growth. Alternatively, their activity and regulation may be tailored for a role in the Calvin cycle. This certainly seems to be true for the inducible FBPase of *N. opaca* and *X. flavus*. The high SBPase activity of this enzyme, and the regulatory differences between the autotrophic and heterotrophic enzymes, indicate that the in vivo role of this enzyme is that of an SBPase, a characteristic enzyme of the Calvin cycle.

The biochemical characterization of the two FBPase forms of *X. flavus* and *N. opaca* indicates that the constitutive enzyme characterized by a high FBPase-to-SBPase ratio plays a role in the first APT unit of the Calvin cycle, whereas the inducible FBPase/SBPase catalyzes the dephosphorylation of sedoheptulose bisphosphate in the second APT unit (Figure 1). In addition to the inducible FBPase/SBPase, the *cbb* operon also frequently encodes transketolase and class II aldolase (see THE GENES AND THEIR LOCATION/ORGANIZATION, below). It is therefore likely that the latter enzymes also operate in the second APT unit.

CARBOXYSOMES

The enzymes of the Calvin cycle are located in the cytoplasm in most autotrophic bacteria. A notable exception is RuBisCO, which in a number of chemoautotrophic bacteria and all cyanobacteria is located in polyhedral inclusion bodies known as carboxysomes.

Occurrence

Inclusion bodies with polygonal profiles have been reported in all cyanobacteria examined, in a very limited number of other photosynthetic prokaryotes, and in many, but not all, chemoautotrophic bacteria (20, 81, 82, 147–149). Presumably, all polyhedral bodies in organisms utilizing CO_2 as a carbon source are carboxysomes.

Polyhedral bodies were first isolated over 25 years ago from *T. neapolitanus*, shown to contain the enzyme ribulose-1,5-bisphosphate carboxylase/oxygenase, and named carboxysomes (144). Since then, carboxysomes have been isolated from a number of organisms, including *Chlorogloeopsis fritschii, Nitrobacter agilis, Nitrobacter hamburgensis, Nitrobacter winogradskyi, Nitrosomonas* sp., *Synechococcus* PCC7942, and *Thiobacillus thyasiris* (10, 20, 30, 57, 58, 81, 82, 120, 147–149). The presence of carboxysomes in chemoautotrophs appears to be limited either to the obligate autotrophs or to facultative types that grow equally well both autotrophically and heterotrophically. Those chemoautotrophs that exhibit high heterotrophic growth yields have little need for the selective advantage gained by possessing carboxysomes.

Structure

In thin section, the carboxysomes of chemoautotrophs are most commonly regular hexagons about 120 nm in diameter, exhibiting granular substructure, and surrounded by a shell 3–4 nm across (57, 115, 145, 146). This size, 120 nm, would also be acceptable for the majority of carboxysomes, but much larger inclusions with polygonal profiles have been reported in some cyanobacteria (20, 147). In rare instances the bodies appear elongated in one dimension (145).

To date, the question of whether RuBisCO fills the carboxysome completely or just lines the inner surface of the carboxysome shell has not been resolved. The appearance of a uniform, granular substructure in a multitude of thin sections of carboxysomes within cells suggests to most researchers that the inclusions are filled with RuBisCO. Immunogold labeling of numerous thin sections of cyanobacterial carboxysomes is even more supportive (98, 157). However, assuming a shell thickness of 3.5 nm, a RuBisCO diameter of 11 nm, and a 60- to 80-nm average thickness for the thin sections, one might argue that obtaining an exact center section would be an unlikely event. Furthermore, the appearance of this center section within the cell might be missed unless one is specifically searching for that particular image. The arrangement of RuBisCO in rows inside of isolated carboxysomes of *T. neapolitanus* as observed in negative stained preparations prompted Holthuijzen and coworkers (57) to hypothesize that RuBisCO lines the inner surface of the shell as a monolayer. In their opinion, this hypothesis was strongly supported by the discovery of a RuBisCO-to-shell peptides ratio of 1:1 and by the extremely tight association of the small subunit of RuBisCO with the shell in broken carboxysomes (57, 58). These authors did not speculate on the possible contents or function, if any, of the resulting center cavity.

Examination of thin sections of purified carboxysomes of *T. neapolitanus* revealed three distinct profiles: hexagons with completely empty centers, hexagons with centers of decreased staining intensity, and hexagons of even staining intensity (145; JM Shively, unpublished observations). These observations seem to suggest that RuBisCO might line the inner surface of the shell. However, these structures might represent different stages of carboxysome breakage or may simply be an artifact created by the isolation and/or electron microscopy preparation procedures. The RuBisCO-lining-the-shell model is attractive because it eliminates the necessity for the extensive diffusion of the enzyme substrates/products in and out of a filled and crowded structure, i.e. the fixation of CO_2 becomes a "surface"—albeit inner surface of the shell—phenomenon. Recent evidence does suggest that the diffusion of the substrate RuBP and the product PGA play a role in the activity of carboxysomal RuBisCO, but this observation neither favors nor discounts either of the structural possibilities (139). An attempt to differentiate between the two structural possibilities using the stoichiometry of the *T. neapolitanus* carboxysome polypeptides elucidated by sodium dodecyl sulfate (SDS)-polyacrylamide gel electrophoresis (PAGE) in conjunction with a variety of other known carboxysome parameters was unsuccessful (148). Finally, it should be noted that the much larger polyhedral bodies found in some cyanobacteria possess an even, granular substructure throughout, thereby providing evidence for a filled structure. One should keep in mind, however, that a polyhedral body is not necessarily a carboxysome; the polyhedral bodies of enteric bacteria (see below) and the carboxysomes of chemoautotrophs provide a perfect example. One might also consider that two polyhedral bodies with RuBisCO, i.e. carboxysomes from chemoautotrophs and cyanobacteria, may not necessarily possess identical structure/function. However, this seems unlikely because carboxysome-possessing chemoautotrophs and cyanobacteria for all practical purposes face the same dilemma, the over-riding presence of O_2.

RuBisCO is the only enzyme whose presence in the carboxysome has been unequivocally established. Carbonic anhydrase (CA) was reported to be present in the carboxysomes of *Synechococcus* PCC7942, but the preparations were grossly contaminated with cytoplasmic/outer membrane, a common cellular location of carbonic anhydrase (120). Attempts to show CA activity in carboxysome fractions of *C. fritschii* were unsuccessful even though the activity was easily demonstrated in whole cells and broken cell fractions (83). Admittedly, the method of analysis used in this study might not detect minute amounts present in the carboxysomes.

Purified carboxysomes of *T. neapolitanus*, three *Nitrobacter* spp., and *C. fritschii* were shown several years ago to possess from 7 to 15 polypeptides (7, 10, 30, 58, 81). The molecular masses of these polypeptides were from 10 to 120 kDa. No additional information has been reported for the carboxysomes of either *C. fritschii* or *Nitrobacter*. Recently the presence of eight polypeptides in the carboxysomes of *T. neapolitanus* has been shown (149). The existence of three additional polypeptides is deemed likely, i.e. a total of 11 (149; JM Shively, unpublished data). Two of the identified polypeptides with molecular masses of 51 and 9 kDa are the L and S subunits of RuBisCO, respectively. Because the six identified polypeptides have not been shown to possess any enzymatic activity, they have been designated with Cso for carboxysome, S for shell, and numbers and letters for further identification (35, 149). These six polypeptides, with molecular masses—as determined by SDS-PAGE—of 130 kDa (CsoS2B), 80 kDa (CsoS2A), 44 kDa (CsoS3), 15 kDa (CsoS1B), and 5 kDa (CsoS1A and CsoS1C), are believed to be shell components and are thought to be glycosylated (10, 35, 58, 149; JM Shively, unpublished data). The established presence of these peptides was based on the determination of the polypeptide composition of purified carboxysomes using SDS-PAGE and on genetic analyses (see below). CsoS1A, B, and C are structural homologs resulting from the transcription/translation of a three-gene repeat (35). Sequencing of the N terminus of CsoS2A and B yielded the same 15 amino acids, indicating they are products of the same gene; at least one and possibly both are the result of posttranslational processing (149; JM Shively, unpublished data). One of the additional polypeptides (38 kDa) appears to be a component of the mature carboxysome, but its corresponding gene is yet to be identified (see below). The other two polypeptides, structural homologs resulting from the transcription/translation of a duplicated gene (ORFA and B), have not been observed by SDS-PAGE of purified carboxysomes. Their presence is inferred from their being part of a putative carboxysome operon (149; see below). One needs to consider that some polypeptides may be required solely for the assembly of the carboxysome. The existence of other quantitatively minor but functionally essential components is always possible. Although a number of putative carboxysome genes have been elucidated in *Synechococcus* PCC7942, the direct identification of polypeptides as components of the purified carboxysome has not been forthcoming.

Hypothesized Function(s)

The regulation of RuBisCO synthesis, as well as its packaging into the carboxysome, via the availability of CO_2 to growing cultures of *T. neapolitanus* strongly suggested an active role for the carboxysome in carbon dioxide fixation (5). Under CO_2 limitation, the cells respond by synthesizing elevated levels of RuBisCO and sequestering a greater percentage of the enzyme into the carboxysome. Under conditions of excess CO_2 the reverse is true: Less RuBisCO is synthesized and a lower percentage is packaged. Furthermore, in the facultative autotroph, *T. intermedius*, both the enzyme and the carboxysomes are, for all practical purposes, totally repressed when the culture is supplied with suitable organic nutrients (148). More importantly, alleviating the repression by transferring the cells to an autotrophic growth environment results in the immediate formation of carboxysomes, i.e. a large pool of enzyme is not accumulated before packaging begins (JM Shively, unpublished data). The availability of carbon is also a controlling factor in the sequestering of RuBisCO into the carboxysomes of cyanobacteria (98, 165). Compelling evidence for the carboxysome's active role in carbon fixation has been garnered via genetic manipulation and analysis. Insertion mutation of carboxysome genes in both *T. neapolitanus* and *Synechococcus* results in a reduction in number, a structural modification, or a total loss of carboxysomes, resulting in a requirement of elevated CO_2 for growth (34, 42, 95, 96, 118, 134; JM Shively, unpublished data). Thus, it is now widely accepted that the carboxysome is not only actively involved in the fixation of CO_2, it somehow enhances the CO_2-fixing ability of its RuBisCO.

The possible presence of CA inside the carboxysomes of the cyanobacterium *Synechococcus* PCC7942 provides the basis for an interesting and attractive hypothesis regarding carboxysome function (69, 119, 129). Essentially, the mechanism for CO_2 fixation enhancement calls for the diffusion of HCO_3^- into the interior of the carboxysome, where it is rapidly converted to CO_2 by the CA. The CO_2 is fixed by RuBisCO before outward diffusion can take place. The uptake of HCO_3^- into the carboxysome would have to be driven by a concentration gradient of some type. This gradient would be created, at least in part, by the rapid conversion of HCO_3^- to CO_2 by CA and the concomitant fixation of the formed CO_2 by RuBisCO. It should be noted that the production of protons by the RuBisCO reaction itself could promote the formation of CO_2 without the necessity of CA. Furthermore, as reported above, the presence of CA has not been unequivocally established. Also, it is somewhat bothersome that the carboxysome shell appears to exhibit little if any selective permeability for the substrates and/or products of RuBisCO (139, 148). O_2 can readily diffuse into the carboxysome. An alternate, and as yet untested, hypothesis to account for the fixation enhancement is the creation of an improved τ value for RuBisCO. This might result from the association of the enzyme with a certain polypeptide(s) in the assembled carboxysome. The very tight association of the small subunit of RuBisCO with components of the carboxysome shell has been reported (57, 58). Chen and coworkers (18) suggest that a nuclear-encoded protein in *Chlamydomonas reinhardtii* might be able to influence the structure (conformation) of RuBisCO, which in turn alters its τ factor.

Relationship of Other Polyhedral Bodies to Carboxysomes

Inclusions with polygonal profiles have recently been described in *Salmonella enterica* (*Salmonella typhimurium*), in *Klebsiella oxytoca*, and in *Escherichia coli* (146), and their presence in other enteric bacteria seems likely. The polyhedral bodies of enteric organisms, also surrounded by a shell or envelope, are commonly more irregular, slightly smaller, and somewhat less dense in appearance than those of their chemoautotrophic counterparts (146). Extensive electron microscopy studies suggest that the bodies of enteric bacteria are filled with a proteinaceous material (HC Aldrich, personal communication). Although these bodies are obviously not carboxysomes, they are definitely related structurally (see below). The inclusions are formed only when the organisms are grown anaerobically with either ethanolamine or propanediol as the energy source (17, 154). *S. typhimurium* with mutations in either the ethanolamine (*eut*) or propanediol (*pdu*) operons are unable to form the inclusions (17, 154). The degradation of both ethanolamine and propanediol proceeds via a coenzyme vitamin B_{12}-dependent pathway, thereby providing a common thread for their occurrence/structure.

No information is available on the peptide composition of these polyhedral bodies. However, genetic studies demonstrate that the probable shell polypeptides PduA and CchA, coded for by genes in the propanediol and ethanolamine operons, respectively, have significant sequence similarity with the CsoS1 carboxysome polypeptides of *T. neapolitanus* and with putative carboxysome polypeptides of *Synechococcus* PCC7942 (146).

Interestingly, it was proposed that the polyhedral bodies of *S. typhimurium* might serve to protect the Pdu (propanediol) enzymes from O_2 (17). Alternatively, the microcompartments might serve to protect the enzymes of ethanolamine and propanediol metabolism from toxic aldehydes (154).

THE GENES AND THEIR LOCATION/ORGANIZATION

Calvin:Benson:Bassham (cbb) Genes

The organization of the Calvin cycle genes of both photo- and chemoautotrophic bacteria has been recently reviewed in detail (80). Thus, our coverage is limited to Figure 2, which shows gene location/organization, some summary statements, and a few comments that provide essential information for other parts of the review. The genes for the bacterial enzymes that function exclusively in the Calvin-Benson-Bassham cycle are given the designation *cbb* (161). (The gene/enzyme designations are explained in Figures 1 and 2.) The *cbb* genes may be plasmid encoded (*Oligotropha carboxidovorans*), chromosome encoded (common), or both (*N. hamburgensis, R. eutropha*). A number of organisms exhibit *cbb* gene duplications. The presence of the *cbbM* gene for RuBisCO is more common than originally believed, with a number of bacteria possessing both *cbbLS* and *cbbM*. The *cbbR* gene, encoding the transcriptional regulator of the *cbb* genes, is present in all but three of the bacteria thus far examined, and in the latter bacteria *cbbR* may be located elsewhere on the genome. The *cbbR* does not have to be directly adjacent to the genes it regulates. With the exception of *cbbR* in *R. rubrum*, *cbbR*, when adjacent to *cbb* genes, is always transcribed in the opposite direction.

Carboxysome Genes

The carboxysome (*cso*) genes of *T. neapolitanus* reside immediately downstream of, and form an operon with, the genes for the large (*cbbL*) and small (*cbbS*) subunits of RuBisCO in the following order: *csoS2, csoS3*, ORFA, ORFB, *csoS1C, csoS1A*, and *csoS1B* (35, 149; JM Shively, unpublished data; see Figure 2). These genes code for their corresponding proteins described in the section on carboxysome structure. The location and function of the products of the gene repeat ORFA and B have not been determined. Probing genomic DNA of several thiobacilli and nitrifying and ammonia oxidizing bacteria with *csoS1A* yielded positive results, indicating that these carboxysome formers all have the well-conserved *csoS1*. The same gene organization observed in

Xanthobacter flavus	R	L	S	X	F	P	T	A	E			
Xanthobacter flavus		gap	pgk									
Ralstonia eutropha-C	R	L	S	X	Y	E	F	P	T	Z	G	K A B
Ralstonia eutropha-P	R	L	S	X	Y	E	F	P	T	Z	G	K A
Nitrobacter vulgaris	R	L	S	G								
Nitrobacter vulgaris	R	F	P	A								
Oligotropha carboxidovorans	R	F	P	K	L	S	X					
Rhodobacter sphaeroides	R	F	P	A	L	S	X	Y	Z			
Pseudomonas hydrogenothermophila		L	S	Q								
Rhodobacter capsulatus	R	L	S	Q								
Chromatium vinosum-1	R	L	S	Q								
Chromatium vinosum-2		L	S									
Thiobacillus denitrificans	R	L	S									
Thiobacillus ferrooxidans-1	R	L	S									
Thiobacillus ferrooxidans-2		L	S									
Thiobacillus intermedius		L	S	Carboxysome								
Thiobacillus neapolitanus		L	S	Carboxysome								
Hydrogenovibrio marinus-1		L	S									
Hydrogenovibrio marinus-2		L	S									
Mn Oxidizer (SI85-9A1)	A	L	S									

Hydrogenovibrio marinus						M	Q	
Thiobacillus intermedius					R	M	Q	
Thiobacillus neapolitanus					R	M	Q	
Thiobacillus denitrificans					R	M		
Rhodobacter sphaeroides		F	P	T	G	A	M	
Rhodobacter capsulatus	R	F	P	T	G	A	M	E
Rhodospirillum rubrum	A	T	P	F	E	R	M	

T. neapolitanus has been elucidated in *T. intermedius* K12 (JM Shively, unpublished data). In *T. denitrificans*, however, the *csoS1* three-gene repeat is located in closer proximity to *cbbM*, the gene for the form II RuBisCO (JM Shively, unpublished data). Obviously, the carboxysome genes will also be located differently in *Nitrobacter vulgaris* (see Figure 2). The putative genes (*ccmK, ccmL, ccmM, ccmN,* and *ccmO*) for the carboxysomes of *Synechococcus* PCC7942 lie upstream (in descending order) from *cbbLS* but constitute a separate operon (42, 121, 134). *Synechocystis* PCC6803 has four copies of *ccmK* and a *ccmO*; none are located close to *cbbLS* (146). As indicated above, the CcmK and CcmO polypeptides have a considerable amount of similarity to the CsoS1 polypeptides of *T. neapolitanus* as well as to the CchA and PduA polypeptides of *S. typhimurium* (146). Watson & Tabita (180) recently reported the presence of a *ccmK* directly upstream and in the same operon with *cbbLS* in a marine cyanobacterium. The product of this *ccmK* has greater homology to the CsoS1A of *T. neapolitanus* than to the CcmK of both *Synechococcus* and *Synechocystis* (146). All of the carboxysome-containing chemoautotrophs thus far examined possess at least one *cbbR*. The location varies (Figure 2) and the regulation of the expression of carboxysome genes by CbbR is yet to be determined.

Phylogeny—Origin of the Calvin Cycle

The phylogeny of RuBisCO and the implications for the origins of eukaryotic organelles was recently extensively reviewed (24, 181), and therefore we only present the major conclusions regarding RuBisCO. A phylogenetic analysis of the large subunit of RuBisCO reveals two groups of form I RuBisCO: green-like and red-like (Figure 3). The green-like RuBisCO proteins are subdivided into a cluster containing proteins of cyanobacteria and of plastids of plants and green algae and into a cluster containing representatives of the α-, β-, and γ-subdivisions of the proteobacteria and cyanobacteria. The red-like RuBisCO proteins are also subdivided into two clusters: One contains proteins of non-green algae and one contains sequences of the α- and β-subdivisions of proteobacteria. Clearly, the RuBisCO phylogeny is at odds with

←———————————————————————————————

Figure 2 Calvin cycle genes and their location/organization. Each letter identifies a *cbb* gene. Gene and product abbreviations: *cbbLS,* the large and small subunits of form I RuBisCO, respectively; *cbbM,* form II RuBisCO; *cbbR,* LysR-like transcriptional activator; *cbbF,* FBPase/SBPase; *cbbP,* PRK; *cbbT,* TKT; *cbbA,* FBA/SBA; *cbbE,* PPE; *cbbG* and *gap,* GAPDH; *cbbK* and *pgk,* PGK; *cbbZ,* PGP (phosphoglycolate phosphatase); *cbbB, X, Y,* and *Q,* unknown. See Figure 1 for enzymes. Gene size is not depicted here. (*arrows*) Operons and transcriptional direction. *R. eutropha-C* and -*P* are chromosome and plasmid, respectively. Carboxysome genes are located downstream of *cbbLS* in *Thiobacillus intermedius* and *Thiobacillus neapolitanus*. (From References 11, 49, 50, 72, 77, 80, 113, 137, 174, 175, 187, 188.)

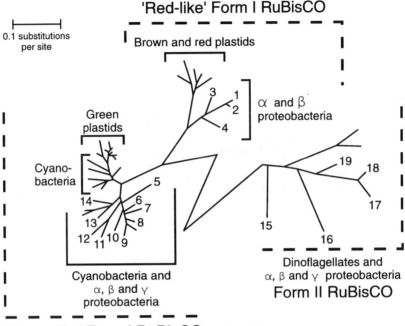

Figure 3 Phylogenetic relationships of form II and the large subunit of form I ribulose-1,5-bisphosphate carboxylase/oxygenase (RuBisCO). The unrooted phylogenetic tree is based on a distance matrix calculated using the Kimura model and constructed with the neighbor joining method as implemented in the PHYLIP 3.5c software package. Insertions and deletions were not taken into account. For clarity, only sequences of proteobacteria are indicated. 1, *Ralstonia eutropha* (chromosomal); 2, *R. eutropha* (plasmid); 3, *Rhodobacter sphaeroides*; 4, *Xanthobacter flavus*; 5, *Hydrogenovibrio marinus* 1; 6, *Thiobacillus neapolitanus*; 7, *Nitrobacter vulgaris*; 8, *Thiobacillus denitrificans*; 9, *Chromatium vinosum* 1; 10, *Thiobacillus ferrooxidans*; 11, *Pseudomonas hydrogenothermophila*; 12, *Rhodobacter capsulatus*; 13, *C. vinosum* 2; 14, *H. marinus* 2; 15, *T. denitrificans*; 16, *H. marinus*; 17, *R. sphaeroides*; 18, *R. capsulatus*; 19, *Rhodospirillum rubrum*.

16S rRNA–based phylogenies, which led Delwiche & Palmer (24) to propose at least four lateral gene transfers to explain the dichotomy between red- and green-like RuBisCO sequences. One of the best examples of lateral gene transfer of form I RuBisCO is the presence of a green-like RuBisCO in *Rhodobacter capsulatus*, whereas *R. sphaeroides* has a red-like protein. In addition, both species contain a closely related form II RuBisCO. This indicates that the *Rhodobacter* ancestral species contained form II RuBisCO and subsequently acquired form I RuBisCO via lateral gene transfer (114). The phylogenetic tree of CbbR has the same topology as the RuBisCO tree, which includes grouping of *R. capsulatus* CbbR, which is encoded upstream from *cbbLS*, with proteins encoded upstream from green-like RuBisCO. This strongly suggests that *cbbR-cbbLSQ* of *R. capsulatus* was obtained by lateral gene transfer from a bacterium containing green-like RuBisCO (114).

The relationship between form I and form II RuBisCO is unclear at present. The simple structure and poor catalytic characteristics of form II RuBisCO suggest that the more complex form I RuBisCO is derived from this protein. In comparison to form I RuBisCO, the phylogenetic distribution of form II RuBisCO is limited. However, this is hardly surprising given the fact that the primitive form II RuBisCO only functions well at low oxygen and high CO_2 concentrations, conditions that reflect the ancient earth atmosphere. The emergence of form I RuBisCO allows CO_2 fixations under present-day atmospheric conditions.

A phylogenetic analysis of class II aldolase proteins shows that the branching order of the aldolase phylogenetic tree is different from those based on 16S rRNA alignments (170). The proteins (CbbA) from the autotrophic proteobacteria *X. flavus*, *R. eutropha*, *R. sphaeroides*, and *N. vulgaris* are closely related to aldolase of the gram-positive bacterium *Bacillus subtilis* but not to proteins of heterotrophic proteobacteria (141, 170). This suggests that the autotrophic aldolase proteins were obtained from a gram-positive bacterium. The *cbbLS* genes are usually clustered with either *cbbX* or *cbbQ*. Interestingly, *cbbX* is only present downstream from the red-like RuBisCO genes, whereas *cbbQ* is always found in conjunction with green-like RuBisCO. In contrast, aldolase of *N. vulgaris* groups with those from *X. flavus*, *R. eutropha*, and *R. sphaeroides*, even though the RuBisCO protein of the former is green-like and that of the latter three red-like (WG Meijer, unpublished results). These data indicate that *cbbQ* and *cbbX* may have been acquired after, and *cbbA* before, the divergence of green- and red-like RuBisCO sequences. In summary, the phylogenies of the *cbb* genes examined to date suggest that extensive lateral gene transfer took place. As a result, bacteria acquired the ability to assimilate CO_2 or, as in the case in *Rhodobacter*, acquired a more efficient RuBisCO, which allows CO_2 fixation to proceed at low CO_2 concentrations or under aerobic conditions.

The selfish operon model proposes that operons gradually assemble following lateral transfer of genes clusters (85). Acquisition of a gene cluster encoding related metabolic functions allows the host to exploit new niches, whereas transfer of individual genes provides no benefit. This theory provides an attractive explanation for the high rate of lateral gene transfer of the highly conserved clusters *cbbLS* and *cbbFP* (Figure 2). Acquisition of these clusters, but not of the individual genes, provides a heterotrophic bacterium with the bare essentials to assimilate CO_2. The variability in genetic organization of other *cbb* genes suggests that these were obtained later (e.g. *cbbX* and *cbbQ*), depending on the metabolic needs of the cell. It is interesting to note that *R. eutropha* has recruited the glyceraldehyde phosphate dehydrogenase and phoshoglycerate kinase genes (*cbbKG*) into the *cbb* operon, whereas the closely related *X. flavus* employs the *gap-pgk* operon for both heterotrophic and autotrophic growth. The *cbb* operon is frequently found on mobile genetic elements of both proteobacteria and gram-positive bacteria, which provides a means for the lateral transfer of autotrophy (31, 43, 179).

REGULATION OF GENE EXPRESSION

Growth Conditions and Expression of Calvin Cycle Genes

The fixation of three molecules of CO_2 by the Calvin cycle requires the investment of nine molecules of ATP and six of NADH to obtain one molecule of glyceraldehyde-3-phosphate. It is obvious that this is an expensive process that has to be regulated carefully. The Calvin cycle is usually not induced during growth of facultatively autotrophic bacteria on substrates supporting fast heterotrophic growth, for example succinate or acetate (26, 45, 51). Growth of *R. eutropha* on fructose, glycerol, or gluconate, substrates that allow only intermediate or low growth rates leads to intermediate or high Calvin cycle enzyme levels (45). However, heterotrophic derepression of the Calvin cycle is only observed in strains harboring a megaplasmid containing one of the two *cbb* operons of *R. eutropha* (9). The facultatively autotrophic bacterium *Pseudomonas oxalaticus* simultaneously uses acetate and formate in batch cultures. Under these growth conditions formate is only used as an ancillary energy source and the Calvin cycle is not induced (26). The repression of the Calvin cycle is less severe during growth under carbon-limiting conditions (25, 44). Addition of formate to the feed of an acetate-limited continuous culture of *P. oxalaticus* resulted in simultaneous and complete utilization of the two substrates. Interestingly, the enzymes of the Calvin cycle remained absent at formate concentrations below 40 mM, whereas at higher concentrations they were induced. Because the bacterial dry mass of the culture increased by 40% when the formate concentration was increased from 0 to 40 mM, it was concluded that the use of formate as an additional energy source allowed a decreased dissimilation (via the citric acid cycle) and an increased assimilation (via the glyoxylate pathway) of acetate. Interestingly, growth of obligately autotrophic bacteria under CO_2 limitation also leads to increased activities of the Calvin cycle (5).

Although growth under carbon limitation conditions alleviates repression of the Calvin cycle, carbon starvation does not lead to the induction of the Calvin cycle (45). Under these conditions, the presence of reduced compounds supporting autotrophic growth are required, which generally stimulates the expression of the Calvin cycle. For example, the Calvin cycle is rapidly induced to autotrophic levels following the addition of formate, methanol, or H_2 to *X. flavus* growing on gluconate (102).

The results of these experiments show that two physiological parameters, the availability of carbon sources (including CO_2) and reduced (in)organic substrates supporting autotrophic growth, control the expression of the Calvin cycle in chemoautotrophic bacteria. This is most likely mediated by metabolites originating from carbon and energy metabolism. A reduced metabolite originating from the oxidation of reduced compounds (e.g. H_2S) is most likely responsible for the induction of the Calvin cycle. Chemostat experiments showed that methanol is a more potent inducer than formate, even though these compounds are metabolized via the same linear pathway (23). However, the oxidation of methanol generates three reducing equivalents whereas formate oxidation only produces one. The importance of reducing equivalents in the regulation of the Calvin cycle is also evident from an interesting series of experiments involving the phototrophs *R. capsulatus* and *R. sphaeroides*. Anaerobic photoheterotrophic growth of these bacteria depends on the Calvin cycle to dispose of excess reducing equivalents. Consequently, the activity of the Calvin cycle is higher during growth on butyrate than on the less-reduced substrate malate. Conversely, the expression of Calvin cycle enzymes is dramatically reduced in the presence of alternative electron acceptors, e.g. dimethyl sulfoxide (54, 131, 178).

The search for the identity of the repressor intermediate derived from carbon metabolism has focused on the end products of the Calvin cycle or closely related compounds. During growth of *R. eutropha* on fructose, the levels of RuBisCO increased fivefold upon addition of sodium fluoride. This compound inhibits, among other things, the activity of enolase, which catalyzes the conversion of 2-phosphoglycerate to phosphoenolpyruvate (61). An increase in RuBisCO activity was also observed in a phosphoglycerate mutase mutant of *R. eutropha* grown on fructose (130). The reduced ability to convert fructose to 2-phosphoglycerate and phosphoenolpyruvate apparently alleviates repression of the Calvin cycle enzymes. Mutants of *X. flavus* that are devoid of either phosphoglycerate kinase or triosephosphate isomerase activity display enhanced repression of the Calvin cycle by gluconeogenic substrates (99, 103). The metabolic block introduced by the *pgk* and *tpi* mutations are likely to cause a buildup of intermediary metabolites during gluconeogenic growth. These data strongly suggest that the concentration of an intermediary metabolite, most likely phosphoenolpyruvate, signals the carbon status of the cell. However, the results of experiments involving an isocitrate lyase mutant of *P. oxalaticus* indicate that a metabolite related to acetyl coenzyme A, rather than phosphoenolpyruvate, fulfills this signaling role (104).

Transcripts

THE *CBB* PROMOTERS   The *cbb* operons of the chemoautotrophs *T. ferrooxidans*, *X. flavus*, and *R. eutropha* are transcribed from a single promoter (77, 78, 101). In contrast to the −10 region, which is poorly conserved, the −35 regions of the *cbb* promoters of these bacteria are virtually identical and resemble the *E. coli* σ^{70} consensus sequence (Figure 4). A characteristic of the *cbb* operon is the close proximity of the divergently transcribed *cbbR* gene (Figure 2). The intergenic region between *cbbR* and the *cbb* operon can be as small as 89 bp, which indicates that the promoters of *cbbR* and the *cbb* operon may have regulatory elements in common. This was recently established for *R. eutropha* (78). Analysis of *cbbR* transcripts and regulation studies involving transcriptional fusions between *lacZ* and *cbbR* showed that *cbbR* of *R. eutropha* is constitutively transcribed during heterotrophic and autotrophic growth, giving rise to a monocistronic transcript of 1.4 kb. Surprisingly, *cbbR* is transcribed from two different σ^{70} promoters, depending on the growth conditions. Promoter P_{Rp} is located 120 bp upstream from the *cbbR* gene, which is within the transcribed region of the *cbb* operon. As a consequence, P_{Rp} is not active during autotrophic growth because of the high activity of the *cbb* operon promoter. Under these conditions, therefore, *cbbR* is transcribed from the alternative promoter P_{Ra}, 75 bp upstream from the *cbbR* gene, which partially overlaps the promoter of the *cbb* operon. The activity of the *cbbR* promoters is only 4% of the promoter of the *cbb* operon, which could be the result of the low similarity to the *E. coli* σ^{70} consensus sequence.

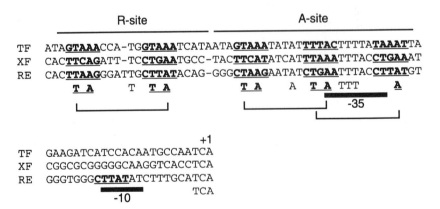

Figure 4 Alignment of the promoter of the *cbb* operon of *Thiobacillus ferrooxidans* (TF), *Xanthobacter flavus* (XF), and *Ralstonia eutropha* (RE). (*brackets*) Position of a LysR-motif (T-N$_{11}$-A). (*bold* and *underlined*) Putative CbbR binding sites. (*bars*) Position of the −35 and −10 regions of the σ^{70} promoter. +1, Transcription initiation site.

MRNA PROCESSING AND TRANSCRIPTION TERMINATION   The insertion of antibiotic resistance markers or transposons in *cbb* genes of *R. eutropha*, *X. flavus*, and *R. sphaeroides* prevents the expression of *cbb* genes downstream from the insertion. These experiments revealed that the *cbb* genes are organized in large operons, which in *R. eutropha* may encompass as much as 15 kb (8, 49, 100, 141, 182). However, although *cbbLS* mRNA is relatively abundant in cells following autotrophic growth, mRNA containing all *cbb* genes has not been detected (3, 36, 59, 77, 101). Analysis of *cbb* transcripts of *X. flavus* showed that *cbbLS* mRNA encompassing the 5′ end of the *cbb* operon is six times as abundant as the 3′ end (101). Similar observations were made in *R. eutropha* and interpreted as a premature transcription termination at a sequence resembling a terminator structure downstream from the *cbbLS* genes (140). Sequences that may form a hairpin structure are also present downstream of the *cbbLS* genes of *T. ferrooxidans*, *T. denitrificans*, and *X. flavus* (56, 123). RuBisCO is usually synthesized to high levels during autotrophic growth because the enzyme is a poor catalyst. Since all *cbb* genes are transcribed from a single promoter, differential *cbb* gene expression in *X. flavus* and *R. eutropha* seems to be achieved by a relative abundance of *cbbLS* mRNA.

The LysR-Type Regulator CbbR

PRIMARY STRUCTURE   To date, *cbbR* genes have been cloned and sequenced from the chemoautotrophs *R. eutropha*, *X. flavus*, *T. denitrificans*, *T. neapolitanus*, *T. intermedius*, *T. ferrooxidans*, and *N. vulgaris* (76, 90, 101, 156, 171; JM Shively, unpublished results) and from the photoautotrophic bacteria *R. rubrum*, *R. capsulatus*, *R. sphaeroides*, and *Chromatium vinosum* (38, 48, 114, 175). The predicted molecular mass of the CbbR proteins ranges from 31.7 to 35.9 kDa. CbbR is a LysR-type transcriptional regulator, a protein family that includes over 70 proteins from gram-positive and gram-negative bacteria (142). LysR-type regulators are dimeric or tetrameric proteins, which contain a conserved amino-terminal DNA binding domain. The central and carboxy terminal domains are involved in ligand binding and multimerization, respectively. Because LysR-type regulators control a wide range of cellular processes, they bind a variety of chemically unrelated compounds. It is therefore not surprising that the central ligand binding domain is not conserved. Interestingly, the CbbR proteins are similar (35% identity) throughout the sequence, indicating that these proteins bind the same or related ligands. The structure of the ligand binding domain of CysB (amino acids 88–324), which is closely related to CbbR, was recently solved by Tyrrell et al (166). The CysB monomer contains two α/β domains enclosing a coinducer binding cavity. One sulfate anion was found in the coinducer binding cavity between the two domains in each monomer. All four threonine residues involved in ligand binding by CysB

```
Conserved
region            1              2              3                    4
KP-CysB100    -THTQA-104    148-ATEA-151    199-FGFTGR-204    243-IASMAVDP-VSDPDLVKLD-260
               ^ *_^            ~                *   *        ^      ^ ^_^     _*^  *_

RE         105-IST-S-108    152-MGRP-155    202-PGSGTR-207    246-LSLHTLGLELRTGEIGLLD-264
XF         102-VST-A-105    149-MGRP-152    200-PGSGTR-205    245-ISAHTVAAEVADGRLRVLE-263
RR         103-VST-A-106    151-MGRP-154    198-PGSGTR-203    242-LSRNTMSLELSVGRLVILD-260
RS         126-VST-G-129    153-MGRP-156    203-EGSGTR-208    248-LSLHVVMDELRFGQLVQLA-266
RC-1       100-VST-A-103    147-LGQP-150    197-PGSGTR-202    241-LSQDTITLEAETGRLAVLD-259
RC-2       106-VST-A-109    153-MGRP-156    202-PGSGTR-207    247-LSLHTVTEELGSGRLVELS-265
CV         100-AST-V-103    147-MGVP-150    197-EGSGTR-202    241-VSLHTIELELETRRLVTLD-259
TF         103-LST-T-106    150-LGQP-153    200-PGSGTR-205    244-LSASTIRAELASGKLAILD-262
            ^**_^           ^*_*             *****            ^*^_*^ ^*^  *_*^^*_

CONSENSUS     VST-^           MGRP            PGSGTR           LS^HT^X^ELXXGRL^^LD
```

Figure 5 Alignment of CysB of *Klebsiella aerogenes* with CbbR of *Ralstonia eutropha* (RE), *Xanthobacter flavus* (XF), *Rhodospirillum rubrum* (RR), *Rhodobacter sphaeroides* (RS), *Rhodobacter capsulatus* 1 (RC-1), *R. capsulatus* 2 (RC-2), *Chromatium vinosum* (CV), and *Thiobacillus ferrooxidans* (TF). Threonine residues forming a complex with sulfate in CysB(88-324) are shown (*boldfaced*). ^, Hydrophobic residue (GALIVFPYWTMS); *, identical residue; ~, charged/hydrophilic residue (HNDKRE); X, any residue. (See References 112, 164 for alignment methods.)

are located in three regions that are conserved in all CbbR proteins (Figure 5). This strongly suggests that these conserved residues also play a role in ligand binding by CbbR. In addition, a fourth region of CysB involved in ligand response and/or multimerization is also conserved in all CbbR proteins.

Because the level of expression of CbbR is too low for biochemical studies, the proteins of *R. eutropha*, *T. ferrooxidans*, *C. vinosum*, and *X. flavus* have been overexpressed in *E. coli*. Only the proteins from *R. eutropha* and *X. flavus* have been purified to homogeneity (79, 173). The CbbR proteins from *R. eutropha* and *X. flavus* are dimers in solution, although the protein from *R. eutropha* forms a tetramer at 4°C (79, 173).

GENES UNDER CBBR CONTROL   *Autoregulation of cbbR gene expression*   The *cbbR* gene of *R. eutropha* is constitutively transcribed from two promoters 75 and 120 bp upstream from the *cbbR* gene. DNA footprinting studies (see below) showed that CbbR binds to a DNA fragment 52–104 bp upstream from the *cbbR* gene; this indicates that CbbR binding may repress transcription from the *cbbR* promoters. Removal of the CbbR binding site nearest to the *cbbR* gene resulted in a 4- and 35-fold increase in *cbbR* expression during autotrophic and heterotrophic growth, respectively (78, 79). This shows that the overlap between CbbR binding sites and promoters creates an autoregulatory circuit in which *cbbR* expression is repressed by CbbR. The CbbR binding sites of *T. ferrooxidans* and *X. flavus* are also adjacent to *cbbR*, indicating that transcription of *cbbR* in these organisms is regulated in a similar manner.

Induction and super induction of Calvin cycle genes   Disruption of *cbbR* by insertion of Tn5 or an antibiotic resistance gene completely abolished the expression of the *cbb* operons of *R. eutropha* and *X. flavus* (171, 183). The fact that both the chromosomal and plasmid operons of *R. eutropha* were affected showed that the activity of both *cbb* promoters is dependent on chromosomally encoded CbbR protein. The similarity in genetic organization and regulation of the *cbb* operons indicates that the dependence on CbbR for transcription of the *cbb* operon is a common characteristic of chemoautotrophic bacteria. In addition to regulating transcription of the *cbb* operon and *cbbR*, CbbR has been shown to control the expression of at least one other transcriptional unit. The *gap-pgk* operon of *X. flavus* is constitutively expressed and superinduced following a transition to autotrophic growth conditions (106). The *cbbR* mutant strain failed to superinduce this operon, indicating that CbbR is required for autotrophic regulation of *gap-pgk*.

MECHANISM OF ACTION   *Interactions between CbbR and the cbb promoter*
Bandshift assays and footprinting experiments were used to analyze binding of CbbR to the promoter region of the *cbb* operon of *R. eutropha*, *X. flavus*, *T. ferrooxidans*, *T. neapolitanus*, *T. denitrificans*, and *T. intermedius* (76, 79, 171, 173, 183; JM Shively, unpublished results). Footprinting experiments showed that CbbR of *T. ferrooxidans* protected nucleotides from position -76 to -14 relative to the *cbbL1* transcriptional start site (76). CbbR from *R. eutropha* and *X. flavus* protected similar regions that are located between -74 and -29 and between -75 and -23 relative to the transcriptional start site of the *cbb* operon, respectively (79, 173). Increasing concentrations of CbbR of *R. eutropha* protected additional nucleotides up to position $+13$. These experiments show that the binding site of CbbR overlaps the -35 region of the *cbb* promoter. This facilitates contact with the α subunit of RNA polymerase, which is essential for transcriptional activation by LysR-type regulators (162).

Bandshift assays using DNA fragments containing segments of the binding sites of *R. eutropha* or *X. flavus* revealed the presence of two subsites (R- and A-site; Figure 4) (79, 169, 171). These experiments showed that CbbR from *R. eutropha* and *X. flavus* have different DNA binding characteristics. CbbR from *R. eutropha* is able to bind to both subsites, regardless of the protein concentration or the presence of a ligand, which is similar to NodD (40). The protein from *X. flavus* behaves like TrpI, which has a high affinity for the promoter distal site (R-site) and only binds to the promoter proximal site (A-site) at higher protein concentrations or in the presence of a ligand (15).

Comparison of the DNA sequences that are protected in footprinting experiments using CbbR of *T. ferrooxidans*, *X. flavus*, and *R. eutropha* reveal a number of interesting similarities (Figure 4). The R-site of the *cbb* operon promoter of *X. flavus* and *T. ferrooxidans* contains a LysR-motif (T-N_{11}-A), which forms the core of an inverted and a direct repeat, respectively. A related motif (T-N_{12}-A) is present in the imperfect inverted repeat of the R-site of *R. eutropha*.

Comparison of these repeats reveals the CbbR consensus sequence $TnA-N_{7/8}$-TnA. Interestingly, the consensus sequence is both a direct and an inverted repeat. CbbR of *X. flavus* and *R. eutropha* is able to bind to a DNA fragment containing only the R-site. Furthermore, mutagenesis of the T in the LysR-motif of *X. flavus* abolishes binding of CbbR, which strongly suggests that the $TnA-N_7-TnA$ sequence represents the CbbR binding site (G van Keulen, WG Meijer, unpublished results).

The A-site of *R. eutropha* and *X. flavus* contains two partially overlapping $TnA-N_7-TnA$ sequences (Figure 4). The A-site of *T. ferrooxidans* has overlapping $TnA-N_7-TnA$ and $TnA-N_7-AnA$ sequences. The sequences containing the CbbR-consensus are related to those of the R-site. Most striking is the conservation of the right half-site of the putative CbbR binding site of the R-site (GTAAA, *T. ferrooxidans*; CTGAA, *X. flavus*; CTTAT, *R. eutropha*). Interestingly, this sequence is repeated in the -10 region of *R. eutropha*, which may account for CbbR binding in this region at high protein concentrations.

The CbbR consensus sequences in the R- and A-sites are separated by two and three turns of the DNA-helix and are located on the same side. Insertion of two additional nucleotides between the R- and A-sites of *R. eutropha* did not abolish DNA binding of CbbR, although the *cbb* promoter was no longer active (79). This indicates that protein-protein contacts, which are dependent on the proper positioning of the two CbbR molecules, are essential for transcriptional activation. DNase I hypersensitive sites were observed between the R- and A-sites at positions -47 and -48 in the *cbb* promoter of *R. eutropha* and *X. flavus*, which could be the result of DNA bending induced by binding of CbbR (79, 173). Bandshift assays using circular permutated DNA fragments showed that CbbR of *X. flavus* induces a 64° DNA bend upon binding (173). Protein-induced DNA bending is frequently observed in LysR-type proteins (142).

Ligands of CbbR   As discussed above, the expression of the Calvin cycle genes in autotrophic bacteria depends on the availability of suitable carbon and energy sources. The discovery that CbbR is a transcriptional regulator of the *cbb* operons in autotrophic bacteria strongly suggests that CbbR is required for the transduction of these signals to the transcription apparatus. Band-shift assays using purified CbbR of *X. flavus* showed that DNA binding of this protein is increased threefold following the addition of 100 μM NADPH to the binding assay. In addition, CbbR-induced DNA bending is decreased by 9° in the presence of NADPH (173). Similar changes in DNA binding characteristics following binding of ligands was previously observed for other LysR-type regulators (e.g. 177).

Although these in vitro experiments do not prove that the in vivo expression of the *cbb* operon is controlled by the intracellular concentration of NADPH, a number of experiments indicate that autotrophic growth is associated with elevated concentrations of NADPH. The transition from heterotrophic to autotrophic growth of *P. oxalaticus* is accompanied by an increase in the NADPH-to-NADP ratio (71). Furthermore, NADP is completely reduced when *R. rubrum* is incubated under anaerobic conditions in the light, when cells normally induce the *cbb* operon. NADPH was subsequently oxidized within 1 min when the cells were exposed to oxygen or incubated in the dark, growth conditions under which *R. rubrum* does not have an active Calvin cycle (62).

IN RELATION TO PHOTOAUTOTROPHIC BACTERIA   The same principles that apply to transcriptional regulation of Calvin cycle genes by CbbR in chemoautotrophs also apply to photoautotrophic bacteria. The expression of the *cbbM* of *R. rubrum* and of the form I *cbb* operon of *R. sphaeroides* depends on the presence of a functional CbbR protein (38, 48). However, these phototrophs are metabolically extremely versatile. It is therefore not surprising that there are some interesting differences with chemoautotrophic bacteria regarding transcriptional regulation of the Calvin cycle genes.

Like *R. eutropha*, *R. sphaeroides* also has two *cbb* operons and only one *cbbR* gene (48). However, in contrast to *R. eutropha*, the transcription of only one *cbb* operon is completely dependent on CbbR. The form II operon was still expressed at 30% of the level found in the wild type in a *cbbR* mutant strain. As a result, photoautotrophic growth was completely abolished. However, photoheterotrophic growth on malate or butyrate was still possible, albeit at a reduced growth rate. Exposure to oxygen completely repressed the synthesis of form II RuBisCO, indicating that in addition to CbbR, other control mechanisms exist in purple-nonsulfur bacteria. Two genes of unknown function, *orfU* and *orfV*, are located upstream from the form II operon of *R. sphaeroides* (186). Using transcriptional fusions with *xylE* it was shown that the expression of these genes was increased in a *cbbR* mutant, indicating that the expression of these two genes is repressed by CbbR.

No biochemical studies of CbbR of phototrophic bacteria have been reported to date. However, the extensive similarities in primary structure between CbbR from chemoautotrophic and photoautotrophic bacteria indicate that the molecular mechanism underlying transcriptional regulation by CbbR are similar in both types of autotrophic bacteria.

Additional Transcriptional Regulators?

REPRESSORS?   There is only circumstantial evidence for the presence of a repressor of the *cbb* operon. The presence of an additional *cbb* operon located on an indigenous megaplasmid or a broad host range plasmid results in heterotrophic derepression of the Calvin cycle in *R. eutropha* and *X. flavus*, respectively (9, 88). Fusions between the *cbb* promoter of *R. eutropha* and *X. flavus* and *lacZ* are also active during heterotrophic growth when present in multiple copies (78, 101). The anomalous activity of the *cbb* promoter under these conditions is likely to be due to titration of a repressor molecule by multiple repressor binding sites.

Tn5 mutagenesis of *R. eutropha* identified a gene that is required for autotrophic growth. The mutant (*a*) failed to induce the Calvin cycle during heterotrophic growth on gluconate or formate, (*b*) displayed reduced glycolytic activity, and (*c*) altered colony morphology. Interestingly, the *cbb* operon was still inducible by formate. Southern hybridization showed that the *aut* gene is also present in *Pseudomonas* sp. and *T. intermedius*, and possibly also in *R. capsulatus* and *Paracoccus denitrificans*. The Aut protein displays similarities with cytidyltransferases; however, its function remains unknown (41).

PHOSPHO-RELAY SYSTEM IN PHOTOAUTOTROPHS   A global signal transduction system that includes the two-component phospho-relay system, RegA-RegB, has been identified in purple nonsulfur bacteria that integrates CO_2 fixation with other important processes such as photosynthesis and nitrogen fixation (67, 125, 143). The RegA-RegB system is involved in the positive regulation of the *cbb* operons of *R. rubrum* and *R. sphaeroides* as well as in the expression of genes important for the alternative CO_2 fixation pathway (67, 125). RegB, the sensor kinase, is autophosphorylated in response to an unidentified (external) signal (37) and is required for the phosphorylation of the response regulator RegA, which in turn controls photosynthetic gene expression. RegA does not have a DNA-binding domain, which indicates that an additional factor must be involved in transducing the signals to the transcription machinery.

Model of Regulation of cbb Gene Expression

The data from physiological experiments strongly suggest that a reduced metabolite derived from the oxidation of autotrophic substrates is responsible for induction of the Calvin cycle. Genetic and biochemical data show that the transcriptional regulator CbbR binds to *cbb* promoters and regulates the expression of Calvin cycle genes. Furthermore it has been shown that CbbR interacts with NADPH, which enhances DNA binding and reduces DNA bending by CbbR. Based on this we propose the model shown in Figure 6 for the regulation of the expression of the Calvin cycle in chemoautotrophic bacteria.

During heterotrophic growth of facultatively autotrophic bacteria, reducing power and ATP is generated via catabolic pathways and oxidative phosphorylation, which is subsequently used for the assimilation of organic substrates into biomass via anabolic pathways. Catabolic and anabolic reactions in the cell are thus balanced. The balance between catabolism and anabolism is disturbed following the transition from heterotrophic to autotrophic growth conditions.

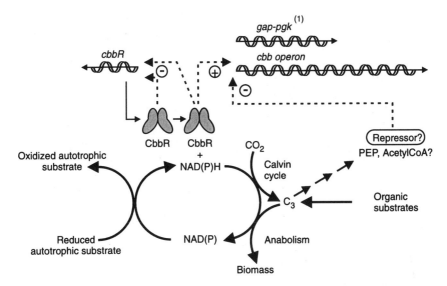

Figure 6   Model for the regulation of expression of the Calvin cycle genes. C_3, Triosephosphate, a metabolite intermediary to the Calvin cycle and intermediary metabolism; PEP, phosphoenolpyruvate; CoA, coenzyme A. The helices represent the *cbb* operon, *cbbR*, and the *gap-pgk* operon. (1) The *gap-pgk* operon has only been identified in *Xanthobacter flavus*.

The oxidation of reduced (in)organic compounds supporting autotrophic growth leads to the production of reducing equivalents. However, since organic carbon sources are not available, anabolic reactions do not proceed, which leads to a buildup of reducing equivalents such as NADPH, which may be formed from NADH by transhydrogenase. The increased intracellular concentration of NADPH in the cell favors the formation of an NADPH-CbbR complex, which activates transcription of the *cbb* operon and in *X. flavus* also of the *gap-pgk* operon. The subsequent fixation of CO_2 via the Calvin cycle provides the cell with a source of carbon, which allows anabolism to proceed and dissipates excess reducing equivalents.

The extent by which reducing power is dissipated depends on the carbon flux through the Calvin cycle. CO_2 limitation reduces the flow of carbon through the Calvin cycle and hence reduces the rate by which reducing equivalents are oxidized. As a result, the NADPH concentration increases, which shifts the equilibrium between CbbR and NADPH-CbbR to the latter. This results in an enhanced transcription of the Calvin cycle.

The available data indicate the presence of a repressor protein that may respond to the intracellular concentration of an intermediary metabolite such as phosphoenolpyruvate. Because actively growing cells contain NADPH, part of the CbbR population will be activated following binding of NADPH, which may lead to a low level expression of the Calvin cycle during heterotrophic growth. The role of the repressor protein may therefore be the prevention of *cbb* transcription during heterotrophic growth.

CONCLUDING REMARKS

A wealth of new information contributing to our understanding of carbon dioxide fixation in chemoautotrophs has been gathered since the previous comprehensive reviews on the subject. However, for every question answered, many others have been unmasked. For example, how does the carboxysome enhance CO_2 fixation? How is the expression of the carboxysome genes regulated? How is the carboxysome assembled? Where are the Calvin cycle genes organized in chemoautotrophic bacteria in which the *cbbLS* is not clustered with other *cbb* genes? Is CO_2 fixation controlled by a phospho-relay system as is the case in *R. sphaeroides*? Does NADPH activate CbbR in vivo? What is the mechanism by which CbbR activates transcription of the Calvin cycle genes? Is CbbR involved in the regulation of carboxysome gene expression? How did the Calvin cycle evolve? The answer to these, as well as to many other questions, will likely be the subject of a future review.

ACKNOWLEDGMENTS

Some of the material in this article is based on work supported by the Cooperative State Research Service, US Department of Agriculture, under agreement No. 92-37306-7663 (JMS); a National Science Foundation Grant, MCB-9513481 (JMS); a grant from the Royal Society (WGM); and a European Union Human Capital and Mobility Institutional Grant (ERBCHBGCT 930293) (WGM). We wish to thank K Verschueren and T Wilkinson of the Department of Chemistry, University of York, for helpful discussions on CysB structural data.

Literature Cited

1. Abdelal ATH, Schlegel HG. 1974. Purification and regulatory properties of phosphoribulokinase from *Hydrogenomonas eutropha* H16. *Biochem. J.* 139:481–89
2. Amachi T, Bowien B. 1979. Characterization of two fructose bisphosphatase isoenzymes from the hydrogen bacterium *Nocardia opaca* 1b. *J. Gen. Microbiol.* 113: 347–56
3. Andersen K, Wilke-Douglas M. 1987. Genetic and physical mapping and ex-
pression in *Pseudomonas aeruginosa* of the chromosomally encoded ribulose bisphosphate carboxylase genes of *Alcaligenes eutrophus*. *J. Bacteriol.* 169:1997–2004
4. Ballard RW, MacElroy RD. 1971. Phosphoenolpyruvate, a new inhibitor of phosphoribulokinase in *Pseudomonas facilis*. *Biochem. Biophys. Res. Commun.* 44: 614–18
5. Beudeker RF, Cannon GC, Kuenen JG,

Shively JM. 1980. Relations between D-ribulose-1,5-bisphosphate carboxylase, carboxysomes, and CO_2 fixing capacity in the obligate chemolithotroph *Thiobacillus neapolitanus* grown under different limitations in the chemostat. *Arch. Microbiol.* 124:185–89

6. Beudeker RF, Gottschal JC, Kuenen JG. 1982. Reactivity versus flexibility in thiobacilli. *Antonie van Leeuwenhoek J. Microbiol. Serol.* 48:39–51

7. Biederman M, Westphal K. 1979. Chemical composition and stability of Nb_1-particles from *Nitrobacter agilis*. *Arch. Microbiol.* 121:187–91

8. Bömmer D, Schäferjohann J, Bowien B. 1996. Identification of $cbbB_c$ as an additional distal gene of the chromosomal *cbb* CO_2 fixation operon from *Ralstonia eutropha*. *Arch. Microbiol.* 166:245–51

9. Bowien B, Friedrich B, Friedrich C. 1984. Involvement of megaplasmids in heterotrophic derepression of the carbon-dioxide assimilating enzyme system in *Alcaligenes* spp. *Arch. Microbiol.* 139:305–10

10. Cannon GC, Shively JM. 1983. Characterization of a homogenous preparation of carboxysomes from *Thiobacillus neapolitanus*. *Arch. Microbiol.* 134:52–59

11. Caspi R, Haygood MG, Tebo BM. 1996. Unusual ribulose-1,5-bisphosphate carboxylase/oxygenase genes from a marine manganese-oxidizing bacterium. *Microbiology* 142:2549–59

12. Cavanaugh CM. 1994. Microbial symbiosis: patterns of diversity in the marine environment. *Am. Zool.* 34:79–89

13. Cavanaugh CM, Levering PR, Maki JS, Mitchell R, Lidstrom ME. 1987. Symbiosis of methylotrophic bacteria and deep-sea mussels. *Nature* 325:346–48

14. Cavanaugh CM, Robinson JJ. 1995. CO_2 fixation in chemoautotroph-invertebrate symbiosis: expression of form I and form II RuBisCO. In *Microbial Growth on C_1 Compounds*, ed. ME Lidstrom, FR Tabita, pp. 285–92. Dordrecht: Kluwer

15. Chang M, Crawford IP. 1991. In vitro determination of the effect of indoleglycerol phosphate on the interaction of purified TrpI protein with its DNA-binding sites. *J. Bacteriol.* 173:1590–97

16. Chen KY, Morris JC. 1972. Kinetics of oxidation of aqueous sulfide by O_2. *Environ. Sci. Technol.* 6:529–37

17. Chen P, Andersson DI, Roth JR. 1994. The control region of the *pdu/cob* regulon in *Salmonella typhimurium*. *J. Bacteriol.* 176:5474–82

18. Chen Z, Green D, Westhoff C, Spreitzer RJ. 1990. Nuclear mutation restores the reduced CO_2/O_2 specificity of ribulose bisphosphate carboxylase/oxygenase in a temperature-conditional chloroplast mutant of *Chlamydomonas reinhardtii*. *Arch. Biochem. Biophys.* 283:60–67

19. Chung SY, Yaguchi T, Nishihara H, Igarashi Y, Kodama T. 1993. Purification of form L_2 RuBisCO from a marine obligately autotrophic hydrogen-oxidizing bacterium. *FEMS Microbiol. Lett.* 109:49–54

20. Codd GA. 1988. Carboxysomes and ribulose bisphosphate carboxylase/oxygenase. In *Advances in Microbial Physiology*, ed. AH Ross, DW Tempest, pp. 115–64. London: Academic

21. Cohen Y, de Jonge I, Kuenen JG. 1979. Excretion of glycolate by *Thiobacillus neapolitanus* grown in continuous culture. *Arch. Microbiol.* 122:189–94

22. Corliss JB, Dymond J, Gordon J, Edmond JM, Herzen RPV, et al. 1979. Submarine thermal springs on the Galapagos Rift. *Science* 203:1073–83

23. Croes LM, Meijer WG, Dijkhuizen L. 1991. Regulation of methanol oxidation and carbon dioxide fixation in *Xanthobacter* strain 25a grown in continuous culture. *Arch. Microbiol.* 155:159–63

24. Delwiche CF, Palmer JD. 1996. Rampant horizontal transfer and duplication of rubisco genes in eubacteria and plastids. *Mol. Biol. Evol.* 13:873–82

25. Dijkhuizen L, Harder W. 1979. Regulation of autotrophic and heterotrophic metabolism in *Pseudomonas oxalaticus* OX1. Growth on mixtures of acetate and formate in continuous culture. *Arch. Microbiol.* 123:47–53

26. Dijkhuizen L, Knight M, Harder W. 1978. Metabolic regulation in *Pseudomonas oxalaticus* OX1. Autotrophic and heterotrophic growth on mixed substrates. *Arch. Microbiol.* 116:77–83

27. Distel DL, Felbeck H, Cavanaugh CM. 1994. Evidence for phylogenetic congruence among sulfur-oxidizing chemoautotrophic bacterial endosymbionts and their bivalve hosts. *J. Mol. Biol.* 38:533–42

28. Distel DL, Lee HKW, Cavanaugh CM. 1995. Intracellular coexistence of methano- and thioautotrophic bacteria in a hydrothermal vent mussel. *Proc. Natl. Acad. Sci. USA* 92:9598–602

29. Dubilier N, Giere O, Distel DL, Cavanaugh CM. 1995. Characterization of chemoautotrophic bacterial symbionts in a gutless marine worm (Oligochaeta, Annelida) by phylogenetic 16S rRNA

sequence analysis and in situ hybridization. *Appl. Environ. Microbiol.* 61:2346–50

30. Ebert A. 1982. *Ribulose-1,5-bisphosphate carboxylase in Nitrobacter.* PhD thesis. Univ. Hamburg, Hamburg, Germany. 84 pp.

31. Ecker C, Reh M, Schlegel HG. 1986. Enzymes of the autotrophic pathway in mating partners and transconjugants of *Nocardia opaca* 1b and *Rhodococcus erythropolis. Arch. Microbiol.* 145:280–86

32. Eisen JA, Smith SW, Cavanaugh CM. 1992. Phylogenetic relationships of chemoautotrophic bacterial symbionts of *Solemya velum* Say (Mollusca: Bivalvia) determined by 16S rRNA gene sequence analysis. *J. Bacteriol.* 174:3416–21

33. Emerson S, Cranston RE, Liss PS. 1979. Redox species in a reducing fjord: equilibrium and kinetic considerations. *Deep-Sea Res.* 26:859–78

34. English RS, Jin S, Shively JM. 1995. Use of electroporation to generate a *Thiobacillus neapolitanus* carboxysome mutant. *Appl. Environ. Microbiol.* 61:3256–60

35. English RS, Lorbach SC, Qin X, Shively JM. 1994. Isolation and characterization of a carboxysome shell gene from *Thiobacillus neapolitanus. Mol. Microbiol.* 12:647–54

36. English RS, Williams CA, Lorbach SC, Shively JM. 1992. Two forms of ribulose-1,5-bisphosphate carboxylase/oxygenase from *Thiobacillus denitrificans. FEMS Microbiol. Lett.* 94:111–19

37. Eraso JM, Kaplan S. 1996. Complex regulatory activities associated with the histidine kinase PrrB in expression of photosynthesis genes in *Rhodobacter sphaeroides* 2.4.1. *J. Bacteriol.* 178:7037–46

38. Falcone DL, Tabita FR. 1993. Complementation analysis and regulation of CO_2 fixation gene expression in a ribulose-1,5-bisphosphate carboxylase/oxygenase deletion strain of *Rhodospirillum rubrum. J. Bacteriol.* 175:5066–77

39. Fisher CR, Brooks JM, Vodenichar JS, Zande JM, Childress JJ, et al. 1993. The co-occurrence of methanotrophic and chemoautotrophic sulfur-oxidizing bacterial symbionts in a deep-sea mussel. *Marine Ecol.* 14:277–89

40. Fisher RF, Long SR. 1993. Interactions of NodD at the *nod* box: NodD binds to two distinct sites on the same face of the helix and induces a bend in the DNA. *J. Mol. Biol.* 233:336–48

41. Freter A, Bowien B. 1994. Identification of a novel gene, *aut*, involved in autotrophic growth of *Alcaligenes eutrophus. J. Bacteriol.* 176:5401–8

42. Friedberg D, Kaplan A, Ariel R, Kessel M, Seijffers J. 1989. The 5′-flanking region of the gene encoding the large subunit of ribulose-1,5-bisphosphate carboxylase oxygenase is crucial for growth of the cyanobacterium *Synechococcus* sp. strain PCC 7942 at the level for CO_2 in air. *J. Bacteriol.* 171:6069–76

43. Friedrich B, Hogrefe C, Schlegel HG. 1981. Naturally occurring genetic transfer of hydrogen-oxidizing ability between strains of *Alcaligenes eutrophus. J. Bacteriol.* 147:198–205

44. Friedrich CG. 1982. Derepression of hydrogenase during limitation of electron donors and derepression of ribulose bisphosphate carboxylase during carbon limitation of *Alcaligenes eutrophus. J. Bacteriol.* 149:203–10

45. Friedrich CG, Friedrich B, Bowien B. 1981. Formation of enzymes of autotrophic metabolism during heterotrophic growth of *Alcaligenes eutrophus. J. Gen. Microbiol.* 122:69–78

46. Gale NL, Beck JV. 1966. Competitive inhibition of phosphoribulokinase by AMP. *Biochem. Biophys. Res. Commun.* 22:792–96

47. Gibson JL, Tabita FR. 1987. Organization of phosphoribulokinase and ribulose bisphosphate carboxylase/oxygenase genes in *Rhodopseudomonas* (Rhodobacter) *sphaeroides. J. Bacteriol.* 169:3685–90

48. Gibson JL, Tabita FR. 1993. Nucleotide sequence and functional analysis of CbbR, a positive regulator of the Calvin cycle operons of *Rhodobacter sphaeroides. J. Bacteriol.* 175:5778–84

49. Gibson JL, Tabita FR. 1996. The molecular regulation of the reductive pentose phosphate pathway in proteobacteria and cyanobacteria. *Arch. Microbiol.* 166:141–50

50. Gibson JL, Tabita FR. 1997. Analysis of the *cbbXYZ* operon in *Rhodobacter sphaeroides. J. Bacteriol.* 179:663–69

51. Gottschal JC, Kuenen JG. 1980. Mixotrophic growth of *Thiobacillus* A2 on acetate and thiosulfate as growth limiting substrates in the chemostat. *Arch. Microbiol.* 126:33–42

52. Grassle JF. 1985. Hydrothermal vent animals: distribution and biology. *Science* 229:713–17

53. Hallbeck L, Stahl F, Pedersen K. 1993. Phylogeny and phenotypic characterization of the stalk-forming and iron-oxidizing bacterium *Gallionella ferruginea. J.*

Gen. Microbiol. 139:1531–35

54. Hallenbeck PL, Lerchen R, Hessler P, Kaplan S. 1990. Roles of CfxA, CfxB, and external electron acceptors in regulation of ribulose-1,5-bisphosphate carboxylase/oxygenase expression in *Rhodobacter sphaeroides*. *J. Bacteriol.* 172:1736–48

55. Hartman FC, Harpel MR. 1994. Structure, function, regulation, and assembly of D-ribulose-1,5-bisphosphate carboxylase/oxygenase. *Annu. Rev. Biochem.* 63:197–234

56. Hernandez JM, Baker SH, Lorbach SC, Shively JM, Tabita FR. 1996. Deduced amino acid sequence, functional expression, and unique enzymatic properties of the form I and form II ribulose bisphosphate carboxylase/oxygenase from the chemoautotrophic bacterium *Thiobacillus denitrificans*. *J. Bacteriol.* 178:347–56

57. Holthuijzen YA, van Breeman JFL, Konings WN, van Bruggen EFJ. 1986. Electron microscopic studies of carboxysomes of *Thiobacillus neapolitanus*. *Arch. Microbiol.* 144:258–62

58. Holthuijzen YA, van Breemen JFL, Kuenen JG, Konings WN. 1986. Protein composition of the carboxysomes of *Thiobacillus neapolitanus*. *Arch. Microbiol.* 144:398–404

59. Husemann M, Klintworth R, Büttcher V, Salnikow J, Weissenborn C, et al. 1988. Chromosomally and plasmid-encoded gene clusters for CO_2 fixation (cfx) genes in *Alcaligenes eutrophus*. *Mol. Gen. Genet.* 214:112–20

60. Igarashi Y, Kodama T. 1995. Genes related to carbon dioxide fixation in *Hydrogenovibrio marinus* and *Pseudomonas hydrogenothermophila*. In *Microbial Growth on C_1 Compounds*, ed. ME Lidstrom, FR Tabita, pp. 88–93. Dordrecht: Kluwer

61. Im D, Friedrich CG. 1983. Fluoride, hydrogen and formate activate ribulose bisphosphate carboxylase formation in *Alcaligenes eutrophus*. *J. Bacteriol.* 154:803–8

62. Jackson JB, Crofts AR. 1968. Energy-linked reduction of nicotinamide adenine dinucleotides in cells of *Rhodospirillum rubrum*. *Biochem. Biophys. Res. Commun.* 32:908–15

63. Jannasch HW, Mottl MJ. 1985. Geomicrobiology of deep-sea hydrothermal vents. *Science* 229:717–25

64. Johnson EJ, MacElroy RD. 1973. Regulation in the chemolithotroph *Thiobacillus neapolitanus*: fructose-1,6-diphosphate.

Arch. Microbiol. 93:23–28

65. Jordan DB, Ogren WL. 1981. Species variation in the specificity of ribulose biphosphate carboxylase/oxygenase. *Nature* 291:513–15

66. Jorgensen BB, Revsbech NP. 1983. Colorless sulfur bacteria, *Beggiatoa* spp. and *Thiovulum* spp. in O_2 and H_2S microgradients. *Appl. Environ. Microbiol.* 45:1261–70

67. Joshi H, Tabita FR. 1996. A global two component signal transduction system that integrates the control of photosynthesis, carbon dioxide assimilation, and nitrogen fixation. *Proc. Natl. Acad. Sci. USA* 93:14515–20

68. Jouanneau Y, Tabita FR. 1986. Independent regulation of synthesis of form I and form II ribulose bisphosphate carboxylase-oxygenase in *Rhodopseudomonas sphaeroides*. *J. Bacteriol.* 165:620–24

69. Kaplan A, Friedberg D, Schwarz R, Ariel R, Seijffers J, et al. 1989. The "CO_2 concentrating mechanism" of cyanobacteria: physiological, molecular and theoretical studies. *Photosynth. Res.* 17:243–55

70. Kiesow LA, Lindsley BF, Bless JW. 1977. Phosphoribulokinase from *Nitrobacter winogradski*: activation by reduced nicotinamide dinucleotide and inhibition by pyridoxal phosphate. *J. Bacteriol.* 130:20–25

71. Knight M, Dijkhuizen L, Harder W. 1978. Metabolic regulation in *Pseudomonas oxalaticus* OX1. Enzyme and coenzyme concentration changes during substrate transition experiments. *Arch. Microbiol.* 116:85–90

72. Kobayashi H, Viale AM, Takabe T, Akazawa T, Wada K, et al. 1991. Sequence and expression of genes encoding the large and small subunits of ribulose-1,5-bisphosphate carboxylase/oxygenase from *Chromatium vinosum*. *Gene* 97:55–62

73. Krieger TJ, Miziorko HM. 1986. Affinity labeling and purification of spinach leaf ribulose-5-phosphate kinase. *Biochemistry* 25:3496–501

74. Krueger DM, Cavanaugh CM. 1997. Phylogenetic diversity of bacterial symbionts of *Solemya* hosts based on comparative sequence analysis of 16S rRNA genes. *Appl. Environ. Microbiol.* 63:91–98

75. Kuenen G, Bos P. 1989. Habitats and ecological niches of chemolitho(auto)trophic bacteria. In *Autotrophic Bacteria*, ed. HG Schlegel, B Bowien, pp. 53–80. Madison: Sci. Tech.

76. Kusano T, Sugawara K. 1993. Specific binding of *Thiobacillus ferrooxidans* RbcR to the intergenic sequence between

the *rbc* operon and the *rbcR* gene. *J. Bacteriol.* 175:1019–25

77. Kusano T, Takeshima T, Inoue C, Sugawara K. 1991. Evidence for two sets of structural genes coding for ribulose bisphosphate carboxylase in *Thiobacillus ferrooxidans. J. Bacteriol.* 173:7313–23

78. Kusian B, Bednardski R, Husemann M, Bowien B. 1995. Characterization of the duplicate ribulose-1,5-bisphosphate carboxylase genes and *cbb* promoters of *Alcaligenes eutrophus. J. Bacteriol.* 177:4442–50

79. Kusian B, Bowien B. 1995. Operator binding of the CbbR protein, which activates the duplicate *cbb* CO$_2$ assimilation operons of *Alcaligenes eutrophus. J. Bacteriol.* 177:6568–74

80. Kusian B, Bowien B. 1997. Organization and regulation of *cbb* CO$_2$ assimilation genes in autotrophic bacteria. *FEMS Microbiol. Rev.* 21:135–55

81. Lanaras T, Codd GA. 1981. Ribulose-1,5-bisphosphate carboxylase and polyhedral bodies of *Chlorogloeopsis fritschii. Planta* 153:279–85

82. Lanaras T, Cook CM, Wood AP, Kelly DP, Codd GA. 1991. Purification of ribulose-1,5-bisphosphate carboxylase/oxygenase and of carboxysomes from *Thiobacillus thyasiris* the putative symbiont of *Thyasira flexuosa* (Montagu). *Arch. Microbiol.* 156:338–43

83. Lanaras T, Hawthornthwaite AM, Codd GA. 1985. Localization of carbonic anhydrase in the cyanobacterium *Chlorogloeopsis fritschii. FEMS Microbiol. Lett.* 26:285–88

84. Lane DJ, Harrison AP, Stahl D, Pace B, Giovannoni S, et al. 1992. Evolutionary relationships among sulfur- and iron-oxidizing bacteria. *J. Bacteriol.* 174:269–78

85. Lawrence JG. 1997. Selfish operons and speciation by gene transfer. *Trends Microbiol.* 5:355–59

86. Leadbeater L, Bowien B. 1984. Control of autotrophic carbon assimilation in *Alcaligenes eutrophus* by the inactivation and reactivation of phosphoribulokinase. *J. Bacteriol.* 157:95–99

87. Leadbeater L, Siebert K, Schobert P, Bowien. B. 1982. Relationship between activities and protein levels of ribulose bisphosphate carboxylase and phosphoribulokinase in *Alcaligenes eutrophus. FEMS Microbiol. Lett.* 14:263–66

88. Lehmicke LG, Lidstrom ME. 1985. Organization of genes necessary for growth of the hydrogen-methanol autotroph *Xan-*

thobacter sp. strain H4-14 on hydrogen and carbon dioxide. *J. Bacteriol.* 162:1244–49

89. Lolis E, Petsko GA. 1990. Crystallographic analysis of the complex between triosephosphate isomerase and 2-phosphoglycolate at 2.5 Å resolution: implication for catalysis. *Biochemistry* 29:6619–25

90. Lorbach SC, Shively JM. 1995. Identification, isolation, and sequencing of the ribulose bisphosphate carboxylase/oxygenase genes (cbbRI and cbbRII) in *Thiobacillus denitrificans. Abstr. Annu. Meet. Am. Soc. Microbiol.,* p. 502 (Abstr.)

91. MacElroy RD, Johnson EJ, Johnson MK. 1968. Characterization of ribulose diphosphate carboxylase and phosphoribulokinase from *Thiobacillus thioparus* and *Thiobacillus neapolitanus. Arch. Biochem. Biophys.* 127:310–16

92. MacElroy RD, Johnson EJ, Johnson MK. 1969. Control of ATP-dependent CO$_2$ fixation in extracts of *Hydrogenomonas facilis:* NADH regulation of phosphoribulokinase. *Arch. Biochem. Biophys.* 131:272–75

93. MacElroy RD, Mack HM, Johnson EJ. 1972. Properties of phosphoribulokinase from *Thiobacillus neapolitanus. J. Bacteriol.* 112:532–38

94. March JJ, Lebherz HG. 1992. Fructose bisphosphate aldolases: an evolutionary history. *Trends Biochem. Sci.* 17:110–13

95. Marco E, Martinez I, Ronen-Tarazi M, Orus I, Kaplan A. 1994. Inactivation of *ccmO* in *Synechococcus* sp. strain PCC 7942 results in a mutant requiring high levels of CO$_2$. *Appl. Environ. Microbiol.* 60:1018–20

96. Martinez I, Orus I, Marco E. 1997. Carboxysome structure and function in a mutant of *Synechococcus* that requires high levels of CO$_2$ for growth. *Plant Physiol. Biochem.* 35:137–46

97. McFadden BA, Shively JM. 1991. Bacterial assimilation of carbon dioxide by the Calvin cycle. In *Variations in Autotrophic Life,* ed. JM Shively, LL Barton, pp. 25–49. London: Academic

98. McKay RML, Gibbs SP, Espie GS. 1993. Effect of dissolved inorganic carbon on the expression of carboxysomes, localization of rubisco and mode of inorganic carbon transport in cells of the cyanobacterium *Synechococcus* UTEX 625. *Arch. Microbiol.* 159:21–29

99. Meijer WG. 1994. The Calvin cycle enzyme phosphoglycerate kinase of *Xanthobacter flavus* required for autotrophic CO$_2$ fixation is not encoded by the *cbb*

operon. *J. Bacteriol.* 176:6120–26

100. Meijer WG. 1996. Genetics of CO₂ fixation in methylotrophs. In *Microbial Growth on C₁ Compounds*, ed. ME Lidstrom, FR Tabita, pp. 118–25. Dordrecht: Kluwer

101. Meijer WG, Arnberg AC, Enequist HG, P Terpstra, Lidstrom ME, et al. 1991. Identification and organization of carbon dioxide fixation genes in *Xanthobacter flavus*. *Mol. Gen. Genet.* 225:320–30

102. Meijer WG, Croes LM, Jenni B, Lehmicke LG, Lidstrom ME, et al. 1990. Characterization of *Xanthobacter* strains H4-14 and 25a and enzyme profiles after growth under autotrophic and heterotrophic growth conditions. *Arch. Microbiol.* 153:360–67

103. Meijer WG, de Boer P, van Keulen G. 1997. *Xanthobacter flavus* employs a single triosephosphate isomerase for heterotrophic and autotrophic metabolism. *Microbiology* 143:1925–31

104. Meijer WG, Dijkhuizen L. 1988. Regulation of autotrophic metabolism in *Pseudomonas oxalaticus* OX1 wild-type and an isocitrate-lyase-deficient mutant. *J. Gen. Microbiol.* 134:3231–37

105. Meijer WG, Enequist HG, Terpstra P, Dijkhuizen L. 1990. Nucleotide sequences of the genes encoding fructosebisphosphatase and phosphoribulokinase from *Xanthobacter flavus* H4-14. *J. Gen. Microbiol.* 136:2225–30

106. Meijer WG, van den Bergh ERE, Smith LM. 1996. Induction of the *gap-pgk* operon encoding glyceraldehyde-3-phosphate dehydrogenase and 3-phosphoglycerate kinase of *Xanthobacter flavus* requires the LysR type transcriptional activator CbbR. *J. Bacteriol.* 178:881–87

107. Miyajima T. 1992. Biological manganese oxidation in a lake: I. Occurrence and distribution of *Metallogenium* sp. and its kinetic properties. *Arch. Hydrobiol.* 124:317–35

108. Moreira D, Amils R. 1997. Phylogeny of *Thiobacillus cuprinus* and other mixotrophic thiobacilli: proposal for *Thiomonas* gen. nov. *Int. J. Syst. Bacteriol.* 47:522–28

109. Morse D, Salois P, Markovic P, Hastings JW. 1995. A nuclear-encoded form II rubisco in dinoflagellates. *Science* 268:1622–24

110. Moyer CL, Dobbs FC, Karl DM. 1995. Phylogenetic diversity of the bacterial community from a microbial mat at an active, hydrothermal vent system, Loihi Seamount, Hawaii. *Appl. Environ. Microbiol.* 61:1555–62

111. Muyzer G, Teske A, Wirsen CO, Jannasch

HW. 1995. Phylogenetic relationships of *Thiomicrospira* species and their identification in deep-sea hydrothermal vent samples by denaturing gradient gel electrophoresis of 16S rDNA fragments. *Arch. Microbiol.* 164:165–72

112. Nicholas KB, Nicholas HB Jr, Deerfield DW II. 1997. Visualization of genetic variation. *Embnet. News* 4:14

113. Paoli GC, Morgan NS, Tabita FR, Shively JM. 1995. Expression of the *cbbLcbbS* and *cbbM* genes and distinct organization of the *cbb* Calvin cycle structural genes of *Rhodobacter capsulatus*. *Arch. Microbiol.* 164:396–405

114. Paoli GC, Soyer F, Shively J, Tabita FR. 1997. *Rhodobacter capsulatus* genes encoding form I ribulose-1,5-bisphosphate carboxylase/oxygenase (*cbbLS*) and neighbouring genes were acquired by a horizontal gene transfer. *Microbiology* 144:219–27

115. Peters KR. 1974. Charakteriserung eines Phagenahnlichen Partikels aus Zellen von *Nitrobacter*. II. Struktur und Grosse. *Arch. Microbiol.* 97:129–40

116. Polz MF, Cavanaugh CM. 1995. Dominance of one bacterial phylotype at a Mid-Atlantic Ridge hydrothermal vent site. *Proc. Natl. Acad. Sci. USA* 92:7232–36

117. Polz MF, Distel DL, Zarda B, Amann R, Felbeck H, et al. 1994. Phylogenetic analysis of a highly specific association between ectosymbiotic, sulfur-oxidizing bacteria and a marine nematode. *Appl. Environ. Microbiol.* 60:4461–67

118. Price GD, Badger MR. 1989. Isolation and characterization of high CO₂-requiring mutants of the cyanobacterium *Synechococcus* PCC 7942: two phenotypes that accumulate inorganic carbon but are apparently unable to generate CO₂ within the carboxysome. *Plant Physiol.* 91:514–25

119. Price GD, Badger MR. 1991. Evidence for the role of carboxysomes in the cyanobacterial CO₂-concentrating mechanism. *Can. J. Bot.* 69:963–73

120. Price GD, Coleman JR, Badger MR. 1992. Association of carbonic anhydrase activity with carboxysomes isolated from the cyanobacterium *Synechococcus* PCC7942. *Plant Physiol.* 100:784–93

121. Price GD, Howitt SM, Harrison K, Badger MR. 1993. Analysis of a genomic DNA region from the cyanobacterium *Synechococcus* sp. strain PCC7942 involved in carboxysome assembly and function. *J. Bacteriol.* 175:2871–79

122. Pronk JT, de Bruyn JC, Bos P, Kuenen JG. 1992. Anaerobic growth of *Thiobacillus*

ferrooxidans. Appl. Environ. Microbiol.
58: 2227–30

123. Pulgar C, Gaete L, Allende J, Orellana O, Jordana X, et al. 1991. Isolation and nucleotide sequence of the *Thiobacillus ferrooxidans* genes for the small and large subunits of ribulose-1,5-bisphosphate carboxylase/oxygenase. *FEBS Lett.* 292: 85–89

124. Purohit K, McFadden BA, Cohen AL. 1976. Purification, quaternary structure, composition, and properties of D-ribulose-1,5-bisphosphate carboxylase from *Thiobacillus intermedius. J. Bacteriol.* 127:505–15

125. Qian Y, Tabita FR. 1996. A global signal transduction system regulates aerobic and anaerobic CO_2 fixation in *Rhodobacter sphaeroides. J. Bacteriol.* 178:12–18

126. Read BA, Tabita FR. 1992. A hybrid ribulose bisphosphate carboxylase/oxygenase enzyme exhibiting a substantial increase in substrate specificity factor. *Biochemistry* 31:5553–60

127. Read BA, Tabita FR. 1992. Amino acid substitutions in the small subunit of ribulose-1,5-bisphosphate carboxylase/oxygenase that influence catalytic activity of the holoenzyme. *Biochemistry* 31: 519–25

128. Read BA, Tabita FR. 1994. High substrate specificity factor ribulose bisphosphate carboxylase/oxygenase from eukaryotic marine algal and properties of recombinant cyanobacterial rubisco containing "algal" residue modifications. *Arch. Biochem. Biophys.* 312:210–18

129. Reinhold L, Kosloff R, Kaplan A. 1991. A model for inorganic carbon fluxes and photosynthesis in cyanobacterial carboxysomes. *Can. J. Bot.* 69:984–88

130. Reutz I, Schobert P, Bowien B. 1982. Effect of phosphoglycerate mutase deficiency on heterotrophic and autotrophic carbon metabolism of *Alcaligenes eutrophus. J. Bacteriol.* 151:8–15

131. Richardson DJ, King GF, Kelly DJ, McEwan AG, Ferguson SJ, et al. 1988. The role of auxiliary oxidants in maintaining redox balance during phototrophic growth of *Rhodobacter capsulatus* on propionate or butyrate. *Arch. Microbiol.* 150:131–37

132. Rindt KP, Ohmann E. 1969. NADH and AMP as allosteric effectors of ribulose-5-phosphate kinase in *Rhodopseudomonas sphaeroides. Biochem. Biophys. Res. Commun.* 36:357–64

133. Rippel S, Bowien B. 1984. Phosphoribulokinase from *Rhodopseudomonas acidophila. Arch. Microbiol.* 139:207–12

134. Ronen-Tarazi M, Lieman-Hurwitz J,

Gabay C, Orus MI, Kaplan A. 1995. The genomic region of *rbcLS* in *Synechococcus* sp. PCC7942 contains genes involved in the ability to grow under low CO_2 concentration and in chlorophyll biosynthesis. *Plant Physiol.* 108:1461–69

135. Rowan R, Whitney SM, Fowler A, Yellowlees D. 1996. Rubisco in marine symbiotic dinoflagellates: form II enzymes in eukaryotic oxygenic phototrophs encoded by a nuclear multigene family. *Plant Cell* 8:539–53

136. Ruby EG, Wirsen CO, Jannasch HW. 1981. Chemolithotrophic sulfur-oxidizing bacteria from the Galapagos Rift hydrothermal vent. *Appl. Environ. Microbiol.* 42:317–24

137. Santiago B, Meyer O. 1997. Purification and molecular characterization of the H_2 uptake membrane-bound NiFe-hydrogenase from the carboxidotrophic bacterium *Oligotropha carboxidovorans. J. Bacteriol.* 179:6053–60

138. Sarbu SM, Kane TC, Kinkle BK. 1996. A chemoautotrophically based cave ecosystem. *Science* 272:1953–55

139. Satoh R, Himeno M, Wadano A. 1997. Carboxysomal diffusion resistance to ribulose-1,5-bisphosphate and 3-phosphoglycerate in the cyanobacterium *Synechococcus* PCC7942. *Plant Cell Physiol.* 38: 769–75

140. Schäferjohann J, Bednarski R, Bowien B. 1996. Regulation of CO_2 assimilation in *Ralstonia eutropha*: premature transcription termination within the *cbb* operon. *J. Bacteriol.* 178:6714–19

141. Schäferjohann J, Yoo J-G, Bowien B. 1995. Analysis of the genes forming the distal parts of the two CO_2 fixation operons from *Alcaligenes eutrophus. Arch. Microbiol.* 163:291–99

142. Schell MA. 1993. Molecular biology of the LysR family of transcriptional regulators. *Annu. Rev. Microbiol.* 47:597–626

143. Sganga MW, Bauer CE. 1992. Regulatory factors controlling photosynthetic reaction center and light-harvesting gene expression in *Rhodobacter capsulatus. Cell* 68:945–54

144. Shively JM, Ball F, Brown DH, Saunders RE. 1973. Functional organelles in prokaryotes: polyhedral inclusions (carboxysomes) in *Thiobacillus neapolitanus. Science* 182:584–86

145. Shively JM, Ball FL, Kline BW. 1973. Electron microscopy of the carboxysomes (polyhedral bodies) of *Thiobacillus neapolitanus. J. Bacteriol.* 116:1405–11

146. Shively JM, Bradburne CE, Aldrich HC, Bobik TA, Mehlman JL, et al. 1998.

Sequence homologs of the carboxysomal polypeptide CsoS1 of the thiobacilli are present in cyanobacteria and enteric bacteria that form carboxysomes/polyhedral bodies. *Can. J. Bot.* 76: In press

147. Shively JM, Bryant DA, Fuller RC, Konopka AE, Stevens SE, et al. 1988. Functional inclusions in prokaryotic cells. *Int. Rev. Cytol.* 113:35–100

148. Shively JM, English RS. 1991. The carboxysome, a prokaryotic organelle: a mini review. *Can. J. Bot.* 69:957–62

149. Shively JM, Lorbach SC, Jin S, Baker SH. 1996. Carboxysomes: the genes of *Thiobacillus neapolitanus.* In *Microbial Growth on C1 Compounds*, ed. ME Lidstrom, FR Tabita, pp. 56–63. Dordrecht: Kluwer

150. Siebert K, Bowien B. 1984. Evidence for an octameric structure of phosphoribulokinase from *Alcaligenes eutrophus. Biochim. Biophys. Acta* 787:208–14

151. Siebert K, Schobert P, Bowien B. 1981. Purification, some catalytic and molecular properties of phosphoribulokinase from *Alcaligenes eutrophus. Biochim. Biophys. Acta* 658:35–44

152. Singer PC, Stumm W. 1970. Acidic mine drainage: the rate limiting step. *Science* 167:1121–23

153. Spreitzer RJ. 1993. Genetic dissection of rubisco structure and function. *Annu. Rev. Plant Physiol. Plant Mol. Biol.* 44:411–34

154. Stojiljkovic I, Baumler AJ, Heffron F. 1995. Ethanolamine utilization in *Salmonella typhimurium*: nucleotide sequence, protein expression, and mutational analysis of the *cchA, cchB, eutE, eutJ, eutG,* and *eutH* gene cluster. *J. Bacteriol.* 177:1357–66

155. Stoner MT, Shively JM. 1993. Cloning and expression of the D-ribulose-1,5-bisphosphate carboxylase/oxygenase form II gene from *Thiobacillus intermedius* in *Escherichia coli. FEMS Microbiol. Lett.* 107:287–92

156. Strecker M, Sickinger E, English RS, Shively JM, Bock E. 1994. Calvin cycle genes in *Nitrobacter vulgaris* T3. *FEMS Microbiol. Lett.* 120:45–50

157. Swift H, Leser GP. 1989. Cytochemical studies on prochlorophytes: localization of DNA and ribulose-1,5-bisphosphate carboxylase-oxygenase. *J. Phycol.* 25:149–52

158. Tabita FR. 1980. Pyridine nucleotide control and subunit structure of phosphoribulokinase from photosynthetic bacteria. *J. Bacteriol.* 143:1275–80

159. Tabita FR. 1988. Molecular and cellular regulation of autotrophic carbon dioxide fixation in microorganisms. *Microbiol. Rev.* 52:155–89

160. Tabita FR. 1995. The biochemistry and metabolic regulation of carbon metabolism and CO_2 fixation in purple bacteria. In *Anoxygenic Photosynthetic Bacteria*, ed. RE Blankenship, MT Madigan, CE Bauer, pp. 885–914. Dordrecht: Kluwer

161. Tabita FR, Gibson JL, Bowien B, Dijkhuizen L, Meijer WG. 1992. Uniform designation for the genes of the Calvin-Benson-Bassham reductive pentose phosphate pathway of bacteria. *FEMS Microbiol. Lett.* 99:107–10

162. Tao K, Fujita N, Ishihama A. 1993. Involvement of the RNA polymerase α subunit C-terminal region in co-operative interaction and transcriptional activation with OxyR protein. *Mol. Microbiol.* 7:859–64

163. Teske A, Alm E, Regan JM, Toze S, Rittmann BE, et al. 1994. Evolutionary relationships among ammonia- and nitrite-oxidizing bacteria. *J. Bacteriol.* 176:6623–30

164. Thompson JD, Higgins DG, Gibson TJ. 1994. Clustal W: improving the sensitivity of progressive multiple sequence alignment through sequence weighting, position-specific gap penalties and weight matrix choice. *Nucl. Acids Res.* 22:4673–80

165. Turpin DH, Miller AG, Canvin DT. 1984. Carboxysome content of *Synechococcus leopoliensis* (Cyanophyta) in response to inorganic carbon. *J. Phycol.* 20:249–53

166. Tyrrell R, Verschueren KH, Dodson EJ, Murshudov GN, Addy C, et al. 1997. The structure of the cofactor-binding fragment of the LysR family member, CysB: a familiar fold with a surprising subunit arrangement. *Structure* 5:1017–32

167. Uemura K, Anwaruzzaman A, Miyachi S, Yokota A. 1997. Ribulose-1,5-bisphosphate carboxylase/oxygenase from thermophilic red algae with strong specificity for CO_2 fixation. *Biochem. Biophys. Res. Commun.* 233:568–71

168. Uemura K, Suzuki Y, Shikanai T, Wadano A, Jensen RG, et al. 1996. A rapid and sensitive method for the determination of relative specificity of RuBisCO from various species by anion-exchange chromatography. *Plant Cell Physiol.* 37:325–31

169. van den Bergh ERE. 1997. *Regulation of CO_2 fixation via the Calvin cycle in the facultative autotroph Xanthobacter flavus.* PhD thesis. Univ. Groningen, Groningen, The Netherlands. 114 pp.

170. van den Bergh ERE, Baker SC, Raggers RJ, Terpstra P, Woudstra EC, et al. 1996.

Primary structure and phylogeny of the Calvin cycle enzymes transketolase and fructosebisphosphate aldolase of *Xantho-bacter flavus. J. Bacteriol.* 178:888–93

171. van den Bergh ERE, Dijkhuizen L, Meijer WG. 1993. CbbR, a LysR-type transcriptional activator, is required for expression of the autotrophic CO_2 fixation enzymes of *Xanthobacter flavus. J. Bacteriol.* 177:6097–104

172. van den Bergh ERE, van der Kooij TAW, Dijkhuizen L, Meijer WG. 1995. Fructosebisphosphatase isoenzymes of the chemoautotroph *Xanthobacter flavus. J. Bacteriol.* 177:5860–64

173. van Keulen G, Girbal L, van den Bergh ERE, Dijkhuizen L, Meijer WG. 1998. The LysR-type transcriptional regulator CbbR controlling autotrophic CO_2 fixation by *Xanthobacter flavus* is an NADPH-sensor. *J. Bacteriol.* 180:1411–17

174. Viale AM, Kobayashi H, Akazawa T. 1989. Expressed genes for plant-type ribulose-1,5-bisphosphate carboxylase/oxygenase in the photosynthetic bacterium *Chromatium vinosum*, which possesses two complete sets of the genes. *J. Bacteriol.* 171:2391–400

175. Viale AM, Kobayashi H, Akazawa T, Henikoff S. 1991. *rbcR*, a gene coding for a member of the LysR family of transcriptional regulators, is located upstream of the expressed set of ribulose-1,5-bisphosphate carboxylase/oxygenase genes in the photosynthetic bacterium *Chromatium vinosum. J. Bacteriol.* 173:5224–29

176. Vlasceanu L, Popa R, Kinkle BK. 1997. Characterization of *Thiobacillus thioparus* LV43 and its distribution in a chemoautotrophically based groundwater ecosystem. *Appl. Environ. Microbiol.* 63:3123–27

177. Wang L, Helmann JD, Winans SC. 1992. The *A. tumefaciens* transcriptional activator OccR causes a bend at a target promoter, which is partially relaxed by a plant tumor metabolite. *Cell* 69:659–67

178. Wang X, Falcone DL, Tabita FR. 1993. Reductive pentose phosphate-independent CO_2 fixation in *Rhodobacter sphaeroides* and evidence that ribulose bis-phosphate carboxylase/oxygenase activity serves to maintain the redox balance of the cell. *J. Bacteriol.* 175:3372–79

179. Warrelmann J, Friedrich B. 1989. Genetic transfer of lithoautotrophy mediated by a plasmid-cointegrate from *Pseudomonas facilis. Arch. Microbiol.* 151:359–64

180. Watson GM, Tabita FR. 1996. Regulation, unique gene organization, and unusual primary structure of carbon fixation genes from a marine phycoerythrin-containing cyanobacterium. *Plant Mol. Biol.* 32:1103–15

181. Watson GM, Tabita FR. 1997. Microbial ribulose-1,5-bisphosphate carboxylase/oxygenase: a molecule for phylogenetic and enzymological investigation. *FEMS Microbiol. Lett.* 146:13–22

182. Windhövel U, Bowien B. 1990. On the operon structure of the *cfx* gene clusters in *Alcaligenes eutrophus. Arch. Microbiol.* 154:85–91

183. Windhövel U, Bowien B. 1991. Identification of *cfxR*, an activator of autotrophic CO_2 fixation in *Alcaligenes eutrophus. Mol. Microbiol.* 5:2695–705

184. Wirsen CO, Jannasch HW. 1980. Deep-sea primary production at the Galápagos hydrothermal vents. *Science* 207:1345–47

185. Wood AP, Kelly DP, Thurston CF. 1977. Simultaneous operation of three catabolic pathways in the metabolism of glucose by *Thiobacillus* A2. *Arch. Microbiol.* 113:265–74

186. Xu HH, Tabita FR. 1994. Positive and negative regulation of sequences upstream of the form II *cbb* CO_2 fixation operon of *Rhodobacter sphaeroides. J. Bacteriol.* 176:7299–308

187. Yaguchi T, Chung SY, Igarashi Y, Kodama T. 1994. Cloning and sequencing of the L_2 form of RuBisCO from a marine obligately autotrophic hydrogen-oxidizing bacterium. *Biosci. Biotech. Biochem.* 58:1733–37

188. Yokoyama K, Hayashi NR, Arai H, Chung SY, Igarashi Y, et al. 1995. Genes encoding RuBisCO in *Pseudomonas hydrogenothermophila* are followed by a novel *cbbQ* gene similar to *nirQ* of the denitrification gene cluster from *Pseudomonas* species. *Gene* 153:75–79

Annu. Rev. Microbiol. 1998. 52:231–86

THE ANTI-SIGMA FACTORS

Kelly T. Hughes

Department of Microbiology, Box 357242, University of Washington, Seattle, Washington 98195; email: hughes@u.washington.edu

Kalai Mathee

Department of Microbiology and Immunology, University of Tennessee, College of Medicine, 858 Madison Avenue, Memphis, Tennessee 38163; email: kmathee@utmem1.utmem.edu

KEY WORDS: transcription, RNA polymerase, anti-anti-sigma, ECF subfamily

ABSTRACT

A mechanism for regulating gene expression at the level of transcription utilizes an antagonist of the sigma transcription factor known as the anti-sigma (anti-σ) factor. The cytoplasmic class of anti-σ factors has been well characterized. The class includes AsiA form bacteriophage T4, which inhibits *Escherichia coli* σ^{70}; FlgM, present in both gram-positive and gram-negative bacteria, which inhibits the flagella sigma factor σ^{28}; SpoIIAB, which inhibits the sporulation-specific sigma factors, σ^{F} and σ^{G}, of *Bacillus subtilis*; RbsW of *B. subtilis*, which inhibits stress response sigma factor σ^{B}; and DnaK, a general regulator of the heat shock response, which in bacteria inhibits the heat shock sigma factor σ^{32}. In addition to this class of well-characterized cytoplasmic anti-sigma factors, a new class of homologous, inner-membrane–bound anti-σ factors has recently been discovered in a variety of eubacteria. This new class of anti-σ factors regulates the expression of so-called extracytoplasmic functions, and hence is known as the ECF subfamily of anti-sigma factors. The range of cell processes regulated by anti-σ factors is highly varied and includes bacteriophage phage growth, sporulation, stress response, flagellar biosynthesis, pigment production, ion transport, and virulence.

CONTENTS

231

0066-4227/98/1001-0231$08.00

INTRODUCTION

The environmental niche occupied by any organism at a given time is defined by the proteins and RNAs encoded in the DNA sequence of its genome and how their levels in the cell are regulated. Adaptation to predictable environmental changes is dependent to a large extent on the ability of an organism's proteins and RNAs to be regulated at the level of gene expression. This review focuses on one aspect of gene regulation carried out in the eubacterial kingdom: inhibition of the σ transcription factor. Transcription in the eubacterial kingdom is facilitated by a complex of proteins called RNA polymerase (for an extensive review, see 145). RNA polymerase has a sequence-specific affinity for the region 5' to coding sequence known as the promoter. The sequence specificity is usually defined by two regions of the 5' noncoding sequence located at about 10 base pairs and 35 base pairs upstream of the mRNA transcriptional start point. These are known as the -10 and -35 regions that define the promoter sequence. There are promoters that lack a good consensus -35 sequence and instead contain an "extended" -10 region that effectively reduces or eliminates the requirement for a -35 binding region (36, 121, 131, 184). Frequently, other 5' sequences, known as upstream activating sequences, are also involved in promoter recognition. They have been found to act in addition to the -10 and -35 promoter sequences or in lieu of the -35 sequence to facilitate promoter recognition. These sequences often bind transcriptional activator proteins that also interact with RNA polymerase to increase promoter recognition under conditions where the transcriptional activator is expressed so as to effect regulation of gene expression.

RNA polymerase is a heteromultimeric complex of five essential protein subunits (reviewed in 145). Four subunits, $\alpha_2\beta\beta'$, copurify as a tightly associated complex known as the core RNA polymerase. Association of core RNA polymerase with one of two families (σ^{70} and σ^{54}) of homologous σ subunits forms the holoenzyme. It is the σ subunit that directs binding and initiation of transcription by RNA polymerase at specific promoter sequences, while either the σ or any of the other subsubunits can interact with transcriptional activator proteins to facilitate promoter recognition. The presence of multiple σ factors provides one mechanism for gene regulation to occur at the level of transcription. A set of genes can be coordinately expressed if their promoter sequences are recognized by a single σ factor. In *Escherichia coli*, the σ^{70} factor is known as the housekeeping σ factor. σ^{70}-holoenzyme is responsible for transcribing

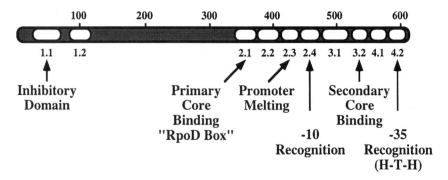

Figure 1   Schematic representation of conserved regions of the σ^{70} family of transcription factors. The conserved regions and their proposed functions are listed.

genes of the essential biosynthetic pathways such as genes required for the biosynthesis of amino acids, nucleotides, enzyme cofactors, cell wall and membrane components, and the genes required for carbon source utilization including enzymes involved in glycolysis, the pentose phosphate shunt and the TCA cycle (reviewed in 82, 92). The remaining σ factors are commonly referred to as alternative sigma factors. These alternative σ factors recognize different -10 and -35 promoter sequences. This confers differential promoter specificity to RNA polymerase allowing sets of genes to be coordinately regulated in response to specific environmental conditions (reviewed in 154, 155). Both the housekeeping and alternative sigma factors share high homology at amino acid level where the conserved regions have similar functions (Figure 1; 86, 154, 163).

Gene regulation in eubacteria has been found to occur at different levels, but the most significant is at the level of transcription initiation. The rate of transcription initiation of a gene is determined both by the intrinsic activity of its promoter and by the interaction of DNA-binding proteins, repressors and activators, with the promoter region. Recently, an additional layer of transcriptional regulation has been uncovered that carries out gene regulation by the inhibition of specific σ factors. An anti-σ factor is defined by the ability to prevent its cognate σ factor to compete for core RNA polymerase. The result of the action of an anti-σ factor is to inhibit transcription from a given set of promoters by inhibiting only the σ factor that is required to recognize that particular promoter sequence. This definition excludes other σ factors that inhibit transcription directed by a given σ factor by simply competing for available core RNA polymerase. The mechanisms used by anti-anti-σ factors to inhibit their specific anti-σ factor vary, ranging from enzymatic modification to export out of the cell. The list of recognized and potential anti-σ factors has more than doubled since the first review in anti-σ factors was written just a few years ago

Table 1   List of anti-sigma factors and their cognate sigma factors

Organism	Sigma factors	Anti-σ factors	Size(aa)	Function
Bacteriophage T4	*E. coli* σ^{70}	AsiA	90	Housekeeping
Bacillus subtilis	σ^F	SpoIIAB	147	Sporulation
	σ^G	SpoIIAB	147	Sporulation
	σ^B	RsbW	160	Stress response
	σ^D	FlgM	97	Flagellar biosynthesis
Salmonella typhimurium	σ^{28}	FlgM	97	Flagellar biosynthesis
Eubacteria	σ^{32}	DnaK	638	Heat shock
Streptomyces coelicolor	σ^{WhiG}	?		Sporulation
ECF subfamily				
Alcaligenes eutrophus	σ^{CnrH}	Orf1	192	Co/Ni resistance
Azotobacter vinelandii	σ^{AlgU}	MucA	194	Alginate biosynthesis
Bacillus subtilis	σ^{SigX}	?		
	σ^{SigV}	Anti-SigV	285	
	σ^{SigZ}	?		
Erwinia amylovora	σ^{22}	?		Virulence factors/ Hypersensitivity
Escherichia coli	σ^E	RseA/MclA	208	Extreme temperature survival
	σ^{FecI}	FecR	317	FeIII-Citrate transport
Haemophilus influenzae	σ^E	RseA/MclA	195	
Myxococcus xanthus	σ^{CarQ}	CarR	221	Caroteroid biosynthesis
Mycobacterium tuberculosis	σ^E	?		
Mycobacterium leprae	σ^E	?		
Salmonella typhimurium	σ^E	?		
Streptomyces coelicolor	σ^E	?		Agarase synthesis
Sulfolobus acidocaldarius	σ^E	?		
Synechocystis sp.	σ^E	?		
Photobacterium SS9	σ^E	Orf2	232	High pressure survival
Pseudomonas aeruginosa	$\sigma^{22(AlgT/AlgU)}$	MucA	194	Alginate biosynthesis
	σ^{PvdS}	?		Pyoverdine biosynthesis
Pseudomonas putida	σ^{PupI}	PupR	316	Pseudobactin transport
Pseudomonas syringae	σ^{HrpL}	?		Virulence factors/ Hypersensitivity
Rhodobacter sphaeroides	σ^E	ChrR	213	Cytochrome c_2 expression

(27). The list of known and potential anti-σ factors and their cognate σ factors is provided in Table 1.

This review will attempt to state what is known about the anti-σ factors discovered to date, the cellular processes they regulate, and how environmental cues determine their fate.

BACTERIOPHAGE T4 AsiA PROTEIN: ANTI-σ^{70} OF *ESCHERICHIA COLI*

Temporal Regulation of T4 Transcription

During growth after infection of *Escherichia coli*, the genes of bacteriophage T4 are differentially transcribed according to when they are needed in phage development. T4 promoters are assigned to one of three classes depending on when in the infection state they are recognized: early, middle, and late. The early promoters are transcribed immediately after infection by the *E. coli* σ^{70}-containing RNA polymerase holoenzyme (E·σ^{70}, reviewed in 264). Two early genes, *mod* and *alt*, encode proteins that modify the α subunits of RNA polymerase by ADP-ribosylation. Two more early genes, *asiA* and *rpbA*, encode proteins, 90 and 100 amino acids in length, respectively, that associate with the modified polymerase (99). At this time, RNA polymerase is unable to transcribe the early promoters, but is able to transcribe from the middle promoters. One of the middle genes, gene 55, encodes an alternative σ factor, gp55, that is specifically required to transcribe late promoters (68). Besides gp55, late transcription requires two other DNA-binding proteins, gp33 and DsbA, and onset of phage replication (267). However, *E. coli* σ^{70}-dependent transcription is needed throughout the T4 infection cycle. Normally, *E. coli* σ^{70} and T4 gp55 compete for RNA polymerase either unmodified or ADP-ribosylated and bound by RbpA (266). The AsiA protein is a key player in T4 development. It is required for the transition both from early to middle and from middle to late promoter transcription (99, 100, 207, 246).

AsiA Is an Anti-σ^{70} Factor

The T4 AsiA protein was identified as a 10 kDa protein that co-purified with *E. coli* σ^{70} from T4-infected cells (243, 244, 245). The AsiA protein was purified and found to inhibit transcription of T4 DNA by *E. coli* RNA polymerase (205). Purified AsiA was treated with trypsin to yield a soluble fragment that could be subject to peptide sequence analysis by Edman degradation (205). The amino acid sequence of the fragment was used to locate a corresponding open reading frame of 90 amino acids in the T4 genome. The open reading frame, normally expressed from a T4 early promoter, was subcloned into a plasmid expression vector and the expressed protein was identified as AsiA by its ability to inhibit *E. coli* σ^{70}-dependent transcription and dissociate σ^{70} from core RNA polymerase (205). An amber mutation was introduced at amino acid position 22 of AsiA (207). The *asiA amS22* mutant phage formed pinpoint plaques on an *E. coli* non-suppressing (su⁻) host. This was due to a defect in transcription from middle promoters and a delay in transcription of late promoters and T4 DNA replication. This behavior was similar to results found with mutations in the *motA* gene of T4 that is essential for phage growth (176). Furthermore, transcription of middle promoters by T4 phage-modified *E. coli* RNA polymerase is dependent on the MotA protein (87, 98, 229).

Role of AsiA in Middle Promoter Activation by MotA

The requirement for AsiA in middle and late promoter transcription suggests a role as a positive transcriptional activator protein in addition to a role as an anti-σ^{70} factor (99, 100, 207, 246). The MotA protein behaves as a classical transcriptional activator protein for T4 middle promoter transcription (87). Upon initial T4 infection, early promoters, which have σ^{70} consensus -10 and -35 binding sites, are transcribed by the unmodified host RNA polymerase. Early genes transcribe the *alt* and *mod* genes that modify RNA polymerase by ADP ribosylation in addition to the *motA* and *asiA* genes, whose protein products bind either modified or unmodified σ^{70} RNA polymerase holoenzyme (264). The middle promoters lack a -35 consensus sequence, but have an "extended -10 sequence" that allows both modified and unmodified σ^{70} holoenzyme to bind. However, based on DNase footprinting analysis, interactions are different between modified and unmodified σ^{70} holoenzyme and middle promoter DNA (100). MotA binds a specific sequence located about 30 base pairs upstream of the translation initiation site, effectively replacing the -35 region of the promoter (100). This would account for MotA-dependent enhancement of the ability of T4-modified and unmodified RNA polymerase to bind middle promoter DNA (100). However, the transition from closed to open complex and subsequent transcription requires both the AsiA and MotA proteins (100, 206). Hinton *et al* (100) have suggested that AsiA could bind the -35 binding region of σ^{70} (region 4.2; Figure 1) substituting for the DNA contacts and assist in the interaction between DNA-bound MotA and RNA polymerase to enhance MotA-dependent transcriptional activation from middle class promoters. Consistent with this hypothesis, AsiA does not affect transcription at extended -10 promoters, nor does AsiA inhibit σ^{70}-dependent transcription from preformed open complex between RNA polymerase and the *lac* promoter DNA (209). Also, a trypsin digested fragment of σ^{70} containing regions 3 and 4 (Figure 1) binds AsiA with high affinity and the specific epitope was localized between amino acids 551 and 608 in region 4.2 (238).

Role of AsiA in Late Promoter Transcription by gp55

The binding of AsiA to σ^{70} holoenzyme would prevent early promoter transcription, and in the presence of MotA allow the transition to middle promoter transcription. What role does AsiA play in the transition from middle to late promoter transcription? Transcription of the late promoters is dependent on the gp55 holoenzyme in the presence of the gp33 late gene-specific transcriptional activator (68). Because *E. coli* σ^{70}-dependent transcription is needed throughout the T4 infection cycle, an AsiA-dependent mechanism has evolved allowing both middle and late transcription to occur. Normally, *E. coli* σ^{70} will outcompete T4 gp55 for RNA polymerase either unmodified or ADP-ribosylated (164, 165, 266). Because AsiA weakens σ^{70}-core RNA polymerase interactions, it would enable gp55 to effectively compete for core to such a degree that both middle and late promoter transcription can simultaneously occur.

Summary

A model for the regulation of T4 temporal gene expression is presented in Figure 2. AsiA is a key player in each stage of T4 development playing dual roles as a positive and a negative regulator of σ^{70}-dependent transcription. Once AsiA and the other early gene products have been produced, AsiA acts as an anti-σ^{70} factor to prevent further early promoter transcription, and AsiA inhibition of σ^{70} enables gp55 sigma factor to compete for core RNA polymerase in late promoter transcription. In middle promoter transcription AsiA acts in combination with MotA, as a transcriptional activator, possibly to allow MotA, bound to its DNA upstream activation site in middle promoters, to stimulate transcription.

THE FlgM ANTI-σ FACTOR AND FLAGELLAR BIOSYNTHESIS

Transcription of the Flagellar Regulon Is Coupled to Flagellar Morphogenesis

Regulation of promoter transcription in the flagella biosynthetic pathway in *Salmonella typhimurium* is similar to bacteriophage T4 in that the flagellar genes are organized into a transcriptional hierarchy of three classes: Class 1, Class 2, and Class 3. Each later class of genes requires that all the genes of the previous classes be functional in order for the next class to be expressed (128, 136). Mutants defective in Class 1 genes do not express Class 2 and Class 3 genes. Mutants in Class 2 genes do not express Class 3 genes.

--→

Figure 2   Model for regulation of temporal gene expression during bacteriophage T4 lytic development. (*A*) The T4 genome is transcribed as a transcriptional hierarchy of three classes of operons needed at different times, early, middle, and late, during the T4 infection cycle. *E. coli* σ^{70} holoenzyme transcribes early promoters, P_E, which transcribe the *mod, alt, asiA, rbpA*, and *mot* genes necessary to activate transcription from middle promoters, P_M. The *gp55* gene, transcribed from a middle promoter, encodes a late-gene specific alternative σ factor, gp55, which associates with modified core RNA polymerase and transcribes the late genes. (*B*) Diagram depicting the cycle of temporal gene expression during T4 lytic development. Upon infection, early promoters are transcribed along with the host genes. Once the *alt* and *mod* gene products are made, they modify the host RNA polymerase by NAD-dependent ADP-ribosylation. Once the anti-σ^{70} factor AsiA is made, it inactivates σ^{70} by direct interaction. The action of AsiA and RNA polymerase modification inactivates host gene transcription. In association with the Mot and RbpA proteins, AsiA-bound σ^{70} directs transcription from the middle promoters. The alternative σ factor, gp55, is expressed from a middle promoter and when produced will outcompete σ^{70} for modified RNA polymerase, and in the presence of the late promoter-specific transcription activator, gp33, directs transcription from the late promoters to complete the T4 infection cycle.

A:

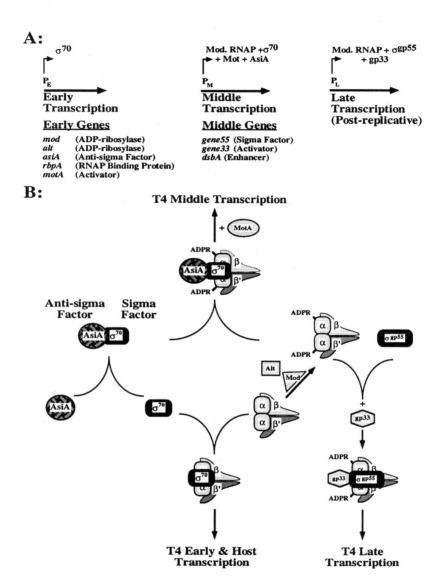

σ^{70}

P_E
Early
Transcription

Early Genes

mod	(ADP-ribosylase)
alt	(ADP-ribosylase)
asiA	(Anti-sigma Factor)
rbpA	(RNAP Binding Protein)
motA	(Activator)

Mod. RNAP + σ^{70}
+ Mot + AsiA

P_M
Middle
Transcription

Middle Genes

gene55	(Sigma Factor)
gene33	(Activator)
dsbA	(Enhancer)

Mod. RNAP + σ^{gp55}
+ gp33

P_L
Late
Transcription
(Post-replicative)

B:

T4 Middle Transcription

+ MotA

Anti-sigma   Sigma
Factor   Factor

T4 Early & Host
Transcription

T4 Late
Transcription

The Class 1 genes, *flhC* and *flhD*, are regulated in response to numerous environmental stimuli such as cAMP levels and temperature (128, 240). The FlhD and FlhC proteins interact, forming a heteromultimeric complex that acts as a transcriptional activator to stimulate transcription from promoters of Class 2 flagellar genes by σ^{70}-containing RNA polymerase (150). Similar to the T4 system involving MotA, the Class 2 promoters in the flagella system lack a conserved -35 sequence. The FlhDC complex binds an $\sim$40 base pair region upstream of the -35 promoter region and is thought to interact with the C-terminal region of the α subunit of RNA polymerase to initiate transcription of the Class 2 promoters (150, 152, 153).

The Class 2 genes encode the proteins needed for the structure and assembly of the basal body-hook structure as well as the *fliA* regulatory gene (136). The *fliA* gene encodes a flagellar-specific alternative σ factor, σ^{28}, necessary for transcription from Class 3 promoters (151, 203). All of the 35 Class 2 genes must be functional to lead to both the completion of the basal body-hook intermediate and Class 3 gene expression. The Class 3 genes encode proteins involved in the final stages of flagellar assembly including the flagellar motor force generators, the flagellin subunit genes (*fliC* and *fljB*) that are polymerized into the long external filament, and the chemotactic signal transduction pathway (Figure 3; 159). While σ^{28} is essential for Class 3 promoter expression, it plays an additional role in both Class 1 and Class 2 gene expression (133, 134, 153).

Role of FlgM in Coupling Gene Expression to Flagellar Morphogenesis

Early work on the regulation of the Class 3 flagellin gene (*fliC*) in *E. coli* revealed that transcription of the *fliC* flagellin gene was coupled to the completion of an earlier stage of flagellar assembly, the hook-basal body structure (HBB). The expression of genes required for later assembly stages is coupled to the

---

Figure 3 Model for the regulation of flagellar expression in *Salmonella typhimurium*. (*A*) Induction of the flagellar regulon leads to expression of genes required for the synthesis and assembly of the hook-basal body complex and the *fliA* and *flgM* regulatory genes. The *fliA* gene encodes an alternative σ factor, σ^{28}, which is specific for genes required late in assembly and genes of the chemosensory system. The *flgM* gene encodes an anti-σ^{28} factor that inhibits σ^{28} activity by direct interaction in the absence of a functional hook-basal body intermediate assembly structure. (*B*) Upon completion of the hook-basal body intermediate structure, a signal is transmitted to the flagellar-specific type III secretion to stop the export of hook subunits and begin export of late assembly structures expressed from σ^{28}-dependent promoters. FlgM carries the late export determinants and is secreted out of the cell in response to hook-basal body completion. (*C*) Secretion of FlgM through the hook-basal body structure relieves σ^{28}-inhibition, transcription from late promoters ensues, and flagellar assembly is completed.

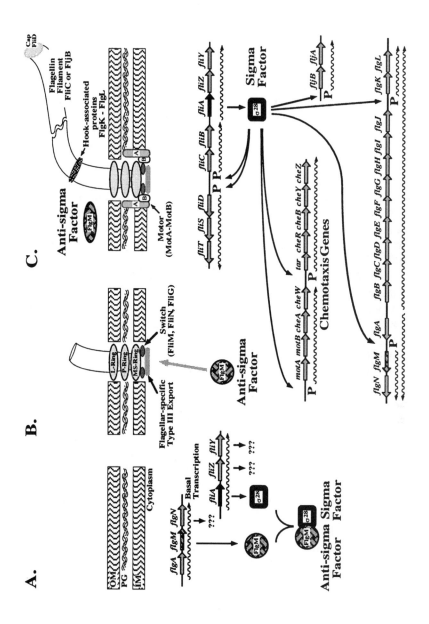

morphogenesis of an intermediate assembly stage. If any of 35 genes necessary for formation of the HBB structure is defective or absent, transcription of the *fliC* gene does not occur (126, 127). The same form of negative regulation was found true in *S. typhimurium* for all Class 3 genes (71, 136). A negative regulatory gene, *flgM*, was identified that prevented Class 3 gene expression in mutants defective in HBB formation (71, 72). Disruption of the *flgM* gene resulted in Class 3 gene expression in mutant strains defective in any of the Class 2 genes required for HBB formation. Thus, FlgM appeared to be the negative regulatory protein responsible for sensing completion of the HBB structure in order for Class 3 promoters to be transcribed. While FlgM of *S. typhimurium* is discussed in this review because it has been best character-ized, other FlgM isolates have been found to exist in the flagellar regulatory systems of *Bacillus subtilis* (34, 63, 185), *Yersinia enterocolitica* (117), *Pro-teus mirabilis* (88), *Shigella species* (2), and *E. coli* (DS Mytelka and MJ Chamberlin, personal communication).

FlgM Is an Anti-σ^{28} Factor

Given that FlgM acted to inhibit Class 3 transcription in HBB defective strains, it seemed likely that FlgM would act on the transcription factor, σ^{28}, that directed RNA polymerase to Class 3 promoters. Purified FlgM protein was able to inhibit Class 3 transcription of the *fliC* promoter in levels stoichiometric to the σ^{28} holoenzyme and was thus labeled an anti-σ^{28} factor (204). When purified FlgM and purified σ^{28} were mixed *in vitro* followed by addition of cross-linking reagent, a hybrid product consisting of both proteins in 1:1 stoichiometry was detected.

Further experiments led to the model that FlgM acts by binding directly to σ^{28} to prevent both its interaction with core RNA polymerase and interaction of free σ^{28} to Class 3 promoter DNA (134). *In vivo*, single amino acid changes at seven positions within the σ^{28} structural gene, *fliA*, were isolated that over-came negative regulation by FlgM (134). These FlgM-insensitive *fliA* mutants mapped to three regions of the *fliA* gene: region 2.1 (amino acid 14), region 3.1 (amino acid 142), and region 4 (amino acids 199, 202, 209, and 213) (Figure 1). Regions of FlgM and σ^{28} that interact with each other were identified by anal-ysis of truncated proteins (110, 134). A fragment of σ^{28} including amino acids 154 through 239 was the smallest stable fragment able to inhibit FlgM func-tion *in vivo*, while fragment 2–213 did not. This suggests that the C-terminal region 4 of σ^{28} can interact with FlgM *in vivo* while regions 2 and 3 alone do not. One conclusion that can be drawn from these results is that the region 4 mutants are defective in direct interactions with FlgM while region 2.1 and 3.1 mutants overcome FlgM inhibition by other mechanisms. For example, region 2.1 is known to interact directly with RNA polymerase. FlgM-insensitive σ^{28} mutants at position His-14 may overcome FlgM inhibition by having a tighter interaction with core. Since the function of region 3.1 is not certain, it is difficult to speculate on how mutants at positions Thr-138 and Gln-142 overcome FlgM inhibition. However, the characterization of the mutants in region 3.1 may provide insight into why this region is conserved among the σ factors.

In FlgM, truncated proteins that include amino acids 26 through 97 retain almost all anti-σ^{28} activity, and fragment 50 through 88 retains about 10% activity. These results suggest that it is the C-terminal portion of FlgM that binds to σ^{28} (110). Deletion analysis identified residues 42 through 88 to contain the anti-σ^{28} domain. Single amino acid changes were isolated that were defective in anti-σ^{28} activity *in vivo* (41). These include changes I58L, L66S, I82T, an insertion of Asn between amino acids 58 and 59, and deletion of Met-77. The belief that these single amino acid changes directly affect anti-σ^{28} activity and not FlgM structure is supported by the finding that FlgM itself lacks any detectable secondary structure (41).

Streptomyces coelicor and *S. aureofaciens* are gram-positive soil bacteria that undergo differentiation and developmental changes in the process of spore formation (38, 129, 182). Late in development, ariel hyphae are produced that develop into chains of mature spores. The *whiG* locus was identified as a block that failed to initiate sporulation and was highly homologous to the σ^{28} equivalent of *B. subtilis* (37). A surprising result was the finding that the *S. coelicor whiG* gene would complement a *fliA* null allele of *S. typhimurium* and that this complementation could be inhibited by the *S. typhimurium* FlgM protein (J Nodwell and R Losick, personal communication). This result leads to the intriguing possibility that an FlgM-like anti-σ factor may be involved in the developmental pathway leading to sporulation in *Streptomyces*.

Native FlgM Exists in an Unfolded State

The interaction between FlgM and σ^{28} was more extensively characterized using multidimensional heteronuclear NMR spectroscopy (41); this finding redefined the current dogma that the biologically significant native structure of a protein should also include the unfolded state (215). NMR analysis showed that FlgM was mostly unstructured by itself in solution. In the presence of σ^{28}, the C-terminal amino acids from residues 47 through 94 of FlgM became structured as evidenced by NMR shifts, but the N-terminal 46 amino acids remained unstructured in complex with σ^{28}. This is in agreement with the genetic and molecular results presented above.

Regulation of FlgM by Export Through the HBB Intermediate Structure

Despite the identification of FlgM as an anti-σ^{28} factor that coupled HBB completion to Class 3 gene expression, it was not clear how a single protein could assess the normal function of all 35 genes required for HBB formation. The answer turned out to be a surprising and elegant mechanism of detecting the assembled HBB structure itself. The Type III flagellar export machinery is highly selective. It not only distinguishes flagellar exported proteins required for organelle assembly, but depending on the stage of assembly it selectively chooses which flagellar proteins are to be exported. Export of Class 2 proteins, such as the hook subunits, is a prerequisite for the export of Class 3 proteins. Upon completion of the HBB structure, the flagellar-specific Type III export machinery is modified by an unknown mechanism through interaction between the FliK and FlhB gene products so that it no longer exports Class 2 proteins, but now exports Class 3 proteins (101, 120, 135, 265). This is the signal recognized by FlgM that the HBB structure is complete and export-competent for flagellin subunits. In the presence of a functional HBB complex, FlgM is found in the spent growth medium presumably by export through this structure (105, 132). In strains defective in any of the genes required for HBB formation, FlgM is found only in the cytoplasm. Once FlgM is removed from the cytoplasm, σ^{28} inhibition is thereby relieved, the Class 3 genes are transcribed, and flagellar assembly is completed. In this manner, FlgM is able to sense the integrity of the hook-basal body structure. If flagella are detached by physical means, such as shearing, FlgM is presumably exported from the cell, thereby relieving inhibition of σ^{28}. This results in the immediate transcription and synthesis of more flagellin. This mechanism provides a system of gene regulation that can sense the development of the flagella and requirements for new filament protein subsequent to HBB completion and following flagellar loss.

Flagellar-Specific Secretion

At least three types of secretory pathways exist for proteins that are destined to be outside the cytoplasm (19, 69, 111, 219). The general secretory (*sec*-dependent) and the type II secretion pathways involve proteins that have N-terminal signal peptides that target them for export. The Type I pathway is *sec*- and signal-peptide–independent and requires signaling from the C-terminal portion of the secreted proteins. Other secreted proteins that are targeted to specific organelles such as the flagella (with the exception of the P- and L-ring structural and assembly proteins) are exported through a Type III secretory pathway that is also *sec*- and signal-peptide–independent. In the case of the flagellar organelle, the exported axial proteins that make up the rod, hook, and filament structure lack a cleavable signal peptide. The axial structure is thought to be hollow and a flagellar-specific export machinery directs the transport of these and the FlgM regulatory protein through the hollow structure. The proteins thought to be involved in Type III flagellar-specific export are the FlhA, B, E, FliE, H, I, J, O, P, Q, and FliR proteins (202). Fundamental questions regarding export through the flagellar organelle remain to be answered. Is there an energy source that drives the reaction? Recent studies in *S. typhimurium* and *Caulobacter cresentus* show that FliI is an ATPase, and it has been postulated as a possible energy source for export (58, 242). Are proteins exported in the order they are assembled? We have found that flagellin when expressed from a *lac* promoter will not be exported with Class 2 proteins (J Karlinsey and K Hughes, unpublished results). Similar results were observed in *C. cresentus* flagellin gene expression (6). Strains carrying null mutations in the *fliK* gene exhibit a "polyhook" phenotype (210, 252). Hook length control is abolished and hooks continue to grow. In order to allow export of specific components, there must exist targeting information on proteins transported through the filament. Several conserved features have been identified near the amino terminus of the axial proteins, including a heptad repeat of hydrophobic residues, an SLG tripeptide, and an ANNLAN-related hexapeptide domain (20). Whether these sequences are export signals or provide a role in the assembly and structure of the axial proteins is not known.

In *C. cresentus*, which shares many of the homologous genes in flagellar biosynthesis (272), deletion analysis of the flagellar hook protein identified amino acids 38–58 as essential for export (129a). This sequence does not include the SLG or ANNLAN sequences and is not conserved among the other flagellar exported proteins. Specific amino acids required for export were not determined in this study. In *S. typhimurium*, individual mutants in either the FliC or FljB flagellin were isolated and shown by Western analysis to accumulate flagellin in vivo (104). These mutations all map to the C-terminal portion of each flagellin, and these mutant proteins are presumed to be defective in export. DNA sequence analysis of the individual mutations was not reported. Deletion analysis of the N-terminus of FlgM revealed that this region is essential for export (110), although detailed characterization between the interaction of this region of FlgM and the Type III export system remains to be elucidated. These studies suggest that flagellar-specific export is complex and may be different in different species, using N-terminal sequences in *Caulobacter* and both N- and C-terminal sequences in *S. typhimurium*.

Summary

A model for the regulation of flagellar gene expression coupled to flagellar assembly is presented in Figure 3. The FlgM story has had far-reaching effects beyond that of regulating flagellar gene expression in eubacteria. Work in the Ohnishi and co-workers lab established that FlgM acted as an anti-σ factor (anti-σ^{28}) (204). This report taken with earlier work on the T4 AsiA anti-σ factor (anti-σ^{70}) and subsequent discoveries of anti-σ factors in *Bacillus subtilis*, RsbW (anti-σ^{B}) and SpoIIAB (anti-SpoIIAA), established the field of gene regulation through the action of an anti-σ factor, and more examples of such a regulatory mechanism are continuing to be uncovered. Recent work has revealed that the native state of FlgM is in an unfolded state, and secondary structure formation requires the interaction with σ^{28} (41). This result is changing the way that the native, biologically significant state of a protein is perceived (215). Finally, the evidence that FlgM could detect that the BBH structure was complete and export-competent for proteins required for later flagellar assembly stages, by itself being a substrate for export, defined a completely novel method for regulating gene expression (105, 132). A similar regulatory scheme was recently discovered for regulation of the expression of virulence factors in *Yersinia pseudotuberculosis* (213). A set of secreted virulence factors, called Yops, are expressed and secreted into the cytosol of a targeted eukaryotic cell. They are secreted through a Type III secretion system. The family of Type-III secretion systems includes the flagellar export pathway of eubacteria. Normally, a negative regulator, LcrQ, prevents expression of the Yop virulence factors. However, upon contact with the eukaryotic cell, LcrQ is secreted out of the cell, and repression of Yop gene expression is relieved. Just as FlgM secretion through the BBH is dependent on the flagellar Type III secretion system in *Salmonella*, secretion of the LcrQ protein is also dependent on the Type III Yop secretion system in *Yersinia* (213). Thus, regulation by export of a negative regulator is no longer unique to the flagellar regulatory system. The characterization of the FlgM protein and its role in regulation has provided groundbreaking results in a number of fields.

THE SpoIIAB ANTI-σ FACTOR IN THE *BACILLUS* SPORULATION REGULATORY CASCADE

Sporulation in Bacillus

Regulation of morphological development through the action of anti-σ activity is a key component of the sporulation pathway in *Bacillus subtilis*. Sporulation in *B. subtilis* proceeds through a series of morphological stages that are regulated by the activity of a cascade of different σ factors, each required for the expression of a set of operons necessary for the transition from one developmental stage to the next (reviewed by 56, 89, 247). Four different σ factors are specifically required for sporulation: σ^E, σ^F, σ^G, and σ^K. Sporulation is activated by known stress signals such as nutrient starvation. The subsequent nuclear division yields daughter nuclei that are separated along the long axis of the cell, and a septum is formed between the daughter nuclei. However, instead of septum formation in the middle of the dividing cell, it forms at an extreme polar position. The two cell types of the diploid cell are known as the mother cell and the forespore. Following asymmetric division is the activation of two cell-specific σ factors, σ^F, which is activated in the forespore, and σ^E, which is activated in the mother cell. Soon after the forespore becomes engulfed within the mother cell, the gene for another σ factor, σ^G, is transcribed by σ^F-associated RNA polymerase within the forespore. Transcription of genes required in the late stages of sporulation within the forespore requires σ^G. Late gene expression within the mother cell is directed by σ^K.

SpoIIAB Is an Anti-σ Factor

The mechanism by which σ^F activity is inhibited in the mother cell has been extensively characterized. The activities of σ^F and σ^G are controlled by the regulatory proteins SpoIIAA, SpoIIAB, SpoIIE, and purine nucleotides (Figure 4). SpoIIAB is an anti-σ factor that inhibits σ^F and σ^G activities by direct interaction (5, 54, 122). Mutations in σ^F were isolated that are insensitive to negative regulation by SpoIIAB mapped to regions 2.1, 3.1, and 4.1 (Figure 1; 43). These are the same regions from which mutations in the *S. typhimurium* σ^{28} factor were obtained that are insensitive to anti-σ factor FlgM (135). Unlike SpoIIAB, only the region 4 peptide from σ^{28} has been shown to interact with FlgM, while all three regions from *B. subtilis* σ^F bind SpoIIAB. Thus the mutations in regions 2.1 and 3.1 may overcome inhibition by different mechanisms in the two systems.

The Anti Anti-σ Factor SpoIIAA

SpoIIAB is regulated in turn by the anti-anti σ factor SpoIIAA, which binds the SpoIIAB-σ^F and the SpoIIAB-σ^G complexes to release σ^F and σ^G in the forespore (5, 49, 53). Even though σ^F and σ^G are only active in the forespore, using immunofluorescence methods σ^F was found to be present in the predivisional cell and after division in both the mother cell and the forespore, but it is inactive in the predivisional cell and the mother cell by the action of SpoIIAB (139, 247). In the mother cell, SpoIIAB has kinase activity that inactivates SpoIIAA by phosphorylation at Ser-58 (161, 183, 193). The characterization of different amino acid substitutions at Ser-58 supports a model that it is the conformation of SpoIIAA that affects its ability to interact with SpoIIAB, and this conformation is determined by a combination of phosphorylation and interaction with ADP and ATP (156). Recently, the dissociation constants of the SpoIIAA-SpoIIAB-ADP, σ^F-SpoIIAB-ADP, and σ^F-SpoIIAB-ATP were determined to examine the effect of ATP and ADP on binding preference of SpoIIAB for SpoIIAA and σ^F (162). The results suggest that SpoIIAB's preference for either SpoIIAA or σ^F is not dependent on the ATP/ADP ratio in the cell.

Compartment-Specific Gene Expression Directed by SpoIIE

In the mother cell, SpoIIAB acts as both an anti-σ factor for σ^F and as an anti-anti-anti-σ factor by phosphorylating its own inhibitor, the anti-anti-σ factor

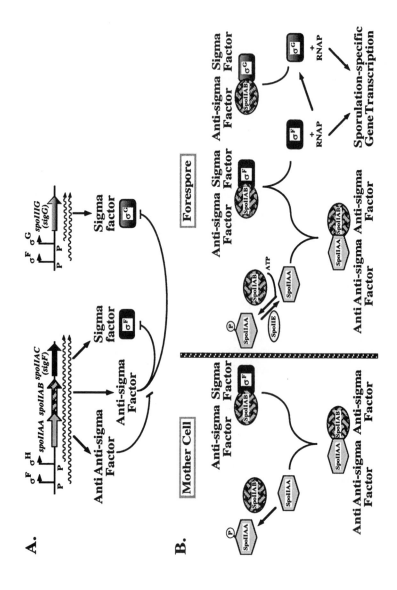

SpoIIAA (Figure 4). The key to σ^F activity in the forespore is the activity of SpoIIE (166). SpoIIE is a bifunctional protein. SpoIIE is required in the normal formation of the septum between the mother cell and the forespore (59). Prior to division, it is visualized at potential septum sites at both ends of the cell, but shortly after polar division between the mother cell and forespore it is only found at the sporulation septum (8, 13). SpoIIE is also a phosphorylase, active only in the forespore, where it is specific for SpoIIAA-PO$_4$ (52). Each activity is genetically separable by mutation (14). Because SpoIIE dephosphorylates SpoIIAA-PO$_4$, it is an antagonist for the SpoIIAB phosphorylation activity on SpoIIAA. Unphosphorylated SpoIIAA accumulates in the forespore, resulting in the release of σ^F from the SpoIIAB-σ^F complex, and σ^F-dependent transcription ensues (139).

Summary

The sporulation pathway of *B. subtilis* continues to provide fundamental knowledge to the process of initiation of distinct cell types following division, a process fundamental to any developmental pathway. A model for the regulation of sporulation-specific gene expression in *B. subtilis* is presented in Figure 4. The SpoIIAB anti-σ factor is responsible for preventing σ^F-dependent transcription of sporulation genes in the mother cell. In the mother cell, SpoIIAB acts as a kinase to phosphorylate, thereby inactivating its antagonist SpoIIAA. This leaves SpoIIAB free to bind to and inhibit σ^F. The key initial step in compartmental-specific gene expression is decided by a single protein, SpoIIE. In the forespore, SpoIIE is active where it dephosphorylates SpoIIAA-PO$_4$ leading to the accumulation of SpoIIAA. SpoIIAA then releases σ^F from the SpoIIAB-σ^F complex, and gene expression leading to development of the spore is initiated by σ^F-bound RNA polymerase.

Figure 4 Model for the regulation of sporulation-specific expression in *Bacillus subtilus*. (A) Induction of sporulation leads to expression of the *spoIIA* operon, *spoIIAA-spoIIAB-spoIIAC*, from a σ^H-dependent promoter. The *spoIIAC* gene encodes an alternative sigma factor, σ^F, which is expressed in both the forespore and in the mother cell. The *spoIIAB* gene encodes an anti-σ^F factor, and the *spoIIAC* gene encodes an anti-SpoIIAB factor (anti-anti-σ^F). σ^F is autoregulatory, and once expressed from the σ^H-dependent promoter, σ^F will transcribe the *spoIIAA-spoIIAB-spoIIAC* operon as well as the *spoIIIG* operon. The *spoIIIG* gene encodes the sporulation-specific alternative sigma factor, σ^G. (B) In the mother cell, SpoIIAB inactivates SpoIIAA by phosphorylation. SpoIIAB is free to bind σ^F to prevent σ^F-dependent gene expression in the mother cell. SpoIIAB will also bind any σ^G expressed in the mother cell and prevent σ^G-dependent transcription. In the forespore, SpoIIE protein is active. SpoIIE is a phosphorylase that dephosphorylates SpoIIAA-phosphate. Unphosphorylated SpoIIAA binds SpoIIAB to prevent its interaction with σ^F. σ^F is free to transcribe sporulation-specific genes, including the σ^G structural gene, *spoIIIG*. σ^G is also free to transcribe sporulation-specific genes.

THE RsbW ANTI-σ FACTOR OF *BACILLUS* REGULATES STRESS RESPONSE

The Anti-σ Factor RsbW and the Anti-Anti-σ Factor RsbV

Like many organisms, *B. subtilis* responds to a variety of environmental stresses in addition to stresses that induce sporulation. Transcription of stress-induced genes throughout the chromosome utilizes a stress-response–specific, alternative σ factor, σ^B (for a review see 91). The transcription factors σ^B and σ^F are closely related, and the regulatory mechanisms affecting their activity are strikingly similar. Just as σ^F activity is regulated by the anti-σ factor SpoIIAB and its antagonists SpoIIAA, σ^B activity is regulated by the anti-σ factor RsbW and its antagonist RsbV (116, 274). The anti-σ factors RsbW and SpoIIAB are themselves similar, with a 27% amino acid identity, suggesting evolution from a common ancestor protein. The same is true for the anti-anti-σ factors RsbV and SpoIIAA, with a 32% amino acid identity (115). Interactions between RsbW and both σ^B and RsbV have been demonstrated using gel filtration chromatography, coimmuneprecipitation, chemical cross-linking, and the yeast two hybrid system (4, 18, 50, 51, 258, 260). The stoichiometry of the complex is $RsbW_2\sigma^B_2$ (4, 50).

Regulation by Phosphorylation of RsbV

How does the RsbW-dependent regulatory system respond to stress? There are two distinct stress response pathways, but in each case the key signal is the phosphorylation state of RsbV (4, 50, 274). Like SpoIIAB, RsbW is a kinase, and during normal exponential growth, RsbW inactivates RsbV by phosphorylation. RsbW is then free to bind σ^B and inhibit σ^B-dependent transcription. Under stress response, RsbV-PO_4 is dephosphorylated, and it is this form of RsbV that binds the RsbW-σ^B complex to release σ^B and allow transcription from σ^B-dependent promoters. Not surprisingly, the input from stress response signals to inhibit RsbW activity is by the dephosphorylation of RsbV and RsbV catalyzed release of σ^B from the RsbW-σ^B complex. RsbW is not subject to regulation by protein degradation (222), a result similar to that observed with the FlgM anti-σ factor in the *S. typhimurium* flagellar system (118). There are a variety of additional regulatory components that respond to environmental stresses and induce dephosphorylation of RsbV-PO_4 leading to σ^B-dependent transcription of stress-response genes.

Response of the RsbU-RsbV-RsbW Regulatory Module to ATP Levels

One important stress signal is a drop in ATP levels in the cell (energy stress signal). This occurs during carbon source limitation and entry into stationary-phase growth, both conditions leading to σ^B-dependent transcription (5, 51,

116). The ability of RsbW to complex with σ^B is unaffected by the addition of ATP, while the presence of ATP had a strong inhibitory effect on the binding of RsbV to RsbW (4). Similarly, the ability of RsbW to inhibit σ^B-dependent transcription in vitro is unaffected by addition of ATP. RsbV inhibited the anti-σ^B activity of RsbW in the absence of ATP or in the presence of the nonhydrolyzable analog of ATP, AMP-PNP. In the presence of ATP, RsbV did not inhibit the anti-σ^B activity of RsbW because of RsbW-dependent conversion of RsbV to RsbV-PO$_4$ (4).

The RsbX-RsbR-RsbS-RsbT Regulatory Module Responds to Environmental Stress

A separate regulatory mechanism allows induction of σ^B-dependent gene expression by stresses that have no direct effect on the cell's ATP levels. These environmental stresses include exposure to elevated temperatures, high NaCl concentrations, addition of ethanol, peroxide, and acid shock (116, 260). In these cases, other regulatory proteins have been identified that affect RsbW and RsbV activities. The σ^B structural gene, sigB, is cotranscribed in an operon with seven other genes (rsbR, S, T, U, V, W, sigB, and rsbX) transcribed from a σ^A-dependent promoter. In addition, the rsbV, W, sigB, and rsbX are also transcribed from a separate σ^B-dependent promoter (115, 268). The gene products from all of these genes are implicated in the control of RsbW/RsbV-dependent regulation of σ^B activity (116, 257–260, 274). RsbU was identified as a positive effector of σ^B-dependent transcription in response to environmental stress (257, 260, 268). RsbU is a phosphatase of RsbV whose activity is in opposition to the kinase activity of RsbW (274). These results led to the model that the role of RsbU was to stimulate release of σ^B from the RsbW-σ^B complex, while RsbU is in turn inhibited from this activity by the action of the RsbX, RsbS, and RsbT regulators (257, 274). RsbU activity is controlled directly by the binding of RsbT (274). Similar to the RsbV antagonistic effect on the RsbW-σ^B complex, the RsbT-RsbU complex is disrupted by unphosphorylated RsbS. The phosphorylation state of RsbS is dependent on the opposing RsbT kinase and RsbX phosphatase activities, just as the phosphorylation state of RsbV is dependent on opposing RsbW kinase and RsbU phosphatase activities. These results led to the proposal that regulation of σ^B activity is under control of two partner switching modules (Figure 5; 116, 274). One module, RsbU-RsbV-RsbW, composed of a phosphatase (U)—antagonist (V)—kinase (W), responds to energy stress through ATP levels, while the other module, RsbX-RsbS-RsbT, also composed of a phosphatase (X)—antagonist (S)—kinase (T), responds to environmental stress. Not surprisingly, there is significant amino acid homology between the RsbU and RsbX phosphatases (257, 268), between the RsbV and RsbS antagonists (274), and between the RsbW and RsbT kinases (274).

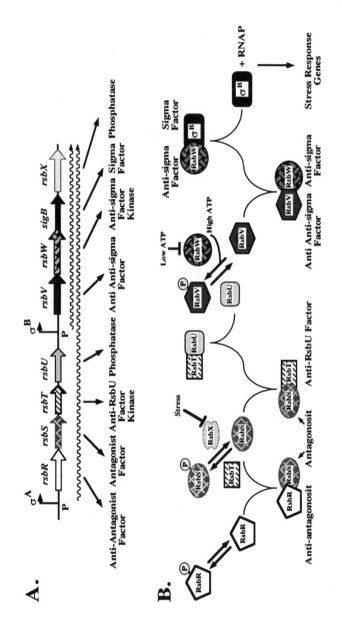

The role of RsbR in modulating the stress response has been recently characterized (1). RsbR is homologous to the antagonist family of homologous *Bacillus* regulators: RsbS, RsbV, and SpoIIAA. A null mutant in *rsbB* was defective in response to salt and heat stress, but unaffected in response to ethanol stress, suggesting it is a positive regulator involved in specific stress responses (1). Because RsbR belongs to the antagonist family, RsbS, RsbV, and SpoIIAA, and because this family of proteins is inactivated by phosphorylation, the homologous site of phosphorylation in RsbR, T205, was mutated by site-directed mutagenesis. Mutants T205A and T205D were defective in RsbR activity, suggesting that the unphosphorylated state of RsbR is active and that it is inactivated by phosphorylation (1). RsbR was shown to interact strongly with RsbS in the yeast two hybrid system and weakly with RsbT and RsbW (1, 259). This led to a model that unphosphorylated RsbR interacts with the RsbS protein to prevent RsbS antagonist action on RsbT. This is consistent with genetic results. The *rsbR* mutant had no phenotype in the absence of stress, whereas the *rsbR rsbX* double mutant had elevated levels of σ^B activity in the absence of stress, but not as high as the *rsbX* single mutant. This is consistent with *rsbR* having an indirect role in modulating the stress response through RsbS. RsbR interaction with RsbS was proposed to make RsbS a poorer substrate for RsbX phosphatase or a better substrate for RsbT kinase (1). This remains to be tested.

Summary

A model for the regulation of σ^B activity is presented in Figure 5. *Bacillus subtilis* appears to have exploited the anti-σ factor control mechanism to three

← _____

Figure 5   Model for the regulation of the stress response alternative sigma factor, σ^B, in *Bacillus subtilis*. (*A*) The σ^B structural gene, *sigB*, is transcribed in an eight-gene operon from a general "housekeeping" σ^A-dependent promoter. It is also autoregulated and will transcribe the *rsbV—rsbW—sigB—rsbX* four-gene operon. RsbW is an anti-σ^B factor and its activity is modulated by the other members of the operon in response to either energy stress (measured as a drop in ATP levels) or to environmental stress. (*B*) Under stress-free conditions, RsbW binds σ^B to prevent σ^B-dependent expression of stress-response genes. RsbW is also a kinase and under stress-free conditions will phosphorylate and inactivate its antagonist, RsbV. Under poor growth conditions, energy stress occurs, ATP levels drop, and unphosphorylated RsbV accumulates and inhibits RsbW. σ^B is free to transcribe stress-response genes.

In the absence of environmental stress, RsbT is free to interact with RsbU phosphorylase to prevent it from dephosphorylating RsbV-phosphate; under these conditions, RsbW is free to inhibit σ^B-dependent expression of stress-response genes. RsbT is also a kinase that will phosphorylate and inactivate its antagonist, RsbS. In the presence of environmental stress, RsbX, a RsbS-PO$_4$ phosphatase, is activated to dephosphorylate RsbS-PO$_4$. In addition, stress also induces dephosphorylation of RsbR. Unphosphorylated RsbS is free to interact with RsbT and/or RsbR, disrupting the RsbT-RsbU complex. The net effect of this interaction is to titrate RsbR, RsbS, and RsbT. This frees RsbU to dephosphorylate RsbV-PO$_4$. Unphosphorylated RsbV is free to interact with RsbW, disrupting the RsbW-σ^B complex. σ^B is free to transcribe stress-response genes.

areas of gene regulation by gene duplication of a key regulatory module: the phosphatase—antagonist—kinase module. In regulation of σ^F-dependent expression of sporulation genes, described earlier, the module includes SpoIIE (phosphatase)-SpoIIAA (antagonist)-SpoIIAB (kinase). In regulation of σ^B-dependent stress-response genes, the modules are RsbU (phosphatase)-RsbV (antagonist)-RsbW (kinase), and RsbX (phosphatase)-RsbS (antagonist)-RsbT (kinase). Furthermore, RsbR modulates responses to specific environmental stresses as an antagonist of RsbS (1). Significant homology was noted between the SpoIIE, RsbU, and RsbX phosphatases (8, 257, 268); between the SpoIIAA, RsbV, RsbS, and RsbR antagonists (1, 115, 274); between the SpoIIAB, RsbW and RsbT kinases (115, 274); and between the σ^F and σ^B transcription factors (154). Given that the gene orders are *rsbR-rsbS-rsbT-rsbU-rsbV-rsbW-sigB-rsbX, spoIIAA-spoIIAB-spoIIAC(σ^F)*, a striking similarity was noted between the *spoIIAA-spoIIAB-spoIIAC(σ^F)* operon and the *rsbV-rsbW-sigB(σ^B)* operon (115). It is tempting to speculate that successive gene duplications occurred. Perhaps the *rsbV-rsbW-sigB-rsbX* operon duplicated to yield *spoIIAA-spoIIAB-spoIIAC(σ^F)-rpoIIE*, where *rpoIIE* was eventually separated to a single gene operon (13), and the *rsbS-rsbT-rsbU* module arose following deletion of the σ factor gene. Both the RsbX-RsbS-RsbT and RsbU-RsbV-RsbW modules affect the ability of the RsbW anti σ factor to inhibit σ^B-dependent transcription, just as the RpoIIE-SpoIIAA-SpoIIAB module affects the ability of the SpoIIAB anti σ factor to inhibit σ^F-dependent transcription. The *rsbR* locus could have arisen by a single tandem duplication of *rsbS* followed by amino acid changes, resulting in a protein that interacts with RsbS under specific environmental stress conditions.

DnaK, AN ANTI-σ FACTOR REGULATOR OF THE HEAT-SHOCK RESPONSE

The Bacterial Heat-Shock Response

The heat-shock response of *E. coli* is the coordinate induction of a set of genes following temperature shift from 30° to 42°C. The heat-shock proteins that are produced act as chaperones to prevent protein turnover by maintaining proper folding, or they are proteases that facilitate protein turnover (70, 77, 93). An alternative σ factor, σ^{32}, is induced following heat shock, and σ^{32} then directs RNA polymerase to transcribe genes encoding heat-shock proteins (84, 85, 251). What is significant about the heat-shock response is the conservation of many of these proteins throughout the biological kingdoms (190, 228, 278). The DnaK (Hsp70 homolog), DnaJ (Hsp40 homolog), and GrpE proteins make up the prokaryotic Hsp70 system. The regulation of the heat-shock response in *Escherichia coli* is the best understood system and has been extensively reviewed (30, 70, 81, 83, 160, 279). This review will only focus on the role of DnaK, in association with DnaJ and GrpE, as an anti-σ factor in the regulation of σ^{32}.

Regulation of the Heat-Shock Response in E. coli

Regulation of the heat-shock response is achieved by changes in the concentration of σ^{32} in the cell, suggesting that posttranscriptional control of σ^{32} levels plays a major role in regulating the heat-shock response (251, 278). One key player in controlling the levels of σ^{32} is the HflB (FtsH) protease. The HflB protease was first identified from a mutant in the *hflB* locus that allowed the high-frequency lysogeny of bacteriophage λ (67). HflB is an ATP-dependent protease for both the lambda cII and cIII proteins that promote transcription from lysogenic promoters in phage λ (12, 94, 96, 124, 239). HflB protease is also the protease responsible for degradation of σ^{32} (95, 255). HflB protease appears specific for σ^{32}; in the presence of purified σ^{32} and σ^{70}, only σ^{32} is degraded (255).

The DnaK, DnaJ, and GrpE Negative Regulators of the Heat-Shock Response

Mutants in the *dnaK*, *dnaJ*, and *grpE* genes were found to have a negative regulatory effect on heat-shock gene expression. A mutation in *dnaK* failed to turn off the heat-shock response after temperature upshift and overproduction of wild-type DnaK protein inhibited the expression of heat-shock proteins at all temperatures (254). It was later shown that mutations in either *dnaK*, *dnaJ*, or *grpE* resulted in failure to shut off the heat-shock response after temperature upshift, and elevated levels of heat-shock proteins at low temperature, while mutations in other heat-shock genes, *lon*, *groEL*, and *groES* had no effect (250). The DnaK, DnaJ, and GroE proteins are themselves heat-shock proteins that function in protein folding, either for the initial folding of nascent polypeptide chains, preventing proteins destined to be translocated across the cytoplasmic membrane from folding prior to translocation, or in the refolding of proteins following stress such at heat shock (216). They also perform a central role in HflB-dependent degradation of σ^{32}.

DnaK as an Anti-σ^{32} Factor

Biochemical evidence suggests a mechanism by which DnaK binds and releases proteins, including σ^{32}, and how this activity is regulated by ATP hydrolysis, DnaJ, and GrpE (119). Purified DnaK has a weak ATPase activity (282) that is modulated by DnaJ and GrpE (114). Binding of ATP to DnaK-substrate complex accelerated the dissociation of bound substrates (177). Binding kinetics suggests that it is the ATP-bound DnaK that initiates substrate binding to form an unstable complex that can be stabilized by ATP hydrolysis in the presence of DnaJ (177, 214, 224, 263). The binding and hydrolysis of ATP induce conformational changes in DnaK that are essential for its chaperone function. The ATPase activity and substrate binding occur in separate but interactive domains (28, 29). If both DnaJ and GrpE are present, ATP hydrolysis is further enhanced, and the presence of GrpE also stimulates substrate release following ATP hydrolysis (3, 142, 143, 177, 281). It is the release of substrates catalyzed by GrpE that frees DnaK to recycle its chaperone activity (137, 253).

What was found for the activities of the DnaK system in its chaperone functions described above is also seen in the interactions of the DnaK system with σ^{32}. DnaK, DnaJ, and GroE proteins bind σ^{32} (65, 140, 141). The interaction between DnaK and σ^{32} and between GrpE and σ^{32} could be disrupted by the addition of ATP, but ATP had no effect on the interaction between DnaJ and σ^{32}. DnaK and DnaJ were able to bind independently to σ^{32}, while the presence of ATP promoted the formation of a ternary complex of σ^{32}, DnaK and DnaJ. DnaK and DnaJ only bind σ^{32} when it is dissociated from RNA polymerase (66). The DnaK-σ^{32} interaction is destabilized by ATP, but if DnaJ is present, it stimulates DnaK hydrolysis of ATP, resulting in the stable triprotein complex bound to ADP. Thus, DnaJ binds σ^{32} and presents σ^{32} to ATP-bound DnaK. GrpE binding to the ADP–DnaK–σ^{32}–DnaJ complex stimulates the release of ADP, allowing ATP to enter and dissociate the complex (208, 224).

In a comprehensive study to determine the substrate specificity of DnaK, 37 known DnaK substrates were divided into libraries of 13 amino acid lengths that overlap by three amino acids and attached to cellulose. By characterizing the segments that bound to DnaK, a binding motif was uncovered. The DnaK binding motif includes a hydrophobic core flanked by regions rich in basic residues that are mostly buried within a folded protein (226). Fusions of different segments of σ^{32} to β-galactosidase were examined for those that were targeted for DnaK-dependent degradation. This identified a segment of σ^{32}, termed region C, that is subject to DnaK-dependent degradation that spans the C-terminal portion of region 2.4 through the N-terminal portion of region 3.1 (Figure 1; 192). A more detailed study using the 13 amino acid peptide library system described above revealed seven peptide regions within σ^{32} that are recognized by DnaK, with region C constituting one of the larger binding regions (178). Region C is particularly attractive as a DnaK binding domain because it is conserved only among the σ^{32} homologs of eubacterial σ factors (154). The substrate binding domain of DnaK resides in the C-terminal 250 amino acids, while the N-terminal portion of DnaK includes the ATPase domain (28, 29, 33, 80, 280). Overexpression of the C-terminal substrate binding fragment was toxic at 42°C, presumably resulting from interactions between the overexpressed fragment and normal DnaK substrates (33). Mutants in the C-terminal fragment were isolated that did not kill the cells when induced at 42°C. This identified 10 amino acid positions at which substitutions (S398F, P419L, P419S, E444K, M408I, G405S, G406D, E402K, A488T, D526N, and G539D) were defective in substrate binding. Two of these substitutions, D526N and G539D, reside in a region of homology between σ^{32} and the flagellar-specific anti-σ factor, FlgM (27).

Summary

A model for the regulation of the heat-shock response of E. coli is presented in Figure 6. Following heat treatment of 51°C for 15 minutes, σ^{70} bound to RNA polymerase becomes inactivated (21). Inactive σ^{70} separates from the active core RNA polymerase complex and forms aggregates. The model for the heat-shock response is that under these conditions, σ^{32} remains active, and after temperature down-shift, associates with free core to direct transcription from σ^{32}-dependent promoters. The DnaK, DnaJ, and GrpE are induced by the heat-shock response and, when returned to normal growth temperatures, will act on the σ^{70} aggregates to refold σ^{70} into active protein. As protein aggregate substrates for the DnaK, DnaJ, and GrpE proteins are removed, they are free to act on σ^{32}. Under normal growth conditions, in the presence of ATP, DnaJ induces DnaK to compete with RNA polymerase for binding σ^{32} and to form a stable DnaJ—σ^{32}—DnaK—ADP complex after ATP hydrolysis (144). GrpE can interact with this complex to release active σ^{32}, which can reassociate with RNA polymerase or present unfolded σ^{32} to the HflB (FtsH) protease for degradation.

THE ECF SUBFAMILY OF σ FACTORS AND THEIR ANTI-σ FACTORS

Members of the ECF Family

A subfamily of eubacterial RNA polymerase s factors activates the expression of extracytoplasmic functions (ECF) (Table 1; 155). This ECF subfamily includes (a) Pseudomonas aeruginosa AlgT (σ^{22}), (b) Myxococcus xanthus CarQ (σ^{CarQ}), (c) E. coli RpoE (σ^{24} or σ^{E}), (d) Photobacterium sp. RpoE, (e) E. coli FecI, (f) P. putida PupI, (g) Streptomyces coelicolor SigE (σ^{28}), and others (Table 1). AlgT controls alginate overproduction in P. aeruginosa (47, 97, 170); M. xanthus CarQ activates carotenoid biosynthesis in response to blue light production (227); RpoE of E. coli controls σ^{32} expression as part of the extreme heat-shock response in E. coli (55, 102, 220, 225); RpoE from the deep-sea Photobacterium activates barometric adaptation (15); S. coelicolor SigE activates extracellular agarase production (155), and FecI of E. coli and PupI of P. putida control citrate-dependent iron transport and pseudobactin transport,

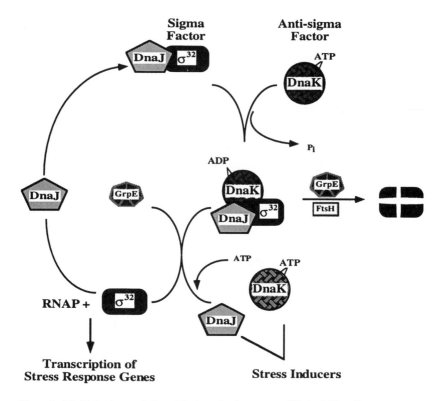

Figure 6   Model for the regulation of the heat-shock response of *Escherichia coli*.

Under normal growth conditions, DnaJ protein binds the heat-shock–specific alternative σ factor, σ^{32}, and presents it to ATP-bound DnaK protein. In the presence of ATP, DnaJ induces DnaK to compete with RNA polymerase for binding σ^{32} and to form a stable DnaJ-σ^{32}-DnaK-ADP complex after ATP autohydrolysis by DnaK. GrpE directs this complex to FtsH protease for degradation of σ^{32} and complex dissociation.

Following heat-shock treatment, σ^{70}-bound to RNA polymerase becomes inactivated, and many cell proteins unfold and are substrates for both the DnaK-DnaJ-GrpE chaperone activity. Inactive σ^{70} separates from the active core RNA polymerase complex and forms aggregates. The DnaK, DnaJ, and GrpE act on the σ^{70} aggregates to refold σ^{70} into active protein. The FtsH protease acts on proteins that have become denatured as a result of heat shock. σ^{32} is unaffected by heat shock and is free to interact with core RNA polymerase and transcribe σ^{32}-dependent operons. As protein aggregate substrates for the DnaK, DnaJ, and GrpE proteins are removed, they are free to act on σ^{32}. As above, in the presence of ATP, DnaJ induces DnaK to compete with RNA polymerase for binding σ^{32} and to form a stable DnaJ-σ^{32}-DnaK-ADP complex after ATP hydrolysis. GrpE can interact with this complex to release active σ^{32} if FtsH is unavailable, because it is acting on other substrates. Free σ^{32} can reassociate with RNA polymerase and transcribe σ^{32}-dependent operons. If Free FtsH is present, GrpE can present unfolded σ^{32} to the FtsH (HflB) protease for degradation.

respectively (7, 130). Many ECF s factors have been found to be cotranscribed with and regulated by their cognate anti-σ factors, which are listed in Table 1.

PSEUDOMONAS AERUGINOSA AlgT (AlgU, σ^{22}) AND MucA

The AlgT and MucA Proteins Are Regulators of Alginate Biosynthesis in Pseudomonas sp.

Pseudomonas aeruginosa is an opportunistic human pathogen that afflicts patients that are immunocompromised by surgery, cytotoxic drugs, or burn wounds as well as those suffering from cystic fibrosis (107, 217). The ability to survive in a wide range of environments is attributed to its ability to produce several extracellular products such as elastase, a staphylolytic protease, alkaline protease, exotoxin A, phospholipase, and pyocyanine. Patients with cystic fibrosis (CF) are frequently afflicted with pulmonary infections by *P. aeruginosa*. Following colonization of the CF respiratory tract, mucoid variants of *P. aeruginosa* emerge, become predominant, and lead to chronic pulmonary disease (79, 199). The appearance of mucoid strains is correlated to inflammation and poor prognosis (108, 109, 211). The mucoid phenotype is due to the production of alginate, a linear polymer composed of O-acetylated-D-mannuronate and L-guluronate (57, 146). The genes controlling alginate production are located to five operons (including a single 18-kb operon) that are coordinately regulated (Figure 7; reviewed in 78, 201). The *algT* gene (also called *algU*) encodes a 22-kDa alternative σ factor (σ^{22}) that directs transcription of the alginate regulon (44, 47, 73, 97, 167, 236, 269, 270). The *algT* gene is a trans-acting positive regulator of alginate biosynthesis located at minute 67.5 on the chromosome (60, 167). Alginate production is negatively regulated by the *mucA* gene product (74, 169). Most CF isolates harbor a mutation in *mucA* gene and this appears to be the primary molecular mechanism for promoting mucoid phenotype (24). The *mucA* locus was originally identified by mutants also mapping at minute 67.5 on the chromosome that resulted in a mucoid phenotype because of induction of the alginate regulon (64, 158). The inactivation of the *mucA* gene in nonmucoid *P. aeruginosa* PAO1 also results in constitutive alginate production (44, 169; K Mathee, O Ciofu, C Sternberg, P Lindum, D Ohman, M Givskov, S Molin, N Høiby, A Kharazmi, unpublished results). This is consistent with MucA acting as a negative regulator. Fine mapping located the *mucA* gene to an open reading frame directly downstream of *algT* (44, 169). The mucoid phenotype can also be induced by the introduction of the *algT* gene present on a multicopy plasmid (60, 61, 74, 158). Conversely, the introduction of the

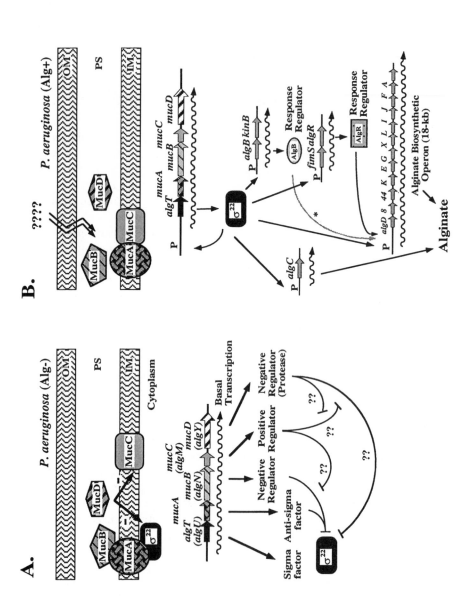

wild type *mucA* gene present on a multicopy plasmid will suppress the mucoid phenotype of CF mutant strains (167).

MucA Is an Anti-σ^{22} Factor

The Alg$^+$ phenotype resulting from mutation in *mucA* is unstable under laboratory growth conditions where the cultures are not well aerated (200). Mutations resulting in reversion from Alg$^+$ to Alg$^-$ were also mapped to 67.5 min and reported to result from mutation in an *algT*-linked locus (60, 61). Subsequent work demonstrated that these mutations were localized within *algT* gene (47, 233). The Alg$^-$ suppressor mutations mapped to the proposed core binding region (47) and to the -10 and -35 promoter recognition regions of σ^{22} (233), suggesting that these mutations were defective in transcription at either the level of binding to core RNA polymerase or the level of promoter recognition. The isolation of *algT* suppressors of Alg$^+$ mutants in *mucA* suggested that MucA might be an anti-σ factor by interacting directly to inhibit σ^{22} activity. Direct physical interaction between these proteins was suggested by cosedimentation of σ^{22} and MucA in a glycerol density gradient, crosslinking studies of σ^{22} and MucA expressed in *E. coli*, and the ability of σ^{22} to bind to immobilized MucA (235, 273). This interaction may affect σ^{22} stability in vivo. In Alg$^+$ strains, σ^{22} is readily detected; in Alg$^-$ strains, σ^{22} is absent even though these strains carry a functional and fully expressed *algT* locus (175). It is possible that binding of MucA to σ^{22} targets σ^{22} for degradation. Direct proof of MucA anti-σ^{22} activity was demonstrated by the ability of purified MucA to directly inhibit σ^{22}-dependent transcription in vitro; however, this inhibition was not stoichiometric, requiring a 10- to 20-fold molar excess of MucA (235, 273). The requirement for a 10- to 20-fold molar excess of purified MucA to inhibit σ^{22}-dependent transcription may reflect conformational changes in MucA resulting from purification that render MucA less active in vitro. MucA is an inner membrane protein (175). Genetic analysis using phoA fusions to different regions of MucA suggests that the N-terminal 83 amino acids of MucA

←——————————————————————————————————

Figure 7   Model for the regulation of alginate biosynthesis in *Pseudomonas aeruginosa*. (A) The σ^{22} structural gene *algT* (*algU*) is transcribed in an operon of five genes. It is postulated that the gene products from the *algT(algU)* operon are made in stoichiometric amounts, and the regulation occurs at the inner membrane. In the absence of signal, σ^{22} made during basal transcription is tightly controlled and kept from activating other alginate genes by the cognate regulators, the anti-σ^{22} factor MucA, MucB, and MucD protease, that are cotranscribed. *Question marks* indicate that the specific targets have not been conclusively demonstrated. (B) The *alg* regulon. Upon induction by an unknown signal (*????*), σ^{22} is released from the control by its cognate regulators, allowing it to activate all the genes (*solid lines*) in the *alg* regulon resulting in alginate production. *Asterisks* indicate indirect control of P$_{algD}$ by AlgB. [Adapted from Ohman et al (201).]

preceding the membrane spanning segment resides in the cytoplasm and the C-terminal 30 amino acids reside in the periplasm (175). Thus, the high molar excess of MucA required to inhibit σ^{22}-dependent transcription may result from structural changes of MucA in the absence of membrane.

The results of these studies suggest two mechanisms by which MucA can act as an anti-σ^{22} factor. One mechanism is similar to DnaK protein of *E. coli*, where MucA binds to σ^{22} and targets it for proteolysis; the second mechanism would be a direct binding to σ^{22} by MucA to inhibit σ^{22}-dependent transcription. Whether MucA acts in vivo by one or both mechanisms remains to be elucidated.

Other Regulators of σ^{22}

The σ^{22} structural gene, *algT (algU)*, is cotranscribed in an operon with four other genes, *algT—mucA—mucB(algN)—mucC(algM)—mucD(algY)* (Figure 7; 23, 44, 60, 74, 168, 201, 276). The *mucB* gene encodes a negative regulator of alginate biosynthesis (74, 168). Induction of alginate synthesis (in an otherwise nonalginate-producing strain PAO1) by the presence of *algT* on a multicopy plasmid is prevented if the plasmid also carries *mucB* (47, 74). Inactivation of *mucB* results in constitutive alginate production and detection of σ^{22} in the cell (74, 168, 175). Purified MucB had no effect on the ability of σ^{22}-dependent transcription in vitro, suggesting that it does not act as an anti-σ^{22} factor (235, 273). The 5'-end of the translated sequence contains a putative sec-dependent signal sequence, and deletion of this sequence resulted in a protein that failed to exert its negative regulatory effect (235). PhoA fusion and cell fractionation analysis also localized MucB to the periplasm (175, 235). This suggests that MucB exerts its negative regulatory effect on σ^{22} from the periplasm.

The *mucC* gene is a positive regulator of alginate biosynthesis (201). The presence of a wild-type *mucC(algM)* gene will overcome the negative regulatory effects of the *mucB* and *mucD* genes on *algT/U* (201). The amino acid sequence predicted for MucC is very hydrophobic, suggesting it may be a transmembrane protein.

Inactivation of *mucD* also results in constitutive alginate production (23, 201). MucD has a high degree of amino acid sequence identity to the membrane-associated HtrA(DegP) periplasmic protease of *E. coli* (23, 147, 201). In addition, Boucher et al identified a cytoplasmic HtrA(DegP)-like protease AlgW that is also involved in alginate production (23).

Summary

A model has been proposed whereby MucB interacts with the periplasmic domain of the anti-σ^{22} factor MucA, altering the conformation of the cyto-plasmic domain of MucA so that it binds σ^{22} and targets it for degradation (Figure 7; 175); however, no direct interaction between MucA and MucB has been demonstrated. In strains defective in *mucA*, the negative regulatory signal from MucB would not be transduced into the cytoplasm destabilizing σ^{22}. If σ^{22} turnover requires the action of specific proteases, either MucD (23, 201) or the cytoplasmic HtrA(DegP)-like protease AlgW (23) could perform this role in the regulation of σ^{22} levels in the cell. A resemblance has been observed between the Kyte-Doolittle hydrophobicity plots of MucC(AlgM) and *B. subtilis* RsbS, the positive regulator of *B. subtilis* σ^{B} activity described earlier. By analogy to RsbS, MucC may serve as an anti-anti-σ factor of MucA by antagonizing the MucA-σ^{22} or protease-σ^{22}.

Many strides have been taken in understanding the molecular mechanism for alginate, in particular σ^{22} regulation, however many fundamental questions remain to be answered. How do σ^{22}, MucA, MucB, MucC, and MucD communicate with each other? How is the signal from the environment perceived and transmitted? Finally, what is the signal in the lungs of CF patients that selects for deregulation of σ^{22}? Mucoid *P. aeruginosa* activates the complement system (112) and the oxidative burst of polymorphonuclear lymphocytes (PMNs) (113); these cells predominate in the inflammatory response in the lungs of CF patients (11, 123). The inflammatory environment of the CF lungs was mimicked by growing a nonalginate producing strain in a biofilm flow-cell and subjecting it repeatedly to low concentrations of hydrogen peroxide or activated PMNs; Alg^{+} variants arose under these conditions that resulted from mutations in *mucA* (K Mathee, O Ciofu, C Sternberg, P Lindum, D Ohman, M Givskov, S Molin, N Høiby, A Kharazmi, unpublished results).

MYXOCOCCUS XANTHUS CarQ (σ^{CarQ}) AND CarR

Carotenoid Pigment Production

Carotenoid pigments provide protection against photochemically generated oxidants such as singlet oxygens by quenching (17, 48, 221). In *Myxococcus xanthus*, carotenoid biosynthesis is induced by blue light, and the photosensitizing molecule appears to be protoporphyrin IX (31, 32, 221). Carotenoids are only produced during exponential growth. If the cells are in stationary phase, exposure to blue light results in cell lysis (known as photolysis) (31, 32). Carotenoids are ubiquitous compounds that have a basic polyene hydrocarbon chain derived from eight isoprene units that can be modified in many different ways (9, 26, 75, 237). *M. xanthus* produces two monoesterified, glycosylated, monocyclic carotenoids called myxobactin and myxobacton (32, 223).

The Carotenoid Regulon

The carotenoid biosynthesis genes of *M. xanthus* compose a regulatory network of four operons (*carQRS*, *carB-A*, *carC*, and *carD*) called the light or *car* regulon (Figure 8; 22, 62, 196, 227). The *carQRS* operon encodes three

A.   **B.**

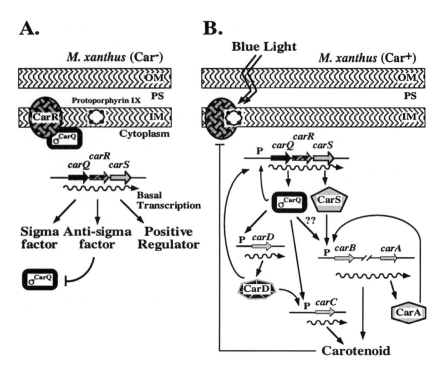

Figure 8   Model for the activation of carotenoid synthesis in *Myxococcus xanthus*. (*A*) The CarQ (σ^{CarQ}) structural gene, *carQ*, is transcribed in an operon of three genes, *carQ-carR-carS*. It is postulated that the gene products of *carQRS* operon are made in stoichiometric amounts, and the regulation occurs at the inner membrane. In this model, σ factor, σ^{CarQ} made in the absence of blue light inducer, is sequestered by anti-σ^{CarQ} factor CarR at the inner membrane. (*B*) The *car* regulon. Upon exposure to blue light, activated protoporphyrin IX releases a signal that results in the degradation of CarR, releasing σ^{CarQ} to activate the *car* regulon, resulting in carotenoid production. Carotenoids quench the light-induced signal and subsequently downregulate carotenoid production. *Question marks* indicate that the real target has not been conclusively demonstrated.

regulatory proteins, two positive activators, CarQ and CarS, and a negative regulator, CarR. CarQ is a member of the ECF subfamily of σ factors (76, 155) required for expression of the *car* regulon (103). CarS is a transacting positive regulator of the *carB-A* operon (76). A negative regulatory locus, *carR*, was originally identified by transposon Tn*5* insertions that resulted in the constitutive light-independent production of carotenoids (a constitutive Car^c phenotype) (172). Further genetic analysis of this region also identified an autoregulating, transacting, positive regulator. These *carR*-linked mutations were defective in the expression of the carotenoid biosynthetic genes either under inducing conditions or in the presence of null mutations in *carR* (Car$^-$) (10, 103, 171–173).

Subsequent sequence analysis revealed the presence of three open reading frames, designated *carQ*, *carR* and *carS*, that are co-transcribed and likely to be translationally coupled (179). The Tn5 insertions that resulted in a Car^c phenotype mapped to *carR* (179). An increase in promoter activity was observed in the presence of light, and there was light-independent expression in *carR* mutants (76, 103, 179). The *carQ* gene was identified as the gene coding for the positive regulator identified from the Car$^-$ mutants linked to *carR* (179).

CarR Is an Anti-σ^{CarQ} Factor

Similar to the Alg^c phenotype of *mucA* mutants in *P. aeruginosa*, the Car^c phenotype of *carR* null mutants is unstable, and spontaneous suppressors arise that are defective in carotenoid production (179). Two suppressor mutations of a *carR* null allele resulted from in-frame deletions in the *carQ* locus encoding the σ factor (179). One of these deletions was mapped to the proposed 20mer box in region 2.4 (*carQ1*), and the other to the RpoD box in region 4.2 (*carQ2*). Both these regions are required for σ subunit interaction with the -10 and -35 regions of the promoters, respectively (155, 179). The occurrence of secondary suppressor mutations of *carR* in *carQ* provided genetic evidence that CarR modulates CarQ activity, and this modulation could be through direct interaction (179). To date, CarQ has not been demonstrated to be a σ factor, nor has a physical interaction between CarQ and CarR been demonstrated. However, based on sequence similarity and analogy to other members of the ECF family, it is referred to as a σ factor, and the negative regulator CarR as the anti-σ factor (76). Sequence analysis of CarR and differential membrane fractionation of a Protein A-CarR fusion peptide suggest that CarR is an integral membrane protein (76, 179). It has been postulated that CarQ interacts with CarR polypeptide that is being translated and results in the sequestering of CarQ by nascent CarR protein (76). The negative regulatory effect of CarR over CarQ can be removed by either translational uncoupling of these two genes or by increasing the copy number of CarQ (76), similar to the σ^{22}-MucA system in *P. aeruginosa*. The requirement for precise stoichiometry suggests a direct interaction between CarQ and CarR proteins.

Other Regulators of σ^{CarQ}

The open reading frame downstream of *carQ* and *carR* is *carS* that is co-transcribed and likely to be translationally coupled (179). Neither the *carQRS* nor the *carC* operons were affected by the absence of CarS protein, but a *carS* deletion mutant failed to show light-dependent induction of the *carB-A* operon (76). An amber mutation in *carS*, which removes 25 amino acids from the C-terminus of the wild-type 111 amino acid protein, resulted in constitutive expression of the *carB-A* promoter, but not the entire *car* regulon (179). CarS expressed from a σ^{CarQ}-independent promoter is sufficient for expression of the *carB-A* operon (76). This indicates that the positive effect of CarQ on *carB-A* expression is an indirect effect of σ^{CarQ}-dependent expression of CarS.

The *carD* operon is part of the *car* or light regulon (196). The CarD protein is required for sensing two completely different environmental cues: exposure to blue light and starvation. The former results in the activation of *car* regulon and the latter in the development of fruiting bodies. However, σ^{CarQ} is not required for the activation of CarD-dependent development-specific promoters (196). The CarD protein is a novel protein that shares homology to class I eukaryotic transcription factors (195). It is required for activation of both *carQRS* and *carC* operons (196). The *carQRS* promoter lacking the CarD protein binding site failed to respond to blue light (195).

Summary

The observation by Burchard and Dworkin (31) that the membrane is the site of light damage in carotenoid nonproducing cells implies that activation of the *car* regulon involves a signal transduction pathway at the membrane similar to the *alg* regulon in *P. aeruginosa*. The observation made by Gorham et al (76) that a CarR-β-galactosidase fusion protein was stable in the dark in strains constitutively expressing carotenoid but not in the light suggests that either translation of *carR* is inhibited or the protein is rapidly degraded in the presence of light. According to the model proposed by McGowan et al (179), CarR protein activity could be inhibited by its interaction with blue light-activated protoporphyrin with or without singlet oxygen, resulting in release of CarQ protein that would then induce the *car* regulon and result in carotenoid production (Figure 8). In the absence of induction, it is plausible that translational coupling ensures that CarQ protein is held inactive as soon as it is made. Quenching of the oxidants by the newly synthesized carotenoid would free CarR protein, restoring its activity and thereby resulting in down regulation of the *car* regulon. Inactivation of *carQRS* transcription resulting from carotenoid production has been demonstrated (103); however, it is not known if the proposed CarQ–CarR interaction destabilizes the σ factor as observed in *P. aeruginosa* σ^{22}-MucA interaction (175).

ESCHERICHIA COLI RpoE (σ^E) AND RseA

Identification of RpoE

The first member of the ECF subfamily identified is σ^E, of *E. coli*. Biochemical analysis revealed that an alternative σ factor, σ^E (Mr = 24,000), is required for high temperature survival or thermotolerance (>40°C) (55). The specificity of σ^E is distinctly different from other known σ factors, including σ^{70} and σ^{32} (55).

σ^E recognizes the P3 promoter for *rpoH* (σ^{32}), which is active at 50°C (83, 262), and the promoter for *htrA*(*degP*), which encodes a periplasmic serine protease (55, 148, 149, 248, 249). Using these two promoters as reporters, several loci were identified that modulated σ^E activity; these map to genes coding for outer membrane proteins (OMPs) (181). These findings led to the hypothesis that the accumulation or misfolding of outer membrane proteins activates σ^E. The structural gene for σ^E, *rpoE*, was later identified by scanning databases with a σ^E oligopeptide sequence (45, 155, 194, 225). Subsequently, the *rpoE* gene was subcloned from the Kohara phage library (225) and also by polymerase chain reaction amplification of *E. coli* chromosomal DNA (277). In addition, mutants that suppressed or had elevated expression of σ^E regulon were used to locate the *rpoE* gene (220). The *rpoE* gene maps to 55.5 minutes on the *E. coli* chromosome (102, 220, 225). Purified RpoE exhibited σ factor activity on σ^E-dependent promoters in vitro (102, 220, 225).

Temperature-Dependent Regulation of ropE

Similar to other ECF σ factors, *rpoE* is autoregulated (102, 220, 225). Temperature upshift causes a 1.2- to 1.5-fold increase in transcription from the *rpoE* P2 promoter that results in a 4- to 8-fold increase in *degP*(*htrA*) expression (220, 225). How temperature upshift is translated into increased σ^E-dependent transcription is not clear. Loss of function mutations in *E. coli rpoE* resulted in lower levels of *htrA* mRNA and HtrA protein. These mutations were mapped to the conserved region 2.1 (L25P) implicated in core binding, and to region 4.2 (S172P and R178G) implicated in −35 recognition (220).

Modulation of σ^E Activity by Protein Status

Over- or underexpression of certain OMPs was shown to affect σ^E-dependent transcription (225) and its activity (181). Increasing the amounts of OMPs, OmpX, OmpT, OmpF, or OmpC, increased σ^E-dependent expression from the *rpoH-P3* and *degP* promoters (181). Also, reduction in either the major outer membrane lipoprotein Lpp or OmpR (the transcriptional activator of OmpC and OmpF) resulted in a decrease in σ^E-dependent transcription (181). *E. coli* strains containing mutations that affect protein folding such as *htrC*, *dsbA*, *dsbC*, *dsbD*, *dsbG*, *trxB*, *surA*, *fkpA*, and *ompH/skp* have elevated levels of σ^E (138, 186, 220). Elevated levels of σ^E were also observed in strains with an altered composition of outer membrane proteins resulting from mutation in the lipopolysaccharide gene *htrM/rfaD* (197, 212, 220). Finally, overexpression of the *dsb* genes, coding for proteins involved in disulfide bond formation, and the *surA* and *fkpA* genes, coding for distinct peptidyl-prolyl isomerases, can compensate for and/or complement some of these mutations for σ^E-dependent transcription (186, 188). However, it is not certain if underexpression of *ompS* affects σ^E-dependent transcription, considering that mutations in genes such as *surA* and *ompH*, which have reduced Omp levels, have higher levels of σ^E-dependent transcription (186).

RseA Is an Anti-σ^E Factor

The *rpoE* locus in both *E. coli* and *P. aeruginosa* is flanked by the *nadB* and *lepA* genes (46, 220, 225; Figure 9; C McPherson, K Mathee, DE Ohman, unpublished results). σ^E and σ^{22} are closely related in amino acid sequence and a plasmid carrying the *E. coli rpoE* gene will complement an *algT* null mutant of *P. aeruginosa* (277). Sequence analysis of the region immediately downstream from *E. coli rpoE* revealed an operon structure, *rpoE-rseA-rseB-rseC* (regulator of σ^E) or *rpoE-mclA-mclB-mclC* (mucA-like protein) homologous to the *algT-mucA-mucB-nucC* genes of *P. aeruginosa* (188, 220, 225, 277). Clones containing the *resA-rseB-rseC* region downregulate σ^E-dependent transcription from *rpoE* P2 and *htrA* promoters (187, 220, 225). Null mutations in *rseA* have increased σ^E activity, suggesting that RseA negatively regulates σ^E (42, 187).

RseA is an inner membrane protein with the N-terminal portion in the cytoplasm and the C-terminal portion in the periplasm (42, 187). Copurification, crosslinking, and coimmunoprecipitation analysis demonstrated that RseA

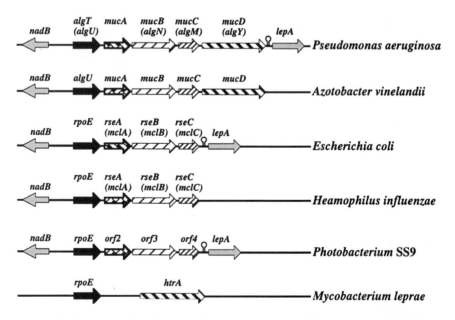

Figure 9 Comparison of *rpoE* locus in different bacterial species. This locus seems to be tightly linked to *nadB*, which codes for L-aspartate oxidase, and in known cases to *lepA* encoding leader peptidase A.

interacts with σ^E (42, 187). Several lines of evidence are consistent with the model that the N-terminus of RseA interacts directly with σ^E: (*a*) deletion of first 28 residues of the cytoplasmic N-terminal region of RseA no longer inhibited σ^E activity in vitro or in vivo (187), (*b*) point mutations in the *rseA* gene defective in σ^E inhibition occurred within the first 32 amino acid residues of RseA (187), and (*c*) overexpression of the cytoplasmic N-terminal domain with or without the membrane spanning region is sufficient to inhibit effect on σ^E activity in vivo (42, 187). In the analogous *P. aeruginosa* system, expression of MucA deleted for its C-terminal periplasmic domain did not inhibit σ^{22} activity in vivo (175). It is possible that the periplasmic domain of MucA is required for the folding of the cytoplasmic region into an active conformation or that MucA, deleted for the periplasmic region, is unstable while the analogous deletion of RseA is not.

Other Regulators of σ^E Activity

Mutation in *rseB* resulted in slight (two- to threefold) increase in σ^E activity, suggesting that RseB is a negative regulator (42, 187). Complete inhibition of σ^E activity in vivo results from overexpression of both RseA and RseB (187). Analogous to *P. aeruginosa* MucB, RseB was localized to periplasm (42, 175, 187, 235). Purified RseB had no inhibitory effect on σ^E activity in vitro (187, 235, 273). RseA and RseB coimmnunoprecipitate using anti-RseA or anti-RseB antibodies (42, 187). RseB also copurified with the C-terminal portion of RseA, which is consistent with the hypothesis that RseB exerts its negative regulatory activity through RseA via interaction with the C-terminal periplasmic region of RseA (42, 187).

The effect of RseC on σ^E activity is not dramatic. In one case, deletion of this gene had no effect on σ^E activity (42), while another group showed that an insertional mutation in *rseC* resulted in a 30% reduction of σ^E activity and overexpression of RseC resulted in a twofold increase in σ^E activity (187). This slight positive effect of RseC on σ^E activity would be consistent with the hypothesis that RseC is an antagonist of RseA activity analogous to MucC in *P. aeruginosa* (201).

Summary

Clearly, *E. coli* σ^E is required for survival at high temperature as well as sensing the state proteins. It is not clear how all these signals are transmitted via the cognate regulators, RseA, RseB, RseC, and RseD. The conservation of these loci among various species argues that the molecular mechanism of their action is likely to be conserved as well. Even though the analysis of this locus in *P. aeruginosa* has been centered around alginate production, it is possible that the role of σ^{22} in *P. aeruginosa* in a normal environment (that is, in wild-type *mucA* gene background) could be similar to *E. coli* σ^E. *P. aeruginosa* σ^{22} was shown to be involved in induction of *rpoH* (σ^{32}) promoter and survival during stress (170, 234). σ^{22} activity could also be modulated by over- or underexpression of OMPs or misfolded proteins as seen in the case of *E. coli* σ^E. Mutation in *dsbB* gene in *P. aeruginosa* resulted in alginate production indicating increased activity of σ^{22} (K Mathee and D Ohman, unpublished results).

PHOTOBACTERIUM SP. STRAIN SS9 *rpoE* LOCUS

Barophiles are deep-sea microorganisms that have adapted to grow optimally at hydrostatic pressure greater than 1 atmosphere (atm); their optimal growth temperature is usually at or below 15°C (241, 275). *Photobacterium* sp. strain SS9 (SS9) is a moderate barophile that grows optimally at 272 atm, but is able to grow at up to 23°C at 1 atm. The ability of SS9 to sense and adapt to high pressure is achieved by modulating the production of several outer membrane proteins (OMP), including OmpH (15, 39). Expression of OmpH, a porin-like outer membrane protein, increases 10- to 100-fold when the pressure is shifted from 1 to 272 atm (15, 16, 40). Insertion mutants, unlinked to the *ompH* gene, were isolated that are defective in the barosensitive regulation of OmpH (15). Subsequent molecular and sequence analysis of one of these regulatory mutants identified an operon of four open reading frames (40). Based on sequence similarity to σ^E, ORF1 presumed to be the RpoE equivalent of SS9 (40). Consistent with SS9 RopE being a member of the ECF σ factor family, the open reading frames immediately downstream of *rpoE, orf2, orf3*, and *orf4* encode proteins homologous to MucA, MucB, and MucC of *P. aeruginosa*, respectively (Figure 9; 40). Insertion mutants in *orf3* have altered levels of OMPs, but showed normal high-pressure induction of OmpH and another barosensitive OMP, OmpI (40). The growth rate of these *orf3* mutants was reduced at 15°C and at high pressure (272 atm), and completely abolished at cold temperatures (5°C) (40). Both of these phenotypes could be suppressed by providing ORF4 in trans. These findings led the authors to postulate that the role of ORF4 is in baroadaptation, whereas that of ORF3 is to influence *omp* expression (40). All the revertants of ORF3 mutants showed several different rearrangements in the *rpoE* locus; these await further characterization. By analogy to RpoE from *E. coli*, the role of RpoE in barosensitive gene regulation in *Photobacterium* SS9 may also depend on the state and abundance of extracytoplasmic proteins (40, 181).

ESCHERICHIA COLI FecI (σ^{FecI})–FecR
AND *PSEUDOMONAS PUTIDA* PupI (σ^{PupI})–PupR

Two of the alternative sigma factors belonging to the ECF subfamily are involved in transport systems, namely *E. coli* FecI for iron-citrate and *P. putida* PupI for pseudobactin transport (7, 155, 130). Both systems seem to involve three analogous proteins, a cytoplasmic sigma factor (FecI and PupI), an inner membrane protein (FecR and PupR), and an outer membrane receptor (FecA and PupA) (reviewed in 25). The current model postulates that the binding of a signal molecule to the receptor results in a conformational change that gets transmitted through the inner membrane protein to signal the sigma factor (90, 198, 271).

$E.$ $coli$ has many siderophore-mediated iron transport systems (reviewed in 25). Of these, substrate-mediated transport has been demonstrated in the case of the iron-citrate system (106). In iron-limiting conditions, the ferric-citrate transport genes (fec) are induced by ferric-citrate without entering the cell by binding to an outer membrane receptor protein called FecA (218, 261). Somehow this signal is transmitted to the cytoplasm to FecI via an inner membrane protein called FecR (218, 271). Recent biochemical analysis has shown that FecI is an alternative sigma factor (7). Unlike other ECF sigma factors, σ^{FecI} is not autoregulated (198). σ^{70}-dependent constitutive transcription is kept very low in the absence of inducer by the iron-loaded Fur repressor (198). Early genetic analysis suggested that FecR is a negative regulator of σ^{FecI}, in which σ^{FecI} activity is kept minimal by FecR in the absence of inducers (256). Deletion of the C-terminal portion of FecR resulted in constitutive expression of the fec transport genes, a result similar to MucA, the anti-sigma factor of σ^{22} in $P.$ $aeruginosa$ (198). Analogous to MucA, FecR is an inner membrane protein with a single transmembrane domain, where the N-terminus interacts with σ^{FecI} and the C-terminus with the N-terminus of FecA (125, 175, 198). However, σ^{FecI} is not activated in $fecR$ mutants under normal induction conditions, suggesting that FecR is a positive regulator and that it is essential for the activation of σ^{FecI} (198). It is possible that FecR is essential for transmitting the signal and by the same token acts as an anti-sigma factor in the absence of signal. This would be the first instance in the ECF subfamily in which an anti-sigma factor is shown to be essential, and an outer membrane component that is able to sense the signal has been identified.

ROLE OF ANTI-σ FACTORS IN AVIRULENCE

Role of FlgM in Salmonella Virulence

The $mviS$ locus of $S.$ $typhimurium$ was identified as a gene, linked to a cluster of flagellar genes (including $flgM$), that when deleted, exhibited reduced virulence in mice (35). Further characterization revealed that a defect in $mviS$ resulted in a decreased survival in mouse macrophages (232). DNA sequence analysis of a complementing clone of the $mviS$ locus gave a sequence identical to that reported for the $flgM$ gene (M Carsiotis, personal communication). Examination of $flgM$ mutant alleles revealed that loss of $flgM$ was also attenuated for mouse virulence and resulted in decreased survival in macrophages (232). This result, and the fact that the original *mviS* allele is complemented by a minimal clone of *flgM* or clones of *flgM* from nonvirulent *Salmonella* sp. (231), provides strong evidence that *mviS* and *flgM* are the same gene. How does FlgM influence virulence? Evidence suggests that FlgM is a positive factor in virulence through its anti-σ^{28} activity. The avirulent phenotype resulting from null alleles of *flgM* can be reversed by a second mutation in the structural gene for σ^{28}, *fliA* (232). Later, it was shown that the loss of flagellin expression also restored virulence in a *flgM* mutant strain (230). This result led to the model that loss of *flgM* results in increased σ^{28}-dependent expression of flagellin that led to the avirulent phenotype of a *flgM* mutant strain.

Role of MucA in Pseudomonas Virulence

Chronic pulmonary infections in patients with CF are often caused by the presence of alginate-producing *P. aeruginosa* in their lungs (79, 199). Alginate overproduction requires the presence of σ^{22}, which is tightly regulated by the anti-σ MucA (described above). The possible role of alginate in pathogenesis has recently been summarized (78). In a recent survey of Alg$^+$ CF isolates, 84% arose by mutation in the *mucA* gene (24). Using a burned mouse model, Luzar and Montie (157) showed that CF isolates were avirulent. With a respiratory infection mouse model, alginate production was shown to be important for bacterial survival in the lungs in a strain- and mouse-specific manner (24). Avirulent phenotype resulting from mutations in *mucA* gene can be reversed by a second mutation in the structural gene for σ^{22}, *algT/algU* (47, 233). These strains overproduce other known virulence factors such as proteases (189). Despite evidence that alginate production provides the bacteria with an advantage in a high-stress environment, attempts to show virulence in acute animal models have failed. Similar to FlgM, it seems that the anti-σ activity of MucA promotes virulence, especially in a non-CF environment.

PERSPECTIVES

Regulation of Transcription

If a mechanism is possible, it is usually waiting to be discovered. It has been 40 years since the discovery of the lac repressor paradigm. The ensuing prejudice biased molecular biology for many years into thinking that only one mode of gene regulation existed; many mechanisms have since been elucidated. The discovery of anti-σ factors provides another mechanism—a posttranslational mechanism—of gene regulation by direct protein:protein interaction between an anti-σ factor and its cognate σ factor to modulate the transcriptional regulation of a set of related genes. By definition, the anti-σ factors all act by a similar mechanism: the binding to the cognate σ factor and prevention of its interaction with RNA polymerase. This definition may yet expand to include those that bind specific holoenzyme complexes only or in addition to free σ. The anti-σ factors themselves are quite different. The non-ECF members, AsiA, FlgM, SpoIIAB, RsbW, and DnaK, all exist as cytoplasmic proteins, while the ECF subfamily appear to have sensor domains in the periplasm and σ factor binding domains in the cytoplasm. However, common themes are also apparent, such as the targeting of σ factor to proteolytic degradation in the DnaK and MucA systems, and the phosphatase-antagonist-kinase modules for anti-σ activity in the SpoIIAB and RsbW systems. Regulation of the anti-σ factors is quite varied, ranging from enzymatic modification by anti anti-σ factors to export out of the cell through a flagellar intermediate structure.

In the ECF subfamily of σ factors, the sigma factors are cotranscribed with their cognate negative regulators, which are known as anti-sigma factors (Table 1). In several species, the sigma factor locus *rpoE* is tightly linked to *nadB* and *lepA* (Figure 9). The ECF anti-σ factors that have been characterized are inner membrane proteins with at least one transmembrane domain. The amino terminus of ECF anti-σ factors is conserved, and evidence from *E. coli* RseA and *P. aeruginosa* MucA suggests that it is this region that interacts directly with the cognate σ factor. Poor homology is observed near the C-terminus of ECF anti-σ factors (except between RseA of *E. coli* and MclA of *H. influenzae*), in line with the possibility that they may respond to different extracytoplasmic signals. The ECF anti-σ factors are also likely to be regulated by another coexpressed protein, an anti-anti-sigma factor, as seen in the *Bacillus* systems. These proteins, MucC(AlgM) homologs, including RsbV, are very hydrophobic and likely to be localized to membranes. Accumulation or stabilization of the ECF σ factors has been proposed as the mechanism by which they are controlled. This mechanism would involve a signal transduction pathway in which an extracytoplasmic signal is transferred to the cytoplasm via proteins in three compartments—the outer membrane, periplasmic space, and inner membrane. Existing data suggest that there must be an outer membrane component that needs to be identified.

Since the initial anti-σ review (27) came out just a few years ago, progress in the field has been made in two main areas. The demonstration that several members of the ECF subfamily do act as anti-σ sigma factors in vitro has put this group solidly in the camp. Given the high degree of homology of the ECF subfamily between species, it is likely that more will be quickly identified and more genome sequences become available. The second major area of progress is the molecular and biochemical detail of the structure function relationships between the anti-σ and cognate σ factors. A crystal structure of regions 1 and 2 of σ^{70} has recently been elucidated (163), and perhaps the rest will come soon. NMR analysis of FlgM gave the surprising discovery that FlgM is unfolded in its native state and only takes on structure when bound to σ^{28} (41). This is different from DnaK and its homologs, which are highly structured without being substrate bound. Also, the crystal structure of the DnaK substrate binding domain complexed with peptide substrate has been determined. It shows the DnaK substrate binding region to include a β-sandwich domain followed by α-helical segments (280). Work on AsiA, FlgM, and SpoIIAB has demonstrated that region 4 of their cognate σ factors is the target site for interaction and inhibition, while DnaK appears to target an area between 2.4 and 3.1 for binding. Both these results probably reflect the different mechanisms of inhibition between DnaK, which targets σ^{32} for degradation, and the others, which appear to simply bind their cognate σ factors and inhibit activity. More anti-σ and σ structures have been solved, more binding constants have been determined, and more specific amino acid interactions have been identified. The detailed molecular mechanisms that transmit the environmental to the effector molecules are being resolved.

ACKNOWLEDGMENTS

The authors thank Arsalan Kharazmi and Søren Molin for making it possible for the authors to meet at the Danish Technical University to complete this review. We thank David Ambruster, Kit Brown, Allan Christensen, Martha Howe, Peter Lindum, Giri Narasimhan, Satish Raina, Phil Youderian, and Daniel Wozniak for critical comments, Dennis Ohman for his continuous support to KM, and the Annual Reviews Board for their patience. We thank M Carsiotis, J Nodwell, R Losick, D Mytelka, and M Chamberlin for sharing unpublished results.

Visit the *Annual Reviews home page* at
http://www.AnnualReviews.org.

Literature Cited

1. Akbar S, Kang CM, Gaidenko TA, Price CW. 1997. Modulator protein RsbR regulates environmental signalling in the general stress pathway of *Bacillus subtilis*. *Mol. Microbiol.* 24:567–78
2. Al Mamun AA, Tominaga A, Enomoto M. 1996. Detection and characterization of the flagellar master operon in the four *Shigella* subgroups. *J. Bacteriol.* 178:3722–26
3. Alfano C, McMacken R. 1989. Heat shock protein-mediated disassembly of nucleoprotein structures is required for the initiation of bacteriophage lambda DNA replication. *J. Biol. Chem.* 264:10709–18
4. Alper S, Dufour A, Garsin DA, Duncan L, Losick R. 1996. Role of adenosine nucleotides in the regulation of a stress-response transcription factor in *Bacillus subtilis*. *J. Mol. Biol.* 260:165–77
5. Alper S, Duncan L, Losick R. 1994. An adenosine nucleotide switch controlling the activity of a cell type-specific transcription factor in *Bacillus subtilis*. *Cell* 77:195–205
6. Anderson DK, Newton A. 1997. Post-transcriptional regulation of *Caulobacter* flagellin genes by a late flagellum assembly checkpoint. *J. Bacteriol.* 179:2281–88
7. Angerer A, Enz S, Ochs M, Braun V. 1995. Transcriptional regulation of ferric citrate transport in *Escherichia coli* K-12. FecI belongs to a new subfamily of σ^{70}-type factors that respond to extracytoplasmic stimuli. *Mol. Microbiol.* 18:163–74

8. Arigoni F, Pogliano K, Webb CD, Stragier P, Losick R. 1995. Localization of protein implicated in establishment of cell type to sites of asymmetric division. *Science* 270:637–40

9. Armstrong GA. 1994. Eubacteria show their true colors: genetics of carotenoid pigment biosynthesis from microbes to plants. *J. Bacteriol.* 176:4795–802

10. Balsalobre JM, Ruiz-Vázquez RM, Murillo FJ. 1987. Light induction of gene expression in *Myxococcus xanthus*. *Proc. Natl. Acad. Sci. USA* 84:2359–62

11. Baltimore RS, Christie CD, Smith GJ. 1989. Immunohistopathologic localization of *Pseudomonas aeruginosa* in lungs from patients with cystic fibrosis. Implications for the pathogenesis of progressive lung deterioration. *Am. Rev. Respir. Dis.* 140:1650–61

12. Banuett F, Hoyt MA, McFarlane L, Echols H, Herskowitz I. 1986. *hflB*, a new *Escherichia coli* locus regulating lysogeny and the level of bacteriophage cII protein. *J. Mol. Biol.* 187:213–24

13. Barak I, Behari J, Olmedo G, Guzman P, Brown DP, et al. 1996. Structure and function of the SpoIIE protein and its localization to sites of sporulation septum assembly. *Mol. Microbiol.* 19:1047–60

14. Barak I, Youngman P. 1996. SpoIIE mutants of *Bacillus subtilis* comprise two distinct phenotypic classes consistent with a dual functional role for the SpoIIE protein. *J. Bacteriol.* 178:4984–89

15. Bartlett D, Wright M, Yayanos AA, Silverman M. 1989. Isolation of a gene regulated by hydrostatic pressure in deep-sea bacterium. *Nature* 342:572–74

16. Bartlett DH, Chi E, Wright ME. 1993. Sequence of the *ompH* gene from the deep-sea bacterium *Photobacterium* SS9. *Gene* 131:125–28

17. Benasson RV, Land EJ, Truscot TG. 1993. *Excited States and Free Radicals in Biology and Medicine*. Oxford, UK: Oxford Univ. Press

18. Benson AK, Haldenwang WG. 1993. *Bacillus subtilis* σ^B is regulated by a binding protein (RsbW) that blocks its association with core RNA polymerase. *Proc. Natl. Acad. Sci. USA* 90:2330–34

19. Binet R, Letoffe S, Ghigo JM, Delepelaire P, Wandersman C. 1997. Protein secretion by Gram-negative bacterial ABC exporters—a review. *Gene* 192:7–11

20. Blair DF. 1995. How bacteria sense and swim. *Annu. Rev. Microbiol.* 49:489–522

21. Blaszczak A, Zylicz M, Georgopoulos C, Liberek K. 1995. Both ambient temperature and the DnaK chaperone machine modulate the heat shock response in *Escherichia coli* by regulating the switch between σ^{70} and σ^{32} factors assembled with RNA polymerase. *EMBO. J.* 14:5085–93

22. Botella JA, Murillo FJ, Ruiz-Vázquez R. 1995. A cluster of structural and regulatory genes for light-induced carotenogenesis in *Myxococcus xanthus*. *Eur. J. Biochem.* 233:238–48

23. Boucher JC, Martinez-Salazar J, Schurr MJ, Mudd MH, Yu H, Deretic V. 1996. Two distinct loci affecting conversion to mucoidy in *Pseudomonas aeruginosa* in cystic fibrosis encode homologs of the serine protease HtrA. *J. Bacteriol.* 178:511–23

24. Boucher JC, Yu H, Mudd MH, Deretic V. 1997. Mucoid *Pseudomonas aeruginosa* in cystic fibrosis: characterization of muc mutations in clinical isolates and analysis of clearance in a mouse model of respiratory infection. *Infect. Immun.* 65:3838–46

25. Braun V. 1997. Surface signaling: novel transcription initiation mechanism starting from the cell surface. *Arch. Microbiol.* 167:325–31

26. Britton G. 1995. Structure and properties of carotenoids in relation to function. *FASEB. J.* 9:1551–58

27. Brown KL, Hughes KT. 1995. The role of anti-sigma factors in gene regulation. *Mol. Microbiol.* 16:397–404

28. Buchberger A, Theyssen H, Schroder H, McCarty JS, Virgallita G, et al. 1996. Nucleotide-induced conformational changes in the ATPase and substrate binding domains of the DnaK chaperone provide evidence for interdomain communication. *J. Biol. Chem.* 270:16903–10

29. Buchberger A, Valencia A, McMacken R, Sander C, Bukau B. 1994. The chaperone function of DnaK requires the coupling of ATPase activity with substrate binding through residue E171. *EMBO. J.* 13:1687–95

30. Bukau B. 1993. Regulation of the *Escherichia coli* heat shock response. *Mol. Microbiol.* 9:671–80

31. Burchard RP, Dworkin M. 1966. Light-induced lysis and carotenogenesis in *Myxococcus xanthus*. *J. Bacteriol.* 91:535–45

32. Burchard RP, Hendricks SB. 1969. Action spectrum for carotenogenesis in *Myxococcus xanthus*. *J. Bacteriol.* 97:1165–68

33. Burkholder WF, Zhao X, Zhu X, Hendrickson WA, Gragerov A, Gottesman ME. 1996. Mutations in the C-terminal fragment of DnaK affect peptide binding. *Proc. Natl. Acad. Sci. USA* 93: 10632–37

34. Caramori T, Barilla D, Nessi C, Sacchi L, Galizzi A. 1996. Role of FlgM in σ^D-dependent gene expression in *Bacillus subtilis*. *J. Bacteriol.* 178:3113–18

35. Carsiotis M, Stocker BA, Weinstein DL, O'Brien AD. 1989. A *Salmonella typhimurium* virulence gene linked to *flg*. *Infect. Immun.* 57:3276–80

36. Chan B, Spassky A, Busby S. 1990. The organization of open complexes between *Escherichia coli* RNA polymerase and DNA fragments carrying promoters either with or without consensus −35 region sequences. *Biochem. J.* 270:141–48

37. Chater KF, Bruton CJ, Plaskitt KA, Buttner MJ, Mendez C, Helmann JD. 1989. The developmental fate of *Streptomyces coelicor* hyphae depends upon a gene product homologous with the motility sigma factor of *Bacillus subtilis*. *Cell* 59:133–43

38. Chater KF, Losick R. 1996. The mycelial life-style of *Streptomyces coelicor* and its relatives. In *Bacteria As Multicellular Organisms*, ed. JH Shapiro, M Dworkin. New York: Oxford Univ. Press

39. Chi E, Bartlett DH. 1993. Use of a reporter gene to follow high-pressure signal transduction in the deep-sea bacterium *Photobacterium* sp. strain SS9. *J. Bacteriol.* 175:7533–40

40. Chi E, Bartlett DH. 1995. An *rpoE*-like locus controls outer membrane protein synthesis and growth at cold temperatures and high pressures in the deep-sea bacterium *Photobacterium* sp. strain SS9. *Mol. Microbiol.* 17:713–26

41. Daughdrill GW, Chadsey MS, Karlinsey JE, Hughes KT, Dahlquist FW. 1997. The C-terminal half of the anti-sigma factor, FlgM, becomes structured when bound to its target σ^{28}. *Nat. Struct. Biol.* 4:285–91

42. De Las Penas A, Connolly L, Gross CA. 1997. The σ^E-mediated response to extracytoplasmic stress in *Escherichia coli* is transduced by RseA and RseB, two negative regulators of σ^E. *Mol. Microbiol.* 24:373–85

43. Decatur A, Losick R. 1996. Three sites of contact between the *Bacillus subtilis* transcription factor σ^F and its antisigma factor SpoIIAB. *Genes Dev.* 10:2348–58

44. Deretic V, Govan JR, Konyecsni WM, Martin DW. 1990. Mucoid *Pseudomonas aeruginosa* in cystic fibrosis: mutations in the *muc* loci affect transcription of the *algR* and *algD* genes in response to environmental stimuli. *Mol. Microbiol.* 4:189–96

45. Deretic V, Schurr MJ, Boucher JC, Martin DW. 1994. Conversion of *Pseudomonas aeruginosa* to mucoidy in cystic fibrosis: environmental stress and regulation of bacterial virulence by alternate sigma factors. *J. Bacteriol.* 176:2773–80

46. DeVries CA, Hassett DJ, Flynn JL, Ohman DE. 1995. Genetic linkage in *Pseudomonas aeruginosa* of *algT* and *nadB*: mutation in *nadB* does not affect NAD biosynthesis or alginate production. *Gene* 156:63–67

47. DeVries CA, Ohman DE. 1994. Mucoid-to-nonmucoid conversion in alginate-producing *Pseudomonas aeruginosa* often results from spontaneous mutations in *algT*, encoding a putative alternative sigma factor, and shows evidence for autoregulation. *J. Bacteriol.* 176:6677–87

48. Di Mascio P, Devasagayam TP, Kaiser S, Sies H. 1990. Carotenoids, tocopherols and thiols as biological singlet molecular oxygen quenchers. *Biochem. Soc. Trans.* 18:1054–56

49. Diederich B, Wilkinson JF, Magnin T, Najafi M, Errington J, Yudkin M. 1994. Role of interactions between SpoIIAA and SpoIIAB in regulating cell-specific transcription factor σ^F of *Bacillus subtilis*. *Genes Dev.* 8:2653–63

50. Dufour A, Haldenwang WG. 1994. Interactions between a *Bacillus subtilis* anti-σ factor (RsbW) and its antagonist (RsbV). *J. Bacteriol.* 176:1813–20

51. Dufour A, Voelker U, Voelker A, Haldenwang WG. 1996. Relative levels and fractionation properties of *Bacillus subtilis* σ^B and its regulators during balanced growth and stress. *J. Bacteriol.* 178:3701–9

52. Duncan L, Alper S, Arigoni F, Losick R, Stragier P. 1995. Activation of cell-specific transcription by a serine phosphatase at the site of asymmetric division. *Science* 270:641–44

53. Duncan L, Alper S, Losick R. 1996. SpoIIAA governs the release of the cell-type specific transcription factor σ^F from its anti-sigma factor SpoIIAB. *J. Mol. Biol.* 260:147–64

54. Duncan L, Losick R. 1993. SpoIIAB is an anti-σ factor that binds to and inhibits transcription by regulatory protein σ^F from *Bacillus subtilis*. *Proc. Natl. Acad. Sci. USA* 90:2325–29

55. Erickson JW, Gross CA. 1989. Identification of the σ^E subunit of *Escherichia coli* RNA polymerase: a second alternate σ factor involved in high-temperature gene expression. *Genes Dev.* 3:1462–71

56. Errington J. 1996. Determination of cell fate in *Bacillus subtilis. Trends Genet.* 12:31–34

57. Evans LR, Linker A. 1973. Production and characterization of the slime polysaccharide of *Pseudomonas aeruginosa. J. Bacteriol.* 116:915–24

58. Fan F, Macnab RM. 1996. Enzymatic characterization of FliI, an ATPase involved in flagellar assembly in *Salmonella typhimurium. J. Biol. Chem.* 271:31981–88

59. Feucht A, Magnin T, Yudkin MD, Errington J. 1996. Bifunctional protein required for asymmetric cell division and cell-specific transcription in *Bacillus subtilis. Genes Dev.* 10:794–803

60. Flynn JL, Ohman DE. 1988. Cloning of genes from mucoid *Pseudomonas aeruginosa* which control spontaneous conversion to the alginate production phenotype. *J. Bacteriol.* 170:1452–60

61. Flynn JL, Ohman DE. 1988. Use of a gene replacement cosmid vector for cloning alginate conversion genes from mucoid and nonmucoid *Pseudomonas aeruginosa* strains: *algS* controls expression of *algT. J. Bacteriol.* 170:3228–36

62. Fontes M, Ruiz-Vázquez R, Murillo FJ. 1993. Growth phase dependence of the activation of a bacterial gene for carotenoid synthesis by blue light. *EMBO. J.* 12:1265–75

63. Fredrick K, Helmann JD. 1996. FlgM is a primary regulator of σ^D activity, and its absence restores motility to a *sinR* mutant. *J. Bacteriol.* 178:7010–13

64. Fyfe JA, Govan JR. 1980. Alginate synthesis in mucoid *Pseudomonas aeruginosa*: a chromosomal locus involved in control. *J. Gen. Microbiol.* 119:443–50

65. Gamer J, Bujard H, Bukau B. 1992. Physical interaction between heat shock proteins DnaK, DnaJ, and GrpE and the bacterial heat shock transcription factor σ^{32}. *Cell* 69:833–42

66. Gamer J, Multhaup G, Tomoyasu T, McCarty JS, Rudiger S, et al. 1996. A cycle of binding and release of the DnaK, DnaJ, and GrpE chaperones regulates activity of the *Escherichia coli* heat shock transcription factor σ^{32}. *EMBO. J.* 15:607–17

67. Gautsch JW, Wulff DL. 1974. Fine structure mapping, complementation, and physiology of *Escherichia coli hfl* mutants. *Genetics* 77:435–48

68. Geiduschek EP. 1991. Regulation of expression of the late genes of bacteriophage T4. *Annu. Rev. Genet.* 25:437–60

69. Genin S, Boucher CA. 1994. A superfamily of proteins involved in different secretion pathways in gram-negative bacteria: modular structure and specificity of the N-terminal domain. *Mol. Gen. Genet.* 243:112–18

70. Georgopoulos C, Liberek K, Zylicz M, Ang D. 1994. Properties of the heat shock proteins of *Escherichia coli* and the autoregulation of the heat shock response. In *The Biology of Heat Shock Proteins and Molecular Chaperones*, ed. RI Morimoto, A Tissieres, C Georgopoulos, pp. 209–49. Cold Spring Harbor, NY: Cold Spring Harbor Lab.

71. Gillen KL, Hughes KT. 1991. Negative regulatory loci coupling flagellin synthesis to flagellar assembly in *Salmonella typhimurium. J. Bacteriol.* 173:2301–10

72. Gillen KL, Hughes KT. 1991. Molecular characterization of *flgM*, a gene encoding a negative regulator of flagellin synthesis in *Salmonella typhimurium. J. Bacteriol.* 173:6453–59

73. Goldberg JB, Dahnke T. 1992. *Pseudomonas aeruginosa* AlgB, which modulates the expression of alginate, is a member of the NtrC subclass of prokaryotic regulators. *Mol. Microbiol.* 6:59–66

74. Goldberg JB, Gorman WL, Flynn JL, Ohman DE. 1993. A mutation in *algN* permits trans activation of alginate production by *algT* in *Pseudomonas* species. *J. Bacteriol.* 175:1303–8

75. Goodwin TW. 1983. Developments in carotenoid biochemistry over 40 years. The third Morton lecture. *Biochem. Soc. Trans.* 11:473–83

76. Gorham HC, McGowan SJ, Robson PR, Hodgson DA. 1996. Light-induced carotenogenesis in *Myxococcus xanthus*: light-dependent membrane sequestration of ECF sigma factor CarQ by anti-sigma factor CarR. *Mol. Microbiol.* 19:171–86

77. Gottesman S, Wickner S, Maurizi MR. 1997. Protein quality control: triage by chaperones and proteases. *Genes Dev.* 11:815–23

78. Govan JR, Deretic V. 1996. Microbial pathogenesis in cystic fibrosis: mucoid *Pseudomonas aeruginosa* and *Burkholderia cepacia. Microbiol. Rev.* 60:539–74

79. Govan JR, Harris GS. 1986. *Pseudomonas aeruginosa* and cystic fibrosis:

unusual bacterial adaptation and pathogenesis. *Microbiol. Sci.* 3:302–8

80. Gragerov A, Zeng L, Zhao X, Burkholder W, Gottesman ME. 1994. Specificity of DnaK-peptide binding. *J. Mol. Biol.* 235: 848–54

81. Gross CA. 1996. Function and regulation of the heat shock proteins. In *Escherichia coli and Salmonella Cellular and Molecular Biology*, ed. FC Neidhardt, pp. 167–89. Washington, DC: Am. Soc. Microbiol. 2nd ed.

82. Gross CA, Lonetto M, Losick R. 1992. Bacterial sigma factors. In *Transcriptional Regulation*, ed. K Yamamoto, S McKnight. Cold Spring Harbor, NY: Cold Spring Harbor Lab Press

83. Gross CA, Straus DB, Erickson JW, Yura T. 1990. The function and regulation of heat shock proteins in *Escherichia coli*. In *Stress Proteins in Biology and Medicine*, ed. RI Morimoto, A Tissieres, C Georgopoulos, pp. 167–89. Cold Spring Harbor, NY: Cold Spring Harbor Lab.

84. Grossman AD, Erickson JW, Gross CA. 1984. The *htpR* gene product of *E. coli* is a sigma factor for heat shock promoters. *Cell* 38:383–90

85. Grossman AD, Straus DB, Walter WA, Gross CA. 1987. σ^{32} synthesis can regulate the synthesis of heat shock proteins in *Escherichia coli*. *Genes Dev.* 1:179–84

86. Gruber TM, Bryant DA. 1997. Molecular systematic studies of eubacteria, using σ^{70}-type sigma factors of group 1 and group 2. *J. Bacteriol.* 179:1734–47

87. Guild N, Gayle M, Sweeney R, Hollingsworth T, Modeer T, Gold L. 1988. Transcriptional activation of bacteriophage T4 middle promoters by the MotA protein. *J. Mol. Biol.* 199:241–58

88. Gygi D, Fraser G, Dufour A, Hughes C. 1997. A motile but non-swarming mutant of Proteus mirabilis lacks FlgN, a facilitator of flagella filament assembly. *Mol. Microbiol.* 25:597–604

89. Haldenwang WG. 1995. The sigma factors of *Bacillus subtilis*. *Microbiol. Rev.* 59:1–30

90. Harle C, Kim I, Angerer A, Braun V. 1995. Signal transfer through three compartments: transcription initiation of the *Escherichia coli* ferric citrate transport system from the cell surface. *EMBO. J.* 14:1430–38

91. Hecker M, Schumann W, Volker U. 1996. Heat-shock and general stress response in *Bacillus subtilis*. *Mol. Microbiol.* 19:417–28

92. Helmann JD, Chamberlin MJ. 1988. Structure and function of bacterial sigma factors. *Annu. Rev. Biochem.* 57:839–72

93. Hendrick JP, Hartl FU. 1993. Molecular chaperone functions of heat-shock proteins. *Annu. Rev. Biochem.* 62:349–84

94. Herman C, Ogura T, Tomoyasu T, Hiraga S, Akiyama Y, et al. 1993. Cell growth and λ phage development controlled by the same essential *Escherichia coli* gene, *ftsH/hflB*. *Proc. Natl. Acad. Sci. USA* 90: 10861–65

95. Herman C, Thevenet D, D'Ari R, Bouloc, P. 1995. Degradation of σ^{32}, the heat shock regulator in *Escherichia coli*, is governed by HflB. *Proc. Natl. Acad. Sci. USA* 92:3516–20

96. Herman C, Thevenet D, D'Ari R, Bouloc, P. 1997. The HflB protease of *Escherichia coli* degrades its inhibitor χ cIII. *J. Bacteriol.* 179:358–63

97. Hershberger CD, Ye RW, Parsek MR, Xie Z, Chakrabarty AM. 1995. The *algT* (*algU*) gene of *Pseudomonas aeruginosa*, a key regulator involved in alginate biosynthesis, encodes an alternative σ factor (σ^{E}). *Proc. Natl. Acad. Sci. USA* 92:7941–45

98. Hinton DM. 1991. Transcription from a bacteriophage T4 middle promoter using T4 MotA protein and phage-modified RNA polymerase. *J. Biol. Chem.* 266: 18034–44

99. Hinton DM, March-Amegadzie R, Gerber J, Sharma M. 1996. Bacteriophage T4 middle transcription system: T4-modified RNA polymerase; AsiA, a σ^{70} binding protein; and transcriptional activator MotA. *Methods Enzymol.* 274:43–57

100. Hinton DM, March-Amegadzie R, Gerber JS, Sharma M. 1996. Characterization of pre-transcription complexes made at a bacteriophage T4 middle promoter: involvement of the T4 MotA activator and the T4 AsiA protein, a σ^{70} binding protein, in the formation of the open complex. *J. Mol. Biol.* 256:235–48

101. Hirano T, Yamaguchi S, Oosawa K, Aizawa SI. 1994. Roles of FliK and FlhB in determination of flagellar hook length in *Salmonella typhimurium*. *J. Bacteriol.* 176:5439–49

102. Hiratsu K, Amemura M, Nashimoto H, Shinagawa H, Makino K. 1995. The *rpoE* gene of *Escherichia coli*, which encodes σ^{E}, is essential for bacterial growth at high temperature. *J. Bacteriol.* 177:2918–22

103. Hodgson DA. 1993. Light induced carotenogenesis in *Myxococcus xanthus*: ge-

netic analysis of the *carR* region. *Mol. Microbiol.* 7:471–88

104. Homma M, Fujita H, Yamaguchi S, Iino T. 1987. Regions of *Salmonella typhimurium* flagellin essential for its polymerization and excretion. *J. Bacteriol.* 169:291–96

105. Hughes KT, Gillen KL, Semon MJ, Karlinsey JE. 1993. Sensing structural intermediates in bacterial flagellar assembly by export of a negative regulator. *Science* 262:1277–28

106. Hussein S, Hantke K, Braun V. 1981. Citrate-dependent iron transport system in *Escherichia coli* K-12. *Eur. J. Biochem.* 117:431–37

107. Høiby N. 1974. Epidemiological investigations of the respiratory tract bacteriology in patients with cystic fibrosis. *Acta Pathol. Microbiol. Scand.* 82:541–50

108. Høiby N. 1977. *Pseudomonas aeruginosa* infection in cystic fibrosis. Diagnostic and prognostic significance of pseudomonas aeruginosa precipitins determined by means of crossed immunoelectrophoresis. A survey. *Acta Pathol. Microbiol. Scand. Suppl.* 262:1–96

109. Høiby N, Jacobsen L, Jorgensen BA, Lykkegaard E, Weeke B. 1974. *Pseudomonas aeruginosa* infection in cystic fibrosis. Occurrence of precipitating antibodies against *Pseudomonas aeruginosa* in relation to the concentration of sixteen serum proteins and the clinical and radiographical status of the lungs. *Acta Paediatr. Scand.* 63:843–48

110. Iyoda S, Kutsukake K. 1995. Molecular dissection of the flagellum-specific anti-sigma factor, FlgM, of *Salmonella typhimurium*. *Mol. Gen. Genet.* 249:417–24

111. Izard JW, Kendall DA. 1994. Signal peptides: exquisitely designed transport promoters. *Mol. Microbiol.* 13:765–73

112. Jensen ET, Kharazmi A, Garred P, Kronborg G, Fomsgaard A, et al. 1993. Complement activation by *Pseudomonas aeruginosa* biofilms. *Microb. Pathog.* 15:377–88

113. Jensen ET, Kharazmi A, Lam K, Costerton JW, Høiby N. 1990. Human polymorphonuclear leukocyte response to *Pseudomonas aeruginosa* grown in biofilms. *Infect. Immun.* 58:2383–85

114. Jordan R, McMacken R. 1995. Modulation of the ATPase activity of the molecular chaperone DnaK by peptides and the DnaJ and GrpE heat shock proteins. *J. Biol. Chem.* 270:4563–69

115. Kalman S, Duncan ML, Thomas SM, Price CW. 1990. Similar organization of the sigB and spoIIA operons encoding alternate sigma factors of *Bacillus subtilis* RNA polymerase. *J. Bacteriol.* 172:5575–85

116. Kang CM, Brody MS, Akbar S, Yang X, Price CW. 1996. Homologous pairs of regulatory proteins control activity of *Bacillus subtilis* transcription factor σ^B in response to environmental stress. *J. Bacteriol.* 178:3846–53

117. Kapatral V, Olson JW, Pepe JC, Miller VL, Minnich SA. 1996. Temperature-dependent regulation of *Yersinia enterocolitica* Class III flagellar genes. *Mol. Microbiol.* 19:1061–71

118. Karlinsey JE, Pease AJ, Winkler ME, Bailey JL, Hughes KT. 1997. The *flk* gene of *Salmonella typhimurium* couples flagellar P- and L-ring assembly to flagellar morphogenesis. *J. Bacteriol.* 179:2389–400

119. Karzai AW, McMacken R. 1996. A bipartite signaling mechanism involved in DnaJ-mediated activation of the *Escherichia coli* DnaK protein. *J. Biol. Chem.* 271:11236–46

120. Kawagishi I, Homma M, Williams AW, Macnab RM. 1996. Characterization of the flagellar hook length control protein *fliK* of *Salmonella typhimurium* and *Escherichia coli*. *J. Bacteriol.* 178:2954–59

121. Keilty S, Rosenberg M. 1987. Constitutive function of a positively regulated promoter reveals new sequences essential for activity. *J. Biol. Chem.* 262:6389–95

122. Kellner EM, Decatur A, Moran CP Jr. 1996. Two-stage regulation of an anti-sigma factor determines developmental fate during bacterial endospore formation. *Mol. Micribiol.* 21:913–24

123. Kharazmi A, Schiøtz PO, Høiby N, Baek L, Doring G. 1986. Demonstration of neutrophil chemotactic activity in the sputum of cystic fibrosis patients with *Pseudomonas aeruginosa* infection. *Eur. J. Clin. Invest.* 16:143–48

124. Kihara A, Akiyama Y, Ito K. 1997. Host regulation of lysogenic decision in bacteriophage lambda: transmembrane modulation of FtsH (HflB), the cII degrading protease, by HflKC (HflA). *Proc. Natl. Acad. Sci. USA* 94:5544–49

125. Kim I, Stiefel A, Plantor S, Angerer A, Braun V. 1997. Transcription induction of the ferric citrate transport genes via the N-terminus of the FecA outer membrane protein, the Ton system and the electrochemical potential of the cy-

toplasmic membrane. *Mol. Microbiol.* 23:333–44

126. Komeda Y. 1986. Transcriptional control of flagellar genes in *Escherichia coli* K-12. *J. Bacteriol.* 168:1315–18

127. Komeda Y, Iino T. 1979. Regulation of expression of the flagellin gene (*hag*) in *Escherichia coli* K-12: analysis of *hag-lac* gene fusions. *J. Bacteriol.* 139:721–29

128. Komeda Y, Suzuki H, Ishidsu JI, Iino T. 1975. The role of cAMP in flagellation of *Salmonella typhimurium. Mol. Gen. Genet.* 142:289–98

129. Kormanec J, Potuckova L, Rezuchova B. 1994. The *Streptomyces aureofaciens* homologue of the whiG gene encoding a putative sigma factor essential for sporulation. *Gene* 143:101–3

129a. Kornacker MG, and Newton A. 1994. Information essential for cell-cycle-dependent secretion of the 591-residue *Caulobacter* hook protein is confined to a 21-amino acid sequence near the N-terminus. *Mol. Microbiol.* 14:73–85

130. Koster M, van Klompenburg W, Bitter W, Leong J, Weisbeek P. 1994. Role for the outer membrane ferric siderophore receptor PupB in signal transduction across the bacterial cell envelope. *EMBO. J.* 13:2805–13

131. Kumar A, Malloch RA, Fujita N, Smillie DA, Ishihama A, Hayward RS. 1993. The minus 35-recognition region of *Escherichia coli* σ^{70} is inessential for initiation of transcription at an "extended minus 10" promoter. *J. Mol. Biol.* 232:406–18

132. Kutsukake K. 1994. Excretion of the anti-sigma factor through a flagellar substructure couples flagellar gene expression with flagellar assembly in *Salmonella typhimurium. Mol. Gen. Genet.* 243:605–12

133. Kutsukake K. 1997. Autogenous and global control of the flagellar master operon, *flhD*, in *Salmonella typhimurium. Mol. Gen. Genet.* 254:440–48

134. Kutsukake K, Iino T. 1994. Role of the FliA-FlgM regulatory system on the transcriptional control of the flagellar regulon and flagellar formation in *Salmonella typhimurium. J. Bacteriol.* 176:3598–605

135. Kutsukake K, Iyoda S, Ohnishi K, Iino T. 1994. Genetic and molecular analyses of the interaction between the flagellum-specific σ and its anti-σ factors in *Salmonella typhimurium. EMBO. J.* 13:4568–76

135a. Kutsukake K, Minamino T, Yokoseki T.

1994. Isolation and characterization of FliK-independent flagellation mutants from *Salmonella typhimurium. J. Bacteriol.* 176:7625–7629

136. Kutsukake K, Ohya Y, Iino T. 1990. Transcriptional analysis of the flagellar regulon of *Salmonella typhimurium. J. Bacteriol.* 172:741–47

137. Langer T, Lu C, Echols H, Flanagan J, Hayer MK, Hartl FU. 1992. Successive action of DnaK, DnaJ and GroEL along the pathway of chaperone-mediated protein folding. *Nature* 356:683–89

138. Lazar SW, Kolter R. 1996. SurA assists the folding of *Escherichia coli* outer membrane proteins. *J. Bacteriol.* 178:1770–73

139. Lewis PJ, Magnin T, Errington J. 1996. Compartmentalized distribution of the proteins controlling the prespore-specific transcription factor, σ^{F} of *Bacillus subtilis. Genes Cells.* 1:881–94

140. Liberek K, Galitski TP, Zylicz M, Georgopoulos C. 1992. The DnaK chaperone modulates the heat shock response of *Escherichia coli* by binding to the σ^{32} transcription factor. *Proc. Natl. Acad. Sci. USA* 89:3516–20

141. Liberek K, Georgopoulos C. 1993. Autoregulation of the *Escherichia coli* heat shock response by the DnaK and DnaJ heat shock proteins. *Proc. Natl. Acad. Sci.* 90:11019–23

142. Liberek K, Georgopoulos C, Zylicz M. 1988. Role of the *Escherichia coli* DnaK and DnaJ heat shock proteins in the initiation of bacteriophage λ DNA replication. *Proc. Natl. Acad. Sci. USA* 85:6632–36

143. Liberek K, Marszalek J, Ang D, Georgopoulos C, Zylicz M. 1991. *Escherichia coli* DnaJ and GrpE heat shock proteins jointly stimulate ATPase activity of DnaK. *Proc. Natl. Acad. Sci. USA* 88:2874–78

144. Liberek K, Wall D, Georgopoulos C. 1995. The DnaJ chaperone catalytically activates the DnaK chaperone to preferentially bind the σ^{32} heat shock transcriptional regulator. *Proc. Natl. Acad. Sci. USA* 92:6224–28

145. Lin ECC, Lynch AS. 1996. *Regulation of Gene Expression in Escherichia Coli.* Austin, TX: Landes

146. Linker A, Jones RS. 1966. A new polysaccharide resembling alginic acid isolated from Pseudomonads. *J. Biol. Chem.* 241: 3845–51

147. Lipinska B, Fayet O, Baird L, Georgopoulos C. 1989. Identification, characterization, mapping of the *Escherichia*

coli htrA gene, whose product is essential for bacterial growth only at elevated temperature. *J. Bacteriol.* 171:1574–84

148. Lipinska B, Sharma S, Georgopoulos C. 1988. Sequence analysis and regulation of the *htrA* gene of *Escherichia coli*: a σ^{32}-independent mechanism of heat-inducible transcription. *Nucleic Acids Res.* 16:10053–67

149. Lipinska B, Zylicz M, Georgopoulos C. 1990. The HtrA(DegP) protein, essential for *Escherichia coli* survival at high temperature, is an endopeptidase. *J. Bacteriol.* 172:1791–97

150. Liu X, Matsumura P. 1994. The FlhD/FlhC complex, a transcriptional activator of the *Escherichia coli* flagellar Class II operons. *J. Bacteriol.* 176:7345–51

151. Liu X, Matsumura P. 1995. An alternative sigma factor controls transcription of flagellar class-III operons in *Escherichia coli*: gene sequence, overproduction, purification and characterization. *Gene* 164:81–84

152. Liu X, Matsumura P. 1995. The C-terminal region of the α subunit of *Escherichia coli* RNA polymerase is required for transcriptional activation of the flagellar level II operons by the FlhD/FlhC complex. *J. Bacteriol.* 177:5186–88

153. Liu X, Matsumura P. 1996. Differential regulation of multiple overlapping promoters in flagellar Class II operons in *Escherichia coli*. *Mol. Microbiol.* 21:613–20

154. Lonetto M, Gribskow M, Gross CA. 1992. The σ^{70} family: sequence conservation and evolutionary relationships. *J. Bacteriol.* 174:3843–49

155. Lonetto MA, Brown KL, Rudd KE, Buttner MJ. 1994. Analysis of the *Streptomyces coelicolor sigE* gene reveals the existence of a subfamily of eubacterial RNA polymerase σ factors involved in the regulation of extracytoplasmic functions. *Proc. Natl. Acad. Sci. USA* 91:7573–77

156. Lord M, Magnin T, Yudkin MD. 1996. Protein conformational change and nucleotide binding involved in regulation of σ^F in *Bacillus subtilis*. *J. Bacteriol.* 178: 6730–35

157. Luzar MA, Thomassen MJ, Montie TC. 1985. Flagella and motility alterations in *Pseudomonas aeruginosa* strains from patients with cystic fibrosis: relationship to patient clinical condition. *Infect. Immun.* 50:577–82

158. MacGeorge J, Korolik V, Morgan AF, Ashe V, Holloway B. 1986. Transfer of a chromosomal locus responsible for mucoid colony morphology in *Pseudomonas aeruginosa* isolated from cystic fibrosis patients to *Pseudomonas aeruginosa* PAO. *J. Med. Microbiol.* 21: 331–36

159. Macnab RM. 1995. Flagella and motility. In Escherichia coli *and* Salmonella typhimurium: *Cellular and Molecular Biology*, ed. FC Neidhart, p. 123–45. Washington, DC: Am. Soc. Microbiol. 2nd ed.

160. Mager WH, de Kruijff AJ. 1995. Stress-induced transcriptional activation. *Microbiol. Rev.* 59:506–31

161. Magnin T, Lord M, Errington J, Yudkin MD. 1996. Establishing differential gene expression in sporulating *Bacillus subtilis*: phosphorylation of SpoIIAA (anti-anti-σ^F) alters its conformation and prevents formation of a SpoIIAA/SpoIIAB/ADP complex. *Mol. Microbiol.* 19:901–7

162. Magnin T, Lord M, Yudkin MD. 1997. Contribution of partner switching and SpoIIAA cycling to regulation of σ^F activity in sporulating *Bacillus subtilis*. *J. Bacteriol.* 179:3922–27

163. Malhotra A, Severinova E, Darst SA. 1996. Crystal structure of a σ^{70} subunit fragment from *Escherichia coli* RNA polymerase. *Cell* 87:127–36

164. Malik S, Goldfarb A. 1988. Late σ factor of bacteriophage T4. Formation and properties of RNA polymerase-promoter complexes. *J. Biol. Chem.* 263: 1174–81

165. Malik S, Zalenskaya K, Goldfarb A. 1987. Competition between sigma factors for core RNA polymerase. *Nucleic Acids Res.* 15:8521–30

166. Margolis P, Driks A, Losick R. 1991. Establishment of cell type by compartmentalized activation of a transcription factor. *Science* 254:562–65

167. Martin DW, Holloway BW, Deretic V. 1993. Characterization of a locus determining the mucoid status of *Pseudomonas aeruginosa*: AlgU shows sequence similarities with a *Bacillus* sigma factor. *J. Bacteriol.* 175:1153–64

168. Martin DW, Schurr MJ, Mudd MH, Deretic V. 1993. Differentiation of *Pseudomonas aeruginosa* into the alginate-producing form: inactivation of *mucB* causes conversion to mucoidy. *Mol. Microbiol.* 9:497–506

169. Martin DW, Schurr MJ, Mudd MH, Govan JR, Holloway BW, Deretic V. 1993.

Mechanism of conversion to mucoidy in *Pseudmonas aeruginosa* infecting cystic fibrosis patients. *Proc. Natl. Acad. Sci. USA* 90:8377–81

170. Martin DW, Schurr MJ, Yu H, Deretic V. 1994. Analysis of promoters controlled by the putative sigma factor AlgU regulating conversion to mucoidy in *Pseudomonas aeruginosa*: relationship to σ^E and stress response. *J. Bacteriol.* 176:6688–96

171. Martínez-Laborda A, Balsalobre JM, Fontes M, Murillo FJ. 1990. Accumulation of carotenoids in structural and regulatory mutants of the bacterium *Myxococcus xanthus*. *Mol. Gen. Genet.* 223:205–10

172. Martínez-Laborda A, Elías M, Ruiz-Vázquez R, Murillo FJ. 1986. Insertion of Tn5 linked to mutations affecting carotenoid synthesis in *Myxococcus xanthus*. *Mol. Gen. Genet.* 205:107–14

173. Martínez-Laborda A, Murillo FJ. 1989. Genic and allelic interactions in the carotenogenic response of *Myxococcus xanthus* to blue light. *Genetics* 122:481–90

174. Deleted in proof

175. Mathee K, McPherson CJ, Ohman DE. 1997. Posttranslational control of the *algT* (*algU*)-encoded sigma 22 for expression of the alginate regulon in *Pseudomonas aeruginosa* and localization of its antagonist proteins MucA and MucB (AlgN). *J. Bacteriol.* 179:3711–20

176. Mattson T, van Houwe G, Epstein RH. 1978. Isolation and characterization of conditional lethal mutations in the *mot* gene of bacteriophage T4. *J. Mol. Biol.* 126:551–70

177. McCarty JS, Buchberger A, Reinstein J, Bukau B. 1995. The role of ATP in the functional cycle of the DnaK chaperone system. *J. Mol. Biol.* 249:126–37

178. McCarty JS, Rudiger S, Schonfeld HJ, Schneider-Mergener J, Nakahigashi K, et al. 1996. Regulatory region C of the *Escherichia coli* heat shock transcription factor, σ^{32}, constitutes a DnaK binding site and is conserved among eubacteria. *J. Mol. Biol.* 256:829–37

179. McGowan SJ, Gorham HC, Hodgson DA. 1993. Light-induced carotenogenesis in *Myxococcus xanthus*: DNA sequence analysis of the *carR* region. *Mol. Microbiol.* 10:713–35

180. Deleted in proof

181. Mecsas J, Rouviere PE, Erickson JW, Donohue TJ, Gross CA. 1993. The activity of σ^E, an *Escherichia coli* heat-inducible σ-factor, is modulated by expression of outer membrane proteins. *Genes Dev.* 7:2618–28

182. Mendez C, Chater KF. 1987. Cloning of whiG, a gene critical for sporulation of *Streptomyces coelicolor* A3(2). *J. Bacteriol.* 169:5715–20

183. Min KT, Hilditch CM, Diederich B, Errington J, Yudkin MD. 1993. σ^F, the first compartment-specific transcription factor of *Bacillus subtilis*, is regulated by an anti-σ factor that is also a protein kinase. *Cell* 74:735–42

184. Minchin S, Busby S. 1993. Location of close contacts between *Escherichia coli* RNA polymerase and guanine residues at promoters either with or without consensus-35 region sequences. *Biochem. J.* 289:771–75

185. Mirel DB, Lauer P, Chamberlin MJ. 1994. Identification of flagellar synthesis regulatory and structural genes in a σ^D-dependent operon of *Bacillus subtilis*. *J. Bacteriol.* 176:4492–500

186. Missiakas D, Betton JM, Raina S. 1996. New components of protein folding in extracytoplasmic compartments of *Escherichia coli* SurA, FkpA and Skp/OmpH. *Mol. Microbiol.* 21:871–84

187. Missiakas D, Mayer MP, Lemaire M, Georgopoulos C, Raina S. 1997. Modulation of the *Escherichia coli* σ^E (RpoE) heat-shock transcription-factor activity by the RseA, RseB and RseC proteins. *Mol. Microbiol.* 24:355–71

188. Missiakas D, Raina S, Georgopoulos C. 1996. Heat shock regulation. In *Regulation of Gene Expression in* Escherichia coli, ed. EC Lin, AS Lynch, pp. 481–501. Austin, TX: Landes

189. Mohr CD, Rust L, Albus AM, Iglewski BH, Deretic V. 1990. Expression patterns of genes encoding elastase and controlling mucoidy—co-ordinate regulation of two virulence factors in *Pseudomonas aeruginosa* isolates from cystic fibrosis. *Mol. Microbiol.* 4:2103–10

190. Morimoto RI, Tissieres A, Georgopoulos C. 1994. *The Biology of Heat Shock Proteins and Molecular Chaperones*. Cold Spring Harbor, NY: Cold Spring Harbor Lab.

191. Deleted in proof

192. Nagai H, Yuzawa H, Kanemori M, Yura T. 1994. A distinct segment of the σ^{32} polypeptide is involved in DnaK-mediated negative control of the heat shock response in *Escherichia coli*. *Proc. Natl. Acad. Sci. USA* 91:10280–84

193. Najafi SM, Willis AC, Yudkin MD.

1995. Site of phosphorylation of SpoI-IAA, the anti-anti-sigma factor for sporulation-specific σ^F of *Bacillus sutilis*. *J. Bacteriol.* 177:2912–13

194. Nashimoto H. 1993. Non-ribosomal proteins affecting the assembly of ribosomes in *Escherichia coli*. In *The Translational Apparatus: Structure, Function, Regulation, Evolution*, ed. KH Nierhaus, F Franceschi, AR Subramanian, pp. 185–96. New York: Plenum

195. Nicolás FJ, Cayuela ML, Martinez-Argudo IM, Ruiz-Vázquez R, Murillo FJ. 1996. High mobility group I(Y)-like DNA-binding domains on a bacterial transcription factor. *Proc. Natl. Acad. Sci. USA* 93:6881–85

196. Nicolás FJ, Ruiz-Vázquez R, Murillo FJ. 1994. A genetic link between light response and multicellular development in the bacterium *Myxococcus xanthus*. *Genes Dev.* 8:2375–87

197. Nikaido H, Vaara M. 1985. Molecular basis of outer membrane permeability. *Microbiol. Rev.* 49:1–32

198. Ochs M, Angerer A, Enz S, Braun V. 1996. Surface signaling in transcriptional regulation of the ferric citrate transport system of *Escherichia coli*: mutational analysis of the alternative sigma factor FecI supports its essential role in *fec* transport gene transcription. *Mol. Gen. Genet.* 250:455–65

199. Ogle JW, Janda JM, Woods DE, Vasil ML. 1987. Characterization and use of a DNA probe as an epidemiological marker for *Pseudomonas aeruginosa*. *J. Infect. Dis.* 155:119–26

200. Ohman DE, Chakrabarty AM. 1981. Genetic mapping of chromosomal determinants for the production of the exopolysaccharide alginate in a *Pseudomonas aeruginosa* cystic fibrosis isolate. *Infect. Immun.* 33:142–48

201. Ohman DE, Mathee K, McPherson CJ, DeVries CA, Ma S, et al. 1996. Regulation of the alginate (*algD*) operon in *Pseudomonas aeruginosa*. In *Molecular Biology of Pseudomonads*, ed. T Nakazawa, K Furukawa, D Haas, S Silver, pp. 472–83. Washington, DC: Amer. Soc. Microbiol.

202. Ohnishi K, Fan F, Schoenhals GJ, Kihara M, Macnab RM. 1997. The FliO, FliP, FliQ, FliR proteins of *Salmonella typhimurium*: putative components for flagellar assembly. *J. Bacteriol.* 179:6092–99

203. Ohnishi K, Kutsukake K, Suzuki H, Iino T. 1990. Gene *fliA* encodes an alternative σ factor specific for flagellar operons in *Salmonella typhimurium*. *Mol. Gen. Genet.* 221:139–47

204. Ohnishi K, Kutsukake K, Suzuki H, Iino T. 1992. A novel transcriptional regulatory mechanism in the flagellar regulon of *Salmonella typhimurium*: an anti-sigma factor inhibits the activity of the flagellum-specific sigma factor, σ^F. *Mol. Microbiol.* 6:3149–57

205. Orsini G, Ouhammouch M, Le Caer JP, Brody EN. 1993. The *asiA* gene of bacteriophage T4 codes for the anti-σ^{70} protein. *J. Bacteriol.* 175:85–93

206. Ouhammouch M, Adelman K, Harvey SR, Orsini G, Brody EN. 1995. Bacteriophage T4 MotA and AsiA proteins suffice to direct *Escherichia coli* RNA polymerase to initiate transcription at T4 middle promoters. *Proc. Natl. Acad. Sci. USA* 92:1451–55

207. Ouhammouch M, Orsini G, Brody EN. 1994. The *asiA* gene product of bacteriophage T4 is required for middle mode RNA synthesis. *J. Bacteriol.* 176:3956–65

208. Packschies L, Theyssen H, Buchberger A, Bukau B, Goody RS, Reinstein J. 1997. GrpE accelerates nucleotide exchange of the molecular chaperone DnaK with an associative displacement mechanism. *Biochemistry* 36:3417–22

209. Pahari S, Chatterji D. 1997. Interaction of bacteriophage T4 AsiA protein with *Escherichia coli* σ^{70} and its variant. *FEBS Lett.* 411:60–62

210. Patterson-Delafield J, Martinez RJ, Stocker BA, Yamaguchi S. 1973. A new *fla* gene in *Salmonella typhimurium*—*flaR*—and its mutant phenotype—superhooks. *Arch. Mikrobiol.* 90:107–20

211. Pedersen SS, Moller H, Espersen F, Sorensen CH, Jensen T, Høiby N. 1992. Mucosal immunity to *Pseudomonas aeruginosa* alginate in cystic fibrosis. *Acta Pathol. Microbiol. Immunol. Scand.* 100:326–34

212. Pegues JC, Chen LS, Gordon AW, Ding L, Coleman WG Jr. 1990. Cloning, expression and characterization of *Escherichia coli* K-12 *rfaD* gene. *J. Bacteriol.* 172:4652–60

213. Pettersson J, Nordfelth R, Dubinina E, Bergman T, Gustafsson M, et al. 1996. Modulation of virulence factor expression by pathogen target cell contact. *Science* 273:1231–33

214. Pierpaoli EV, Sandmeier E, Baici A, Schonfeld HJ, Gisler S, Christen P. 1997. The power stroke of the

DnaK/DnaJ/GrpE molecular chaperone system. *J. Mol. Biol.* 269:757–68

215. Plaxco KW, Groß M. 1997. Cell biology. The importance of being unfolded. *Nature* 836:657

216. Polissi A, Goffin L, Georgopoulos C. 1995. The *Escherichia coli* heat shock response and bacteriophage lambda development. *FEMS Microbiol. Rev.* 17: 159–69

217. Pollack M. 1995. *Pseudomonas aeruginosa.* In *Principles and Practice of Infectious Diseases*, ed. GL Mandell, JE Bennett, R Dolin, Vol. 2, pp. 1980–2003. New York: Churchill Livingstone. 4th ed.

218. Pressler U, Staudenmaier H, Zimmermann L, Braun V. 1988. Genetics of the iron dicitrate transport system of *Escherichia coli. J. Bacteriol.* 170:2716–24

219. Pugsley AP, Francetic O, Possot OM, Sauvonnet N, Hardie KR. 1997. Recent progress and future directions in studies of the main terminal branch of the general secretory pathway in gram-negative bacteria—a review. *Gene* 192:13–19

220. Raina S, Missiakas D, Georgopoulos C. 1995. The *rpoE* gene encoding the σ^E (σ^{24}) heat shock sigma factor of *Escherichia coli. EMBO. J.* 14:1043–55

221. Rau W. 1988. Functions of carotenoids other than photosynthesis. In *Plant Pigments*, ed. T Goodwin, pp. 231–55. London: Academic

222. Redfield AR, Price CW. 1996. General stress transcription factor σ^B of *Bacillus subtilis* is a stable protein. *J. Bacteriol.* 178:3668–70

223. Reichenbach H, Kleinig H. 1984. Pigments of myxobacteria. In *Myxobacteria: Development and Cell Interaction*, ed. E Rosenberg, p. 127–37. New York: Springer-Verlag

224. Reid KL, Fink AL. 1996. Physical interactions between members of the DnaK chaperone machinery: characterization of the DnaK.GrpE complex. *Cell Stress Chaperones* 1:127–37

225. Rouviere PE, De Las Penas A, Mecsas J, Lu CZ, Rudd KE, Gross CA. 1995. *rpoE*, the gene encoding the second heat-shock sigma factor, σ^E, in *Escherichia coli. EMBO. J.* 14:1032–42

226. Rudiger S, Germeroth L, Schneider-Mergener J, Bukau B. 1997. Substrate specificity of the DnaK chaperone determined by screening cellulose-bound peptide libraries. *EMBO. J.* 16:1501–7

227. Ruiz-Vázquez R, Fontes M, Murillo FJ. 1993. Clustering and co-ordinated activation of carotenoid genes in *Myxococcus xanthus* by blue light. *Mol. Microbiol.* 10:25–34

228. Schlesinger MJ. 1996. Heat shock in vertebrate cells. *Cell Stress Chaperones* 1:213–14

229. Schmidt RP, Kreuzer KN. 1992. Purified MotA protein binds the -30 region of a bacteriophage T4 middle-mode promoter and activates transcription in vitro. *J. Biol. Chem.* 267:11399–407

230. Schmitt CK, Darnell SC, O'Brien AD. 1996. The attenuated phenotype of a *Salmonella typhimurium flgM* mutant is related to expression of FliC flagellin. *J. Bacteriol.* 178:2911–15

231. Schmitt CK, Darnell SC, O'Brien AD. 1996. The *Salmonella typhimurium flgM* gene, which encodes a negative regulator of flagella synthesis and is involved in virulence, is present and functional in other *Salmonella* species. *FEMS Microbiol. Lett.* 135:281–85

232. Schmitt CK, Darnell SC, Tesh VL, Stocker BA, O'Brien AD. 1994. Mutation of *flgM* attenuates virulence of *Salmonella typhimurium*, mutation of *fliA* represses the attenuated phenotype. *J. Bacteriol.* 176:368–77

233. Schurr MJ, Martin DW, Mudd MH, Deretic V. 1994. Gene cluster controlling conversion to alginate-overproducing phenotype in *Pseudomonas aeruginosa*: functional analysis in a heterologous host and role in the instability of mucoidy. *J. Bacteriol.* 176:3375–82

234. Schurr MJ, Yu H, Boucher JC, Hibler NS, Deretic V. 1995. Multiple promoters and induction by heat shock of the gene encoding the alternative sigma factor AlgU (sigma E) which controls mucoidy in cystic fibrosis isolates of *Pseudomonas aeruginosa. J. Bacteriol.* 177:5670–79

235. Schurr MJ, Yu H, Martinez-Salazar JM, Boucher JC, Deretic V. 1996. Control of AlgU, a member of the σ^E-like family of stress sigma factors, by the negative regulators MucA and MucB and *Pseudomonas aeruginosa* conversion to mucoidy in cystic fibrosis. *J. Bacteriol.* 178:4997–5004

236. Schurr MJ, Yu H, Martinez-Salazar JM, Hibler NS, Deretic V. 1995. Biochemical characterization and posttranslational modification of AlgU, a regulator of stress response in *Pseudomonas aeruginosa. Biochem. Biophys. Res. Commun.* 216:874–80

237. Senger H, ed. 1987. *Blue Light Response: Phenomena and Occurrence*

in *Plants and Microorganisms*, Vol. 2. Boca Raton, FL: CRC

238. Severinova E, Severinov K, Fenyo D, Marr M, Brody EN, et al. 1996. Domain organization of the *Escherichia coli* RNA polymerase σ^{70} subunit. *J. Mol. Biol.* 263: 637–47

239. Shotland Y, Koby S, Teff D, Mansur N, Oren DA, et al. 1997. Proteolysis of the phage lambda CII regulatory protein by FtsH (HflB) of *Escherichia coli*. *Mol. Microbiol.* 24:1303–10

240. Silverman M, Simon M. 1974. Characterization of *Escherichia coli* flagellar mutants that are insensitive to catabolite repression. *J. Bacteriol.* 120:1196–203

241. Somero G. 1991. Hydrostatic pressure and adaptations to the deep sea. In *Environmental and Metabolic Animal Physiology*, ed. CL Prosser, pp. 167–68. New York: Wiley-Liss

242. Stephens C, Mohr C, Boyd C, Maddock J, Gober J, Shapiro L. 1997. Identification of the *fliI* and *fliJ* components of the *Caulobacter* flagellar type III protein secretion system. *J. Bacteriol.* 179:5355–65

243. Stevens A. 1972. New small polypeptides associated with DNA-dependent RNA polymerase of *Escherichia coli* after infection with bacteriophage T4. *Proc. Natl. Acad. Sci. USA* 69:603–7

244. Stevens A. 1973. An inhibitor of host sigma-stimulated core enzyme activity that purifies with DNA-dependent RNA polymerase of *Escherichia coli* following T4 phage infection. *Biochem. Biophys. Res. Commun.* 54:488–93

245. Stevens A. 1976. A salt-promoted inhibitor of RNA polymerase isolated from T4 phage-infected *Escherichia coli*. In *RNA Polymerase*, ed. R Losick, M Chamberlin. Cold Spring Harbor, NY: Cold Spring Harbor Lab.

246. Stitt B, Hinton D. 1994. Regulation of middle-mode transcription. In *Molecular Biology of Bacteriophage T4*, ed. JD Karam, pp. 142–60. Washington, DC: Am. Soc. Microbiol.

247. Stragier P, Losick R. 1996. Molecular genetics of sporulation in *Bacillus subtilis*. *Annu. Rev. Genet.* 30:297–341

248. Strauch KL, Beckwith J. 1988. An *Escherichia coli* mutation preventing degradation of abnormal periplasmic proteins. *Proc. Natl. Acad. Sci. USA* 85: 1576–80

249. Strauch KL, Johnson K, Beckwith J. 1989. Characterization of *degP*, a gene required for proteolysis in the cell envelope and essential for growth of *Es-*

cherichia coli at high temperatures. *J. Bacteriol.* 171:2689–96

250. Straus D, Walter W, Gross CA. 1990. DnaK, DnaJ, GrpE heat shock proteins negatively regulate heat shock gene expression by controlling the synthesis and stability of σ^{32}. *Genes Dev.* 4:2202–9

251. Straus DB, Walter WA, Gross CA. 1987. The heat shock response of *E. coli* is regulated by changes in the concentration of σ^{32}. *Nature* 329:348–51

252. Suzuki T, Iino T. 1981. Role of the *flaR* gene in flagellar hook formation in *Salmonella* spp. *J. Bacteriol.* 148:973–79

253. Szabo A, Langer T, Schroder H, Flanagan J, Bukau B, Hartl FU. 1994. The ATP hydrolysis-dependent reaction cycle of the *Escherichia coli* Hsp70 system DnaK, DnaJ, GrpE. *Proc. Natl. Acad. Sci. USA* 91:10345–49

254. Tilly K, McKittrick N, Zylicz M, Georgopoulos C. 1983. The DnaK protein modulates the heat-shock response of *Escherichia coli*. *Cell* 34:641–46

255. Tomoyasu T, Gamer J, Bukau B, Kanemori M, Mori H, et al. 1995. *Escherichis coli* FtsH is a membrane-bound, ATP-dependent protease which degrades the heat-shock transcription factor σ^{32}. *EMBO. J.* 14:2551–60

256. Van Hove B, Staudenmaier H, Braun V. 1990. Novel two-component transmembrane transcription control: regulation of iron dicitrate transport in *Escherichia coli* K-12. *J. Bacteriol.* 172:6749–58

257. Voelker U, Dufour A, Haldenwang WG. 1995. The *Bacillus subtilis rsbU* gene product is necessary for RsbX-dependent regulation of σ^{B}. *J. Bacteriol.* 177:114–22

258. Voelker U, Voelker A, Haldenwang WG. 1996. Reactivation of the *Bacillus subtilis* anti-σ^{B} antagonist, RsbV, by stress- or starvation-induced phosphatase activities. *J. Bacteriol.* 178:5456–63

259. Voelker U, Voelker A, Haldenwang WG. 1996. The yeast two-hybrid system detects interactions between *Bacillus subtilis* σ^{B} regulators. *J. Bacteriol.* 178: 7020–23

260. Voelker U, Voelker A, Maul B, Hecker M, Dufour A, Haldenwang WG. 1995. Separate mechanisms activate σ^{B} of *Bacillus subtilis* in response to environmental and metabolic stresses. *J. Bacteriol.* 177:3771–80

261. Wagegg W, Braun V. 1981. Ferric citrate transport in *Escherichia coli* requires outer membrane receptor protein *fecA*. *J. Bacteriol.* 145:156–63

262. Wang QP, Kaguni JM. 1989. A novel sigma factor is involved in expression of the *rpoH* gene of *Escherichia coli*. *J. Bacteriol.* 171:4248–53

263. Wawrzynow A, Banecki B, Wall D, Liberek K, Georgopoulos C, Zylicz M. 1995. ATP hydrolysis is required for the DnaJ-dependent activation of DnaK chaperone for binding to both native and denatured protein substrates. *J. Biol. Chem.* 270:19307–11

264. Wilkens K, Rüger W. 1994. Transcription from early promoters. In *Molecular Biology of Bacteriophage T4*, ed. JD Karam, pp. 132–41. Washington, DC: Am. Soc. Microbiol.

265. Williams AW, Yamaguchi S, Togashi F, Aizawa SI, Kawagishi I, Macnab RM. 1996. Mutations in *fliK* and *flhB* affecting flagellar hook and filament assembly in *Salmonella typhimurium*. *J. Bacteriol.* 178:2960–70

266. Williams KP, Kassavetis GA, Geiduschek EP. 1987. Interaction of the bacteriophage T4 gene 55 product with *Escherichia coli* RNA polymerase. Competition with *E. coli* σ^{70} and release from late T4 transcription complexes following initiation. *J. Biol. Chem.* 262:12365–71

267. Williams KP, Kassavetis GA, Herendeen DR, Geiduschek EP. 1994. Regulation of late-gene expression. In *Molecular Biology of Bacteriophage T4*, ed. JD Karam, pp. 161–75. Washington, DC: Am. Soc. Microbiol.

268. Wise AA, Price CW. 1995. Four additional genes in the *sigB* operon of *Bacillus subtilis* that control activity of the general stress factor σ^{B} in response to environmental signals. *J. Bacteriol.* 177:123–33

269. Wozniak DJ, Ohman DE. 1993. Involvement of the alginate *algT* gene and integration host factor in the regulation of the *Pseudomonas aeruginosa algB* gene. *J. Bacteriol.* 175:4145–53

270. Wozniak DJ, Ohman DE. 1994. Transcriptional analysis of the *Pseudomonas aeruginosa* genes *algR*, *algB* and *algD* reveals a hierarchy of alginate gene expression which is modulated by *algT*. *J. Bacteriol.* 176:6007–14

271. Wriedt K, Angerer A, Braun V. 1995. Transcriptional regulation from the cell surface: conformational changes in the transmembrane protein FecR lead to altered transcription of the ferric citrate transport genes in *Escherichia coli*. *J. Bacteriol.* 177:3320–22

272. Wu J, Newton A. 1997. Regulation of the *Caulobacter* flagellar gene hierarchy; not just for motility. *Mol. Microbiol.* 24:233–39

273. Xie ZD, Hershberger CD, Shankar S, Ye RW, Chakrabarty AM. 1996. Sigma factor-anti-sigma factor interaction in alginate synthesis: inhibition of AlgT by MucA. *J. Bacteriol.* 178:4990–96

274. Yang X, Kang CM, Brody MS, Price CW. 1996. Opposing pairs of serine protein kinases and phosphatases transmit signals of environmental stress to activate a bacterial transcription factor. *Genes Dev.* 10:2265–75

275. Yayanos AA. 1986. Evolutional and ecological implications of the properties of deep-sea barophilic bacteria. *Proc. Natl. Acad. Sci. USA* 83:9542–46

276. Yu H, Schurr M, Boucher J, Martinez-Salazar J, Martin D, Deretic V. 1996. Molecular mechanism of conversion to mucoidy in *Pseudomonas aeruginosa*. In *Molecular Biology of Pseudomonads*, ed. T Nakazawa, K Furukawa, D Haas, S Silver, pp. 384–97. Washington, DC: Amer. Soc. Microbiol.

277. Yu H, Schurr MJ, Deretic V. 1995. Functional equivalence of *Escherichia coli* σ^{E} and *Pseudomonas aeruginosa* AlgU: *E. coli rpoE* restores mucoidy and reduces sensitivity to reactive oxygen intermediates in *algU* mutants of *P. aeruginosa*. *J. Bacteriol.* 177:3259–68

278. Yura T. 1996. Regulation and conservation of the heat-shock transcription factor σ^{32}. *Genes Cells.* 1:277–84

279. Yura T, Nagai H, Mori H. 1993. Regulation of the heat shock response in bacteria. *Annu. Rev. Microbiol.* 47:321–50

280. Zhu X, Zhao X, Burkholder WF, Gragerov A, Ogata CM, et al. 1996. Structural analysis of substrate binding by the molecular chaperone DnaK. *Science* 272:1606–14

281. Zylicz M, Ang D, Liberek K, Georgopoulos C. 1989. Initiation of λDNA replication with purified host- and bacteriophage-encoded proteins: the role of the DnaK, DnaJ, GrpE heat shock proteins. *EMBO. J.* 8:1601–8

282. Zylicz M, LeBowitz JH, McMacken R, Georgopoulos C. 1983. The DnaK protein of *Escherichia coli* possesses an ATPase and autophosphorylating activity and is essential in an in vitro DNA replication system. *Proc. Natl. Acad. Sci. USA* 80:6431–35

Annu. Rev. Microbiol. 1998. 52:287–331

DEVELOPMENT OF HYBRID STRAINS FOR THE MINERALIZATION OF CHLOROAROMATICS BY PATCHWORK ASSEMBLY

Walter Reineke

Bergische Universität–Gesamthochschule Wuppertal, D-42097 Wuppertal, Germany; e-mail: reineke@uni-wuppertal.de

KEY WORDS: chlorocatechol, conjugative gene transfer, enzyme and effector specificity, clustering of genes, modified *ortho* pathway

ABSTRACT

The persistence of chloroaromatic compounds can be caused by various bottlenecks, such as incomplete degradative pathways or inappropriate regulation of these pathways. Patchwork assembly of existing pathways in novel combinations provides a general route for the development of strains degrading chloroaromatics. The recruitment of known complementary enzyme sequences in a suitable host organism by conjugative transfer of genes might generate a functioning hybrid pathway for the mineralization of some chloroaromatics not degraded by the parent organisms. The rational combination uses (*a*) peripheral, funneling degradation sequences originating from aromatics-degrading strains to fulfill the conversion of the respective analogous chloroaromatic compound to chlorocatechols as the central intermediates; (*b*) a central chlorocatechol degradation sequence, the so-called modified *ortho* pathway, which brings about elimination of chlorine substituents; and (*c*) steps of the 3-oxoadipate pathway to reach the tricarboxylic acid cycle. The genetic organization of these pathway segments has been well characterized. The specificity of enzymes of the xylene, benzene, biphenyl, and chlorocatechol pathways and the specificity of the induction systems for the chlorinated substrates are analyzed in various organisms to illustrate eventual bottlenecks and to provide alternatives that are effective in the conversion of the "new" substrate. Hybrid pathways are investigated in "new" strains degrading chlorinated benzoates, toluenes, benzenes, and biphenyls. Problems

287

0066-4227/98/1001-0287$08.00

occurring after the conjugative DNA transfer and the "natural" solution of these are examined, such as the prevention of misrouting into the *meta* pathway, to give a functioning hybrid pathway. Some examples clearly indicate that patchwork assembly also happens in nature.

CONTENTS

INTRODUCTION

Microorganisms have been used for more than 100 years to degrade organic substances in the biological treatment of wastewater streams. The entirety of existing microbes has the capacity to degrade nearly all naturally occurring substances. Without this degradative capability, circulation of compounds would not be possible in nature, and biological waste treatment would not exist.

Bacteria are well adapted for degrading natural substances. However, mineralization takes place only when environmental conditions (such as water activity, presence or absence of oxygen, temperature, and pH value) are suitable. Degradation is likely to fail when a target compound is either very concentrated or very diluted as it confronts microorganisms. Some synthetic chemicals have been found to be persistent. Xenobiotics, which resemble naturally occurring compounds, usually do not lead to problems concerning degradation. On the other hand, structures that strongly deviate from natural patterns are often recalcitrant.

In addition to other classes, chlorinated aromatics belong to this group of compounds. Well-known examples are polychlorinated biphenyls (PCBs), chlorinated dibenzo-*p*-dioxins, and dibenzofurans. Chlorobenzenes and chloro-phenols are of more importance concerning pattern of use and amounts released into the environment.

For a successful biological cleanup of waste or soil contaminated with such compounds, aerobic microorganisms that can use chloroaromatics as a source of carbon and energy are helpful. However, the dechlorinating potential of anaerobic microbial populations can also be useful in cleanup processes.

The following approaches can be used to obtain active cultures with the desired aerobic biodegradative capabilites:

1. enrichment of aerobic organisms from soil or water samples able to use chloroaromatic compounds as the growth substrate;

2. creation of hybrid pathways by patchwork assembly of existing pathway segments through conjugative gene transfer, sometimes termed in vivo construction; and

3. tailoring of pathways by genetic engineering techniques.

This review focuses mostly on the in vivo approach. The biochemical aspects and genetic basis of the development of hybrid organisms by cell-cell contact are discussed.

THE CONCEPT OF PATCHWORK ASSEMBLY

Incubation of aromatic-degrading bacteria with chlorinated analogous substrates often leads to black-colored products—autoxidation products of dead-end metabolites. In most cases, these products result because of the absence of enzyme sequences able to transform chlorocatechols, which occur during the degradation of the chlorinated aromatics. The establishment of a complete sequence for the degradation of chloroaromatics in one strain should lead to a new strain able to use these compounds as growth substrates. This idea is summarized in Figure 1. A hybrid pathway for chloroaromatics leading to chloride, CO_2, and biomass results from the combination of two pathways in a suitable host organism bearing a 3-oxoadipate pathway: (*a*) a peripheral, funneling sequence taken from an aromatic-degrading strain, which is able to transform a chlorinated aromatic into a chlorocatechol, and (*b*) a chlorocatechol-degrading central sequence, the so-called modified *ortho* pathway.

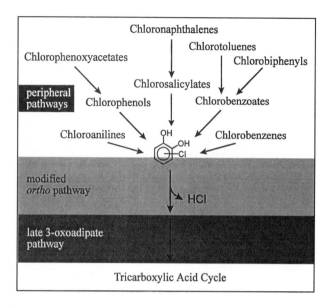

Figure 1 Combination of pathway segments to develop a hybrid pathway for the mineralization of chloroaromatic compounds.

THE DEGRADATIVE PATHWAY SEGMENTS

Peripheral, Funneling Sequences

The peripheral, funneling pathways are shown in Figure 2. The degradation of aromatic compounds comprises many reaction steps prior to the stage of catechol, which can be divided into few types of reactions. As has been demonstrated for the degradation of mononuclear aromatics such as aniline, benzene, benzoate, salicylate, and toluene, and moreover for binuclear compounds such as biphenyl, dibenzofuran, dibenzo-*p*-dioxin, and naphthalene, analogous degradative sequences are present.

---→

Figure 2 Peripheral degradation sequences for aromatics with relevance for the development of hybrid strains using chloroaromatics. 1, naphthalene; 2, dibenzofuran; 3, dibenzo-*p*-dioxin; 4, benzene; 5, phenol; 6, aniline; 7, biphenyl; 8, toluene; 9, salicylate; and 10, benzoate. Path 4 will also be used for toluene and ethylbenzene while xylenes are degraded via path 8 so that methyl- or ethyl-substituted metabolites result, respectively. Catechol or the methyl- and ethyl-substituted analogs, the central metabolites, will be degraded via the *meta* pathway. The cleavage of carbon-carbon bonds is marked by *a line of stars*. This information was compiled from the following sources: 6, 10, 19–21, 41, 47, 48, 56, 58–60, 65, 106, 123, 135, 136, 153, 176, 177.

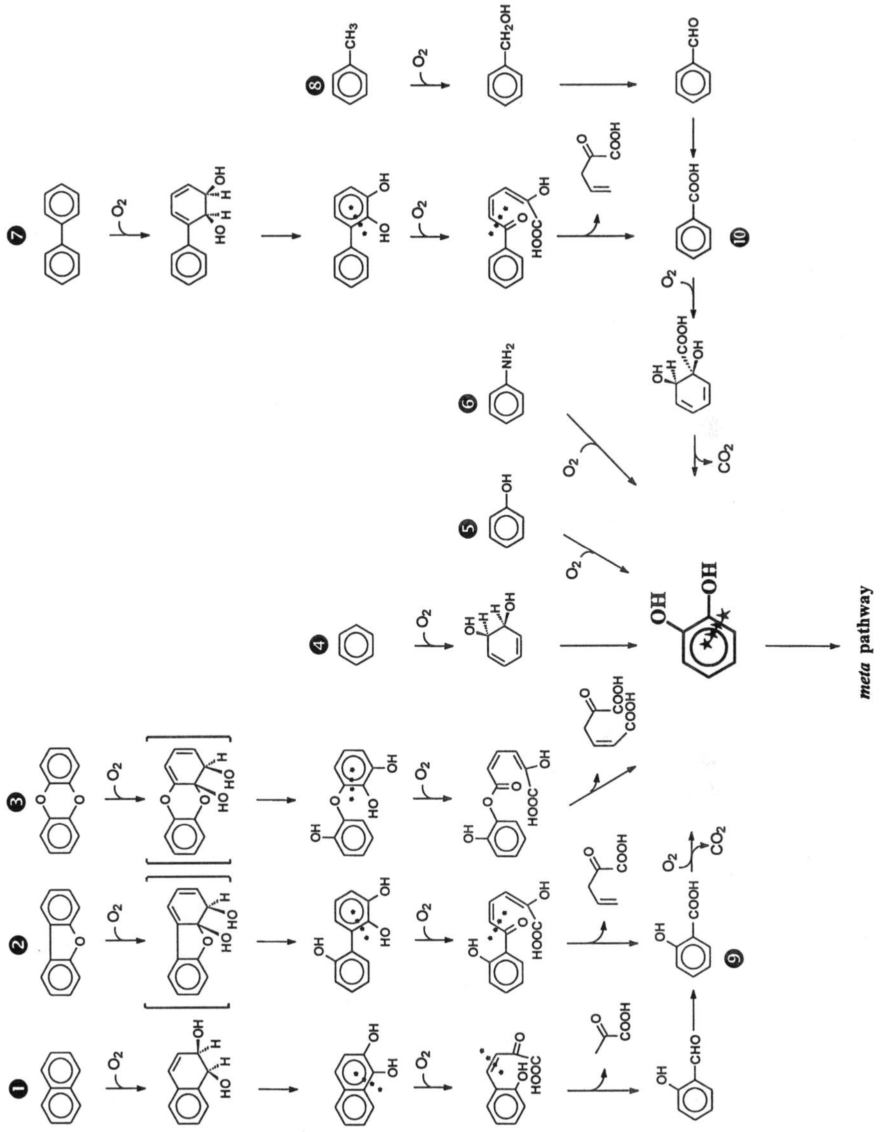
meta pathway

In these pathways, an aromatic compound without a hydroxyl group is activated through a multicomponent dioxygenase such as benzene dioxygenase or biphenyl 2,3-dioxygenase using molecular oxygen. Subsequently, the resulting *cis*-dihydrodiol is converted by a dehydrogenase to give a 1,2-diphenolic structure, regaining the aromatic state. The resulting diphenol is substrate for ring cleavage, often *meta* cleavage, which requires another mole of molecular oxygen. In the case of mononuclear aromatics, the ring-cleaving dioxygenation belongs to the central pathway.

During aniline degradation, the stage of the diphenol is reached by the action of an aniline dioxygenase eliminating NH_3 at the same time. In this case no dehydrogenase is needed. Phenol will be converted by a monooxygenase reaction to produce catechol.

In the degradation of xylenes, three oxidative reactions—xylene monooxygenase, benzylalcohol, and benzaldehyde dehydrogenase—initiate the pathway prior to the mentioned ring activation by the toluate dioxygenase.

During the degradation of binuclear aromatics, the formation and cleavage of the 1,2-diphenol take place twice. The first *cis*-dihydrodiols, which result during degradation of dibenzo-*p*-dioxin and dibenzofuran, are unstable hemiacetals. Therefore, the ether bond is cleaved spontaneously, and a dehydrogenase is not necessary to reach the 1,2-diphenolic structure. *Meta* cleavage by a dioxygenase yields a yellow product that is subject to hydrolytic shortening of the side chain. Pyruvate and compounds of the central *meta* pathway, such as 4-oxalocrotonate and 2-oxopent-4-enoate, are the products of the hydrolase reaction. The aromatic products of the reaction are benzoate, salicylaldehyde, and catechol, respectively.

The peripheral degradation of a multitude of aromatics converges at the stage of catechol. Chlorocatechols are formed via the mentioned sequences from chlorinated compounds.

Central Degradative Sequences

MODIFIED *ORTHO* PATHWAY   The reaction steps of the central degradation pathway of chlorocatechols are shown in Figure 3. Ring cleavage by chlorocatechol 1,2-dioxygenase with consumption of molecular oxygen results in the formation of the corresponding chloro-*cis,cis*-muconate. Further degradation by a chloromuconate cycloisomerase leads to a lactone, which eliminates chlorine as HCl from position 4 or 5, bringing about the formation of an exocyclic double bond. In this way, a (chloro)dienelactone is formed. It is transformed to the corresponding (chloro)maleylacetate via hydrolytic cleavage by (chloro)dienelactone hydrolase. A maleylacetate reductase forms 3-oxoadipate, an intermediate shared with the degradation of various aromatic compounds. Chlorinated derivatives are handled the same way to give chlorosubstituted

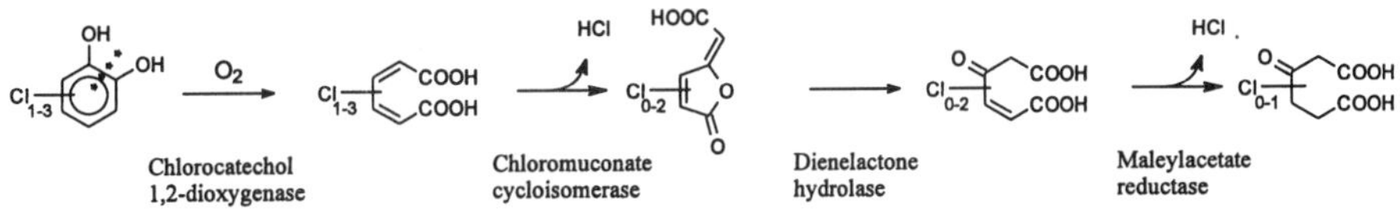

Figure 3 Simplified presentation of the modified *ortho* pathway. The pathway for the degradation of di- and trichlorocatechols includes two dechlorinating steps. The *ortho* cleavage is marked by *a line of stars*. Information was compiled from the following sources: 35, 36, 44, 88, 89, 115, 138, 141, 143, 144, 159, 167–169.

3-oxoadipates. Besides functioning to reduce the carbon-carbon double bond, the maleylacetate reductase is able to eliminate a possible chlorine substituent from position 2 so that 3-oxoadipate results from 2-chloromaleylacetate or 3-chloro-4-oxoadipate from 2,3-dichloromaleylacetate.

3-OXOADIPATE PATHWAY   The use of chlorocatechols as a source of carbon and energy by bacterial cells requires the synthesis of the last two enzymes of the well-known 3-oxoadipate (β-ketoadipate) pathway—3-oxoadipate:succinyl-CoA transferase and 3-oxoadipyl-CoA thiolase—to reach the Krebs cycle (Figure 4). Thus modified *ortho* pathways depend on the presence of the 3-oxoadipate pathway in a host strain in order to operate. Various benzoate- or *p*-hydroxybenzoate-degrading bacteria fulfill this requirement. In a classic study on the taxonomy of pseudomonads, Stanier et al (151) tested 267 strains for their ability to grow with any of 146 different organic compounds, including benzoate and *p*-hydroxybenzoate. Only 65% of the 175 strains of the fluorescent group (*Pseudomonas aeruginosa*, *Pseudomonas fluorescens*, *Pseudomonas putida*) were able to use benzoate as the growth substrate, whereas

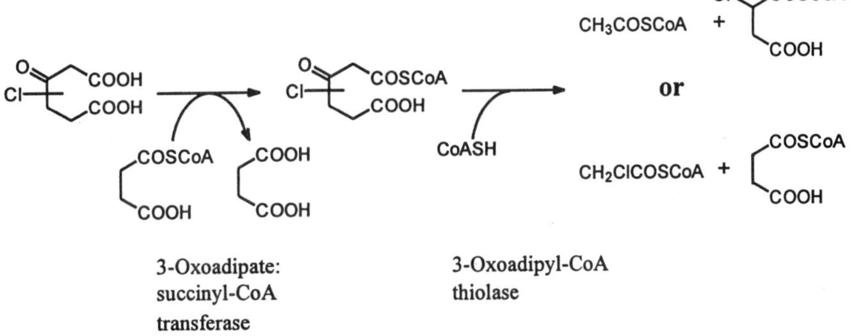

Figure 4 Proposed reactions of the late 3-oxoadipate pathway for chlorinated 3-oxoadipates.

most widely utilized was *p*-hydroxybenzoate with 95% positive strains. Therefore, a benzoate-negative character does not necessarily indicate the absence
of a 3-oxoadipate pathway. However, the *p*-hydroxybenzoate-positive phenotype is not an absolute proof that the degradation takes place via 3-oxoadipate.
Two species of the acidovorans group, *Pseudomonas acidovorans* and *Pseudomonas testosteroni* (today *Comamonas testosteroni*) were characterized by
meta cleavage of protocatechuate, the intermediate in the degradation of *p*-
hydroxybenzoate. Suitability of a strain for the patchwork assembly process is
indicated by growth of the strain with *p*-hydroxybenzoate and the detection of
3-oxoadipate as an intermediate in the pathway by the Rothera reaction.

ENZYME SPECIFICITY OF DEGRADATIVE SEQUENCES FOR CHLORINATED COMPOUNDS

I use a number of examples to answer the question of whether the enzymes of the
degradative sequences are unspecific enough for the degradation of a multitude
of chloroaromatic compounds or whether bottlenecks are present that have to
be eliminated through an isoenzyme from a different source. It is no matter of
course that all enzymes of a strain that has been enriched with a nonchlorinated
aromatic show activity with chlorinated compounds. First to be discussed is
information on the specificity of enzymes of the xylene degradative pathway
for chlorotoluenes and the biphenyl degradation pathway for chlorobiphenyl
congeners. Second, enzymes of a peripheral pathway with a known former
function in the degradation of chloroaromatics are compared with those without
any experience with chloroaromatics, i.e. the first steps in benzene degradation.
Third, the enzymes of the central chlorocatechol pathway are analyzed with
respect to chloroaromatic history in comparison with isofunctional steps of the
3-oxoadipate pathway.

Peripheral Pathways Without a History for Chloroaromatics

Pathways evolved for the degradation of methyl-substituted growth substrates
can often deal with substrates bearing chlorine substituents because methyl
and chlorine substituents are of similar size. The van der Waals radius of a
chlorine substituent is 1.8 Å, whereas a methyl group is slightly larger, 2.0 Å.
This preadaptation avoids steric hindrance and formation of a bottleneck for a
chlorinated substrate.

The oxygenation of different chlorotoluenes has been investigated in *P. putida*
PaW1, a strain using *m*- and *p*-xylene as growth substrate. The degree of trans-
formation of chlorotoluenes depends on the position of the chlorine substituent.
The substrate analogs 3-chloro- and 4-chlorotoluene are transformed at high

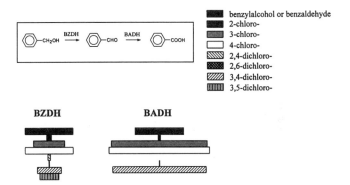

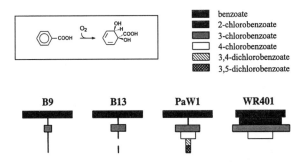

Figure 5 Dehydrogenation of benzylalcohols and benzaldehydes by *Pseudomonas putida* PaW1, *m*-xylene–grown (15). The activity was determined by the nonsubstituted substrate (100%) and compared with activity determined by substituted substrate in a cell-free assay.

rates, whereas no or only low activity has been found with other chlorotoluenes. Substituents in the *ortho* position impaired substrate binding.

Benzylalcohol and benzaldehyde dehydrogenase show broader specificities (Figure 5). But again, a chlorine substituent in *ortho* position leads to a drastic decrease in the substrate conversion.

Specificities are known for three types of enzymes that dioxygenate chlorinated benzoates (Figure 6). *Ralstonia eutropha* (formerly *Alcaligenes eutrophus*), a benzoate-degrading strain, and *Pseudomonas* sp. B13, a 3-chlorobenzoate–degrading strain, possess similar benzoate 1,2-dioxygenases with a narrow specificity. In benzoate and 3-chlorobenzoate–grown cells of strain B13,

Figure 6 Dioxygenation of benzoates by strains *Ralstonia eutropha* (formerly *Alcaligenes eutrophus*) benzoate-grown, *Pseudomonas* sp. B13 3-chlorobenzoate–grown, *Pseudomonas putida* PaW1 *p*-toluate–grown, and *Burkholderia cepacia* (formerly *Pseudomonas* sp.) WR401 *o*-toluate–grown (40, 129). The activity was determined by the nonsubstituted substrate (100%) and compared with activity determined by substituted substrate in a whole-cell assay.

the specificity of the benzoate 1,2-dioxygenase did not change, indicating the absence of significant adaptation. In addition to benzoate, only 3-chlorobenzoate is converted by the B13-enzyme. In contrast, the toluate 1,2-dioxygenase of *P. putida* PaW1, whose natural function is the conversion of *m*- and *p*-toluate, transforms all tested chlorobenzoates with the exception of 2-chlorobenzoate. 4-Chlorobenzoate is a usable substrate because of its structural analogy to *p*-toluate (4-methylbenzoate). The corresponding isoenzyme from *Burkholderia cepacia* (formerly *Pseudomonas* sp.) WR401, an *o*-toluate degrader, is not a bottleneck even for 2-chlorobenzoate.

These data indicate that electronic effects, which might have been expected from chlorosubstituents, are not responsible for the varying activities found with different chlorobenzoates. Instead, preadaptation for growth with different methyl-substituted substrates has eliminated the steric hindrance for a respective chloroanalogous compound.

The dihydrodihydroxybenzoate dehydrogenases occurring in *R. eutropha*, *Pseudomonas* sp. B13, and *P. putida* PaW1 are characterized by broad substrate specificity and the ability to transform all dihydrodiols tested to the corresponding catechols (Figure 7).

In the biphenyl pathways, the above-mentioned preadaptation has not taken place for the conversion of chlorobiphenyls.

Concerning the 2,3-dioxygenation of chlorobiphenyls, congeners chlorinated only in one ring were found to be transformed at a high rate by biphenyl-degrading strains *P. putida* BN10 and JHR. On the other hand, substrates carrying chlorine substituents in both rings, such as 2,2'-dichloro- and 4,4'-dichlorobiphenyl, are converted at low rates only (H Mokroß & W Reineke, unpublished results).

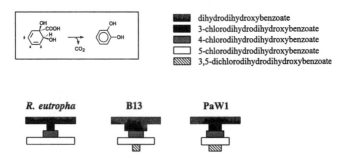

Figure 7 Dehydrogenation of dihydrodihydroxybenzoates by *Ralstonia eutropha* (formerly *Alcaligenes eutrophus*), benzoate-grown; *Pseudomonas* sp. B13, 3-chlorobenzoate–grown, and *Pseudomonas putida* PaW1, *m*-toluate–grown (130; A Stolz & W Reineke, unpublished results). The activity was determined by the nonsubstituted substrate (100%) and compared with activity determined by substituted substrate in a cell-free assay or by purified enzyme.

B. cepacia (formerly *Pseudomonas* sp.) strain LB400 and *Pseudomonas pseudoalcaligenes* strain KF707, the best-studied biphenyl-degrading strains, have been tested with a mixture of PCB congeners toward the potential for transformation (57). The defined congener depletion assay does not allow the calculation of rates, but this elegant procedure is highly sensitive. Overall, both strains were capable of degrading a broad spectrum of PCB congeners, including some hexachlorobiphenyls in the case of strain LB400. Biphenyl-grown cells of strain LB400 oxidized a much wider range of chlorinated biphenyls than did *P. pseudoalcaligenes* KF707. The different substrate specificity in both strains was due to relatively few differences in *bphA*, the gene coding the large subunit of the terminal biphenyl dioxygenase (43). Site-directed mutagenesis of the LB400 *bphA* gene resulted in an enzyme combining the broad congener specificity of strain LB400 with increased activity against congeners oxidized especially by the KF707-dioxygenase.

At present, only a small number of chlorinated phenylcatechols are available. Therefore, data on the specificity of the following enzymes of the biphenyl-degrading pathway should be considered fragmentary. 2'-Chloro-, 3'-chloro-, and 4'-chlorophenylcatechol are metabolized at high rates, while a substituent located in the dihydroxylated ring (see 4-chloro- and 5-chlorophenylcatechol) leads to a drastic decrease in the transformation rate. Substrate specificity turns out to be similar in biphenyl-degrading strains such as *P. putida* BN10 and JHR and in a bibenzyl-degrading strain (H Mokroß, SR Kaschabek, W Reineke, unpublished results).

The rate of the hydrolase reaction is slowed by a chlorine substituent next to the position of cleavage (see 3-chloro-2-hydroxy-6-oxo-6-phenylhexa-2,4-diennoate). An unexpected finding was that 3',5'-dichloro-2-hydroxy-6-oxo-6-phenylhexa-2,4-dienoate, which carries substituents only in the ring not participating in the reaction, is hydrolized at a low rate, whereas the same substituents in the positions 3' and 4' are of only minor influence (H Mokroß, SR Kaschabek, W Reineke, unpublished results).

Data on the conversion of chlorobiphenyls through the biphenyl pathway have provided a conclusive explanation for the observation of a bottleneck concerning effective degradation of PCB congeners chlorinated in both rings. It remains unclear, however, to what extent other properties of these congeners, such as solubility or vapor pressure, are responsible for the reduction of the transformation rate.

Peripheral Pathways with a History for Chloroaromatics

Elucidation of the first steps in the degradation of chlorobenzenes—i.e. the (chloro)benzene 1,2-dioxygenase and the (chloro)benzene *cis*-1,2-dihydrodiol dehydrogenase—showed the following data on their specificity (Figures 8 and 9). The benzene 1,2-dioxygenase of *P. putida* F1, a benzene-, toluene-, and

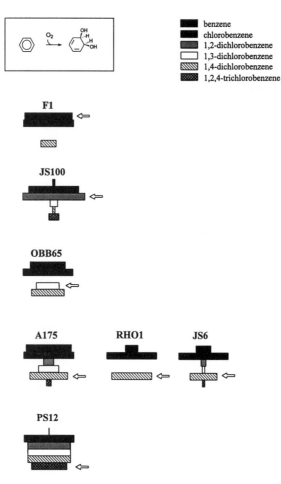

Figure 8 Dioxygenation of benzenes. Maximum velocities are given as percentages with respect to the reaction with chlorobenzene (= 100%) determined in a whole-cell assay. The respective growth substrate is indicated by an *arrow*. *Missing boxes* indicate that data have not been determined. The data were compiled from the following sources: *Alcaligenes* sp. A175 1,4-dichlorobenzene–grown (142); *Pseudomonas* sp. P12, 1,2,4-trichlorobenzene–grown (138); *Pseudomonas* sp. JS6, 1,4-dichlorobenzene–grown (147); *Pseudomonas* sp. JS100, 1,2-dichlorobenzene–grown (62); *Pseudomonas putida* F1 (original substrates benzene, toluene, or ethylbenzene) in WR1323 (117); *Pseudomonas aeruginosa* RHO1, 1,4-dichlorobenzene–grown (117); *Alcaligenes* sp. OBB65, 1,3-dichlorobenzene–grown (31).

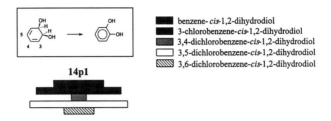

14p1

■ benzene-*cis*-1,2-dihydrodiol
■ 3-chlorobenzene-*cis*-1,2-dihydrodiol
▨ 3,4-dichlorobenzene-*cis*-1,2-dihydrodiol
☐ 3,5-dichlorobenzene-*cis*-1,2-dihydrodiol
▧ 3,6-dichlorobenzene-*cis*-1,2-dihydrodiol

Figure 9 Dehydrogenation of dihydrodihydroxybenzenes by *Xanthobacter flavus* (148). The activity was determined by the nonsubstituted substrate (100%) and compared with activity determined by substituted substrate in a cell-free assay.

ethylbenzene-degrading strain, shows high activity with benzene and chlorobenzene. However, the enzymes in all other strains are strongly adapted to the conversion of the respective chlorinated benzene used as the enrichment substrate, so they are more appropriately termed chlorobenzene 1,2-dioxygenases (Figure 8). With respect to the substrates benzene and chlorobenzene, the enzymes of strain OBB65, enriched with 1,3-dichlorobenzene, and strain A175, enriched with 1,4-dichlorobenzene, are similar to the F1-enzyme. In contrast, in the other strains the potential for conversion of benzene is decreased or lost. For that reason the relative activities can be calculated only through chlorobenzene as the compound with 100% activity. It is interesting to find that the enzyme of strain JS100 is efficiently adapted for the conversion of 1,2-dichlorobenzene, the enrichment substrate, while exhibiting higher specificity toward the other dichlorobenzenes. The chlorobenzene 1,2-dioxygenase of strain P12, enriched with 1,2,4-trichlorobenzene, transforms all chlorobenzenes at similarly high rates and is the dioxygenase with the broadest substrate spectrum for chlorinated benzenes.

The (chloro)benzene *cis*-1,2-dihydrodiol dehydrogenase of *Xanthobacter flavus* 14p1, the only enzyme tested toward specificity for chlorinated substrates, is characterized by broad substrate specificity and the ability to transform all dihydrodiols tested to the corresponding catechols (Figure 9). However, 3,4-dichloro- and 3,6-dichlorobenzene *cis*-1,2-dihydrodiol, the latter the metabolite in the degradation of 1,4-dichlorobenzene, were degraded more slowly than the other substrates.

Central Pathway for Chlorocatechols

In Figures 10–12, data are given on the specificity of enzymes of the modified *ortho* pathway—the pathway used for the degradation of chlorocatechols—and on the specificity of enzymes that degrade nonchlorinated aromatics through the 3-oxoadipate pathway. Chlorocatechol 1,2-dioxygenases as well as chloromuconate cycloisomerases catalyze reactions that are isofunctional to those of catechol 1,2-dioxygenase and muconate cycloisomerase. Both enzyme steps, ring cleavage and lactonization, have often been studied toward specificity with chlorosubstituted substrates to search for some adaptation to the growth with chloroaromatic compounds and to characterize the events that have taken place during adaptation.

In general, the main difference between catechol and chlorocatechol 1,2-dioxygenases can be found in the high affinities and activities of the latter with chlorocatechols (Figure 10). Catechol 1,2-dioxygenases cleave catechol at a high rate and accept 4-chlorocatechol as a poor substrate, but they fail to show ring cleavage with other chlorinated catechols. In contrast, chlorocatechol 1,2-dioxygenases prefer to use chlorocatechols as the substrate. Some kind of adaptation is visible, especially to the respective chlorocatechol; it occurs during degradation of the enrichment substrate. Thus the specificity of about five types of chlorocatechol 1,2-dioxygenases can be distinguished. The adaptation is shown best with the substrates 3,4-dichloro- and 3,5-dichlorocatechol occurring as the central metabolites during growth with 1,2-dichlorobenzene and 2,4-D, respectively. The enzymes of the strains JS100 and P51, which grow with 1,2-dichlorobenzene, prefer 3,4-dichlorocatechol as a substrate, while the chlorocatechol 1,2-dioxygenases of the other strains bring about poor cleavage of this compound. 3,5-Dichlorocatechol is the best substrate for the enzymes of the 2,4-D–degrading strains CSV90, JMP134, and HV3. Overall, the enzyme of strain PS12 has been characterized with the widest set of chlorinated substrates and was found to cleave them all at high rates.

Similarly, chloromuconate cycloisomerase accepts 2-chloro- and 3-chloromuconate in addition to *cis,cis*-muconate, whereas muconate cycloisomerase

———————————————————————————————————→

Figure 10   Ring cleavage of catechols by catechol or chlorocatechol 1,2-dioxygenases. Activity was determined by the nonsubstituted substrate taken as 100% and was compared with activity determined by substituted substrate. The respective catechol occurring during the degradation is indicated by an *arrow*. *Missing boxes* indicate that data have not been obtained. The data were compiled from the following sources: *Pseudomonas* sp. B13, benzoate- or 3-chlorobenzoate–grown (35, 36, 117, 132); *Alcaligenes eutrophus* JMP134, benzoate- or 2,4-dichlorophenoxyacetate–grown (122); *Pseudomonas putida* AC866, 3-chlorobenzoate–grown (16); *Pseudomonas acidovorans* CA28, 3-chloroaniline–grown (72); *Alcaligenes denitrificans* BRI 6011, 2,5-dichlorobenzoate–grown (113); *Pseudomonas cepacia* CSV90, 2,4-dichlorophenoxyacetate–grown (13); *Pseudomonas* sp. HV3, 2-methyl-4-chlorophenoxyacetate–grown (92); *Pseudomonas* sp. PS12, 1,2,4-trichlorobenzene–grown (138); strain WR1306, chlorobenzene–grown (133); *Pseudomonas* sp. P51, 1,2-dichloro- or 1,2,4-trichlorobenzene–grown (163, 165); *Pseudomonas* sp. JS6, 1,4-dichlorobenzene–grown (63, 147); *Pseudomonas* sp. JS100, 1,2-dichlorobenzene–grown (62); *Pseudomonas aeruginosa* RHO1, 1,4-dichlorobenzene–grown (117); *Alcaligenes* sp. A175, 1,4-dichlorobenzene–grown (142); and *Xanthobacter flavus* 14p1, 1,4-dichlorobenzene–grown (149).

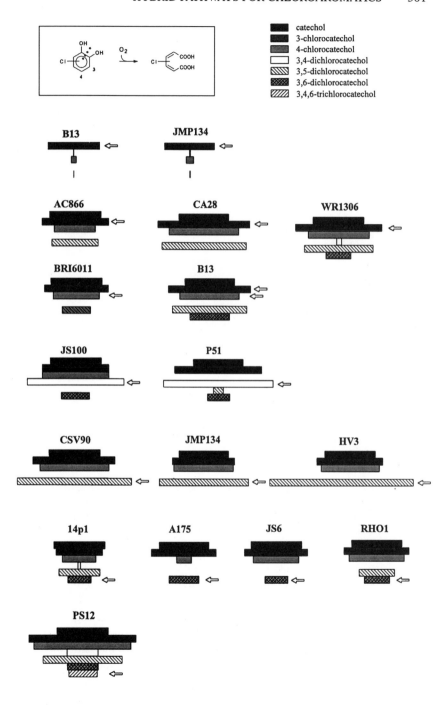

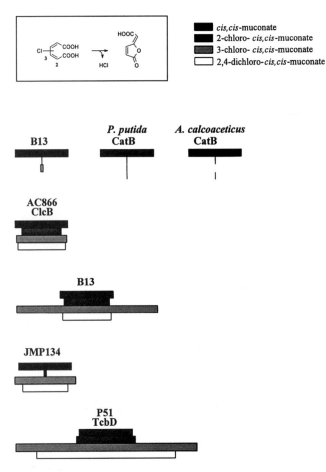

Figure 11 Lactonization of muconates by *Pseudomonas* sp. B13, *Ralstonia eutropha* (formerly *Alcaligenes eutrophus*) JMP134 (96, 141), *Pseudomonas putida* (CatB), *Acinetobacter calcoaceticus* (CatB), *Pseudomonas putida* AC866 (ClcB), and *Pseudomonas* sp. P51 (TcbD) (140). The activity was determined by the nonsubstituted substrate (100%) and compared with activity determined by substituted substrate.

shows very low affinities and activities toward chlorinated substrates (Figure 11). The chloromuconate cycloisomerase of 2,4-D–degrading strains develops some activity toward *cis,cis*-muconate but only a very low affinity. This enzyme is highly specific to 2,4-dichloromuconate occurring during 2,4-D degradation.

Dienelactone hydrolase is not able to transform the corresponding metabolite of the 3-oxoadipate pathway, 3-oxoadipate enollactone. In the same way, the 3-oxoadipate enollactone hydrolase fails to convert *cis*- and *trans*-dienelactone.

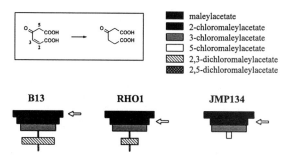

Figure 12   Reduction of maleylacetates by strains *Pseudomonas* sp. B13, *Pseudomonas aeruginosa* RHO1, and *Ralstonia eutropha* (formerly *Alcaligenes eutrophus*) JMP134 (89, 115). The substrate resulting in the degradation of the growth substrate is marked by the *arrow*. The activity was determined by the nonsubstituted substrate (100%) and compared with activity determined by substituted substrate.

Therefore, the dienelactone hydrolase cannot be considered simply a 3-oxoadipate enollactone hydrolase with low substrate specificity.

Maleylacetate reductase has no equivalent in the 3-oxoadipate pathway. No specialization has been found, although the strains have been enriched with different chloroaromatics (Figure 12). While maleylacetate is the natural substrate in the degradation of 3-chlorobenzoate in strain B13, 2-chloromaleylacetate arises from 3,5-dichloro- or 3,6-dichlorocatechol in strain JMP134 or RHO1. These enzymes transform many chlorinated substrates. However, a chlorine substituent in position 5, as it is in 5-chloro- and 2,5-dichloromaleylacetate, leads to a drastic decrease in the transformation rate.

No data are available characterizing the potential of the 3-oxoadipate:succinyl-CoA transferase and 3-oxoadipyl-CoA thiolase, the enzymes of the 3-oxoadipate pathway, toward chlorosubstituted substrates. But both steps have to be productive to finalize the degradation of trichlorocatechols and to reach the Krebs cycle.

GENE TRANSFER BY PLASMIDS

For production of the necessary enzymes, the genetic information must be transferred from one strain to another. Table 1 shows that numerous degradative sequences are encoded on plasmids. Of the total number of strains on the list, three quarters are *Pseudomonas*. Nearly half of these are strains of *P. putida*, and the next most abundant group is composed of anonymous *Pseudomonas* species for which not enough information exists to support their assignment to the genus as presently defined. The majority of the presented plasmids are transferable via conjugation. Conjugational intra- and intergeneric gene transfers have been

Table 1 Degradative plasmids with relevance to the construction of chloroaromatics-degrading hybrid strains[a]

Plasmid	Size (kb)	Conjugative	Incompatibility group	Substrate	Host	Reference
Peripheral pathways						
TOL	117	+	P-9	Xylenes, toluene, toluate	*Pseudomonas putida*	9, 100–102, 172
NAH7	83	+	P-9	Naphthalene via salicylate	*Pseudomonas putida*	9, 37, 101, 139, 180, 181
pWW60-1	87	+	P-9	Naphthalene via salicylate	*Pseudomonas* sp.	17
pDTG1	83	+	P-9	Naphthalene via salicylate	*Pseudomonas putida*	145
SAL1	85	+	P-9	Salicylate	*Pseudomonas putida*	9, 22, 101, 180, 181
pKF1	82	–	ND	Biphenyl via benzoate	*Acinetobacter* sp. (reclassified as *Rhodococcus globerulus*)	5, 53
pWW100	~200	–	ND	Biphenyl via benzoate methylbiphenyls via toluates	*Pseudomonas* sp.	105
pWW110	>200	ND	ND	Biphenyl via benzoate methylbiphenyls via toluates	*Pseudomonas* sp.	18
pCIT1	100	ND	ND	Aniline	*Pseudomonas* sp.	2
pEB	253	ND	ND	Ethylbenzene	*Pseudomonas fluorescens*	12
pRE4	105	ND	ND	Isopropylbenzene	*Pseudomonas putida*	38, 39
pWW174	200	+	ND	Benzene	*Acinetobacter calcoaceticus*	175
pHMT112	112	ND	ND	Benzene	*Pseudomonas putida*	157
pEST1005	44	ND	ND	Phenol	*Pseudomonas putida*	94
pVI150	mega	+	P-2	Phenol, cresols, 3,4-Dimethylphenol	*Pseudomonas* sp.	8, 146
Central pathways						
pAC25	117	+	P-9	3-Chlorobenzoate	*Pseudomonas putida*	26
pJP4	77	+	P-1	3-Chlorobenzoate, 2,4-D	*Ralstonia eutropha* (formerly *Alcaligenes eutrophus*)	33
pBR60	85	+	ND	3-Chlorobenzoate	*Alcaligenes* sp.	179
pRC10	45	ND	ND	2,4-D	*Flavobacterium* sp.	28
pP51	100	–	ND	1,2,4-Trichlorobenzene	*Pseudomonas* sp.	165
pMAB1	90	ND	ND	2,4-D	*Burkholderia* (formerly *Pseudomonas*) *cepacia*	14

[a]ND, not determined; 2,4-D, 2,4-dichlorophenoxyacetate.

demonstrated. The plasmids contain genes for peripheral pathways as well as for central ones. The detection of plasmids may be difficult in some cases; however, even then, transfer by conjugation can be demonstrated. Transfer of a plasmid from one bacterium to another can have rates of 10^{-2} for R plasmid RP4 of *P. aeruginosa* (30, 152), $1-10^{-2}$ for TOL plasmid derivatives (127), and $10^{-5}-10^{-7}$ for a *Pseudomonas* sp. strain B13 conjugable element (134) per cell-cell contact. In some bacterial strains, the genes for certain degradative sequences are located on the chromosome (such as the genes coding the toluene/benzene degradation of *P. putida* F1 or genes coding the biphenyl degradation of *P. pseudoalcaligenes* KF707) or on nonconjugative plasmids. With a few exceptions, these strains can be used only as acceptors for additional DNA. In a typical assay for conjugative gene transfer, the participating organisms are brought into close contact, usually being on a solid complex medium, which enables them to exchange genetic material. Afterward, the population is transferred onto selective media to develop the desired properties.

In some cases, insertion (IS) elements and transposons are associated with genes for degradative pathways; they have been important in the formation and distribution of new catabolic pathways (162, 178). The movement of IS elements and transposons can even lead to capture and mobilization of other genes. Transposons may transpose to plasmids and undergo plasmid-mediated transfer. So genes located on a nontransferable plasmid can be mobilized to another host by RP4, the broad host range *Inc*P1 R plasmid. Recently, Springael et al (150) reported the transfer of chromosomally located pathway by means of prime plasmid formation. Clustered genes of a catabolic pathway such as those coding a biphenyl pathway were found to be amenable to the in vivo cloning and transfer of genes.

GENE ORGANIZATION

In this discussion, the genes of biphenyl, xylene, and benzene degradation have been chosen to serve as examples for peripheral pathways because of their importance in the degradation of xenobiotics as well as the multitude of data available for these pathways. Furthermore, information on the gene organization of the central pathway for chlorocatechol degradation is presented.

Biphenyl Genes

The organization of the genes of biphenyl degradation located on the chromosome, usually denoted *bph*, has been clarified in various laboratories. Figure 13 shows 12 structural genes of the peripheral pathway of biphenyl degradation lying side by side with the central *meta* pathway. As far as is known, one transcript is formed from the *bph* gene cluster. There are many similarities but as

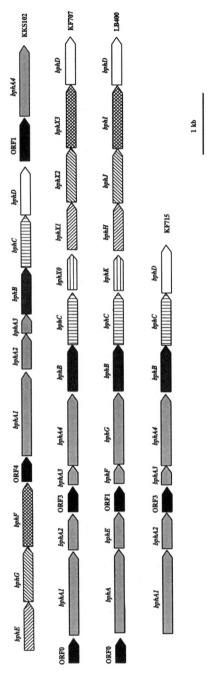

Figure 13   Operon structure of the genes of the biphenyl degradation pathway in strains KKS102, KF707, LB400, and KF715. Each *arrow* corresponds to a gene or an open reading frame (ORF). Homologous genes are *shaded* the same way. The genes of the biphenyl 2,3-dioxygenase appear *light gray*: terminal dioxygenase, large subunit (*bphA1*, *bphA*); terminal dioxygenase, small subunit (*bphA2*, *bphE*); ferredoxin (*bphA3*, *bphF*); and ferredoxin reductase (*bphA4*, *bphG*). Genes of biphenyldihydrodiol dehydrogenase (*bphB*) are colored *dark gray*, and those of phenylcatechol 1,2-dioxygenase (*bphC*) are marked with *stripes*. Genes that code hydrolases (*bphD*) are *without coloration*. The genes of strains KKS102, KF707, LB400, and JHR22 that code enzymes of the *meta* pathway are the following: 2-hydroxypenta-2,4-dienoate hydratase (*bphE*, *bphX1*, *bphH*); acetaldehyde dehydrogenase (acylating) (*bphG*, *bphX2*, *bphI*); 4-hydroxy-2-oxovalerate aldolase (*bphF*, *bphX3*, *bphJ*); and glutathione S-transferase (*bphX0*, *bphK*). Information was compiled from the following sources: 42, 51, 52, 54, 70, 74, 75, 90, 91, 93, 154, 155.

well some differences in the organization of the operons. Differences exist, for example, between the composition of the clusters of well-investigated strains such as *Pseudomonas* sp. strain KSS102 and the composition of the clusters of strains like *B. cepacia* (formerly *Pseudomonas* sp.) LB400 and *P. pseudoalcaligenes* KF707. The order of the structural genes in the strains LB400, KF707, and *B. cepacia* JHR22, a chlorobiphenyl-degrading hybrid strain, is identical (D Springael, unpublished results). Difficulties arise from the different nomenclature of the genes.

The regulation of biphenyl-degrading pathways in the above-mentioned strains is not known. Lau et al (99) reported that a two-component signal transduction system regulates the biphenyl degradation pathway in *Rhodococcus globerulus* strain M5.

Xylene Genes

Among the genes for peripheral pathways, the *xyl* genes, lying on the TOL plasmid pWW0 of *P. putida* strain PaW1, are the best characterized. These genes, which are essential for the total degradation of toluene and xylenes, form two functional units, the upper and the *meta* operon. A simplified model is given in Figure 14. The upper operon *xylCMABN* encodes three enzymes that oxidize toluene and xylenes to benzoate and toluates, respectively. The function of the protein encoded by *xylN* is unknown. The promoter of the upper pathway genes, Pu, is regulated positively by the regulatory gene *xylR*. This gene contains the information for a protein, which enhances transcription after the binding of an inducing molecule (toluene, *m*-xylene, and the respective benzoates). XylR belongs to the NtrC family of σ^{54}-dependent transcriptional regulators (1, 81). Subsequently, benzoate and toluate are transformed into the respective catechols, which are mineralized by the enzymes of the *meta* pathway.

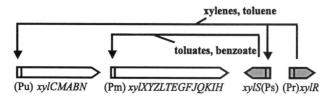

Figure 14   Simplified model of the regulation of *xyl* gene operons (*xyl*, xylene). The upper operon (*xylCMABN*) codes xylene monooxygenase, benzylalcohol, and benzaldehyde dehydrogenase, while 13 enzymes are encoded by the *meta* operon (*xylXYZLTEGFJQKIH*), including toluate1,2-dioxygenase (XylXYZ), toluate dihydrodiol dehydrogenase (XylL), and catechol 2,3-dioxygenase (XylE). The regulator genes *xylS* and *xylR* are shown in *gray*. The promoter regions are marked by *small boxes*. The *arrows* indicate induction by XylR and XylS regulatory protein in concert with the respective aromatic effectors. Data from 64, 77, 79, 80, 112, 124, 126.

These enzymes are encoded by the genes of the *meta* operon. Misunderstandings are caused by the nonuniform use of the term *meta*, which can refer to the operon or the biochemical pathway. The *meta* operon contains the genes *xylXYZ* and *xylL*. These genes encode the enzymes toluate 1,2-dioxygenase and toluate dihydrodiol dehydrogenase, which belong to the peripheral pathway. The central, *meta*, pathway starts with the degradation of catechol. No fewer than nine enzymes are expressed in this polycistronic operon coordinately from the promoter Pm. The product of the *xylS* gene is needed for an efficient expression of the *meta* operon by the promoter Pm. Inducing compounds such as benzoate and toluate bind to XylS, which belongs to the AraC family of transcriptional regulators (55, 125), bringing about an activated form.

Both operons and regulatory genes are indispensable for the transformation of toluenes to catechols. All the necessary genes are jointly transferred during the plasmid exchange. Problems might arise, however, because when the genes of peripheral sequences are transferred, information for pathways disturbing the degradation of chloroaromatics (for example, the central *meta* pathway) is also transferred.

Toluene/Benzene/Chlorobenzene Genes

The best-studied degradative system for toluene/benzene in terms of the enzymology involved is that of *P. putida* strain F1 (61). The 10 genes coding for the enzymes for the total degradation of toluene into Krebs cycle intermediates are arranged in the order *todFC1C2BADEGIH* (98, 111, 182)—the gene arrangement does not reflect the order of biochemical reactions (Figure 15). In terms of gene organization and the nucleotide sequence, the *tod* genes are similar to corresponding *bph* genes coding the biphenyl pathway of *P. pseudoalcaligenes* strain KF707 (52, 73, 155). The *todC1C2BAD* genes showed significant homology to the nucleotide sequence for benzene dioxygenase and benzene *cis*-dihydrodiol dehydrogenase from *P. putida* strain 136R-3 (82, 182). Studies on the regulation of *tod* genes indicated that at least the *todFC1C2BADE* genes are arranged in an operon, and these genes are coordinately induced by toluene. The expression of *tod* genes was found to be controlled by a two-component strain-specific regulatory system coded by two genes, designated *todS* and *todT*, situated downstream of *todH* (99). Upstream of *todF*, the gene *todX* was found coding a membrane protein (170). A toluene-inducible promoter was localized in front of *todX*, which is probably responsible for the expression of all *tod* structural genes. A truncated LysR-type regulator, which seems to have a negligible role in *tod* gene regulation and is coded by *todR*, is oriented in the direction opposite the other *tod* genes. Homology comparisons indicated that *tcbAaAbAcAbB* genes of strain P51 coding chlorobenzene dioxygenase and chlorobenzene *cis*-dihydrodiol dehydrogenase

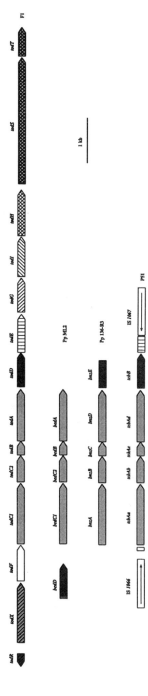

Figure 15   Operon structure of the genes of the benzene/toluene/chlorobenzene degradation pathway in strains F1, ML2, 136-R3, and P51. Homologous genes are *shaded* the same way: terminal dioxygenase, large subunit (*todC1, bedC1, bnzA, tcbAa*); terminal dioxygenase, small subunit (*todC2, bedC2, bnzB, tcbAb*); ferredoxin (*todB, bedB, bnzC, tcbAc*); ferredoxin reductase (*todA, bedA, bznD, tcbAd*); benzene *cis*-dihydrodiol dehydrogenase (*todD, bedD, bznE, tcbB*); catechol 2,3-dioxygenase (*todE*); 2-hydroxy-6-hepta-2,4-dienoate hydrolase (*todF*); 2-hydroxypenta-2,4-dienoate hydratase (*todG*); acetaldehyde dehydrogenase (acylating) (*todI*); 4-hydroxy-2-oxovalerate aldolase (*todH*); membrane protein (*todX*); regulator (*todS*); regulator (*todT*); and truncated LysR-type activator (*todR*). Information was compiled from the following sources: 46, 82, 98, 99, 111, 156–158, 166, 170, 171, 182.

are closely related to those of toluene (*todC1C2BAD* of *P. putida* strain F1) and benzene degradation (*bedDC1C2BA* of *P. putida* strain ML2 and *bnz ABCD* of *P. putida* strain 136-R3). The *tcbAaAbAcAbB* gene block of strain P51, however, is no longer integrated with the genes for the *meta* pathway, but cut out by the action of two insertion elements (IS*1066* and IS*1067*) and moved to a new position near the genes for the chlorocatechol pathway. Surprisingly, the organization of the *tod*, *tcb*, and *bzn* genes on one side and *ben* on the other is quite different. The dehydrogenase gene *benD* of strain ML2 lies upstream of the dioxygenase cluster and is flanked by 42-bp direct repeats, each containing a 14-bp sequence identical to the inverted repeat of IS*26*. From this configuration one can assume that the *benD* gene has been transferred to be juxtaposed with the dioxygenase gene cluster. The different G+C codon usage of the *benD* and *benC1C2BA* genes indicated that they derive from different origins.

Just recently, the organization of the chlorobenzene dioxygenase gene cluster *tecA1A2A3A4* of *Burkholderia* sp. PS12, a strain using 1,2,4-trichloro- and 1,2,4,5-tetrachlorobenzene as the growth substrate, was found to resemble the equivalent *tcb* and *tod* genes (11).

Chlorocatechol Genes

Plasmids that contain the genes for the degradative pathway of chlorocatechols, the modified *ortho* pathway, have been investigated well. The structures of the corresponding operons are nearly identical, in spite of the different origins of the bacteria (Figure 16). The *clc* and *tcb* genes, for example, are organized

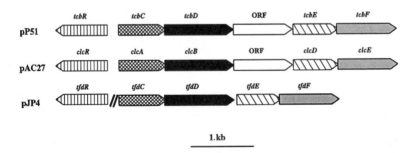

Figure 16 Operon structure of the genes of the modified *ortho* pathway with adjacent regulatory genes (*tcb*, trichlorobenzene; *clc*, chlorocatechol; *tfd*, two-four-D). Each *arrow* corresponds to a gene or an open reading frame (ORF). Homologous genes are marked the same way. The genes of the chlorocatechol 1,2-dioxygenases (*tcbC*, *clcA*, *tfdC*) are marked with *crossed diagonals*, those of the chloromuconate cycloisomerases (*tcbD*, *clcB*, *tfdD*) are colored *black*, those of the dienelactone hydrolases (*tcdE*, *clcD*, *tfdE*) are marked with *diagonals*, and those of the maleylacetate reductase (*tcdF*, *clcE*, *tfdF*) are colored *gray*. The regulator genes *tcbR*, *clcR*, and *tfdR* are marked with *vertical stripes*. *tfdR* is located several kilobases upstream of *tfdCDEF*. *Arrows without tips* denote an overlap of the reading frames. Information was compiled from the following sources: 29, 49, 85, 86, 103, 108, 121, 163, 164.

identically, and the *tfdCDEF* operon differs from these genes only in the absence of an open reading frame between *tfdD* and *tfdE*. The regulatory genes *tfdR*, *clcR*, and *tcbR* are oriented in the opposite direction. The operons of the *clc*, *tcb*, and *tfd* genes have similar structure. Furthermore, a comparison of the sequences shows a high identity between the corresponding enzymes of the different bacteria. The gene products of the regulatory genes, TfdR, ClcR, and TcbR, which are members of the LysR family, act as positive regulators. It has long remained unclear which substances are the effectors, activating the regulatory proteins. Because of several analogies to the regulation of the normal 3-oxoadipate pathway with *cis,cis*-muconate as the effector of the *cat* genes, chlorinated muconates are believed to play this role in the modified *ortho* pathway. Clear-cut proof has recently been obtained of 2,4-dichloro-*cis,cis*-muconate being the inducing metabolite of the enzymes of the modified *ortho* pathway coded on the plasmid pJP4 by use of transposon mutants (45). Just recently, 2-chloro-*cis,cis*-muconate was demonstrated to be the inducer of the *clcABCDE* operon (110).

Oxoadipate Genes

Molecular studies of the 3-oxoadipate pathway carried out with *P. putida* PRS2000 have shown that the chromosomal *pca* genes are relatively scattered with at least four separated transcriptional units (Figure 17). These genes code the protocatechuate branch and the set of enzymes to complete the conversion of β-ketoadipate enol-lactone—the conversion point of catechol and protocatechuate degradation—to tricarboxylic acid cycle intermediates. The enzymes coded by *pca* genes in *P. putida* are induced by 3-oxoadipate (118) through intermediacy of the transcriptional activator encoded by *pcaR*. PcaR is a protein that belongs to a newly described family of regulatory proteins called the PobR family (32). The *pcaR* gene has been shown to lie about 15 kbp upstream from the *pcaBDC* operon (78). The *pcaI* and *pcaJ* genes, which encode the two subunits of β-ketoadipate:succinyl-CoA transferase, constitute a separate operon (119, 120). The last enzyme in the pathway, β-ketoadipyl-CoA thiolase, is encoded by the *pcaF* gene, which is part of the *pcaRKF* operon (67, 137). The *cat* genes encode the catechol branch of the 3-oxoadipate pathway. But analogous cat genes coding transferase and thiolase are absent in *P. putida*.

INDUCTION POTENTIAL OF CHLORINATED COMPOUNDS

The induction of enzymes, in addition to their specificity, can be a bottleneck during the degradation of chlorinated aromatics. Overall, only a few data exist on the effector specificity of degradative pathways. Figure 18 schematically summarizes the information about the induction of the upper and *meta* operons, which are encoded on the TOL plasmid, by different chlorinated substrate

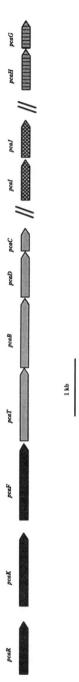

Figure 17 Transcriptional units of *pca* genes scattered on the chromosome of *Pseudomonas putida* PRS2000, which codes the protocatechuic acid degradation pathway. The following gene products are formed. *pcaRKF* operon (*dark gray*): activator protein, 4-hydroxybenzoate transporter protein, 3-oxoadipyl-CoA thiolase. *pcaHG* operon (*stripes*): protocatechuate 3,4-dioxygenase. *pcaTBDC* operon (*light gray*): dicarboxylic acid transporter protein, β-carboxy-*cis,cis*-muconate cycloisomerase, γ-carboxymuconolactone decarboxylase, β-ketoadipate enol-lactone hydrolase. *pcaIJ* operon (*crosshatches*): two subunits of 3-oxoadipate:succinyl-CoA transferase. Information was compiled from the following sources: 50, 67, 68, 78, 119, 120, 137, 174.

1 kb

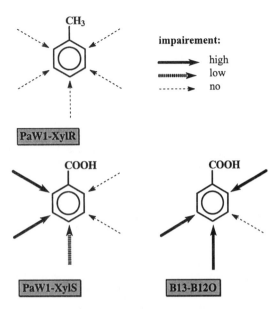

Figure 18 Schematic representation of impairment of induction of the upper and *meta* pathway operon of strain PaW1 and the initial benzoate pathway of strain B13 by chlorosubstituents in toluenes or benzoates (1, 15, 126; A Stolz & W Reineke, unpublished results). No, low, and high impairments of efficiency as effector by substituents in certain positions are marked by *various arrows.*

analogs. Furthermore, the induction of the first enzyme of the 3-chlorobenzoate degradation pathway has been investigated toward effector specificity.

Two different methods, the reporter gene technique (Pu::*lacZ* and Pm::*lacZ* fusion) and induction with whole cells, lead to similar results concerning the effector specificity of the proteins XylR and XylS. All chlorotoluenes tested (mono- and dichlorinated ones with substitution in position 2, 3, 4, 5, 6) were effectors for XylR. This result indicates a broad effector specificity of XylR. It is remarkable that 2-chlorotoluene, a bad substrate for xylene monooxygenase, is the best effector.

In the case of the XylS protein, a different degree of substitution in positions 2, 3, and/or 4 is possible without loss of effector activity. The binding site of XylS seems to exclude a disubstitution with one substituent placed in *ortho* position. Simultaneous substitution of positions 3 and 5 negatively influences the effector activity, too.

Only 3-chlorobenzoate is a good inducer of the peripheral pathway of strain B13, whereas 4-chloro- and 3,5-dichlorobenzoate do not function as inducing compounds (A Stolz & W Reineke, unpublished results).

The benzene dioxygenase and benzene *cis*-dihydrodiol dehydrogenase of *P. putida* strain F1 are induced by chloro-, 1,2-dichloro-, 1,3-dichloro-, 1,4-dichloro-, and 1,2,4-trichlorobenzene (DT Gibson, V Subramanian, W Reineke, unpublished observations).

Data are not available that allow an assessment of whether some chloroaromatic isomeric compounds or their metabolites, chlorosubstituted *cis,cis*-muconates, are such poor effectors that the induction of the central modified *ortho* pathway is a main bottleneck. It is unknown whether the PcaR protein, the regulator of the 3-oxoadipate pathway, is able to recognize chlorosubstituted 3-oxoadipates besides 3-oxoadipate as effectors.

PRODUCTION OF A FUNCTIONING HYBRID PATHWAY

Some examples of the development of chlorobenzoate-, chlorotoluene-, chlorobenzene-, and chlorobiphenyl-degrading strains answer the question of whether changes in the genetic material are necessary, and if so what type of changes are necessary, after transfer of existing DNA into the 3-oxoadipate–degrading host organism to obtain a functioning hybrid pathway for the chloroaromatics.

Chlorobenzoate-Degrading Strains

A multitude of information resulted from the experiments with *Pseudomonas* sp. strain B13 and *P. putida* strain PaW1, which led to hybrid strains able to grow with 4-chloro- and 3,5-dichlorobenzoate (Figure 19). Both parent strains harbor a 3-oxoadipate pathway. Strain B13 was enriched with 3-chlorobenzoate (34) and converts this substrate to 3-chloro- and 4-chlorocatechol, which are mineralized via the modified *ortho* pathway. Strain B13 is not able to use 4-chloro- and 3,5-dichlorobenzoate as growth substrates because its benzoate 1,2-dioxygenase has a narrow specificity and does not convert these compounds (see Figure 6). In addition, both compounds fail to induce the initial enzymes in strain B13 (see Figure 18). However, the expected metabolites, 4-chloro- and 3,5-dichlorocatechol, can be mineralized by strain B13. The toluate 1,2-dioxygenase of *P. putida* PaW1, which is encoded on the TOL plasmid, has a broader substrate specificity than the corresponding B13 enzyme and accepts 4-chloro- and 3,5-dichlorobenzoate as substrates. Transconjugants of the mating between *Pseudomonas* sp. B13 as the acceptor (a streptomycin-resistant derivative was used) and *P. putida* PaW1 as the donor showed the phenotype of strain B13 with the additional ability to grow with *m*- and *p*-toluate (strain WR211). However, this hybrid strain was not able to utilize 4-chlorobenzoate. Mutants like strain WR216 that are able to grow with 4-chlorobenzoate could easily be obtained by using solid media with 4-chlorobenzoate as the sole source

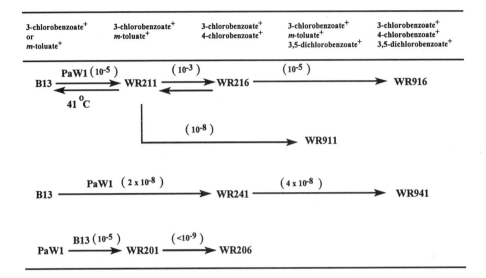

Figure 19 Genealogy of hybrid strains degrading various chlorobenzoates from the matings between streptomycin-resistant B13 and PaW1 and streptomycin-resistant PaW1 and B13. The phenotypes given in the top lines are valid for all strains in the respective columns below. The frequency given in *parentheses* is defined as the number of transconjugants per donor cell or as spontaneous mutation on the selective substrate (131; A Stolz & W Reineke, unpublished results).

of carbon. It was surprising to find this strain unable to grow with the toluates. These results clearly show that plasmid transfer is necessary but not sufficient for the development of a new degradative pathway.

The TOL plasmid underwent various rearrangements in the host strain B13 to give the 4-chlorobenzoate–degrading strain WR216. A 56-kb segment of TOL DNA is integrated into the chromosome, as has been detected in the intermediate strain WR211. A deletion of a 39-kb segment from TOL (leaving a cryptic plasmid) and an insertion of a DNA segment of about 3 kb into the *xylE* gene of the plasmid encoding the catechol 2,3-dioxygenase was observed in strain WR216 (84).

Transconjugants like WR241 were obtained by mating the wild-type strains B13 and PaW1 when selected directly for growth with 4-chlorobenzoate. They occurred at a frequency similar to the product built from the frequencies of the single steps during selection of the strains WR211 and WR216. Investigation of the substrates used by strains like WR241 showed that strain B13 had functioned as the acceptor in the mating.

On the other hand, strain B13 can also serve as the donor for the genes of the modified *ortho* pathway. This function was shown in a mating between strain

B13 and a streptomycin-resistant derivative of strain PaW1. Transconjugants like strain WR201 occurred at a frequency similar to that observed in the
matings leading to strains like WR211 when the media contained 3-chloroben-
zoate for selection of the genes encoding the chlorocatechol degradation and
additionally streptomycin for counterselection against the donor strain B13.
These strains showed the phenotype of strain PaW1 as well as the ability to
grow with 3-chlorobenzoate. However, no growth with 4-chlorobenzoate could
be observed.

Strains of the type WR206, which acquired the potential to use 4-chloroben-
zoate as the growth substrate, occurred at very low frequency. This finding
explains why only transconjugants have been obtained with strain B13 as the
acceptor of the TOL plasmid when strains B13 and PaW1 have been mated
without use of a resistance selecting against the donor strain.

In addition to the bottlenecks concerning the turnover of substrates, the induction of a pathway by the novel compound sometimes does not take place. One
example of this is 3,5-dichlorobenzoate. While 3,5-dichlorobenzoate fails to induce the toluate 1,2-dioxygenase in the strains WR211, WR216, and WR241,
the compound is an effector in the strains WR911, WR916, and WR941, which
have occurred on solid media containing 3,5-dichlorobenzoate from the respective origins. Strains such as WR911 retained a m-toluate$^+$ phenotype.

At this point it can be ascertained that the following phenotypes are compatible: 3-chlorobenzoate$^+$, 3,5-dichlorobenzoate$^+$, and m-toluate$^+$. 4-Chloro-
benzoate$^+$ and m-toluate$^+$ phenotypes seem to exclude each other in the system
investigated.

The conclusions above may be interpreted as follows: Because the desired
peripheral sequences as well as the *meta* pathway are encoded on plasmids,
misrouting of chlorocatechols into the *meta* pathway has to be avoided in the
hybrid strains. This is discussed for the ring-cleavage substrates 3-chloro-,
4-chloro-, and 3,5-dichlorocatechol (Figure 20). The critical element for the
degradation through the *ortho* pathway in the hybrid strains is the presence or
absence of both ring-cleaving dioxygenases as well as the activity and affinity to
the chlorocatechols. The *meta*-cleaving catechol 2,3-dioxygenase is present at
a high basic level and will be induced by substrates such as toluate in *P. putida*
PaW1. In contrast, product induction by *cis,cis*-muconate has been demonstrated for the *ortho*-cleaving catechol 1,2-dioxygenase. 2,4-Dichloro-*cis,cis*-
muconate has been determined to induce the enzymes of the *tfd* operon (45).
McFall et al (110) just recently demonstrated by use of in vitro transcription
assay and *lacZ* transcription fusions in vivo that 2-chloromuconate is the inducer of the *clcABCDE* operon coded by plasmid pAC27. One can presume
that the same product induction by 2-chloro-*cis,cis*-muconate might function
in strain B13 and its TOL derivatives.

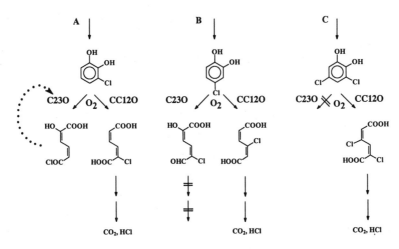

Figure 20   Divergent pathways for (*A*) 3-chlorocatechol, (*B*) 4-chlorocatechol, and (*C*) 3,5-dich-lorocatechol in hybrid strains resulting from the mating of strains B13 and PaW1 (7, 36, 132; W Reineke, unpublished results). The *pointed arrow* illustrates the reaction of the reactive acylchloride with the catechol 2,3-dioxygenase. *Left: meta* cleavage by catechol 2,3-dioxygenase (C23O). *Right: ortho* cleavage by chlorocatechol 1,2-dioxygenase (CC12O) and total degradation with elimination of chloride.

Misrouting of 3-chlorocatechol into the *meta* pathway can be avoided. Both ring-cleaving dioxygenases exhibit activity with 3-chlorocatechol. Catechol 2,3-dioxygenase forms an acylchloride from 3-chlorocatechol, which reacts in the catalytic site of the enzyme, resulting in the nonreversible so-called suicide inactivation of the enzyme. For this reason, further conversion of 3-chloro-catechol into the *meta* pathway is prevented, and thus induction of 1,2-dioxygenase can occur.

A different mechanism is necessary for 4-chlorocatechol to avoid misrouting. In this case no acylchloride is formed; therefore the induction of catechol 2,3-dioxygenase causes a constant misrouting of 4-chlorocatechol into the *meta* pathway. Consequently, no inducer of the *ortho*-cleaving chlorocatechol 1,2-dioxygenase is formed. But the *meta* pathway can be prevented on the gene level because the gene of the catechol 2,3-dioxygenase, *xylE*, can be inactivated by an insertion (see strain WR216) or a point mutation (see strain WR206) (84; PA Williams, personal communication).

In analogy to the suicide inactivation mechanism observed for 3-chlorocate-chol, an acylchloride was also expected to be formed from 3,5-dichlorocatechol via catechol 2,3-dioxygenase. However, misrouting of 3,5-dichlorocatechol into the *meta* pathway did not occur because the catechol 2,3-dioxygenase was found to exhibit low activity and affinity with the compound. In contrast, chlorocatechol 1,2-dioxygenase showed high activity and affinity toward 3,5-dichlorocatechol.

Natural mechanisms eliminate the unsuitable *meta* pathway. The utilization of whole plasmids, which leads to the transfer of all genes coding a peripheral pathway inclusive of the essential regulatory proteins, does not create a problem, although genes encoding the *meta* pathway are cotransferred.

Gene transfer of the chlorocatechol-degradative genes of strain B13 into *B. cepacia* strain WR401, capable of utilizing salicylate and methylsalicylates but not chlorosalicylate, allowed the development of cells that utilize both salicylate and chlorosalicylate (134). Because the host organism degrades *o*-toluate by use of a dioxygenase via 3-methylcatechol (40), the transconjugants such as JH230 were found to be able to grow with 2-chlorobenzoate via 3-chlorocatechol (66). The hybrid organism has acquired the potential to use 2-chloro-, 3-chloro-, 4-chloro-, 2,4-dichloro-, and 3,5-dichlorobenzoate as the growth substrate.

Chlorotoluene-Degrading Strains

By mating strains PaW1 and B13 with selection for growth with 3-chloro- and 4-chlorotoluene, strains such as WR1412 or WR1441 occurred at high frequency (15). Strains such as WR1426 that have the ability to grow slowly with 3,5-dichlorotoluene were obtained from the same mating. The investigation of the enzymes clearly indicates that the TOL plasmid–coded enzymes, including the toluate 1,2-dioxygenase and those of the modified *ortho* pathway, are used by the hybrid strains for growth with the chlorotoluenes.

Because strain JH230 is able to grow with 2-chlorobenzoate, it was thought to be the right partner for a mating with strain PaW1 to obtain 2-chlorotoluene–degrading hybrid strains. 3-Chlorotoluene–positive strains such as WR233 occurred at low frequency from the mating between strains PaW1 and JH230. However, strain WR233 fails to use 2-chlorotoluene as the growth substrate. Considering that 2-chlorotoluene was shown to be an effective inducer of the upper pathway and the property of the organism to grow with 2-chlorobenzylalcohol, the xylene oxygenase seems to be the sole bottleneck to be overcome.

Chlorobiphenyl-Degrading Strains

In contrast to the examples mentioned above, for some chloroanalogous substrates the respective peripheral pathway does not bring about the conversion to the chlorocatechols. In this case more than just the mating of two strains is necessary: a novel pathway has to be constructed in one organism by segments from at least three organisms (see Figure 21).

Strains able to mineralize 3-chlorobiphenyl were obtained from the mating of *P. putida* strain BN10 with strain B13. Both organisms were found to function

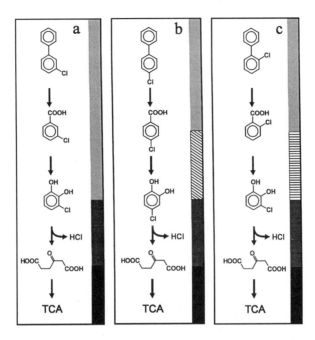

Figure 21   Hybrid pathways for the mineralization of chlorobiphenyls in hybrid strains: (*a*) *Pseudomonas putida* strain BN210, (*b*) *Pseudomonas putida* strain KE210, and (*c*) *Burkholderia cepacia* strain JHR22. The *color* or the *hatch* on the right side of the pathway characterizes the origin of the respective pathway segment in the hybrid strains.
Peripheral pathway segment 1:
   ▨, biphenyl-degrading *Pseudomonas putida* strains BN10 (a, b) or JHR (c).
Peripheral pathway segment 2:
   ▧, *p*-toluate–degrading *P. putida* strain PaW1 or
   ▤, *o*-toluate–degrading *B. cepacia* strain WR401.
Modified *ortho* pathway segment:
   ■, 3-chlorobenzoate–degrading *Pseudomonas* sp. strain B13.
Late 3-oxoadipate pathway segment:
   ■, *P. putida* strain BN10 (a, b) or *B. cepacia* strain WR401 (c).
   Information was compiled from the following sources: 69, 114; K Engelberts & W Reineke, unpublished results.

as donor and recipient in this type of mating (114). Conjugational transfer of the chlorocatechol-degradative ability from strain B13 to strain BN10 resulted at a frequency 10^{-6} in transconjugants such as strain BN210 that are able to grow with 3-chlorobiphenyl with stoichiometric release of chloride. Transfer of the genes coding enzymes for the degradation of biphenyl from strain BN10 to strain B13 was observed to occur at a lower rate to give strain B131. The benzoate 1,2-dioxygenases of both transconjugants fail to use 4-chlorobenzoate

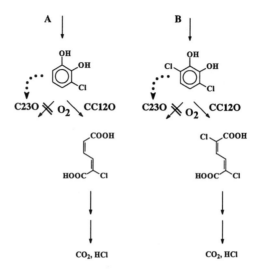

Figure 22 Divergent pathways for (*A*) 3-chlorocatechol and (*B*) 3,6-dichlorocatechol in hybrid strains resulting from the mating of strains B13 and F1. The *pointed arrow* illustrates the inactivation of the catechol 2,3-dioxygenase by the respective chlorocatechol. *Left: meta* cleavage by catechol 2,3-dioxygenase (C23O). *Right: ortho* cleavage by chlorocatechol 1,2-dioxygenase (CC12O) and total degradation with elimination of chloride. Information was compiled from the following sources: 36, 95, 132; SR Kaschabek & W Reineke, unpublished results.

as the substrate. However, the introduction of the TOL plasmid into strain BN210 allowed the isolation of 4-chlorobiphenyl–positive derivatives, such as *P. putida* strain KE210 (K Engelberts & W Reineke, unpublished results). To obtain 2-chlorobiphenyl–degrading organisms, the biphenyl-degrading *P. putida* strain JHR was mated with *B. cepacia* strain JH230, a strain with the potential to degrade chlorobenzoates including *ortho*-substituted isomers (see above). The biphenyl-positive derivative of strain JH230, strain JHR2, was able to grow with 3-chloro- and 4-chlorobiphenyl but failed to use 2-chloro-, 2,4-dichloro, and 3,5-dichlorobiphenyl as the growth substrate. However, growth and stoichiometric elimination of chloride occurred with 2-chloro- and 3,5-dichlorobenzoate. Further selection for growth with 2-chlorobiphenyl allowed the isolation of strain JHR22, which was able to grow with 2-chloro-, 3-chloro-, 4-chloro-, 2,4-dichloro-, and 3,5-dichlorobiphenyl with stoichiometric elimination of chloride (69). The nature of some events that have taken place during the development is unknown.

Chlorobenzene-Degrading Strains

Strain F1, enriched with ethylbenzene and identified as *P. putida* biotype B (59), was found to be a suitable partner for the development of chlorobenzene-degrading strains. The protocatechuate branch of the 3-oxoadipate pathway seems to be present, because 90% or more of the *P. putida* biotype B strains are able to use *p*-hydroxybenzoate as the growth substrate (151). The conjugational transfer of the genes coding the degradation of chlorocatechols from strain B13 into strain F1 resulted in chlorobenzene-degrading strains such as WR1313 (117). The presence of the catechol 2,3-dioxygenase is not a disadvantage, because 3-chlorocatechol is able to inactivate the enzyme by chelating the ferrous ion (95); therefore a genetic elimination of the *todE* gene is not necessary (Figure 22). Strain WR1323, which was able to grow with 1,4-dichlorobenzene, was derived from strain WR1313 after growth on chlorobenzene for 18 months in the presence of 1,4-dichlorobenzene. Catechol 2,3-dioxygenase was observed in 1,4-dichlorobenzene–grown cells of strain WR1323, indicating that the presence of the *meta* pathway is not a problem for productive degradation of 3,6-dichlorocatechol through the modified *ortho* pathway.

CONCLUDING REMARKS

The judicious combination of peripheral and central pathway sequences in a suitable host organism can provide complete hydrid pathways for various mono- and disubstituted chloroaromatics such as chlorinated benzenes, benzoates, biphenyls, naphthalenes, phenols, salicylates, and toluenes (1, 15, 23–25, 27, 66, 69, 97, 104, 114, 117, 128, 131, 134, 143, 160). The procedure using the DNA transfer with whole degradative plasmids is superior because of its technical simplicity and effective positive selection. The enrichment of chloroaromatics-degrading strains isolated from contaminated material often takes a couple of months. Hybrid strains, however, can be developed by well-aimed mating within 1 to 4 weeks because the organisms with the suitable pathway segments are inoculated together on solid media, making a gene transfer easy. However, the selection steps have to be done in the right order. It is important to establish the chlorocatechol-degradative sequence in the beginning of the development of the hybrid strains, because otherwise the accumulated chlorocatechols will harm the cells.

In principle, similar strains can also be designed through genetic engineering techniques. The major advantage of this approach is the potential to combine genetic material from completely different sources. In contrast, the strains involved in the in vivo process described here are taxonomically related organisms, i.e. pseudomonads, although the assignment to the genus as currently defined has not been successful for various strains like strain B13. The major disadvantage of the sophisticated in vitro approach is the huge mass of research necessary prior to the construction experiment leading to the hybrid stains. Cloning of structural as well as regulatory genes has to be done, followed by establishment in a suitable host.

It is highly probable that strains, which have been obtained from enrichment cultures able to use a chloroaromatic as the growth substrate, are the product of patchwork assembly. This result has been shown for a strain degrading 1,2,4-trichlorobenzene that has been enriched from sediment of the Rhine River. Van der Meer et al (166) found that a transposon carries genes that code for the two initial enzymes in the degradation of 1,2,4,-trichlorobenzene: chlorobenzene 1,2-dioxygenase and chlorobenzene *cis*-dihydrodiol dehydrogenase. This element is part of the plasmid p51, so in combination with the chlorocatechol sequence, a hybrid pathway was formed. Other examples illustrating the natural selection for novel combination are given by Ogawa & Miyashita (116) for the degradation of 3-chlorobenzoate and by Fong et al (46) for the degradation of benzene. Surprisingly, the benzene pathway is encoded on a plasmid instead of being localized on the chromosome, which normally carries the genes of the *ortho* pathway. The different G+C codon usage of the different segments of the benzene pathway allows the assumption that different organisms are the source of the genetic material used for the hybrid pathway. *Burkholderia* (formerly *Pseudomonas*) *cepacia* strain AC1100, degrading 2,4,5-trichlorophenoxyacetate through the chlorohydroxyhydroquinone pathway, provides an example of an organism that appears to have evolved rapidly and acquired genes originally present in a consortium. Genes specifying a portion of the pathway are associated with the IS element RS1100 (161), which was not detected in other *B. cepacia* strains. These examples clearly indicate that patchwork assembly also happens in nature (see also the discussion by Williams & Sayers, 173).

At present the dibenzofuran and dibenzo-*p*-dioxin degradation sequences mentioned in Figure 2 have not been used successfully to develop hybrid strains. An additional future task should be the degradation of higher-chlorinated biphenyls. New strains that can be used for in vivo construction should harbor a peripheral pathway with the ability to convert biphenyls chlorinated in both rings effectively. But these strains must also be the right partner in matings with strains such as B13.

Another future issue might be the use of an alternative or addition to the modified *ortho* pathway segment in the development of chloroaromatics-degrading strains. The above-mentioned data seem to support the generalization that the *meta* pathway is unsuitable for the degradation of chloroaromatics. One reason has been found in the formation of a suicide product from 3-chlorocatechol by the catechol 2,3-dioxygenase (7) inactivating the ring-cleavage enzyme. However, some recent publications gave evidence that compounds can be degraded via the *meta* pathway when the resulting chlorocatechol is substituted in the 4 position (3, 4, 71, 76, 83, 109). Just recently, the use of the *meta* cleavage pathway for the mineralization of chlorobenzene with 3-chlorocatechol as the central metabolite has been determined (87, 107). So the above-mentioned suicide inactivation is not a mechanism that should be generalized. An involvement of the *meta* pathway for the creation of hybrid pathways by patchwork assembly remains to be seen.

ACKNOWLEDGMENTS

This review is a modified, updated, and more comprehensive version of a recent article on the same subject (68a). I thank all my collaborators for their enthusiastic participation in chloroaromatic degradation research. The work on the development of chloroaromatics-degrading strains in my laboratory was generously supported by the Deutsche Forschungsgemeinschaft; the Bundesministerium für Bildung, Wissenschaft, Forschung und Technologie; and the European Commission.

> Visit the *Annual Reviews home page* at
> http://www.AnnualReviews.org.

Literature Cited

1. Abril M-A, Michan C, Timmis KN, Ramos JL. 1989. Regulator and enzyme specificities of the TOL plasmid-encoded upper pathway for degradation of aromatic hydrocarbons and expansion of the substrate range of the pathway. *J. Bacteriol.* 171:6782–90
2. Anson JG, Mackinnon G. 1984. Novel *Pseudomonas* plasmid involved in aniline degradation. *Appl. Environ. Microbiol.* 48:868–69
3. Arensdorf JJ, Focht DD. 1994. Formation of chlorocatechol *meta* cleavage products by a pseudomonad during metabolism of monochlorobiphenyls. *Appl. Environ. Microbiol.* 60:2884–89
4. Arensdorf JJ, Focht DD. 1995. A *meta* cleavage pathway for 4-chlorobenzoate, an intermediate in the metabolism of 4-chlorobiphenyl by *Pseudomonas cepacia* P166. *Appl. Environ. Microbiol.* 61:443–47
5. Asturias JA, Moore E, Yakimov MM, Klatte S, Timmis KN. 1994. Reclassification of the polychlorinated biphenyldegraders *Acinetobacter* sp. strain P6 and *Corynebacterium* sp. strain MB1 as *Rhodococcus globerulus*. *System. Appl. Microbiol.* 17:226–31
6. Bachofer R, Lingens F. 1975. Conversion of aniline into pyrocatechol by a *Nocardia* sp.: incorporation of oxygen-18. *FEBS Lett.* 50:288–90

7. Bartels I, Knackmuss H-J, Reineke W. 1984. Suicide inactivation of catechol 2,3-dioxygenase from *Pseudomonas putida* mt-2 by 3-halocatechols. *Appl. Environ. Microbiol.* 47:500–5
8. Bartilson M, Nordlund I, Shingler V. 1990. Location and organization of the dimethylphenol catabolic genes of *Pseudomonas* CF600. *Mol. Gen. Genet.* 220: 294–300
9. Bayley SA, Morris DW, Broda P. 1979. The relationship of degradative and resistance plasmids of *Pseudomonas* belonging to the same incompatibility group. *Nature* 280:338–39
10. Bayly RC, Wigmore GJ. 1973. Metabolism of phenol and cresols by mutants of *Pseudomonas putida*. *J. Bacteriol.* 113:1112–20
11. Beil S, Happe B, Timmis KN, Pieper DH. 1997. Genetic and biochemical characterization of the broad spectrum chlorobenzene dioxygenase from *Burkholderia* sp. strain PS12. Dechlorination of 1,2,4,5-tetrachlorobenzene. *Eur. J. Biochem.* 247: 190–99
12. Bestetti G, Galli E. 1984. Plasmid-coded degradation of ethylbenzene and 1-phenylethanol in *Pseudomonas fluorescens*. *FEMS Microbiol. Lett.* 21:165–68
13. Bhat MA, Ishida T, Horiike K, Vaidyanathan CS, Nozaki M. 1993. Purification of 3,5-dichlorocatechol 1,2-dioxygenase,

a nonheme iron dioxygenase and a key enzyme in the biodegradation of a herbicide, 2,4-dichlorophenoxyacetic acid (2,4-D), from *Pseudomonas cepacia* CSV90. *Arch. Biochem. Biophys.* 300:738–46

14. Bhat MA, Tsuda M, Horiike K, Nozaki M, Vaidyanathan CS, Nakazawa T. 1994. Identification and characterization of a new plasmid carrying genes for degradation of 2,4-dichlorophenoxyacetate from *Pseudomonas cepacia* CSV90. *Appl. Environ. Microbiol.* 60:307–12

15. Brinkmann U, Reineke W. 1992. Degradation of chlorotoluenes by in vivo constructed hybrid strains: problems of enzyme specificity, induction and prevention of *meta*-pathway. *FEMS Microbiol. Lett.* 96:81–88

16. Broderick JB, O'Halloran TV. 1991. Overproduction, purification, and characterization of chlorocatechol dioxygenase, a non-heme iron dioxygenase with broad substrate tolerance. *Biochemistry* 30:7349–58

17. Cane PA, Williams PA. 1982. The plasmid-coded metabolism of naphthalene and 2-methylnaphthalene in *Pseudomonas* strains: phenotypic changes correlated with structural modification of the plasmid pWW60-1. *J. Gen. Microbiol.* 128:2281–90

18. Carrington B, Lowe A, Shaw LE, Williams PA. 1994. The lower pathway operon for benzoate catabolism in biphenyl-utilizing *Pseudomonas* sp. strain IC and the nucleotide sequence of the *bphE* gene for catechol 2,3-dioxygenase. *Microbiology* 140:499–508

19. Catelani D, Colombi A. 1974. Metabolism of biphenyl. Structure and physicochemical properties of 2-hydroxy-6-oxo-6-phenylhexa-2,4-dienoic acid, the *meta*-cleavage product from 2,3-dihydroxybiphenyl by *Pseudomonas putida*. *Biochem. J.* 143:431–34

20. Catelani D, Colombi A, Sorlini C, Treccani V. 1973. Metabolism of biphenyl. 2-Hydroxy-6-oxo-6-phenylhexa-2,4-dienoate: the *meta*-cleavage product from 2,3-dihydroxybiphenyl by *Pseudomonas putida*. *Biochem. J.* 134:1063–66

21. Catelani D, Sorlini C, Treccani V. 1971. The metabolism of biphenyl by *Pseudomonas putida*. *Experientia* 27:1173–74

22. Chakrabarty AM. 1972. Genetic basis of the biodegradation of salicylate in *Pseudomonas*. *J. Bacteriol.* 112:815–23

23. Chapman PJ. 1988. Constructing microbial strains for degradation of halogenated aromatic hydrocarbons. In *Environmental Biotechnology: Reducing Risks from Environmental Chemicals Through Biotechnology*, ed. GS Omenn, *Basic Life Sciences* 45:81–95. New York: Plenum

24. Chatterjee DK, Chakrabarty AM. 1982. Genetic rearrangements in plasmids specifying total degradation of chlorinated benzoic acids. *Mol. Gen. Genet.* 188:279–85

25. Chatterjee DK, Kellogg ST, Furukawa K, Kilbane JJ, Chakrabarty AM. 1981. Genetic approaches to the problems of toxic chemical pollution. In *Recombinant DNA*, ed. AG Walton, pp. 199–212. Amsterdam: Elsevier.

26. Chatterjee DK, Kellogg ST, Hamada S, Chakrabarty AM. 1981. Plasmid specifying total degradation of 3-chlorobenzoate by a modified *ortho* pathway. *J. Bacteriol.* 146:639–46

27. Chatterjee DK, Kellogg ST, Watkins DR, Chakrabarty AM. 1981. Plasmids in the biodegradation of chlorinated aromatic compounds. In *Molecular Biology, Pathogenicity, and Ecology of Bacterial Plasmids*, ed. SB Levy, RC Clowes, EL Koenig, pp. 519–28. New York: Plenum

28. Chaudhry GR, Huang GH. 1988. Isolation and characterization of a new plasmid from a *Flavobacterium* sp. which carries the genes for degradation of 2,4-dichlorophenoxyacetate. *J. Bacteriol.* 170:3897–902

29. Coco WM, Rothmel RK, Henikoff S, Chakrabarty AM. 1993. Nucleotide sequence and initial functional characterization of the *clcR* gene encoding a LysR family activator of the *clcABD* chlorocatechol operon in *Pseudomonas putida*. *J. Bacteriol.* 175:417–27

30. Datta N, Hedges RW, Shaw EJ, Sykes RB, Richmond MH. 1971. Properties of an R factor from *Pseudomonas aeruginosa*. *J. Bacteriol.* 108:1244–49

31. De Bont JAM, Vorage MJAW, Hartmans S, van den Tweel WJJ. 1986. Microbial degradation of 1,3-dichlorobenzene. *Appl. Environ. Microbiol.* 52:677–80

32. DiMarco AA, Averhoff B, Ornston LN. 1993. Identification of the transcriptional activator *pobR* and characterization of its role in the expression of *pobA*, the structural gene for *p*-hydroxybenzoate hydroxylase in *Acinetobacter calcoaceticus*. *J. Bacteriol.* 175:4499–506

33. Don RH, Pemberton JM. 1981. Properties of six pesticide degradation plasmids isolated from *Alcaligenes eutrophus*. *J. Bacteriol.* 145:681–86

34. Dorn E, Hellwig M, Reineke W, Knackmuss H-J. 1974. Isolation and characterization of a 3-chlorobenzoate degrading

pseudomonad. *Arch. Microbiol.* 99:61–70

35. Dorn E, Knackmuss H-J. 1978. Chemical structure and biodegradability of halogenated aromatic compounds. Two catechol 1,2-dioxygenases from a 3-chlorobenzoate-grown pseudomonad. *Biochem. J.* 174:73–84

36. Dorn E, Knackmuss H-J. 1978. Chemical structure and biodegradability of halogenated aromatic compounds. Substituent effects on 1,2-dioxygenation of catechol. *Biochem. J.* 174:85–94

37. Dunn NW, Gunsalus IC. 1973. Transmissible plasmid coding early enzymes of naphthalene oxidation in *Pseudomonas putida. J. Bacteriol.* 114:974–79

38. Eaton RW, Timmis KN. 1986. Characterization of a plasmid-specified pathway for catabolism of isopropylbenzene in *Pseudomonas putida* RE204. *J. Bacteriol.* 168:123–31

39. Eaton RW, Timmis KN. 1986. Spontaneous deletion of a 20-kilobase DNA segment carrying genes specifying isopropylbenzene metabolism in *Pseudomonas putida* RE204. *J. Bacteriol.* 168:428–30

40. Engelberts K, Schmidt E, Reineke W. 1989. Degradation of *o*-toluate by *Pseudomonas* sp. strain WR401. *FEMS Microbiol. Lett.* 59:35–38

41. Engesser KH, Strubel V, Christoglou K, Fischer P, Rast HG. 1989. Dioxygenolytic cleavage of aryl ether bonds: 1,10-dihydro–1,10-dihydroxyfluoren–9-one, a novel arene dihydrodiol as evidence for angular dioxygenation of dibenzofuran. *FEMS Microbiol. Lett.* 65:205–10

42. Erickson BD, Mondello FJ. 1992. Nucleotide sequencing and transcriptional mapping of the genes encoding biphenyl dioxygenase, a multicomponent polychlorinated-biphenyl-degrading enzyme in *Pseudomonas* strain LB400. *J. Bacteriol.* 174:2903–12

43. Erickson BD, Mondello FJ. 1993. Enhanced biodegradation of polychlorinated biphenyls after site-directed mutagenesis of a biphenyl dioxygenase gene. *Appl. Environ. Microbiol.* 59:3858–62

44. Evans WC, Smith BSW, Fernley HN, Davies JI. 1971. Bacterial metabolism of 2,4-dichlorophenoxyacetate. *Biochem. J.* 122:543–51

45. Filer K, Harker AR. 1997. Identification of the inducing agent of the 2,4-dichlorophenoxyacetic acid pathway encoded by plasmid pJP4. *Appl. Environ. Microbiol.* 63:317–20

46. Fong KP, Goh CB, Tan HM. 1996. Characterization and expression of the plasmid-borne *bedD* gene from *Pseudomonas putida* ML2, which codes for a NAD$^+$-dependent *cis*-benzene dihydrodiol dehydrogenase. *J. Bacteriol.* 178:5592–601

47. Fortnagel P, Harms H, Wittich R-M, Krohn S, Meyer H, et al. 1990. Metabolism of dibenzofuran by *Pseudomonas* sp. strain HH69 and the mixed culture HH27. *Appl. Environ. Microbiol.* 56:1148–56

48. Fortnagel P, Wittich R-M, Harms H, Schmidt S, Franke S, et al. 1989. New bacterial degradation of the biaryl ether structure. Regioselective dioxygenation prompts cleavage of ether bonds. *Naturwissenschaften* 76:523–24

49. Frantz B, Chakrabarty AM. 1987. Organization and nucleotide sequence determination of a gene cluster involved in 3-chlorocatechol degradation. *Proc. Natl. Acad. Sci. USA* 84:4460–64

50. Frazee RW, Livingston DM, LaPorte DC, Lipscomb JD. 1993. Cloning, sequencing, and expression of the *Pseudomonas putida* protocatechuate 3,4-dioxygenase genes. *J. Bacteriol.* 175:6194–202

51. Fukuda M, Yasukouchi Y, Kikuchi Y, Nagata Y, Kimbara K, et al. 1994. Identification of the *bphA* and *bphB* genes of *Pseudomonas* sp. strain KKS102 involved in degradation of biphenyl and polychlorinated biphenyls. *Biochem. Biophys. Res. Commun.* 202:850–56

52. Furukawa K, Arimura N, Miyazaki T. 1987. Nucleotide sequence of the 2,3-dihydroxybiphenyl dioxygenase gene of *Pseudomonas pseudoalcaligenes. J. Bacteriol.* 169:427–29

53. Furukawa K, Chakrabarty AM. 1982. Involvement of plasmids in total degradation of chlorinated biphenyls. *Appl. Environ. Microbiol.* 44:619–26

54. Furukawa K, Kimura N, Iwakiri R, Nishi A, Suyama A. 1996. Construction of hybrid operons conferring expanded capability for degrading aromatic hydrocarbons and chlorinated compounds. In *Molecular Biology of Pseudomonads*, ed. T Nakazawa, K Furukawa, D Haas, S Silver, pp. 81–93. Washington, DC: ASM

55. Gallegos M-T, Michan C, Ramos JL. 1993. The XylS/AraC family of regulators. *Nucleic Acids Res.* 21:807–10

56. Gibson DT, Cardini GE, Maseles FC, Kallio RE. 1970. Incorporation of oxygen-18 into benzene by *Pseudomonas putida. Biochemistry* 9:1631–35

57. Gibson DT, Cruden DL, Haddock JD, Zylstra GJ, Brand JM. 1993. Oxidation of polychlorinated biphenyls by *Pseudo-*

monas sp. strain LB400 and *Pseudomonas pseudoalcaligenes* KF707. *J. Bacteriol.* 175:4561–64

58. Gibson DT, Hensley M, Yoshioka H, Mabry RJ. 1970. Formation of (+)-*cis*–2,3-dihydroxy–1-methylcyclohexa–4,6–diene from toluene by *Pseudomonas putida. Biochemistry* 9:1626–30

59. Gibson DT, Koch JR, Kallio RE. 1968. Oxidative degradation of aromatic hydrocarbons by microorganisms. I. Enzymatic formation of catechol from benzene. *Biochemistry* 7:2653–62

60. Gibson DT, Roberts RL, Wells MC, Kobal VM. 1973. Oxidation of biphenyl by a *Beijerinckia* species. *Biochem. Biophys. Res. Commun.* 50:211–19

61. Gibson DT, Zylstra GJ, Chauhan S. 1990. Biotransformations catalyzed by toluene dioxygenase from *Pseudomonas putida* F1. In *Pseudomonas. Biotransformations, pathogenesis, and evolving biotechnology*, ed. S Silver, AM Chakrabarty, B Iglewski, S Kaplan, pp. 121–32. Washington, DC: ASM

62. Haigler BE, Nishino SF, Spain JC. 1988. Degradation of 1,2-dichlorobenzene by a *Pseudomonas* sp. *Appl. Environ. Microbiol.* 54:294–301

63. Haigler BE, Spain JC. 1989. Degradation of *p*-chlorotoluene by a mutant of *Pseudomonas* sp. strain JS6. *Appl. Environ. Microbiol.* 55:372–79

64. Harayama S, Leppick RA, Rekik M, Mermod N, Lehrbach PR, Timmis KN. 1986. Gene order of the TOL catabolic plasmid upper pathway operon and oxygenation of both toluene and benzyl alcohol by the *xylA* product. *J. Bacteriol.* 167:455–61

65. Harms H, Wittich R-M, Sinnwell V, Meyer H, Fortnagel P, Francke W. 1990. Transformation of dibenzo-*p*-dioxin by *Pseudomonas* sp. strain HH69. *Appl. Environ. Microbiol.* 56:1157–59

66. Hartmann J, Engelberts K, Nordhaus B, Schmidt E, Reineke W. 1989. Degradation of 2-chlorobenzoate by in vivo constructed hybrid pseudomonads. *FEMS Microbiol. Lett.* 61:17–22

67. Harwood CS, Nichols NN, Kim M-K, Ditty JL, Parales RE. 1994. Identification of the *pcaRKF* gene cluster from *Pseudomonas putida*: Involvement in chemotaxis, biodegradation, and transport of 4-hydroxybenzoate. *J. Bacteriol.* 176:6479–88

68. Harwood CS, Parales RE. 1996. The β-ketoadipate pathway and the biology of self-identity. *Annu. Rev. Microbiol.* 50:553–90

68a. Havel J, Kasberg T, Kaschabek SR, Mokroß H, Müller D, Reineke W. 1996. Development of hybrid strains for the mineralization of chloroaromatics. In *Molecular Biology of Pseudomonas*, ed. T Nakazawa, K Furukawa, D Haas, S Silver, pp. 94–120. Washington, DC: ASM

69. Havel J, Reineke W. 1991. Total degradation of various chlorobiphenyls by co-cultures and in vivo constructed hybrid pseudomonads. *FEMS Microbiol. Lett.* 78:163–70

70. Hayase N, Taira K, Furukawa K. 1990. *Pseudomonas putida* KF715 *bphABCD* operon encoding biphenyl and polychlorinated biphenyl degradation: cloning, analysis, and expression of soil bacteria. *J. Bacteriol.* 172:1160–64

71. Higson FK, Focht DD. 1992. Utilization of 3-chloro–2-methylbenzoic acid by *Pseudomonas cepacia* MB2 through the *meta* fission pathway. *Appl. Environ. Microbiol.* 58:2501–4

72. Hinteregger C, Loidl M, Streichsbier F. 1992. Characterization of isofunctional ring-cleaving enzymes in aniline and 3-chloroaniline degradation of *Pseudomonas acidovorans* CA28. *FEMS Microbiol. Lett.* 97:261–66

73. Hirose J, Suyama A, Hayashida S, Furukawa K. 1994. Construction of hybrid biphenyl (*bph*) and toluene (*tod*) genes for functional analysis of aromatic ring dioxygenase. *Gene* 138:27–33

74. Hofer B, Backhaus S, Timmis KN. 1994. The biphenyl/polychlorinated biphenyl-degradation locus (*bph*) of *Pseudomonas* sp. LB400 encodes four additional metabolic enzymes. *Gene* 144:9–16

75. Hofer B, Eltis LD, Dowlings DN, Timmis KN. 1993. Genetic analysis of a *Pseudomonas* locus encoding a pathway for biphenyl/polychlorinated biphenyl degradation. *Gene* 130:47–55

76. Hollender J, Dott W, Hopp J. 1994. Regulation of chloro- and methylphenol degradation in *Comamonas testosteroni* JH5. *Appl. Environ. Microbiol.* 60:2330–38

77. Holtel A, Abril M-A, Marques S, Timmis KN, Ramos JL. 1990. Promoter-upstream activator sequences are required for expression of the *xylS* gene and upper-pathway operon on the *Pseudomonas* TOL plasmid. *Molec. Microbiol.* 4:1551–56

78. Hughes EJ, Shapiro MK, Houghton JE, Ornston LN. 1988. Cloning and expression of *pca* genes from *Pseudomonas putida* in *Escherichia coli. J. Gen. Microbiol.* 134:2877–87

79. Inouye S, Ebina Y, Nakazawa A; Naka-

zawa T. 1984. Nucleotide sequence surrounding transcription initiation site of *xylABC* operon on TOL plasmid of *Pseudomonas putida*. *Proc. Natl. Acad. Sci. USA* 81:1688–91

80. Inouye S, Nakazawa A, Nakazawa T. 1984. Nucleotide sequence of the promoter region of the *xylDEGF* operon on TOL plasmid of *Pseudomonas putida*. *Gene* 29:323–30

81. Inouye S, Nakazawa A, Nakazawa T. 1988. Nucleotide sequence of the regulatory gene *xylR* of the TOL plasmid from *Pseudomonas putida*. *Gene* 66:301–6

82. Irie S, Doi S, Yorifuji T, Takai M, Yano K. 1987. Nucleotide sequencing and characterization of the genes encoding benzene oxidation enzymes of *Pseudomonas putida*. *J. Bacteriol.* 169:5174–79

83. Janke D, Fritsche W. 1979. Dechlorierung von 4-Chlorphenol nach extradioler Ringspaltung durch *Pseudomonas putida*. *Z. Allgem. Mikrobiol.* 19:139–41

84. Jeenes DJ, Reineke W, Knackmuss H-J, Williams PA. 1982. TOL plasmid pWW0 in constructed halobenzoate-degrading *Pseudomonas* strains: enzyme regulation and DNA structure. *J. Bacteriol.* 150:180–87

85. Kasberg T, Daubaras DL, Chakrabarty AM, Kinzelt D, Reineke W. 1995. Evidence that operons *tcb*, *tfd*, and *clc* encode maleylacetate reductase, the fourth enzyme of the modified *ortho* pathway. *J. Bacteriol.* 177:3885–89

86. Kasberg T, Seibert V, Schlömann M, Reineke W. 1997. Cloning, characterization, and sequence analysis of the *clcE* gene encoding maleylacetate reductase of *Pseudomonas* sp. strain B13. *J. Bacteriol.* 179:3801–3

87. Kaschabek SR, Kasberg T, Müller D, Mars AE, Janssen DB, Reineke W. 1998. Degradation of chloroaromatics: purification and characterization of a novel type of chlorocatechol 2,3-dioxygenase of *Pseudomonas putida* GJ31. *J. Bacteriol.* 180:296–302

88. Kaschabek SR, Reineke W. 1992. Maleylacetate reductase of *Pseudomonas* sp. strain B13: dechlorination of chloromaleylacetates, metabolites in the degradation of chloroaromatic compounds. *Arch. Microbiol.* 159:412–17

89. Kaschabek SR, Reineke W. 1995. Maleylacetate reductase of *Pseudomonas* sp. strain B13: specificity of substrate conversion and halide elimination. *J. Bacteriol.* 177:320–25

90. Kikuchi Y, Nagata Y, Hinata M, Kimbara K, Fukuda M, et al. 1994. Identification of the *bphA4* gene encoding ferredoxin reductase involved in biphenyl and polychlorinated biphenyl degradation in *Pseudomonas* sp. strain KKS102. *J. Bacteriol.* 176:1689–94

91. Kikuchi Y, Yasukochi Y, Nagata Y, Fukuda M, Takagi M. 1994. Nucleotide sequence and functional analysis of the *meta*-cleavage pathway involved in biphenyl and polychlorinated biphenyl degradation in *Pseudomonas* sp. strain KKS102. *J. Bacteriol.* 176:4269–76

92. Kilpi S, Backström V, Korhola M. 1983. Degradation of catechol, methylcatechols and chlorocatechols by *Pseudomonas* sp. HV3. *FEMS Microbiol. Lett.* 18:1–5

93. Kimbara K, Hashimoto T, Fukuda M, Koana T, Takagi M, et al. 1989. Cloning and sequencing of two tandem genes involved in degradation of 2,3-dihydroxybiphenyl to benzoic acid in the polychlorinated biphenyl-degrading soil bacterium *Pseudomonas* sp. strain KKS102. *J. Bacteriol.* 171:2740–47

94. Kivisaar M, Horak R, Kasak L, Heinaru A, Habicht J. 1990. Selection of independent plasmids determining phenol degradation in *Pseudomonas putida* and the cloning and expression of genes encoding phenol monooxygenase and catechol 1,2-dioxygenase. *Plasmid* 24:25–36

95. Klečka GM, Gibson DT. 1981. Inhibition of catechol 2,3-dioxygenase from *Pseudomonas putida* by 3-chlorocatechol. *Appl. Environ. Microbiol.* 41:1159–65

96. Kuhm AE, Schlömann M, Knackmuss H-J, Pieper DH. 1990. Purification and characterization of dichloromuconate cycloisomerase from *Alcaligenes eutrophus* JMP134. *Biochem. J.* 266:877–83

97. Latorre J, Reineke W, Knackmuss H-J. 1984. Microbial metabolism of chloroanilines: enhanced evolution by naturally genetic exchange. *Arch. Microbiol.* 140:159–65

98. Lau PCK, Bergeron H, Labbe D, Wang Y, Brousseau R, Gibson DT. 1994. Sequence and expression of the *todGIH* genes involved in the last three steps of toluene degradation by *Pseudomonas putida* F1. *Gene* 146:7–13

99. Lau PCK, Wang Y, Labbe D, Bergeron H, Garnon J. 1996. Two-component signal transduction systems regulating toluene and biphenyl-polychlorinated biphenyl degradations in a soil pseudomonad and an actinomycete. In *Molecular Biology of Pseudomonads*, ed. T Nakazawa, K Furukawa, D Haas, S Silver, pp. 176–87. Washington, DC: ASM

100. Lehrbach PR, Jeenes DJ, Broda P. 1983.

328   REINEKE

Characterization by molecular cloning of insertion mutants in TOL catabolic functions. *Plasmid* 9:112–25

101. Lehrbach PR, McGregor I, Ward JM, Broda P. 1983. Molecular relationships between *Pseudomonas* Inc P-9 degradative plasmids TOL, NAH, and SAL. *Plasmid* 10:164–74

102. Lehrbach PR, Ward JM, Meutien P, Broda P. 1982. Physical mapping of TOL plasmids pWW0 and pND2 and various R plasmid-TOL derivatives from *Pseudomonas* spp. *J. Bacteriol.* 152:1280–83

103. Leveau JHJ, de Vos WM, van der Meer JR. 1994. Analysis of the binding site of the LysR-type transcriptional activator TcbR on the *tcbR* and *tcbC* divergent promoter sequences. *J. Bacteriol.* 176:1850–56

104. Liu T, Chapman PJ. 1983. Degradation of halogenated aromatic acids and hydrocarbons by *Pseudomonas putida*. *Annu. Meet. Am. Soc. Microbiol.* K211, p. 212 (Abstr.)

105. Lloyd-Jones G, de Jong C, Ogden RC, Duetz WA, Williams PA. 1994. Recombination of *bph* (biphenyl) catabolic genes from plasmid pWW100 and their deletion during growth on benzoate. *Appl. Environ. Microbiol.* 60:691–96

106. Lunt D, Evans WC. 1970. The microbial metabolism of biphenyl. *Biochem. J.* 118:54p

107. Mars AE, Kasberg T, Kaschabek SR, van Agteren MH, Janssen DB, Reineke W. 1997. Microbial degradation of chloroaromatics: use of the *meta*-cleavage pathway for mineralization of chlorobenzene. *J. Bacteriol.* 179:4530–37

108. Matrubutham U, Harker AR. 1994. Analysis of duplicated gene sequence associated with *tfdR* and *tfdS* in *Alcaligenes eutrophus* JMP134. *J. Bacteriol.* 176:2348–53

109. McCullar MV, Brenner V, Adams RH, Focht DD. 1994. Construction of a novel polychlorinated biphenyl-degrading bacterium: utilization of 3,4′-dichlorobiphenyl by *Pseudomonas acidovorans* M3GY. *Appl. Environ. Microbiol.* 60:3833–39

110. McFall SM, Parsek MR, Chakrabarty AM. 1997. 2-Chloromuconate and ClcR-mediated activation of the *clcABD* operon: in vitro transcriptional and DNase I footprint analyses. *J. Bacteriol.* 179:3655–63

111. Menn F-M, Zylstra GJ, Gibson DT. 1991. Location and sequence of the *todF* gene encoding 2-hydroxy–6-oxohepta–2,4-dienoate hydrolase in *Pseudomonas putida* F1. *Gene* 104:91–94

112. Mermod N, Lehrbach PR, Reineke W, Timmis KN. 1984. Transcription of the TOL plasmid toluate catabolic pathway operon of *Pseudomonas putida* is determined by a pair of co-ordinately and positively overlapping promoters. *EMBO J.* 3:2461–66

113. Miguez CB, Greer CW, Ingram JM. 1993. Purification and properties of chlorocatechol 1,2-dioxygenase from *Alcaligenes denitrificans* BRI 6011. *Can. J. Microbiol.* 39:1–5

114. Mokross H, Schmidt E, Reineke W. 1990. Degradation of 3-chlorobiphenyl by in vivo constructed hybrid pseudomonads. *FEMS Microbiol. Lett.* 71:179–86

115. Müller D, Schlömann M, Reineke W. 1996. Maleylacetate reductases in chloroaromatic-degrading bacteria using the modified *ortho* pathway: comparison of catalytic properties. *J. Bacteriol.* 178:298–300

116. Ogawa N, Miyashita K. 1995. Recombination of a 3-chlorobenzoate catabolic plasmid from *Alcaligenes eutrophus* NH9 mediated by direct repeat elements. *Appl. Environ. Microbiol.* 61:3788–95

117. Oltmanns RH, Rast HG, Reineke W. 1988. Degradation of 1,4-dichlorobenzene by constructed and enriched strains. *Appl. Microbiol. Biotechnol.* 28:609–16

118. Ornston LN. 1966. The conversion of catechol and protocatechuate to β-ketoadipate by *Pseudomonas putida*. IV. Regulation. *J. Biol. Chem.* 241:3800–10

119. Parales RE, Harwood CS. 1992. Characterization of the genes encoding β-ketoadipate:succinyl-coenzyme A transferase in *Pseudomonas putida*. *J. Bacteriol.* 174:4657–66

120. Parales RE, Harwood CS. 1993. Regulation of *pcaIJ* genes for aromatic acid degradation in *Pseudomonas putida*. *J. Bacteriol.* 175:5829–38

121. Perkins EJ, Gordon MP, Caceres O, Lurquin PF. 1990. Organization and sequence analysis of the 2,4-dichlorophenol hydroxylase and dichlorocatechol oxidative operons of plasmid pJP4. *J. Bacteriol.* 172:2351–59

122. Pieper DH, Reineke W, Engesser K-H, Knackmuss H-J. 1988. Metabolism of 2,4-dichlorophenoxyacetic acid, 4-chloro–2-methylphenoxyacetic acid and 2-methylphenoxyacetic acid by *Alcaligenes eutrophus* JMP 134. *Arch. Microbiol.* 150:95–102

123. Pothuluri JV, Cerniglia CE. 1994. Microbial metabolism of polycyclic aromatic hydrocarbons. In *Biological Degradation and Bioremediation of Toxic Chemicals,*

ed. GR Chaudhry, pp. 92–124. Portland, OR: Dioscorides

124. Ramos JL, Mermod N, Timmis KN. 1987. Regulatory circuits controlling transcription of TOL plasmid operon encoding *meta*-cleavage pathway for degradation of alkylbenzoates by *Pseudomonas*. *Mol. Microbiol.* 1:293–300

125. Ramos JL, Michan C, Rojo F, Dwyer D, Timmis KN. 1990. Signal-regulator interactions. Genetic analysis of the effector binding site of XylS, the benzoate-activated positive regulator of *Pseudomonas* TOL plasmid *meta*-cleavage pathway operon. *J. Mol. Biol.* 211:373–82

126. Ramos JL, Stolz A, Reineke W, Timmis KN. 1986. New effector specificities in regulators of gene expression: TOL plasmid *xylS* mutants and their use to engineer expansion of the range of aromatics degraded by bacteria. *Proc. Natl. Acad. Sci. USA* 83:8467–71

127. Ramos-Gonzalez M-I, Duque E, Ramos JL. 1991. Conjugational transfer of recombinant DNA in cultures and in soils: host range of *Pseudomonas putida* TOL plasmids. *Appl. Environ. Microbiol.* 57: 3020–27

128. Reineke W, Jeenes DJ, Williams PA, Knackmuss H-J. 1982. TOL plasmid pWW0 in constructed halobenzoate-degrading *Pseudomonas* strains: prevention of meta pathway. *J. Bacteriol.* 150:195–201

129. Reineke W, Knackmuss H-J. 1978. Chemical structure and biodegradability of halogenated aromatic compounds. Substituent effects on 1,2-dioxygenation of benzoic acid. *Biochim. Biophys. Acta* 542: 412–23

130. Reineke W, Knackmuss H-J. 1978. Chemical structure and biodegradability of halogenated aromatic compounds. Substituent effects on dehydrogenation of 3,5-cyclohexadiene–1,2-diol–1-carboxylic acid. *Biochim. Biophys. Acta* 542: 424–29

131. Reineke W, Knackmuss H-J. 1979. Construction of haloaromatics utilising bacteria. *Nature* 277:385–86

132. Reineke W, Knackmuss H-J. 1980. Hybrid pathway for chlorobenzoate metabolism in *Pseudomonas* sp. B13 derivatives. *J. Bacteriol.* 142:467–73

133. Reineke W, Knackmuss H-J. 1984. Microbial metabolism of haloaromatics: isolation and properties of a chlorobenzene-degrading bacterium. *Appl. Environ. Microbiol.* 47:395–402

134. Reineke W, Wessels SW, Rubio MA, Latorre J, Schwien U, et al. 1982. Degra-

dation of monochlorinated aromatics following transfer of genes encoding chlorocatechol catabolism. *FEMS Microbiol. Lett.* 14:291–94

135. Reiner AM. 1971. Metabolism of benzoic acid by bacteria: 3,5-cyclohexadiene–1,2-diol–1-carboxylic acid is an intermediate in the formation of catechol. *J. Bacteriol.* 108:89–94

136. Reiner AM, Hegeman GD. 1971. Metabolism of benzoic acid by bacteria. Accumulation of (−)-3,5-cyclohexadiene–1,2-diol–1-carboxylic acid by a mutant strain of *Alcaligenes eutrophus*. *Biochemistry* 10:2530–36

137. Romero-Steiner S, Parales RE, Harwood CS, Houghton JE. 1994. Characterization of the *pcaR* regulatory gene from *Pseudomonas putida*, which is required for the complete degradation of *p*-hydroxybenzoate. *J. Bacteriol.* 176: 5771–79

138. Sander P, Wittich R-M, Fortnagel P, Wilkes H, Francke W. 1991. Degradation of 1,2,4-trichloro- and 1,2,4,5-tetrachlorobenzene by *Pseudomonas* strains. *Appl. Environ. Microbiol.* 57:1430–40

139. Schell MA. 1983. Cloning and expression in *Escherichia coli* of the naphthalene degradation genes from plasmid NAH7. *J. Bacteriol.* 153:822–29

140. Schlömann M. 1995. *Die Evolution des bakteriellen Chloraromaten-Abbaus: Untersuchungen zum Chlorbrenzkatechin-Zweig des 3-Oxoadipat-Weges.* Habilitation thesis. University of Stuttgart, Stuttgart, Germany

141. Schmidt E, Knackmuss H-J. 1980. Chemical structure and biodegradability of halogenated aromatic compounds. Conversion of chlorinated muconic acids into maleoylacetic acid. *Biochem. J.* 192: 339–47

142. Schraa G, Boone ML, Jetten MSM, van Neerven ARW, Colberg PJ, Zehnder AJB. 1986. Degradation of 1,4-dichlorobenzene by *Alcaligenes* sp. strain A175. *Appl. Environ. Microbiol.* 52:1374–81

143. Schwien U, Schmidt E. 1982. Improved degradation of monochlorophenols by a constructed strain. *Appl. Environ. Microbiol.* 44:33–39

144. Schwien U, Schmidt E, Knackmuss H-J, Reineke W. 1988. Degradation of chlorosubstituted aromatic compounds by *Pseudomonas* sp. strain B13: fate of 3,5-dichlorocatechol. *Arch. Microbiol.* 150:78–84

145. Serdar CM, Gibson DT. 1989. Isolation and characterization of altered plasmids in mutant strains of *Pseudomonas*

putida NCIB 9816. *Biochem. Biophys. Res. Commun.* 164:764–71

146. Shingler V, Franklin FCH, Tsuda M, Holroyd D, Bagdasarian M. 1989. Molecular analysis of a plasmid-encoded phenol hydroxylase from *Pseudomonas* CF600. *J. Gen. Microbiol.* 135:1083–92

147. Spain JC, Nishino SF. 1987. Degradation of 1,4-dichlorobenzene by a *Pseudomonas* sp. *Appl. Environ. Microbiol.* 53:1010–19

148. Spiess E, Görisch H. 1996. Purification and characterization of chlorobenzene *cis*-dihydrodiol dehydrogenase from *Xanthobacter flavus* 14p1. *Arch. Microbiol.* 165:201–5

149. Spiess E, Sommer C, Görisch H. 1995. Degradation of 1,4-dichlorobenzene by *Xanthobacter flavus* 14p1. *Appl. Environ. Microbiol.* 61:3884–88

150. Springael D, van Thor J, Goorissen H, Ryngaert A, de Baere R, et al. 1996. RP4::Mu3A-mediated *in vivo* cloning and transfer of a chlorobiphenyl catabolic pathway. *Microbiology* 142:3283–93

151. Stanier RY, Palleroni NJ, Doudoroff M. 1966. The aerobic *Pseudomonas*: a taxonomic study. *J. Gen. Microbiol.* 43:159–271

152. Stanisich VA, Richmond MH. 1975. Gene transfer in the genus *Pseudomonas*. In *Genetics and Biochemistry of Pseudomonas*, ed. PH Clarke, MH Richmond, pp. 163–90. London: Wiley

153. Strubel V, Engesser KH, Fischer P, Knackmuss H-J. 1991. 3-(2-Hydroxyphenyl)catechol as substrate for proximal *meta* ring cleavage in dibenzofuran degradation by *Brevibacterium* sp. strain DPO 1361. *J. Bacteriol.* 173:1932–37

154. Taira K, Hayase N, Arimura N, Yamashita S, Miyazaki T, Furukawa K. 1988. Cloning and nucleotide sequence of the 2,3-dihydroxybiphenyl dioxygenase gene from the PCB-degrading strain of *Pseudomonas paucimobilis* Q1. *Biochemistry* 27:3990–96

155. Taira K, Hirose J, Hayashida S, Furukawa K. 1992. Analysis of *bph* operon from the polychlorinated biphenyl-degrading strain of *Pseudomonas pseudoalcaligenes* KF707. *J. Biol. Chem.* 267:4844–53

156. Tan HM, Fong KP. 1993. Molecular analysis of the plasmid-borne *bed* gene cluster from *Pseudomonas putida* ML2 and cloning of the *cis*-benzene dihydrodiol dehydrogenase gene. *Can. J. Microbiol.* 39:357–62

157. Tan H-M, Mason JR. 1990. Cloning and expression of the plasmid-encoded benzene dioxygenase genes from *Pseudomonas putida* ML2. *FEMS Microbiol. Lett.* 72:259–64

158. Tan H-M, Tang H-Y, Joannou CL, Abdel-Wahab NH, Mason JR. 1993. The *Pseudomonas putida* ML2 plasmid-encoded genes for benzene dioxygenase are unusual in codon usage and low in G+C content. *Gene* 130:33–39

159. Tiedje JM, Duxbury JM, Alexander M, Dawson JE. 1969. 2,4-D metabolism: pathway of degradation of chlorocatechols by *Arthrobacter* sp. *J. Agric. Food Chem.* 17:1021–26

160. Timmis KN, Lehrbach PR, Harayama S, Don RH, Mermod N, et al. 1985. Analysis and manipulation of plasmid encoded pathways for the catabolism of aromatic compounds by soil bacteria. In *Plasmids in Bacteria*, ed. DR Helinski, SN Cohen, DB Clewell, DA Jackson, A Hollaender, pp. 719–39. New York: Plenum

161. Tomasek PH, Frantz B, Sangodkar UMX, Haugland RA, Chakrabarty AM. 1989. Characterization and nucleotide sequence determination of a repeat element isolated from a 2,4,5-T degrading strain of *Pseudomonas cepacia*. *Gene* 76:227–38

162. van der Meer JR, de Vos WM, Harayama S, Zehnder AJB. 1992. Molecular mechanisms of genetic adaptation to xenobiotic compounds. *Microbiol. Rev.* 56:677–94

163. van der Meer JR, Eggen RIL, Zehnder AJB, de Vos WM. 1991. Sequence analysis of the *Pseudomonas* sp. strain P51 *tcb* gene cluster, which encodes metabolism of chlorinated catechols: evidence for specialization of catechol 1,2-dioxygenases for chlorinated substrates. *J. Bacteriol.* 173:2425–34

164. van der Meer JR, Frijters ACJ, Leveau JHJ, Eggen RIL, Zehnder AJB, de Vos WM. 1991. Characterization of the *Pseudomonas* sp. strain P51 gene *tcbR*, a LysR-type transcriptional activator of the *tcbCDEF* chlorocatechol oxidation operon, and analysis of the regulatory region. *J. Bacteriol.* 173:3700–8

165. van der Meer JR, van Neerven ARW, de Vries EJ, de Vos WM, Zehnder AJB. 1991. Cloning and characterization of plasmid-encoded genes for the degradation of 1,2-dichloro-, 1,4-dichloro-, and 1,2,4-trichlorobenzene of *Pseudomonas* sp. strain P51. *J. Bacteriol.* 173:6–15

166. van der Meer JR, Zehnder AJB, de Vos WM. 1991. Identification of a novel composite transposable element, Tn*5280*, carrying chlorobenzene dioxygenase genes of *Pseudomonas* sp. strain P51. *J. Bacteriol.* 173:7077–83

167. Vollmer MD, Fischer P, Knackmuss H-J, Schlömann M. 1994. Inability of muconate cycloisomerases to cause dehalogenation during conversion of 2-chloro-*cis,cis*-muconate. *J. Bacteriol.* 176:4366–75

168. Vollmer MD, Schlömann M. 1995. Conversion of 2-chloro-*cis,cis*-muconate and its metabolites 2-chloro- and 5-chloromuconolactone by chloromuconate cycloisomerase of pJP4 and pAC27. *J. Bacteriol.* 177:2938–41

169. Vollmer MD, Stadler-Fritzsche K, Schlömann M. 1993. Conversion of 2-chloromaleylacetate in *Alcaligenes eutrophus* JMP134. *Arch. Microbiol.* 159:182–88

170. Wang Y, Rawlings M, Gibson DT, Labbe D, Bergeron H, et al. 1995. Identification of a membrane protein and a truncated LysR-type regulator associated with the toluene degradation pathway in *Pseudomonas putida* F1. *Mol. Gen. Genet.* 246:570–79

171. Werlen C, Kohler H-PE, van der Meer JR. 1996. The broad substrate chlorobenzene dioxygenase and *cis*-chlorobenzene dihydrodiol dehydrogenase of *Pseudomonas* sp. strain P51 are linked evolutionarily to the enzymes for benzene and toluene degradation. *J. Biol. Chem.* 271:4009–16

172. Williams PA, Murray K. 1974. Metabolism of benzoate and methylbenzoates by *Pseudomonas putida* (*arvilla*) mt-2: evidence for the existence of a TOL plasmid. *J. Bacteriol.* 120:416–23

173. Williams PA, Sayers JR. 1994. The evolution of pathways for aromatic hydrocarbon oxidation in *Pseudomonas*. *Biodegradation* 5:195–217

174. Williams SE, Woolridge EM, Ransom SC, Landro JA, Babbitt PC, Kozarich JW. 1992. 3-Carboxy-*cis,cis*-muconate lactonizing enzyme from *Pseudomonas putida* is homologous to the class II fumarase family: a new reaction in the evolution of a mechanistic motif. *Biochemistry* 31:9768–76

175. Winstanley C, Taylor SC, Williams PA. 1987. pWW174: a large plasmid from *Acinetobacter calcoaceticus* encoding benzene catabolism by the β-ketoadipate pathway. *Mol. Microbiol.* 1:219–27

176. Wittich R-M, Wilkes H, Sinnwell V, Francke W, Fortnagel P. 1992. Metabolism of dibenzo-*p*-dioxin by *Sphingomonas* sp. strain RW1. *Appl. Environ. Microbiol.* 58:1005–10

177. Worsey MJ, Williams PA. 1975. Metabolism of toluene and xylenes by *Pseudomonas putida* (*arvilla*) mt-2: evidence for a new function of the TOL plasmid. *J. Bacteriol.* 124:7–13

178. Wyndham RC, Cashore AE, Nakatsu CH, Peel MC. 1994. Catabolic transposons. *Biodegradation* 5:323–42

179. Wyndham RC, Singh RK, Straus NA. 1988. Catabolic instability, plasmid gene deletion and recombination in *Alcaligenes* sp. BR60. *Arch. Microbiol.* 150:237–43

180. Yen K-M, Gunsalus IC. 1983. Plasmid gene organization: naphthalene/salicylate oxidation. *Proc. Natl. Acad. Sci. USA* 79:874–78

181. Yen K-M, Sullivan M, Gunsalus IC. 1983. Electron microscope heteroduplex mapping of naphthalene oxidation genes on the NAH7 and SAL1 plasmids. *Plasmid* 9:105–11

182. Zylstra GJ, Gibson DT. 1989. Toluene degradation by *Pseudomonas putida* F1. Nucleotide sequence of the *todC1C2-BADE* genes and their expression in *Escherichia coli*. *J. Biol. Chem.* 264:14940–46

Annu. Rev. Microbiol. 1998. 52:333–60
Copyright © 1998 by Annual Reviews. All rights reserved

VIRULENCE GENES OF
CLOSTRIDIUM PERFRINGENS

Julian I. Rood

Department of Microbiology, Monash University, Clayton 3168, Australia;
e-mail: Julian.Rood@med.monash.edu.au

KEY WORDS: genome, toxin, pathogenesis, regulation, two-component signal transduction

ABSTRACT

Clostridium perfringens causes human gas gangrene and food poisoning as well as several enterotoxemic diseases of animals. The organism is characterized by its ability to produce numerous extracellular toxins including α-toxin or phospholipase C, θ-toxin or perfringolysin O, κ-toxin or collagenase, as well as a sporulation-associated enterotoxin. Although the genes encoding the α-toxin and θ-toxin are located on the chromosome, the genes encoding many of the other extracellular toxins are located on large plasmids. The enterotoxin gene can be either chromosomal or plasmid determined. Several of these toxin genes are associated with insertion sequences. The production of many of the extracellular toxins is regulated at the transcriptional level by the products of the *virR* and *virS* genes, which together comprise a two-component signal transduction system.

CONTENTS

0066-4227/98/1001-0333$08.00

INTRODUCTION

The genus *Clostridium* consists of a diverse group of primarily Gram-positive anaerobic bacteria distinguished by their ability to form heat-resistant endospores (113). These organisms cause diseases such as botulism, tetanus, gas gangrene, and pseudomembranous colitis—diseases generally characterized by the involvement of potent extracellular toxins (1, 48, 115).

Clostridium perfringens is a normal inhabitant of the gastrointestinal tract of humans and animals and is commonly found in the soil. It causes gas gangrene and food poisoning in humans and several enterotoxemic diseases in domestic animals (93, 111). It is characterized by its ability to produce numerous extracellular toxins and enzymes including α-toxin, β-toxin, ε-toxin, θ-toxin, κ-toxin, λ-toxin, ι-toxin, μ-toxin, and sialidase. Individual isolates can be divided into five distinct toxin types, A to E, based on the toxins they produce (93, 111). *C. perfringens* is the paradigm species for genetic studies on the pathogenic clostridia, primarily because of its oxygen tolerance, relatively fast growth rate, and ability to be genetically manipulated (92). This review focuses primarily on the genetic organization and regulation of the genes encoding extracellular toxin production in *C. perfringens*, but it starts with a brief review of the pathogenesis of the major diseases caused by this organism.

PATHOGENESIS OF INFECTIOUS DISEASES CAUSED BY *C. PERFRINGENS*

Clostridial Myonecrosis or Gas Gangrene

Gas gangrene is one of the classical bacterial diseases of humans. The pathogenesis of the disease involves the entry of vegetative cells or spores of *C. perfringens* type A into the body cavity via a major traumatic injury or a surgical wound. These organisms originate from soil that has contaminated the wound or from leakage of intestinal contents as a result of damage to the gastrointestinal tract. Vascular damage resulting from the injury facilitates establishment of the anaerobic conditions required for germination of *C. perfringens* spores and growth of the resultant vegetative cells. Growth is accompanied by the production of extracellular toxins and leads to gas production, extensive necrosis, and tissue damage. If not controlled quickly by a combination of antibiotic treatment and surgical intervention, the patient rapidly develops systemic toxemia and shock; death results from the effects of the toxins on the hemodynamic systems of the body (7, 67, 115).

The primary toxin implicated in the disease is the α-toxin, the first bacterial toxin demonstrated to have enzymatic activity (66). *C. perfringens* α-toxin is a lethal dermonecrotic toxin that has both phospholipase C and sphingomyelinase activities (119, 120). Although for many years it was believed to be the major toxin implicated in gas gangrene, definitive proof of its role was obtained only recently (2, 135). Vaccine studies showed that it was possible to protect mice against experimental infection with lethal doses of *C. perfringens* type A by immunization with a recombinant *Escherichia coli*–derived protein comprising the C-terminal domain of the α-toxin (135). Furthermore, genetic studies carried out in this laboratory have shown that a *C. perfringens* mutant in which the α-toxin structural gene, *plc* (or *cpa*), had been insertionally inactivated by homologous recombination, was unable to cause clostridial myonecrosis in mice (2). Virulence was restored when the *plc* mutation was complemented by a plasmid carrying the wild-type *plc* gene, indicating that α-toxin is essential for the disease process. Although there is no definitive evidence, other toxins including the θ-toxin (perfringolysin O), κ-toxin (collagenase), and μ-toxin (hyaluronidase), have been implicated in the disease (93). To date, additional defined mutants have been constructed only for the *pfoA* gene, which encodes the θ-toxin. These mutants are still virulent in the mouse myonecrosis model, although the data do suggest that θ-toxin may play a role in the disease process (2, 116).

Food Poisoning

C. perfringens is also one of the major bacterial causes of human food poisoning. Improper heating, storage, or reheating of meat, or food containing meat, leads to the germination of contaminating *C. perfringens* spores and the rapid growth of the resultant vegetative cells. Ingestion of this contaminated food leads to infection of the gastrointestinal tract (48). The *C. perfringens* isolates that cause human food poisoning carry the *cpe* gene, which encodes a sporulation-associated enterotoxin (74). Sporulation in the intestine leads to the production of the enterotoxin, which causes fluid secretion into the lumen and subsequent diarrhea. The disease is usually fairly self-limiting and in healthy individuals resolves within one or two days. Elderly or debilitated patients may be affected more severely. Not all isolates of *C. perfringens* can cause food poisoning because not all isolates carry the *cpe* gene and, therefore, have the ability to produce enterotoxin (48, 74).

The enterotoxin is a single 35.3-kDa heat-labile polypeptide that acts on epithelial cells of the gastrointestinal tract to cause fluid and electrolyte loss. It is different from cholera and *E. coli* enterotoxins in that it does not affect cAMP levels. The toxin is cytotoxic, damaging intestinal epithelial cells at the tips of the villi (56, 73).

Enteritis Necroticans

C. perfringens type C is responsible for a very rare form of enteritis known as enteritis necroticans (61, 78). The disease is characterized by vomiting, diarrhea, severe abdominal pain, and the presence of blood in the stools. Death may occur from intestinal obstruction or systemic toxemia. The disease was known as Darmbrand in Germany at the end of World War II and as the endemic children's disease Pig Bel in Papua New Guinea in the 1950s and 1960s. It has been reported more recently in Vietnam. Nutritional and social factors are very important in the epidemiology of the disease. Enteritis necroticans is associated with the ingestion of high protein meat meals by undernourished or protein-deficient individuals. The sudden change of diet results in the rapid growth of *C. perfringens* type-C cells present in either the endogenous flora or in the meat, and concomitant extracellular toxin production (59, 61).

The toxin implicated in enteritis necroticans is the *C. perfringens* β-toxin, an extracellular toxin that is inactivated rapidly in the gastrointestinal tract by trypsin. Low-protein diets lead to a reduction in the level of pancreatic enzymes such as trypsin and, therefore, to reduced β-toxin inactivation. In addition, the sweet potato is the dietary staple in endemic Pig Bel areas of Papua New Guinea and is often consumed with pig meat in traditional feasts. Sweet potatoes contain large amounts of a trypsin inhibitor. The presence of this inhibitor, combined with the reduced levels of trypsin in the gastrointestinal tract, prevents the normal inactivation of any β-toxin that might be produced (59, 61). Although its precise mode of action is not known, β-toxin is a lethal necrotizing toxin that causes hemorrhagic necrosis and significant destruction of the intestinal villi. The susceptibility of the toxin to tryptic hydrolysis explains why predisposing dietary conditions are required if β-toxin-producing strains of *C. perfringens* are to cause enteritis necroticans (60, 101).

Diseases of Domestic Animals

C. perfringens strains of most toxin types cause a variety of economically significant diseases in domestic animals (110, 111). These diseases include several enteric syndromes such as fowl necrotic enteritis, bovine and ovine enterotoxemia, and lamb dysentery. Both α-toxin and β-toxin have been implicated in necrotic enteritis. Although the precise pathogenic mechanisms are not known, β-toxin is likely to be of importance in infections caused by type-C strains. The sudden death syndrome known as ovine enterotoxemia, or pulpy kidney disease, is caused by type-D strains of *C. perfringens* and is associated with the change from a low to a high protein feed. The disease is mediated by the production of ε-toxin, which, after tetanus and botulinum toxins, is the most potent toxin produced by the clostridia (84). The toxin is produced as an inactive prototoxin and is activated by proteolysis in the gastrointestinal tract. Although its mode of action is not known, active ε-toxin appears to be a potent neurotoxin that causses edema in the brain and also affects vascular permeability (84).

GENETIC ORGANIZATION OF EXTRACELLULAR TOXIN GENES

Organization of the C. perfringens *Chromosome*

C. perfringens was the first Gram-positive bacterium for which a genetic map was elucidated (11). Pulsed-field gel electrophoresis revealed that strain CPN50 had a single circular 3.6-Mb chromosome and led to the mapping of 24 genes or gene regions. In an arrangement similar to that of *Bacillus subtilis*, the 10 ribosomal RNA operons are located on either side of the origin of chromosomal replication, *oriC*, and are confined to only one-third of the chromosome (11, 17). The latest published map of strain CPN50 has approximately 100 mapped markers (54) (Figure 1). Analysis of this map revealed that several of the genes encoding extracellular toxins and enzymes are clustered within a 250-kb region near *oriC*. These genes include *plc, pfoA, colA* (which encodes the κ-toxin), and *nagH* (the putative μ-toxin gene). Subsequent studies showed that the *pfoA* and *colA* genes are located within 10 kb of each other, although the intervening genes do not appear to be involved in virulence (82). The *nagH* gene is located on the same 30-kb *ApaI-SmaI* restriction fragment as *colA* (54). The *nanH* and *nanI* genes, which encode distinct sialidases of unknown importance in virulence, are located within 200 kb of each other at 0.8 kb and 1.0 kb, respectively, on the CPN50 map (17, 54). Another 200 kb from *nanI*, there is a gene, *pspA*, a homologue of which encodes a putative surface protein that is a virulence factor in *Streptococcus pneumoniae*.

Other potential virulence genes that have been mapped were identified by genome scanning. These genes include *pfoS*, which has similarity to *pfoR*, a gene located immediately upstream of *pfoA* and thought to encode a *pfoA* activator, and *copR*, which has similarity to a putative transcriptional activator from *Pseudomonas syringae* (54). There is no evidence that *pfoS, copR*, or *pspA* are involved in virulence in *C. perfringens*. Note that the *virRS* operon, the

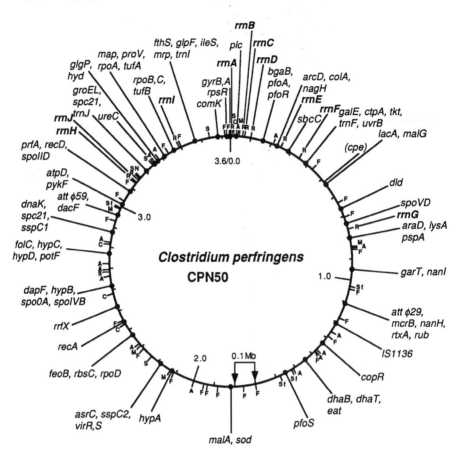

Figure 1 Detailed physical and genetic map of the 3.6-Mb *C. perfringens* strain CPN50 chromosome. Cleavage sites for the enzymes *Apa*I (A), *Sac*II (C), *Fsp*I (F), *Mlu*I (M), *Nru*I (N), the repeated 2*Sac*II-*Sma*I-*Nru*I-2*Sma*I-*Nru*I-I-*Ceu*I locus (R), *Sma*I (S), and *Sfi*I (Sf) are shown. The relative positions of genes located on the same fragment are arbitrary. Reproduced from Katayama and coworkers (54) with the permission of the authors and the American Society for Microbiology.

products of which are involved in the regulation of extracellular toxin production and virulence, is located near the putative replication terminus, on the opposite side of the chromosome to the virulence gene region (54).

Similar methods were used to compare the genomic organization of the virulence gene regions of 10 diverse *C. perfringens* isolates, representing all 5 toxin types (13). Three hypervariable regions were identified. One of these regions spans the *plc* gene locus; this gene is present at the same site in all strains, but the size of the region between the *plc* gene and *rrnA* varies. Similarly, variation was observed in the *pfoA-nagH* region, although their map positions remained unaltered. It is not known why the *plc, pfoA, colA*, and *nagH* genes are located in more variable chromosomal regions. The final hypervariable region encompasses the *cpe* gene, which encodes the enterotoxin. Only one of the strains tested produced enterotoxin; the remaining strains did not carry the *cpe* gene. One of the major conclusions of this study was that the five toxin types of *C. perfringens* have a very similar genomic organization and that the ability to produce different toxins reflects the acquisition or loss of specific toxin genes (11).

Location of Several Toxin Genes on Extrachromosomal Elements

Preliminary evidence suggested that both the ε-toxin structural gene, *etx*, and the β-toxin structural gene, *cpb*, were located on large plasmids (13, 25). Subsequent reports indicated that many of the extracellular toxin genes are plasmid determined (53). These studies involved digestion with the intron-encoded endonuclease I-*Ceu*I, which cuts only within rRNA operons found on the chromosome. The analysis of I-*Ceu*I digests of 16 *C. perfringens* isolates confirmed that the *plc, pfoA, colA,* and *nagH* genes are invariably located on the chromosome. By contrast, the *etx, cpb, iap/ibp* (ι-toxin), lam (λ-toxin), and urease genes are located on large extrachromosomal elements that do not contain I-*Ceu*I sites (26, 53). Clearly, there is a need for further studies that involve the detailed analysis of such plasmids.

The differentiation of *C. perfringens* type-A isolates from isolates of types B to E is dependent on the ability of the latter isolates to produce various combinations of β-toxin, ε-toxin, and ι-toxin (93). These differences in toxin type appear to reflect the acquisition or loss of extrachromosomal elements that contain the structural genes encoding these extracellular toxins (53).

The α-Toxin Structural Gene, plc

Different *C. perfringeens* isolates produce different amounts of α-toxin, with type-A strains generally producing higher levels (10, 55, 129). The *plc* gene was cloned originally from the type-A strains NCTC8237 (63, 83, 122, 128) and NCTC8798 (100). More recent studies have involved the cloning and sequence analysis of *plc* genes from other *C. perfringens* toxin types (29, 55, 129). As expected these *plc* genes, and their deduced α-toxin sequences, are very similar. Comparison of the nucleotide sequences of 10 *plc* genes, from all 5 toxin types, has enabled a phylogenetic tree to be constructed (129).

The relationship between these *plc* genes does not depend on the parent strain's toxin type (129). Purification of the recombinant α-toxin proteins derived from several of these strains revealed that they have very similar specific

```
CTAAGTTAAA   ACCTGTTTTT   GATTGAAAAT  TTTTATTATC  CATATTAAAA   50
 E  L  N  F   G  T  K      S  Q  F    N  K  N  D   M   ORF2
                                                           ◄────

TCCTTTGCCT   TATAATTTAT   TTCAAATTTT  ATTCCATCCC  TTATATTATG  100

TGTAAAAATT   CTTATTAAAT   TAAAAAACAA  GATTTAACTT  ATTATAGCAC  150

TAATAATTGT   AAATTTTCAT   ATTAAAAATA  AGTTTAACAA  TTTAGAGTGG  200
  -10                       -35
GTAAGGTTAG   ATATGTTTAA   TTGAAATTTG  AATTGTATTC  AAAAATATTT  250
                                      ───────────────►
TAAAAAAATAT  TCAAAAATTT   AGTGAGCTTA  TGGTAATTAT  ATGGTATAAT  300
───────────────────►       -35                         -10
TTCAGTGCAA   GTGTTAATCG   TTATCAAAAA  AGGGGAGATT  AATACTTGAA  350
    └►
AAAAATTAAC   GGGGGATATA   AAAAATGAAA  AGAAAG
              plc          M   K   R  K
              ───►
```

Figure 2 The upstream *plc* gene region. The sequence of the region located between ORF2 and the *plc* gene in strain NCTC8237 (Accession No. X62825) is shown (55, 125). The amino acid sequences of the N-termini of both gene products are also shown. The putative −35 and −10 promoter sequences are underlined. The transcriptional start-point of the *plc* gene is shown by the bent arrow. The directly repeated sequence is indicated by the long arrows. The poly(A) tracts are shown by large bold letters.

activities, which implies that the observed differences in α-toxin levels of the respective parent strains reflect differences in the regulation of *plc* gene expression rather than amino acid sequence changes. This conclusion was supported by quantitative mRNA determinations, which showed that the two strains that produce higher levels of α-toxin produce more *plc*-specific mRNA (129). Other workers have drawn similar conclusions from experiments in which a shuttle plasmid carrying the *plc* gene from strain NCTC8237 was introduced into the type-A strain 13 and the type-C strain NCTC8533. Sixteen-fold more α-toxin was produced in the type-A strain (10).

ORF2 is located upstream of the *plc* gene but is transcribed in the opposite orientation (55, 100, 125) (Figure 2). Because ORF2 was also found upstream of a low-level α-toxin-producing type-C strain it was concluded that it is not involved in the regulation of the *plc* gene (55). The putative ORF2 product has significant amino acid sequence similarity to the f478 gene product (38% identity) from *E. coli* (6), the SPAC20G.05c protein (34% identity) from *Schizosaccharomyces pombe* (Accession No. G2330761), and the sll1464 protein (28% identity) from *Synechocystis* sp. (51). Although these hypothetical proteins are closely related and probably share a common conserved function, that function remains unknown.

Titball and coworkers (122) observed a directly repeated sequence (5′ TATTCAAAAAT 3′) located upstream of the −35 box of the *plc* promoter and suggested that it might be involved in the regulation of the *plc* gene. These repeats form part of a 77-bp AT-rich region containing three $d(A)_{5-6}$ tracts (Figure 2). In *E. coli*, deletion of this region leads to a significant increase in α-toxin production, suggesting that the poly(A) tracts represent a region of inherent DNA bending, which inhibits transcription of the *plc* gene in *cis* (125). Subsequent studies compared the equivalent upstream regions from a low-level type-C α-toxin producer and the high-level type-A strain NCTC 8237 (55). Although significantly higher *plc*-specific mRNA levels were found in the type-A strain, the poly(A) tracts were present in both upstream regions.

Gel retardation studies showed that the fragment carrying the upstream region and the promoter are retarded by crude extracts from both the type-A and type-C strains. However, only the type-A extract retards a fragment from within the *plc* gene. On the basis of these results it was suggested that the observed differences in α-toxin production are related not to the region of DNA curvature but to the presence of an unidentified DNA-binding protein in the type-A strain (55). This putative protein, which we have designated PlcR (94), remains to be identified.

Specific deletion of the poly(A) tracts progressively decreases DNA curvature (69). Plasmids containing three, two, or one poly(A) tract were introduced into a *C. perfringens* strain 13–derived *plc* mutant and α-toxin levels were measured. In contrast to the results previously obtained in *E. coli*, deletion of these regions progressively decreased α-toxin production and transcription of the *plc* gene. Insertion of 5 to 21 nucleotides between the first poly(A) tract and the -35 region had a similar effect. Furthermore, the poly(A) tracts were responsible for the temperature dependence of α-toxin production, with more toxin being produced at lower temperatures. These results indicated that optimal rotational placement of the poly(A) tracts is important for maximal *plc* gene expression and that the intrinsic DNA curvature induced by these tracts contributes to the thermoregulation of the *plc* gene (69).

The *C. perfringens* α-toxin is a zinc metalloenzyme that degrades both lecithin and sphingomyelin (119). Comparative analysis of the deduced amino acid (aa) sequence of the 398-aa α-toxin precursor polypeptide revealed that it has three distinct domains (122, 123). The first 28 amino acids represent a classical signal peptide sequence, as expected for a secreted protein (63, 83, 122). The second domain, from aa 1 to 248 of the mature protein, has sequence similarity (29% identity) to the smaller (245-aa) nontoxic phospholipase C from *Bacillus cereus* and clearly contains the phospholipase C domain (63, 122, 128). The crystal structure of the phospholipase C from *B. cereus* has been determined (40). The five histidine residues proposed to be involved in binding three separate zinc ions are conserved in the *C. perfringens* α-toxin and the phospholipase C from *Clostridium bifermentans* (63, 124). The third domain, from aa 249 to 370, has sequence similarity (29% identity) to human arachadonate-5′-

lipoxygenase, an enzyme involved in the arachadonic acid cascade (123). Mice can be protected from clostridial myonecrosis by immunization with a recombinant protein consisting of aa 247 to 370 (135). This domain is not present in the *B. cereus* enzyme, and although it has no known intrinsic enzymatic activity, it is necessary for the hemolytic, lethal, and sphingomyelinase activities of the α-toxin (121). The latter activity appears to involve the same active site responsible for phospholipase C activity. It has been suggested that the C-terminal domain of α-toxin is responsible for recognition of sphingomyelin (123), a postulate supported by recent studies suggesting that this C-terminal domain mediates Ca^{2+}-dependent recognition of membrane phospholipids (30).

Genetic methods have been used in structure-function studies on the α-toxin. Site-directed mutagenesis was used to replace each of nine α-toxin histidine residues and the resultant proteins purified and analyzed (80). The results showed that the His-68, His-126, His-136, and His-148 residues are critical for phospholipase C, sphingomyelinase, hemolytic, and lethal activity. Three of these proteins, H68G, H126G, and H136G, like the wild-type enzyme, have two bound zinc ions per protein molecule but have significantly reduced capacity to bind $^{62}Zn^{2+}$. Active α-toxin has two tightly bound zinc ions and one zinc, cobalt, or manganese ion that can be removed by EDTA treatment. The data suggest that the His-68, His-126, and His-136 residues are involved in the binding of this latter, more easily removed, metal ion (80). The H148G protein contains only one bound zinc ion, indicating that this residue is involved in binding one of the tightly bound ions (80). This conclusion is consistent with those drawn from comparison with the crystallographic data (40).

Other studies have shown that all five histidine residues predicted by comparison with the *B. cereus* phospholipase C structure to be involved in zinc binding are essential for the biological activity of α-toxin (31). In this work H11S, H68S, H126S, H136S, and H148S mutants were constructed and shown to have dramatically reduced biological activity, as did mutations at several other sites, including Asp-56. These studies (31) and other mutagenesis studies (79) indicate that a single active site is responsible for all of the biological activities of the α-toxin and that Asp-56 is required for catalytic activity. Random PCR mutagenesis has shown that Thr-74 is required for activity (81).

Recent studies have involved mutagenesis of residues in the C-terminal domain (30). Alignment of this domain with human arachidonate-5'-lipoxygenase revealed the presence of four conserved tyrosine residues (30), which appear to be involved in substrate recognition (102, 121, 123). Alteration of two of these residues resulted in enzymes that were still hemolytic but had somewhat reduced phospholipase C activity (30). The Y275F and Y277F enzymes, and mutant enzymes generated by replacement of the Asp-269 and Asp-336 residues, had reduced phospholipase C activity under conditions of both optimal and

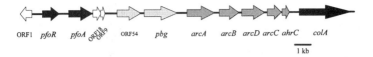

ORF1 pfoR pfoA ORF49 ORF54 pbg arcA arcB arcD arcC ahrC colA

1 kb

Figure 3 Genetic organization of the 20.7-kb *pfoA/colA* gene region. The genes with known or potential functions include *pfoA*, the θ-toxin structural gene; *pfoR*, the putative *pfoA* regulatory gene; ORF54, a putative lactose permease gene; *pbg*, which encodes a β-galactosidase; *arcABDC* and *ahrC*, genes involved in arginine metabolism; and *colA*, the κ-toxin structural gene. Genes involved in extracellular toxin production (*black*), lactose metabolism (*light gray*), and arginine metabolism (*darker gray*) are shown as indicated. ORFs of unknown function are in white. The genetic map of this gene region was compiled from published sequence data (82, 106, 107).

suboptimal Ca^{2+} concentrations. These results are consistent with the finding that calcium ions are required for the binding of α-toxin to lipid films (77). Substitution of Thr-272 with proline also significantly affected biological activity (30). On the basis of these studies and structural comparisons, it was concluded that the C-terminal domain of the α-toxin is a member of the C_2 β-barrel domain family and interacts with membrane phospholipids in a calcium-dependent manner.

The pfoA *(θ-Toxin) and* colA *(κ-Toxin) Gene Region*

The genes encoding the θ-toxin and κ-toxin are located within an approximately 10-kb region on the *C. perfringens* chromosome (82). These genes, *pfoA* and *colA*, which encode a hemolysin (perfringolysin O) and a collagenase, respectively, are separated by the *pbg* gene, which encodes a β-galactosidase, and an operon whose products have sequence similarity to proteins involved in arginine metabolism (Figure 3). A putative regulatory gene, *pfoR*, is located immediately upstream of *pfoA*. Deletion derivatives that lack the *pfoR-pfoA-colA* gene region have been isolated by Tn916 mutagenesis (3).

The *colA* gene from a *C. perfringens* type-C strain has been cloned and sequenced (70). It encodes a 126-kDa preprocollagenase, which is cleaved to produce a 116-kDa extracellular collagenase that is closely related (36.4% identity) to the collagenase from *Clostridium histolyticum* (136). In addition to containing a typical signal peptide, the 86-aa leader peptide contains a collagenase target sequence, PLGP, suggesting that the precursor is also cleaved by an autocatalytic process. As expected for a zinc metalloprotease, the mature collagenase protein contains a consensus HEXXH zinc-binding domain (70). The role of the collagenase in virulence remains to be determined.

The *pfoR* gene is located 591-bp upstream of *pfoA* and encodes a 343-aa basic protein that contains several SPXX-like domains and a putative helix-turn-helix motif, suggesting that it is involved in gene regulation (107). Deletion of *pfoR* from a recombinant plasmid that carries the *pfoR* and *pfoA* genes leads to reduced hemolytic activity in *E. coli*, indicating that *pfoR* is a positive regulator of *pfoA* expression. In-frame deletion of an internal fragment that encodes the helix-turn-helix motif has a similar effect on *pfoA* expression. However, since these studies were carried out exclusively in *E. coli*, confirmation of the significance of the regulatory role of *pfoR* awaits genetic studies in *C. perfringens*.

The *pfoA* gene encodes a polypeptide that consists of a 27-aa signal peptide and the mature 500-aa perfringolysin O protein (107, 130, 131). Sequence analysis indicates that perfringolysin O has similarity (40–70% identity) with a family of cholesterol-binding cytolysins such as streptolysin O, listeriolysin O, and pneumolysin (131, 133). The region of similarity includes the conserved tryptophan-rich ECTGLAWEWWR motif, which probably represents the cholesterol-binding domain (97, 133). Chemical modification of the cysteine residue within this domain, which is the only cysteine residue in perfringolysin O, inactivates the hemolysin (45), although this cysteine does not appear to be essential for toxin function (132, 133). It appears that this modification disturbs the structure of the tryptophan-rich domain, whereas replacement of this cysteine residue with an amino acid of similar size retains an active conformation (97).

The crystal structure of perfringolysin O has been determined (97). The protein has an unusually elongated shape and contains four discontinuous domains. Biochemical studies have suggested that aggregation of perfringolysin O on the membrane is essential for hemolytic activity (37). On the basis of the crystallographic data, a model for the formation and insertion of these aggregates into the mammalian cell membrane has been proposed. This model involves the formation of a pore by the insertion of domain 4 of the perfringolysin O aggregates into the membrane and is consistent with the biological data (97).

Although perfringolysin O may modulate the virulence of *C. perfringens* strains it is not an essential virulence factor since *pfoA* mutants still cause clostridial myonecrosis in mice (2, 116). By contrast, the closely related cytolysin listeriolysin O is an essential virulence factor that is responsible for the release of the intracellular pathogen *Listeria monocytogenes* from the phagocytic vacuole. It is the only member of this cytolysin group that has an intracellular rather than extracellular role. Although replacement of the *hly* gene, which encodes listeriolysin O, with *pfoA* still enabled *L. monocytogenes* cells to be released from the host vacuole, perfringolysin O could not complement listeriolysin O for virulence because it was toxic and killed the host cell (49). However, by selecting for mutants more suited to a biological role within mammalian cells, perfringolysin O derivatives that were functionally more like listeriolysin O were selected. Interestingly, these compensatory mutations were all within conserved regions, including the tryptophan-rich region, and all decreased rather than increased the sequence identity between perfringolysin O and listeriolysin O (50). Some of these single amino acid changes reduced the specific activity or intracellular half-life of the toxin. In another mutant a leucine-to-phenylalanine mutation within the tryptophan-rich region altered the pH optimum, reducing its relative activity at neutral pH so that like listeriolysin O it was more active at the acidic pH of the phagocytic vacuole. It was concluded that single mutations can eliminate host cell toxicity without eliminating its capacity to lyse cell membranes (50). Now that the crystal structure of perfringolysin O is available it would be interesting to analyze the predicted structural effects of these mutations.

The ε-Toxin Gene, etx

Type-B and type-D strains of *C. perfringens* are characterized by their ability to produce ε-toxin, which is secreted as an inactive prototoxin and is activated in the gastrointestinal tract by proteolytic cleavage (84). The ε-toxin structural gene, *etx*, is encoded on an extrachromosomal element, but little is known about this large host plasmid (53).

An insertion sequence, IS*1151*, which has significant similarity to IS*231* from *Bacillus thuringiensis*, is located 96 bp upstream of the *etx* gene (22, 68) as well as near other toxin genes from *C. perfringens* (53). Immediately upstream of IS*1151* there is an ORF (Accession No. X60694) with significant similarity to the transposase-encoding *tnpA* gene from the Tn*3* family of transposons, including the cryptic Tn*4430* from *B. thuringiensis* (68). In *B. thuringiensis*, IS*231* preferentially inserts at the ends of Tn*4430* (32). Although further sequence analysis of the region upstream of the *etx* gene is needed, the presence of IS*1151* and a region with similarity to Tn*4430* suggests that a similar situation may exist in *C. perfringens*. Downstream of *etx* there is a region with some sequence similarity (42) to the transposon Tn*4001* from *Staphylococcus aureus*. This region appears to encode a fusion protein with similarity to transposases from *IS1201* and IS*905* from *Lactococcus helveticus* (117) and *Lactococcus lactis* (24), respectively. These IS elements belong to the Mutator family of insertion sequences (27). However, the functional significance of the *etx*-associated sequences is unknown.

The *etx* gene encodes a single polypeptide that has a 32-aa signal peptide sequence; trypsin cleaves another 13 amino acids from the secreted prototoxin to form an active 283-aa toxin (38, 42). The prototoxin can also be activated by cleavage with *C. perfringens* λ-toxin (76). Little is known about the mode of action of ε-toxin. Structure-function studies have led to the identification of several essential amino acids, including the sole tryptophan residue (84, 103, 104). The toxin is lethal and dermonecrotic; it primarily appears to increase vascular permeability in the brain, kidneys, and intestine, which is consistent with the pathology observed in both type-B and type-D enterotoxemia (84). ε-toxin is toxic for MDCK cells, where it appears to form a large membrane complex and increase K^+ efflux (87). Recent studies have shown that ε-toxin has 20–27% sequence identity with the mosquitocidal toxins Mtx2 and Mtx3 from *Bacillus sphaericus* (64, 118) and with a parasporal crystal protein, c53, from *B. thuringiensis* (Accession No. X98616). The similarity of both the ε-toxin and its upstream region to sequences found in *B. thuringiensis* may be fortuitous, but it could indicate that these gene regions have a common evolutionary origin.

The β-Toxin Gene, cpb

β-toxin is produced by type-B and type-C strains of *C. perfringens* (93). Like ε-toxin it is encoded on a large uncharacterized plasmid that also carries IS*1151* (25, 53). The structural gene, *cpb*, has been cloned and shown to encode a single polypeptide with a 27-aa signal sequence, which upon secretion is cleaved to produced a 309-aa trypsin-sensitive extracellular toxin (41, 114). β-toxin has significant sequence similarity to other toxins including *S. aureus* α-toxin and components of the γ-toxin and leukocidins from *S. aureus*. Although these toxins are all classified as pore-forming cytolysins, β-toxin is not cytolytic and its cytotoxic mode of action remains unknown (41, 62). Recent studies have shown that Arg-212 is important for the biological activity of β-toxin (114a). A distinct toxin known as Beta2 toxin has been identified from a porcine type C strain of *C. perfringens* (28a). This toxin does not have amino acid sequence similarity to *C. perfringens* β-toxin. The Beta2 toxin gene, *cpb2*, is also found on a large plasmid but does not hybridize with the *cpb* gene.

The ι-Toxin Genes, iap and iab

The relatively rare type-E strains of *C. perfringens* produce ι-toxin, which unlike all other extracellular *C. perfringens* toxins consists of two separate polypeptides. The active 47.5-kDa component Ib is encoded by the *iap* gene and has actin-specific ADP-ribosyltransferase activity, whereas the 71.5-kDa binding component Ib is encoded by the *ibp* gene (14, 28, 109). The Ia component has the conserved ADP-ribosylation site found in other ADP-ribosylating toxins and a conserved actin-binding motif (28). In addition, Ib has 33% sequence identity with the protective antigen component of anthrax toxin, including the central transmembrane domain that is involved in cell membrane translocation, although the mode of action of these proteins is not identical (85). The ι-toxin is very closely related to the binary toxin from *Clostridium spiroforme*, to the extent that hybrid molecules are functional (14). A similar binary toxin is also produced by *Clostridium difficile* (86). The precise role in virulence of the ι-toxin family remains to be determined.

The plasmid-determined (28, 53, 85) *iap* and *ibp* genes are separated by 40 bp and appear to be transcribed from a single promoter located upstream of *iap* (85). A large 140-kb plasmid that carries the *iap/ibp* operon, the *cpe* gene, and IS*1151* has been identified in two *C. perfringens* type-E strains (28). The equivalent *sas* and *sbs* genes from *C. spiroforme* (28), and the *cdtA* and *cdtB* genes from *C. difficile* (86), all of which hybridize with their respective *iap* and *ibp* homologs, have a similar genetic organization but are chromosomally determined. These operons appear to have a common evolutionary origin (28, 86).

The Sialidase Genes, nanH *and* nanI

Like many pathogens, *C. perfringens* produces extracellular sialidases or neuraminidases, which have the ability to hydrolyze the sialic acid residues located on many mammalian cell membranes (90). Two distinct *C. perfringens* sialidase genes, *nanH*, which encodes the small sialidase (15, 89), and *nanI*, which encodes the large sialidase (127), have been cloned and sequenced. The *nanH* gene encodes a 382-aa or 42.8-kDa polypeptide that does not have a typical hydrophobic leader sequence (89). The resultant small NanH sialidase is closely related to sialidases from *Clostridium sordellii* (98) and *Clostridium septicum* (99), although the latter enzyme is significantly larger.

Comparative sequence analysis revealed that NanH contains four repeats of a 12-aa motif that is common to other bacterial and viral sialidases (88). This motif includes a highly conserved SXDXGXTW sequence known as an Asp box. About 37 amino acids upstream of the first Asp box is a short conserved sequence, FRIP. Site-directed mutagenesis showed that replacement of the conserved Arg-37 residue of this FRIP motif with lysine significantly increases the K_M and decreases the V_{MAX} of the resultant NanH enzyme, indicating that the conserved arginine residue is involved in substrate binding (91). Other mutagenesis studies indicated that additional conserved residues, namely Asp-62, Asp-100, and Glu-230, are involved in catalysis. In addition, comparative homology modeling of the NanH structure, using the known crystal structure of the *Salmonella typhimurium* sialidase, revealed that these residues, and Arg-37, can be superimposed on the active site pocket of the latter enzyme (16). Mutations of other residues located within the Asp boxes do not alter the kinetic parameters, suggesting that these residues play a structural rather than a catalytic role.

The *nanI* gene encodes a 694-aa polypeptide that includes a 41-aa hydrophobic signal sequence, leading to secretion of the mature 73.0-kDa large sialidase enzyme (127). This enzyme, which is often called the large NanH sialidase, has 26% sequence identity to the small *C. perfringens* sialidase and is most closely related to the large *C. septicum* enzyme. The *C. perfringens* sialidases differ in their pH and temperature optima and substrate specificity. The large enzyme has four appropriately spaced Asp boxes and an upstream YRIP motif (127).

Recent database searches revealed that upstream and downstream of the *nanI* gene are ORFs with sequence similarity to a spore photoproduct lyase from *B. subtilis* and to various hypothetical bacterial proteins, respectively. The latter ORF does not encode a sialic acid permease as initially suggested (126, 127). Note that both the *nanH* and the *nanI* genes are chromosomally determined (17).

Genes Encoding Other Extracellular Toxins or Enzymes

C. perfringens also encodes other extracellular enzymes and toxins whose roles in virulence are unknown (93). Some of these genes have been cloned and analyzed. The *nagH* gene is located within 30 kb of *colA* on the *C. perfringens* chromosome and encodes a 114-kDa secreted endo-β-N-acetylglucosaminidase, or hyaluronidase, which is also known as the μ-toxin (12). The deduced protein contains several repeated motifs, which is typical for a carbohydrate-binding protein.

A caseinase gene, *lam*, which encodes a 36-kDa thermolysin-like zinc metalloprotease, has been cloned from a type-B strain and designated as the λ-toxin (47). The *lam*-encoded protease appears to be secreted as a proenzyme that contains a long 200-aa precursor sequence. The mature protein not only contains the expected HEXXH zinc-binding motif but also has consensus sequences of the thermolysin protease family. Although the *lam* gene is clearly located on a 70-kDa plasmid in this strain, other studies have reported the presence of a much smaller plasmid that also carries a caseinase-encoding gene (5). Injection of the λ-toxin intradermally into mice revealed that it can degrade conective tissue (47), but determination of its role in virulence awaits the construction and virulence testing of *lam* mutants. λ-Toxin may have a role in the activation of ε-Toxin (76).

GLOBAL REGULATION OF EXTRACELLULAR TOXIN PRODUCTION IN *C. PERFRINGENS*

Isolation of Pleiotropic Toxin Mutants

Early studies involving chemical mutagenesis of *C. perfringens* cells have led to the isolation of mutants altered in their ability to produce several extracellular toxins or enzymes (39, 96). These studies suggested that global regulatory systems may operate in *C. perfringens*. The θ-toxin mutants could be divided into two groups, a and b, neither of which could produce θ-toxin (39). However, when these strains were cross-streaked on sheep blood agar, the group a strains complemented the mutation in the group b mutants and hemolytic activity was detected (39, 44). Similar observations were made when protease, κ-toxin, and hemagglutinin activities were examined (44). Subsequent studies showed that the group a mutants secreted an as-yet-uncharacterized small-molecular-weight compound called substance A, which had the ability to induce toxin production in the group b mutants (43). These results can be explained by postulating that extracellular toxin production in *C. perfringens* is controlled by the production of substance A, which activates genes that regulate the expression of the toxin structural genes. The group a mutants presumably produce substance A but have mutations in specific toxin structural genes, whereas the group b mutants have wild-type toxin genes but are unable to produce substance A and therefore are unable to produce wild-type toxin levels (108).

VirR and VirS: a Two-Component Signal Transduction System That Regulates Extracellular Toxin Production

There is convincing evidence suggesting the presence of a global regulatory network that controls extracellular toxin production in *C. perfringens* (65, 105). Complementation of an NTG-derived pleiotropic mutant was used to identify the *virR* regulatory gene (105). This mutant was unable to produce θ-toxin and produced reduced levels of κ-toxin and hemagglutinin. In another study, Tn916 mutagenesis was used to isolate a mutant that produced no θ-toxin and reduced levels of α-toxin, protease, and sialidase (65). A 4.3-kb fragment that could complement this mutation was cloned and sequenced and shown to encode several ORFs. The product of one of these ORFs, ORF10C, has very significant amino acid sequence similarity (64%) to ORF10 from the *C. perfringens* bacteriocin plasmid pIP404, but its function remains unknown. Downstream of ORF10, and on the same strand, are two regulatory genes, *virR* and *virS*, which encode a two-component signal transduction system. The Tn916 insertion site is located within *virS* (65).

The *virS* gene product has a relatively hydrophobic N-terminus that contains six putative transmembrane domains, suggesting that as in most sensor histidine kinases it is located in the cell membrane (65). The C-terminal domain of VirS appears to be cytoplasmic and contains three of the four motifs that are common to sensor histidine kinases; the fourth motif is located in a putative cytoplasmic loop between the N-terminal transmembrane domains 4 and 5. These motifs include the His-255 residue that is equivalent to the site of autophosphorylation in other sensor kinases (65).

The N-terminal domain of VirR has sequence similarity to conserved motifs found in other response regulators (65, 105). These amino acids include the conserved aspartate residue that accepts the phosphate group from the cognate sensor kinase. The available data (65, 105) indicate that the VirR and VirS proteins comprise a two-component signal transduction system that regulates extracellular toxin production and virulence in *C. perfringens*. Evidence that virulence is affected was obtained by experiments whereby the *virS::Tn916* mutant and a complemented derivative containing the cloned wild-type *virRvirS* operon were examined in a mouse myonecrosis model. The results showed that mice injected with the *virS* mutant have reduced muscle destruction and necrosis compared to mice injected with the complemented derivative (65).

The VirR/VirS system regulates toxin production at the level of transcription (4), as expected for a classical two-component regulatory pathway. However, it has differential effects on the production of the various toxins. No θ-toxin is produced in either a *virR* or a *virS* mutant (65, 105). By contrast, α-toxin, κ-toxin, protease, and sialidase production in these mutants is reduced, not eliminated. Primer extension studies have identified the VirR-regulated promoters of several of the toxin structural genes (4). The *pfoA* gene has a major promoter, expression from which is totally dependent on the VirR/VirS system. Similarly, the *plc* gene is expressed from a single promoter. However, transcription from this promoter is only reduced in the *virS* mutant, explaining why α-toxin activity is still observed, albeit at lower levels, in this derivative. Finally, κ-toxin production is dependent upon two distinct promoters, only one of which is VirR dependent, again explaining why reduced levels of collagenase activity are observed in the various mutants (4).

There are no common DNA sequences located upstream of the *pfoA*, *plc*, or *colA* genes. Therefore, it is not possible to identify the VirR-binding site by comparative sequence analysis. Recent studies in this laboratory have involved the purification of a functional VirR protein and gel retardation studies on the putative target genes. The results have shown that VirR binds specifically to a 182-bp region that includes the *pfoA* promoter but that VirR does not bind to regions upstream of the *plc, colA* or *pfoR* genes (JK Cheung & JI Rood, unpublished results). It is concluded that the VirR/VirS system regulates θ-toxin production by directly modulating the transcription of the *pfoA* gene and does not act via the putative regulatory gene, *pfoR*, which is located immediately upstream of *pfoA*. The precise role of *pfoR* therefore remains to be determined. By contrast, since VirR does not bind to the other toxin structural genes, it presumably regulates the expression of as-yet-unidentified regulatory genes, which in turn regulate the expression of specific toxin structural genes.

The current model for the regulation of extracellular toxin production in *C. perfringens* can be summarized as follows (see Figure 4): in conditions favorable for toxin production, an unknown environmental or growth phase signal, which could be substance A, is thought to bind to the N-terminal domain of the VirS sensor kinase, which is located in the cell membrane. The resultant conformational change enables the autophosphorylation of the His-255 residue located in the cytoplasmic C-terminal domain of VirS. The phosphorylated VirS protein then acts as a phosphodonor and facilitates the phosphorylation of the appropriate aspartate residue of the cognate response regulator, VirR. The activated VirR protein then is able to bind to the promoter region of *pfoA* and activate transcription.

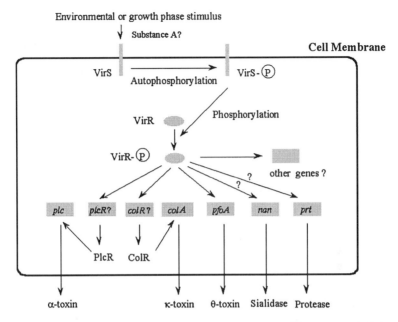

Figure 4 Model for the regulation of extracellular toxin production. The current model is based on previous diagrams (65, 94) and incorporates the latest VirR-binding data (JK Cheung & JI Rood, unpublished data). The putative protease gene is designated as the *prt* gene. Putative regulatory genes whose existence has not been proven are designated with a question mark.

Phosphorylated VirR must also activate the transcription of secondary regulatory genes, such as the putative *plcR* and *colR* genes. The products of these genes then activate transcription of the target structural genes, *plc* and *colA*, respectively. Regulation of extracellular toxin production in *C. perfringens* therefore involves a VirR/VirS-dependent regulatory cascade, only part of which has been identified. Superimposed upon this cascade may be the effects of other regulatory elements such as the *pfoR* gene product, the poly(A) region located upstream of *plc*, and genes involved in the synthesis of substance A.

GENETIC ORGANIZATION OF THE ENTEROTOXIN STRUCTURAL GENE

The cpe *Gene Can Be Found on the Chromosome or on a Plasmid*

The enterotoxin is the only known *C. perfringens* toxin that is not secreted from vegetative bacterial cells but is produced during sporulation. It is a cytotoxic protein that kills susceptible mammalian cells by altering membrane

Figure 5   Genetic organization of the putative enterotoxin transposon. The organization of the chromosomal *cpe* gene region is shown. ORFs are indicated by the arrows. The IS elements are shown by the gray boxes. The scale is shown in kb. Based on previous data (8, 9, 19).

permeability (71). The enterotoxin structural gene, *cpe*, was first cloned as separate segments (36, 46, 134) and then in its entirety (21). The gene encodes a 319-aa polypeptide with a deduced molecular size of 35,317. Less than 6% of *C. perfringens* isolates carry the *cpe* gene and produce enterotoxin (23, 58, 112, 134).

In strains NCTC8798 and NCTC8239 the *cpe* gene is located on the chromosome, between *rrnF* and *rrnG* (Figure 1), within 0.5 Mb of the region known to encode other toxin genes (13). However, in other isolates the *cpe* gene is located on a 100–120-kb plasmid (19). Sequence analysis has revealed that the *cpe* gene is closely associated with several IS sequences and in some chromosomal strains appears to be located on a transposable genetic element (8, 9). The enterotoxin-producing isolates examined to date contain a 789-bp IS*200*-like element, IS*1469*, located about 1.2 kb upstream of the *cpe* gene (Figure 5). Immediately upstream of IS*1469* in the chromosomal *cpe* strain NCTC8239 is another insertion sequence, IS*1470*, which is a member of the IS*30* family. This strain contains a second copy of IS*1470* 1.1 kb downstream of the *cpe* gene. These elements therefore appear to comprise a 6.3-kb compound transposon that contains an internal IS*1469* element and the *cpe* gene (Figure 5). At this time there is no genetic evidence that this putative transposon is capable of transposition.

In *C. perfringens* strains isolated from food poisoning outbreaks the *cpe* gene is chromosomally determined and appears to be located on the same restriction fragment, implying that it is located on the same compound transposon. However, in enterotoxin-producing *C. perfringens* isolates of animal origin, the *cpe* gene is plasmid determined (18, 19). In these strains the *cpe* gene is associated with IS*1469* in the same manner as before, but not with IS*1470*. These plasmids also carry IS*1151*, and in at least one strain, 945P, the *cpe* plasmid also carries the *etx* gene. In strain F3686, IS*1151* is located approximately 260 bp downstream of the *cpe* gene, whereas in strain 945P, IS*1151* appears to be linked closely to the *etx* gene but not to the *cpe* gene (19). Finally, recent studies have shown that the *cpe* gene is also plasmid encoded in *C. perfringens* strains associated with nonfoodborne human gastrointestinal infections (18).

Regulation of Enterotoxin Production

Enterotoxin is produced in significant quantities only in sporulating *C. perfringens* cells. However, the nature of the regulatory mechanisms that prevent *cpe* expression in vegetative cells and activate expression in sporulating cells is not known. The most likely explanation is that *cpe* transcription is dependent upon a sporulation-specific sigma factor or transcriptional activator (72, 74). Indeed, SigE- and SigK-like promoters were recently identified upstream of the *cpe* gene (137). In *B. subtilis*, SigE and SigK are sporulation-specific factors that are active in the mother cell component during sporulation. Sequence analysis of regions upstream of the *cpe* gene has revealed that although the *cpe* genes of the six strains examined have the same transcriptional start points, the strains can be divided into two classes, one of which has a 45-bp insertion upstream of the promoter region. It is not known if this insert has any regulatory or functional role, although expression vector studies have shown that the different upstream regions do lead to somewhat different expression patterns (75).

These studies (75) and those of other workers (20) have shown that *cpe* expression is regulated at the transcriptional level. Introduction of a recombinant *cpe* gene into *cpe*-negative *C. perfringens* strains resulted in sporulation-specific enterotoxin production, indicating that the as-yet-unidentified elements involved in the regulation of *cpe* expression are present in such strains (20). Clearly, we know very little about how enterotoxin production is regulated in *C. perfringens*; much exciting research remains to be done.

Enterotoxin Structure-Function Studies

The mode of action of *C. perfringens* enterotoxin has not been elucidated precisely but entails an ordered process that involves the binding of the enterotoxin to a receptor on the surface of the mammalian cell and the formation of a small complex; insertion of the bound enterotoxin complex into the membrane; formation of a large enterotoxin-containing complex, and resultant cell membrane permeability changes (56). The enterotoxin binds to one of two related membrane proteins (52). Initial studies showed that the C-terminal half of the enterotoxin (aa 171–319) contains the receptor-binding domain but is nontoxic (33, 36). The receptor-binding domain was localized subsequently to the terminal 30 amino acids of the enterotoxin protein (34). By contrast, although the removal of the N-terminal 25 amino acids by limited trypsin digestion does not affect receptor binding, it does lead to a two- to threefold increase in cytotoxic activity (35).

Subsequent genetic studies have involved the construction and functional analysis of a series of *cpe*-deletion derivatives that encode sequential N-terminal and C-terminal truncations of the enterotoxin protein (57). These proteins were then analyzed for their ability to carry out the successive processes previously identified as being essential for enterotoxin activity. The results showed that deletion of the first 44 amino acids increases toxic activity by resulting in the formation of increased amounts of large complex in the mammalian cell membrane. However, removal of the first 53 amino acids results in a protein that is not cytotoxic because it cannot carry out the post-binding insertion step. Residues located between amino acids 44 and 53 are therefore essential for cytotoxic activity. Deletion of even five amino acids from the C-terminus eliminates receptor-binding activity (57).

FUTURE PERSPECTIVES

The elucidation of reliable techniques for the genetic manipulation of *C. perfringens* (92) has led to exciting advances in our understanding of both virulence mechanisms and the structure and function of virulence genes of *C. perfringens*. However, much remains to be done. In particular, although we now know that the α-toxin is the major toxin involved in gas gangrene, we do not know exactly how its cellular effects are mediated. We also do not know the precise role of the other extracellular toxins in the disease process. Similarly, the mechanisms of pathogenesis of the enterotoxemic diseases caused by the other *C. perfringens* toxin types are unknown.

Our knowledge of how extracellular toxin production in *C. perfringens* is regulated has advanced significantly in recent years (94, 108). But, despite the fact that the VirR/VirS two-component signal transduction system has been identified and partially characterized, not enough is known about this important regulatory network. In particular, we do not know the environmental conditions under which this system is activated or how this activation is mediated. The secondary regulatory genes that must be involved in this cascade also remain to be identified. Finally, we need to elucidate the mechanism by which *C. perfringens* enterotoxin production is regulated in a sporulation-specific manner. The answer to many of these questions regarding the identification and functional characterization of the genes involved in *C. perfringens* virulence awaits the determination of the sequence of the *C. perfringens* genome and its detailed analysis.

ACKNOWLEDGMENTS

I thank my colleagues Stewart Cole, Michel Popoff, Roland Schauer, Mike Parker, Bruce McClane, and Per Einar Granum for providing copies of their papers prior to publication. The work carried out in my laboratory was supported by grants from the Australian National Health and Medical Research Council.

Literature Cited

1. Arnon SS. 1997. Human tetanus and human botulism. See Ref. 95, pp. 95–115
2. Awad MM, Bryant AE, Stevens DL, Rood JI. 1995. Virulence studies on chromosomal α-toxin and θ-toxin mutants constructed by allelic exchange provide genetic evidence for the essential role of α-toxin in *Clostridium perfringens*-mediated gas gangrene. *Mol. Microbiol.* 15:191–202
3. Awad MM, Rood JI. 1997. Isolation of α-toxin, θ-toxin and κ-toxin mutants of *Clostridium perfringens* by Tn916 mutagenesis. *Microb. Pathogen.* 22:275–84
4. Ba-Thein W, Lyristis M, Ohtani K, Nisbet IT, Hayashi H, et al. 1996. The *virR/virS* locus regulates the transcription of genes encoding extracellular toxin production in *Clostridium perfringens*. *J. Bacteriol.* 178:2514–20
5. Blaschek HP, Solberg M. 1981. Isolation of a plasmid responsible for caseinase activity in *Clostridium perfringens* ATCC 3626B. *J. Bacteriol.* 147:262–66
6. Blattner FR, Plunkett G III, Bloch CA, Perna NT, Burland V, et al. 1997. The complete genome sequence of *Escherichia coli* K-12. *Science* 277:1453–74
7. Bryant AE, Stevens DL. 1997. The pathogenesis of gas gangrene. See Ref. 95, pp. 185–96
8. Brynestad S, Iwanejko LA, Stewart GSAB, Granum PE. 1994. A complex array of Hpr consensus DNA recognition sequences proximal to the enterotoxin gene in *Clostridium perfringens* type A. *Microbiology* 140:97–104
9. Brynestad S, Synstad B, Granum PE. 1997. The *Clostridium perfringens* enterotoxin gene is on a transposable genetic element in type A human food poisoning strains. *Microbiology* 143:2109–15
10. Bullifent HL, Moir A, Awad MM, Scott PT, Rood JI, et al. 1996. The level of expression of α-toxin by different strains of *Clostridium perfringens* is dependent upon differences in promoter structure and genetic background. *Anaerobe* 2:365–71
11. Canard B, Cole ST. 1989. Genome organization of the anaerobic pathogen *Clostridium perfringens*. *Proc. Natl. Acad. Sci. USA* 86:6676–80
12. Canard B, Garnier T, Saint-Joanis B, Cole ST. 1994. Molecular genetic analysis of the *nagH* gene encoding a hyaluronidase of *Clostridium perfringens*. *Mol. Gen. Genet.* 243:215–24
13. Canard B, Saint-Joanis B, Cole ST. 1992. Genomic diversity and organization of virulence genes in the pathogenic anaerobe *Clostridium perfringens*. *Mol. Microbiol.* 6:1421–29
14. Carmen RJ, Perelle S, Popoff MR. 1997. Binary toxins from *Clostridium spiroforme* and *Clostridium perfringens*. See Ref. 95, pp. 359–67
15. Chien CH, Huang YC, Chen HY. 1997. Small neuraminidase gene of *Clostridium perfringens* ATCC 10543: cloning, nucleotide sequence, and production. *Enzyme Microb. Technol.* 20:277–85
16. Chien CH, Shann YJ, Sheu SY. 1996. Site-directed mutations of the catalytic and conserved amino acids of the neuraminidase gene, *nanH*, of *Clostridium perfringens* ATCC 10543. *Enzyme Microb. Technol.* 19:267–76
17. Cole ST, Canard B. 1997. Structure, organization and evolution of the genome of *Clostridium perfringens*. See Ref. 95, pp. 49–63
18. Collie RE, McClane BA. 1998. Evidence that the enterotoxin gene can be episomal in *Clostridium perfringens* isolates associated with nonfoodborne human gastrointestinal disease. *J. Clin. Microbiol.* 36:30–36
19. Cornillot E, Saint-Joanis B, Daube G, Katayama S-I, Granum PE, et al. 1995. The enterotoxin gene (*cpe*) of *Clostridium perfringens* can be chromosomal or plasmid-borne. *Mol. Microbiol.* 15:639–47
20. Czeczulin JR, Collie RE, McClane BA. 1996. Regulated expression of *Clostridium perfringens* enterotoxin in naturally *cpe*-negative type A, B, and C isolates of *C. perfringens*. *Infect. Immun.* 64:3301–9
21. Czeczulin JR, Hanna PC, McClane BA. 1993. Cloning, nucleotide sequencing, and expression of the *Clostridium perfringens* enterotoxin gene in *Escherichia*

coli. Infect. Immun. 61:3429–39

22. Daube G, Simon P, Kaeckenbeeck A. 1993. IS*1151*, an IS-like element of *Clostridium perfringens*. *Nucleic Acids Res.* 21:352

23. Daube G, Simon P, Limbourg B, Manteca C, Mainil J, et al. 1996. Hybridization of 2,659 *Clostridium perfringens* isolates with gene probes for seven toxins (alpha, beta, epsilon, iota, theta, mu, and enterotoxin) and for sialidase. *Am. J. Vet. Res.* 57:496–501

24. Dodd HM, Horn N, Gasson MJ. 1994. Characterization of IS*905*, a new multicopy insertion sequence identified in lactococci. *J. Bacteriol.* 176:3393–96

25. Duncan CL, Rokos EA, Christenson CM, Rood JI. 1978. Multiple plasmids in different toxigenic types of *Clostridium perfringens*: possible control of beta toxin production. In *Microbiology 1978*, ed. D Schlessinger, pp. 246–48. Washington, DC: Am. Soc. Microbiol.

26. Dupuy B, Daube G, Popoff MR, Cole ST. 1997. *Clostridium perfringens* urease genes are plasmid borne. *Infect. Immun.* 65:2313–20

27. Eisen JA, Benito MI, Walbot V. 1994. Sequence similarity of putative transposases links the maize Mutator autonomous element and a group of bacterial insertion sequences. *Nucleic Acids Res.* 22:2634–36

28. Gibert M, Perelle S, Daube G, Popoff M. 1997. *Clostridium spiroforme* toxin genes are related to *C. perfringens* iota toxin genes but have a different genomic localization. *Syst. Appl. Microbiol.* 20:337–47

28a. Gibert M, Jolivet-Renaud C, Popoff MR. 1997. Beta2 toxin, a novel toxin produced by *Clostridium perfringens*. *Gene* 203:65–73

29. Ginter A, Williamson ED, Dessy F, Coppe P, Bullifent H, et al. 1996. Molecular variation between the alpha-toxins from the type strain (NCTC 8237) and clinical isolates of *Clostridium perfringens* associated with disease in man and animals. *Microbiology* 142:191–98

30. Guillouard I, Alzari PM, Saliou B, Cole ST. 1997. The carboxy-terminal C$_2$-like domain of the α-toxin from *Clostridium perfringens* mediates calcium-dependent membrane recognition. *Mol. Microbiol.* 26:867–76

31. Guillouard I, Garnier T, Cole ST. 1996. Use of site-directed mutagenesis to probe structure-function relationships of alpha-toxin from *Clostridium perfrin-*

gens. Infect. Immun. 64:2440–44

32. Hallet B, Rezsohazy R, Mahillon J, Delcour J. 1994. IS*231A* insertion specificity: consensus sequence and DNA bending at the target site. *Mol. Microbiol.* 14:131–39

33. Hanna PC, McClane BA. 1991. A recombinant C-terminal toxin fragment provides evidence that membrane insertion is important for *Clostridium perfringens* enterotoxin cytotoxicity. *Mol. Microbiol.* 5:225–30

34. Hanna PC, Mietzner TA, Schoolnick GK, McClane BA. 1991. Localization of the receptor-binding region of *Clostridium perfringens* enterotoxin utilizing cloned toxin fragments and synthetic peptides. The 30 C-terminal amino acids define a functional binding region. *J. Biol. Chem.* 266:11037–43

35. Hanna PC, Wieckowski EU, Mietzner TA, McClane BA. 1992. Mapping of functional regions of *Clostridium perfringens* type A enterotoxin. *Infect. Immun.* 60:2110–14

36. Hanna PC, Wnek AP, McClane BA. 1989. Molecular cloning of the 3′ half of the *Clostridium perfringens* enterotoxin gene and demonstration that this region encodes receptor-binding activity. *J. Bacteriol.* 171:6815–20

37. Harris RW, Sims PJ, Tweten RK. 1991. Kinetic aspects of the aggregation of *Clostridium perfringens* θ-toxin on erythrocyte membranes. *J. Biol. Chem.* 266:6936–41

38. Havard HL, Hunter SE, Titball RW. 1992. Comparison of the nucleotide sequence and development of a PCR test for the epsilon toxin gene of *Clostridium perfringens* type B and D. *FEMS Microbiol. Lett.* 76:77–81

39. Higashi Y, Chazono M, Inoue K, Yanagase Y, Amano T, et al. 1973. Complementation of θ-toxinogenicity between mutants of two groups of *Clostridium perfringens*. *Biken J.* 16:1–9

40. Hough E, Hansen LK, Birkness B, Jynge K, Hansen S, et al. 1989. High resolution (1.5 Å) crystal structure of phospholipase C from *Bacillus cereus*. *Nature* 338:357–60

41. Hunter SEC, Brown JE, Oyston PCF, Sakurai J, Titball RW. 1993. Molecular genetic analysis of beta-toxin of *Clostridium perfringens* reveals sequence homology with alpha-toxin, gamma-toxin, and leukocidin of *Staphylococcus aureus*. *Infect. Immun.* 61:3958–65

42. Hunter SEC, Clarke IN, Kelly DC, Titball RW. 1992. Cloning and nucleotide

sequencing of the *Clostridium perfringens* epsilon-toxin gene and its expression in *E. coli. Infect. Immun.* 60:102–10

43. Imagawa T, Higashi Y. 1992. An activity which restores theta toxin activity in some theta toxin-deficient mutants of *Clostridium perfringens. Microbiol. Immunol.* 36:523–27

44. Imagawa T, Tatsuki T, Higashi Y, Amano T. 1981. Complementation characteristics of newly isolated mutants from two groups of strains of *Clostridium perfringens. Biken J.* 24:13–21

45. Iwamoto M, Ohno-Iwashita Y, Ando S. 1987. Role of the essential thiol group in the thiol-activated cytolysin from *Clostridium perfringens. Eur. J. Biochem.* 167:425–30

46. Iwanejko LA, Routledge MN, Stewart GSAB. 1989. Cloning in *Escherichia coli* of the enterotoxin gene from *Clostridium perfringens* type A. *J. Gen. Microbiol.* 135:903–9

47. Jin F, Matsushita O, Katayama S-I, Jin S, Matsushita C, et al. 1996. Purification, characterization, and primary structure of *Clostridium perfringens* lambda-toxin, a thermolysin-like metalloprotease. *Infect. Immun.* 64:230–37

48. Johnson S, Gerding DN. 1997. Enterotoxemic infections. See Ref. 95, pp. 117–40

49. Jones S, Portnoy DA. 1994. Characterization of *Listeria monocytogenes* pathogenesis in a strain expressing perfringolysin O instead of listeriolysin O. *Infect. Immun.* 62:5608–13

50. Jones S, Preiter K, Portnoy DA. 1996. Conversion of an extracellular cytolysin into a phagosome-specific lysin which supports the growth of an intracellular pathogen. *Mol. Microbiol.* 21:1219–25

51. Kaneko T, Sato S, Kotani H, Tanaka A, Asamizu E, et al. 1996. Sequence analysis of the genome of the unicellular cyanobacterium *Synechocystis* sp. strain PCC6803. II. Sequence determination of the entire genome and assignment of potential protein-coding regions. *DNA Res.* 3:109–36

52. Katahira J, Sugiyama H, Inoue N, Horiguchi Y, Mastsuda M, et al. 1997. *Clostridium perfringens* enterotoxin utilizes two structurally related membrane proteins as functional receptors *in vivo. J. Biol. Chem.* 272:26652–58

53. Katayama S-I, Dupuy B, Daube G, China B, Cole ST. 1996. Genome mapping of *Clostridium perfringens* strains with I-*Ceu*I shows many virulence genes to be plasmid-borne. *Mol. Gen. Genet.* 251:720–26

54. Katayama S-I, Dupuy B, Garnier T, Cole ST. 1995. Rapid expansion of the physical and genetic map of the chromosome of *Clostridium perfringens. J. Bacteriol.* 177:5680–85

55. Katayama S-I, Matsushita O, Minami J, Mizobuchi S, Okabe A. 1993. Comparison of the alpha-toxin genes of *Clostridium perfringens* type A and C strains: evidence for extragenic regulation of transcription. *Infect. Immun.* 61:457–63

56. Kokai-Kun JF, McClane BA. 1997. The *Clostridium perfringens* enterotoxin. See Ref. 95, pp. 325–57

57. Kokai-Kun JF, McClane BA. 1997. Deletion analysis of the *Clostridium perfringens* enterotoxin. *Infect. Immun.* 65:1014–22

58. Kokai-Kun JF, Songer JG, Czeczulin JR, Chen F, McClane BA. 1994. Comparison of Western immunoblots and gene detection assays for identification of potentially enterotoxigenic isolates of *Clostridium perfringens. J. Clin. Microbiol.* 32:2533–39

59. Lawrence G. 1979. The pathogenesis of pig-bel in Papua New Guinea. *Papua New Guinea Med. J.* 22:39–49

60. Lawrence G, Walker PD. 1976. Pathogenesis of enteritis necroticans in Papua-New Guinea. *Lancet* 1:125–26

61. Lawrence GW. 1997. The pathogenesis of enteritis necroticans. See Ref. 95, pp. 197–207

62. Leary SEC, Titball RW. 1997. The *Clostridium perfringens* β-toxin. See Ref. 95, pp. 243–50

63. Leslie D, Fairweather N, Pickard D, Dougan G, Kehoe M. 1989. Phospholipase C and haemolytic activities of *Clostridium perfringens* alpha-toxin cloned in *Escherichia coli*: sequence and homology with *Bacillus cereus* phospholipase C. *Mol. Microbiol.* 3:383–92

64. Liu JW, Porter AG, Wee BY, Thanabalu T. 1996. New gene from nine *Bacillus sphaericus* strains encoding highly conserved 35.8-kilodalton mosquitocidal toxins. *Appl. Environ. Microbiol.* 62:2174–76

65. Lyristis M, Bryant AE, Sloan J, Awad MM, Nisbet IT, et al. 1994. Identification and molecular analysis of a locus that regulates extracellular toxin production in *Clostridium perfringens. Mol. Microbiol.* 12:761–77

66. MacFarlane MG, Knight BCJG. 1941. The biochemistry of bacterial toxins. I. The lecithinase activity of *Cl. welchii*

toxins. *Biochem. J.* 35:884–902

67. MacLennan JD. 1962. The histotoxic clostridial infections of man. *Bact. Rev.* 26:177–276

68. Mahillon J, Rezsohazy R, Hallet B, Delcour J. 1994. IS*231* and other *Bacillus thuringiensis* transposable elements: a review. *Genetica* 93:13–26

69. Matsushita C, Matsushita O, Katayama S-I, Minami J, Takai K, et al. 1996. An upstream activating sequence containing curved DNA involved in activation of the *Clostridium perfringens plc* promoter. *Microbiology* 142:2561–66

70. Matsushita O, Yoshihara K, Katayama S-I, Minami J, Okabe A. 1994. Purification and characterization of a *Clostridium perfringens* 120-kilodalton collagenase and nucleotide sequence of the corresponding gene. *J. Bacteriol.* 176:149–56

71. McClane BA. 1994. *Clostridium perfringens* enterotoxin acts by producing small molecule permeability alterations in plasma membranes. *Toxicology* 87:43–67

72. McClane BA. 1998. New insights into the genetics and regulation of expression of *Clostridium perfringens* enterotoxin. *Curr. Top. Microbiol. Immunol.* 225:37–55

73. McClane BA. 1996. An overview of *Clostridium perfringens* enterotoxin. *Toxicon* 34:1335–43

74. Melville SB, Collie RE, McClane BA. 1997. Regulation of enterotoxin production in *Clostridium perfringens*. See Ref. 95, pp. 471–87

75. Melville SB, Labbe R, Sonenshein AL. 1994. Expression from the *Clostridium perfringens cpe* promoter in *C. perfringens* and *Bacillus subtilis*. *Infect. Immun.* 62:5550–58

76. Minami J, Katayama S-I, Matsushita O, Matsushita C, Okabe A. 1997. Lambda-toxin of *Clostridium perfringens* activates the precursor of epsilon-toxin by releasing its N- and C-terminal peptides. *Microbiol. Immunol.* 41:527–35

77. Moreau H, Pieroni G, Joilivet-Reynaud C, Alouf JE, Verger R. 1988. A new kinetic approach for studying phospholipase C (*Clostridium perfringens* α-toxin) activity on phospholipid monolayers. *Biochemistry* 27:2319–23

78. Murrell TGC. 1989. Enteritis necroticans. In *Anaerobic Infections in Humans*, ed. SM Finegold, WL George, pp. 639–59. New York: Academic

79. Nagahama M, Nakayama T, Michiue K, Sakurai J. 1997. Site-specific mutagenesis of *Clostridium perfringens* alpha-toxin: replacement of Asp-56, Asp-130, or Glu-152 causes loss of enzymatic and hemolytic activities. *Infect. Immun.* 65:3489–92

80. Nagahama M, Okagawa Y, Nakayama T, Nishioka E, Sakurai J. 1995. Site-directed mutagenesis of histidine residues in *Clostridium perfringens* alpha-toxin. *J. Bacteriol.* 177:1179–85

81. Nagahama M, Sakurai J. 1996. Threonine-74 is a key site for the activity of *Clostridium perfringens* alpha-toxin. *Microbiol. Immunol.* 40:189–93

82. Ohtani K, Bando M, Swe T, Banu S, Oe M, et al. 1997. Collagenase gene (*colA*) is located in the 3′-flanking region of the perfringolysin O (*pfoA*) locus in *Clostridium perfringens*. *FEMS Microbiol. Lett.* 146:155–59

83. Okabe A, Shimizu T, Hayashi H. 1989. Cloning and sequencing of a phospholipase C gene of *Clostridium perfringens*. *Biochem. Biophys. Res. Commun.* 160:33–39

84. Payne D, Oyston P. 1997. The *Clostridium perfringens ε-toxin*. See Ref. 95, pp. 439–47

85. Perelle S, Gibert M, Boquet P, Popoff MR. 1993. Characterization of *Clostridium perfringens* iota-toxin genes and expression in *Escherichia coli* (Authors correction, 1995; 63:4967). *Infect. Immun.* 61:5147–56

86. Perelle S, Gibert M, Bourlioux P, Corthier G, Popoff M. 1997. Production of a complete binary toxin (actin-specific ADP-ribosyltransferase) by *Clostridium difficile* CD196. *Infect. Immun.* 65:1402–7

87. Petit L, Gibert M, Gillet D, Laurent-Winter C, Boquet P, Popoff MR. 1997. *Clostridium perfringens* epsilon-toxin acts on MDCK cells by forming a large membrane complex. *J. Bacteriol.* 179:6480–87

88. Roggentin P, Rothe B, Kaper JB, Galen J, Lawrisuk L, et al. 1989. Conserved sequences in bacterial and viral sialidases. *Glycoconj. J.* 6:349–53

89. Roggentin P, Rothe B, Lottspeich F, Schauer R. 1988. Cloning and sequencing of a *Clostridium perfringens* sialidase gene. *FEBS Lett.* 238:31–34

90. Roggentin P, Schauer R. 1997. Clostridial sialidases. See Ref. 95, pp. 423–37

91. Roggentin T, Kleineidam RG, Schauer R, Roggentin P. 1992. Effects of site-specific mutations on the enzymatic properties of a sialidase from *Clostridium perfringens*. *Glycoconj. J.* 9:235–40

92. Rood JI. 1997. Genetic analysis in *C. perfringens*. See Ref. 95, pp. 65–72
93. Rood JI, Cole ST. 1991. Molecular genetics and pathogenesis of *Clostridium perfringens*. *Microbiol. Rev.* 55:621–48
94. Rood JI, Lyristis M. 1995. Regulation of extracellular toxin production in *Clostridium perfringens*. *Trends Microbiol.* 3:192–96
95. Rood JI, McClane BA, Songer JG, Titball RW. 1997. *The Clostridia: Molecular Biology and Pathogenesis*. London: Academic. 533 pp.
96. Rood JI, Wilkinson RG. 1975. Isolation and characterization of *Clostridium perfringens* mutants altered in both hemagglutinin and sialidase production. *J. Bacteriol.* 123:419–27
97. Rossjohn J, Feil SC, McKinstry WJ, Tweten RK, Parker MW. 1997. Structure of a cholesterol-binding, thiol-activated cytolysin and a model of its membrane form. *Cell* 88:685–92
98. Rothe B, Roggentin P, Frank R, Blöcker H, Schauer R. 1989. Cloning, sequencing and expression of a sialidase gene from *Clostridium sordellii* G12. *J. Gen. Microbiol.* 135:3087–96
99. Rothe B, Rothe B, Roggentin P, Schauer R. 1991. The sialidase gene from *Clostridium septicum*: cloning, sequencing, expression in *Escherichia coli* and identification of conserved sequences in sialidases and other proteins. *Mol. Gen. Genet.* 226:190–97
100. Saint-Joanis B, Garnier T, Cole ST. 1989. Gene cloning shows the alpha toxin of *Clostridium perfringens* to contain both sphingomyelinase and lecithinase activities. *Mol. Gen. Genet.* 219:453–60
101. Sakurai J, Duncan CL. 1978. Some properties of beta-toxin produced by *Clostridium perfringens*. *Infect. Immun.* 21:678–80
102. Sakurai J, Fujii Y, Torii K, Kobayashi K. 1989. Dissociation of various biological activities of *Clostridium perfringens* alpha toxin by chemical modification. *Toxicon* 27:317–23
103. Sakurai J, Nagahama M. 1985. Role of one tryptophan residue in the lethal activity of *Clostridium perfringens* epsilon toxin. *Biochem. Biophys. Res. Commun.* 128:760–66
104. Sakurai J, Nagahama M. 1987. Histidine residues in *Clostridium perfringens* epsilon toxin. *FEMS Microbiol. Lett.* 41:317–19
105. Shimizu T, Ba-Thein W, Tamaki M, Hayashi H. 1994. The *virR* gene, a member of a class of two-component response regulators, regulates the production of perfringolysin O, collagenase, and hemagglutinin in *Clostridium perfringens*. *J. Bacteriol.* 176:1616–23
106. Shimizu T, Kobayashi T, Ba-Thein W, Ohtani K, Hayashi H. 1995. Sequence analysis of flanking regions of the *pfoA* gene of *Clostridium perfringens*: beta-galactosidase gene (*pbg*) is located in the 3'-flanking region. *Microbiol. Immunol.* 39:677–86
107. Shimizu T, Okabe A, Minami J, Hayashi H. 1991. An upstream regulatory sequence stimulates expression of the perfringolysin O gene of *Clostridium perfringens*. *Infect. Immun.* 59:137–42
108. Shimizu T, Okabe A, Rood JI. 1997. Regulation of toxin production in *C. perfringens*. See Ref. 95, pp. 451–70
109. Simpson L, Stiles B, Zepeda H, Wilkins T. 1987. Molecular basis for the pathological action of *Clostridium perfringens* iota toxin. *Infect. Immun.* 55:118–22
110. Songer JG. 1997. Clostridial diseases of animals. See Ref. 95, pp. 153–82
111. Songer JG. 1996. Clostridial enteric diseases of domestic animals. *Clin. Microbiol. Rev.* 9:216–34
112. Songer JG, Meer RR. 1996. Genotyping of *Clostridium perfringens* by polymerase chain reaction is a useful adjunct to diagnosis of clostridial enteric disease in animals. *Anaerobe* 2:197–203
113. Stackebrandt E, Rainey FA. 1997. Phylogenetic relationships. See Ref. 95, pp. 3–19
114. Steinthórsdóttir V, Fridriksdóttir V, Gunnarsson E, Andrésson OS. 1995. Expression and purification of *Clostridium perfringens* beta-toxin glutathione S-transferase fusion protein. *FEMS Microbiol. Lett.* 130:273–78
114a. Steinthórsdóttir V, Fridriksdóttir V, Gunnarsson E, Andrésson OS. 1998. Site-directed mutagenesis of *Clostridium perfringens* beta-toxin: expression of wild-type and mutant toxins in *Bacillus subtilis*. *FEMS Microbiol. Lett.* 158:17–23
115. Stevens DL. 1997. Necrotizing clostridial soft tissue infections. See Ref. 95, pp. 141–51
116. Stevens DL, Tweten R, Awad MM, Rood JI, Bryant AE. 1997. Clostridial gas gangrene: evidence that alpha and theta toxins differentially modulate the immune response and induce acute tissue necrosis. *J. Infect. Dis.* 176:189–95
117. Tailliez P, Ehrlich SD, Chopin MC.

1994. Characterization of IS*1201*, an insertion sequence isolated from *Lactobacillus helveticus*. *Gene* 145:75–79

118. Thanabalu T, Porter AG. 1996. A *Bacillus sphaericus* gene encoding a novel type of mosquitocidal toxin of 31.8 kDa. *Gene* 170:85–89

119. Titball RW. 1993. Bacterial phospholipases C. *Microbiol. Rev.* 57:347–66

120. Titball RW. 1997. Clostridial phospholipases. See Ref. 95, pp. 223–42

121. Titball RW, Fearn AM, Williamson ED. 1993. Biochemical and immunological properties of the C-terminal domain of the alpha-toxin of *Clostridium perfringens*. *FEMS Microbiol. Lett.* 110:45–50

122. Titball RW, Hunter SEC, Martin KL, Morris BC, Shuttleworth AD, et al. 1989. Molecular cloning and nucleotide sequence of the alpha-toxin (phospholipase C) of *Clostridium perfringens*. *Infect. Immun.* 57:367–76

123. Titball RW, Leslie DL, Harvey S, Kelly D. 1991. Hemolytic and sphingomyelinase activities of *Clostridium perfringens* alpha-toxin are dependent on a domain homologous to that of an enzyme from the human arachadonic acid pathway. *Infect. Immun.* 59:1872–74

124. Titball RW, Rubidge T. 1990. The role of histidine residues in the α-toxin of *Clostridium perfringens*. *FEMS Microbiol. Lett.* 68:261–66

125. Toyonaga T, Matsushita O, Katayama S-I, Minami J, Okabe A. 1992. Role of the upstream region containing an intrinsic DNA curvature in the negative regulation of the phospholipase C gene of *Clostridium perfringens*. *Microbiol. Immunol.* 36:603–13

126. Traving C, Roggentin P, Schauer R. 1997. Cloning, sequencing and expression of the acylneuraminate lyase gene from *Clostridium perfringens* A99. *Glycoconj. J.* 14:821–30

127. Traving C, Schauer R, Roggentin P. 1994. Gene structure of the 'large' sialidase isoenzyme from *Clostridium perfringens* A99 and its relationship with other clostridial nanH proteins. *Glycoconj. J.* 11:141–51

128. Tso JY, Siebel C. 1989. Cloning and

expression of the phospholipase C gene from *Clostridium perfringens* and *Clostridium bifermentans*. *Infect. Immun.* 57:468–76

129. Tsutsui K, Minami J, Matsushita O, Katayama S-I, Taniguchi Y, et al. 1995. Phylogenetic analysis of phospholipase C genes from *Clostridium perfringens* types A to E and *Clostridium novyi*. *J. Bacteriol.* 177:7164–70

130. Tweten RK. 1988. Cloning and expression in *Escherichia coli* of the perfringolysin O (theta toxin) gene from *Clostridium perfringens* and characterization of the gene product. *Infect. Immun.* 56:3228–34

131. Tweten RK. 1988. Nucleotide sequence of the gene for perfringolysin O (theta toxin) from *Clostridium perfringens*: significant homology with the genes for streptolysin O and pneumolysin. *Infect. Immun.* 56:3235–40

132. Tweten RK. 1995. Pore-forming toxins in gram-positive bacteria. In *Virulence Mechanisms of Bacterial Pathogens*, ed. JA Roth, CA Bolin, KA Brogden, C Minion, MJ Wannemuehler, pp. 207–29. Washington, DC: Am. Soc. Microbiol.

133. Tweten RK. 1997. The thiol-activated clostridial toxins. See Ref. 95, pp. 211–21

134. Van Damme-Jongsten M, Wernars K, Notermans S. 1989. Cloning and sequencing of the *Clostridium perfringens* enterotoxin gene. *Anton. van Leeuwen.* 56:181–90

135. Williamson ED, Titball RW. 1993. A genetically engineered vaccine against the alpha-toxin of *Clostridium perfringens* protects against experimental gas gangrene. *Vaccine* 11:1253–58

136. Yoshihara K, Mastushita O, Minami J, Okabe A. 1994. Cloning and nucleotide sequence analysis of the *colH* gene from *Clostridium histolyticum* encoding a collagenase and a gelatinase. *J. Bacteriol.* 176:6489–96

137. Zhao Y, Melville SB. 1998. Identification and characterization of sporulation-dependent promoters upstream of the enterotoxin gene (*cpe*) of *Clostridium perfringens*. *J. Bacteriol.* 180:136–42

Annu. Rev. Microbiol. 1998. 98:361–95

TOUR DE PACLITAXEL:
Biocatalysis for Semisynthesis

Ramesh N. Patel

Department of Microbial Technology, Bristol-Myers Squibb Pharmaceutical Research
Institute, PO Box 191, New Brunswick, New Jersey 08903;
e-mail: ramesh_n._patel@cc mail.bms.com.

KEY WORDS: C-13 taxolase, C-10 deacetylase, C-7 xylosidase, paclitaxel C-13 side chain
             synthons, paclitaxel semisynthesis

ABSTRACT

In collaboration with the National Cancer Institute, Bristol-Myers Squibb has
developed paclitaxel for treatment of various cancers; it has been approved by
the Food and Drug Administration for the treatment of ovarian and metastatic
breast cancer. Originally paclitaxel was isolated and purified from the bark of
Pacific yew trees. This source of paclitaxel was considered to be economically
and ecologically unsuitable as it required the destruction of the yew trees. This
review article describes alternate methods for the production of paclitaxel, specifically, a semisynthetic approach and the application of biocatalysis in enabling
the semisynthesis of paclitaxel. Three novel enzymes were discovered in our
laboratory that converted the variety of taxanes to a single molecule, namely
10-deacetylbaccatin III (paclitaxel without C-13 side chain and C-10 acetate), a
precursor for paclitaxel semisynthesis. These enzymes are C-13 taxolase (catalyzes the cleavage of C-13 side chain of various taxanes), C-10 deacetylase
(catalyzes the cleavage of C-10 acetate of various taxanes), and C-7 xylosidase
(catalyzes the cleavage of C-7 xylose from various xylosyltaxanes). Using a biocatalytic approach, paclitaxel and a variety of taxane in extracts of a variety of
Taxus cultivars were converted to a 10-deacetylbaccatin III. The concentration of
10-deacetylbaccatin III was increased by 5.5- to 24-fold in the extracts treated
with the enzymes, depending upon the type of *Taxus* cultivars used. Biocatalytic
processes have also been described for the preparation of C-13 paclitaxel side
chain synthons. The chemical coupling of 10-deacetylbaccatin III or baccatin
III to C-13 paclitaxel side chain has been summarized to prepare paclitaxel by
semisynthesis.

0066-4227/98/1001-0361$08.00

CONTENTS

INTRODUCTION: The Discovery of Taxol

In December 1992, Taxol® (paclitaxel) (Compound 1) was approved for marketing for the treatment of refractory ovarian cancer. It took almost 30 years of work that began with the collection of *Taxus brevifolia* in the Pacific Northwest in Washington state in 1962.

In the early 1950s, the major cancer drug discovery and development program in the United States was at the Sloan-Kettering Research Institute. Most of the compounds evaluated were at the Sloan-Kettering; however, capacity was very low for the needs of the scientific community. In 1955, Congress directed the National Cancer Institute (NCI) to organize and support a program for research in cancer drug discovery and treatment (97). This led to the formation of the Cancer Chemotherapy National Service Center. It has the mission of supporting research in cancer drug discovery and treatment and of developing the standardization service so that material received from various national laboratories could all be tested under identical conditions and in a similar test system. Screening of plant extracts from various sources for the plant screening program was initiated at the NCI under the direction of Dr. Jonathan Hartwell. An agreement was negotiated with the United States Department of Agriculture (USDA) to collect and identify materials for the screening program (76). In addition to collecting new plant samples, attempts were made to secure access to many existing collections of plant extracts made for other screening purposes. Thousands of ethanolic extracts of plants prepared at the

USDA Eastern Regional Research Laboratory under guidance of Dr. Monroe Wall were submitted for initial testing in mice using an adenocarcinoma tumor model. The plant collection agreement between the NCI and USDA resulted in many collecting trips by botanists to a variety of locations in the US for a broad taxonomic cross-section of plants with morphological diversity and the unique variety of chemicals these plants produced as secondary metabolites (83). A USDA collection team under the direction of Arthur Barclay went to Washington state to the Gifford Pinchot National Forest and collected samples of the Pacific yew, *Taxus brevifolia*, family *Taxaceae*. The *Taxaceae* is a family with only one genus, *Taxus*. Two samples were collected by the team, one from stems and fruit and the other from stems and bark. Both samples were dried, extracted, and evaluated against carcinoma of the nasopharynx (KB) cells. Only the latter sample (stems and bark) was active and demonstrated cytotoxic activity. The fractionation of plant extracts was subsequently carried out in Dr. Wall's laboratories, at the Research Triangle Institute in North Carolina, from 30 pounds of bark of the Pacific yew, *Taxus brevifolia*. Fractions from these extracts were found to be active in vivo against P-1534 leukemia, Walker 256 carcinosarcoma, and P388 leukemia. Paclitaxel was isolated by following the bioactivity of the compound using cytotoxicity to KB cells and in vivo antitumor activity in the above systems (90, 91).

In 1966, paclitaxel was isolated in pure form (it took about two years to accomplish this). Isolation of paclitaxel proved to be a challenging task because it was present in very small amounts in the bark. The yield of paclitaxel was 0.02% from the dried bark of *T. brevifolia*. It was also reported to be present in other *Taxus* species such as *T. baccata* and *T. cuspidata* (57). The structure of paclitaxel was determined by a combination of NMR and X-ray crystallographic techniques. The paclitaxel was poorly soluble in many solvents and readily crystallized; however, the needle-shaped crystals were not suitable for X-ray crystallography. The solvolysis of paclitaxel in methanol gave a tetraol characterized by X-ray crystallography as its bisiodoacetate (Compound 2) (Figure 1).

A methylester (Compound 3) derived from the side chain (Figure 1) was also characterized by X-ray crystallography by its *p*-bromobenzoate ester and methyl acetate. These data plus additional ^{1}H-NMR studies were used to deduce the complex structure of paclitaxel (Compound 1) (Figure 1) with a molecular formula of $C_{47}H_{51}NO_{14}$ (92). Evidence was also presented demonstrating that both the paclitaxel nucleus, also called Baccatin III (Compound 4), and the C-13 side chain were essential for antitumor activity of paclitaxel.

DEVELOPMENT OF PACLITAXEL

Paclitaxel showed antitumor activity in the L1210 leukemia and the Walker 256 carcinosarcoma models. Activity was also demonstrated in the P1534 and

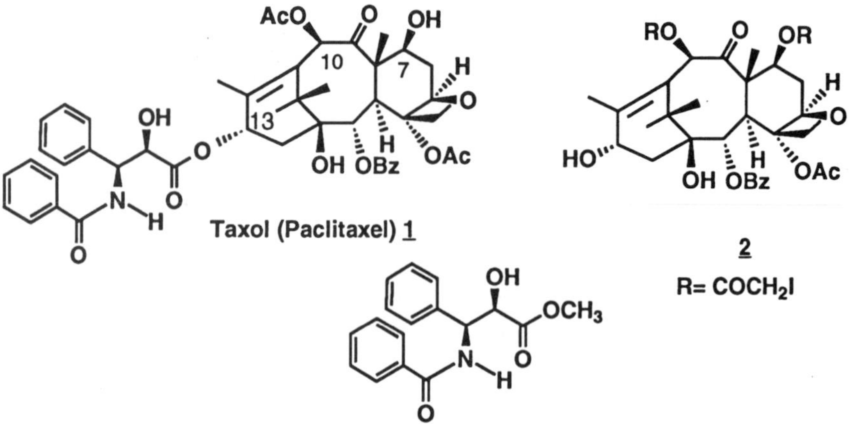

Taxol (Paclitaxel) 1

2

R= $COCH_2I$

C-13 Side-chain methyl ester 3

Figure 1 Structure of Taxol (Paclitaxel).

P388 leukemia models; however, it was only weakly active (85). The biological activity of paclitaxel was not particularly impressive in comparison to that of other known and novel anticancer agents such as vinblastine, colchicine, and vincristine. Paclitaxel thus met minimal activity in all of these models but did not meet NCI criteria for development in any of them. It was recognized in the early 1970s that the best-studied models, such as the L1210 leukemia system for antitumor activity, were valuable for drug discovery, but that the clinical spectrum of agents was mainly against rapidly growing tumors such as leukemias and lymphomas. Further, the need was recognized to develop new models against solid tumors, which led to the introduction of B16 melanoma as a secondary screen for compounds active in the primary leukemia screen. The first test of paclitaxel against B16 melanoma was done in 1974, and the data showed an increase in life span (ILS) of up to 126% at a dose of 10 mg/Kg per intraperitonial injection on a daily dose schedule for nine days against implanted tumor. This was clearly more than the ILS of 50% needed for consideration in the B16 melanoma model (88). Paclitaxel was presented to the NCI decision-making committee in 1977 as a development candidate and was later approved. Still, there were significant hurdles to development of paclitaxel owing to its low solubility and poor potency, relatively difficult purification from bark of the yew tree, and need for an appropriate formulation to supply paclitaxel.

Formulation Studies

The solubility of paclitaxel in water and other aqueous-based systems is poor. The potency of paclitaxel is three- to ten-fold less than that of other natural-product-derived anticancer compounds such as vinblastine, doxorubicin, and daunorubicin. Thus, the formulation problem was due to the high dose and low aqueous solubility of the compound. Because paclitaxel lacks functional groups that are ionizable in a pharmaceutically useful range, the manipulation of pH does not enhance solubility. Further approaches to improve solubility by the addition of charged complexing agents or by producing salts of the drug were not feasible. Various attempts were carried out using emulsions, mixed solvent, and liposome formulations without much success. Various prodrug approaches that retain pharmacological activity were also considered (15, 55, 59). [A prodrug is defined as a chemically modified form of a drug that undergoes in vivo transformation to its active form (38).] The chemical derivatization at either 2′ or the 7 position appeared to give the best prospect for prodrug development (8, 10, 24, 25, 29, 31). Various succinate and glutarate half-esters and salts of these acids were found to have better activity. A number of amino-acid derivatives at the 2′-position were also prepared. (Several chemical derivatives at the 7-position have also been prepared, but they turned out to be more stable and do not undergo hydrolysis to paclitaxel at a suitable rate.) Many prodrugs showed only marginal improvements in solubility or were unstable. Some prodrugs were stable with relatively higher solubility, but the regeneration of paclitaxel from prodrugs was very slow. An alternative to the synthesis of paclitaxel derivatives (prodrugs) is the use of solvents or carriers to provide sufficient solubility for human administration and suitable pharmaceutical properties. Various agents such as ethanol, dimethylsulfoxide, Tween 80 (polysorbate-80), carboxymethyl cellulose (CMC), hydroxypropyl cellulose (HPC), polyethylene glycols (PEG-400, PEG-hydroxystearate), soybean oil, lecithin, soy lipids, and polyoxyethylated castor oil (Cremophor® EL), alone or in combination were evaluated in formulation studies (1, 74, 80, 84). Finally, the formulation selected for clinical development consisted of paclitaxel solubilized at a concentration of 6 mg/mL in 50% polyoxyethylated castor oil and anhydrous ethanol. One common packaging of formulated paclitaxel is in vials containing 5 mL of this solution. For adminstration, the drug is diluted 5- to 20-fold in saline or 5% dextrose to a final concentration of 0.3 to 1.2 mg/mL.

Mechanism of Action

Microtubules are a major component of the spindle fibers and the cytoskeleton of eukaryotic cells that participate in cellular functions such as mitosis, morphogenesis, motility, and intracellular organelle transport. Microtubules are assembled mainly from tubulin, a dimeric protein of about 110,000 MW, consisting of two polypeptides chains (α and β) of identical molecular weight (6). Microtubules contain tubulin and several copurified proteins referred to as microtubule-associated proteins (MAPs) that are components of the mitotic spindle and other specialized cellular structure and are important for microtubule stability. MAPs are also important in both the nucleation and elongation phases of the assembly reaction. Tubulin has two sites for binding guanine nucleotides; one is exchangeable and the other is not. GTP binds at the exchangeable site, initiates the assembly of microtubules, and is hydrolyzed to GDP (7). Tubulin also has several sites for drug binding (95). Antimicrotubule agents are among the most important anticancer drugs. They arrest cell division at metaphase by interfering with normal spindle formation.

Antimicrotubule drugs can be divided into two classes based on their effect on the microtubule. Most compounds, including vinca alkaloids, colchicine, podophyllotoxin, and nocodazole, are involved in destabilization of the microtubule polymer and interfere with the assembly of tubulin (95). Paclitaxel is unique in its mechanism of inhibition: It interferes with microtubule function by binding to the polymer and stabilizing. Paclitaxel inhibits depolymerization of microtubules back to tubulin (77). Paclitaxel promotes the assembly of tubulin even under conditions that generally do not support polymerization (such as the absence of MAPs or exogenous GTP, low temperature, and mild alkalinity). Paclitaxel promotes both the nucleation and elongation phases of the polymerization process and can reduce the critical concentration of tubulin to almost zero (64). Researchers using several photoaffinity analogues of paclitaxel demonstrated that paclitaxel binds preferentially to β-tubulin (48, 73).

Antitumor Studies

Paclitaxel had no substantive activity against the B16 melanoma when the drug was given by the oral or intravenous route against the intraperitoneally (i.p.)-implanted tumor; marginal activity occurred when paclitaxel was administered i.p. against the subcutaneous implanted tumor. Paclitaxel had good activity against the B16 melanoma only when administered i.p. against i.p.-implanted tumor. Similar results were obtained by examination of data for the L1210, P388, and P1534 leukemias (85). In 1976, NCI introduced the human tumor xenograft. In conjunction with the discovery of "nude" mouse (immunosuppressed from the lack of thymus gland), this led to a breakthough in the development of human tumor models to screen potential anticancer drugs (19). Three models were introduced, representing major types of human solid tumors, (a) LX-1 lung xenograft, (b) MX-1 breast xenograft, and (c) CX-1 colon xenograft. Paclitaxel caused regression of the implanted tumor in the MX-1 model, and 80–90% inhibition of tumor growth in the LX-1 and CX-1 models (28). Subsequent to completion of toxicology studies in 1982, an Investigational New Drug (IND) application was filed with the Food and Drug Adminstration (FDA), and the FDA approved paclitaxel for clinical trials in 1984. Early Phase-I clinical trials were conducted on a variety of schedules including 1, 3, 6, or 24-hour infusions every 21 days, and 1- to 6-hour infusions daily for five days. Throughout the Phase-I clinical trials and the early Phase-II trials, paclitaxel was in short supply, and the number of studies that could be initiated was restricted; consequently, many important tumor types were not studied. The early supply of paclitaxel (500 g to 1 Kg) was prepared from 10,000 pounds of Pacific yew bark by Polysciences, Inc (84). After the Johns Hopkins group discovered the important activity of paclitaxel in treatment of refractory ovarian cancer in 1989 (56), the NCI put forward a request for applications for a Cooperative Research and Development Agreement (CRADA) to expand and supply extensive clinical trials leading to marketing of paclitaxel in cancer treatment. The next major clinical development was the discovery that paclitaxel had remarkable activity in primary metastatic breast cancer, with a response rate of 56% (39). The CRADA competition was won by Bristol-Myers Squibb Company (BMS) in 1991; as a part of this agreement, BMS was responsible for increasing the production of paclitaxel from bark and for developing an alternative supply of paclitaxel. BMS signed two cooperative agreements with the Department of Agriculture's US Forest Service (USFS) and the Department of the Interior's Bureau of Land Management (BLM) for the production of paclitaxel from Pacific yew bark. Hauser Chemicals, the authorized collector of bark for BMS, harvested approximately 73,000 pounds in 1990, 1.6 million pounds in 1991, and 1.5 million pounds in 1992. Hauser Chemical Research began to supply GMP-grade paclitaxel to BMS through a production process that involved bark collection, drying, chipping, grinding, and chemical purification of paclitaxel, including column chromatography (13). About 16,000 pounds of bark from approximately 2000–2500 yew trees were required to prepare 1 Kg of paclitaxel. BMS provided about 16 Kg of formulated paclitaxel for human treatment in vials in 1992 (13).

DIVERSITY AND DISTRUBUTION OF PACLITAXEL

Taxus is widespread, although rarely abundant, in North America and Eurasia. *Taxus* generally occurs in moist, temperate forests in the Northern Hemisphere but also occurs in subtropical and tropical areas of Southeast Asia and Central America. *Taxus* occurs in the southern hemisphere in Sumatra and Celebes. In North and Central America, four native species of *Taxus* are recognized by their geographical location. The Pacific yew (western yew), *T. brevifolia*, is widely distributed in western parts of Canada and the United States. The Pacific yew grows from Alaska to the Sierra Nevada mountains in California, especially in the Cascade Ranges in Washington and Oregon. In the Rocky Mountains, it grows from British Columbia through Idaho and Montana. The Mexican yew, *T. globosa*, is native to Mexico, El Salvador, Honduras, and Guatamala. The

Florida yew, *T. floridana*, is endemic in northern Florida. The Canadian yew, *T. canadensis*, grows from Newfoundland to Manitoba and south from Iowa to North Carolina. The English yew, *T. baccata*, is native to Europe and north Africa.

T. cuspidata, Japanese yew, is native to eastern Asia (China, Japan, Korea and parts of Russia). The Himalayan yew, *T. wallichiana*, grows from eastern Afghanistan to Tibet and China (3, 72). Variation in the taxane content of the *Taxus* plant [paclitaxel (Compound 1), 10-deacetyltaxol (Compound 6), cephalomannine (Compound 7), baccatin III (Compound 4), 10-deacetylbaccatin III (Compound 5), and taxol C (Compound 8); Figures 1 and 2] was also observed relative to geographical locations, environmental, agricultural influence, harvesting, and storage conditions (89, 94).

Although the paclitaxel content of Pacific yew bark ranges from 0.0001 to 0.08%, the average isolated yield is in the range of 0.014–0.017% (82, 89). About 16,000 pounds of bark is required to produce 1 Kg paclitaxel (13). Discovery of this fact led to development programs for alternative sources of paclitaxel. The least active (but very abundant) compound in the dried leaves of *T. baccata* (14, 30) was Compound 5, or 10-deacetylbaccatin III (10-DAB),

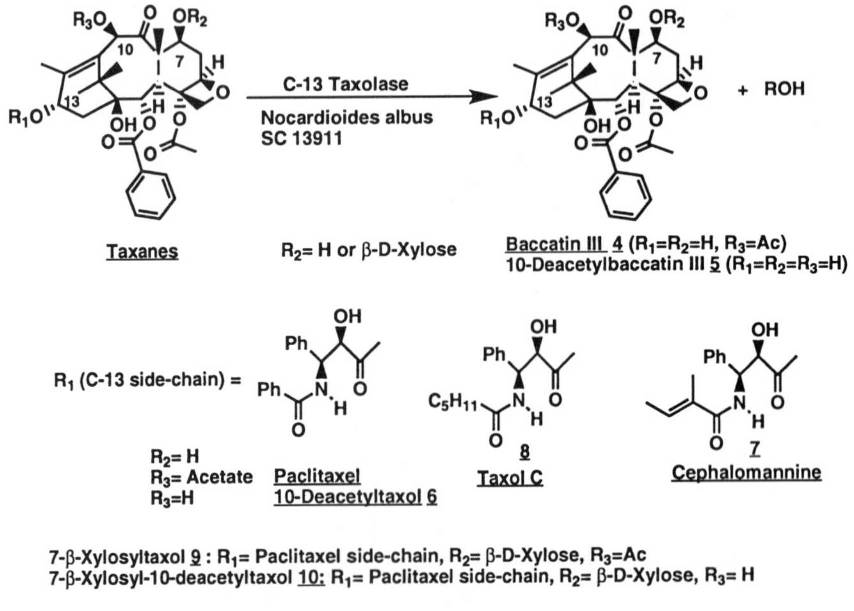

Figure 2 Enzymatic hydrolysis of the C-13 side chain from taxanes by C-13 taxolase from *N. albus* SC 13911.

a taxane without the C-10 acetate and C-13 side chain of paclitaxel. This recognition led to the use of renewable resources of yew leaves to develop a process for paclitaxel synthesis (40, 41).

NURSERY PRODUCTION OF TAXUS PLANTS

The benefits of cultivation of medicinal plants include increasing the amount of material available, stabilizing supply, increased quality control, and assurance of desired botanical sources, in addition to increased production of the desired chemicals by genetic improvement, agronomic manipulation, and postharvesting handling (62). For maximum plant growth and enhanced production of secondary metabolites, a complete agronomic system should include (a) plant line selection; (b) cultivation at optimum sites for climate and soil type; (c) optimum plant spacing; (d) fertilization and irrigation protocol; (e) weed, disease, and insect controls; and (f) harvesting, drying, and storage protocols to maximize the desired chemical content for extraction and purification (87).

A commercial-scale nursery plantation of *Taxus* for paclitaxel and other taxanes was established by the Weyerhaeuser Company with the support of BMS. The American nursery industry is the world's largest producer of cultivated *Taxus*, with over 30 million yews in cultivated fields that annually generate surplus biomass from clippings, root pruning, and cull plants. A collaborative effort between USDA, NCI, and BMS with Zelneka Nursery led to the collection of over 40,000 pounds of dried clippings and leaves (stripped from dried whole plants of *T. hicksii*) for extraction and isolation of paclitaxel (70). Hauser Chemical Research Inc. also studied laboratory-scale isolation of paclitaxel and was able to produce paclitaxel in 98% purity in 0.01% isolation yield (5). The nursery production in Europe of high-yielding *Taxus* for 10-DAB is being supported by Rhone-Poulenc Rorer (17).

SEMISYNTHESIS OF PACLITAXEL

Extracts of *Taxus* cultivars contain a complex mixture of taxanes, with paclitaxel usually constituting a very low proportion of the total taxanes. The most valuable material in this mixture for semisynthesis is the taxane "nucleus" component of baccatin III (Compound 4) and 10-DAB (Compound 5). Conversion of taxanes to 10-DAB by cleavage of the C-10 acetate and the C-13 paclitaxel side chain is a very attractive approach to increase the concentration of this valuable compound in yew extracts. Cleavage of paclitaxel at C-10 or C-13 has been described; however, epimerization at C-7 was observed (9). Magri et al (54) reduced paclitaxel (Compound 1), cephalomannine (Compound 7), or a mixture of the two compounds to give baccatin III (Compound 4), but application of this method to yew extracts was not demonstrated. We reported the enzymatic conversion of a complex mixture of taxanes to a single compound (10-DAB, Compound 5) by treatment of extracts prepared from a variety of yew cultivars (32–34, 58, 69). Details of this process are described below.

Biocatalytic Hydrolysis of Taxanes to 10-Deacetylbaccatin III

By using selective enrichment techniques, we have isolated two strains of *Nocardioides* that contain the novel enzymes C-13 taxolase and C-10 deacetylase (32–34, 58, 69). The extracellular C-13 taxolase derived from filtrate of fermentation broth of *Nocardioides albus* SC 13911 catalyzed the cleavage of the C-13 side chain from paclitaxel and related taxanes such as taxol C (Compound 8), cephalomannine (Compound 7), 7-β-xylosyltaxol (Compound 9), 7-β-xylosyl-10-deacetyltaxol (Compound 10), and 10-deacetyltaxol (Compound 6) (Figure 2).

The kinetics of the C-13 taxolase reaction were demonstrated using an ethanolic concentrate of *T. hicksii* cultivars. The reaction mixture contained 120 mL of ethanolic concentrate, 880 mL of 25 mM potassium phosphate buffer (pH 7.0), and ten units of C-13 taxolase. The reaction was carried out at 25°C and 200 RPM agitation.

At the end of the 8-hr reaction, the concentration of baccatin III was increased fourfold from 25 mmoles/L to 110 mmoles/L. The total taxane concentration remained the same (Figure 3).

The reaction yield was close to 100% for conversion of taxanes to baccatin III and 10-DAB. High pressure liquid chromatography (HPLC)/thermospray mass spectroscopy was used to confirm the identity of hydrolysis products (Table 1). A fermentation process was developed for growth of *N. albus* SC 13911 to produce C-13 taxolase and was scaled up to 5000-L batches. In a batch-fermentation process, the growth of *N. albus* SC 13911 was completed in 10 hr.

The C-13 taxolase production started at 8 hr and was completed 28 hr after inoculation. Most C-13 taxolase is produced upon completion of the growth period. The enzyme activity is not related to growth of the organism (Figure 4). An 11 U/L (mmoles of baccatin III produced from taxanes/min/mL) activity was obtained. The extracellular enzyme, after a cell-removal step, was precipitated by salt (60% saturation) and recovered by centrifugation after addition of 0.6% SolkaFloc. An overall 90% recovery of C-13 taxolase was obtained from the fermentation broth.

The intracellular C-10 deacetylase derived from fermentation of *Nocardioides luteus* SC 13912 catalyzed the cleavage of C-10 acetate from paclitaxel, related taxanes, and baccatin III to yield 10-DAB (Figure 5).

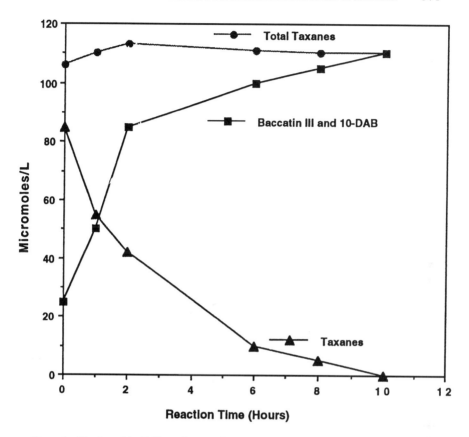

Figure 3   Kinetics of the C-13 taxolase reaction.

The kinetics of the C-10 deacetylase reaction are shown in Figure 6. The reaction mixture after C-13 taxolase treatment was used. To a 1L reaction mixture, 15 units of C-10 deacetylase (whole cells) were added, and the reaction was continued. A reaction yield of 100% was obtained in a 6-hr reaction time when catalyzed at pH 7.0 and 25°C.

The fermentation process for growth of *N. luteus* SC 13912 was developed and also scaled up to 4000-L batches. Growth of the organism was completed in 16 hours. The intracellular C-10 deacetylase production was started upon completion of growth. Five U/L (mmoles of 10-DAB produced from baccatin III/min/mL) of activity was obtained (Figure 4). C-10 deacetylase was recovered as wet cell paste (40 Kg) in 94% overall yield from the fermentation broth.

A biotransformation process was demonstrated for the conversion of paclitaxel and related taxanes in extracts of *Taxus* plant cultivars to a single compound

Table 1 HPLC/thermospray mass spectrometry of products of C-13 taxolase and C-10 deacetylase reactions

Enzyme	Substrate	Products	(M+H)	(M+NH^4)	(M+Na)	(M+K)
C-13	Paclitaxel	Baccatin III	587	604		
		C-13 Side-chain	286			
	Cephalomannine	Baccatin III	587	604		
	7-Xylosyltaxol	7-Xylosylbaccatin III	719	736		
	7-Xylosyl-10-deacetyltaxol	7-Xylosyl-10-deacetyl baccatin III		694	699	
	10-Deacetyltaxol	10-Deactylbaccatin III	545	562		
C-10	Baccatin III	10-DAB	545	562	567	583
	Paclitaxel	10-Deacetyltaxol	812		834	850
C-13 & C-7	7-Xylosyl-10-deacetyltaxol	10-DAB	545	562	567	583
C-13 & C-7	7-Xylosyltaxol	Baccatin III	587	604		

10-DAB by using both enzymes. In the bioconversion process, ethanolic extracts of the whole young plant (dried, milled, and extracted with ethanol) of five different cultivars of *Taxus* independently were first treated with a crude preparation of the C-13 taxolase to give complete conversion of measured taxanes to baccatin III and 10-DAB in 6 hr.

N. luteus SC 13192 whole cells were then added to the reaction mixture to give complete conversion of baccatin III to 10-DAB. The concentration of 10-DAB was increased by 5.5- to 24-fold in the extracts treated with the two enzymes. The bioconversion process was also applied to extracts of the bark of *T. brevifolia* to give a nearly 12-fold increase in 10-DAB concentration (Table 2). The enhancement of 10-DAB concentration in yew extracts was very useful in increasing the amount and purification of this key precursor (10-DAB).

The C-13 taxolase (47,000 MW) from the filtrate of fermentation broth of *N. albus* SC 13911 and C-10 deacetylase (40,000 MW) from the cell extract of *N. luteus* SC 13912 were purified to homogeneity. Both N-terminal and internal sequences of each enzyme were determined to enable cloning (32, 58).

Biocatalytic Hydrolysis of Xylosyltaxanes to 10-DAB

Using enrichment culture techniques, organisms capable of hydrolyzing 7-β-xylosyltaxanes were isolated (35). Soil samples were plated on an enriched medium containing 1% birchwood xylan. After a week of incubation at room temperature, plates were overlaid with 0.8% agarose containing 1 mM 4-methyl-umbelliferyl β-D-xylose. Colonies that showed fluorescence under a longwave

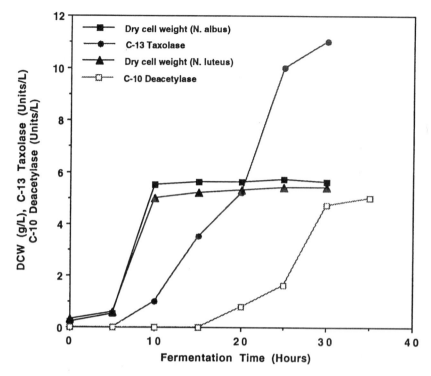

Figure 4 Fermentation of *N. albus* (C-13 taxolase) and *N. luteus* (C-10 deacetylase) in a 5000-L batch process.

ultraviolet lamp were isolated and further characterized. About 125 strains with xylosidase activity (as indicated by hydrolysis of 4-methyl-umbelliferyl-β-D-xyloside to the fluorescent 4-methyl-umbelliferone) were isolated; nine isolates produced 10-deacetyltaxol (Compound 6) and 10-DAB (Compound 5) from 7-β-xylosyl-10-deacetyltaxol (Compound 10), and 7-β-xylosyl-10-DAB (Compound 11), respectively (Figure 7). Both products were identified by HPLC retention times and also by LC/MS. The best culture that catalyzed the cleavage of xylose from 7-xylosyltaxol and 7-xylosyl-10-deacetyltaxol was identified as a strain of *Morexella* sp. Production of xylosidase was scaled up from a 15-L to a 500-L batch fermentation process.

The maximum xylosidase activity of 0.3 U/L was observed after 46 hours growth. From 500 L of broth, 5 Kg of wet cell paste was collected and used in the biotransformation process. Cell suspensions of *Moraxella* sp. in 50 mM phosphate buffer (pH 7.0) gave complete conversion of 7-xylosyl-10-deacetyltaxol to 10-deacetyltaxol and 7-xylosyl-10-DAB to 10-DAB (Figure 8).

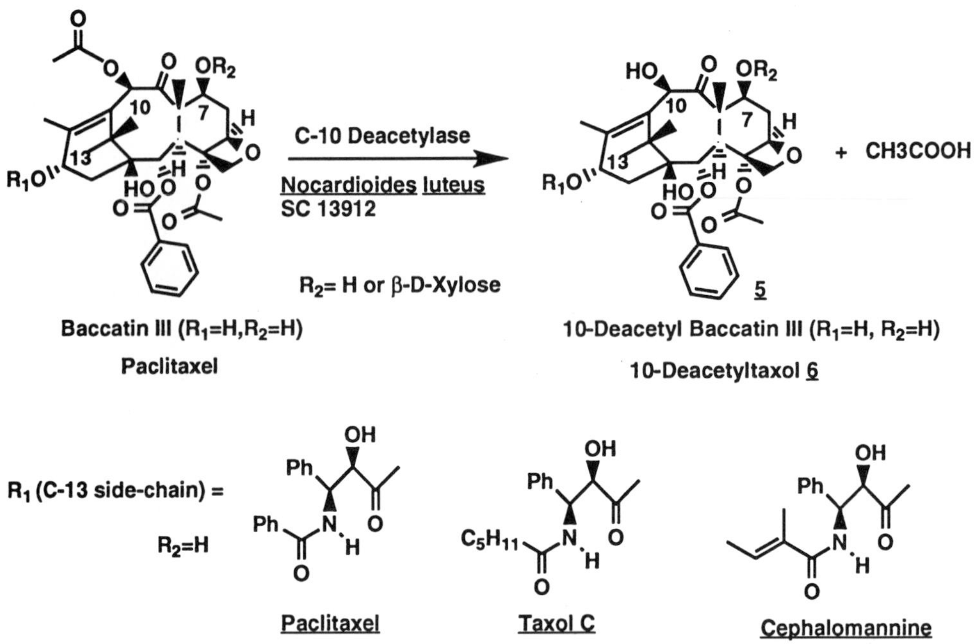

Figure 5 Enzymatic hydrolysis of C-10 acetate from taxanes by C-10 deacetylase from *N. luteus* SC 13192.

The C-13 side chains of 7-xylosyltaxol (Compound 9) and 7-xylosyl-10-deacetyltaxol (Compound 10) were removed by C-13 taxolase from *N. albus* SC 13911. The corresponding products were converted quantitatively to baccatin III (Compound 4) and 10-DAB (Compound 5) by treatment with xylosidase. HPLC/thermospray mass spectroscopy was used to confirm the identity of hydrolysis products (Table 1). Xylosidase activity in extracts of *Moraxella* sp. was found in both the soluble and particulate fractions. About 80% of the total activity was found in the pellet fraction obtained after centrifugation of the soluble fraction at 101,000 X g. Various xylosyltaxanes (7-β-xylosyltaxol (Compound 9), 7-β-xylosylcephalomannine (Compound 12), 7-β-xylosyl-10-deacetyltaxol (Compound 10), 7-β-xylosyl-10-deacetylcephalomannine (Compound 13), 7-β-xylosyl-10-DAB (Compound 11), and 7-β-xylosylbaccatin III (Compound 14) were converted to 10-DAB by treatment with three enzymes: xylosidase (*Moraxella* sp.), C-13 taxolase (*N. albus*), and C-10 deacetylase (*N. luteus*) from microbial sources.

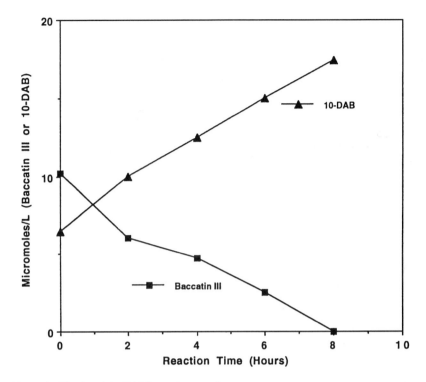

Figure 6 Kinetics of the C-10 deacetylase reaction.

Biocatalytic Preparation of Paclitaxel Side-Chain Intermediates

Two different processes were developed for the preparation of C-13 paclitaxel side-chain intermediates. In the first process, the stereoselective enzymatic
resolution of racemic acetate (Compound 15, cis-3-acetyloxy-4-phenyl-2-azeti-
dinone) to yield the desired (3R-cis)-acetate (Compound 16) and (3S-cis)-hy-
droxy-4-phenyl-2-azetidinone (Compound 17) by lipases has been demonstrated (Figure 9). Product (3R-cis)-acetate (Compound 16) is a key C-13
paclitaxel side-chain intermediate required for semisynthesis of paclitaxel
(66, 68).

Among commercially available enzymes evaluated, lipases from *Pseudomonas cepacia* (PS-30 from Amano International Enzyme Co.), *Geotrichum can-
didum* (GC-20), *Candida cylindraceae* (AY-30), and porcine pancreatic lipase catalyzed the hydrolysis of the undesired enantiomer in racemic acetate
(Compound 15) to the corresponding S(−)-alcohol (Compound 17). The desired

Table 2 Bioconversion of taxanes by C-13 taxolase and C-10 deacetylase

	Reaction time (hours)	Reaction volume (mL)	Cephalomannine (mg)	10-Deacetyltaxol (mg)	Paclitaxel (mg)	Baccatin III (mg)	10-DAB (mg)	10-DAB (X)
T. Adams (120 ml)	0	876	6.1	17.5	14	7	6.1	
	6		0	0	0	27.1	20.1	
	22		0	0	0	0	41.1	6.7
T. Runyani (200 ml)	0	1460	4.4	10.2	13.2	2.3	1.4	
	6		0	0	0	21.9	13.1	
	22		0	0	0	0	33.8	24
T. Dark Green Spreader (150 ml)	0	1170	9.3	8.2	12.8	3.51	3.5	
	6		0	0	0	28.1	8.2	
	22		0	0	0	0	44.4	12.6
T. Hicksii (120 ml)	0	936	16.3	15.9	23.4	2.8	12.2	
	6		0	0	0	42.1	27.1	
	22		0	0	0	2.4	66.4	5.5
T. Caspidata (150 ml)	0	1170	11.2	10.5	10.6	8.2	5.8	
	6		0	0	0	19.2	18.7	
	22		0	0	0	0	41	7
T. Brevifolia (150 mL)	0	1220	15.8	31.7	12.2	8.5	10.8	
	6		0	0	0	82.9	41.4	
	22		0	0	0	0	128	11.8

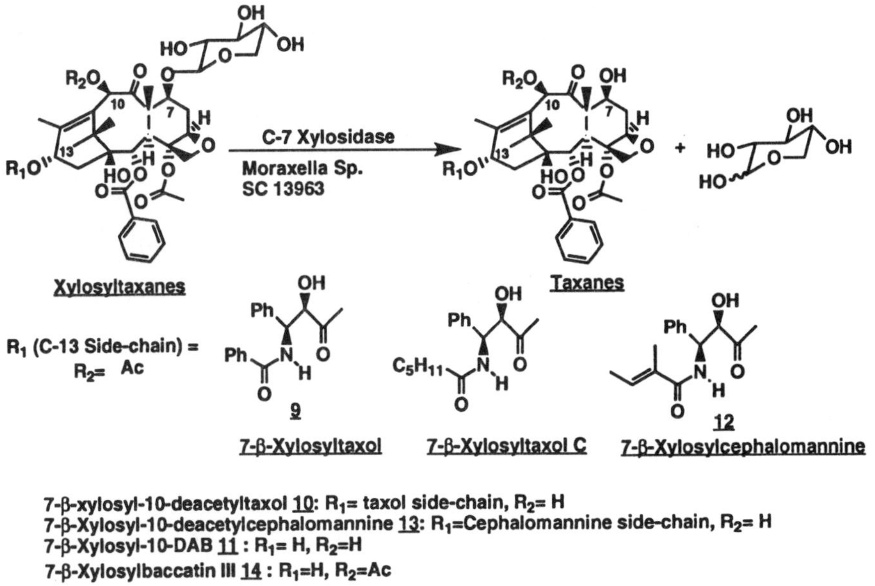

Xylosyltaxanes

R₁ (C-13 Side-chain) =
R₂= Ac

R_1 (C-13 Side-chain) =
R_2= Ac

Taxanes

9
7-β-Xylosyltaxol

7-β-Xylosyltaxol C

12
7-β-Xylosylcephalomannine

7-β-xylosyl-10-deacetyltaxol **10**: R₁= taxol side-chain, R₂= H
7-β-Xylosyl-10-deacetylcephalomannine **13**: R₁=Cephalomannine side-chain, R₂= H
7-β-Xylosyl-10-DAB **11** : R₁= H, R₂=H
7-β-Xylosylbaccatin III **14** : R₁=H, R₂=Ac

Figure 7   Enzymatic hydrolysis of C-7 xylose from xylosyltaxanes by C-7 xylosidase from *Morexella* sp. SC 13963.

enantiomer (3R-cis) acetate (Compound 16) remained unreacted. The reaction yield (40–96%) and the optical purity (94–99.6%) of chiral acetate (Compound 16) obtained depended upon the lipase used in the reaction mixture (Table 3).

For an in-house source of enzyme (BMS lipase), a lipase fermentation and recovery process was developed using *Pseudomonas* sp. SC 13865. The highest lipase activity achieved in a fed-batch fermentation process was 1500 units/mL. During fermentation, most of the glucose and soybean oil was consumed during the first 36 hours, which corresponded to the period of rapid cell growth and extracellular lipase production. Crude BMS lipase (1.7 kg) containing 140,000 units/g of lipase activity was recovered from the filtrate of fermentation broth by ethanol precipitation. BMS lipase was immobilized on Accurel polypropylene with 98% adsorption efficiency. The immobilized BMS lipase was evaluated in the resolution of racemic acetate (Compound 15). Substrate was used at 10 g/L, and immobilized enzyme was used at 3 g/L concentration. A reaction yield of 96 M% and the optical purity of 99.5% was obtained for (3R-cis) acetate (Compound 16) after 40 hours reaction time (Table 4). For an alternative enzyme source, lipase PS-30 was also immobilized on Accurel polypropylene. The kinetics of hydrolysis of racemic acetate (Compound 15) was investigated

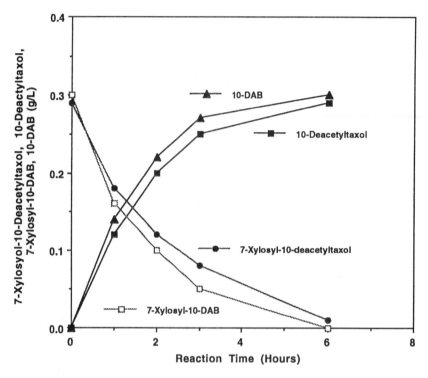

Figure 8 Kinetics of the C-7 xylosidase reaction.

independently using immobilized BMS lipase and lipase PS-30. The reaction
was conducted at 10 g/L substrate and 3 g/L enzyme concentrations.

As shown in Table 4, the initial rate (0–24 hr) of hydrolysis was faster than
the later rate of hydrolysis (24–40 hr). The immobilized BMS lipase and lipase
PS-30 were reused for ten cycles without any loss of activity, productivity, or
the optical purity of the product. Results of reusability studies using lipase
PS-30 are shown in Table 5. An average reaction yield of 94% and an optical
purity of 99.3% were obtained for chiral acetate (Compound 16). About 3.1
grams of Compound 16 was isolated from each cycle in an overall 62% yield.
The rate of hydrolysis of Compound 15 remained constant over ten cycles
(0.12 g/L/h). Similar results were obtained in reusability studies using BMS
lipase, except the reaction was completed in 36 hours.

The enzymatic resolution process was scaled up to 75-L and 150-L using
immobilized BMS lipase and lipase PS-30, respectively. After 38 hr reaction
time, 94 M% reaction yield and 99.4% optical purity of Compound 16 were
obtained in each batch. From the reaction mixtures, 331 and 675 g of chiral

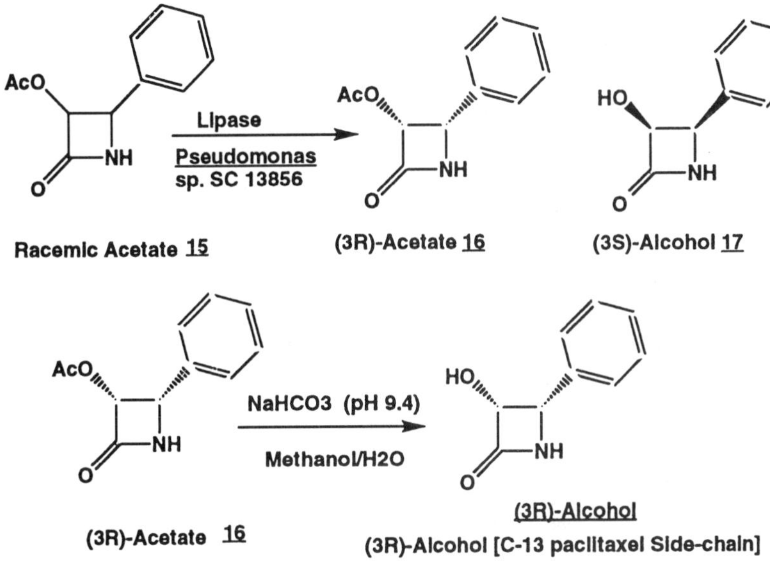

Figure 9   Enzymatic resolution of paclitaxel C-13 side chain intermediates by lipases.

acetate (Compound 16) were isolated (overall yield of 88 M% and 90 M%, respectively) from the two batches. The isolated acetate (Compound 16) ($[\alpha]_D = -15.6$) from both batches gave an optical purity of 99.5% by chiral HPLC and 99.9% area % purity by HPLC.

In the second process, the enzymatic synthesis of paclitaxel side-chain synthon by the stereoselective microbial reduction of Compound 18 [2-keto-3-(N-benzoyl-amino)-3-phenyl propionic acid ethyl ester] to yield Compound 19a [(2R,3S)-(−)-N-benzoyl-3-phenyl isoserine ethyl ester] was demonstrated (65, 67). The reduction of Compound 18 could result in the formation of four possible alcohol diastereomers (Compounds 19a–d) (Figure 10). Various microbial cultures were evalutated for stereoselective reduction of Compound 18. The reaction yield and stereoselectivity using the best positive cultures were dependent upon the microorganism used during the reduction of Compound 18 to Compound 19a (Table 6).

Organisms from genus *Nocardia, Candida, Rhodococcus, Mortierella, Saccharomyces,* and *Hansenula* predominantly converted ketone (Compound 18) to the desired alcohol (Compound 19a) in high optical purity (>91%), while organisms from genus *Pullularia* and *Trichoderma* (data not shown) gave lower optical purity (75–88%) of the desired alcohol (Compound 19a). *H. polymorpha* SC 13865 and *H. fabianii* SC 13894 effectively reduced Compound 18 to

Table 3 Enzymatic hydrolysis of *cis*-3-acetyloxy-4-phenyl-2-azetidinone[a]

Enzyme	Reaction time (hours)	Yield of compound 16 (%)	Optical purity of compound 16 (%)
BMS Lipase P	42	95	99.4
Pseudomonas cepacia (lipase PS-30)	48	96	99.6
Geotrichum candidum (lipase GC-20)	48	84	99
Candida cylindraceae (lipase AY-30)	48	90	94
Porcine pancreatic lipase	48	90	99.2
Pseudomonas sp. (lipase AK)	48	74	98.3

[a]Reaction mixure in 10 mL of 100 mM potassium phosphate buffer (pH 7.0) contained 100 mg of substrate compound 15 and 200 mg of crude lipase powder. Reaction was carried out in a pH state at 30°C with mixing. Reaction yields and optical purities were determined by HPLC.

Compound 19*a*. Reaction yields of >80% and optical purities of >95% were observed for these bioreductions.

Cells of *H. polymorpha* SC13865 and *H. fabianii* SC13894 were grown in a 500-L fermentor containing glucose as the carbon source. Cells were collected by centrifugation, suspended in buffer (20% w/v, wet cells), and used in the reduction of Compound 18. Semipreparative-scale reductions of ketone (Compound 18) to the desired alcohol (Compound 19*a*) were carried out in a 5-L

Table 4 Kinetics of hydrolysis of *cis*-3-acetyloxy-4-phenyl-2-azetidinone by lipase PS-30 and BMS lipase[a]

Reaction time (hours)	(3R)-Acetate (g/L)	(3S)-Acetate (g/L)	(3S)-Alcohol (g/L)	(3R)-Alcohol (g/L)	Conversion (%)	Optical purity (3R)-Acetate (%)
Lipase PS-30						
0.5	5	4.5	0.4	0	8	54
16	4.95	2.5	2.5	0	50	75
24	4.92	1.2	3.8	0.013	72	86
40	4.82	0	4.95	0.14	96.4	99.6
BMS Lipase						
0.5	5	4.2	0.8	0	14	57
16	4.98	2.48	2.49	0.02	50	75
24	4.96	1.5	3.6	0.04	66	86
40	4.8	0	4.92	0.16	96	99.5

[a]The reaction mixture in 1L of 25 mM potassium phosphate buffer (pH 7.0) contained 10 g of substrate Compound 15 and 3 g of immobilized lipase. Reactions were carried out at 30°C, 200 RPM. Reaction yields and optical purities were determined by HPLC.

Table 5 Enzymatic hydrolysis of *cis*-acetyloxy-4-phenyl-2-azetidinone: reusability of immobilized lipase PS-30[a]

Cycle #	Reaction time (hours)	Reaction rate (g/L/hr)	Yield of Compound 16 (%)	Yield of Compound 16 (grams)	Optical purity of Compound 16 (%)
1	40	0.123	94	3.1	99.4
3	36	0.12	96	3.05	99.5
5	37	0.15	94	3.1	99.5
7	40	0.124	93	3.12	99.4
9	42	0.12	95	3.12	99.2
10	42	0.121	95	3.1	99.3

[a]The reaction mixture in 750 mL of 25 mM potassium phosphate buffer (pH 7.0) contained 7.5 g of substrate15, and 2.2 g of immobilized lipase PS-30 or BMS lipase. Reactions were carried out at 29°C, 250 RPM agitation at pH 7.0.

fermentor using cell suspensions of *H. polymorpha* SC 13865 and *H. fabianii* SC 13894 in independent experiments. In both batches, a reaction yield of 85–90% and an optical purity of >95% were obtained for compound 19a (Table 7). From one batch (*H. polymorpha* SC 13865), 5.2 gram of Compound 19a was isolated in 65% overall yield. The isolated compound gave 99.6% optical purity and 99.8% gas chromotography (GC) area purity. A specific rotation of −21.7 was obtained for Compound 19a in chloroform.

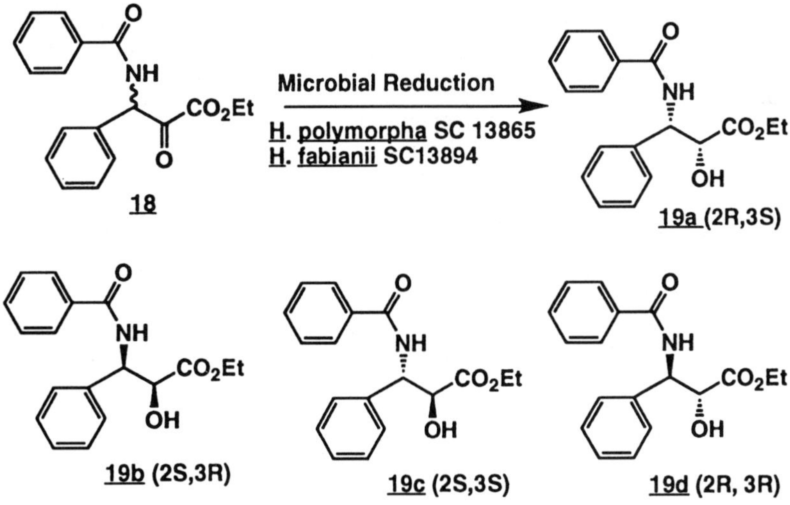

Figure 10 The stereoselective reduction process to prepare paclitaxel C-13 side chain intermediates.

Table 6 Stereoselective microbial reduction of ketone compound 18 to alcohol compound 19a[a]

Microorganisms	Reaction yield Compound 19a (%)	Optical purity Compound 19a (%)
Candida guilliermondi ATCC 20318	31	95
Rhodococcus erythropolis ATCC 4277	39	96
Saccharomyces cerevisiae ATCC 24702	35	94
Hansenula polymorpha SC 13865	80	99
Pseudomonas putida ATCC 11172	32	94
Mortierella rammanianna ATCC 38191	35	97
Hansenula fabianii SC 13894	85	95
Pichia methanolica ATCC 58403	80	26
Nocardia salmonicolor SC 6310	45	99

[a]Cells of various microbial cultures were suspended in 50 mM potassium phosphate buffer (pH 6.0) at 20% (w/v, wet cells) cell concentration and supplemented with 2 mg/mL of substrate compound 18 and 35 mg/mL of glucose. Reduction was carried out at 28°C, and 280 RPM. The reaction yields and the optical purity were determined by HPLC.

A single-stage fermentation/bioreduction process was developed for conversion of Compound 18 to Compound 19a with cells of *H. fabianii* SC 13894. Cells were grown in a 25-L fermentor, and after a 48-hour growth cycle the bioreduction process was initiated by addition of substrate and glucose and continued for a 48-hr period. A reaction yield of 88% and an optical purity of 95% were obtained for the desired alcohol (Compound 19a) (Figure 11). From the 2-L fermentation broth after bioreduction, 2.4 grams of Compound 19a were isolated. The isolated compound gave 99.5% optical purity and 99.5% GC area purity. Paclitaxel side chains prepared by either the resolution or reduction process could be coupled chemically to 10-DAB or baccatin III to yield paclitaxel, as described in the following section (44, 45).

Table 7 Semi-preparative-scale reduction of compound 18 by cell suspensions of *Hansenula* strains: two-stage process[a]

Microorganisms	Reaction time (hours)	Anti diastereomers (19c, 19d) (%)	Syn diastereomers (19a, 19b) (%)	Optical purity desired syn 19a (%)
Hansenula polymorpha SC 13865	72	20	80	99
Hansenula fabianii SC 13894	48	10	90	94

[a]Cells were suspended in 4 L of 50 mM phosphate buffer (pH 6) at 20% (w/v) concentration. Cell-suspensions were supplemented with 2 g/L of substrate compound 18 and 35 g/L of glucose. The bioreduction of compound 18 was carried out at 28°C, 250 RPM in a 5-L NB Bioflo fermentor.

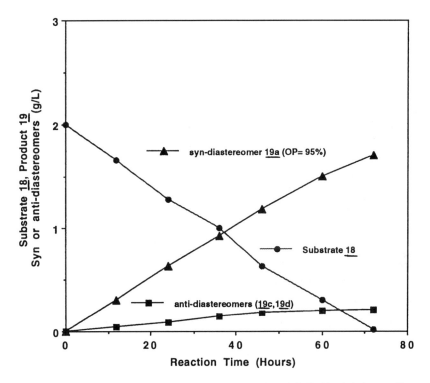

Figure 11   The single-stage process for preparation of paclitaxel C-13 side chain intermediates by *H. fabianii* SC 13894.

Chemistry of C-13 Paclitaxel Side-Chain Intermediates

The initial synthesis of paclitaxel published by the Potier group required the esterification of cinnamic acid to baccatin, followed by hydroxylation of the C-13 side chain; it gave very low yields and was not commercially feasible (30). Subsequently, they demonstrated the synthesis of paclitaxel from the esterification of baccatin III with the phenylisoserine side chain protected at the 2'-OH position. This process gave only about 40% yield. Subsequent practical and efficient syntheses were developed by Holton (41) and others (e.g. by BMS) using different side-chain intermediates and different reaction conditions, and providing much higher yields without any formation of epimers. β-lactams can serve as excellent precursors of paclitaxel side chains (42). Methods for preparation of racemic β-lactam and subsequent conversion to paclitaxel side chains are outlined in Figure 12. N-acetyl-β-lactams were prepared by Holton for the purpose of exploring their utility as direct acylating agents (40).

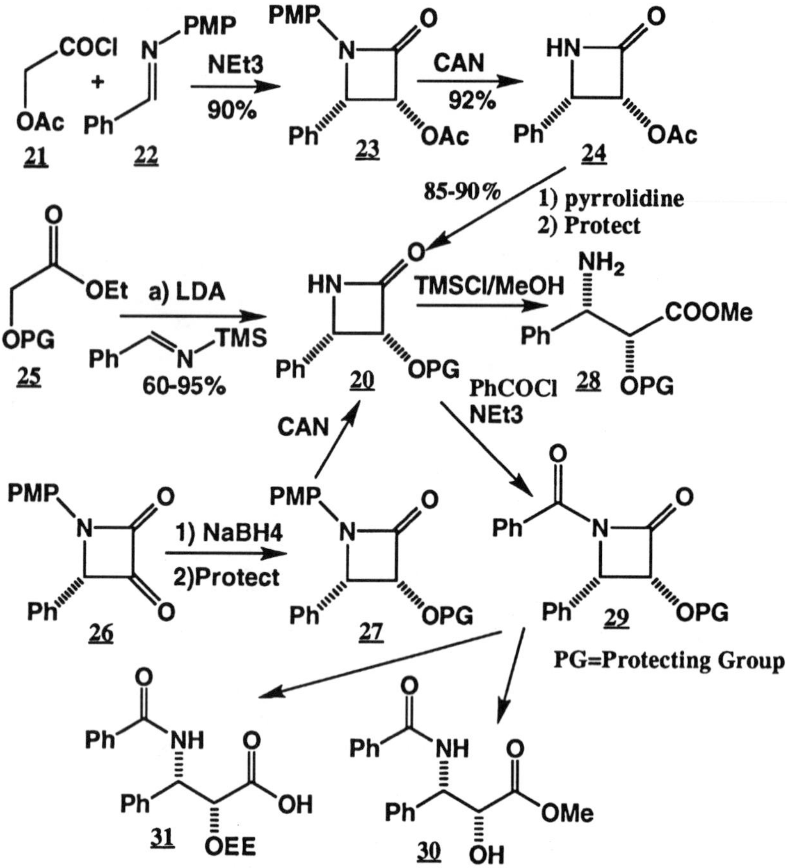

Figure 12   Chemical synthesis for preparation of β-lactams (precursors of the paclitaxel C-13 side chain).

In contrast, Palomo and coworkers synthesized β-lactam intermediates for preparing suitably protected phenylisoserines (63). Racemic β-lactam (Compound 20) can be prepared in >75% yield by various routes as shown in Figure 12 (42). The Staudinger reaction of acetyl glycolyl chloride (Compound 21) with imine (Compound 22) in the presence of triethylamine gives cis-β-lactam (Compound 23) in 90% yield. The removal of p-methoxyphenyl (PMP) substituent with ceric ammonium nitrate then provides Compound 24 in 92% yield. The acetyl group can be selectively removed by treatment with pyrrolidine, and the resulting alcohol can be protected with a variety of groups to yield Compound 20 in an overall yield of 85–90%. In an alternate route, the enolate of glycolate ester (Compound 25), protected with a silyl group, an acetal, or one of several other nonacyl groups, reacts with a trimethylsilyl aldimine to provide Compound 20 directly in 60–95% yield depending upon the protecting group used in the reaction. Palomo et al described the reduction of 3-keto azetidin-2-ones such as Compound 26 to provide exclusively cis-β-lactam. The β-lactam can then be protected to give Compound 27, which upon dearylation gives Compound 20. The β-lactam (Compound 20, protecting group = methyl) undergoes ring opening in the presence of trimethylsilyl chloride/methanol to give amino ester (Compound 28), or acylation to give N-acetyl-β-lactam (Compound 29) in high yield. Treatment of Compound 29 with trimethylsilyl chloride/methanol then provides methyl ester (Compound 30) or in the presence of NaOH, Compound 29 is converted to the protected side chain acid (Compound 31, protecting group = EE or other acetal). The enzymatic synthesis of the chiral paclitaxel side chain from β-lactam [cis-3-acetyloxy-4-phenyl-2-azetidinone] could be prepared as described above by a resolution process using lipases (66, 68). Brieva et al (4) have studied the enzymatic kinetic resolution of β-lactams and have achieved the resolution of Compound 23 in 49% yield and >99.5% optical purity. They have also demonstrated the resolution of Compound 24 and Compound 29 in 39% yield and 96% enantiomeric excess and 46% yield and 99.8% ee, respectively (4). Asymmetric syntheses of β-lactams via the Staudinger reaction (23), cycloaddition (43), and the ester enolate-imine cyclocondensation (61) have been demonstrated.

Coupling of C-13 Paclitaxel Side-Chain Intermediates to Baccatin III Derivatives

Holton et al (41, 42) first recognized N-acyl-β-lactams such as Compound 29a–d as suitable acylating agents for baccatin III derivatives. The hindered 7-triethylsilyl baccatin III (Compound 32) reacted with 5 equivalents of Compound 29a in pyridine to give 2'-ethoxyethyl-7-triethylsilyl taxol (Compound 33a) in 96% yield at 96% conversion (Figure 13). Deprotection of Compound 33a gave paclitaxel in 91% yield (42). However, hydrolysis of the ethoxy ethyl protecting group was slow. Holton et al subsequently carried out the esterification of 7-triethylsilyl baccatin III (Compound 32) by forming a C-13 metal alkoxide. Addition of n-butyllithium to an approximately 0.1 M solution of Compound 32 in THF at −45°C produced alkoxide (Compound 34, M = lithium) which was followed by the addition of Compound 29a (0.2 M in THF), and the mixture was warmed to 0°C for 2 hr to give Compound 33a in nearly quantitative yield (44). Under these conditions, β-lactams (Compound 29b–d) were equally effective. Deprotection of Compound 33a was carried out using HF/pyridine in acetonitrile solution to give paclitaxel in >98% yield (45, 46).

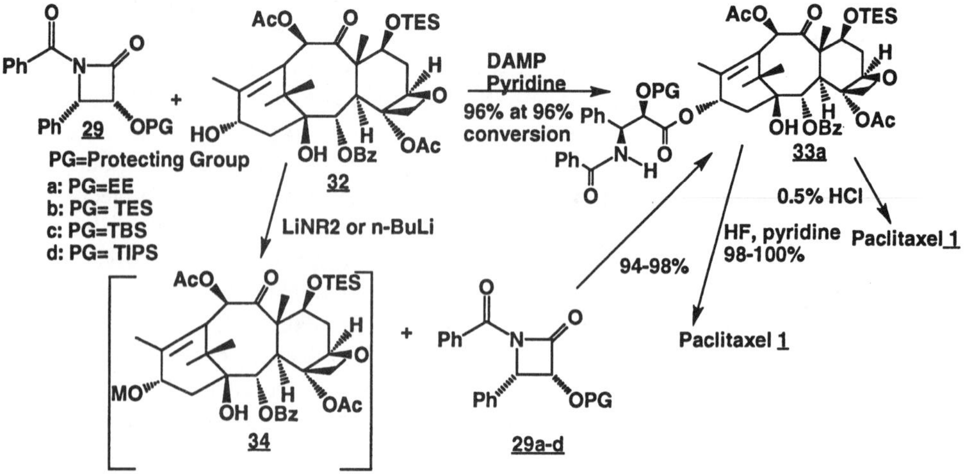

Figure 13   Coupling of the C-13 paclitaxel side chain to baccatin III derivatives.

AN ALTERNATE SOURCE OF PACLITAXEL

Plant-Cell Culture for Taxane Synthesis

Plant-cell culture is expected be a future supply of paclitaxel. Cell culture can possibly lead to exploration of the genetic diversity within the genera that might be useful for selecting higher-yielding variants for cell-culture processes. Cultures can also be selected continuously for elite variants and are amenable to genetic manipulation (26). At present, the biosynthetic pathway for paclitaxel and taxanes is being elucidated and will further identify the rate-limiting enzymes for further manipulation in plant-cell cultures. Plant-cell and tissue cultures are established by selecting and isolating living plant tissues, called explants, separate from intact plants. The environments of explants are strictly controlled under aseptic conditions with respect to composition of media and physical conditions (temperature, light, and humidity). Both undifferentiated-culture and differentiated-culture techniques have been developed for *Taxus* cell cultures.

A callus is an unorganized, proliferating mass of undifferentiated cells. The selection of the explant tissue for use in initiating a callus is guided by the source tissue from which the secondary metabolite of interest is synthesized. The induction of a callus requires a growth medium containing plant-growth regulators and an environment in which some of the cells can divide and proliferate. Callus cultures have been successfully established from bark, young stems, and needles of various species of *Taxus* (11, 16, 18, 20, 21, 27, 52, 71, 78, 96). Plant-cell suspension cultures are usually induced from established callus cultures by transferring callus into an appropriate liquid medium favoring suspension-culture growth. Suspension-cultures are advantageous, since the system is amenable to scale-up. The tissue of *T. brevifolia* has been successfully cultured to produce paclitaxel and related taxanes from both callus and suspension culture (11, 16, 18, 20, 21, 52, 71, 78, 96). Differentiated cultures have also been evaluated for the paclitaxel synthesis by cell cultures. Differentiated cultures contain organized tissues that maintain distinct cellular, tissue, and organ interactions similar to an intact plant. A differentiated culture may be advantageous for secondary product formation, considering that it may retain the temporal and spatial expression of the genes essential for the synthesis of the desired compound. Differentiated cell cultures using embryos from mature seeds, root, and hairy root (derived from plant tissue that has been genetically transformed by *Agrobacterium rhizogene*) have been developed using *Taxus* species (16, 18, 20, 21, 71). The first report of tissue culture production of paclitaxel from *T. brevifolia* was from the USDA, and a patent has been obtained for it. Phyton Catalytic Company has licenced this technology and is developing a cell-culture process for the production of paclitaxel with support from BMS (16). A second company, ESCA Genetics has also developed tissue-culture technology. Although details of their technology for paclitaxel production are not available, it has been speculated that it uses hairy-root cultures. Paclitaxel production from callus cultures of *T. cuspidata* and *T. canadensis* has been demonstrated on a laboratory scale, with 0.02% yield (16). Suspension cultures of *T. cuspidata* have also been established and immobilized on glass fiber mats, yielding paclitaxel in 0.012% yield (16). Phenylalanine in the growth medium promoted paclitaxel biosynthesis in callus cultures. The production of paclitaxel from *T. brevifolia* has also been reported by a Japanese group (75) with a reported yield of 0.05% (dry weight basis). Acetate and mevalonate have been reported to be biosynthetic precursors of paclitaxel in plant cell cultures (96). Paclitaxel production using free and immobilized cells of *T. cuspidata* has been reported at specific production rate of 0.3 mg/g dry cell weight per day for up to 40 days (78). Because plant cells require different conditions for cell proliferation and secondary product formation, a two-stage process is preferred. The growth phase allows sufficient biomass to be accumulated, followed by a second stage that triggers and extends the secondary metabolite synthesis. Media can be optimized for each stage of the process. Plant-cell cultures also produce other taxanes besides paclitaxel such as 7-xylosyl-10-deacetytaxol, 10-deacetyltaxol, cephalomannine, 7-xylosylcephalomannine, and many other taxanes. Semisynthetic processes for production of paclitaxel by converting a variety of taxanes produced by plant cell cultures to 10-DAB by microbial enzymes (C-13 taxolase, C-10 deacetylase, and C-7 xylosidase) will be promising for future supplies of paclitaxel.

Fungal Production of Paclitaxel

In 1993, Stierle et al reported the production of paclitaxel by the fungus *Taxomyces andeanae*, which was found growing on a specimen of *T. brevifolia* (79). The yield of paclitaxel was very low (25–50 nanogram/L). It is likely that strain improvement by mutagenesis or genetic engineering will result in yield improvement; however, the yield improvement required will be orders of magnitude higher than in any other process accomplished so far.

Total Synthesis of Paclitaxel

The baccatin III molecule of paclitaxel is a complex tetracyclic diterpene with many functional groups and stereochemical features that represented a tremendous challenge for synthetic organic chemists. Many groups were involved in the total chemical synthesis of paclitaxel (2, 46, 60, 93). The first total synthesis of paclitaxel was achieved in 1994 by Holton and his coworkers (46) and by Nicolaou et al (60). Although total synthesis of paclitaxel represents an enormous synthetic accomplishment, it is not yet practical for paclitaxel production. The semisynthetic approach described above is much more economically attractive at this time.

BIOSYNTHESIS OF PACLITAXEL

In the long term, it would be desirable to produce paclitaxel and analogs thereof by a process that does not depend on the extraction of plant material, but which can be carried out under controlled conditions. Such a biotechnological process may be based on a plant-cell culture fermentation or a microbial process with a genetically engineered organism in combination with some chemical/enzymatic steps. The development of such a process would be greatly benefited by an understanding of the biosynthesis of paclitaxel in plants, and by characterization of enzymes catalyzing various reactions and identification of genes coding the key enzymes. Identification of rate limiting enzymes would be essential to increase paclitaxel synthesis.

Biosynthesis of the Diterpene Moiety

Based on its structure and from general biochemistry knowledge, the taxane ring system can arise by a cyclization of geranyl-geranyl pyrophosphate (GGPP) (Compound 35) to give, ultimately, a hydrocarbon precursor, 4(20),11(12)-taxadiene. Early attempts to experimentally verify the isoprenoid origin of the taxane ring system by feeding radiolabeled mevalonic acid to *T. baccata* were unsuccessful (22). Recently, Zamir et al (96) fed [5-^{3}H]mevalonate to ground, new-growth leaves of *T. canadensis* for three weeks and obtained incorporation of radioactivity into paclitaxel. Stierle et al (79) fed precursors to aseptically prepared pieces of the inner bark of *T. brevifolia* and observed 0.1% specific incorporation of [2-^{14}C] mevalonate and 0.38% of [1-^{14}C] acetate. They also reported the incorporation of ^{14}C-labeled acetate and phenylalanine, but not benzoate and leucine, into paclitaxel and baccatin III by the fungus *Taxomyces andreanae* (79). The incorporation is extremely small because of the extremely low level (25–50 nanog/L) of paclitaxel production in this system. Croteau et al (12) have investigated the transformation of radioactive GGPP in cell-free extracts of *Taxus brevifolia*. They observed the formation of a labeled hydrocarbon compound with a molecular ion of m/z 272, as expected for $C_{20}H_{32}$. The elimination product of pyrophosphate from GGPP will have the same molecular weight; however, in the absence of a reference standard, it is hard to derive any conclusive information from this data. Recently, Lin et al (53) employed [1-^{2}H$_2$, 20-^{2}H$_3$] and [20-^{2}H$_3$]GGPP as substrates with the partially purified taxadiene synthase from *Taxus brevifolia* stems to examine the possibility of a preliminary cyclization to taxa-4-(20),11(12)-diene, followed by isomerization to the more stable endicyclic double-bond isomer. From the results, Lin et al (53) proposed a stereochemical mechanism for the taxadiene synthase reaction involving the initial cyclization of GGPP (Compound 35) to a transient verticillyl cation intermediate, with transfer of the C11 α-proton to C7 to initiate transannular B/C-ring closure to the taxenyl cation, followed by deprotonation at C5 to yield the taxa-4(5),11(12)-diene product (Compound 36) (Figure 14). They ruled out the possibility of formation of taxa-4(20),11(12)-diene. Later studies on taxadiene synthase from *Taxus canadensis*, which catalyzes the cyclization of GGPP to taxa-4(5),11(12)-diene, have also been demonstrated to be chromatographically, electrophoretically, and kinetically similar to taxadiene synthase from *Taxus brevifolia* (37).

Following the generation of the taxane ring system, the large number of functionalities present in paclitaxel must be introduced successively into the diterpenoid ring system. These functionalizations include hydroxylations, acylations of hydroxy groups, oxidation to a ketone, and generation of the oxetane ring system. No experimental data are available on the order in which these reactions take place; however, a comparison of the structures of all taxanes (22) provides some information. The least functionalized compounds isolated carry at least two oxygens, namely on C-5 and C-10. Thus, these two oxygens must be introduced first in the reaction sequence. Oxygens at C-2 and C-9 are next in abundance, followed by C-13. The acetylation of some of the existing oxygens may precede the introduction of new ones. Oxygens at C-7 and C-1 are probably introduced last. Epoxidation, oxetane ring formation, the acylation of C-13 with a side chain precursor, and the oxidation of C-9 to the ketone are late steps (50). Recently, Hefner et al (36) have demonstrated the cytochrome P-450–catalyzed hydroxylation of taxa-4(5),11(12)-diene to

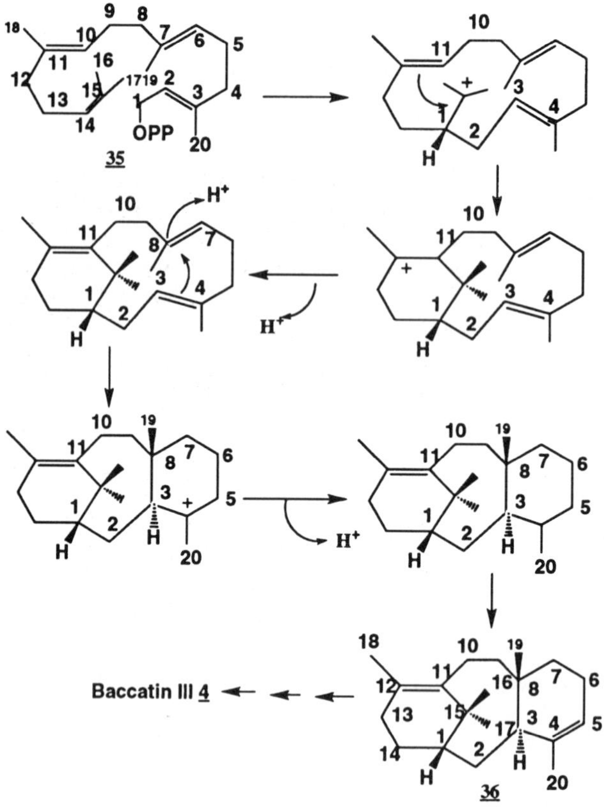

Figure 14   Biosynthesis of the diterpene moiety of paclitaxel.

taxa-4(20),11(12)-dien-5α-ol by microsomal enzymes from the *Taxus* stem and cultured cells. Selective deacetylation and hydroxylation of 2α,5α,10β,14β-tetra-acetoxy-4(20)11-taxadiene by cells of *Cunninghaemella echinulata* has also been demonstrated by Hu et al (49).

Biosynthesis of C-13 Paclitaxel Side-Chains

In a feeding experiment with D,L-[3-C^{14}]phenylalanine, Leete & Bodem (51) demonstrated after careful degradation of the product that 92% of the radioactivity was located in C-3 of (3R)-N,N-dimethyl-β-phenylalanine. Incorporation of radiolabeled phenylalanine into the phenylisoserine side chain of paclitaxel was reported by Zamir et al (96) and Strobel et al (81) by a feeding experiment using ground, new-growth leaves of *T. canadensis* and aseptically prepared pieces of the inner bark of *T. brevifolia*, respectively. Fleming et al (20) synthesized

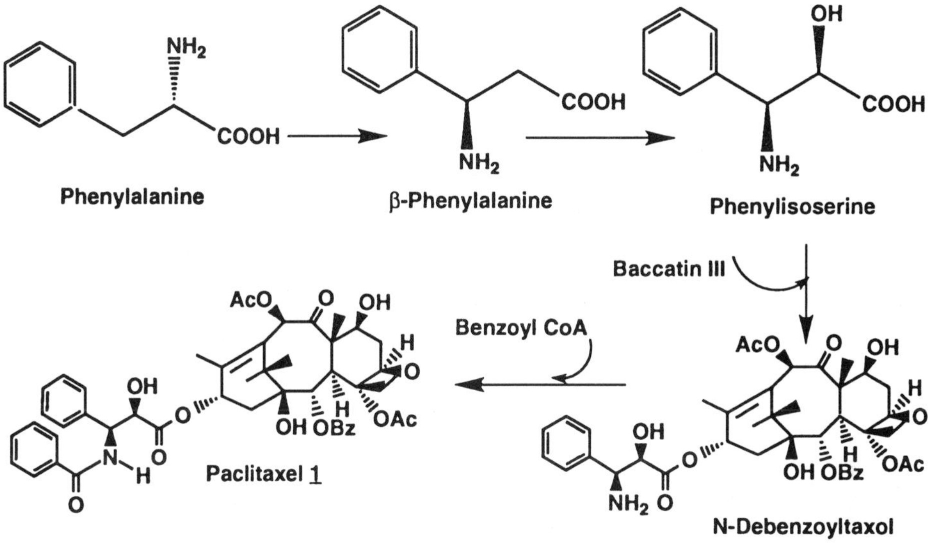

Figure 15   Biosynthesis of the C-13 paclitaxel side chain and synthesis of paclitaxel from diterpene and the side chain.

and fed ring-pentadeuterated precursors to inner bark pieces of *T. brevifolia* to investigate the sequence of reactions by which the side chain is assembled. The resulting paclitaxel and cephalomannine samples were analyzed by electrospray ES-MS/MS in the presence of methylammonium ions. From these experimental results, they concluded that the phenylisoserine side chain is formed from phenylalanine via β-phenylalanine (Figure 15). Both phenylisoserine and the entire paclitaxel side chain are specifically incorporated into paclitaxel.

Feeding experiments with [7-^{14}C] benzoic acid and [^{14}C] benzoylphenyliso-serine added to cut twigs of *T. brevifolia* demonstrated that both the phenylisoser-ine moiety and the benzoate moiety of the paclitaxel side chain must have been formed from β-phenylalanine and phenylisoserine (20).

Biosynthesis of Paclitaxel from Diterpene and Side-Chain

The final issue is the synthesis of paclitaxel from the diterpene moiety and the paclitaxel side chain. The simplest assumption on the basis of the existing natural taxanes would be that the diterpene is fully functionalized before the C-13 side chain is attached. [13-^{3}H] baccatin III was prepared and fed to cut twigs of *T. brevifolia*. Labeled paclitaxel was obtained after crystallization to constant specific radioactivity. Upon reductive cleavage of the C13 ester bond, 95% of the radioactivity was recovered in the diterpene moiety, whereas the side chain fragment was not labeled. These results were further confirmed by feeding tetradeuterated baccatin III to inner bark pieces and demonstrating by ES-MS/MS the incorporation of deuterium into paclitaxel. These experiments demonstrated the presence of an enzyme in *T. brevifolia* tissue that is capable of coupling the side chain to C-13 of baccatin III (22). N-debenzoyltaxol carrying five atoms of deuterium in the benzene ring of the phenylisoserine side chain and three in the 10-acetyl group was synthesized and fed. This precursor gave very good incorporation, giving paclitaxel that contained eight atoms of deuterium. Thus, unlike N-benzoylphenylisoserine, this precursor was incorporated intact, indicating that the side chain is attached to baccatin III at or prior to the stage of phenylisoserine (22). On the basis of available information, the late stages of the paclitaxel biosynthesis can be summarized in Figure 15.

Visit the *Annual Reviews home page* at
http://www.AnnualReviews.org.

Literature Cited

1. Adams JD, Flora KP, Goldspiel BR, Wilson JW, Finley R. 1993. Taxol: a history of pharmaceutical development and current pharmaceutical concerns. *J. Natl. Cancer Inst. Monogr.* 15:141–44

2. Begley MJ, Jackson CG, Pattenden G. 1990. Total synthesis of verticillene. A biomimetric approach to the taxane family of alkaloids. *Tetrahedron* 46:4907–12

3. Bolsinger CL, Jaramillo AE. 1990. *Taxus brevifolia* Nutt., Pacific yew. In *Silvics of North America*, USDA Forest Service, Agric. Handb. 654, ed. RM Bruns, BH Honkala, pp. 573–75. Washington DC: US Dept. Agric.

4. Brieva R, Crich JZ, Sih CJ. 1993. Chemoenzymatic synthesis of the C–13 side chain of taxol: optically active 3-hydroxy–4-phenyl β-lactam derivatives. *J. Org. Chem.* 58:1068–72

5. Bristol-Myers Squibb. 1993. Feasibility study on the isolation and purification of taxol from yew-leaf derived resin provided by the NCI. Washington, DC: USDA

6. Bruns RG, Surridge CD. 1994. Tubulin: conservation and structure. In *Microtubules*, ed. JS Hyams, CW Lloyd, pp. 85–109. New York: Wiley-Liss

7. Carlier M-F, Didry D, Pantaloni D. 1987. Microtubule elongation and GTP hydrolysis. Role of Guanine nucleotides in microtubule dynamics. *Biochemistry* 26:4428–33

8. Chaudhary AG, Gharpure MM, Rimoldi JM, Chordia MD, Gunatilaka AAL, et al. 1994. Unexpectedly facile hydrolysis of the 2-benzoale group of taxol and syntheses of analogs with increased activities. *J. Am. Chem. Soc.* 116:4097–102

9. Chen S-H, Haung S, Wei J, Farina V. 1993. The chemistry of taxanes: reaction of taxol and baccatin III derivatives with Lewis acids in aprotic and protic media. *Tetrahedron* 49:2805–28

10. Chen S-H, Farina V, Wei J-M, Long B, Fairchild C, et al. 1994. Structure-activity relationships of taxol: synthesis and biological evaluation of C-2 taxol analogue. *Bioorg. Med. Chem. Lett.* 4:479–82

11. Christen AA, Gibson DM, Bland J. 1991. *U.S. Patent No. 5019504*

12. Croteau R, Hezari M, Hefner J, Koepp A, Lewis NG. 1995. Paclitaxel biosynthesis. Early steps. In *Toxic Cancer Agents: Basic Science and Current Status*, ed. GI George, TT Chen, I Ojima, DT Vyas, pp. 72–80. Washington, DC: AM. Chem. Soc.

13. Croom EM. 1995. Taxus for taxol and taxoids. In *Taxol: Science and Applications*, ed. M Suffness, pp. 37–70. Boca Raton, FL: CRC

14. Denis J-N, Greene AE, Guenard D, Gueritte-Vogelin F, Mangatal L, et al. 1988. Highly efficient, practical approach to natural taxol. *J. Am. Chem. Soc.* 110:5917–21

15. Deutsch HM, Glinski JA, Hernandez M, Haugwitz RD, Narayanan VL, et al. 1989. Synthesis of congeners and prodrugs. III:

Water-soluble prodrugs of taxol with potent antitumor activity. *J. Med. Chem.* 32: 788–92

16. Edgington SM. 1991. Plant cell culture for taxol production. *Bio/Technology* 9:479–82
17. Fabre JL. 1993. *Taxotere supply strategy.* Presented at Stony Brook Symp. Taxol Taxotere, State Univ. NY, Stony Brook
18. Fett-Neto AG, DiCosmo F, Reynold WF, Sakata K. 1992. Cell culture of *Taxus* as a source of the antineoplastic drug taxol and related taxanes. *Bio/Technology* 10:1572–75
19. Flanagan SP. 1966. "Nude", a new hairless gene with pleitropic effects in the mouse. *Genet. Res.* 8:295–99
20. Fleming PE, Mocek U, Floss HG. 1993. Biosynthesis studies on taxol. *J. Am. Chem. Soc.* 115:805–7
21. Flores T, Wagner LJ, Flores HE. 1993. Embryo culture and taxane production in *Taxus* sp. *In Vitro* 29:160–64
22. Floss HG, Mocek U. 1995. Biosynthesis of taxol. In *Taxol: Science and Applications*, ed. M Suffness, pp. 191–208. Boca Raton, FL: CRC
23. Georg GI, Mashava PM, Akgun E, Milstead MW. 1991. Asymmetric synthesis of β-lactam and N-benzoyl-3-phenylisoserine via the Staudinger reaction. *Tetrahedron Lett.* 32:3151–55
24. Georg GI, Cheruvallath ZS, Himes RH, Mejillano MR, Burke CT. 1992. Synthesis of biologically active taxol analogues with modified phenylisoserine side chains. *J. Med. Chem.* 35:4230–35
25. Georg GI, Boge TC, Cheruvallath ZS, Clowers JS, Harriman GCB, et al. 1995. The medicinal chemistry of taxol. In *Taxol: Science and Applications*, ed. M Suffness, pp. 318–75. Boca Raton, FL: CRC
26. Gibson DM, Ketchum EB, Hirasuna TJ, Shuler ML. 1995. Potential of plant cell culture for taxane production. In *Taxol: Science and Applications*, ed. M Suffness, pp. 71–95. Boca Raton, FL: CRC
27. Gibson DM, Ketchum REB, Vance NC, Christen AA. 1993. Initiation and growth of cell lines of *Taxus brevifolia* (Pacific yew). *Plant Cell Rep.* 12:479–84
28. Giovanella BC, Yim SO, Stewhlin JS, Williams LJ. 1972. Development of invasive tumors in the nude mouse after injection of cultured human melanoma cells. *J. Natl. Cancer Inst.* 48:1531–34
29. Guenard D, Gueritte-Vogelein F, Potier P. 1993. Taxol and taxotere: discovery, chemistry, and structure-activity relationships. *Acc. Chem. Res.* 26:160–66
30. Gueritte-Voegelein F, Senilh V, David B,

Guenard D, Potier P. 1986. Chemical studies of 10-deacetyl baccatin III. Hemisynthesis of taxol derivatives. *Tetrahedron* 42: 4451–60
31. Gueritte-Voegelein F, Guenard D, Lavelle F, Le Goff M-T, Mangatal L, et al. 1991. Relationships between the structure of taxol analogues and their antitumor activity. *J. Med. Chem.* 34:992–96
32. Hanson RL, Wasylyk JM, Nanduri VB, Cazzulino DL, Patel RN, et al. 1994. Site-specific enzymatic hydrolysis of taxanes at C-10 and C-13. *J. Biol. Chem.* 35:22145–49
33. Hanson RL, Patel RN, Szarka LJ. 1996. *U.S. Patent No. 5516676*
34. Hanson RL, Patel RN, Szarka LJ. 1996. *U.S. Patent No. 5523219*
35. Hanson RL, Howell JM, Brzozowski BD, Sullivan SA, Patel RN, et al. 1997. Enzymatic hydrolysis of 7-xylosyltaxanes by xylosidase from *Moraxella* sp. *Biotechnol. Appl. Biochem.* 26:152–58
36. Hefner J, Rubenstein SM, Ketchum REB, Gibson DM, Williams RM, et al. 1996. Cytochrome P450-catalyzed hydroxylation of taxa-4(5),11(12)-diene to taxa-4(20),11(12)-dien-5a-ol: the first oxygenation step in taxol biosynthesis. *Chem. Biol.* 3:479–89
37. Hezari M, Ketchum REB, Gibson DM, Croteau R. 1997. Taxol production and taxadiene synthetase activity in *Taxus canadensis* cell suspension cultures. *Arch. Biochem. Biophys.* 337:185–90
38. Higuchi T, Stella V, eds. 1975. *Prodrugs as Novel Drug Delivery Systems.* Washington, DC: Am. Chem. Soc.
39. Holmes FA, Walters RS, Thierault RL, Forman AD, Newton LK, et al. 1991. Phase II trial of taxol, an active drug in the treatment of metastatic breast cancer. *J. Natl. Cancer Inst.* 83:1797–1802
40. Holton RA. 1991. *U.S. Patent No. 5015744*
41. Holton RA. 1992. *U.S. Patent No. 5175315*
42. Holton RA. 1993. *U.S. Patent No. No. 5254703*
43. Holton RA, Liu JH. 1993. A novel asymmtric synthesis of cis-3-hydroxy-4-aryl azetidin-2-ones. *Bioorg. Med. Chem. Lett.* 3:2475–78
44. Holton RA. 1993. *U.S. Patents 5229526, 5274124*
45. Holton RA, Biediger RJ. 1993. *U.S. Patent No. 5243045*
46. Holton RA, Somoza C, Kim HB, Liang F, Biediger RJ, et al. 1994. First total synthesis of taxol. 1. Functionalization of the B ring. *J. Am. Chem. Soc.* 116:1597–60
47. Holton RA, Biediger RJ, Boatman PD. 1995. Semisynthesis of taxol and taxotere.

In *Taxol: Science and Applications*, ed. M Suffness, pp. 3–25. Boca Raton, FL: CRC

48. Horwitz SB. 1992. Mechanism of action of taxol. *Trends Pharmacol. Sci.* 13:134–40

49. Hu S, Tian X, Zhu W, Fang Q. 1996. Microbial transformation of taxoids: selective deacetylation and hydroxylation of $2\alpha,5\alpha,10\beta,14\beta$-tetra-acetoxy-4(20),11-taxadiene by the fungus *Cunninghamella echinulata*. *Tetrahedron* 52:8739–46

50. Kingston GDI. 1995. Natural taxoids: Structure and chemistry. In *Taxol: Science and Applications*, ed. M Suffness, pp. 287–315. Boca Raton, FL: CRC

51. Leete E, Bodem GB. 1966. The biosynthesis of 3-dimethylamino-3-phenylpropanoic acid in yew. *Tetrahedron Lett.* 3925–55

52. LePage MT. 1968. Mise en evidence d'une dormance associée à une immaturité de l' embryons chez *Taxus baccata*. *C. R. Acad. Sci.* 266:482–84

53. Lin X, Hezari M, Koepp AE, Floss HG, Croteau R. 1996. Mechanism of taxadiene synthase, a diterpene cyclase that catalyzes the first step of taxol biosynthesis in Pacific yew. *Biochemistry* 35:2968–77

54. Magri NF, Kingston DGI, Jitrangsri C, Piccareillo T. 1989. Modified taxol III. Preparation and acylation of baccatin III. *J. Org. Chem.* 51:3239–42

55. Mathews AE, Mejillano MR, Nath JP, Himes RH, Stella VJ. 1992. Synthesis and evaluation of some water-soluble prodrugs and derivatives of taxol with antitumor activity. *J. Med. Chem.* 35:145–50

56. McGuire WP, Rowinsky EK, Rosenheim NB, Grumbine FC, Ettinger DS, et al. 1989. Taxol: a unique antineoplastic agent with significant activity in advanced ovarian epithelial neoplasm. *Ann. Intern. Med.* 111:273–76

57. Miller RW. 1980. A brief survey of *Taxus* alkaloids and other taxane derivatives. *J. Nat. Prod.* 43:425–30

58. Nanduri VB, Hanson RL, LaPorte TL, Ko R, Patel RN, et al. 1995. Fermentation and isolation of C-10 deacetylase for the production of 10-DAB from baccatin III. *Biotech. Bioeng.* 48:547–50

59. Nicolau KC, Riemer C, Kerr MA, Rideout D, Wrasidlo W. 1993. Design, synthesis and biological activity of protaxols. *Nature* 364:464–67

60. Nicolaou KC, Yang Z, Liu JJ, Ueno H, Nantermet PG, et al. 1994. Total synthesis of taxol. *Nature* 367:630–34

61. Ojima I, Habus I. 1990. Asymmetric synthesis of β-lactams by chiral ester enolate-imine condensation. *Tetrahedron Lett.* 31:4289–92

62. Palevitch D. 1991. Agronomy applied to medicinal plant conservation. In *Herbs, Spices, and Medicinal Plants*, ed. O Akerele, V Heywood, H Synge, pp. 167–80. New York: Cambridge Univ. Press

63. Palomo C, Arrieta A, Cossio FP, Aizpurua JM, Mielgo A, et al. 1990. Highly stereoselective synthesis of α-hydroxy β-amino acids through β-lactams: application to the synthesis of taxol and bestatin side chains and related systems. *Tetrahedron Lett.* 31:6429–35

64. Parness J, Horwitz SB. 1981. Taxol binds to polymerized tubulin in vitro. *J. Cell Biol.* 91:479–87

65. Patel RN, Banerjee A, Howell JM, McNamee CG, Brzozowski D, et al. 1993. Microbial synthesis of (2R, 3S)-(−)-N-benzoyl-3-phenyl isoserine ethyl ester—a taxol side-chain synthon. *Tetrahedron: Asymmetry* 4:2069–84

66. Patel RN, Banerjee A, Ko R, Howell JM, Li W-S, et al. 1994. Enzymatic preparation of (3R-cis)-3-(acetyloxy)-4-phenyl-2-azetidinone: a taxol side-chain synthon. *Biotechnol. Appl. Biochem.* 20:23–33

67. Patel RN, Banerjee A, McNamme CG, Thottathil JK, Szarka LJ. 1995. *U.S. Patent No. 5420337*

68. Patel RN, Szarka LJ, Partyka R. 1996. *U.S. Patent No. 5567614*

69. Patel RN. 1997. Stereoselective biotransformations in synthesis of some pharmaceutical intermediates. *Adv. Appl. Microbiol.* 43:91–140

70. Piesch RF, Wyant VP. 1993. Intensive cultivation of yew species: Weyerhaeuser's contribution to the taxol supply dilemma. *Proc. Int. Yew Resources Conf.: Yew (Taxus) Conserv. Biol. Interact.*, Univ. Calif. Native Yew Conserv. Counc., World Conserv. Union, Berkeley, CA, pp. 27–35

71. Plaut-Carcasson YY, Benkrima L, Dawkins M, Wheeler N, Yanchuk A, et al. 1993. *Proc. Int. Yew Resources Conf.* World Conserv. Union, Berkeley, CA, pp. 45–48

72. Price RA. 1990. The genera of Taxaceae in the southeastern United States. *J. Arnold Arb.* 71:69–75

73. Rao S, Horwitz SB, Ringel I. 1992. Direct photoaffinity labeling of tubulin with taxol. *J. Natl. Cancer Inst.* 84:785–88

74. Rose WC. 1992. Taxol: a review of its preclinical in vivo antitumor activity. *Anti-Cancer Drugs* 3:311–20

75. Saito K, Ohashi H, Hibi M, Tahara M. 1992. Production of taxol by cell culture. *Chem. Abstr.* 117:190280w

76. Scheparta SA. 1976. History of the NCI and the plant screening program. *Cancer Treat. Rep.* 60:975–78

77. Schiff PB, Fant J, Horwitz SB. 1979. Promotion of microtubule assembly in vitro by taxol. *Nature* 227:9479–90

78. Seki M, Ohzora C, Takeda M, Furusaki S. 1997. Taxol (paclitaxel) production using free and immobilized cells of *Taxus cuspidata*. *Biotechnol. Bioeng.* 53:214–19

79. Stierle A, Strobel GA, Stierle D. 1993. Taxol and taxane production by *Taxomyces andreanae*, an endophytic fungus of Pacific yew. *Science* 260:214–16

80. Straubinger RM. 1995. Biopharmaceutics of paclitaxel (taxol): formulation, activity, and pharmacokinetics. In *Taxol: Science and Applications*, ed. M Suffness, pp. 237–58. Boca Raton, FL: CRC

81. Strobel GA, Stierle A, van Kuijk FJGM. 1992. Factors influencing the in vitro production of radiolabeled taxol by Pacific yew, *Taxus brevifolia*. *Plant Sci.* 84:65–68

82. Stull DP, Jans NA. 1992. *Current taxol production from yew bark and future production strategies*. Proc. NCI Workshop Taxol Taxus, 2nd, Alexandria, VA

83. Suffness M. 1989. Development of antitumor natural products at the NCI. *Gann Monogr. Cancer Res.* 36:21–23

84. Suffness M. 1993. Taxol: from discovery to therapeutic use. *Annu. Rep. Med. Chem.* 28:305–15

85. Suffness M, Wall ME. 1995. Discovery and development of taxol. In *Taxol: Science and Applications*, ed. M Suffness, pp. 3–25. Boca Raton, FL: CRC

86. US House of Representatives. 1993. *Pricing of Drugs Codeveloped by Federal Laboratories and Private Companies*, Serial No. 103-2. Hearing before Subcommittee on Regulation, Business Opportunities, and Technology, Washington, DC

87. Vanhaelen M, Lejoly J, Hanocq M, Molle L. 1991. Climatic and geographical aspects of medicinal plant constituents. In *The Medicinal Plant Industry*, ed. ROB Wijesekera, pp. 33–47. Boca Raton, FL: CRC

88. Venditti JM, Wesley RA, Plowman J. 1984. Current NCI preclinical antitumor screening in vivo: results of tumor panel screening, 1976–1982, and future directions. *Adv. Pharmacol. Chemother.* 20:1–20

89. Vidensek N, Lim P, Campbell A, Carlson C. 1990. Taxol content in bark, wood, root, leaf, twig, and seedling from several *Taxus* species. *J. Nat. Prod.* 53:1609–15

90. Wall ME, Wani MC, Cook CE, Palmer CE, McPhail AT, et al. 1966. Plant antitumor agents. I: The isolation and structure of camtothecin, a novel alkaloidal leukemia and tumor inhibitor from *Camptotheca acuminata*. *J. Am. Chem. Soc.* 88:3888–94

91. Wall ME, Wani MC, Taylor HL. 1976. Isolation and chemical characterization of antitumor agents from plants. *Cancer Treat Rep.* 60:1011–14

92. Wani MC, Taylor HL, Wall ME, Coggon P, McPhail AT. 1971. VI: The isolation and structure of taxol, a novel antileukemic and antitumor agent from *Taxus brevifolia*. *J. Am. Chem. Soc.* 93:2325–27

93. Wender PA, Natchus MG, Shuker AJ. 1995. Towards the total synthesis of taxol and its analogues. In *Taxol: Science and Applications*, ed. M Suffness, pp. 123–87. Boca Raton, FL: CRC

94. Wheeler NC, Jech K, Masters S, Brobst SW, Alvarado AB, Hoover AJ, et al. 1992. Effect of genetic, epigenetic and environmental factors on taxol content in *Taxus brevifolia* and related species. *J. Nat. Prod.* 55:432–37

95. Wilson A, Jordan MA. 1994. Phamacological probes of microtubule function. In *Microtubules*, ed. JS Hyams, CW Lloyd, pp. 59–83. New York: Wiley-Liss

96. Zamir LO, Nedea ME, Garneau FX. 1992. Taxol and related taxanes by plant cell cultures. *Tetrahedron Lett.* 33:5235–36

97. Zubrod CG, Schepartz SA, Leiter J, Endicott KM, Carrese LM, et al. 1966. The chemotherapy program of the NCI: history, analysis, and plans. *Cancer Chemother. Rep.* 50:349–55

Annu. Rev. Microbiol. 1998. 52:397–421

VIROCRINE TRANSFORMATION:
The Intersection Between Viral Transforming Proteins and Cellular Signal Transduction Pathways

Daniel DiMaio, Char-Chang Lai, and Ophir Klein
Department of Genetics, Yale University School of Medicine, 333 Cedar Street, New Haven, Connecticut 06510; email: daniel.dimaio@yale.edu

KEY WORDS: viral oncogenes, bovine papillomavirus E5 protein, spleen focus-forming virus gp55, polyomavirus middle T antigen, Epstein-Barr virus LMP-1

ABSTRACT

This review describes a mechanism of viral transformation involving activation of cellular signaling pathways. We focus on four viral oncoproteins: the E5 protein of bovine papillomavirus, which activates the platelet-derived growth factor β receptor; gp55 of spleen focus forming virus, which activates the erythropoietin receptor; polyoma virus middle T antigen, which resembles an activated receptor tyrosine kinase; and LMP-1 of Epstein-Barr virus, which mimics an activated tumor necrosis factor receptor. These examples indicate that diverse viruses induce cell transformation by activating cellular signal transduction pathways. Study of this mechanism of viral transformation will provide new insights into viral tumorigenesis and cellular signal transduction.

CONTENTS

397

0066-4227/98/1001-0397$08.00

INTRODUCTION

Many tumor viruses stimulate the proliferation of their host cells. The analysis of viral transforming proteins has revealed several strategies by which viruses achieve this end. In some cases, viruses have transduced cellular genes involved in the control of cell growth and differentiation. Other viruses encode proteins that induce the synthesis of cellular DNA replication proteins, thereby mobilizing the cellular replication machinery so that it can support replication of the viral DNA. Activation of the replication machinery in this manner frequently stimulates cellular as well as viral DNA replication. Viruses have evolved two general strategies to induce this replicative state. Several groups of DNA viruses encode proteins that inactivate cellular tumor suppressor proteins such as p53 and p105^{RB} that inhibit cell proliferation. This releases the brakes on the cell cycle, thereby resulting in increased proliferation. Viruses also can activate cellular pathways that normally stimulate cell proliferation. Several RNA and DNA viruses encode homologues of cellular growth factors, growth factor receptors, or downstream components of growth factor signaling pathways, which act in a similar manner to their cellular counterparts. This review describes another class of viral transforming proteins that activate growth factor signaling pathways, even though these viral proteins bear no obvious resemblance to cellular proteins. These viral proteins either activate components of the signaling pathway or mimic the structure of an activated component. Here, we focus on viral proteins that either activate a particular growth factor receptor in the absence of its normal ligand or mimic an activated receptor. We propose the term "virocrine transformation" to encompass this mechanism of viral transformation (36).

A BRIEF INTRODUCTION TO GROWTH FACTOR RECEPTOR FUNCTION

Peptide growth factors or cytokines are often required for cellular survival and proliferation. These proteins are usually soluble and initiate signaling by binding to specific receptors at the cell surface, resulting in receptor activation. In general, receptors are composed of an extracellular ligand binding domain, a transmembrane segment, and a cytoplasmic domain required for signaling. Signaling is initiated by ligand binding. For most receptor tyrosine kinases, including the platelet-derived growth factor (PDGF) receptor, ligand binding induces receptor dimerization (Figure 1) (56). This stimulates the intrinsic tyrosine kinase activity of the receptor and induces receptor autophosphorylation

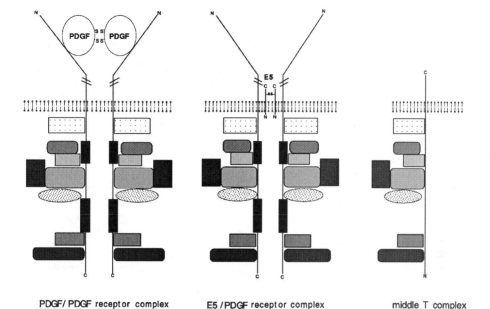

PDGF/PDGF receptor complex   E5/PDGF receptor complex   middle T complex

Figure 1 Schematic diagram of activated receptor tyrosine kinase complexes and the related middle T complex. Left, a complex containing dimeric PDGF β receptor activated by PDGF bound to the extracellular domain of the receptor. Middle, a complex containing dimeric PDGF β receptor activated by the BPV E5 protein bound to the transmembrane region of the receptor. Right, a complex containing polyomavirus middle T antigen. The cell membrane is shown with the cytoplasmic domain of the receptor or mT extending down from the membrane, bound to various cellular signaling proteins represented by *shaded boxes*. The split kinase domain of the PDGF β receptor is depicted as *black boxes*. The spectrum of cellular proteins bound under these various situations is not fully defined and may not be the same in all three cases. [Modified from (36).]

and tyrosine phosphorylation of other proteins. Autophosphorylation of the receptor on tyrosine residues plays an essential role in receptor signaling by generating specific binding sites for cellular signaling proteins containing *src* homology 2 (SH2) or phosphotyrosine binding (PTB) domains (22). Cellular proteins that stably bind receptors in such a phosphotyrosine-dependent fashion include *src* family tyrosine kinases, phospholipase Cγ (PLCγ), phosphoinositol 3' kinase (PI3K), and Shc (22). Once recruited to the tyrosine-phosphorylated receptor, these cellular proteins are activated by a variety of mechanisms, including conformational changes, direct phosphorylation by the receptor, and subcellular relocalization. The signal initiated by ligand binding and receptor activation is then propagated through a cascade of cytoplasmic and nuclear signal transducers, including the *ras*-MAP kinase pathway (22).

Unlike receptor tyrosine kinases, many cytokine receptors, such as the erythropoietin receptor, do not possess intrinsic tyrosine kinase activity. Nevertheless, these receptors utilize a related mechanism of signaling (63, 133). Upon ligand binding, cytokine receptors dimerize and form complexes with cytoplasmic tyrosine kinases, such as the JAK family kinases (63, 96). This results in activation of these kinases, which in turn catalyze tyrosine phosphorylation of the receptor and recruitment of other proteins to the receptor, thereby activating the downstream signaling cascade.

The tumor necrosis factor (TNF) receptor can also signal cell proliferation in some situations. Like cytokine receptors, the TNF receptor is not an enzyme and seems to function by mobilizing intracellular proteins. TNF and related proteins bind to TNF receptors and induce aggregation of the receptor. This results in the aggregation of proteins that bind TNF receptors, including tumor necrosis factor receptor-associated factors (TRAFs) that are constitutively bound to the cytoplasmic tail of the TNF receptor 2 (112, 128). TRAF aggregation in some cell types including B lymphocytes is believed to activate cytoplasmic serine/threonine kinases, which in turn activate the transcription factor NFκB and stimulate proliferation (12). The activated TNF receptor 1 binds directly to other proteins such as TNF receptor-associated death domain protein (TRADD), which can itself bind TRAF2, leading to NFκB activation and proliferation (61a, 128). Alternatively, TRADD can initiate apoptosis by activating a different signaling cascade.

Viral proteins interact with each of the cellular signaling pathways outlined above. Table 1 lists the viral proteins described in this review and indicates the points at which these viral proteins intersect with cellular signaling pathways.

Table 1 Viral proteins and their actions

Viral proteins that activate cellular receptors	
Bovine papillomavirus E5 protein	PDGF β receptor
Spleen focus forming virus gp55	Erythropoietin receptor
Viral proteins that mimic cellular receptors	
Polyomavirus middle T antigen	Receptor tyrosine kinase
Epstein-Barr virus LMP-1	TNF receptor 1 and 2
Viral proteins that engage downstream signaling components	
Herpesvirus saimiri Tip	$pp56^{c\text{-}lck}$
Herpesvirus saimiri STP-C488	$p21^{c\text{-}ras}$
Herpesvirus saimiri STP-A11	*src* family tyrosine kinases
Epstein-Barr virus LMP-2	*c-lyn* and *c-syk* tyrosine kinases
Hepatitis B virus HBx	$pp60^{c\text{-}src}$

DIRECT ACTIVATION OF GROWTH FACTOR RECEPTORS BY VIRAL PROTEINS THAT DO NOT RESEMBLE NORMAL LIGANDS

Some oncogenic retroviruses and DNA tumor viruses encode proteins that closely resemble the normal cellular ligands of growth factor receptors (see 10, 35, 132). These proteins evidently bind the receptors and initiate signaling by the same mechanisms that are utilized by their cellular counterparts. In many cases, these viral proteins can act in an autocrine fashion, i.e. they are synthesized in the same cell as the receptor. Recent studies of viral transformation have revealed a related but distinct mechanism of receptor activation.

The BPV E5 Protein and the PDGF β Receptor

The E5 gene of bovine papillomavirus type 1 (BPV) can cause stable and acute transformation of cultured fibroblasts. This gene encodes the E5 protein, a very unusual hydrophobic transforming protein of only 44 amino acids that is primarily localized to the membranes of the endoplasmic reticulum (ER) and Golgi apparatus in transformed cells (19, 20, 117). The E5 protein is thought to be a type II transmembrane protein with its carboxyl-terminal third in the lumen of the ER and Golgi apparatus (19). The predominant form of the E5 protein in transformed cells is a disulfide-linked homodimer, but monomeric E5 protein is also present (17).

In transformed fibroblasts, the BPV E5 protein is present in a stable complex with mature and immature intracellular forms of the endogenous PDGF β receptor (105, 106) (Figure 1). Strikingly, the PDGF β receptor in cells expressing the E5 protein is constitutively activated as assessed by several criteria: It displays increased tyrosine-kinase activity, exhibits increased levels of tyrosine phosphorylation, and is constitutively bound to several cellular signal transduction proteins (37, 106, 107). Receptor activation occurs rapidly and in a dose-dependent fashion upon acute expression of the E5 protein (107). These findings suggest the simple model that the E5 protein directly induces activation of the PDGF β receptor, and that the activated receptor then initiates a sustained mitogenic signal similar to that induced by the normal ligand, resulting in growth transformation.

Several studies indicate that the PDGF β receptor can indeed mediate transformation by the E5 protein. Variant fibroblast cell lines with decreased ability to respond to PDGF are relatively resistant to E5 transformation (111). In addition, E5 mutants unable to bind and activate the PDGF receptor are transformation-defective (61, 91, 101, 119). [There is some controversy about this point, but the most recent and complete experiments examining this issue revealed an excellent correlation between the abilities of various E5 mutants to bind and activate the receptor and to transform C127 cells (101, 121; O Klein, GW Polack, D DiMaio, unpublished observations).] The strongest evidence for an essential role for the PDGF β receptor in E5 transformation is provided by gene transfer experiments in which cells that do not express endogenous PDGF β receptor become susceptible to E5-induced proliferation or tumorigenesis only when an exogenous PDGF β receptor is expressed (37, 51, 102, 123). In these heterologous systems, transformation is accompanied by E5/PDGF β receptor complex formation and by E5-induced tyrosine phosphorylation of the receptor. Other receptor tyrosine kinases, including the closely related PDGF α receptor and the epidermal growth factor (EGF) receptor, are unable to mediate E5-induced transformation (51, 102, 123).

Further insight into the role of the PDGF β receptor in transformation by the E5 protein is provided by studies in which PDGF β receptor mutants have been analyzed for their ability to bind the E5 protein and support E5 transformation. The intrinsic tyrosine-kinase activity of the receptor is required for mitogenic signaling, suggesting that E5 transformation occurs through activation of the receptor signal transduction pathway (37). This interpretation is supported by the finding that treatment of E5-transformed cells with specific inhibitors of PDGF receptor tyrosine kinase activity reverses receptor tyrosine phosphorylation and the transformed phenotype (O Klein, P Irusta, D DiMaio, unpublished observations). In contrast, deletion of most of the extracellular ligand binding domain of the receptor prevented receptor activation by PDGF but had no effect on complex formation with the E5 protein and did not impair E5 induced receptor activation and mitogenic activity (37, 123). Therefore, E5 transformation proceeds in the absence of activation of the receptor by PDGF, indicating that it is ligand-independent.

PDGF is soluble, hydrophilic, and much larger than the E5 protein, and it binds the extracellular domain of the receptor. Thus, the biochemical mechanism of PDGF β receptor activation by PDGF is likely to differ from that utilized by the E5 protein, which is very small, hydrophobic, and does not interact with the extracellular domain of the receptor. The molecular interactions responsible for complex formation and receptor activation are still under investigation, but currently available evidence favors the model that the interaction between these two proteins is direct and not mediated by other cellular proteins (101, 108, 121). Analysis of PDGF receptor mutants and chimeras indicates that the transmembrane region of the PDGF receptor is essential for complex formation and activation, suggesting that the transmembrane domains of these two proteins may line up with one another in the cell membrane (25, 52, 108, 123). In fact, specific amino acids in the transmembrane and juxtamembrane domains of both proteins have been identified that are required for complex formation and receptor activation (101, 108). The ability of the E5 protein to distinguish between the related PDGF α and β receptors also maps to the transmembrane/juxtamembrane region of the receptor (123). Strikingly, the specific residues of the PDGF β receptor required for complex formation with the E5 protein are absent from the α receptor.

The E5 protein, like PDGF, induces receptor dimerization (CC Lai, C Henningson, D DiMaio, unpublished information). Because the E5 protein exists in a dimeric form in transformed cells and dimerization-defective E5 mutants do not activate the receptor nor induce cell transformation (61, 91, 101), complex formation between two molecules of the PDGF β receptor and a dimeric E5 protein can in principle be sufficient to cause dimerization of the receptor, which in turn activates the receptor tyrosine kinase. On the basis of computational searches and other considerations, we have recently proposed a model in which an E5 monomer is unable to bind the PDGF β receptor, because the site for receptor binding is generated by dimerization of the E5 protein (T Surti, O Klein, K Ascheim, D DiMaio, SO Smith, submitted for publication). According to this model, one receptor molecule binds to one face of the E5 dimer and a second receptor molecule binds to the other face of the dimer (Figure 2). The finding that dimerization-defective E5 mutants are unable to bind the PDGF β receptor is consistent with this model (101). However, it should be noted that the stoichiometry of the E5 protein and the PDGF receptor in the complex has not been established, and there is no high-resolution structural information available.

The results summarized above provide compelling evidence that the E5 protein activates the PDGF β receptor by means of biochemical interactions that are distinct from those utilized by the normal ligand, and that receptor activation drives cell transformation. However, cellular proteins other than the PDGF β receptor can bind to the E5 protein. These proteins include the 16-kd pore-forming subunit of vacuolar H^+-ATPase, a 125-kd protein related to α-adaptin, and the EGF receptor (36). The role of these proteins in E5-induced transformation of fibroblasts remains unclear (121), although it is possible that these interactions are important for other biological activities of the BPV E5 protein, such as NFκB induction, modulation of EGF receptor stability, and as-yet-poorly-defined effects in epithelial cells (20, 75, 90).

The E5 proteins of other papillomaviruses, including the high-risk human papillomaviruses (HPV), also appear to modulate signal transduction pathways. Early experiments suggested that growth factor signaling was enhanced by the HPV16 E5 protein (79, 109). This may be due, at least in part, to effects on recycling of the EGF receptor (124). More recently, it has been shown that the E5 proteins of HPV16 and the related rhesus monkey papillomavirus can activate *ras*, MAP kinase, and PI3K pathways in cells (48, 53). However, unlike the case for the BPV E5 protein, the mechanisms involved are far from

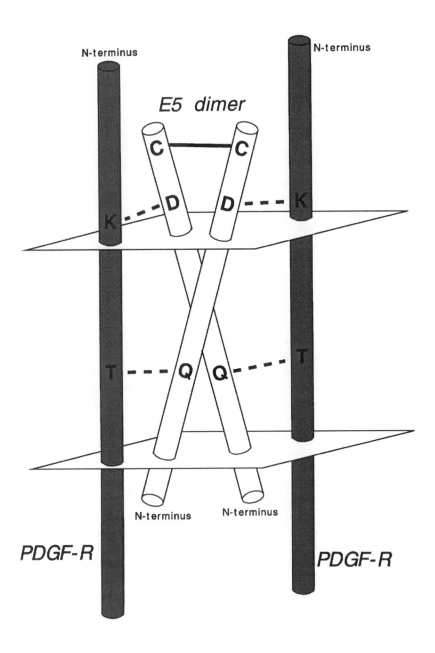

clear. Although HPV E5 proteins can form stable complexes with growth factor receptors in cell systems that overexpress the viral protein and the receptor, such complexes have not been observed with endogenous growth factor receptors in epithelial cells, the normal host cells for HPV replication and tumorigenesis (26, 62).

Friend Erythroleukemia Virus gp55 and the Erythropoietin Receptor

The spleen focus forming virus (SFFV) is a replication-defective retrovirus that induces erythroleukemia in infected mice. The viral protein gp55 is responsible for the initial proliferative stage of erythroleukemia (2, 3, 24, 31, 115). Most acutely transforming retroviruses carry a transduced cellular oncogene. In contrast, the SFFV gp55 gene is an intrinsic viral gene that encodes a dimeric transmembrane viral envelope glycoprotein (31, 50, 136).

Normal erythroblast proliferation results from the interaction between erythropoietin (EPO) and its cell surface receptor, a member of the cytokine receptor superfamily (133). EPO binding induces homodimerization of the EPO receptor, activation of cytoplasmic tyrosine kinases, and recruitment and activation of cellular signal transducing proteins. Leukemic erythroblasts isolated from SFFV-infected mice can proliferate in an EPO-independent manner, and similar effects can be elicited by SFFV infection of erythroblast cell lines, suggesting that gp55 might activate the EPO receptor signaling pathway (59, 114, 133). This hypothesis was confirmed in cell lines that normally require interleukin-3 (IL-3) for continued survival and proliferation. Coexpression of gp55 and murine EPO receptor in these cells allows sustained proliferation in the absence of the IL-3, whereas expression of neither protein alone supports proliferation in the absence of growth factors (59, 83). The ability of gp55 plus EPO receptor to substitute for IL-3 suggests that gp55 expression results in constitutive activation of the EPO receptor, which provides the necessary survival and proliferative signals (31, 133). This interpretation is substantiated by the

←———————————————————————————————

Figure 2   Model for the interaction between the BPV E5 protein and the PDGF β receptor. A complex between a dimer of the E5 protein (open rods) and the transmembrane and juxtamembrane domains of two PDGF receptor molecules (grey rods) is shown. The space bounded by the planes represents the cell membrane, with the cytoplasm at the bottom. The transmembrane domains of the E5 protein and the PDGF receptor are oriented in an antiparallel fashion relative to each other. Solid line, a disulfide bond between two carboxy-terminal cysteines in the E5 protein; dotted lines, putative non-covalent bonds between residues that have been shown by mutational analysis to be important for complex formation. A central feature of this model is that each PDGF receptor molecule forms contacts with both monomers of the E5 protein in the E5 dimer.

observation that a dominant-negative EPO receptor blocks gp55 transformation, and by the correlation between pathogenicity of various gp55 mutants and their ability to confer EPO resistance (5, 45, 130a). In addition, in cells expressing gp55, JAK kinases and other proteins downstream of the EPO receptor appear constitutively activated (84a, 102a, 120, 137).

Biochemical studies demonstrate that gp55 and the EPO receptor exist in the cells as a stable complex with at least a fraction of these complexes at the cell surface (45, 81, 83, 141). However, as is the case with BPV E5 and PDGF, gp55 bears no amino acid sequence similarity to EPO. Thus, the nature of the interaction between gp55 and EPO receptor is quite different from the inter- action between EPO and the EPO receptor. The biochemical mechanism of EPO receptor activation by gp55 has not been established, and it has not been clearly shown that gp55 and EPO receptor contact one another directly, nor that gp55 induces dimerization of the EPO receptor (76, 127). Nevertheless, genetic studies have identified the domains required for the productive interac- tion between these two proteins. Both the extracellular ligand binding domain and transmembrane domain of the EPO receptor are required for signaling by gp55 (31, 142). Similarly, the transmembrane and extracytoplasmic domains of gp55 are required for complex formation with the EPO receptor and signaling (31, 119a, 122, 133), and biological studies show that these same two domains are required for leukemogenicity (24, 123, 131).

The nature of the interaction between the EPO receptor and gp55 has been probed in more detail by using two different variants of gp55, gp55A and gp55P. Both proteins induce leukemia and bind the EPO receptor, but only gp55P is able to cooperate with the EPO receptor to confer EPO-independent proliferation of hematopoietic cell lines (115). This difference between gp55A and gp55P, as well as differences in their pathogenicity in animals, maps to their transmembrane domains, again highlighting the importance of this portion of gp55 in mediating its interaction with EPO receptor (24, 115). It has recently been shown that gp55A is able to support erythroid differentiation in murine fetal liver cells, a process that is normally mediated by EPO (27). Strikingly, gp55A had no such effect on cells isolated from EPO receptor knock-out mice, establishing that this effect is mediated by the EPO receptor (27). Thus, gp55A as well as gp55P can activate the EPO receptor, although the details of the cellular response depend on the particular system used. Taken together, analysis of gp55 demonstrates that it mimics EPO by binding and activating the EPO receptor, which in turn delivers a proliferative signal.

Other cytokine receptors may also serve as the targets of retroviral envelope proteins. The gp70 proteins of some mink cell focus forming (MCF) viruses associate with the interleukin 2 (IL-2) receptor, resulting in receptor activation (82). A small hydrophobic protein encoded by human T cell leukemia virus 1

(HTLV-1), p12^I also appears to interact with the IL-2 receptor (99). Thus, activation of cytokine receptors by retroviral proteins may be a strategy commonly employed by these viruses to induce cell proliferation.

VIRAL PROTEINS THAT MIMIC ACTIVATED GROWTH FACTOR RECEPTORS

The analysis of BPV E5 and gp55 summarized in the preceding section revealed that viruses can activate growth factor receptors and transform cells without synthesizing proteins that resemble the normal ligands of the receptor. As summarized in this section, viruses have evolved another mechanism to activate growth factor signaling pathways, namely the synthesis of proteins that resemble not a growth factor, but an activated receptor itself.

Polyomavirus Middle T Antigen

The middle T antigen (mT) of mouse polyomavirus and related viruses can cause growth transformation of established lines of rodent fibroblasts and is required for virus-induced tumorigenicity in rodents (14, 34). mT is a 421-amino acid protein anchored to the plasma membrane by a carboxyl-terminal segment of hydrophobic amino acids, with most of the protein extending into the cytoplasm (Figure 1). Although mT has no known intrinsic catalytic activity, in transformed cells it forms a stable complex with pp60^{c-src}, the product of the c-src cellular proto-oncogene, and other kinases of the src family including c-fyn and c-yes (29, 60). The association of c-fyn with hamster polyomavirus mT appears to be mediated by the ability of the c-fyn SH2 domain to recognize a specific phosphotyrosine on mT, but SH2/phosphotyrosine interactions do not mediate the association of mouse polyomavirus mT with src family members (13, 40, 41). Complex formation with mT stimulates the tyrosine kinase activity of pp60^{c-src} and related kinases (11, 28). The mechanism of kinase activation by mT is unclear, but it may involve removal of an inhibitory phosphate group from a carboxyl-terminal tyrosine in pp60^{c-src} and stabilization of the enzyme in an active conformation.

The importance of the mT/src interaction has been assessed by using a variety of approaches. Studies of mT mutants indicate that the association of mT with src family members is necessary for transformation, a conclusion supported by the inhibitory effects of dominant negative alleles of src family members on mT transformation (14, 15, 34, 49). However, activation of src family members is not sufficient for transformation, because some transformation-deficient mT mutants still display elevated levels of associated tyrosine kinase activity (84). Studies in animals unable to express specific src family members and in cells derived from such animals indicate that no individual src family member is required for mT activity (73, 74, 126). Rather, these studies indicate that mT induced tumorigenesis and cell transformation is a complex process involving redundant and/or alternative cellular signaling pathways. In the case of mouse mT-induced hemangioma formation, *c-yes* appears to play a more important role than the other *src* family members tested (73).

The primary biochemical role of the *src* kinases in mT transformation appears to be the phosphorylation of mT itself on tyrosine residues. Tyrosine phosphorylation of mT generates binding sites for a variety of other cellular signaling proteins including PI3K, PLC$_\gamma$, and Shc (21, 33, 125, 134, 139). As is the case for the association between these same signaling proteins and growth factor receptors, the association with mT appears to be mediated by the ability of the SH2 and PTB domains of these proteins to recognize specific phosphotyrosine residues on the viral protein (21, 33, 139). Once Shc and PLC$_\gamma$ are recruited to the phosphorylated mT complex, they also undergo tyrosine phosphorylation, an event that is also probably catalyzed by *src* kinases (9, 21, 33). Phosphorylated Shc bound to mT recruits Grb2 to the mT complex, which in turn activates the *ras* protooncogene product and eventually the MAP kinase signaling pathway (9, 21, 22). In addition, association between mT and PI3K results in increased PI3K enzymatic activity and the elevation of PI3K products in the cells (84). Thus, the interactions between cellular signaling molecules and mT mobilize signaling pathways that are similar to those elicited by growth factor treatment. Indeed, analysis of mT mutants indicates that efficient transformation depends on the assembly of these complexes and the consequent activation of the cellular proteins (9, 33, 49, 134). Similarly, a dominant negative mutant of Shc blocked mT transformation (9), and *ras* activity is essential for mT transformation (15, 110). In animals, the efficiency of tumor formation or the spectrum of tumors formed is affected by mutations in mT that inhibit association with the cellular signaling proteins (17, 138). In addition to proteins known to be involved in propagating mitogenic signals, several other interesting cell proteins are present in the mT complex and may play roles in transformation. The 14-3-3 proteins also form a complex with mT and may activate the raf kinase (44, 46, 64, 103). mT also forms a complex with phosphoprotein phosphatase 2A, an interaction proposed to be important for the assembly or the correct subcellular distribution of the mT signaling complex (13, 49, 104, 129).

Thus, the structure and activity of the complex of mT and its associated cellular proteins closely resemble that of the transmembrane and cytoplasmic domains of an activated growth factor receptor bound to proteins involved in signal transduction (34). However, unlike ligand-activated growth factor receptors, mT appears to exist in activated kinase complexes as a monomer, and dimerization of mT is not required for cell transformation (23, 118). Because mT is devoid of intrinsic tyrosine kinase activity but associates with and activates *src* kinases, mT is most similar to activated cytokine receptors that are tyrosine phosphorylated by the action of associated cytoplasmic kinases such as JAK kinases. These results indicate that the mT protein mimics an activated growth factor receptor and thereby delivers a constitutive mitogenic signal to cells.

Epstein-Barr Virus LMP-1

Epstein-Barr Virus (EBV) encodes latent membrane protein 1 (LMP-1), a 62 kDa integral membrane protein with six transmembrane domains that is constitutively aggregated in the plasma membrane (57). LMP-1 is expressed in B lymphocytes latently infected with EBV, and it is required for B cell transformation by the virus (57, 71). In addition, LMP-1 is sufficient to induce stable growth transformation of cultured rodent fibroblasts (4, 130). In B cells, LMP-1 induces the expression of a variety of cellular genes including B cell activation markers and Bcl-2 (85, 88, 113). LMP-1 expression also results in activation of the transcription factor NFκB (55, 58, 95).

The structure of LMP-1 is somewhat reminiscent of G-protein coupled receptors, leading to the suggestion that LMP-1 resembles a cell surface receptor (89). This notion has been confirmed by biochemical and genetic studies. To identify cellular proteins that might mediate LMP-1 transformation, Mosialis et al (98) used the yeast two-hybrid system to identify B lymphocyte proteins able to bind to a membrane-proximal segment of the cytoplasmic domain of LMP-1 required for transformation of primary B lymphocytes (72). In both yeast and B cells, it was found that this segment of LMP-1 binds to TRAFs, the factors that associate with activated tumor necrosis factor (TNF) receptor 2 and mediate its proliferative signal (98, 112). By binding various TRAFs, LMP-1 induces their constitutive aggregation in the absence of ligand and therefore appears to mimic the structure of the ligand-activated TNF receptor 2 complex (Figure 3) (98). In B-lymphocytes, the primary TNF receptor is CD40, which like LMP-1 engages multiple TRAFs and induces NFκB activation. Dominant negative mutants of TRAF2 and TRAF3 partially inhibit LMP-1-induced activation of NFκB, thus providing a genetic link between the viral protein and activation of the transcription factor (32, 70). Genetic results also suggest that the ability of LMP-1 to aggregate and bind TRAFs is responsible for NFκB activation, the initial proliferative stages of B-lymphocyte transformation, and fibroblast transformation (32, 65, 70, 95, 97, 116). The ability of LMP-1 to engage TRAFs also appears responsible for its effects in epithelial cells, including activation of NFκB, synthesis of interleukin-6, and induction of EGF receptor expression (42, 43, 94).

The study of the mechanism of LMP-1–induced lymphocyte transformation is complicated by the ability of a segment of LMP-1 downstream of the

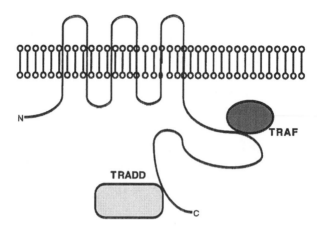

LMP-1   complex

Figure 3 A model for the EBV LMP-1 signaling complex. LMP-1 is present in the plasma membrane with six membrane-spanning domains. The membrane proximal segment of the cytoplasmic carboxy-terminal tail associates with TRAFs, and the membrane-distal segment of the tail associates with TRADD. Although for simplicity only a single LMP-1 molecule is shown, LMP-1 exists in membranes in a constitutively aggregated form, so the associations represented here result in aggregation of the cellular signaling molecules.

TRAF binding site to activate NFκB by a second, independent mechanism (16, 32, 70, 116). Recently, this segment of LMP-1 has been shown to associate with another TNF receptor-associated protein, TRADD, which is involved in mediating TNF-induced NFκB activation and apoptosis (66). Mutations in the carboxyl-terminal tail of LMP-1 that block TRADD binding also impair B cell transformation and NFκB activation (66). Thus, the LMP-1/TRADD interaction appears important for the ability of LMP-1 to induce high level NFκB activation and efficient long-term outgrowth of lymphoblastoid cells. The ability of LMP-1 to engage productively two different classes of TNF receptor-associated signaling molecules, and the strikingly similar response of B lymphocytes to LMP-1 expression and TNF treatment, provides compelling reason to regard LMP-1 as a constitutively activated mimic of the TNF receptor (12, 98). It is interesting to note that upon TNF addition, TNF receptor 1 directly engages TRADD but not TRAFs, whereas TNF receptor 2 engages TRAFs but not TRADD (128). Therefore, LMP-1, which directly engages both TRADD and TRAFs, may deliver a composite signal different from that delivered by either TNF receptor 1 or TNF receptor 2 alone.

VARIATIONS ON A THEME, AND IMPLICATIONS

Growth factors are powerful mitogenic stimuli, and cells have evolved elaborate signaling cascades to transmit proliferative signals from the cell surface to the nucleus. The examples reviewed here illustrate that viral activation of these cellular growth factor signaling pathways can result in cell transformation. These examples differ in some regards from the classically described mechanisms of growth factor receptor activation, which involve circuits composed of authentic growth factors and growth factor receptors or structurally related molecules. We suggest the term "virocrine transformation" to capture the central features of the viral mechanism described here (36). This term emphasizes the viral origin of the transforming protein and indicates a mechanism of action involving activation of signal transduction pathways. The viral protein acts either by activating a cellular signaling molecule such as a growth factor receptor, as in the case of BPV E5 or SFFV gp55, or through mimicry of an activated cellular signaling molecule, as in the case of polyomavirus mT or EBV LMP-1. Importantly, viral proteins that activate signaling pathways do not necessarily bear any obvious sequence similarity to cellular signaling molecules. In fact, studies on BPV E5 and gp55 have established the principle that growth factor receptors can be activated by proteins that do not resemble natural ligands. In those cases where the viral proteins deviate markedly from the structure of the normal ligands, the nature of the interactions that drive receptor binding may be very different from the interactions responsible for binding of the normal ligand to the receptor. Virocrine transforming proteins could even activate a signaling pathway by using biochemical mechanisms that are entirely different from those utilized by the normal ligand. It is also possible that receptors and pathways activated by different means may deliver qualitatively different signals to cells. This is suggested by the finding that EPO receptor activated by EPO or by gp55 elicits different biological responses and displays differential association with cellular proteins (2, 127, 137).

The virocrine mechanism of transformation differs in some crucial aspects from the other strategy commonly employed by DNA tumor viruses, namely inactivation of cellular tumor suppressors. Tumor-suppressor proteins and the cell cycle machinery they control are widely expressed in many cell types, whereas there is considerable specificity to the signal transduction pathways that are present in various cells. Therefore, a virus utilizing a virocrine strategy can target its proliferative stimulus to a specific cell type that expresses a particular receptor or set of downstream signaling molecules. Because growth factors can promote survival or differentiation as well as induce proliferation, viral proteins that signal through these pathways can influence many aspects of cell behavior. Finally, transforming proteins that activate positive signaling pathways function in a dominant fashion since they do not have to overcome inhibitory signals. As a consequence, low concentrations of a positive factor may be sufficient to drive proliferation. For example, it appears that a relatively small fraction of the PDGF receptor is in a complex with BPV E5, and mT binds only a small fraction of pp60$^{c\text{-}src}$ (C Henningson, D DiMaio, unpublished observations; 23). This is in contrast to the situation of viral proteins that inactivate tumor suppressor proteins, because in that case the viral protein must neutralize a substantial fraction of its cellular target to release the brakes on cell growth.

Although the examples cited in this review involve activation of the initial steps of signaling pathways, viral proteins also interact with and activate more downstream components. For example, the STP-C488 oncoprotein of herpesvirus saimiri (HVS) subgroup C forms a complex with cellular p21^{ras}, resulting in ras activation (67). Point mutations in STP that interfere with *ras* binding and activation also inhibit cell transformation, suggesting that STP induces cell transformation by activating *ras* (67). In contrast, the weak transforming protein STP A11 from HVS subgroup A forms a stable complex with pp60$^{c\text{-}src}$ (78). Although this association does not result in increased *src* activity, STP-A11-tyrosine phosphorylated by pp60$^{c\text{-}src}$ is able to bind to *lck* and *fyn*, which are *src* family members specifically expressed in T lymphocytes, the natural host cell for HVS transformation. These interactions appear mediated by the SH2 domain of *src* family members recognizing a phosphotyrosine on STP-A11. These findings raise the possibility that STP-A11 may transform cells by binding to pp60$^{c\text{-}src}$ which catalyzes tyrosine phosphorylation of the viral protein, thereby allowing it to interact with and activate other *src* family members (78). *Src* family members may also be the crucial cellular targets of the HBx protein of hepatitis B virus. This protein activates the *ras-raf*-MAP kinase signaling pathway in mammalian cells, resulting in cell cycling (6, 7, 30, 100). Recent evidence indicates that HBx activates *src* family members, which are then responsible for activation of the downstream signaling pathway (77). The mechanism of *src* activation by HBx has not been established.

There are also viral proteins that bind cellular signaling proteins and inhibit their activity, thereby modulating the response of cells to the normal ligands. An attractive mechanism for such an effect is that a viral protein may act in a dominant negative fashion by interacting with a subset of signaling proteins that is not sufficient to initiate a signal or by interacting with the entire complement of signaling molecules without being able to induce the next step in the signaling cascade. This may be the case for LMP-2, an EBV-encoded membrane protein that is tyrosine-phosphorylated and forms a complex with the *src* family protein tyrosine kinases c-lyn and c-syk in B lymphocytes transformed by EBV (18, 86). However, unlike polyoma mT, which activates *src* family members, LMP-2 inhibits them (86, 92). Specifically, LMP-2 inhibits the effects of surface immunoglobulin cross-linking on tyrosine kinase activity, activation of downstream signaling events such as substrate tyrosine phosphorylation and calcium mobilization, and reactivation of latent EBV (92, 93). Considering that the normal B-lymphocyte response to immunoglobulin cross-linking is activation of *src* kinases, LMP-2 may be acting as a dominant negative decoy by binding to these signaling proteins without having the ability to activate them. In fact, the specific short amino-acid sequences on LMP-2 required for kinase association and inhibition are the same motifs used by normal B-cell signaling proteins to assemble active kinase complexes (47). Thus, LMP-2 appears to be a negative regulator of the *src* family kinases in B lymphocytes, thereby playing a role in the maintenance of the latent state.

Another viral protein that regulates tyrosine kinase activity in lymphocytes is the Tip protein of various strains of herpesvirus saimiri, a protein that cooperates with STP in inducing T-lymphocyte transformation (38). In transformed cells, Tip is tyrosine-phosphorylated and in a complex with the T cell specific *src* family kinase $p56^{c\text{-}lck}$ (8, 68, 87). $p56^{c\text{-}lck}$ can tyrosine phosphorylate Tip in vitro (8), but the biological consequences of $Tip/p56^{c\text{-}lck}$ complex formation are unclear. Some reports maintain that this interaction results in stimulation of *lck* tyrosine kinase activity, but other evidence indicates that Tip downregulates the activity of $p56^{c\text{-}lck}$ and inhibits *lck*-mediated signal transduction in T cells (39, 54, 69, 135). HVS encoding a Tip mutant unable to bind *lck* causes more severe neoplastic disease in animals than does wild type HVS, lending further support to the conclusion that the Tip-*lck* interaction is not essential for oncogenicity and indeed may inhibit transformation (39). If Tip, in fact, inhibits tyrosine kinase signaling in T cells, it would seem quite analogous to EBV LMP-2, which has similar effects in B cells. A major role of Tip, as appears to be the case for LMP-2, may be in establishing or maintaining the latent viral state. Tip also interacts with another cellular protein, Tap, and this interaction appears to affect NFκB activity and other aspects of the T cell response to Tip (140).

Small viral proteins such as BPV E5 may also serve as models for the rational design of small proteins and other molecules that can influence cellular signaling pathways. The tendency of membrane spanning regions to adopt an α-helical conformation and the availability of powerful computational methods to explore the energetics of potential interhelical interactions suggests that the transmembrane domains of growth factor receptors may be particularly suitable targets for such an approach (1, 80). Proteins that modulate growth factor signaling pathways may have important research and clinical uses. For example, the ability to activate growth factor receptors in the central nervous system or to inactivate the erbB2 growth factor receptor in breast carcinoma may have great clinical utility. The targeted expression of such proteins has an advantage over systemic delivery of drugs that modulate receptor activity, because the effects of these proteins may be restricted to the cell in which they are expressed, thereby providing a mechanism to fine-tune the response of particular cells to a circulating growth factor. The specificity of the BPV E5 protein suggests that it may even be possible to design molecules that display greater selectivity than do the natural ligands of the receptor. Similarly, the ability of LMP-1 to bind directly both TRAFs and TRADD suggests that proteins with novel signaling properties can be constructed.

As the mechanisms by which viral proteins transform cells continue to be elucidated, additional examples of virocrine transformation will be identified. There is a great diversity of signaling molecules and pathways in cells. Therefore, it is likely that viral transforming proteins will show a comparable diversity in their mode of action. In fact, it is possible that study of such viral proteins will identify novel cellular signaling proteins and pathways, much as study of SV40 large T antigen resulted in the identification of p53. Thus, we can be confident that studies on viral transforming proteins will provide new insights into cellular signaling pathways and may suggest novel approaches for the manipulation of these pathways.

ACKNOWLEDGMENTS

We thank colleagues for access to unpublished information. The research carried out in the authors' laboratory was supported by grants from the National Institutes of Health (CA16038 and CA37157). We thank J Zulkeski for assistance in preparing this manuscript.

Visit the *Annual Reviews home page* at
http://www.AnnualReviews.org.

Literature Cited

1. Adams PD, Engelman DM, Brünger AT. 1996. Improved prediction for the structure of the dimeric transmembrane domain of glycophorin A obtained through global searching. *Proteins* 26:257–61; 27:132, Erratum

2. Ahlers N, Hunt N, Just U, Laker C, Ostertag W, Nowock J. 1994. Selectable retrovirus vectors encoding Friend virus gp55 or erythropoietin induce polycythemia with different phenotypic expression and disease progression. *J. Virol.* 68:7235–43

3. Aizawa S, Suda Y, Furuta Y, et al. 1990. Env-derived gp55 gene of Friend spleen focus forming virus specifically induces neoplastic proliferation of erythroid progenitor cells. *EMBO J.* 9:2107–16

4. Baichwal VR, Sugden V. 1988. Transformation of Balb/3T3 cells by the *BNLF-1* gene of Epstein-Barr virus. *Oncogene* 4:67–74

5. Barber DL, DeMartino JC, Showers MO, D'Andrea AD. 1994. A dominant negative erythropoietin (EPO) receptor inhibits EPO-dependent growth and blocks F-gp55-dependent transformation. *Mol. Cell. Biol.* 14:2257–65

6. Benn J, Schneider RJ. 1994. Hepatitis B virus HBx protein activates Ras-GTP complex formation and establishes a Ras, Raf, MAP kinase signalling cascade. *Proc. Natl. Acad. Sci. USA* 91: 10350–54

7. Benn J, Schneider RJ. 1995. Hepatitis B virus HBx protein deregulates cell cycle

checkpoint controls. *Proc. Natl. Acad. Sci. USA* 92:11215–19

8. Biesinger B, Tsygankov AY, Fickenscher H, Enimrich F, Fleckenstein B, et al. 1995. The product of the herpesvirus saimiri open reading frame 1 (tip) interacts with T cell-specific kinase p56*lck* in transformed cells. *J. Biol. Chem.* 270:4729–34

9. Blaikie PA, Foumier E, Dilworth SM, Birnbaum D, Borg JP, Margolis B. 1997. The role of the Shc phosphotyrosine interaction/phosphotyrosine binding domain and tyrosine phosphorylation sites in polyoma middle T antigen-mediated cell transformation. *J. Biol. Chem.* 272:20671–71

10. Blomquist MC, Hunt LT, Barker WC. 1984. Vaccinia virus 19-kilodalton protein: relationship to several mammalian proteins, including two growth factors. *Proc. Natl. Acad. Sci. USA* 81:7363–67

11. Bolen JB, Thiele CJ, Israel MA, Yonemoto W, Lipisch LA, Brugge JS. 1984. Enhancement of cellular *src* gene product associated tyrosyl kinase activity following polyoma virus infection and transformation. *Cell* 38:767–77

12. Boussiotis VA, Nadler LM, Strominger JL, Goldfeld AE. 1994. Tumor necrosis factor alpha is an autocrine growth factor for normal human B cells. *Proc. Natl. Acad. Sci. USA* 91:7007–11

13. Brewster CEP, Glover HR, Dilworth SM. 1997. pp60$^{c\text{-}src}$ binding to polyomavirus middle T-antigen (MT) requires residues 185 to 210 of the MT sequence. *J. Virol.* 71:5512–20

14. Brizuela L, Olcese LM, Courtneidge SA. 1994. Transformation by middle T antigens. *Semin. Virol.* 5:381–89

15. Brizuela L, Ulug ET, Jones MA, Courtneidge SA. 1995. Induction of interleukin-2 transcription by the hamster polyomavirus middle T antigen: a role for Fyn in T cell signal transduction. *Eur. J. Immunol.* 25:385–93

16. Brodeur SR, Cheng G, Baltimore D, Thorley-Lawson DA. 1997. Localization of the major NF-κB-activating site and the sole TRAF3 binding site of LMP-1 defines two distinct signaling motifs. *J. Biol. Chem.* 272:19777–84

17. Bronson R, Dawe C, Carroll J, Benjamin T. 1997. Tumor induction by a transformation-defective polyoma virus mutant blocked in signaling through Shc. *Proc. Natl. Acad. Sci. USA* 94:7954–58

18. Burkhardt AL, Bolen JB, Kieff E, Longnecker R. 1992. An Epstein-Barr virus transformation-associated membrane protein interacts with src family tyrosine kinases. *J. Virol.* 66:5161–67

19. Burkhardt A, Willingham M, Gay C, Jeang K-T, Schlegel R. 1989. The E5 oncoprotein of bovine papillomavirus is oriented asymmetrically in Golgi and plasma membranes. *Virology* 170:334–39

20. Burnett S, Jareborg N, DiMaio D. 1992. Localization of bovine papillomavirus type 1 E5 protein to transformed basal keratinocytes and permissive differentiated cells in fibropapilloma tissue. *Proc. Natl. Acad. Sci. USA* 89:5665–69

21. Campbell KS, Ogris E, Burke B, Su W, Auger KR, et al. 1994. Polyoma middle tumor antigen interacts with SHC protein via the NPTY (asn-pro-thr-tyr) motif in middle tumor antigen. *Proc. Natl. Acad. Sci. USA* 91:6344–48

22. Cantley LC, Auger KR, Carpenter C, Duckworth B, Graziani A, et al. 1991. Oncogenes and signal transduction. *Cell* 64:281–302

23. Cheng SH, Espino PC, Marshall J, Harvey R, Smith AE. 1990. Stoichiometry of cellular and viral components in the polyomavirus middle-T antigentyrosine kinase complex. *Mol. Cell. Biol.* 10:5569–74

24. Chung S-W, Wolff L, Ruscetti SK. 1989. Transmembrane domain of the envelope gene of a polycythemia-inducing retrovirus determines erythropoietin-independent growth. *Proc. Natl. Acad. Sci. USA* 86:7957–60

25. Cohen BD, Goldstein DJ, Rutledge L, Vass WC, Lowy DR, et al. 1993. Transformation-specific interaction of the bovine papillomavirus E5 oncoprotein with the platelet-derived growth factor receptor transmembrane domain and the epidermal growth factor receptor cytoplasmic domain. *J. Virol.* 67:5303–11

26. Conrad M, Goldstein D, Andresson T, Schlegel R. 1994. The E5 protein of HPV-6, but not HPV-16, associates efficiently with cellular growth factor receptors. *Virology* 200:796–800

27. Constantinescu SN, Wu H, Liu X, Beyer W, Fallon A, Lodish HF. 1997. The anemic friend virus gp55 envelope protein induces erythroid differentiation in fetal liver CFU-Es. *Blood* 91:1163–72

28. Courtneidge SA. 1985. Activation of pp60$^{c\text{-}src}$ kinase by middle T-antigen binding or by dephosphorylation, *EMBO J.* 4:1471–77

29. Courtneidge SA, Smith AE. 1983. Polyoma virus transforming protein

associates with the product of the c-*src* gene. *Nature (London)* 303:433–38

30. Cross JC, Wen P, Rutter WJ. 1993. Transactivation by hepatitis B virus X protein is promiscuous and dependent on mitogen activated cellular serine/threonine kinases. *Proc. Natl. Acad. Sci. USA* 90:8078–82

31. D'Andrea AD. 1992. The interaction of the erythropoietin receptor and gp55. *Cancer Surv.* 15:19–36

32. Devergne O, Hatzivassiliou E, Izumi KM, Kaye KM, Kleijnen MF, et al. 1996. Association of TRAF1, TRAF2, and TRAF3 with an Epstein-Barr virus LMP1 domain important for B-lymphocyte transformation: role in NF-κB activation. *Mol. Cell. Biol.* 16:7098–7108

33. Dilworth SM, Brewster CEP, Jones MD, Lanfrancone L, Pelicci G, Pelicci PG. 1994. Transformation by polyoma virus middle T antigen involves the binding and tyrosine phosphorylation of Shc. *Nature* 367:87–90

34. Dilworth SM. 1995. Polyoma virus middle T antigen: meddler or mimic? *Trends Microbiol.* 3:31–35

35. Doolittle RF, Hunkapiller MW, Hood LE, Devare SG, Robbins KC, et al. 1983. Simian sarcoma virus *onc* gene, v-*sis*, is derived from the gene (or genes) encoding a platelet-derived growth factor. *Science* 221:275–77

36. Drummond-Barbosa D, DiMaio D. 1997. Virocrine transformation. *Biochim. Biophys. Acta* 1332:M1–17

37. Drummond-Barbosa D, Vaillancourt RR, Kazlauskas A, DiMaio D. 1995. Ligand-independent activation of the platelet-derived growth factor β receptor: requirements for bovine papillomavirus E5-induced mitogenic signaling. *Mol. Cell. Biol.* 15:2570–81

38. Duboise SM, Guo J, Czajak S, Desrosiers RC, Jung JU. 1998. STP and tip are essential for herpesvirus saimiri oncogenicity. *J. Virol.* 72:1308–13

39. Duboise SM, Lee H, Guo J, Choi J-K, Czajak S, et al. 1997. Mutation of the lck-binding motif of tip enhances lymphoid cell activation by herpesvirus saimiri. *J. Virol.* 72:2607–14

40. Dunant NM, Messerschmitt AS, Ballmer-Hofer K. 1997. Functional interaction between the SH2 domain of Fyn and tyrosine 324 of hamster polyomavirus middle-T antigen. *J. Virol.* 71:199–206

41. Dunant NM, Senften M, Ballmer-Hofer K. 1996. Polyomavirus middle-T antigen associates with the kinase domain

of Src-related tyrosine kinases. *J. Virol.* 70:1323–30

42. Eliopoulos AG, Dawson CW, Mosialos G, Floettmann JE, Rowe M, et al. 1996. CD40-induced growth inhibition in epithelial cells is mimicked by Epstein-Barr Virus-encoded LMP1: involvement of TRAF3 as a common mediator. *Oncogene* 13:2243–54

43. Eliopoulos AG, Stack M, Dawson CW, Kaye KM, Hodgkin L, et al. 1997. Epstein-Barr virus-encoded LMP1 and CD40 mediate IL-6 production in epithelial cells via an NF-κB pathway involving TNF receptor-associated factors. *Oncogene* 14:2899–916

44. Fantl WJ, Muslin AJ, Kikuchi A, Martin JA, MacNicol AM, et al. 1994. Activation of Raf-1 by 14-3-3 proteins. *Nature* 371:612–14

45. Ferro Jr FE, Kozak SL, Hoatlin ME, Kabat D. 1993. Cell surface site for mitogenic interaction of erythropoietin receptors with the membrane glycoprotein encoded by Friend erythroleukemia virus. *J. Biol. Chem.* 268:5741–47

46. Freed E, Symons M, Macdonald SG, McCormick F, Ruggieri R. 1994. Binding of 14-3-3 proteins to the protein kinase Raf and effects on its activation. *Science* 265:1713–16

47. Fruehling S, Longnecker R. 1997. The immunoreceptor tyrosine-based activation motif of Epstein-Barr virus LMP2A is essential for blocking BCR-mediated signal transduction. *Virology* 235:241–51

48. Ghai J, Ostrow RS, Tolar J, McGlennen RC, Lemke TD, et al. 1996. The E5 gene product of rhesus papillomavirus is an activator of endogenous Ras and phosphatidylinositol-3'-kinase in NIH 3T3 cells. *Proc. Natl. Acad. Sci. USA* 93:12879–84

49. Glenn GM, Eckhart W. 1993. Mutation of a cysteine residue in polyomavirus middle T antigen abolishes interactions with protein phosphatase 2A, pp60$^{c\text{-}src}$, and phosphatidylinositol-3 kinase, activation of c-fos expression, and cellular transformation. *J. Virol.* 67:1945–52

50. Gliniak BC, Kozak SL, Jones RT, Kabat D. 1991. Disulfide bonding controls the processing of retroviral envelope glycoproteins. *J. Biol. Chem.* 266:22991–97

51. Goldstein DJ, Li W, Wang L-M, Heidaran MA, Aaronson S, et al. 1994. The bovine papillomavirus type 1 E5 transforming protein specifically binds and activates the β-type receptor for the platelet-derived growth factor but not

other related tyrosine kinase-containing receptors to induce cellular transformation. *J. Virol.* 68:4432–41

52. Goldstein DJ, Andresson T, Sparkowski JJ, Schlegel R. 1992. The BPV-1 E5 protein, the 16 kDa membrane poreforming protein and the PDGF receptor exist in a complex that is dependent on hydrophobic membrane interactions. *EMBO J.* 11:4851–59

53. Gu Z, Matlashewski G. 1995. Effect of human papillomavirus type 16 oncogenes on MAP kinase activity. *J. Virol.* 69:8051–56

54. Guo J, Duboise M, Lee H, Li M, Choi J-K, et al. 1997. Enhanced downregulation of lck-mediated signal transduction by a Y_{114} mutation of herpesvirus saimiri tip. *J. Virol.* 71:7092–96

55. Hammarskjold ML, Simurda MC. 1992. Epstein-Barr virus latent membrane protein transactivates the human immunodeficiency virus type 1 long terminal repeat through induction of NF-κB activity. *J. Virol.* 66:6469–6501

56. Heldin C-H. 1995. Dimerization of cell surface receptors in signal transduction. *Cell* 80:213–23

57. Hennessy K, Fennewald S, Hummel M, Cole T, Kieff E. 1984. A membrane protein encoded by Epstein-Barr virus in latent growth-transforming infection. *Proc. Natl. Acad. Sci. USA* 81:7201–11

58. Herrero JA, Mathew P, Paya CV. 1995. LMP-1 activates NF-κB by targeting the inhibitory molecule I kappa B alpha. *J. Virol.* 69:2168–74

59. Hoatlin ME, Kozak SL, Lilly F, Chakraborti A, Kozak CA, Kabat D. 1990. Activation of erythropoietin receptors by Friend viral gp55 and by erythropoietin and down-modulation by the murine *Fv-2ʳ* resistance gene. *Proc. Natl. Acad. Sci. USA* 87:9985–89

60. Horak ID, Kawakami T, Gregory F, Robbins KC, Bolen JB. 1989. Association of p60*fyn* with middle tumor antigen in murine polyomavirus-transformed rat cells. *J. Virol.* 63:2343–47

61. Horwitz BH, Burkhardt AL, Schlegel R, DiMaio D. 1988. 44-amino-acid E5 transforming protein of bovine papillomavirus requires a hydrophobic core and specific carboxyl terminal amino acids. *Mol. Cell. Biol.* 8:4071–78

61a. Hsu HL, Shu HB, Pan MG, Goeddel DV. 1996. TRADD-TRAF2 and TRADD-FADD interactions define two distinct TNF receptor 1 signal transduction pathways. *Cell* 84:299–308

62. Hwang E-S, Nottoli T, DiMaio D. 1995.

The HPV16 E5 protein: expression, detection, and stable complex formation with transmembrane proteins in COS cells. *Virology* 211:227–33

63. Ihle JN. 1995. Cytokine receptor signalling. *Nature* 377:591–94

64. Irie K, Gotoh Y, Yashar BM, Errede B, Nishida E, Matsumoto K. 1994. Stimulatory effects of yeast and mammalian 14-3-3 proteins on the Raf protein kinase. *Science* 265:1716–19

65. Izumi KM, Kaye KM, Kieff ED. 1997. The Epstein-Barr virus LMP1 amino acid sequence that engages tumor necrosis factor receptor associated factors is critical for primary B lymphocyte growth transformation. *Proc. Natl. Acad. Sci. USA* 94:1447–52

66. Izumi KM, Kieff ED. 1997. The Epstein-Barr virus oncogene product latent membrane protein 1 engages the tumor necrosis factor receptor-associated death domain protein to mediate B lymphocyte growth transformation and activate NF-κB. *Proc. Natl. Acad. Sci. USA* 94:12592–97

67. Jung JU, Desrosiers RC. 1995. Association of the viral oncoprotein STP-C488 with cellular ras. *Mol. Cell. Biol.* 15:6506–12

68. Jung JU, Lang SM, Friedrich U, Jun T, Robert TM, et al. 1995. Identification of Lck-binding elements in tip of herpesvirus saimiri. *J. Biol. Chem.* 270: 20660–67

69. Jung JU, Lang SM, Jun T, Roberts TM, Veillette A, Desrosiers RC. 1995. Downregulation of Lck-mediated signal transduction by tip of herpesvirus saimiri. *J. Virol.* 69:7814–22

70. Kaye KM, Devergne O, Harada JN, Izumi KM, Yalamanchili R, et al. 1996. Tumor necrosis factor receptor associated factor 2 is a mediator of NF-κB activation by latent infection membrane protein 1, the Epstein-Barr virus transforming protein. *Proc. Natl. Acad. Sci. USA* 93:11085–90

71. Kaye KM, Izumi KM, Kieff E. 1993. Epstein-Barr virus latent membrane protein 1 is essential for B-lymphocyte growth transformation. *Proc. Natl. Acad. Sci. USA* 90:9150–54

72. Kaye KM, Izumi KM, Mosialos G, Kieff E. 1995. The Epstein-Barr virus LMP1 cytoplasmic carboxyl terminus is essential for B lymphocyte transformation: fibroblast co-cultivation complements a critical function within the terminal 155 residues. *J. Virol.* 69:675–83

73. Kiefer F, Anheuser 1, Soriano P, Aguzzi

A, Courtneidge SA, Wagner EF. 1994. Endothelial cell transformation by polyomavirus middle T antigen in mice lacking Src-related kinases. *Curr. Biol.* 4:100–9

74. Kiefer F, Courtneidge SA, Wagner EF. 1994. Oncogenic properties of the middle T antigens of polyomaviruses. *Adv. Cancer Res.* 64:125–57

75. Kilk A, Talpsepp T, Vali U, Ustav M. 1996. Bovine papillomavirus oncoprotein E5 induces the NF-κB activation through superoxide radicals. *Biochem. Mol. Biol. Int.* 40:689–97

76. Kishi A, Chiba T, Sugiyama M, Machide M, Nagata Y, et al. 1993. Erythropoietin receptor binds to Friend virus gp55 through other membrane components. *Biochem. Biophys. Res. Commun.* 192:1131–38

77. Klein NP, Schneider RJ. 1997. Activation of Src family kinases by hepatitis B virus HBx protein and coupled signaling to ras. *Mol. Cell. Biol.* 17:6427–36

78. Lee H, Tremble IJ, Yoon D-W, Regier D, Desrosiers RC, Jung JU. 1997. Genetic variation of herpesvirus saimiri subgroup A transforming protein and its association with cellular src. *J. Virol.* 71:3817–25

79. Leechanachai P, Banks L, Moreau F, Matlashewski G. 1992. The E5 gene from human papillomavirus type 16 is an oncogene which enhances growth factor-mediated signal transduction to the nucleus. *Oncogene* 7:19–25

80. Lemmon MA, Engelman DM. 1994. Specificity and promiscuity in membrane helix interactions. *Q. Rev. Biophys.* 27:157–218

81. Li J-P, Hu H-O, Niu Q-T, Fang C. 1995. Cell surface activation of the erythropoietin receptor by Friend spleen focus-forming virus gp55. *J. Virol.* 69:1714–19

82. Li J-P, Baltimore D. 1991. Mechanism of leukemogenesis induced by the MCF murine leukemia viruses. *J. Virol.* 65:2408–14

83. Li J-P, D'Andrea AD, Lodish HF, Baltimore D. 1990. Activation of cell growth by binding of Friend spleen focus-forming virus gp55 glycoprotein to the erythropoietin receptor. *Nature* 343:762–64

84. Ling LE, Druker BJ, Cantley LC, Roberts TM. 1992. Transformation-defective mutants of polyomavirus middle T antigen associate with phosphatidylinositol 3-kinase (PI 3-kinase) but are unable to maintain wild-type levels of PI

3-kinase products in intact cells. *J. Virol.* 66:1702–8

84a. Linnekin D, Evans GA, D'Andrea A, Farrar WL. 1992. Association of the erythropoietin receptor with protein tyrosine kinase activity. *Proc. Natl. Acad. Sci. USA* 89:6237–41

85. Longnecker R, Kieff E, Rickinson A. 1991. Induction of bcl-2 expression by Epstein-Barr virus latent membrane protein 1 protects infected B cells from programmed cell death. *Cell* 65:1107–15

86. Longnecker R, Miller CL. 1996. Regulation of Epstein-Barr virus latency by latent membrane protein 2. *Trends Microbiol.* 4:38–42

87. Lund T, Medveczky MM, Neame PJ, Medveczky PG. 1996. A herpesvirus saimiri membrane protein required for interleukin-2 independence forms a stable complex with p56lck. *J. Virol.* 70:600–6

88. Martin JM, Veis D, Korsmeyer SJ, Sugden B. 1993. Latent membrane protein of Epstein-Barr virus induces cellular phenotypes independently of expression of Bcl-2. *J. Virol.* 67:5269–78

89. Martin J, Sugden B. 1991. The latent membrane protein oncoprotein resembles growth factor receptors in the properties of its turnover. *Cell Growth Dim.* 2:653–60

90. Martin P, Vass WC, Schiller JT, Lowy DR, Velu TJ. 1989. The bovine papillomavirus E5 transforming protein can stimulate the transforming activity of EGF and CSF-1 receptors. *Cell* 59:21–32

91. Meyer AM, Xu Y-F, Webster MK, Smith AE, Donoghue DJ. 1994. Cellular transformation by a transmembrane peptide: structural requirements for the bovine papillomavirus E5 oncoprotein. *Proc. Natl. Acad. Sci. USA* 91:4634–38

92. Miller CL, Burkhardt AL, Lee JH, Stealey B, Longnecker R, et al. 1995. Integral membrane protein 2 of Epstein-Barr virus regulates reactivation from latency through dominant negative effects on protein-tyrosine kinases. *Immunity* 2:155–66

93. Miller CL, Lee JH, Kieff E, Longnecker R. 1994. An integral membrane protein (LMP2) blocks reactivation of Epstein-Barr virus from latency following surface immunoglobulin crosslinking. *Proc. Natl. Acad. Sci. USA* 91:772–76

94. Miller WE, Mosialos G, Kieff E, Raab-Traub N. 1997. Epstein-Barr virus LMP1 induction of the epidermal growth

factor receptor is mediated through a TRAF signaling pathway distinct from NF-κB activation. *J. Virol.* 71:586–94

95. Mitchell T, Sugden B. 1995. Stimulation of NF-κB-mediated transcription by mutant derivatives of the latent membrane protein of Epstein-Barr virus. *J. Virol.* 69:2968–76

96. Miura O, Ihle JN. 1993. Dimer- and oligomerization of the erythropoietin receptor by disulfide bond formation and significance of the region near the WSXWS motif in intracellular transport. *Arch. Biochem. Biophys.* 306:200–8

97. Moorthy RK, Thorley-Lawson DA. 1993. All three domains of the Epstein-Barr virus-encoded latent membrane protein LMP-1 are required for transformation of Rat-1 fibroblasts. *J. Virol.* 67:1638–46

98. Mosialos G, Birkenbach M, Yalamanchili R, Van Arsdale T, Ware C, Kieff E. 1995. The Epstein-Barr virus transforming protein LMP1 engages signaling proteins for the tumor necrosis factor receptor family. *Cell* 80:389–99

99. Mulloy JC, Crowley RW, Fullen J, Leonard WJ, Franchini G. 1996. The HTLV-I p12^I protein binds the IL-2 receptor β and γ$_c$ chains and affects their expression on the cell surface. *J. Virol.* 70:3599–3605

100. Natoli G, Avantaggiati ML, Chirillo P, Purl PL, Ianni A, et al. 1994. Ras- and raf-dependent activation of c-jun transcriptional activity by the hepatitis B virus transactivator pX. *Oncogene* 9:2837–43

101. Nilson LA, DiMaio D. 1993. Platelet-derived growth factor receptor can mediate tumorigenic transformation by the bovine papillomavirus E5 protein. *Mol. Cell. Biol.* 13:4137–45

102. Nilson LA, Gottlieb RL, Polack GW, DiMaio D. 1995. Mutational analysis of the interaction between the bovine papillomavirus E5 transforming protein and the endogenous β receptor for platelet-derived growth factor in mouse C127 cells. *J. Virol.* 69:5869–74

102a. Ohashi T, Masuda M, Ruscetti SK. 1995. Induction of sequence-specific DNA-binding factors by erythropoietin and the spleen focus-forming virus. *Blood* 85:1454–62

103. Pallas DC, Fu H, Haehnel LC, Weller W, Collier RJ, Roberts TM. 1994. Association of polyomavirus middle tumor antigen with 14-3-3 proteins. *Science* 265:535–37

104. Pallas DC, Shahrik LK, Martin BL, Jaspers S, Miller TB, et al. 1990. Polyoma small and middle T antigens and SV40 small t antigen form stable complexes with protein phosphatase 2A. *Cell* 60:167–76

105. Petti L, DiMaio D. 1992. Stable association between the bovine papillomavirus E5 transforming protein and activated platelet-derived growth factor receptor in transformed mouse cells. *Proc. Natl. Acad. Sci. USA* 89:6736–40

106. Petti L, DiMaio D. 1994. Specific interaction between the bovine papillomavirus E5 transforming protein and the β receptor for platelet-derived growth factor in stably transformed and acutely transfected cells. *J. Virol.* 68:3582–92

107. Petti L, Nilson L, DiMaio D. 1991. Activation of the platelet-derived growth factor receptor by the bovine papillomavirus E5 protein. *EMBO J.* 10:845–55

108. Petti LM, Reddy V, Smith SO, DiMaio D. 1997. Identification of amino acids in the transmembrane and juxtamembrane domains of the platelet-derived growth factor receptor required for productive interaction with the bovine papillomavirus E5 protein. *J. Virol.* 71:7318–27

109. Pim D, Collins M, Banks L. 1992. Human papillomavirus type 16 E5 gene stimulates the transforming activity of the epidermal growth factor receptor. *Oncogene* 7:27–32

110. Raptis L, Marcellus R, Corbley MJ, Krook A, Whitfield J, et al. 1991. Cellular ras gene activity is required for full neoplastic transformation by polyomavirus. *J. Virol.* 65:5203–10

111. Riese DJ II, DiMaio D. 1995. An intact PDGF signaling pathway is required for efficient growth transformation of mouse C127 cells by the bovine papillomavirus E5 protein. *Oncogene* 10:1431–39

112. Rothe M, Wong SC, Henzel WJ, Goeddel DV. 1994. A novel family of putative signal transducers associated with the cytoplasmic domain of the 75 kDa tumor necrosis factor receptor. *Cell* 78:681–92

113. Rowe M, Peng-Pilon M, Huen DS, Hardy R, Croom-Carter D, et al. 1994. Upregulation of bcl-2 by the Epstein-Barr virus latent membrane protein LMP1: a B-cell specific response that is delayed relative to NF-κB activation and to induction of cell surface markers. *J. Virol.* 68:5602–12

114. Ruscetti SK, Janesch NJ, Chakraborti A, Sawyer ST, Hankins WD. 1990. Friend

spleen focus-forming virus induces factor independence in an erythropoeitin-dependent erythroleukemia cell line. *J. Virol.* 43:1057–62

115. Ruscetti S, Wolff L. 1985. Biological and biochemical differences between variants of spleen focus-forming virus can be localized to a region containing the 3′ end of the envelope gene. *J. Virol.* 56:717–22

116. Sandberg M, Hammerschmidt W, Sugden B. 1997. Characterization of LMP-1's association with TRAF1, TRAF2, and TRAF3. *J. Virol.* 71:4649–56

117. Schlegel R, Wade-Glass M, Rabson MS, Yang YC. 1986. The E5 transforming gene of bovine papillomavirus encodes a small, hydrophobic polypeptide. *Science* 233:464–67

118. Senften M, Dilworth S, Ballmer-Hofer K. 1997. Multimerization of polyomavirus middle-T antigen. *J. Virol.* 71:6990–95

119. Settleman J, Fazeli A, Malicki J, Horwitz BH, DiMaio D. 1989. Genetic evidence that acute morphologic transformation, induction of cellular DNA synthesis, and focus formation are mediated by a single activity of the bovine papillomavirus E5 protein. *Mol. Cell. Biol.* 9:5563–72

119a. Showers MO, DeMartino JC, Saito Y, D'Andrea AD. 1993. Fusion of the erythropoietin receptor and the Friend spleen focus-forming virus gp55 glycoprotein transforms a factor-dependent hematopoietic cell line. *Mol. Cell. Biol.* 13:739–48

120. Showers MO, Moreau J-F, Linnekin D, Druker B, D'Andrea AD. 1992. Activation of the erythropoietin receptor by the Friend spleen focus-forming virus gp55 glycoprotein induces constitutive protein tyrosine phosphorylation. *Blood* 80:3070–78

121. Sparkowski J, Mense M, Anders J, Schlegel R. 1996. E5 oncoprotein transmembrane mutants dissociate fibroblast transforming activity from 16-kilodalton protein binding and platelet-derived growth factor receptor binding and phosphorylation. *J. Virol.* 70:2420–30

122. Srinivas RV, Kilpatrick DR, Tucker S, Rui Z, Compans RW. 1991. The hydrophobic membrane-spanning sequences of the gp52 glycoprotein are required for the pathogenicity of Friend spleen focus-forming virus. *J. Virol.* 65:5272–80

123. Staebler A, Pierce JH, Brazinski S, Heidaran MA, Li W, et al. 1995. Mutational analysis of the β-type platelet-derived growth factor receptor defines the site of interaction with the bovine papillomavirus type 1 E5 transforming protein. *J. Virol.* 69:6507–17

124. Straight SW, Hinkle PM, Jewers RJ, McCance DJ. 1993. The E5 oncoprotein of human papillomavirus type 16 transforms fibroblasts and effects the downregulation of the epidermal growth factor receptor in keratinocytes. *J. Virol.* 67:4521–32

125. Su W, Liu W, Schaffhausen BS, Roberts TM. 1995. Association of *Polyomavirus* middle tumor antigen with phospholipase C-γ1. *J. Biol. Chem.* 270:12331–34

126. Thomas JE, Aguzzi A, Soriano P, Wagner EF, Brugge JS. 1993. Induction of tumor formation and cell transformation by polyoma middle T antigen in the absence of Src. *Oncogene* 8:2521–29

127. Tam K, Watowich SS, Longmore GD. 1997. Cell surface organization of the erythropoietin receptor complex differs depending on its mode of activation. *J. Biol. Chem.* 272:9099–107

128. Vandenabeele P, Declercq W, Beyaert R, Fiers W. 1995. Two tumor necrosis factor receptors: structure and function. *Trends Cell Biol.* 5:392–99

129. Walter G, Ruediger R, Slaughter C, Mumby M. 1990. Association of protein phosphatase 2A with polyoma virus medium tumor antigen. *Proc. Natl. Acad. Sci. USA* 87:2521–25

130. Wang D, Liebowitz D, Kieff E. 1985. An EBV membrane protein expressed in immortalized lymphocytes transforms established rodent cells. *Cell* 43:831–40

130a. Wang Y, Kayman SC, Li JP, Pinter A. 1993. Erythropoietin receptor (EpoR)-dependent mitogenicity of spleen focus-forming virus correlates with viral pathogenicity and processing of env protein but not with formation of gp52-EpoR complexes in the endoplasmic reticulum. *J. Virol.* 67:1322–27

131. Watanabe N, Nishi M, Ikawa Y, Amanuma H. 1991. Conversion of Friend mink cell focus-forming virus to Friend spleen focus-forming virus by modification of the 3′ half of the env gene. *J. Virol.* 65:132–37

132. Waterfield MD, Scarce GT, Whittle N, Stroobant P, Johnsson A, et al. 1983. Platelet-derived growth factor is structurally related to the putative transforming protein p18^sis of simian sarcoma virus. *Nature* 304:35–39

133. Watowich SS, Wu H, Socolovsky M, Klingmuller U, Constantinescu SN,

Lodish HF. 1996. Cytokine receptor signal transduction and the control of hematopoietic cell development. *Annu. Rev. Cell Dev. Biol.* 12:91–128

134. Whitman M, Kaplan DR, Schaffhausen B, Cantley L, Roberts TM. 1985. Association of phosphatidylinositol kinase activity with polyoma middle T competent for transformation. *Nature* 315:239–42

135. Wiese N, Tsygankov AY, Klauenberg U, Bolen JB, Fleischer B, Broker BM. 1996. Selective activation of T cell kinase p56*lck* by herpesvirus saimiri protein tip. *J. Biol. Chem.* 271:847–52

136. Wolff L, Scolnick E, Ruscetti S. 1983. Envelope gene of the Friend spleen focus-forming virus. *Proc. Natl. Acad. Sci. USA* 80:4718–22

137. Yamamura Y, Senda H, Kageyama Y, Matsuzaki T, Noda M, Ikawa Y. 1998. Erythropoietin and friend virus gp55 activate different JAK/STAT pathways through the erythropoietin receptor in erythroid cells. *Mol. Cell. Biol.* 18:1172–80

138. Yi X, Peterson J, Freund R. 1997. Transformation and tumorigenic properties of a mutant polyomavirus containing a middle T antigen defective in Shc binding. *J. Virol.* 71:6279–86

139. Yoakim M, Hou W, Liu Y, Carpenter CL, Kapeller R, Schaffhausen BS. 1992. Interactions of polyomavirus middle T with the SH2 domains of the pp85 subunit of phosphatidylinositol-3-kinase. *J. Virol.* 66:5485–91

140. Yoon D-W, Lee H, Seol W, DeMaria M, Rosenzweig M, Jung JU. 1997. Tap: A novel cellular protein that interacts with Tip of herpesvirus saimiri and induces lymphocyte aggregation. *Immunity* 6:571–82

141. Yoshimura A, D'Andrea AD, Lodish HF. 1990. Friend spleen focus-forming virus glycoprotein gp55 interacts with the erythropoietin receptor in the endoplasmic reticulum and affects receptor metabolism. *Proc. Natl. Acad. Sci. USA* 87:4139–43

142. Zon LI, Moreau J-F, Koo J-W, Mathey-Prevot B, D'Andrea AD. 1992. The erythropoietin receptor transmembrane region is necessary for activation by the Friend spleen focus-forming virus gp55 glycoprotein. *Mol. Cell. Biol.* 12:2949–57

Annu. Rev. Microbiol. 1998. 52:423–52

NEW PERSPECTIVES ON MICROBIAL DEHALOGENATION OF CHLORINATED SOLVENTS:
Insights from the Field

M. D. Lee
Terra Systems Inc, 1035 Philadelphia Pike, Suite E, Wilmington, Delaware 19809;
e-mail m_d_lee@msn.com

J. M. Odom
DuPont Company, Life Sciences, Experimental Station (E328/B47),
Wilmington, Delaware, 19898;
e-mail: j-martin.odom@usa.dupont.com

R. J. Buchanan Jr.
DuPont Company, Engineering, BMP27/2122, Lancaster Pike & Rt. 41,
Wilmington, Delaware, 19805; e-mail: ron.j.buchanan@usa.dupont.com

KEY WORDS: natural attenuation, reductive dechlorination, in situ, intrinsic bioremediation, anaerobic

---

ABSTRACT

A variety of microbial dechlorination mechanisms have been demonstrated in laboratory microcosms, pure cultures, and in situ sedimentary environments. New perspectives on in situ processes from these efforts allow the design of more realistic bioremediation strategies that complement natural processes regardless of whether the strategy used is one of engineered accelerated bioremediation or natural attenuation. Since 1994 the scientific community has acquired considerable knowledge regarding natural attenuation of organochlorine compounds. Natural attenuation of chlorinated solvents has been documented at a number of field sites. Reductive dechlorination driven by co-contaminants or naturally occurring organics as substrates in combination with aerobic or co-metabolic degradation contains certain chlorinated solvent plumes. Although natural attenuation is not a panacea, at sites where it is applicable, it offers a scientifically sound,

0066-4227/98/1001-0423$08.00

cost-effective method to remediate groundwater contaminated with chlorinated solvents.

CONTENTS

INTRODUCTION

The understanding of the fate of natural and man-made organochlorine compounds in the environment has changed dramatically over the past two decades. The shift has been from viewing the carbon-chlorine bond as generally recalcitrant in biological systems to the realization that diverse natural, as well as man-made, organochlorine compounds are microbiologically degraded by known bacteria and novel microorganisms. A major factor in this change has been the combined efforts of academic, government, and industry laboratories to understand the fate of chlorinated solvents in the biosphere in general and, in particular, their fate in aquifer ecosystems.

This review focuses on studies of the biological fate of chlorinated solvents comprising chlorinated alkenes and alkanes in aquifer sediments and groundwaters, where most of this type of pollution is found. Perchloroethene (PCE), trichloroethene (TCE), dichloroethene (DCE), vinyl chloride (VC), carbon tetrachloride (CT), chloroform (CF), dichloromethane (DCM), and chloromethane (CM) are the most common of these pollutants. (A hypothetical PCE plume, which is transported down-gradient and is spatially dispersed in the aquifer, is shown in Figure 1.) Site characterization and groundwater-monitoring wells provide information to determine the extent of contamination. Traditionally, groundwater has been extracted through extraction wells and treated ex situ via air stripping or some other physical-chemical technology.

This review compares engineered approaches to naturally occurring chlorinated solvent biodegradation in aquifer systems. Engineered approaches are generally predicated on a single mechanism approach: anaerobic or aerobic biodegradation. Naturally occurring processes are often the result of reductive and oxidative biochemistries operating in concert. Because groundwaters and sediment microenvironments are frequently limited in oxygen, it is generally believed that reductive dehalogenation is a key initial biological step to achieve the biodegradation of highly chlorinated compounds in these environments. This review also discusses reductive dehalogenation in the natural environment, where intrinsic heterogeneity allows for mixtures of oxidative and reductive processes to occur either spatially or temporally separated, resulting in branched biodegradation pathways.

The demonstration of in situ microbial processes resulting in the biodegradation of chlorinated solvents is the biological component of a remediation strategy termed natural attenuation. The strategy is formally defined by the United States Environmental Protection Agency (USEPA) as follows:

> Naturally occurring processes in soil and groundwater environments that act without human intervention to reduce the mass, toxicity, mobility, volume, or concentration of contaminants in those media. These in situ processes include biodegradation, dispersion, dilution, adsorption, volatilization, and chemical or biological stabilization or destruction of contaminants (89).

The biological component of this strategy is useful only in cases where solvent concentration, the presence of dechlorinating bacteria, and the absence of impacted receptors allow for this destruction/containment approach.

Chlorinated solvents are produced on a very large scale and are typically used in metal degreasing operations and textile processing. These solvents are more dense than water and only sparingly soluble. These properties contribute to the occurrence and persistence of these solvents in groundwater. Improper disposal and accidental spills of these compounds clearly have contributed to groundwater pollution; however, in many cases the presence of more lightly chlorinated solvents (i.e. cis-DCE, VC, CF, DCM, and CM) results not from accidents or improper disposal but rather from microbial dehalogenation of more highly chlorinated solvents (9, 69). If such lightly chlorinated solvents

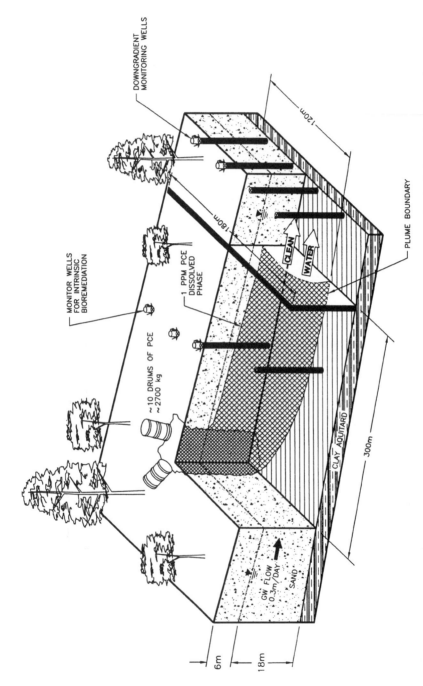

Figure 1 Template site showing a spill of perchloroethene (PCE) contaminating the groundwater with a dissolved plume. Monitoring wells are placed above and below the treatment zone. TCE, trichloroethene; DCE, dichloroethene; VC, vinyl chloride.

are not used at a particular site, the occurrence of more reduced compounds can be viewed as evidence for reductive dehalogenation.

The potential cost of this pollution in terms of environmental damage or loss of usable water cannot be quantified precisely; however, the estimated monetary cleanup costs are enormous [on the order of one-half to one trillion dollars for the United States alone (74)]. The extent of groundwater contamination in the United States is significant; groundwater contamination by hazardous chemicals may exist at from 33,000 sites (88) to 400,000 sites (63). The USEPA has already placed more than 1400 of these sites on the National Priorities List (NPL) under the federal Comprehensive Environmental Response, Cleanup, and Liability Act (CERCLA) or Superfund program. These are sites that, according to the USEPA, pose an unacceptable risk to human health or the environment and, thus, require some form of remediation. Of these sites, over 85% have groundwater contamination (68). Moreover, 1700 additional hazardous waste facilities governed under the Resource Conservation and Recovery Act (RCRA) and located throughout the United States are suspected of contributing to groundwater contamination. TCE is the most frequently detected chemical contaminant in groundwater at NPL sites, and PCE is number three on the USEPA's list (63).

Appreciation of the magnitude of the financial and environmental cost to clean up contaminated aquifers using known technologies has accelerated research in the microbiology of dehalogenation over the past 15 years. In addition to the known ability of many aerobic bacteria to oxidize the more reduced halogenated compounds for carbon and energy, early reports of microbial reductive dehalogenation of haloaromatics (85) and haloaliphatics (9, 69) were landmark discoveries. A significant body of literature documents the fact that many types of naturally occurring bacteria can cause the release of chloride from aliphatic and aromatic organochlorines to produce nontoxic metabolites. One of the most significant findings, from a practical perspective, was that bacteria can derive biologically useful energy from the complete reductive dechlorination of chlorinated solvents (e.g. PCE) in the absence of oxygen and, in the process, yield chloride and ethene, ethane, or carbon dioxide as sole degradation products (23, 35, 46, 58, 64, 76, 94). The term dehalorespiration has been used to describe this process, whereby the cell uses the solvent as an electron acceptor for growth under dark, anaerobic conditions.

MICROBIAL DECHLORINATION PATHWAYS

The evolutionary origin of dehalorespiring bacteria or other microbial types that appear to be well adapted to growth on chlorinated solvents is uncertain. The existence of such dehalorespiring organisms is intriguing when one considers the relatively recent (in the past 50 years) and very localized dispersal of these solvents into the environment. However, from a global perspective, a number of naturally produced organochlorine compounds have been in the environment over geologic time. Plants and algae produce over 2000 compounds containing either chlorine, bromine, or iodine (37). Chloromethane is produced by marine algae, fungi, and certain species of evergreen trees in amounts that dwarf human emissions of this compound to the atmosphere. TCE and PCE are emitted during volcanic eruptions (43).

In light of these facts, it is less surprising that microbial enzyme systems exist that are specialized for degrading organochlorine compounds. Therefore, microbial dehalogenation need not be viewed as an oddity or accident but rather as another microbial adaptation to available carbon and energy sources. Microbial metabolism of chlorinated aliphatics can be categorized broadly into four areas, as described below. These areas have been reviewed extensively by others (30, 44, 93) and, therefore, are discussed only briefly here.

Energy-Yielding Solvent Oxidations

A wide range of chlorinated compounds, including chloromethanes, chloroethanes, chloroalcohols, chloroalkanoic acids, and chloroalkenoic acids, support microbial growth by acting as the sole source of carbon and energy (71). VC, a product of reductive dehalogenation, is mineralized to carbon dioxide by aerobic bacteria such as *Mycobacterium* sp. (39), *Rhodococcus* sp. (57), *Actinomycetales* sp. (70), or *Nitrosomonas* sp. (91). Davis & Carpenter (21) used radiolabeled VC to demonstrate direct oxidation of VC to carbon dioxide in aerobic aquifer microcosms. The role of anaerobic iron reduction in the oxidation of more reduced organochlorine species is currently being resolved, but evidence suggests that iron-reducing species such as *Geobacter* sp. play a role in the oxidation of hydrocarbons and perhaps the more reduced chloroethenes such as DCE and VC (12, 55). Anaerobic dichloromethane degradation by methanogenic microbial consortia have also been reported (84).

Co-Metabolic Oxidations

Methanotrophic bacteria containing monoxygenase and dioxygenase enzymes are widespread in nature, including aquifer environments. The utility of these organisms for oxidative, co-metabolic destruction of chloroethenes via formation of chloroethene epoxides has been investigated widely and applied to aquifer environments. The inducible oxygenases oxidatively degrade partially chlorinated solvents such as TCE, *cis*-DCE, or VC during normal oxidation of hydrocarbons such as toluene, phenol, methane, or propane (30, 93). However, the fully chlorinated ethene PCE is resistant to degradation via this mechanism. Another limitation of this approach for aquifer environments is the oxygen requirement, which presents problems in terms of delivery and aquifer plugging and fouling. There is also a need for a co-substrate such as methane or an aromatic compound (25). This oxidative, co-metabolic process has been used successfully in situ to degrade solvents where sufficient oxygen and a hydrocarbon source can be delivered to indigenous microflora (65).

Energy-Yielding Reductions: Dehalorespiration

Isolation of novel, dehalorespiring strains capable of using PCE, TCE, or chlorobenzoates as electron acceptors for biologically useful energy generation has been demonstrated in several laboratories (44). Known examples of these strains include *Desulfomonile tiedjei*, (38), *Dehalococcoides ethenogenes*, (58), *Desulfitobacterium dehalogenans* (90), *Desulfitobacterium chlororespirans* (75), *Dehalobacter restrictus* (77), and *Dehalospirillum multivorans* (64). These species are distinct from the anaerobic, co-metabolic dechlorinators found among the sulfate-reducers and methanogens. Depending on the species, these bacteria may produce *cis*-DCE as a final end product or may carry out complete dechlorination to ethene. Facultative, enterobacterial isolates have been found that can derive energy from partial dechlorination of PCE to *cis*-DCE (79).

Microorganisms catalyzing dehalogenation of haloaromatics appear to be a distinct group from the haloalkane or haloalkene respirers and are typified by *Desulfitobacterium chlororespirans* (75), *Desulfomonile tiedjei* (18, 38), or *Desulfitobacterium dehalogenans* (90). Laboratory demonstrations of energy-yielding anaerobic reductive dehalogenation is of great practical significance for the technology of bioremediation. These microbial processes show that oxygen need not be supplied to effect complete mineralization and suggest further that because of the energy-yielding nature of the reactions, chlorinated solvent plumes may be self-enriching for dehalogenating bacteria. More recent field data strongly supports the idea that these bacteria are not laboratory oddities but are more widespread in nature and play a role in the natural decay process of chlorinated solvents.

Co-Metabolic Reductive Dehalogenation Processes

PCE is fully chlorinated and does not serve as an electron donor for aerobic or anaerobic bacteria. However, PCE and TCE can be reductively dehalogenated by many types of anaerobic bacteria, including certain species of methanogens, sulfate-reducing bacteria (3), and novel microbial types that do not fall into either category (59). The reaction carried out by these types of bacteria is not thought to be energy-yielding but rather co-metabolic because only a small fraction of the total reducing equivalents derived from the oxidation of electron donors is used to reduce the solvent (3). Solvent reduction appears to be a minor side-reaction in these cases. However, in environmental situations where high levels of organics and intense methanogenic or sulfidogenic respiration are found (i.e. in wetlands, landfills, and landfill leachates), partial co-metabolic dechlorination of solvents can be significant.

DECHLORINATION MECHANISMS IN COMPLEX AQUIFER ENVIRONMENTS

The fact that bacteria are present in significant numbers in aquifer environments is unequivocal. Microbes have been detected at least 800 m below the surface of the earth, and their numbers generally decline with depth (45). Most contaminated aquifers are relatively close to the surface and harbor bacterial populations at densities from 10^4 to 10^7 bacteria/g (dry weight) of soil (48). Aquifers may be characterized macroscopically as reduced environments, denoting the lack of oxygen and presence of indicators of anaerobic microbial activity such as methane and high concentrations of dissolved iron. Alternatively, oxidizing environments are associated with high concentrations of dissolved oxygen and the absence of anaerobic metabolites. Accordingly, there is a tendency to ascribe a single dominant dechlorination pathway to an aquifer based on bulk phase macroscopic parameters. However, many aquifers do not fit either macroscopic characterization but rather are complex heterogeneous mixtures of aerobic and anaerobic microenvironments that result from differences in sediment permeability, channeling of water flow, and, frequently, proximity to sources of organic contaminants such as fuel hydrocarbons or landfill leachates. The proximity of aerobic and anaerobic zones could vary over a range of millimeters to meters. Measuring bulk phase parameters in flowing groundwater can lead to a correct macroscopic or averaged view of the aquifer but may underestimate the true diversity of microbial capabilities and lead to incorrect expectations of the types of solvent degradation products to be found within the aquifer.

Environmentally relevant aerobic and anaerobic process are complementary in terms of the chlorinated species degraded. Highly chlorinated species such as PCE will be more rapidly degraded anaerobically, whereas less chlorinated species such as DCE may be longer lived in anaerobic environments and more readily degraded by aerobic processes.

Multiple dechlorination pathways are likely to operate in heterogeneous aquifers where there is some influx of oxygen and organic carbon into a physically dynamic and heterogeneous milieu. Figure 2 depicts how juxtaposed aerobic and anaerobic mechanisms could allow all four mechanisms described above to participate in the total biodegradation of PCE. Methane generation in anaerobic zones can provide carbon and energy for methanotrophic oxidation of TCE, DCE, or VC in subsequent aerobic zones. Direct oxidation of VC or DCE with molecular oxygen or ferric iron could also occur in zones at higher redox potential. Thus in the natural environment PCE biodegradation is

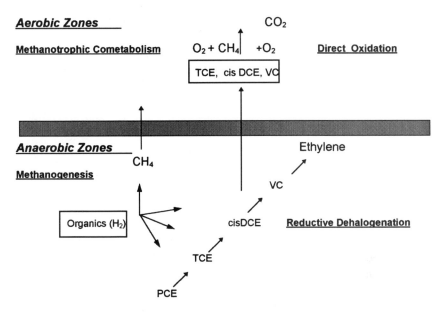

Figure 2 The interplay between different biochemical mechanisms operating within a continuum of aquifer conditions from aerobic to strictly anaerobic. Where oxygen is present, current data suggest that vinyl chloride and perhaps dichloroethene can be oxidized directly to carbon dioxide and chloride. Alternatively, at the interface between aerobic and anaerobic micro-environments where methane and oxygen are co-incident, co-metabolic oxidations convert chlorinated ethenes to carbon dioxide and chloride. Within strictly anaerobic environments where organic electron donors or hydrogen are present, reductive dehalogenation is the predominant mechanism yielding ethene and chloride. PCE, perchloroethene.

likely to be a multimechanism process with both ethene and carbon dioxide as ultimate end products. This scheme does not preclude complete dechlorination by dehalorespiring bacteria, but neither does it require their presence.

The application and utility of these microbial processes has, until recently, been viewed in terms of engineering or enhancing bioremediation present in a contaminated area via addition of cultured bacteria or nutritional amendments or both to the contaminated aquifer. Implicit in these approaches is the assumption that natural solvent decay, which occurs in the absence of any human intervention, is of limited value for groundwater cleanup; however, field evidence indicates that natural processes can play a significant role in the containment and eventual cleanup of contaminated groundwater. The strategy of natural attenuation must be applied appropriately and may be combined with other engineered technologies at sites where contaminant mobility exceeds contaminant destruction, contaminant concentrations are inhibitory to microbial growth, or the contaminant plume adversely affects receptors.

IN SITU ENGINEERED APPROACHES

In Situ Conditions for Reductive Dehalogenation

Reductive dehalogenation holds considerable promise as an engineered, in situ bioremediation technology because of its potential for completely dechlorinating fully chlorinated solvents such as PCE. In the absence of any other mechanism, this process could achieve total dechlorination without the need for molecular oxygen and the associated costs and risks. However, this process is not fully understood, and the ability to enhance or control the process is more uncertain.

To anaerobically degrade chlorinated solvents in situ, the following conditions are necessary:

1. A microbial consortium capable of degrading the chlorinated solvent contaminants must be present (or added by bioaugmentation).

2. Contaminant concentrations must be within an acceptable range that the microbes can biodegrade.

3. An appropriate electron donor, or substrate, must be supplied to achieve reducing conditions in the aquifer and to meet the demand for organic carbon for cell growth and metabolism.

4. Requisite electron acceptors and nutrients must be provided, and environmental conditions must be favorable (e.g. appropriate pH and the absence of toxicants).

Chlorinated Solvent Biodegradation in the Presence of Different Electron Acceptors

Many chlorinated solvents are biodegraded anaerobically. Hydrogen produced by the breakdown of complex materials and intermediates is used by the dechlorinating organisms to reduce chlorinated solvents (61). Dechlorinating organisms compete for electrons in hydrogen with organisms using other electron acceptors. Whether dechlorination will occur at a particular site depends on the contaminant, the microbial consortia present, and the relative affinity of the microbial consortia to use the available hydrogen for dechlorination rather than the affinity for other electron-accepting reactions.

In general, complete dechlorination occurs under methanogenic conditions when competing electron acceptors such as oxygen, nitrate, nitrite, and sulfate have been depleted (60). However, reductive dehalogenation can occur in microenvironments even in a bulk aerobic environment, as Enzien et al (31) demonstrated in a column study where PCE and TCE were converted to

DCE. Facultative organisms such as the MS-1 organism isolated by Sharma & McCarty (79) can be grown aerobically and degrade PCE to DCE under anaerobic conditions. The more highly chlorinated compounds such as CT, TCE, and PCE are more energetically favorable electron acceptors than are nitrate, sulfate, or carbon dioxide (93). Compounds such as CT can be degraded under redox conditions associated with nitrate reduction (10).

Sulfate can be inhibitory to reductive dehalogenation (42), whereas in other laboratory studies, dechlorination can proceed despite substantial sulfate concentrations. Klecka et al (49) demonstrated 1,1,1-trichloroethane (1TCA) biodegradation with sulfate concentrations as high as 770 mg/liter (mg/L). Iron reducers biodegrade some chlorinated solvents, including DCE and VC (13). Although reductive dechlorination is often associated with methanogenic conditions, the dechlorinators may be a population separate from the methanogens (4). Many of the dehalorespirers can outcompete the methanogens for the available electrons at low hydrogen concentrations (4).

Where reductive dehalogenation results in incomplete dechlorination, aerobic oxidation can completely mineralize residual, partially dehalogenated daughter products. Compounds such as PCE or CT can be dechlorinated partially to TCE and DCE or CF, respectively, under reducing conditions and then directly metabolized or co-metabolically degraded by aerobic organisms (32). This treatment strategy should eliminate concerns about VC generation at sites with PCE or TCA contamination and would be beneficial at sites where organisms capable of complete reductive dehalogenation are not present.

Distribution of Dechlorinators

Microbial populations capable of completely dechlorinating chlorinated solvents have been isolated from such diverse environments as anaerobic digester sludges from municipal wastewaters (24), a winery wastewater (16), aquifer solids downgradient of a landfill or a waste lagoon (42), river sediments and sugar beet wastewater sludge (23), an anaerobic filter (95), and a bioreactor treating groundwater contaminated with DCM and trace TCE (101). In a survey of six sites with PCE- or TCE-contaminated groundwaters, Odom et al (67) found evidence at four sites in field and/or laboratory microcosms of the biodegradation of PCE or TCE to VC and ethene. The potential for complete dechlorination varied from site to site and within locations at a single site.

Microbial populations from reducing environments with relatively high concentrations of dissolved organics are more likely to have organisms capable of complete reductive dehalogenation (99). Typically, the presence of a microbial population capable of complete dechlorination can be inferred from field measurements of the daughter products and laboratory microcosm studies. At sites where a microbial population capable of complete dechlorination is not present, it may be possible to bioaugment with a dehalogenating culture. An interesting approach at the Schoolcraft, Michigan, site involved adjusting the aquifer pH to favor a dechlorinating strain of denitrifiers (*Pseudomonas stutzeri* strain KC) (27). Strain KC degrades CT to carbon dioxide and nonvolatile products with minimal CF production (20). Although strongly reducing conditions do not guarantee reductive dehalogenation, the process is favored by low redox potential environments where sulfate reduction, methogenesis, or acetogenesis occurs.

Maximum Solvent Concentrations

High concentrations of some solvents or intermediates may inhibit dechlorination. Smatlak et al (82) observed PCE dechlorination at up to 550 μM (91 mg/L) in a culture acclimated to high PCE levels. PCE dechlorination to *cis*-DCE by *Dehalospirillum multivorans* was inhibited by PCE concentrations greater than 300 μM or 50 mg/L (64). Concentrations of *cis*-DCE above 83 μM (8.0 mg/L) were inhibitory to the dechlorinating culture isolated from a Victoria, Texas, site (42). VC levels between 30 and 60 μM (1.9 to 3.8 mg/L) inhibited VC dechlorination (24). DCM concentrations of 100 μM (8.5 mg/L) were completely removed in a bioreactor fed 0.5 mg/L yeast extract with as short as 0.25-day hydraulic retention time, but 780 μM DCM (66 mg/L) was not completely removed in a 2-day hydraulic retention time (33). High concentrations of 750 μM (100 mg/L) 1TCA inhibited growth of *Clostridium* sp. (34).

Several strategies have been attempted for treating soils or groundwater with very high concentrations of solvents, including dense nonaqueous phase liquids (DNAPLs). DiStefano et al (24) selected for PCE-tolerant bacterial strains by increasing PCE dosages. After a pilot demonstration of surfactant flushing of a CT DNAPL–affected site in Corpus Christi, Texas, Lee et al (50) documented that CT losses were influenced by (*a*) the biodegradation of the CT DNAPL using the two nonionic surfactants as substrate and sulfate as the electron acceptor and (*b*) surfactant flushing. After the surfactant flushing pilot was completed, the following changes were observed over a 19-month period: 74% reduction in average CT concentrations, 98% reduction in surfactant concentrations in groundwater, and increases in CF, DCM, and carbon disulfide (CDS) concentrations. Weathers & Parkin (96) evaluated a coupled metallic iron-enhanced biotransformation system where the metallic iron could directly degrade the DNAPL solvent and supply the dechlorinating bacteria with hydrogen.

Electron Donor for Reductive Dehalogenation

Electron donors are required for reductive dehalogenation. A number of organic substrates such as benzoate, lactate, butyrate, ethanol, methanol, acetate, and hydrogen support this process (52). The question of whether these substrates act directly or through molecular hydrogen as a key intermediate is being evaluated. Molecular hydrogen is an effective substrate for reductive dehalogenation.

Some dechlorinating bacteria can outcompete methanogens or sulfate-reducing bacteria at extremely low ambient hydrogen concentrations (4, 82); however, not all substrates are effective in every case.

For example, although acetate and methanol did not stimulate dechlorination in laboratory microcosms from a mixed PCE-fuel site in Traverse City, Michigan (36), these substrates have supported dechlorination at other sites. Metabolic intermediates such as benzoate, lactate, propionate, and acetate may target dechlorinators more efficiently (82). Substrates that generate hydrogen slowly (e.g. butyrate) may result in sustained dechlorination and enrichment for dechlorinating bacteria. Ethanol, lactate, and butyrate, which are fermented directly to hydrogen without the production of methane, may also promote dechlorination.

The use of inexpensive substrates reduces anaerobic bioremediation costs. Inexpensive, complex substrates such as molasses; a wastewater containing formate, acetate, and propionate; cheese whey permeate (a waste product from the manufacture of cheese); corn steep liquor (an inexpensive product used for fermentation and that is produced by steeping corn in water); and the dissolved organic fraction of chicken manure were shown to support reductive dechlorination of PCE to VC in microcosm studies (52). Yeast extract, a more expensive substrate, also supported dechlorination. In a laboratory microcosm study comparing dechlorinating activity at six sites with various substrates, complex substrates such as yeast extract gave higher rates of dechlorination than simpler substrates (67). Degradation rates increased for enrichment cultures repeatedly fed complex substrates compared to enrichment cultures fed simple compounds such as acetate and hydrogen (73). However, complex substrates may result in the growth of numerous competing anaerobes that are not involved in the dechlorination process.

It appears that if a dechlorinating population is present at a site, almost any fermentable substrate can be effective in stimulating its activity. However, because of the variable responses to substrates by organisms from different sites, laboratory or small-scale field studies are necessary to confirm that a particular substrate will support dechlorination at the site. No substrate that reliably supports complete dechlorination at all sites has been identified to date.

Although the theoretical electron demand for 1 g of PCE is 0.4 g of chemical oxygen demand (COD), many times more substrate is actually required because of competition for the electron donors by the organisms carrying out the reductive dechlorination and other organisms in the consortium (60). A minimum of about 60 mg/L of total organic carbon (TOC; mg C/L) of any of the inexpensive substrates or yeast extract was necessary to support dechlorination beyond DCE in microcosm studies with the Victoria, Texas, soils (52). Other researchers have found that varying concentrations of electron donors are necessary for dechlorination depending on the microbial consortia, substrate, chlorinated solvent concentrations, and relative efficiency of conversion of the substrate to drive reductive dechlorination.

Bouwer (8) estimated substrate requirements for the complete dechlorination of 10 liters (16.3 kg) of PCE in a cubic meter of soil. The relatively inefficient stoichiometry provided by de Bruin et al (23) was used with lactate as the substrate and the more efficient kinetics from DiStefano et al (24) was used with methanol as the electron donor. The estimates ranged from 13.6 kg for methanol with a very efficient dechlorinating culture to 970 kg for lactate with a less efficient microbial culture. Skeen et al (80) estimated a substrate requirement of 35 kg of methanol for another culture to dechlorinate 300 mg of PCE or the equivalent of 1.9 million kg of methanol to treat the 16.3 kg of PCE in Bouwer's example (8). Clearly, this poor stoichiometry would not be acceptable for a field project (80).

Vitamins and Micronutrients

Nitrogen and phosphorous are often limiting in aquifer environments and are added to ensure that nutrient availability does not limit microbial activity. The general rule of thumb for biodegradation of hydrocarbons to cell mass is to add nitrogen and phosphorus at a ratio of carbon to nitrogen to phosphorus of 100:10:1 (65). At this addition rate, the nutrients may be present in excess of the true demand, particularly for anaerobic systems where less biomass is produced than in aerobic systems. Although other minerals and trace metals are often included in anaerobic media used in the laboratory (92), the true requirements for these nutrients by organisms in the environment is unknown. Some anaerobic cultures require the addition of vitamins and trace minerals for optimal degradation of chlorinated solvents. For example, the pure culture of *Dehalococcus ethenegenes* requires vitamin B_{12} and anaerobic digester sludge supernatant or cell extracts to dechlorinate PCE (58).

Corrinoids such as vitamin B_{12} are a factor in biological reductive dehalogenation by pure strains such as *Dehalospirillum multivorans* (64). Additions of corrinoids such as vitamin B_{12} increase the extent and rate of biodegradation of chlorinated solvents such as CT (66) and CF (5). Additions of vitamin B_{12} to a DCM-degrading enrichment culture increased the extent of biodegradation of CT tenfold, increased the oxidation to carbon dioxide, and minimized accumulations of DCM and CM compared to the culture without the vitamin (41). The lowest dosage of cyanocobalamin (a commercially available form of corrinoid) that supported the complete dechlorination of CT was 5 μM (40). Dechlorination of CT to carbon monoxide can be stimulated by vitamin B_{12} even in an organism such as *Shewanella alga* strain BrY, which does not normally degrade CT (104). Vitamin B_{12} can catalyze abiotic dechlorination when an external reductant is added. Additions of vitamin B_{12} in the presence of titanium citrate can support the complete abiotic dechlorination of PCE DNAPL (53).

IN SITU ENGINEERED EXPERIENCES

Field Applications of Reductive Dehalogenation

Only a limited number of field pilots have been conducted for the anaerobic reductive dehalogenation of chlorinated solvents. Table 1 presents a summary of these trials. As shown in Table 1, field pilots have been conducted under denitrifying conditions, sulfate-reducing conditions, methanogenic conditions, and combined anaerobic-aerobic conditions. In situ pilots ranged from simple tests in which a single borehole was dosed with substrate to systems where the groundwater was recirculated. Geologic environments included sandy aquifers to fractured bedrock systems. Substrates included acetate, benzoate, casamino acids, yeast extract, methanol, ethanol, glucose, sucrose, and surfactants.

DENITRIFYING CONDITIONS   A pilot study to biodegrade CT under denitrifying conditions was conducted in 1989 at the Moffett Field Air Force Base in Mountain View, California (78). The recirculating pilot consisted of an injection well, three sampling wells, and an extraction well 6 m from the injection well in the confined sandy aquifer. CT was added to the groundwater continually at 0.26 μM along with short pulses of acetate, providing average acetate concentrations of 25–46 mg/L. With the addition of acetate to groundwater containing 22 mg/L nitrate, the maximum CT removal was about 80% with 55–67% of the CT transformed to CF. When nitrate was removed in an aboveground bioreactor, an average of 96% of the CT was transformed in the first 4 m of the cell with a residence time of about 28 h. Less CF, 30–40% of the transformed CT was produced during this period. During the period with no nitrate additions, the groundwater contaminants were reduced by an average of 72% for fluorotrichloromethane (Freon-11), 18% for 1,1,2-trichloro-1,2,2-trifluoroethane (Freon-113), and 9% for 1TCA in the 4 m from the injection point to the third sampling well.

   Hooker et al (46a) conducted another pilot under denitrifying conditions to degrade CT at the Department of Energy site at Hanford, Washington. Under continuous operating conditions with pulsed additions of acetate and nitrate, CT concentrations were reduced by 30% with minimal production of CF. When an excess of acetate was added and the nitrate depleted, CF was produced in greater quantities. An increase in the biomass of an estimated one order of magnitude resulted in aquifer plugging and reduced recirculation rates.

SULFATE-REDUCING OR METHANOGENIC CONDITIONS   Beeman et al (7) were the first to demonstrate the complete reductive dechlorination of chlorinated ethenes in a field experiment. A recirculating pilot was constructed downgradient of a landfill in a sandy aquifer at the DuPont Victoria, Texas, plant where the groundwater was contaminated with PCE and benzene. Benzene was degraded under initial aerobic conditions, but little degradation of PCE was observed.

Table 1 Reductive dechlorination pilots

Site	Contaminant concentration in μM (% reduction)	Type of pilot/ geology	Electron donors/electron acceptors/others	Reference
Moffett Field Air Station, CA 1,200 hr.	0.26 CT (80); *CF (−50); 0.38 1TCA (5); 0.032 Freon 113 (16); 0.022 Freon 11 (56)	recirc./sand	acetate/nitrate	78
1,450–1,550 hr.	0.26 CT (96); *CF (−27); 0.38 1TCA (5); 0.032 Freon 113 (19); 0.022 Freon 11 (72)	recirc./sand	acetate/sulfate	78
Hanford, WA	12.5 CT (30); 0.010 CF (−40)	dual stage recirc./silty sandy gravel	acetate/nitrate	46a, 81
Schoolcraft, MI Day 142	0.29 CT (73)	single pass + bioaugment/glacial outwash sand	acetate/nitrate/phosphate, Pseudomonas stutzeri strain KC	27
Victoria, TX Day 426	0.86 PCE (>97); 0.79 TCE (>95); 0.66 DCE (>92); <0.16 VC (0); 0.75 ETE (−95)	recirc./fine sand	benzoate/sulfate methane	7
Niagara Falls, NY Day 187	0.63 PCE (81); 51 TCE (96); 5.9 1,2-DCE (96); 1.5 ETE (89); 0.69 1TCA (>78); 0.68 1DCA (>71); 0.40 ETA (40); 0.42 CF (−119); 0.94 DCM (49)	borehole/fractured bedrock	yeast extract/sulfate	14

FAA Facility, OK	0.028-0.038 mg/liter CAH (77 PCE, detect DCE)	recirc./fractured bedrock	methanol, sucrose/ unknown EA/nitrate	17
TX, Gulf Coast	Average 170 TCE (76); 12 DCE (−42); 0.24 VC (−220)	recirc./interbedded sands & clays	methanol/unknown EA/nitrate	54
Corpus Christi, TX	860 CT (74); *CF (−17); *DCM (−4); *DCS (−1)	recirc./interbedded sands and clays	Tergitol 15-S-12, Witconol 2722/sulfate	50
Breda, Netherlands	Day 48 80% PCE (94); 5% TCE (80); 10% DCE (−800); 3% VC (0); Day 350 Well A3b 400 DCE (100); 60 VC (100)	recirc. anaerobic to aerobic single pass/ medium to fine sand	methanol/EA unknown phenol/oxygen	2, 83
Gulf Coast	Biocell B3 1,700 2DCA (78); 970 CA (63); 450 2TCA (73); 490 CF (87); 270 DCE (73); 440 mg/L CAH (77)	static cells in pond bottom/fine sand, silt, and clay	cas-amino acids, acetate/ methane, oxygen	32
Crosby, TX Well S1-106	~50 VC (100); ~100 1DCA (100); ~20 2DCA (100); ~90 CF (100)	recirc./interbedded clay, silt, and sand	none/nitrate, oxygen	22, 86
Central NJ Days 114–181	3.3-62 1TCA (95 to >99)	bioreactor/organic rich loam + GAC	glucose + ethanol/EA unknown/nitrogen, phosphorus	11

*Initial concentration not known, CA = chloroethane, CAH = chlorinated aliphatic hydrocarbons, CDS = carbon disulfide, CF = chloroform, CT = carbon tetrachloride, d = day, 1DCA = 1,1-dichloroethane, 2DCA = 1,2-dichloroethane, DCE = 1,2-dichloroethene, DCM = dichloromethane, EA = electron acceptor, ETE = ethane, ETE = ethane, GAC = granular activated carbon, PCE = tetrachloroethene, recirc. = recirculating, 1TCA = 1,1,1-trichloroethane, 2TCA = 1,1,2-trichloroethane, TCE = trichloroethene, VC = vinyl chloride, >x = removal to below detection limit.

Alternate additions of 38 mg/L sodium benzoate or 56 mg/L magnesium sulfate began to drive the groundwater anaerobic. Within 35 days, PCE and TCE concentrations at the nearest well began to drop and DCE concentrations increased. Over the next 100 days, DCE concentrations peaked and began to fall. VC was detected once sulfate concentrations were below 60 mg/L (6). A second phase of the pilot was initiated three months after the first anaerobic phase had ended (7). A mass balance calculation on day 426 showed the conversion of 0.86 μM PCE, 0.79 μM TCE, and 0.66 μM DCE in the influent wells to nondetectable levels by the well 19 m downgradient. Ethene, ethane, and chloroethane (CA) were produced at this site. Beeman (6) received a patent as a result of his demonstration of the complete reductive dechlorination pathway for PCE in the field.

Buchanan et al (14) conducted borehole pilot demonstrations at a Niagara Falls, New York, site contaminated with several chloroethenes and chloroethanes. A recirculating pilot could not be established at this site because of uncertainties in the direction of groundwater flow in the fractured bedrock units. Yeast extract and sulfate were added up to three times per week to the water in the well. Over the 187 days of substrate additions, PCE concentrations decreased by 81%, TCE by 96%, DCE by 96%, ethene by 89%, 1TCA by more than 78%, 1,1-dichloroethane (1DCA) by more than 71%, ethane by 40%, and DCM by 49%. CF concentrations increased by 119%.

A full-scale system for treating TCE-contaminated groundwater was conducted at a site on the Gulf Coast of Texas (54). Groundwater amended with methanol was circulated between a series of injection and extraction trenches in an aquifer comprised of relatively low hydraulic conductivity interbedded sands and clays. The average concentrations of TCE in wells within the area being treated dropped 69% during the five-month period in which methanol and nutrients were added. An estimated one pore volume of groundwater was treated. A portion of the TCE removal was due to natural dilution as chloride declined from 147 to 99 mg/L. The increase in DCE and VC concentrations, 70 and 225%, respectively, suggests that reductive dechlorination occurred faster than dilution.

Christopher et al (17) described a recirculation pilot for a fractured bedrock site owned by the Federal Aviation Administration (FAA) in Oklahoma. The extent of dechlorination at this site was unclear.

COMBINED ANAEROBIC-AEROBIC CONDITIONS   The successful application of the anaerobic-aerobic process has been demonstrated in the field. A full-scale system for treating CF, 1DCA, 1,2-dichloroethane, VC, and petroleum hydrocarbons, including benzene, toluene, and ethylbenzene, was conducted at the French Limited Superfund site (22, 86). Pure oxygen and nitrate were injected under pressure as electron acceptors. Concentrations of 1DCA, CF, VC, and 2DCA decreased by more than 99.9%. No substrate other than the hydrocarbon contaminants was added to support degradation of the chlorinated compounds. The half-life of VC in the test area was reported to be 61 days once active bioremediation began.

Another pilot demonstrated the potential for applying combined anaerobic-aerobic treatment of highly chlorinated solvents in pond sediments and aerobic treatment of the less chlorinated contaminants near the surface of the pond (32). After removing most of the chlorinated solvents discarded into two surface impoundments on a Gulf Coast site, residual contamination remained in the pond sediments. Static biocells were constructed with 0.61-m diameter steel pipes and were fed a single dose of substrate. In the biocell B3 amended with 0.2% casamino acid plus 0.2% acetate, the following percent removals were achieved during the 12-week study: 78% (2DCA), 63% (CA), 73% (1,1,2-trichloroethane or 2TCA), 87% (CF), and 73% (total DCEs). Increases in daughter products were not observed, probably as a result of aerobic co-metabolic degradation in the upper water column with the methane produced in the deeper zone.

Microorganisms from a site in Breda, The Netherlands, could degrade 35 μM PCE to ethene with either glucose or formate in laboratory columns (2). Up to 250 μM TCE, *cis*-DCE, and VC were degraded with 0.1 mM/day phenol and oxygen in another soil column. Based on these results, a field pilot was designed that included an anaerobic recirculation loop fed methanol followed by an aerobic loop where phenol was added and oxygen introduced by air sparging. About 1100 kg of methanol was distributed throughout the site, with groundwater concentrations ranging from 3.8 to 26 mg/L (83). PCE concentrations in the anaerobic zone decreased by 94%. DCE and VC concentrations increased, and ethene was produced. When the aerobic portion of the test was in operation, DCE and VC concentrations decreased from 400 μM and 60 μM, respectively, in the anaerobic zone to nondetectable levels within the aerobic loop. The phenol was degraded completely in the aerobic loop. All of the PCE and daughter products were degraded biologically in this in situ aquifer pilot, thus successfully combining anaerobic and aerobic treatments.

NATURAL ATTENUATION APPROACHES

In contrast to the engineered approaches predicated on a single microbial mechanism, the strategy of natural attenuation may involve primarily anaerobic or aerobic processes or a combination of the two. Reductive dehalogenation, oxidative aerobic biodegradation, co-metabolic biodegradation, sorption, volatilization, dispersion, and abiotic reactions act upon the contaminant plumes. Natural attenuation has been documented for chlorinated ethenes at a number of sites (see Table 2). Biodegradation half-lives were estimated using several methods, including mass fluxes along groundwater plume transects (97), modeling efforts using the USEPA Multimedia Exposure Assessment Model (51), estimates using chloride or trimethylbenzene (TMB) as tracers (100), a method with a semilog plot of individual well analyses versus time of transport (28), and a graphical extrapolation method that assumes the plume is at steady state and calculates the time for the water to move between points along the center line of the plume (28).

Wiedemeier et al (99) have documented chlorinated solvent contaminated sites demonstrating natural attenuation with different sources of organic carbon to support reductive dechlorination. Such sources include petroleum hydrocarbons, other organic contaminants, or natural organic carbon. Still other sites are aerobic and reductive dehalogenation is generally not an important process. Mixed sites with combinations of these types are also possible.

A knowledge base has been developing that helps researchers recognize and define natural attenuation of chlorinated solvent plumes in the field. This knowledge is derived from the following lines of evidence (99):

1. reduction in contaminant concentrations and total mass downgradient through the groundwater flow field;

2. geochemical and biochemical indicators through the groundwater flow field;

3. microbial laboratory microcosm studies.

The combination of the first two lines of evidence indicates that natural attenuation of organochlorine contaminants is occurring in the ambient environment. Definition of attenuation rates, geo- and biochemical indicators, and mass flux through a groundwater flow field have been demonstrated. Laboratory microcosm studies may be performed but only when necessary or if the first two lines of evidence are nonexistent. Such microcosm studies are costly to implement and time consuming (taking many months to perform because of inherent biotransformation rates).

NATURAL ATTENUATION FIELD EXPERIENCES

Natural Attenuation with Co-Contaminants

Sites where fuels or other organic contaminants support reductive dehalogenation include the St. Joseph, Michigan, Superfund site (1, 62, 97, 98); Necco

Table 2 Chlorinated solvent natural attenuation demonstrations

Site	Initial average concentration μM (% reduction)	RT (yr.)	1st Order rate half-life (yr.)	Electron donor/electron acceptors	Reference
Co-contaminant Driven Sites					
St. Joseph, MI	56 TCE (>99)	18	3.2 net	undefined COD/CO$_2$,	1, 62,
2nd transect to lake	94 cDCE (>99)	18	4.1 net	SO$_4$, Fe^{III}, O$_2$	97, 98
	7.4 tDCE (>99)	18			
	3.5 1DCE (>99)				
	16 VC (>99)	18	5.8 net		
Necco Park, NY	27 PCE (91)			TOC/CO$_2$, SO$_4$,	51, 103
Wells 146F to 156F	220 TCE (90)			Fe^{III}, O$_2$	
	100 cDCE (49)				
	22 tDCE (68)				
	10 1DCE (80)				
	22 VC (−82)				
	3.5 ETE (−166)				
	14 TeCA (92)				
	47 2TCA (79)				
	2DCA (−4)				
	140 CF (98);				
	73 mg/L TVOC (81)	1.6	1.0		
Plattsburgh Air Force	190 TCE (99)	18	1.9–3.6	BTEX/Fe^{III}, CO$_2$,	100
Base, NY MW-02-	530 DCE (99)		0.8–22	SO$_4$, O$_2$	
108 to 35PLTW13	23 VC (99)		1.6–6.3		
Tibbetts Road, NH	5.4 TCE (99)	10	1.3	BTEX/Fe^{III}	102
	2.3 DCE (99)		1.6		
Naval Air Station	0.16 PCE (ND)	1.2	0.1–0.2	fuels/Fe^{III}, SO$_4$, CO$_2$	15, 103
Cecil Field, FL	0.34 cDCE (ND)				
	0.05 VC (ND)		0.1–0.2		
Eielson Air Force	680 TCE (99)	6.3	0.7 (0.3–3.2)	hydrocarbon/NA	26
Base, AK site 45/57					
to end of plume					
Naturally Occurring Organic-Driven Sites					
Pictatinny Arsenal,	14 TCE (59)	3.1	2.2	DOC/Fe^{III}, SO$_4$	47, 103
NJ source to discharge	NATCE	4.2	0.6		
	NA DCE		>1.8		
Mixed Behavior Sites					
Sacramento, CA	11 TCE (99)	1.8	0.6	BOD/SO$_4$, CO$_2$, O$_2$	19, 103
Wells 1727 to 3054	29 DCE (98)		0.8		
	22 VC ((>99)		0.2		
	14 1DCA (99)				
	1.3 CA (>88)				
Dover Air Force	150 TCE (ND)	51	2.8–4.3 TCE>DCE	NA/CO$_2$, SO$_4$, Fe^{III},	29
Base, DE	100 DCE (NA)		1.4–2.8 DCE>VC	NO$_3$, O$_2$	
Entire Plume	16 VC (NA)		1.8–22 VC>ETE		

BOD = biochemical oxygen demand, BTEX = benzene, toluene, ethylbenzene, and xylenes, CA = chloroethane, cDCE = *cis*-1,2-dichloroethene, CF = chloroform, CO$_2$ = carbon dioxide, COD = chemical oxygen demand, 1DCA = 1,1-dichloroethane, 2DCA = 1,2-dichloroethane, 1DCE = 1,1-dichloroethene, DOC = dissolved organic carbon, ETE = ethene, Fe^{III} = iron reducing, NA = not available, ND = non-detect, NO$_3$ = nitrate, O$_2$ = oxygen, PCE = tetrachloroethene, RT = residence time, SO$_4$ = sulfate, 1TCA = 1,1,1-trichloroethane, 2TCA = 1,1,2-trichloroethane, TCE = trichloroethene, tDCE = *trans*-1,2-dichloroethene, TeCA = 1,1,2,2-tetrachloroethane, TVOC = total volatile organic compounds, VC = vinyl chloride, >x = removal to below detection limit.

Park, New York (51, 103); Plattsburgh Air Force Base, New York (100); Tibbetts Road, New Hampshire (102); Naval Air Station Cecil Field, Florida (15, 103); and Eielsen Air Force Base, Arkansas (26). The electron donors for these sites include undefined COD, TOC, benzene, toluene, ethylbenzene, xylenes (BTEX), fuels, and hydrocarbons. As much as 25% of the 190 mg/L reduction in COD across the St. Joseph, Michigan, site was estimated to result from methanogenesis and 18% from TCE dechlorination (62). The first-order half-lives for these sites are shown in Table 2. Estimated degradation half-lives of 0.1 to 0.2 years were highest at the Naval Air Station Cecil Field, Florida, where unknown fuels and warm temperatures support rapid reductive dechlorination rates (15). Important electron acceptors at these sites include oxygen, nitrate, iron (Fe^{III}), sulfate, and carbon dioxide. Iron reduction may be an important terminal electron acceptor process at St. Joseph, Michigan, and aerobic degradation of VC may occurr there (1). Zones of sulfate-reduction and iron reduction surrounded the methanogenic zones at sites such as Naval Air Station Cecil Field, Florida (15).

Natural Attenuation with Naturally Occurring Organics

The Building 24 site of the Picatinny Arsenal, New Jersey, was thought to be an example of a site where natural organics supported natural attenuation (47). Between 5 and 14 mg/L of dissolved organic carbon, consisting primarily of humic and fulvic acids, may be the electron donors for this site. Humic acids can support reductive dechlorination (56). Estimates for the TCE half-life ranged from 0.6 years (103) to 2.2 years (47). The DCE half-life at this site was more than 1.8 years. Both Fe^{III} and sulfate were documented as electron acceptors.

The plume discharges to a brook at a rate of an estimated 50 kg TCE per year. A pump-and-treat system was installed and has been in operation since 1992 to provide hydraulic control and prevent the movement of the plume into the brook. TCE desorption into the groundwater was estimated to contribute 550 kg TCE per year to the plume with an unknown quantity of TCE from DNAPL dissolution. An estimated 360 kg TCE is biotransformed anaerobically each year, and another 50 kg TCE per year is volatilized during advection. Lateral dispersion, diffusion-driven volatilization, and sorption are each thought to contribute less than 1 kg of TCE losses per year.

Natural Attenuation with Sequential Anaerobic and Aerobic Behaviors

The Sacramento, California, site (19) and the Dover Air Force Base, Delaware, plume (29) were both thought to exhibit sequential behaviors (i.e. an anaerobic portion followed by an aerobic portion). At the Sacramento site, contamination levels decreased 99.4% for TCE, 98% for DCE, more than 99.3% for VC,

99.6% for 1DCA, and more than 88% for CA as groundwater moved from wells 1727 to 3054, with an estimated residence time of 1.8 years (19). During this same interval, a reduction in 4 mg/L BOD to less than 1 mg/L was noted. This undefined organic may support some reductive dechlorination by consuming sulfate and producing methane. In the downgradient areas, the redox potential becomes positive and dissolved oxygen can be used for co-metabolic biodegradation of the TCE, DCE, and VC. First-order biodegradation half-lives of 0.6, 0.8, and 0.2 years were estimated for TCE, DCE, and VC, respectively (103).

The chlorinated solvent plume at Dover Air Force Base, which has multiple sources of solvent contamination, extends for approximately 2700 m and is 910 m wide (29). The plumes of TCE, DCE, VC, and ethene are spatially segregated, suggesting that the more mobile daughter products are degraded before they move away. Dissolved oxygen, nitrate, sulfate, iron, and carbon dioxide are all present and could be potential electron acceptors at the site. Microcosm studies with soils from the site indicate the direct aerobic mineralization of ^{14}C-DCE (29). Half-lives for TCE to DCE, DCE to VC, and VC to ethene were estimated to be 2.8 to 4.2 years, 1.4 to 2.8 years, and 1.8 to 2.2 years, respectively.

ECONOMICS

The economics of remediation systems are highly variable and depend in large part on the type of contaminant, site hydrogeology, and the type of remediation technology applied. In general, the monetary costs for installing and operating an ex situ technology such as pump-and-treat is relatively high, while the costs for systems that rely on natural attenuation tend to be much lower.

Chlorinated solvent groundwater contamination clearly plays an important role in the overall environmental burden of polluted groundwater. Thus the stake from an impaired groundwater resource standpoint is high. The costs associated with the remediation of sites with groundwater contamination will be large and will escalate.

In a landmark study, Russell et al (74) estimated the cost of cleaning up all hazardous waste sites in the United States over the next 30 years. Depending on the assumptions and scenario utilized, cleanup costs ranged from a half trillion dollars under a less stringent cleanup policy scenario to over one trillion dollars if the USEPA were to shift to a more stringent policy. Table 3 shows the range of these cost estimates. Although such estimates contain significant assumptions and uncertainty, they illustrate the order of magnitude of potential site cleanup costs nationwide.

Table 3 Thirty-year estimate of hazardous waste site cleanup[a]

Estimated costs (Billions $)	Policy scenario	Scenario type
752	Current	Circa 1988–1991
484	Less stringent	Containment
1,177	More stringent	Treatment/removal

[a]Adapted from Reference 74.

Remediation Technology Cost and Performance

Quinton et al (72) used a present value model to provide a consistent basis for cost and performance comparisons. The method uses a template site with a PCE plume 300 m long by 120 m wide and includes the ability to use varying depths (up to 27 m deep). Figure 1 presents a schematic of the template site. Input variables include remediation duration, estimated engineering and flow/transport modeling costs, equipment costs, operations and maintenance costs, and monitoring costs. In this evaluation, the total present value life-cycle cost of a remediation technology is the sum of the present value for capital costs plus the annual operating costs throughout the remediation period. The following equation illustrates the mathematical algorithm used for the model:

$$PV = \sum_{(i=1 \text{ to } n)} Yi/(1+r)^i,$$

where PV is present value, Yi represents the dollars spent in the ith year, r is the discount rate, and n is the number of years of remediation.

In applying this model, a combined discount rate, r', that accounted for inflation was used in place of the general discount rate, r. Four primary technologies were compared for remediating a PCE groundwater contaminant plume, including in situ bioremediation (ISB) involving substrate-enhanced anaerobic bioremediation (both source area recirculating system and downgradient biobarrier), intrinsic bioremediation/natural attenuation (IBR/NA), in situ permeable reactive barrier (PRB; zero valent iron wall technology), and a simple pump-and-treat (P&T) system (air stripping with vapor- and liquid-phase activated carbon). Cost metrics were calculated for total system cost, cost per unit of contaminant removed, and cost per unit volume of groundwater treated using the discounted cash flow analysis. This model also accounted for the volume of groundwater flowing through the template site during the remediation period. The results of this analysis indicated that the per unit costs for removing PCE contaminant significantly increased in the following order: IBR/NA < ISB < PRB < P&T. Examples of the associated treatment technology costs are shown in Table 4 ($i = 30$ years; $r' = 8.7\%$).

Table 4 Template site remediation cost comparison[a]

Metric	IBR/NA	ISB recirc	ISB barrier	PRB	P&T
$/1000 liters treated	0.32	0.48	1.10	1.40	2.35
$/kg PCE removed	330	480	1,100	1,400	3,500

[a]Adapted from Reference 72. IBR/NA, intrinsic bioremediation/natural attenuation; ISB, in situ bioremediation; PRB, permeable reactive barrier; P&T, Pump-and-treat; PCE, perchloroethene.

These authors demonstrated the usefulness of a template site against which a variety of remedial technologies can be evaluated on a consistent basis using lowest cost present value analysis. The pump-and-treat technologies evaluated for the template site provided the highest remediation costs when compared to the other technologies, and under assumptions of equivalent protectiveness, intrinsic bioremediation (natural attenuation) provided the lowest cost.

CONCLUSIONS

Chlorinated solvents are among the most common groundwater contaminants. Microbes have a number of mechanisms for biodegrading these solvents, including energy-yielding oxidations, co-metabolic oxidations, energy-yielding reductive dehalogenation, and co-metabolic reduction. Reductive dehalogenation is a key step for the breakdown of highly chlorinated compounds such as PCE. Other microbes such as dehalorespiring organisms may derive energy from these reactions.

Several trials of in situ engineered approaches of chlorinated solvent biodegradation have been implemented successfully, including the complete anaerobic degradation of the solvents, combined anaerobic-aerobic processes, and bioaugmentation (where the microbial populations capable of completely degrading the contaminants were not present). Less chlorinated daughter products can be biodegraded subsequently in aerobic zones. Natural attenuation via both anaerobic and aerobic processes can act in combination to biodegrade chlorinated solvent plumes. Reductive dehalogenation of chlorinated solvents occurs naturally at a number of sites supported by the oxidation of co-contaminants or naturally occurring organics in sequence with aerobic or co-metabolic degradation.

Natural attenuation is not a panacea. However, at sites where natural attenuation is protective, and where the physical, chemical, and biochemical reactions are relatively complete, it offers a scientifically sound, cost-effective solution to chlorinated solvent contaminant remediation.

Literature Cited

1. Adriaens P, Lendvay J, McCormick ML, Dean SM. 1997. Biogeochemistry and dechlorination potential at the St. Joseph Aquifer-Lake Michigan interface. In *Papers from the 4th Int. In Situ and On Site Bioremediation Symp., New Orleans, LA*, 3:173–78. Columbus, OH: Battelle

2. Alphenaar A, Gerritse J, Kloetstra G, Spuij P, Urlings L, et al. 1995. In-situ bioremediation of chloroethene-contaminated soil. In *Contaminated Soil '95*. Vol. II. *Proc. 5th Int. FZK/TNO Conf. on Contaminated Soil, Masstrictht, Netherlands*, ed. WJ van Den Brink, R Bosman, F Arendt, pp. 833–41. Dordrecht, The Netherlands: Kluwer

3. Bagley DM, Gossett JM. 1989. Tetrachloroethene transformation to trichloroethene and cis–1,2-dichloroethene by sulfate-reducing enrichment culture. *Appl. Environ. Microbiol.* 56(8):2511–16

4. Ballapragada BS, Stensel HD, Puhakka JA, Ferguson JF. 1997. Effect of hydrogen on reductive dechlorination of chlorinated ethenes. *Environ. Sci. Technol.* 31 (6):1728–34

5. Becker JG, Freedman DL. 1994. Use of cyanocobalamin to enhance anaerobic biodegradation of chloroform. *Environ. Sci. Technol.* 28(11):1942–49

6. Beeman RE. 1994. *U.S. Patent No. 5,277,815*

7. Beeman RE, Shoemaker SH, Howell JE, Salazar EA, Buttram JR. 1994. A field evaluation of in situ microbial reductive dehalogenation by the biotransformation of chlorinated ethenes. In *Bioremediation of Chlorinated and Polycyclic Aromatic Hydrocarbon Compounds*, ed. RE Hinchee, A Leeson, L Semprini, SK Ong, pp. 14–27. Boca Raton, FL: Lewis

8. Bouwer EJ. 1994. Bioremediation of chlorinated solvents using alternate electron acceptors. See Ref. 65a, pp. 149–75

9. Bouwer EJ, McCarty PL. 1983. Transformations of 1- and 2-carbon halogenated aliphatic organic compounds under methanogenic conditions. *Appl. Environ. Microbiol.* 45(4):1286–94

10. Bouwer EJ, McCarty PL. 1983. Transformations of halogenated organic compounds under denitrification conditions. *Appl. Environ. Microbiol.* 45(4):1295–99

11. Boyer JD, Ahlert AC, Kosson DS. 1988. Pilot plant demonstration of in-situ biodegradation of 1,1,1-trichloroethane. *J. Water Pollut. Control Fed.* 60(10): 1843–49

12. Bradley PM, Chapelle FH. 1996. Anaerobic mineralization of vinyl chloride in Fe(III)-reducing aquifer systems. *Environ. Sci. Technol.* 30(6):2084–86

13. Bradley PM, Chapelle FH. 1997. Kinetics of DCE and VC mineralization under methanogenic and Fe(III) reducing conditions. *Environ. Sci. Technol.* 31(9):2692–96

14. Buchanan RJ Jr, Ellis DE, Odom JM, Mazierski PF, Lee MD. 1995. Intrinsic and accelerated anaerobic biodegradation of perchloroethylene in groundwater. See Ref. 42a, pp. 245–52

15. Chapelle FH. 1996. *Identifying redox conditions that favor the natural attenuation of chlorinated ethenes in contaminated ground-water systems*. Presented at Int. Bus. Commun. 2nd Annu. Int. Sym. on Intrinsic Bioremediation: Natural Attenuation, Annapolis, MD, pp. 1–10

16. Chiu Y-C, Lu C-J, Huang S-Y. 1997. Anaerobic dechlorination of tetrachloroethylene to ethene using winery microbial consortium. In *Papers from the 4th Int. In Situ and On Site Bioremediation Symp., New Orleans, LA*, 5:51–56. Columbus, OH: Battelle

17. Christopher M, Litherland ST, O'Cleirigh D, Vaughn C. 1997. Anaerobic bioremediation of chlorinated organics in a fractured shale environment. In *Papers from the 4th Int. In Situ and On Site Bioremediation Symp., New Orleans, LA*, 5:507–12. Columbus, OH: Battelle

18. Cole JR, Cascarelli AL, Mohn WW, Tiedje JM. 1994. Isolation and characterization of a novel bacterium growing via reductive dehalogenation on 2-chlorophenol. *Appl. Environ. Microbiol.* 60(10):3536–42

19. Cox E, Edwards E, Lehmicke L, Major D. 1995. Intrinsic biodegradation of trichloroethene and trichloroethane in a sequential anaerobic-aerobic aquifer. See Ref. 42a, pp. 223–31

20. Criddle CS, DeWitt JT, Grbic-Galic D, McCarty PL. 1990. Transformation of

carbon tetrachloride by *Pseudomonas* sp. Strain KC under denitrification conditions. *Appl. Environ. Microbiol.* 56(11):3240–47

21. Davis JW, Carpenter CL. 1990. Aerobic biodegradation of vinyl chloride in groundwater samples. *Appl. Environ. Microbiol.* 56(12):3878–80

22. Day MJ, Thomson JAM, O'Hayre AP. 1993. *In situ bioremediation of DNAPL-impacted groundwater and subsoils using alternate electron acceptors at the French Limited Superfund Site, Texas.* I-EC Special Symp. ACS, Atlanta, GA, pp. 819–22. (Extended Abstr.)

23. de Bruin WP, Kotterman MJJ, Posthumus MA, Schraa G, Zehnder AJB. 1992. Complete biological reductive transformation of tetrachloroethene to ethane. *Appl. Environ. Microbiol.* 58(6):1996–2000

24. DiStefano TD, Gossett JM, Zinder SH. 1991. Reductive dehalogenation of high concentrations of tetrachloroethene to ethene by an anaerobic enrichment culture in the absence of methanogenesis. *Appl. Environ. Microbiol.* 57(8):2287–92

25. Dolan ME, McCarty PL. 1995. Methanotrophic chloroethene transformation capacities and 1,1-dichloroethene transformation product toxicity. *Environ. Sci. Technol.* 29:2741–47

26. DuPont RR, Gorder K, Sorensen DL, Kemblowski MW, Haas P. 1996. Case study: Eielson Air Force Base, Alaska. See Ref. 88a, pp. 104–9

27. Dybas MJ, Bezborodinikov S, Voice T, Wiggert DC, Davies S, et al. 1997. Evaluation of bioaugmentation to remediate an aquifer contaminated with carbon tetrachloride. In *Papers from the 4th Int. In Situ and On Site Bioremediation Symp. New Orleans, LA,* 4:507–12. Columbus, OH: Battelle

28. Ellis DE. 1996. Intrinsic remediation in the industrial marketplace. See Ref. 88a, pp. 120–23

29. Ellis DE, Lutz EJ, Klecka GM, Pardieck DL, Salvo JJ, et al. 1996. Remediation Technology Development Forum intrinsic remediation project at Dover Air Force Base, Delaware. See Ref. 88a, pp. 93–97

30. Ensley BD. 1991. Biochemical diversity of trichloroethylene metabolism. *Annu. Rev. Microbiol.* 45:283–99

31. Enzien MV, Picarel F, Hazen TC, Arnold RG, Fliermans CB. 1994. Reductive dechlorination of trichloroethylene and tetrachloroethylene under aerobic conditions in a sediment column. *Appl. Environ. Microbiol.* 60(6):2200–4

32. Fathepure BZ, Youngers GA, Richter DL,

Downs CE. 1995. In situ bioremediation of chlorinated hydrocarbons under field aerobic-anaerobic environments. See Ref. 42b, pp. 169–86

33. Freedman DL, Gossett JM. 1991. Biodegradation of dichloromethane in a fixed film reactor under methanogenic conditions. In *On-Site Bioreclamation Processes for Xenobiotic and Hydrocarbon Treatment,* ed. RE Hinchee, RF Olfenbuttel, pp. 113–33. Stoneham, MA: Butterworth-Heinemannn

34. Galli R, McCarty PL. 1989. Biotransformation of 1,1,1-trichloroethane, trichloromethane, and tetrachloromethane by a *Clostridium* sp. *Appl. Environ. Microbiol.* 55(4):837–44

35. Gerritse J, Renard V, Pedro-Gomes TM, Lawson PA, Collins MD, et al. 1996. *Desulfitobacterium* sp. Strain PCE1, an anaerobic bacterium that can grow by reductive dechlorination of tetrachloroethene or ortho-chlorinated phenols. *Arch. Microbiol.* 165:132–40

36. Gibson SA, Sewell GW. 1992. Stimulation of reductive dechlorination of tetrachloroethene in anaerobic aquifer microcosms by additions of short-chain organic acids or alcohols. *Appl. Environ. Microbiol.* 58(4):1392–93

37. Gribble GW. 1994. The natural production of chlorinated compounds. *Environ. Sci. Technol.* 28(7):310–19

38. Griffith GD, Cole JR, Quensen JF, Tiedje JM. 1992. Specific deuteration of dichlorobenzoate during reductive dehalogenation by *Desulfomonile tiedjei* in deuterium oxide. *Appl. Environ. Microbiol.* 58(1):409–11

39. Hartmans S, De Bont JAM, 1992. Aerobic vinyl chloride metabolism in *Mycobacterium aurum* L1. *Appl. Environ. Microbiol.* 58(4):1220–26

40. Hashsham SA, Freedman DL. 1997. Enhanced biotransformation of carbon tetrachloride by an anaerobic enrichment culture. In *Papers from the 4th Int. In Situ and On Site Bioremediation Symp. New Orleans, LA,* 4:465–70. Columbus, OH: Battelle

41. Hashsham SA, Scholze R, Freedman DL. 1995. Cobalamin-enhanced anaerobic biotransformation of carbon tetrachloride. *Environ. Sci. Technol.* 29(11):2856–63

42. Haston ZC, Sharma PK, Black JNP, McCarty PL. 1994. Enhanced reductive dechlorination of chlorinated ethenes. In *Abstr., Symp. on Bioremediation of Hazardous Wastes: Research, Development, and Field Evaluation.* U.S. EPA/600/R–

94/075, pp. 11–14. Washington, DC: USEPA Office of Research and Development

42a. Hinchee RE, Wilson JT, Downey DC, ed. 1995. *Intrinsic Bioremediation*. Columbus, OH: Battelle

42b. Hinchee RE, Leeson A, Semprini L, ed. 1995. *Bioremediation of Chlorinated Solvents*. Columbus, OH: Battelle

43. Hoekstra EJ, De Leer EWB. 1995. Organohalogens: the natural alternatives. *Chem. Britain*. Feb:127–31

44. Holliger C. 1995. The anaerobic microbiology and biotreatment of chlorinated ethenes. *Curr. Opin. Biotechnol.* 6:347–51

45. Holliger C, Gaspard S, Glod G, Heijman C, Schumacher W, et al. 1997. Contaminated environments in the subsurface and bioremediation: organic contaminants. *FEMS Microbiol. Rev.* 20:517–23

46. Holliger C, Schumacher W. 1994. Reductive dehalogenation as a respiratory process. *Antonie van Leeuwenhoek* 66:239–46

46a. Hooker BS, Skeen RS, Truex MJ, Johnson CD, Peyton BM, et al. 1998. In situ bioremediation of carbon tetrachloride: field test results. *Bioremediation J.* 1:181–82

47. Imbrigiotta TE, Ehlke TA, Wilson BH, Wilson JT. 1996. Case study: natural attenuation of a trichloroethene plume at Picatinny Arsenal, New Jersey. See Ref. 88a, pp. 83–89

48. Jones RE, Beeman RE, Suflita JM. 1988. Anaerobic metabolic processes in the deep terrestrial subsurface. *Geomicrobiol. J.* 7:117–30

49. Klecka GM, Gonsior SG, Markham DA. 1990. Biological transformation of 1,1,1-trichloroethane in subsurface soils and ground water. *Environ. Toxicol. Chem.* 9(12):1437–51

50. Lee MD, Gregory GE III, White DG, Fountain JF, Shoemaker SH. 1995. Surfactant-enhanced anaerobic bioremediation of a carbon tetrachloride DNAPL. See Ref. 42b, pp. 147–51

51. Lee MD, Mazierski PF, Buchanan RJ Jr, Ellis DE, Sehayek LS. 1995. Intrinsic in situ anaerobic biodegradation of chlorinated solvents at an industrial landfill. See Ref. 42a, pp. 205–22

52. Lee MD, Quinton GE, Beeman RE, Biehle AA, Liddle RL, et al. 1997. Scale-up issues for in situ anaerobic tetrachloroethene bioremediation. *J. Ind. Microbiol. Biotechnol.* 18(2/3):106–15

53. Lesage S, Brown D, Millar K. 1996. Vitamin B_{12}-catalyzed dechlorination of perchloroethylene present as residual DNAPL. *Ground Water Monit. Rev.* Fall: 76–85

54. Litherland ST, Anderson DW. 1997. Full-scale bioremediation at a chlorinated solvent site. In *Papers from the 4th Int. In Situ and On Site Bioremediation Symp., New Orleans, LA*, 5:425–30. Columbus, OH: Battelle

55. Lovley DR. 1997. Microbial Fe(III) reduction in subsurface environments. *FEMS Microbiol. Rev.* 20:305–13

56. Lyon WG, West CC, Osborn ML, Sewell GW. 1995. Microbial utilization of vadose zone organic carbon for reductive dechlorination of tetrachloroethene. *J. Environ. Sci. Health A* 30(7):1629–39

57. Malachowsky KJ, Phelps TJ, Teboli AB, Minnikin DE, White DC. 1994. Aerobic mineralization of trichloroethylene, vinyl chloride, and aromatic compounds by *Rhodococcus* species. *Appl. Environ. Microbiol.* 60(2):542–48

58. Maymo-Gatell X, Chien Y-t, Gossett JM, Zinder SH. 1997. Isolation of a bacterium that reductively dechlorinates tetrachloroethene to ethene. *Science* 276: 1568–71

59. Maymo-Gatell X, Tandoi V, Gossett JM, Zinder SH. 1995. Characterization of an H_2-utilizing enrichment culture that reductively dechlorinates tetrachloroethene to vinyl chloride and ethene in the absence of methanogenesis and acetogenesis. *Appl. Environ. Microbiol.* 61(11):3928–33

60. McCarty PL. 1996. Biotic and abiotic transformations of chlorinated solvents in ground water. See Ref. 88a, pp. 5–9

61. McCarty PL. 1997. Breathing with chlorinated solvents. *Science* 276:1521–22

62. McCarty PL, Wilson JT. 1992. Natural anaerobic treatment of a TCE plume, St. Joseph, Michigan, NPL Site. In *Bioremediation of Hazardous Wastes*, EPA/600/R–92/126, pp. 47–50. Cincinnati, OH: USEPA Center for Environmental Research Information

63. National Research Council. 1994. *Alternatives for Ground Water Cleanup*. Washington, DC: National Academy

64. Neumann A, Scholz-Muramuatsu H, Diekert G. 1994. Tetrachloroethene metabolism of *Dehalospirillum multivorans*. *Arch. Microbiol.* 162:295–301

65. Norris RD. 1994. In-situ bioremediation of soils and ground water contaminated with petroleum hydrocarbons. See Ref. 65a, pp. 17–37

65a. Norris RD, Hinchee RE, Brown R, McCarty PL, Semprini L, et al, eds. 1994.

Handbook of Bioremediation. Boca Raton, FL: Lewis

66. Odom JM, Nagel E, Tabinowski J. 1995. Chemical-biological catalysis for in situ anaerobic dehalogenation of chlorinated solvents. See Ref. 42b, pp. 35–43

67. Odom JM, Tabinowski J, Lee MD, Fathepure BZ. 1995. Anaerobic biodegradation of chlorinated solvents: comparative laboratory study of aquifer microcosms. See Ref. 42b, pp. 17–24

68. Olsen RL, Kavanaugh MC. 1993. Can groundwater restoration be achieved? *Water Environ. Technol.* 5(3):42–47

69. Parsons F, Wood PR, DeMarco J. 1984. Transformation of tetrachloroethylene and trichloroethylene in microcosms and groundwater. *J. Am. Water Works Assoc.* 76:56–59

70. Phelps TJ, Malachowsky K, Schram RM, White DC. 1991. Aerobic mineralization of vinyl chloride by a bacterium of the order *Actinomycetales*. *Appl. Environ. Microbiol.* 57(4):1252–54

71. Pries F, van der Ploeg JR, Dolfing J, Janssen DB. 1994. Degradation of halogenated aliphatic compounds: the role of adaptation. *FEMS Microbiol. Rev.* 15:279–95

72. Quinton GE, Buchanan RJ Jr, Ellis DE, Shoemaker SH. 1997. A method to compare ground water cleanup technologies. *Remediation* Autumn:7–16

73. Rasmussen G, Komisar SJ, Ferguson JF. 1994. Transformation of tetrachloroethene to ethene in mixed methanogenic cultures: effect of electron donor, biomass levels, and inhibitors. See Ref. 42b, pp. 309–13

74. Russell M, Colglazier E, English MR. 1991. *Hazardous Waste Remediation: The Task Ahead*. Knoxville: Univ. Tennessee, Waste Manage. Res. Educ. Inst.

75. Sanford RA, Cole JR, Loeffler FE, Tiedje JM. 1996. Characterization of *Desulfitobacterium chlororespirans* sp. nov., which grows by coupling the oxidation of lactate to the reductive dechlorination of 3-chloro-4-hydroxybenzoate. *Appl. Environ. Microbiol.* 62(10):3800–8

76. Scholz-Muramatsu H, Neumann A, Mebmer M, Moore E, Diekert G. 1995. Isolation and characterization of *Dehalospirillum multivorans* gen. nov., sp. nov., a tetrachloroethene-utilizing, strictly anaerobic bacterium. *Arch. Microbiol.* 163:48–56

77. Schumacher W, Holliger C. 1996. The proton/electron ratio of the menaquinone-dependent electron transport from dihydrogen to tetrachloroethene in

Dehalobacter restictus. *J. Bacteriol.* 178(8):2328–33

78. Semprini L, Hopkins GD, Roberts PV, McCarty PL. 1992. In situ transformation of carbon tetrachloride and other halogenated compounds resulting from biostimulation under anoxic conditions. *Environ. Sci. Technol.* 26(12):2454–61

79. Sharma PK, McCarty PL. 1997. Isolation and characterization of a facultatively aerobic bacterium that reductively dehalogenates tetrachloroethene to cis–1,2-dichloroethene. *Appl. Environ. Microbiol.* 62(3):761–65

80. Skeen RS, Gao J, Hooker BS. 1995. Kinetics of chlorinated ethylene dehalogenation under methanogenic conditions. *Biotechnol. Bioeng.* 45:659–66

81. Skeen RS, Lutrell SP, Brouns TM, Hooker BS, Petersen JN. 1993. In situ bioremediation of Hanford groundwater. *Remediation* Summer:353–67

82. Smatlak C, Gossett JM, Zinder SH. 1996. Comparative kinetics of hydrogen utilization for reductive dechlorination of tetrachloroethene and methanogenesis in an anaerobic enrichment culture. *Environ. Sci. Technol.* 30(9):2850–58

83. Spuij F, Alphenaar A, de Wit H, Lubbers R, v/d Brink K, et al. 1997. Full-scale application of in situ bioremediation of PCE-contaminated soil. In *Papers from the 4th Int. In Situ and On Site Bioremediation Symp., New Orleans, LA,* 5:431–37. Columbus, OH: Battelle

84. Stromeyer SA, Winkelbauer W, Kohler H, Cook AM, Leisinger T. 1991. Dichloromethane utilized by an anaerobic mixed culture: acetogenesis and methanogenesis. *Biodegradation* 2:129–37

85. Suflita JM, Horowitz A, Shelton DR, Tiedje JM. 1982. Dehalogenation: a novel pathway for the anaerobic biodegradation of haloaromatic compounds. *Science* 218:1115–17

86. Thomson JAM, Day MJ, Sloan RL, Collins ML. 1995. In situ aquifer bioremediation at the French Limited Superfund Site. In *Applied Bioremediation of Petroleum Hydrocarbons*, ed. RE Hinchee, JA Kittel, HJ Reisinger, pp. 453–59. Columbus, OH: Battelle

87. Deleted in proof

88. USEPA. 1993. *Cleaning Up The Nation's Waste Sites: Markets And Technology Trends*. EPA 542-R–92–020. Washington, DC: USEPA

88a. USEPA. 1996. *Symposium on Natural Attenuation of Chlorinated Organics in Ground Water, Dallas, TX.* EPA/540/R–96/509. Washington, DC: USEPA, Office of Research and Development

89. USEPA. 1997. *Memorandum: draft interim final OSWER monitored natural attenuation policy (OSWER directive 9200.4-17)*, p. 2. Washington, DC: USEPA, Office of Solid Waste and Emergency Response

90. Utkin I, Woese C, Wiegel J. 1994. Isolation and characterization of *Desulfitobacterium dehalogenans* gen. nov., sp. nov., an anaerobic bacterium which reductively dechlorinates chlorophenolic compounds. *Int. J. Syst. Bacteriol.* 44(4):612–19

91. Vanelli T, Logan M, Arciero DM, Hooper AB. 1990. Degradation of halogenated aliphatic compounds by the ammonia-oxidizing bacterium *Nitrosomonas europaea*. *Appl. Environ. Microbiol.* 56(4):1169–71

92. Vogel TM. 1994. Natural bioremediation of chlorinated solvents. See Ref. 65a, pp. 201–25

93. Vogel TM, Criddle CS, McCarty PL. 1987. Transformations of halogenated aliphatic compounds. *Environ. Sci. Technol.* 21(8):722–36

94. Vogel TM, McCarty PL. 1985. Biotransformation of tetrachloroethylene to trichloroethylene, dichloroethylene, vinyl chloride, and carbon dioxide under methanogenic conditions. *Appl. Environ. Microbiol.* 49(5):1080–83

95. Vogel TM, McCarty PL. 1987. Abiotic and biotic transformation of 1,1,1-trichloroethane under methanogenic conditions. *Environ. Sci. Technol.* 21(12):1208–13

96. Weathers LJ, Parkin GF. 1995. Metallic iron-enhanced biotransformation of carbon tetrachloride and chloroform under methanogenic conditions. See Ref. 42b, pp. 117–22

97. Weaver JW, Wilson JT, Kampbell DH. 1996. Case study of natural attenuation of trichloroethene at St. Joseph, MI. See Ref. 88a, pp. 65–68

98. Weaver JW, Wilson JT, Kampbell DH. 1996. Extraction of degradation rate constants from the St. Joseph, Michigan, trichloroethene site. See Ref. 88a, pp. 69–73

99. Wiedemeier TH, Swanson MA, Moutoux DE, Wilson JT, Kampbell DH, et al. 1996. Overview of the technical protocol for natural attenuation of chlorinated aliphatic hydrocarbons in ground water under development for the U.S. Air Force Center for Environmental Excellence. See Ref. 88a, pp. 35–59

100. Wiedemeier TH, Wilson JT, Kampbell DH. 1996. Natural attenuation of chlorinated aliphatic hydrocarbons at Plattsburgh Air Force Base, New York. See Ref. 88a, pp. 74–82

101. Wild AP, Winhelbauer W, Leisinger T. 1995. Anaerobic dechlorination of trichloroethene, tetrachloroethene, and 1,2-dichloroethane by an acetogenic mixed culture in a fixed bed reactor. *Biodegradation* 6:309–18

102. Wilson BH, Wilson JT, Luce D. 1996. Design and interpretation of microcosm studies for chlorinated compounds. See Ref. 88a, pp. 21–28

103. Wilson JT, Kampbell DH, Weaver JW. 1996. Environmental chemistry and kinetics of biotransformation of chlorinated organic compounds in ground water. See Ref. 88a, pp. 124–27

104. Workman DJ, Woods SL, Gorby YA, Fredrickson JK, Truex MJ. 1997. Microbial reduction of vitamin B_{12} by *Shewanella alga* Strain BrY with subsequent transformation of carbon tetrachloride. *Environ. Sci. Technol.* 31(8):2292–97

Annu. Rev. Microbiol. 1998. 52:453–90

NOSOCOMIAL OUTBREAKS/PSEUDO-OUTBREAKS CAUSED BY NONTUBERCULOUS MYCOBACTERIA

Richard J. Wallace, Jr.[1,2], *Barbara A. Brown*[1], *David E. Griffith*[2]

[1]University of Texas Health Center, Department of Microbiology and [2]Center for Pulmonary Infectious Disease Control, Post Office Box 2003, Tyler, Texas 75710; e-mail: melanie@uthct.edu

KEY WORDS: rapidly growing mycobacteria, sternal wound infections, postinjection abscesses, bronchoscope contamination, mycobacteria and water

---

ABSTRACT

Nosocomial outbreaks and pseudo-outbreaks caused by the nontuberculous mycobacteria (NTM) have been recognized for more than 20 years and continue to be a problem. Most of these outbreaks have involved the rapidly growing mycobacterial species *Mycobacterium fortuitum* and *M. abscessus*. The reservoir for these outbreaks is generally municipal and (often separate) hospital water supplies. These mycobacterial species and others are incredibly hardy, able to grow in municipal and distilled water, thrive at temperatures of 45°C or above (*M. xenopi* and *M. avium* complex), and resist the activity of organomercurials, chlorine, 2% concentrations of formaldehyde and alkaline glutaraldehyde, and other commonly used disinfectants. Disease outbreaks usually involve sternal wound infections, plastic surgery wound infections, or postinjection abscesses. Pseudo-outbreaks most commonly relate to contaminated bronchoscopes and endoscopic cleaning machines (*M. abscessus*) and contaminated hospital water supplies (*M. xenopi*). Knowledge of the reservoir of these species, their great survival capabilities within the hospital, and newer molecular techniques for strain comparison have helped control and more quickly identify current nosocomial outbreaks or pseudo-outbreaks caused by the NTM.

453

0066-4227/98/1001-0453$08.00

CONTENTS

INTRODUCTION

The first clinical and laboratory description of *Mycobacterium fortuitum* was reported by da Costa Cruz in 1936 (26). This was also the first case of nosocomial disease proved to be caused by the nontuberculous mycobacteria (NTM), because the patient described by da Costa Cruz had a postinjection abscess that followed an intramuscular vitamin injection. In the 1960s, several clusters of postinjection abscesses were reported from the Congo, the Netherlands, and Texas (8, 46, 64, 89) that were due to rapidly growing mycobacteria and were the first NTM disease outbreaks to be reported. In 1975 a cluster of sternal wound infections caused by *M. abscessus* occurred in a hospital in North Carolina (44, 73). Although the source of the outbreak was never found, this was the first surgical wound infection outbreak traceable to the NTM to be reported.

Since that time, outbreaks and pseudo-outbreaks caused by NTM—especially the rapidly growing mycobacteria—have become relatively common (see Tables 1 and 2). In a summary of outbreaks or pseudo-outbreaks investigated by the Centers for Disease Control and Prevention (CDC) between 1980 and 1990, 4% were caused by mycobacteria (47). Much has been learned about the hospital epidemiology and reservoirs of the NTM and the mechanism of disease and pseudo-disease outbreaks. This study reviews these outbreaks and what they have taught us about the nosocomial involvement of the environmental species comprising the NTM.

Table 1 Outbreak or clustered nosocomial diseases caused by nontuberculous mycobacteria

Clinical disease	Number of reported outbreaks
Surgical wound infections	
Cardiac surgery	7
Plastic surgery	4
Vein stripping	(1)
Augmentation mammaplasty and blepharoplasty	(1)
Nasal reconstruction	(1)
Liposuction	(1)
Laparoscopy	1
Laminectomy	1
Postinjection abscesses	11
Dialysis related infections	3
Otitis media (post tympanostomy tube)	1
Disseminated disease	1
Lung disease	1
Totals	30

NOSOCOMIAL EPIDEMIOLOGY AND PATHOGENESIS

NTM in Municipal Water Systems

Municipal water supplies are now recognized as a major haven for the NTM and as the major reservoir for most nosocomial outbreaks and pseudo-outbreaks caused by these organisms. A good review of the history of mycobacteria in piped and treated waters is provided by Collins et al (22). Carson et al (16) found NTM in municipal water supplies or treated water supplies from 95 of 115 (83%) of dialysis centers throughout the United States. More than 50% of the municipal water samples were positive (16). Representative species included *M. mucogenicum, M. chelonae* (no separation of *M. chelonae* versus *M. abscessus*), *M. fortuitum, M. gordonae, M. avium* complex, and *M. terrae* complex. Approximately two thirds of the isolates were rapidly growing species. In a study of potable water supplies in Los Angeles, California, *M. avium* complex isolates were recovered from 42 of 108 (32%) tested locations that included homes, hospitals, commercial buildings, and reservoirs (6). Other NTM have been described in municipal water systems, especially *M. kansasii* (3, 22, 30, 61, 76).

Biofilms

Understanding of the microbial dynamics within municipal (piped) water systems has been expanded significantly by recognition and understanding of the

Table 2 Nontuberculous mycobacterial species associated with nosocomial disease outbreaks or pseudo-outbreaks

Disease outbreaks	Number
Rapidly growing mycobacteria	
M. abscessus	14
M. fortuitum	7
M. chelonae	3
M. mucogenicum	1
Uncertain species	5[a]
Total	30
Slowly growing mycobacteria	
M. xenopi	2
M. avium complex	1
Total	3

Pseudo-outbreaks	Number
Rapidly growing mycobacteria	
M. abscessus	7
M. fortuitum group	4
M. chelonae	0
M. mucogenicum	0
Total	11
Slowly growing mycobacteria	
M. xenopi	3
M. avium complex	4
M. terrae complex	1
M. gordonae	2
M. simiae	1
Total	11

[a]Outbreaks occurred before 1980 when species identification was uncertain. All involved postinjection abscesses.

concept of biofilms. Biofilms are the filmy layer at the solid (pipe) and liquid (water) interface. They have come to be recognized as a frequent localization site for mycobacteria (76) as well as other bacterial species. In one study of 50 biofilm samples from Germany, 90% of the sampled biofilms within a variety of piped water systems contained mycobacteria (76). Identified species included *M. fortuitum*, *M. chelonae*, *M. flavescens*, *M. gordonae*, *M. kansasii*, and *M. terrae* complex. This film appears to be present in almost all collection and piping systems and likely provides the nutritional support for the organisms. The presence of the organisms in the free-flowing tap water likely reflects the number and concentration of species within the biofilm. Currently, municipal water systems are considered the major environmental reservoirs for *M. kansasii*, *M. xenopi*, and (possibly) *M. simiae*, species that are rarely recovered from other environmental sources.

NTM in Hospital Water Systems

THERMOPHILIC NTM   Several mycobacterial species are thermophilic and survive and grow well at 45°C. This includes *M. xenopi*, *M. smegmatis*, and *M. avium* complex. The decision to reduce hospital hot water temperatures from 52°C to 43°C likely has facilitated the presence of these species, although these species are capable of growing in hospital water kept at temperatures as high as 55°C (28). Both *M. xenopi* and *M. avium* complex are found in hospital hot water systems, both in the holding tank and at the faucet (61, 67, 92).

NONTHERMOPHILIC NTM   Other mycobacterial species do not appear to tolerate temperatures of 43°C or above and are generally found only in cold water systems. This includes *M. kansasii*, *M. gordonae*, *M. fortuitum*, *M. chelonae* (does not tolerate temperatures much above 30°C, although will grow at 35°C), *M. abscessus*, and *M. mucogenicum*. The presence of *M. kansasii* in hospital water systems has been reported worldwide, including reports from England (61, 99), France (70), the Netherlands (30), Canada (59), and the United States (3). In one such circumstance, the organism was concentrated in long-term, low-flow storage tanks (54a). *M. avium* complex grows well at both high and low temperatures, so it can be found in either system within the hospital.

Some nonthermophilic slow-growing NTM species such as *M. simiae*, *M. terrae* complex, *M. haemophilum*, *M. genavense*, and *M. malmoense* have not been studied extensively or are fastidious and difficult to study. Their involvement with municipal and hospital water systems is relatively unknown.

Virulence Factors

ABILITY TO GROW IN DISTILLED WATER   Nontuberculous mycobacteria appear able to survive and grow not only in tap water, but also in distilled water (see Table 3). The assumption that distilled water sources will remain sterile because of inhospitable growth conditions is clearly untrue (14, 97). Presumably, carbon sources and other nutrients leach from the water container into the water and provide sufficient nutritional support to allow for organism growth. The rapidly growing mycobacteria, especially *M. abscessus* and *M. mucogenicum*, have been shown to do this experimentally, with isolates reaching concentrations of 10^6 cfu/ml which were maintained over a one-year period (14). These species have been associated with outbreaks and pseudo-outbreaks, with contaminated distilled water as the reservoir (75, 97).

Table 3 Virulence factors that enhance the risk of nosocomial disease caused by the nontuberculous mycobacteria

Growth in municipal water	Resistance to standard disinfectants
Growth in distilled water	Resistance to chlorine
Resistance to 2% formaldehyde,	Growth in biofilms
alkaline glutaradehyde	Growth at 45°C
Resistance to inorganic, organic	(*M. xenopi, M. avium* complex)
mercurials (merbromin, etc)	Resistance to gentian violet (1%)

RESISTANCE TO CHLORINE   Many NTM species are relatively chlorine resistant (14, 16), and tolerate the lower chlorine concentrations (0.05 to 0.2 μg/ml of free chlorine) found at the tap. They are 20 to 100 times more resistant to chlorine than are coliforms. This includes species such as *M. avium* complex, *M. fortuitum, M. abscessus, M. kansasii,* and *M. mucogenicum* (formerly *M. chelonei*-like organism or MCLO) (4, 14, 16).

RESISTANCE TO GLUTARALDEHYDE AND FORMALDEHYDE   A number of other features of the organisms allow for their survival in commercial products and hospital equipment. Formaldehyde, and more recently, alkaline glutaraldehyde have been used for equipment sterilization. Slow-growing NTM, including *M. avium, M. intracellulare,* and *M. gordonae,* appear to be relatively resistant to the compounds, surviving 2% concentrations for more than 10 min (23). Water-adapted strains of the rapidly growing mycobacterium *M. mucogenicum* have been shown to survive 2% concentrations of aqueous formaldehyde for 24 h (14, 42), and for up to 10% concentrations for relatively short periods. The genetic mechanism for this relative resistance is unknown, but a formaldehyde inactivating enzyme (formaldehyde dehydrogenase) has been identified in multiple species of *Enterobacteriaceae* and *Pseudomonas* (50). Resistance to these agents used as disinfectants has been associated with several disease outbreaks, including one involving reusable hemodialysis equipment (7).

RESISTANCE TO MERCURY   Organomercurial compounds such as phenylmercuric acetate, thimerosal, and merbromin have been used for many years as antiseptics and to maintain sterility of medical solutions. Mercury resistance in rapidly growing mycobacteria varies from species to species but is present in 83% of random clinical strains of *M. mucogenicum* and 20% of random clinical strains of *M. abscessus* (81a). Mercury resistance in these species appears to be broad spectrum and includes the cytoplasmic enzyme mercuric reductase that converts inorganic mercury to elemental mercury, and organomercurial lyase that disrupts the carbon-mercury linkage (81a). Such resistance allows these species to survive and thrive if introduced into multidosing vials or solutions that rely on one of the organomercurial compounds as a bacteriostatic preservative. Mercury resistance has been shown to be present in a number of slow-growing environmental species as well, including *M. scrofulaceum*, *M. intracellulare*, and *M. avium* (31). At least one multihospital disease outbreak has been reported that was related to the presence of *M. abscessus* in a commercial solution of merbromin used as a skin disinfectant in patients undergoing vein stripping (32).

RESISTANCE TO STANDARD DISINFECTANTS   NTM are known to be relatively resistant to many commercial routine disinfectants, more so than most bacterial species. This presumably relates to their lipid-rich cell wall, which is relatively impermeable to organic and inorganic compounds. Several NTM outbreaks have been related to the use of these routine commercial products as disinfectants instead of FDA-registered disinfectants (57), including at least one hemodialysis outbreak (56). One outbreak following the use of a jet injector for lidocaine occurred because the disinfectant (a quaternary ammonium) was only slowly (over hours) mycobactericidal (97).

RESISTANCE TO GENTIAN VIOLET   Gentian violet (mixtures of hexamethyl-, pentamethyl-, and tetramethyl-pararosaniline hydrochloride) has been in use for more than 100 years as a topical antiseptic. It is still used for this purpose, but more recently it has gained popularity as a skin-marking agent for surgical incisions (75). A recent study showed that some strains of *M. abscessus* are able to survive and grow in 1% solutions prepared in distilled water; one such contaminated solution was responsible for an outbreak of plastic surgery wound infections when it was used as a skin marking agent (15, 75).

LABORATORY ASPECTS OF NTM ASSOCIATED WITH NOSOCOMIAL OUTBREAKS

Microbiology

The NTM associated with nosocomial outbreaks or pseudo-outbreaks are generally easily recovered in the laboratory. Most species will grow well at 35°C in Middlebrook 7H10 or 7H11 agar, Lowenstein Jensen agar, and BACTEC 12B broth. Other newer culture systems such as the MIGIT system (Becton-Dickinson, Sparks, MD) and the ESP System (Difco, Detroit, MI) also seem to support NTM growth well. Exceptions exist, however, with optimal growth temperatures other than 35°C being the major parameter change needed for recovery of some organisms. Select species are relatively heat intolerant and can often be recovered on primary isolation at (or below) 30°C. This includes *M. chelonae*, *M. abscessus* (some strains), *M. marinum*, and *M. haemophilum*.

Other species are thermotolerant (*M. avium*, *M. smegmatis*) or, in the case of *M. xenopi*, grow optimally at 45°C and poorly at 35°C.

Organism identification currently utilizes several rapid systems: (*a*) commercial RNA-DNA probes (*M. avium* complex, *M. kansasii*, *M. gordonae*) (Accuprobe, GenProbe Inc.); (*b*) high-performance liquid chromatography (HPLC), which assesses patterns of extracted long-chain fatty acids (12, 48); and most recently, (*c*) PCR restriction fragment length polymorphism analysis (PRA) of a 439-bp fragment of the 65-kDa heat-shock protein gene (82, 86). Traditional biochemical methodologies are still employed in a number of laboratories, and provide satisfactory, albeit slower, identification. Separation of the rapidly growing mycobacteria into its multiple species (*M. abscessus*, *M. chelonae*, *M. fortuitum*, *M. peregrinum*, and sometimes *M. mucogenicum*) cannot be done by HPLC, which only groups these species into the *M. fortuitum* complex. Use of biochemicals (citrate, mannitol, inositol, and sorbitol; nitrate reduction; iron uptake; 3-day arylsulfatase activity) (77) and PRA (82, 86) are the only well-established methods for identification at the species level.

Susceptibility-testing of the rapidly growing mycobacteria has been useful in strain comparison for most outbreaks involving these organisms. This has usually involved a broth microdilution method in Mueller-Hinton broth (84).

METHODS FOR STRAIN COMPARISON   Early outbreaks used standard biochemical methods to determine the cause of the outbreak. For the rapidly growing mycobacteria, this meant the exact species by current taxonomic standards is unknown, because species identification prior to 1981 often reported no laboratory methodologies, did not separate *M. abscessus* from *M. chelonae* (both were generally included as *M. chelonei*), did not accurately separate *M. fortuitum* from *M. chelonae*, and did not recognize other species such as *M. mucogenicum*.

Recent times have seen better and more accurate NTM species identification. This has not been sufficient to investigate outbreaks, because the frequent presence of these species in the hospital environment meant that their recovery (especially from tap water) did not establish this as the outbreak source. Initially a variety of other phenotypic methodologies were used to characterize outbreak strains, most with rapidly growing mycobacteria. This included phage-typing, antibiogram patterns, heavy-metal resistance (including mercury), beta lactamase patterns by isoelectric focusing, and multilocus enzyme electrophoresis.

NTM outbreak investigation with accurate strain comparison did not come of age until the introduction of genetic strain comparisons—specifically pulsed-field gel electrophoresis (PFGE). This methodology has become the standard method for strain comparison with the NTM and has been used for investigation of outbreaks resulting from *M. chelonae*, *M. abscessus*, *M. xenopi*, *M. fortuitum*, *M. kansasii*, and *M. avium* complex (see Figure 1). It potentially could be used for any species of NTM that have not been investigated to date.

Some limitations for PFGE do exist. PFGE is technically difficult and labor intensive, and the electrophoresis equipment is expensive. Hence only a few laboratories currently perform this testing. Approximately 50% of isolates of *M. abscessus* produce broken or digested DNA with standard genomic DNA preparation and cannot be tested by this method (they produce only a smear on the electrophoresis gel) (95). Finally, some species show minimal genetic differences between unrelated isolates, so defining epidemic strain relatedness can be difficult (e.g. *M. xenopi* and *M. kansasii*).

PCR-based technologies for bacterial strain comparison are being used with increasing frequency for outbreak investigations. The use of arbitrarily primed (AP) or random-amplified polymorphic DNA (RAPD) PCR has been described with *M. malmoense* (49) and *M. abscessus* (101), including outbreak strains that could not be compared by PFGE (101). This technology is much easier to perform but is much more method-dependent and requires carefully controlled conditions with a need for multiple primers, and the strains for comparison generally run on the same gel (101). An example of this technology with isolates of *M. abscessus* is shown in Figure 2.

Another PCR-based method involves amplification of a 439-bp sequence of the 65-kDa heat shock protein gene followed by restriction enzyme digestion (82, 86). This results in primarily species-specific patterns but sometimes subspecies-specific patterns. Picardeau et al were able to divide clinical and environmental isolates of *M. kansasii* into five groups on the basis of this technique, but ultimate individual strain comparison still required PFGE (70).

The use of repetitive insertional sequences such as IS*6110* with *M. tuberculosis* complex that are randomly inserted in the chromosome has also been studied. Although repetitive or insertional elements have been described with several NTM species, most studies have involved *M. avium*, where four repetitive elements—IS*901*, IS*110*, IS*1245*, and IS*1311*—have been described (41, 71). Southern hybridization and PCR, which amplify the space between insertional elements, have both been used to type strains of this species (41, 71).

NOSOCOMIAL DISEASE OUTBREAKS

Nosocomial disease outbreaks traceable to the NTM have been recognized since 1975. Most involve wound infections following cardiac surgery or plastic surgical procedures, and postinjection abscesses. Sporadic disease of an identical nature is also recognized. Although the surgical wound infection outbreaks are relatively rare in the 1990s, they still occur and are a reminder of the remarkable tolerance of the NTM to survive the seeming inhospitable environment surrounding modern-day surgery.

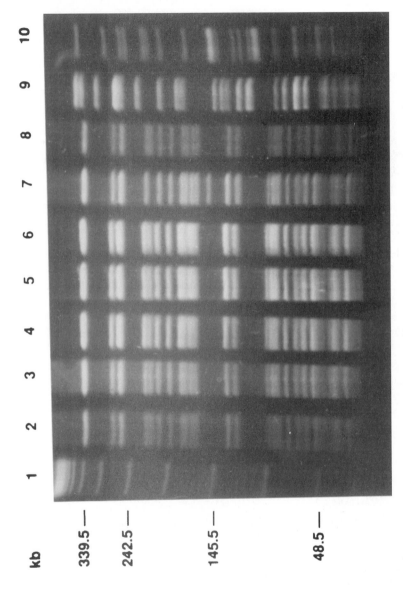

Figure 1 AsnI (LRF) patterns of genomic DNA from isolates of *M. abscessus* in an outbreak related to contaminated bronchoscopes. *Lanes 2 to 8*, epidemic isolates; *lane 9*, random isolate of *M. abscessus*; *lane 10*, *M. abscessus* ATCC 35749; *lane 1*, lambda DNA concatemer standard. Note the two band differences in the epidemic strain in *lane 7*. [Reprinted with permission from *J. Clin. Microbiol.* (95).]

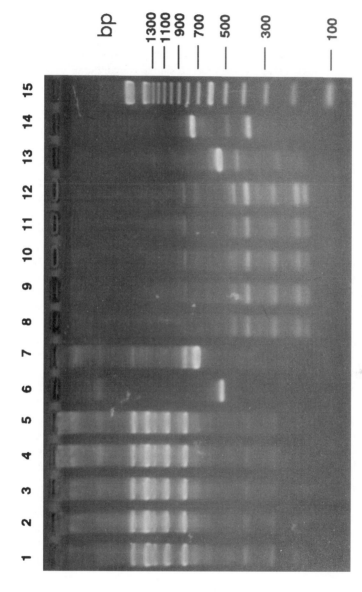

Figure 2 RAPD-PCR profile of five epidemic isolates of *M. abscessus*, obtained by two methods of DNA extraction and primer OPA-18. *Lanes 1–7* contain the band patterns from the five epidemic strains (*lanes 1–5*) and two unrelated control strains (*lanes 6 and 7*), with DNA extracted with phenol and chloroform, while *lanes 8–14* contain the band patterns from the same five epidemic strains (*lanes 8–12*) and unrelated controls (*lanes 13 and 14*), with DNA extracted by boiling for 30 min; *lane 15*, commercial linear DNA standards. [Reprinted with permission from *Clin. Infect. Dis.* (91).]

Cardiac Surgery

Between 1975 and 1981, five cardiac surgery–related outbreaks occurred in the United States, and one occurred in Hungary. Five of these were surgical wound outbreaks (usually involving the sternum), and one involved contaminated porcine prosthetic valves. In only one of these six outbreaks were environmental isolates identical to the outbreak strain recovered, but recovery of these environmental isolates and subsequent molecular strain comparison in one surgical wound outbreak in Texas turned out to be the key to understanding these outbreaks.

The one outbreak associated with environmental isolates involved six patients who underwent surgery in a single hospital in Corpus Christi, Texas, in 1981; the outbreak was extensively investigated by the CDC. It involved both *M. fortuitum* and *M. abscessus*. The outbreak strain of *M. fortuitum* and one of two outbreak strains of *M. abscessus* were recovered from nonsterile ice, made from tap water, used to cool the cardioplegia solution in the cardiovascular operating room. [It was postulated that contamination of the operative field and/or operative personnel occurred when the bag of cardioplegia solution was brought to the operating table and given to the anesthesiologist (51).] The outbreak strain of *M. fortuitum* was also recovered from the cold water tap in the operating room, from an ice machine in the hospital, from an additional patient operated on for a noncardiac procedure (laparotomy), and, finally, from municipal water coming into the hospital (51). Identity of the *M. fortuitum* isolates was proven by plasmid profiles (94a), PFGE (43), and multilocus enzyme electrophoresis (MEE) (94a). One of two outbreak strains of *M. abscessus* was recovered from a pair of scissors taped to a stand in the anesthesia area and one was recovered from nonsterile cooling solution (51). The identity of the two environmental strains of *M. abscessus* to patient isolates was ultimately demonstrated by RAPD-PCR (101). Following the CDC investigation, with recovery of *M. abscessus* and *M. fortuitum* from the tap water ice, sterile water/ice was used to cool the cardioplegia solution and no further cases occurred (51). Similar changes (i.e. discontinued use of tap water as a coolant) in other operating rooms may explain the ending of these outbreaks in the United States in the early 1980s.

Only one cardiovascular outbreak has been reported in the more than 15 years since the recognition of these first six outbreaks, and it occurred between 1987 and 1989 in Hong Kong (100). Molecular analysis has been performed on isolates from all six sternal wound outbreaks by a variety of methods, including plasmid profiles, PFGE, and ribotyping (43, 94a, 95, 100, 101). These studies, combined with phenotypic evaluations, demonstrated that multiple strains and/or species were involved in three out of six of the total outbreaks, which would support the likelihood of environmental spread/contamination rather than contamination of some medical product or patient-to-patient spread of the disease.

CARDIAC OUTBREAK DETAILS   Details of the seven cardiac surgery–related outbreaks are as follows. The first cardiac surgery wound infection outbreak occurred over a ten-week period in 1975 in North Carolina (44, 73). Nineteen of 80 persons (24%) operated on over this period developed sternal wound infections from *M. abscessus*, of whom five (26%) died of their disease. A second *M. abscessus* outbreak occurred in 1977 in Budapest, Hungary. Six patients were infected over a one-week period, with three deaths (50% mortality) (85). Plasmid analysis subsequently performed on isolates from North Carolina and Hungary showed the presence of a single pattern for each outbreak (94a).

Contaminated porcine valves used as prosthetic valves (made by Hancock Lab, Inc.) were recognized in a nationwide outbreak in the United States in 1976. Small tissue tags provided as a culture control grew an organism ultimately identified as *M. chelonae*. Twenty-five valves from 24 patients and ten medical centers were involved. Only two patients (two out of 24, or 8%) were known to have developed clinical disease because of the contamination (18). Relative resistance to the 2% formaldehyde used for valve sterilization was ultimately shown to be the cause of the outbreak (18, 52).

The first outbreak to involve *M. fortuitum* occurred in 1976 in a hospital in Colorado. Four of nine persons who underwent cardiovascular surgery over a two-week period became infected, with no deaths. Another single case caused by *M. fortuitum* occurred six months earlier, but the isolate was phenotypically and genetically unrelated to the outbreak strain (43, 94a). A settle-plate isolate of *M. fortuitum* was obtained during epidemiologic investigation, and it was similar to the outbreak strain morphologically and by multilocus enzyme electrophoresis (MEE) (94a). However, it was subsequently shown to be unique (different) when studied by PFGE (43). (The four outbreak strains were identical morphologically, by MEE, and by PFGE.)

The second outbreak triggered by *M. fortuitum* occurred in 1981 in Nebraska and involved six patients. Two strains of *M. fortuitum* were involved, as demonstrated by MEE (94a), plasmid profiles (94a), and PFGE (43). Details of this outbreak have never been published (it was reported only in abstract form) (72).

The only outbreak to involve both *M. fortuitum* and *M. abscessus* was the above-discussed outbreak in Corpus Christi, Texas, in 1981 (51). Six of 51 patients (11.3%) operated on over a six-month period became infected, with sternal wound infections (four cases), prosthetic valve endocarditis (one case), and a saphenous vein graft-site infection (one case) (51). Two of the six patients (33%) died as a result of their infections.

The last and most recent cardiac surgery outbreak occurred in Hong Kong between 1987 and 1989. Twenty-one cases of wound infections were identified. The majority of cases were due to *M. peregrinum* (12 of 21, or 57%), but seven cases were due to *M. fortuitum* (seven of 21, or 33%), and two cases to *M. fortuitum* third-biovariant complex (two of 21, or 10%) (100). Further grouping was evident with susceptibility testing and ribotyping, which suggested that as many as seven different strains were involved.

Postinjection Abscesses

After cardiac surgery outbreaks and contaminated bronchoscopes, postinjection abscesses are perhaps the best known of the nosocomial diseases caused by the NTM. The first postinjection abscess outbreak occurred in 1961 and 1962 (64, 89), and the last as recently as 1997 (19). A total of 11 such outbreaks have been reported over this more than 30-year period. By organism, these outbreaks can be divided into two periods. Six outbreaks occurred before 1980 (8, 38, 46, 64, 69, 89). Because of limited taxonomic methods, the exact species involved in these outbreaks is uncertain, although some isolates were identified to the species level. The taxonomy of the causative agent was well established in the five most recent outbreaks (all since 1985) (13, 19, 36, 62, 97), the analysis of which benefited from modern taxonomic methods: three were *M. abscessus*, one was *M. chelonae*, and one was *M. fortuitum*.

Two clinical settings were seen for these 11 outbreaks. One was the use of multidose vials (8, 38, 64, 69) or contaminated biologicals (13, 19) that were used for injections (nine outbreaks), with contamination of the fluids the suspected source of the outbreak. Injected materials included vaccines (8, 64, 46), lidocaine (13), sterile saline water (38), allergens (36, 36), and adrenal cortical extracts (19). Inadequate skin disinfectation (36) was considered a contributing factor in one of these outbreaks. In two of the more recent outbreaks (13, 19), the same organism has been recovered from the vials used for injections. The second setting involved inadequate sterilization of equipment used for injections or noninjecting needles (two outbreaks), with contaminated water used for cleaning or rinsing the equipment the likely source of the outbreak (62, 97). In one of these outbreaks (97), the causative organism was recovered from distilled water used for disinfection.

VACCINE RELATED OUTBREAKS   The first group of four outbreaks all related to clinic injections of vaccines. A series of 50 among 1200 patients reported from Texas received injections, most occurring in a six-month period in 1961. Thirty-six patients received influenzae vaccine (36 of 50), and reusable syringes were used for some patients (64).

Borghaus & Stanford reported an outbreak of 50 postinjection abscesses that occurred over a six-month period in 1969 at a health clinic in the Netherlands. Multiple dose containers, reusable needles, and use of a common needle for vaccine extraction were all potential causes. The organism belonged to the *M. chelonae/abscessus* group (8).

Inman et al reported 12 patients with local abscesses that occurred over a 14-month period in 1996–1997 in an ENT clinic in England. No source was identified for the outbreak, but the multidose vials of histamine were considered a probable source (46). The organism appeared to belong to the *M. chelonae/abscessus* group.

The fourth vaccine-related outbreak occurred in an allergy clinic in Georgia in 1990. The causative organism was *M. chelonae*. No source of the outbreak was found, but the use of a quatenary ammonium with tap water and soaked cotton balls (used as skin disinfectant) was the possible source (cotton fibers are likely to result in inactivation of the quaternary) (36).

CONTAMINATED BIOLOGICS FOR ALTERNATIVE MEDICINE   Two of the outbreaks related to administration of biologics as part of alternative medicine practices. In both, the infecting organism was recovered from the biologic fluid as well as from patients. In one of these outbreaks (91), identity of patient isolates was demonstrated by RAPD-PCR.

The largest outbreak of postinjection abscesses to date was reported from Barranquilla, Colombia, in 1996 and 1997 (13, 91). Over a five-month period, 350 of approximately 2000 lidocaine-injected patients in a single bioenergetic (practitioner of alternative medicine) office developed localized cutaneous abscesses or cellulitis. Of 210 abscess samples cultured, 205 grew *M. abscessus*. Five isolates available for testing were identical in drug and heavy-metal resistance patterns and DNA RAPD-PCR profiles (91). Macroscopic examination of the lidocaine vials revealed deteriorated rubber, with several punctures and different liquid volumes, suggesting refilling, repeated use, and sterilization (13). *M. abscessus* was isolated from one of the lidocaine vials that had the same biochemical characteristics as the patient isolates (13).

In a similar outbreak from Denver, Colorado, in 1996, 47 out of 69 patients who were injected with adrenal cortex extract (ACE), (a product not approved by the Food and Drug Administration) as part of a weight-loss regimen developed abscesses at the site of injection. Five additional patients who were treated in Wyoming also developed abscesses following injection of the preparation, and two other Denver area patients treated the previous year were reported. Specimens from 11 Colorado patients grew rapidly, growing mycobacteria "most consistent with *M. abscessus*." Four of eleven patient isolates were confirmed as *M. abscessus*, and additionally *M. abscessus* was isolated from one unopened and three opened vials of ACE. This outbreak is still under investigation by the CDC (19).

INADEQUATE STERILIZATION NEEDLES/INJECTION MATERIALS   Two additional outbreaks of postinjection abscesses involved not contaminated biologics but inadequate sterilization of the equipment used for needle insertion.

An outbreak of infections caused by *M. fortuitum* was described in six patients who received electromyography (EMG) at a facility in Washington (62). *M. fortuitum* was isolated from five patients. Standard procedure for EMG included the use of reusable needle electrodes disinfected with 2% glutaraldehyde and subsequently rinsed in tap water. Tap water samples and glutaraldehyde stock solution samples were cultured and found to be sterile. However, further investigation revealed that the procedures used for processing the disinfectant and tap water specimens for culture may not have been optimal. Institution of autoclaving in place of manual disinfection eliminated further cases.

Similarly, an outbreak from Florida in 1988 occurred when eight persons who had undergone invasive podiatric procedures were diagnosed with *M. abscessus* foot infections (97). Upon investigation, it was discovered that the jet injector used to administer lidocaine was held between procedures in a mixture of distilled water and disinfectant, as recommended by the manufacturer. Inoculation of patients with the mycobacteria by the jet injector may have occurred because of slow killing of the *M. abscessus* by the disinfectant. Additionally, isolates of *M. abscessus* with the same antimicrobial susceptibility pattern as the patients' strains were cultured from distilled water in a reusable, nonsterilized container.

MISCELLANEOUS OUTBREAKS   One outbreak of subcutaneous abscesses in Texas involved nursing students who repeatedly gave themselves practice injections from a single bag of normal saline. The bag was noted to have become cloudy, but students continued to use it. The abscesses grew *M. abscessus*. The container of normal saline was not cultured (38).

The first reported outbreak of postinjection abscesses occurred between 1960 and 1965 and involved 100 patients in Kinshasa, Congo. The organism from the abscesses was identified as *M. fortuitum* by the CDC, but no details of the injections were provided (89).

Plastic Surgery Wound Outbreaks

Sporadic wound infections caused by the NTM following plastic surgery are well known. Most have followed augmentation mammaplasty and involved *M. fortuitum* (21, 94). Four NTM disease outbreaks have been described that involved plastic surgery procedures. This includes augmentation mammaplasty, face lifting with blepharoplasty, nasal reconstruction, liposuction, and vein stripping. All involved rapidly growing mycobacteria. No unique mechanism for these outbreaks was seen, but judgment errors like those seen with other surgical wound infections occurred with these outbreaks—i.e. patients exposed to tap water or distilled water assumed to be sterile that in fact contained mycobacteria. Two outbreaks occurred in the United States and two occurred outside the United States. None of the outbreaks to date have been studied by molecular techniques.

The first reported NTM plastic surgery–related outbreak occurred in 1974–1975 in Barcelona, Spain, and involved contaminated commercial merbromin (an organomercurial) used for presurgical antisepsis of varicose veins that were to be excised. The outbreaks occurred in two hospitals and involved *M. abscessus* (32). Additional isolates were recovered from additional hospitals using merbromin, but without recorded outbreaks, and ultimately from commercial bottles of merbromin themselves. The outbreaks stopped when use of the merbromin was discontinued.

The first plastic surgery–related outbreak in the United States occurred in Florida in 1985. It involved eight women undergoing augmentation mammaplasty or face-lift blepharoplasty, and the infections were related to a 1% aqueous solution of gentian violet contaminated with *M. abscessus* used for skin marking (75). The causative strain was subsequently shown to grow in 1% solutions of gentian violet prepared in distilled water (15).

The third outbreak involved nasal reconstructive surgery and occurred in Mexico in 1987–1988. The organism was identified as *M. abscessus*, but its susceptibility pattern (very drug susceptible) was highly suggestive of *M. mucogenicum* (79). The causative organism was recovered from hospital water systems and other environmental sources, but molecular strain comparison was not performed.

The last outbreak involved patients undergoing liposuction in California in 1997. The outbreak was traced to a rapidly growing organism, with *M. chelonae* isolated from both patients and tap water, but details of the outbreak have not been reported (DF Moore, H Meyers, J Curry, personal communication).

Dialysis Related Outbreaks

Sporadic cases of peritonitis that resulted from NTM in patients undergoing peritoneal dialysis is well recognized. Of the three reported NTM outbreaks involving dialysis, one involved automated peritoneal dialysis (4); the other two involved hemodialysis (7, 56). In all three outbreaks, the causative organism was recovered from hospital tap water, and the combination of inadequate sterilization of reusable dialysis equipment and exposure to contaminated tap water was the major disease mechanism. All three outbreaks involved rapidly growing mycobacteria (one *M. mucogenicum*, two *M. abscessus*). Molecular analysis was performed in one of these outbreaks involving hemodialysis (7), demonstrating the identity of environmental water strains and disease strains of *M. abscessus* (101). These outbreaks demonstrated the frequent presence of NTM in community tap water, and the relative resistance of these species to 2% formaldehyde and alkaline glutaraldehyde and even newer antiseptics such as renalin (56). Current dialysis procedures employ more effective sterilization, and no dialysis-related outbreak has occurred in the past ten years.

The first reported NTM dialysis outbreak involved *M. mucogenicum* (designated initially as an *M. chelonae*-like organism or MCLO) and automated peritoneal dialysis; cases occurred in ten patients who were receiving intermittent chronic peritoneal dialysis at one of two centers (4). It is the only known nosocomial outbreak traceable to this species.

The first hemodialysis NTM outbreak occurred in 1982 in a dialysis center in Louisiana (7). Twenty-seven (19%) of 140 patients being treated at the center became infected. Causative organisms were *M. abscessus* (26 cases) and *M. mucogenicum* (one case). Widespread contamination with *M. abscessus* and *M. mucogenicum* was evident in the center, including processed water used for rinsing dialyzers (mean number of cfu in the water was 9.0×10^3 per ml) and from the patients' side of 5 of 31 hemodialyzers that had been processed and were ready for use. Of the 27 infected patients, 14 (51%) died of their disease. Clinical and environmental strains were subsequently shown to be identical by PFGE (95, 101) and RAPD-PCR (101).

The second NTM hemodialysis outbreak involving high-flux hemodialysis occurred in California in 1987–1988. The dialysis center had adopted a commercial antiseptic called renalin (hydrogen peroxide/peracetic acid-based disinfectant) in 1986 to replace 4% formaldehyde for reprocessing dialyzers, and this proved to be inadequate at the concentration used (2.5%) to sterilize the hemodialyzers (56). Infection was again caused by *M. abscessus*. Five patients were infected: four of the five had infected arteriovenous grafts, and two patients (40%) died. *M. abscessus* was recovered from the patients as well as from the municipal water and the hose of the water spray device used for reprocessing the high-flux but not from the regular dialyzers (use of high-flux dialysis was seen with 100% of disease cases but only 30% of controls). Despite increasing use of renalin in place of formaldehyde or glutaraldehyde, no further hemodialysis outbreaks have been reported (34, 56).

Disseminated M. avium Disease in AIDS

Von Reyn et al (92) published a provocative study of 36 AIDS patients from Boston and New Hampshire with disseminated *M. avium*. Isolates were subjected to PFGE. Three clusters of isolates were seen. Three AIDS patients treated at hospital A had one strain, with the same strain recovered from the hot-water system of the hospital. A second group of three AIDS patients from hospital B had another strain, again with the same strain recovered from the hot-water system of hospital B. The study clearly suggests that outbreaks of disseminated *M. avium* disease can occur with hospital hot water supplies as their source. Because of the infrequent nature of NTM environmental studies and the complexity of PFGE, failure to recognize similar outbreaks may reflect the fact that PFGE studies of environmental and clinical isolates of this nature have not been done.

Miscellaneous Outbreaks

Disease outbreaks have occurred following a wide variety of surgical procedures, but clearly with less frequency than those following cardiac surgery, plastic surgery, and local intramuscular injections. Implicated procedures include laparoscopy, surgical irrigation of chronic otitis media, and spinal disk removals (diskectomy).

An outbreak of wound infections following laparoscopy occurred in one hospital in Mississippi in 1985. Sixteen cases were identified, with the causative organism (*M. abscessus*) recovered from four of them. Following an extensive CDC investigation, *M. abscessus* with the same antimicrobial susceptibility pattern was recovered from mineral oil used to lubricate the laparoscopy and from multiple specimens of tap water within the hospital. The outbreak stopped when a sterile aqueous-based solution was used in place of the mineral oil, and when laparoscopes underwent cleaning without exposure to tap water. Unfortunately, molecular analysis of these strains was never performed (34).

One outbreak of chronic otitis media was described in the office of an ear, nose, and throat surgeon in Louisiana (56). Seventeen cases of otitis media caused by *M. abscessus* were detected among patients in 1987. All the patients had a tympanostomy tube or tubes in place or had one or more tympanic-membrane perforations. Thirteen of the 14 ear isolates of *M. abscessus* had the same unusual high-level resistance to aminoglycosides. None of the isolates had detectable plasmids. The original source of the epidemic isolate was thought to be tap water. Case control studies demonstrated an increased risk of infection associated with suctioning and multiple ear examinations. Contact with contaminated tap water may have occurred during suctioning, and the chance of exposure to contaminated instruments increased with repeated ear examinations. The fact that most patients were infected by the strains with this identical resistance pattern suggested patient-to-patient transfer via contaminated instruments rather than mere exposure to contaminate municipal tap water. The source of the outbreak was unknown, but a patient with a known *M. abscessus* infection who had been examined prior to the onset of the outbreak was suspected. In a recent study by Zhang et al (101), selected isolates of tap water, suction equipment, and patients were studied. By RAPD-PCR of ten selected environmental and clinical isolates, a single strain of *M. abscessus* was identified among the outbreak isolates, but the original source of the outbreak remains unknown because the isolate from the suspected patient had a different pattern. Because of the unusually high resistance to aminoglycosides of the epidemic strain, the belief remains that a previously infected patient was the probable source of this outbreak. The ultimate source of infection for the source patient, however, was probably from contaminated municipal tap water (101).

Between 1988 and 1992, *M. xenopi* was reported in 16 of 4306 patients who had undergone diskectomies in the same surgical center in France. Epidemiological investigation implicated the water supply. Hot water tanks and tap filters delivering a "so-called sterile water" used to rinse the material for diskectomy after glutaraldehyde disinfection yielded large numbers of *M. xenopi*. Pulsed field gel electrophoresis patterns and hybridization patterns with the repetitive sequence IS*1081* on 12 clinical and 18 environmental isolates were identical.

Summary

In summary, nosocomial disease outbreaks became a major concern beginning in 1975, and they remain a clinical issue 20 years later. The recognition of the risk of contaminated tap water and distilled water has reduced the incidence of cardiac surgery–related disease, but the frequent need and use of water with various surgical procedures make subsequent disease outbreaks almost a certainty, despite what has been learned about environmental sources of these organisms and their resistance to sterilization and disinfection procedures. Molecular techniques have greatly improved our ability to investigate these outbreaks.

NOSOCOMIAL PSEUDO-OUTBREAKS

Nosocomial pseudo-outbreaks or pseudo-epidemics can be defined as clustering of false infections or artifactual clustering of real infections (96). With false infections, patients by definition are not infected and hence do not exhibit evidence of disease. However, because of administration of unnecessary therapeutic measures, they may be at risk for complications related to these procedures.

Bronchoscopy Related Pseudo-Outbreaks

Pseudoinfection related to bronchoscopy (see Table 4) should be suspected with the repeated isolation of unusual rarely pathogenic mycobacterial species (i.e. any NTM). Pseudoinfection can occur at any stage of the procedure with sources of contamination that include defective bronchoscopes, inadequate disinfection of the bronchoscopes, solutions of local anesthetics, tap water, disinfectant/cleaner, and antimicrobial solution used in culture medium. The source of organisms can be essentially any fluid reservoir that is not routinely and effectively decontaminated and sterilized. Several mycobacterial species can be involved; however, *M. abscessus* is the most common. It is important to maintain surveillance of the culture results from the bronchoscopy suite as almost all NTM pseudo-outbreaks were identified by an unusual clustering of NTM isolates.

CONTAMINATED TOPICAL ANESTHESIA   Steere et al (81) reported *M. gordonae* isolates recovered from 52 patients (out of 699) obtained at bronchoscopy over a 2.5-year period. The source of the positive cultures was traced to a contaminated bottle of green dye (a food dye derived from coal tars, containing tartrazine, brilliant blue, citric acid, and sodium benzoate) that was added to the preoperative topical anesthetic usually by a single bronchoscopist. Cultures of the dye yielded *M. gordonae*.

CONTAMINATED AND/OR MALFUNCTIONING BRONCHOSCOPES   In four of the following studies, pseudoinfection occurred because an instrument component was not adequately sterilized and in one because instrument damage created a space for contaminated fluid to pool. In another report, contaminated tap water used for specimen collection was responsible for the pseudoinfection. It is noteworthy that in the two examples where the bronchoscope itself was contaminated, the instruments became essentially permanently contaminated.

Pappas et al (65) reported 72 patients in an 11-month period of time with positive respiratory cultures for "*M. chelonei*" (presumably *M. abscessus*). Eighty-seven percent of the isolates were from bronchoscopic specimens and 13% from postbronchoscopy sputum specimens. Only two patients were felt to have significant mycobacterial disease. Bronchoscopes had been cleaned with glutaraldehyde, but with the *M. chelonei* outbreak the bronchoscopes were sterilized with ethylene oxide. Despite this change, specimens continued to be positive for *M. chelonei*. Two bronchoscopes subsequently submitted for repair were found to have punctured suction channels. A portion of the interior of each bronchoscope, usually sealed, was grossly contaminated with "slimy, proteinaceous material" that was culture positive for *M. chelonei*.

Wheeler et al (97a) identified three separate episodes of mycobacterial contamination of specimens following bronchoscopy, two involving *M. avium* complex. The contamination occurred despite the adherence to an accepted cleaning and disinfection protocol. The most likely source of contamination was the bronchoscope suction valve, a component that cannot be disinfected by commonly used commercial products. In related experiments, bronchoscopes

Table 4 Bronchoscope related pseudo-outbreaks caused by nontuberculous mycobacteria (NTM)[a]

Reference	Location	Year	Organism	Source	Incidence (%)	Environmental source (+)
Rapidly growing mycobacteria						
58	Kentucky	1992	*M. abscessus*	Contaminated automated washer	17/1009 (2)	+
63	England	1989	*M. abscessus*	Tap water rinse (manual disinfection)	8/58 (15)	+
40	Switzerland	1989	*M. abscessus*	Contaminated automated washer	16/46 (35)	+
65	Illinois	1981	*M. abscessus*	Two bronchoscopes with broken suction channels (manual disinfection)	72/195 (37)	−
33	Missouri	1989–90	*M. abscessus*	Contaminated automated washer	14/1270 (1)	+
Unpublished	Maryland	1990–91	*M. abscessus*	Manual/automated	12/unknown	−
Unpublished	Pennsylvania	1993	*M. abscessus*	Automated washer	30/unknown	−
68a	Michigan	1993	*M. abscessus*	Contaminated automated washer	19 (7.6)	+
Unpublished	Florida	1993	*M. fortuitum* third biovar	Automated washer	4/unknown	−
Slowly growing mycobacteria						
97a	Tennessee	1986	MAC (*M. avium*)	Contaminated suction valve with inadequate decontamination (automated disinfection)	2/unknown	−

26a	Queensland, Australia	1980	MAC (M. intracellulare)	Contaminated plastic tubing (inadequate disinfection of bronchoscopes)	3/unknown	+
68	North Carolina	1990–93	MAC	Contaminated tap water used for terminal rinsing (automated disinfection)	3/unknown	+
9	Pennsylvania	1996	MAC	Contaminated tap water used for rinsing (unknown disinfection method)	16/80 (20)	+
83	Illinois	1983–84	M. gordonae	Contaminated tap water used for rinsing (manual disinfection)	22/70 (31)	+
81	Connecticut	1975–77	M. gordonae	Contaminated green food dye added to the topical anesthesia at bronchoscopy	52/699 (7)	+
40	Switzerland	1989	M. gordonae	Contaminated deionized tap water used for rinsing (automated disinfection)	3/unknown	+
5a	Michigan	1988–91	M. xenopi	Contaminated tap water used for rinsing (manual disinfection)	21/60 (35)	+

[a]MAC, Mycobacterium avium complex. M. fortuitum third biovar, Mycobacterium fortuitum third biovariant complex.

and related equipment were exposed to a saline suspension of *M. fortuitum*. Bronchoscopes were readily sterilized by routine cleaning and disinfection procedures, but the spring-operated suction valves remained contaminated, even after a 30-min exposure to 2% glutaraldehyde or after passage through a commercial bronchoscope washer. Suction valves that have been heavily contaminated with mycobacterial organisms cannot be reliably disinfected with glutaraldehyde.

Dawson et al (26a) reported positive cultures for the same serovar of *M. intracellulare* from three bronchoscopy specimens within a one-week period. One patient was likely infected by the organism, but the other two patients did not have any clinical evidence of invasive mycobacterial disease. The source of the contamination proved to be the plastic tubing through which the washings were aspirated into the sterile collection tube. Although the bronchoscope was sterilized with glutaraldehyde, the collection tubing was not. These investigators also demonstrated that 2% glutaraldehyde could be a rapid and highly effective disinfectant against mycobacteria if the organisms were adequately exposed to the glutaraldehyde.

Stine et al (83) found 22 patients (out of a total of 70) with positive AFB smears over an eight-month period of time who did not have a clinical picture consistent with mycobacterial disease. *Mycobacterium gordonae* was cultured from six patient specimens and from water used by a pathology laboratory and the bronchoscopy suite on two floors of the same hospital building. Clinical contamination apparently occurred while obtaining the bronchoscopic specimens and during specimen processing in the pathology lab. In addition to *M. gordonae*, *M. avium* complex and *M. scrofulaceum* were identified in the contaminated water. The pseudo-outbreak was controlled by inserting a polymer filter in the outflow tubing of the pathology tap water–deionizing unit and the tap water source in the bronchoscopy suite.

CONTAMINATED TERMINAL RINSE WATER (TAP WATER)   These four scenarios illustrate that pseudoinfection can occur if the instrument is rinsed with contaminated tap water, even if effective sterilization steps preceded the rinse procedure. One involved *M. abscessus*, one *M. xenopi*, and two *M. avium* complex. Contamination by this route is perhaps predictable, given the potential for tap water contamination at other stages of the procedure discussed above. One study (involving *M. xenopi*) used PFGE on a limited basis to support the environmental source of the pseudo-outbreak.

Nye et al (63) recovered "*M. chelonei*" (presumably *M. abscessus*) from eight bronchoalveolar lavage (BAL) fluid specimens from seven patients (out of 58 BAL specimens) over a six-month period. The bronchoscopes had been manually disinfected and then rinsed in tap water. "*M. chelonei*" was isolated from rinsing fluid after disinfection and from the hospital water supply. The pseudo-outbreak was interrupted when the bronchoscopes were disinfected with ethylene oxide and use of tap water in rinsing was abandoned.

Bennett et al investigated an apparent outbreak of *M. xenopi*, an NTM rarely isolated in the United States, in a northern Michigan hospital (5a). Over a 37-month period, 21 of 60 (35%) mycobacterial isolates from respiratory specimens were identified as *M. xenopi*. No other hospitals in the region had isolated *M. xenopi* in the same time-frame. Three patients met criteria for *M. xenopi* disease, accounting for four *M. xenopi* isolates. The 17 remaining *M. xenopi* isolates included 13 bronchoscopy specimens. Almost all water sources in the hospital that were sampled yielded at least one mycobacterial species (including *M. avium* complex, *M. gordonae*, and/or *M. xenopi*). *M. xenopi* was cultured from tap water in the outpatient building where the endoscopy unit was located. Because *M. xenopi* could be cultured from bronchoscopes after disinfection, it appeared that *M. xenopi* in the tap water contaminated the bronchoscope during cleaning. Bronchoscopes had been manually disinfected in a 0.13% glutaraldehyde-phenate and tap-water bath, and then were rinsed in tap water. Water from the hot water tank supplying this area yielded *M. xenopi*; however, cultures of the disinfectant liquid did not yield *M. xenopi*. Limited investigation of the pseudo-outbreak by PFGE analysis confirmed an environmental source for at least four *M. xenopi* isolates associated with pseudoinfection.

Two additional brief reports involve MAC. Bourbeau et al (9) recognized the new isolation of 16 mucoid MAC isolates from the bronchoscopy suite, while no mucoid MAC was isolated from elsewhere in the hospital. Subsequent cultures of the tap water in the bronchoscopy suite and from bronchoscopes, after the standard cleaning process, also yielded mucoid MAC. Following ethylene oxide sterilization of the bronchoscopes, no further mucoid MAC isolates were recovered. Pentony et al (68) noted an elevenfold increase in MAC isolates from bronchoscopic specimens in a one-year span. Filters had been placed in the tap water system and cultures from pre- and post-filter water as well as from the filter itself were culture positive for MAC. Changing the final rinse to sterile water decreased the number of MAC culture positive bronchoscopy specimens.

CONTAMINATED AUTOMATED ENDOSCOPE WASHERS Contamination by this route is perhaps the most surprising. As with the contaminated bronchoscopes, pseudoinfection on this basis can be refractory to sterilization efforts and may require replacement of cleaning equipment. These pseudo-outbreaks involved the Olympus EW 10 and EW 20 and the Keymed auto disinfector II (34). Because of these outbreaks, the United States Food and Drug Administration issued a class II recall in 1990 that prohibited further sale of the Olympus EW 10 and EW 20 (34). Modification of these machines was also required to try to eliminate the contamination problem (34). Of the four reported pseudo-outbreaks related to the machines, at least one has occurred since these modifications were made (58). All four outbreaks involved *M. abscessus*, and two of the four used PFGE not only to confirm the pseudo-outbreak but also to identify (or confirm) the source of the pseudoinfection.

In a brief report in 1993, Petersen et al (68a) noted that 18 cultures from bronchoscopies were positive for *M. abscessus* over a four- to five-month period. No cultures from any other part of the hospital were positive for *M. abscessus* during that time. *Mycobacterium abscessus* was isolated from sites from all four endoscope cleaning machines tested. After instituting manual cleaning of the bronchoscopes, no further *M. abscessus* isolates were identified.

Gubler et al (40) described a pseudoepidemic of NTM triggered by contamination of the water tank of a machine used to clean and disinfect bronchoscopes. Over a six-month period, 16 out of 46 bronchoscopic specimens were AFB-smear and/or culture positive. Two specimens, including one positive for *M. tuberculosis*, were felt to represent true positives, and in four patients, only the AFB smears were positive. In seven patients, the cultures were positive for "*M. chelonei*" (presumably *M. abscessus*) and three patients had specimens positive for *M. gordonae*. None of these ten patients had clinical evidence of mycobacterial disease. The source of two out of three of the *M. gordonae* isolates was found to be a contaminated antimicrobial solution used in the BACTEC mycobacterial culture system. Additionally, both *M. chelonae* and *M. gordonae* were cultured from tap water passed through the suction channel of the bronchoscope, and *M. gordonae* was cultured from the storage tank of rinsing water within the machine used for cleaning and disinfecting the bronchoscopes.

Fraser et al (33) reported the isolation of *M. abscessus* from bronchoscopic washings from 13 patients on a single clinical service over an approximately ten-month period. None of these patients had evidence of invasive *M. abscessus* disease. These 13 isolates had the same antimicrobial susceptibility pattern, with an unusually high degree of resistance to cefoxitin (MIC > 256 μg/ml). None of the 10 control isolates of *M. abscessus* obtained from elsewhere in the hospital matched the drug susceptibility pattern of the outbreak strain. However, an *M. abscessus* strain with the same pattern of high-degree resistance to cefoxitin was isolated from the rinse water collected from the bronchoscope disinfecting machine. Subsequent analysis demonstrated that the patient isolates had identical PFGE patterns and were also identical to an isolate recovered from the automated bronchoscope disinfection machine (95). Aggressive infection control measures on the disinfecting machine, including use of sterile water in the wash and rinse cycles, increasing the 2% alkaline glutaraldehyde exposure time, frequent replacement of the glutaraldehyde and disinfection of the machine, failed to eradicate the *M. abscessus*, presumably because of the presence of an infected biofilm inside the machine. Rinsing the scopes with 70% alcohol after automated disinfection finally eliminated the outbreak strain. This study demonstrated that automatic bronchoscope disinfecting machines may become heavily contaminated with mycobacteria that resist usual disinfection, resulting in a recurring source of bronchoscope contamination.

Maloney et al (58) reported on 15 out of 47 patients undergoing bronchoscopy over a three-month period with positive cultures for *M. abscessus* following bronchoscopy without clinical evidence of invasive mycobacterial disease. *M. abscessus* was recovered from the automated washer, the inlet water feeding the washer, and a bronchoscope. All three environmental isolates matched all 15 pseudo-infection isolates by RFLP analysis.

SUMMARY   In summary, bronchoscope NTM pseudoinfections can present the clinician with difficult patient management problems. Although most NTM species have relatively low virulence, some (such as *M. abscessus*) can cause invasive lung disease. Especially in an area endemic for NTM lung disease, it may be difficult—short of an invasive procedure such as a biopsy—to be certain that an NTM isolate is not clinically significant. Conversely, in the midst of a pseudo-outbreak, patients with clinical syndromes that are compatible with mycobacterial disease may have their disease attributed to the mycobacterium causing the pseudo-outbreak, simply because there is no plausible alternative explanation. AFB smear-positive bronchoscopy specimens caused by pseudo-infection can be especially pernicious, because they may lead to the initiation of potentially toxic antituberculous therapy for presumed tuberculosis.

Some recommended procedures for cleaning bronchoscopes, when ethylene oxide sterilization is not feasible, include (*a*) careful adherence to manufacturer's protocol for machine disinfection (with the awareness that colonization of the washer holding tanks may occur despite the use of the manufacturer's recommended disinfection protocol); (*b*) immediate mechanical cleaning, with wiping of outside and brushing all channels; (*c*) autoclaving of removable, heat-stable parts; (*d*) cleaning with fresh detergent solution, rinsing with high quality tap water at minimum; (*e*) high-level disinfection, rinsing with sterile deionized water; (*f*) thorough drying; (*g*) storage in a sterile environment; and (*h*) consideration of terminal rinsing with 70% alcohol immediately before use. Active surveillance should be maintained, and if infection or pseudoinfection are suspected, cultures of tap water, automated washer tanks, and endoscopes should be obtained to assess possible washer contamination.

Non-Bronchoscopy Related Pseudo-Outbreaks

From 1969 to the present (1998), at least 17 non-bronchoscopy related pseudo-outbreaks related to the NTM have been described (see Table 5). The majority

Table 5 Non-bronchoscope related pseudo-outbreaks caused by nontuberculous mycobacteria (NTM)[a]

Reference	Location	Year	Organism	Culture site	Incidence	Source	Typing
54	Worcester, MA	1993–94	*M. abscessus*	Respiratory (+) other	14/19	Contaminated (inhouse sterilized) distilled water	AP PCR
2	New Jersey	1994, Apr.–Dec.	*M. abscessus*	Blood cultures	23	Contaminated vial of Septi-chek AFB supplement contaminated "sterilized" water	MEE
53	Albany, NY	1985, Oct.–Nov.	*M. peregrinum*	Respiratory	5–40	Contaminated ice from ice machine	ND
45	Houston, TX	1985, Feb.–Mar.	*M. fortuitum*	Bone marrow	4	Syringe with marrow aspirate plunged into ice for transport to lab (viral culture requested)	ND
5	Connecticut	1988–89, Jan.–Dec.	*M. fortuitum*	Respiratory	90	Contaminated hospital tap water	ND
37	Virginia	1983–85, Jul.–Nov.	MAC	Urine	29	Contaminated hospital tap water used to prepare phenol red solution for urine processing	Serotype
20	Alabama	1994	MAC	Sterile body	8	Contaminated manufactured broth media and supplement	RFLP
90	Missouri	1971	MAC	Blood	3	Malfunction of needle sterilization in BACTEC cultures	ND

Reference	Organism	Specimen	Number	Source	Typing	Location	Year
24, 39	*M. xenopi*	Respiratory, other	>900	Contaminated hot water taps	ND	Connecticut	1969–77
55	*M. terrae*	Respiratory, other	131 (163 specimens)	Contaminated hospital potable water, ice machine, shower, hot water, patient sink	ND	Michigan	1986
25	*M. simiae*	Respiratory and stool	56	Unknown	MEE	New Mexico	1991–93
74	*M. simiae*	Respiratory	22/26	Unknown	Unknown	Arizona	1995
35	*M. gordonae*	Induced sputum	7/50	Contaminated tap water used to induce sputum samples	ND	Texas	1976
87	*M. gordonae*	Respiratory, other	34 patients, 46 samples	Contaminated PANTA solution (BACTEC)	ND	Michigan	1989 (eight weeks)
10	*M. kansasii*	Respiratory	50 patients, 83 isolates	Contaminated ice ingested by patients before sputa collection	Unknown	Ohio	1973–74?
55	*M. kansasii*	Respiratory	Unknown	Contaminated tap water used for intubations	ND	Paris, France	1979–80
59	*M. kansasii*	Respiratory	17	Contaminated tap water used for lab preparation of N-acetyl cysteine sodium hydroxide decontamination processing of samples	Unknown	Manitoba, Canada	1973–74?

[a]RFLP, Restriction fragment length polymorphism; MEE, multilocus enzyme electrophoresis; AP PCR, arbitrarily primed PCR; ND, typing not done; MAC, *M. avium* complex.

of these outbreaks have involved rapidly growing mycobacteria (RGM) and *Mycobacterium avium* complex (MAC). Other species, including *M. gordonae* and *M. terrae*, and more recently *M. xenopi*, *M. simiae*, and *M. kansasii*, however, are being recognized with increasing frequency. Because many species of NTM reside normally in tap water, the source of the outbreaks can most often be traced to a contaminated water supply somewhere in the hospital setting.

RAPIDLY GROWING MYCOBACTERIA   Five pseudo-outbreaks have been reported traceable to RGM, with contaminated solutions, water, or ice being incriminated or suspected in all of the outbreaks. In one pseudo-outbreak from a 370-bed public hospital, 90 positive cultures of *M. fortuitum* were reported between January 1988 and December 1989. Only one isolate of this species had been recovered in the year prior to the apparent outbreak, and none of the patients had disease. Systematic sampling of municipal and on-site hospital water reservoirs incriminated the on-site reservoir. After recommendation to close this reservoir and to avoid tap water use with equipment and solutions for respiratory or invasive procedures, no new isolates were identified (5). DNA fingerprinting of the isolates was not performed.

Water sampling and other environmental culturing does not always, however, readily identify the source of a pseudo-outbreak. In a large single hospital laboratory, *Mycobacterium abscessus* was responsible for low-level specimen contamination for almost five years before it was resolved (54). Although contaminated water was assumed to be the cause of the pseudo-outbreak, efforts to identify any environmental source were unsuccessful until the occurrence of the second outbreak, when a laboratory quality-control sample of the autoclaved distilled water grew the same strain of *M. abscessus*. These two separate pseudo-outbreaks were traced to the same strain [as confirmed by random amplified polymorphic DNA (RAPD)-PCR of genomic DNA] (101) and were ultimately traced to contaminated distilled water used to process the samples in the mycobacteriology laboratory. Intermittent inadequate sterilization of the water (decreased autoclave time or inappropriate load capacity) was hypothesized to be the cause of the contamination (54).

A third pseudo-outbreak caused by RGM was reported in 1997, when 23 blood cultures from HIV-infected patients were reported positive for *M. abscessus*. Several bacterial species and *M. abscessus* were cultured from an opened multidose culture supplement vial (BBL, Septi-Chek AFB supplement) that had been used for mycobacterial cultures. The finding that multiple, unopened vials were sterile suggested extrinsic contamination of the previously opened vial. Observations of laboratory procedures suggested several points of entry for contamination, including manual removal of the injection cap of the bottle, repeated entry of the injection cap bottle, multiple use of a common sterile water source, and lapses in sterile technique during specimen processing (2).

In the last two pseudo-outbreaks involving the *M. fortuitum* group, contaminated ice was found to be the source of the organism. One of these first pseudo-outbreaks involving sputum cultures occurred in 1985 in a New York hospital, where an ice machine serving patients in one medical ward had become colonized with *M. peregrinum* (formerly *M. fortuitum* biovar *peregrinum*). When the ice machine was turned off (one or two times daily on each shift), a large amount of water accumulated and allowed replication of the strain already in the water supply. Recovery of *M. peregrinum* from sputum was associated with recent consumption of ice water from bedside pitchers, melted ice, and ice with medications, but not with the use or consumption of other potable water sources (53). A second similar pseudo-outbreak involving ice contaminated with *M. fortuitum* occurred that same year at a hospital in Texas. Four patients on the same hospital floor who underwent bone marrow aspiration grew *M. fortuitum* from their aspirates. None of the patients had evidence of disease resulting from *M. fortuitum*. Each of the patients had viral cultures performed on their bone marrow aspirates, which necessitated plunging the syringe containing the bone marrow aspirate into ice for transport. Samples of ice from the ice machine on the patient floor were positive for *M. fortuitum*, while samples taken from other floors were free of mycobacteria (45).

MYCOBACTERIUM AVIUM COMPLEX   Of the 17 described pseudo-outbreaks traceable to the NTM, three have involved MAC. One of these clusters involved hospital water supplies. A second water-related MAC pseudo-outbreak occurred in Virginia in which the deionized tap water used to prepare a phenol red solution used in the processing of urine samples was contaminated with MAC. The same MAC strain (by serotype) was isolated in both the ward and the laboratory water samples (37).

The remaining two MAC pseudo-outbreaks involved breakdowns in commercial culture methods for mycobacteria. One pseudo-outbreak reported in 1994 was associated with a contaminated manufactured broth media and supplement (20). Growth of MAC occurred in two separate lot numbers of the growth supplement, and DNA fingerprint analysis showed that the suspect specimen isolates and the isolates from the commercial supplement were identical. The second cluster of cases related to commercial culture systems involved an outbreak of pseudobacteremia from a malfunction in the heating block used to sterilize the needle of the BACTEC 460 TB system (90). Three cases of cross-contamination occurred as a consequence. When the defective heating block was replaced, no further problems were encountered (90).

MYCOBACTERIUM XENOPI   *M. xenopi* has been involved in several outbreaks, at least one of which was determined to be a non-endoscope-related pseudo-outbreak. The largest number of isolates reported in the United States was reported by Costrini and colleagues at the West Haven, Connecticut, Veterans Administration Hospital during an eleven-year period (1969–1980) (24). More than 900 isolates from sputum, urine, bronchial washing, and lung tissue were recovered. Of these, only 19 patients had clinical evidence of disease caused by *M. xenopi*. (At the time of the publication, lung disease caused by *M. xenopi* had been reported in only five persons within the United States.) The remainder of the isolates recovered from over 600 patients were thought to be contaminants from the hospital hot water supply. Environmental sampling of the water taps and hot water generators in the hospital were positive for *M. xenopi*, often with almost confluent growth. Efforts to determine the source of contamination outside of the hospital substantiated that only the hospital water reservoir was colonized (24, 39). Although DNA fingerprinting was not performed, it is presumed that all of these isolates were a single strain.

MYCOBACTERIUM SIMIAE   Another slowly growing NTM, *M. simiae* has also been implicated in at least two pseudo-outbreaks, although it has yet to be recovered environmentally in either case. Clinical data in both outbreaks indicated that there was no disease traceable to *M. simiae* in any of the patients from whom the organism was recovered. The local water supply was the presumed source of the organism in both cases (25, 74).

MYCOBACTERIUM GORDONAE   Another long-recognized water-related environmental species, *M. gordonae*, has been associated with at least two non-bronchoscopy pseudo-outbreaks in the United States (81, 83). Tokars et al published the first report of pseudo-infection with *M. gordonae* caused by contamination of a commercial product when they reported that two lot numbers of the antimicrobial solution PANTA PLUS (Becton-Dickinson Instrument Systems) shipped to 173 laboratories were contaminated with *M. gordonae* (87). Finally, *M. gordonae* was also isolated from a large number of sputa samples when patients rinsed and gargled with tap water before sputa induction (35).

MYCOBACTERIUM TERRAE COMPLEX   Less commonly, pseudo-outbreaks whose causative agents were other NTM such as *M. terrae* have been investigated. In the first report defining the epidemiological aspects of *M. terrae* contaminating clinical specimens, Lockhart and coworkers described 163 positive cultures for *M. terrae* from a variety of clinical specimens. The investigators documented the hospital's potable water system as the reservoir of *M. terrae*. The species was recovered from skin, a shower, an ice machine, and potable water samples (35).

MYCOBACTERIUM KANSASII   In 1979, 75% of the isolates of NTM in a hospital in Paris, France, were identified as *M. kansasii*. In searching for the etiology, investigators discovered that nonsterilized tap water was being used to perform sputum induction and that the *M. kansasii* was localized in the water compressors (54a). Two other pseudo-outbreaks traceable to *M. kansasii* were described earlier in connection with contaminated ice ingested before sputa collection (10), and in Canada when the N-acetyl cysteine sodium hydroxide measuring containers used in the laboratory for decontamination and digestion of AFB samples were rinsed with tap water contaminated with *M. kansasii* (59).

SUMMARY   In summary, nosocomial pseudo-outbreaks unrelated to bronchoscopy may be problematic in that (*a*) inappropriate therapy may result, causing risks of drug-related adverse events; (*b*) patients and physicians may develop concern for a disease that does not exist; and (*c*) unnecessary expense may be incurred by the patient and the hospital. The clinical laboratory must expend time and resources to identify spurious isolates, determine their source, and notify the physician to discover clinical significance, if any, of the isolates recovered. False positive culture results delay the ordering of tests to confirm an alternative diagnosis. Furthermore, mycobacterial pseudo-outbreaks present unique challenges because of the long periods needed for growth and identification and evaluation of the isolates. Prevention of pseudo-outbreaks depends on careful surveillance techniques, appropriate quality-control measures, and continuing education for all hospital personnel. Additionally, good communication between laboratory personnel and clinicians may ensure proper interpretation of culture results to rule out insignificant laboratory results. Newer technologies including molecular techniques continue to expedite recognition and subsequent investigation of pseudo-outbreaks caused by the NTM.

Literature Cited

1. Alvarado C, Stolz S, Maki D. 1991. Nosocomial infections after contaminated endoscope washer: a flawed automated endoscope washer. An investigation using molecular epidemiology. *Am. J. Med.* 91:2772–80S

2. Ashford DA, Kellerman S, Yakrus M, Brim S, Good RC, et al. 1997. Pseudo-outbreak of septicemia due to rapidly growing mycobacteria associated with extrinsic contamination of culture supplement. *J. Clin. Microbiol.* 35:2040–42

3. Bailey RK, Wyles S, Dingley M, Hesse F, Kent GW. 1970. The isolation of high catalase *Mycobacterium kansasii* from tap water. *Am. Rev. Respir. Dis.* 101:430–31

4. Band JD, Ward JI, Fraser DW, Peterson NJ, Silcox VA, Good RC, et al. 1982. Peritonitis due to a *Mycobacterium chelonae*-like organism associated with intermittent chronic peritoneal dialysis. *J. Infect. Dis.* 145:9–17

5. Bendaña N, Glover N, Skolnick S, Barba

D, Mascola L, Yakrus M. 1991. Pseudoepidemic of *Mycobacterium fortuitum* associated with a contaminated on-site hospital water reservoir. *Am. J. Infect. Cont.* 19(2):106 (Abstr.)

5a. Bennett SN, Peterson DE, Johnson DR, Hall WN, Robinson-Dunn B, Dietrich S. 1994. Bronchoscopy-associated *Mycobacterium xenopi* pseudoinfections. *Am. J. Respir. Crit. Care Med.* 150:245–50

6. Boian MG, Aronson T, Holtzman A, Bishop NH, Tran T, et al. 1997. *A comparison of clinical and potable water isolates of* Mycobacterium avium *using PCR of genomic sequences between insertion elements.* Presented at Gen. Meet. Am. Soc. Microbiol., 97th. (Abstr. U-161, p. 571)

7. Bolan G, Reingold AL, Carson CA, Silcok VA, Woodley CL, et al. 1985. Infections with *Mycobacterium chelonei* in patients receiving dialysis and using processed hemodialyzers. *J. Infect. Dis.* 152:1013–19

8. Borghaus JGA, Stanford JL. 1973. *Mycobacterium chelonei* in abscesses after injection of diphtheria-tetanus-polio vaccine. *Am. Rev. Respir. Dis.* 107:1–8

9. Bourbeau P, Kline B, Leberfinger M, Bross, Pfaller M. 1997. *Pseudoinfection of* Mycobacterium avium *complex associated with the use of bronchoscopes rinsed with tap water.* Presented at Annu. Meet. Intersci. Conf. Antimicrob. Agents Chemother., 37th. (Abstr. J-141, p. 314)

10. Brust RA, Ayers LW. 1974. *"Epidemic" contamination of patients' sputa by an environmental* Mycobacterium kansasii *contaminating ingested ice.* Presented at Annu. Meet. Am. Soc. Microbiol., 74th. (Abstr. M-245, p. 107)

11. Burns DN, Wallace RJ Jr, Schultz ME, Zhang YS, Zubairi SQ, et al. 1991. Nosocomial outbreak of respiratory tract colonization with *Mycobacterium fortuitum*: demonstration of the usefulness of pulsed-field gel electrophoresis in an epidemiologic investigation. *Am. Rev. Respir. Dis.* 144:1153–59

12. Butler WR, Jost KC Jr, Kilburn JO. 1991. Identification of mycobacteria in high-performance liquid chromatography. *J. Clin. Microbiol.* 29:2468–72

13. Carmago D, Saad C, Ruiz F, Ramirez ME, Lineros M, et al. 1996. Iatrogenic outbreak of *M. chelonae* skin abscesses. *Epidemiol. Infec.* 117:113–19

14. Carson LA, Peterson NJ, Favero MS, Agnero SM. 1978. Growth characteristics of atypical mycobacteria in water and their comparative resistance to disinfectants. *Appl. Environ. Microbiol.* 36:839–46

15. Carson LA, Aguero SM, Safranek TJ, Jarvis WR. 1987. Growth characteristics of *Mycobacterium chelonei* in aqueous gentian violet solutions. Presented at Annu. Meet. Am. Soc. Microbiol., 87th. (Abstract L-14, p. 406)

16. Carson LA, Bland LA, Cusick LB, Favero MS, Bolan GA, et al. 1988. Prevalence of nontuberculous mycobacteria in water supplies of hemodialysis centers. *Appl. Environ. Microbiol.* 54:3122–25

17. Centers for Disease Control. 1991. Nosocomial infection and pseudoinfection from contaminated endoscopes and bronchoscopes—Wisconsin and Missouri. *Morbid. Mortal. Wkly. Rep.* 40:675–78

18. Centers for Disease Control. 1978. Follow-up on mycobacterial contamination of porcine heart valve prosthesis—United States. *Morbid. Mortal. Wkly. Rep.* 27:92, 97–98

19. Centers for Disease Control. 1996. Infection with Mycobacterium abscessus associated with intramuscular injection of adrenal cortex extract—Colorado and Wyoming, 1995–1996. *Morbid. Mortal. Wkly. Rep.* 45:713–15

20. Chapin KC, McDonald CL. 1994. *Pseudoepidemic of* M. avium-intracellulare *complex associated with manufactured product.* Presented at Annu. Meet. Intersci. Conf. Antimicrob. Agents Chemother., 34th. J256:243

21. Clegg HW, Foster MT, Sanders WE Jr, et al. 1983. Infection due to organisms of the *Mycobacterium fortuitum* complex after augmentation mammaplasty: clinical and epidemiological features. *J. Infect. Dis.* 147:427–33

22. Collins CH, Grange JM, Yates MD. 1984. Mycobacteria in water. *J. Appl. Bacteriol.* 57:193–211

23. Collins FM. 1986. Bactericidal activity of alkaline glutaraldehyde solution against a number of atypical mycobacterial species. *J. Appl. Bacteriol.* 61:247–51

24. Costrini AM, Mahler DA, Gross WM, Hawkins JE, Yesner R, D'Esopo ND. 1981. Clinical and roentgenographic features of nosocomial pulmonary disease due to *Mycobacterium xenopi. Am. Rev. Respir. Dis.* 123:104–9

25. Crossey MJ, Yakrus MA, Cook MB, Rasmussen SK, McEntee TM, et al. 1994. *Isolation of* Mycobacterium simiae *in a Southwestern hospital and typing by multilocus enzyme electrophoresis.* Presented at Annu. Meet. Am. Soc. Microbiol., 94th. U38:179

26. da Costa Cruz J. 1938. *"Mycobacterium*

fortuitum", um novo bacilo acido-resistente patogenico para o homen. *Acta Med.* 1:297–301

26a. Dawson DJ, Armstrong JG, Blacklock ZM. 1982. Mycobacterial cross-contamination of bronchoscopy specimens. *Am. Rev. Respir. Dis.* 126:1095–97

27. Desplaces N, Picardeau M, Dinh V, Leonard PH, Mamoudy P, et al. 1995. *Spinal infections due to Mycobacterium xenopi after discectomies.* Presented at Annu. Meet. Intersci. Conf. Antimicrob. Agents Chemother., 35th, J-162. (Abstr.)

28. du Moulin GC, Stottmeier KD, Pelletier PA, Tsang AY, Hedley-Whyte J. 1988. Concentration of *Mycobacterium avium* by hospital hot water systems. *J. Am. Med. Assoc.* 260:1599–1601

29. Elston RA, Hay AJ. 1991. Acid-fast bacillus contamination of a bronchoscope washing machine [letter]. Comment on: *J. Hosp. Infect. 1990;* 16:257–61; *J. Hosp. Infect.* 19:72–73

30. Engel HWB, Berwald LG, Havelaar AH. 1980. The occurrence of *Mycobacterium kansasii* in tap water. *Tubercle* 61:21–26

31. Falkinham JO III, George KL, Parker BC, Gruft H. 1984. In vitro susceptibility of human and environmental isolates of *Mycobacterium avium*, *M. intracellulare*, and *M. scrofulaceum* to heavy-metal salts and oxyanions. *Antimicrob. Agents Chemother.* 25:137–39

32. Foz A, Roy C, Jurado J, Arteago E, Ruiz JM, et al. 1978. *M. chelonae* iatrogenic infections. *J. Clin. Microbiol.* 7:319–21

33. Fraser VJ, Jones M, Murray PR, Medoff G, Zhang Y, et al. 1992. Contamination of flexible fiberoptic bronchoscopes with *Mycobacterium chelonae* linked to an automated bronchoscope disinfection machine. *Am. Rev. Respir. Dis.* 145:853–55

34. Fraser V, Wallace RJ Jr. 1996. Nontuberculous mycobacteria. In *Hospital Epidemiology and Infection Control*, ed. C Glen Mayhall. Baltimore: Williams & Wilkins. 1224 pp.

35. Gangadharem PRJ, Lockhart JA, Awe RJ, Jenkins DE. 1976. Mycobacterial contamination through tap water. *Am. Rev. Respir. Dis.* 113:894

36. Georgia Department of Human Resources. 1990. Abscesses in an allergy practice due to *M. chelonae. Georgia Epidemiol. Rep.* 6:2

37. Graham LJR, Warren NG, Tsang AY, Dalton HP. 1988. *Mycobacterium avium* complex pseudobacteriuria from a hospital water supply. *J. Clin. Microbiol.* 26:1034–36

38. Gremillion DH, Mursch SB, Lerner RJ. 1983. Injection sites abscesses caused by *M. chelonae. Infect. Control* 4:25–28

39. Gross WN, Hawkins JE, Murphy DB. 1976. Origin and significance of *M. xenopi* in clinical specimens. *Bull. Internat. Union Tuberc.* 51:267–69

40. Gubler JGH, Salfinger M, von Graevenitz A. 1992. Pseudoepidemic of nontuberculous mycobacteria due to a contaminated bronchoscope cleaning machine. Report of an outbreak and review of the literature. *Chest* 101:1245–49

41. Guerrero C, Bernasconi C, Burki D, Bodmer T, Telenti A. 1995. A novel insertion element from *Mycobacterium avium*, IS*1245*, is a specific target for analysis of strain relatedness. *J. Clin. Microbiol.* 33:304–7

42. Hayes PS, McGiboney DL, Band JD, Feeley JC. 1982. Resistance of *Mycobacterium chelonei*-like organisms to formaldehyde. *Appl. Environ. Microbiol.* 43:722–24

43. Hector JS, Pang Y, Mazurek GH, Zhang Y, Brown BA, et al. 1992. Large restriction fragment patterns of genomic *Mycobacterium fortuitum* DNA as strain-specific markers and their use in epidemiologic investigation of four nosocomial outbreaks. *J. Clin. Microbiol.* 30:1250–55

44. Hoffman PC, Fraser DW, Robicsek F, O'Bar P, Mauneg CV. 1981. Two outbreaks of sternal wound infections due to organisms of the *Mycobacterium fortuitum* complex. *J. Infect. Dis.* 143:533–42

45. Hoy J, Rolston K, Hopfer RL. 1987. Pseudoepidemic of *Mycobacterium fortuitum* in bone marrow biopsies. *Am. J. Infect. Control* 15:268–71

46. Inman PM, Beck A, Brown AE, Stanford JL. 1969. Outbreak of infection abscesses due to *M. abscessus. Arch. Dermatol.* 100:141–47

47. Jarvis WR. 1991. Nosocomial outbreaks: the Centers for Disease Control's Hospital Infections Program Experience, 1980–1990. *Am. J. Med.* 91:101–6S (Suppl. 3B)

48. Jost KC Jr, Dunbar D. 1992. *Automated identification of mycobacteria by high-performance liquid chromatography using computer-aided pattern recognition algorithms.* Presented at Annu. Meet. Am. Soc. Microbiol., 92nd. U69:177

49. Kauppinen J, Mäntyjärvi, Katila M-L. 1994. Random amplified polymorphic DNA genotyping of *Mycobacterium malmoense. J. Clin. Microbiol.* 32:1827–29

50. Kümmerle N, Feucht H-H, Kaulfers

P-M. 1996. Plasmid-mediated formaldehyde resistance in *Escherichia coli*: characterization of resistance gene. *Antimicrob. Agents Chemother.* 40:2276–79

51. Kuritsky JN, Bullen MG, Broome CV, Silcox VA, Good RC, et al. 1983. Sternal wound infections and endocarditis due to organisms of the *M. fortuitum* complex. *Ann. Intern. Med.* 98:938–39

52. Laskowski LF, Marr JJ, Spernoga JF, Frank NJ, Barner HB, et al. 1977. Fastidious mycobacteria grown from porcine prosthetic heart valve cultures. *N. Engl. J. Med.* 297:101–2

53. Laussucq S, Baltch A, Smith R, Smithurck R, Davis B, et al. 1988. Nosocomial *M. fortuitum* colonization from a contaminated ice machine. *Am. Rev. Respir. Dis.* 138:891–94

54. Lai KK, Brown BA, Westerlin JA, Fontecchio SA, Melvin ZS, et al. 1998. Long-term laboratory contamination due to *Mycobacterium abscessus* resulting in two pseudo outbreaks: recognition using random amplified polymorphic DNA (RAPD) PCR *Clin. Infect. Dis.* In press

54a. Levy-Frebault V, David HL. 1983. *Mycobacterium kansasii*: contaminant du réseau d'eau potable d'un hôpital. *Rev. Epidém. Santé Publ.* 31:11–20

55. Lockwood WW, Friedman C, Bus N, Pierson C, Gaynes R. 1989. An outbreak of *Mycobacterium terrae* in clinical specimens associated with a hospital potable water supply. *Am. Rev. Respir. Dis.* 140:1614–17

56. Lowry PW, Beck-Sague CM. 1990. *M. chelonae* infection among patients receiving hi-flux dialysis in a hemodialysis clinic in California. *J. Infect. Dis.* 161:85–90

57. Lowry PW, Jarvis WR, Oberle AD, Bland LA, Silberman R, et al. 1988. *M. chelonae* causing otitis media in an ear-nose-and-throat practice. *N. Engl. J. Med.* 319:978–82

58. Maloney S, Welbel S, Daves B, Adams K, Becker S, et al. 1994. *Mycobacterium abscessus* pseudoinfection traced to an automated endoscope washer: utility of epidemiologic and laboratory administration. *J. Infect. Dis.* 169:166–69

59. Maniar AC, Vanbuckenhout LR. 1976. *Mycobacterium kansasii* from an environmental source. *Can. J. Public Health.* 67:59–63

60. Mazurek GH, Chin DP, Hartman S, Reddy V, Horsburgh CR Jr, et al. 1997. Genetic similarity among *Mycobacterium avium* isolates from blood, stool, and sputum of persons with AIDS. *J. Infect. Dis.* 176:976–83

61. McSwiggan DA, Collins CH. 1974. The isolation of *M. kansasii* and *M. xenopi* from water systems. *Tubercle* 55:291–97

62. Nolan CM, Hashisaki PA, Dundas DF. 1991. An outbreak of soft-tissued infections due to *Mycobacterium fortuitum* associated with electromyography. *J. Infect. Dis.* 163:1150–53

63. Nye K, Chadha DK, Hodgkin P, Bradley C, Hancox J, et al. 1990. *Mycobacterium chelonei* isolation from broncho-alveolar lavage fluid and its practical implications. *J. Hosp. Infect.* 16:257–61

64. Owen M, Smith A, Coultras J. 1963. Granulomatous lesions occurring at site of injections of vaccines and antibiotics. *South. Med. J.* 56:949–52

65. Pappas SA, Schaff DM, Dicostango MB, King FW, Sharp JJ. 1983. Contamination of flexible fiberoptic bronchoscopes. *Am. Rev. Respir. Dis.* 127:381–92

66. Pattyn SR, Vandepitte J, Portaels F, DeMuynck A. 1971. Cases of *M. borstelense* and *M. abscessus* infection observed in Belgium. *J. Med. Microbiol.* 4:145–49

67. Pelletier PA, du Moulin GC, Stottmeier KD. 1988. Mycobacteria in public water supplies: comparative resistance to chlorine. *Microbiol. Sci.* 5:147–48

68. Pentony T, Haywood H, Smith G, Hayes A, Ingram C. 1994. *Pseudo-outbreak of Mycobacterium avium complex (MAC) in bronchoscopy specimens due to contaminated potable water supply.* Annu. Meet. Soc. Hosp. Epidemiol. Am., 4th. S33:30

68a. Petersen K, Bus N, Walter V, Chenoweth C. 1994. Pseudoepidemic of Mycobacterium abscessus associated with bronchoscopy. *Infect. Cont. Hosp. Epid.* S32: 30

69. Pettini B, Hellstrand P, Erickson M. 1980. Infection with *M. chelonae* following injections. *Scand. J. Infect. Dis.* 12:237–38

70. Picardeau M, Prod'Hom G, Raskine L, LePennec MP, Vincent V. 1997. Genotypic characterization of five subspecies of *Mycobacterium kansasii*. *J. Clin. Microbiol.* 35:25–32

71. Picardeau M, Vincent V. 1996. Typing of *Mycobacterium avium* isolates by PCR. *J. Clin. Microbiol.* 34:389–92

72. Preheim LC, Bittner MJ, Giger DK, Sanders WE Jr. 1982. Mycobacterium fortuitum *(Mf) sternotomy infections treated with amikacin (Am), cefoxitin (C), and rifampin (R): serum static (S) and Killing (K) titers.* Presented at Annu. Meet. Intersci. Conf. Antimicrob. Agents Chemother., 22nd. 564:165

73. Robicsek F, Daugherty HK, Cook JW,

Selle JG, Masters TN, et al. 1978. *My-cobacterium fortuitum* epidemics after open heart surgery. *J. Thorac. Cardio-vasc. Surg.* 75:91–96

74. Rynkiewicz DL, Ampel NM. 1994. *Lack of clinical significance of* Mycobacterium simiae. Presented at Annu. Meet. Infect. Dis. Soc. Am., 34th. 305:92

75. Safranek TJ, Jarvis WR, Carson CA, Cusick LB, Bland LA, et al. 1987. *M. chelonae* wound infections after plastic surgery employing contaminated gentian violet making solution. *N. Engl. J. Med.* 317:197–201

76. Schulze-Röbbecke R, Janning B, Fis-cheder R. 1992. Occurrence of mycobac-teria in biofilm samples. *Tuberc. Lung Dis.* 73:141–44

77. Silcox VA, Good RC, Floyd MM. 1984. Identification of clinically significant *M. fortuitum* complex isolates. *J. Clin. Mi-crobiol.* 14:686–91

78. Sniadack DH, Ostroff SM, Karlix MA, Smithwick RW, Schwartz B, et al. 1993. A nosocomial pseudo-outbreak of *My-cobacterium xenopi* due to a contami-nated potable water supply: lessons in prevention. *Infect. Cont. Hosp. Epide-miol.* 14:636–41

79. Soto LE, Bobadilla M, Villalobos Y, Si-fuentes J, Avelar J, et al. 1991. Post-surgical nasal cellulitis outbreak due to *Mycobacterium chelonae. J. Hosp. Infect.* 19:99–106

80. Spach DH, Silverstein FE, Stamm WE. 1993. Transmission of infection by gas-trointestinal endoscopy and broncho-scopy. *Ann. Intern. Med.* 118:117–28

81. Steere AC, Corrales J, von Graevenitz A. 1979. A cluster of *Mycobacterium gor-donae* isolates from bronchoscopy speci-mens. *Am. Rev. Respir. Dis.* 120:214–16

81a. Steingrube VA, Wallace RJ Jr, Steele LC, Nash DR. 1991. Mercuric reductase activ-ity and evidence of broad spectrum mer-cury resistance among clinical isolates of rapidly growing mycobacteria. *Antimi-crob. Agents Chemother.* 35:819–23

82. Steingrube VA, Gibson JL, Brown BA, Zhang Y, Wilson RW, et al. 1995. PCR amplification and restriction endonucle-ase analysis of a 65-kilodalton heat shock protein gene sequence for taxonomic sep-aration of rapidly growing mycobacteria. *J. Clin. Microbiol.* 33:149–53

83. Stine TM, Harris AA, Levin S, Rivera N, Kaplan RL. 1987. A pseudoepidemic due to atypical mycobacteria in a hospital wa-ter supply. *J. Am. Med. Assoc.* 258:809–11

84. Swenson JM, Thornsberry C, Silcox VA. 1982. Rapidly growing mycobacteria:

testing of susceptibility to 34 antimicro-bial agents by broth microdilution. *An-timicrob. Agents Chemother.* 22:186–92

85. Szabo I, Sarkozi K. 1980. *M. chelonae* endemy after heart surgery with fatal con-sequences. *Am. Rev. Respir. Dis.* 121:607 (Letter)

86. Telenti A, Marchesi F, Balz M, Bally F, Böttger EC, Bodmer T. 1993. Rapid iden-tification of mycobacteria to the species level by polymerase chain reaction and restriction enzyme analysis. *J. Clin. Mi-crobiol.* 31:175–78

87. Tokars JI, McNeil MM, Tablan OC, Chapin-Robertson K, Patterson JE, et al. 1990. *Mycobacterium gordonae* pseu-doinfection associated with a contami-nated antimicrobial solution. *J. Clin. Mi-crobiol.* 28:2765–69

88. Tyras DH, Kaiser GC, Barner HB, Laskowski LF, Marr JJ. 1978. Atypical mycobacteria and the xenograft valve. *J. Thorac. Cardiovasc. Surg.* 75:331–37

89. Vandepitte J, Dessmyter J, Gatti F. 1969. Mycobacteria, skins, and needles. *Lancet* 4:691

90. Vannier AM, Tarrand JJ, Murray PR. 1988. Mycobacterial cross contamination during radiometric culturing. *J. Clin. Mi-crobiol.* 26:1867–68

91. Villaneuva A, Calderon RV, Vargas BA, Ruiz F, Aguero S, et al. 1997. Report on an outbreak of postinjection abscesses due to *Mycobacterium abscessus*, includ-ing management with surgery and clar-ithromycin therapy and comparison of strains by random amplified polymorphic DNA polymerase chain reaction. *Clin. In-fect. Dis.* 24:1147–53

92. von Reyn CF, Maslow JN, Barber TW, Falkinham JO III, Arbeit RD. 1994. Per-sistent colonisation of potable water as a source of *Mycobacterium avium* infection in AIDS. *Lancet* 343:1137–41

93. Wallace RJ. 1987. Nontuberculous my-cobacteria and water: a love affair with in-creasing clinical importance. *Infect. Dis. Clin. North Am.* 3:677–86

94. Wallace RJ Jr, Steele LC, Labidi A, Silcox VA. 1989. Heterogeneity among isolates of rapidly growing mycobacteria respon-sible for infections following augmenta-tion mammaplasty despite case clustering in Texas and other southern coastal States. *J. Infect. Dis.* 160:281–88

94a. Wallace RJ Jr, Musser JM, Hull SI, Silcox VA, Steele LC, et al. 1989. Diversity and sources of rapidly growing mycobacteria associated with infections following car-diac surgery. *J. Infect. Dis.* 159:708–16

95. Wallace RJ Jr, Zhang Y, Brown B, Fraser

V, Mazurek GH, et al. 1993. DNA large restriction fragment patterns of sporadic and epidemic nosocomial strains of *M. chelonae* and *M. abscessus*. *J. Clin. Microbiol.* 31:2697–701

96. Weinstein RA, Stamm WE. 1977. Pseudoepidemics in hospital. *Lancet* 2:862–64

97. Wenger JD, Spika JS. 1990. Outbreak of *M. chelonae* infection associated with use of jet injectors. *J. Am. Med. Assoc.* 264:373–76

97a. Wheeler PW, Lancaster D, Kaiser AB. 1989. Bronchopulmonary cross-colonization and infection related to mycobacterial contamination of suction valves of bronchoscopes. *J. Infect. Dis.* 159:954–58

98. Wolinsky E. 1979. Nontuberculous mycobacterium and associated diseases. *Annu. Rev. Respir. Dis.* 119:107–59

99. Wright EP, Collins CH, Yates MD. 1985. *Mycobacterium xenopi* and *Mycobacterium kansasii* in a hospital water supply. *J. Hosp. Infect.* 6:175–78

100. Yew W-w, Wong P-c, Woo H-s, Yip C-w, Chan C-y, et al. 1993. Characterization of *Mycobacterium fortuitum* isolates from sternotomy wounds by antimicrobial susceptibilities, plasmid profiles, and ribosomal ribonucleic acid gene restriction patterns. *Diagn. Microbiol. Infect. Dis.* 17:111–17

101. Zhang Y, Rajagopalan M, Brown BA, Wallace RJ Jr. 1997. Randomly amplified polymorphic DNA PCR for comparison of *Mycobacterium abscessus* strains from nosocomial outbreaks. *J. Clin. Microbiol.* 35:3132–39

Annu. Rev. Microbiol. 1998. 52:491–532

THE HIV-1 REV PROTEIN

Victoria W. Pollard

Department of Microbiology, University of Pennsylvania School of Medicine,
Philadelphia, Pennsylvania 19104-6148; e-mail: pollard@hhmi.upenn.edu

Michael H. Malim

Departments of Microbiology and Medicine and Howard Hughes Medical Institute,
University of Pennsylvania School of Medicine, Philadelphia, Pennsylvania
19104-6148; e-mail: malim@hhmi.upenn.edu

KEY WORDS: RRE, RNA export, NES, exportin 1

ABSTRACT

The nuclear export of intron-containing HIV-1 RNA is critically dependent on
the activity of Rev, a virally encoded sequence-specific RNA-binding protein.
Rev shuttles between the nucleus and the cytoplasm and harbors both a nuclear
localization signal and a nuclear export signal. These essential peptide motifs have
now been shown to function by accessing cellular signal-mediated pathways for
nuclear import and nuclear export. HIV-1 Rev therefore represents an excellent
system with which to study aspects of transport across the nuclear envelope.

CONTENTS

0066-4227/98/1001-0491$08.00

INTRODUCTION

A molecular understanding of viral functions frequently provides fundamental insights into basic cellular mechanisms. This philosophy has been richly rewarded in the study of the human immunodeficiency virus type-1 (HIV-1) Rev protein. It is generally accepted that Rev functions by activating the nuclear export of unspliced (intron-containing) viral mRNAs. The analysis of Rev has therefore become one of the preferred model systems with which one can study the signal-mediated export of macromolecules from the nucleus to the cytoplasm. For simplicity, this review is organized chronologically, starting with the discovery of Rev in 1986, proceeding through its assignment as a *trans*-activator of RNA nuclear export, and culminating with descriptions of interacting cellular proteins and a molecular model for its mechanism of action. In addition, the fact that Rev is essential for HIV-1 replication makes it an attractive target for antiviral approaches. The development of novel methods for preventing and controlling HIV-1 infections remains a critical objective in HIV research as the number of people infected worldwide continues to escalate.

RETROVIRAL GENE EXPRESSION

Overview of the Retroviral Life Cycle

HIV-1 is regarded as the prototype member of the lentivirus subfamily of retroviruses. The other two subfamilies are the comparatively obscure foamy viruses (also known as spumaviruses) and the long-studied oncoretroviruses (RNA tumor viruses); examples of the latter subfamily are avian leukosis viruses (ALVs) and murine leukemia viruses (MLVs). The major features of the lentivirus and oncoretrovirus life cycles are well conserved and have been reviewed in detail (20, 27, 62, 105, 128). Infectious virions initially bind to cellular receptors on the surface of susceptible cells via envelope (Env) glycoproteins. Fusion of the viral and cellular lipid membranes ensues, and the viral core (nucleocapsid) enters the cytoplasm. The linear single-stranded RNA genome of the virus is then copied into a double-stranded linear DNA molecule by the viral enzyme reverse transcriptase (RT); this conversion of RNA to DNA gives retroviruses their name.

At a point thereafter, the DNA enters the nucleus as a nucleic acid–protein complex (the preintegration complex) and is incorporated into the cell's genome by the action of a second viral enzyme, integrase (IN). The covalently integrated form of viral DNA, which is defined as the provirus, serves as the template for viral transcription. Retroviral RNAs are synthesized, processed, and then transported to the cytoplasm, where they are translated to produce the viral proteins. The proteins that form the viral core, namely the products of the *gag* and *pol* genes, initially assemble into immature nucleocapsids together with two copies of full-length viral RNA (these become the viral genome). As these structures bud through the plasma membrane, they become encapsulated by a layer of membrane that also harbors the viral Env glycoproteins. Coincident with budding, a third viral enzyme known as protease (PR) cleaves the core proteins into their final forms. This final step primes the viral particles for the next round of infection and is termed maturation. Although this general scheme was also believed to reflect foamy virus replication (59), recent findings have suggested that these viruses may represent an evolutionary intermediate between retroviruses and the hepadnaviruses (230)—a family of DNA viruses that includes hepatitis B virus (HBV) and that also uses reverse transcription to replicate its genome (71).

Alternative Splicing and the RNA Nuclear Export Problem

Examination of a typical proviral organization for each retroviral subfamily immediately reveals a critical gene expression problem that must be surmounted by these viruses (Figure 1): specifically, how to express multiple genes from a single proviral unit? Indeed, many viruses have to deal with this issue, and a variety of solutions have evolved. Three different strategies for generating an array of mRNAs from a single template include alternative sites of transcriptional initiation, alternative splicing, and alternative sites of 3'-end formation.

All retroviruses utilize the 5'-long terminal repeat (LTR) as the promoter to direct synthesis of the full-length viral RNA. For oncoretroviruses and lentiviruses, this is the only promoter present in the provirus; in contrast, foamy viruses also utilize an internal promoter to express one of their subgenomic mRNAs (Figure 1) (19, 125). Although retroviral transcription is mediated by RNA polymerase II and a combination of cellular basal and promoter (enhancer) specific factors, several viruses, for example HIV-1, additionally encode their own transcriptional activator. Importantly, all retroviruses utilize alternative splicing of the full-length transcript to produce the diversity of viral mRNAs required to position the initiator codons of all genes in close enough proximity to the 5'-cap to ensure appropriately efficient translation. As discussed below, retroviral splicing is mediated entirely by cellular pre-mRNA splicing factors and is, by necessity, relatively inefficient. There is no evidence that retroviruses exploit alternative 3'-end formation as all mRNAs are polyadenylated in the 3'-LTR.

The least complex genetic organization of a provirus is exemplified by the oncoretrovirus ALV (Figure 1). Here, only the three genes that are common to all replication-competent retroviruses—*gag*, *pol*, and *env*—are present. The Gag polyprotein is expressed from the ~7.5-kb unspliced full-length transcript by conventional translation. Pol is also expressed from this mRNA, but as a Gag-Pol fusion protein that is generated by ribosomal frameshifting in the region of overlap between *gag* and *pol*. The Env glycoprotein precursor is expressed

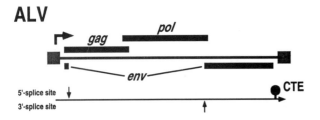

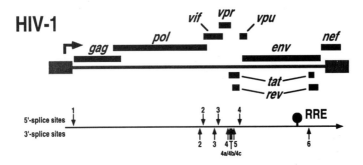

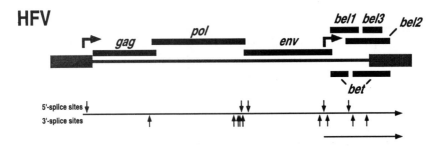

Figure 1   Genetic structure of retroviruses. The provirus organizations of an oncoretrovirus [avian leukosis virus (ALV)], a lentivirus (HIV-1), and human foamy virus (HFV) are shown. (*Solid boxes*) The positions of the various genes. (*Gray boxes*) The 5'- and 3'-long terminal repeats (LTRs). (*Horizontal arrows*) The sites of transcription initiation. The primary transcripts of each virus are shown below the proviruses together with the positions of known 5'-splice sites (*downward-pointing arrows*), known 3'-splice sites (*upward-pointing arrows*), the constitutive transport element (CTE) of ALV, and the Rev response element (RRE) of HIV-1. The numbering of the HIV-1 splice sites is as previously described (173).

from a processed ~2.2-kb mRNA that is the product of a unique splice that utilizes the unique 5'- and 3'-splice sites of the virus and excises the bulk of *gag* and *pol*. In contrast, the complexity of HIV-1 is markedly greater than that of ALV (Figure 1). In addition to *gag*, *pol*, and *env*, there are six regulatory and/or accessory genes to be expressed; the specific functions of the encoded Tat, Nef, Vif, Vpr, and Vpu proteins are discussed elsewhere (30, 46, 107, 122, 193). More than 30 different viral mRNA species can be present in HIV-1 infected cells, and at least 4 different 5'-splice sites and 8 different 3'-splice sites are utilized during their biogenesis (53, 77, 173, 179, 189). These RNA species include ~9-kb unspliced full-length transcripts that encode Gag and Gag-Pol; many different ~4-kb singly (partially) spliced mRNAs that lack the *gag-pol* region and encode Env, Vif, Vpr, or Vpu; and various ~2-kb multiply (fully) spliced mRNAs that lack *gag*, *pol*, and much of *env* and encode Rev, Tat, or Nef (Figures 2,3). The splicing pattern employed by human foamy virus (HFV) is

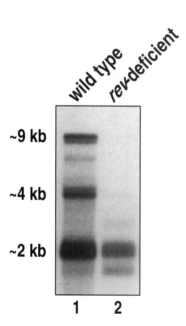

Figure 2 Rev is required for the cytoplasmic expression of intron-containing HIV-1 mRNAs. Northern analysis of cytoplasmic RNA isolated from human T cells harboring wild-type (*lane 1*) or *rev*-deficient (*lane 2*) HIV-1 proviruses (135). In cells expressing Rev, all intron-containing (~9-kb and ~4-kb) and fully spliced (~2-kb) mRNAs accumulate to high levels in the cytoplasm, whereas only the ~2-kb mRNAs are detected in the cytoplasms of cells lacking Rev. Purified RNAs were fractionated by denaturing agarose gel electrophoresis and then transferred to nitrocellulose, hybridized to a [^{32}P]-labeled long terminal repeat probe, and visualized by autoradiography.

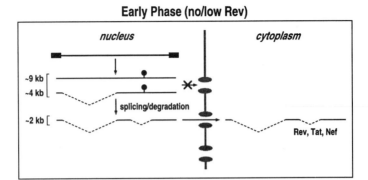

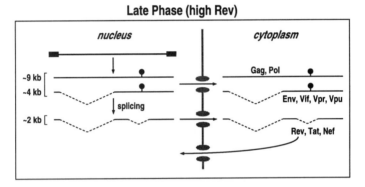

Figure 3 Early and late phases of HIV-1 mRNA expression. Full-length unspliced ~9-kb RNA, partially spliced ~4-kb mRNA, and fully spliced ~2-kb mRNA species are constitutively expressed in the nucleus. In the absence of Rev (*upper panel*), or when the concentration of Rev is below the threshold necessary for function, the ~9-kb and ~4-kb transcripts are excluded from the cytoplasm and either spliced or degraded. In contrast, the fully processed ~2-kb mRNA are constitutively exported to the cytoplasm and used to express Rev, Tat, and Nef. When the levels of Rev in the nucleus are sufficiently high (*lower panel*), the nuclear export of ~9-kb and ~4-kb RNAs is activated and the translation of all viral proteins ensues. (*Left*) The lengths of the RNAs. (*Broken lines*) The approximate regions of RNA that are excised as introns to generate the ~4-kb and ~2-kb mRNAs. (*Ball and stick*) The Rev response element. The ~4-kb mRNAs lack *gag* and *pol* and encode Env, Vif, Vpr, and Vpu, whereas the ~2-kb mRNAs lack this region and most of *env*.

at least as complex as for HIV-1 (Figure 1) (59, 150); as mentioned already, this virus also creates additional transcript diversity by virtue of an internal promoter.

The full-length transcripts of retroviruses have three distinct functions: (*a*) they constitute the genomes of these viruses, (*b*) they serve as templates for translation (Gag and Gag-Pol), and (*c*) they function as precursor RNAs (pre-mRNAs) for the production of diverse subgenomic mRNAs. Because the first two of these functions depend upon localization to the cytoplasm, a mechanism (or mechanisms) must exist for exporting these unspliced RNAs out of the nucleus. Importantly, the translocation of RNAs that harbor functional introns (which, by definition, these RNAs do, as they function as pre-mRNAs) to the cytoplasm is highly unusual. Under normal circumstances, intron-containing pre-RNAs are retained in the nucleus by the interaction of splicing factors (sometimes referred to as commitment factors) until they are either spliced to completion or degraded (21, 124, 153). The same export problem also confronts the incompletely spliced transcripts of HIV-1 and HFV; all these RNAs contain at least one functional intron but must enter the cytoplasm to act as templates for the synthesis of proteins such as Env.

Retroviruses exploit two different, but potentially related, posttranscriptional mechanisms by which their intron-containing mRNAs circumvent nuclear sequestration and are exported to the cytoplasm. The first to be discovered is typified by the Rev *trans*-activator of HIV-1 and is the major subject of this review. Here, a virally encoded protein, Rev, interacts directly with a *cis*-acting target, the Rev response element (RRE), which is present in all incompletely spliced viral mRNAs and induces their nuclear export (Figure 1). This mechanism is employed by all known lentiviruses (21, 47, 54, 80, 91, 113, 132, 137, 184, 187, 201, 214) as well as the human oncoretroviruses, human T-cell leukemia viruses type-I and -II (HTLV-I and -II) (83, 88); in HTLV-I and -II, the protein is known as Rex and the RNA target is termed the Rex response element (RxRE). The second mechanism was initially discovered in Mason-Pfizer monkey virus (M-PMV) (17) and is thought to be utilized by the remaining oncoretroviruses, including ALV (158), as well as the foamy viruses. It comprises a *cis*-acting RNA sequence that functions independently of any virally encoded protein; because these sequences must interface directly with cellular RNA export factors, they have been termed constitutive transport elements (CTEs). In contrast to all retroviral full-length transcripts and the partially spliced mRNAs of HIV-1 and HFV, the various completely spliced viral mRNAs are presumed to exit the nucleus by the same pathway or pathways that are utilized by fully processed cellular mRNAs.

REV FUNCTION

The Essential Role of Rev in HIV-1 mRNA Nuclear Export

As noted above, HIV-1 encodes six regulatory and/or accessory genes in addition to *gag*, *pol*, and *env*. One of the first to be identified was *tat* (an acronym for transcription *trans*-activator), and it was rapidly established that the 86 amino acid Tat protein potently *trans*-activates the expression of all virus genes (106, 107). During the genetic delineation of *tat*, it was noted that disruption of neighboring sequences could give rise to viruses that still expressed Tat but were unable to express Gag, Pol, or Env and were therefore unable to replicate (52, 195). A gene that overlaps with *tat* in the +1 reading frame and encodes a protein of 116 amino acids was immediately recognized; in accordance with its initially described function, this essential gene was termed *rev* (an acronym for regulator of expression of virion proteins).

Most initial Rev studies were performed by using cells transiently transfected with a Rev-responsive reporter plasmid (either proviral or subgenomic) and a Rev cDNA expression vector. By analyzing such cultures with subcellular fractionation and in situ hybridization techniques, a number of groups demonstrated that Rev is absolutely required for the cytoplasmic accumulation of intron-containing ~9-kb and ~4-kb HIV-1 mRNAs (Figure 2) (21, 47, 54, 80, 99, 137, 236). In contrast, the spliced ~2-kb mRNAs are found in the cytoplasm both in the presence and in the absence of Rev. Given that the incompletely spliced ~9-kb and ~4-kb mRNAs encode Gag, Pol, and Env, these findings provided a straightforward explanation of why Rev is essential for virion structural protein expression and therefore for viral replication (summarized in Figure 3) (52, 78, 118, 135, 195). A natural corollary of Rev-regulated mRNA expression is that the ~2-kb mRNAs (which encode Tat and Nef in addition to Rev) accumulate in the cytoplasm prior to the ~9-kb and ~4-kb mRNAs. In other words, HIV-1 gene expression is biphasic in nature and can be segregated into early (Rev-independent) and late (Rev-dependent) stages (112, 171). This hallmark feature of lentivirus replication may have important consequences for in vivo infections and, in particular, for the ability of HIV-1 to establish and maintain a latent state in certain cell populations.

The establishment of assay systems for measuring Rev function enabled many groups to define the sequences within an RNA that render it Rev-responsive and to map the functional domains of Rev (discussed in detail below). Two distinct features of the RNA are essential: (*a*) the *cis*-acting sequence-specific target for Rev, the RRE, and (*b*) elements that ensure that a sufficiently accessible pool of RNA is available in the nucleus to serve as a substrate for Rev. This latter requirement is illustrated by the high levels of Rev-responsive intron-containing HIV-1 transcripts that are detected in the nuclei of transfected (or infected) cells in the presence or absence of Rev (Figure 3) (54, 135, 137). Importantly, the nuclear accumulation of pre-mRNAs is not typical of cellular genes. Most pre-mRNAs are rapidly spliced to completion and efficiently exported to the cytoplasm as fully processed mRNAs (102, 153). The examination of a series of RRE-containing β-globin pre-mRNAs that harbored wild-type or altered splice sites provided insight into how intron-containing viral RNAs escape this fate (21). Here, it was demonstrated that an intermediate (suboptimal) level of splicing efficiency critically underlies Rev's ability to activate the cytoplasmic expression of unspliced RNA. If splicing is too rapid, intron-containing RNAs fail to accumulate in the nucleus and therefore cannot serve as targets for Rev. On the other hand, if splicing is too inefficient (presumably because of a lack of recognition by splicing factors), the primary transcripts are equivalent to intronless RNAs and are able to enter the cytoplasm constitutively—in other words, independently of Rev. It has been proposed, therefore, that the Rev-responsive RNAs present in HIV-1 infected cells are, like cellular pre-mRNAs, retained in the nucleus by interacting splicing factors (21, 81, 126, 205).

Based on Rev's ability to activate the cytoplasmic expression of unspliced RNAs, it was proposed that Rev's principal function was either to induce the nuclear export of incompletely spliced HIV-1 transcripts or, alternatively, to inhibit the splicing of those RNAs. According to the first model, Rev would directly target an RNA for export and, as a result, override nuclear sequestration. In contrast, the second model suggests that Rev would act to remove splicing factors from an RNA such that it would then be recognized by cellular export factors and efficiently transported to the cytoplasm. Although the splicing inhibition model was supported by Rev's ability to prevent splicing in vitro (116), the inability to segregate the effects of splicing inhibition from export activation with a high degree of confidence in transfection-based experiments made it difficult to obtain definitive evidence in favor of one model over the other.

Microinjected *Xenopus laevis* oocytes offer a number of important advantages over transfected cells for studying the nuclear export of RNA, proteins, and complexes thereof (termed ribonucleoproteins, RNPs). First, purified substrates that are biochemically defined are introduced directly into the nucleus. Second, the experiments are performed over a relatively short time frame (minutes or hours), which reduces the likelihood of indirect effects confounding the outcome. Third, it is relatively straightforward to separate oocytes into nuclear and cytoplasmic fractions. Accordingly, a number of distinct signal-mediated RNA export pathways have been operationally defined in *Xenopus* oocytes (104). Using this system, it was demonstrated that the coinjection of recombinant Rev protein with RRE-containing pre-mRNAs resulted in their export, together with lariat RNAs, out of the nucleus (58). Although this result was consistent with earlier transfection studies, a more significant finding was that a synthetic RNA that lacked recognizable introns or splice sites and was a poor substrate for export because of its AppppG 5'-cap (mRNAs have m^7GpppN-caps at their 5'-termini) (79) was also rendered export-competent by the RRE in the presence of Rev. More recently, two small nuclear U-rich RNAs (U snRNAs) that are both localized exclusively to the nucleus, namely U3 and U6, were also shown to be sensitive to Rev-mediated export when expressed as RRE chimeras (57, 165). Because the U3 and U6 RNAs are retained in the nucleus by mechanisms that are distinct from the splicing factor-mediated retention of pre-mRNAs (11, 212, 213), it has been concluded that Rev can function independent of splicing and that Rev is, therefore, a bona fide activator of RNA nuclear export.

In certain experimental configurations, Rev has been shown to have additional effects on unspliced viral RNAs. First, some transfection studies have shown that the fold effect of Rev on reporter gene expression exceeds Rev's effect on unspliced RNA distribution; it has been suggested, therefore, that Rev enhances the translation of Rev-responsive RNAs (5, 32). Second, it has been demonstrated that the expression of Rev dramatically stabilizes RRE-containing pre-mRNAs in the nucleus (54, 135); for example, Rev increases the nuclear half-life of the ~9-kb RNA from ~10 min to more than 6 h in provirus-containing human T cells. In our view, however, these effects should be considered secondary to Rev's primary function as a sequence-specific nuclear export factor.

One additional feature that is often attributed—in our opinion, inappropriately—to Rev is the ability to downregulate its own expression as well as that of the other early viral gene products. This idea stems from initial transient transfection studies where the unspliced RNAs were comparatively stable and were, presumably, exported at the expense of splicing. In contrast, experiments that utilized full-length proviruses have shown that spliced RNA is also, in fact, more abundant in the presence of Rev. This increase in spliced RNA abundance is presumably a consequence of the dramatic increase in precursor RNA stability (135). The latter situation is more reflective of virus infection; thus we consider it unlikely that Rev functions to suppress early gene expression.

Features of Rev-Responsive RNAs

Rev-responsive RNAs contain two features required for regulation: (a) the RRE and (b) sequences that determine nuclear retention in the absence of Rev. The RRE is a large RNA structure that resides within the env intron [between 5'-splice site 4 and 3'-splice site 6 (Figure 1)] and is, therefore, present in all ~9-kb and ~4-kb mRNAs. Although it was initially described as a ~210-nucleotide (nt) sequence, computer-assisted RNA folding analyses indicated that this region coincided with a complex 234-nt stem-and-loop structure (Figure 4A) (137). Importantly, the validity of this structure has been confirmed by in vitro and in vivo structure-probing analyses using a variety of chemical and enzymatic methodologies (22, 114). Of note, more recent studies have now extended the full RRE to 351 nt (141).

The specific sequences within the RRE that determine Rev responsiveness are surprisingly limited; in fact, recognition by Rev appears to be imparted by the presence of a single high-affinity binding site. A combination of RRE mutagenesis (39, 86, 89, 139), in vitro bi.. :ing (29), chemical modification interference (115, 216), and iterative in vitro genetic selection assays (6) have mapped this Rev binding site to stems IIB and IID (Figure 4A; refer to legend for details). Nuclear magnetic resonance (NMR) studies of a Rev-RRE complex that consisted of a 34-nt RNA hairpin bound to a 23 amino acid Rev peptide that contained residues 34 to 50 have subsequently elucidated the structure of the protein-bound stem IIB/IID region and identified the points of contact between Rev and the RRE (7). Although this stem region is helical in nature, the local structure is substantially distorted by the formation of two non-Watson-Crick purine-purine base pairs (G47-A73 and G48-G71) (6). The resulting widening of the helix's major groove by ∼5 Å allows the α-helical RNA-binding domain (RBD) of Rev to enter the major groove and contact specific nucleotides. It has been proposed that the G-G base pair, although not actually in direct contact with Rev, is primarily responsible for establishing this "open" conformation.

Even though the RRE harbors only a single high-affinity Rev binding site, in vitro binding and footprinting studies using full-length Rev and larger RRE fragments have demonstrated that multiple Rev molecules bind to single RNA (29, 33, 86, 114, 139). The mechanism that underlies multimerization as well as its functional significance is discussed in the following sections. An important question regarding RRE structure is, Why is it so complex? One likely explanation is that its free energy of formation ($\Delta G = -115.1$ kCal/mol^{-1} for the 234-nt element) ensures appropriate folding and presentation of the high-affinity Rev binding site; it is also possible that the remaining stems and loops provide preferred secondary Rev binding sites (114). In summary, the RRE serves as the docking site for Rev on RNAs that are confined to the nucleus. This conclusion is supported by two lines of evidence: First, the RRE is the only HIV-1 sequence that needs to be transferred to a heterologous RNA for Rev responsiveness to be acquired (21, 57, 58, 165, 205); second, the RRE can be functionally replaced by heterologous RNA sequences provided that Rev is also appended to the appropriate cognate RNA-binding domain (143, 219).

Although the *Xenopus* oocyte experiments described above demonstrate that Rev-activated RNA export is not dependent on one mode of nuclear retention, many groups have described elements in intron-containing HIV-1 mRNAs that render cytoplasmic expression Rev-dependent. These sequence elements fall into two basic categories. First, there are the splice sites themselves; as discussed earlier, these are utilized inefficiently by the cellular splicing machinery such that appropriate pools of unspliced RNAs are able to accumulate in the nucleus (21, 126, 205). As with the regulation of cellular pre-mRNA splicing (10, 222), the features of viral RNAs that dictate splicing efficiency are likely to be both complex and variable, particularly because the assorted 5'- and 3'-sites are used with vastly differing efficiencies (173). For example, the branchpoint

A

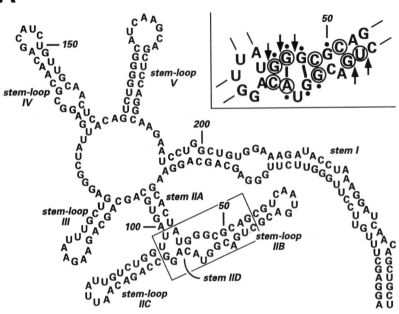

B

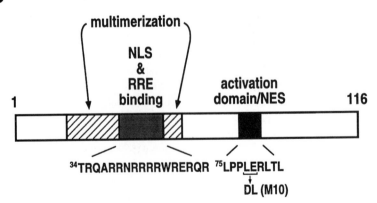

sequence as well as the polypyrimidine tract of the 3′-splice site that borders the 5′-terminus of the second coding exons of *tat* and *rev* are both suboptimal (197). Furthermore, as has been shown for many cellular splice sites, the activity of HIV-1 splice sites is also modulated by the combined influences of exonic splicing enhancer and silencer elements (3, 198).

Sequence elements of the second category have been termed *cis*-acting repressive sequences (CRSs) and/or instability sequences (INSs) (26, 131, 180, 186, 188, 190). When appended in *cis* to heterologous reporter genes, these elements inhibit expression of those genes in a manner that is itself reversed by the action of Rev/RRE. CRSs and INSs are distributed throughout the regions of HIV-1 that form introns and are frequently adenine-uracil (AU)-rich in character. The contributions that these sequences make to HIV-1 gene expression in the context of viral infection are difficult to evaluate, as the experimental RNA substrates that have been used to identify them have lacked functional introns and are therefore quite dissimilar to Rev-responsive viral pre-mRNAs. Nevertheless, the identification of host cell factors that interact with these sequences (160) should shed light on the possible mechanisms by which they may assist in the nuclear sequestration and degradation of RNA. Interestingly, AU-rich sequences have also been shown to be important determinants of instability in a number of short-lived cellular mRNAs, for example, those encoding a number of proto-oncogenes or cytokines (23, 103). However, unlike the ∼9-kb and ∼4-kb HIV-1 transcripts, these mRNAs are degraded in the cytoplasm as fully processed transcripts. Whether there is a mechanistic link between the rapid turnover of these processed mRNAs and the marked nuclear degradation of intron-containing HIV-1 RNAs in the absence of Rev remains to be determined.

←

Figure 4   Domain structures of (*A*) the Rev response element (RRE) and (*B*) Rev from HIV-1$_{HXB}$. (*A*) The RRE is drawn as a 234-nucleotide (nt) structure using the generally accepted nomenclature for stems and loops. The primary Rev binding site is *boxed* and shown in greater detail in the *inset*. Watson-Crick base pairs are represented with the bases in close proximity (for example, C49-G70). The two noncanonical purine-purine base pairs (G47-A73 and G48-G71) are drawn with *lines* between the bases; these lines also reflect the widening of the major groove of the helix in this region. The bases that are *circled* were invariant in iterative in vitro genetic selection experiments (6). Chemical modification of the purines marked with *dots* resulted in severely reduced Rev binding in vitro (115, 216), as did modification of the sugar-phosphate backbone at the positions indicated (with *arrows*) (115). (*B*) The 116 amino acid Rev protein harbors regions that mediate RNA (RRE) binding and nuclear localization (*gray box*), are required for protein multimerization (*hatched boxes*) and function as the NES/activation domain (*solid box*). The amino acid sequences of the arginine-rich and leucine-rich domains (residues 34 to 50 and 75 to 83, respectively) are indicated *below*, together with the changes at positions 78 and 79 that are present in the RevM10 mutant.

Domain Structure of Rev

The 116 amino acid HIV-1 Rev protein has a number of attributes that are essential for function in vivo. In particular, this ~18-kDa phosphoprotein is capable of being imported into the nucleus, binding specifically to the RRE, forming multimers, and directing the nuclear export of large RRE containing RNP complexes. The Rev protein can be divided into two relatively discrete domains that have been delineated by extensive mutagenesis (Figure 4B). The amino-terminal domain contains an arginine-rich sequence that serves both as the nuclear localization signal (NLS) (15, 120, 133, 167) and as the RNA-binding domain (RBD) (15, 33, 91, 134, 159, 231a, 232). It is flanked on both sides by less well defined sequences that are required for multimerization (12, 130, 134, 159, 232). The leucine-rich carboxy-terminal domain (90, 133, 138, 144, 159, 220, 223), originally known as the activation (or effector) domain, contains the nuclear export signal (NES) (56, 146, 225). The NLS and NES each function by accessing cellular pathways for nuclear import and export.

Although indirect immunofluorescence experiments show that Rev accumulates predominantly in the nucleoli of transfected cells (31, 54, 133, 145), its role as an activator of RNA export requires that it be, in actuality, in dynamic equilibrium between several subnuclear and cytoplasmic sites. The arginine-rich region that spans residues 34 to 50 mediates both nuclear import and nucleolar targeting; the demonstration that the transfer of this sequence to a heterologous protein resulted in the acquisition of import capacity established this domain as a bona fide NLS (120, 167). Importantly, this motif has also been shown to bind specifically to the RRE (115, 209). The NMR-based studies discussed above demonstrated that four arginine residues (at positions 35, 39, 40, and 44) participate in base-specific contacts with the high-affinity binding site in stems IIB and IID of the RRE, whereas other residues (the threonine at position 34 and the arginines at positions 38, 41–43, 46, and 48) contact the sugar-phosphate backbone (Figure 4A) (7). However, the contributions that these amino acids make to binding in vivo have not been fully evaluated through assays that measure Rev function. In fact, it has been shown that single substitutions of each of these specific residues for leucine or lysine do not inhibit Rev activity in transient transfection assays (82).

Through in vitro RRE-binding assays, it has been shown that Rev can bind to its high-affinity binding site as a monomer (28, 29, 134, 216) and that this binding results in some localized melting of the RNA structure (36, 174, 210) as well as stabilization of the protein helix (209, 210). Additional Rev molecules then bind and multimerize via a combination of cooperative protein-protein and protein-RNA interactions (34, 235) such that eight or more Revs are bound to a single RRE (33, 114, 141). Mutant Rev proteins that are deficient for multimerization are still able to occupy the high-affinity site and form a binary complex but are unable to assemble into higher-order complexes (130, 134, 216). Although the order of Rev binding has not been determined in vivo, structure-probing studies using intact yeast (*Saccharomyces cerevisiae*) cells treated with dimethyl sulfate (DMS) have shown that a multimerization-deficient mutant can still bind to the high-affinity site but that a conformational change in stem IIA that is induced by wild-type Rev binding does not occur (22). In addition to the regions of Rev indicated in Figure 4B, multimerization in vivo can be influenced by the integrity of the NLS/RBD and NES domains (12, 130). It has also been shown that multimerization can occur in the absence of the RRE both in vitro (28, 59) and in vivo (12, 92, 159). The possible contribution of bridging factors to multimerization in vivo should not be overlooked, however.

The RNP complex that is formed by the interaction of multiple Revs with an RRE-containing RNA is targeted for export by the activation domain. Early mutagenesis experiments demonstrated that disruption of this leucine-rich domain (residues 75 to 83) yields proteins (for example, RevM10) that localize to the nucleus, bind RRE-containing RNA, multimerize, and yet do not facilitate the export of those RNAs from the nucleus (35, 93, 133, 134, 159, 216, 220, 232). Because such mutants also exhibit a dominant negative (*trans*-dominant) phenotype (133, 144, 220) and because repeats of similar leucine-rich motifs are involved in protein-protein interactions between diverse proteins (119), it was proposed that this domain interacts with cellular proteins required for Rev function (138). Rev has been shown to shuttle rapidly between the nucleus and the cytoplasm in heterokaryon experiments (145) and also to accumulate in the cytoplasm in the absence of RNA polymerase II transcription (109, 145, 175, 199, 227); this latter phenomenon is poorly understood but has also been described for some shuttling hnRNP proteins (168). Importantly, mutations in the activation domain that inhibit Rev function also abolish both shuttling and cytoplasmic accumulation in response to RNA polymerase II inhibition (145, 199, 206, 227). More recently, it has been demonstrated that the activation domain is, in fact, a nuclear export signal (NES) in that it will promote the export of a heterologous protein when appended to it (56, 146, 225). Indeed, Rev together with the inhibitor of cAMP-dependent protein kinase (PKI) were the first proteins shown to harbor bona fide NESs. A consensus sequence for a functional leucine-rich NES has been determined for HTLV-I Rex by using an in vivo randomization-selection assay (13). These studies showed that although there is no strict sequence requirement, a loose consensus of relatively evenly spaced bulky hydrophobic (frequently leucine) residues could be discerned (Figure 5A). It is now known that many cellular and viral proteins harbor hydrophobic motifs similar to the NESs of Rev and Rex and that these elements are important for nuclear export (these are discussed in more detail below).

A Leucine-rich NESs

HIV-1 Rev	L P P · L E R · L T L
HTLV-I Rex	L S A Q L Y S S L S L
visna Rev	M V G · M E N · L T L
E4-34 kDal	M V · L T R E E L V L
RanBP1	V A E K L E A · L S V
Gle1p	L P · · L G K · L T L
TAP (Mex67p) repeat	L X X · L X · L X_2 N X L
IκBα	L G Q · L T · L E N L
PKI	L A L K L A G · L D I
Leu-rich consensus:	L X_{2-3} λ X_{2-3} L X L/I

FIV Rev	K A F K K M M T D L E D R F R K L F G S P S K D E Y T
EIAV Rev	G P L E S G Q W C R V L R Q S L P E

B Arginine-rich NLSs

HIV-1 Rev	T R Q A R R N R R R W R E R Q R
HTLV-I Rex	M P K T R R R P R R S Q R K R P P T

IBB consensus:
RL--FKNKG-----E-RRRR-EV-VELRKAKKDEQ--KRRNV---EED--SP

Figure 5 Amino acid sequences of leucine-rich nuclear export signals (NESs) and arginine-rich nuclear localization signals (NLSs). (*A*) The following (*from top to bottom*) are aligned with a consensus leucine (Leu)-rich NES sequence that was derived in vivo (13): the leucine-rich transferable NESs of four viral proteins [HIV-1 Rev (56, 146, 225), human T-cell leukemia viruses type-I (HTLV-I) Rex (111, 162), visna Rev (146), and adenoviral E4-34 kDal (40)]; one cellular transport factor [RanBP1 (177)]; two RNA-binding proteins [Gle1p (151) and the TAP (human Mex67p) consensus of four repeat regions (192)]; and two cellular proteins [IκBα (4) and the inhibitor of cAMP-dependent protein kinase (PKI) (225)]. The *dots* in the sequences indicate gaps introduced for the purpose of alignment, X represents any amino acid, and λ indicates amino acids with bulky hydrophobic side chains. The relatively conserved hydrophobic residues are shaded in *gray boxes*. The non-leucine-rich NESs of FIV Rev (140) and EIAV Rev (146) are shown for the sake of comparison. (*B*) The arginine-rich NLSs of Rev (127, 167) and Rex are shown (194) together with the importin-β-binding (IBB) domain consensus sequence (74, 224). The *dashes* in the consensus sequence indicate nonconserved residues.

The NLS/RBD and NES domains of Rev have been shown to be interchangeable with analogous domains from other proteins. In other words, Rev has a modular structure that is reminiscent of many transcription and RNA processing factors. For example, chimeric proteins made by fusing Rev to the RNA-binding sequences of the MS2 bacteriophage coat protein have been shown to export RNAs in which the RRE has been replaced with the MS2 operator (143, 219). As expected, these chimeric proteins tolerate point mutations in Rev's NLS/RBD, which have been shown to disrupt RBD function. Mutations that disrupt nucleolar accumulation are also well tolerated, indicating that nucleolar localization may not be essential for Rev function. Surprisingly, even though the sequence of the arginine-rich domain can be scrambled with no noticeable loss of function relative to fusion proteins carrying the wild-type sequence, its deletion generates proteins that are only marginally functional (143). This implies that the spacing or the overall basic character of this region, rather than its primary sequence, is essential for activity in the context of the chimeras. Because RNA binding and nuclear localization are imparted by the MS2 sequences, one possibility is that the arginine-rich domain participates in multimerization (as noted above). It was also found that these chimeric proteins are more active when two or more copies of the operator sequence are present in the RNA (143); this also seems to underscore the importance of Rev multimerization for RNP export.

As with the alteration in RNA target specificity by introduction of heterologous RNA-binding domains, substitutions of the Rev NES with the leucine-rich NESs from other proteins also reconstitutes functional proteins. For instance, the leucine-rich elements of other Rev proteins (66, 72, 140, 215), Rex (90, 223), PKI (64), transcription factor IIIA (TFIIIA) (65), the α-inhibitor of NF-κB (IκBα) (67), and the fragile-X mental retardation protein-1 (FMR-1) (63) are fully functional when substituted for the Rev NES. While these assorted chimera-based experiments demonstrate that Rev has a modular structure, matters are not as simple as they appear. In particular, it has not been possible to construct a "Rev-less" Rev that is composed entirely of heterologous protein sequences. This suggests that other attributes of Rev itself are critical for biological activity. In addition to our speculation of the requirement for arginine residues in the Rev-MS2 chimeras, we speculate that essential aspects of multimer formation cannot be accurately recreated by non-Rev sequences.

The discovery and definition of peptide NESs that mediate rapid, sequence-specific nuclear export resolved a long-standing controversy regarding the translocation of proteins and RNPs from the nucleus to the cytoplasm. One school of thought had been that export from the nucleus was a default pathway that could be overcome by nuclear retention (185). The alternative school of thought was that export was an active, signal-mediated process somewhat analogous to the nuclear import of proteins and RNPs. The continued analysis of Rev has therefore been instrumental in establishing the latter model as a fundamental principle that underlies the nucleocytoplasmic transport of many molecules and complexes.

Pathways of Nuclear Transport: A General Overview

Given that Rev shuttles between the nucleus and the cytoplasm (145), a comprehensive description of Rev function requires some discussion of cellular nuclear import and export pathways. As is already evident, nuclear export cannot, in fact, be discussed without engaging the Rev issue. Although there are numerous recent reviews of nuclear transport and nuclear pore structure (37, 38, 42, 73, 75, 152, 155, 217), we provide a brief overview of the current models for nuclear transport. Importantly, many of the essential features of import and export appear to be well conserved among all eukaryotes.

Transport into and out of the nucleus is highly regulated for most molecules and macromolecular complexes and occurs via nuclear pore complexes (NPCs). These ~125-MDa gated structures contain at least 50 different proteins, termed nucleoporins, which are often characterized by domains that contain numerous repeats of the di-peptide phenylalanine-glycine (FG-repeat domains) (38, 42). The architecture of the NPC is essentially that of a wheel with spokes surrounding a central transporter structure. Filaments extend from the NPC into the cytoplasm, while the nuclear face exhibits a basketlike structure. Signal-mediated protein import into the nucleus takes place in a number of discrete steps (73, 75, 155). Initially, the molecule to be imported is recognized in the cytoplasm by one of an emerging family of ~100-kDa transport receptors. The first of these receptors to be described was importin-β (also known as karyopherin-β or PTAC97). This ~97-kDa protein acts together with a ~60-kDa protein, importin-α, to mediate the import of proteins that contain "classical" basic-type NLSs. It is now recognized that importin-α is a bridging factor that connects basic-type import substrates with importin-β and that importin-β interacts directly with the NPC. More recently, a number of additional import receptors that each share sequence similarity with importin-β and that participate in diverse transport pathways have been identified (217). In contrast to the classical pathway in which importin-α serves as a bridge between the substrate and importin-β, these receptors have been shown to bind directly to the NLSs of their respective import substrates. Once the receptor/substrate complex is formed, it docks at the cytoplasmic filaments of the NPC via a direct interaction between the import receptor and the filament proteins. Although currently ill understood, translocation through the pore is also thought to be facilitated by sequential direct interactions between the import receptor and various nucleoporins (42, 75).

Other essential components of the nuclear import pathway include NTF-2 and the GTPase Ran (73). NTF-2 is an essential protein in yeast that binds to nucleoporins, importin-β, and RanGDP and is proposed to play a role in translocation through the NPC. The small GTPase Ran is also an essential component of nuclear import. Although the exact mechanisms of Ran function are still being elucidated, it is thought that Ran plays at least two distinct roles during import. The first is to provide the energy, via GTP hydrolysis, that is necessary for translocation through the pore; the details of how this energy is coupled to locomotion are, however, still mysterious. Ran also provides directionality to the transport process by virtue of the asymmetry of RanGTP/RanGDP that exists across the nuclear membrane (73, 100). This gradient is created as a consequence of the subcellular distributions of the guanine nucleotide exchange factor (GEF), RCC1 (regulator of chromosomal condensation 1), the Ran GTPase-activating protein (RanGAP), and Ran binding protein 1 (RanBP1). Because RCC1 is a chromatin-associated nuclear protein, GDP is exchanged for GTP only in the nucleus. In contrast, RanGAP and RanBP1 (which potentiates RanGAP function) are both localized predominantly in the cytoplasm and are associated with the cytoplasmic filaments of the NPC; consequently, most GTP hydrolysis occurs at the cytoplasmic face of the NPC. All known import receptors bind to RanGTP with high affinity. However, binding is mutually exclusive with the binding of import substrates or, in the case of the classical pathway, of importin-α. Thus, Ran that is in the cytoplasm is bound to GDP and is not inhibitory to receptor-substrate interactions, whereas nuclear Ran, which is bound to GTP, induces the efficient dissociation of such complexes. RanGTP is therefore responsible for terminating import reactions and for liberating the receptors in readiness for export back to the cytoplasm.

Export from the nucleus was historically perceived as an RNA-specific issue that represented a default event that occurred in the absence of complete nuclear retention. Contrary to this view, saturation studies using microinjected *Xenopus* oocytes demonstrated that mRNA, tRNA, U snRNA, and 5S rRNA require specific factors and distinct pathways for export (104, 152, 153). Because RNAs leave the nucleus in the context of large RNP complexes rather than as naked RNA (37, 152, 153), it is now thought that signals that determine RNA export probably reside within the associated proteins. This concept, together with the discovery that many viral and cellular RNA-binding proteins shuttle between the nucleus and the cytoplasm, prompted numerous studies, which have culminated in the definition of a variety of peptide NESs.

These NESs can be classified in accordance with their primary amino acid sequences. First, and as discussed already, an ever-increasing number of proteins have been shown to contain leucine-rich NESs. Many of these proteins have been implicated in RNA metabolism, for example, Rev (56, 146, 225), TFIIIA (65), the *S. cerevisiae* GLFG-lethal 1 protein (Gle1p) (151), FMR-1 (63), and the *S. cerevisiae* ~67-kDa mRNA export factor (Mex67p) (192). Several others, including IκBα (4, 67), PKI (225), and MAP kinase kinase (MAPKK) (70), are involved in the regulation of other critical cellular functions. In addition to leucine-rich motifs of this type, several other sequences that can function as NESs have been identified. Oddly, the NES sequences of the Rev proteins from FIV (140) and EIAV (66, 140, 146) bear little resemblance to the leucine-rich NES (Figure 5). Unlike the leucine-rich domains, the critical sequence determinants of these elements have not been characterized extensively. Two other types of NESs, the hnRNP A/B M9 sequence (148) and the hnRNP K KNS (K nuclear shuttling) sequence (149), that also bear no primary sequence resemblance to leucine-rich NESs have been described. In addition to determining export, these sequences also function as NLSs; as such, they are therefore involved in both legs of the shuttling cycles of these hnRNP proteins.

From the outset, it was considered likely that NESs, like the peptide NLSs that determine import, would function by interacting with proteins that would specifically access receptor-mediated export pathways. Initial genetic evidence in *S. cerevisiae* suggested that such export receptors could be members of the importin-β superfamily (191). This notion has subsequently been proven with the identification of CRM1/exportin 1 as the export receptor for Rev, MAPKK, IκBα, PKI, and DIS 1 kinase (Dsk 1) (60, 69, 154, 161, 196) (discussed in detail below) and CAS/Cse (cellular apoptosis susceptibility/chromosome segregation gene 1) as the export receptor for importin-α (121). Importantly, binding of RanGTP to the export receptor is required for formation of the export receptor/substrate complex. The events that take place following complex assembly are currently under investigation and may vary for the export of RNPs and between different RNA-free NESs. In particular, RanGAP-induced hydrolysis of GTP by Ran is not required for the export of a leucine-rich NES (100, 176) but does appear to be important for both RNP export (100) and for the postexport disassembly of importin-α from CAS at the cytoplasmic filaments of the NPC (121).

Cellular Proteins That Interact with the Rev NES

For the reasons expounded above, there has been a great deal of interest in cellular proteins that interact with the NES (activation domain) of HIV-1 Rev. Those that have been described are, in chronological order, eIF-5A (181), several FG-repeat-containing nucleoporins including Rip/Rab (14, 67, 68, 202, 204), and CRM1/exportin 1 (60, 69, 154). The interaction between the ~19-kDa protein eIF-5A (eukaryotic initiation factor 5A) and the NES was identified by cross-linking proteins from cellular extracts to synthetic peptides and then sequencing a specifically bound protein (181). The significance of this interaction was evaluated through an assay system that lacks endogenous eIF-5A, namely *Xenopus* oocytes. By using DNA expression vectors to produce Rev and the reporter RNA, coinjection of eIF-5A was shown to be required for Rev-mediated *trans*-activation. These data are, therefore, in contrast to other oocyte experiments where purified Rev efficiently activated the export of nuclear-injected RNAs in the absence of added eIF-5A (58, 165, 182). Some potential explanations for these discrepancies can be envisaged. First, the oocytes utilized by the different groups could have been at different developmental stages, thus giving rise to conflicting results. Second, the experiments with plasmid DNA templates are more complicated since RNA production, protein synthesis, and Rev nuclear import must all occur as prerequisites for Rev function. It is therefore plausible that indirect effects could have had an impact on these experiments; indeed, it has been shown that Rev itself is imported into oocyte nuclei inefficiently (202). These possible caveats aside, more recent results further support the conclusion that eIF-5A has a role in Rev function. In particular, eIF-5A mutant proteins that retain their ability to bind Rev both inhibit Rev nuclear export in somatic cell microinjection experiments and can block HIV-1 replication (9). Interestingly, although originally implicated in translational initiation, the normal function of eIF-5A is unknown (164). eIF-5A is the only protein known to contain the amino acid hypusine; this posttranslational modification of a lysine residue is not only conserved in all eukaryotes but is also essential for eukaryotic cell proliferation.

FG-repeat–containing nucleoporins have also been implicated in the export of Rev. This idea originally emerged from the identification of Rip/Rab (Rev interacting protein/Rev activation domain–binding protein) as an NES interacting protein in yeast two-hybrid screening experiments (14, 68, 204). Importantly, the interaction between the ~60-kDa Rip/Rab and Rev correlated precisely with the functionality of an extensive series of NES mutations. Indeed, the ability to interact with Rip/Rab was used to define the consensus leucine-rich NES described earlier (13) (Figure 5A). Rip/Rab has been shown to be an inessential protein in *S. cerevisiae*, to exhibit nucleoplasmic as well as NPC localization, and to modestly enhance Rev function when overexpressed in transfected cells (14, 68). It was dubbed a nucleoporin by virtue of its FG-repeat sequences and the fact that its closest relatives (by amino acid sequence similarity) in yeast and mammals are the FG-nucleoporins Nup153p and CAN1, respectively. Because Rev is still functional, albeit to a lesser extent, in an *S. cerevisiae* mutant strain that lacks Rip/Rab, it has been concluded that Rip/Rab may be involved in, but cannot be essential for, Rev function (204). Recently, however, this protein has been shown to be essential for the nuclear export of *S. cerevisiae* heat shock mRNAs (183, 203). Interestingly, overexpression of Rev (but not of inactive Rev NES mutants) partially blocks the export of heat shock mRNAs, a finding that implies that these mRNAs and the Rev NES share some export pathway components—possibly including Rip/Rab. Rev also interacts by two-hybrid analysis with the FG-repeat regions of a number of additional distinct yeast and vertebrate nucleoporins (67, 202); moreover, microinjection of a subset of these FG-repeats into *Xenopus* oocytes can modestly inhibit Rev export (202). Recent experiments have demonstrated, however, that the interactions between the NES and these FG-nucleoporins are indirect and are bridged by the export receptor CRM1/exportin 1 (154). Specifically, these interactions are not detected in *S. cerevisiae* strains lacking the CRM1 gene.

Exportin 1, an ~112-kDa member of the importin-β family of transport receptors, was originally named CRM1 (chromosome region maintenance gene 1) because conditional mutations in this gene produced *Schizosaccharomyces pombe* strains that exhibited deformed chromosomes in the nonpermissive conditions (1). This mutant phenotype is probably a secondary consequence of this protein's role in nuclear transport. CRM1 was implicated in transport, rather circuitously, by studies involving the antibiotic leptomycin B (LMB). In the course of screening low-molecular-weight compounds for the ability to interfere with Rev function, LMB was shown to be an inhibitor of both Rev-mediated RNA export and HIV-1 replication (228). Prior to this, studies of LMB toxicity in *S. pombe* had revealed that one class of mutants that conferred insensitivity to LMB mapped to the CRM1 gene (156). Further clues that CRM1 was a transport receptor came from (*a*) its association with CAN/Nup214 in vivo (61); (*b*) its localization to the nucleoplasm, nucleolus, and NPC (1, 61); and (*c*) the finding that NPC localization can be competitively inhibited by the FG-repeat domain of CAN/Nup214 (61). Most significantly, it was noted that the CRM1 protein shared sequence similarity with the importin-β superfamily of transport receptors (61).

These assorted facts as well as the notion that importin-β family members might be involved in RNA export (191) prompted several different groups, each utilizing different assay systems, to test whether CRM1 could mediate the nuclear export of leucine-rich NESs. Exportin 1 was shown to shuttle between the nucleus and the cytoplasm (196) and to be required for Rev function in *S. cerevisiae* (154). As would be predicted for a bona fide export receptor, exportin 1 was shown to bind in vitro to NESs in the context of rabbit reticulocyte lysates and bacterial extracts (60, 161), in the yeast two-hybrid system (154, 196), and using purified recombinant proteins (69). This protein also interacts in vitro with Ran (60, 196), but only in the presence of an NES and when Ran is in its GTP-bound state; the nuclear localization of RanGTP presumably ensures that formation of an exportin 1/NES/RanGTP complex can occur only in the nucleus (60). Once assembled, this complex is then thought to be competent for export through the NPC. By analogy with protein import and because exportin 1 interacts with nucleoporins in the two-hybrid system (154), translocation is likely to be facilitated by the binding of exportin 1 to the NPC. As discussed above, the two-hybrid interactions of leucine-rich NESs with nucleoporins are indirect and are ablated in *S. cerevisiae* strains that lack exportin 1 (154).

Experiments performed using LMB have helped confirm the role of exportin 1 in nuclear export. LMB associates with the exportin 1 proteins of humans and *S. pombe* (but not of *S. cerevisiae*) and induces conformational changes that culminate in the disruption of the exportin 1/NES/RanGTP complex and, hence, the inhibition of leucine-rich NES nuclear export (60). Importantly, the inhibitory effects of LMB on leucine-rich NES export in the human (161), *Xenopus* (60), and *S. pombe* (69) systems have been shown to be overcome by overexpression of exportin 1 (60). Because LMB inhibits Rev NES and U snRNA export in *Xenopus* oocytes, it can be inferred that these substrates may both utilize exportin 1-mediated pathway(s) of nuclear export (60). In contrast, mRNA, tRNA, and importin-α are all insensitive to the action of LMB and are likely, therefore, to access alternative pathways of export. As noted above, importin-α utilizes another importin-β-related protein, CAS, as its export receptor. Interestingly, the involvement of exportin 1 in mRNA export in *S. cerevisiae* is less clear as temperature-sensitive null strains rapidly accumulate polyA mRNA in the nucleus following shift to the nonpermissive temperature (196).

The Rev Nuclear Transport Cycle

As stated earlier, the Rev shuttling cycle is dependent on the utilization of cellular nuclear import and export pathways. We propose that the model shown in Figure 6 reflects many of the critical aspects of this cycle. Following translation, the cycle is initiated by nuclear import. Recent studies have implicated importin-β as the receptor that mediates the import of both Rev and its HTLV-I counterpart, Rex (87, 163). Specifically, Rev binds directly to importin-β in vitro and Rex is efficiently imported into the nucleus in the absence of importin-α in permeabilized cell import assays (2). Of note, sequence comparisons between the arginine-rich NLSs of Rev and Rex and the importin-β-binding (IBB) domain of importin-α reveal that these domains are similar (Figure 5B). We suggest, therefore, that Rev and Rex (like the IBB domain itself) utilizes importin-β in the absence of a bridging factor. Following translocation into the nucleus, the interaction of importin-β with RanGTP presumably induces disassembly of the importin-β/Rev complex and releases Rev into the nucleoplasm (87, 163). Because the arginine-rich NLS of Rev also functions as the RBD (Figure 4B), dissociation results in Rev becoming available for binding to the RRE.

Rev initially binds to its primary site in stems IIB and IID. Additional copies of Rev are then bound to secondary sites within the RRE by a series of cooperative protein-protein and protein-RNA interactions. Precisely when RRE binding commences in terms of HIV-1 RNA biogenesis is unclear. One could predict that binding would be preceded by recognition of the viral RNAs by the splicing commitment factors that impart nuclear retention. Interestingly, it has been shown by using somatic cells that Rev uses only nascent transcripts as substrates for export (97). This finding implies that Rev binding as well as interactions with retention factors may occur cotranscriptionally. This notion is consistent with current thinking regarding the intimate coupling of transcription with RNA processing and fate determination in

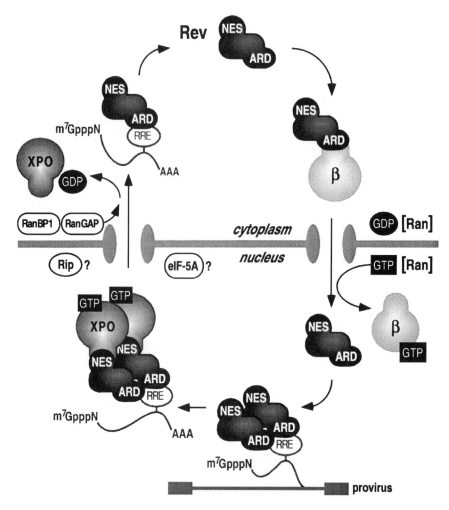

Figure 6   The Rev nuclear transport cycle. This model reflects various critical interactions between Rev, RNA, and cellular proteins. Rev binds to importin-β (β) or the Rev response element (RRE) via its arginine-rich domain (ARD), to exportin 1 (XPO) via its nuclear export signal (NES), as well as to itself (multimerization). Ran is shown in its GDP-bound (*oval*) and GTP-bound (*rectangle*) states. The m^{7}GpppN 5′-cap and the 3′-polyA tail of the RNA are both indicated. The precise molecular roles of Rip/Rab and eIF-5A in the transport cycle remain to be defined.

eukaryotes (142, 200, 231). It should be pointed out, however, that nascent transcripts are not required for a Rev response in all systems as *Xenopus* oocytes support the Rev-mediated export of microinjected RNA substrates (58, 165, 182).

Multimerization, while unequivocally essential for Rev function, remains poorly understood at the molecular level. Even though monomeric leucine-rich NESs apparently still serve as substrates for exportin 1–mediated protein export (60, 69, 161, 196), the binding of a single Rev to the RRE is incapable of activating RNP export in somatic cells (133, 134). This characteristic strongly implies that multiple export receptors need be recruited to with the Rev/RRE RNP for export to be activated. The observation that coexpression of wild-type and NES-deficient Rev proteins results in severely inhibited RNP export relative to the expression of wild-type Rev alone is entirely consistent with this conclusion (58, 133, 220). We speculate that exportin 1 multiplicity may be required either for facilitating multiple (simultaneous) interactions with the NPC during RNP translocation or, alternatively, for overcoming the interactions that are otherwise responsible for nuclear retention. It is also possible that Rev multimerization is required for the formation of stable Rev/RRE complexes in vivo. However, the DMS-mediated modification experiments performed in yeast tend to argue against this latter notion because a multimerization-deficient Rev mutant was found to occupy the primary Rev binding site as efficiently as wild-type Rev (22).

Once the appropriate Rev/RRE/exportin 1/RanGTP complexes are assembled, they are presumably targeted for efficient export through the NPC. In view of the similarities between nuclear import and export, the steps involved in this process presumably include migration to the nuclear face of the NPC, docking, interactions between exportin 1 and FG-repeat nucleoporins (including Rip/Rab, as described above), translocation, and disassembly of the complex in the cytoplasm (termination). Given that it was only recently discovered that exportin 1 is the export receptor for Rev, it can be expected that new information concerning these processes will be forthcoming. As discussed earlier and in more detail in the following section, studies in *Xenopus* oocytes have shown that the export of Rev-responsive RNAs (as well as other classes of RNA) is inhibited by a mutant derivative of Ran (RanQ69L) that is unable to hydrolyze GTP (117). It is also assumed that Rev is released from the RRE in the cytoplasm and then imported back into the nucleus in preparation for further rounds of export. Whether dissociation takes place as exportin 1 is liberated, as the RNA is translated or packaged, or by another means is currently unknown.

A number of additional proteins are also capable of interacting with HIV-1 Rev. First and foremost is eIF-5A; although this protein was identified based on its ability to bind to Rev's NES (181), it is currently unclear what its precise role in the proposed Rev cycle is. However, genetic experiments are consistent with the idea that eIF-5A function is involved at some level (9, 108). Two proteins that interact with the arginine-rich NLS/RBD of Rev have also been described. Interaction with a ~32-kDa protein (p32) normally associated with the cellular splicing factor SF2/ASF was detected in yeast two-hybrid analyses (129), whereas binding to the nucleolar shuttling protein B23 was initially identified by using cellular extracts and Rev affinity columns (50). In addition, SF2/ASF itself has also been shown to interact with the RRE in vitro in a Rev-dependent manner (172). It has been proposed that Rev's interactions with various proteins involved in splicing may participate in the removal of splicing factors from intron-containing RNAs. The relevance of the interaction between B23 and Rev also remains largely unexplored.

The inherent nature of the Rev cycle may have important sequelae for HIV-1 replication and pathogenesis. In particular, the use of a virally encoded protein (Rev) as a *trans*-activator of ~9-kb and ~4-kb mRNA nuclear export results in there being two basic stages of viral gene expression—early and late (Figure 3) (112). The distinction between these phases is thought to be accentuated by Rev's requirement for multimerization (134). Thus, when Rev expression is low, Rev's ability to form Rev/RRE complexes of sufficiently high multiplicity is diminished. This predicts that Rev function is nonlinear with respect to the intracellular concentration of Rev. This expectation is supported by dose-response experiments performed with transfection-based assays (169) and infected cell lines in various states of transcriptional activation (18, 147, 171). It is possible that the nonlinearity of Rev function could, first, delay the expression of viral antigens (in particular Gag and Env) in infected cells in vivo and thereby maximize eventual virus production by allowing such cells to temporarily escape the processes that would otherwise result in their rapid demise (51, 157). Second, latent proviruses could be established in the following manner. Infected cells that encounter unfavorable transcriptional conditions that result in low levels of Rev accumulation would disproportionately reduce or, in some cases, end the nuclear export of unspliced viral RNA. The ensuing lack of viral antigen expression could then result in the avoidance of cell turnover. In the event that viral transcription subsequently stopped altogether, a state of proviral latency would have been attained. Although perceptions regarding the contributions that latency makes to HIV-1 pathogenesis have ebbed and flowed over the years, recent results obtained using samples derived from patients undergoing HAART (highly active antiretroviral therapy) have indicated that CD4+ memory T cells do indeed harbor latent proviruses (55, 229). It seems plausible that Rev's mode action may contribute to this critical phenomenon.

Unresolved Issues and Future Directions

During the past few years, the assignment of NES activity to the activation domain of Rev and the identification of exportin 1 as the Rev export receptor have answered key questions regarding the mechanism of Rev function. However, several significant and overlapping issues remain unresolved. These include, but are certainly not limited to, the role of phosphorylation in Rev function, the biochemical characterization of Rev-containing RNP complexes, the molecular basis for the role of the multimerization requirement during Rev-mediated RNP export, the elucidation of the precise roles of Rip/Rab and eIF-5A, the mechanism by which Rev overcomes nuclear (and nucleolar) retention, the mechanism of translocation to and through the NPCs, and the role of GTP hydrolysis and Ran function in RNP nuclear export.

Rev is phosphorylated at more than one serine residue by a nuclear kinase or nuclear kinases (24, 85). No insight, however, regarding the potential significance of phosphorylation has been forthcoming. In fact, disruption of the major sites of Rev phosphorylation (the serines at positions 92 and 99) does not appear to affect Rev function in transfected cells (25, 133). However, given that phosphorylation is a well-known regulator of protein-protein interactions, the potential significance of this modification in terms of HIV-1 infection should not be discounted. For instance, phosphorylation (or dephosphorylation) may govern protein activity when the levels of Rev protein are low and/or limiting (for example, as might be encountered during establishment of, or emergence from, latency).

The detailed biochemical characterization of Rev RNP complexes will be essential for identifying the full complement of cellular factors that are involved in Rev-mediated gene regulation. Nascent RNA polymerase II transcripts are coated with proteins, including hnRNP proteins and RNA processing factors, that determine the fate of these RNAs (153, 200). Obviously, HIV-1 RNAs will also be complexed with these cellular proteins; as discussed earlier, interactions with splicing factors have been shown to be important for the nuclear sequestration of unspliced viral transcripts and, therefore, for the Rev response. In contrast to factors involved in retention, several factors that interact with nascent RNAs have also been found to exert positive effects on the nuclear export of diverse RNPs (41, 45, 94–96). Examples of specific proteins involved in RNP export include the cap binding complex (CBC), which has been shown to facilitate the export of mRNA and U snRNA (101, 221), and the shuttling hnRNP A1 protein, which, when injected into *Xenopus* oocytes, was found to inhibit mRNA export (101). How these and other assorted proteins function in the Rev-mediated export pathway remains to be determined. Moreover, the extent to which various export pathways overlap with each other and with the leucine-rich NES (Rev) pathway remains ill defined at the molecular level. For example, injection experiments in the *Xenopus* system have revealed that saturating amounts of a wild-type Rev NES peptide inhibit not only Rev export but also 5S rRNA and U snRNA export, indicating that these export pathways share at least some common components (56, 165, 166). This effect is not a general one, however, as excess NES peptide has little effect on the export of tRNAs or ribosomes. Experiments that have addressed Rev NES competition with mRNA export have yielded variable results; whereas one group could demonstrate inhibition (166), a second one could not (56). Consistent with the latter result is the observation that LMB, presumably by selectively disrupting exportin 1 function, is unable to inhibit mRNA export in *Xenopus* oocytes (60); alternatively, successful blocking of mRNA export by the Rev NES may reflect competition at a step downstream of exportin 1 binding. Thus, the identification of the proteins that are associated with Rev/RRE RNPs, as well as with other export RNPs, at all stages of export should provide critical insight into the mechanisms of nuclear export and nuclear retention.

At least two lines of evidence suggest that RNP export and leucine-rich NES-mediated protein export are mechanistically different. First, a multimerization-deficient mutant of Rev (RevM7) that appears to be export competent (145) is incapable of activating the export of RRE-containing RNAs (133); importantly, in vitro and in vivo assays indicate that this is not due to an RNA-binding defect (130, 134). As discussed above, we hypothesize that this difference devolves from the necessity for an RNP to be associated with multiple export receptors. Second, the nuclear export of leucine-rich NES containing export substrates (for example, Rev), importin-α, importin-β, or transportin is not inhibited by mutant forms of Ran that bind, but cannot hydrolyze, GTP (100, 176). In contrast, the nuclear export of Rev-dependent RNAs, mRNA, or U snRNA is markedly less efficient in the presence of such mutant Ran proteins in *Xenopus* oocytes (57, 100, 182). Although the inhibition of RNA export could be due to indirect effects such as a block in the import of an essential export factor, the rapidity with which inhibition is observed tends to argue for a direct effect on export itself.

A number of speculative (nonexclusive) proposals can be made concerning why GTP hydrolysis by Ran may be required for RNP export but not for RNA-free protein export. First, perhaps the hydrolysis of GTP is required to release nuclear retention factors from RNAs (and RNPs) prior to export. This notion requires that Ran-mediated GTP hydrolysis occurs in the nucleus; although a nuclear GAP has recently been described (84), there is currently no evidence that it activates GTP hydrolysis in vivo. Second, GTP hydrolysis may be required for the translocation of comparatively large RNPs to the nuclear envelope, through the NPC and/or for the final step(s) of export. An inability to hydrolyze Ran-bound GTP could result in prolonged tethering at the cytoplasmic filaments of the NPC and, potentially, in a broad inhibition of subsequent export; this would be reminiscent of the general inhibition of protein nuclear import that is observed with mutant forms of importin-β that are unable to bind RanGTP (76). Third, it is possible that energy consumption differs among different export pathways and that leucine-rich NES-mediated protein export does not involve GTP hydrolysis by Ran. Fourth, RanQ69L may saturate a factor that is required for RNP export but not for RNA-free protein export.

In summary, the past 10 years have demonstrated that Rev provides a tractable system for examining these issues. It is likely therefore that the continued study of Rev will provide exciting information regarding both HIV-1 replication and intracellular trafficking. In particular, Rev will assuredly remain one of the standard systems with which to study processes related to nuclear export. As with many biological issues, improved structural data for the full-length Rev protein (both on and off the RRE) will undoubtedly provide crucial fresh insight into many of these areas.

CONSTITUTIVE TRANSPORT ELEMENTS (CTEs)

In contrast to the lentiviruses and HTLVs, the remaining retroviruses do not encode *trans*-acting factors that mediate the nuclear export of unspliced viral RNAs. Instead, these viruses rely solely on *cis*-acting RNA elements, known as CTEs, which, together with host cell factors, target their unspliced transcripts to the cytoplasm. Although initially reported for M-PMV (17), CTEs have now been described in the closely related simian retrovirus type-1 (SRV-1) (237) as well as in ALV (158) and a mouse intracisternal A-particle (IAP) (207). In each case, the CTE is located close to the site of polyadenylation (Figure 1) and, at least for M-PMV and SRV-1, has been shown to be essential for virus replication (48). CTEs, like the RRE, can also impart nuclear export when transferred to heterologous RNA substrates and tested in mammalian or *Xenopus* assay systems (165, 182). Moreover, they can also restore replicative capacity, albeit somewhat inefficiently, to *rev*-deficient mutants of HIV-1 (17, 237).

The primary sequences of the M-PMV and SRV-1 CTEs are over 90% identical and consist of two ~70-nt imperfect direct repeats that are loosely palindromic (49, 208). These two repeats have been shown, by a combination of chemical modification and mutational analyses, to form one elongated stem structure with a 9-nt terminal loop and two internal 16-nt loops (49, 208). The sequences of the internal loops are especially well conserved and share significant sequence similarity with the CTE of IAPs and the putative CTEs of

SRV-2, HFV, and simian foamy virus (SFV) (49). Mutations that have been introduced into the loop sequences of the M-PMV or SRV-1 CTEs, as well as into the adjacent stem sequences, markedly inhibit CTE function. Because stem mutations can be deleterious even when restorative complementary mutations are introduced (49, 165, 208), it has been suggested that the single-stranded loop sequences, together with the surrounding stem structure, interact with cellular factors required for CTE function.

Competition experiments performed in microinjected *Xenopus* oocytes have shown that the CTEs of M-PMV or SRV-1 inhibit mRNA export but not Rev/RRE or tRNA export (165, 182). This suggests that the CTE and mRNA pathways share a common factor. This factor has recently been identified as TAP, the human homolog of the *S. cerevisiae* mRNA export factor Mex67p (76a). This protein binds to the loop sequences of the CTE and promotes CTE export in Xenopus oocytes. Excess TAP also overcomes CTE saturation of mRNA export in oocytes. The only cellular protein of known identity that has been reported to bind to a CTE is RNA helicase A (211). Interestingly, RNA helicase A relocalizes from the nucleus to the cytoplasm in cells that overexpress CTE-containing mRNA; however, it has yet to be determined whether this protein helps promote RNA export or, rather, is exported by virtue of being associated with CTE-containing RNPs. It has also been noted that the CTE of M-PMV functions only when located in close proximity to the site of polyadenylation (178); this suggests, perhaps, that factors involved in CTE export and 3′-end formation may interact with each other.

In contrast to the Rev/RRE system of HIV-1, CTEs (as the name implies) function constitutively and, therefore, facilitate the simultaneous synthesis of all virally encoded proteins. A possible consequence of this mode of gene expression may be that the decision between productive infection and viral latency cannot be influenced by posttranscriptional regulatory events for these viruses. Consistent with this notion, we are unaware of instances where CTE-containing retroviruses have been shown to establish latent infections that are capable of spontaneous reactivation.

REV AS A THERAPEUTIC TARGET

Based on the essential and specific role of Rev in HIV-1 replication in all cultured cell systems and in vivo (52, 98, 195), and an urgent need for additional antiretroviral approaches, Rev likely represents an attractive target for the development of novel therapeutic strategies. In addition to more general approaches that can be directed against *rev* or the RRE, such as ribozymes and antisense nucleic acids, several more specific anti-Rev strategies are either being developed or can be envisioned (170). It is likely that approaches aimed at inhibiting the action of leucine-rich NESs will lack selectivity. In particular, now that it appears that the exportin 1–mediated pathway of nuclear export is utilized by important cellular proteins, disruption of this pathway may well be broadly toxic to cells. In contrast, interference with the binding of Rev to its highly specialized target RNA (the RRE) or to itself (multimerization) appears to be more attractive in terms of specificity. Indeed, aminoglycoside antibiotics and aromatic heterocyclic compounds have been shown to inhibit Rev/RRE binding in vitro and HIV-1 structural gene expression in cultured cells (233, 234). Related small-molecule inhibitors may therefore represent good candidates for further development.

At various points during this review, inactive mutant proteins that exhibit dominant negative (*trans*-dominant) phenotypes with respect to their wild-type counterparts have been described. In the case of Rev itself, such mutations are within the NES and are exemplified by the RevM10 protein (133). It has been proposed that such proteins bind to the RRE and assemble into mixed multimers together with wild-type Rev (134); as discussed above, these RNPs presumably have a limited ability to access the required number of export receptors and, as a result, fail to be transported out of the nucleus. In a somewhat similar manner, dominant negative forms of eIF-5A are thought to bind to Rev/RRE complexes but then be incapable of interacting with required export factors (9). In both of these cases, stable expression of the dominant negative proteins has been demonstrated to inhibit HIV-1 replication effectively in cultured cell systems (8, 9, 16, 108, 136, 218). Moreover, the introduction of RevM10 into the peripheral CD4+ lymphocytes of HIV-1-infected individuals has been shown to provide those cells with a survival advantage over cells treated with a control gene (226). Protection is thought to occur because RevM10-expressing cells do not synthesize new virions following infection and are not therefore cleared by the processes that eliminate productively infected cells in vivo.

Two additional gene-based approaches have also been shown to suppress replication in culture. Here, the objective is to create a "sink" for Rev within cells such that Rev is diverted away from its RRE-containing RNA targets. In one example, Rev-specific single-chain variable antibody fragments (SFvs) have been used to sequester Rev in the cytoplasm (43, 44). In the other, multiple stable copies of the RRE are expressed in the absence of other viral sequences (123); these RNAs have been termed RNA decoys. Importantly, and as with all genetic-based therapies, the major challenge with each of these four versions of "intracellular immunization" is to deliver the inhibitory sequences to the critical cell targets with sufficient simplicity and effectiveness (110).

Visit the *Annual Reviews home page* at
http://www.AnnualReviews.org.

Literature Cited

1. Adachi Y, Yanagida M. 1989. Higher order chromosome structure is affected by cold-sensitive mutations in a *Schizosaccharomyces pombe* gene crm1$^+$ which encodes a 115-kD protein preferentially localized in the nucleus and its periphery. *J. Cell Biol.* 108:1195–207

2. Adam SA, Marr RS, Gerace L. 1990. Nuclear protein import in permeabilized mammalian cells requires soluble cytoplasmic factors. *J. Cell Biol.* 111:807–16

3. Amendt BA, Hesslein D, Chang L-J, Stoltzfus CM. 1994. Presence of negative and positive *cis*-acting RNA splicing elements within and flanking the first *tat* coding exon of human immunodeficiency virus type 1. *Mol. Cell. Biol.* 14: 3960–70

4. Arenzana-Seisdedos F, Turpin P, Rodriguez M, Thomas D, Hay RT, et al. 1997. Nuclear localization of IκBα promotes active transport of NF-κB from the nucleus to the cytoplasm. *J. Cell Sci.* 110:369–78

5. Arrigo SJ, Chen ISY. 1991. Rev is necessary for translation but not cytoplasmic accumulation of HIV-1 *vif*, *vpr*, and *env/vpu2* RNAs. *Genes Dev.* 5:808–19

6. Bartel DP, Zapp ML, Green MR, Szostak JW. 1991. HIV-1 regulation involves recognition of non-Watson-Crick base pairs in viral RNA. *Cell* 67:529–36

7. Battiste JL, Mao H, Rao NS, Tan R, Muhanidram DR, et al. 1996. α helix-RNA major groove recognition in an HIV-1 Rev peptide-RRE RNA complex. *Science* 273:1547–51

8. Bevec D, Dobrovnik M, Hauber J, Böhnlein E. 1992. Inhibition of human immunodeficiency virus type 1 replication in human T cells by retroviral-mediated gene transfer of a dominant-negative Rev trans-activator. *Proc. Natl. Acad. Sci. USA* 89:9870–74

9. Bevec D, Jaksche H, Oft M, Wöhl T, Himmelspach M, et al. 1996. Inhibition of HIV-1 replication in lymphocytes by mutants of the Rev cofactor eIF-5A. *Science* 271:1858–60

10. Black DL. 1995. Finding splice sites within a wilderness of RNA. *RNA* 1: 763–71

11. Boelens WC, Palacios I, Mattaj IW. 1995. Nuclear retention of RNA as a mechanism for localization. *RNA* 1:273–83

12. Bogerd H, Greene WC. 1993. Dominant negative mutants of human T-cell leukemia virus type 1 Rex and human immunodeficiency virus type 1 Rev fail to multimerize in vivo. *J. Virol.* 67:2496–502

13. Bogerd HP, Fridell RA, Benson RE, Hua J, Cullen BR. 1996. Protein sequence requirements for function of the human T-cell leukemia virus type 1 Rex nuclear export signal delineated by a novel in vivo randomization-selection assay. *Mol. Cell. Biol.* 16:4207–14

14. Bogerd HP, Fridell RA, Madore S, Cullen BR. 1995. Identification of a novel cellular cofactor for the Rev/Rex class of retroviral regulatory proteins. *Cell* 82:485–94

15. Böhnlein E, Berger J, Hauber J. 1991. Functional mapping of the human immunodeficiency virus type 1 Rev RNA binding domain: new insights into the domain structure of Rev and Rex. *J. Virol.* 65:7051–55

16. Bonyhadi ML, Moss K, Voytovich A, Auten J, Kalfoglou C, et al. 1997. RevM10-expressing T cells derived in vivo from transduced human hematopoietic stem-progenitor cells inhibit human immunodeficiency virus replication. *J. Virol.* 71:4707–16

17. Bray M, Prasad S, Dubay JW, Hunter E, Jeang K-T, et al. 1994. A small element from the Mason-Pfizer monkey virus genome makes human immunodeficiency virus type 1 expression and replication Rev-independent. *Proc. Natl. Acad. Sci. USA* 91:1256–60

18. Butera ST, Roberts BD, Lam L, Hodge T, Folks TM. 1994. Human immunodeficiency virus type 1 RNA expression by four chronically infected cell lines indicates multiple mechanisms of latency. *J. Virol.* 68:2726–30

19. Campbell M, Renshaw-Gegg L, Renne R, Luciw PA. 1994. Characterization of the interal promoter of simian foamy viruses. *J. Virol.* 68:4811–20

20. Cann AJ, Chen ISY. 1996. Human T-cell leukemia virus types I and II. In *Fields Virology*, ed. BN Fields, DM Knipe, PM Howley, 3:1849–80. Philadelphia: Lippincott-Raven

21. Chang DD, Sharp PA. 1989. Regulation by HIV Rev depends upon recognition of splice sites. *Cell* 59:789–95

22. Charpentier B, Stutz F, Rosbash M. 1997. A dynamic *in vivo* view of the HIV-1 Rev-RRE interaction. *J. Mol. Biol.* 266:950–62

23. Chen CY, Shyu AB. 1995. AU-rich elements: characterization and importance in mRNA degradation. *Trends Biochem. Sci.* 20:465–70

24. Cochrane A, Kramer R, Ruben S, Levine J, Rosen CA. 1989. The human immunodeficiency virus *rev* protein is a nuclear phosphoprotein. *Virology* 171:264–66

25. Cochrane AW, Golub E, Volsky D, Ruben S, Rosen CA. 1989. Functional significance of phosphorylation to the human immunodeficiency virus rev protein. *J. Virol.* 63:4438–40

26. Cochrane AW, Jones KS, Beidas S, Dillon PJ, Skalka AM, Rosen CA. 1991. Identification and characterization of intragenic sequences which repress human immunodeficiency virus structural gene expression. *J. Virol.* 65:5305–13

27. Coffin JM. 1996. *Retroviridae*: the viruses and their replication. In *Fields Virology*, ed. BN Fields, DM Knipe, PM Howley, 3:1767–847. Philadelphia: Lippincott-Raven

28. Cole JL, Gehman JD, Shafer JA, Kuo LC. 1993. Solution oligomerization of the rev protein of HIV-1: implications for function. *Biochemistry* 32:11769–75

29. Cook KS, Fisk GJ, Hauber J, Usman N, Daly TJ, Rusche JR. 1991. Characterization of HIV-1 REV protein: binding stoichiometry and minimal RNA substrate. *Nucleic Acids Res.* 19:1577–83

30. Cullen BR. 1996. HIV-1: Is Nef a PAK animal? *Curr. Biol.* 6:1557–59

31. Cullen BR, Hauber J, Campbell K, Sodroski JG, Haseltine WA, Rosen CA. 1988. Subcellular localization of the human immunodeficiency virus *trans*-acting *art* gene product. *J. Virol.* 62:2498–501

32. D'Agostino DM, Felber BK, Harrison JE, Pavlakis GN. 1992. The rev protein of human immunodeficiency virus type 1 promotes polysomal association and translation of *gag/pol* and *vpu/env* mRNAs. *Mol. Cell. Biol.* 12:1375–86

33. Daly TJ, Cook KS, Gray GS, Maione TE, Rusche JR. 1989. Specific binding of HIV-1 recombinant Rev protein to the Rev-responsive element in vitro. *Nature* 342:816–19

34. Daly TJ, Doten RC, Rennert P, Auer M, Haksche H, et al. 1993. Biochemical characterization of binding of multiple HIV-1 Rev monomeric proteins to the Rev responsive element. *Biochemistry* 32:10497–505

35. Daly TJ, Rennert P, Barry JK, Dundas M, Rusche JR, et al. 1993. Perturbation of the carboxy terminus of HIV-1 Rev

affects multimerization on the Rev responsive element. *Biochemistry* 32:8945–54

36. Daly TJ, Rusche JR, Maione TE, Frankel AD. 1990. Circular dichroism studies of the HIV-1 Rev protein and its specific RNA binding site. *Biochemistry* 29:9791–95

37. Daneholt B. 1997. A look at messenger RNP moving through the nuclear pore. *Cell* 88:585–88

38. Davis LI. 1995. The nuclear pore complex. *Annu. Rev. Biochem.* 64:865–96

39. Dayton ET, Konings DAM, Powell DM, Shapiro BA, Butini L, et al. 1992. Extensive sequence-specific information throughout the CAR/RRE, the target sequence of the human immunodeficiency virus type 1 Rev protein. *J. Virol.* 66:1139–51

40. Dobbelstein M, Roth J, Kimberly WT, Levine AJ, Shenk T. 1997. Nuclear export of the E1B 55-kDa and E4 34-kDa adenoviral oncoproteins mediated by a rev-like signal sequence. *EMBO J.* 16:4276–84

41. Donello JE, Beeche AA, Smith GJ III, Lucero GR, Hope TJ. 1996. The hepatitis B virus posttranscriptional regulatory element is composed of two subelements. *J. Virol.* 70:4345–51

42. Doye V, Hurt E. 1997. From nucleoporins to nuclear pore complexes. *Curr. Opin. Cell Biol.* 9:401–11

43. Duan L, Bagasra O, Laughlin MA, Oakes JW, Pomerantz RJ. 1994. Potent inhibition of human immunodeficiency virus type 1 replication by an intracellular anti-Rev single-chain antibody. *Proc. Natl. Acad. Sci. USA* 91:5075–79

44. Duan L, Zhu M, Ozaki I, Zhang H, Wei DL, Pomerantz RJ. 1997. Intracellular inhibition of HIV-1 replication using a dual protein- and RNA-based strategy. *Gene Ther.* 4:533–43

45. Eckner R, Ellmeier W, Birnstiel ML. 1991. Mature mRNA 3′ end formation stimulates RNA export from the nucleus. *EMBO J.* 10:3513–22

46. Emerman M. 1996. HIV-1, Vpr and the cell cycle. *Curr. Biol.* 6:1096–103

47. Emerman M, Vazeux R, Peden K. 1989. The *rev* gene product of the human immunodeficiency virus affects envelope-specific RNA localization. *Cell* 57:1155–65

48. Ernst RK, Bray M, Rekosh D, Hammarskjöld M-L. 1997. A structured retroviral RNA element that mediates nucleocytoplasmic export of intron-containing RNA. *Mol. Cell. Biol.* 17:135–44

49. Ernst RK, Bray M, Rekosh D, Hammarskjöld M-L. 1997. Secondary structure and mutational analysis of the Mason-Pfizer monkey virus RNA constitutive transport element. *RNA* 3: 210–22

50. Fankhauser C, Izaurralde E, Adachi Y, Wingfield P, Laemmli UK. 1991. Specific complex of human immunodeficiency virus type 1 Rev and nucleolar B23 proteins: dissociation by the Rev response element. *Mol. Cell. Biol.* 11:2567–75

51. Fauci AS. 1996. Host factors and the pathogenesis of HIV-induced disease. *Nature* 384:529–34

52. Feinberg MB, Jarrett RF, Aldovini A, Gallo RC, Wong-Staal F. 1986. HTLV-III expression and production involve complex regulation at the levels of splicing and translation of viral RNA. *Cell* 46:807–17

53. Felber BK, Drysdale CM, Pavlakis GN. 1990. Feedback regulation of human immunodeficiency virus type 1 expression by the Rev protein. *J. Virol.* 64:3734–41

54. Felber BK, Hadzopoulou-Cladaras M, Cladaras C, Copeland T, Pavlakis GN. 1989. Rev protein of human immunodeficiency virus type 1 affects the stability and transport of the viral mRNA. *Proc. Natl. Acad. Sci. USA* 86:1495–99

55. Finzi D, Hermankova M, Pierson T, Carruth LM, Buck C, et al. 1997. Identification of a reservoir for HIV-1 in patients on highly active antiretroviral therapy. *Science* 278:1295–300

56. Fischer U, Huber J, Boelens WC, Mattaj IW, Lührmann R. 1995. The HIV-1 Rev activation domain is a nuclear export signal that accesses an export pathway used by specific cellular RNAs. *Cell* 82:475–83

57. Fischer U, Pollard VW, Lührmann R, Teufel M, Michael WM, Dreyfuss G, Malim MH. 1998. Rev-mediated nuclear export of RNA is dominant over nuclear retention and is coupled to the Ran-GTPase cycle. Submitted

58. Fischer U, Meyer S, Teufel M, Heckel C, Lührmann R, Rautmann G. 1994. Evidence that HIV-1 Rev directly promotes the nuclear export of unspliced RNA. *EMBO J.* 13:4105–12

59. Flügel RM. 1993. The molecular biology of the human spumavirus. In *Human Retroviruses*, ed. BR Cullen, pp.193–214. Oxford: Oxford Univ. Press

60. Fornerod M, Ohno M, Yoshida M, Mattaj IW. 1997. CRM1 is an export receptor for leucine-rich nuclear export signals. *Cell* 90:1051–60

61. Fornerod M, van Deursen J, van Baal S, Reynolds A, Davis D, et al. 1997. The human homologue of yeast CRM1 is in a dynamic subcomplex with CAN/Nup214 and a novel nuclear pore component Nup88. *EMBO J.* 16:807–16

62. Frankel AD, Young JAT. 1998. HIV-1: fifteen proteins and an RNA. *Annu. Rev. Biochem.* In press

63. Fridell RA, Benson RE, Hua J, Bogerd HP, Cullen BR. 1996. A nuclear role for the fragile X mental retardation protein. *EMBO J.* 15:5408–14

64. Fridell RA, Bogerd HP, Cullen BR. 1996. Nuclear export of late HIV-1 mRNAs occurs via a cellular protein export pathway. *Proc. Natl. Acad. Sci. USA* 93:4421–24

65. Fridell RA, Fischer U, Lührmann R, Meyer BE, Meinkoth JL, et al. 1996. Amphibian transcription factor IIIA proteins contain a sequence element functionally equivalent to the nuclear export signal of human immunodeficiency virus type 1 Rev. *Proc. Natl. Acad. Sci. USA* 93:2936–40

66. Fridell RA, Partin KM, Carpenter S, Cullen BR. 1993. Identification of the activation domain of equine infectious anemia virus rev. *J. Virol.* 67:7317–23

67. Fritz CC, Green MR. 1996. HIV Rev uses a conserved cellular protein export pathway for the nucleocytoplasmic transport of viral RNAs. *Curr. Biol.* 6:848–54

68. Fritz CC, Zapp ML, Green MR. 1995. A human nucleoporin-like protein that specifically interacts with HIV Rev. *Nature* 376:530–33

69. Fukuda M, Asano S, Nakamura T, Adachi M, Yoshida M, et al. 1997. CRM1 is responsible for intracellular transport mediated by the nuclear export signal. *Nature* 390:308–11

70. Fukuda M, Gotoh I, Gotoh Y, Nishida E. 1996. Cytoplasmic localization of mitogen-activated protein kinase directed by its NH2-terminal, leucine-rich short amino acid sequence, which acts as a nuclear export signal. *J. Biol. Chem.* 271:20024–28

71. Ganem D. 1996. *Hepadnaviridae*: the viruses and their replication. In *Fields Virology*, ed. BN Fields, DM Knipe, PM Howley, 3:2703–37. Philadelphia: Lippincott-Raven

72. Garrett ED, Cullen BR. 1992. Comparative analysis of Rev function in human

immunodeficiency virus types 1 and 2. *J. Virol.* 66:4288–94

73. Görlich D. 1997. Nuclear protein import. *Curr. Opin. Cell Biol.* 9:412–19

74. Görlich D, Henklein P, Laskey RA, Hartmann E. 1996. A 41 amino acid motif in importin-αβ confers binding to importin-β and hence transit into the nucleus. *EMBO J.* 15:1810–17

75. Görlich D, Mattaj IW. 1996. Nucleocytoplasmic transport. *Science* 271: 1513–18

76. Görlich D, Panté N, Kutay U, Aebi U, Bischoff FR. 1996. Identification of different roles for RanGDP and RanGTP in nuclear protein import. *EMBO J.* 15:5584–94

76a. Grüter P, Tabernero C, von Kobbe C, Schmitt C, Saavedra C, et al. 1998. TAP, the human homolog of Mex67p, mediates CTE-dependent RNA export from the nucleus. *Mol. Cell* 1:649–59

77. Guatelli JC, Gingeras TR, Richman DD. 1990. Alternative splice acceptor utilization during human immunodeficiency virus type 1 infection of cultured cells. *J. Virol.* 64:4093–98

78. Hadzopoulou-Cladaras M, Felber BK, Cladaras C, Athanassopoulos A, Tse A, Pavlakis GN. 1989. The *rev* (*trs/art*) protein of human immunodeficiency virus type 1 affects viral mRNA and protein expression via a *cis*-acting sequence in the *env* region. *J. Virol.* 63:1265–74

79. Hamm J, Mattaj IW. 1990. Monomethylated cap structures facilitate RNA export from the nucleus. *Cell* 63:109–18

80. Hammarskjöld M-L, Heimer J, Hammarskjöld B, Sangwan I, Albert L, Rekosh D. 1989. Regulation of human immunodeficiency virus *env* expression by the *rev* gene product. *J. Virol.* 63:1959–66

81. Hammarskjöld M-L, Li H, Rekosh D, Prasad S. 1994. Human immunodeficiency virus *env* expression becomes Rev-independent if the *env* region is not defined as an intron. *J. Virol.* 68:951–58

82. Hammerschmid M, Palmeri D, Ruhl M, Jakshe H, Weichselbraun I, et al. 1994. Scanning mutagenesis of the arginine-rich region of the human immunodeficiency virus type 1 Rev *trans* activator. *J. Virol.* 68:7329–35

83. Hanly SM, Rimsky LT, Malim MH, Kim JH, Hauber J, et al. 1989. Comparative analysis of the HTLV-I Rex and HIV-1 Rev *trans*-regulatory proteins and their RNA response elements. *Genes Dev.* 3:1534–44

84. Hattori M, Tsukamoto N, Nur-e-Kamal MS, Rubinfeld B, Iwai K, et al. 1995. Molecular cloning of a novel mitogen-inducible nuclear protein with a RNA GTPase-activating domain that affects cell cycle progression. *Mol. Cell. Biol.* 15:552–60

85. Hauber J, Bouvier M, Malim MH, Cullen BR. 1988. Phosphorylation of the *rev* gene product of human immunodeficiency virus type 1. *J. Virol.* 62:4801–4

86. Heaphy S, Dingwall C, Ernberg I, Gait MJ, Green SM, et al. 1990. HIV-1 regulator of virion expression (Rev) protein binds to an RNA stem-loop structure located within the Rev response element region. *Cell* 60:685–93

87. Henderson BR, Percipalle P. 1997. Interactions between HIV Rev and nuclear import and export factors: the Rev nuclear localisation signal mediates specific binding to human importin-beta. *J. Mol. Biol.* 274:693–707

88. Hidaka M, Inoue J, Yoshida M, Seiki M. 1988. Post-transcriptional regulator (*rex*) of HTLV-1 initiates expression of viral structural proteins but suppresses expression of regulatory proteins. *EMBO J.* 7:519–23

89. Holland SM, Ahmad N, Maitra RK, Wingfield P, Venkatesan S. 1990. Human immunodeficiency virus Rev protein recognizes a target sequence in Rev-responsive element RNA within the context of RNA secondary structure. *J. Virol.* 64:5966–75

90. Hope TJ, Bond BL, McDonald D, Klein NP, Parslow TG. 1991. Effector domains of human immunodeficiency virus type 1 Rev and human T-cell leukemia virus type I Rex are functionally interchangeable and share an essential peptide motif. *J. Virol.* 65:6001–7

91. Hope TJ, Huang X, McDonald D, Parslow TG. 1990. Steroid-receptor fusion of the human immunodeficiency virus type 1 Rev transactivator: mapping cryptic functions of the arginine-rich motif. *Proc. Natl. Acad. Sci. USA* 87:7787–91

92. Hope TJ, Klein NP, Elder ME, Parslow TG. 1992. *trans*-Dominant inhibition of human immunodeficiency virus type 1 Rev occurs through formation of inactive protein complexes. *J. Virol.* 66: 1849–55

93. Hope TJ, McDonald D, Huang X, Low J, Parslow TG. 1990. Mutational analysis of the human immunodeficiency virus type 1 Rev transactivator: essential residues near the amino terminus. *J. Virol.* 64:5360–66

94. Huang J, Liang TJ. 1993. A novel hepatitis B virus (HBV) genetic element with *rev* response element-like properties that is essential for expression of HBV gene products. *Mol. Cell. Biol.* 13:7476–86

95. Huang Y, Carmichael GG. 1996. Role of polyadenylation in nucleocytoplasmic transport of mRNA. *Mol. Cell. Biol.* 16:1534–42

96. Huang Z-M, Yen TSB. 1995. Role of the hepatitis B virus post-transcriptional regulatory element in export of intronless trancripts. *Mol. Cell. Biol.* 15:3864–69

97. Iacampo S, Cochrane A. 1996. Human immunodeficiency virus type 1 Rev function requires continued synthesis of its target mRNA. *J. Virol.* 70:8332–39

98. Iversen AKN, Shpaer EG, Rodrigo AG, Hirsch MS, Walker BD, et al. 1995. Persistence of attenuated *rev* genes in a human immunodeficiency virus type 1-infected asymptomatic individual. *J. Virol.* 69:5743–53

99. Ivey-Hoyle M, Rosenberg M. 1990. Rev-dependent expression of human immunodeficiency virus type 1 gp160 in *Drosophila melanogaster* cells. *Mol. Cell. Biol.* 10:6152–59

100. Izaurralde E, Kutay U, von Kobbe C, Mattaj IW, Görlich D. 1997. The asymmetric distribution of the constituents of the Ran system is essential for transport into and out of the nucleus. *EMBO J.* 16:6535–47

101. Izaurralde E, Lewis J, Gamberi C, Jarmolowski A, McGuigan C, Mattaj IW. 1995. A cap-binding protein complex mediating U snRNA export. *Nature* 376:709–12

102. Izaurralde E, Mattaj IW. 1995. RNA export. *Cell* 81:153–59

103. Jacobson A, Peltz SW. 1996. Interrelationships of the pathways of mRNA decay and translation in eukaryotic cells. *Annu. Rev. Biochem.* 65:693–739

104. Jarmolowski A, Boelens WC, Izaurralde E, Mattaj IW. 1994. Nuclear export of different classes of RNA is mediated by specific factors. *J. Cell Biol.* 124:627–35

105. Joag SV, Stephens EB, Narayan O. 1996. Lentiviruses. In *Fields Virology*, ed. BN Fields, DM Knipe, PM Howley, 3:1977–96. Philadelphia: Lippincott-Raven

106. Jones K, Peterlin B. 1994. Control of RNA initiation and elongation at the HIV-1 promoter. *Annu. Rev. Biochem.* 63:717–43

107. Jones KA. 1997. Taking a new TAK on Tat transactivation. *Genes Dev.* 11:2593–99

108. Junker U, Bevec D, Barske C, Kalfoglou C, Escaich S, et al. 1996. Intracellular expression of cellular eIF-5A mutants inhibits HIV-1 replication in human T cells: a feasibility study. *Hum. Gene Ther.* 7:1861–69

109. Kalland K-H, Szilvay AM, Brokstad KA, Sçtrevik W, Haukenes G. 1994. The human immunodeficiency virus type 1 Rev protein shuttles between the cytoplasm and nuclear compartments. *Mol. Cell. Biol.* 14:7436–44

110. Kay MA, Liu D, Hoogerbrugge PM. 1997. Gene therapy. *Proc. Natl. Acad. Sci. USA* 94:12744–46

111. Kim FJ, Beeche AA, Hunter JJ, Chin DJ, Hope TJ. 1996. Characterization of the nuclear export signal of human T-cell lymphotropic virus type 1 rex reveals that nuclear export is mediated by position-variable hydrophobic interactions. *Mol. Cell. Biol.* 16:5147–55

112. Kim S, Byrn R, Groopman J, Baltimore D. 1989. Temporal aspects of DNA and RNA synthesis during human immunodeficiency virus infection: evidence for differential gene expression. *J. Virol.* 63:3708–13

113. Kiyomasu T, Miyazawa T, Furuya T, Shibata R, Sakai H, et al. 1991. Identification of feline immunodeficiency virus *rev* gene activity. *J. Virol.* 65:4539–42

114. Kjems J, Brown M, Chang DD, Sharp PA. 1991. Structural analysis of the interaction between the human immunodeficiency virus Rev protein and the Rev response element. *Proc. Natl. Acad. Sci. USA* 88:683–87

115. Kjems J, Calnan BJ, Frankel AD, Sharp PA. 1992. Specific binding of a basic peptide from HIV-1 Rev. *EMBO J.* 11:1119–29

116. Kjems J, Frankel AD, Sharp PA. 1991. Specific regulation of mRNA splicing in vitro by a peptide from HIV-1 Rev. *Cell* 67:169–78

117. Klebe C, Bischoff FR, Ponstingl H, Wittinghofer A. 1995. Interaction of the nuclear GTP-binding protein Ran with its regulatory proteins RCC1 and RanGAP1. *Biochemistry* 34:639–47

118. Knight DM, Flomerfelt FA, Ghrayeb J. 1987. Expression of the art/trs protein of HIV and study of its role in viral envelope synthesis. *Science* 236:837–40

119. Kobe B, Deisenhofer J. 1994. The leucine-rich repeat: a versatile binding motif. *Trends Biochem. Sci.* 19:415–21

120. Kubota S, Siomi H, Satoh T, Endo S-I, Maki M, Hatanaka M. 1989. Functional

similarity of HIV-1 *rev* and HTLV-1 *rex* proteins: identification of a new nucleolar-targeting signal in *rev* protein. *Biochem. Biophys. Res. Commun.* 162:963–70

121. Kutay U, Bischoff FR, Kostka S, Kraft R, Görlich D. 1997. Export of importin α from the nucleus is mediated by a specific nuclear transport factor. *Cell* 90:1061–71

122. Lamb RA, Pinto LH. 1997. Do Vpu and Vpr of human immunodeficiency virus type 1 and NB of influenza B virus have ion channel activities in the viral life cycles. *Virology* 229:1–11

123. Lee S-W, Gallardo HF, Gilboa E, Smith C. 1994. Inhibition of human immunodeficiency virus type 1 in human T cells by a potent Rev response element decoy consisting of the 13-nucleotide minimal Rev-binding domain. *J. Virol.* 68:8254–64

124. Legrain P, Rosbash M. 1989. Some *cis*-and *trans*-acting mutants for splicing target pre-mRNA to the cytoplasm. *Cell* 57:573–83

125. Löchelt M, Muranyi W, Flügel RM. 1993. Human foamy virus genome possesses an internal, Bel-1-dependent and functional promoter. *Proc. Natl. Acad. Sci. USA* 90:7317–21

126. Lu X, Heimer J, Rekosh D, Hammarskjöld M-L. 1990. U1 small nuclear RNA plays a direct role in the formation of a rev-regulated human immunodeficiency virus *env* mRNA that remains unspliced. *Proc. Natl. Acad. Sci. USA* 87:7598–602

127. Lu Y, Touzjian N, Stenzel M, Dorfman T, Sodroski JG, Haseltine WA. 1990. Identification of *cis*-acting repressive sequences within the negative regulatory element of human immunodeficiency virus type 1. *J. Virol.* 64:5226–29

128. Luciw PA. 1996. Human immunodeficiency viruses and their replication. In *Fields Virology*, ed. BN Fields, DM Knipe, PM Howley, 3:1881–952. Philadelphia: Lippincott-Raven

129. Luo Y, Yu H, Peterlin BM. 1994. Cellular protein modulates effects of human immunodeficiency virus type 1 Rev. *J. Virol.* 68:3850–56

130. Madore SJ, Tiley LS, Malim MH, Cullen BR. 1994. Sequence requirements for Rev multimerization *in vivo*. *Virology* 202:186–94

131. Maldarelli F, Martin MA, Strebel K. 1991. Identification of posttranscriptionally active inhibitory sequences in human immunodeficiency virus type 1

RNA: novel level of gene regulation. *J. Virol.* 65:5732–43

132. Malim MH, Böhnlein S, Fenrick R, Le S-Y, Maizel JV, Cullen BR. 1989. Functional comparison of the Rev transactivators encoded by different primate immunodeficiency virus species. *Proc. Natl. Acad. Sci. USA* 86:8222–26

133. Malim MH, Böhnlein S, Hauber J, Cullen BR. 1989. Functional dissection of the HIV-1 Rev *trans*-activator-derivation of a *trans*-dominant repressor of Rev function. *Cell* 58:205–14

134. Malim MH, Cullen BR. 1991. HIV-1 structural gene expression requires the binding of multiple Rev monomers to the viral RRE: implications for HIV-1 latency. *Cell* 65:241–48

135. Malim MH, Cullen BR. 1993. Rev and the fate of pre-mRNA in the nucleus: implications for the regulation of RNA processing in eukaryotes. *Mol. Cell. Biol.* 13:6180–89

136. Malim MH, Freimuth WW, Liu J, Boyle TJ, Lyerly HK, et al. 1992. Stable expression of transdominant Rev protein in human T cells inhibits human immunodeficiency virus replication. *J. Exp. Med.* 176:1197–201

137. Malim MH, Hauber J, Le S-Y, Maizel JV, Cullen BR. 1989. The HIV-1 *rev* trans-activator acts through a structured target sequence to activate nuclear export of unspliced viral mRNA. *Nature* 338:254–57

138. Malim MH, McCarn DF, Tiley LS, Cullen BR. 1991. Mutational definition of the human immunodeficiency virus type 1 Rev activation domain. *J. Virol.* 65:4248–54

139. Malim MH, Tiley LS, McCarn DF, Ruche JR, Hauber J, Cullen BR. 1990. HIV-1 structural gene expression requires binding of the Rev *trans*-activator to its RNA target sequence. *Cell* 60:675–83

140. Mancuso VA, Hope TJ, Zhu L, Derse D, Phillips T, Parslow TG. 1994. Posttranscriptional effector domains in the Rev proteins of feline immunodeficiency virus and equine infectious anemia virus. *J. Virol.* 68:1998–2001

141. Mann DA, Mikaélian I, Zemmel RW, Green SM, Lowe AD, et al. 1994. Cooperative Rev binding to stem I of the Rev-response element modulates human immunodeficiency virus type-1 late gene expression. *J. Mol. Biol.* 241:193–207

142. McCracken S, Fong N, Yankulov K, Ballantyne S, Pan G, et al. 1997. The C-terminal domain of RNA polymerase II

couples mRNA processing to transcription. *Nature* 385:357–61

143. McDonald D, Hope TJ, Parslow TG. 1992. Posttranscriptional regulation by the human immunodeficiency virus type 1 Rev and human T-cell leukemia virus type I Rex proteins through a heterologous RNA binding site. *J. Virol.* 66:7232–38

144. Mermer B, Felber BK, Campbell M, Pavlakis GN. 1990. Identification of *trans*-dominant HIV-1 rev protein mutants by direct transfer of bacterially produced proteins into human cells. *Nucleic Acids Res.* 18:2037–44

145. Meyer BE, Malim MH. 1994. The HIV-1 Rev *trans*-activator shuttles between the nucleus and the cytoplasm. *Genes Dev.* 8:1538–47

146. Meyer BE, Meinkoth JL, Malim MH. 1996. Nuclear transport of the human immunodeficiency virus type 1, Visna virus, and equine infectious anemia virus Rev proteins: identification of a family of transferable nuclear export signals. *J. Virol.* 70:2350–59

147. Michael NL, Morrow P, Mosca J, Vahey MA, Burke DS, Redfield RR. 1991. Induction of human immunodeficiency virus type 1 expression in chronically infected cells is associated primarily with a shift in RNA splicing patterns. *J. Virol.* 65:1291–303

148. Michael WM, Choi M, Dreyfuss G. 1995. A nuclear export signal in hnRNP A1: a signal-mediated, temperature-dependent nuclear protein export pathway. *Cell* 83:415–22

149. Michael WM, Eder PS, Dreyfuss G. 1997. The K nuclear shuttling domain: a novel signal for nuclear import and nuclear export in the hnRNP K protein. *EMBO J.* 16:3587–98

150. Muranyi W, Flügel RM. 1991. Analysis of splicing patterns of human spumaretrovirus by polymerase chain reaction reveals complex RNA structures. *J. Virol.* 65:727–35

151. Murphy R, Wente SR. 1996. An RNA-export mediator with an essential nuclear export signal. *Nature* 383:357–60

152. Nakielny S, Dreyfuss G. 1997. Nuclear export of proteins and RNAs. *Curr. Opin. Cell Biol.* 9:420–29

153. Nakielny S, Fischer U, Michael WM, Dreyfuss G. 1997. RNA transport. *Annu. Rev. Neurosci.* 20:269–301

154. Neville M, Stutz F, Lee L, Davis LI, Rosbash M. 1997. The importin-beta family member Crm1p bridges the interaction between Rev and the nuclear pore complex during nuclear export. *Curr. Biol.* 7:767–75

155. Nigg EA. 1997. Nucleocytoplasmic transport: signals, mechanisms and regulation. *Nature* 386:779–87

156. Nishi NK, Yoshida M, Fujiwara D, Nishikawa M, Horinouchi S, Beppu T. 1994. Leptomycin B targets a regulatory cascade of crm1, a fission yeast nuclear protein, involved in control of higher order chromosome structure and gene expression. *J. Biol. Chem.* 269:6320–24

157. O'Brien WA, Pomerantz RJ. 1996. HIV infection and associated diseases. In *Viral Pathogenesis*, ed. N Nathanson, R Ahmed, F Gonzalez-Scarano, DE Griffin, KV Holmes, et al, pp. 815–36. Philadelphia: Lippincott-Raven

158. Ogert RA, Lee LH, Beemon KL. 1996. Avian retroviral RNA element promotes unspliced RNA accumulation in the cytoplasm. *J. Virol.* 70:3834–943

159. Olsen HS, Cochrane AW, Dillon PJ, Nalin CM, Rosen CA. 1990. Interaction of the human immunodeficiency virus type 1 Rev protein with a structured region in *env* mRNA is dependent on multimer formation mediated through a basic stretch of amino acids. *Genes Dev.* 4:1357–64

160. Olsen HS, Cochrane AW, Rosen C. 1992. Interaction of cellular factors with intragenic *cis*-acting repressive sequences within the HIV genome. *Virology* 191:709–15

161. Ossareh-Nazari B, Bachelerie F, Dargemont C. 1997. Evidence for a role of CRM1 in signal-mediated nuclear protein export. *Science* 278:141–44

162. Palmeri D, Malim MH. 1996. The human T-cell leukemia virus type 1 posttranscriptional trans-activator Rex contains a nuclear export signal. *J. Virol.* 70:6442–45

163. Palmeri D, Malim MH. 1998. Nuclear import of HTLV-I Rex requires importin-β but not importin-α. Submitted

164. Park MH, Wolff EC, Folk JE. 1993. Is hypusine essential for eukaryotic cell proliferation? *Trends Biochem. Sci.* 18: 476–79

165. Pasquinelli AE, Ernst RK, Lund E, Grimm C, Zapp ML, et al. 1997. The constitutive transport element (CTE) of Mason-Pfizer monkey virus (MPMV) accesses a cellular mRNA export pathway. *EMBO J.* 16:7500–10

166. Pasquinelli AE, Powers MA, Lund E, Forbes D. 1997. Inhibition of mRNA export in vertebrate cells by nuclear export

signal conjugates. *Proc. Natl. Acad. Sci. USA* 94:14394–99

167. Perkins A, Cochrane AW, Ruben SM, Rosen CA. 1989. Structural and functional characterization of the human immunodeficiency virus *rev* protein. *J. Acquired Immune Defic. Syndr.* 2:256–63

168. Piñol-Roma S, Dreyfuss G. 1992. Shuttling of pre-mRNA binding proteins between nucleus and cytoplasm. *Nature* 355:730–32

169. Pomerantz RJ, Seshamma T, Trono D. 1992. Efficient replication of human immunodeficiency virus type 1 requires a threshold level of Rev: potential implications for latency. *J. Virol.* 66:1809–13

170. Pomerantz RJ, Trono D. 1995. Genetic therapies for HIV infections: promise for the future. *AIDS* 9:985–93

171. Pomerantz RJ, Trono D, Feinberg MB, Baltimore D. 1990. Cells nonproductively infected with HIV-1 exhibit an aberrant pattern of viral RNA expression: a molecular model of latency. *Cell* 61:1271–76

172. Powell DM, Amaral MC, Wu JY, Maniatis T, Greene WC. 1997. HIV Rev-dependent binding of SF2/ASF to the Rev response element: possible role in Rev-mediated inhibition of HIV RNA splicing. *Proc. Natl. Acad. Sci. USA* 94:973–78

173. Purcell DFJ, Martin MA. 1993. Alternative splicing of human immunodeficiency virus type 1 mRNA modulates viral protein expression, replication, and infectivity. *J. Virol.* 67:6365–78

174. Renwick SB, Critchley AD, Adams CJ, Kelly SM, Price NC, Stockley PG. 1995. Probing the details of the HIV-1 Rev-Rev-responsive element interaction: effects of modified nucleotides on protein affinity and conformational changes during complex formation. *Biochem. J.* 308:447–53

175. Richard N, Iacampo S, Cochrane A. 1994. HIV-1 Rev is capable of shuttling between the nucleus and cytoplasm. *Virology* 204:123–31

176. Richards SA, Carey KL, Macara IG. 1997. Requirement of guanosine triphosphate-bound ran for signal-mediated nuclear protein export. *Science* 276:1842–47

177. Richards SA, Lounsbury KM, Carey KL, Macara IG. 1996. A nuclear export signal is essential for the cytosolic localization of the Ran binding protein, RanBP1. *J. Cell Biol.* 134:1157–68

178. Rizvi TA, Schmidt RD, Lew KA. 1997.

Mason-Pfizer monkey virus (MPMV) constitutive transport element (CTE) functions in a position-dependent manner. *Virology* 236:118–29

179. Robert-Guroff M, Popovic M, Gartner S, Markham P, Gallo RC, Reitz MS. 1990. Structure and expression of *tat*-, *rev*-, and *nef*-specific transcripts of human immunodeficiency virus type 1 in infected lymphocytes and macrophages. *J. Virol.* 64:3391–98

180. Rosen CA, Terwilliger E, Dayton A, Sodroski JG, Haseltine WA. 1988. Intragenic *cis*-acting art gene-responsive sequences of the human immunodeficiency virus. *Proc. Natl. Acad. Sci. USA* 85:2071–75

181. Ruhl M, Himmelspach M, Bahr GM, Hammerschmid F, Jaksche H, et al. 1993. Eukaryotic initiation factor 5A is a cellular target of the human immunodeficiency virus type 1 Rev activation domain mediating *trans*-activation. *J. Cell Biol.* 123:1309–20

182. Saavedra C, Felber B, Izaurralde E. 1997. The simian retrovirus-1 constitutive transport element, unlike the HIV-1 RRE, uses factors required for cellular mRNA export. *Curr. Biol.* 7:619–28

183. Saavedra CA, Hammell CM, Heath CV, Cole CN. 1997. Yeast heat shock mRNAs are exported through a distinct pathway defined by Rip1p. *Genes Dev.* 11:2845–56

184. Sakai H, Shibata R, Sakuragi J-I, Kiyomasu T, Kawamura M, et al. 1991. Compatibility of *rev* gene activity in the four groups of primate lentiviruses. *Virology* 184:513–20

185. Schmidt-Zachmann MS, Dargemont C, Kühn LC, Nigg EA. 1993. Nuclear export of proteins: the role of nuclear retention. *Cell* 74:493–504

186. Schneider R, Campbell M, Nasioulas G, Felber BK, Pavlakis GN. 1997. Inactivation of the human immunodeficiency virus type 1 inhibitory elements allows Rev-independent expression of Gag and Gag/protease and particle formation. *J. Virol.* 71:4892–903

187. Schoborg RV, Saltarelli MJ, Clements JE. 1994. A Rev protein is expressed in caprine arthritis encephalitis virus (CAEV)-infected cells and is required for efficient viral replication. *Virology* 202:1–15

188. Schwartz S, Campbell M, Nasioulas G, Harrison J, Felber BK, Pavlakis GN. 1992. Mutational inactivation of an inhibitory sequence in human immunodeficiency virus type 1 results in

Rev-inedependent *gag* expression. *J. Virol.* 66:7176–82

189. Schwartz S, Felber BK, Benko DM, Fenyö E-M, Pavlakis GN. 1990. Cloning and functional analysis of multiply spliced mRNA species of human immunodeficiency virus type 1. *J. Virol.* 64:2519–29

190. Schwartz S, Felber BK, Pavlakis GN. 1992. Distinct RNA sequences in the *gag* region of human immunodeficiency virus type 1 decrease RNA stability and inhibit expression in the absence of Rev protein. *J. Virol.* 66:150–59

191. Seedorf M, Silver PA. 1997. Importin/karyopherin protein family members required for mRNA export from the nucleus. *Proc. Natl. Acad. Sci. USA* 94:8590–95

192. Segref A, Sharma K, Doye V, Hellwig A, Huber J, et al. 1997. Mex67p, a novel factor for nuclear mRNA export, binds to both poly(A)$^+$ RNA and nuclear pores. *EMBO J.* 16:3256–71

193. Simon JHM, Miller DL, Fouchier RAM, Soares MA, Peden KWC, Malim MH. 1998. The regulation of primate immunodeficiency virus infectivity by Vif is cell species restricted: a role for Vif in determining virus host range and cross-species transmission. *Embo J.* 17:1259–67

194. Siomi H, Shida H, Nam SH, Nosaka T, Maki M, Hatanaka M. 1988. Sequence requirements for nucleolar localization of human T cell leukemia virus type I pX protein, which regulates viral RNA processing. *Cell* 55:197–209

195. Sodroski J, Goh WC, Rosen C, Dayton A, Terwilliger E, Haseltine W. 1986. A second post-transcriptional *trans*-activator gene required for HTLV-III replication. *Nature* 321:412–17

196. Stade K, Ford CS, Guthrie C, Weis K. 1997. Exportin 1 (Crm1p) is an essential nuclear export factor. *Cell* 90:1041–50

197. Staffa A, Cochrane A. 1994. The *tat/rev* intron of human immunodeficiency virus type 1 is inefficiently spliced because of suboptimal signals in the 3′ splice site. *J. Virol.* 68:3071–79

198. Staffa A, Cochrane A. 1995. Identification of positive and negative splicing regulatory elements within the terminal tat-rev exon of human immunodeficiency virus type 1. *Mol. Cell. Biol.* 15:4597–605

199. Stauber R, Gaitanaris GA, Pavlakis GN. 1995. Analysis of trafficking of Rev and transdominant Rev proteins in living cells using green fluorescent protein

fusions: Transdominant Rev blocks the export of Rev from the nucleus to the cytoplasm. *Virology* 213:439–49

200. Steinmetz EJ. 1997. Pre-mRNA processing and the CTD of RNA polymerase II: the tail that wags the dog? *Cell* 89:491–94

201. Stephens RM, Derse D, Rice NR. 1990. Cloning and characterization of cDNAs encoding equine infectious anemia virus Tat and putative Rev proteins. *J. Virol.* 64:3716–25

202. Stutz F, Izaurralde E, Mattaj IW, Rosbash M. 1996. A role for nucleoporin FG repeat domains in export of human immunodeficiency virus type 1 Rev protein and RNA from the nucleus. *Mol. Cell. Biol.* 16:7144–50

203. Stutz F, Kantor J, Zhang D, McCarthy T, Neville M, Rosbash M. 1997. The yeast nucleoporin Rip1p contributes to multiple export pathways with no essential role for its FG-repeat region. *Genes Dev.* 11:2857–68

204. Stutz F, Neville M, Rosbash M. 1995. Identification of a novel nuclear pore-associated protein as a functional target of the HIV-1 Rev protein in yeast. *Cell* 82:495–506

205. Stutz F, Rosbash M. 1994. A functional interaction between Rev and yeast pre-mRNA is related to splicing complex formation. *EMBO J.* 13:4096–104

206. Szilvay AM, Brokstad KA, Kopperud R, Haukenes G, Kalland K-H. 1995. Nuclear export of the human immunodeficiency virus type 1 nucleocytoplasmic shuttle protein Rev is mediated by its activation domain and is blocked by transdominant negative mutants. *J. Virol.* 69:3315–23

207. Tabernero C, Zolotukhin AS, Bear J, Schneider R, Karsenty G, Felber BK. 1997. Identification of an RNA sequence within an intracisternal-A particle element able to replace Rev-mediated post-transcriptional regulation of human immunodeficiency virus type 1. *J. Virol.* 71:95–101

208. Tabernero C, Zolotukhin AS, Valentin A, Pavlakis GN, Felber BK. 1996. The posttranscriptional control element of the simian retrovirus type 1 forms an extensive RNA secondary structure necessary for its function. *J. Virol.* 70:5998–6011

209. Tan R, Chen L, Buettner JA, Hudson D, Frankel AD. 1993. RNA recognition by an isolated α helix. *Cell* 73:1031–40

210. Tan R, Frankel AD. 1994. Costabilization of peptide and RNA structure in

an HIV Rev peptide-RRE complex. *Biochemistry* 33:14579–85

211. Tang H, Gaietta GM, Fischer WH, Ellisman MH, Wong-Staal F. 1997. A cellular cofactor for the constitutive transport element of type D retrovirus. *Science* 276:1412–15

212. Terns MP, Dahlberg JE. 1994. Retention and 5′ cap trimethylation of U3 snRNA in the nucleus. *Science* 264:959–61

213. Terns MP, Grimm C, Lund E, Dahlberg JE. 1995. A common maturation pathway for small nucleolar RNAs. *EMBO J.* 14:4860–71

214. Tiley LS, Brown PH, Le S-Y, Maizel JV, Clements JE, Cullen BR. 1990. Visna virus encodes a post-transcriptional regulator of viral structural gene expression. *Proc. Natl. Acad. Sci. USA* 87:7497–501

215. Tiley LS, Malim MH, Cullen BR. 1991. Conserved functional organization of the human immunodeficiency virus type 1 and visna virus Rev proteins. *J. Virol.* 65:3877–81

216. Tiley LS, Malim MH, Tewary HK, Stockley PG, Cullen BR. 1992. Identification of a high-affinity RNA-binding site for the human immunodeficiency virus type 1 Rev protein. *Proc. Natl. Acad. Sci. USA* 89:758–62

217. Ullman KS, Powers MA, Forbes DJ. 1997. Nuclear export receptors: from importin to exportin. *Cell* 90:967–70

218. Vandendriessche T, Chuah MK, Chiang L, Chang HK, Ensoli B, Morgan RA. 1995. Inhibition of clinical human immunodeficiency virus (HIV) type 1 isolates in primary CD4+ T lymphocytes by retroviral vectors expressing anti-HIV genes. *J. Virol.* 69:4045–52

219. Venkatesan S, Gerstberger SM, Park H, Holland SM, Nam Y. 1992. Human immunodeficiency virus type 1 Rev activation can be achieved without Rev-responsive element RNA if Rev is directed to the target as a Rev/MS2 fusion protein which tethers the MS2 operator RNA. *J. Virol.* 66:7469–80

220. Venkatesh LK, Chinnadurai G. 1990. Mutants in a conserved region near the carboxy-terminus of HIV-1 identify functionally important residues and exhibit a dominant negative phenotype. *Virology* 178:327–30

221. Visa N, Izaurralde E, Ferreira J, Daneholt B, Mattaj IW. 1996. A nuclear cap-binding complex binds Balbiani ring pre-mRNA cotranscriptionally and accompanies the ribonucleoprotein particle during nuclear export. *J. Cell Biol.* 133:5–14

222. Wang J, Manley JL. 1997. Regulation of pre-mRNA splicing in metazoa. *Curr. Opin. Genet. Dev.* 7:205–11

223. Weichselbraun I, Farrington GK, Rusche JR, Böhnlein E, Hauber J. 1992. Definition of the human immunodeficiency virus type 1 Rev and human T-cell leukemia virus type I Rex protein activation domain by functional exchange. *J. Virol.* 66:2583–87

224. Weis K, Ryder U, Lamond AI. 1996. The conserved amino-terminal domain of hSRP1 alpha is essential for nuclear protein import. *EMBO J.* 15:1818–25

225. Wen W, Meinkoth JL, Tsien RY, Taylor SS. 1995. Identification of a signal for rapid export of proteins from the nucleus. *Cell* 82:463–73

226. Woffendin C, Ranga U, Yang Z-Y, Xu L, Nabel GJ. 1996. Expression of a protective gene prolongs survival of T cells in human immunodeficiency virus-infected patients. *Proc. Natl. Acad. Sci. USA* 93:2889–94

227. Wolff B, Cohen G, Hauber J, Meshcheryakova D, Rabeck C. 1995. Nucleocytoplasmic transport of the Rev protein of human immunodeficiency virus type 1 is dependent on the activation domain of the protein. *Exp. Cell Res.* 217:31–41

228. Wolff B, Sanglier J-J, Wang Y. 1997. Leptomycin B is an inhibitor of nuclear export: inhibition of nucleo-cytoplasmic translocation of the human immunodeficiency virus type 1 (HIV-1) Rev protein and Rev-dependent mRNA. *Chem. Biol.* 4:139–47

229. Wong JK, Hezareh M, Günthard HF, Havlir DV, Ignacio CC, et al. 1997. Recovery of replication-competent HIV despite prolonged suppression of plasma viremia. *Science* 278:1291–95

230. Yu SF, Baldwin DN, Gwynn SR, Yendapalli S, Linial ML. 1996. Human foamy virus replication: a pathway distinct from that of retroviruses and hepadnaviruses. *Science* 271:1579–82

231. Yurev A, Patturajan M, Litingtun Y, Joshi RV, Gentile C, et al. 1996. The C-terminal domain of the largest subunit of RNA polymerase II interacts with a novel set of serine/arginine-rich proteins. *Proc. Natl. Acad. Sci. USA* 93: 6975–80

231a. Zapp ML, Green MR. 1989. Sequence specific RNA binding by the HIV-1 Rev protein. *Nature* 342:714–16

232. Zapp ML, Hope TJ, Parslow TG, Green MR. 1991. Oligomerization and RNA binding domains of the type 1 human immunodeficiency virus Rev protein: a

dual function for an arginine-rich binding motif. *Proc. Natl. Acad. Sci. USA* 88:7734–38

233. Zapp ML, Stern S, Green MR. 1993. Small molecules that selectively block RNA binding of HIV-1 Rev protein inhibit Rev function and viral production. *Cell* 74:969–78

234. Zapp ML, Young DW, Kumar A, Singh R, Boykin DW, et al. 1997. Modulation of the Rev-RRE interaction by aromatic hetercyclic compounds. *Bioorg. Med. Chem.* 5:1149–55

235. Zemmel RW, Kelley AC, Karn J, Butler PJG. 1996. Flexible regions of RNA structure facilitate co-operative Rev assembly on the Rev-response element. *J. Mol. Biol.* 258:763–77

236. Zhang G, Zapp ML, Yan G, Green MR. 1996. Localization of HIV-1 RNA in mammalian nuclei. *J. Cell Biol.* 135:9–28

237. Zolotukhin AS, Valentin A, Pavlakis GN. 1994. Continuous propagation of RRE(−) and Rev(−)RRE(−) human immunodeficiency virus type 1 molecular clones containing a *cis*-acting element of simian retrovirus type 1 in human peripheral blood lymphocytes. *J. Virol.* 68:7944–52

Annu. Rev. Microbiol. 1998. 52:533–60

AGING IN *SACCHAROMYCES CEREVISIAE*

David Sinclair, Kevin Mills, and Leonard Guarente

Department of Biology, Massachusetts Institute of Technology, Cambridge, Massachusetts 02139; e-mail: leng@mit.edu, davids@mit.edu

KEY WORDS: yeast, senescence, longevity, ERC, extrachromosomal, rDNA, RAS, UTH, SGS1

ABSTRACT

The budding yeast *Saccharomyces cerevisiae* divides asymmetrically, giving rise to a mother cell and a smaller daughter cell. Individual mother cells produce a finite number of daughter cells before senescing, undergoing characteristic changes as they age such as a slower cell cycle and sterility. The average life span is fixed for a given strain, implying that yeast aging has a strong genetic component. Genes that determine yeast longevity have highlighted the importance of such processes as cAMP metabolism, epigenetic silencing, and genome stability. The recent finding that yeast aging is caused, in part, by the accumulation of circular rDNA molecules has unified many seemingly disparate observations.

CONTENTS

0066-4227/98/1001-0533$08.00

INTRODUCTION

The budding yeast *Saccharomyces cerevisiae* has been used as a model for replicative senescence and aging since 1959, when Mortimer & Johnston (84) first noted that individual yeast cells have a finite life span. Since then, much of yeast aging research remained descriptive, and many aging models have been proposed, then discounted. The 1990s have seen a concerted effort to understand the yeast aging process at a molecular level, driven in part by rapid advances in genetic and biochemical techniques. One of the most important advances in the field was the development of techniques to isolate large quantities of old cells, a requirement for any biochemical analysis of yeast aging.

Over the past few years, great progress has been made in identifying key genes in the yeast aging process, including those involved in metabolism, cell cycle control, and epigenetic silencing. One recent discovery has implicated DNA stability in the aging process and, in so doing, may have validated the yeast aging model that invokes a molecular aging clock timing the accumulation of a senescence factor.

AGING THEORIES: An Overview

Studies of model organisms such as *Drosophila*, *Caenorhabditis elegans*, and *S. cerevisiae* have revealed that many biological themes are highly conserved. Many eukaryotes experience an exponential decline in fitness and fecundity

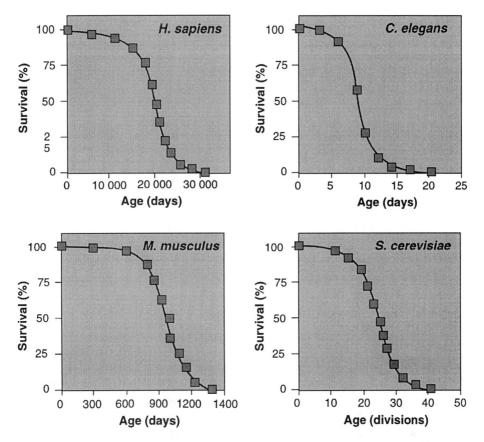

Figure 1   Mortality curves for various species. Human, worm, mouse, and yeast curves redrawn from human census data (US women 1980; 107) and References 129, 122, and 114, respectively.

over time, most evident after the reproductive phase of life. In the absence of external influences that affect the mortality rate (such as parasitic disease or predation), a genetic component of aging may be seen. Many eukaryotic organisms have rates of survival that are fixed for a genetic background, and these rates often follow a pattern predicted by Gompertz-Makeham kinetics (30, 37). Figure 1 shows graphs of survival by age, known as mortality curves, for various species.

Two aging theories exist currently. The first is the multi-hit theory, by which organisms are subject to damage from their internal and external environments, leading to various types of lesions at the cellular level. The other theory suggests that aging is directed by genetic influences that are intimately linked to the rate of metabolism or the cell cycle. These two theories are not mutually exclusive—it is possible, for example, that oxidative damage leads to the initiation of a genetic program—but for the purposes of this review the theories will be described separately.

MULTI-HIT THEORIES

The gradual accumulation of innumerable random defects and toxic metabolic products may lead to the death of individual cells, damage to tissues, and eventual loss of homeostasis. One prominent multi-hit hypothesis postulates that aging is due in part to the accumulation of irreversible molecular oxidative damage leading to senescence (43, 94). Molecular oxygen can generate a number of partially reduced molecules and secondary reactive oxygen metabolites (ROMs) that lead to extensive damage, including DNA modification, lipid peroxidation, and increased protein degradation or inactivation (reviewed in 124). The exponential increase in mortality with replicative age has been used as an argument against a simple stochastic aging mechanism (54, 97).

PROGRAM THEORIES AND GENETIC INFLUENCES

An alternate aging hypothesis postulates that age-associated decreases in fitness are determined genetically. The strong view is that organisms are programmed to senesce after reproducing, by a process that is an extension of development. One form of the theory predicts genetic clocks that govern the timing of age-associated phenotypes. This theory also encompasses the idea that genes involved in the maintenance of cellular function eventually become disregulated, ultimately resulting in the breakdown of the organism (18, 28, 70, 80, 127, 131).

Telomeres

Telomere length has been proposed to influence many biological processes: the finite replicative capacity of cultured human diploid fibroblasts (2, 42), somatic differentiation (17), tumorigenesis (20), and human aging (reviewed in 40). In contrast to most normal somatic cells, transformed cell lines and cells derived from human tumors often have rejuvenated telomerase activity. This observation has led to the appealing theory that the replicative capacity of mammalian cells is determined by telomerase activity and telomere length. Recent results, however, complicate such a model (138). Introduction of telomerase into two telomerase-negative normal human cells increased the number of in vitro cell doublings by at least twofold, implying that replicative senescence is caused, in part, by a barrier imposed by telomere length (7a). This result is complicated by the fact that some immortalized cell lines maintain telomere length without expressing telomerase (10), and the early generations of mice lacking telomerase do not exhibit aging phenotypes (7). The role of telomerase in human aging, in vivo, remains to be determined.

Genetic Influences

A corollary of the genetic theories of aging is that life span may be manipulated genetically (128), and some progress has been made in this direction (reviewed in 103; 121). Long-lived *Drosophila* strains have been generated by selective breeding (103, 104), whereas *drop-dead* strains exhibit age-associated phenotypes and die soon after emergence from the pupa (47, 102). Studies using *C. elegans* have identified genes that appear to extend life span by slowing the overall metabolic rate (23, 29, 63, 65). Screens for longevity genes in *S. cerevisiae* have taken two approaches. First, a differential hybridization screen identified genes whose expression is regulated in an age-dependent manner (14, 21). Second, as discussed before, a search for mutants with increased stress resistance identified several genes that regulate longevity (59).

Stochastic Trigger and Senescence Factor

Jazwinski (51) first suggested that "the mechanisms governing [yeast] aging and senescence [may] comprise a stochastic trigger and a deterministic effector." The stochastic trigger would set in motion a genetic program in which the deterministic effector would cause aging. A stochastic element could explain the variability in the life spans of genetically identical (87) cells within a population.

The dominance of aging characteristics and the ability of old mother cells to generate prematurely old daughter cells has been seen as evidence for a cytoplasmic senescence factor (27, 58). As cells age, the senescence factor may accumulate in old mother cells to levels that allow leakage into daughter cells. Restoration of youth may occur by diluting the senescence factor over successive generations or degrading it in daughter cells. Recent findings may have identified a molecule with properties resembling a senescence factor. Self-replicating extrachromosomal DNA circles, discussed in greater detail below, may be a direct cause of yeast aging.

PHENOMENOLOGY OF YEAST AGING

A key requirement for tracing aging in an organism is the ability to distinguish and follow that individual over time. This is easy to do in multicellular eukaryotes, such as flies, mice, and humans. In microbial organisms the task is more

Table 1 Life spans of various haploid yeast strains

Strain	Average	Maximum	Reference
	Life span[1]		
BKy1-14c (*uth4-14c*)	16	25	60
SP1	18	28	15
DBY747	22	32	59
PSY316	23	34	83
X2180-1A	24	39	26
X30	24	42	84
W303-1A	25	43	114
S 288C	28	47[2]	88
PSY142	29	55	59

[1]Life spans determined on complete glucose media. Life spans are unaffected by ploidy (87), richness of media (113), or temperature (89).
[2]Max. life span of this strain on ethanol medium was 67 (88).

difficult. The problem is compounded by the fact that many microorganisms divide symmetrically, giving rise to two cells that are visually indistinguishable from their progenitor. One attractive feature of *S. cerevisiae* is that the progenitor cell is easily distinguished from its descendants. Yeast cells divide asymmetrically by budding: the original cell gives rise to a smaller, visibly and molecularly distinct cell generating mother cells that are older by one division and daughter cells that have a full life-span potential.

Experiments by Mortimer & Johnston (84) first demonstrated that yeast cells are mortal. By micromanipulating daughter cells away from larger mother cells they showed that cells undergo a fixed number of divisions, which define the replicative life span (4, 84). The average life span is relatively constant for a given strain (56, 84, 87), suggesting that the yeast life span is influenced strongly by genetic background (Table 1). These observations have set the stage for the utilization of yeast as a system for the genetic study of aging and senescence.

Chronological vs Replicative Life Span

Life span in almost all multicellular organisms is expressed in units of time.[1] Yeast life span can also be measured chronologically, but no strong correlation exists between chronological life span and replicative life span. For instance, yeast chronological life span can be increased dramatically without markedly altering replicative life span by cooling the cells, briefly blocking cell division, or slowing the budding rate (89). Although some evidence suggests that nonreplicative aging may occur, most yeast aging research has focused on the

[1]For many simple organisms, life span is influenced by external factors such as temperature (64, 66); thus for those organisms, chronological life span is not an absolute number.

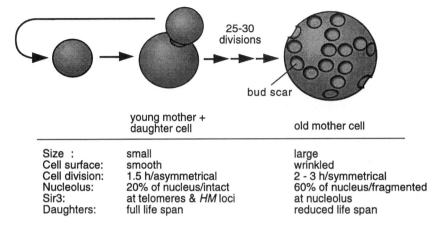

	young mother + daughter cell	old mother cell
Size :	small	large
Cell surface:	smooth	wrinkled
Cell division:	1.5 h/asymmetrical	2 - 3 h/symmetrical
Nucleolus:	20% of nucleus/intact	60% of nucleus/fragmented
Sir3:	at telomeres & *HM* loci	at nucleolus
Daughters:	full life span	reduced life span

Figure 2   The biomarkers of yeast aging. *S. cerevisiae* undergoes asymmetric division producing a mother cell and a smaller daughter each cell cycle. This process results in the formation on the mother's cell surface of a permanent chitinous bud scar that can be used as a measure of age.

replicative constituent of life span. Yeast cells undergo many characteristic changes as they age. Some changes are manifest visibly, whereas others reflect more subtle effects, including alterations in gene expression patterns. Figure 2 shows the major age-related changes associated with yeast aging.

Nonreplicative Life Span

Under a variety of conditions, yeast cell division may be halted without killing the cell. Under conditions of nutrient starvation, *S. cerevisiae* cells may enter stationary phase (reviewed in 134). During this nonreplicative phase, cells undergo characteristic changes, over a period ranging from weeks to months, before they senesce. In the laboratory, stationary-phase cells are encountered most commonly in response to glucose starvation. When glucose has been depleted from the growth medium, fermentation stops, growth slows, and respiratory metabolism begins. When cells cease respiratory growth, they enter stationary phase, where they become refractile, exhibit altered cell wall structures, display thermotolerance, and undergo changes in transcriptional and translational activities. These changes allow for long-term survival under nutrient-depleted conditions. However, cells in stationary phase do not remain viable indefinitely—after prolonged maintenance of cells in this phase, the efficiency of plating to nutrient-rich medium declines for reasons that are unclear. However, rather than entering a state of dormancy, stationary-phase cells are thought to actively promote their nonreplicative longevity. For example, superoxide dismutase (SOD) is essential for stationary-phase survival, indicating that stationary cells still generate and detoxify reactive oxygen metabolites (74). Interestingly, expression of the human anti-apoptotic gene *Bcl-2* in yeast restores stationary-phase survival to SOD mutant cells, possibly revealing a stationary-phase senescence mechanism (73).

Nonreplicating cells can also be produced by incubating temperature-sensitive cell cycle mutants at a restrictive temperature. Cell cycle–arrested cells differ from stationary-phase cells in several ways: They are more metabolically active than stationary-phase cells; they do not acquire characteristics of stationary-phase cells, such as thickened cell wall or increased thermotolerance; and they lose viability much more rapidly than stationary-phase cells. If *cdc28-ts* mutants are incubated at the restrictive temperature they arrest prior to entry into the S phase of the cell cycle. Cells held in the arrested state continue to grow for 24 h, after which they begin to degrade their own DNA, RNA, and many proteins, in a process known as autophagic death (86). Because autophagic death requires protein synthesis, a mechanism appears to exist that responds to arrest conditions by activating an autophagic pathway. Several other temperature-sensitive cell cycle mutants also exhibit autophagic death, further indicating that nonreplicative senescence mechanisms may become engaged.

Bud Scar Accumulation

When a yeast cell replicates, it does so by localizing newly replicated nuclear and cytoplasmic components to the newly formed bud. Until cytokinesis, the bud remains attached to the mother cell and continues to grow in volume. To separate the bud from the mother cell, the boundary between the two cells constricts, leaving behind on the mother cell's surface a circular chitin-containing remnant termed the bud scar. Bud scars remain permanently deposited on the surface of the mother cell, each occupying an estimated 1% of the available cell surface. As a mother cell goes through successive rounds of cell division, bud scars accumulate on the cell surface, serving as a convenient marker for the number of divisions realized by a single cell. As a cell continues dividing, previous sites of budding are seldom reused.

Although it has been hypothesized that the accumulation of bud scars may impose a theoretical upper limit on a cell's replicative potential (12), evidence indicates that it does not result in senescence of wild-type strains. First, bud scars typically occupy approximately 1% of the available cell surface, giving a theoretical upper limit of at least 100 divisions, but most lab strains have an average life span of 20–30 divisions (53, 56, 58, 89). Second, increasing cell size and thus increasing available budding sites, does not extend life span (88). Third, it is possible to artificially elevate the deposition of chitin by briefly incubating cells containing a temperature-sensitive allele of *CDC24* at the nonpermissive temperature (117, 118). At this temperature *cdc24-ts* cells arrest in an unbudded state and randomly accumulate cell wall chitin. Elevated chitin deposition does not adversely affect life span (27), further demonstrating that bud scar accumulation is likely not a cause of either senescence or aging.

Increased Cell Size

Another visible characteristic accrued by aging yeast is cell size. After three or four divisions mother cells are easily distinguished from daughter cells by inspection. It has been proposed that cells senesce because they reach a critical upper size limit (84). Theoretically, increased cell volume could impose a replicative limit if the rate of nutrient diffusion to disparate parts of a large cell becomes limiting, or if metabolic activity decreases as a result of diminishing cell surface-to-volume ratio. A careful test of this hypothesis has been complicated by the occurrence of other age-related changes in gene expression and protein synthesis. However, a simple increase in cell size appears insufficient to halt division since larger parental diploid cells do not live longer than their haploid counterparts (88). Perhaps more convincingly, cell size can be modulated by arresting the cell cycle for various lengths of time during the G_1 phase (119). Cells arrested at this phase remain metabolically active and continue to grow in size. After release from G_1 arrest, cells can be greatly increased in size but have an identical life span potential to that of untreated cells (58). Cells may experience a metabolic decline as they age, but that decline is probably not related to size or surface-to-volume ratio.

Metabolic and Physiologic Changes

Although cell size does not limit life span, for other reasons cells may become less metabolically or physiologically robust until they are unable to support the process of cell division. Such incapacity may result from accumulation of DNA or protein damage or from a reduction in protein synthesis, among other causes.

Evidence suggests that the incidence of unrepaired DNA damage probably does not result in cell death in *S. cerevisiae*. First, cells heterozygous for different nutritional marker genes do not display an increased rate of auxotrophy when clones were derived from old cells (87). Second, if DNA damage resulted in restricted life span, then this trait should be heritable and all cells derived from old mother cells should exhibit a short life span. This is clearly not the case: The life span of cells derived from old mother cells is restored to normal within a few generations, indicating that there has been no permanent loss of genetic information (56, 59). However, contribution of reversible epigenetic changes to senescence cannot be ruled out (59, 115).

Another factor that may contribute to age-related physiologic incapacity is a decrease in the efficiency of protein synthesis. If protein synthesis declines over time, then the cell may be unable to support essential processes, stop dividing, and die. To test this hypothesis, the rate of in vitro protein synthesis in a cell-free system was examined using ribosomes from aged yeast cells. The synthetic rate using old cell-derived ribosomes was about 40% less than that of young cell controls (85). Also, this observation may indicate that translational activity decreases over time. It is not clear if decreased translational activity results from old, deteriorating ribosomes or from newly synthesized, but less active, ribosomes in old cells. Additionally, total protein content was elevated in aged cells, despite a decreased synthetic rate, suggesting that cells may somehow compensate for a depressed rate of synthesis. Decreased efficiency of translational activity in old cells may be terminal if the rate of synthesis of essential factors does not meet cellular demand or is exceeded by the rate of degradation.

Loss of Asymmetry

As yeast cells age they grow dramatically larger but continue to give rise to small daughter cells throughout most of their life span. However, very old mother cells tend to produce large, short-lived daughter cells (56, 58, 84). At or near the final division daughter cells often do not separate from the mother until both cells are similar in size (56, 59). Loss of asymmetry in old cells is intriguing because it provides another argument for a diffusible senescence factor. When divisions cease to be asymmetrical, faithful discrimination between mother and daughter is lost and a diffusible senescence factor would be more equally distributed to both cells. This could account for the decreased replicative potential of daughter cells derived from old mother cells.

Changes in Gene Expression

If aging has a genetic component, then patterns of gene expression will likely change over time, as age-related genes become activated or inactivated. Gene expression changes in cultured fibroblasts are well documented and involve a variety of cellular processes, from cell cycle control to cellular architecture (reviewed in 128). The causal role of such gene expression changes in senescence has not been established, but such changes likely are a consequence, rather than a cause, of in vitro aging. In yeast, the identification of genes regulated according to age may provide molecular explanations for both the cessation of division in old mother cells and the restoration of full division potential to daughter cells. To investigate the age-related alterations in *S. cerevisiae* gene expression, Egilmez et al (25) employed a differential hybridization approach to identify genes whose transcript level changed with replicative age. This screen identified six genes that exhibited differential gene expression with age. Two of these (*LAG1* and *LAG2*) have been characterized partially and are described below. The identification of other genes displaying agedependent expression patterns will no doubt shed further light on how the cell responds to advancing age and perhaps identify key determinants of life span.

Loss of Fertility and Silencing

S. cerevisiae can exist in either a haploid or diploid state. When two fertile haploid cells of opposite mating type encounter one another they mate to form a diploid zygote. Young yeast cells are normally fertile since the two repositories of mating-type information, *HMR* and *HML*, remain in a transcriptionally silent state (reviewed in 69). To switch between the two mating types (*MAT*a and *MAT*α), a cell transposes the opposite silent information to the mating-type locus where it is expressed. The silent state at *HM* loci is maintained, in part, by the Sir2/3/4p silencing complex.

When cells grow old they become sterile (88), and this phenotype remains one of the most reliable external marker of yeast aging. Sterility in old cells is caused by a loss of transcriptional silencing at the cryptic mating-type loci, *HMR*a and *HML*α, resulting in simultaneous expression of both a and α information (120). To demonstrate that age-dependent sterility results from a loss of silencing, rather than interruption of other steps of the mating pathway, *HMR*a was deleted from a *MAT*α strain. The resulting strain exhibited the same characteristic life span as wild type but no longer became sterile, even in very old cells. In addition to a loss of mating-type silencing, old cells lose silencing near telomeres (62). Loss of transcriptional silencing is caused by relocalization of the silent information regulator Sir3p (and perhaps Sir4p) to the nucleolus in old cells (60).

Nucleolar Fragmentation

Another biomarker for aging in yeast is fragmentation of the nucleolus. The nucleolus is a nuclear structure containing the ribosomal RNA genes (rDNA) and other components required for ribosome assembly (81, 109, 112). Yeast rDNA, located on chromosome XII, contains 100–200 tandemly repeated copies of a 9.1-kb unit (96, 105). In young yeast cells, the nucleolus forms a crescent-shaped structure retained near the nuclear periphery (110). However, in old cells the nucleolus becomes enlarged and fragmented into multiple, rounded structures (115).

Fragmentation of the nucleolus and relocalization of Sir3p may be related. Relocalization of Sir proteins to the nucleolus may delay or slow damage that eventually leads to fragmentation. Numerous observations are consistent with this hypothesis: rDNA has been identified as a native target of silencing (10, 123); deletion of *SIR2*, *SIR3*, or *SIR4* dramatically shortens life span (59); and Sir proteins have been implicated recently in repair of double-stranded DNA breaks (130).

Telomere Length

Yeast *est1* mutants defective in telomere elongation display a clonally senescent phenotype that is evident after approximately 85 generations (76). Although the effects of the *est1* mutation on aging have not been examined, the phenotypes are very different from aging in wild-type cells, suggesting that the mutation does not accelerate a normal aging pathway. Strains manipulated genetically to have short telomeres actually have longer life span potentials than those with longer telomeres (3), and this life span extension is dependent on the Sir3 silencing/DNA repair protein that resides at telomeres. In strains with shorter telomeres, more Sir3p may be liberated to silence or repair other sites in the genome, leading to life span extension (41). In a wild-type cell, release of Sir3 from telomeres probably requires an active process because telomere length remains constant throughout the cell's lifetime (22, 119).

SIMILARITIES BETWEEN YEAST AND MAMMALIAN AGING

Dominant Aging Substance

Daughter cells of old yeast cells often exhibit many characteristics of old age, indicating that age can be inherited in a dominant fashion (27, 58). The dominance of age has been supported further by yeast mating experiments in which the aged phenotype of the mother is seen in the resulting diploid (88). The senescent phenotype of mammalian cells is also dominant in heterokaryons formed between young and old cells, leading to arrest of DNA synthesis at the G_1-S boundary (95; reviewed in 92). For both yeast and mammals, this phenomenon has been interpreted as evidence for the existence of a cytoplasmic senescence factor (27). Identification of such a factor in yeast would provide much insight into the processes that cause senescence.

Life Span

Yeast cells undergo a limited number of divisions that are independent of chronological age (84, 89). Many mammalian cells in culture also have a restricted, chronologically independent, replicative potential (43). For many years the asymmetry in yeast cell division was thought to demonstrate a fundamental difference between yeast and mammalian aging. However, recent studies have identified mechanisms that underlie self-renewing asymmetric divisions, including those that regulate cytoskeletal organization, cell-cycle progression, and mitotic orientation (reviewed in 71). In terms of the whole organism, the proliferative capacity of many stem-cell populations undergo age-dependent changes (6). Major heterochromatic changes in myeloid and lymphoid stem cells occur because of a stochastic loss of original stem cell lineages (33). Some evidence also exists for age-dependent alterations in skin stem cells (82).

Gene Expression and Heterochromatin

Loss of the fidelity of gene regulation has been proposed as a causal factor in the aging process in many species (45). Yeast gene expression has a temporal component, and Egilmez et al (25) have identified at least six genes that are regulated in this way. Although only two of these genes (*LAG1* and *LAG2*) have been characterized further and their molecular functions are unknown, expression of specific genes has been observed to change with time. This observation lends credence to the hypothesis that yeast cells, like mammalian cells, contain a mechanism to mark the passage of successive cell cycles.

Differential gene expression during the life span has been examined in *Drosophila*, mice, humans, and yeast, as well as cultured fibroblasts (reviewed in 128). The irreversible arrest of mammalian cells at the G_1-S boundary of the cell cycle during cellular senescence is associated with numerous changes in transcript levels. The expression levels of cyclin-dependent kinase inhibitors p21 and p16 are increased in late stages of cellular senescence (1, 136). p21 and p16 may influence growth rate by modulating the activity of the retinoblastoma gene product (Rb) and E2F activity, respectively. The expression levels of several matrix-associated structural proteins, including stromelysin, fibronectin, α1-collagen, and interstitial collagenase, are also altered during senescence (11, 49).

A common theme between mammalian and yeast aging is the redistribution of regulatory proteins resulting in age-dependent heterochromatic changes. Yeast aging is associated with redistribution of the Sir3 silencing protein to the nucleolus, perhaps to silence rRNA genes. Human collagenase expression is increased in late passages of human diploid fibroblasts, concomitant with dissociation of the transcriptional repressor Oct-1 from the nuclear periphery (48). Other studies show that silencing of a beta-globin/*lacZ* transgene in mouse erythrocytes decreases in an age-dependent manner when inserted at various independent loci (101). A fatty acid–binding protein/human growth hormone fusion undergoes age-related reductions in expression in transgenic mice aged 1–10 months (16). Although the results for both yeast and mammals are preliminary, they demonstrate that heterochromatic changes are likely to be fundamental to the aging process.

Genome Stability

The yeast *SGS1* gene, isolated as a slow growth suppressor in a *top3* deletion mutant, encodes an ATP-dependent DNA helicase that interacts with both topoisomerases I and II (34, 133). Mutations in *SGS1* result in increased genome-wide instability, including a sevenfold increase in recombination at the repeated rDNA locus (34, 132). Strains deleted for *SGS1* have a short life span and prematurely exhibit yeast aging phenotypes including sterility and nucleolar fragmentation (115; see below).

A human progeroid syndrome, Werner's syndrome, is characterized by a premature appearance of many symptoms seen in normal aged individuals including predisposition to cancer, gray hair, atherosclerosis, and loss of skin plasticity (reviewed in 107). The gene responsible for the disease, *WRN*, encodes a protein with similarity to the yeast Sgs1 protein (39, 137; see below). However, the exact function of the *WRN* protein in vivo is not known. One of the first clues to the nature of the disease came from the in vivo study of Werner's fibroblasts, which are severely reduced in division potential and exhibit a 10-fold increase in spontaneous mutation rate (32). Translocation frequency was also found to be elevated in Werner's lymphocytes (107). These results strengthen the notion that genome stability is a crucial determinant of life span.

METHODS FOR ISOLATING OLD CELLS

Any systematic molecular analysis of aging requires the isolation of aged individuals. For *S. cerevisiae*, the determination of life span may be performed on a relatively small number of single cells. However, a study of the physical or biochemical changes that accompany aging requires the separation of large quantities of old individuals from young. This prospect is complicated in single-celled, exponentially growing organisms because the fraction of old cells in the population is vanishingly small. Consider a yeast strain with an average life span of 20 divisions. Because the population doubles with each cell division, 50% of the population will be virgin cells, 25% will be one division old, 12.5% will be 2 divisions old, and only 1 in 2 million cells will have realized 20 divisions. The techniques for accomplishing this separation are reviewed here.

Yeast "Baby Machine"

One technique for separating mother cells from daughter cells has been described as a "baby machine" (44). Cells are attached to a membrane and allowed to divide. Mother cells remain attached to the membrane, but daughter cells are continuously eluted and collected by washing the membrane. Although this technique efficiently produces virgin cells, it has not been used successfully to isolate old cells.

Sucrose Gradient Centrifugation

After only a few divisions, mother cells become substantially larger than the daughter cells they generate. To generate an age-matched population of cells, a saturated yeast culture is separated on a 10–30% sucrose gradient generating two distinct bands of cells, one of which is composed mostly of virgin cells (27). Virgin cells are synchronized in the cell cycle by exposure to mating pheromone and then allowed to proceed through as many as three divisions. Daughter cells are separated from mother cells by repeating the sucrose gradient step except that aged rather than virgin cells are collected. By repeating cycles of synchronization, growth, and separation, it is possible to isolate a population of cells that are age matched up to twenty generations and are contaminated by less than 10% young cells. One drawback of this technique is that it requires many rounds of manipulation to obtain old cells. However, it is advantageous because it allows large-scale isolation of a relatively pure population of cells of advanced age. Using this technique, researchers have been able to prepare old cells and use them to characterize various biomarkers of yeast aging (26).

Centrifugal Elutriation

A variation on the cell-size-dependent separation scheme relies on centrifugal elutriation to continuously separate daughter cells from mother cells (135). Cells are grown in the chamber of an elutriation rotor and eluted under centrifugation. Daughter cells are collected and subjected to a second round of elutriation. During the second elutriation, daughter cells are discarded and mother cells retained. Continuous growth and elution over 15 divisions can produce a population 10 generations or older, with 70% purity. Use of this strategy over longer times results in greater heterogeneity in the final separated population. Although the fidelity of this technique is compromised at high numbers of divisions, elutriation allows the rapid, manipulation-free purification of age-matched, middle-aged yeast cells.

Fluorescence-Activated Cell Sorting

When yeast cells divide, they synthesize a bud that will eventually become the daughter cell. The cell surface of the bud is derived de novo, with no contribution from previously synthesized mother-cell components. This process allows mother cells to be "tagged" and then distinguished from daughter cells, even when they are greatly outnumbered in culture. Proteins on the cell surface are first conjugated to biotin and then cells are cultured in broth for up to 15 generations. By adding fluorochrome-conjugated avidin to the final population of cells, old cells become specifically labeled and may be separated using fluorescence-activated cell sorting (FACS) (118). This technique allows the rapid purification of an old cell population that has greater than 99% purity, although yield is compromised for the sake of purity. Typically, this method allows procurement of 10^4 old cells.

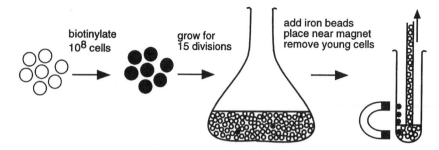

Figure 3   Magnetic sorting of old cells. Young cells are biotinylated and allowed to grow for up to 15 divisions. Cells are harvested and washed with paramagnetic iron beads that adhere to old cells. A magnet is then used to separate old cells from young cells, resulting in a million-fold enrichment.

Magnetic Sorting

A superior variation on the biotinylation method relies on magnetic rather than fluorescent discrimination of mother cells from the unbiotinylated daughter cells (118). Cells are biotinylated and allowed to divide as in FACS sorting (Figure 3). Avidin-coated paramagnetic iron beads are then added to the final population. The beads coat the old cells, which are then separated from the bulk population by placing the culture near a magnet. Several rounds of magnetic sorting and resuspension allow the high-fidelity separation of old cells. Magnetic sorting is the most useful of the methods described here because it combines the purity of biotin-dependent sorting with the ability to rapidly sort large numbers of cells. Magnetic sorting can easily yield 10^8 old cells. A high-yield, pure population of old cells is critical for studies of age-dependent molecular changes. Magnetic sorting has been used extensively to demonstrate a variety of age-associated molecular events (60, 115, 120).

GENES INVOLVED IN YEAST AGING

To isolate genes or mutations that affect yeast aging, researchers have taken either physical or genetic approaches. Both of these approaches have been successful, resulting in the identification of eight genes that affect yeast longevity (summarized in Table 2). The characterization of these genes, as well as the identification of other genes that affect aging, will permit further molecular dissection of the aging process.

The ras Genes

The *ras* genes have attracted attention because of their importance in malignant transformation of human cells. Activated *ras* genes are found in 5–40% of

Table 2   Genes that affect yeast longevity

Gene/Allele	Function of wt allele	Effects on mean life span		Parental strain	Reference
		Deletion	Overexpression		
v-Ha-*RAS*	Mouse oncogene	—	70% increase	SP1	14
RAS1	c-AMP/cell-cycle control	23% increase	no effect	SP1	124
RAS2	c-AMP/cell-cycle/ cell size	23% decrease	20–40% increase	SP1	124
UTH2/ SIR4-42	Silencing, DNA repair	30% decrease	45% increase	BKy1-14c	60
UTH4	Silencing	30% decrease	20% increase	PSY142	61
LAG1[1]	Putative membrane protein	50% decrease	60% increase	SP1	21, 53
LAG2	Putative membrane protein	50% decrease	36% increase	SP1	15
SGS1	Genome stability	40% decrease	20% decrease[2]	W303-1A	114

[1]The *LAG1* homolog, *LAC1*, is a longevity gene presented in a review by Jazwinski (54).
[2]Overexpression of *SGS1* at moderate levels (under *ADH*p rather than *GAL1*p control) does not affect life span.

all tumors examined. *S. cerevisiae* contains two *ras* homologues, *RAS1* and *RAS2*, and both genes are functionally interchangeable with mammalian *ras* genes (19, 57). Although no growth phenotype is observed upon deletion of either *RAS1* or *RAS2*, the double deletion is lethal. Ras forms part of the A complex, which integrates metabolism with cell cycle control via adenylate cyclase activity (36).

Two studies have examined the role of the *RAS* genes in yeast longevity (13, 125). Expression of the Harvey murine sarcoma virus, v-Ha-*RAS*, at moderate levels extended yeast life span by 70% and markedly increased cell size early in life (13). The life-extending properties of v-Ha-*RAS* may stem from an inhibition of the hypothetical senescence factor (13, 51).

Following this promising result, the effects of the yeast *RAS* genes on life span were examined (125). The two genes have reciprocal roles in yeast longevity: Deletion of *RAS1* extended life span 20–40%, but overexpression of *RAS2* had a similar effect. Surprisingly, overexpression of a mutant *RAS2* extended life span to the same extent as the wild-type allele, indicating that the effect may be independent of the well-characterized cAMP pathway (125). Perhaps the *RAS* genes exert their influence through an unknown signal transduction pathway (125).

LAG1, LAC1, and LAG2

A hybridization screen using labeled poly(A)$^+$ mRNA from young and old cells was used to isolate genes differentially expressed over the life span (25).

Six genes were identified that displayed age-dependent changes in expression level, ranging from 1.8- to 7-fold. Repeated screening produced a total of 14 genes that display differential patterns of expression, five of which have been sequenced. Two of the genes preferentially expressed in young cells, *LAG1* and *LAG2* (for *longevity assurance gene*), were chosen for further analyses.

LAG1 encodes a 47-kDa protein with two putative transmembrane domains (21). Deletion of *LAG1* shortens life span by 40% (21). Interestingly, expression of *LAG1* at high levels from a *GAL1* (galactose) promoter results in the senescence of many young cells, but those that survive often live longer than do the wild type. Under more moderate levels of *LAG1* expression controlled by the glucocorticoid receptor response element, the life span of the parent strain increases 60% (21). Because the presence of glucocorticoid in the medium reduces yeast life span, the possibility remains that *LAG1* expression ameliorates the effect of glucocorticoid (21).

LAG2 encodes a 78-kDa protein with a single putative transmembrane domain and a potential mitochondrial import sequence (14). Deletion of *LAG2* results in caffeine sensitivity and a 50% reduction in life span. Overexpression of a single copy of *LAG2* using the *GAL10* promoter extends life span by 20%. Induction of *LAG2* expression at generation 12 extends mean life span even further, implying that overexpression of *LAG2* in young cells is deleterious (14). The functions of *LAG1* and *LAG2* are not known.

The UTH *Genes and* SIR4-42

A major obstacle to the isolation of yeast longevity mutants is that life span can (currently) be determined for individual cells only by micromanipulation. To circumvent this problem, Kennedy et al (59) capitalized on the correlation between stress resistance and longevity. A genetic screen for starvation-resistant mutants was performed, and each mutant was tested for increased longevity. Four complementation groups were obtained (*UTH1–4*) that increased life span by 20–55%. Two of these, *UTH2* and *UTH4*, have been characterized in further detail.

The strain used in the genetic screen, BKy-14c, was chosen in part for its short life span. The shorter life span was the result of a frame-shift mutation in the 2.5-kb *UTH4/MPT5/HTR1* gene (61) that truncated the gene product after 207 residues (designated as the *uth4-14c* allele). One of the long-lived strains isolated in the screen restored the reading frame of the *uth4-14c* allele, producing a full-length Uth4 protein. *UTH4* encodes a protein with homology to two other yeast genes, *YGL023* and *YLL014*. Together these genes constitute a family of yeast genes encoding proteins showing similarity to the *Drosophila* RNA-binding protein *pumilio*. The functions of *UTH4*, *YGL023*, and *YLL014* are unknown; however, they have been implicated in the control of transcriptional silencing. Deletion of *UTH4* strengthens silencing at telomeres and weakens silencing at rDNA, whereas overexpression of *UTH4* has the opposite effect (K Mills, D Sinclair, & L Guarente, unpublished results). In addition, *UTH4* overexpression extends the life span of long-lived strains, showing that it has relevance to yeast longevity (60). The putative RNA-binding activity of Uth4p may target it to the RNA-rich nucleolus, where it may ameliorate some damage (60).

UTH2, isolated as an allele-specific suppressor of the *uth4-14c* mutation, is allelic to *SIR4* (59). Curiously, the mutation is recessive for sterility but dominant for stress resistance and life-span extension. The mutation creates a stop codon removing 121 residues from the C-terminus of Sir4. The allele, designated *SIR4-42*, results in the constitutive relocalization of Sir4-42p from *HM* loci and telomeres to the nucleolus by a process that requires both *UTH4* and *YGL023* (38, 60).

SGS1

Human progeroid syndromes such as Werner's syndrome and Hutchinson-Guilford syndrome are diseases that produce many symptoms resembling premature aging (reviewed in 107). The gene for Werner's syndrome, *WRN*, encodes an ATP-dependent DNA helicase with homology to the *Escherichia coli RecQ* family of proteins (137). Patients mutant for *WRN* have a higher rate of genome instability (78). One interesting aspect of the disease is that many cell types cultured from Werner's patients divide approximately half as many times as those from normal individuals (108), suggesting that there may be a link between replicative senescence and organismal aging.

The yeast homolog of *WRN*, *SGS1*, isolated by two groups studying topoisomerase (*TOP*) function (34, 133), encodes a DNA helicase that interacts with Top2p and Top3p. Mutation of *SGS1* suppresses the slow growth exhibited by a *top3* mutant strain (34, 75). Sgs1p is concentrated in the nucleolus (115), and mutant *sgs1* strains have a higher rate of genomic instability, particularly at rDNA (34, 132), and a life span 40% shorter than wild type (115). *sgs1* cells senesce owing to the rapid progression of a normal aging process, as indicated by the early onset of age-specific markers such as sterility and nucleolar fragmentation. Because both the human *WRN* and yeast *SGS1* genes forestall the onset of aging phenotypes, these helicases may be involved in a conserved aging mechanism.

Since Sgs1p is concentrated in the nucleolus, its absence may predispose the organelle to DNA damage. In young *sgs1* cells, the nucleolus is a compact structure identical to the nucleolus in a young wild-type cell, occupying about one fifth of the nucleus. In old *sgs1* cells, however, the nucleolus becomes enlarged and fragments into numerous mini-nucleoli that form an array around the nuclear periphery (115). Old wild-type nucleoli undergo similar changes, demonstrating that this phenomenon is not an aberration of the *sgs1* mutation.

EXTRACHROMOSOMAL CIRCULAR DNA

Extrachromosomal circular DNA (eccDNA) molecules have been found in the majority of eukaryotic cells thus far examined, including trypanosomatids, tobacco, *Xenopus laevis spp.*, *Drosophila*, chicken, mouse, rat, hamster, monkey, human tissue, cultured human fibroblasts, filamentous fungi, and yeast (reviewed in 35). The number and composition of eccDNAs vary during development and aging, both in vitro and in vivo (31, 75, 77). A role for these molecules in senescence has been demonstrated for filamentous fungi and, more recently, for yeast, providing impetus for the careful examination of the effects of eccDNA in higher organisms.

Senescence in Podospora anserina

The filamentous fungus *Podospora anserina* has been used extensively as a model for cellular senescence. Each strain has a specific life span that allows it to grow a characteristic distance before senescing. The senescent state is inherited maternally through sexual crosses (79, 100) and is infectious through anastomoses between mycelia in the absence of nuclear migration. Senescence is coincidental with the accumulation of extrachromosomal circular DNA derived from the mitochondrial genome called senDNA (5, 50). Three distinct groups of senDNA (α, β, γ) originate from separate regions of the mitochondrial genome (reviewed in 24). Among these, α-senDNA (a group II intron) occurs most frequently in senescing cultures (93), and this sequence is absent from the genomes of many long-lived strains (111). Double-stranded DNA breaks may be a trigger for circle formation since the formation of some senDNAs is preceded by DNA double-stranded breaks at the 3' end of the *cox1*-e1 and *cox1*-e4 exons (106). Mutations that specifically prevent the accumulation of α-senDNA do not significantly affect the rate of senescence, implying that intron α-senDNA may not be the only major determinant of senescence (113).

Clonal Senescence of nib1 *Mutants*

The 2μ plasmid is found in virtually all strains of *S. cerevisiae*. The plasmid contains five open reading frames, four of which maintain the plasmid copy number at a stable 50–100 copies per cell (90, 98). In most strains, the 2μ plasmid confers no phenotype: *cir*$^+$ cells are indistinguishable from *cir*$^-$ cells, and both form round, smooth colonies. In contrast, strains mutant for *nib1* (for *nibbled colony morphology*) exhibit runaway amplification of the 2μ plasmid in a subpopulation of cells (<10%) that eventually senesce clonally. The phenotypes of *nib1* mutants are similar to those of aging cells: a slower S phase, cell enlargement, and the breakdown of asymmetry in older cells (46). An undefined extrachromosomal element can also confer the *Nib1⁻* phenotype in the absence of 2μ DNA; however, its identity is currently unknown (126).

Extrachromosomal rDNA Circles (ERCs) and Yeast Aging

The rDNA locus is comprised of 100–200 tandem repeats of a 9.1-kb unit. Each repeat unit encodes the 5S and 35S rRNAs and is sufficient to act as a focus for nucleolus formation (91). The nontranscribed spacer (NTS) of each repeat contains three autonomously replicating sequences (*ARS*s) that can support the replication of recombinant plasmids (67). However, less than one third of repeats are used as origins of replication during each S phase (8, 72).

As cells age, the normally compact structure of the nucleolus becomes enlarged and eventually fragments, and this event may represent a primary cause of yeast aging (discussed above). Yeast *sgs1* mutants experience rapid aging phenotypes, leading to a dramatic increase in the size and number of nucleoli within only 7–12 divisions (115). Occasionally, homologous recombination between adjacent rDNA repeats produces 3-μm circular rDNA molecules (15, 68). These molecules can be detected as a background band on a Southern blot of young cell DNA, at a level below one copy per cell (115). In contrast, the majority of the rDNA of old cells is present in the form of extrachromosomal rDNA molecules, called ERCs, that are comprised of multimers of the 9.1-kb repeat (114; reviewed in 116). Old cells may contain more than 500 ERCs, equaling the DNA content of the entire yeast genome. These molecules are the likely cause of the enlarged and fragmented nucleoli observed in old cells. Consistent with the notion that *sgs1* mutants age rapidly, young *sgs1* cells begin life with low levels of ERCs, but they accumulate ERCs at twice the rate of wild-type cells. Figure 4 shows a model for yeast senescence by ERC accumulation.

Two simple properties of ERCs can explain their abundance in old cells: replication and asymmetrical inheritance. Once excised from the rDNA locus, an ERC has the ability to replicate once per S phase, but both copies are likely to be retained within the mother cell (114). The replication event may be the molecular clock that times aging of the cell. If the sequence of replication then segregation to the mother cell is reiterated 15 times, then the number of ERC molecules in that cell will be approximately 500. This estimate fits well with observed ERC numbers in old cells (114).

Using a Cre-*loxP* system to release an ERC from a centromeric (*ARS-CEN*) sequence that maintained the plasmid as a single copy per cell, it was demonstrated that excision of a single ERC shortens life span (114). Once liberated, the ERC was free to segregate asymmetrically and accumulate to high levels. The average life span of a strain after ERC release was half that of the wild type.

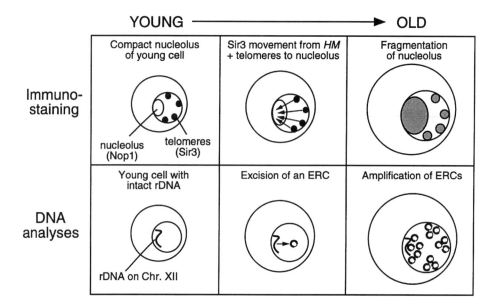

Figure 4   The role of the nucleolus in yeast aging. The nucleolus of a young yeast cell is a compact crescent-shaped structure on the nuclear periphery that contains the rDNA and other components for ribosome synthesis. As cells age, the nucleolus becomes enlarged and fragmented, possibly owing to an extrachromosomal rDNA circle (ERC) that amplifies exponentially in mother cells. Sir proteins may relocalize to the nucleolus as a response to damage to rDNA or to the increase in rDNA copy number.

ERC-induced senescence is correlated with premature sterility, demonstrating that ERCs can cause aging in yeast. The life span of a strain carrying an *ARS1* plasmid is also shorter than control strains, suggesting that any circular molecule that can replicate and segregate asymmetrically can reduce life span. The asymmetrical inheritance pattern of ERCs breaks down in old cells resulting in the leakage of ERCs into daughter cells, thus explaining how young cells from old mother cells become prematurely old themselves. ERCs may be diluted over successive generations, restoring the full life-span potential of those cells. It seems likely that ERCs are the hypothetical senescence factor proposed by Egilmez & Jazwinski (27).

CONCLUSIONS AND PERSPECTIVES

Beginning in 1950 with the discovery that yeast cells are mortal, the phenomenon of yeast aging has been characterized steadily, first descriptively then genetically. We are now in a period of rapid growth in yeast aging research.

Genetic and differential hybridization screens have identified many genes that affect aging, and their further characterization will no doubt aid in the elucidation of other aging mechanisms. Even at this early stage, yeast aging is clearly multifaceted, with influences from such processes as cAMP metabolism, epigenetic silencing, and genomic stability.

One of the more notable discoveries has been identification of a putative senescence factor. These molecules, called ERCs, explain many aspects of yeast aging including the dominance in cell fusion experiments and the kinetics of yeast mortality. Research is under way to determine the processes that govern ERC formation and proliferation. Perhaps the most exciting finding is the role of genome stability in yeast and human aging, implying that some aging processes may be common to both organisms. If so, the progress being made in yeast research will only further the endeavor to understand aging in more complex eukaryotes.

ACKNOWLEDGMENTS

We would like to thank Brad Johnson, Tod Smeal, Heidi Tissenbaum, Shinichiro Imai, David McNabb, and other members of the Guarente laboratory for advice and manuscript preparation. DAS is supported by a Helen-Hay Whitney Foundation Fellowship; KM is supported by an NIH training grant; the Guarente lab is supported by NIH grant AG11119.

> Visit the *Annual Reviews home page* at
> http://www.AnnualReviews.org.

Literature Cited

1. Alcorta DA, Xiong Y, Phelps D, Hannon G, Beach D, Barrett JC. 1996. Involvement of the cyclin-dependent kinase inhibitor p16 (INK4a) in replicative senescence of normal human fibroblasts. *Proc. Natl. Acad. Sci. USA* 93:13742–47

2. Allsopp RC, Vaziri H, Patterson C, Goldstein S, Younglai EV, et al. 1992. Telomere length predicts replicative capacity of human fibroblasts. *Proc. Natl. Acad. Sci. USA* 89:10114–18

3. Austriaco NR, Guarente L. 1997. Changes of telomere length cause reciprocal changes in the lifespan of mother cells in *Saccharomyces cerevisiae. Proc. Natl. Acad. Sci. USA* 94:9768–72

4. Barton AA. 1950. Some aspects of cell division in *Saccharomyces cerevisiae. J. Gen. Microbiol.* 4:84–86

5. Belcour L, Begel O. 1980. Life-span and senescence in *Podospora anserina*: effect of mitochondrial genes and functions. *J. Gen. Microbiol.* 119:505–15

6. Bergman RJ, Gazit D, Kahn AJ, Gruber H, McDougall S, Hahn TJ. 1996. Age-related changes in osteogenic stem cells in mice. *J. Bone Miner. Res.* 11:568–77

7. Blasco MA, Lee HW, Hande MP, Samper E, Lansdorp PM, et al. 1997. Telomere shortening and tumor formation by mouse cells lacking telomerase RNA. *Cell* 91:25–34

7a. Bodnar AG, Guellette M, Frolkis M, Hott SE, Chiu C, et al. 1998. Extension of life span by introduction of telomerase into normal human cells. *Science* 279:349–52

8. Brewer BJ, Fangman WL. 1988. A replication fork barrier at the 3' end of yeast ribosomal RNA genes. *Cell* 55:637–43

9. Bryan TM, Englezou A, Gupta J,

Bacchetti S, Reddel RR. 1995. Telomere elongation in immortal human cells without detectable telomerase activity. *EMBO J.* 14:4240–48

10. Bryk M, Banjeree M, Murphy M, Knudsen KE, Garkinkle DJ, Curcio MJ. 1997. Transcriptional silencing of Ty1 elements in the EDN locus of yeast. *Genes Dev.* 11:255–69

11. Burke EM, Horton WE, Pearson JD, Crow MT, Martin GR. 1994. Altered transcriptional regulation of the human interstitial collagenase in cultured skin fibroblasts from older donors. *Exp. Gerontol.* 29:37–53

12. Cabib E, Ulane R, Bowers B. 1974. A molecular model for morphogenesis: the primary septum of yeast. *Curr. Top. Cell. Regul.* 8:1–32

13. Chen JB, Sun J, Jazwinski SM. 1990. Prolongation of the yeast life span by the v-Ha-*RAS* oncogene. *Mol. Microbiol.* 4:2081–86

14. Childress AM, Franklin D, Pinswasdi C, Kale S, Jazwinski SM. 1996. *LAG2*, a gene that determines yeast longevity. *Microbiology* 142:2289–97

15. Clark-Walker GD, Azad AA. 1980. Hybridizable sequences between cytoplasmic ribosomal RNAs and 3 micron circular DNAs of *Saccharomyces cerevisiae* and *Torulopsis glabrata*. *Nucleic Acids Res.* 8:1009–22

16. Cohn S, Roth KA, Birkenmeier EH, Gordon JI. 1991. Temporal and spatial patterns of transgene expression in aging adult mice provide insights about the origins, organization, and differentiation of the intestinal epithelium. *Proc. Natl. Acad. Sci. USA* 88:1034–38

17. Cooke HJ, Smith BA. 1986. Variability at the telomeres of human X/Y pseudo autosomal region. *Cold Spring Harbor Symp. Quant. Biol.* L1:213–19

18. Cutler RG. 1991. Antioxidants and aging. *Am. J. Clin. Nutr.* 53:373S–79S

19. DeFeo-Jones D, Tatchel K, Robinson LC, Sigal IS, Vass WC, et al. 1985. Mammalian and yeast *ras* gene products: biological function in their heterologous systems. *Science* 228:179–84

20. de Lange T. 1994. Activation of telomerase in a human tumor. *Proc. Natl. Acad. Sci. USA* 91:2882–85

21. D'mello NP, Childress AM, Franklin DS, Kale SP, Pinswasdi C, Jazwinski SM. 1994. Cloning and characterization of *LAG1*, a longevity-assurance gene in yeast. *J. Biol. Chem.* 269:15451–59

22. D'mello NP, Jazwinski SM. 1991. Telomere length constancy during aging of *Saccharomyces cerevisiae*. *J. Bacteriol.* 173:6709–13

23. Dorman JB, Albinder B, Shroyer T, Kenyon C. 1995. The age–1 and daf–2 genes function in a common pathway to control the lifespan of *Caenorhabditis elegans*. *Genetics* 141:1399–406

24. Dujon B, Belcour L. 1989. Mitochondrial instabilities and rearrangements in yeasts and fungi. In *Mobile DNA*, ed. DE Berg, MM Howe, pp. 861–78. Washington, DC: *Am. Soc. Microbiol.*

25. Egilmez NK, Chen JB, Jazwinski SM. 1989. Specific alterations in transcript prevalence during the yeast life span. *J. Biol. Chem.* 264:14312–17

26. Egilmez NK, Chen JB, Jazwinski SM. 1990. Preparation and partial characterization of old yeast cells. *J. Gerontol.* 45:B9–B17

27. Egilmez NK, Jazwinski SM. 1989. Evidence for the involvement of a cytoplasmic factor in the aging of the yeast *Saccharomyces cerevisiae*. *J. Bacteriol.* 171:37–42

28. Ermini M, Moret ML, Reichlmeier K, Dunne T. 1978. Age-dependent structural changes in human neuronal chromatin. *Aktuelle Gerontol.* 8:675–80

29. Ewbank JJ, Barnes TM, Lakowski B, Lussier M, Bussey H, Hekimi S. 1997. Structural and functional conservation of the *Caenorhabditis elegans* timing gene clk-1. *Science* 275:980–83

30. Finch CE. 1990. *Longevity, Senescence, and the Genome*. Chicago, IL: Univ. Chicago Press

31. Flores SC, Sunnerhagen P, Moore TK, Gaubatz JW. 1988. Characterization of repetitive sequence families in mouse heart small polydisperse circular DNA: age-related studies. *Nucleic Acids Res.* 16:3889–906

32. Fukuchi K, Martin GM, Monnat RJ. 1989. Mutator phenotype of Werner syndrome is characterized by extensive deletions. *Proc. Natl. Acad. Sci. USA* 86:5893–97

33. Gale RE, Fielding AK, Harrison CN, Lynch DC. 1997. Acquired skewing of X-chromosome inactivation patterns in myeloid cells of the elderly suggests stochastic clonal loss with age. *Br. J. Haematol.* 98:512–19

34. Gangloff S, McDonald JP, Bendixen C, Arthur L, Rothstein R. 1994. The yeast type I topoisomerase Top3 interacts with Sgs1, a DNA helicase homolog: a potential eukaryotic reverse gyrase. *Mol. Cell. Biol.* 14:8391–98

35. Gaubatz JW, Flores SC. 1990. Tissue-specific and age-related variations in

repetitive sequences of mouse extrachromosomal circular DNAs. *Mutat. Res.* 237:29–36

36. Gibbs JB, Marshal MS. 1989. The *ras* oncogene—an important regulatory element in lower eucaryotic organisms. *Microbiol. Rev.* 53:171–85

37. Gompertz B. 1825. On the nature of the function expressive of the law of human mortality, and on a new mode of determining life contingencies. *Philos. Trans. R. Soc.* 115:513–85

38. Gotta M, Strahl-Bolsinger S, Renauld H, Laroche T, Kennedy BK, et al. 1997. Localization of Sir2p: the nucleolus as a compartment for silent information regulators. *EMBO J.* 16:3243–55

39. Gray MD, Shen JC, Kamath-Loeb AS, Blank A, Sopher BL, et al. 1997. The Werner syndrome protein is a DNA helicase. *Nat. Genet.* 17:100–3

40. Greider CW. 1990. Telomeres, telomerase, and senescence. *BioEssays* 12:363–69

41. Guarente L. 1997. Link between aging and the nucleolus. *Genes Dev.* 11:2449–55

42. Harley CB, Futcher AB, Greider CW. 1990. Telomeres shorten during ageing of human fibroblasts. *Nature* 345:458–60

43. Hayflick L, Moorhead P. 1961. The serial cultivation of human diploid cell strains. *Exp. Cell Res.* 25:585–621

44. Helmstetter CE. 1991. Description of a baby machine for *Saccharomyces cerevisiae*. *New Biol.* 3:1089–96

45. Holliday R. 1987. The inheritance of epigenetic defects. *Science* 238:163–70

46. Holm C. 1982. Clonal lethality caused by the yeast plasmid 2μ DNA. *Cell* 29:585–94

47. Hotta Y, Benzer S. 1972. Mapping of behaviour in *Drosophila* mosaics. *Nature* 240:527–35

48. Imai S, Nishibayashi S, Takao K, Tomifuji M, Fujino T, et al. 1997. Dissociation of Oct-1 from the nuclear peripheral structure induces the cellular aging-associated collagenase gene expression. *Mol. Cell. Biol.* 8:2407–19

49. Imai S, Takano T. 1992. Loss of collagenase gene expression in immortalized clones of SV-40 T antigen-transformed human diploid fibroblasts. *Biochem. Biophys. Res. Commun.* 189:148–53

50. Jamet-Vierny C, Begel O, Belcour L. 1980. Senescence in Podospora anserina: amplification of a mitochondrial DNA sequence. *Cell* 21:189–94

51. Jazwinski SM. 1990. Aging and senescence of the budding yeast *Saccharomyces cerevisiae*. *Mol. Microbiol.* 4:337–43

52. Jazwinski SM. 1990. An experimental system for the molecular analysis of the aging process: the budding yeast *Saccharomyces cerevisiae*. *J. Gerontol.* 45:B68–74

53. Jazwinski SM. 1993. The genetics of aging in the yeast *Saccharomyces cerevisiae*. *Genetica* 91:35–51

54. Jazwinski SM. 1996. Longevity, genes, and aging. *Science* 273:54–59

55. Jazwinski SM, Egilemez NK, Chen JB. 1989. Replication control and cellular life span. *Exp. Gerontol.* 24:423–36

56. Johnston JR. 1966. Reproductive capacity and mode of death of yeast cells. *Antonie van Leeuwenhoek* 32:94–98

57. Kataoka T, Powers S, Cameron S, Fasano O, Strathern J, et al. 1985. Functional homology of mammalian and yeast RAS genes. *Cell* 40:19–26

58. Kennedy BK, Austriaco NR, Guarente L. 1994. Daughter cells of *Saccharomyces cerevisiae* from old mothers display a reduced life span. *J. Cell Biol.* 127:1985–93

59. Kennedy BK, Austriaco NR, Zhang J, Guarente L. 1995. Mutation of the silencing gene SIR4 can delay aging in S. cerevisiae. *Cell* 80:485–96

60. Kennedy BK, Gotta M, Sinclair DA, Mills K, McNabb DS, et al. 1997. Redistribution of silencing proteins from telomeres to the nucleolus is associated with extension of life span in S. cerevisiae. *Cell* 89:381–91

61. Kikuchi Y, Oka Y, Kabayashi M, Uesono Y, Toh-e A, Kikuchi A. 1994. A new yeast gene, *HTR1*, required for growth at high temperature, is needed for recovery from mating pheromone-induced G1 arrest. *Mol. Gen. Genet.* 245:107–16

62. Kim S, Villeponteau B, Jazwinski SM. 1996. Effect of replicative age on transcriptional silencing near telomeres in *Saccharomyces cerevisiae*. *Biochem. Biophys. Res. Commun.* 219:370–76

63. Kimura KD, Tissenbaum HA, Liu Y, Ruvkun G. 1997. daf–2, an insulin receptor-like gene that regulates longevity and diapause in *Caenorhabditis elegans*. *Science* 277:942–46

64. Klass MR. 1977. Aging in the nematode Caenorhabditis elegans: major biological and environmental factors influencing life span. *Mech. Ageing Dev.* 6:413–29

65. Lakowski B, Hekimi S. 1996. Determination of life-span in *Caenorhabditis elegans* by four clock genes. *Science* 272:1010–13

66. Deleted in proof

67. Larionov V, Kouprina N, Karpova T. 1984. Stability of recombinant plasmids containing the *ARS* sequence of yeast extrachromosomal rDNA in several strains of *Saccharomyces cerevisiae*. *Gene* 28: 229–35

68. Larionov VL, Grishin AV, Smirnov MN. 1980. 3 micron DNA—an extrachromosomal ribosomal DNA in the yeast *Saccharomyces cerevisiae*. *Gene* 12:41–49

69. Laurenson P, Rine J. 1992. Silencers, silencing, and heritable states. *Microbiol. Rev.* 56:543–60

70. Lezhava TA. 1984. Heterochromatinization as a key factor in aging. *Mech. Ageing Dev.* 28:279–87

71. Lin H, Schagat T. 1997. Neuroblasts: a model for the asymmetric division of stem cells. *Trends Genet.* 13:33–39

72. Linskens MH, Huberman JA. 1988. Organization of replication of ribosomal DNA in *Saccharomyces cerevisiae*. *Mol. Cell. Biol.* 8:4927–35

73. Longo VD, Ellerby LM, Breesen DE, Valentine JS, Gralla EB. 1997. Human Bcl-2 reverses survival defects in yeast lacking superoxide dismutase and delays death of wild-type yeast. *J. Cell Biol.* 137: 1581–88

74. Longo VD, Gralla EB, Valentine JS. 1996. Superoxide dismutase activity is essential for stationary phase survival in *Saccharomyces cerevisiae*. *J. Biol. Chem.* 271: 12275–80

75. Lu J, Mullen JR, Brill SJ, Kleff S, Romeo AM, Sternglanz R. 1996. Human homologues of yeast helicase. *Nature* 383:678–79

76. Lundblad V, Szostak JW. 1989. A mutant with a defect in telomere elongation leads to senescence in yeast. *Cell* 57:633–43

77. Maciera-Coelho A, Garcia-Giralt E, Adrian M. 1971. Changes in lysosomal associated structures in human fibroblasts kept in resting phase. *Proc. Soc. Exp. Biol. Med.* 138:712–18

78. Maciera-Coelho A, Puvion-Dutilleul F. 1985. Genome reorganization during aging of dividing cells. *Adv. Exp. Med. Biol.* 190:391–420

79. Marcou D. 1961. *Notion de longevité et nature cytoplasmique du déterminant de sénescence chez quelques champignons*. *Ann. Sci. Natur. Bot.* 11:653–764

80. Mays-Hoopes LL, Brown A, Huang RC. 1983. Methylation and rearrangement of mouse intracisternal A particle genes in development, aging, and myeloma. *Mol. Cell. Biol.* 3:1371–80

81. Melese T, Xue Z. 1995. The nucleolus: an organelle formed by the act of building a ribosome. *Curr. Opin. Cell Biol.* 7:319–24

82. Michel M, Török N, Godbout M-J, Lussier M, Gaudreau P, et al. 1996. Keratin 19 as a biochemical marker of skin stem cells in vivo and in vitro. *J. Cell Sci.* 109:1017–28

83. Deleted in proof

84. Mortimer RK, Johnston JR. 1959. Life span of individual yeast cells. *Nature* 183: 1742–51

85. Motizuki M, Tsurugi K. 1992. The effect of aging on protein synthesis in the yeast *Saccharomyces cerevisiae*. *Mech. Ageing Dev.* 64:235–45

86. Motizuki M, Yakota S, Tsurugi K. 1995. Autophagic death after cell-cycle arrest at the restrictive temperature in temperature-sensitive cell division cycle and secretory mutants of the yeast *Saccharomyces cerevisiae*. *Eur. J. Cell Biol.* 68:275–87

87. Muller I. 1971. Experiments on ageing in single cells of *Saccharomyces cerevisiae*. *Arch. Mikrobiol.* 77:20–25

88. Muller I. 1985. Parental age and the life-span of zygotes of *Saccharomyces cerevisiae*. *Antonie van Leeuwenhoek* 51:1–10

89. Muller I, Zimmerman M, Becker D, Flomer M. 1980. Calendar life span *versus* budding life span of *Saccharomyces cerevisiae*. *Mech. Ageing Dev.* 12:47–52

90. Murray JA, Scarpa M, Rossi M, Cesareni G. 1987. Antagonistic controls regulate copy number of the yeast 2μ plasmid. *EMBO J.* 6:4205–12

91. Nierras CR, Liebman SW, Warner JR. 1997. Does *Saccharomyces* need an organized nucleolus? *Chromosoma* 105:444–51

92. Norwood TH. 1978. Somatic cell genetics in the analysis of in vitro senescence. In *The Genetics of Aging*, ed. EL Schneider. New York: Plenum

93. Osiewacz HD, Esser K. 1984. The mitochondrial plasmid of *Podospora anserina*: a mobile intron of a mitochondrial gene. *Curr. Genet.* 8:299–305

94. Pacifici RE, Davies KJ. 1991. Protein, lipid and DNA repair systems in oxidative stress: the free-radical theory of aging revisited. *Gerontology* 37:166–80

95. Pereira-Smith OM, Smith JR. 1983. Evidence for the recessive nature of cellular immortality. *Science* 221:964–66

96. Petes TD. 1979. Yeast ribosomal DNA genes are located on chromosome XII. *Proc. Natl. Acad. Sci. USA* 76:410–14

97. Pohley HJ. 1987. A formal mortality analysis for populations of unicellular

organisms (*Saccharomyces cerevisiae*). *Mech. Ageing. Dev.* 38:231–43

98. Reynolds AE, Murray AW, Szostak JW. 1987. Roles of the 2μm gene products in stable maintenance of the 2μm plasmid of *Saccharomyces cerevisiae*. *Mol. Cell. Biol.* 7:3566–73

99. Rittling SR, Brooks KM, Cristofalo VJ, Baserga R. 1986. Expression of cell cycle–dependent genes in young and senescent WI-38 fibroblasts. *Proc. Natl. Acad. Sci. USA* 83:3316–20

100. Rizet G. 1952. *Les phénomènes de barrage chez Podospora anserina. I. Analyse genetique des barrages entre les souches S et s. Rev. Cytol. Biol. Veget.* 13:51–92

101. Robertson G, Garrick D, Wilson M, Martin DI, Whitelaw B. 1996. Age-dependent silencing of globin transgenes in the mouse. *Nucleic Acids Res.* 24:1465–71

102. Rogina B, Benzer S, Halfland SL. 1997. Drosohila drop-dead mutations accelerate the time course of age-related markers. *Proc. Natl. Acad. Sci. USA* 94:6303–6

103. Rose MR, Archer MA. 1996. Genetic analysis of mechanisms of aging. *Curr. Opin. Genet. Dev.* 6:366–70

104. Rose MR, Charlesworth B. 1981. Genetics of life history in *Drosophila melanogaster*. II. Exploratory selection experiments. *Genetics* 97:187–96

105. Rustchenko EP, Sherman F. 1994. Physical constitution of ribosomal genes in common strains of *Saccharomyces cerevisiae*. *Yeast* 10:1157–71

106. Sainsard-Chanet A, Begel O, Belcour L. 1994. DNA double-strand break *in vivo* at the 3′ extremity of exons located upstream of group II introns. Senescence and circular DNA introns in *Podospora* mitochondria. *J. Mol. Biol.* 242:630–43

107. Salk D. 1982. Werner's syndrome: a review of recent research with an analysis of connective tissue metabolism, growth control of cultured cells, and chromosomal aberrations. *Hum. Genet.* 62:1–5

108. Salk D, Bryant E, Hoehn H, Johnston P, Martin GM. 1985. Growth characteristics of Werner syndrome cells in vitro. *Adv. Exp. Med. Biol.* 190:305–12

109. Scheer U, Benavente R. 1990. Functional and dynamic aspects of the mammalian nucleolus. *BioEssays* 12:14–21

110. Schimmang T, Tollervey D, Kern H, Frank R, Hurt EC. 1989. A yeast nucleolar protein related to mammalian fibrillarin is associated with small nucleolar RNA and is essential for viability. *EMBO J.* 8:4015–24

111. Schulte E, Kuck U, Esser K. 1988. Extrachromosomal mutants from *Podospora anserina*: permanent vegetative growth in spite of multiple recombination events in the mitochodrial genome. *Mol. Gen. Genet.* 211:342–39

112. Shaw PJ, Highett MI, Beven AF, Jordan EG. 1995. The nucleolar architecture of polymerase I transcription on processing. *EMBO J.* 14:2896–906

113. Silar P, Koll F, Rossignol M. 1997. Cytosolic ribosomal mutations that abolish accumulation of circular intron in the mitochondria without preventing senescence of *Podospora anserina*. *Genetics* 145:697–705

114. Sinclair DA, Guarente L. 1997. Extrachromosomal rDNA circles—a cause of aging in yeast. *Cell* 91:1033–42

115. Sinclair DA, Mills K, Guarente L. 1997. Accelerated aging and nucleolar fragmentation in yeast *sgs1* mutants. *Science* 277:1313–16

116. Sinclair DA, Mills K, Guarente L. 1998. Molecular mechanisms of yeast aging. *Trends Biochem. Sci.* 24:In press

117. Sloat BF, Adams A, Pringle JR. 1981. Roles of the *CDC24* gene product in cellular morphogenesis during *Saccharomyces cerevisiae* cell cycle. *J. Cell. Biol.* 89:395–405

118. Sloat BF, Pringle JR. 1978. A mutant of yeast defective in cellular morphogenesis. *Science* 200:1171–73

119. Deleted in proof

120. Smeal T, Claus J, Kennedy B, Cole F, Guarente L. 1996. Loss of transcriptional silencing causes sterility in old mother cells of S. cerevisiae. *Cell* 84:633–42

121. Smeal T, Guarente L. 1997. Mechanisms of cellular senescence. *Curr. Opin. Gen. Dev.* 7:281–87

122. Smith GS, Walford RL. 1977. Influence of the main histocompatibility complex on ageing in mice. *Nature* 270:727–29

123. Smith JS, Boeke JD. 1997. An unusual form of transcriptional silencing in yeast ribosomal DNA. *Genes Dev.* 11:241–54

124. Sohal RS, Weindruch R. 1996. Oxidative stress, caloric restriction, and aging. *Science* 273:59–63

125. Sun J, Kale SP, Childress AM, Pinswasdi C, Jazwinski SM. 1994. Divergent roles of *RAS1* and *RAS2* in yeast longevity. *J. Biol. Chem.* 269:18638–45

126. Sweeney R, Zakian VA. 1989. Extrachromosomal elements cause a reduced division potential in *nib1* strain of *Saccharomyces cerevisiae*. *Genetics* 122:749–57

127. Swisshelm K, Disteche CM, Thorvaldsen

J, Nelson A, Salk D. 1990. Age-related increase in methylation of ribosomal genes and inactivation of chromosome-specific rRNA gene clusters in mouse. *Mutat. Res.* 237:131–46

128. Thakur MK, Oka T, Natori Y. 1993. Gene expression and aging. *Mech. Ageing Dev.* 66:283–98

129. Tissenbaum HA. 1997. *Genetic and molecular analysis of genes controlling diapause and longevity in* Caenorhabditis elegans. Ph.D. thesis, Harvard University

130. Tsukamoto Y, Kato J, Ikeda H. 1997. Silencing factors participate in DNA repair and recombination in *Saccharomyces cerevisiae. Nature* 388:900–3

131. Vanyushin BF, Nemirovsky LE, Klimenko VV, Vasiliev VK, Belozersky AN. 1973. The 5-methylcytosine in DNA of rats. Tissue and age specificity and the changes induced by hydrocortisone and other agents. *Gerontologia* 19:138–52

132. Watt PM, Hickson ID, Borts RH, Louis EJ. 1996. *SGS1*, a homologue of the Bloom's and Werner's syndrome genes, is required for maintenance of genome

stability in *Saccharomyces cerevisiae. Genetics* 144:935–45

133. Watt PM, Louis EJ, Borts RH, Hickson ID. 1995. Sgs1: a eukaryotic homolog of E. coli RecQ that interacts with topoisomerase II in vivo and is required for faithful chromosome segregation. *Cell* 81:253–60

134. Werner-Washburne M, Braun E, Johnston GC, Singer RA. 1993. Stationary phase in the yeast *Saccharomyces cerevisiae. Microbiol. Rev.* 57:383–401

135. Woldringh CL, Fluiter K, Huls PG. 1995. Production of senescent cells of *Saccharomyces cerevisiae* by centrifugal elutriation. *Yeast* 11:361–69

136. Yang ZY, Simari RD, Perkins ND, San H, Gordon D, et al. 1996. Role of the p21 cyclin-dependent kinase inhibitor in limiting intimal cell proliferation in response to arterial injury. *Proc. Natl. Acad. Sci. USA* 93:7905–10

137. Yu CE, Oshima J, Fu YH, Wijsman EM, Hisama F, et al. 1996. Positional cloning of the Werner's syndrome gene. *Science* 272:258–62

Annu. Rev. Microbiol. 1998. 52:561–90
Copyright © 1998 by Annual Reviews. All rights reserved

METABOLIC CHANGES OF THE MALARIA PARASITE DURING THE TRANSITION FROM THE HUMAN TO THE MOSQUITO HOST

N. Lang-Unnasch and A. D. Murphy

Division of Geographic Medicine, Department of Medicine, University of Alabama at Birmingham, Birmingham, AL 35294-2170; e-mail: nlang-unnasch@geomed.dom.uab.edu

KEY WORDS: *Plasmodium falciparum*, carbohydrate metabolism, gametocytogenesis, nucleic acid biosynthesis, protein synthesis

ABSTRACT

Plasmodium falciparum is an obligate human parasite that is the causative agent of the most lethal form of human malaria. Transmission of *P. falciparum* to a new human host requires a mosquito vector within which sexual replication occurs. *P. falciparum* replicates as an intracellular parasite in man and as an extracellular parasite in the mosquito, and it undergoes multiple developmental changes in both hosts. Changes in the environment and the activities of parasites in these various life-cycle stages are likely to be reflected in changes in the metabolic needs and capabilities of the parasite. Most of our knowledge of the metabolic capabilities of *P. falciparum* is derived from studies of the asexual erythrocytic cycle of the parasite, the portion of the parasite life cycle found in infected humans that is responsible for malarial symptoms. Efforts to control transmission and to understand the sometimes unique biology of this parasite have led to information about the metabolic capabilities of sexual and/or sporogonic stages of these parasites. This review focuses on comparing and contrasting the carbohydrate, nucleic acid, and protein synthetic capabilities of asexual erythrocytic stages and sexual stages of *P. falciparum*.

561

0066-4227/98/1001-0561$08.00

CONTENTS

INTRODUCTION

Like that of all malaria parasites, the life cycle of *Plasmodium falciparum* requires two hosts. When *P. falciparum* enters its human host, it first undergoes exoerythrocytic replication in liver cells and then initiates the asexual erythrocytic cycle (Figure 1). Parasites in the asexual erythrocytic cycle are responsible for the clinical symptoms of malaria. A portion of these asexual parasites become committed to sexual development and undergo gametocytogenesis (3). *P. falciparum* gametocyte maturation takes longer than that of other malaria parasites, about 10 days in vivo or in vitro (126), and five morphologically distinguished intermediate stages have been identified in this process (44). Sexual dimorphism of the gametocytes can first be detected ultrastructurally at stage III of gametocytogenesis (126). Mature female gametocytes, the macrogametocytes, are readily distinguished from male gametocytes, the microgametocytes, at the light microscopic level. *P. falciparum* gametocytes have a long half life (2.4 days) and period of infectivity as compared to gametocytes of other *Plasmodium* species (127). The rest of the *Plasmodium* sexual cycle, including gametogenesis, fertilization, and meiosis, takes place in the mosquito (Figure 1). Gametogenesis can be induced in vitro by a drop in temperature and an increase in pH, but in vivo it may be triggered by a mosquito factor (8a, 37, 127). The ookinete stage of the parasite produces a chitinase that enables the parasite to penetrate the peritrophic membrane surrounding the blood meal (103). It then migrates through the midgut epithelium, stopping and rounding up at the basal lamina, where the oocyst develops (91). The development of sporozoites within the oocyst, their migration to the salivary gland (131), and their injection into a human complete the parasite's life cycle.

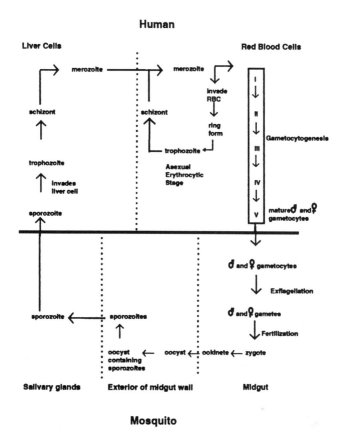

Figure 1 The life cycle of *Plasmodium falciparum*. (*lower case*) Developmental stages of the parasite, localized by the host in which they occur (human stages above *thick line* and mosquito stages *below*) and by the compartment (*dotted vertical lines*) within each host in which they reside.

Transmission of the parasite between hosts results in dramatic differences in its environment. Human stages of the parasite are intracellular, while those in the mosquito are extracellular. The temperature in the human host is 37°C, while that in the mosquito is ambient. The rich milieu of the blood in the human host is modified by trypsin and other mosquito factors in the blood meal, and is completely absent outside the midgut at the site of oocyst and sporozoite development. Such differences suggest that the metabolic requirements and capabilities of the parasite are likely to change over the course of this complex life cycle.

This review focuses on three areas of *P. falciparum* metabolism that have been investigated in both asexual erythrocytic and in sporogonic stages of the life cycle: carbohydrate metabolism, nucleic acid synthesis, and protein synthesis. Some aspects of these metabolic pathways may prove to be species specific. However, in the interest of developing a comprehensive model for the changes in *P. falciparum*, we have included information about the metabolism of other *Plasmodium* species. It is hoped that this model will form the basis for further studies in *P. falciparum*.

CARBOHYDRATE METABOLISM

Mitochondrial Changes

The asexual erythrocytic *P. falciparum* parasite has a single mitochondrion (133), which contains few if any cristae (77). Fluorescent cationic probes, such as Rhodamine 123 and 3,3'-dihexyloxacarbocyanine [$DIOC_6(3)$], which specifically stain areas of high negative membrane potential, have been used to follow the development of the mitochondria throughout the parasite life cycle. Such studies indicate that, over the course of gametocytogenesis, there are changes in mitochondrial shape, in the mitochondrial membrane potential, and in sensitivity to mitochondrial inhibitors (25, 62, 135).

During the third morphological stage of *P. falciparum* gametocyte formation (Figure 1), macrogametocytes begin to show a marked increase in cristate mitochondria (77, 126). Interestingly, this is the same stage of *P. falciparum* gametocyte development in which the drug sensitivity of the parasite changes. For the most part, late-stage gametocytes (stages IV–V) are more resistant to antimalarial drugs (15) and metabolic inhibitors (130) than are early-stage gametocytes or asexuals. The one drug to which the late-stage gametocytes are more sensitive than asexual stages is primaquine (8, 108). Primaquine's mechanism of action is not yet clear, but ^{3}H-primaquine selectively accumulates in *Plasmodium* mitochondria (2). Other studies indicate it causes swelling in mitochondria of several *Plasmodium* species (6, 10, 50), including the gametocytes of *P. falciparum* (79). Evidence that *P. falciparum* mitochondria change in terms of both ultrastructure and chemotherapeutic susceptibility suggests that there may be related biochemical changes in the mitochondrially directed metabolic activities of the parasite.

Glycolysis

Asexual erythrocytic-stage *P. falciparum* require glucose for growth, and glucose cannot be replaced by ribose, mannose, fructose, or galactose (39). *P. falciparum*–infected RBCs use up to 100 times more glucose than uninfected cells (112, 117), and after 22 h of growth, virtually all of the glucose can be accounted for by the production of lactate (117). Furthermore, *P. falciparum* ATP levels appear to be minimally affected or unaffected by mitochondrial inhibitors, though ATP levels are greatly reduced by exposure of the parasite to inhibitors of glucose transport (36). Together with data, described below, indicating that asexual erythrocytic-stage parasites lack a functional tricarboxylic acid (TCA) cycle, these studies strongly suggest that the primary source of asexual parasite energy is ATP generated via fermentation of glucose.

P. falciparum–infected RBCs have increased activity of the glycolytic enzymes and of lactate dehydrogenase as compared to uninfected RBCs (111). Similarly, the specific activity of the glycolytic enzymes and lactate dehydrogenase from isolated P. falciparum trophozoites is higher than that from uninfected RBCs (148). Most of the P. falciparum enzymes used for fermentation of glucose during asexual erythrocytic growth have now been characterized antigenically, enzymatically, and/or genetically (Figure 2).

It is generally presumed that the glycolytic enzymes are present in the mosquito stages of Plasmodium, but despite the current availability of antibody and/or nucleotide probes, there are no published studies supporting this presumption. A parasite-specific lactate dehydrogenase isoenzyme has been detected in Plasmodium yoelii–infected Anopheles stephensi and Anopheles gambiae (107). These results suggest that this rodent malaria parasite utilizes at least some glucose by a fermentative pathway. The relevance of these findings to other Plasmodium parasites, including P. falciparum, needs to be confirmed. During asexual erythrocytic stages of development, P. falciparum appear to be more exclusively dependent on fermentation of glucose to lactate than are Plasmodium species causing avian, rodent, and monkey malarias (116, 114). Further studies are likely to highlight other species-specific metabolic differences.

Tricarboxylic Acid Cycle

Electron microscopic studies show that the mitochondria of erythrocytic-stage P. falciparum have few cristae, while those of macrogametocytes are cristate (77, 126, 128). Asexual erythrocytic-stage avian malaria parasites such as Plasmodium lophurae are cristate (1), while those of rodent and most other mammalian malaria parasites are acristate (74). Erythrocytic-stage avian parasites appear to use some respiratory metabolism, but mammalian parasites of this stage, including P. falciparum, appear to depend on fermentative metabolism (116, 121). Thus, it has been proposed that the development of cristate mitochondria is associated with the presence of an active TCA cycle and respiration (48, 77).

Energy generation via the TCA cycle requires the presence of all eight enzymes, even though biosynthetic pathways may continue to function if only some of the enzymes are present. Although it is difficult to prove the absence of an activity, efforts to identify an activity for three TCA cycle enzymes in asexual erythrocytic-stage P. falciparum have been unsuccessful. No parasite-

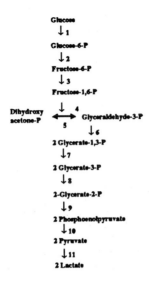

Glucose
↓ 1
Glucose-6-P
↓ 2
Fructose-6-P
↓ 3
Fructose-1,6-P

Dihydroxy   ↓ 4
acetone-P ←——→ Glyceraldehyde-3-P
              5   ↓ 6

2 Glycerate-1,3-P
↓ 7
2 Glycerate-3-P
↓ 8
2-Glycerate-2-P
↓ 9
2 Phosphoenolpyruvate
↓ 10
2 Pyruvate
↓ 11
2 Lactate

Figure 2 Pathways by which glucose is used to generate energy. Parasite-specific properties of most of the *Plasmodium falciparum* enzymes used for fermentation of glucose have been determined. A parasite isoenzyme of hexokinase [1] has been visualized (111), parasite-specific properties of this enzyme reported (110), the gene for it has been cloned, and specific antibodies generated (96). The glucose phosphate isomerase [2] protein (136) has been characterized, antibodies were generated (136), and the gene has been cloned (61). A parasite-specific isoenzyme of phosphofructokinase [3] has been visualized (111). The *P. falciparum* aldolase [4] gene has been cloned (66, 67). Antibodies to aldolase were used to localize it in a soluble cytoplasmic form and in a membrane-associated, inactive form (67). Some evidence suggests that aldolase may associate with other glycolytic enzymes by association with the cytoskeleton (51). The triose phosphate isomerase [5] gene has been cloned (105) and its crystalline structure determined (152). A higher specific activity of glyceraldehyde-3-phosphate dehydrogenase [6] and phosphoglycerate mutase [8] was measured in purified trophozoite-stage *P. falciparum* than in uninfected RBCs. The phosphoglycerate kinase [7] gene encodes a protein with 60% homology to the human homolog (46). The enolase [9] gene shows phylogenetic similarity to those of plants (106). A parasite isoenzyme of pyruvate kinase [10] has been visualized (111). *P. falciparum* lactate dehydrogenase [11] has enzymatic (149) and antigenic properties (64) that distinguish it from the human homolog. The lactate dehydrogenase gene (11) has been characterized, as has the structure of the protein (27).

specific activity could be detected for α-ketogluterate dehydrogenase (9), while the parasite-specific activities of malate dehydrogenase (75, 76) and isocitrate dehydrogenase (150) appear to be cytoplasmically localized and to participate in other metabolic pathways.

Malate dehydrogenase activity detected in asexual erythrocytic-stage *P. falciparum* was thought to possibly represent a TCA cycle enzyme (151). Investigation of this activity resulted in the identification, purification, and characterization of a single *P. falciparum*–specific malate dehydrogenase (75). Typically, eukaryotic cells contain two genetically distinct, cytoplasmic and mitochondrial compartmentalized isoenzymes of malate dehydrogenase. Only the mitochondrially located isoenzyme functions as part of the TCA cycle. Based on characteristics of the purified enzyme (75), on subcellular fractionation studies of *P. falciparum* (76), and on the inability of malate to act as a substrate for mitochondrial oxidation using purified *P. falciparum* mitochondria (34), it appears that the one malate dehydrogenase isoenzyme present in asexual erythrocytic-stage parasites does not function in the TCA cycle.

Efforts to identify additional parasite-specific malate dehydrogenase in gametocyte stages of *P. falciparum* were inconclusive (Lang-Unnasch & Murphy, unpublished). However, late-stage *P. falciparum* gametocytes do appear to have 2–3 times more malate dehydrogenase enzyme and enzyme activity than do asexual parasites (Lang-Unnasch, unpublished).

An isocitrate dehydrogenase that is dependent on NADP is found at 20-fold higher specific activity in *P. falciparum* than in uninfected RBCs (150). Partial purification and characterization of an NADP-specific isocitrate dehydrogenase from *Plasmodium knowlesi* indicated that it is localized in the cytosolic fraction (115). Assuming that the *P. falciparum* and *P. knowlesi* NADP-specific isocitrate dehydrogenases are the same, it is unlikely that they function in the TCA cycle. Starch gel electrophoresis of *Plasmodium berghei* suggested that neither an NAD- nor an NADP-dependent, parasite-specific isoenzyme of isocitrate dehydrogenase was present in infected RBCs (49).

Although no parasite-specific isoenzymes of isocitrate dehydrogenase were found in asexual erythrocytic-stage *P. berghei*, oocysts could be shown to contain parasite-specific isoenzymes of both an NADP- and an NAD-dependent isocitrate dehydrogenase (49). If the NAD-dependent isocitrate dehydrogenase is localized in oocyst mitochondria, this would provide concrete evidence for an increased biosynthetic and/or energetic role of mitochondria in oocysts as compared to erythrocytic-stage *P. berghei*.

Succinate dehydrogenase is a TCA cycle enzyme and is referred to as Complex II of the electron transport chain because it contains a flavin adenine dinucleotide (FAD) by which electrons are transferred directly to ubiquinone. Succinate dehydrogenase activity has been demonstrated in subcellular fractions enriched for the mitochondria of asexual erythrocytic-stage *P. falciparum* or *P. yoelii*, by succinate-dependent stimulation of oxygen respiration (34). Genes for the flavoprotein and iron-sulphur–containing subunits of succinate dehydrogenase of *P. falciparum* have been sequenced (accession nos. D86573 and D86574). RNA for the iron-sulphur subunit of the *P. falciparum* succinate dehydrogenase was detected by reverse transcriptase–mediated polymerase chain reaction (RT-PCR) amplification with gene-specific oligonucleotide primers indicating that the gene for this subunit is transcribed in asexual parasites

(N Lang-Unnasch, unpublished). This result was predicted based on the presence of succinate dehydrogenase activity in asexual erythrocytic-stage parasites. It has been suggested that in asexual erythrocytic-stage *P. falciparum*, succinate dehydrogenase together with an NADH-fumarate reductase might carry out a cyclic interconversion of succinate and fumarate to help reoxidize NADH (33). As yet, no NADH-fumarate reductase activity has been identified in *Plasmodium*, and fumarate does not stimulate respiration by *P. falciparum* or *P. yoelii* mitochondria (33). Thus, the role of succinate dehydrogenase in the asexual cycle of these parasites remains unclear.

In *P. berghei*, no succinate dehydrogenase activity was detected by a histochemical assay of asexual erythrocytic- or gametocyte-stage parasites, but it was found in oocysts and developing sporozoites (48). The apparent difference between the activity of succinate dehydrogenase in asexual erythrocytic-stage *P. berghei* and in *P. yoelii* may reflect the relative insensitivity of the histochemical assay used by Howells (48) as compared to the biochemical assay used by Fry (34). This suggests that further study will show that succinate dehydrogenase is more highly expressed in oocysts than in asexual stages of these rodent malaria parasites and probably also in *P. falciparum*. Given the current availability of the gene sequence for two subunits of succinate dehydrogenase, quantitative RT-PCR methods could readily be used to delineate the expression of these genes throughout human and mosquito stages of the parasite life cycle.

Classical Electron Transport

The classical electron transport or respiratory chain consists of NADH dehydrogenase (Complex I), succinate dehydrogenase (Complex II), ubiquinone, and the cytochromes c, $b + c_1$ (Complex III) and $a + a_3$ (Complex IV). Translocation of a H^+ ion can occur at Complexes I, III, and IV. In the presence of F_1F_0-ATPase, this H^+ translocation is usually coupled to ATP synthesis. Despite the absence of a TCA cycle, there is substantial evidence that asexual erythrocytic-stage *P. falciparum* have a functional electron transport chain. The first evidence for its existence in *Plasmodium* centered on Complex IV, or cytochrome c oxidase. Activity for *P. berghei* cytochrome c oxidase was localized in the whorled cytoplasmic membrane structures that are now recognized as the mitochondria of this rodent malaria (141). Shortly thereafter, cytochrome oxidase activity was also found in asexual erythrocytic-stage *P. falciparum* (118). More recently, the mitochondria of the rodent malaria parasites *P. yoelii* (34) and *P. berghei* (71) and the human malaria parasite *P. falciparum* (34) have been purified and characterized. Low-temperature difference spectra indicated the presence of cytochromes $a + a_3$ as well as b, c, and c_1. Cytochrome c oxidase purified from *P. berghei* mitochondria catalyzes the oxidation of reduced cytochrome c, contains cytochrome a, and is as sensitive to inhibition by cyanide or azide as is mouse liver cytochrome oxidase (71). The genes for the *P. falciparum* cytochrome c oxidase subunits I and III have been identified on the 6-kb extrachromosomal DNA element of *P. falciparum* (28) and other *Plasmodium* species (4, 145). Cytochrome c oxidase activity has been used as a marker for *P. falciparum* mitochondrial purification (70).

In *P. berghei*, cytochrome oxidase activity has also been found in the mitochondria of young oocysts and in developing sporozoites (47). Cytochrome oxidase in *P. falciparum* gametocytes has been investigated indirectly by assaying the sensitivity of mitochondrial membrane potential to inhibition by cytochrome oxidase inhibitors. Using Rhodamine 123 as a fluorescent probe of parasite mitochondrial membrane potential, membrane potential in *P. falciparum* trophozoites (25) but not late-stage gametocytes (62) was diminished by 1 mM cyanide or 1 mM azide.

The cytochrome b gene of *P. falciparum* and other *Plasmodium* species is localized on the 6-kb mitochondrial DNA element. Genes for the nuclear encoded components of Complex III have not yet been identified. The activity and spectra of Complex III have been studied in purified *P. falciparum* and *P. yoelii* mitochondria (34). Biochemical assays suggest that the respiratory activity of mitochondria from asexual erythrocytic *P. falciparum* is 10- to 100-fold less sensitive to Complex III inhibitors such as antimycin A ($EC_{50} = 2$ υM) and myxothiazol ($EC_{50} = 30$ μM) (34) than the respiratory activity of mammalian mitochondria (142). This difference in sensitivity may be due to nontypical amino acids in 2 of 5 sites critical for inhibitor sensitivity in the Q_1 binding site of the cytochrome b protein (146). In contrast, Complex III of *P. falciparum* is 1000 times more sensitive to inhibition by the hydroxynapthoquinone atovaquone than are mammalian mitochondria (35).

Comparisons of Complex III activity in asexual and sexual-stage *P. falciparum* have been based on assays of mitochondrial membrane potential and on the sensitivity of different life-cycle stages to atovaquone. Both asexual erythrocytic parasites (26) and late-stage gametocytes (62) have diminished mitochondrial membrane potential following exposure to 1 μM antimycin A as indicated by staining with rhodamine 123. In contrast, the number of asexual and early-stage gametocytes but not late-stage gametocytes of *P. falciparum* is reduced by atovaquone (31). Pretreatment of late-stage gametocytes with atovaquone also did not inhibit exflagellation (31). This may mean that although Complex III is present and inhibitable by atovaquone in late-stage gametocytes, the activity of Complex III is not required for the survival or exflagellation of these gametocytes. Studies of the atovaquone sensitivity of mosquito stages of *P. berghei* also suggest the insensitivity of mosquito stages of the life cycle prior to ookinete formation (32).

Based on these studies, one might expect to find transcripts for the cytochrome *b* gene in late-stage gametocytes as well as in asexual erythrocytic-stage *P. falciparum* (29). In fact, the steady-state level of cytochrome *b* RNA in early-stage gametocytes is several-fold higher than that in asexual erythrocytic-stage *P. falciparum* (Figure 3) (59); and in late-stage gametocytes is 7-fold higher than in asexual erythrocytic *P. falciparum* as measured using quantitative RT-PCR (100). It will be interesting to determine in which, if any, life-cycle stages the increased cytochrome *b* mRNA may be reflected as an increase in Complex III or electron transport activity. Thus far, studies of Complex III suggest that increased activity probably does not occur prior to ookinete formation. Evidence for an increase in Complex II (succinate dehydrogenase) activity in oocysts as compared to asexual erythrocytic-stage parasites is discussed above.

Biochemical evidence suggests that asexual erythrocytic *Plasmodium* mitochondria do not contain any Complex I (NADH-ubiquinone reductase) activity. Purified *P. falciparum* and *P. yoelii* mitochondria do not oxidize NAD^+-linked substrates, such as α-ketoglutarate, malate, or pyruvate (34). Furthermore, mitochondrial respiration by these parasites is insensitive to the Complex I inhibitor rotenone (34). These observations are further corroborated by assays in *P. yoelii* demonstrating that rotenone has no significant effect on mitochondrial membrane potential or oxygen consumption (135).

Alternative Electron Transport

In the alternative respiratory pathway of plants and some fungi, an alternative oxidase transfers electrons directly from ubiquinone to oxygen, resulting in cyanide-resistant respiratory activity (90). Despite evidence that *P. berghei* cytochrome oxidase is sensitive to 25 μM of cyanide (71), polarographic assays of *P. falciparum* trophozoites indicate that 2 mM of cyanide inhibits only about three quarters of the oxygen consumed by these parasites (93). This cyanide-resistant oxygen consumption is sensitive to both salicylhydroxamic acid (SHAM) and propyl gallate, two well-characterized alternative oxidase inhibitors (93). One indication that the cyanide-resistant, SHAM and propyl gallate–sensitive oxygen consumption is important for parasite growth is that seven different alternative oxidase inhibitors, including SHAM and propyl gallate, inhibit asexual parasite growth in vitro with IC_{50}s ranging from

$\longrightarrow$

Figure 3 Northern blot analysis of the *Plasmodium falciparum* cytochrome *b* gene. Total RNA (10 μg/lane) from asexual erythrocytic-stage (2) and early-stage gametocytes (1) was electrophoresed on a 1.5% agarose formaldehyde gel, blotted, and hybridized with a radiolabeled PCR product from the cytochrome *b* gene. A control experiment with a probe for the 18s rRNA indicated comparable loading in the two lanes.

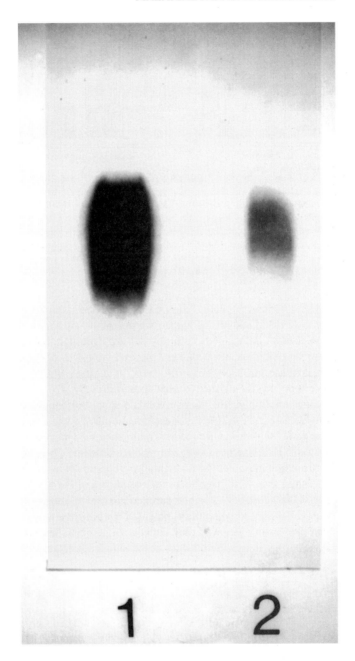

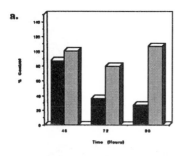
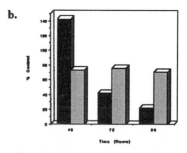

Figure 4 Effect of alternative oxidase inhibitors on *Plasmodium falciparum* cultures containing gametocytes. A single culture was divided into three flasks that were fed daily with medium containing 100 μM propyl gallate, (b)1 μM SHAM, or (c) no inhibitor. Giemsa-stained slides made at each time point for each culture were examined by light microscopy. The fraction of red blood cells infected with asexual parasites (*dark bars*) or late-stage gametocytes (*shaded bars*) in inhibitor-treated cultures was compared to the medium alone control For each slide, at least 100 gametocytes were counted. Preliminary experiments showed that SHAM and propyl gallate had no effect on survival of blood cells kept under the same conditions.

2 to 247 μM (93). Consistent with evidence for a branched respiratory pathway in *P. falciparum*, SHAM and propyl gallate can potentiate the classical respiratory pathway inhibitor, atovaquone (Murphy & Lang-Unnasch, unpublished).

The relative susceptibility of asexual and gametocyte stages of *P. falciparum* to alternative oxidase inhibitors has been tested in culture. Both SHAM and propyl gallate reduced the number of asexual parasites by 70–75%, whereas under the same conditions, the number of late-stage gametocytes was reduced by only 25% (Figure 4). Mitochondrial membrane potential in late-stage *P. falciparum* gametocytes is more readily diminished by 1 mM of SHAM or a combination of SHAM and cyanide than by cyanide alone (94). In contrast, mitochondrial membrane potential of asexual *P. falciparum* is unaffected by SHAM, suggesting that there may be an increased role of the alternative respiratory pathway in gametocytes as compared to asexual parasites (94). Since SHAM affects mitochondrial membrane potential but not gametocyte survival, the alternative respiratory pathway may increase in gametocytes but be more functionally important in mosquito stages of the parasite life cycle.

NUCLEIC ACID SYNTHESIS

Pyrimidine and Folate Biosynthesis

One role of the electron transport pathway of *Plasmodium* mitochondria is the de novo synthesis of pyrimidines (119). The enzymatic link between electron transport and pyrimidine biosynthesis is dihydroorotate dehydrogenase

(DHODase), which is believed to donate electrons to the electron transport chain via ubiquinone (43). DHODase is sensitive to a number of inhibitors of classical electron transport, including cyanide (43, 53), monoctone (43), and atovaquone (52). After at least 24 h of asexual *P. falciparum* exposure to tetracycline, this antibiotic also inhibits DHODase activity in such a manner that the activity can be rescued by the addition of electron carriers to the reaction (102). DHODase is also sensitive to inhibition by the alternative respiration inhibitor, SHAM (43). As with atovaquone (52), SHAM inhibited DHODase activity at a much lower concentration (IC_{50} = 0.5 μM) (43) than that required to inhibit parasite growth (IC_{50} of 247 μM) (93). These results suggest that inhibition of the classical and/or alternative electron transport pathways may inhibit asexual parasite growth because of the resulting inhibition of pyrimidine biosynthesis.

Plasmodium pyrimidine biosynthesis can also be inhibited by inhibiting the parasite pathway for tetrahydrofolate production. Tetrahydrofolate functions in the transfer of methyl units and is an essential cofactor in pyrimidine biosynthesis. The sulfonamides and sulfones act as competitive inhibitors of para-aminobenzoic acid (PABA) for binding to dihydropterate synthase, thereby inhibiting the formation of dihydropterate, a precursor of dihydrofolate (92, 160). In *P. falciparum*, dihydrofolate can also be synthesized by salvage of intact or degraded folic acid in the medium (72, 92, 155, 160). Parasite dihydrofolate synthesized by either the de novo or folate salvage pathways is the substrate for dihydrofolate reductase, the enzyme that converts dihydrofolate to tetrahydrofolate (162). The *Plasmodium* dihydrofolate reductase is specifically inhibited by pyrimethamine, cycloguanil, and WR99210 (30, 164).

Pyrimethamine inhibits survival of early- but not late-stage gametocytes of *P. falciparum* (15, 16, 138). Similarly, atovaquone (31) and SHAM (Murphy & Lang-Unnasch, unpublished) inhibit survival of early- but not late-stage *P. falciparum* gametocytes. These results suggest that de novo pyrimidine biosynthesis is not essential for late-stage gametocytes and that any pyrimidines needed in *P. falciparum* gametocytes are synthesized in the early stages of development.

The sporontocidal action of pyrimethamine appears to occur during the first few days of parasite development in the mosquito. Introduction of 100-nM or higher concentrations of pyrimethamine at the time of feeding mosquitos with *P. falciparum* or *P. vivax* gametocytes resulted in reduction or elimination of oocyst formation (16, 125, 139). Microscopic studies following pyrimethamine treatment indicated that development proceeds through the point of ookinete penetration of the gut wall, but the formation of the rounded-up ookinete is inhibited (125). Introduction of even 10 μM pyrimethamine 4 days after infection of the mosquito with gametocytes (after oocysts were established beneath the basal lamina of the midgut epithelium) did not result in a reduction in oocyst formation or development (16). Consistent with these results, the number of oocysts formed by *P. berghei* in mosquitos pretreated with pyrimethamine and/or sulphormethoxine is reduced compared to that in untreated controls (104). Similarly, studies show that *P. berghei* oocyst formation is greatly reduced by the addition of electron transport inhibitors that are known to affect DHODase activity. Atovaquone, when added in high concentrations (50 nM or higher) to *P. berghei* gametocyte cultures just prior to or just following exflagellation, inhibits ookinete formation in vitro (32). Higher concentrations of atovaquone (200 nM), when added following meiotic DNA replication, 6 h after exflagellation, also inhibited ookinete development. Similarly, monoctone, when added to *P. berghei* gametocytes at the time of mosquito infection, results in reduction or elimination of ookinete formation (17). It appears that, as in other stages of development, *Plasmodium* oocysts do not take up pyrimidines and are therefore dependent on de novo synthesized pyrimidines for nucleic acid synthesis (22). It is quite possible that the electron transport inhibitors inhibit more than just DHODase. However, at a minimum, the *P. berghei* results suggest pyrimidine biosynthesis does not occur during gametogenesis or zygote formation, and probably not prior to meiotic DNA replication. Taken together, these results suggest that pyrimidine synthesis in the mosquito is accomplished late in ookinete development or shortly after formation of the oocyst.

As indicated above, PABA is a precursor of dihydrofolate, which in turn is required for parasite pyrimidine biosynthesis. Sporogonic development of *P. falciparum* is not stimulated by PABA (7), unlike the sporogonic development of avian (140) and rodent (99) malaria parasites. Both the *P. falciparum* and *P. berghei* experiments were done in *Anopheles stephensi*, so it is unlikely that the difference simply represents the availability of PABA in the mosquito. Interestingly, the development of *P. berghei* ookinetes in vitro was not increased by the addition of PABA to the medium (129), suggesting that stimulation of sporogonic development in the mosquito occurs just after oocyst formation, a time consistent with the presumed time of pyrimidine formation. As indicated above, asexual erythrocytic-stage *P. falciparum* appear to use a folate salvage pathway as well as synthesis from PABA. The possible use of a folate salvage pathway by other *Plasmodium* species has not been investigated. One explanation of the differing effects of PABA on sporogonic development by different *Plasmodium* species is that folate salvage might be the primary pathway for dihydrofolate production in sporogonic *P. falciparum* but not sporogonic *P. berghei*, *P. yoelii*, or *P. gallinaceum*.

RNA Synthesis

A variety of antigens specific for sexual-stage parasites have been identified in avian, rodent, and human malarias (60). In most cases, the genes for sexual-stage antigens are transcribed in a developmentally regulated fashion, consistent with a major role for transcriptional control of the expression of the gene product. For instance, the expression of the genes for *P. falciparum* gametocyte-stage antigens Pfs16 (23), Pfs230 (163), Pfs48/45 (153), and Pfg27 (85) all seem to be regulated primarily at the level of transcription initiation, although Pfs230 also undergoes stage-specific processing (163). Similarly, the Pfs25 antigen is expressed primarily in macrogametes, zygotes, and developing ookinetes and only at a low level in gametocytes (153), and the RNA for this antigen is also synthesized at very low levels in gametocytes and at higher levels in macrogametes and zygotes (45).

In asexual *P. falciparum*, transcription occurs in three compartments, the nucleus, the mitochondrion, and the plastid (29, 80). Actinomycin D is a general inhibitor of RNA synthesis, presumably acting in all three compartments. At low concentrations ($IC_{50} = 0.8$ nM), actinomycin D inhibits growth of asexual erythrocytic stage-*P. falciparum* (40). At the lowest concentration tested (50 nM), actinomycin D also inhibits early- and, to a lesser extent, late-stage *P. falciparum* gametocytes. In short-term experiments (90 min), inhibition of exflagellation by *P. yoelii* microgametocytes required similar actinomycin D concentrations (40 nM). Synthesis of transcripts for the 25-kDa surface antigen on macrogametes and zygotes can be inhibited by actinomycin D at a concentration (4 nM) similar to that used to inhibit asexual parasite transcription (45). Together, these studies suggest that RNA synthesis is essential for all of these parasite stages, but less so in late-stage gametocytes and during gametogenesis.

Asexual-stage *P. falciparum* have an IC_{50} for rifampin of 0.3 μM (40). Evidence that genes on the *P. falciparum* 35-kb DNA plastid genome encode polypeptides with homology to subunits of a bacterial-type RNA polymerase led to the hypothesis that a 35-kb DNA-encoded RNA polymerase is the target of the antimalarial action of rifampin (38). Most efforts to confirm this have been inconclusive (137 and Lang-Unnasch, unpublished). Recent studies supporting this hypothesis are based on the observation that plastid RNAs, but not nuclear RNAs, are selectively eliminated by 6-h treatment of asexual *P. falciparum* with rifampin (87). Early-stage *P. falciparum* gametocytes are comparably susceptible to rifampin, but the number of late-stage gametocytes is reduced only by drug concentrations that are 25-fold higher. These results are similar to those indicating that higher concentrations of actinomycin D are needed to inhibit survival of late gametocytes than are needed to inhibit survival of early gametocytes. Three possible explanations of the sensitivity of late-stage gametocytes only to higher concentrations of RNA synthesis inhibitors are that the late-stage gametocytes are less accessible to these drugs, that RNA synthesis is less essential during this stage of developmental, or that the high concentrations of these drugs are acting on targets other than RNA synthesis.

DNA Synthesis

The most extensive studies of DNA synthesis in sexual-stage parasites have been carried out in rodent malaria parasites, especially *P. berghei*. These studies have included analyses of ultrastructure (129) and cytology (12, 18, 19, 55, 57), and the effects of DNA synthesis inhibitors (56, 57, 143). Early inferences about DNA synthesis in sexual-stage parasites have had to be reconsidered as problems with feulgen staining methods and the inhibitor mitomycin C became apparent in the mid 1980s (18, 56). It now appears that some DNA synthesis occurs in *P. berghei* gametocytes (55, 57) such that individual *P. berghei* gametocytes partially replicate or selectively amplify some part of their DNA (macrogametocytes have 1.4 x the haploid DNA content and microgametocytes have 1.8 x the haploid DNA content) (55, 57).

In *P. falciparum*, mature gametocytes have a full diploid DNA content consistent with replication (54). Some stage II *P. falciparum* gametocytes contain an intranuclear spindle, and it has been suggested that these nuclei may undergo a single division at this time (126).

P. berghei microgametocytes undergo a 5-fold increase in DNA content during the approximately 10 min between activation and exflagellation (55, 57). The final DNA content of the microgametocyte is slightly more than octoploid, while the free microgametes each had a haploid DNA content. In contrast, the DNA content in mature macrogametocytes does not change in macrogametes. Janse and coworkers (55, 57) also used aphidicolin, a specific inhibitor of eukaryotic DNA polymerase-α, to investigate DNA synthesis in sexual-stage parasites. Aphidicolin inhibits the increase in DNA in the activated microgametocyte and inhibits exflagellation, indicating that DNA synthesis during exflagellation is essential.

When *P. berghei* macrogametes were fertilized, feulgen staining indicated that the resulting zygotes had slightly more than a diploid DNA content (2.0–2.4 x) (57). Replication to an almost tetraploid DNA content (3.7 x) occurred between 1 and 3 h post activation, within the first 2 h after fertilization. Addition of aphidicolin after fertilization resulted in the development of ookinetes that at 24 h appeared to be morphologically normal and mature but that contained only a diploid DNA content. Ultrastructural evidence showing a synaptonemal complex in the zygote suggests that the first meiotic division may occur shortly after the increase in DNA content to tetraploid (129).

PROTEIN SYNTHESIS

Amino Acids

There are three sources of amino acids for erythrocytic-stage malaria parasites (121): through degradation of ingested hemoglobin, by import from the serum, and by de novo biosynthesis. Experiments using radiolabeled amino acids demonstrate that many different *Plasmodium* species have the ability to incorporate amino acids from the environment (24, 120). Using a defined culture medium containing dialyzed human serum (24) or Albumax (73, 95), growth of asexual erythrocytic-stage *P. falciparum* requires only seven exogenously supplied amino acids (isoleucine, methionine, cystine or cysteine, glutamate, glutamine, proline, and tyrosine). The requirement for exogenously supplied glutamate is interesting because this amino acid is only poorly incorporated into *P. falciparum* proteins (24). It seems likely that the exogenously supplied glutamate is utilized by parasite glutamate dehydrogenase in the following reversible reaction:

$$\text{glutamate} + \text{NADP}^+ + \text{H}_2\text{O} \rightleftharpoons \alpha\text{-ketoglutarate} + \text{NH}_4^+ + \text{NADPH} + \text{H}^+.$$

The *P. falciparum* glutamate dehydrogenase enzyme appears to be a hexamer of identical 49-kDa subunits (69, 84). Its activity has been used as a marker for the parasite compartment of the infected red blood cells (151). It has been shown to have a strong cofactor preference for NADP over NAD (69). The enzyme has a higher specific activity in the direction of glutamate synthesis (69, 150), but the direction of the reaction may be controlled by the availability of substrates. For instance, *P. falciparum* glutamate dehydrogenase has a high K_m (8 mM) for ammonia, so this may favor the breakdown of glutamate and the net gain of NADPH by the parasite. The breakdown of glutamate by glutamate dehydrogenase is believed to be a primary source of parasite NADPH (113). NADPH is a cofactor required for a variety of enzymes, including glutathione reductase, a key enzyme for the parasite's ability to respond to oxidative stress (69, 84, 113).

Erythrocytic-stage *P. falciparum* (9), *P. knowlesi* (101, 123), and *P. lophurae* (122) have been shown to be able to fix radiolabeled CO_2 into small amounts of the amino acids glutamate, aspartate, and alanine. Presumably, this amino acid biosynthesis reflects carbon dioxide fixation via phosphoenolpyruvate carboxylase or carboxykinase (132) into α-ketoglutarate followed by the action of glutamate dehydrogenase, aspartate amino transferase, and alanine amino transferase.

There are two sources of amino acids for mosquito stages of malaria parasites: import from the environment and de novo biosynthesis. *P. relictum* oocysts (5) and *P. berghei* ookinetes (161) and oocysts (147) are capable of uptake of radiolabeled leucine, but similar studies have not been extended to other *Plasmodium* species or other amino acids. More recently, methods have been developed for the growth in culture of *P. gallinaceum* (14, 63), *P. berghei* (58, 129), and *P. falciparum* (13) ookinetes and of *P. gallinaceum* oocysts (156). These systems should provide means of testing both for incorporation of radiolabeled amino acids into parasite protein and for testing to determine which exogenously supplied amino acids are essential for growth.

The increasing availability of genomic sequence data from *P. falciparum* may also help to elucidate biochemical pathways that may be specifically relevant to mosquito stages of the parasite. For instance, a partial sequence available through the TIGER genomic database (gnl|TIGR|PF2HA56R) appears to encode a protein with high homology to the flavin mononucleotide and the 3Fe-4S binding sites of plant and bacterial glutamate synthases (98). This enzyme carries out the reaction:

$$\text{glutamine} + \alpha\text{-ketoglutarate} + \text{NAD(P)H} + \text{H}^+ \rightleftharpoons 2\,\text{glutamate} + \text{NAD(P)}^+.$$

It is very unlikely that such a glutamate synthase functions in the asexual erythrocytic stages of *P. falciparum*, since growth in culture requires that both glutamate and glutamine be supplied exogenously (24). If glutamate synthase were present in these parasites, glutamine should be able to act as the precursor for de novo biosynthesis of glutamate, thereby eliminating the need for an exogenous source. Thus, it is possible that the *P. falciparum* putative glutamate synthase is expressed only in mosquito stages of the parasite and represents an increased biosynthetic capacity in these stages.

Effects of Protein Synthesis Inhibitors

Inhibition of asexual erythrocytic *P. falciparum* protein synthesis with 100 μM cycloheximide, an inhibitor of eukaryotic protein synthesis, eliminates more than 98% of incorporation of radiolabeled amino acids into protein (41) and results in only a few electrophoretically distinguishable proteins being synthesized (65). The cycloheximide-resistant protein synthesis may represent protein synthetic activity in the mitochondrion and/or the plastid. Cycloheximide, emetine, and puromycin, all inhibitors of eukaryotic protein synthesis, readily inhibit the growth of asexual erythrocytic and early gametocyte stages of *P. falciparum* (40, 130). The survival of late-stage *P. falciparum* gametocytes is also reduced by these inhibitors, though comparable effects require about a 5-fold higher concentration than needed to reduce early-stage gametocytes (130). These protein synthesis inhibitors also block exflagellation by *P. yoelii* microgametocytes, but this requires about 100-fold higher concentration of cycloheximide over the short duration (90 min) of this inhibition study (143).

Ribosomes

The eukaryotic rRNA genes of *Plasmodium* species differ from those of other eukaryotes in terms of copy number, gene dispersal, genetic variability, and developmentally regulated isoforms (for recent reviews, see 82, 89, 157). Of particular interest in the context of this review is the developmental regulation of distinct rRNA isoforms and the potential implications for a novel form of translational regulation in this parasite.

The copy number of the ribosomal operon, which encodes the 28S, 5.8S, and 18S rRNAs, is 4–8 in *Plasmodium* species (21, 20, 78, 144), which is much lower than for other eukaryotic organisms (86). As described below, only a portion of these rRNA genes are expressed in a particular developmental stage. Thus, 2–3 *Plasmodium* rRNA genes must generate the necessary rRNAs for the parasite ribosomes. To maintain the rapid growth of the parasite, there must be a high transcription rate for each of the rRNA genes. Furthermore, sequence differences in individual rRNA genes may have profound effects on ribosomal structure and/or function.

All of the studies of *Plasmodium* rRNA genes have indicated the unusually heterogenous character of these genes. Most studies have focused on the 18S RNA genes, which can differ by 3.5% to 17.5% (20, 42, 81, 83, 159). In *P. falciparum*, representatives of two classes of 28S RNA genes have been sequenced and differ from each other by about 20% (88, 109). Restriction digest and nucleotide sequence data suggest that there are at least five different *P. falciparum* 5.8S rDNAs that differ by 2 to 24% (124).

DEVELOPMENTAL REGULATION OF DISTINCT RIBOSOMAL RNA CLASSES   It is now well established that *Plasmodium* species have at least two developmentally regulated rDNA transcription units (20, 42, 81, 83, 158, 159, 165). Initially, it was assumed that the developmental regulation of the different classes of rRNA genes in each *Plasmodium* species would be essentially the same, but differences are emerging (Figure 5). One common feature of all of the *Plasmodium* species investigated thus far is the class of rRNA transcription unit expressed in sporozoites [S gene(s)] differs from the predominant rRNA transcription unit expressed during the asexual cycle (A gene(s)) (20, 42, 83, 159). Interestingly, a third rRNA gene class (O gene(s)) has been identified in *Plasmodium vivax* (81). The O gene rRNAs are found in ookinetes and oocysts but not in sporozoites (81).

In *P. berghei*, during the asexual erythrocytic cycle the A rRNA is the predominant mature rRNA, accumulating at least 20-fold more than S rRNA (42). Expression of the S rRNA genes in these asexual cultures appears to be limited to a transient burst, probably during early gametocytogenesis (158). Mature macrogametocytes express little if any S rRNA, but do continue to express A rRNAs, while mature microgametocytes appear to transcribe no rRNAs (158). Within the mosquito, parasite expression of A rRNA stops almost completely within 2 h of fertilization. S rRNA expression begins during the same interval and increases dramatically during ookinete development (158).

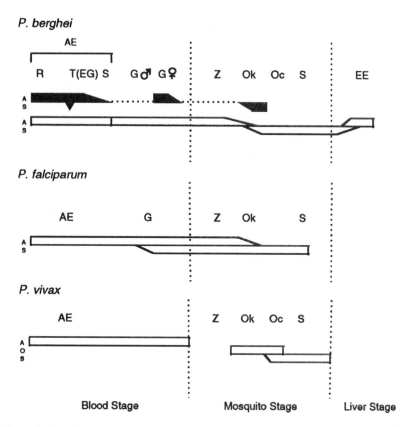

Figure 5 Developmental changes in rRNA gene expression in *Plasmodium* species. Studies of steady-state (*open bars*) and transcriptional (*solid bar*) expression of A rRNA (*A*), S rRNA (*S*), and O rRNA (*O*) in *Plasmodium berghei* (42, 158, 165), *Plasmodium falciparum* (109, 159), and *Plasmodium vivax* (81, 83) are diagrammatically summarized. Life-cycle stages in mammals include the asexual erythrocytic cycle (*AE*), which is further delineated as ring (*R*), trophozoite (*T*) and early gametocyte (*EG*), and schizont (*S*) stages, the male and female gametocyte (*G*) stages, and the exoerythrocytic (*EE*) stages. Life-cycle stages in the mosquito include the zygote (*Z*), ookinete (*Ok*), oocyst (*Oc*), and sporozoite (*S*).

In *Plasmodium falciparum*, the transition from A rRNA to S rRNA during sexual development differs in some detail from that which occurs in *P. berghei* (Figure 5) (159). The predominant 18S rRNA in gametocytes and zygotes is the mature A rRNA. About 20-fold less of a partially processed S rRNA was detected in gametocytes. The amount of this partially processed S rRNA increased in zygotes, where the mature form of the S rRNA could also be detected. During ookinete formation, there was an increased proportion of mature form S rRNA relative to the partially processed form, and there was a decrease in the steady-state levels of A rRNA. The decrease in A rRNA resulted from specific processing events within conserved rRNA regions associated with ribosomal functions.

An important difference between the developmental regulation of rRNAs in *P. berghei* and *P. falciparum* is the accumulation of a partially processed form of the S rRNA in the gametocytes of *P. falciparum* and the apparent absence of such precursors in *P. berghei* gametocytes. It is not clear if the transient expression of S rRNA in early gametocytes of *P. berghei* results in the accumulation of mature S rRNA containing ribosomes in *P. berghei* gametocytes. If so, zygotes of both species could be expected to contain a preponderance of A ribosomes and a small fraction of S type ribosomes. The mechanism of achieving this would clearly differ between the species, but this difference may reflect the species-specific factors in gametocyte development and longevity. In particular, *P. falciparum* gametocytes develop much more slowly and remain infective for much longer durations than do those of *P. berghei* (126, 134). The absence of mature type S ribosomes in *P. falciparum* gametocytes might help prevent the premature triggering of zygote- or ookinete-specific functions.

In *P. vivax*, the developmental pattern of rRNA gene expression differs from that in both *P. falciparum* and *P. berghei* in that the S rRNAs are not produced until oocyst formation (83) (Figure 5). Neither A nor S rRNAs are found in ookinetes, but a new class of rRNA, called O rRNA, is expressed in ookinete and oocyst stages (81). Both O rRNA and S rRNA are expressed at high levels in oocysts, but only the S rRNA is found in sporozoites, suggesting that there is a mechanism by which different ribosome classes are segregated in these cells. It is not clear if any of the rRNA gene units from *P. falciparum* or *P. berghei* currently designated as an S rRNA gene has a pattern of expression consistent with O rRNA type regulation.

POTENTIAL DIFFERENCES IN THE FUNCTION OF DIFFERENTIALLY EXPRESSED RIBOSOMES   Most of the sequence differences between the 18S A and S rRNA classes are clustered in the nonconserved portions of these rRNAs, and so their possible influence on RNA structure or function are difficult to infer (20, 42, 88). Recent studies of an 18S O rRNA gene from *P. vivax* indicates numerous differences that are expected to affect translational accuracy (81). Similarly, differences near the GTPase site in the 28S A and S rRNA genes of *P. falciparum* suggest that these rRNAs might interact differently with the large subunit ribosomal protein L11 or have differing rates of GTP hydrolysis (109). Studies in which the GTPase sites of *P. falciparum* A and S ribosomes have been substituted into yeast genes indicate that the observed sequence differences in these rRNAs do result in phenotypically different ribosomes (T McCutchan & S Liebman, personal communication). Potentially, phenotypic changes in ribosomes could facilitate global regulatory change by favoring the translation of particular subsets of mRNAs or by enhancing or decreasing the overall rate of translation.

To date, there has been no mRNA whose translation has been shown to be controlled by developmentally regulated ribosomes. However, the expression of the *P. berghei* 21-kDa ookinete surface protein (Pbs21) is regulated translationally in such a way that it might represent an example of control by developmentally regulated ribosomes. The mRNA for Pbs21 accumulates in the cytoplasm of female gametocytes, zygotes, and ookinetes, but the protein is translated only in ookinetes (97, 154). It is not known why the translation of Pbs21 mRNA is repressed prior to the ookinete stage of parasite development, but it is suggestive that synthesis of the S rRNAs begins in the ookinete (158). It has not been reported whether the *P. falciparum* homolog of Pbs21 (Pfs28) also undergoes translational regulation.

CONCLUSIONS

There are clear indications that the metabolic activities of *P. falciparum* differ between those developmental stages carried out in the mosquito host and those carried out in the human host. In some mosquito stages of development, the parasites may have more mitochondrial activity than do asexual erythrocytic-stage parasites, as suggested by an increase in the activity or expression of the classical electron transport pathway enzymes, succinate dehydrogenase (48) and cytochrome *b* (59, 100). Similarly, there are hints of a possible increase in an alternative electron transport pathway (94). Such an increase in mitochondrial activity in mosquito-stage parasites could include the development of a TCA cycle (49), which is almost certainly absent from asexual erythrocytic-stage parasites (9, 75, 76, 150). Pyrimidine biosynthesis appears to occur in asexual erythrocytic-stage parasites, in early gametocytes, and in late-stage ookinetes or oocysts. Late-stage gametocytes, gametes, zygotes, and ookinetes presumably use preformed pyrimidines. In most parasite stages, inhibition of electron transport only affects development if pyrimidine biosynthesis is also inhibited. Ookinetes are exceptional in that their development can be blocked by inhibitors of electron transport (32) but not by inhibitors of pyrimidine biosynthesis (125), suggesting an important new role for the electron transport chain in these mosquito-stage parasites.

Amino acids may be less available to mosquito stages than to human stages of the parasite. Here we have suggested the possibility that mosquito stages of *P. falciparum* may express enzymes required to synthesize amino acids such as glutamate that are acquired from other sources in asexual erythrocytic-stage parasites. A qualitative difference between protein synthesis in human and mosquito stages is suggested by the shift from A ribosomes that are used by *P. falciparum* while in the human host to S ribosomes that are used in ookinete through sporozoite stages in the mosquito host (159). The potential of using this ribosomal change to globally alter the translational patterns in these parasites has not yet been explored.

New opportunities exist to greatly advance our understanding of the sexual and sporogonic development of malaria parasites in general and of *P. falciparum* in particular. Advances in culturing techniques for mosquito-stage parasites (13, 14, 58, 63, 129); in molecular methods, such as RT-PCR and in situ hybridization, that are very specific and require only very small amounts of material (83, 100, 158); and in the availability of molecular and immunological probes for metabolic enzymes expressed in the asexual erythrocytic stages of the parasite life cycle (e.g. Figure 2 and www.ncbi.nlm.nih.gov/Malaria/index.html) have barely been exploited thus far. Furthermore, our mind set about what metabolic pathways might exist in these parasites has expanded due to recent evidence that an ancient progenitor of Plasmodium and other apicomplexans may have acquired a plastid by engulfing a photosynthetic algae (68; N Lang-Unnasch, M Reith, J Munholland, and J Barta, unpublished). Evidence for a glycolytic enzyme that is phylogenetically most closely related to those of plants (106) supports the hypothesis that some of the algal genes for basic metabolic enzymes may have been incorporated into that ancestor of *Plasmodium.*

The new reagents, methods, and mind set may lead to advances in our understanding of the metabolic capabilities of *P. falciparum* throughout its life cycle. Such knowledge would be invaluable for the development of chemotherapeutic, immunological, and biological interventions that could reduce the transmission as well as the symptoms of malaria.

ACKNOWLEDGMENTS

We thank David Aiello for assistance with the figures. Supported by National Institutes of Health NIH grant AI-38329.

Visit the *Annual Reviews home page* at
http://www.AnnualReviews.org.

Literature Cited

1. Aikawa M. 1966. The fine structure of the erythrocytic stages of three avian malarial parasites, *Plasmodium fallax*, *P. lophurae* and *P. cathemerium. Am. J. Trop. Med. Hyg.* 15:449–71
2. Aikawa M, Beaudouin RL. 1970. *Plasmodium fallax*: high resolution autoradiography of exoerythrocytic stages treated with primaquine in vitro. *Exp. Parasitol.* 27:454–63
3. Alano P, Carter R. 1990. Sexual differ-

entiation in malaria parasites. *Annu. Rev. Microbiol.* 44:429–49
4. Aldritt SM, Joseph JT, Wirth DF. 1989. Sequence identification of cytochrome b in *Plasmodium gallinaceum. Mol. Cell. Biol.* 9:3614–20
5. Ball GH, Chao J. 1976. Use of amino acids by *Plasmodium relictum* oocysts in vitro. *Exp. Parasitol.* 39:115–18
6. Beaudoin RL, Aikawa M. 1968.

Primaquine-induced changes in morphology. *Science* 160:1233–34

7. Beier MS, Pumpuni CB, Beier JC, Davis JR. 1994. Effects of para-aminobenzoic acid, insulin, and gentamicin on *Plasmodium falciparum* development in anopheline mosquitoes (Diptera: Culicidae). *J. Med. Entomol.* 31:561–65

8. Bhasin VK, Trager W. 1987. Gametocytocidal effects *in vitro* of primaquine and related compounds on *Plasmodium falciparum*. In *Primaquine: pharmacokinetics, metabolism, toxicity and activity.*, ed. PI Trigg, pp. 145–53. New York: Wiley & Sons

8a. Bilker O, Lindo V, Panico M, Etienne AE, Paxton T, et al. 1998. Identification of xanthurenic acid as the putative inducer of malaria development in the mosquito. *Nature* 392:289–92

9. Blum JJ, Ginsberg H. 1984. Absence of α-ketogluterate dehydrogenase activity and presence of CO2-fixing activity in *Plasmodium falciparum* grown in vitro in human erythrocytes. *J. Protozool.* 31:167–69

10. Boulard Y, Landau I, Miltgen F, Ellis DS, Peters W. 1983. The chemotherapy of rodent malaria, XXXIV. Causal prophylaxis Part III: ultrastructural changes induced in exo-erythrocytic schizonts of *Plasmodium yoelii yoelii* by primaquine. *Ann. Trop. Med. Parasitol.* 77:555–68

11. Bzik DJ, Fox BA, Goyner K. 1993. Expression of *Plasmodium falciparum* lactate dehydrogenase in *Escherichia coli*. *Mol. Biochem. Parasitol.* 59:155–66

12. Canning EU, Sinden RE. 1975. Nuclear organization in gametocytes of *Plasmodium* and *Hepatocystis*: a cytochemical study. *Z. Parasitenkunde* 46:297–99

13. Carter R, Graves PM, Quakyi IA, Good MF. 1989. Restricted or absent immune responses in human populations to *Plasmodium falciparum* gamete antigens that are targets of malaria transmission-blocking antibodies. *J. Exp. Med.* 169:135–47

14. Carter R, Kaushal DC. 1984. Characterization of antigens on mosquito stages of *Plasmodium gallinaceum*. III. Changes in zygote surface proteins during the transition to mature ookinete. *Mol. Biochem. Parasitol.* 13:235–41

15. Chutmongkonkul M, Maier WA, Steitz HM. 1992. A new model for testing gametocytocidal effects of some antimalarial drugs on *Plasmodium falciparum* in vitro. *Ann. Trop. Med. Parasitol.* 86:207–15

16. Chutmongkonkul M, Maier WA, Steitz HM. 1992. *Plasmodium falciparum*: effect of chloroquine, halofantrine and pyrimethamine on the infectivity of gametocytes for *Anopheles stephensi* mosquitos. *Ann. Trop. Med. Parasitol.* 86:103–10

17. Coleman RE, Clavin AM, Milhous WK. 1992. Gametocytocidal and sporontocidal activity of antimalarials against *Plasmodium berghei* anka in ICR mice and *Anopheles stephensi* mosquitos. *Am. J. Trop. Med. Hyg.* 50:646–53

18. Cornelissen AWCA, Overdulve JP, Ploeg MV. 1984. Cytochemical studies on nuclear DNA of four eucoccidian parasites, *Isospora (Toxoplasma) gondii*, *Eimeria tenella*, *Sarcocystis cruzi* and *Plasmodium berghei*. *Parasitology* 88:13–25

19. Cornelissen AWCA, Overdulve JP, Ploeg MV. 1984. Determination of nuclear DNA of five Eucoccidian parasites, *Isospora (Toxoplasma) gondii*, *Sarcocystis cruzi*, *Eimeria tenella*, *E. acervulina* and *Plasmodium berghei*, with special reference to gamontogenesis and meiosis in *I. (T.) gondii*. *Parasitology* 88: 531–53

20. Corredor V, Enea V. 1994. The small subunit RNA isoforms in *Plasmodium cynomolgi*. *Genetics* 136:857–65

21. Dame JB, McCutchan TF. 1983. The four ribosomal DNA units of the malaria parasite *Plasmodium berghei*. *J. Biol. Chem.* 258:6984–90

22. Davies EE, Howells RE. 1973. Uptake of ³H-adenosine and ³H-thymidine by oocysts of *Plasmodium berghei berghei*. *Trans. R. Soc. Trop. Med. Hyg.* 67:20

23. Dechering KJ, Thompson J, Dodemont HJ, Eling W, Konings RNH. 1997. Developmentally regulated expression of pfs16, a marker for sexual differentiation of the human malaria parasite *Plasmodium falciparum*. *Mol. Biochem. Parasitol.* 89:235–44

24. Divo AA, Geary TG, Davis NL, Jensen JB. 1985. Nutritional requirements of *Plasmodium falciparum* in culture. I. Exogenously supplied dialyzable components necessary for continuous growth. *J. Protozool.* 32:59–64

25. Divo AA, Geary TG, Jensen JB, Ginsburg H. 1985. The mitochondrion of *Plasmodium falciparum* visualized by rhodamine 123 fluorescence. *J. Protozool.* 32:442–46

26. Divo AA, Patton CL, Sartorelli AC. 1993. Evaluation of rhodamine 123 as a probe for monitoring mitochondrial membrane potential in *Trypanosoma*

brucei spp. *J. Eukaryotic Microbiol.* 40:329–35

27. Dunn C, Banfield M, Barker J, Higham C, Moreton K, et al. 1996. The structure of lactate dehydrogenase from *Plasmodium falciparum* reveals a new target for anti-malarial design. *Nat. Struct. Biol.* 3:912–15

28. Feagin JE. 1992. The 6-kb element of *Plasmodium falciparum* encodes mitochondrial cytochrome genes. *Mol. Biochem. Parasitol.* 52:145–48

29. Feagin JE, Drew ME. 1995. *Plasmodium falciparum:* alterations in organelle transcript abundance during the erythrocytic cycle. *Exp. Parasitol.* 80:430–40

30. Fidock DA, Wellems TE. 1997. Transformation with human dihydrofolate reductase renders malaria parasites insensitive to WR99210 but does not affect the intrinsic activity of proguanil. *Proc. Natl. Acad. Sci. USA* 94:10931–36

31. Fleck SL, Pudney M, Sinden RE. 1996. The effect of atovaquone (566C80) on the maturation and viability of *Plasmodium falciparum* gametocytes *in vitro*. *Trans. R. Soc. Trop. Med. Hyg.* 90:309–12

32. Fowler RE, Sinden RE, Pudney M. 1995. Inhibitory activity of the anti-malarial atovaquone (566C80) against ookinetes, oocysts, and sporozoites of *Plasmodium berghei*. *J. Parasitol.* 81:452–58

33. Fry M. 1991. Mitochondria of *Plasmodium*. In *Biochemical Protozoology*, ed. GH Coombs, MJ North, pp. 154–67. Washington, DC: Taylor & Francis

34. Fry M, Beesley JE. 1991. Mitochondria of mammalian *Plasmodium* spp. *Parasitology* 102:17–26

35. Fry M, Pudney M. 1992. Site of action of the antimalarial hydroxynapthoquinone, 2-[trans-4-(4'-chlorphenyl) cyclohexyl]-3-hydroxy-1,4-napthoquinone (566C80). *Biochem. Pharmacol.* 43:1545–53

36. Fry M, Webb E, Pudney M. 1990. Effect of mitochondrial inhibitors on adenosinetriphosphate levels in *Plasmodium falciparum*. *Comp. Biochem. Physiol.* 96:775–82

37. Garcia GE, Wirtz RA, Rosenberg R. 1997. Isolation of a substance from the mosquito that activaties *Plasmodium* fertilization. *Mol. Biochem. Parasitol.* 88:127–35

38. Gardner MJ, Williamson DH, Wilson RJM. 1991. A circular DNA in malaria parasites encodes an RNA polymerase like that of prokayotes and chloroplasts. *Mol. Biochem. Parasitol.* 44:115–24

39. Geary TG, Divo AA, Bonanni LC, Jensen JB. 1985. Nutritional requirements of *Plasmodium falciparum* in culture. III. Further observations on essential nutrients and antimetabolites. *J. Protozool.* 32:608–13

40. Geary TG, Divo AA, Jensen JB. 1989. Stage specific actions of antimalarial drugs on *Plasmodium falciparum* in culture. *Am. J. Trop. Med. Hyg.* 40:240–44

41. Gershon PD, Howells RE. 1986. Mitochondrial protein synthesis in *Plasmodium falciparum*. *Mol. Biochem. Parasitol.* 18:37–43

42. Gunderson JH, Sogin ML, Hollingdale GW, Cruz VF, Waters AP, McCutchan TF. 1987. Structurally distinct ribosomes occur in *Plasmodium*. *Science* 238:933–37

43. Gutteridge WE, Dave D, Richards WH. 1979. Conversion of dihydroorotate to orotate in parasitic protozoa. *Biochim. Biophys. Acta* 582:390–401

44. Hawking F, Wilson ME, Gammage K. 1971. Evidence for cyclic development and short lived maturity in the gametocytes of *Plasmodium falciparum*. *Trans. R. Soc. Trop. Med. Hyg.* 65:549–59

45. Hella CW, Marieke BAC, Lamers AC, Deursen J, Ponnudurai T, Meuwissen JHET. 1990. Biosynthesis of the 25-kDa protein in the macrogametes/zygotes of *Plasmodium falciparum*. *Exp. Parasitol.* 71:229–35

46. Hicks KE, Read M, Holloway SP, Sims PF, Hyde JE. 1991. Glycolytic pathway of the human malaria parasite *Plasmodium falciparum:* primary sequence analysis of the gene encoding 3-phospoglycerate kinase and chromosomal mapping studies. *Gene* 100:123–29

47. Howells RE. 1970. Cytochrome oxidase activity in a normal and some drug-resistant strains of *Plasmodium berghei*– a cytochemical study. II Sporogonic stages of a drug-sensitive strain. *Ann. Trop. Med. Parasitol.* 64:223–25

48. Howells RE. 1970. Mitochondrial changes during the life cycle of *Plasmodium berghei*. *Ann. Trop. Med. Parasitol.* 64:181–87

49. Howells RE, Maxwell L. 1973. Further studies on the mitochondrial changes during the life cycle of *Plasmodium berghei:* electrophoretic studies on isocitrate dehydrogenases. *Ann. Trop. Med. Parasitol.* 67:279–83

50. Howells RE, Peters W, Fullard J. 1970. The chemotherapy of rodent malaria, XIII. Fine structural changes observed in the erythrocytic stages of *Plasmodium*

berghei berghei following exposure to primaquine and menoctone. *Ann. Trop. Med. Parasitol.* 64:203–07

51. Itin C, Burki Y, Certa U, Dobeli H. 1993. Selective inhibition of *Plasmodium falciparum* aldolase by a tubulin derived peptide and identification of the binding site. *Mol. Biochem. Parasitol.* 58:135–44

52. Ittarat I, Asawamahasakda W, Meshnick SR. 1994. The effects of antimalarials on the *Plasmodium falciparum* dihydroorotate dehydrogenase. *Exp. Parasitol.* 79:50–56

53. Ittarat I, Webster HK, Yuthavong Y. 1992. High-performance liquid chromatographic determination of dihydroorotate dehydrogenase of *Plasmodium falciparum* and effects of antimalarials on enzyme activity. *J. Chromatogr.* 582:57–64

54. Janse C, Ponnudurai T, Lensen AHW, Meuwissen JHET, Ramesar J. 1988. DNA synthesis in gametocytes of *P. falciparum. Parasitology* 96:1–7

55. Janse CJ, Klooster PF, Kaay HJ, Ploeg M, Overdulve JP. 1986. DNA synthesis in *Plasmodium berghei* during asexual and sexual development. *Mol. Biochem. Parasitol.* 20:173–82

56. Janse CJ, Klooster PF, Kaay HJ, Ploeg M, Overdulve JP. 1986. Mitomycin-C is an unreliable inhibitor for study of DNA synthesis in *Plasmodium. Mol. Biochem. Parasitol.* 21:33–36

57. Janse CJ, Klooster PF, Kaay HJ, Ploeg M, Overdulve JP. 1986. Rapid repeated DNA replication during microgametogenesis and DNA synthesis in young zygotes of *Plasmodium berghei. Trans. R. Soc. Trop. Med. Hyg.* 80:154–57

58. Janse CJ, Mons B, Rouwenhorst RJ, Klooster PF, Overdulve JP, Kaay HJ. 1985. *In vitro* formation of ookinetes and functional maturity of *Plasmodium berghei* gametocytes. *Parasitology* 91:19–29

59. Johnson P, Lang-Unnasch N. 1993. Expression of cytochrome b in *Plasmodium falciparum.* Presented at Woods Hole Parasitology meeting, Woods Hole, MA

60. Kaslow DC. 1993. Transmission-blocking immunity against malaria and other vector-borne diseases. *Curr. Opin. Immunol.* 5:557–65

61. Kaslow DC, Hill S. 1990. Cloning metabolic pathway genes by complementation in *Escherichia coli. J. Biol. Chem.* 265:12337–41

62. Kato M, Tanabe K, Miki A, Ichimori K, Waki S. 1990. Membrane potential of *Plasmodium falciparum* gametocytes monitored with rhodamine 123. *FEMS Microbiol. Lett.* 69:283–88

63. Kaushal DC, Carter R, Russel JH, McAuliffe FL. 1983. Characterization of antigens on mosquito midgut stages of *Plasmodium gallinaceum.* I. Zygote surface antigens. *Mol. Biochem. Parasitol.* 8:53–69

64. Kaushal DC, Watts R, Haider S, Singh N, Kaushal NA, Dutta GP. 1988. Antibodies to lactate dehydrogenase of *Plasmodium knowlesi* are specific to *Plasmodium* species. *Immunol. Inv.* 17:507–16

65. Kiatfuengfoo R, Suthiphongchai T, Prapunwattana P, Yuthavong Y. 1989. Mitochondria as the site of action of tetracycline on *Plasmodium falciparum. Mol. Biochem. Parasitol.* 34:109–16

66. Knapp B, Hundt E, Kupper HA. 1989. A new blood stage antigen of *Plasmodium falciparum* transported to the erythrocyte surface. *Mol. Biochem. Parasitol.* 37:47–56

67. Knapp B, Hundt E, Kupper HA. 1990. *Plasmodium falciparum* aldolase: gene structure and localization. *Mol. Biochem. Parasitol.* 40:1–12

68. Kohler S, Delwiche CF, Denny PW, Tilney LG, Webster P, et al. 1997. A plastid of probable green algal origin in apicomplexan parasites. *Science* 275:1485–89

69. Krauth-Siegel RL, Muller JG, Lottspeich F, Schirmer RH. 1996. Glutathione reductase and glutamate dehydrogenase of *Plasmodium falciparum*, the causative agent of tropical malaria. *Eur. J. Biochem.* 235:345–50

70. Krungkrai J. 1995. Purification, characterization and localization of mitochondrial dihydroorotate dehydrogenase in *Plasmodium falciparum*, human malaria parasite. *Biochim. Biophys. Acta* 1243:351–60

71. Krungkrai J, Krungkrai SR, Bhumiratana A. 1993. *Plasmodium berghei*: partial purification and characterization of the mitochondrial cytochrome c oxidase. *Exp. Parasitol.* 77:136–46

72. Krungkrai J, Webster HK, Yuthavong Y. 1989. De novo and salvage biosynthesis of pteroylpentaglutamates in the human malaria parasite, *Plasmodium falciparum. Mol. Biochem. Parasitol.* 32:25–37

73. LaBlanc SB. 1995. Studies on pyrimidine biosynthesis and linkage with mitochondrial respiration in *Plasmodium falciparum.* Univ. Ala., Birmingham. 148 pp.

74. Ladda RL. 1969. New insights into the fine structure of rodent malarial parasites. *Mil. Med.* 134:825–65

75. Lang-Unnasch N. 1992. Purification and properties of *Plasmodium falciparum* malate dehydrogenase. *Mol. Biochem. Parasitol.* 50:17–26

76. Lang-Unnasch N. 1995. *Plasmodium falciparum:* antiserum to malate dehydrogenase. *Exp. Parasitol.* 80:357–59

77. Langreth SG, Jensen JB, Reese RT, Trager W. 1978. Fine structure of human malaria in vitro. *J. Protozool.* 25:443–52

78. Langsley G, Hyde JE, Goman M, Scaife JG. 1983. Cloning and characterization of the rRNA genes from the human malaria parasite *Plasmodium falciparum. Nucleic Acids Res.* 11:8703–17

79. Lanners HN. 1991. Effect of the 8-aminoquinoline primaquine on culture-derived gametocytes of the malaria parasite *Plasmodium falciparum. Parasitol. Res.* 77:478–81

80. Lanzer M, Bruin DD, Ravitch JV. 1992. Transcription mapping of a 100 kb locus of *Plasmodium falciparum* identifies a region in which transcription terminates and reinitiates. *EMBO J.* 11:1949–55

81. Li J, Gutell RR, Damberger SH, Wirtz RA, Kissinger JC, Rogers MJ, Sattabongkot J, McCutchan TF. 1997. Regulation and trafficking of three distinct 18S ribosomal RNAs during development of the malaria parasite. *J. Mol. Biol.* 269:203–13

82. Li J, McConkey GA, Rogers JM, Waters AP, McCutchan TR. 1994. *Plasmodium:* the developmentally regulated ribosome. *Exp. Parasitol.* 78:437–41

83. Li J, Wirtz RA, McConkey GA, Sattabongkot J, McCutchan TF. 1994. Transition of *Plasmodium vivax* ribosome types corresponds to sporozoite differentiation in the mosquito. *Mol. Biochem. Parasitol.* 65:283–89

84. Ling IT, Cooksley S, Bates PA, Hempelmann E, Wilson RJM. 1986. Antibodies to the glutamate dehydrogenase of *Plasmodium falciparum. Parasitology* 92:313–24

85. Lobo C-A, Konings RNH, Kumar N. 1994. Expression of early gametocyte-stage antigens Pfg27 and Pfs16 in synchronized gametocytes and non-gametocyte producing clones of *Plasmodium falciparum. Mol. Biochem. Parasitol.* 68:151–54

86. Long EO, Dawid IB. 1980. Repeated genes in eukaryotes. *Annu. Rev. Biochem.* 49:727–64

87. McConkey GA, Rogers MJ, McCutchan TF. 1997. Inhibition of *Plasmodium falciparum* protein synthesis: targeting the plastid-like organelle with thiostrepton. *J. Biol. Chem.* 272:2046–49

88. McCutchan TF, Cruz VFdl, Lal AA, Gunderson JH, Elwood HJ, Sogin ML. 1988. Primary sequence of two small subunit ribosomal RNA genes from *Plasmodium falciparum. Mol. Biochem. Parasitol.* 28:63–68

89. McCutchan TF, Li J, McConkey GA, Rogers MJ, Waters AP. 1995. The cytoplasmic ribosomal RNAs of *Plasmodium* spp. *Parasitol. Today* 11:134–38

90. McIntosh L. 1994. Molecular biology of the alternative oxidase. *Plant Physiol.* 105:781–86

91. Meis JFGM, Pool G, Lensen AHW, Ponnudurai T, Meuwissen JHET. 1989. *Plasmodium falciparum* ookinetes migrate intercellularly through *Anopheles stephensi* midgut epithelium. *Parasitol. Res.* 76:13–19

92. Milhous WK, Weatherly NF, Bowdre JH, Desjardins RE. 1985. In vitro activities and mechanisms of resistance to antifol antimalarial drugs. *Antimicrob. Agents Chemother.* 27:525–30

93. Murphy AD, Doehler J, Lang-Unnasch N. 1997. *Plasmodium falciparum:* cyanide-resistant oxygen consumption. *Exp. Parasitol.* 87:112–20

94. Murphy AD, Unnasch NL. 1994. Mitochondrial activity in *Plasmodium falciparum* gametocytes. *Am. J. Trop. Med. Hyg.* 51S:205

95. Ofulla AVO, Okoye VCN, Khan V, Githure JI, Roberts CR, Johnson AJ, Martin SK. 1993. Cultivation of *Plasmodium falciparum* parasites in a serum-free medium. *Am. J. Trop. Med. Hyg.* 49:335–40

96. Olafsson P, Certa U. 1994. Expression and cellular localization of hexokinase during bloodstage development of *Plasmodium falciparum. Mol. Biochem. Parasitol.* 63:171–74

97. Paton MG, Barker GC, Matsuoka H, Ramesar J, Janse CJ, et al. 1993. Structure and expression of a conserved and post-transcriptionally regulated gene encoding a surface antigen of *Plasmodium berghei. Mol. Biochem. Parasitol.* 59:263–76

98. Pelanda R, Vanoni MA, Perego M, Piubelli L, Galizzi A, et al. 1993. Glutamate synthase genes of the diazotroph *Azospirillum brasilense.* Cloning, sequencing, and analysis of functional domains. *J. Biol. Chem.* 268:3099–106

99. Peters W, Ramkaran AE. 1980. The chemotherapy of rodent malaria, XXXII. The influence of p-aminobenzoic acid on the transmission of *Plasmodium yoelii* and *P. berghei* by *Anopheles stephensi*. *Ann. Trop. Med. Parasitol.* 74:275–82

100. Petmitr S, Krungkrai J. 1995. Mitochondrial cytochrome b gene in two developmental stages of the human malarial parasite *Plasmodium falciparum*. *S. E. Asian J. Trop. Med. Public Health* 26:600–5

101. Polet H, Brown ND, Angel CR. 1969. Biosynthesis of amino acids from 14C-U-glucose, pyruvate, and acetate by erythrocytic forms of *P. knowlesi* in vitro. *Proc. Soc. Exp. Biol. Med.* 131:1215–18

102. Prapunwattana P, O'Sullivan WJ, Yuthavong Y. 1988. Depression of *Plasmodium falciparum* dihydroorotate dehydrogenase in *in vitro* culture by tetracycline. *Mol. Biochem. Parasitol.* 27: 119–24

103. Ramasamy MS, Kulasekera R, Srikrishnaraj KA, Ramasamy R. 1996. Different effects of modulation of mosquito (Diptera: Culicidae) trypsin activity on the infectivity of two human malaria (Hemosporidia: Plasmodidae) parasites. *J. Med. Entomol.* 33:777–82

104. Ramkaran AE, Peters W. 1969. The chemotherapy of rodnt malaria, VIII. The action of some sulphonamides alone or with folic reductase inhibitors against malaria vectors and parasites, part 3: the action of sulphormethoxine and pyrimethamine on the sporogonic stages. *Ann. Trop. Med. Parasitol.* 63:449–54

105. Ranie J, Kumar VP, Balaram H. 1993. Cloning of the triosephosphate isomerase gene of *Plasmodium falciparum* and expression in *Escherichia coli*. *Mol. Biochem. Parasitol.* 61:159–69

106. Read M, Hicks KE, Sims PF, Hyde JE. 1994. Molecular characterisation of the enolase gene from the human malaria parasite *Plasmodium falciparum*. Evidence for ancestry within a photosynthetic lineage. *Eur. J. Biochem.* 220:513–20

107. Riandey MF, Sannier C, Peltre G, Monteny N, Cavaleyra M. 1996. Lactate dehydrogenase as a marker of *Plasmodium* infection in malaria vector *Anopheles*. *J. Am. Mosq. Control Assoc.* 12:194–98

108. Rieckmann KH, McNamara JV, Kass L, Powell RD. 1969. Gametocytocidal and sporontocidal effects of primaquine upon two strains of *Plasmodium falciparum*. *Mil. Med.* 134:802–19

109. Rogers MJ, Gutell RR, Damberger SH, Li J, McConkey GA, et al. 1996. Structural features of the large subunit rRNA expressed in *Plasmodium falciparum* sporozoites that distinguish it from the asexually expressed large subunit rRNA. *RNA.* 2:134–45

110. Roth EF. 1987. Malarial parasite hexokinase and hexokinase dependent glutathione reduction in the *Plasmodium falciparum* infected erythrocyte. *J. Biol. Chem.* 262:15678–82

111. Roth EF, Calvin M-C, Max-Audit I, Rosa J, Rosa R. 1988. The enzymes of the glycolytic pathway in erythrocytes infected with *Plasmodium falciparum* malaria parasites. *Blood* 72:1922–25

112. Roth EF, Raventos-Suarez C, Perkins M, Nagel RL. 1982. Glutathione stability and oxidative stress in *P. falciparum* infection in vitro: response of normal and G6PD deficient cells. *Biochem. Biophys. Res. Commun.* 109:355–62

113. Roth EF, Schulman S, Vanderberg J, Olson J. 1986. Pathways for the reduction of oxidized glutathione in the *Plasmodium falciparum*-infected erythrocyte: Can parasite enzymes replace host red cell glucose–6–phosphate dehydrogenase? *Blood* 67:827–30

114. Roth EJ. 1990. *Plasmodium falciparum* carbohydrate metabolism: a connection between host cell and parasite. *Blood Cells* 16:453–60

115. Sahni SK, Saxena N, Puri SK, Dutta GP, Pandey VC. 1992. NADP-specific isocitrate dehydrogenase from the simian malaria parasite *Plasmodium knowlesi*: partial purification and characterization. *J. Protozool.* 39:338–42

116. Scheibel LW. 1988. *Plasmodium* metabolism and related organellar function during various stages of the life-cycle: carbohydrates. See Ref. 161a, 1:171–217

117. Scheibel LW, Adler A, Trager W. 1979. Tetraethylthiuram disulfide (Antabuse) inhibits the human malaria parasite *Plasmodium falciparum*. *Proc. Natl. Acad. Sci. USA* 76:5303–7

118. Scheibel LW, Pflaum WK. 1970. Cytochrome oxidase activity in platelet free preparations of *Plasmodium falciparum*. *J. Parasitol.* 56:1054

119. Scheibel LW, Sherman IW. 1988. *Plasmodium* metabolism and related organellar function during various stages of the life-cycle: proteins, lipids, nucleic acids and vitamins. See Ref. 161a, 1: 219–52

120. Sherman IW. 1977. Transport of amino acids and nucleic acid precursors in malarial parasites. *Bull. WHO* 55:211–25

121. Sherman IW. 1979. Biochemistry of *Plasmodium* (malarial parasites). *Microbiol. Rev.* 43:453–95

122. Sherman IW, Ting IP. 1966. Carbon dioxide fixation in malaria (*Plasmodium lophurae*). *Nature* 212:1387–88

123. Sherman IW, Ting IP. 1968. Carbon dioxide fixation in malaria. II. *Plasmodium knowlesi* (monkey malaria). *Comp. Biochem. Physiol.* 24:639–42

124. Shippen-Lentz D, Afroze T, Vezza AC. 1990. Heterogeneity and expression of the *Plasmodium falciparum* 5.8S ribosomal RNA genes. *Mol. Biochem. Parasitol.* 38:113–20

125. Shute PG, Maryon M. 1954. The effect of pyrimidine (daraprim) on the gametocytes and oocysts of *Plasmodium falciparum* and *Plasmodium vivax. Trans. R. Soc. Trop. Med. Hyg.* 48:50–63

126. Sinden RE. 1982. Gametocytogenesis of *Plasmodium falciparum* in vitro: an electron microscopic study. *Parasitology* 84:1–11

127. Sinden RE, Butcher GA, Billker O, Fleck SL. 1996. Regulation of infectivity of *Plasmodium* to the mosquito vector. *Adv. Parasitol.* 38:54–117

128. Sinden RE, Canning EU, Bray RS, Smalley ME. 1978. Gametocyte and gamete development in *Plasmodium falciparum. Proc. R. Soc. London Ser. B* 201:375–99

129. Sinden RE, Hartley RH, Winger L. 1985. The development of *Plasmodium* ookinetes in vitro: an ultrastructural study including a description of meiotic division. *Parasitology* 91:227–44

130. Sinden RE, Smalley ME. 1979. Gametocytogenesis of *Plasmodium falciparum* in vitro: the cell-cycle. *Parasitology* 79:277–96

131. Sinden RE, Strong K. 1978. An ultrastructural study of the sporogonic development of *Plasmodium falciparum* in *Anopheles gambiae. Trans. R. Soc. Trop. Med. Hyg.* 72:477–91

132. Siu PM. 1967. Carbon dioxide fixation in *Plasmodia* and the effect of some antimalarial drugs on the enzyme. *Comp. Biochem. Physiol.* 23:785–95

133. Slomianny C, Prensier G. 1986. Application of the serial sectioning and tridimensional reconstruction techniques to the morphological study of the *Plasmodium falciparum* mitochondrion. *J. Parasitol.* 72:595–98

134. Smalley ME, Sinden RE. 1977. *Plasmodium falciparum* gametocytes, their longevity and infectivity. *Parasitology* 74:1–8

135. Srivastava IK, Rottenberg H, Vaidya AB. 1997. Atovaquone, a broad spectrum antiparasitic drug, collapses mitochondrial membrane potential in a malaria parasite. *J. Biol. Chem.* 272:3961–66

136. Srivastava IK, Schmidt M, Grall M, Certa U, Garcia AM, Perrin LH. 1992. Identification and purification of glucose phosphate isomerase of *Plasmodium falciparum. Mol. Biochem. Parasitol.* 54:153–64

137. Strath M, Scott-Finnigan T, Gardner M, Williamson D, Wilson I. 1993. Antimalarial activity of rifampicin in vitro and in rodent models. *Trans. R. Soc. Trop. Med. Hyg.* 87:211–16

138. Strickland GT, Fox E, Sarwar M, Khaliq AA, MacDonald M. 1986. Effects of chloroquine, amodiaquine and pyrimethamine-sulfadoxine on *Plasmodium falciparum* gametocytemia. *Am. J. Trop. Med. Hyg.* 35:259–62

139. Taklehaimanot A, Nguyen-Dinh P, Collins WE, Barber AM, Campbell CC. 1985. Evaluation of the sporontocidal compounds using *Plasmodium falciparum* gametocytes produced in vitro. *Am. J. Trop. Med. Hyg.* 34:429–34

140. Terzian LA, Stahler N, Ward PA. 1952. The effect of antibiotics and metabolites on the immunity of mosquitoes to malaria infection. *J. Infect. Dis.* 90:115–23

141. Theakston RDG, Howells RE, Fletcher KA, Peters W, Fullard J, Moore GA. 1969. The ultrastructural distribution of cytochrome oxidase activity in *Plasmodium berghei* and *P. gallinaceum. Life Sci.* 8:521–29

142. Thierbach G, Reichenbach H. 1981. Myxothiazol, a new inhibitor of the cytochrome b-c_1 segment of the respiratory chain. *Biochim. Biophys. Acta* 638:282–89

143. Toye PJ, Sinden RE, Canning EU. 1977. The action of metabolic inhibitors on microgametogenesis in *Plasmodium yoelii nigeriensis. Z. Parasitenkunde* 53:133–41

144. Unnasch TR, Wirth DF. 1983. The avain malaria parasite *Plasmodium lophurae* has a small number of heterogenous ribosomal RNA genes. *Nucleic Acids Res.* 11:8443–59

145. Vaidya AB, Akella R, Suplick K. 1989. Sequences similar to genes for two mitochondrial proteins and portions of

ribosomal RNA in tandemly arrayed 6-kilobase-pair DNA of a malarial parasite. *Mol. Biochem. Parasitol.* 35:97–108

146. Vaidya AB, Lashgari MS, Pologe LG, Morrisey J. 1993. Structural features of *Plasmodium* cytochrome *b* that may underlie susceptibility to 8-aminoquinolines and hydroxynapthoquinones. *Mol. Biochem. Parasitol.* 58:33–42

147. Vanderberg J, Rdodin J, Yoeli M. 1967. Electron microscopic and histological studies of sporozoite formation in *Plasmodium berghei*. *J. Protozool.* 14:82–103

148. VanderJagt DL, Hunsaker LA, Campos NM, Baack BR. 1990. D-Lactate production in erythrocytes infected with *Plasmodium falciparum*. *Mol. Biochem. Parasitol.* 42:277–84

149. VanderJagt DL, Hunsaker LA, Heidrich JE. 1981. Partial purification and characterization of lactate dehydrogenase from *Plasmodium falciparum*. *Mol. Biochem. Parasitol.* 4:255–64

150. VanderJagt DL, Hunsaker LA, Kibirige M, Campos NM. 1989. NADPH production by the malarial parasite *Plasmodium falciparum*. *Blood* 74:471–74

151. VanderJagt DL, Intress C, Heidrich JE, Mrema JE, Reickmann KH, Heidrich HG. 1982. Marker enzymes of *Plasmodium falciparum* and human erythrocytes as indicators of parasite purity. *J. Parasitol.* 68:1068–71

152. Velanker SS, Ray SS, Gokhale RS, Balaram H, Balaram P, Murthy MR. 1997. Triosephosphate isomerase from *Plasmodium falciparum*: The crystal structure provides insights into antimalarial drug design. *Structure* 5:751–61

153. Vermeulen AN, Durawn J, Brakenhoff RH, Lensen THW, Ponnundurai T, Meuwissen JHET. 1986. Characterization of *Plasmodium falciparum* sexual stage antigens and their biosynthesis in synchronized gametocyte cultures. *Mol. Biochem. Parasitol.* 20:155–63

154. Vervenne RAW, Dirks RW, Ramesar J, Waters AP, Janse CJ. 1994. Differential expression in blood stages of the gene coding for the 21-kilodalton surface protein of *Plasmodium berghei* as detected by RNA in situ hybridization. *Mol. Biochem. Parasitol.* 68:259–66

155. Wang P, Read M, Sims PFG, Hyde JE. 1997. Sulfadoxine resistance in the human malarial parasite *Plasmodium falciparum* is determined by mutations in dihydropteroate synthetase and an additional factor associated with folate utilization. *Mol. Microbiol.* 23:979–86

156. Warburg A, Miller LH. 1992. Sporogonic development of a malaria parasite in vitro. *Science* 255:448–50

157. Waters AP. 1994. The ribosomal RNA genes of *Plasmodium*. *Adv. Parasitol.* 34:34–79

158. Waters AP, Spaendonk RML, Ramesar J, Vervenne RAW, Dirks RW, et al. 1997. Species-specific regulation and switching of transcription between stage-specific ribosomal RNA genes in *Plasmodium berghei*. *J. Biol. Chem.* 272:3583–89

159. Waters AP, Syin C, McCutchan TF. 1989. Developmental regulation of stage-specific ribosome populations in *Plasmodium*. *Nature* 342:438–40

160. Watkins WM, Sixsmith DG, Chulay JD, Spencer HC. 1985. Antagonism of sulfadoxine and pyrimethamine antimalarial activity by p-aminobenzoic acid, p-aminobenzoylglutamic acid and folic acid. *Mol. Biochem. Parasitol.* 14:55–61

161. Weiss MM, Vanderberg JP. 1976. Studies on *Plasmodium* ookinetes. I. Isolation and concentration from mosquito midguts. *J. Protozool.* 23:547–51

161a. Wernsdorfer WH, McGregor SI, eds. 1988. *Malaria Principles and Practice of Malariology*, Vols. 1 & 2. New York: Churchill Livingstone

162. Wernsdorfer WH, Trigg PI. 1988. Recent progress of malaria research: chemotherapy. See Ref. 161a, 2:1569–1674

163. Williamson KC, Fujioka H, Aikawa M, Kaslow DC. 1996. Stage-specific processing of Pfs230, a *Plasmodium falciparum* transmission-blocking vaccine candidate. *Mol. Biochem. Parasitol.* 78:161–69

164. Wooden JM, Hartwell LH, Vasquez B, Sibley CH. 1997. Analysis in yeast of antimalarial drugs that target the dihydrofolate reductase of *Plasmodium falciparum*. *Mol. Biochem. Parasitol.* 85:25–40

165. Zhu J, Waters AP, Appiah A, McCutchan TF, Lal AA, Hollingdale MR. 1990. Stage-specific ribosomal RNA expression switches during sporozoite invasion of hepatocytes. *J. Biol. Chem.* 265:12740–44

Annu. Rev. Microbiol. 1998. 52:591–625

THYMINE METABOLISM AND THYMINELESS DEATH IN PROKARYOTES AND EUKARYOTES

S. I. Ahmad[1], S. H. Kirk[1], and A. Eisenstark[2]

[1]Department of Life Sciences, Nottingham Trent University, Clifton Lane, Nottingham NG11 8NS, England; e-mail:lif3ahmadsi@NTU.AC.UK, sandra.kirk@ntu.ac.uk.
[2]Cancer Research Center, 3501 Berrywood Drive, Columbia, Missouri 65201; e-mail: abe@biosci.mbp.missouri.edu

KEY WORDS: TLD, DNA damage, DNA repair, thymidylate synthetase, mutation, apoptosis

ABSTRACT

For many years it has been known that thymine auxotrophic microorganisms undergo cell death in response to thymine starvation [thymineless death (TLD)]. This effect is unusual in that deprivation of many other nutritional requirements has a biostatic, but not lethal, effect. Studies of numerous microbes have indicated that thymine starvation has both direct and indirect effects. The direct effects involve both single- and double-strand DNA breaks. The former may be repaired effectively, but the latter lead to cell death. DNA damaged by thymine starvation is a substrate for DNA repair processes, in particular recombinational repair. Mutations in *recBCD* recombinational repair genes increase sensitivity to thymineless death, whereas mutations in RecF repair protein genes enhance the recovery process. This suggests that the RecF repair pathway may be critical to cell death, perhaps because it increases the occurrence of double-strand DNA breaks with unique DNA configurations at lesion sites. Indirect effects in bacteria include elimination of plasmids, loss of transforming ability, filamentation, changes in the pool sizes of various nucleotides and nucleosides and in their excretion, and phage induction. Yeast cells show effects similar to those of bacteria upon thymine starvation, although there are some unique features. The mode of action of certain anticancer drugs and antibiotics is based on the interruption of thymidylate metabolism and provides a major impetus for further studies on TLD. There are similarities between TLD of bacteria and death of

591

eukaryotic cells. Also, bacteria have "survival" genes other than *thy* (thymidylate synthetase), and this raises the question of whether there is a relationship between the two. A model is presented for a molecular basis of TLD.

CONTENTS

Introduction

Cohen & Barner (53) discovered that a thymine auxotrophic mutant (*thyA*) of *Escherichia coli*, when starved of thymine, loses colony-forming ability, i.e. thymine starvation causes cell death. This so-called unbalanced growth or thymineless death (TLD) (24) is a unique biological phenomenon, since starvation of bacteria of other growth factors, such as amino acids or vitamins, stops growth but does not cause cell death (23).

Despite the publication of more than 250 papers on TLD in several organisms, from prokaryotes to eukaryotes (Table 1), the exact molecular mechanism of TLD is unclear. Many molecular events have been implicated in cell death, but the obvious lack of any compelling direct relationship(s) has so far precluded agreement on a definitive hypothesis. Although it may not exclude other possibilities, a model based on the detailed experiments of Nakayama et al (159) may be a basis for further research.

Most information on TLD has been obtained using *E. coli;* hence this organism provides a basic paradigm for explaining the essential phenomena and molecular mechanism(s) of TLD. However, findings in other microbes, including yeasts, are also described. With regard to TLD in mammalian cells, information is increasingly becoming available (59, 99, 105, 107–109, 133). Studies of TLD are important from a clinical standpoint, as certain widely used chemotherapeutic agents mimic TLD in eukaryotes by inhibiting thymidylate

Table 1 Thymine auxotrophic microorganisms

Determinant	Reference
Phages	
T4	121
Gram-negative bacteria	
Aerobacter aerogenes	97
Escherichia coli	22
Klebsiella aerogenes	176
Mycoplasma laidlawii	201
Pseudomonas acidovorans	122
Pseudomonas stutzeri	51
Proteus mirabilis	167
Salmonella typhimurium	67, 168
Sulfolobus acidocaldarius	94
Gram-positive bacteria	
Bacillus anthracis	117
Bacillus megatarium	218
Bacillus subtilis	73
Bacillus thuringiensis	61
Diplococcus pneumoniae	36
Enterococcus	94
Lactobacillus acidophilus	183
Micrococcus radiodurans	137
Pneumococcus	81
Staphylococcus aureus	144, 145
Streptococcus faecalis	10, 33, 82
Fungi	
Saccharomyces cerevisiae	20, 91
Mammalian cells	222

metabolism: Methotrexate inhibits dihydrofolate reductase and thereby interferes with thymidylate synthesis; fluorodeoxyuridine (FUdR), in addition to exerting a variety of toxic effects, inhibits thymidylate synthetase (100); and sulfa drugs (analogues of *p*-aminobenzoic acid) can impede de novo folate synthesis and thereby decrease the supply of reduced folates participating in the methylation of deoxyuridine monophosphate (dUMP) (138).

Although TLD in bacteria was a focal subject between the time of its discovery in 1954 until 1994, very few relevant papers have appeared since. The gene *thy* is unique because it codes for the only nutritional component needed for survival. More recent studies have focused on survival of bacteria in stationary phase (SPD–stationary-phase death) (66–69, 101, 102, 114, 124, 200, 225).

The concepts, experimental approaches, and reported mechanisms are not identical to those proposed for TLD, but it may be worthwhile to use the molecular genetic approaches in study of SPD for renewed studies of TLD.

Thymine Metabolism

An understanding of metabolic events upon thymine starvation may be helpful in identifying the initial lesion that culminates in TLD. The first step in deoxythymidine triphosphate (dTTP) synthesis is the conversion of deoxyuridine monophosphate (dUMP) to deoxythymidine monophosphate (dTMP) (Figure 1, reaction 1) catalyzed by thymidylate synthetase (encoded by thyA) (13). This step involves the transfer of a C-1 unit from N^5, N^{10}-methylene tetrahydrofolate to the 5 position of dUMP, tetrahydrofolate being consumed in the reaction (80). dTMP is sequentially phosphorylated to the corresponding diphosphate and triphosphate (Figure 1, reactions 2 and 3) by dTMP kinase (161) and dTDP kinase (135).

In thymine auxotrophic E. coli, thyA cells have no active thymidylate synthetase. Consequently no dTMP, dTDP, or dTTP is produced endogenously. If thymine is present in the growth medium, in contrast to wild-type cells that cannot use exogenous thymine, thy mutants can incorporate it via a salvage pathway. The reaction involves deoxyribose-1-phosphate (dRib-1-P) and thymine being converted to thymidine (dTR) and subsequently to dTMP (Figure 1, reactions 4 and 5). Wild-type cells normally cannot use exogenous thymine because they lack a dRib-1-P pool (48). This raises the question of where the dRib-1-P comes from in thyA mutants

In a thyA cell, there is no intracellular pool of dRib-1-P. Also, there is a decline in the dTTP pool due to continuing DNA synthesis. This decline leads to derepression of nucleoside diphosphate reductase (Figure 1, reaction 6) (35). The dCTP pool increases (Figure 1, reactions 6 and 3) and is then deaminated to dUTP (Figure 1, reaction 7) (165). dUTP is subsequently degraded to dUMP (Figure 1, reaction 8). In the absence of thymidylate synthetase, dUMP is prevented from being converted to dTMP and is degraded to deoxyuridine and subsequently to uracil and dRib-1-P (Figure 1, reactions 9 and 4). Thus, the pool of dRib-1-P is generated from dUMP. Further evidence for the importance of dRib-1-P in promoting the incorporation of thymine comes from the observation that incorporation of exogenous fluorouracil is increased if the compound is accompanied by a source of dRib-1-P, such as deoxyadenosine (181).

Wild-type strains of E. coli and Salmonella typhimurium can also incorporate the nucleoside moiety of dTMP derived from exogenous sources. However, the nucleotide must first be dephosphorylated and subsequently rephosporylated to dTMP without being broken down by thymidine kinase. The enzyme that cleaves the nucleotide is located within the periplasmic space (190).

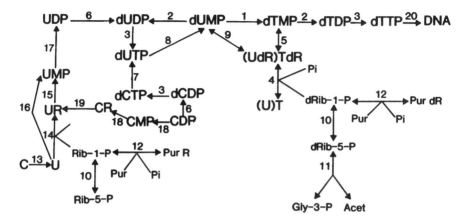

Figure 1   Nucleotide metabolism in *E. coli.*

Numbers represent enzymes catalysing individual steps, and these are:

1. Thymidylate synthetase
2. Deoxyribonucleotide kinase
3. Deoxynucleotide diphosphate kinase
4. Thymidine phosphorylase
5. Thymidine kinase
6. Nucleotide diphosphate reductase
7. dCTP deaminase
8. Deoxyuridine triphosphatase
9. dUMP phosphatase
10. Phosphodeoxyribo mutase
11. Phosphodeoxyribo aldolase
12. Purine nucleoside phosphorylase
13. Cytosine deaminase
14. Uridine phosphorylase
15. Uridine kinase
16. UMP pyrophosphorylase
17. UMP kinase
18. Nucleotidase
19. Cytidine deaminase
20. DNA polymerase

Abbreviations used: Acet, acetaldehyde; C, cytosine; CR, cytidine; CMP, CDP, and CTP, cytidine mono-, di-, and triphosphate; dCDP and dCTP, deoxcytidine di and triphosphate; dRib-1-P, deoxyribose-1-phosphate; dTMP, dTDP, and dTTP, deoxythymidine mono-, di-, and triphosphate; dUMP and dUDP, deoxyuridine mono- and diphosphate; Gly-3-P, glyceraldehyde-3-phosphate; Pur, purine; Pur dR, purine deoxyriboside; Pur R, purine riboside; Rib-1-P, ribose-1-phosphate; TdR, deoxythymidine; U, uracil; UR, UMP, and UDP, uridine, uridine mono-, and diphosphate; UdR, deoxyuridine.

An alternative pathway for the formation of pyrimidine deoxyribosides is via deoxyribosyl transferase action; it is observed in other bacteria, including *Lactobacilli* and *Corynebacterium* species (38, 189).

Selection of Thy *Bacteria and Phages*

Most *thy* mutants have been isolated as cells resistant to folate antagonists (e.g. aminopterin, trimethoprim) in the presence of thymine or thymidine (29, 168, 206). The isolation of *thyA* mutants is more efficient with trimethoprim than with aminopterin, the former having a better penetration into bacterial cells (206). Thymine auxotrophic mutants have been isolated from various Gram-positive and Gram-negative bacterial species as well as from *Saccharomyces cerevisiae* (Table 1).

In *Pseudomonas acidovorans* and *Diplococcus pneumoniae*, thymidylate synthetase mutants require thymidine to grow because they lack thymidine phosphorylase (Figure 1, reaction *4*) and cannot convert exogenous thymine to thymidine (81, 122). A similar situation can exist in *E. coli* mutants lacking thymidine phosphorylase and thymidylate synthetase. Even though *Bacillus anthracis* lacks thymidine phosphorylase, *thy* mutants of this bacterium can utilize exogenous thymine, presumably due to deoxyribosyl transferase activity (117). *Pseudomonas stutzeri* and *Pseudomonas aeruginosa*, unlike *P. acidovorans*, lack thymidine kinase in addition to thymidine phosphorylase, and as a result thymine auxotrophs are unable to utilize exogenous thymidine and hence are difficult to select. Carlson et al (51) were able to clone *tdk* (thymidine kinase) from *E. coli* in a variety of *Pseudomonas* species. Subsequently, it was possible to isolate a thymidine auxotroph of *P. stutzeri*. Thymine auxotrophic mutants of *Bacillus thuringiensis* requiring low levels of thymine (61) and thymidylate-requiring mutants of *Enterococcus*, *Streptococcus*, *Micrococcus*, and *Staphylococcus* have been isolated (10, 94, 137, 144), as have thymine-requiring mutants of phage T4 (121). It is interesting to note that *Micrococcus radiodurans*, which shows unusually high resistance to UV light, is not hyper-resistant to TLD (137). A likely explanation is that the nature of damage to DNA by these two diverse agents is different and that the damage is amenable to separate repair pathways.

In *Bacillus subtilis*, two genes (*thyA* and *thyB*) code for different thymidylate synthetases, and both must be mutated to produce a *thy* phenotype similar to that of *E. coli* (163). *thyA* encodes a heat-resistant enzyme that contributes 95% of wild-type thymidylate synthetase activity; the heat-sensitive *thyB* product contributes the remaining activity.

An interesting observation concerning thymine auxotrophy in *E. coli* is that the mutants show reduced translational fidelity (215). This is manifested as suppression of nonsense and frameshift mutations. Such a phenomenon is of obvious advantage to thymineless cells.

Classes of Thymine Auxotrophic Mutants

All primary mutations for thymine auxotrophy are in *thyA*. Secondary thymine auxotrophic mutants can be isolated and are classed according to how much thymine they require for growth. This depends on the endogenous concentration of dRib-1-P; the higher the dRib-1-P pool, the lower the requirement for exogenous thymine.

The endogenous pool of dRib-1-P depends on three inducible enzymes: phosphopentomutase (Figure 1, reaction *10*), deoxyribose-phosphate aldolase (Figure 1, reaction *11*), and thymidine phosphorylase (Figure 1, reaction *4*), encoded respectively by *deoB*, *deoC*, and *deoA* (5, 13).

Auxotrophs that carry only a *thyA* mutation require about 20 μg of thymine per ml to grow. If their dRib-1-P pool is taken as a standard and low thymine–requiring auxotrophic mutants are isolated, these auxotrophs must carry an additional mutation, either in *deoB* or *deoC* (Figure 1, reactions *10* and *11*). Such mutations prevent the catabolism of dRib-1-P and dRib-5-P, respectively, resulting in an enhanced intracellular pool of the former, and the bacteria require only 2 to 5 μg of thymine per ml in the medium to grow. The next group of thymine auxotrophs are super low thymine requirers (0.2 to 0.5 μg per ml). In addition to *thyA* and *deoB* or *deoC*, they have a mutation in *deoR* (the gene that codes for the repressor of the *deo* operon). As a result of mutation in *deoR*, constitutive synthesis of thymidine phosphorylase (and other *deo* encoded enzymes) occurs. This leads to enhanced degradation of dUMP to dUR (deoxyuridine) and then to dRib-1-P plus uracil. Due to mutation in *deoB* or *deoC*, the level of dRib-1-P is increased significantly and can trap very low levels of thymine present in the medium (6). The final class of thymine auxotrophs comprises the super high thymine requirers. These are obtained from the super low thymine requirers by the reversion of *deoC* or *deoB* mutations to their respective wild-type phenotypes. As a result, the double mutant *thyA deoR* synthesizes the *deoB* and *deoC* enzymes constitutively and rapidly depletes the dRib-1-P pool, and hence requires a higher level of thymine (6). Mutants with low and high thymine requirements have been isolated in other bacterial species.

Techniques for Thymine Starvation

Thymine starvation can be induced by (*a*) transferring cells from a thymine-supplemented medium to a medium devoid of thymine after centrifugation or filtration; (*b*) adding a ribonucleoside to the growth medium as an inhibitor of thymidine phosphorylase such that conversion of added thymine to thymidine cannot occur (39, 48, 74, 221); (*c*) inhibiting thymidylate synthetase with FUdR; or (*d*) inhibiting folate metabolism, with sulfonamides, which block de novo folate synthesis, with aminopterin or trimethoprim, which inhibit dihydrofolate reductase (212), or by adding cytosine arabinoside (11, 56) or showdomycin (119).

For TLD to occur, carbon and energy sources are needed while DNA synthesis is in progress. Stationary cells or cells between normal rounds of DNA replication are immune to TLD (23, 58, 95, 142, 154).

Bresler et al (44) observed that TLD does not occur in mutants of *E. coli* and *B. subtilis* if cells are plated at moderate density on solid media for up to 70 h. This effect may be due to the thymidylate released from dead cells within individual colonies becoming available more effectively to remaining cells than is possible in liquid medium.

Growth media seem to play a role in determining the slope of the inactivation curve, particularly with *recA* mutant bacteria; the cells die faster when the incubation medium contains casamino acid. This effect was shown to be more pronounced with *recA13* mutant bacteria, and it has been proposed that an increased rate of growth may be the reason for enhanced cell death by thymine starvation in the rich-nutrient medium (1, 157).

Strain Specificity and Kinetics of TLD

In most bacterial strains, the time course of TLD can be divided into two phases: (*a*) the lag period, during which there is little or no apparent decline in viable count; and (*b*) the death period, represented by an exponential decline in viable cell population. The slope is used to determine the kinetics of cell death.

Variabilities in the lag period and death rate may be attributed to (*a*) differences in the endogenous dTTP pools at the onset of thymine starvation—the lower the pool of dTTP, the shorter the expected lag period (different dTTP pools have been observed in *E. coli* strains supplemented with varying concentrations of thymine in the medium) (25, 166); (*b*) the presence of inducible prophage or bacteriocins—in *E. coli* harboring prophages, TLD has been shown to take place immediately after the removal of thymine (71, 125, 149, 152, 195); (*c*) the DNA repair efficiencies of the cells for damage caused by thymine starvation; (*d*) "leakiness" of thymidylate synthetase mutations. An important difference between various strains of *E. coli* with regard to TLD is that some of them carry prophage in their genomes while others do not (see SOS response and prophage induction, below).

E. coli strains B (a UV- and TLD-sensitive mutant), B/r, K-12 *rec21*, and K-12 *lon⁻*, and certain other derivatives of *E. coli* B such as B_{s-1}, B_{s-2}, B_{s-3}, B_{s-8}, and B_{s-11}, have no lag period, and TLD starts immediately after removal of thymine from the medium (*thy* mutations in the *lon* gene) (see cell wall synthesis and filamentation) (58).

The rate of TLD in *B. thuringiensis* is much slower than that observed in a typical *thy* mutant of *E. coli*, but the process is otherwise similar (61). In *Streptococcus faecalis* subsp. *liquefaciens*, the lag period is longer and the death rate is slower than in *E. coli* (82).

Biochemical and Physiological Changes
upon Thymine Deprivation

A major consequence of thymine starvation in *E. coli* 15T⁻ and its derivatives is alteration in endogenous deoxyribonucleoside triphosphate (dNTP) pools resulting from catabolic reactions that degrade dATP, dCTP, and dGTP (35, 164, 166). In summary, not only does thymine starvation disrupt the endogenous thymidylate pool, but the levels of several other nucleotides are also affected. The level of dATP rises 4- to 5-fold within 30 min and then remains constant. In contrast, the dCTP pool remains steady for the first 30 min and then increases linearly 10- to 15-fold within 90 min. The dGTP level remains constant. As expected, the level of dTTP drops dramatically within 30 min (to 10% of its initial concentration), and that of dTMP drops within 90 min. These changes are related to alterations in the rates of nucleotide synthesis during thymine starvation (164). The rate of dATP synthesis remains constant for at least 75 min, that of dCTP increases sharply over the same time period, and that of dGTP decreases to a very low level. These effects obviously reflect the intimate relationships between biosynthesis and catabolism of these metabolites.

Thymine starvation results in excretion of more than twenty compounds, including hypoxanthine, uracil, and orotic acid (53). Deoxyribose and uracil are excreted in notably high amounts, confirming that dUMP is the chief source of dRib-1-P (40, 162, 164).

Loss of thymine from DNA in thymine-starved *E. coli* (41) and in *B. subtilis* (186) has been shown, and it has been proposed that this is due to incorporation of dUMP in place of dTMP in newly synthesized regions.

One intriguing aspect observed during thymine starvation is filamentation of bacterial cells (24). This phenomenon has been shown to be strain specific. For example, *E. coli* K-12 *rec21* does not filament (110), whereas strains B, B_{s-3}, and B_{s-12} do (57). Donachie (65) was able to demonstrate an association between cell division and DNA replication, and therefore it is not surprising that in the thymine-starved condition when DNA synthesis is impeded, septa formation cannot take place, although cell growth continues, leading to filamentation (24).

At the molecular level, cellular filamentation upon thymine starvation has been associated with induction of *sulA*, an SOS gene also known as *sfi* and which is responsible for synthesizing an inhibitor of septum formation (112, 115). As a result of high concentrations of the inhibitor, septa cannot form and cells elongate. The Sul protein has been shown to be unstable and can be degraded by Lon protease. Hence filamentation is more pronounced in *lon* mutants of *E. coli* K-12 and in wild-type *E. coli* B (a naturally occurring *lon* mutant unable to degrade septum-inhibitor protein). A mutation invoking constitutive expression of *sulA* leads to lethal filamentation, a situation observed in cases of TLD (88).

An interesting observation is that *LexA* mutants of *E. coli* (with deregulated SOS repair) divide more frequently under conditions of thymine starvation compared to *lexA*+ strains, the mutant bacteria giving rise to cells that lack DNA (111). These authors suggest that this reflects a role of the *lexA* product in inactivation of an inhibitor of cell division, and that this effect is exacerbated in the absence of thymine.

Bisaillion et al (33) showed that D-cycloserine increased the TLD rate of a *S. faecalis* thymine auxotrophic strain. However, other antibiotics, including bacitracin, penicillin, and vancomycin, decreased it. Another variant of *S. faecalis*, var. *zymogens*, behaved similarly, except in this case penicillin also increased the rate of TLD. These results suggest that thymine starvation affects cell wall synthesis in this bacterium.

Molecular Effects of Thymine Starvation

When *thy* mutants of *E. coli* are deprived of thymine, synthesis of RNA and protein continue at a rapid rate, followed by a rapid decline (147, 148). RNA, but not protein, synthesis may be a prerequisite for TLD (85, 95). A direct involvement of protein synthesis in TLD is unlikely. For example, as Ganguli & Bhattacharjee (87) found, chloramphenicol treatment reduces the rate of TLD but does not abolish it in a multiple auxotrophic mutant of *E. coli* 15 T⁻. However, in various strains of *E. coli*, inhibition of RNA synthesis can prevent cell death (95, 173, 187). RNA synthesized during TLD in *E. coli* B3 and K-12 has the same base composition as that synthesized in the presence of thymine (85, 197), indicating that TLD is not caused by the synthesis of grossly abnormal species of RNA, but rather that the process of DNA unwinding involved in transcription is implicated. In support, it has been shown that low concentrations of phenethyl alcohol, which interfere with mRNA synthesis, prevent TLD (187).

As expected, in various *E. coli* species the rate of mRNA synthesis decreases exponentially during thymine starvation because of the inability of DNA to serve as a template for RNA polymerase. Indeed, purification of DNA from thymine-starved cells increases the priming efficiency of RNA polymerase, possibly because damaged templates are removed (141).

In *thy* mutants of *B. subtilis*, the situation appears similar in that actinomycin (at concentrations sufficient to inhibit RNA but not DNA synthesis) blocks TLD. By studying cells at different times after removal of thymine, Rolfe (186) concluded that in this bacterium, TLD involved two distinct steps: one requiring protein and RNA synthesis, and a subsequent step requiring continuous thymine starvation but independent of RNA and protein synthesis.

Induced synthesis of a number of proteins has been associated with thymine starvation; thus in *E. coli* B and B/r, two proteins, one of 80 kDa and another of 88 kDa (likely components of the SOS system), are synthesized in increased quantities during thymine starvation (60), as are thymidine phosphorylase and nucleoside diphosphate reductase (35, 40), an endonuclease (78), and a DNA masking protein (209). The latter is able to bind (and inactivate) the single-stranded phage (øX174) and is likely to be an SSB (single-strand binding) protein induced during thymine starvation. Increased production of enzymes involved in DNA repair and thymidylate metabolism is understandable under these conditions, and their induced synthesis is not directly responsible for TLD.

Tiganos & Herrington (215) noted that lack of thymidylate synthetase activity alters the fidelity of translation in *E. coli*. This alteration takes the form of suppression of nonsense and frameshift mutations, providing a useful survival mechanism.

Thymine starvation provokes substantial DNA damage: There is (*a*) difficulty in isolating high-molecular-weight DNA from bacteria after they have been thymine starved (146), (*b*) loss of transforming activity of DNA (153), (*c*) induction of mutation (55), (*d*) induction of prophage and bacteriocins (125, 151, 195), and (*e*) lack of assembly of Okazaki fragments (77).

Single-strand DNA breaks within genomic DNA occur in *thy* mutants of *E. coli* B, K-12, B/r, and 15T⁻ upon thymine starvation (30, 36, 41, 76, 103, 184). The extent of DNA breaks is strain dependent, the breaks being especially apparent in strains deficient in DNA polymerase I (155). However, despite the sensitivity of *polA* mutants to TLD (27), no DNA strand breaks could be observed in W3110, a *thy⁻ polA1⁻* strain, after thymine starvation (15, 191). Interestingly, in those *polA⁻* strains susceptible to TLD, although addition of thymine to the medium resulted in nearly complete restoration of the DNA sedimentation profile, it did not restore cell survival, suggesting that strand integrity is not the principal reason for recovery from TLD (155). It is possible that nonrecoverable damage is mediated by one of the repair processes (possibly *recF*) that, in the absence of thymine, induces double-strand breaks as part of an homologous recombination repair mechanism. Indeed, by using a neutral density gradient assay, Yoshinaga (223) was able to detect double-strand DNA breaks in thymine-starved *E. coli*. These double-strand breaks potentially become lethal if not efficiently repaired. The configuration of the DNA strands at the break may be unique for thymine starvation. Hence it is the type of damage and the process, rather than the extent of damage, that determine lethality.

The incidence of DNA strand breaks may be DNA type dependent; thus Baker & Hewitt (15) and Sedgwick & Bridges (191) were unable to detect chromosomal strand breaks in *E. coli* C *thy*-321 after 150 min of thymine starvation, whereas Freifelder (76) detected breaks in F and F-prime in *E. coli* K-12. In support, Shuster (192) observed when using an F^+ derivative of *E. coli* CR34 *thy⁻*, lysogenic for a noninducible lambda prophage, that plasmid but not prophage DNA is lost on thymine starvation. Because F DNA requires a membrane-bound protein during its replication, it has been suggested that this protein is responsible for introducing nicks in F DNA when it passes through the membrane (76) that are inefficiently repaired under conditions of thymine starvation, contributing to episome loss. Supporting evidence for this came from the observations of Hill & Fangman (103) that, upon thymine starvation, the rate of occurrence of single-strand DNA breaks in F-*lac* was higher than that observed in the genome.

Assembly of Okazaki fragments, which normally is completed fairly rapidly, is inefficient in thymine-starved cells (77), although DNA ligase is not inhibited (78). This is probably due to the incorporation of uracil into the fragments, which is removed by uracil glycosylase, generating an increased numbers of gaps.

Similar results have been shown as regards single-strand DNA breaks for *Diplococcus pneumoniae* and *B. subtilis* (36, 50, 185). Both single-strand DNA breaks—measured in *B. subtilis* by loss of transforming activity of DNA from behind the replication fork (182)—and cell death start within 10 min of the initiation of thymine starvation, and it is likely that prophage induction is the cause of DNA breakdown. Buick & Harris (49) also implicated hypermethylation as causal in formation of strand breaks. No double-strand DNA breaks were detected by Bousque & Sicard (36) in *D. pneumoniae* upon prolonged thymidine starvation. In the same way as for *E. coli*, it is suggested that thymine starvation in these bacteria leads to TLD via initial damage to newly synthesised DNA.

6-Methyl aminopurine (MAP), which is usually present in small amounts in *E. coli* DNA, is incorporated at increased levels if *thy* mutants of *E. coli* 15T⁻ are grown in the absence or presence of subnormal concentrations of thymine or in the presence of the thymine antagonists 5-aminouracil or 2-thiothymine. The increase is not universal; thymine auxotrophs of *E. coli* B and *H. influenzae* grown in the presence of FUdR or 5-aminouracil do not show any increase in the MAP content of their DNA (211). Initially, increased methylation was thought to be due to production of a novel methylase during thymine starvation. However, Yudelevich & Gold (224) discovered that the methylase responsible for the results with 15T⁻ appeared only after 3.5 h of starvation and subsequent addition of thymine associated with cell lysis. They associated the activity with a phage-coded "late enzyme" that could not be produced during thymine starvation, during which time phage DNA synthesis was prevented. In further support of this, a mutant derivative, 15T⁻R, isolated as hyperresistant to mitomycin C (MC) and TLD—and which may be similar to SA236 isolated by Ahmad et al, (2, 3)—showed no increase in DNA methylase activity upon thymine starvation (149, 150). Hypermethylation of DNA per se in *E. coli* has little to do with TLD (75, 141), but may be a consequence of phage activation during thymine starvation.

On the contrary, in *B. subtilis*, Buick & Harris (49) observed an involvement of abnormal methylation in TLD. Their evidence shows, in methylase activity on thymine starvation, an initial rise that subsequently reduces to normal levels. They propose that such methylases preferentially act at the sites of single-strand breaks that accumulate early following thymine deprivation. This abnormal methylation pattern is recognized by cellular exonucleases that remove five MC residues and several associated bases, enlarging the single-strand gaps. When replication occurs, such single-stranded regions convert to double-strand breaks, which are lethal. Buick & Harris (49) proposed that this involvement of methylation exacerbates TLD rather than being its primary cause.

DNA Repair Mutations Affecting TLD

Examination of mutants sensitive to TLD (*polA*, *lig*, *uvrD*, *recBC*, or *rep*) assists in understanding the molecular events involved. Since many DNA replication and repair mutants are hypersensitive to TLD, the effect is expected to be mediated directly or indirectly at the level of DNA replication and repair (1, 9, 27, 62, 89, 158, 172, 191).

The sensitivity of *lig* mutants to TLD can be explained on the basis that in *E. coli* cells surviving thymine starvation, DNA molecules are ligated at a late stage in recovery, which cannot occur in the mutant. Such damage, however, could not be detected in the presence of rifampicin, and cell lethality was significantly reduced (172), reflecting an involvement of both transcription and DNA repair in TLD.

Berg & O'Neill (27) compared the graphs of TLD of *polA1⁻* and *pol⁺* strains of *E. coli* W3110 and concluded that the curves are identical in shape but different in slope. *PolA⁺* cells showed a shoulder in the early phase of TLD while *polA⁻* did not, and the commencement of TLD took place soon after starvation began. In these studies, it was surprising to find the same level of DNA strand breaks in both *polA⁻* and *polA⁺* strains. The authors suggested that in wild-type *E. coli*, initial damage to DNA caused by thymine starvation may be repaired by DNA polymerase I, but that when the number of lesions exceeds the number of available enzyme molecules, damage accumulates and TLD ensues. Further TLD studies of *pol* mutants and their respective *pol⁺* derivatives have confirmed these findings (155); e.g. in a comparison of MM384 and MM383, the former (a low thymine–requiring mutant of *E. coli* K-12) was shown to exhibit little, if any, change in the sedimentation profile of DNA on thymine starvation at 30°C for up to 150 min compared to cells supplemented with thymine, but showed considerable broadening and shift of the peak at 42°C. However, in MM383 (a *polA⁻ts* mutant of MM384), a change in sedimentation pattern was observed at 30°C and at 42°C that was much greater than in MM384. This clearly demonstrates a role for DNA polymerase I in the repair of thymine starvation–induced DNA damage. Further evidence for an involvement of polymerase I in TLD comes from the finding that a *uvh* (TLD hyperresistant) strain of *E. coli* synthesizes the enzyme constitutively (7).

The *E. coli rep* gene encodes DNA helicase III, a 67-kDa protein that contains an ATPase activity stimulated by single-stranded DNA regions and that is not essential for cell survival. Some impairment of DNA repair and replication has been noted in *rep⁻* mutant strains (62, 208). It is not surprising to observe TLD sensitivity and lack of support of øX174 growth when thymine is removed from the growth medium of *rep* mutants (62).

Mutants of *E. coli* deficient in *uvrD* (also known as *dar-2*, *mutU*, *pdeB*, *recL*, *uvrE*, and *uvr502*, encoding helicase II) are sensitive to TLD. This mutant (originally isolated as *mutU4*) shows enhanced mutator activity and sensitivity to UV (199). Indeed, DNA helicase II and helicase III double mutants are inviable, suggesting complementary actions of the two proteins (210).

The addition of coumermycin or nalidixic acid (inhibitors of DNA gyrase) to growth media sensitizes cells to TLD, suggesting that DNA gyrase is also involved in the repair of DNA damage generated by TLD (158). Further support comes from the recent work of Ahmad (2), in which an *E. coli* mutant with enhanced resistance to TLD has been shown to be resistant to nalidixic acid.

E. coli mutants blocked in *ruv* are sensitive to TLD (118) as well as to mitomycin C, UV, and ionizing radiation (139, 170). It is suggested that *ruvA* and *ruvB* gene products are involved in recombinational repair and that this repair is thus operative during thymine starvation.

Conflicting reports have been published concerning the sensitivities of *recBC* and *recA* mutants of *E. coli* to TLD (9, 157); e.g. the *recB⁻C⁻* strain SDB1318 is no more sensitive to TLD than its parent strain, whereas measurement of TLD in a number of different strains of *E. coli* having mutations in *recB21*, *recC22*, *recB21 C22*, *recB21 C22 sbcB15*, and *recB21 C22* and *xonA* show that all are sensitive to TLD (9). Subsequently, Ahmad (1) showed that SDB1318 was also sensitive to TLD. Aizenberg et al (8) interpreted results obtained with *recB58* as indicating a reduced sensitivity to TLD, although a careful analysis of the data suggests that this may not be correct and that the *recB* mutant is also sensitive to TLD. Additionally, the *recB⁻ recC⁻* mutant, when suppressed by a *sbc* mutation, was as sensitive to TLD as its parent strain (157). *Xon* mutations, which suppress UV sensitivity in the *recBC* mutant but not its recombination deficiency, exert no effect on TLD in this strain (132). From these studies, it can be argued that most, if not all, *recB⁻C⁻* mutants of *E. coli* are sensitive to

TLD. Some controversy may be due to differences in experimental conditions and strains employed.

Mutants of *E. coli* showing enhanced resistance to TLD have also been isolated, including *dnaB, infA, groE, psiB, recJ, recO, recF, recQ, ruv,* and a UV-hyperresistant mutant, *uvh* (2, 3, 14, 64, 156, 157; for *psiB*: personal communication from R Devoret). Also, certain *recA* mutants of *E. coli* have been found to be more resistant to TLD than their respective parent $recA^+$ strain (1).

The observation that *recQ1* mutants are resistant to TLD but sensitive to UV and deficient in conjugational recombination on a *recBCD sbcB* background (158) is particularly revealing. The *recQ* gene encodes a DNA helicase and is involved in inhibiting illegitimate recombination (94a). Mutations in *recJ* and *recO* also enhance resistance (160). In more recent studies, Nakayama's group have proposed a mechanism for involvement of *recF* in TLD. Their studies of the hyperresistance phenotype of *thy recF* (*thyA⁻ recF143, thyA⁻ recF144,* and *thyA⁻ recF145*) to TLD (106, 157) imply that thymine starvation induces lesions in DNA that are substrates for RecF repair systems and that the process leads to the formation of irreparable, lethal structures. Subsequently, Nakayama et al (159) identified these probably lethal structures as nonmigrating DNA (nmDNA) comprising DNA with regions of single strands internally or as tails and branched structures.

TLD of temperature-sensitive *dnaA* and *dnaCD* mutants is prevented at nonpermissive temperatures (37). Since these are initiation-defective mutants, it is not surprising that they are resistant to TLD; lack of initiation of replication prevents the accumulation of strand breaks and protects against TLD. With regard to studies of *dnaB* mutants, the results of Sicard & Bouvier (194) show that prevention of DNA unwinding at the replication fork protects against TLD. However, the presence of a *dnaG dnaB* temperature-sensitive double mutation leads to loss of protection. It has been proposed that this effect is due to the occurrence of stable replication fork–associated single-stranded regions in the DNA of the double mutant at the permissive temperature.

Ahmad et al (3) isolated an *E. coli* mutant called *uvh* (selected originally as hyperresistant to UVC) hyperresistant to TLD. This mutant synthesizes DNA polymerase I, endonuclease I, and exonuclease III at high concentrations. Overexpression of certain DNA repair enzyme and resultant hyperrepair activity is the suggested cause for the resistance to various damaging agents, including TLD (7). This is likely to be due to increased repair efficiency, which extends the lag phase of TLD. The mutant is similar to *infA*, which is also resistant to X-rays, UV irradiation, and thymine starvation (14). The *infA* mutants are also defective for phage induction, and induction of phage or colicinogenic factors is known to increase sensitivity to TLD in *E. coli* (88) and *B. subtilis* (72).

There are DNA metabolism mutants that have little or no effect on TLD. The value of examining these is that their activities may be excluded in developing a mechanistic scheme for TLD.

Despite important roles of uracil DNA glycosylase in removing uracil from DNA, *ung* mutants of *E. coli* show normal sensitivity to TLD, suggesting that the incorporation of uracil into DNA is not a reason for TLD (111, 128, 158). However, an *ung dut* double mutant incorporates high levels of uracil into DNA, and this is lethal (70). Thus, a likely explanation for the insensitivity of *ung* mutants to TLD is that in this strain, due to high *dut* gene product activity (catabolizing dUTP to dUMP), the incorporation of uracil into DNA is small. Nakayama (158) showed that *xth* (exonuclease III) mutants of *E. coli* are no more sensitive to TLD than are wild type, supporting the above explanation. In contrast to *E. coli*, *ung* mutants of *B. subtilis* show a modest degree of resistance to TLD (143), suggesting that during TLD there is perhaps less uracil incorporated into DNA in this bacterium, which may be due to stronger expression of *dut*.

recN mutants of *E. coli* are as sensitive to TLD as their $recN^+$ parent strains, indicating no significant role of this component in the repair process in TLD (154, 160). The sensitivities of a variety of DNA repair/replication mutants clearly point to involvement of transcription, replication, and repair in TLD. These processes are intimately involved with one another within the cell, involving helix unwinding and incorporation of nucleotides into DNA strands, ligation, etc. In the absence of thymine, the levels of nucleotide pools are disrupted, repair and replication are interrupted, and TLD ensues. Mutations in genes encoding enzymes involved in replication and repair further exacerbate this effect. In addition, a variety of reports are available that suggest overlap between exposures to UV, X-rays, ^{32}P, thymidine analogues, and subsequent death by TLD (16, 31, 84, 87, 169). Thus, for example, Balgavy et al (18) showed that *E. coli* B/r, if starved of thymine, become more sensitive to UV for a short period, the surviving fraction subsequently showing resistance. Likewise, UV-irradiated cells are more sensitive to TLD. The results suggest some overlap in the mechanisms used to repair damage induced by UV and thymine starvation.

The SOS response (also known as the DNA damage tolerance mechanism, trans-lesion DNA synthesis, inducible repair, and error-prone repair) is the process in which more than twenty genes in a network are induced in response to damage to DNA (96, 220). Some of the genes involved in this and other repair processes and which have been implicated in TLD are shown in Table 2. Indirect evidence for an involvement of the SOS response in TLD comes from the facts that they share common phenomena: (*a*) A number of genes belonging to SOS regulons influence the response to thymine starvation–mutation in these

Table 2   DNA rapair genes: functions and effects of mutation on TLD

Gene	Function or effect
Mutants resistant to TLD	
dnaA	Initiation of DNA replication and part of SOS response
dnaA/dnaB ts	Same, but not LexA regulated
dnaB	DNA chain elongation
groE	Stress protein
(mop)	
infA	Protein chain initiation factor 1
psiB	Inhibitor of SOS induction
recA	DNA recombination and repair
recF	DNA recombination and repair
recJ	DNA recombination and repair
recO	Conjugational recombinational and repair (DNA helicase)
recQ	Conjugational recombinational and repair
ung in *B. subtilis*	Uracil DNA glycosylase
Mutants sensitive to TLD	
recBC	Exonuclease V
lig	DNA ligase
polA	DNA polymerase I
rep	Replicase
uvrA	UV endonuclease
uvrD*	DNA-dependent ATPase and helicase II
dnaC/D	Initiation and chain elongation
dnaG	Primase
DNA repair mutants showing no effect	
lexA	Repressor of SOS
recN	Recombination and repair
ung	Uracil DNA glycosylase

genes may confer hyperresistance or hypersensitivity; (*b*) both induce cell filamentation; (*c*) thymine starvation induces DNA damage, which in turn induces the SOS system; (*d*) both enhance mutagenicity; (*e*) both cause prophage and colicin induction (116, 152, 195).

Prophage and colicin induction are noted upon thymine starvation in lysogenic and Col-carrying strains of *E. coli* (71, 150), in *Staphylococcus* (188), and in *B. subtilis*. In a PBSX lysogenic *B. subtilis*, extensive DNA degradation has been shown to occur upon thymine starvation, leading to faster cell death. This DNA degradation has been associated with a phage-encoded nucleolytic activity (72).

Thymine Starvation Effects on Conjugation and Genetic Recombination

Using *thy⁻* HfrC and HfrB1 strains, Hart (98) demonstrated a difference between them in terms of chromosome transfer during conjugation to a *thy⁺* F⁻ strain. Although the transfer gradient of HfrC was normal for a period following removal of thymine, that of HfrB1 was markedly reduced. It was implied that the difference was due to variations in the modification of thymine starvation–induced DNA lesions between the strains, but that DNA damage accumulates in both in response to thymine starvation, interfering with conjugative transfer.

By using pulsed-field gel electrophoresis combined with cell lysis in agarose gels, peculiar DNA electrophoretic profiles have been discovered among cells starved of thymine (159). Nonmigrating DNA appears enriched in such structures as single-stranded tails or gaps and branching with single-stranded arms. A functional *recA* gene and certain *recF* family genes are required for such structures to occur, suggesting an involvement of recombinational repair in their formation. Indeed, homologous recombination between sister duplexes has been proposed to occur upon thymine starvation.

Thymine starvation increases the frequency of chromosomal recombination in λ (83) and in merodiploid *E. coli* (83). An explanation for the enhanced genetic recombination is that damage induced by thymine starvation acts as a nucleation site for recombination.

Thymine Starvation and Mutagenesis

Since the discovery in 1956 by Coughlin & Adelberg of a 100-fold increased mutation frequency upon thymine starvation (55), a variety of evidence has been uncovered linking thymine starvation with an increased mutagenesis in *E. coli*, *B. subtilis*, and *S. typhimurium*.

Deutch & Pauling (63) reported reversion of *arg⁻* to *arg⁺* in *E. coli* TAU⁻ immediately after the start of thymine starvation, reaching a plateau during the lag phase, with an increase of 10- to 15-fold (63). This has been confirmed for *E. coli* B/r by Balgavy & Turek (17). Deutch & Pauling (63) found that plating on minimal medium lacking both thymine and arginine increased the frequency of reversion. This is to be expected, as the absence of arginine will provide a selective pressure for *arg⁺* revertants, and thymine deprivation increases the mutation rate. On the other hand, when starved of thymine and uracil, this strain showed reduction in both thymineless mutagenesis and TLD, suggesting that uracil incorporation is central to both of these effects.

In a study of *E. coli*, *exr⁺* and *exr⁻* (lexA) strains (46) showed an association between error-prone repair and thymineless mutagenesis, the rate of mutation in the former strain being higher than in the latter on thymine starvation. In support, no thymineless mutagenesis in *E. coli recA*⁻ mutants was observed by Witkin (220). Pons & Mennigmann (180) have also implicated error-prone repair in mutagenesis of *E. coli* B/r strains.

Pauling (171) demonstrated that incubating *E. coli* B3 with 2-aminopurine (AP) before thymine starvation increased the frequency of mutation considerably. On the other hand, loading with 5-bromouracil (BU) had no effect. A strong interaction between thymine starvation and AP substitution in DNA is therefore possible, AP pairing errors occurring during residual DNA synthesis. Thus, AT/GC transition occurs according to the following: A:TÆAP:TÆAP: CÆG:C. In support, Holmes & Eisenstark (104) have presented evidence in *S. typhimurium* that the increase in number of mutants arising in medium devoid of thymine is not due to selective pressure, and that the mutants are true revertible, nonrevertible, and mutagen stable. In addition, in AT/GC transitions in phage T4 (205) and GC/AT transitions, all possible transversions, deletions, and frameshift mutations are induced on thymine starvation in *E. coli* (45, 46, 128, 171) and in *B. subtilis* (44, 45).

Mutagenesis in *ung*⁻ strains of *E. coli* showed a different profile on thymine starvation compared to that in wild-type strains. Although the overall frequency of mutation was unaffected, the frequencies of individual types varied, e.g. nonsense mutations were reduced by 30%, indicating a role of uracil DNA-glycosylase in thymineless mutagenesis. Hence it is likely that the overall effects of thymine starvation on mutagenesis are the result of a summation of several effects, including those of error-prone repair and uracil DNA-glycosylase activities.

The rate of thymineless mutagenesis is also increased if *B. subtilis thy*⁻ mutants are exposed to low concentrations of thymine (0.2–0.3 μg per ml) rather than being totally deprived of it (44, 45). At these concentrations, mutations were induced in all regions of the chromosome in 50% of the cell population within 50–60 h of plating on solid medium, and the frequency was 1000 to 10,000 times the background mutation frequency. This finding was reflected in *E. coli* as determined by Zaritsky & Pritchard (226): They showed that the velocity of DNA replication influences thymineless mutagenesis. At subnormal thymine concentrations, when the rate of DNA replication is reduced, thymineless mutagenesis is increased.

As indicated above, thymine deprivation is also mutagenic for phage T4 (28, 205). In fact, Bernstein et al (28) have shown that both thymine deficiency and its presence in high concentrations are mutagenic, and that these effects are independent of host thymidylate synthetase.

Biological Effects of Thymine Starvation

It is not surprising to note that most kinds of bacterial plasmids, including F, F′, F′lac, colI, colE, and R plasmids of N compatibility groups, are eliminated from host *E. coli*, *K. aeruginosa*, and *S. typhimurium* cells upon thymine starvation (32, 76, 177). Also, treatment of cells that interferes with DNA synthesis, such as by addition of hydroxyurea or cytosine arabinoside, and with agents that affect thymidylate synthetase, such as FUdR and trimethoprim, eliminates plasmids (176, 180). Since addition of chloramphenicol to growth media abolishes this phenomenon, it is suggested that protein synthesis is essential (179). R factor 1818, which is not curable by acridines, ethidium bromide, or showdomycin, is eliminated during thymineless incubation, suggesting that two different elimination processes are possible (177).

Possible reasons for the elimination of certain plasmids include introduction of single-stranded DNA breaks (76) due to the synthesis of certain R-mediated nucleases. For those plasmids that are refractory to curing, it has been suggested that a plasmid-specific ligase is involved that reduces the number of strand breaks (177).

Chromosomal DNA isolated from thymine-starved *B. subtilis* and *D. pneumoniae* shows reduced transforming activity (of 85% or more for the latter) compared to that from wild-type organisms (36, 47, 153, 193). If, however, the DNA is purified by phenol treatment, its transforming activity is regained; a likely explanation is that purification eliminates damaged and inactive DNA fragments (193). Lack of transforming ability of DNA appears not to be caused by decreases in DNA uptake (198). It seems likely that reduced transforming capacity is a result of DNA strand breaks. Ephrati-Elizur et al (72) showed that a *B. subtilis* strain cured of prophage did not lose the transforming activity of DNA whether the cells were starved of thymine or not. It is therefore likely that some loss of transforming activity (at least in lysogenic bacteria) is due to the degradation of DNA by phage-coded nuclease activities.

A DNase (endo- and exonuclease)-deficient mutant of *D. pneumoniae* was tested to verify that loss in transforming ability is not due to these nuclease activities within the nonmutant recipient cells. No difference between the mutants and the wild-type strains was noted (36, 47), and single- (rather than double-) strand damage to DNA was implicated with the lack of transforming activity. DNA strands that break by other means (such as X-rays) have been demonstrated to reduce transforming activity (214). However, use of a DNase-deficient mutant does not preclude the possibility that other, as yet unidentified, nucleases are induced in this organism and participate in strand nicking.

Thymidylate Biosynthesis and Selection of dTMP Mutants in S. cerevisiae

As in most bacterial species, the only biosynthetic pathway for dTMP in *Saccharomyces cerevisiae* is via methylation of dUMP to dTMP (34). A major source of dUMP in the cell is dUTP, which is converted to dUMP by deoxyuridine triphosphatase (Figure 1, reaction 8). However, unlike most bacterial thymine auxotrophs that can grow on media supplemented with thymine or thymidine, mutants of *S. cerevisiae* deficient in thymidylate synthetase cannot use thymidylate because this organism lacks thymidine kinase (Figure 1, reaction 5) (91).

For dTMP-requiring mutants to be viable, therefore, it is necessary that permease mutations (*tup*) be introduced to allow free access of exogenous dTMP to the cells (136). dTMP auxotrophs generated from *tup* mutants were found to fall into two categories, those defective in TS activity (*tmp1*), and those defective in folate metabolism (*fol*). The latter strains are multiple auxotrophs that rely on addition of compounds such as methionine, adenine, etc (86, 136, 138). An alternative approach was presented in 1976, when Brendel published a method for the isolation of permease mutants of a thermosensitive *tmp1* strain (42).

In the absence of dTMP, *S. cerevisiae* mutants display behavior similar to TLD in *thy⁻* bacteria: Cellular growth takes place for one generation, and then viability decreases exponentially for about 20 h (34, 42, 43). In a methionine- and dTMP-requiring double mutant, starvation of dTMP for 24 h in the presence of methionine resulted in 0.5% survival, but when both compounds were omitted, a significant fraction was found to escape cell death (20). This may be due to lack of cell growth in the absence of methionine, preventing TLD, which appears to rely on active DNA metabolism. In the presence of methionine, DNA replication, repair, and transcription continue to an extent sufficient enough to allow the manifestation of TLD. Barclay et al (19) demonstrated substantial DNA strand breaks in cells undergoing thymidylate deprivation and suggested that these arise as a result of incorporation of uracil in DNA and its subsequent excision followed by mitotic recombination.

Other manifestations of thymine starvation in dTMP mutants of *S. cerevisiae* are changes in cell size and development of buds, cells appearing swollen and becoming highly structured (21). Upon reaching the size of the mother cell, buds do not detach. This is analogous to findings in *thy⁻ lon⁻* bacteria in which thymine starvation results in elongation of cells but an inability to form septa and dissociate into daughter cells. RNA and protein synthesis are also affected, the rate of the former dropping significantly after 90 min, and of the latter after 2 h, during which period no DNA synthesis occurs.

Certain dTMP mutants of *S. cerevisiae* are converted to petite forms upon thymine starvation, and this appears to correlate with accumulation of mutation in and loss of mitochondrial DNA, which leads to the production of respiratory-defective mutants (*mit⁻*). Clear evidence of nuclear single- and double-strand DNA breaks has also been obtained (21, 86, 129).

dTMP starvation is not mutagenic in terms of accumulation of point mutations in the yeast nuclear genome (20), although it does result in increased recombination (131). Although mutants auxotrophic for methionine, lysine, leucine, histidine, cytosine, and uracil have been isolated from cells surviving one or two cycles of dTMP starvation, this is thought to be the result of spontaneous rather than dTMP starvation-induced mutation. However, induction of mutation is prominent in mitochondrial DNA. After 20 h of dTMP starvation, a 16-fold increase in erythromycin-resistant and an 8-fold increase in chloramphenicol-resistant mutants is observed, due to mitochondrial DNA mutations (21, 123).

As for *E. coli*, cell culture density affects TLD; at low density (approximately 10 cells/ml), cells starved of dTMP have a 10% survival rate after 24 h, whereas at high cell density (10^3 cells/ml), there is 95% survival, showing a substantial protection against killing (21). As in bacteria, this effect is likely to be due to decreased growth rate of more dense cultures and/or to dTMP becoming available from dead cells.

Thymidylate starvation is also recombinagenic at the nuclear level (19, 127, 129–131). Studies on thymidylate stress in dTMP auxotrophs or cells treated with methotrexate or FdUMP showed a high level of mitotic recombination in the forms of both gene conversion and crossing over (19, 127). In studies of *rad* mutants, considered to be involved in error-prone DNA repair, there is clear evidence of links between thymidylate stress, DNA strand breaks, and enhanced recombination. (129, 134). Subsequent studies have shown direct evidence for homologous recombination between chromosomes specifying different mating types (130), and for gene conversion in a mutant containing two copies of a particular *leu2* allele on the same chromosome separated by a short *E. coli* plasmid sequence (131).

A possible explanation for the recombinagenic effects of dTMP starvation, although no evidence has been presented, is that uracil is incorporated into DNA in place of thymine. It has been hypothesized that such uracil may be excised by uracil DNA glycosylase, generating single-strand DNA breaks or gaps that are inefficiently repaired due to thymine deprivation and then act as substrates for genetic recombination (127, 136). This hypothesis allows for the differential effects of dTMP starvation on mitochondrial and nuclear genomes, if it can be assumed that the activities of uracil DNA glycosylase vary between mitochondria and nuclei. If enzyme activity is higher in the nucleus, this would prevent accumulation of point mutations but would lead to enhanced levels of strand breaks and hence recombination (136). Although an interesting hypothesis, as mentioned above, no supporting evidence has so far been reported.

Kohalmi & Kunz (123) have recently shown that dTMP starvation severely diminishes the dTTP pool and elevates those of dATP, dGTP, and dCTP. Thymine starvation also increases the frequency of mutation in a tRNA gene, such mutants largely comprising point changes.

Relationship to Mammalian TLD

Although the past five years have seen an explosion of research dealing with the mechanism of mammalian cell death, for decades clinicians have used a major cancer therapeutic method to cheat cells of thymine. Thus, inducing thymineless death, particularly by loading cells with the analog 5-fluorouracil and with the folic acid antagonist, Methotrexate, has been a rational form of treatment (113, 217, 219). It is now known that specific genes program normal cell death, a phenomenon known as apoptosis (133, 217, 219). Thus, it is not surprising that the knowledge gained from TLD in bacteria furnished a basis for study of the biology of the cancerous cell. Note that a hallmark of DNA structure in apoptosis is similar to that of DNA structure in bacterial TLD, i.e. single-stranded DNA. In apoptosis, the DNA fragments form a ladder in agarose electrophoretic gels. Single-stranded structures can also be detected by immunochemical labeling of single-stranded DNA, as well as by end-labeling of DNA breaks in DNA strands in situ (12). Houghton et al (107) studied the signal gene products that induce thymineless stress, particularly in colon carcinoma cells. In particular, they demonstrated that certain gene products involved in Fas-FasL interaction link DNA damage induced by thymineless stress to the apoptotic machinery of colon carcinoma cells. Actually, although numerous drugs are being tested as pragmatic anticancer agents, many of these are effective because they simulate thymine starvation.

A particularly interesting example of a relationship between apoptosis and oxidative stress is that described by Steinman (207). He found that expression of the bcl-2 oncogene, which exerts both antiapoptotic and antioxidant action when transformed into superoxide dismutase-defective *E. coli* cells, results in increased transcription of *katG*, a catalase gene, and resistance to H_2O_2. In addition, mutation rate is increased. Bcl-2 may control entry into apoptosis. Both *katG* and its regulator, *oxyR*, are required for aerobic survival of cells containing Bcl-2. Steinman's data indicate that Bcl-2 influences levels of reactive oxygen intermediates that induce endogenous cellular antioxidants. The influence of reactive oxygen species has also been studied in apoptosis. This information, together with that presented by Houghton et al (107–109), suggests that a relationship between TLD and oxidative damage should be explored for common molecular mechanisms.

Hypothesis for TLD

The studies reported in this review indicate that an easily identifiable mechanism for TLD has yet to be fully elucidated. Part of the difficulty is that TLD shows different characteristics among bacterial species and even among members of the same species. This is apparent in that the kinetics, death slopes, and phases of TLD vary between the microbes tested. Nevertheless, examination of the most plausible mechanism may be a way to resolve uncertainties.

Factors influencing TLD can be divided into three main groups: (*a*) genetic backgrounds of strains employed, (*b*) pathways of thymidylate metabolism and DNA repair of the organisms studied, and (*c*) physiological conditions and growth media used.

Despite the observed differences in manifestation of TLD, common factors associated with it are DNA damage (of both single- and double-strand type) and associated induction of DNA repair system(s), impairment of transcriptional activity, and alterations in nucleotide pool sizes. Other manifestations of thymine starvation specific to certain bacterial strains include induction of prophage in lysogenic cells and cell filamentation.

Let us consider what may be happening immediately after actively growing cells are transferred to media devoid of thymidylate. Some cells may just have completed DNA replication and have entered the D, or division, phase. In the absence of thymine, it is unlikely that these cells will start the next round of replication, and hence they may be immune to TLD. Those cells that have started or are in the process of DNA synthesis will continue normal metabolism in the presence of residual intracellular thymidylate. This will lead to a fall in the dTTP pool. When the intracellular level of thymidylate falls below that required for DNA synthesis, due to continued replication, uridylate moieties may be incorporated into the DNA. Incorporated uracil becomes a substrate for uracil DNA glycosylase, producing AP sites that in turn become substrates for AP endonuclease. This will produce a nicked DNA strand, which may become a substrate for DNA polymerase activity and be repaired, leading to a viable cell. If, on the other hand, repair is incomplete due to depletion of dTTP, damage will accumulate in the DNA of cells in which synthesis, transcription, and repair are ongoing. Thus, DNA polymerase will either leave gaps in the replicating strand opposite A (adenine) residues or insert incorrect bases.

Gaps left in the newly synthesized strand induce cellular DNA repair systems, including the SOS response, and appear to be associated with recombinational events. This is evidenced by the fact that *recA* mutants fail to convert short nascent DNA strands into high-molecular-weight products (202). A number of genes have been implicated in the tolerance to DNA damage associated with discontinuities and gap-filling of daughter-strand DNA (96, 203, 204). Others, e.g. *recB*, *recC*, *recJ*, *recQ*, *recA*, *recF*, *recO*, *recR*, *ssb*, *ruvABC*, and *recG*, have been associated with recombinational repair; and because mutations in certain of these genes have been shown to change the kinetics of TLD, it is likely that this process plays an important role in TLD.

It is likely that gaps within the DNA become substrates for exonucleases and are widened, particularly in transcriptionally activated regions. Damage of this kind may induce *recBC* and *recF* recombinational repair processes; when attended by the former, repair is achieved more successfully (judged by the sensitivity of *recBC* mutants to TLD) than when attended by the latter, in which case double-strand DNA breaks and unusual lethal DNA structures containing regions with extensive single-strand tails and branched molecules occur (159). Double-strand DNA breaks are therefore likely to be the main cause of TLD. Although various recombinational repair proteins have been implicated (213), in the absence of knowledge of the precise roles of these genes (Table 2), it is difficult to propose an exact mechanism for TLD.

As discussed above, it appears that induction of colicins and prophage have additional effects on TLD in relevant strains.

Possible Relationship Between TLD and Death from Mutation in Other Genes

Almost all of the research dealing with TLD in bacteria was performed before the 1990s. Considerable research since then has focused on survival strategies and the death process, particularly identification of structural and regulatory genes needed for survival of bacteria. A question arises whether the newer information might contribute to our understanding of TLD. The answer can be only speculative, since many new experiments would need to be performed to identify a relationship between TLD and strategies for survival against other stress conditions. In particular, the new experiments would require new constructs of numerous strains with presence and absence of critical genes that might be involved in survival. The new constructs might involve genes associated with stationary-phase death, as well as repair and recombination genes.

Many recent studies of genes needed for survival of *E. coli* under starvation and other environmental stresses involve the role of *rpoS*, a regulatory sigma factor that codes for genes whose products protect cells from death while in stationary phase. Wild-type *E. coli* cells may remain alive for long periods. Indeed, cultures that have been sealed and stored for 40 years contain live cells that can be recovered (67–69) and that grow readily upon subculturing. This is not true for cells that contain certain mutations (e.g. *rpoS*, or certain genes regulated by RpoS) (140). Almost the entire population of *rpoS* cells in a culture die after about 10 days of dormancy. Indeed, the only survivors may be suppressor, secondary mutations of *rpoS*. It is obvious that certain genes regulated by *rpoS* are needed for protection, whereas wild-type cells remain dormant for long periods. Studies show that many (if not all) of the same "survival" genes protect cells against low fluences of near-UV and other oxidative events (66). This hypothesis is based on assays of transcription of *rpoS*, *katG*, *katE*, *dps*, and *xthA*, using *lacZ* fusions of each.

Although considerable information is accumulating concerning "survival genes," i.e. gene products that are needed for survival of dormant cells under extremely adverse conditions, it has yet to be seen whether the pathways of SPD and TLD have anything in common.

Conclusion and Perspectus

Despite studies carried out over 30 years, precise molecular steps in TLD remain unclear. Evidence exists that TLD leads to a variety of damage to DNA and that DNA repair processes are activated to rectify the damage. It has been shown that mutations in various DNA repair genes may either reduce or enhance the rate of TLD, and from these studies it can be hypothesized that recombinational repair systems are principally involved. A working model might be that stretches of single-strand DNA, subsequent to repair and recombination events, result in a folded configuration that is lethal. Perhaps more recent knowledge of DNA damage and repair, especially the larger number of DNA repair mutants of *E. coli* now available, could clarify the roles of the enzymes encoded. Among gene products recently described that could have a major role would be recQ helicase (94a).

It is now known that mutations in other genes, particularly in the stationary-phase regulator, rpoS, can be lethal. Perhaps a comparision of the death processes of SPD and TLD could be made to note whether or not the two processes have molecular steps in common.

Because the mode of action of a variety of chemotherapeutic agents (both antibiotics and anticancer agents) is based on TLD, an understanding of this phenomenon may lead to further applications in medicine.

The results of further studies of TLD in bacteria may be directly or indirectly applicable to work on mammalian cells, particularly as applied to the phenomenon of apoptosis. Of interest are the observations that Bloom's and Werner's syndromes involve defects in helicases with functional homologies to E. coli RecQ helicases (121a, 223a). Further information on DNA repair, mutation, and recombination mechanisms may allow development of drugs that have increased efficacy.

ACKNOWLEDGMENTS

The authors are most indebted to Drs. Devoret, Hanawalt, and Nakayama for their input in the planning of the review and in preparation of the manuscript. SIA is grateful for funding from the USPH grant 2R01 FD00658 and SIA and SHK for HEFCE, UK suppport. Research of A. E. was made possible by grant ES04889 from the National Institute of Environmental Health

Science, NIH, USA. We thank Shawna McGilvray and Marnie Tutt for secretarial assistance.

Visit the *Annual Reviews home page* at
http://www.AnnualReviews.org.

Literature Cited

1. Ahmad SI. 1980. Thymineless death in recombination deficient mutants of *Escherichia coli*. *Z. Naturforsch. Teil C* 35: 279–83

2. Ahmad SI. 1996. A mutant of *Escherichia coli* hyper-resistant to a number of DNA damaging agents: location of the mutational site. *J. Photochem. Photobiol. B*36:47–53

3. Ahmad SI, Atkinson A, Eisenstark A. 1980. Isolation and characterization of a mutant of *Escherichia coli* synthesizing DNA polymerase I and endonuclease I constitutively. *J. Gen. Microbiol.* 117: 419–22

4. Deleted in proof

5. Ahmad SI, Pritchard RH. 1969. A map of four gene specifying enzymes involved in the catabolism of nucleosides and deoxynucleosides in *Escherichia coli*. *Mol. Gen. Genet.* 104:351–59

6. Ahmad SI, Pritchard RH. 1971. A regulatory mutant affecting the synthesis of enzymes involved in the catabolism of nucleosides in *Escherichia coli*. *Mol. Gen. Genet.* 111:77–83

7. Ahmad SI, Van Sluis CA. 1987. Inducible DNA polymerase I synthesis in a UV hyper-resistant mutant of *Escherichia coli*. *Mutat. Res.* 190:77–81

8. Aizenberg OA, Samoylenko II, Fradkin GE. 1977. Role of *recB*⁺ gene product in thymineless death and characteristics of this phenomenon on *thy⁻ recB⁻* mutant of *Escherichia coli* K12. *Genetika* 13:1988–96

9. Anderson JA, Barbour SD. 1973. Effect of thymine starvation on deoxyribonucleic acid repair systems of *Escherichia coli* K12. *J. Bacteriol.* 113:114–21

10. Andrew MHE. 1973. Use of trimethoprim to obtain thymine requiring mutants of *Streptococcus faecalis*. *J. Gen. Microbiol.* 74:195–99

11. Atkinson C, Stacey K. 1968. Thymineless death induced by cytosine arabinoside. *Biochim. Biophys. Acta* 166:705–7

12. Ayusawa D, Shimizu K, Koyana H, Takeishi K, Seno T. 1983. Accumulation of DNA strand breaks during thymineless death in thymidylate synthase-negative mutants of mouse FM3A cells. *J. Biol. Chem.* 258:12448–54

13. Bachmann BJ. 1990. Linkage map of *Escherichia coli* K12, Edn. 8. *Microbiol. Rev.* 54:130–97

14. Bailone A, Blanco M, Devoret R. 1975. *E. coli K12 inf:* a mutant deficient in prophage λ induction and cell filamentation. *Mol. Gen. Genet.* 136:291–307

15. Baker NL, Hewitt RR. 1971. Influence of thymine starvation on the integrity of deoxyribonucleic acid in *Escherichia coli*. *J. Bacteriol.* 105:733–38

16. Balgavy P, Masek F, Turek R. 1975. Thymine dimer excision and DNA degradation after thymine starvation and ultraviolet irradiation in *Escherichia coli*. *Biochim. Biophys. Acta* 390:24–27

17. Balgavy P, Turek R. 1976. Influence of thymine starvation on uv-mutability of *Escherichia coli*. *Biologia* (Bratislava) 31:401–10

18. Balgavy P, Turek R, Vlcek D. 1977. Effect of a thymine starvation on reparation processes in cells *Escherichia coli* after UV irradiation. *Acta FRN Univ. Comen. Genetica* 8:23–37

19. Barclay BJ, Kunz BA, Little JG, Haynes RH. 1982. Genetic and biochemical consequences of thymidylate stress. *Can. J. Biochem.* 60:172–94

20. Barclay BJ, Little JG. 1977. Selection of yeast auxotrophs by thymidylate starvation. *J. Bacteriol.* 132:1036–37

21. Barclay BJ, Little JG. 1978. Genetic damage during thymidylate starvation in *Saccharomyces cerevisiae*. *Mol. Gen. Genet.* 160:33–40

22. Barner HD, Cohen SS. 1954. The induction of thymine synthesis by T2 infection of a thymine requiring mutant of *Escherichia coli*. *J. Bacteriol.* 68:80–88

23. Barner HD, Cohen SS. 1957. The isolation and properties of amino acid requiring mutants of a thymineless bacterium. *J. Bacteriol.* 74:350–55

24. Bazill CW. 1967. Lethal unbalanced growth in bacteria. *Nature* 226:346–49

25. Beacham IR, Beacham K, Zaritsky A, Pritchard RH. 1971. Intracellular thymidine triphosphate concentrations in wild type and in thymine requiring mutants of *Escherichia coli* 15 and K12. *J. Mol. Biol.* 60:75–86
26. Deleted in proof
27. Berg CM, O'Neill JM. 1973. Thymineless death in *polA*⁺ and *polA*⁻ strains of *Escherichia coli. J. Bacteriol.* 115:707–8
28. Bernstein C, Bernstein H, Mufti S, Strom B. 1972. Stimulation of mutation in phage T4 by lesions in gene 32 and by thymidine imbalance. *Mutat. Res.* 16:113–19
29. Bertino JB, Stacey KA. 1966. A suggested mechanism for the selective procedure for isolating thymine requiring mutants of *Escherichia coli. Biochem. J.* 101:320–21
30. Bhattacharjee SB, Das J. 1973. Strand breaks in deoxyribonucleic acid of thymine starved bacteria. *Int. J. Radiat. Biol.* 23:167–74
31. Bhattacharjee SB, Samanta HK. 1978. Radiation response of thymine starved bacteria. *Radiat. Res.* 74:144–51
32. Birks JH, Pinney RJ. 1975. Correlation between thymineless elimination and absence of *hsp*II (*Eco*RII) specificity in N-group R factors. *J. Bacteriol.* 121:1208–10
33. Bisaillon JG, de Repentigny J, Mathiau LG. 1974. Effects of wall inhibitors on thymineless *Streptococcus faecalis. Can. J. Microbiol.* 20:1185–87
34. Bisson L, Thorner J. 1977. Thymidine 5′-monophosphate requiring mutants of *Saccharomyces cerevisiae* are deficient in thymidylate synthetase. *J. Bacteriol.* 132: 44–50
35. Biswas C, Hardy J, Beck WS. 1965. Release of repressor control of ribonucleotide reductase by thymine starvation. *J. Biol. Chem.* 240:3631–40
36. Bousque JL, Sicard N. 1976. Size and transforming activity of deoxyribonucleic acid in *Diplococcus pneumoniae* during thymidine starvation. *J. Bacteriol.* 128:540–48
37. Bouvier F, Sicard N. 1975. Interference of *dnats* mutations of *Escherichia coli* with thymineless death. *J. Bacteriol.* 124:1198–204
38. Boyce RP, Setlow RB. 1962. A simple method of increasing the incorporation of thymidine into the deoxyribonucleic acid of *Escherichia coli. Biochim. Biophys. Acta* 61:618–20
39. Boyle JV, Jones NE. 1970. Effects of ribonucleosides on thymidine incorporation: selective reversal of the inhibition of deoxyribonucleic acid synthesis in thymineless auxotrophs of *Escherichia coli. J. Bacteriol.* 104:264–71
40. Breitman TR, Bradford RM. 1964. The induction of thymidine phosphorylase and excretion of deoxyribose during thymine starvation. *Biochem. Biophys. Res. Commun.* 17:786–91
41. Breitman TR, Maury PB, Toal JN. 1972. Loss of deoxyribonucleic acid-thymine during thymine starvation of *Escherichia coli. J. Bacteriol.* 112:646–48
42. Brendel M. 1976. A simple method for the isolation and characterization of thymidylate uptaking mutants in *Saccharomyces cerevisiae. Mol. Gen. Genet.* 147:209–15
43. Brendel M, Langjahr UG. 1974. Thymineless death in a strain of *Saccharomyces cerevisiae* auxotrophic for deoxythymidine–5′-monophosphate. *Mol. Gen. Genet.* 131:351–58
44. Bresler S, Mosevitsky M, Vyacheslavov L. 1970. Complete mutagenesis in a bacterial population induced by thymine starvation on solid media. *Nature* 225:764–66
45. Bresler S, Mosevitsky M, Vyacheslavov L. 1973. Mutations as possible replication errors in bacteria growing under conditions of thymine deficiency. *Mutat. Res.* 19:281–93
46. Bridges BA, Law J, Munson RJ. 1968. Mutagenesis in *Escherichia coli*. II. Evidence for a common pathway for mutagensis by ultraviolet light, ionizing radiation and thymine deprivation. *Mol. Gen. Genet.* 103:266–73
47. Brunel F, Sicard AM, Sicard N. 1971. Transforming ability of bacterial deoxyribonucleic acid in relation to the marker efficiencies in *Diplococcus pneumoniae* during thymidine starvation. *J. Bacteriol.* 106:904–7
48. Budman DR, Pardee AB. 1967. Thymidine and thymine incorporation into deoxyribonucleic acid: inhibition and repression by uridine of thymidine phosphorylase of *Escherichia coli. J. Bacteriol.* 94:1546–50
49. Buick RN, Harris WJ. 1975. Aberrant DNA methylation under conditions of thymine deprivation in *B. subtilis. J. Gen. Microbiol.* 90:347–54
50. Buick RN, Harris WJ. 1975. Thymineless death in *Bacillus subtilis. J. Gen. Microbiol.* 88:115–22
51. Carlson CA, Stewart JG, Ingraham JL.

1985. Thymidine salvage in *Pseudomonas stutzeri* and *Pseudomonas aeruginosa* provided by heterologous expression of *Escherichia coli* thymidine kinase gene. *J. Bacteriol.* 163:291–93

52. Clowes RC, Moody EEM, Pritchard RH. 1965. The elimination of extrachromosomal elements in thymineless strains of *E. coli* K12. *Genet. Res.* 6:147–52

53. Cohen SS, Barner HD. 1954. Studies on unbalanced growth in *Escherichia coli. Proc. Natl. Acad. Sci. USA* 40:885–93

54. Cohen SS, Barner HD. 1956. Studies on the induction of thymine deficiency and on the effects of thymine and thymidine analogues in *Escherichia coli. J. Bacteriol.* 71:588–97

55. Coughlin CA, Adelberg EA. 1956. Bacterial mutation induced by thymine starvation. *Nature* 178:531–32

56. Cummings DJ. 1969. Thymineless death induced by cytosine arabinoside. *Biochim. Biophys. Acta* 179:237–38

57. Cummings DJ, Kusy AR. 1969. Thymineless death in *Escherichia coli*: inactivation and recovery. *J. Bacteriol.* 99: 558–66

58. Cummings DJ, Kusy AR. 1970. Thymineless death in *Escherichia coli*: deoxyribonucleic acid replication and the immune state. *J. Bacteriol.* 102:106–17

59. Curtin NJ, Aherne AL. 1991. Mechanism of cell death following thymidylate synthetase inhibition: 2′-deoxyuridine triphosphate accumulation, DNA damage, and growth inhibition following exposure to CB3717 and dipyridamole. *Cancer Res.* 51:2346–52

60. Dankberg F, Cummings DJ. 1973. Protein synthesis and deoxyribonucleic acid membrane-attachment during thymineless death in *Escherichia coli. J. Bacteriol.* 113:711–17

61. De Barjac H. 1970. Thymine requiring mutants of the insect pathogen *Bacillus thuringiensis. J. Invertebr. Pathol.* 16:321–24

62. Denhardt DT, Iwaya M, Larison LL. 1972. The *rep* mutation. II. Its effects on *Escherichia coli* and on the replication of bacteriophage ϕX174. *Virology* 49:486–96

63. Deutch CE, Pauling C. 1974. Thymineless mutagenesis in *Escherichia coli. J. Bacteriol.* 119:861–67

64. Devoret R, Blanco M. 1970. Mutants of *Escherichia coli* K12 (λ)$^+$ non-inducible by thymine deprivation. I. Method of isolation and classes of mutants obtained. *Mol. Gen. Genet.* 107:272–80

65. Donachie WD. 1969. Control of cell division in *Escherichia coli*: experiments with thymine starvation. *J. Bacteriol.* 100:260–68

66. Eisenstark A, Calcutt MJ, Becker-Hapak M, Ivanova A. 1996. Role of *Escherichia coli rpoS* and associated genes in defense against oxidative damage. *Free Radic. Biol. Med.* 21:975–93

67. Eisenstark A, Eisenstark R, Cunningham S. 1968. Genetic analysis of thymineless mutant in *Salmonella typhimurium. Genetics* 58:493–506

68. Eisenstark A, Miller C, Jones J, Leven S. 1992. *Escherichia coli* genes involved in cell survival during dormancy: role of oxidative stress. *Biochem. Biophys. Res. Commun.* 188:1054–59

69. Eisenstark A, Yallaly P, Ivanova A, Miller C. 1995. Genetic mechanisms involved in cellular recovery from oxidative stress. *Arch. Insect Biochem. Physiol.* 29:159–73

70. El Hajj HH, Wang L, Weiss B. 1992. Multiple mutant of *Escherichia coli* synthesising virtually thymineless DNA during limited growth. *J. Bacteriol.* 174: 4450–56

71. Endo H, Ayabe K, Amako K, Takeya K. 1965. Inducible phage of *E. coli* 15. *Virology* 25:469–71

72. Ephrati-Elizur E, Yosuv D, Shmueli E, Horowitz A. 1974. Thymineless death in *Bacillus subtilis*: correlation between cell lysis and deoxyribonucleic acid breakdown. *J. Bacteriol.* 119:36–43

73. Farmer JL, Rotham F. 1965. Transformable thymine-requiring mutant of *Bacillus subtilis. J. Bacteriol.* 89:262–63

74. Freifelder D. 1965. Technique for starvation of *Escherichia coli* of thymine. *J. Bacteriol.* 90:1153–54

75. Freifelder D. 1967. Lack of relation between deoxyribonucleic acid methylation and thymineless death in *Escherichia coli. J. Bacteriol.* 93:1732–33

76. Freifelder D. 1969. Single-strand breaks in bacterial DNA associated with thymine starvation. *J. Mol. Biol.* 45:1–7

77. Freifelder D, Katz G. 1971. Persistence of small fragments of newly synthesised DNA in bacteria following thymidine starvation. *J. Mol. Biol.* 57:351–54

78. Freifelder D, Levine E. 1972. Stimulation of nuclease activity by thymine starvation. *Biochem. Biophys. Res. Commun.* 46:1782–87

79. Freifelder D, Maaløe O. 1964. Energy requirement for thymineless death in cells of *Escherichia coli. J. Bacteriol.* 88:987–90

80. Friedkin M, Kornberg A. 1957. The

enzymatic conversion of deoxyuridylic acid to thymidylic acid and the participation of tetrahydrofolic acid. In *A Symposium on the Chemical Basis of Heredity*, ed. WD McElroy, B Glass, p. 609. Baltimore, MD: John Hopkins Univ. Press

81. Friedman LR, Ravin AW. 1972. Genetic and biochemical properties of thymidine-dependent mutants of *Pneumococcus*. *J. Bacteriol.* 109:459–61

82. Fu KP, Kimble EF, Coldrek RJ, Konopka EA. 1984. Antimicrobial susceptibility growth: kinetic and pathogenicity of thymidine requiring streptococcus species. *Chemotherapy* 30:373–78

83. Gallant J, Spottswood T. 1965. The recombinogenic effect of thymidylate starvation in *Escherichia coli* merodiploids. *Genetics* 52:107–18

84. Gallant J, Suskind SR. 1961. Relationship between thymineless death and ultraviolet inactivation in *Escherichia coli*. *J. Bacteriol.* 82:187–94

85. Gallant J, Suskind SR. 1962. Ribonucleic acid synthesis and thymineless death. *Biochim. Biophys. Acta* 55:627–38

86. Game JC. 1976. Yeast cell cycle mutant cdc21 is a temperature sensitive thymidylate auxotroph. *Mol. Gen. Genet.* 146:313–15

87. Ganguli K, Bhattacharjee SB. 1969. Interaction of x-ray ultraviolet light, and thymineless incubation in killing 5-bromodeoxyuridine bacteria. *Radiat. Res.* 39:126–34

88. Gherardi M, Sicard N. 1970. Induction of filament formation and thymineless death in *Escherichia coli* K12. *J. Bacteriol.* 102:293–95

89. Gilchrist CA, Denhardt DT. 1987. *Escherichia coli* rep gene: sequence of the gene, the encoded helicase, and its homology with *uvrD*. *Nucleic Acids Res.* 15:465–75

90. Gray MD, Shen JC, Kamath-Loeb AS, Blank A, Sopher BL. 1997. The Werner syndrome protein is a DNA helicase. *Nat. Genet.* 17:100–3

91. Grivell AR, Jackson JF. 1968. Thymidine kinase: evidence for its absence from *Neurospora crassa* and some other microorganisms and the relevance of this to the specific labelling of deoxyribonucleic acid. *J. Gen. Microbiol.* 54:307–17

92. Gupta S. 1997. Microbial survival: *mutations that confer a competitive advantage during starvation*. MA thesis. Cambridge, MA: Harvard Univ

93. Deleted in proof

94. Haltimer RC, Migneau PC, Robertson RG. 1980. Incident of thymidine dependent Enterococci detected on Muellen-Hinton agar with low thymidine content. *Antimicrob. Agents Chemother.* 18:365–68

94a. Hanada K, Ukita T, Kohno Y, Saito K, Kato J, Ikeda H. 1997. *RecQ* DNA helicase is a suppressor of illegitimate recombination in *Escherichia coli*. *Proc. Natl. Acad. Sci. USA* 94:3860–65

95. Hanawalt PC. 1963. Involvement of synthesis of RNA in thymineless death. *Nature* 198:286

96. Hanawalt PC, Cooper PK, Ganesan A. 1979. DNA repair in bacteria and mammalian cells. *Annu. Rev. Biochem.* 48:783–836

97. Harrison AP. 1965. Thymine incorporation and metabolism by various classes of thymineless bacteria. *J. Gen. Microbiol.* 41:321–33

98. Hart MGR. 1966. Thymine starvation and genetic damage in *Escherichia coli*. *J. Gen. Microbiol.* 45:489–96

99. Harwood FG, Frazier MW, Krajewski S, Reed JC, Houghton JA. 1996. Acute and delayed apoptosis induced by thymidine deprivation correlates with expression of p53 and p53-regulated genes in colon carcinoma cells. *Oncogene* 12:(10) 2057–67

100. Heidelberg C, Chaudhuri NK, Danneberg PB, Mooren D, Griesbach L, et al. 1957. Fluorinated pyrimidine, a new class of tumor inhibitory compounds. *Nature* 179:663

101. Hengge-Aronis R. 1993. Survival of hunger and stress: the role of rpoS in early stationary phase gene regulation of *E. coli*. *Cell* 72:165–68

102. Hengge-Aronis R. 1996. Back to log phase-sigma(s) as a global regulator in the osmotic control of gene-expression in *Escherichia coli*. *Mol. Microbiol.* 21:887–93

103. Hill WE, Fangman WL. 1973. Single-strand breaks in deoxyribonucleic acid and viability loss during deoxyribonucleic acid synthesis inhibition in *Escherichia coli*. *J. Bacteriol.* 116:1329–35

104. Holmes AJ, Eisenstark A. 1968. The mutagenic effect of thymine-starvation on *Salmonella typhimurium*. *Mutat. Res.* 5:15–21

105. Horii TA, Ayusawa D, Shimizu K, Koyama E, Seno T. 1984. Chromosome breakage induced by thymidylate stress in thymidylate synthase-negative mutants of mouse FM3A cells. *Cancer Res.* 44:703–9

106. Horii ZI, Clark AJ. 1973. Genetic analysis of RecF pathway to genetic recombination in *Escherichia coli* K12. Isolation and characterization of mutants. *J. Mol. Biol.* 80:327–44

107. Houghton JA, Harwood FG, Tillman DM. 1997. Thymineless death in colon carcinoma cells is mediated via Fas signaling. *Proc. Natl. Acad. Sci. USA* 94:8144–49

108. Houghton JA, Harwood FG, Frazier MW, Krajewski S, Reed C. 1996. Acute and delayed thymineless death correlates with expression of p53 and p53-regulated genes in colon carcinoma (cc) cells. *Proc. Am. Assoc. Cancer Res. Annu. Meet.* 37:14–15

109. Houghton JA, Harwood FG, Houghton PJ. 1994. Cell cycle control processes determine cytostasis or cytotoxicity in thymineless death of colon cancer cell. *Cancer Res.* 54:4967–73

110. Howard-Flanders P, Simson E, Theriot L. 1964. A locus that controls filament formation and sensitivity to radiation in *Escherichia coli* K12. *Genetics* 49:237–46

111. Howe WE, Mount WD. 1975. Production of cells without deoxyribonucleic acid during thymidine starvation of lexA⁻ cultures of Escherichia coli K12. *J. Bacteriol.* 124:1113–21

112. Howe WE, Mount WD. 1979. Distribution of cell lengths in cultures of a *lexA* mutant of *Escherichia coli* K12. *J. Bacteriol.* 138:273–74

113. Huang EY, Mohler AM, Rohlman CE. 1997. Protein expression in response to folate stress in Escherichia coli. *J. Bacteriol.* 179:17 5648–53

114. Huisman GW, Siegele DH, Zambrano MM, Kolter R. 1996. Morphological and physiological changes during stationary phase. In *Escherichia coli and Salmonella: Cellular and Molecular Biology*, ed. FC Neidhardt, R Curtiss, JL Ingraham, ECC Lin, KB Low, B Magasanik, WS Resnikoff, M Riley, M Schaechter, HE Umbarger, pp. 1672–82. Washington DC: Am. Soc. Microbiol.

115. Huisman O, D'Ari R. 1981. An inducible replication cell division coupling mechanism in *Escherichia coli*. *Nature* 290:797–99

116. Ishibashi M, Hirota Y. 1965. Hybridisation between *Escherichia coli* K12 and 15T⁻ and thymineless death of their derivatives. *J. Bacteriol.* 90:1496–97

117. Ivanovics G, Dobozy A, Pal L. 1969. Incorporation of thymine into prototrophic and thymine-dependent mutants of *Bacillus anthracis. J. Gen. Microbiol.* 59:337–49

118. Iyehara-Ogawa H, Otsuji N. 1984. Induction of prophage λ in ultraviolet light sensitive *ruv* mutant of *Escherichia coli*. *J. Gen. Genet.* 59:593–99

119. Kalman TI. 1972. Inhibition of thymidylate synthetase by showdomycin and its 5′-phosphate. *Biochem. Biophys. Res. Commun.* 49:1007–13

120. Deleted in proof

121. Kaneko T, Kuno S. 1979. Selective isolation of a thymine-requiring mutant of phage T4. *Virology* 93:275–76

121a. Karow JK, Chakraverty RK, Hickson ID. 1997. The Bloom's syndrome gene product is a 3′–5′ DNA helicase. *J. Biol. Chem.* 272:30,611–14

122. Kelln RA, Warren RAJ. 1971. Obligate thymidine auxotrophs of *Pseudomonas acidovorans*. *J. Bacteriol.* 113:510–11

123. Kohalmi SE, Kunz BA. 1993. Mutational specificity of thymine nucleotide depletion in yeast. *Mutat. Res.* 289:73–81

124. Kolter R, Siegele D, Tormo A. 1993. The stationary phase of the bacterial life cycle. *Annu. Rev. Microbiol.* 47:855–74

125. Korn D, Weissbach A. 1962. Thymineless induction in *Escherichia coli* (λ). *Biochim. Biophys. Acta* 61:775–90

126. Kunz BA. 1996. Inhibitor of thymine nucleotide biosynthesis: antimetabolites that provoke genetic change via primary non-DNA target. *Mutat. Res.* 355:485–87

127. Kunz BA, Barclay BJ, Game JC, Little JG, Haynes RH. 1980. Induction of mitotic recombination in yeast by starvation for thymine nucleotides. *Proc. Natl. Acad. Sci. USA* 37:6057–61

128. Kunz BA, Glickman BW. 1985. Mechanism of mutation by thymine starvation in *Escherichia coli*: clues from mutagenic specificity. *J. Bacteriol* 162:895–64

129. Kunz BA, Haynes RH. 1982. DNA repair and the genetic effects of thymidylate stress in yeast. *Mutat. Res.* 93:353–75

130. Kunz BA, Taylor GR, Haynes RH. 1985. Mating type switching in yeast is induced by thymine nucleotide depletion. *Mol. Gen. Genet.* 99:540–42

131. Kunz BA, Taylor GR, Haynes RH. 1986. Induction of intrachromosomal recombination in yeast by inhibition of thymidylate biosynthesis. *Genetics* 114:375–92

132. Kushner SR, Nagaishi H, Clark AJ. 1972. Indirect suppression of *recB*

and *recC* mutations by exonuclease I deficiency. *Proc. Natl. Acad. Sci. USA* 69:1366–70

133. Kyprianou N, Bains AK, Jacobs SC. 1994. Induction of apoptosis in androgen independent human prostate cancer cells undergoing thymineless death. *Prostate* 25:66–75

134. Lawrence CW, Christsen R. 1979. UV mutagenesis in radiation sensitive strains of yeast. *Genetics* 82:207–32

135. Lehman IR, Bessman MJ, Simms ES, Kornberg A. 1958. Enzymatic synthesis of deoxyribonucleic acid. I. Preparation of substrate and partial purification of an enzyme from *Escherichia coli*. *J. Biol. Chem.* 233:163–67

136. Little JG. 1985. Genetic and biochemical effects of thymidylate stress in yeast. *Basic Life Sci.* 31:211–31

137. Little JG, Hanawalt PC. 1973. Thymineless death and ultraviolet sensitivity in *Micrococcus radiodurans*. *J. Bacteriol.* 113:233–40

138. Little JG, Haynes RH. 1979. Isolation and characterisation of yeast mutants auxotrophic for 2′-deoxythymidine 5′-monophosphate. *Mol. Gen. Genet.* 168:141–51

139. Lloyd RG, Benson FE, Shurvinton CE. 1984. Effect of *ruv* mutations on recombination and DNA repair in *Escherichia coli*. *Mol. Gen. Genet.* 194:303–9

140. Loewen PC, Hengge-Aronis R. 1994. The role of the sigma-factor sigma(s) (katF) in bacterial global regulation. *Annu. Rev. Microbiol.* 48:53–80

141. Luzzati D. 1966. Effect of thymine starvation on messenger ribonucleic acid synthesis in *Escherichia coli*. *J. Bacteriol.* 92:1435–46

142. Maaløe O, Hanawalt PC. 1961. Thymine deficiency and the normal DNA replication cycle. I. *J. Mol. Biol.* 3:144–55

143. Makino F, Munakata N. 1978. Deoxyuridine residues in DNA of thymine *B. subtilis* strains with defective N-glycosidase activity for uracil containing DNA. *J. Bacteriol.* 134:24–29

144. Maskell R, Okubadejo OA, Hayne RH. 1977. Human infections with thymine requiring bacteria. *J. Med. Microbiol.* 11:33–45

145. Mathieu LG, De Repentigny J, Turgeon S, Sonea S. 1968. Thymineless death of *S. aureus* and formation of its alpha toxin. *Can. J. Microbiol.* 14:983–87

146. McFall E, Magasanik B. 1962a. The relation of enzyme synthesis to the ribonucleic acid level of normal and of thymine-starved *Escherichia coli*. *Biochim. Biophys. Acta* 55:909–19

147. McFall E, Magasanik B. 1962b. The effects of thymine deprivation on the synthesis of protein in *Escherichia coli*. *Biochim. Biophys. Acta* 55:920–28

148. Medoff G. 1972. Nucleic acid and protein synthesis during thymineless death in lysogenic and nonlysogenic thymine auxotrophs. *J. Bacteriol.* 109:462–64

149. Medoff G, Overholt S. 1970. Thymineless death in *Escherichia coli* 15T⁻ and recombinants of 15T⁻ and *Escherichia coli* K12. *J. Bacteriol.* 102:213–16

150. Medoff G, Swartz MN. 1969. Induction of a defective phage and DNA methylation in *Escherichia coli* 15T⁻. *J. Gen. Virol.* 4:15–21

151. Melechen NE, Skaar PD. 1962. The provocation of an early step of induction by thymine deprivation. *Virology* 16:21–29

152. Menningmann HD. 1964. Induction in *Escherichia coli* 15 of the colicinogenic factor by thymineless death. *Biochem. Biophys. Res. Commun.* 16:373

153. Menningmann HD, Szybalski W. 1962. Molecular mechanism of thymineless death. *Biochem. Biophys. Res. Commun.* 9:398–404

154. Nakayama H. Couch JL. 1973. Thymineless death in *Escherichia coli* in various assay systems: viability determined in liquid medium. *J. Bacteriol.* 114:228–32

155. Nakayama H, Hanawalt P. 1975. Sedimentation analysis of deoxyribonucleic acid from thymine-starved *Escherichia coli*. *J. Bacteriol.* 121:537–47

156. Nakayama H, Nakayama K, Nakayama R, Irino N, Nakayama Y, et al. 1984. Isolation and genetic characterisation of a thymineless death-resistant mutant of *Escherichia coli* K12: identification of a new mutation (*recQ1*) that blocks the RecF recombination pathway. *Mol. Gen. Genet.* 195:474–80

157. Nakayama H, Nakayama K, Nakayama R, Nakayama Y. 1982. Recombination deficient mutations and thymineless death in *Escherichia coli* K12: reciprocal effects of *recBC* and *recF* and indifference of *recA* mutations. *Can. J. Microbiol.* 28:425–30

158. Nakayama K. 1984. The genetic study of the mechanism for thymineless death in *Escherichia coli*. *Fukuoka Igaku Zasshi.* 75:89–106

159. Nakayama K, Kusano K, Irino N, Nakayama H. 1994. Thymine starvation induced structural changes in *Escherichia coli* DNA. *J. Mol. Biol.* 243:611–20

160. Nakayama K, Shiota S, Nakayama H. 1988. Thymineless death in *Escherichia coli* mutants deficient in RecF recombinant pathway. *Can. J. Microbiol.* 34: 905–7

161. Nelson DJ, Carter CE. 1969. Purification and characterisation of thymidine 5′-monophosphate kinase from *Escherichia coli* B. *J. Biol. Chem.* 244: 5254–62

162. Neuhard J. 1966. Studies on the acid soluble nucleotide pool in thymine requiring mutants of *Escherichia coli* during thymine starvation. *Biochim. Biophys. Acta* 129:104–15

163. Neuhard J, Price AR, Schack L, Thomassen E. 1978. Two thymidylate synthetases in *Bacillus subtilis*. *Proc. Natl. Acad. Sci. USA* 75:1194–98

164. Neuhard J, Thomassen E. 1971. Turnover of the deoxyribonucleoside triphosphates in *Escherichia coli* 15T during thymine starvation. *Eur. J. Biochem.* 20: 36–43

165. O'Donovan GA, Edlin G, Fuchs JA, Neuhard J, Thomassen E. 1971. Deoxycytidine triphosphate deaminase: characterisation of an *Escherichia coli* mutant deficient in the enzyme. *J. Bacteriol.* 105:666–72

166. Ohkawa T. 1975. Studies of intracellular thymidine nucleotides: thymineless death and the recovery after re-addition of thymine in *Escherichia coli* K12. *Eur. J. Biochem.* 60:57–66

167. Okabadeju OA, Maskell R. 1973. Thymine requiring mutant of *Proteus mirabilis* selected by trimoxazole. *J. Gen. Microbiol.* 77:533–35

168. Okada T, Homma J, Sonohara H. 1962. Improved method for obtaining thymineless mutants of *Escherichia coli* and *Salmonella typhimurium*. *J. Bacteriol.* 4:602–3

169. Okagaki H, Tsubota Y, Sibatani A. 1960. Unbalanced growth and bacterial death in thymine-deficient and ultraviolet irradiated *Escherichia coli*. *J. Bacteriol.* 80:762–71

170. Otsuji N, Iyehara H, Hideshima Y. 1974. Isolation and characterisation of an *Escherichia coli ruv* mutant which forms non-septate filaments after low doses of ultraviolet light irradiation. *J. Bacteriol.* 117:337–44

171. Pauling C. 1968. The specificity of thymineless mutagenesis. In *Struct. Chem. Mol. Biol. Conf.*, pp. 383–98. New York: Freeman

172. Pauling C, Beck LA, Wilczynski SP. 1976. Properties of a DNA ligase mutant of *Escherichia coli*: introduction of strand breaks in DNA. *J. Gen. Microbiol.* 94:297–304

173. Pauling C, Hanawalt PC. 1965. Nonconservative DNA replication in bacteria after thymine starvation. *Proc. Natl. Acad. Sci. USA* 54:1728–35

174. Deleted in proof

175. Pedersen-Lane J, Maley GF, Chu E, Maley F. 1997. High level expression of human thymidylate synthetase. *Protein Expr. Purif.* 10:256–62

176. Pinney RJ, Hernadi F, Smith JT. 1978. Curing of an R factor from *Escherichia coli* by hydroxyurea and cytosine arabinoside. *Chemotherapy* 24:240–48

177. Pinney RJ, Smith JT. 1971. R-factor elimination by thymine starvation. *Genet. Res.* 18:173–77

178. Pinney RJ, Smith JT. 1972. R factor elimination during thymine starvation: effects of inhibition of protein synthesis and readdition of thymine. *J. Bacteriol.* 111:361–67

179. Pinney RJ, Smith JT. 1973. Curing of an R factor from *Escherichia coli* by trimethoprim. *Antimicrob. Agents Chemother.* 3:670–76

180. Pons FW, Menningmann HD. 1979. Mutation induction by thymine deprivation in *Escherichia coli* B/R II. Mutation in the repressed and derepressed tryptophan operon. *Mutat. Res.* 60:271–78

181. Pritchard RH, Ahmad SI. 1971. Fluorouracil and the isolation of mutants lacking uridine phosphorylase in *Escherichia coli*: location of the gene. *Mol. Gen. Genet.* 111:84–88

182. Ramareddy G, Reiter H. 1970. Sequential loss of loci in thymine starved *Bacillus subtilis* 168 cells. *J. Mol. Biol.* 50:525–32

183. Reich D, Soska J. 1973. Thymineless death in *Lactobacillus acidophilus* strain R–26. II. Factors determining the rate of the reproductive inactivation. *Folia Microbiol.* 18:361–67

184. Reichenbach DL, Schaiberger GE, Sallman B. 1971. The effect of thymine starvation on chromosomal structure of *Escherichia coli* JG–151. *Biochem. Biophys. Res. Commun.* 42:23–30

185. Reiter H, Ramareddy G. 1970. Loss of DNA behind the growing point of thymine-starved *Bacillus subtilis* 168. *J. Mol. Biol.* 50:533–48

186. Rolfe R. 1967. On the mechanism of thymineless death in *Bacillus subtilis*. *Proc. Natl. Acad. Sci. USA* 57:114–21

187. Rozenkranz HS, Carr HS, Rose HM. 1965. Phenethyl alcohol. II. Effect on

thymine-requiring *Escherichia coli. J. Bacteriol.* 89:1370–73

188. Rudin L, Lindberg M. 1975. Thymineless bacteriophage induction in *Staphylococcus aureus.* I. High frequency transduction with lysates containing a bacteriophage related to bacteriophage φ11. *J. Virol.* 16:1357–66

189. Sakai T, Tochikura T, Ogata K. 1966. Metabolism of nucleasides in bacteria. *Agr. Biochim.* 30:245

190. Schwan H, Holldorf AW. 1975. Effective utilization of exogenous deoxythymidine 5′-monophosphate for DNA synthesis in enterobacteria. *FEBS Lett.* 57:179–82

191. Sedgwick SG, Bridges BA. 1971. Alkaline sucrose gradient sedimentation of chromosomal deoxyribonucleic acid from *Escherichia coli polA*⁺ and *polA*⁻ strains during thymine starvation. *J. Bacteriol.* 108:1422–23

192. Shuster RC. 1973. Influence of thymine starvation on the integrity of episomal and chromosomal deoxyribonucleic acids in *Escherichia coli* CR34 (lambda ind-). *J. Bacteriol.* 116:1067–70

193. Sicard N, Anagnastopoulos C. 1964. Activité transformante de l'acide désoxyribonucleique de *Bacillus subtilis* lors de la carence en thymine. *Compt. Rend. Acad. Sci. Paris* 259:4173–76

194. Sicard N, Bouvier F. 1977. Thymineless death in *Escherichia coli dnaB* mutants and in a *dnaB dnaG* double mutant. *J. Bacteriol.* 132:779–83

195. Sicard N, Devoret R. 1962. Effects de la carence en thymine sur des souches d'*Escherichia coli* lysogenes K12 T⁻ et colicinogenes 15T. *Compt. Rend. Acad. Sci. Paris* 255:1417–19

196. Sicard N, Devoret R. 1962. Effets de la carence en thymine sur des souches d'*Escherichia coli* lysogenes K12T⁻ et colicinogene 15T⁻. *Compt. Rend. Acad. Sci. Paris* 255:1417–19

197. Sicard N, Simonnet G, Astrachan L. 1967. Base composition of rapidly-labelled RNA in *E. coli* undergoing thymineless death. *Biochem. Biophys. Res. Commun.* 26:532–38

198. Sicard N, Venema G. 1969. Penetration of thymine-starved bacterial DNA during transformation of *B. subtilis* 168 T⁻. *Biochem. Biophys. Res. Commun.* 36:647–50

199. Siegal EC. 1973. Ultraviolet-sensitive mutator strain of *Escherichia coli* K12. *J. Bacteriol.* 1132:145–60

200. Siegele DA, Kolter R. 1992. Life after Log. *J. Bacteriol.* 174:345–48

201. Smith DW, Hanawalt PC. 1968. Macromolecular synthesis and thymineless death in *Mycoplasma laidlawii* B. *J. Bacteriol.* 96:2066–76

202. Smith KC, Meun DHC. 1970. Repair of radiation induced DNA in *Escherichia coli.* I. Effects of *rec* mutation in postreplication repair of damage done to UV irradiation. *J. Mol. Biol.* 51:459–72

203. Smith KC, Wang TC. 1987. RecA dependent DNA repair in UV irradiated *Escherichia coli. J. Photochem. Photobiol.* 1:1–11

204. Smith KC, Wang TC. 1989. RecA dependent DNA repair process. *Bioassay* 10:12–16

205. Smith MD, Green RR, Ripley LS, Drake JW. 1973. Thymineless mutagenesis in bacteriophage T4. *Genetics* 74:393–403

206. Stacey KA, Simson E. 1965. Improved method for the isolation of thymine-requiring mutants of *Escherichia coli. J. Bacteriol.* 90:554–55

207. Steinman HM. 1995. The Bcl–2 oncoprotein fuctions as a pro-oxidant. *J. Biol. Chem.* 270:3487–90

208. Takahashi S, Hours C, Chu A, Denhardt DT. 1979. The *rep* mutant VI. purification and properties of the *Escherichia coli rep* protein DNA helicase III. *Can. J. Biochem.* 57:855–66

209. Taketo A, Kuno S. 1972. Accumulation of a "DNA-masking protein" in *E. coli* under inhibitory conditions of DNA synthesis. *J. Biochem.* 71:497–505

210. Taucher-Schloz G, Monem MA, Hoffman-Berling H. 1983. DNA helicases in repair. In *Mechanisms of DNA Replication and Recombination,* ed. NR Cozarelli, pp. 65–67. New York: Liss

211. Theil EC, Zammenhof S. 1963. Studies on 6-methylaminopurine (6-methyladenine) in bacterial deoxyribonucleic acid. *J. Biol. Chem.* 238:3058–64

212. Then R, Angehrn P. 1973. Sulfonamide induced thymineless death in *Escherichia coli. J. Gen. Microbiol.* 76:255–63

213. Thoms B, Wackernagel W. 1987. Regulatory role of RecF in the SOS response of *Escherichia coli*: impaired induction of SOS genes by UV irradiation and nalidixic acid in recF mutant. *J. Bacteriol.* 169:1731–36

214. Thorsett GO, Hutchinson F. 1971. Effects on bacterial transformation of single-strand breaks in DNA produced by deoxyribonuclease I and y rays. *Biochim. Biophys. Acta* 238:67–74

215. Tiganos E, Harrington MB. 1993.

Kasugamycin inhibition of non-sense suppression by thymine requiring strains of *Escherichia coli. Can. J. Microbiol.* 39:448–50

216. Deleted in proof

217. Wadler S, Horowitz R, Mao X, Schwartz EL. 1996. Effect of interferon on 5-fluorouracil-induced perturbation in pools of deoxynucleotide triphosphates and DNA strand breaks. *Can. Chemother. Pharmacol.* 38:529–35

218. Waschsman JT, Kemp S, Hogg L. 1964. Thymineless death in *Bacillus negaterium. J. Bacteriol.* 87:1079–86

219. Wei X, McLeod HL, McMurrough J, Gonzalez FJ, Fernandez-Salguero P. 1996. Molecular basis of the human dihydropyrimidine dehydrogenase deficiency and fluorouracil toxicity. *J. Clin. Invest.* 98:610–15

220. Witkin E. 1976. Ultraviolet mutagenesis and inducible DNA repair in *Escherichia coli. Bacteriol. Rev.* 40:869–907

221. Womack JE. 1977. Inhibition of thymidine phosphorylase *in vivo* provides a rapid method for switching DNA labelling. *Mol. Gen. Genet.* 158:11–15

222. Yamao F, Nagai Y, Kaneda S, Yoshida S, Seno T. 1993. Conditional resistance to thymineless death predominantly selects DNA synthesis-deficient mutants of mammalian cells. *Mutat. Res.* 289:83–89

223. Yoshinaga K. 1973. Double strand scission of DNA involved in thymineless death of *Escherichia coli* 15TAU. *Biochim. Biophys. Acta* 294:204–13

223a. Yu CE, Oshima J, Wijsman EM, Nakura J, Miki T, et al. 1997. Mutations in the consensus helicase domains of the Werner syndrome gene. Werner syndrome collaborative group. *Am. J. Hum. Genet.* 60:330–41

224. Yudelevich A, Gold M. 1969. A specific DNA methylase induced by bacteriophage 15. *J. Mol. Biol.* 40:77–91

225. Zambrano M, Kolter R. 1996. Gasping for life in stationary phase. *Cell* 86:181–84

226. Zaritsky A, Pritchard RH. 1971. Replication time of the chromosome in thymineless mutants of *Escherichia coli. J. Mol. Biol.* 60:65–74

Annu. Rev. Microbiol. 1998. 52:627–86

HOW DO ANIMAL DNA VIRUSES GET TO THE NUCLEUS?

H. Kasamatsu and A. Nakanishi

Molecular, Cell and Developmental Biology and Molecular Biology Institute, University of California at Los Angeles, 405 Hilgard Avenue, Los Angeles, California 90095

KEY WORDS: nuclear locus for infection, viral protein NLS, viral protein nuclear import, nuclear targeting of infecting DNA, virion structure-function

ABSTRACT

Genome and pre-genome replication in all animal DNA viruses except poxviruses occurs in the cell nucleus (Table 1). In order to reproduce, an infecting virion enters the cell and traverses through the cytoplasm toward the nucleus. Using the cell's own nuclear import machinery, the viral genome then enters the nucleus through the nuclear pore complex. Targeting of the infecting virion or viral genome to the multiplication site is therefore an essential process in productive viral infection as well as in latent infection and transformation. Yet little is known about how infecting genomes of animal DNA viruses reach the nucleus in order to reproduce. Moreover, this nuclear locus for viral multiplication is remarkable in that the sizes and composition of the infectious particles vary enormously. In this article, we discuss virion structure, life cycle to reproduce infectious particles, viral protein's nuclear import signal, and viral genome nuclear targeting.

CONTENTS

627

INTRODUCTION

Viral infection of animal hosts is characterized by a series of distinct and obligatory stages: entry into susceptible host cells through cell-surface receptor(s), targeting to the reproductive site, uncoating, virus replication and maturation, spreading within the host, shedding of infectious virions outside the host, and finally transmission to the next host, repeating the infection cycle. Viruses enter an animal host, a multicellular organism with great structural complexity, through a single portal tissue, and then they disseminate to a few target tissues. Not all tissues support virus replication, primarily because each virus needs different cellular factors and machineries for virion reproduction; nor can all animal species be infected by a single virus species. Clearance of virions by host defense mechanisms also helps disseminate viruses within the host. These subjects have been thoroughly discussed recently (161, 260).

Table 1 Virus multiplication location

Virus	Genome replication	Nucleocapsid formation	Virion maturation
Poxviruses	CYT	CYT	CYT
Herpesviruses	NUC	NUC	NUC & CYT
Adenoviruses	NUC	NUC	NUC
Papovaviruses	NUC	NUC	NUC
Parvoviruses	NUC	NUC	NUC
Hepadnaviruses	NUC*	CYT	MEM

NUC, CYT, or MEM represents the nucleus, the cytoplasm, or the membrane, respectively.
Nuc*: the site of pre-genome RNA multiplication.

Viral infection in cell culture is much simpler than in the multicellular animal because multiplication and spread of the virus take place in the same culture population; yet the study of viral infection in cultured cells can lead to an understanding of some aspects of the basic viral reproduction processes. Whether in animals or in cultured cells, all animal DNA viruses, except poxviruses, multiply entirely—or at least in the early stages—within the cell nucleus; thus, nuclear targeting of the incoming viral genome is a prerequisite for initiating productive infection, latent infection, and viral transformation.

For adenoviruses, papovaviruses, and parvoviruses, the nucleus is the site of genome replication, nucleocapsid formation, and virion maturation. For herpesviruses, part of virion maturation occurs in the cytoplasm following nuclear reproduction. Hepadnaviruses, which convert of virion DNA genome from RNA pre-genome, also use the nucleus for pre-genome RNA multiplication. Among animal RNA viruses, orthomyxoviruses and deltaviruses use the nucleus for genome replication and nucleocapsid formation as well as for retroviruses' genome multiplication.

Although little is known about how the viral genome reaches its reproductive site, the specific size, constituency, and structure of each virus appear to determine its nuclear entry route. DNA may enter the nucleus in at least three ways: DNA may be (*a*) associated with nuclear localization signal (NLS)-bearing structural proteins, which facilitate entry; (*b*) packaged into virion along with a virally encoded protease that disassembles the virion by shedding structural proteins, exposing NLS that has been tightly or covalently bound to the DNA and so facilitates entry; or (*c*) packaged with cellular proteins that modify viral proteins to create functional NLS for the DNA's nuclear entry.

The cycle of DNA viruses that replicate in the nucleus, except hepadnaviruses, is divided into early and late phases separated by the onset of viral DNA replication. When the infecting virus interacts with a susceptible host cell, it adsorbs to the cell, penetrates into the cytoplasm, and immigrates toward the nucleus. After the viral genome enters the nucleus, transcription and translation of a set of early genes occur. These gene products target cellular gene products, causing physiological changes that favor the production of progeny virions. Certain viruses can become cryptic at this stage and establish latent infection in which only a subset of early gene products (for example, expression of E1 and E2 in papillomavirus-infected basal cells) or latency-associated transcripts (LAT) (for example, expression of LAT in herpes simplex virus-infected neurons) can be detected. When the papillomavirus-infected cells differentiate or physiological change of the virus-infected neuron occurs, productive infection is induced from the latent virus. During normal productive infection, viral early gene products also initiate viral DNA replication, which triggers expression of a new set of late viral genes for progeny virion assembly. One reproduction cycle initiated by a single virion produces many more progeny virion particles.

The life cycle of hepadnaviruses is particularly remarkable in that having gained entry to the nucleus, the incoming genome could maintain copies of transcription-competent and covalently closed circular viral DNAs through semiconservative DNA replication in the nucleus; instead, nuclear viral DNAs are copied by reverse transcription from viral RNA in the cytoplasm as part of progeny virion formation and are then transported back into the nucleus.

DNA VIRUSES THAT REPLICATE IN THE NUCLEUS

Size of the Particles

The size, composition, and gene organization of the DNA viruses vary enormously. The relative sizes of the capsids or core of the nuclear replicating DNA viruses are contrasted in Figure 1 with that of the nuclear pore complex (NPC). Diameters of HSV nucleocapsid in Figure 1, *panel 1* (34), Ad2 in Figure 1, *panel 2* (349), polyomavirus in Figure 1, *panel 3* (141, 296), HBV-core in Figure 1, *panel 4* (36, 69), and human parvovirus B19 in Figure 1 (see color insert at end of volume), *panel 5* (55), are 125 nm, 90 nm, 50 nm, 30 nm, and 26 nm, respectively. Their genome sizes vary according to the size of the shells, from about 150 Kbp in herpesviruses to 2.5 Kbp in hepatitis B virus, and their DNA takes different physical forms: double-stranded linear DNA (herpesviruses and adenoviruses), single-stranded linear DNA (parvoviruses), double-stranded circular DNA (papovaviruses), and partially double-stranded circular DNA (hepadnaviruses). Even within the *Herpesviridae* family, genome size varies from about 80 Kbp to 250 Kbp. The total number of virion proteins encoded by the individual genome generally corresponds to its complexity and is estimated to range from 34 to 39 for herpesviruses and from 4 to 6 for hepadnaviruses, parvoviruses, and papovaviruses. Both herpesviruses and hepadnaviruses are families of enveloped viruses whose envelopes contain virally encoded proteins as well as host glycoproteins with mature glycans characteristic of the maturation site, acquired during maturation. Virion capsids, the protein shells of the viruses, shown in Figure 1, are composed from one to six structural proteins.

Reliance on Nuclear Function

As intracellular parasites, these viruses have evolved to multiply in the nucleus because they depend on its cellular DNA replication machineries and/or the transcription machinery. The smaller DNA viruses, papovaviruses and parvoviruses, rely on the host cell DNA polymerases for genome propagation, whereas more complex and larger viruses, herpesviruses and adenoviruses, use

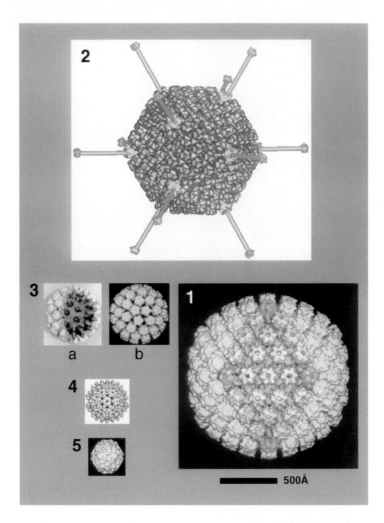

Figure 1 Photographs of virions and NPC. Three-dimensional reconstitution images of herpes simplex type 1 (HSV-1) capsid (*panel 1*), intact adenovirus type 2 (*panel 2*), intact polyomavirus (*panel 3b*), intact polyomavirus cut open to reveal the internal core (*panel 3a*), in vitro assembled human hepadnavirus core (*panel 4*), B19 empty capsid (*panel 5*), and NPC (*panel 6*). *Panel 6*: en face (*a*), oblique (*b*), edge view (*c*), edge view cut open to reveal internal organization (*d*), or consensus model of the functional NPC—the basic NPC framework with appending structural components (*e*). *Panel 6f*; an electron micrograph showing en face views of many NPCs. Scale bar below *panel 1*, 500Å or 50 nm, applies to 3-D images in *panels 1-5* and *6a-6d*. Color coding is as follows. *Panel 1*: *blue*, Vp5 hexon; *turquoise*, Vp5 penton; *orange*, triplex adjacent to hexon; *red*, triplex adjacent to penton. *Panel 2*: *blue*, hexon ; *green*, fiber;

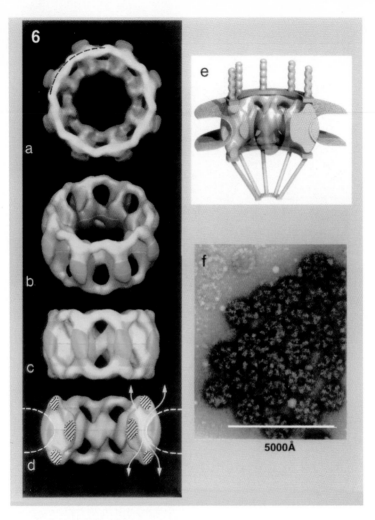

yellow, penton; *red*, RGD sequence. *Panel 6 a-d*: *yellow*, outer rings; *green*, inner annular subunit; *tan*, columnar subunit; *blue*, lumenal subunit; *dash-lined*, the path of nuclear membrane; *double-headed arrows*, passive nucleo-cytoplasmic exchange route; *shaded area*, cut end of rings and inner annulus. *Panel 6e*: *brown*; nuclear envelope; *pink*, basic NPC framework; *opaque gray*, central transporter; *gray*, cytoplasmic filaments with cytoplasmic ring; *orange*, nuclear basket with nuclear ring.

  (Photographs were reproduced with permission from Conway et al. 1996. and Steven and Spear 1997. for *panel 1* ; from Rayment et al. 1982. and Griffith et al. 1992. for *panel 3* ; from Conway et al. 1997 for *panel 4* ; from Chipman et al. 1996. *panel 5* ; from Hinshaw et al. 1992. for *panel 6a-d* and *f* ; from Pante and Aebi 1995. for *panel 6e*. *Panel 2*, courtesy of Stewart PL.)

virally encoded DNA polymerases. In fact, the herpesvirus genome encodes a large array of proteins involved in nucleic acid metabolism and DNA synthesis, but DNA unwinding during the viral DNA replication requires host topoisomerase II. The hepadnaviruses, however, encapsidate within the virion DNA polymerase/reverse transcriptase, which converts partially single-stranded circular DNA into fully duplexed circular DNA in the nucleus following the genome's nuclear targeting. Hepadnavirus polymerase also converts the genomic viral RNA into DNA genome in the cytoplasm as the RNA is being packaged to produce progeny virions, some of which are retargeted to the nucleus. All these DNA viruses use cellular RNA polymerase II—and in some instances RNA polymerase III—and cellular transcription factors for viral transcription. A variation seen in hepadnaviruses—the packaging of DNA polymerase/reverse transcriptase that is covalently attached to one end of a full length, single-stranded DNA—appears to reflect a unique feature of the conversion of pre-genomic RNA into the DNA genome during virion maturation in the cytoplasm.

Unlike nuclearly replicating DNA viruses, poxviruses have evolved to multiply entirely in the cytoplasm independently of the host cell nucleus. Virally encoded RNA polymerases packaged in the virion (see 247) ensure that the viral genome will be transcribed when it enters the cytoplasm. Its large genome (about 240 Kbp) encodes enough proteins to provide for its own transcription and DNA replication machinery.

Nuclear Targeting Events

Two nuclear targeting events occur during the life cycle of all DNA viruses that replicate in the nucleus: nuclear targeting of both the infecting viral genome and the protein products encoded by it. The two events that occur during reproductive cycle of SV40 are shown in Figure 2.

For one incoming genome to target the nucleus, the infecting viral DNA must be transported to the nucleus, perhaps as the particle itself or together with the particle's protein components (Figure 2, arrows). Two constraints on this targeting are the low specific infectivity of most animal viruses and the size of the virion particles. Because infection is initiated by a single virion, one genome's targeting should theoretically be sufficient to launch an infection. Because many potentially infectious particles fail to reproduce—i.e., the ratio of virion particles to infectious units is high—excessive particles are usually applied per cell to probe the targeting event, and the probing method must be able to distinguish one productive virion from many nonproductive ones. The size of virion or capsid particles can range from 26 nm to 125 nm; yet they must pass through the gating channel embedded in the nuclear envelope that opens up to 28 nm (109). How does each virus adopt its pathway into the nucleus?

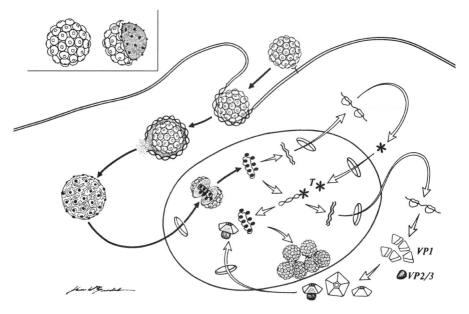

Figure 2 Nuclear targeting events in reproductive SV40 cycle.
The targeting routes of an infecting SV40 are marked by *solid arrows*, while those of the synthetic events are marked by *open arrows*. Symbols: Y, a receptor; double-woven line, caveolae; double-line; plasma membrane; circular, beads-on-a string, minichromosome; two wavy lines in the nucleus, transcripts; a star, LT; and a wavy line with two ribosomes, translation site. In *insert* at the upper left: an intact SV40 and that revealing the internal structure of Vp2/3 and a minichromosome. A hypothetical structural alteration exposes NLS, shown at the left center in the cytoplasm.

It has been recently shown that infecting SV40 DNA enters through the NPC in association with capsid proteins (254). If we assume that these viral DNAs enter the nucleus through NPCs, do the entries of large and small viruses differ?

The second nuclear targeting event—the nuclear import of virus-encoded transcription factors and transcription activators—occurs after the viral genome has reached the nucleus and been transcribed (Figure 2, open arrows). Because this targeting is tied to protein synthesis, its accomplishment permits many molecules to begin entering the nucleus, in turn enabling multiplication of progeny viral genomes. For herpesviruses, adenoviruses, papovaviruses, and parvoviruses, whose assembly occurs entirely or partly in the nucleus, the second targeting event includes the import of structural proteins as well as of proteins needed for assembly and genome packaging. The resident protein domain presumably selected for this function is an NLS. Protein-protein interactions or protein modification patterns are known to influence this synthetic targeting event. The NLS of many viral proteins has been identified in natural or surrogate host cells (Table 2). For hepadnaviruses, which mature in the cytoplasm, the second targeting event is the nuclear entry of a fraction of the progeny viral DNAs that have been assembled in the cytoplasm.

The two nuclear-targeting events for DNA and proteins are programmed precisely according to the life cycle of individual viruses and are clearly separated by the export of newly synthesized viral RNAs from the nucleus into the cytoplasm and its essential protein translation machinery. All three nuclear transport processes—the import of incoming viral genomes, the export of mRNAs, and the import of newly synthesized proteins—take place through a double membrane of the nuclear envelope (NE).

NUCLEAR ENTRY MACHINERY OF THE CELL

Nuclear Envelope and NPC

In eukaryotic cells the NE separates the nucleus from the cytoplasm. The NE has few structurally distinct components: (*a*) the outer nuclear membrane (ONM), similar in composition to the endoplasmic reticulum (ER) and physically contiguous with it; (*b*) the inner nuclear membrane (INM), the chemical composition of which differs from that of the ONM and is replete with inositol-containing phospholipids that modulate the NTPase, which is in turn important for macromolecular translocation processes (4, 337); (*c*) the perinuclear cisterna, the enclosed space between the ONM and INM; and (*d*) the NPC.

The nuclear pore complex (NPC), the key structure in nucleocytoplasmic transport, is disassembled and reassembled during mitosis. The number of NPCs per nucleus varies with the cell's physiological condition; in general, actively dividing cells have more NPCs than non-dividing cells (264). The NPC structure (158) (Figure 1, *panel 6f*) is generally 90 to 100 nm in external diameter. Its cylindrical structure, with an eight-fold rotational symmetry in the plane of the NE, connects the ONM and INM. NPC consists of a highly symmetric basic framework within which eight peripheral channels and a central channel have been imaged (Figure 1, *panel 6*). The three-dimensional structure of detergent-released NPC shows highly symmetrical, interconnected subunits (Figure 1, *panel 6a–d*). The columnar subunit, located centrally (tan), attaches to the two rings (yellow) on both sides of the NPC and to the inner annular subunit (green) as well as to the outer lumenal subunit (blue), which is embedded in the double nuclear membrane (Figure 1, *panel 6d*). The basic framework is devoid of fibers and a cage attaching to the cytoplasmic side and nucleoplasmic side as well as components of a central transporter, all of which function in the active transport of large molecules (Figure 1, *panel 6e*). Eight fibrils extend from the cytoplasmic ring into the cytoplasm, and eight fibrils connect the

Table 2 NLS of viral proteins

Species	Gene (product)	Amino acids	Number of NLS sequence	Reference
MONOPARTITE NLS				
Herpesviruses				
Herpes simplex virus 1	α0(ICP0)	500–508	VRPRKRRGS	248
	α4(ICP4)	726–732	GRKRKSP	248
Herpes simplex virus 2[a]	UL3	188–192	RKPRK	396
Merk's disease virus[a]	MEQ	30–35	RRKKRK	214
Equine herpesvirus[a]	α22(ICP22)	273–278	KKRRKR	159
Varicella-zoster virus[a]	gene 61	387–392	RGAKRR	346
Human cytomegalovirus[a]	IE2*	321–328	PRKKKSKR	288
		144–155	SPRKKPRK	
Epstein-Barr virus[a]	EBNA-1	379–387	LKRPRSPSS	7
Herpesvirus papio[a]	EBNA-2**	478–485	SKRPRPS	212
		341–355	TPEEETVRKRVSRPRQATLRKPRP	
Epstein-Barr virus[a]	EBNA-3a	144–153	RDRRRNPASR	201
Adenoviruses				
Adenovirus	E1a	235–243	DLSCKRPRP	135
Ademovirus 2	pTP	362–373	RLPVRRRRVP	412
	Fiber***	1–5	MKRAR	163
		1–11	MKRARPSEDTF	
Adenovirus 5	pIVa	434–446	DRDRWSRAYRARK	218
Papillomaviruses				
Bovine papillomavirus 1	E2***	105–116	PKRCFKKGAR	336
		340–353	KCYRFRVKKNHRHR	
Human papilloma virus16	E7	17–41	PETTDLYCYEQLNDSSEEEDEIDGP	120
	L1	525–530	KRKKKR	413
	L2***	5–11	RARRRKR	355
		444–449	RKRRKR	

Polyomaviruses				
SV40	Large T	126–132	PKKKRKV	181, 198
	Vp2/3	199–206	PNKKKRKL	63, 400
Murine polyomavirus	Vp1	2–6	APKRK	245
	Vp2/3	196–203	PQKKKRKL	51, 52
Avian polyomavirus	Vp2	308–317	VPKRKRKLPT	303
BKV	Large T	128–134	PKKKRKV	257
Hepadnaviruses				
Hepatitis virus B	Core Protein***	145–156	ETTVVRRRGRSP	101
		172–183	RRRRSQSRESQC	
BIPARTITE NLS				
Herpesviruses				
Herpes simplex virus 1	α27(ICP27)	111–127	RRPSCSPERHGGKVARL	236
	UL29(ICP8)	1170–1188	RKRAFHGDDPFGEGPPDKK	122
	UL9	793–804	KREFAGARFKLR	225
Herpes simplex virus 2[a]	α27(ICP27)	111–127	RRSASPREPHGGKVARI	236
Merk's disease virus[a]	MEQ	62–78	RRRKRNRDAARRRRRKQ	214
Human cytomegalovirus[a]	UL83(pp65)	537–561	PKRRRHRQDALPGPCIASTPKKHRG	323
Epstein-Barr virus[a]	EB-1(ZBRA)	178–204	KRYKNRVASRKCRAKFK	240
Adenoviruses				
Adenovirus 2	Pol	8–46	RARR-(X)$_{13}$-RRRVR-(X)$_{11}$-RARRRR	411
Adenovirus 5	DBP	42–89	PPKKR-(X)$_{37}$-PKKKKK	246
Papillomaviruses				
Bovine papilloma virus 1	E1	84–108	KRK-(X)$_{16}$-PVKRRK	203
Human papilloma virus 16	E6	121–147	KKQRF-(X)$_{18}$-RTRR	331
Polyomaviruses				
SV40	Vp1	5–19	KRKGSCPGAAPKKPK	172
Murine polyomavirus	Large T	88–285	VSRKRPRU-(X)$_{184}$-PKKARED	301
Parvoviruses				
Minute virus of mice	NS1	194–216	KKDYTKCVLFGNMIAYYFLTKKK	272

[a]Viruses not discussed in the text; *, either sequence can be an NLS; **, 478/485 NLS is the stronger of the two NLS; ***, see text; (X)n, n-amino acids between NLSs.

nuclear ring and a small ring to form a basket-like structure (Figure 1, *panel 6e*) (133, 279). The NPCs are supported by a regular meshwork of filaments that appear to be distinct from the nuclear lamina (133).

An NPC is believed to be composed of about 100 different proteins (20). Some of them are integral membrane proteins anchored in and associated with the nuclear membranes, constituents of the basic NPC framework, whereas others are peripheral membrane proteins, collectively called the nucleoporins (20). A fraction of these proteins have been identified, and the amino acid sequences of some NPC proteins deduced from respective cDNA sequences show a few distinct features. Some have O-linked N-acetyl-glucosamine residues and can bind an O-linked glycoprotein-specific lectin, wheat germ agglutinin (WGA), and several pentapeptides represented by the motif XFXFG (X indicates any amino acid). In yeast, either XFXFG or GLFG motif without the O-linked glycosylation is found in each NPC protein (20, 81).

The characteristic feature of integral membrane NPC proteins is the presence of transmembrane domains. That of the peripheral NPC proteins is the presence, in addition to the O-linked glycosylation and the pentapeptide motif, of either the cysteine-rich zinc-finger motif, capable of binding nucleic acids, the leucine zipper motif, or the α-helical coiled-coil domain. These domains and motifs are thought to be important for NPC assembly and the active transport of proteins and RNAs in and out of the nucleus (for additional references, see 279).

Protein Nuclear Entry

Because the incoming viral genome enters the nucleus through the NPC, and because the incoming viral genome's targeting may rely on karyophilic proteins that are components of the virion particle, we discuss below the size limitation for NPC passages, a few known examples of regulated protein nuclear transport in the cell, the signal-mediated import pathways and the known cellular proteins and NPC components that participate in the NLS-mediated nuclear entry. The regulated nuclear entry of macromolecules is important for cell architecture and physiology, particularly for signal transduction, development, and differentiation. The nuclear transport machinery, the NPC proteins' subunit organization, and the functions of these proteins in nuclear import have been extensively described elsewhere (20, 81, 93, 99, 110, 134, 177, 256, 269, 281) and continue to be elucidated in this rapidly expanding field.

An NPC possesses two types of aqueous channels (Figure 1, *panel 6*), eight peripheral channels, and a central channel, which can serve to diffuse small molecules such as ions, metabolites, and small macromolecules up to 9 nm diameter. Protein can accumulate in the nucleus by being diffused through the diffusional channel of the NPCs and by being selectively retained via interaction with nuclear-residing macromolecules such as the nuclear skeletal lamin network, nucleic acids, and transcription factors. Reversal of this diffusion, the nuclear exit of the protein, can also occur, leading to shuttling of the protein between the two compartments. Most proteins, however, are transported rapidly through the central gated channel by a specific signal-mediated mechanism. In fact, macromolecules up to 28 nm in diameter can enter the nucleus through the central channel if they harbor at least one NLS (109).

Protein must contain the nuclear localization signal (NLS) in order for it to pass through the central transporter channel in signal-mediated protein import. The signal consists of predominantly basic amino acids appearing in either one (simple type or monopartite NLS) or two (bipartite NLS) short clusters. Even though the signal must be exposed to and recognized by the cell's transport machinery, an NLS-bearing protein may nevertheless be held in the cytoplasmic membrane through a transmembrane domain until proteolytic cleavage releases the truncated protein for nuclear routing, as is known to occur for protein kinase A (see 268 and reference therein). Phosphorylation, or dephosphorylation, can change the protein's conformation, thus exposing the NLS (185, 268). Such modification may also alter protein-protein interaction, thereby affecting the import of a protein through its binding with an NLS-harboring protein.

Three pathways of signal-mediated import have been identified. One is the nuclear import of classical NLS-containing proteins (223 and reviewed by 137, 237, 269, 282); the other two involve protein components of heterogeneous nuclear ribonucleoproteins (hnRNPs) (239, 289), in particular hnRNP A1 and hnRNP K, that shuttle continuously between the nucleus and cytoplasm, thus harboring both the nuclear export signal (NES) and the NLS. The difference between these three processes lies in the soluble cytoplasmic carrier proteins that act as receptors for NES and NLS during nucleo-cytoplasmic trafficking. Importin (also known as karyopherin and SRP1) recognizes the classical NLS-protein, and transportin recognizes a glycine-rich M9 domain of the A1 protein. The NES and NLS differ within the 38-amino-acid M9 domain (239). The receptor for hnRNP K has not been identified but appears to be different from importin or transportin (239), and the K nuclear shuttling domain (KNS) that confers bi-directional transport has been recently identified. Both the M9 domain and the KNS are unlike the prototype-monopartite NLS identified first in SV40 large tumor antigen (181, 197) and the bipartite NLS identified in nucleoplasmins (94, 307). Other transport processes are likely to be identified in the future.

The well-characterized nuclear import of importin-mediated NLS proteins involves five conserved soluble factors: (a) importin α, which recognizes NLS and forms the NLS protein-importin α complex; (b) importin β, which interacts with the complex through the importin-binding domain residing in the amino-terminus of importin α and forms a trimeric importin β-importin α-NLS protein complex; (c) heat shock protein, hsp70, which is known to participate in conformational changes and interacts with NLS-proteins; (*d*) NTF2, which binds preferentially to Ran-GDP, thereby elevating the concentration of Ran-GTP; and (*e*) GTPase Ran, operating as a molecular switch by cycling between its GTP- and GDP-bound conformations. Ran-GTP specifically interacts with the cytosolic Ran binding protein 1 (Ran BP1), importin-β, and a protein component of the NPC (Ran BP2 or Nup 358, the putative docking site of the import substrate). Consequently, the trimeric NLS-protein importin-α/β complex arrives at the periphery of the NPC. The Ran BP2/Nup 358 located at fibrils emanating from the cytoplasmic periphery of the NPC (223 and reference therein) forms a stable complex with GTPase-associating protein (RanGAP1), which activates GTP hydrolysis. The association of RanGAP1 with the NPC component Ran BP2/Nup 358 requires ATP-dependent posttranslational conjugation of RanGAP1 with a small ubiquitin-related modifier (SUMO-1). Ran BP2/Nup 358-Ran GAP1-SUMO-1 complex then is able to import the NLS-protein complex (223). In short, we can speculate that the importin α/β/NLS-protein complex interacts with Ran BP2/NUP358, a component of the cytoplasmic fibril of NPC, together with Ran-GTP; consequently, modified Ran GAP1-SUMO-1 interacts with Ran BP2/Nuc 358, leading to the import of the NLS-protein with presumably coordinated NFT2, Ran BP1, and GTP hydrolysis.

Although import pathways in the cell and molecules involved in the active macromolecule transport have been well described, the overall flow of molecular movement toward the nucleus has not. How does a newly translated viral protein, for example, move within the highly viscous environment of the cytoplasm, filled with cytoskeletal fibrils? The exciting recent finding is that the modified RanGAP1 interacts with the NPC component RanBP2/Nup 358 attached to the fibrils, which indicates the active role of fibrils in importation. How do these fibers and the import substrate such as virion particles or viral genomes move to the distant central NPC channel? Even less clear is a scenario for the flow of molecules such as animal viruses or signal molecules, which enter cells by endocytosis and move toward the nucleus. What mediates intracytoplasmic movement? It is presently difficult to understand how the endocytosed virus moves toward the nucleus or how large viruses or their DNA genomes enter it. Structural proteins of SV40 have been proven to function in genome nuclear targeting, and thus their NLSs may be essential to it. The resident NLSs for virion components may also play a part in targeting the infecting genome.

PAPOVAVIRUSES

The name *Papovaviridae*, papovavirus, is derived from three characteristics of these viruses: They can cause small neoplastic tumors of the skin (*pa*pilloma) and multifocal tumors (*po*lyomas), and they can induce vacuoles (simian *va*cuolating virus) in infected cells. Papovaviruses are non-enveloped DNA tumor viruses that replicate in the nucleus of permissive host cells and have similar chemical composition and structure. Each papovavirus forms an icosahedral virion particle containing structural proteins and a double-stranded covalently closed-circular genome, which is condensed by the four core histones into a minichromosome. The structure of the icosahedrally symmetric capsid, the protein shell of the virus, is determined by the major coat protein, L1 for papillomaviruses and Vp1 for polyomaviruses. Five L1 or Vp1 proteins form a pentameric capsomer, and 72 capsomers form the respective capsid.

In contrast to these common properties, the papillomaviruses have a different genome organization and replicative cycle from the murine polyomavirus, SV40, and others. The *Papovaviridae* have therefore been subclassified into two subfamilies, polyomaviruses and papillomaviruses. The differences include the following: (*a*) papillomaviruses are larger (55 to 60 nm in diameter) than the polyomaviruses (50 nm); (*b*) papillomaviral genome (8 Kbp), encoding six to eight regulatory proteins (E1 through E8) and two structural proteins (L1 and L2), is larger than that of the polyomaviral genome (5 Kbp), which encodes two to three regulatory proteins (large, middle, and small tumor antigen) and three structural proteins (Vp1, Vp2, and Vp3); (*c*) genetic information is on one strand of the duplex DNA in papillomaviruses, but on both strands of the polyomaviruses; and (*d*) transcription of the papillomaviruses is complex because of many promoters scattered within the half of the genome and the presence of multiple splicing patterns, but transcription of polyomaviruses is relatively simple and occurs from two promoters 400 bp apart. Details on the control of viral transcription, viral DNA replication, and the life cycle of papovaviruses have been described (58, 66, 166).

The mode of DNA replication during productive infection is also different in the two subfamilies. After entering the nucleus, the infecting virion's DNA is transcribed and replicated according to a temporal program specific to each subfamily. For polyomaviruses, replication of viral DNA leads to a programmed switch in transcription from early to late genes and the production of progeny virion particles. For papillomaviruses, virions are not necessarily produced in the infected cell.

Papillomavirus Structure and Replication Cycle

The three-dimensional structure of bovine papillomavirus type 1 (BPV1) capsid, determined to 9 Å resolution, closely resembles that of BVP1, human papillomavirus type 1 (HPV1), and cottontail rabbit papillomavirus (CRPV) previously determined to moderate resolution (367 and reference therein). A few distinct differences exist between the structure of polyomaviruses and BPV1. The diameter of BPV1 is 60 nm, ~10 nm larger than that of polyomaviruses, primarily due to the length of the L1 protein chain. Only one interhexavalent capsomer linkage is seen in the SV40 structure (209, 406), whereas two cross-bridges have been seen in BPV1 (367). Although L1 domains engaged in the multiple cross-linkages have not been identified, residues spanning amino acids 451 to 471, known to be involved in the formation of the BPV1 capsid from capsomer subunits, could account for one of them (278). As in polyomaviruses, 60 hexavalent pentamers and 12 pentavalent pentamers compose the icosahedron of BPV1, but each of the BPV1's hexavalent capsomers has a $\sim$25 Å hole at the center, while the center of the pentavalent capsomer is occluded. Virion-like particles (VLPs) of many papillomaviruses, bovine as well as human, can form in surrogate hosts or in vitro by expressing L1, L2, or both (see reference in 367). The pentavalent pentamer's center occlusion has been hypothesized to be part of the L2 protein (367), and a recent observation that amino acids 61 through 123 of BPV1 L2 are likely to be exposed at the surface (215) agrees with structural observation. L1 or L2 or both appear to interact differently with viral DNA among HPV types, and this difference could influence nuclear targeting of the infecting viral genome. While HPV16 L1 does not bind DNA, and the DNA binding domain at the amino-terminal-12 residues of HPV 16 L2 mediates the binding (414), HPV 11 L1 does bind DNA and has a DNA binding domain at the very end (204). Such a difference has also been reported among structural proteins of SV40 and murine polyomavirus.

The early region of papillomaviral genomes contains clusters of six to eight open reading frames. Many more early gene products, compared to those of polyomaviruses, function in the regulation of viral DNA replication (E1 and E2) and early gene expression (E2), in viral egress via cytokeratin (E4), and in cell transformation (E6 and E7). The late region encodes L1 and L2 capsid proteins. Some aspects of papillomavirus biology are difficult to study in cell culture because of the presence of the strict species- and tissue-tropism, and the association of virion production with a differentiation program of infected epithelial cells [see recent articles (58, 166) detailing papillomavirus biology and pathogenesis].

Each papilloma virus exhibits tropism for mucosal or cutaneous epithelium. The mucosotropic papillomaviruses produce small numbers of virions, some of which are known to be associated with malignancy, whereas the cutaneous papillomaviruses cause benign lesions leading to substantial quantities of virions in warts.

In order to produce progeny virions, an infected epithelial cell must differentiate and switch DNA replication from plasmid to vegetative mode. During papilloma infection, the virion must first reach the basal cells inside the epithelial cell layers, perhaps by entering the cell through an epithelial lesion during wound healing. A recent study identified integrin α6 as a component of the cellular receptor for HPV 6b LI-VLP (104). The proposed model that HPV 6b binds to $\alpha6\beta4$ integrin during wound healing and is endocytosed during the epithelial cell's migration to cover the wound site is consistent with the observed binding of VLPs to many cell types in culture (250, 311). Furthermore, the $\alpha6\beta1$ heterodimer is expressed in mesenchymal tissues, whereas $\alpha6\beta4$ integrin heterodimer is expressed exclusively in the basal layer of stratified squamous epithelium and is known to be turned off at the onset of cell differentiation (104). Initial infection—including virion uptake, endocytosis, nuclear entry of the viral DNA, and uncoating—occurs in a basal cell in the inner portion of the epithelial layer. The viral DNA in that cell is maintained as a stable multicopy plasmid—replicated an average of once per cell cycle and distributed equally to the daughter cells (200, 258). This type of papillomaviral DNA replication, or plasmid replication, is noted in persistent and latent infections, and the transcription of viral E1 and E2 genes is necessary for episomal replication (196, 373). Recent results suggest that the phosphorylation of BPV1 E2 controls proper segregation of plasmid DNA into daughter cells: The distribution into the daughter cells of mutant BPV-1 DNAs, in which the E2 phosphorylation sites are mutated, is unequal (D Lehman and MR Botchan, unpublished data). When the basal cell becomes committed to the differentiation pathway, the terminally differentiated keratinocyte ceases to divide, and a burst of viral DNA synthesis occurs along with late region transcription and capsid protein production. This type of DNA replication is referred to as vegetative.

Very little is known about how one incoming viral DNA replicates to attain a certain plasmid copy number in the infected cell in order to initiate plasmid replication, how the plasmid copy number in daughter basal cells is maintained, or what regulates the switch from plasmid to vegetative viral DNA replication in a keratinocyte. The kinds of transcripts expressed during the differentiation-specific productive processes have begun to be identified using the raft culture system programmed to differentiate (97, 169, 277). This intriguing aspect of PV virus biology is likely to be characterized soon.

Although very little is known about nuclear targeting of the infecting genome, a role for the microfilament network in intracellular transportation of the virion to the nucleus has been proposed, based on the physical proximity of $\alpha6\beta4$ integrin to the network (104).

Nuclear Import of Papillomavirus Proteins

The first early gene products following the viral DNA's entry into the nucleus and gene expression are the E1 and E2 proteins required for replication. The E1-NLS of BPV1 E1 is located within 34 amino acids 84 to 115 (203), necessary and sufficient for the nuclear import of non-nuclear proteins (401), and is a bipartite NLS (84-KRK-86 and 103-PVKRRK-108, Table 2) (203). E2 is also a transcription transactivator for early promoters; an E2-related protein with amino-terminal truncation acts as a transcription repressor. The two proteins use different NLS—106-PKRCFKKGAR-115 within the transactivation domain and 340-KCYRFRVKKNHRHR-353 within the DNA binding- and E1 interacting-domain, even though the latter NLS is present in the full-length E2 but presumably masked in that protein, and the first NLS is responsible for the protein's nuclear localization (336).

The nuclear localization of HPV16 E6 oncoprotein is mediated by a bipartite NLS and resides in the carboxyl domain, 121-KKQRF-125 and 144-RTFRR-147 (331) (Table 2), but a single amino acid substitution of cysteine-66 also affects its nuclear localization (182). The HPV16 E7 oncoprotein appears to use an unusual NLS (17-PETTDLYCYEQLNDSSEEEDEIDGP-41) and has been proposed to enter the nucleus by binding through that domain with an unidentified nuclear protein (120). Both oncoproteins are expressed during the vegetative replication cycle, and their nuclear residence should force the cell to enter a cycling state.

The NLSs of structural proteins L1 and L2 have also been identified (Table 2). HPV16 L1-NLSs consist of either amino acids 525 to 530, 525-KRKKRK-530, or a bipartite NLS with two stretches of basic residues appearing in amino acids 510 to 526, 510-KRK-512 and 525-KR-526 (413). The putative NLS at the end of HPV 11 L1 is a DNA binding domain (204). HPV6B L2 has two functional NLSs, (5-RARRRKR-11) and (444-RKRRKR-449), at each end of the protein, but an internal signal (286-IA SRRGLVRYSRIGQRGSMH-306) influences L2 localization (355). The latter sequence has been proposed to function in nuclear retention of the protein. Again, the amino terminal NLS overlaps with the DNA binding domain.

Polyomavirus Structure and Life Cycle

Polyomaviruses, like papilloma viruses, are small, non-enveloped DNA tumor viruses with restricted host ranges. Twelve members of the subgroup polyomavirinae (eleven mammalian and one avian polyomavirus) have been identified and described in detail elsewhere (66).

The structure of virion particles of both SV40 and murine polyomavirus (Py) is known at an atomic resolution and is determined by the amino acids of the Vp1 polypeptide chain (209, 343), five of which form a pentameric capsomere. A total of 72 capsomeres, 12 pentavalent and 60 hexavalent, form the icosahedron (Figure 1, *panel 3b*). The virion contains 72 copies of the minor coat proteins Vp2 and Vp3 (210), which are inferred to form prongs radiating from the viral minichromosome into the central cavities of the 72 Vp1 pentamers (Figure 1, *panel 3a*) (141) and which therefore would contact the viral DNA as well as the Vp1 pentamers. Neither protein's amino acids could be visualized in X-ray crystallographic analysis of the virus structure (209). Since Vp2 is identical in sequence to Vp3 but contains added amino-terminal amino acids, the interaction of both proteins with the pentamers is thought to occur through common residues. In fact, a Vp1-interactive determinant for both viruses is located at or near the carboxy-terminal region of the proteins, although its exact location and amino acid sequence context differ between them (18, 130). The location of Vp2 unique residues of both viruses, which are myristylated at the amino-terminal glycine (350), relative to Vp1 and the minichromosome is not known. Part of the SV40 Vp2 unique region can be deleted without loss of viability (65). A Py mutant lacking Vp2 myristylation is less pathogenic in mice; in culture, however, the mutation does not affect the viron's encapsidation and stability (193, 318), but it does appears to affect uncoating (318).

The two viruses differ in several respects, although their overall structures are quite similar. The differences among the five Vp1 chains that come together at the central axis of a pentamer have been discussed elsewhere in detail (125) and may reflect the way the two viruses interact with respective cell surface receptors and the viral minichromosome. All structural proteins of SV40 bind DNA (61, 83, 339), but only Vp1 of Py binds DNA. The Vp1-DNA binding domain is within the amino-terminal of Vp1s, the first 5 residues of Py Vp1 (50, 244) and the first 19 residues of SV40 Vp1 (205). The same respective domain functions as the Vp1-NLS: The Py Vp1-NLS is a simple NLS, whereas that of SV40 is bipartite (205). SV40 Vp2/3 proteins are 28 residues longer at the carboxyl end than Py counterparts. The carboxyl 13 residues of SV40 Vp 2/3 function in DNA binding (61, 83), Vp1 interaction (83, 129, 130), and cooperative binding with a transcription factor SP1 to an SV40 promoter region (136).

The DNA of polyomaviruses encodes two to three early gene products—large tumor (LT) and small tumor antigen and an additional middle tumor antigen for Py that act in viral DNA replication, transcription, transformation and in cell proliferation—and three structural proteins—Vp1, Vp2, and Vp3. SV40, human JCV and BKV, and human isolates of SV40 also contain viral gene-encoding agnoprotein, which in SV40 is a 61-amino-acid protein encoded in the late leader region. The coding sequence of SV40 agno is well conserved in human polyomaviruses BK and JC and in SV40 isolates from human child's brain tumors (347), and it appears to be essential for infection in natural hosts. In cultured cells, agno mutants are viable but form small plaques, in contrast to wild-type SV40 (329). The mutant's small-plaque phenotype is attributed to reduced production of virion particles as late mRNA and capsid protein levels fall in the mutant infected cells (341). Reversion to normal plaque phenotype accompanies second-site mutations within the Vp1 coding sequence (17, 227); thus we can infer that agno plays a role in VP1 metabolism. Subcellular localization of agno has been shown to be, in general, cytoplasmic (179, 270), although a fraction is found to associate with minichromosome late in infection (176). Such a small protein would be expected to diffuse into and out of the nucleus. While mutations that eliminate agno protein affect Vp1 nuclear localization in BSC-1 (47, 265, 299), a similar mutation has no effect on the nuclear localization in CV-1 and TC7 cells (172). Currently the role of agno function in Vp1 metabolism remains unknown.

Virions bind to their respective cell surface receptors and are internalized into the cytoplasm. Binding or infection of Py, human polyomavirus BK, and monkey B-lymphotropic papovavirus depends on cell surface sialic acids (153, 154, 189, 342, 343), whereas in SV40 it depends on MHC class I molecules (37). Although endocytosis through uncoated vesicles has been described as an entry route for both SV40 and Py (186, 221, 233), virion internalization following initial binding is not well characterized, except for SV40. SV40 entry has been shown to use caveolae (Figure 2, top—see color insert at end of volume), which reside in specialized regions of plasma membrane, can form flask-shaped invaginations with diameter varying from 50 to 100 mm (316), and have unique protein and lipid composition (11, 316). The caveolae-mediated internalization requires an intact actin network and is regulated by kinase activity (283). SV40 Vp1 associates with caveolin, a major component of caveolae, and virion internalization is inhibited by reagents that disrupt the integrity of caveolae (10). Although primary recognition occurs through the MHC class I molecule, the receptor appears not to be internalized (271). How the receptor binding guides the virion to caveolae is not known. Unlike clathrin-dependent adenovirus internalization, which takes no more than 10 min, the caveolae-mediated SV40 internalization takes about 2 hr (10). The binding activates an intracellular signaling pathway, is dependent on a tyrosine kinase and a Ca^{++}-independent PKC (80), and upregulates c-myc, c-jun, and c-sis. Because treatment with an inhibitor of tyrosine-specific protein kinase blocks SV40 entry specifically and reversibly, the transmembrane signal activation also promotes SV40 internalization (80). Although Py uses distinctly different receptors, i.e. membrane glycoproteins having sialic residues, it may also employ caveolae-mediated endocytosis—that is, binding of Py activates the same intracellular signaling pathway (132, 420).

Although SV40 cell entry is much simpler than that of adenovirus, little is known about how SV40 gets from endocytic vesicles to the nuclear envelope. SV40 and adenovirus have different intracytoplasmic trafficking times: adenovirus reaches the nucleus within two hours of infection, but SV40 takes eight hours or more (405). Intracellular transport of virions in infected cells has been reported to require intact microtubules (333), but when virions are introduced directly into the cytoplasm of cells treated with a microtubule-destabilizing reagent, virion-derived proteins localize efficiently to the nucleus (M Yamada, unpublished data). Microtubules may therefore be needed to move vesicles containing the virus rather than the virus itself. However, biologically active SV40 must presumably be released into the cytoplasm prior to nuclear entry.

When the infecting virion DNA enters the nucleus, it is transcribed and replicated according to a specific temporal program. Approximately 10–12 hr into infection, early genes are expressed, LT starts to appear in the nucleus, and viral DNA replication begins (reviewed in 66). A programmed switch in transcription from early to late genes occurs to enable the assembly of progeny virion particles. Late proteins include the three structural proteins and agnoprotein for SV40. Within the nucleus, structural proteins are believed to arrange sequentially on the minichromosomes (8, 30, 124), probably through a general DNA-binding domain (61, 83, 339). The packaging of SV40 DNA requires a *cis*-acting packaging signal, *ses*, present near the origin (275, 276). By the end of the reproductive cycle, many virions are formed in the nucleus. The precise mechanism of virion release from infected cells is not known—infectious particles could be released from the cell either by cell lysis or by non-lytic secretion.

Nuclear Import of Polyomavirus Proteins

The NLSs of both SV40-LT and Py-LT have been mapped (67, 181, 198, 301, Table 2). Human BKV resembles SV40 in DNA sequence, genomic organization, and nature of the virion-encoded proteins (325), and the BKV-LT NLS has been inferred to be identical to SV40-LT NLS (257). The SV40-LT NLS (126-PKKKRKV-132) is prototype for a simple-type NLS, while the Py-LT NLS (88-VSRKRPR-194 and 279-PKKARED-285) is of a bipartite class of NLSs identified in *Xenopus* nucleoplasmin (95, 96, 307).

SV40-LT NLS has played a key role in elucidating nuclear transport pathways in general as well as identifying protein components of the transport machinery. These subjects have been extensively discussed (81, 93, 99, 110, 134, 177, 264, 281). Both LTs are phosphorylated immediately upstream of the respective NLSs (31, 321). The rate of nuclear import of the carrier protein/gold particles harboring SV40-LT NLS increases if the upstream amino acids 111 to 124, which include phosphorylation sites for casein II kinase and cyclin-dependent p34^{cdc2} kinase, are included with the NLS (CcN) (178, 304), or if the kinase activity unique to the cytosolic fraction of SV40 transformed cells is supplemented (111). Transport-enhancing kinase activity enlarges NPCs (111), affecting cellular nuclear transport apparatus but not the NLS per se (111, 112). Polyomaviruses, therefore, contain a gene whose product alters import machinery to favor the nuclear entry of virus-encoded gene products.

The observation that the phosphorylation site upstream of SV40 LT NLS, or CcN (177), binds importin α/β complex in vitro better than the NLS alone in an ATP-, and GTP-dependent manner (102, 168) agrees with an import-enhancing effect observed in the cytosol of transformed cells (111, 112). In Py-LT, two localization signals behave independently, and thus the effect of phosphorylation is more complex than that of nuclear localization of the SV40-LT (165). Additional phosphorylation might occur once LTs reach the nuclear compartment (165). All in all, these studies illuminate the central importance of both types of resident NLSs in the protein's nuclear entry and the unique strategies viruses employ to benefit their life cycle.

NLSs of the structural proteins have been mapped (Table 2). The Vp1-NLS for SV40 is a bipartite NLS residing within the first 19 residues (1-MAPTKRKGSCPGAAPKKPK-19), in which two clusters of basic residues (5-KRK-7 and 16-KKPK-19) are independently important for nuclear localization (171). Contrary to original descriptions, the Vp1-NLS of SV40 is not a simple-type (399), although the Vp1-NLS for Py is, with 2APKRK6 at the amino terminus (51, 245). The Vp1-NLSs of both viruses (360 in all) function as their DNA binding domains, are located near the minichromosome, and are hidden deep inside the capsid structure (50, 205, 244).

The Vp 2/3-NLSs for both viruses are identical, P N/QKKKRKL (N for SV40 and Q for Py), locating at the very end of the Py proteins (Py Vp3, 196 to 203) and 36 residues inward from the end of the SV40 proteins (SV40 Vp3, 199 to 206) (51, 63, 400). Nuclear import of Vp3 in vitro is signal- and energy-dependent (82). These NLSs are very similar to that of SV40 LT. The Vp2-NLS of avian polyomavirus, the Budgerigar Fledgling disease virus, has also been mapped near the carboxy-terminus (308-VPKRKRKLPT-317, Table 2) (303).

Although in experimental models each viral protein can be targeted to the nucleus individually using its own intrinsic NLS, heterotypic complexes of Vp1, Vp2, and Vp3 are transported into the nucleus during infection for the assembly of virions. Co-transport of SV40 structural proteins has been genetically approached by creating an SV40 variant and by testing for nuclear accumulation of a viral protein lacking an NLS. Because the coding sequence for the carboxy-terminal 38 residues of SV40 Vp2/3 overlaps that for the amino-terminus Vp1 and contains both NLSs as well as other signals, a viable nonoverlapped SV40 (NO-SV40) in which the Vp2/3 and Vp1 genes are separated has been constructed to study the effect of individual NLS mutations on each protein's nuclear accumulation (173). A mutant form of Vp1 lacking an NLS cannot enter the nucleus on its own but can do so when wild-type Vp2 and Vp3 are expressed in the same cell. Similarly, an NLS-negative Vp3 mutant protein can be taken into the nucleus if wild-type Vp1 is expressed in the same cell. This result agrees with that obtained by the biochemical analysis (210), and indicates that virion assembly in the nucleus is controlled by the formation of the complex in the cytoplasm (Figure 2, lower left; see color insert at end of volume).

In insect cells, the intrinsic Vp2/3-NLS of Py (51) appears not to be recognized and results in cytoplasmic localization (87). However, Vp 2/3 can interact with Py Vp1 in the cytoplasm upon cotransfection of plasmids encoding these proteins, and all the proteins are cotransported and form Py capsids (87, 115).

Nuclear Import of SV40 Genome

SV40 is an inefficient producer of infection: approximately 100 physical particles must be applied to a cell in order for one virion particle to reach the reproductive site and to lead to productive infection. When the membrane penetration steps are bypassed by injecting virions directly into the cytoplasm or the nucleus, the efficiency of infection increases: 10 virions will lead to a productive infection when injected into the cytoplasm, and a single virion microinjected directly into the nucleus is sufficient (92, 128). Low rate of infection has slowed investigation of how SV40 enters the nucleus, since morphological and biochemical observations of an entire virion population may reveal little about the rare virion that actually leads to an infection.

To circumvent this low infection rate, injection assays are used to introduce isolated virion particles directly into the cytoplasm in the presence of neutralizing extracellular antibodies, in order to follow the initial fate of SV40 nuclear entry as well as the nuclear entry of the viral genome. Protein components of introduced virions accumulate in the nucleus within 2 hr of injection, and SV40 LT appears in the nucleus, after expression of the incoming viral genome, within 3 to 4 hr of injection (64). The injected virion can complete lytic infection (405). Thus, specific temporal changes in gene expression in injected cells are the same as in infected cells (405 and A Nakanishi, unpublished data), but LT is expressed earlier. The time difference is due to a slow intracytoplasmic trafficking step, which includes the internalization process (405), and is in agreement with the slow process of caveolae-mediated SV40 internalization (271). Nevertheless, results from the injection assay must be verified in natural SV40 infections. The measured endpoint in all of the viral nuclear entry experiments is the LT expression.

The first question regarding SV40's entry into the nucleus is whether the genome enters by itself or as an intact virion. Electron-dense particles have been observed in the nuclei of SV40-infected cells soon after infection (170), as well as in the nucleus after cytoplasmic virion injection (64, 405). Because empty particles can enter the nucleus (254), virion-like particles seen in the nucleus may not contain the genome but may be empty particles produced by the dissociation of virions in the cytoplasm. Experiments to determine the fraction of SV40 [^{3}H]thymidine-labeled DNAs taken up by the nucleus indicate that less than 3% of silver grains are found in the nucleus; thus, most

SV40 genomes remain in the cytoplasm. By contrast, virion Vp1, detected immunologically, is present only in the nucleus (62).

Involvement of NPC in Nuclear Transport of SV40 Genome

Does a virion particle (with or without its genome) use cellular machinery such as the NPC? To test whether SV40 and virion proteins enter the nucleus through the NPC, virions were injected into cells together with antibodies against a nucleoporin, or with wheat-germ agglutinin. Both of these inhibitors of nuclear import completely block the nuclear accumulation of virion proteins and LT expression (64, 254, 405), as does chilling the cells, which blocks facilitated passage through the NPC (405). Antibodies against the NPC proteins also block the nuclear entry of virion particles during normal infection. Furthermore, virion particles and Vp1 can be visualized by electron microscopy (EM) and immuno-EM, respectively, around and in NPCs (64, 405). These results indicate that SV40 enters the nucleus through the NPC (Figure 2, center; see color insert).

If SV40 dissociates in the cytoplasm, would structural proteins accompany SV40 DNA into the nucleus?. Further evidence for a role of Vp1 and Vp2/3 in the nuclear targeting of infecting DNA comes from antibody interception and in vitro reconstitution experiments. In the interception experiments, in which antibodies against Vp1 or Vp3 are present in the cytoplasm during a normal infection, antibodies quantitatively block LT expression. Furthermore, anti-Vp3 antibody injected into the nucleus, but not anti-Vp1 antibody injected identically, also blocks LT expression, the endpoint detected in these nuclear entry assays. When anti-Vp3 antibody and DNA are coinjected into the nucleus, the antibody fails to directly block LT expression driven from SV40 DNA (254), in contrast to the inhibition observed in injection experiments, suggesting that Vp3 accompanies viral DNA into the nucleus. Whether the major capsid protein Vp1 is also involved at this stage is not known. Moreover, reconstitution experiments indicate that viral structural proteins play an important role in genome nuclear targeting. Microinjected empty particles, with or without the addition of viral DNA, enter the nucleus efficiently, whereas naked SV40 DNA or histone-complexed minichromosomes fail to do so (254). It is not understood why NLS-possessing histones do not promote efficient nuclear import of the minichromosome. Thus, Vp1 or Vp3, or both, are necessary and sufficient for the targeting or passage of DNA into the nucleus, or for both (254).

Are the NLSs of structural proteins needed for the nuclear transport of SV40 DNA? Our preliminary results suggest that Vp2/3-NLS plays a role in viral DNA targeting. The importance of the Vp2/3-NLS in the reproductive cycle was probed by reconstituting mutant viral genomes (*NO-SV40 Vp3-NTS mt*) in which one, two, or all five basic residues of the NLS were altered and their viabilities determined. In the mutant genomes, all regions other than the Vp2/3 NLS are the same as in wild-type NO-SV40. All mutant Vp2/3's, regardless of the number of amino-acid alterations, could localize to the nucleus, presumably piggy-backed by Vp1 (173), and the extent of mutant DNA replication was similar to that of wild-type NO-SV40. Mutants with the one or two substitutions were viable, but the one with 5 amino acids substituted ($Vp3\text{-}NTS_{null}$) was not. Virion particles were found in the lysate of the mutant-DNA transfected cells as judged from the sedimentation profile, but the lysate contained no infectious virions (255). The $Vp3\text{-}NTS_{null}$ mutation may therefore influence an early step of the SV40 infection cycle—at nuclear entry of the virion—although it could have also altered adsorption and internalization of the mutant virion as well as its nuclear uptake.

The Vp1-NLSs are located inside the virion particle (209), and location of Vp2/3-NLSs in the virion structure is not known. If any of the combined 432 NLSs is to be recognized by the cellular transport machinery, the structure must be altered before the genome enters the nuclears. Although the disassembly process of SV40 following internalization during infection is not understood, in vitro studies in which redox and ionic conditions (e.g. low Ca^{2+} concentrations) present in the cytoplasm are sufficient to disassemble SV40 particles (68) suggest structural alteration of the virion. The NLSs of the import-competent virion can then be recognized by the cellular machinery. The SV40 minichromosome is about 34 nm wide (14, 141), larger than the maximal diameter of the gated NPC channel: thus, either the import-competent virion or the minichromosome with a few structural proteins attached must change shape as it passes through the NPC. In SV40, therefore, structural proteins facilitate the infecting DNA's entry into the nucleus.

PARVOVIRUSES

Three genera of parvoviruses infect vertebrate species: *Parvovirus* (autonomous parvoviruses), *Erythrovirus* (human parvovirus, B19), and *Dependovirus* (adeno-associated virus, AAV) (27). Autonomous parvoviruses infect a wide range of mammalian and avian hosts (minute virus of mice, MVM; canine parvovirus, CPV; feline parvovirus, FPV; Aleutian mink disease virus; and H-1 parvovirus). The human parvovirus causes the childhood fifth disease and a rheumatic-like syndrome in adults (38). The AAV infects mammals, but it neither propagates by itself nor causes any clinical symptoms. Propagation in the host requires co-infection with adenovirus or any type of herpesvirus (29), or genotoxic stimuli (403, 404).

Structure and Life Cycle

Parvovirus is a small non-enveloped virus, 18 to 26 nm in diameter, with a linear, single-stranded DNA approximately 5 Kb in length. The capsid structure of CPV, FPV, MVM, and B19 is known at an atomic resolution (1, 2, 217, 368, 397, 402) and shows an icosahedral symmetry consisting of 60 subunits, or 60 structural proteins. A three-dimensional structure of B19 (26 nm) is shown in Figure 1, *panel 5*. Each subunit or protein forms a β-barrel core, a motif readily found in other viral capsid structures, with four large loops extending from it. The configuration of the large loops varies among parvovirus species and is known to determine virion antigenisity (3). Each single-protein subunit can be either Vp1, Vp2, or Vp3—three related proteins whose amino-terminal regions vary in length. Most parvoviruses contain abundant Vp3, which is produced by proteolytic shortening of Vp2 amino-terminus, except B19 and Aleutian mink disease virus, which have only one major capsid protein, Vp2. Vp1 is a minor capsid protein for all parvoviruses.

The viral DNA packaged in CPV has been partially visualized (53, 368). Interaction of capsid proteins with DNA consists of protein-base hydrogen bonds and places the DNA's phosphate backbone inside the nucleocapsid. The genomic DNA has palindromic sequences at both termini. Each palindrome's length varies by species and can fold back on itself to form a duplex hairpin structure between the self-complementary sequences. The termini function as primers for viral DNA replication and in packaging reaction (77, 390, 391). Generally, the genome of autonomous parvovirus is of single polarity, complementary to mRNA, while that of both polarities is found in AAV (21, 26).

Parvoviruses propagate in the nucleus of actively dividing cells, and viral genome replication takes place only in the S-phase (25, 357, 380). Without helper virus or some type of cell stimuli, productive infection does not occur in AAV, and the viral genome is integrated into the host chromosome. This integration is by nonhomologous recombination and is partially position-specific; almost two thirds of integrated AAV proviruses are located at the site of human chromosome *19q13.3qter* (192, 319). When cells harboring integrated viral genome are exposed to the helper virus, the latent AAV can be rescued and then resume productive cycle as in autonomous parvovirus.

Cell surface attachment, viral entry, internalization, and intracellular routing to the nucleus appear to be unique to individual viruses. For MVM, sialic acid moieties of cell surface glycoproteins act as the receptor (76). For B19, one glycolipid, globotetraosylceramide, or globocide, is the receptor (39, 55). Following binding, internalization of MVM has been reported to occur via the coated pits (213), whereas that of CPV occurs via endosomes (19, 376a). For MVM and AAV, the virus is shown to reach the nucleus within 30–60 min following attachment (224, 292). Although the mechanism by which the infecting genome targets the nucleus is not known, it is tempting to speculate that it enters the nucleus as an intact nucleocapsid, for several reasons. First, parvoviruses are small enough to pass through the NPC intact. Second, the presumed entry signals are exposed at virion surface (370), and, if so, they can recognize the cell's transport machinery. Third, a Vp1 deleted mutant appears to show a phenotype characteristic of import/uncoating defect (see below) (370).

Once the DNA reaches the nucleus, transcription and replication occur as programmed for the parvovirus life cycle. Parvovirus biology, including DNA replication and gene expression, has been detailed in recent articles (27, 29, 72, 287). The newly synthesized viral proteins are transported to the compartments where they function. The nonstructural protein in autonomous parvovirus, NS1, localizes mostly to the nucleus (272) and functions in viral DNA replication and genome packaging (75, 272, 369, 392). The smaller nonstructural protein, NS2, is mostly cytoplasmic and is required for efficient translation of viral mR-NAs (206, 207), but some may be transported to the nucleus to participate in genome accumulation and capsid assembly (74, 252, 253). In AAV, the functional counterpart of NS1 or NS2 are Rep68 and Rep78 or Rep40 and Rep52, respectively (54, 266). Rep68/78 has other regulatory functions in nonproductive infection, including negative regulation of gene expression (285, 385) and host chromosome integration (131, 211).

The subunit-subunit interaction between structural proteins to form multimers occurs among newly synthesized capsid proteins prior to their nuclear import (70, 251, 393, 394). In both AAV2 and H-1, the empty capsid forms in the nucleolus (5, 335, 393). Empty capsid recognizes the hairpin structure and packages the genome. The observation that the mature virion is redistributed from the nucleolus into the nucleoplasm (117, 171, 381) suggests that nucleolus function, perhaps cellular chaperone protein(s), is needed for genome packaging. Thus, although parvovirus' structure and composition are simple, its packaging pattern appears to have evolved along with that of adenovirus. The mature MVM virion has one molecule of NS 1 covalently attached to the 5′ end of the MVM genome, with the NS 1 outside of nucleocapsid (70, 73). The infecting AAV2 also associates with Rep68/78 (194, 290).

Nuclear Import of Viral Proteins

The 83-kDa full-length MVM NS1 has a bipartite NLS (Table 2) (272), but the truncated 65 kDa protein, known as NS 1*, which lacks the carboxyl-terminal portion of the protein, appears to expose the NLS more effectively and is able to localize more efficiently to the nucleus than full-length NS1 (71). An NS 1 oligomer, known to form in the cytoplasm (272, 293), functions in DNA replication in the nucleus. For H-1 parvovirus, another cellular component present in SV40 transformed cells, but not in nontransformed cells, modulates NS 1's nuclear localization (300). The cytosol of SV40 transformed cells contains a protein kinase activity that enlarges the NPC and expedites protein nuclear import. Therefore, protein-protein interaction and protein modification could work in concert with the protein's resident NLS for the effective function of NS 1 and nuclear localization.

The nucleocapsids of all parvoviruses contain Vp1, and its unique domain, in which a sequence resembling polyomavirus Vp1-NLS seems to present, is exposed on the virion surface. In cells infected with a mutant of MVM in which Vp1 is deleted, progeny particles are produced but are defective in the subsequent infection cycle (370). The defect lies in a post-entry process, not in the initial entry process. Because so-called "uncoating" mutants map in the coding region of the capsid protein (292), further analysis could unveil a unique nuclear entry route for the smallest DNA virus.

HERPESVIRUSES

The herpesviruses are enveloped DNA virus of a large, defined group of more than 100 species. They are classified into three types (313):(*a*) the alphaherpesviruses (Herpes simplex virus type 1, 2), which are neurotropic and have a broad host and cell range; (*b*) the betaherpesviruses (cytomegalovirus, human herpesviruses 6 and 7); and (*c*) the gammaherpesviruses (Epstein-Barr virus, Kaposi's sarcoma herpesvirus). The latter two groups differ in genome size and structure and are restricted to cells of glandular or lymphatic origin.

This review focuses on Herpes simplex virus type 1 (HSV-1), the most extensively studied of all herpesvirus families. HSV-1 causes cold sores, fever blisters, and eye and brain infection. Prevalent with HSV-1, latent infections are without signs of disease, characterized by establishment of latency in trigeminal ganglia of sensory neurons and reactivation to cause recurrent lesions close or identical to the initial infected size.

Structure and Life Cycle

The HSV-1 virion consists of four morphologically distinct components. A dense core or nucleoid is covered by an icosahedral capsid composed of 162 capsomers. This capsid is usually surrounded by an amorphous coat called the tegument, which in turn is covered by a typical lipid bilayer envelope with short glycoprotein spikes on its surface. The diameter of the nucleocapsid is about 125 nm (Figure 1, *panel 1*), and that of the enveloped particle is 200 to 300 nm.

More than 39 different viral proteins form the virion (144, 160). The envelope contains 12 proteins; 10 of them are glycoproteins. The tegument, however, contains 65% of the virion's mass in its 14 or so proteins (324). Different components of this ill-defined structure function in many cellular compartments. For example, one major component, Vp16 (a-TIF), regulates the initiation of immediate early genes by effectively recruiting cellular transcription factors (273);

tegument Vp1/2 appears to deliver the genome into the nucleus (22, 234, 235); and other tegument proteins, UL41 gene product, degrade mRNAs in the cytoplasm (195, 297, 298).

The icosahedral nucleocapsid is composed of four structural proteins, Vp26, Vp23, Vp19, and Vp5, and protease Vp24. The 150-kDa Vp5, the major structural subunit of the nucleocapsid, forms either a hexon capsomer of six interacting Vp5 molecules (Figure 1, *panel 1*, blue) or the penton capsomer of five interacting molecules (Figure 1, *panel 1*, turquoise) (69a, 261, 344). The capsid contains 150 hexons and 12 pentons. Each hexon is appended by a 12-kDa Vp26 (35, 366, 415). One molecule of Vp19C and two molecules of Vp23 make a heterotrimer referred to as the triplex (Figure 1, *panel 1*, red and orange) because of its location at the base of capsomers that form a local three-fold symmetry (261, 365). There are 320 triplexes in the capsid. Mature nucleocapsid also contains the protease Vp24.

The double-stranded DNA of the HSV-1 genome, approximately 152 Kbp, is linear but appears to be toroidal in the capsid (34, 324). Unlike viral DNAs in the capsids of papovavirus and adenovirus to which histones and adenovirus-protein pVII, respectively, associate to form a beads-on-a-string structure, herpesviral DNA appears to be free of proteins, resembling the internal organization of bacteriophage (34). The DNA molecule consists of unique regions and repeat sequences, such that two unique regions, unique long (UL) and unique short (US), are flanked by repeat sequences (314, 378). Of more than 78 genes identified, half are dispensable for propagating virus in cultured cells.

HSV-1 attaches to cells through interaction of the virion's glycoproteins with surface receptors and then enters cells by fusing with the plasma membrane (340). Several cell surface receptors involved in the initial receptor recognition have been identified, including highly sulfated glycosaminoglycans such as heparan sulfate and dermatan sulfate, which bind glycoprotein C (and possibly B) (15, 100, 142, 152, 332, 398). Mannose-6-phosphate receptors (41, 42) and the complement receptor C3b (119) appear to participate in initial binding. Direct fusion between the viral envelope and the plasma membrane internalizes the tegument-covered nucleocapsid into the cytoplasm, and the fusion is known to occur independent of pH (340). Virion-cell fusion requires interactions of viral envelope glycoproteins gB, gD, gH, and gL with several cell surface proteins, including one new member of the tumor necrosis factor receptor family (243, 388) that can mediate HSV-1 entry into human T-cells. Because many cell surface proteins in different cell types are known to mediate HSV-1 entry, HSV is likely to use more than one attachment and entry pathway, described in recent articles on herpesvirus biology (238, 314, 340, 344).

The internalized incoming viruses traverse closer to the nucleus in association with microtubule and dynein-like structure (338). Integrity of the microtubule network is important, for a disorganized network can block this movement. Capsid structure appears to be maintained upon arriving near the NPC, but most, if not all, tegument proteins are shed during the process. "Empty" capsids devoid of electron-dense interior cores are often found closely associated with NPCs and have been thought to result from the genome's having delivered its viral DNA into the nucleus (22, 23, 338, 361). By this stage, tegument proteins dissociate from the nucleocapsid and are probably directed to the various cellular compartments in which they function. For example, nuclear import of the tegument protein Vp16 (α-TIF), the transactivator of α genes (or immediate early genes), is needed to initiate viral gene expression (273).

In the nucleus, temporally coordinated gene expression begins. Expression of the first set of viral genes, immediate early (α) genes, promotes expression of the second set, the early (β) genes, which leads to the production of viral proteins required for viral DNA replication, which soon follows. The onset of replication coincides with or is followed by expression of the last set of genes, the late (γ) genes, which encode virion structural proteins: tegument proteins, envelope glycoproteins, and proteins necessary for nucleocapsid assembly and genome packaging. Parental viral DNA in the nucleus is hypothesized to replicate first via the theta-form intermediates, then by switching to a rolling circle mechanism that produces a long head-to-tail concatemer (33), part of which will be packaged into the preassembled capsid using a head-full rule first identified in the packaging of bacteriophage DNA (377). Although replication intermediates conforming to the rolling circle model have been detected in infected cells, theta intermediates have not yet been detected (314). HSV biology, including viral genes and their expression, DNA replication, assembly, and envelopment, is detailed in elsewhere (33, 160, 273, 314, 344).

Capsid assembly of HSV-1 is a multistep process initiated in the cytoplasm and then completed in the nucleus. Results from studies of cell-free capsid assembly and HSV-1 capsid mutant have shown that capsid formation is controlled by proper protein-protein interactions among structural proteins and between structural and scaffold protein in the cytoplasm, as well as by subsequent nuclear import of protein multimers. Among capsid proteins or their precursor proteins, preVp22a and Vp19C have been shown to be karyophilic, while Vp5, Vp23, and Vp26 are not (267, 306). Interactions of Vp5 with preVp22a and of Vp19C with Vp23 are largely responsible for the nuclear import of Vp5 and Vp23 (188, 267, 306). The domain of preVp22a responsible for interacting with Vp5 resides in the carboxy-terminal end (162, 231, 358), but the Vp23-interaction domain of Vp19C is not known. Vp26 does not interact directly with Vp19C or preVp22a, but its indirect interaction via its affinity to Vp5 suggests a role of preVp22a in nuclear transport (306). These insights reveal how the formation of heterotypic complexes dominates nuclear transport and subsequent assembly. Piggy-back nuclear transports also help to achieve the proper stoichiometry of virion proteins (88).

A series of protein-protein interactions in the nucleus also produces the procapsid, a spherical rather than icosahedral structure whose composition is identical to that of mature capsid's outer shell (262, 365). The procapsid contains abundant Vp22a in the core, where the carboxy-terminal Vp5 interaction domain is cleaved off by viral protease Vp24, which is also incorporated into the procapsid core. The procapsid is then ready for genome packaging. Packaging and cleavage of viral DNA occurs simultaneously, with the large head-to-tail concatemeric replicated DNA, cleaved at the specific site within terminal *a* sequences, present at both termini and in the internal repeated sequences (85, 86, 376). At the time of genome packaging, the scaffold protein Vp22a is expelled, and only the protease (Vp24) remains inside the mature capsid (123). The nucleocapsid then acquires the tegument and a lipid envelope containing immature forms of glycoproteins at the inner nuclear membrane, and these particles transiently accumulate in the perinuclear space. Glycoproteins of the perinuclear virions are immature glycans characteristic of those in the endoplasmic reticulum and differ from those of the mature virions processed through the Golgi apparatus (16, 180, 327, 364). This difference in glycoprotein composition has invited much discussion on the process of HSV-1 envelopment (314, 344). How the virion acquires the mature envelope remains unknown.

In animals, HSV infects and replicates in both epithelial cells and neurons, which have different membrane composition and intracellular protein trafficking. In general, invasion of a sensory neuron follows primary HSV productive infection in the epidermis (379). In the neuron, the infecting HSV is transported from the axon ending to the cell body, mediated by microtubules (219, 363), and establishes latency in sensory ganglia. In the infected neuron, viral DNA is maintained in the nucleus as a nucleosomal, circular episome with a copy number ranging from 10 to 100 (156, 310, 379). Transcription is restricted to the latency associated transcript (LAT), which is retained in the nucleus (91, 345). The LAT's function is not known, but it may be involved in HSV reactivation (157).

Nuclear Import in HSV-1 Life Cycle

The nuclear localization signal of ICP8 resides in the carboxy-terminus of this 1196-amino-acid, single-strand-DNA binding protein. Although the NLS sequence, 1170-KKRAFH-(X)11-KK-1188 (Table 2), alone is sufficient for localizing heterologous protein pyruvate kinase to the nucleus, the presence of an additional ICP8 domain—residues 582 to 1082, the internal domain that suppresses the NLS function—interferes with nuclear localization of the fusion protein (122). If the amino-terminal half of ICP8 is added to the bipartite NLS, this interference is overcome and the fusion protein is now nuclearly localized.

Because exposure of the ICP8-NLS is influenced by the surrounding ICP8 domains even in the fusion protein, protein folding might dominate ICP8 nuclear localization in infected cells. The NLS of the UL9 gene product, an origin binding protein, is also a bipartite NLS (225, Table 2).

During HSV-1 infection, localization of ICP0 within the nucleus changes from the spherical punctate region early in infection to uniform and diffuse nuclear localization at late stage (106). The domain responsible for the punctate ICP0 distribution maps to its amino-terminal portion (105), including potential casein kinase II phosphorylation sites. Yet the domain's function appears to be overridden when other immediate early proteins are present (94, 190, 248, 249, 417–419), since interactions between them are known to enhance or reduce each other's nuclear localization as well as that of other HSV-1 proteins. For example, ICP4 promotes the nuclear localization of ICP0 and the DNA binding protein ICP8 (418, 419). The nuclear localization of the transcriptional activator ICP0 is increased when ICP4 is present, but is reduced when ICP27 is present (417, 418, 419).

The ND10 (nuclear domain 10) or PODs (PML oncogenic domains) are compartmentalized sites within the nucleus that are thought to function in transcription and locate near the splicing sites (12). These sites are rearranged after expression of ICP0 and ICP27 in HSV-1 infected cells (106, 232), and ICP8 has been reported to localize near ND10 in the early phase of infection (84). HSV-1 DNA replication is also known to occur in distinct nuclear compartments (174) where viral DNA polymerase and viral helicase, together with cellular replication machinery, co-localized to form a replication complex (286), although their identities with those of ND10 domains remain speculative.

Because transcription of viral genes, replication of HSV-1 viral genome, and assembly of nucleocapsids occur entirely in the nucleus, the viral proteins required for these activities—more than 30 gene products—must be imported into the nucleus. This import is likely to be controlled at many levels, including posttranslational modifcation, protein-protein interaction, and the proteins' resident NLSs. Nuclear localization of HSV-1 protein will be better understood after the proteins' functional domains are identified and dissected, and then related to the events during HSV-1 infection.

How the infecting genome enters the nucleus is not known. The study of one temperature-sensitive mutant, *tsB7*, may provide some insight (22, 23). This mutant is defective in releasing DNA into the nucleus, and at nonpermissive temperature leaves the DNA-containing nucleocapsid in close association with the NPC. The *tsB7* mutation is within the coding sequence of a large tegument phosphoprotein, ICP1/2 (234, 235). It is present in the virion at approximately 120 to 150 copies (151) with an apparent weight of 270 kDa. The newly synthesized protein localizes to both nucleus and cytoplasm (234). Although the precise function of ICP1/2 is not known, its tight association with the major capsid protein Vp5 and with the viral genome through the *a* repeat sequence, together with a 140-kDa protein of unknown origin (57, 234, 235), suggests its role in viral DNA packaging. HSV-1 appears to adopt the strategy of introducing its DNA alone, leaving an apparently intact capsid shell behind.

ADENOVIRUSES

Adenoviruses infect humans and many other mammals and birds, causing mild respiratory infections and eye infections as well as acute hemorrhagic cystitis. More than 40 human strains, or serotypes, have been serologically identified and described. Each has a slightly different surface which elicits an immune response and leads to the production of type-specific neutralization antibody— for example, adenovirus type 2 (Ad 2) is resistant to neutralization by antisera to adenovirus type 5 (Ad 5).

Structure and Life Cycle

Adenoviruses are non-enveloped virions and are very much larger and more complex than papovaviruses. The adenovirus capsid is 90 nm in diameter with a trimeric fiber protein protruding outwards from each of 12 icosahedral vertices (Figure 1, *panel 2*) (348, 349). The capsid is composed of 240 hexon capsomeres (Figure 1, *panel 2*, blue), each of which contains three tightly associating polypeptide II molecules, and 12 penton capsomeres (Figure 1, *panel 2*, yellow), each of which contains five polypeptide III. Each penton is located at a vertex of the icosahedral particle and houses the trimeric fiber protein polypeptide IV (Figure 1, *panel 2*, green). The amino-termini of the trimeric fiber insert into the penton base, and the carboxy-termini forming a knob mediate virus attachment to a host cell receptor. Hexon-hexon contact is stabilized by association with polypeptides IX and IIIa, and the ring of hexons surrounding a penton capsomere, called peripentonal hexons, interacts with polypeptides VI and VIII, which anchor the capsid and the core. Adenovirus structure has been discussed elsewhere in detail (45).

The adenovirus core is composed of a linear duplex DNA, ranging from 20 to 24×10^6 Da among different serotypes, and four viral encoded proteins— polypeptide V, VII, mu, and the terminal protein (TP). Polypeptide VII is known to condense viral DNA in the same way histones condense eukaryotic DNA (241, 275). The TP is covalently attached to each 5′ end of the duplex DNA. Direct visualization of the core by EM and chemical cross-linking experiments indicate that the viral genome is organized into eight or so supercoiled domains (40, 263, 395) condensed by DNA associating proteins, rather than in a linear structure of 11 μm to 13 μm. One thus can imagine either a rosette structure with eight 45-nm loops connected at the center or a serpentine structure with a stack of eight 45-nm loops. Although the core's presumed size in expanded form on a plane is larger than an NPC opening (28 nm), another hypothetical higher-order compaction could significantly shrink it for nuclear entry. Regardless of the size and shape of the core, experiments have shown that the programmed and step-wise disassembly of the virion structure following internalization is required for the nuclear entry of adenovirus DNA (139).

The adenovirus genome encodes many proteins that mediate viral gene expression, DNA replication, and assembly as well as structural proteins and proteins that reprogram cells to effectively produce a large quantity of virions. The genome carries five early transcription units (E1A, E1B, E2, E3, and E4), two delayed early units (IX and 1Va2), and one late unit, all transcribed by RNA polymerase II and processed by splicing to yield several mRNAs per transcription unit, with each coding for a protein. The genome also carries one or two VA genes that are transcribed by RNA polymerase III.

After reaching the nucleus, infecting DNA associates with the nuclear matrix through the covalently attached TP (32, 118, 320). The initiation of viral gene expression also appears to be programmed by the TP: mutations within the TP of Ad5 (320) are deficient in matrix attachment, and their DNAs are less efficiently transcribed. It is not known which structural features or motifs of the nuclear matrix the TP can recognize. Transcription of the first E1A gene begins soon afterwards, and the E1 proteins activate other early genes according to the programmed adenovirus replication cycle. Gene products for E2, E3, E4, a 140-kDa DNA polymerase (Pol), an 80-kDa pre-terminal protein (pTP), and a 72-kDa single-stranded DNA binding protein (DBP) are made prior to adenovirus DNA replication, and coupled to it is the switch from early to late transcription. A change in viral chromatin conformation, viral gene product IVa2, and cellular transcription factors all contribute to late gene transcription and the production of mRNA molecules that encode viral structural proteins and the proteins needed for either protein processing or assembly. After viral proteins are synthesized, structural proteins are assembled into hexon capsomeres and penton capsomeres in the cytoplasm prior to their nuclear entry. Adenovirus biology and processes involved in adenovirus gene expression, DNA replication, and assembly have recently been described (330).

During virion assembly in the nucleus, the encapsidation of the viral chromosome into a pre-formed empty capsid occurs through the cooperation of the packaging sequence, a *cis*-acting DNA element located near the left end, and a scaffold protein, the L1-coded 55-kDa protein (138, 145–147, 149, 359). Although the point at which the internal proteins VI, VII and VIII are incorporated into the virion is not known, they and the DNA-bound pTP must be processed at defined sites from pVI, pVII, pVIII, and pTP by the virally encoded 23-kDa

L3/23K protease (226, 360, 384). Mutants lacking this functional protease can form particles with unprocessed proteins, resulting in non-infectious virions. The production of infectious particle capable of entering the nucleus involves the encapsidation of 10 copies of this virally encoded functional protease. Thus, the correct assembly of virion particles sets the stage for the uncoating and nuclear targeting of the next infection cycle.

Entry and Uncoating

The entry of adenovirus into susceptible cells is sequential. The virion's fiber knob first attaches to a cell surface receptor. Fiber-mediated attachment of Ad2 and Ad5 has recently been shown to occur through major histocompatibility complex class I and the coxackievirus-adenovirus receptor (CAR) (24, 164, 362). Ad2 internalization requires the interaction of five conserved RGD motifs (Figure 1, *panel 2*, red) of the penton base with specific members of a family of heterodimeric membrane components, $\alpha v \beta 3$ and $\alpha v \beta 5$ vitronectin-binding integrins (13, 113, 389).

The precise domain of integrins that mediate adenovirus internalization has not been identified. The length of the fiber shaft between the fiber knob, a globular carboxy-terminal domain that binds the primary receptor(s), and the amino-terminal domain that interacts with the penton base, varies from 12 to 13 nm depending on the serotypes (59, 90, 317, 334). It is far from the RGD sites that trigger penton-integrin mediated internalization. The virion surface contains 12 sets of penton-fiber structures. Would the penton-integrin interaction use the same penton that is already attached to the receptor-bound fiber? Or is the primary fiber-receptor interaction a means for bringing any penton closer to integrin?

The virus is endocytosed via coated pits and vesicles and enters the cytoplasmic compartment within 5 to 10 min following absorption (48, 49, 114, 374). Contrary to the previously proposed acid-triggered lysis mechanism (284, 291, 328, 356), a recent study has shown that exposure to a low endosomal pH per se is not important for any early event during adenovirus infection including the release of virions from endosomal membrane into the cytosol (312). The next step, dissociation by L3/p23 protease of protein VI linking the viral core to the peripentonal hexon, consistently does not rely on pH change (139); rather, the arrangement of viral as well as host macromolecules appears to be more important (312). Nonetheless, virions are disassembled by the functional L3/p23 cysteine protease, packaged in the virions (9, 78, 382), that requires free cysteins and a disulfide-linked peptide in order to function (384). Sulfhydryl alkylation of virions in vitro renders them noninfectious without grossly affecting the virions' internalization, penetration, and accumulation near the NPCs (140). Protease-defective virions cannot degrade protein VI and thus cannot disassemble the particles by releasing a DNA binding protein, protein VII, from the viral DNA at the cytoplasmic side of the NPC.

Recognition by the cellular receptor and internalization is believed to trigger activation of the L3/p23 protease. Partial disassembly of virions occurs in the cytoplasm as the protease degrades protein IX, which stabilizes hexon-hexon interaction, and protein VI, which associates with peripentonal hexon on one side and the core on the other (139, 140). This programmed disassembly is required for adenovirus nuclear entry: Ad2 *ts1* mutant, which has a defective protease and encapsidates unprocessed proteins within it, is unable to proceed with productive infection (78, 140). The extent of structural alteration during these steps is not clear, but morphologically recognizable electron dense particles can be seen near the NPC (140); thus, partial disassembly may expose several internal signals for the subsequent steps involved in nuclear entry. Retrograde movement of microtubule could be involved in intracytoplasmic trafficking from the attachment and internalization site closer to the NE, since the particle adheres to microtubule through hexon in vitro (216), and Ad infection is inhibited by treatment with microtubule depolymerizing agent (79). Whether the microtubule acts as a guide by directly interacting with the structural component(s) of virion or whether it provides some motive force with other cellular proteins is not known. Endocytosis and intracytoplasmic trafficking in Ad2 infection occur quickly, and many virions can be seen near the nucleus within 40 min of infection (139). One morphological difference is that electron-dense virion-like particles are absent from the nucleus in the early stages of Ad2 infection, but virion-like particles have been observed in the nucleus in SV40-infected cells (170). Ad virion therefore disassembles its structure to a point at which viral DNA can enter the nucleus, transcribe, and replicate. How an infecting adenovirus genome enters through NPC is not well understood, but the NLS of the covalently attached TP is believed to guide the viral DNA nuclear entry (138a, 412). The identity of other proteins accompanying the viral DNA is not known.

Nuclear Import of Adenovirus Proteins

In contrast to the number of known viral gene products, most of which function in the nucleus, very few NLSs have so far been identified. Among early and immediately early proteins, E1a, DBP, Pol, and pTP have been shown to have individual NLSs (Table 2). E1B-55 kDa lacks an NLS but can be cotransported to the nucleus by E4-34 kDa protein, though its NLS has not yet been identified (135).

The E1a-NLS is at the extreme carboxy-terminus, 235-DLSCKRPRP-243 (Table 2) (220), and the proteins' nuclear localization is influenced by immediate upstream residues whose presence impairs the signal function (98).

Ad5 DBP-NLS is a bipartite NLS (42-PPKKR-46 and 84-PKKKKK-89; Table 2) (246) whose parts can independently act in the protein's nuclear localization; yet, the nuclear localization defect can be complemented probably by interaction of the mutant protein with either another virus-encoded protein or with virus-induced cellular proteins (60, 246). The bipartite NLS near the amino-terminal region appears to represent most of the serotypes (383).

Ad2 Pol has an unusual bipartite NLS, with three clusters of basic residues within the amino-terminal 48 residues (8-RARR-11, 25-RRRVR-29, and 41-RARRRR-46; Table 2). A combination of all clusters is required for the nuclear localization of the polymerase; yet the first and last two clusters can function in the nuclear targeting of heterologous β-galactosidase (411). Despite the presence of NLS, Ad2 Pol is distributed in both cytoplasmic and nuclear compartments when expressed in isolation from pTP. Nuclear transport of Ad2 Pol thus is facilitated by interaction with pTP (412). The pTP-NLS is $RLPV(R)_6VP$ (Table 2) (412), located between amino acid residues 360 and 421.

Ad5 pIVa is a 449-residue protein that is essential for the activation of the major late promoter (242) either as a homodimer or as a heterodimer with other IVa2 gene products. The NLS resides in the carboxy-terminus, 434-DRDRWSRAYRARK-446 (Table 2), and partly overlaps a sequence-specific DNA binding domain at the very end (AYRARKTPK) (218).

The Ad2 fiber NLS (amino acids 1 to 5, MKRAR) is required for nuclear localization and with additional downstream residues (1-MKRARPSEDTF-11) promotes the nuclear localization of heterologous β-galactosidase (163).

HEPADNAVIRUSES

The *Hepadnaviridae* family consists of hepatotropic, species-specific viruses that infect certain mammalian and avian species, including the human hepatitis virus (HBV), the woodchuck hepatitis virus (WHV), the ground squirrel hepatitis virus (GSHV), the duck hepatitis virus (DHBV), and other viruses isolated from other animals. Hepadnavirus biology has several unusual features: 1. The infected liver produces three types of virus-related particles—two subviral lipoprotein particles containing only viral envelope proteins differing in shape (20-nm spheres and filaments 20-nm in diameter), and the infectious virion particle known as the Dane particle for HBV. 2. Production of all types of virion particles in the hepatocytes, either in cultured cells or in animals, does not kill the infected cells and often leads to persistent infection. 3. The infectious particle, which is also enveloped, contains a partially duplexed circular DNA and a covalently linked polymerase, which functions in the synthesis of hepadnavirus DNA from viral pregenomic RNA (pgRNA). Primary hepatocytes, terminally differentiated epithelial cells no longer undergoing cell division, which can only be obtained from liver explants, support productive infection. No established cell lines, however, support virion infection, even though some well-differentiated hepatoma cell lines can produce virions upon transfection with cloned viral DNA. Nonetheless, primary and established cell lines have contributed to the understanding of hepadnavirus reproduction. Recent articles provide details of hepadnavirus biology and immunopathogenesis (56, 121).

Structure and Life Cycle

The infectious virion of HBV, a 42-nm particle, consists of an inner nucleocapsid or core 30 or 34 nm in diameter, surrounded by a lipid envelope containing virally encoded surface proteins. Within the core is a 3.2-Kbp viral DNA (309), circular in overall structure but noncontiguous on both strands, and a polymerase (184, 295, 308). HBV and GSHV cores show protein kinase activity (6, 108, 127). In DHBV, heat shock proteins Hsp 90 and p23 are also packaged (167). The amino-terminal 150 residues of the 183 amino-acid HBV core protein are sufficient for capsid assembly, and the remaining 33 residues are required for packaging RNA (pgRNA). Two types of icosahedral shells obtained in vitro with the purified protein expressed in *E. coli*, composed of either 180 or 240 truncated proteins (36, 69), resemble the native core of the virus (Figure 1, *panel 4*). They reveal a largely helical fold of the truncated polypeptide and a clustering between two molecules to form a dimer spike on the virion surface. Although the arrangement of domains for capsid assembly and RNA packaging is conserved, length of the core polypeptide chain varies among hepadnaviruses. For example, DHBV core protein has 262 amino acids, and HBV has 183. The carboxyl-terminal region has three functions: general nucleic acid binding (148, 259), phosphorylation (208, 409, 410), and nuclear localization for the HBV protein (101, 407). The lipid envelope contains three viral envelope glycoproteins, one named after surface protein (S protein), and two comprising the entire S domain plus varying lengths of amino-terminal residues, the M-chain and L-chain, (150), which are required for virion formation and infectivity (43, 44, 222, 354, 372).

After enveloped virion binds to the cell through hepadnavirus receptors, DHBV is endocytosed and then penetrates the membrane in a pH-independent manner (191, 302); afterwards, the core moves closer to the nucleus and the viral genome enters the nucleus. Human hepatoma-derived cell lines, but not fibroblasts, support HBV binding, internalization, and intracytoplasmic movement toward the nucleus; but the virion particles remain associated with, but outside of, the nuclear envelope, and HBV-DNAs do not reach the nucleoplasm (294). If transfected by cloned HBV-DNA (326), these cultured cells can support an infection cycle like primary human hepatocytes infected by the virion (274, 305). Thus the established cell-lines are believed to lack a cellular factor(s) that converts an infecting virion into one capable of nuclear entry.

The conversion of incoming viral DNA into covalently closed circular DNA (cccDNA) and its appearance in the nucleus are among the earliest events in hepadnavirus replication (228, 229, 387). Fifty or more copies of cccDNA are present in each hepatocyte nucleus in addition to the incoming cccDNA genome. This cccDNA amplification in natural infection is confined to the early phase of infection prior to particle envelopment, and these cccDNA molecules serve as templates for viral transcription in the nucleus to produce discrete sets of genomic and subgenomic RNA molecules (reviewed in 121). From all transcripts that serve as the mRNAs for protein production, the smallest of the heterogeneous 3.5-Kbp transcripts (103) is selected for packaging into a capsid whose assembly requires core-dimer subunit interaction with the polymerase-bound pgRNA (199, 416). Encapsidated pgRNA undergoes reverse transcription in the cytoplasm (351). Processes involved in the protein-primed minus-sense DNA synthesis, the hydrolysis of the pgRNA, followed by cap-primed plus-sense DNA synthesis and viral DNA circularization, have been detailed elsewhere (121).

For DHBV, both capsid assembly and reverse transcription require a multi-component chaperone complex, which includes the heat shock protein Hsp90, p23, and possibly other proteins, and encapsidated in the virion (167). The length of the plus-sense DNA varies from long in the secreted and extracellular virion particles to short in intracellular and cytoplasmic core particles, and recent reports indicate a link between envelopment and the extent of DNA synthesis (126, 352, 386). Progeny cores bud into the intracellular membrane, into which virally encoded two- to three-envelope proteins are already inserted, and then acquire the glycoprotein envelope. A posttranslational modification of the L protein, the myristylation of the amino-terminal glycine residue, is essential for the production of infectious virion particles; mutants affecting L myristylation can assemble and secrete particles, but they are not infectious (43, 222, 354). The stage at which modification affects infection (i.e., adsorption, entry, or uncoating) is not yet known.

An important feature of hepadnavirus infection is the establishment of persistent infected hepatocytes that produce virions without being killed. For persistent infection, the size of the cccDNA amplification must be controlled. A fraction of progeny DNA genomes are known to be targeted to the nucleus to produce enough cccDNA to maintain the persistent state.

Nuclear Import of Hepadnavirus Proteins

Subcellular localization studies on the core proteins of HBV and DHBV indicate that different species seem to use different mechanisms to localize core proteins to the nucleus. During natural infection, these proteins primarily act in the cytoplasm for virion assembly, except for initial targeting of infecting genome and—assuming that core proteins accompany viral DNA to the nucleus—for the few capsids that need be routed to the nucleus to replenish the pool of viral cccDNAs. Nuclear localization of HBV core protein, which is karyophilic, depends on the cell cycle, enhanced during the Go/G phase but suppressed during the S phase (408), probably reflecting the hepatocyte's physiological state.

In cultured hepatocytes, internalized capsids accumulate in association with, but outside of, the nuclear membrane (294). In transgenic mice in which core proteins are expressed from the integrated HBV gene, core particles are found in the nucleus of hepatocytes (107, 143). Such particles driven to the cytoplasm during the division of the hepatocytes cannot re-enter the nucleus (143). Viral cccDNAs, however, are absent from the nucleus (107). These two results imply that assembled nucleocapsids cannot traverse the nuclear membrane.

The NLS of HBV core protein is located within the carboxy-terminal 34 amino acids, in which two sequences (145-ETTVVRRRGRSP-156 and 172-RRRRSQSRESQC-183) are important for activity (Table 2) (101, 407). The same residues contain four poly-arginine clusters, of which three, located between amino acids 157 to 183, have an SPRRR motif. The serine residues, in particular the ones penultimate to proline, are phosphorylated in vivo (315). Mutant proteins in which serines in the motif are changed to alanine are non-phosphorylated and localize to the nucleus regardless of the phase of the cell cycle, indicating that phosphorylation influences the protein's nuclear localization (208). The core protein-NLS overlaps with activities for DNA binding and pgRNA encapsidation (148, 259).

In DHBV core protein, however, similar serine-rich domains, whose sequence does not include poly-arginine but is juxtaposed with proline, are present in the carboxy 245 to 262 amino acids and are natural phosphorylation sites (410). Yet the serine-substituted mutant proteins are cytoplasmic, as is the wild-type protein (409). The phosphorylation state of the DHBV protein, in which 3 serines and 1 threonine are differentially phosphorylated, influences the following distinct steps of viral replication: (a) proteins phosphorylated at positions 245 and 259 stimulate DNA synthesis; (b) proteins dephosphorylated at 257 stimulate cccDNA synthesis; (c) proteins dephosphorylated at both 257 and 259 stimulate virus production; and (d) the presence of phosphoserine at 259 then initiates infection. These observations indicate the protein's key role in infection, including the genome's nuclear targeting, through modulating carboxyl residues by phosphorylation. The dynamics of phosphorylation as well as selective and cooperative subunit binding to pgRNA could control the amount of the core subunits destined to be delivered to the nucleus. In the presence of pgRNA during natural infection, the majority of core-dimers are driven for capsid formation, and only a small fraction of cytoplasmic cores containing relaxed viral circular DNA are targeted to the nucleus at any given time. This scarcity of core protein in the nucleus could escape cytochemical detection.

The NLS of HBV polymerase has been reported to reside in the amino-terminal portion of the protein (116, 322). This NLS appears to be hidden in the folded protein, but it becomes recognizable after being truncated from the enzymatic/catalytic domain. This truncated protein distributes in both the cytoplasm and nucleus.

A small preS2/S truncated protein consisting of 102 amino acids has been demonstrated to enter the nucleus by diffusion (155). Found in hepatocarcinoma cells (187), it can transactivate a number of cellular genes. HBV integration into host chromosomal DNA seems to cause truncation of the surface envelope protein gene, and the resulting gene product with altered protein function can distribute evenly within the cell.

Nuclear Import of Hepadnavirus DNA

Recent work implicates WHV-DNA bound polymerase in the nuclear targeting of WHV-DNA. Polymerase-bound viral DNA, but not polymerase-free DNA, was effectively imported into the nucleus in an energy-dependent manner in vitro (183). The proposed model is that following an alteration in its phosphorylation state, nucleocapsid disassembles at the cytoplasmic side of the NPC prior to the nuclear import of viral DNA, and the infecting DNA enters the nucleus via the presumed NLS of the associated polymerase. Although small in quantity, the nuclear import of core-encapsidated DNA was also observed. As stated by Kann et al (183), this result does not rule out the auxiliary role of either monomeric or dimeric core protein in DNA nuclear entry, but it does exclude the possibility that integrity of the virus-core structure must be maintained for the genome's transport. The notion that WHV core binding to the NPC depends on phosphorylation argues in favor of the role of core protein in viral DNA targeting.

How would modification modulate nuclear entry of the infecting genome? For HBV, all core protein NLSs/DBDs in a virion are inside the virion structure, and these NLSs must be unmasked in order to function, perhaps by phosphorylation. For DHBV, mature capsids are underphosphorylated (410), but phosphorylation at 259, the very residue whose dephosphorylation drives the assembly process, is essential for initiating infection, presumably for dissociating the nucleocapsid from the envelope at the viral penetration site (409), which thus requires 259 to be modified. This phosphorylation in turn could begin to alter capsid conformation. Unphosphorylated serine 257 is harbored in the nucleocapsid that is a direct precursor of cccDNA in the nucleus (409), favoring entry of at least some capsid protein together with the infecting DNA.

How would the incoming core be phosphorylated in a unique, sequence-specific manner? In what subcellular compartment would modification begin? In addition to polymerase, HBV and GSHB particles package a protein kinase activity (6, 108, 127), and DHBV particles package heat shock protein Hsp90 and p23 (167). The latter two proteins are normally required for the protein priming reaction of DNA synthesis, and the pgRNA RNP formations for packaging are found tightly associated with polymerase. Packaged cellular proteins could participate in modifying specific amino acids, in step with the virion's interaction with the cytoplasmic components as the nucleocapsid traverses membrane and cytoplasmic compartments, or in step with its interaction with the NPC.

In addition to the primary goal of targeting the infecting genome to the nucleus, progeny viral DNAs must be re-routed to the hepatocyte nucleus to maintain a steady-state level of cccDNA in each nucleus. This maintenance is regulated by the presence of both the capsid protein and the pre-S envelope protein (202, 353, 354, 371). The number of cccDNA copies per nucleus seems to be controlled by differentially phosphorylating a population of progeny core for re-routing to the nucleus as well as for envelopment. Other capsids of DNA viruses discussed here are composed of more than one capsid proteins that may perform different functions in virion assembly and in other stages of the productive infection cycle. Posttranslational modifications of a single hepadnavirus core protein have been postulated to be an alternative to encoding multiple capsid proteins by separate genes (409).

SUMMARY AND PERSPECTIVE

Viral proteins display all the hallmarks of the nuclear import of cellular proteins, including an NLS, the modulation of protein nuclear localization by amino-acid phosphorylation adjacent to the NLS, and tandem, piggy-back transport as a result of protein-protein interaction. Diffusional entry of small viral protein is also observed.

As expected for the proteins' function in assembly and gene expression through DNA-protein interactions, the NLS and the DNA- or RNA-binding domains are often found to overlap, suggesting that the dual signals in one segment of the protein have been adapted to offer proper folding, routing, and functioning.

The cytosol of SV40-transformed cells contains a kinase activity able to enlarge the functional nuclear pore size for gold particles harboring the minimal SV40 LT-NLS. Also, heterologous proteins coupled to a longer SV40 LT-NLS including the phosphorylation sites bind importin α/β, the signal recognizing receptor complex, better than those proteins harboring the minimal NLSs.

These observations clearly indicate the way SV40 selectively orchestrates the routing of the viral proteins to the nucleus.

Most NLSs identified so far in proteins encoded by the genomes of animal DNA viruses (Table 2) are of two types: simple, consisting of a few basic residues (SV40 LT-NLS), and bipartite (Py LT-NLS), consisting of two short stretches of basic residues separated by a spacer of variable length. These prototypes have been seen in many cellular proteins. Because NLSs are domains in the properly folded proteins or protein multimers, and must function in the protein's environment, they must have evolved under influence of the cytoplasmic receptors.

Although we know much about how viral gene products are targeted to the nucleus, little is known still about how infecting viral genomes reach their reproductive site. Clearly, protein components of the virion facilitate nuclear entry of the viral genome, which appears to be determined by differences in size, protein constituents, and structure of each virus.

Structural proteins associate with the SV40 genome and facilitate its nuclear entry through the protein's resident NLSs. The DNA binding domains of the respective proteins ensure their association with the SV40 genome. The complete dissociation of Vp1, Vp2, and Vp3 from the genome leads to the formation of import-incompetent genome. Then energy- and temperature-dependent genome nuclear import occurs through the NPC, and gene expression and virion assembly follow. Genetic evidence that $Vp2/3\text{-}NLS_{null}$ mutant arise noninfectious, supports Vp2/3-NLS's function in genome nuclear targeting. Yet, the internally located NLSs must become exposed to the cell's nuclear transport machinery. In order to accommodate for the 28-nm NPC opening, the 50-nm virion must be structurally altered to liberate some or all NLSs during cell entry and cytoplasmic trafficking. The processes converting SV40 into an import-competent form remain uncharacterized.

For WHV, the terminal protein covalently attached to the genome facilitates the energy-dependent entry of the viral DNA in vitro, and the disassembly of the 30-nm virion, similar to events in the cytoplasm of adenovirus infected cells, has been proposed. For DHBV, heat shock proteins are packaged, and they may participate in genome targeting. Genetic evidence suggests a role of core proteins in the formation of nuclear cccDNA and in genome targeting.

Because the TP of adenovirus is covalently attached to DNA, the TP-DNA complex has been postulated to be capable of nuclear entry. Assisted by a viral protease, the large adeno virions ($\sim$90 nm) undergo programmed disassembly to shed many of the structural proteins during internalization and cytoplasmic trafficking. At the cytoplasmic side of NPC, complete disassembly appears to occur, and the entry-competent genome enters the nucleus.

For parvovirus, the smallest animal DNA virus of 26 nm, the virion could pass through the NPC without much structural alteration if the presumed Vp1-NLSs exposed at the virion's surface remain available for the cell's nuclear import machinery. The MVM mutant with a deletion in the Vp1 region shows a post-entry defect, which is consistent with the interpretation that the loss of the putative NLS's function led to the phenotype.

For HSV, large (200–300 nm) enveloped virion enters the cell by fusion, and the wild-type function of the ICP 1/2 protein, a component of the tegument, appears to be needed for injecting HSV-1 DNA from the 125-nm nucleocapsid at the NPC into the nucleus. The observation that the particles of the *tsB7* mutant with a mutation within the ICP 1/2 coding sequence can reach the NPC with the virion's core still inside the nucleocapsid implies that the interaction of some tegument components with the nucleocapsid plays a key role in the genome nuclear targeting.

Clearly the mature virion of all animal DNA viruses that reproduce in the nucleus contain protein(s) that route(s) the infecting viral genome to the nucleus. Many more experiments must be performed to elucidate each virus's unique strategy for delivering viral DNA into the host nucleus to initiate infection. The protein's resident NLS can function in the genome's targeting through its association with DNA and may require exposure to the cell's import machinery. In particular, when the signal is embedded in the native protein, or is inside the virion particle, the virion's structure may need to be altered in the cytoplasm, either by a protease packaged in the virion or by cell's internalization. Virions can package cellular proteins that can route the protein-associated genome to the nucleus by modifying the protein, then allowing it to board the cell's import receptor and import machinery. Proper assembly of all these components in the mature virion particles sets the stage for nuclear targeting in the next infection cycle.

In the future, we anticipate that the protein domains that specify nuclear localization, DNA binding, and protein-protein interaction will be mapped for many or all proteins packaged in individual virions. We will also learn more about virion proteins' modification patterns. These domains and proteins modifications are likely to cooperate in genome targeting. The study of viral mutants, testing the effect of a domain's mutation when other intact viral genes are present, will perhaps be a key to a deeper understanding of the domain's role in the nuclear targeting.

In the past, DNA viruses have contributed to a molecular understanding of many fundamental cellular processes, including chromatin structure and its organization, gene expression, transcriptional and posttranscriptional regulation, initiation of translation, and protein nuclear transport. In addition, several viral early gene products have led to the identification of cellular proteins that control cell growth. In order to understand how viruses spread, we must understand what mediates the infecting viral genome's entry into the nucleus. Without such basic knowledge, we are still far from developing useful interventions in virus infections and devising effective nuclear delivery methods indispensable for gene therapy.

ACKNOWLEDGMENTS

We thank J Broadale and R Sallaberry for drawing Figure 2, P Sun for help in preparing Table 2, P Li and T Kasamatsu for helpful comments, and C McLaughlin for editing. This work was supported by National Institutes of Health Grant CA50574.

Visit the *Annual Reviews home page* at
http://www.AnnualReviews.org.

Literature Cited

1. Agbandje M, Kajigaya S, McKenna R, Young NS, Rossmann MG. 1994. The structure of human parvovirus B19 at 8 A resolution. *Virology* 230:106–15
2. Agbandje M, McKenna R, Rossmann MR, Strassheim ML, Parrish CR. 1993. Structure determination of feline panleukopenia virus empty capsids. *Proteins: Struc. Func. Genet.* 16:155–71
3. Agbandje M, Parrish CR, Rossmann MG. 1995. The structure of parvovirus. *Sem. Virol.* 6:299–309
4. Agutter PS, Suckling KE. 1982. The fluidity of the nuclear envelope lipid does not affect the rate of nucleocytoplasmic RNA transport in mammalian liver. *Biochim. Biophys. Acta* 696:308–14
5. Al-Lami F, Ledinko N, Toolan H. 1969. Electron microscope study of human NB and SMH cells infected with the parvovirus, H-1: involvement of the nucleolus. *J. Gen. Virol.* 5:485–92
6. Albin C, Robinson WS. 1980. Protein kinase activity in hepatitis B virions. *J. Virol.* 34:297–302
7. Ambinder RF, Mullen M, Chang Y-N, Hayward GS, Hayward D. 1991. Functional domains of Epstein-Barr virus nuclear antigen EBNA-1. *J. Virol.* 65:1466–78
8. Ambrose C, Blasquez V, Bina M. 1986. A block in initiation of simian virus 40 assembly results in the accumulation of minichromosomes, an exposed regulatory region. *Proc. Natl. Acad. Sci. USA* 83:3287–91
9. Amin M, Mirza A, Weber J. 1979. Uncoating of adenovirus type 2. *J. Virol.* 30:462–71
10. Anderson HA, Chen Y, Norkin C. 1996. Bound simian virus 40 translocates to caveolin-enriched membrane domains, and its entry is inhibited by drugs that selectively disrupt caveolae. *Mol. Biol. Cell* 7:1826–34
11. Anderson RGW, Kamen BA, Rothenberg KG, Lacey SW. 1992. Potocytosis: sequestration and transport of small molecules by caveolae. *Science* 255:410–11
12. Ascoli CA, Maul GG. 1991. Identification of a novel nuclear domain. *J. Cell Biol.* 112:785–95
13. Bai M, Harfe B, Freimuth P. 1993. Mutations that alter an Arg-Gly-Asp (RGD) sequence in the adenovirus type 2 penton base protein abolish its cell-rounding activity and delay virus reproduction in flat cells. *J. Virol.* 67:5198–205
14. Baker TS, Drak J, Bina M. 1988. Reconstruction of the three-dimensional structure of simian virus 40 and visualization of the chromatin core. *Proc. Natl. Acad. Sci. USA* 85:422–26
15. Banfield BW, Leduc Y, Esford K, Schubert K, Tufaro F. 1995. Evidence for an interaction of herpes simplex virus with chondroitin sulfate proteoglycans during infection. *Virology* 208:531–39
16. Banfield BW, Tufaro F. 1990. Herpes simplex virus particles are unable to traverse the secretory pathway in the mouse

L-cell gro29. *J. Virol.* 64:5716–29

17. Barkan A, Welch RC, Mertz JE. 1987. Missense mutations in the Vp1 gene of a simian virus 40 that compensate for defects caused by deletions in the viral agnogene. *J. Virol.* 61:3190–98

18. Barouch DH, Harrison SC. 1994. Interactions among the major and minor coat proteins of polyomavirus. *J. Virol.* 68:3982–89

19. Basak S, Turner H. 1992. Infectious entry pathway for canine parvovirus. *Virology* 186:368–76

20. Bastos R, Panté N, Burke B. 1995. Nuclear pore complex proteins. *Int. Rev. Cytol.* 162B:257–302 (Suppl.)

21. Bates R, Snyder CE, Banerjee PT, Mitra S. 1984. Autonomous parvovirus LuIII encapsidates equal amounts of plus and minus DNA strands. *J. Virol.* 49:319–24

22. Batterson W, Furlong D, Roizman B. 1983. Molecular genetics of HSV VIII. Further characterization of a temperature-sensitive mutant defective in release of viral DNA and in other stages of the viral reproductive cycle. *J. Virol.* 45: 397–407

23. Batterson W, Roizman B. 1983. Characterization of the HSV virion-associated factor responsible for the induction of alpha genes. *J. Virol.* 46:371–77

24. Bergelson JM, Cunningham JA, Droguett G, Kurt-Jones EA, Krithivas A, et al. 1997. Isolation of a common receptor for coxsackie B viruses and adenoviruses 2 and 5. *Science* 275:1320–23

25. Berns K, Labow MA. 1987. Parvovirus gene regulation. *J. Gen. Virol.* 68:601–14

26. Berns K, Rose JA. 1970. Evidence for a single-stranded adenovirus-associated virus genome: isolation and separation of complementary single strands. *J. Virol.* 5:693–99

27. Berns KI. 1996. Parvoviridae: the viruses and their replication. In *Fundamental Virology*, ed. BN Fields, DM Knipe, PM Howley, et al. 31:1017–41. Philadelphia: Lippincott-Raven. 3rd ed.

28. Berns KI, Bergoin M, Bloom M, Lederman M, Muzyczka N, et al. 1994. Parvoviridae, IV: the report of international committee on taxonomy of viruses. In *Virus Taxonomy, Arch. Virol. Suppl. 10.* ed. FA Murphy, CM Fauquet, DHL Bishop, et al.

29. Berns KI, Giraud C. 1996. Biology of adeno-associated virus. *Adeno-Associated Virus (AAV) Vectors in Gene Therapy*, ed. KI Berns, C Giraud. *Curr. Top. Microbiol. Immunol.* 218:1–23 (Suppl.)

30. Blasquez V, Stein A, Ambrose C, Bina M. 1986. Simian virus 40 protein Vp1 is involved in spacing nucleosomes in minichromosomes. *J. Mol. Biol.* 191:97–106

31. Bockus BJ, Schaffhausen B. 1987. Localization of the phosphorylations of polymavirus large T antigen. *J. Virol.* 61: 1155–63

32. Bodnar JW, Hanson PI, Polvino-Bodnar M, Zempsky W, Ward DC. 1989. The terminal regions of adenovirus and minute virus of mice DNA's are preferentially associated with the nuclear matrix in infected cells. *J. Virol.* 63:4344–53

33. Boehmer PE, Lehman IE. 1997. HSV DNA replication. *Annu. Rev. Biochem.* 66:347–84

34. Booy FP, Newcomb WW, Trus BL, Brown JC, Baker TS, Steven AC. 1991. Liquid-crystalline, phage-like packing of encapsidated DNA in HSV. *Cell* 64: 1007–15

35. Booy FP, Trus BL, Newcomb WW, Brown JC, Conway JF, Stevens AS. 1994. Finding a needle in a haystack: detection of a small protein (the 12 kD Vp26) in a large complex (the 200MD capsid of HSV). *Proc. Natl. Acad. Sci. USA* 91:5652–56

36. Böttcher B, Wynne SA, Crowther RA. 1997. Determination of the fold of the core protein of hepatitis B virus by electron cryomicroscopy. *Nature* 386:88–91

37. Breau WC, Atwood WJ, Norkin LC. 1992. Class I major histocompatibility proteins are an essential component of the simian virus 40 receptor. *J. Virol.* 66: 2037–45

38. Brown KE, Young NS. 1997. Parvovirus B19 in human disease. *Annu. Rev. Med.* 48:59–67

39. Brown KI, Anderson SM, Young NS. 1993. Erythrocyte P antigen: cellular receptor for B19 parvovirus. *Science* 262: 114–17

40. Brown DT, Westphal M, Burlingham BT, Winterhof U, Doerfler W. 1975. Structure and composition of adenovirus type 2 core. *J. Virol.* 16:366–87

41. Brunetti CR, Burke RL, Hoflack B, Ludwig T, Dingwell KS, Johnson DC. 1995. Role of mannose-6-phosphate receptors in herpes simplex virus entry into cells and cell-to-cell transmission. *J. Virol.* 69:3517–28

42. Brunetti CR, Burke RL, Kornfield S, Gregory W, Masiarz KS, et al. 1995. Herpes simplex virus glycoprotein D acquires mannose-6-phosphate residues

and binds to mannose-6-phosphate receptors. *J. Biol.Chem.* 269:17067–74

43. Bruss V, Ganem D. 1991. The role of envelope proteins in hepatitis B virus assembly. *Proc. Natl. Acad. Sci. USA.* 88:1059–63

44. Bruss V, Thomssen R. 1994. Mapping a region of the large envelope protein required for hepatitis B virion maturation. *J. Virol.* 68:1643–50

45. Burnett RM. 1997. The structure of adenovirus. In *Structural Biology of Viruses,* ed. W Chiu, RM Burnett, RL Garcea, 8:209–38. New York: Oxford Univ. Press

46. Calvert J, Summers J. 1994. Two regions of an avian hepadnavirus RNA pregenome are required in cis for RNA encapsidation. *J. Virol.* 68:2084–90

47. Carswell S, Alwine JC. 1986. Simian virus 40 agnoprotein facilitates perinuclear-nuclear localization of Vp1, the major capsid protein. *J. Virol.* 60:1055–61

48. Chadonnet Y, Dales S. 1970. Early events in the interaction of adenoviruses with Hela cells. I. Penetration of type 5 and intracellular release of the DNA genome. *Virology* 40:462–77

49. Chadonnet Y, Dales S. 1970. Early events in the interaction of adenovirus with Hela cells. II. Comparative observations on the penetration of type 1, 5, 7, and 12. *Virology* 40:478–85

50. Chang D, Cai X, Consigli RA. 1993. Characterization of the DNA binding properties of polyomavirons capsid proteins. *J. Virol.* 67:6327–31

51. Chang D, Haynes JI, Brady JN, Consigli RA. 1992. Identification of a nuclear localization sequence in the polyomavirus capsid protein Vp2. *Virology* 191:978–83

52. Chang D, Haynes JI, Brady JN, Consigli RA. 1992. The use of additive and subtractive approaches to examine the nuclear localization sequence of the polyomavirus major capsid protein Vp1. *Virology* 189:821–27

53. Chapman MS, Rossmann MG. 1995. Single-stranded DNA-protein interactions in canine parvovirus. *Structure* 3:151–62

54. Chejanovsky N, Carter BJ. 1989. Mutagenesis of an AUG codon in the adeno-associated virus rep gene: effects on viral DNA replication. *Virology* 173:120–28

55. Chipman PR, Agbandje-McKenna M, Kajigaya S, Brown KE, Young NS, et al. 1996. Cryo-electron microscopy studies of empty capsids of human parvovirus B19 complexed with its cellular receptor. *Proc. Natl. Acad. Sci. USA* 93:7502–6

56. Chisari FV, Ferrari C. 1997. Viral hepatitis. In *Viral Pathogenesis,* ed. N Nathanson et al, 31:745–78. Philadelphia: Lippincott-Raven. 940 pp.

57. Chou J, Roizman B. 1989. Characterization of DNA sequence-common and sequence specific proteins binding to cis-acting sites for cleavage of the terminal a sequence of the HSV1 genome. *J.Virol.* 63:1059–68

58. Chow LT, Broker TR. 1997. Small DNA tumor viruses. In *Viral Pathogenesis,* ed. N Nathanson, et al, 12:267–301. Philadelphia: Lippincott-Raven

59. Chraboczek J, Ruigrok RWH, Cusak S. 1995. Adenovirions fiber. In *Current Topics in Microbiology and Immunology, The Molecular Repertoire of Adenovirus I,* ed. W Doenfleu, P Böhm. Vol. 199/I, pp. 164–200. Berlin: Springer-Verlag

60. Cleghon VG, Voelkerdig K, Morin N, Delsert C, Klessig DF. 1989. Isolation and characterization of a viable adenovirus mutant defective in nuclear transport of the DNA-binding protein. *J. Virol.* 63:2289–99

61. Clever J, Dean DA, Kasamatsu H. 1993. Identification of a DNA binding domain in simian virus 40 capsid proteins Vp2 and Vp3. *J. Biol. Chem.* 268:20,877–83

62. Clever J. 1992. A structure-function analysis of SV40 structural protein Vp3. PhD thesis. Univ. Calif., Los Angeles. 259 pp.

63. Clever J, Kasamatsu H. 1991. Simian virus 40 V p2/3 small structural proteins harbor their own nuclear transport signal. *Virology* 181:78–90

64. Clever J, Yamada M, Kasamatsu H. 1991. Import of simian virus 40 virions through nuclear pore complexes. *Proc. Natl. Acad. Sci. USA* 88:7333–37

65. Cole CN, Landers T, Goff SP, Maneuil-Brutlag S, Berg P. 1977. Physical and genetic characterization of deletion mutants of simian virus 40 constructed in vitro. *J. Virol.* 24:277–94

66. Cole CN. 1996. Polyomavirinae: The viruses and their replication. In *Fundamental Virology,* ed. BN Field, DM Knipe, PM Howley, 28:917–45. Philadelphia: Lippincott-Raven. 1340 pp. 3rd ed.

67. Colledge WH, Richardson WD, Edge MD, Smith AE. 1986. Extensive mutagenesis of the nuclear location signal

of simian virus 40 large T-antigen. *Mol. Cell. Biol.* 6:4136–39

68. Colomar MC, Degoumois-Sahli C, Beard P. 1993. Opening and refolding of simian virus 40 and in vitro packaging of foreign DNA. *J. Virol.* 67:2779–86

69. Conway JF, Cheng N, Zlotnick A, Wingfield PT, Stahl SJ, Steven AC. 1997. Visualization of a 4-helix bundle in the hepatitis B virus capsid by cryo-electron microscopy. *Nature* 386:91–94

69a. Conway JF, Trus BL, Booy FP, Newcomb WW, et al. 1996. Visualization & three-dimensional density maps reconstructed from cryoelectron micrographs of viral capsids. *J. Struc. Biol.* 116:200–8

70. Cotmore S, Tattersall P. 1988. The NS 1 polypeptide of minute virus of mice is covalently attached to the 5' termini of duplex replicative-form DNA and progeny single strand. *J. Virol.* 62:851–60

71. Cotmore S, Tattersall P. 1990. Alternate splicing in a parvoviral nonstructural gene links a common amino terminal sequence to down stream domains which confer radically different localization and turnover characteristics. *Virology* 177:477–87

72. Cotmore S, Tattersall P. 1995. DNA replication in the autonomous parvoviruses. *Sem. Virol.* 6:271–81

73. Cotmore SF, Tattersall P. 1989. A genome-linked copy of the NS1 polypeptide is located on the outside of infectious parvovirus particles. *J. Virol.* 63:3902–11

74. Cotmore SF, D'Abramo AM Jr, Carbonell LF, Bratton J, Tattersall P. 1997. The NS 2 polypeptide of parvovirus MVM is required for capsid assembly in murine cells. *Virology* 231:267–80

75. Cotmore SF, Nuesch JPF, Tattersall P. 1992. In vitro excision and replication of 5' telomeres of minute virus of mice DNA from cloned palindromic concatemer junctions. *Virology* 190:365–77

76. Cotmore SF, Tattersall P. 1987. The autonomously replicating parvoviruses of vertebrates. *Adv. Virus Res.* 33:91–174

77. Cotmore SF, Tattersall P. 1996. Parvovirus DNA replication. In *DNA Replication in Eukaryotic Cells*, ed. De Phamphilis, pp. 799–813. New York: Cold Spring Harbor Lab.

78. Cotten M, Weber JH. 1995. The adenovirus protease is required for virus entry into host cells. *Virology* 213:494–502

79. Dales S, Chardonnet Y. 1973. Early events in the interaction of adenovirus with Hela cells. IV. Association with

microtubules and the nuclear pore complex during vectorial movement in the inoculum. *Virology* 56:465–83

80. Dangoria NS, Breau WC, Anderson HA, Cishek DM, Norkin LC. 1996. Extracellular simian virus 40 induces an ERK/MAP kinase-independent signaling pathway that activates primary response genes and promotes virus entry. *J. Gen. Virol.* 77:2173–82

81. Davis LI. 1995. The nuclear pore complex. *Annu. Rev. Biochem.* 64:865–96

82. Dean DA, Kasamatsu H. 1994. Signal- and energy-dependent nuclear transport of Vp3 by isolated nuclei. *J. Biol. Chem.* 269:4910–16

83. Dean DA, Li PP, Lee LM, Kasamatsu H. 1995. Essential role of the Vp2 and Vp3 DNA-binding domain in simian virus 40 morphogenesis. *J. Virol.* 69:1115–21

84. DeBruyn Kops A, Knipe DM. 1994. Pre-existing nuclear architecture defines the intranuclear location of herpes virus DNA replication structures. *J. Virol.* 68:3512–26

85. Deiss LP, Chou J, Frenkel N. 1986. Functional domains within the a sequence involved in the cleavage-packaging of herpes simplex virus DNA. *J. Virol.* 59:605–18

86. Deiss LP, Frenkel M. 1986. Herpes simplex virus amplicon: cleavage of concatemeric DNA is linked to packaging and involves amplification of the terminally reiterated a sequence. *J. Virol.* 57:933–41

87. Delos SE, Montross L, Morland RB, Garcea RL. 1993. Expression of the polyomavirus Vp2 and Vp3 proteins in insect cells: co-expression with the major capsid protein Vp1 alters Vp2/Vp3 subcellular localization. *Virology* 194:393–98

88. Desai P, Watkins SC, Person S. 1994. The size and symmetry of B capsids of HSV1 are determined by the gene products of the UL26 ORF. *J. Virol.* 68:5365–74

89. Deshmane SL, Frase NW. 1989. During latency, herpes simplex virus type 1 DNA is associated with nucleosome in a chromatin structure. *J. Virol.* 63:943–47

90. Devaux C, Caillet-Boudin M-L, Jacrot B, Boulanger P. 1987. Crystallization, enzymatic cleavage, and polarity of the adenovirus type 2 fiber. *Virology* 161:121–28

91. Devi-Rao GB, Goodart SA, Hecht LB, Rochford R, Rice MK, Wager EK. 1991. The relationship between polyadenylated and non-polyadenylated herpes

simplex virus type 1 latency-associated transcripts. *J. Virol.* 65:2179–90

92. Diacumakos EG, Gershey EL. 1977. Uncoating and gene expression of simian virus 40 in CV-1 cell nuclei inoculated by microinjection. *J. Virol.* 24:903–06

93. Dingwall C, Laskey RA. 1986. Protein import into the cell nucleus. *Annu. Rev. Cell Biol.* 2:367–90

94. Dingwall C, Laskey RA. 1991. Nuclear targeting sequences—a consensus? *Trends Biochem. Sci.* 16:478–81

95. Dingwall C, Robbins J, Dilworth SM, Roberts B, Richardson WD. 1988. The nucleoplasmin nuclear location sequence is larger and more complex than that of SV40 large T-antigen. *J. Cell Biol.* 107:841–49

96. Dingwall C, Sharmick SV, Laskey RA. 1982. The nucleoplasmin nuclear location sequence is larger and more complex than that of SV40 large T-antigen. *Cell* 30:449–58

97. Dollard SC, Wilson JL, Demeter LM, Bonnez W, Reichman RC, et al. 1992. Production of human papillomavirus and modulation of the infectious program in epithelial raft cultures. *Genes Dev.* 6:1131–42

98. Douglas JL, Quinlam MP. 1996. Efficient nuclear localization and immortalizing ability, two functions dependent on the adenovirus type 5 (Ad5) E1A second exon, are necessary for cotransformation with Ad5 E1B but not with T24 ras. *Virology* 220:339–49

99. Doye V, Hurt EC. 1995. Genetic approaches to nuclear pore structure and function. *Trends Genet.* 11:235–41

100. Dyer AP, Banfield BW, Martindale D, Spannier D-M, Tufaro F. 1997. Dextran sulfate can act as an artificial receptor to mediate a type-specific HSV infection via glycoprotein B. *J. Virol.* 71:191–98

101. Eckhardt SG, Milich DR, McLachlan A. 1991. Hepatitis B virus core antigen has two nuclear localization sequences in the arginine-rich carboxyl terminus. *J. Virol.* 65:575–82

102. Efthymiadis A, Shao H, Hübner S, Jans DA. 1997. Kinetic characterization of the human retinoblastoma protein bipartite nuclear localization sequence (NLS) *in vivo* and *in vitro*. A comparison with the SV40 large T-antigen NLS. *J. Biol. Chem.* 272:22134–39

103. Enders GH, Ganem D, Varmus HE. 1987. 5′-terminal sequences influence the segregation of ground squirrel hepatitis virus RNAs into polyribosomes

and viral core particles. *J. Virol.* 61:35–41

104. Evander M, Frazer IH, Payne E, Qi YM, Hengst K, McMillan NA. 1997. Identification of the $\alpha6$ integrin as a candidate receptor for papillomaviruses. *J. Virol.* 71:2449–56

105. Everette RD. 1988. Analysis of the functional domains of herpes simplex virus type 1 immediate-early polypeptide Vmw 110. *J. Mol. Biol.* 202:87–96

106. Everette RD, Maul GG. 1994. HSV-1 1E protein Vmw 110 causes redistributions of PML. *EMBO J.* 13:5062–69

107. Farza H, Hadchouel M, Scotto J, Tiollais P, Babinet C, Pourcel C. 1988. Replication and gene expression of hepatitis B virus in a transgenic mouse that contains the complete viral genome. *J. Virol.* 62:4144–52

108. Feitelson MA, Marion P, Robinson WC. 1986. Core particles of hepatitis B virus and ground squirrel hepatitis virus II: characterization of the protein kinase reaction associated with ground squirrel hepatitis virus and hepatitis B virus. *J. Virol.* 43:74–748

109. Feldherr CM, Akin D. 1990. The permeability of the nuclear envelope in dividing and nondividing cell culture. *J. Cell Biol.* 111:1–8

110. Feldherr CM, Akin D. 1994. Role of nuclear trafficking in regulating cellular activity. *Int. Rev. Cytol.* 151:183–228

111. Feldherr C, Akin D. 1995. Stimulation of nuclear import by simian virus 40–transformed cell extracts is dependent on protein kinase activity. *Mol. Cell. Biol.* 15:7043–49

112. Feldherr C, Lanford R, Akin D. 1992. Signal-mediated nuclear transport in simian virus 40 transformed cells is regulated by large tumor antigen. *Proc. Natl. Acad. Sci. USA* 89:11002–05

113. Felding-Habermann B, Mueller BM, Romerdahl CA, Cheresh DA. 1992. Involvement of integrin alpha v gene expression in human melanoma tumorigenicty. *J. Clin. Invest.* 89:2018–22

114. FitzGerald DJP, Padmanabhan R, Pastan I, Willingham MC. 1983. Adenovirus-induced release of epidermal growth factor and pseudomonas toxin into the cytosol of kB cells during receptor-mediated endocytosis. *Cell* 32:607–17

115. Forstová J, Krauzewicz N, Wallace S, Street AJ, Dilworth SM, et al. 1993. Cooperation of structural proteins during late events in the life cycle of polyomavirus. *J. Virol.* 67:1405–13

116. Foster GR, Ackrill AM, Goldin RD, Kerr IM, Thomas HC, Stark GR. 1991. Expression of the terminal protein region of hepatitis B virus inhibits cellular responses to interferons α and γ double-stranded RNA. *Proc. Natl. Acad. Sci. USA* 88:2888–92

117. Fox E, Moen PT Jr, Bodner JW. 1990. Replication of minute virus of mice DNA in adenovirus-infected or adeno-virus-transformed cells. *Virology* 176:403–12

118. Fredman JN, Engler JA. 1993. Adenovirus precursor to terminal protein interacts with the nuclear matrix in vivo and in vitro. *J. Virol.* 67:3384–95

119. Friedman HM, Glorioso JC, Cohen GH, Hastings JC, Harris CH, et al. 1986. Binding of complement component C3b to glycoprotein C of herpes simplex virus type 1: mapping of gC-binding sites and demonstration of conserved C3b binding in low passage clinical isolates. *J. Virol.* 60:470–75

120. Fujikawa K, Furuse M, Uwabe K-I, Maki H, Yoshie O. 1994. Nuclear localization and transforming activity of human papillomavirus type 16 E7-β-galactosidase fusion protein: characterization of the nuclear localization sequence. *Virology* 204:789–93

121. Ganem D. 1996. Hepadnaviridae and their replication. In *Fundamental Virology*, ed. BN Fields, DM Knipe, PM Howley, 35:1199–234. Philadelphia: Lippincott-Raven, 1340 pp. 3rd ed.

122. Gao M, Knipe D. 1992. Distal protein sequences can affect the function of a nuclear localization signal. *Mol. Cell. Biol.* 12:1330–39

123. Gao M, Matusick-Kumar L, Hurlburt W, Ditusa SF, Newcomb WM, et al. 1994. The protease of HSV1 is essential for functional capsid formation and viral growth. *J. Virol.* 68:3702–12

124. Garber EA, Seidman MM, Levine AJ. 1980. Intracellular SV40 nucleoprotein complexes: synthesis to encapsidation. *Virology* 107:398–401

125. Garcea RL, Liddington RC. 1997. Structural biology of polyomaviruses. In *Structural Biology of Viruses*, ed. W. Chiu, RM Burnett, RL Garcea, 7:187–208. New York: Oxford Univ. Press

126. Gerelsaikhan T, Tavis JE, Bruso V. 1996. Hepatitis B virus nucleocapsid envelopment does not occur without genomic DNA synthesis. *J. Virol.* 70:4269–74

127. Gerlich W, Goldman U, Muller R, Stibbe W, Wolff W. 1982. Specificity and localization of the hepatitis B virus-associated protein kinase. *J. Virol.* 42:761–66

128. Gershey EL, Diacumakos EG. 1978. Simian virus 40 production after viral uncoating in the CV-1 cell nucleus. *J. Virol.* 28:415–16

129. Gharakhanian E, Kasamatsu H. 1990. Two independent signals, a nuclear localization signal and a Vp1-interactive signal, reside within the carboxy-35 amino acids of SV40 Vp3. *Virology* 178:62–71

130. Gharakhanian E, Takahashi J, Clever J, Kasamatsu H. 1988. *In vitro* assay for protein-protein interaction: carboxyl-terminal 40 residues of simian virus 40 structural protein Vp3 contain a determinant for interaction with Vp1. *Proc. Natl. Acad. Sci. USA* 85:6607–11

131. Giraud C, Winocour E, Berns KI. 1995. Recombinant junctions formed by site-specific integration by adeno-associated virus into an episome. *J. Virol.* 69:6917–24

132. Glenn GJ, Eckhart W. 1990. Transcriptional regulation of early response genes during polyomavirus infection. *J. Virol.* 64:2193–201

133. Goldberg MW, Allen TD. 1992. High resolution scanning electron microscopy of the nuclear envelope: demonstration of a new, regular, fibrous lattice attached to the baskets of the nucleoplasmic face of the nuclear pores. *J. Cell Biol.* 119:1429–40

134. Goldfarb D. 1997. Whose finger is on the switch? *Science* 276:1814–16

135. Goodman FD, Shenk T, Ornelles DA. 1996. Adenovirus early region 4 34-kilodalton protein directs the nuclear localization of the early region 1B 55-kilodalton protein in primate cells. *J. Virol.* 70:6323–35

136. Gordon-Shaag A, Ben-Nun-Shaul O, Kasamatsu H, Oppenheim AB, Oppenheim A. 1998. The SV40 capsid protein Vp3 cooperates with the cellular transcription factor Sp1 in DNA-binding and in regulating viral promoter activity. *J. Mol. Biol.* 275:187–95

137. Görlich D, Mattaj IW. 1996. Nucleocytoplasmic transport. *Science* 271:1513–18

138. Grable M, Hearing P. 1992. *cis* and *trans* requirements for the selective packaging of adenovirus type 5 DNA. *J. Virol.* 66:723–31

138a. Greber U, Suomalainen X, Stidwill RP, Boucke K, Ebersold MW, Helenius A. 1997. The role of the nuclear pore complex in adenovirus DNA entry. *EMBO J.* 16:5998–6007

139. Greber UF, Willetts M, Webster P, Helenius A. 1993. Stepwise dismantling of adenovirus 2 during entry into cells. *Cell* 75:477–86

140. Greber UF, Webster P, Weber J, Helenius A. 1996. The role of the adenovirus protease in virus entry into cells. *EMBO J.* 15:1766–77

141. Griffith JP, Griffith DL, Rayment I, Murakami WT, Casper DLD. 1992. Inside polyomavirus at 25 Å resolution. *Nature* 355:652–54

142. Gruenheid S, Gatzke L, Meadows H, Tufaro F. 1993. Herpes simplex virus infection and propagation in a mouse L-cell mutant lacking heparan sulfate proteoglycans. *J. Virol.* 67:93–100

143. Guidotti LG, Martinez V, Loh Y-T, Ragler CE, Chisari FV. 1994. Hepatitis B virus nucleocapsid particles do not cross the hepatocyte nuclear membrane in transgenic mice. *J. Virol.* 68:5469–75

144. Haarr LS, Skulstad S. 1994. The herpes simplex virus type 1 particle: structure and molecular functions. *APMIS* 102:321–46

145. Hammarskjold ML, Winberg G. 1980. Encapsidation of adenovirus 16 DNA is directed by a small DNA sequence at the left end of the genome. *Cell* 20:787–95

146. Hasson TB, Ornelles DA, Shenk T. 1992. Adenovirus L1 52- and 55-kilodalton proteins are present within assembling virions and co-localize with nuclear structures distinct from replication centers. *J. Virol.* 66:6133–42

147. Hasson TB, Soloway PD, Ornelles DA, Doefler W, Shenk T. 1989. Adenovirus L1 52- and 55-kiladalton proteins are required for assembly of virions. *J. Virol.* 63:3612–21

148. Hatton T, Zhou S, Stranding DN. 1992 RNA- and DNA-binding activities in hepatitis B virus capsid protein: a model for their roles in viral replication. *J. Virol.* 66:5232–41

149. Hearing P, Samulski R, Wishart W, Shenk T. 1987. Identification of a repeated sequence element required for efficient encapsidation of the adenovirus type 5 chromosome. *J. Virol.* 61:2555–58

150. Heermann KH, Goldmann U, Schwartz W, Seyffarth T, Baumgarten H, Gerlich WH. 1984. Large surface proteins of hepatitis B virus containing the pre-S sequence. *J. Virol.* 52:396–402

151. Heine JW, Honess RW, Cassai E, Roizman B. 1974. Proteins specified by HSV. XII. The virion polypeptides of type 1 strains. *J. Virol.* 14:640–51

152. Herold BC, Visalli RJ, Susmarski N, Brandt CR, Spear PG. 1994. Glycoprotein C-independent binding of herpes simplex virus to cells requires cell surface heparan sulfate and glycoprotein B. *J. Gen. Virol.* 75:1211–22

153. Herrmann M, Oppenländer M, Pawlita M. 1995. Fast and high-affinity binding of B-lymphotropic papovirus to human B-lymphonia cell lines. *J. Virol.* 69:5797–804

154. Herrmann M, von der Lieth CW, Stehling P, Rutter W, Pawlita M. 1997. Consequences of a subtle sialic acid modification on the murine polyomavirus receptor. *J. Virol.* 71:5922–31

155. Hildt E, Urban S, Hofschneider PH. 1995. Characterization of essential domains for the functionality of the MHBs^t transcriptional activator and identification of a minimal MHBs^t activator. *Oncogene* 11:2055–66

156. Hill JM, Gebhardt BM, Wen RJ, Bouterie AM, Thompson RJ, et al. 1996. Quantitation of herpes simplex virus type 1 DNA and latency-associated transcripts in rabbit trigeminal ganglia demonstrates a stable reservoir of viral nucleic acids during latency. *J. Virol.* 70:3137–41

157. Hill JM, Sedarati F, Javier RT, Wager EK, Stevens JG. 1990. Herpes simplex virus latent phase transcription facilitates in vitro reactivation. *Virology* 174:117–25

158. Hinshaw JE, Carragher BO, Milligan RA. 1992. Architecture and design of the nuclear pore complex. *Cell* 69:1133–41

159. Holden VR, Caughman GB, Zhao Y, Harty RN, O'Callaghan DJ. 1994. Identification and characterization of the ICP22 protein of equine herpesvirus 1. *J. Virol.* 68:4329–40

160. Homa FL, Brown JC. 1997. Capsid assembly and DNA packaging in herpes simplex virus. *Rev. Med. Virol.* 7:107–22

161. Homes KV. 1997. Localization of viral infection. In *Viral Pathogenesis*, ed. N Nathanson, pp. 3:35–53. Philadelphia: Lippincott-Raven. 940 pp.

162. Hong Z, Beaudet-Miller M, Durkin J, Zhang R, Kwong AD. 1996. Identification of a minimal hydrophobic domain in the HSV1 scaffolding protein which is required for interaction with the major capsid protein. *J. Virol.* 70:533–40

163. Hong JS, Engler JA. 1991. The amino terminus of the adenovirus fiber protein encodes the nuclear localization signal. *Virology* 185:758–67

164. Hong SS, Karayan L, Tournier J, Curiel DT, Boulanger PA. 1997. Adenovirus type 5 fiber knob binds to MHC class I α 2 domain at the surface of human epithelial and B lymphoblastoid cells. *EMBO J.* 16:2294–306

165. Howes SH, Bockus BJ, Schaffhausen BS. 1996. Genetic analysis of polyomavirus large T nuclear localization: nuclear localization is required for productive association with pRb family members. *J. Virol.* 70:3581–88

166. Howley PM. 1996. Papillomavirinae: the viruses and their replication. In *Fundamental Virology*, ed. BN Field, DM Knipe, PM Howley, 29:947–78. Philadelphia: Lippincott-Raven. 1340 pp. 3rd ed.

167. Hu J, Toft DO, Seeger C. 1997. Hepadnavirus assembly and reverse transcription require a multi-component chaperon complex which is incorporated into nucleocapsid. *EMBO J.* 16:59–68

168. Hübner S, Xiao C-Y, Jans DA. 1997. The protein kinase CK2 site (Ser$^{111/112}$) enhances recognition of the simian virus 40 large T-antigen nuclear localization sequence by importin. *J. Biol. Chem.* 272: 17191–95

169. Hummel M, Lim HB, Laimins LA. 1995. Human papillomavirus type 31b late gene expression is regulated through protein kinase C-mediated changes in RNA processing. *J. Virol.* 69:3381–88

170. Hummeler K, Tomassini N, Sokol F. 1970. Morphological aspects of the uptake of simian virus 40 by permissive cells. *J. Virol.* 6:87–93

171. Hunter LA, Samulski RJ. 1992. Colocalization of adeno-associated virus type 2 rep and capsid proteins in the nuclei of infected cells. *J. Virol.* 66:317–24

172. Ishii N, Minami N, Chen EY, Medina AL, Chico MM, Kasamatsu H. 1996. Analysis of a nuclear localization signal of simian virus 40 major capsid protein Vp1. *J. Virol.* 70:1317–22

173. Ishii N, Nakanishi A, Yamada M, Macalalad MH, Kasamatsu H. 1994. Functional complementation of nuclear targeting-defective mutants of simian virus 40 structural proteins. *J. Virol.* 68:8209–19

174. Ishov AM, Maul GG. 1996. The periphery of nuclear domain 10 (ND10) as site of DNA virus deposition. *J. Cell Biol.* 134:815–26

175. Deleted in proof

176. Jackson V, Chalkley R. 1981. Use of whole cell fixation to visualize replication and maturing simian virus 40. Identification of new viral gene product. *Proc. Natl. Acad. Sci. USA.* 78:6081–85

177. Jans DA, Hübner S. 1996. Regulation of protein transport to the nucleus: central role of phosphorylation. *Physiol. Rev.* 76:651–85

178. Jans DA, Jans P. 1994. Negative charge at the casin kinase II site flanking the nuclear localization signal of the SV40 large T-antigen is mechanistically important for enhanced nuclear import. *Oncogene* 9:2961–68

179. Jay G, Nomura S, Anderson CW, Khoury G. 1981. Identification of the SV40 agnogene product: a DNA binding protein. *Nature* 291:346–49

180. Johnson DC, Spear PG. 1983. O-linked oligosaccharides are acquired by herpes simplex virus glycoproteins in the Golgi apparatus. *Cell* 32:987–97

181. Kalderon D, Richardson W, Markham AF, Smith AE. 1984. Sequence requirements for nuclear localization of simian virus 40 large-T antigen. *Nature* 311:33–38

182. Kanda T, Watanabe S, Zanma S, Sato H, Furuno A, Yoshiike K. 1991. Human papillomavirus type 16 E6 proteins with glycine substitution for cysteine in the metal-binding motif. *Virology* 185:536–43

183. Kann M, Bischof A, Gerlich WH. 1997. In vitro model for the nuclear transport of the hepadnavirus genome. *J. Virol.* 71:1310–6

184. Kaplan PM, Greenman RL, Gerin JL, Purcell RH, Robinson WS. 1973. DNA polymerase associated with human hepatitis B antigen. *J. Virol.* 12:995–1005

185. Karin M, Hunter T. 1995. Transcriptional control by protein phosphorylation: signal transmission from the cell surface to the nucleus. *Curr. Biol.* 5:747–57

186. Kartenbeck J, Stukenbrok H, Helenius A. 1989. Endocytosis of simian virus 40 into the endoplasmic reticulum. *J. Cell Biol.* 109:2721–29

187. Kekule A, Lauer U, Meyer M, Caselmann WH, Hofschneider PH, Koshy R. 1990. The pre S2/S region of integrated hepatitis B virus encodes a transcriptional transactivator. *Nature* 343:457–61

188. Kennard J, Rixon FJ, McDougall IM, Tatman JD, Preston VG. 1995. The 25 amino acid residues at the C-terminus

of the HSV1 UL26.5 protein are required for the formation of the capsid shell around the scaffold. *J. Gen. Virol.* 76:1611–21

189. Keppler OT, Stehling P, Herrmann M, Kayser H, Grunow D, Rutter W, Pawlita M. 1995. Biosynthetic modulation of sialic acid-dependent virus-receptor interactions of two primate polyoma viruses. *J. Biol. Chem.* 270:1308–14

190. Knipe DM, Smith JL. 1986. A mutant herpesvirus protein leads to a block in nuclear localization of other viral proteins. *Mol. Cell. Biol.* 6:2371–81

191. Köck J, Borst E-M, Schlicht H-J. 1996. Uptake of duck hepatitis B virus into hepatocytes occurs by endocytosis but does not require passage of the virus through an acidic intracellular compartment. *J. Virol.* 70:5827–31

192. Kotin R, Siniscalco M, Samulski RJ, Zhu XD, Hunter L, et al. 1990. Site-specific integration by adeno-associated virus. *Proc. Natl. Acad. Sci. USA* 87:2211–15

193. Krauzewicz N, Strueli CH, Stuart-Smith N, Jones MD, Wallace S, Griffin BE. 1990. Myristylated polyomavirus Vp2: role in life cycle of the virus. *J. Virol.* 64:4414–20

194. Kube D, Ponnazhagan S, Srivastava A. 1997. Encapsidation of adeno-associated virus type 2 rep proteins in wild type and recombinant progeny virions: repmediated growth inhibition of primary human cells. *J. Virol.* 71:7361–71

195. Kwong AD, Kruper JA, Frenkel N. 1988. HSV virion host shutoff function. *J. Virol.* 62:912–21

196. Lambert PF. 1991. Papillomavirus DNA replication. *J. Virol.* 65:3417–20

197. Lanford RE, Kanda P, Kennedy RC. 1986. Induction of nuclear transport with a synthetic peptide homologous of the SV40 T-antigen transport signal. *Cell* 46:575–82

198. Lanford RE, Butel JS. 1984. Construction and characterization of an SV40 mutant defective in nuclear transport of T antigen. *Cell* 37:801–3

199. Lavine J, Hirsch R, Ganem D. 1989. A system for studying the selective encapsidation of hepadnaviral RNA. *J. Virol.* 63:4257–63

200. Law M-F, Lowy DR, Dvoretzky I, Howley PM. 1981. Mouse cells transformed by bovine papillomavirus contain only extrachromosomal viral DNA sequences. *Proc. Natl. Acad. Sci. USA* 78:2727–31

201. Le Roux A, Berebbi M, Moukaddem M,

Perricaudet M, Joab I. 1993. Identification of a short amino acid sequence essential for efficient nuclear targeting of the Epstein-Barr virus nuclear antigen 3A. *J. Virol.* 67:1716–20

202. Lenhoff R, Summers J. 1994. Coordinate regulation of replication and virus assembly by the large envelope protein of an avian hepadnavirus. *J. Virol.* 68:4565–71

203. Lentz MR, Pak D, Mohr I, Botchan MR. 1993. The E1 replication protein of bovine papillomavirus type 1 contains an extended nuclear localization signal that includes a p34^{cdc2} phosphorylation site. *J. Virol.* 67:1414–23

204. Li M, Cripe TP, Estes PA, Lyon MK, Rose RC, Garcea RL. 1997. Expression of the human papillomavirus type 11 L1 capsid protein in *Escherichia coli*: characterization of protein domains involved in DNA binding and capsid assembly. *J. Virol.* 71:2988–95

205. Li PP, Sun C-K, Miyao A, Kasamatsu H. 1997. Identification of SV40 Vp1 DNA binding domain and its importance for SV40 morphogenesis. *Ann. Meet. Am. Soc. Virol.* 8(1):107 (Abstr.)

206. Li X, Rhode III SL. 1993. The parvovirus H1 NS2 protein affects viral gene expression through sequences in the 3' untranslated region. *Virology* 194:10–19

207. Li X, Rhode III SL. 1991. Nonstructural protein NS2 of parvovirus H1 is required for efficient virion protein synthesis and virus production in rat cells in vivo and in vitro. *Virology* 184:117–30

208. Liao W, Ou J-H. 1995. Phosphorylation and nuclear localization of the hepatitis B virus core protein: significance of serine in the three repeated SPRRR motif. *J. Virol.* 69:1025–29

209. Liddington RC, Yan Y, Moulai J, Sahli R, Benjamin TL, Harrison SC. 1991. Structure of simian virus 40 at 3.8 Å resolution. *Nature* 354:278–84

210. Lin W, Hata T, Kasamatsu H. 1984. Subcellular distribution of viral structural proteins during simian virus 40 infection. *J. Virol.* 50:363–71

211. Linden RM, Winocour E, Berns KI. 1996. The recombination signals for adeno-associated virus site-specific recombination. *Proc. Natl. Acad. Sci. USA* 93:7696–75

212. Ling PD, Ryon JJ, Hayward SD. 1993. EBNA-2 of herpesvirus papio diverges significantly from the type A and type B EBNA-2 proteins of Epstein-Barr virus but retains an efficient transactivation

domain with a conserved hydrophobic motif. *J. Virol.* 67:2990–3003

213. Linster P, Armentrout RW. 1978. Binding of minute virus of mice to cells in culture. In *Replication of Mammalian Parvoviruses*, ed. DC Ward, P Tattersall, pp. 151–60. New York: Cold Spring Harbor Lab.

214. Liu J-L, Lee LF, Ye Y, Qian Z, Kung H-J. 1997. Nucleolar and nuclear localization properties of a herpesvirus bZIP oncoprotein MEQ. *J. Virol.* 71:3188–96

215. Liu WJ, Gissmann L, Sun X-Y, Kanjanahaluethai A, Müller M, et al. 1997. Sequence close to the N-terminus of Õ2 protein is displayed on the surface of bovine papillomavirus type 1 virions. *Virology* 227:474–83

216. Lkuftig RB, Weighing RR. 1975. Adenovirus binds to rat brain microtubules in vitro. *J. Virol.* 16:696–706

217. Llamas-Saiz AL, Agbandje-McKenna M, Wikoff WR, Bratton J, Tattersal P, Rossmann MG. 1997. Structure determination of minute virus of mice. *Acta Crystallogr.* D53:93–102

218. Lutz P, Puvion-Dutilleul F, Lutz Y, Kedinger C. 1996. Nucleoplasmic and nucleolar distribution of the adenovirus IVa2 gene product. *J. Virol.* 70:3449–60

219. Lycke E, Hamark B, Johansson M, Krotochwil A, Lycke J, Svennerholm B. 1988. HSV infection of the human sensory neuron: an electron microscopy study. *Arch. Virol.* 101:87–104

220. Lyons RH, Ferguson BQ, Rosenberg M. 1987. Pentapeptide nuclear localization signal in adenovirus E1a. *Mol. Cell. Biol.* 7:2451–56

221. MacKay R, Consigli RA. 1976. Early events in polyomavirus infection: attachment, penetration, and nuclear entry. *J. Virol.* 19:620–36

222. Macrae D, Bruss V, Ganem D. 1991. Myristylation of duck hepatitis B virus envelope protein is essential for infectivity but not for virus assembly. *Virology* 181:359–63

223. Mahajan R, Delphin C, Guan T, Gerace L, Melchior F. 1997. A small ubiquitin-related polypeptide involved in targeting RanGAP1 to nuclear pore complex protein RanBP2. *Cell* 88:97–107

224. Majaniemi I, Siegl G. 1984. Early events in the replication of parvovirus LuIII. *Arch. Virol.* 81:285–302

225. Malik AK, Shao L, Shanley JD, Weller SK. 1996. Intracellular localization of herpes simplex virus type-1 origin binding protein. *Virology* 224:380–89

226. Mangel WF, McGrath, WJ, Toledo DL, Anderson CW. 1993. Viral DNA and a viral peptide can act as cofactors of adenovirus virion proteinase activity. *Nature* 361:274–75

227. Margolskee RF, Hathans D. 1983. Suppression of a Vp1 mutant of simian virus 40 by missense mutations in serine codons of the viral agnogene. *J. Virol.* 48:405–9

228. Mason W, Halpern M, England J, Seal G, Egan J, Coates L, Aldrich C, Summers J. 1983. Experimental transmission of duck hepatitis B virus. *Virology* 131:375–84

229. Mason WS, Aldrich C, Summers J, Taylor JM. 1982. Asymmetric replication of duck hepatitis B virus DNA in liver cells: free minus strand DNA. *Proc. Natl. Acad. Sci. USA* 79:3997–4001

230. Matusick-Kumar L, Hurlburt W, Weinheimer SP, Newcomb WW, Brown JC, et al. 1994. Phenotype of the HSV1 protease substrate ICP35 mutant virus. *J. Virol.* 68:5384–94

231. Matusick-Kumar L, Newcomb WW, Brown JC, McCann III PJ, Hurlburt W, et al. 1995. The C-terminal 25 amino acids of the protease and its substrate ICP35 of HSV1 are involved in the formation of sealed capsids. *J. Virol.* 69:4347–56

232. Maul GG, Everette RD. 1994. The nuclear location of PML, a cellular member of the C3HC4 zinc-finger domain protein family, is rearranged during herpes simplex virus infection by C3HC4 viral protein ICPO. *J. Gen. Virol.* 75:1223–33

233. Maul GG, Rovera G, Vorbrodt A, Abramczuk J. 1978. Membrane fusion as a mechanism of simian virus 40 entry into different cellular compartments. *J. Virol.* 28:936–44

234. McNabb DS, Courtney RJ. 1992. Characterization of the large tegument protein (ICP1/2) of HSV1. *Virology* 190:221–32

235. McNabb DS, Courtney RJ. 1992. Analysis of the UL36 ORF encoding the large tegument protein (ICP1/2) of HSV1. *J. Virol.* 66:7581–84

236. Mears WE, Lam V, Rice SA. 1995. Identification of nuclear and nucleolar localization signals in the herpes simplex virus regulatory protein ICP27. *J. Virol.* 69:935–47

237. Melchior F, Geraco L. 1995. Mechanisms of nuclear protein import. *Curr. Opin. Cell Biol.* 7:310–18

238. Mettenleiter TC. 1994. Initiation and

spread of Alpha-herpesvirus infection. *Trends Microbiol.* 2:2–3

239. Michael WM, Eder PS, Dreyfuss G. 1997. The K nuclear shuttling domain: a novel signal for nuclear import and nuclear export in the hnRNP K protein. *EMBO J.* 16:3587–98

240. Mikaelian I, Drouet E, Marechal V, Denoyel G, Nicolas J-C, Sergeant A. 1993. The DNA-binding domain of two bZIP transcription factors, the Epstein-Barr virus switch gene product EB1 and Jun, is a bipartite nuclear targeting sequence. *J. Virol.* 67:734–42

241. Mirza MA, Weber J. 1982. Structure of adenovirus chromatin. *Biochem. Biophys. Acta* 696:76–86

242. Mondesert G, Tribouley C, Kedinger C. 1992. Identification of a novel downstream binding protein implicated in late-phase-specific activation of the adenovirus major late promoter. *Nucleic Acids Res.* 20:3881–89

243. Montgomery RI, Warner MS, Lum BJ, Spear PG. 1996. HSV1 entry into cells mediated by a novel member of the TNF/NGF receptor family. *Cell* 87:427–36

244. Moreland RB, Montross L, Garcea RL. 1991. Characterization of the DNA-binding properties of the polyomavirus capsid protein Vp1. *J. Virol.* 65:1168–76

245. Moreland RB, Garcea RL. 1991. Characterization of a nuclear localization sequence in the polyomavirus capsid protein Vp1. *Virology* 18:513–18

246. Mortin N, Delsert C, Klessig DF. 1989. Nuclear localization of the adenovirus DNA-binding protein: requirement for two signals and complementation during viral infection. *Mol. Cell. Biol.* 9:4372–80

247. Moss B. 1996. Poxviridae: the viruses and their replication. In *Fundamental Virology*, ed. BN Field, DM Knipe, PM Howley, et al, pp. 2079–111. Philadelphia: Lippincott-Raven

248. Mullen M-A, Ciufo DM, Hayward GS. 1994. Mapping of intracellular localization domains and evidence for colocalization interactions between the IE110 and IE175 nuclear transactivator proteins of HSV. *J. Virol.* 68:3250–66

249. Mullen M-A, Ciufo DM, Mosca JD, Hayward GS. 1995. Evaluation of colocalization interactions between the IE110, IE175, and IE63 transactivator proteins of herpes simplex virus within subcellular punctate structures. *J. Virol.* 69:476–91

250. Müller M, Gissmann L, Cristiano RJ, Sun X-Y, Frazer IH, et al. 1995. Papillomavirus capsid binding and uptake by cells from different tissues and species. *J. Virol.* 69:948–54

251. Myers W, Carter BJ. 1980. Assembly of adeno-associated virus. *Virology* 102:71–82

252. Naeger LK, Cater J, Pintel DJ. 1990. The small non-structural protein (NS2) of the parvovirus minute virus of mice is required for efficient DNA replication and infectious virus production in a cell type-specific manner. *J. Virol.* 64:6166–75

253. Naeger LK, Salome N, Pintel DJ. 1993. NS2 is required for efficient translation of viral mRNA in minute virus of mice-infected murine cells. *J. Virol.* 67:1034–43

254. Nakanishi A, Clever J, Yamada M, Li PP, Kasamatsu H. 1996. Association with capsid proteins promotes nuclear targeting of simian virus 40. *Proc. Natl. Acad. Sci. USA* 93:96–100

255. Nakanishi A, Li PP, Sun PC-K, Kasamatsu H. 1997. *Role of nuclear localization signal and DNA binding domain of SV40 capsid proteins in morphogenesis.* Presented at Annu. Tumor Virus Meet. on Papovaviruses, Papillomavirus and Adenoviruses. Cambridge, England

256. Nakielny S, Fischer U, Michael WM, Dreyfuss G. 1994. RNA transport. *Annu. Rev. Neurosci.* 20:269–98

257. Nakshatri H, Pater MM, Pater A. 1988. Functional role of BK virus tumor antigens in transformation. *J. Virol.* 62:4613–21

258. Nallaseth FS, DePamphilis ML. 1994. Papillomavirus contains *cis*-acting sequences that can suppress but not replicate origin of DNA replication. *J. Virol.* 68:3051–64

259. Nassal M. 1992. The arginine-rich domain of the hepatitis B virus core protein is required for pregenome encapsidation and productive viral positive-strand DNA synthesis but not for virus assembly. *J. Virol.* 66:4107–16

260. Nathanson N, Tyler KL. 1997. Entry, dissemination, shedding, and transmission of viruses. I. In *Viral Pathogenesis*, ed. N Nathanson, 2:13–33. Philadelphia: Lippincott-Raven. 940 pp.

261. Newcomb WW, Trus BL, Booy FP, Steven AC, Wall JS, et al. 1993. Structure of the Herpes Simplex Virus capsid. Molecular composition of the pentons and triplex. *J. Mol. Biol.* 232:499–511

262. Newcomb WW, Homa FL, Thomsen DR, Booy FP, Trus BL, et al. 1996.

Assembly of the HSV capsid: characterization of intermediates observed during cell-free capsid formation. *J. Mol. Biol.* 263:432–46

263. Newcomb WF, Boring JW, Brown JC. 1984. Ion etching of human adenovirus 2: structure of the core. *J. Virol.* 51:52–56

264. Newport JW, Forbes DJ. 1987. The nucleus: structure, function, and dynamics. *Annu. Rev. Biochem.* 56:535–65

265. Ng S-C, Mertz JE, Sanden-Will S, Bina M. 1985. simian virus 40 maturation in cells harboring mutants deleted in the agnogene. *J. Biol. Chem.* 260:1127–32

266. Ni T, Zhou X, McCarty DM, Zolotukhin I, Muzyczka N. 1994. In virto replication of adeno-associated virus DNA. *J. Virol.* 68:1128–38

267. Nicholson P, Addison C, Cross AM, Kennard J, Preston VG, et al. 1994. Localization of the HSV1 major capsid protein Vp5 to the cell nucleus requires the abundant scaffold protein Vp22a. *J. Gen. Virol.* 75:1091–99

268. Nigg EA. 1990. Mechanisms of signal transduction to the cell nucleus. *Adv. Cancer Res.* 55:271–310

269. Nigg EA. 1997. Nucleocytoplasmic transport: signals, mechanisms and regulation. *Nature* 386:779–87

270. Nomura S, Khoury G, Jay G. 1983. Subcellular localization of the simian virus 40 agnoprotein. *J. Virol.* 45:428–33

271. Norkin LC, Anderson HA. 1996. Multiple stages of virus-receptor interactions as shown by simian virus 40. In *Toward Anti-Adhesion Therapy for Microbial Diseases*, ed. I Kahane, I Ofek, pp. 159–67. New York: Plenum

272. Nuesch JPF, Tattersall P. 1993. Nuclear targeting of the parvoviral replicator molecule NS1. Evidence for self-association prior to nuclear transport. *Virology* 196:637–51

273. O'Hare P. 1993. The virion transactivator of herpes simplex virus. *Sem. Virol.* 4:145–55

274. Ochiya T, Tsurimoto T, Ueda K, Okubo K, Shiozawa M, Matsubara K. 1989. An in vitro system for infection with hepatitis B virus that uses primary human fetal hepatocytes. *Proc. Natl. Acad. Sci. USA* 86:1875–79

275. Oppenheim A, Sandalon Z, Peleg A, Shaul O, Nicolis S, Ottolenghi S. 1992. A *cis*-acting DNA signal for encapsidation of simian virus 40. *J. Virol.* 66:5320–28

276. Oppenheim A, Siani M, Sandalon Z,

Mengerisky G. 1994. Dynamics of the nucleoprotein structure of simian virus 40 regulatory region during viral development. *J. Mol. Biol.* 238:501–513

277. Ozbun MA, Meyers C. 1997. Characterization of late gene transcripts expressed during vegetative replication of human papillomavirus type 31b. *J. Virol.* 71:5161–72

278. Paintsil J, Müller M, Picken M, Gissmann L, Zhou J. 1996. Carboxyl terminus of bovine papillomavirus type-1 L1 protein is not required for capsid formation. *Virology* 223:238–44

279. Panté N, Aebi U. 1994. Toward the molecular details of the nuclear pore complex. *J. Struct. Biol.* 113:179–89

280. Panté N, Aebi U. 1995. Toward a molecular understanding of the structure and function of the nuclear pore complex. *Int. Rev. Cytol.* 162B:225–55 (Suppl.)

281. Panté N, Aebi U. 1996. Toward the molecular dissection of protein import into nuclei. *Curr. Opin. Cell Biol.* 8:397–406

282. Panté N, Aebi U. 1996b. Sequential binding of import ligands to distinct nucleopore regions during their nuclear import. *Science* 273:1729–32

283. Parton RG, Joggerst B, Simons K. 1994. Regulated internalization of caveolae. *J. Cell Biol.* 127:1199–215

284. Patan I, Seth P, FitzGerald D, Willingham M. 1986. In *Adenovirus Entry into Cells: Some New Observations on an Old Problem*, ed. AL Notkins, MBA Oldstone, pp. 141–146. New York: Springer-Verlag

285. Pereira D, McCarty DM, Muzyczka N. 1997. The adeno-associated virus (AAV) rep protein acts as both a repressor and an activator to regulate AAV transcription during productive infection. *J. Virol.* 71:1079–88

286. Phelan A, Dunlop J, Patel AH, Stow ND, Clements JB. 1997. Nuclear sites of HSV1 DNA replication and transcription colocalize at early times postinfection and are largely distinct from RNA processing factors. *J. Virol.* 71:1124–32

287. Pintel DJ, Gersappe A, Haut D, Pearson J. 1995. Determinants that govern alternative splicing of parvovirus pre-mRNAs. *Sem. Virol.* 6:283–90

288. Pizzorno MC, Mullen M-A, Chang Y-N, Hayward GS. 1991. The functionally active IE2 immediate-early regulatory proteins of human cytomegalovirus is an 80-kilodalton polypeptide that contains two distinct activator domains and a

duplicated nuclear localization signal. *J. Virol.* 65:3839–52

289. Pollard VW, Michael M, Nakielny S, Siomi MC, Wang F, Dreyfuss G. 1996. A novel receptor-mediated nuclear protein import pathway. *Cell* 86:985–94

290. Prasad KM, Trempe JP. 1995. The adeno-associated virus Rep78 proteins is covalently linked to a viral DNA in a preformed virion. *Virology* 214:7769–80

291. Prchla E, Plank C, Wagner E, Blaas D, Fuchs R. 1995. Virus-mediated release of endosomal content in vitro: different behavior of adenovirus and rhinovirus serotype 2. *J. Cell Biol.* 131:111–123

292. Previsani N, Fontana S, Hirt B, Beard P. 1997. Growth of the parvovirus minute virus of mice MVMp3 in EL4 lymphocytes is restricted after cell entry and before viral DNA amplification: cell-specific differences in virus uncoating in vitro. *J. Virol.* 71:7769–80

293. Pujol A, Deleu L, Nuesh JPF, Cziepluch C, Jauniau J-C, Rommelaere J. 1997. Inhibition of parvovirus minute virus of mice replication by a peptide involved in the oligomerization of non-structural protein NS1. *J. Virol.* 71:7393–403

294. Qiao M, Macnaughton TB, Gowans E. 1994. Adsorption and penetration of hepatitis B virus in a nonpermissive cell line. *Virology* 201:356–63

295. Radziwill G, Zentgraf H, Schaller H, Bosch V. 1988. The duck hepatitis B virus DNA polymerase is tightly associated with the viral core structure and unable to switch to an exogenous template. *Virology* 163:123–32

296. Rayment I, Baker TS, Casper DLD, Murakami WT. 1982. Polyoma virus capsid structure at 22.5 Å resolution. *Nature* 295:110–15

297. Read GS, Karr BM, Knight K. 1993. Isolation of a HSV1 mutant with a deletion in the virion host shutoff gene and identification of multiple forms of the vhs (UL41) polypeptide. *J. Virol.* 67:7149–60

298. Read GS, Frenkel N. 1983. Herpes simplex virus mutants defective in the virion-associated shutoff of host polypeptide synthesis and exhibiting abnormal synthesis of alpha immediate-early viral polypeptides. *J. Virol.* 46:498–512

299. Resnick J, Shenk T. 1986. Simian virus 40 agnoprotein facilitates normal nuclear location of the major capsid polypeptide and cell-to-cell spread of virus. *J. Virol.* 60:1098–106

300. Rhode III SL, Paradiso PR. 1989. Parvovirus replication in normal and trans-

formed human cells correlates with the nuclear translocation of the early protein NS1. *J. Virol.* 63:349–55

301. Richardson WD, Roberts BL, Smith AE. 1986. Nuclear location signals in polyoma virus large-T. *Cell* 44:77–85

302. Rigg R, Schaller H. 1992. Duck hepatitis B virus infection of hepatocytes is not dependent on low pH. *J. Virol.* 66:2829–36

303. Rihs H-P, Peters R, Hobom G. 1991. Nuclear localization of budgerigar fledgling disease virus capsid protein Vp2 is conferred by residues 308–317. *FEBS Lett.* 291:6–8

304. Rihs HP, Jans DA, Fan H, Peters R. 1991. The rate of nuclear cytoplasmic protein transport is determined by the case in kinase II site flanking the nuclear localization sequence of the SV40 T-antigen. *EMBO J.* 10:633–39

305. Rijntes PJM, Moshage HJ, Yap SH. 1988. In vitro infection of primary cultures of cryopreserved adult human hepatocytes with hepatitis B virus. *Virus Res.* 10:95–100

306. Rixon FJ, Addison C, McGregor A, Macnab SJ, Nicholson P, et al. 1996. Multiple interactions control the intracellular localization of the HSV1 capsid proteins. *J. Gen. Virol.* 77:2251–60

307. Robbins J, Dilworth SM, Laskey RA, Dingwall C. 1991. Two interdependent basic domains in nucleoplasmin nuclear targeting sequence: identification of a class of bipartite nuclear targeting sequence. *Cell* 64:615–23

308. Robinson WS, Grannman R. 1974. DNA polymerase in the core of the human hepatitis B virus candidate. *J. Virol.* 13:1231–36

309. Robinson WS, Clayton DA, Greenman RL. 1974. DNA of a human hepatitis B virus candidate. *J. Virol.* 14:384–91

310. Rock DL, Fraser NW. 1985. Latent herpes simplex virus type 1 DNA contains 2 copies of the virion joint region. *J. Virol.* 62:3820–26

311. Roden RBS, Kirnbauer R, Jenson AB, Lowy DR, Schiller JT. 1994. Interaction of papillomaviruses with the cell surface. *J. Virol.* 68:7260–66

312. Rodriguez E, Everitt E. 1996. Adenovirus uncoating and nuclear establishment are not affected by weak base amines. *J. Virol.* 70:3470–77

313. Roizman B. 1982. The family herpesviridae: general description, taxonomy, and classification. In *The Herpesviruses*, ed. B Roizman, 1:1–23. New York: Plenum

314. Roizman B, Sears A. 1996. Herpes simplex virus and their replication. In *Fundamental Virology*, ed. BN Fields, DM Knipe, PM Howley, et al, 32:1043–62. Philadelphia: Lippincott-Raven. 1340 pp. 3rd ed.

315. Roossinck MJ, Siddigqui A. 1987. In vitro phosphorylation and protein analysis of hepatitis B virus core antigen. *J. Virol.* 61:955–61

316. Rothenberg KG, Heuser JE, Donzell WC, Ying Y-S, Glenney JR, Anderson RGW. 1992. Caveolin, a protein component of caveolae membrane coat. *Cell* 68:673–82

317. Ruigrok RWH, Barge A, Albizes-Rizo C, Dayau S. 1990. Structure of adenovirus fiber. II. Morphology of single fibers. *J. Mol. Biol.* 215:589–96

318. Sahli R, Freund R, Dubensky T, Garcea R, Bronson R, Benjamin T. 1993. Defect in entry and altered pathogenicity of a polyomavirus mutant blocked in Vp2 myristylation. *Virology* 192:142–53

319. Samulski RJ, Zhu X, Xiao X, Brook JD, Housman DE, et al. 1991. Targeted integration of adeno-associated virus (AAV) into human chromosome 19. *EMBO J.* 10:3941–50

320. Schaack J, Ho WY-W, Freimuth П, Shenk T. 1990. Adenovirus terminal protein mediates both nuclear matrix association and efficient transcription of adenovirus DNA. *Genes Dev.* 4:1197–208

321. Scheidtmann K-H, Echle B, Walter G. 1982. Simian virus 40 large T-antigen is phosphorylated at multiple sites clustered in two separate regions. *J. Virol.* 44:116–33

322. Schilling R, Will H. 1996. A fraction of the hepatitis B virus (HBV) P-Protein is transported into the nucleus independent of other viral proteins and its covalent linkage to the viral genome. *Eur. J. Cell Biol.* 69:n. (Suppl. 42)

323. Schmolke S, Drescher P, Jahn G, Plachter B. 1995. Nuclear targeting of the tegument pp65 (UL83) of human cytomegalovirus; an unusual bipartite nuclear localization signal functions with other portions of the protein to mediate its efficient nuclear transport. *J. Virol.* 69:1071–78

324. Schrag JD, Prasad BVV, Rixon FJ, Chiu W. 1989. Three-dimensional structure of the HSV1 nucleocapsid. *Cell* 56:651–60

325. Seif I, Khoury G, Dhar R. 1979. The genome of human papovavirus BKV. *Cell* 18:963–77

326. Sells MA, Chen ML, Acs G. 1987. Production of hepatitis B virus particles in Hep G2 cells transfected with cloned hepatitis B virus DNA. *Proc. Natl. Acad. Sci. USA* 84:1005–9

327. Serafini-Cessi F, Dall'Olio F, Scannavini M, Campedelli-Fiume G. 1983. Processing of herpes simplex virus-1 glycans in cells defective in glycosyl transferases of the Golgi system: relationship to cell fusion and virion egress. *Virology* 131:59–70

328. Seth P, FitzGerald DJ, Willingham MC, Patan I. 1984. Role of a low-pH environment in adenovirus enhancement of the toxicity of a Pseudomonas exotoxin-epidermal growth factor conjugate. *J. Virol.* 51:650–55

329. Shenk T. 1996. Adenoviridae: The viruses and their replication. In *Fundamental Virology*, ed. BN Fields, DM Knipe, PM Howley, 30:979:1016. Philadelphia: Lippincott-Raven. 1340 pp. 3rd ed.

330. Shenk TE, Carbon J, Berg P. 1976. Construction and analysis of viable deletion mutants of simian virus 40. *J. Virol.* 18:664–72

331. Sherman L, Schlegel R. 1996. Serum- and calcium-induced differentiation of human keratinocytes is inhabited by the E6 oncoprotein of human papillomavirus type 16. *J. Virol.* 70:3269–79

332. Shieh M-T, WuDunn D, Montogomery RI, Esko JD, Spear PG. 1992. Cell surface receptors for herpes simplex virus are heparan sulfate proteoglycans. *J. Cell Biol.* 116:1273–81

333. Shimura H, Umeno Y, Kimura G. 1987. Effects of inhibitors of the cytoplasmic structures and functions on the early phase of infection of cultured cells with simian virus 40. *Virology* 158:34–43

334. Signäs C, Akusjärvi G, Pettersson U. 1985. Adenovirus 3 fiber polypeptide genes: implications for the structure of the fiber protein. *J. Virol.* 53:672–78

335. Singer II, Toolan HW. 1975. Ultrastructural study of H-1 parvovirus replication: I. Cytopathology produced in human NB epithelial cells and hamster embryo fibroblast. *Virology* 65:40–54

336. Skiadopoulos MH, McBride AA. 1996. The bovine papillomavirus type 1 E2 transactivator and repressor proteins use different nuclear localization signals. *J. Virol.* 70:1117–24

337. Smith CD, Wells WW. 1984. Solubilization and reconstitution of a nuclear envelope associated ATPase: synergistic activation by RNA and polyphosphoinositides. *J. Bio. Chem.* 259:11890–94

338. Sodeik B, Ebersold MW, Helenius A. 1997. Microtubule-mediated transport of incoming HSV1 capsids to the nucleus. *J. Cell Biol.* 136:1007–21

339. Soussi T. 1986. DNA-binding properties of the major structural protein of simian virus 40. *J. Virol.* 59:740–42

340. Spear PG. 1993. Entry of alphaherpesvirus into cells. *Sem. Virol.* 4:167–80

341. Stacy T, Chamberlain M, Cole CN. 1989. Simian virus 40 host range/helper function mutations cause multiple defects in viral late gene expression. *J. Virol.* 63:5280–15

342. Stehle T, Harrison SC. 1996. Crystal structure of murine polyomavirus in complex with straight-chain and branched-chain sialyloligosaccharide receptor fragments. *Structure* 2:183–84

343. Stehle T, Yan Y, Benjamin TL, Harrison SC. 1994. Structure of murine polyomavirus complexed with an oligosaccharide receptor fragment. *Nature* 369:160–63

344. Steven AC, Spear PG. 1997. Herpesvirus capsid assembly and envelopment. In *Structural Biology of Viruses*, ed. W Chiu, RW Burnett, RL Garcea, 12:312–51. New York: Oxford Univ. Press

345. Stevens JG, Wagner EK, Devi-Rao GB, Cook ML, Feldman LT. 1987. RNA complementary to a herpesvirus alpha gene mRNA is prominent in latently infected neurons. *Science* 235:1056–59

346. Stevenson D, Colman KL, Davidson AJ. 1994. Delineation of a sequence required for nuclear localization of the proteins encoded by varicella-zoster virus gene 61. *J. Gen. Virol.* 75:3229–23

347. Stewart AR, Lednicky JA, Butel JS. 1997. Sequence comparison of SV40 isolates from monkeys and humans. *Am. Soc. Virol.* 50(5):188 (Abstr.)

348. Stewart PL, Burnett RM, Cyrklaff M, Fuller SD. 1991. Image reconstitution reveals the complex molecular organization of adenovirus. *Cell* 67:145–54

349. Stewart PL, Fuller SD, Burnett RM. 1993. Difference imaging of adenovirus: bridging the resolution gap between x-ray crystallography and electron microscopy. *EMBO J.* 12:2589–99

350. Streuli CH, Griffin BE. 1987. Myristic acid is coupled to a structural protein of polyomavirus and SV40. *Nature* 326:619–22

351. Summers J, Mason WS. 1982. Replication of the genome of a hepatitis B-like virus by reverse transcription of an RNA intermediate. *Cell* 29:403–15

352. Summers J, O'Connell A, Millman I. 1975. Genome of hepatitis B virus: restriction enzyme cleavage and structure of DNA extracted from Dane particles. *Proc. Natl. Acad. Sci. USA* 72:4597–601

353. Summers J, Smith P, Horwich A. 1990. Hepadnavirus envelope proteins regulate covalently closed circular DNA amplification. *J. Virol.* 64:2819–24

354. Summers J, Smith P, Huang M, Yu M. 1991. Morphogenetic and regulatory effects of mutations in the envelope proteins of an avian hepadnavirus. *J. Virol.* 65:1310–17

355. Sun X-Y, Frazer I, Müller M, Gissmann L, Zhou J. 1995. Sequences required for the nuclear targeting and accumulation of human papillomavirus type 6B L2 protein. *Virology* 213:321–27

356. Svensson U. 1985. Role of vesicles during adenovirus 2 internalization into Hela cells. *J. Virol.* 55:442–49

357. Tattersall P. 1972. Replication of the parvovirus MVM. I. Dependence of virus multiplication and plaque formation on cell growth. *J. Virol.* 10:586–90

358. Thomsen DR, Newcomb WW, Brown JC, Homa FL. 1995. Assembly of the HSV capsid: requirement for the C-terminal 25 amino acids of the proteins encoded by the UL26 and UL26.5 genes. *J. Virol.* 69:3690–703

359. Tibbetts C. 1980. Viral DNA sequences from incomplete particles of human adenovirus type 7. *Cell* 12:243–49

360. Tihanyi K, Bourbonniere M, Houde A, Rancourt C, Weber JM. 1993. Isolation and properties of adenovirus type 2 proteinase. *J. Biol. Chem.* 268:1780–85

361. Tognon M, Furlong D, Conley AJ, Roizman B. 1981. Molecular genetics of HSV. V, Characterization of a mutant defective in ability to form plaques at low temperatures and in a viral function which prevents accumulation of coreless capsids at nuclear pores late in infection. *J. Virol.* 40:870–80

362. Tomko RP, Xu R, Philipson L. 1997. HCAR and MCAR: the human and mouse cellular receptors for subgroup C adenovirus and group B coxsachievirus. *Proc. Natl. Acad. Sci. USA* 94:3352–56

363. Topp KS, Meade LB, LaVail JH. 1994. Microtubule polarity in the peripheral process of trigeminal ganglion cells: relevance for the retrograde transport of HSV. *J. Neurosci.* 14:318–25

364. Torrisi MR, Di Lazzaro C, Pavan A, Pereira L, Campadelli-Fiume G. 1992. Herpes simplex virus envelopment and maturation studied by fracture label. *J. Virol.* 66:554–61

365. Trus BL, Booy FP, Newcomb WW, Brown JC, Homa FL, et al. 1996. The herpes simplex virus procapsid: structure, conformational changes upon maturation, and roles of the triplex proteins Vp19C and Vp23 in assembly. *J. Mol. Biol.* 263:447–62

366. Trus BL, Homa FL, Booy FP, Newcomb WW, Thomsen DR, et al. 1995. HSV capsids assembled in insect cells infected with recombinant baculoviruses: structural authenticity and localization of Vp26. *J. Virol.* 69:7362–66

367. Trus BL, Roden RBS, Greenstone HL, Vrhel M, Schiller JT, Booy FP. 1997. Novel structural features of bovine papillomavirus capsid revealed by a three-dimensional reconstruction of 9 Å resolution. *Nature Struc. Biol.* 4:413–20

368. Tsao J, Chapman MS, Agbandje M, Keller W, Smith K, et al. 1991. The three dimensional structure of canine parvovirus and its functional implications. *Science* 251:1456–64

369. Tullis G, Labieniec-Pintel L, Clemens KE, Pintel D. 1988. Generation and characterization of a temperature-sensitive mutation in the NS-1 gene of the autonomous parvovirus minute virus of mice. *J. Virol.* 62:2736–44

370. Tullis GE, Burger LR, Pintel DJ. 1993. The minor capsid protein Vp1 of the autonomous parvovirus minute virus of mice is dispensable for encapsidation of progeny single-stranded DNA but is required for infectivity. *J. Virol.* 67:131–41

371. Tuttleman J, Pourcel C, Summers J. 1986. Formation of the pool of covalently closed circular DNA in hepadnavirus-infected cells. *Cell* 47:451–60

372. Ueda K, Tsurimoto T, Matsubara K. 1991. Three envelope proteins of hepatitis B virus: large S, middle S, and major S proteins needed for the formation of Dane particles. *J. Virol.* 65:3251–59

373. Ustav M, Stenlund A. 1991. Transient replication of BPV-1 requires two viral polypeptides encoded by the E1 and E2 open reading frames. *EMBO J.* 10:449–57

374. Varga MJ, Weibull C, Everitt E. 1991. Infectious entry pathway of adenovirus type 2. *J. Virol.* 65:6061–70

375. Vayda ME, Flint SJ. 1987. Isolation and characterization of adenovirus core nucleoprotein subunits. *J. Virol.* 16:3335–9

376. Vermuza SL, Smiley JR. 1985. Signals for site-specific cleavage of HSV DNA: Maturation involves two separate cleav-age events at sites distal to the recognition site. *Cell* 41:792–802

376a. Vihinen-Ranta M, Kalela A, Makinen P, Kakkola L, Marjomaki V, Vuento M. 1998. Intracellular route of canine parvovirus entry. *J. Virol.* 72:802–6

377. Vlazny DA, Kwong A, Frenkel N. 1982. Site-specific cleavage/packaging of herpes simplex virus DNA and the selective maturation of nucleocapsids containing full-length viral DNA. *Proc. Natl. Acad. Sci. USA* 79:1423–27

378. Wadsworth S, Jacob RJ, Roizman B. 1975. Anatomy of herpes simplex virus DNA. II. Size, composition, and arrangement of inverted terminal repetitions. *J. Virol.* 15:1487–97

379. Wagner EK, Bloom DC. 1997. Experimental investigation of herpes simplex virus latency. *Clin. Micro. Rev.* 10:416–43

380. Walter S, Richards R, Armentrout RW. 1980. Cell-cycle dependent replication of the DNA of minute virus of mice a parvovirus. *Biochim. Biophys. Acta* 607:420–31

381. Walton TH, Moen PT, Fox E, Bodnar JW. 1989. Interactions of minute virus of mice and adenovirus with host nucleoli. *J. Virol.* 63:3651–60

382. Weber J. 1976. Genetic analysis of adenovirus type 2. III. Temperature sensitivity of processing of viral proteins. *J. Virol.* 17:462–71

383. Weber JM, Cai F, Horvath J, Guillemette JG. 1994. Predicted structure of the adenovirus DNA binding protein. *Virus Genes* 9:171–75

384. Webster A, Hay RT, Kemp G. 1993. The adenovirus protease is activated by a virus-coded disulphide-linked peptide. *Cell* 72:97–104

385. Weger S, Wistuba A, Grimm D, Kleinschmidt A. 1997. Control of adeno-associated virus type 2 cap gene expression: relative influence of helper virus, terminal repeats, and rep proteins. *J. Virol.* 71:8437–47

386. Wei Y, Tavis JE, Ganem D. 1996. Relationship between viral DNA synthesis and virion envelopment in hepatitis B viruses. *J. Virol.* 70:6455–58

387. Weiser B, Ganem D, Seeger C, Varmus HE. 1983. Closed circular viral DNA and asymmetrical heterogeneous forms in livers from animals infected with ground squirrel hepatitis virus. *J. Virol.* 48:1–9

388. Whitbeck JC, Peng C, Lou H, Xu R, Willis RH, et al. 1997. Glycoprotein D of HSV binds directly to HVEM, a member

of the tumor necrosis factor receptor superfamily and a mediator of HSV entry. *J. Virol.* 71:6083–93

389. Wickham TJ, Mathias P, Cheresh DA, Nemerow GR. 1993. Integrins alpha v beta 3 and alpha v beta 5 promote adenovirus internalization but not virus attachment. *Cell* 73:309–19

390. Willwand K, Hirt B. 1991. The minute virus of mice capsid specifically recognizes the 3′ hairpin structure of the viral replicative-form DNA: Mapping of the binding site by hydroxyl radical footprinting. *J. Virol.* 65:4629–35

391. Willwand K, Hirt B. 1993. The major capsid protein VP2 of minute virus of mice (MVM) can form particles which bind to the 3′-terminal hairpin of MVM replicative form DNA and package single strand viral progeny DNA. *J. Virol.* 67:5660–63

392. Wilson GM, Jindal HK, Yeung DE, Chew W, Astell CR. 1991. Expression of minute virus of mice non-structural proteins in insect cells: purification and identification of ATPase and helicase activity. *Virology* 185:90–98

393. Wistuba A, Kern A, Weger S, Grimm D, Kleinschumidt JA. 1997. Subcellular compartmentalization of adeno-associated virus type 2 assembly. *J. Virol.* 71:1341–52

394. Wistuba A, Weger S, Kern A, Kleinschmidt JA. 1995. Intermediates of adeno-associated virus type 2 assembly: identification of soluble complexes containing Rep and Cap proteins. *J. Virol.* 69:5311–19

395. Wong ML, Hsu MT. 1989. Linear adenovirus DNA is organized into super coiled domains in virus particles. *Nucleic Acids Res.* 17:3535–50

396. Worrad DM, Caradonna S. 1993. The herpes simplex virus type 2 U13 open reading frame encodes a nuclear localizing phosphoprotein. *Virology* 195:364–76

397. Wu H, Rossmann MG. 1993. The canine parvovirus empty capsid structure. *J. Mol. Biol.* 233:231–44

398. WuDunn D, Spear PG. 1989. Initial interaction of herpes simplex virus with cells is binding to heparan sulfate. *J. Virol.* 63:32–58

399. Wychowski C, Benichou D, Girard M. 1986. A domain of SV40 capsid polypeptide Vp1 that specified migration into the cell nucleus. *EMBO J.* 5:2569–76

400. Wychowski C, Benichou D, Girard M. 1987. The intranuclear location of simian virus 40 polypeptides Vp2 and

Vp3 depends on a specific amino acid sequence. *J. Virol.* 61:3862–69

401. Xiao XL, Wilson VG. 1994. Genetically defined nuclear localization signal sequence of bovine papillomavirus E1 protein is necessary and sufficient for the nuclear localization of E1-β-galactosidase fusion proteins. *J. Gen Virol.* 75:2463–67

402. Xie Q, Chapman MS. 1996. Canine parvovirus capsid structure, analyzed at 2.9 Å resolution. *J. Mol. Biol.* 264:497–520

403. Yakinoglu A, Helibronn R. 1988. DNA amplification of adeno-associated virus as a response to cellular genotoxic stress. *Cancer Res.* 48:3123–29

404. Yakobson B, Koch T, Winocour E. 1987. Replication of adeno-associated virus in synchronized cells without a addition of a helper virus. *J. Virol.* 61:972–81

405. Yamada M, Kasamatsu H. 1993. Role of nuclear pore complex in simian virus 40 nuclear targeting. *J. Virol.* 67:119–30

406. Yan Y, Stehle T, Liddington RC, Zhao H, Harrison SC. 1996. Structure determination of simian virus 40 and murine polyomavirus by a combination of 30-fold and 5-fold electrondensity averaging. *Structure* 4:157–64

407. Yeh C-T, Liaw Y-F, Ou JH. 1990. The arginine-rich domain of hepatitis B virus precore and core protein contains a signal for nuclear transport. *J. Virol.* 64:6141–47

408. Yeh C-T, Wong SW, Fung Y-K, Ou J-H. 1993. Cell cycle regulation of nuclear localization of hepatitis B virus core protein. *Proc. Natl. Acad. Sci. USA* 90:6459–63

409. Yu M, Summers J. 1994. Multiple functions of capsid protein phosphorylation in duck hepatitis B virus replication. *J. Virol.* 68:4341–48

410. Yu M, Summers J. 1994. Phosphorylation of the duck hepatitis B virus capsid protein associated with conformational changes in the C terminus. *J. Virol.* 68:2964–69

411. Zhao L-J, Padmanabhan R. 1991. Three basic regions in adenovirus DNA polymerase interact differentially depending on the protein context to function as bipartite nuclear localization signals. *New Biol.* 3:1074–88

412. Zhao LJ, Padmanabhan R. 1988. Nuclear transport of adenovirus DNA polymerase is facilitated by interaction with preterminal protein. *Cell* 55:1005–15

413. Zhou J, Doorbar J, Sun X-Y, Crawford LV, McLean CS, Frazer IH. 1991. Identification of the nuclear localization

signal of human papillomavirus type 16 L1 protein. *Virology* 185:625–32

414. Zhou J, Sun X-Y, Louis K, Frazer IH. 1994. Interaction of human papillomavirus (HPV) type 16 capsid proteins with HVP DNA requires an intact L2 N-terminal sequence. *J. Virol.* 68:619–25

415. Zhou ZH, He J, Jakana J, Tatman JD, Rixon FJ, Chiu W. 1995. Assembly of Vp26 in HSV-1 inferred from structures of wild type and recombinant capsids. *Nature Struct. Biol.* 2:1026–30

416. Zhou S, Yang SQ, Strandring DN. 1992. Characterization of hepatitis B virus capsid particle assembly in *Xenopus* oocytes. *J. Virol.* 66:3086–92

417. Zhu Z, Cai W, Schaffer PA. 1994. Cooperativity among HSV 1 immediate-early regulatory proteins: ICP4 and ICP27 affect the intracellular localization of ICP0. *J. Virol.* 68:3027–40

418. Zhu Z, DeLuca NA, Schaffer PA. 1996. Overexpression of the HSV1 immediate-early regulatory protein, ICP27, is responsible for the aberrant localization of ICP0 and mutant forms of ICP4 in ICP4 mutant virus infected cells. *J. Virol.* 70:5346–56

419. Zhu Z, Schaffer PA. 1995. Intracellular localization of the herpes simplex virus type 1 major transcriptional regulatory protein, ICP4, is affected by ICP27. *J. Virol.* 69:49–59

420. Zullo J, Stiles CD, Garcea RL. 1987. Regulation of c-*myc* and c-*fos* mRNA levels by polyomavirus: distinct roles for the capsid protein Vp1 and the viral early proteins. *Proc. Natl. Acad. Sci. USA* 84:1210–14

NOTE ADDED IN PROOF

Note that photographs for Figure 1 were reproduced with permission from Conway et al 1996 and Steven & Spear 1997 for panel 1.

Annu. Rev. Microbiol. 1998. 52:687–744
Copyright © 1998 by Annual Reviews. All rights reserved

CELL POLARITY
AND MORPHOGENESIS
IN BUDDING YEAST

Kevin Madden and Michael Snyder

Department of Molecular, Cellular and Developmental Biology, Yale University, New Haven, Connecticut 06520-8103; e-mail: michael.snyder@yale.edu

KEY WORDS: cytoskeleton, actin, septins, Cdc42p, Rho1p, cell cycle, MAP kinase, checkpoint, cell fate, asymmetric cell division

ABSTRACT

Eukaryotic cells respond to intracellular and extracellular cues to direct asymmetric cell growth and division. The yeast *Saccharomyces cerevisiae* undergoes polarized growth at several times during budding and mating and is a useful model organism for studying asymmetric growth and division. In recent years, many regulatory and cytoskeletal components important for directing and executing growth have been identified, and molecular mechanisms have been elucidated in yeast. Key signaling pathways that regulate polarization during the cell cycle and mating response have been described. Since many of the components important for polarized cell growth are conserved in other organisms, the basic mechanisms mediating polarized cell growth are likely to be universal among eukaryotes.

CONTENTS

687

0066-4227/98/1001-0687$08.00

INTRODUCTION

Polarized cell growth and directional cell division are fundamental processes that are essential for the development of eukaryotes. Polarized cell growth involves asymmetric growth from one region of a cell to form particular cell structures or shapes. The specialized structures that result are critical for the function of particular cell types and can help mediate diverse cellular interactions during development. Nutrient absorption by the microvilli of epithelial cells (217), plant fertilization (19), and the interaction of helper T cells and antigen-presenting B cells are examples of cellular processes that rely on the formation of polarized cell structures (166).

Directional cell division is a process in which cells divide along specific cleavage planes. Directional divisions can mediate appropriate cell-cell contacts, partition cytoplasmic components asymmetrically during division, and help establish polarized cellular structures. Directional cell divisions occur during the life cycle of many organisms, including during early embryogenesis in *Caenorhabditis elegans* (132), neurogenesis in *Drosophila* (162), spore development in *Bacillus subtilus* (276), and the development of the snail body plan (98). The mechanisms for selecting sites for polarized growth and division as well as for directing growth toward these sites are only beginning to be understood.

The budding yeast *Saccharomyces cerevisiae* undergoes polarized growth during several stages of its life cycle, and growth occurs at defined positions on the cell surface (Figure 1). In the presence of ample nutrients, yeast grow by budding, and the position where the bud forms ultimately determines the plane of cell division. The location of the bud site depends on the mating locus

Vegetative growth

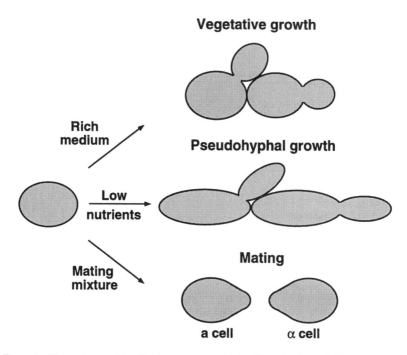

Rich medium

Pseudohyphal growth

Low nutrients

Mating mixture

Mating

a cell   **α cell**

Figure 1 Three phases of the *Saccharomyces cerevisiae* life cycle that exhibit polarized cell growth. Cells grown in a rich medium are round or oval and have defined budding patterns. When exposed to a low-nutrient medium, cells elongate and bud from the distal end to form pseudohyphae. Haploid cells exposed to pheromone from cells of the opposite mating type arrest in G1 and extend a projection toward their mating partner. (Reproduced with permission from *Trends Cell Biol.*)

and pedigree of the cell (49, 100, 128, 287). Bud emergence initiates during the late G1 stage of the cell cycle and the bud continues to grow, first at the tip (apical growth) and then throughout the bud (isotropic growth), until late nuclear division and cytokinesis (180). A second form of polarized growth in yeast occurs when access to specific nutrients, such as nitrogen, is decreased. Under these conditions, diploid yeast cells initiate pseudohyphal growth in which cells elongate, bud primarily from one pole of the cell, and form chains of connected cells that can spread across a substrate and invade a solid medium (110, 165). A related response occurs in haploids, termed haploid invasive growth, in which daughter cells bud to form chains of cells that penetrate an agar medium (258). Specific mechanisms exist to ensure that defined sites for growth are used during budding and the pseudohyphal response.

A third form of polarized growth in yeast occurs during the mating response. Haploid yeast are of two cell types, *MATa* and *MATα*. After exposure to pheromone from cells of the opposite mating type, cells arrest in late G1 and form an elongated mating projection (61, 289). Growth usually occurs at a location on the cell surface nearest the mating partner; this site is not predetermined through intrinsic signals (195). Thus, in contrast to bud site selection in which defined sites for growth are used, during mating external factors influence the selection of sites for polarized growth.

The purpose of polarized cell growth and divisions in yeast is at least three-fold. Budding of haploid cells and mating projection formation are thought to promote cell-cell contact. Budding can also allow cells to spread throughout a medium to forage for nutrients or search for mates. Finally, the production of a bud results in formation of a separate compartment with a distinct cytoplasm from that of the mother cell. This compartmentalization allows segregation of cytosolic factors and thus allows mother and daughter cells to acquire distinct cell fates (see below).

Although vegetative growth, pseudohyphal growth, and the mating response are distinct cellular processes, in each instance a similar series of polarized growth events must take place (193). First, a position on the cell surface is established as the site for growth. Next, the cytoskeleton must be polarized to the site chosen for growth; the cytoskeleton directs membrane deposition and polarized secretion to that site. Cellular organelles such as the nucleus and mitochondria must then be segregated into the newly formed growth structure. Finally, each of these events must be coordinated with cell-cycle progression and cellular signals.

In recent years much insight has been gained into the components that function in polarized growth in yeast and into the molecular and cellular mechanisms that direct and regulate the polarized growth response (reviewed in 29, 164, 178, 182, 190, 248, 261). Many of these components and mechanisms are conserved in a wide variety of eukaryotes. This review describes many aspects of polarized growth in yeast, including selection of sites during budding and mating, bud formation and growth, specification of yeast cell types, and the coordination of cell-cycle control and MAP kinase signaling with polarized growth during budding and mating. (Other recent reviews on these topics can be found in References 28, 127, 190, 261.)

SELECTION OF SITES FOR POLARIZED GROWTH DURING BUDDING

Polarized cell divisions occur by budding at specific sites that are determined by the mating locus of the cell and whether it was last a mother or daughter (Figure 2). Haploid cells and diploid cells that are homozygous at the mating locus (*MATa/MATa* and *MATα/MATα* cells) bud axially (i.e. at proximal sites);

Haploid - axial budding

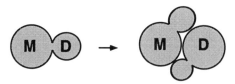

Diploid - bipolar budding

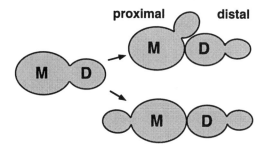

Figure 2   Yeast cells exhibit defined budding patterns. Haploid *MAT*a and *MAT*α cells undergo axial budding in which both mother (M) and daughter (D) cells bud adjacent to the previous site of cytokinesis (i.e. proximal sites). Diploid *MAT*a/*MAT*α cells exhibit a bipolar pattern; mother cells bud at either pole, but daughter cells preferentially form buds opposite the site of cytokinesis (i.e. distal sites). (Reproduced with permission from *Trends Cell Biol.*)

the mother cell forms a bud adjacent to the previous bud site, and the daughter cell buds next to the birth site. Diploid *MAT*a/*MAT*α cells have a different budding pattern. Diploid daughters bud opposite the birth site (distal site), and diploid mother cells bud in a bipolar pattern, i.e. either adjacent (proximal) or opposite to the previous bud site (49, 100, 128, 287). New mother cells usually bud at proximal sites, whereas cells that have undergone one or more divisions exhibit an increased preference for distal sites.

The different budding patterns are thought to fulfill different purposes. The haploid axial budding pattern is believed to facilitate the production of diploid cells (Figure 3) (225). In nature, *MAT*a and *MAT*α yeast cells switch their mating type (i.e. *MAT*a cells become *MAT*α cells and vice versa); mother cells switch using a differentiation pathway that depends on cell polarity (see below). The axial pattern therefore positions *MAT*a and *MAT*α cells in close proximity to one another, which should accelerate diploid formation. Since diploid cells are expected to be more resistant to environmental stress and DNA damaging agents than are haploid cells, the formation of diploids should enhance survival of the species (109). The diploid bipolar pattern is thought to maximize spreading of

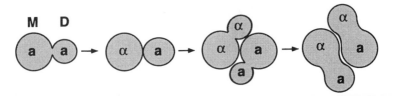

Figure 3 Diploid formation may be promoted by the axial pattern. Mother cells switch their mating type in late G1. Budding at proximal sites will position cells in close proximity to facilitate diploid formation.

a growing microcolony to nutrients (109, 110, 193). This is particularly evident in pseudohyphal cells in which long filamentous chains extend across a surface (110).

Three distinct classes of proteins important for bud site selection have been identified. One class of bud site selection proteins required for axial budding does not appear to affect the bipolar pattern (47, 48, 96, 102, 119, 259, 268). Mutations in genes encoding any of these proteins result in bipolar budding in haploid cells, but they do not disrupt diploid budding patterns. A second class of proteins is not required for haploid axial budding but instead affects the bipolar budding pattern of diploid cells. Thus, axial and bipolar budding are controlled by two separate pathways because either pattern can be altered without affecting the other (18, 46, 287, 325). A final class of proteins is required for both axial and bipolar budding (21, 46, 47, 237).

Analysis of the different proteins involved in bud site selection has led to two important insights. First, a hierarchy of bud sites exists. In haploid cells, loss of axial components through mutation or other means (see below) leads to selection of bipolar sites (49, 195, 288). Thus, in haploid cells a mechanism exists to preferentially select axial sites; bipolar sites (especially distal sites) are used by default. A second insight gained from studies of bud site selection proteins is that several different types of components exist. These include (a) cortical molecules (or tags) critical for selecting the new sites, (b) a GTPase module thought to guide molecules to or help assemble molecules at the specified site, and (c) components assembled at the new site that participate in formation of the new bud. A description of the different components and mechanisms important for bud site selection for both axial and bipolar patterns follows.

The Axial Budding Pattern

Cells that exhibit axial budding utilize a cytokinesis tag in which a component at the cytokinesis site persists into the next cell cycle and directs formation of a new bud at an adjacent site. This tag likely acts as a template for directing assembly of bud formation components at the new bud site (47, 193, 288). The yeast septin proteins (Cdc3p, Cdc10p, Cdc11p, Cdc12p) (48, 96), Bud3p (47, 48), Bud4p (47, 268), and Axl2p (119, 259) may be components of this cortical tag (see below).

SEPTINS PARTICIPATE IN AXIAL BUDDING   *CDC3*, *CDC10*, *CDC11*, and *CDC12* encode related proteins, called septins, that localize to the mother-bud neck and are needed for both cytokinesis and axial bud site selection (reviewed in 190). A temperature-sensitive mutation in any one of these genes causes cytokinesis defects, resulting in the formation of elongated chains of connected cells (3, 121). Cdc3p, Cdc10p, Cdc11p, and Cdc12p localize as a ring at the incipient bud site, and this ring remains at the mother-bud neck during bud formation (Figure 4). During cytokinesis, the septins localize as a double ring at the neck, and importantly, after septation both mother and daughter cells display ring staining throughout much of the G1 phase of the next cell cycle (97, 115, 151; HB Kim, BK Haarer, JR Pringle, unpublished information). The localization patterns—together with the mutant phenotypes—suggest that the septins may be components of the 10-nm neck filaments that lie just underneath the plasma membrane at the neck between the mother cell and bud (37, 39). These filaments, as well as the localization of each septin protein at the neck, are absent in *cdc3*, *cdc10*, *cdc11*, and *cdc12* mutants grown at the restrictive temperature (40, 97, 115, 151). Although the yeast septins have not been shown to form filaments in vitro, *Drosophila* homologs of these proteins form highly ordered filaments in vitro (94). These septins all contain predicted GTP-binding motifs (96; reviewed in 190), and *Drosophila* septin proteins hydrolyze GTP in vitro (94); thus it has been speculated that the neck filaments assemble from nucleotide binding protein monomers, analogous to actin or tubulin filaments (96).

In addition to their role in cytokinesis, the septins are critical for axial bud site selection. Strains containing temperature-sensitive mutations in any of the four septins mentioned are defective in choosing axial sites when incubated at semipermissive temperatures; instead cells choose distal sites at a high frequency (48, 96). Thus, the yeast neck filament system is a cortical structure required for axial bud site selection; recognition of this complex by components important for bud emergence would promote polarized growth events at proximal sites, resulting in the axial pattern (48, 96, 259, 268; see below).

BUD3P, BUD4P, AXL1P, AXL2P, AND CELL WALL SYNTHESIS COMPONENTS ARE IMPORTANT FOR AXIAL BUD SITE SELECTION   Four additional proteins specifically required for axial budding—Bud3p, Bud4p, Axl2p, and Axl1p—have been identified through genetic studies. Loss of any of these four proteins results in

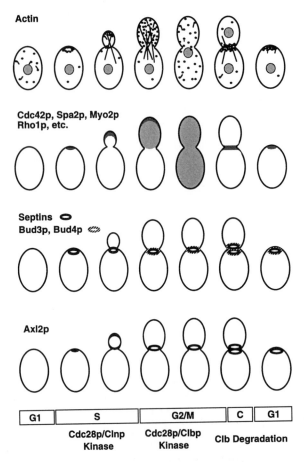

Figure 4   Localization patterns of many proteins important for cell morphogenesis in yeast. (*top row*) Actin and many actin-associated proteins (e.g. Abp1p, Sac6p) localize as patches in a ring at the incipient bud site, at the bud tip of small budded cells, and at the neck in cells undergoing cytokinesis. Actin cables that migrate throughout the mother and bud are also present. For simplicity, the nucleus is shown only once (*first row*). (*second row*) Many proteins important for bud emergence and growth (Cdc42p, Rho1p, Rom2p, Bem1p) and/or other aspects of cell polarization (Spa2p, Pea2p, Bud6p) localize as a diffuse patch at the growth sites mentioned above. In some cases (e.g. Cdc42p, Rho1p, Bem1p) staining at the tip of large budded cells is not detected, nor is the diffuse staining found throughout the bud in these cells. This may be a function of the sensitivity of reagents used for detecting these proteins. (*third row*) Septins (*dark rings*) localize as a ring at the incipient bud site and at the neck of budded cells; this staining persists into the next cell cycle. Bud3p and Bud4p (*shaded rings*) also localize at the neck, but they do not appear until activation of the G2 kinase. (*fourth row*) Axl2p has a localization pattern that is a combination of those described for the second and third rows.

a bipolar budding pattern (2, 47, 48, 102, 268). Bud3p and Bud4p were identified in a screen for mutants defective in bud site selection (47); in addition to
their role in axial budding (48, 268), *bud3*Δ mutants also exhibit a modest defect
in cytokinesis (259). The Bud3p predicted sequence is unique, whereas Bud4p
contains a potential GTP-binding motif near its carboxy terminus (48, 268);
this motif is contained within a domain that is also present in the *Candida albicans* integrin-like protein Int1p (104). Bud3p and Bud4p exhibit interesting
localization patterns that in many respects resemble that of the septins. Both
Bud3p and Bud4p localize as a double ring at the mother-bud neck during mitosis, the double ring persists until cytokinesis, and a single ring is present in
both the mother and daughter cells during much of the G1 phase of the next
cell cycle (48, 268). However, unlike ring staining in the septins, staining of
Bud3p and Bud4p is not detected at the G1/S transition or during S phase but
instead appears later in the cell cycle after activation of the Cdc28p/Clb G2
cyclin-dependent kinase (268). The localization of both proteins to the neck
region is dependent on the putative neck filament proteins (48, 268); there is no
evidence for septin localization requiring Bud3p or Bud4p. These observations
indicate that Bud3p and Bud4p assemble at the septin complex and that they
help select cortical sites for axial budding.

AXL2 encodes a single-spanning integral membrane glycoprotein whose
localization pattern also is consistent with a role in axial bud site selection
(119, 259). Axl2p localizes as a patch at the incipient bud site and at the bud
periphery of small-budded cells (259). This pattern resembles that of many
proteins involved in polarized cell growth (see below). However, unlike other
polarity proteins, Axl2p is also present as a ring at the neck in medium- and
large-budded cells (119, 259), similar to the septins, Bud3p, and Bud4p. This
staining persists after cytokinesis into the G1 stage of the next cell cycle. Axl2p
localizes normally in a *bud3*Δ strain, and Bud4p localization does not require
AXL2 (259). Although Axl2p contains a cytoplasmic domain and an extracellular domain, it does not appear to anchor the septins to the plasma membrane;
Cdc3p still localizes normally in an *axl2*Δ strain (259).

There are two possible mechanisms by which Axl2p may participate in bud
site selection (119, 259). Axl2p may function with the septin proteins, Bud3p,
and/or Bud4p as part of the cytokinesis tag that marks the site of cell division
and recruits other proteins required for the establishment of polarized growth
(119, 259). Alternatively, Axl2p might be involved in tag recognition (259).
For example, Axl2p-containing vesicles may interact with the neck filament
complex and target growth components to the site adjacent to the former bud
site. This model is similar to v-SNARE/t-SNARE vesicle targeting during
protein secretion in which vesicles containing membrane-associated v-SNARE
proteins interact with target membranes containing membrane-bound t-SNARE

proteins (92, 259). Consistent with the hypothesis that Axl2p plays a role in tag recognition are the observations that it localizes to the incipient bud site and that $axl2\Delta$ mutants contain "droopy" buds, a phenotype suggestive of a role in the early steps of bud formation (259).

The localization pattern of Axl1p, a fourth protein required for axial bud site selection (2, 102), has not been reported. Axl1p is a protease for the a-factor mating pheromone, but the protease activity is not required for axial bud site selection (2). *AXL1* is particularly interesting, because its expression is specific to haploid cells (2, 102). Moreover, overexpression of *AXL1* in diploid cells promotes axial budding (102). However, this result must be interpreted cautiously, because *AXL1* was highly overexpressed and only a modest (twofold) increase in budding at the proximal pole was observed. Nevertheless, Axl1p is a candidate for a haploid-specific gene product that specifies axial budding in haploid cells.

Finally, at least two components important for cell wall synthesis—Kre9p, which is involved in $(1 \rightarrow 6)$ β-glucan synthesis, and Hkr1p, a cell surface protein—are also important for axial bud site selection (32, 320). Deletion of either the *KRE9* or *HKR1* gene in haploid cells results in an increase in budding events at random sites. It is likely that these mutations affect either the localization or function of a cortical tag protein, such as Axl2p, which has an extracellular domain. Alternatively, these mutations might disturb underlying cytoskeletal components to disrupt bud site selection. Although not examined, it is likely that these mutations also disrupt the diploid budding pattern. Thus, a combination of intracellular cortical proteins, transmembrane proteins, and extracellular proteins is important for axial bud site selection in yeast.

RSR1P, BUD2P, AND BUD5P CONSTITUTE A GTPASE SIGNALING MODULE INVOLVED IN SELECTING BUD SITES   A GTPase module is important for both axial and bipolar budding patterns and is thought to help direct bud formation components to the selected site of growth. Mutations in *RSR1*, *BUD2*, and *BUD5* cause random budding patterns in both haploid and diploid cells but no other obvious morphological defects (21, 46, 47, 237). *RSR1* encodes a Ras-related GTPase, and several lines of evidence indicate that *BUD2* and *BUD5* encode a GTPase-activating protein (GAP) (20, 237) and a guanine-nucleotide exchange factor (GEF) (46, 244, 328) for the Rsr1p GTPase, respectively. Rsr1p has been localized throughout the cell cortex during all stages of the cell cycle (213); it is predicted that spatially restricted activity of either Bud2p or Bud5p at the future bud site may regulate Rsr1p and thus mediate proper bud site selection.

The Rsr1p GTPase signaling module is thought to direct bud formation components to cortical tags at future bud sites (46, 47, 127, 213, 237). The Rsr1p GTPase module interacts with Cdc42p, Cdc24p, and Bem1p, polarity-establishment components important for bud formation (see below). Cdc42p is a Rac/Rho-type GTPase (144) whose activity is regulated by the (GEF) Cdc24p (329) (Table 1). Genetic evidence suggests an interaction between the Rsr1p and Cdc42p GTPase modules (20, 21, 263). GTP-bound Rsr1p has been shown to bind Cdc24p and GDP-bound Cdc42p (236, 328), whereas GDP-bound Rsr1p binds Bem1p (236), another protein involved in the establishment of yeast cell polarity (46, 52). One possible function of Rsr1p might be to recruit polarity-establishment factors to proper growth sites (236). Since both Cdc24p and Bem1p associate with the plasma membrane in the absence of Rsr1p (213), Rsr1p is not required for their localization at buds but rather is needed for positioning them at the correct sites. How the GTPase-module/polarity-establishment components interact with cortical tags at the selected site is not known.

MODEL FOR AXIAL BUDDING   Based on the observations described above, a "cytokinesis tag" model for axial budding in haploid cells can be summarized as follows (Figure 5) (46, 48, 96, 288): A landmark comprised of the neck filament complex, septins, Bud3p, Bud4p, and perhaps Axl2p can target growth to an adjacent position on the cell cortex in the next budding cycle in both mother and daughter cells (48, 96) [or possibly even during late mitosis/cytokinesis (see 50)]. One or more components at this site are recognized by a complex containing the Rsr1p GTPase (46, 47, 48, 259, 268). GTPase activation plays a role in targeting or assembling bud formation components at the incipient bud site and may help initiate the bud formation process (213, 236, 328).

The Bipolar Budding Pattern

The bipolar budding pattern of diploids is more complex than axial budding in haploids; daughter cells bud at distal sites and mother cells choose either proximal or distal sites (Figure 2). Although a number of components important for the bipolar pattern have been identified using genetic approaches, much less is known about the molecular mechanisms by which proximal and distal sites are selected. A current model is that "growth or polarity" components deposited at the cell surface during early bud formation and growth can serve as cortical tags for selecting distal sites in daughters (47, 48, 193, 325). A distinct set of proteins that appear at the neck during cytokinesis may be important for selecting proximal sites (47, 48, 288, 325). As for axial budding, recognition of these components is thought to require the Rsr1p GTPase module.

THE ACTIN CYTOSKELETON AND ACTIN-BINDING PROTEINS ARE REQUIRED FOR BIPOLAR BUDDING   The actin cytoskeleton is essential for both polarized growth and bud site selection. Consistent with its role in these processes, actin localizes at the incipient bud site, at sites of polarized cell growth, and at the neck in cells undergoing cytokinesis (3, 150, 220, 279; see below). A large number of

Table 1   Genes encoding rho proteins and their regulators[a]

Gene	Features/phenotypes	Localization	Reference
CDC42	Essential; ts mutant arrest as large unbudded cell; depolarized actin	Incipient bud site; small bud tips; neck at cytokinesis	6, 142, 144, 33, 333
GTPase activating protein genes			
BEM3	In vitro Cdc42p-GAP; nonessential		329, 330, 290
RGA1	GAP homology domain; negative regulator of pheromone signaling; 2-hybrid interaction with Cdc42p		
RGA2	Homolog of *RGA1*		290
Other negative regulators			
ZDS1	Deletion mild effect on cell shape; *zds1Δ zds2Δ* double mutant has abnormal buds	Incipient bud site; small bud tips	24
ZDS2	No phenotype; *zds1Δ zds2Δ* double mutant as above	NR	24
Guanine exchange factor gene			
CDC24	Essential; ts mutant arrest as large unbudded cell; depolarized actin, Spa2p	Entire cell periphery	190, 248, 286
Geranyl-geranyl transferase gene			
CDC43	Essential; ts mutant arrest as large unbudded cell; depolarized actin	NR	6, 95
Possible targets Cdc42p: Rsr1p, Cdc24p, Ste20p, Gic1p, Gic2p Cdc24p: Rsr1p			
RHO1	Essential; ts mutants lyse as small budded cells	Incipient bud site; small bud tips; neck at cytokinesis	322

Table 1   (*Continued*)

Gene	Features/phenotypes	Localization	Reference
GTPase activating protein genes			
BEM2	Nonessential; deletion ts and arrests as large unbudded cells; random budding at permissive temp; hypersensitive to benomyl	NR	152, 242, 308
SAC7	Nonessential; defects in actin localization	NR	271
BAG7	Homolog of *SAC7*	NR	271
RDI1	Rho-GDP dissociation inhibitor; deletion no phenotype	Cytosol	156
Guanine exchange factor gene			
ROM1	*rom1*Δ viable; *rom2*Δ inviable	NR	233, 271
ROM2	Deletion viable; ts in some strains; morphology defects	Incipient bud site; small bud tips	199, 233, 271
Possible targets Rho1p: Glucan Synthase (Fks1,2), Pkc1p, Bni1p, Bem4p			
RHO2	Nonessential; deletion increase sensitivity to benomyl	NR	198, 199
GTPase activating protein genes			
BEM2	As above	As above	As above
Guanine exchange factor gene			
ROM1, ROM2	As above	As above	As above
RHO3	Nonessential; deletion slow growing; *rho3*[ts] *rho4*Δ arrests as budded cells with depolarized actin	NR	134, 209
RHO4	Nonessential; deletion, no phenotype, but lethal with *rho3*Δ	NR	209
Possible targets Rho3p & Rho4p: Bem4p, Bni1p, Sec4p			

[a]NR, Not reported.

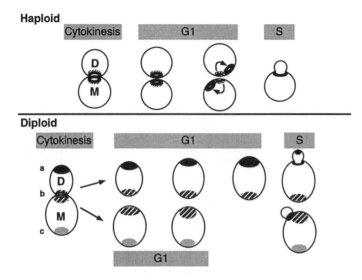

Figure 5   Models for haploid and diploid budding patterns. (*top*) A cytokinesis tag directs axial budding in haploid cells. The neck filament system (*dark rings*) serves as a template for assembly of the axial bud site components Bud3p, Bud4p, and Axl2p (*shaded rings*) at the neck region at the end of the G2 phase. This complex directs a new axial budding event, including the formation of a new neck filament ring, adjacent to the old site before its disassembly (*speckled ring*). (*arrow*) The general bud site selection proteins, Rsr1p, Bud2p, and Bud5p, recognizing this landmark and communicating its position to the polarity-establishment components (e.g. Cdc24p, Cdc42p, Bem1p) that help initiate polarized growth at the adjacent site during the late G1 stage of the next cell cycle. (*bottom*) Bipolar budding pattern of diploid cells. A diploid daughter cell possesses positional information both at the previous site of cell growth (distal site, *a*) (*dark patch*) and adjacent to the site of cytokinesis (*hatched patch*). The distal site is preferred for growth, and Bud8p is required for the function of the distal landmark. After completion of the budding process in the daughter cell, a newborn mother cell contains marks from the two previous cytokinesis events. Many components, including actin, actin-binding proteins, Spa2p, Pea2p, and possibly Bud9p, are required for the function and/or recognition of these cortical landmarks (*hatched patch*). Subsequent budding events can initiate at spatial cues from either pole depending on the age of the cell. (Reproduced in part with permission from *Trends Cell Biol.*)

mutations have been identified in the *ACT1* gene (which encodes actin), many of which affect bipolar budding in diploid cells (78, 323). These mutations do not affect the budding pattern of daughter cells but instead cause diploid mother cells to bud more and more randomly with each cell division. These mutations have little or no effect on axial budding. Interestingly, the amino acids identified as important for bud site selection all map to a specific domain on the actin protein (146, 312, 323), which suggests that this region of Act1p may recognize bipolar-specific proteins or cues.

Mutations in genes encoding several actin-associated proteins (e.g. Sac6p, Srv2p, Sla2p, Rvs167p) or genes that disrupt actin organization (e.g. Sla1p, Rvs161p) cause defects in bipolar budding similar to those of the *act1* mutations (4, 9, 18, 65, 77, 79, 99, 130, 211, 284, 306, 325). One or more of these proteins might interact with the domain of actin required for bipolar budding. The actin cytoskeleton at the incipient bud site and at the neck might help to localize cortical bud site selection tags to their proper location; alternatively, actin filaments might be important for shuttling new components to these tags.

Although most actin-associated proteins are not important for the distal tip budding of diploid daughter cells, one exception to this rule is Bni1p, a member of the highly conserved formin protein family found in *S. cerevisiae* (90), *Schizosaccharomyces pombe* (44, 241), mouse (207), and *Drosophila* (43, 83). Bni1p associates with actin in two-hybrid assays and with regulators of the actin cytoskeleton (Cdc42p, profilin) in coimmunoprecipitation or in vitro binding experiments (90). Diploid *bni1*Δ mutants bud randomly both in the first division and in subsequent divisions (325), but *bni1*Δ haploid cells bud normally. Thus, in addition to its role in bipolar budding of mother cells, Bni1p might also play an important role in the establishment of the distal tag in diploid daughter cells.

OTHER PROTEINS IMPORTANT FOR DIPLOID BUD SITE SELECTION   A variety of other components have been identified that participate in bipolar bud site selection; some of these may also regulate the actin cytoskeleton. Spa2p, Pea2p, and Bud6p comprise a class of proteins important for bipolar bud site selection (10, 287, 304, 325). Like actin mutants, *spa2*, *pea2*, and *bud6* strains are not important for selection of distal sites in diploid daughter cells, but the number of cells that choose random sites increases with successive cell divisions (10, 287, 304, 325). Spa2p, Pea2p, and Bud6p each localize to sites of polarized growth. These proteins localize as a patch at the incipient bud site, to the tip of the growing bud, and to the mother-bud neck region prior to and during cytokinesis; the staining at cytokinesis persists into the unbudded stage of the next cell cycle (10, 90, 287, 288, 304). *spa2*, *pea2*, and *bud6* strains also display defects in mating projection formation, mating efficiency, and pseudohyphal growth (90, 107, 219, 260, 287, 304, 324). Spa2p and Pea2p form a complex, as demonstrated by coimmunoprecipitation and two-hybrid experiments, and Spa2p, Pea2p, and Bud6p migrate together as a 12S complex in sucrose gradients (278). Bud6p has been found to interact with actin by the two-hybrid system (9). Perhaps Spa2p, Pea2p, and Bud6p interact with the actin cytoskeleton in order to promote bipolar budding. Alternatively, because Spa2p and Bud6p interact by the two-hybrid system with components of two mitogen-activated protein (MAP) kinase signaling cascades that regulate polarized cell growth in yeast (see below) (278), it is possible that Spa2p, Pea2p, and Bud6p serve to link bud site selection components to signaling pathways required for polarized growth.

SUR4, *FEN1*, and *BUD7* encode additional proteins required for the bipolar budding pattern; less is known about the function of these proteins. *SUR4* and *FEN1* encode homologous and functionally redundant proteins (254). *sur4* and *fen1* mutants show similar phenotypes, including extragenic suppression of the *rvs161* and *rvs167* mutations that affect the actin cytoskeleton, an overall decrease in phospholipid content, and resistance to an immunosuppressive agent and to ergosterol biosynthesis inhibitors (69, 254). Mutations in *SUR4* and *FEN1* also cause a randomization of the bipolar budding pattern, although this defect has not been analyzed in detail (80, 254). A mutation in *BUD7* causes heterogeneous defects in bipolar budding (325); a null allele of *BUD7* has not been reported. Further analysis of *SUR4*, *FEN1*, and *BUD7* may provide additional insights into mechanisms for recognition of sites during bipolar budding.

BUD8P AND BUD9P   A visual screen to identify components involved in the bipolar budding pattern identified two interesting mutants, *bud8* and *bud9*, that bud exclusively at one pole as evidenced by Calcofluor staining (325). *bud8* cells bud at the proximal pole of the daughter cell, whereas *bud9* cells choose distal sites for growth in the first division of the mother cell and proximal sites in subsequent divisions (325). Bud8p and Bud9p are both predicted to be transmembrane proteins with a serine-rich extracellular domain and a short cytoplasmic domain. Thus, these molecules are candidates to be cortical tags for bud site selection.

bud8 bud9 mutants bud at proximal sites; *bud6 bud9*, *act1 bud9*, and *sla2 bud9* cells display an overall random budding pattern but maintain preference for distal sites during division in daughter buds. *bud6 bud8*, *act1 bud8*, and *sla2 bud8* strains select random sites even in daughter cell division (323, 325). Bud8p is likely to be involved in establishing distal tags in daughter cells; Bud9p functions by establishing proximal sites during daughter cell divisions (see 325 for a more detailed discussion).

MODELS FOR BIPOLAR BUDDING   Diploid cells use positional cues at both proximal and distal sites (Figure 5) (47, 49, 195, 288, 304, 323, 325). Newborn daughter cells choose distal sites for new growth, for several reasons. First, components present from the previous budding process may persist from the bud tip and serve as cortical tags in the next cell cycle. Also, daughter cells have a longer G1 phase of the cell cycle than do mother cells because daughter cells take longer to reach the critical size required for initiation of budding (249); the proximal tag may be more labile than the distal site marker, and this would contribute to the preference for distal sites (96). *BUD8* is required to establish the distal tag and may be a component of this tag (325).

Following distal budding and completion of the first cell cycle, a new mother cell will possess spatial cues from cytokinesis. It is expected that "polarity components" residing at the neck during cytokinesis, such as actin, actin-binding proteins, Spa2p, Pea2p, and perhaps Bud9p, provide information that helps establish tags or interacts with tags at the cytokinesis sites (78, 288, 304, 323, 325). Other proteins, such as Bni1p and possibly Bud7p, may function to place or stabilize both proximal and distal tags (325). If the cytokinesis tags are labile, then as mother cells get older they would be more likely to use distal tags (96); older mother cells have a longer G1 phase. As with axial budding, the different tags are likely to be recognized by Rsr1p complexes to select sites and promote bud emergence.

In summary, a major difference between axial budding in haploid cells and bipolar budding in diploid cells is that diploid cells use a unique spatial cue at distal sites. As described above, it is likely that a haploid-specific gene product (e.g. Axl1p) or set of products functions to make proximal sites the preferred site for growth in haploid cells.

Budding Patterns Can Be Influenced by Environmental Conditions

As noted above, loss of axial-specific budding components (e.g. Bud3p, Bud4p, Axl2p, and Axl1p) often results in an increased frequency of distal budding in haploids (47, 48, 102, 259, 268). A similar phenomenon also occurs when yeast cells are exposed to certain environments.

When starved, haploid cells also lose the proximal tag and, upon reinitiating growth in a rich medium, preferentially select distal sites for bud formation (49, 195). Cells exposed to very low levels of mating pheromone also switch from budding at proximal sites and begin budding at distal sites (see below). In both instances these cells are likely to have a prolonged G1 and thereby may lose their proximal tags. It is likely that the use of distal sites in these circumstances allows the cells to spread efficiently (195). Cells that return to growth after starvation can become more readily exposed to fresh nutrient sources; cells exposed to mating pheromone can spread efficiently to find potential mates.

It has also been reported that when starved for nitrogen, diploid cells on the edge of a colony undergo pseudohyphal growth and convert from a bipolar pattern, in which both ends of the cell are used, to a unipolar pattern, in which all buds emanate from a single pole (110, 165). This interpretation must be interpreted cautiously, however, as analysis of cells from the entire colony suggests that overall, most pseudohyphal cells use the bipolar pattern (260).

Regardless, the primary spreading mechanism of yeast cells is due to the distal budding of daughter cells; this occurs in cells undergoing either bipolar or unipolar budding.

BUD FORMATION AND GROWTH

Bud formation begins when cells traverse the G1/S transition, and bud formation and growth involve polarization of the actin cytoskeleton, secretion, and cell wall synthesis. Bud growth occurs in two phases: Growth occurs initially at the tip of the bud, in what is termed apical growth; later, once the bud reaches about one third of its final size (180), bud growth becomes isotropic, with cell wall deposition occurring uniformly throughout the bud. During both of these phases, there is little growth in the mother cell.

Three lines of evidence indicate that bud formation probably occurs in conjunction with bud site selection. First, mutations in certain components required for bud formation (e.g. Cdc24p, Bem2p) exhibit defects in bud site selection (152, 285). Second, several bud formation proteins and bud site selection components interact physically (236, 242). Finally, a strain containing disruptions of *CLN1* and *CLN2*, which encode two G1 cyclins thought to promote bud formation and apical growth in yeast, is viable, but an additional mutation in *BUD2* results in lethality (23, 66). One interpretation of this result is that Bud2p contributes to the bud formation process such that compromising both Cln function and Bud2p might result in a bud formation defect. Thus, these different observations suggest that bud site selection components work with bud formation proteins to form a new bud.

A variety of proteins important for bud formation and growth have been identified. These include polarity-establishment proteins and Rho-type GTPases, both of which are thought to regulate cytoskeletal components such as actin and septins, secretory and cell wall proteins, and signaling pathways.

The Role of Polarity-Establishment Components

A set of proteins critical for bud formation in yeast are the polarity-establishment proteins. These include Cdc42p, a GTPase most closely related to members of the Rho family, and its GEF, Cdc24p (6, 286, 329). Cells containing temperature-sensitive mutations in either of these genes fail to form buds and form large, round, unbudded cells with multiple nuclei (6, 93, 285, 286). At the restrictive temperature, these strains fail to properly localize many polarized components important for yeast budding, including Spa2p, actin patches, and septins (3, 6, 144, 288, 332).

In mammalian cells, Cdc42p interacts with the PAK protein kinase to help mediate cell polarization (200, 205). Yeast cells contain three PAK kinase homologs, Ste20p, Cla4p, and Skm1p (67, 206). Strains containing either *ste20Δ* or *cla4Δ* mutations are viable, whereas *ste20Δ cla4* double mutant strains are not, which suggests overlap in function between these two kinases (67). A *ste20Δ cla4* temperature-sensitive strain arrests with aberrant cell morphology and defects in cytokinesis (67). *skm1Δ* cells do not exhibit any apparent defects, and increased defects are not observed when the *skm1Δ* mutation is combined with either *cla4Δ* or *ste20Δ* (206). Thus, the normal function of Skm1p is not known.

Ste20p and Cla4p interact physically with Cdc42p, and this interaction is important for the function(s) of these proteins (67, 171, 240, 283). Cdc42p and Ste20p each localize to sites of polarized cell growth in yeast: a patch at the putative bud site in unbudded cells and the bud tip in small budded cells (171, 240, 333). Thus, an attractive model is that Cdc42p may facilitate targeting of Ste20p to polarized growth sites and thereby direct its activity to those sites (171, 240). Consistent with this model, a Ste20p variant unable to bind Cdc42p does not localize to sites of polarized cell growth. This mutation does not cause defects in pheromone signaling (see below), but it is unable to restore growth to *ste20 cla4* cells and causes defects in pseudohyphal growth (171, 240). Thus, the Cdc42p binding site and the proper localization of this protein are important for its function. Although Cdc42p is found at polarized growth sites in yeast, Cdc24p, its GEF, localizes over the entire cell periphery (248). Therefore, either Cdc24p functions only at polarized growth sites where Cdc42p accumulates or it has additional targets besides Cdc42p.

Other components that genetically interact with Cdc42p and Cdc24p have been identified (Table 1). These include Bem3p, a Rho-GAP homolog that serves as a GTPase activating protein for Cdc42p in vitro (290, 329, 330); Rga1p and Rga2p, two Rho-GAP homologs that may serve as GAPs for Cdc42p in vivo (290); and Zds1p and Zds2p, two proteins that are not GAP homologs but that appear to down-regulate Cdc42p in vivo (24). Mutations in another polarity-establishment gene, *BEM1*, are colethal with *MSB1*, a high-copy suppressor of both *cdc24* and *cdc42* (22). Finally, two potential targets of Cdc42p, Gic1p and Gic2p, have been described recently; Gic1p and Gic2p interact genetically with Cdc42p and contain a CRIB domain, characteristic of many Cdc42p interacting proteins (34, 50). Bem1p, Gic1p, Gic2p, and the Zds proteins are all important for cell polarity in yeast, and each of these proteins except Zds2p has been localized to sites of polarized cell growth, similar to Cdc42p (24, 34, 50, 248).

Although strict criteria for defining polarity-establishment components have not been presented, it is important to note that many other components, including cytoskeletal proteins such as actin and type V myosin (Myo2p), are important for bud formation (145, 228). In addition, many other proteins, including Spa2p, Pea2p, Bud6p, and Bni1p, are important for the proper shape of yeast cells

(10, 90, 107, 287, 304). It is likely that many of these proteins function together to promote polarized cell growth. Consistent with this hypothesis, many of these proteins physically associate with one another (90, 278).

The Role of the Cytoskeleton

Three cytoskeletal systems have been well characterized in yeast: microtubules, actin, and septins. At least two of these systems, actin and septins, are important for cell morphogenesis.

ACTIN   Actin is essential for polarized cell growth in yeast and is present in two forms. Actin patches localize at the cortex, primarily in the bud, and actin cables run longitudinally along the length of the cell and often intersect actin patches at their ends (3, 114, 150, 220). Electron microscopic studies indicate that the cortical actin patches are sites of plasma membrane invaginations (220). Actin patches may represent sites of membrane and cell wall deposition (Figure 6) (220). Consistent with this hypothesis, temperature-sensitive actin mutants exhibit a cell wall defect when incubated at the restrictive temperature (103).

Temperature-sensitive mutations in the single actin gene of yeast, *ACT1*, exhibit defects in bud formation and growth, and cells arrest as either large unbudded cells or small budded cells with an enlarged mother cell (228). These cells display delocalized chitin staining, cell lysis defects, and sensitivity to high osmolarity and high and/or low temperatures (228, 229, 280, 312). Many of these phenotypes are probably caused by defects in polarized secretion. Temperature-sensitive *act1* mutants and strains containing mutations in type V myosin (*myo2-1*) and tropomyosin (*tpm1Δ*) accumulate vesicles (113, 145, 187). Furthermore, mutations in *SAC1*, which encodes an actin-binding protein, can

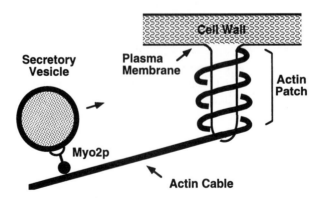

Figure 6   Secretory vesicles containing Myo2p myosin motors may move along actin cables to deliver them to actin patches that are sites of plasma membrane invaginations (220).

suppress defects in *SEC14*, *SEC9*, and *SEC6*, which are involved in late secretory steps in yeast (56). Thus, the actin cytoskeleton participates in secretion to help mediate proper cell wall synthesis.

Actin exhibits a very dynamic localization during the cell cycle (Figure 4) (3, 150). Actin localizes as a ring at the incipient bud site and at the bud tip in apically growing cells. During G2, actin is still concentrated in the bud but localizes isotropically. During mitosis, actin is found throughout both the mother and daughter cell, and at cytokinesis, it localizes at the mother-bud neck. These dynamic changes are regulated by the cell cycle–dependent kinases (180; see below).

In addition to a dynamic localization throughout the cell cycle, actin patches continuously move at a rate of 1 μm per s throughout the bud and even through the neck into the mother cell (75, 307). Given the preferential distribution of actin in the bud, it is likely that actin assembly occurs primarily at the cortex of the bud (similar to growth cones in neuronal cells) and/or that polymerized actin is preferentially disassembled in the mother cell (168).

A large number of proteins that interact physically or genetically with actin have been identified (for more extensive reviews see 28, 114). Many of these components have homologs in other eukaryotes [for example myosin (types I, II, and V), fimbrin, tropomyosin, and cofilin (145, 185, 214, 311)]. Many of these proteins are not essential for cell viability, but mutations in several components often lead to lethality (28). Given the large number of actin-interacting proteins in yeast and other eukaryotes, it is likely that redundant mechanisms mediate actin organization and function.

Of special mention are two classes of actin-binding proteins, myosins and cofilin. There are five myosins in yeast, Myo1p through Myo5p (reviewed in 35). Myo1p encodes a conventional myosin homolog important for proper cytokinesis in yeast (311), and Myo3p and Myo5p are redundant myosin type 1 homologs required for proper polarization of the actin cytoskeleton and endocytosis (108, 112). Myo2p and Myo4p are myosin type V homologs (145). Myo4p localizes to polarized growth sites and is required for cell fate determination (see below). Myo2p also localizes to the incipient bud site and tips of small buds (184) and plays a critical role in bud formation. Like polarity-establishment proteins, cells containing temperature-sensitive mutations in *MYO2* fail to form buds at the restrictive temperature, resulting in the accumulation of large unbudded cells (145). Thus, Myo2p acts early in the budding process. Myo2p may serve as an actin-based motor that translocates secretory vesicles to growth sites (Figure 6). However, despite extensive effort, no direct evidence for this model exists (114). An alternative model is that Myo2p helps localize and organize actin at growth sites, perhaps serving as a cortical tether for actin filaments.

A second type of actin binding protein of particular interest is cofilin (168, 214). Cofilin mediates actin depolymerization in vitro, and temperature-sensitive cofilin mutants accumulate an abundance of actin patches in the mother cell (168). Thus, cofilin may be important for actin depolymerization in the mother cell.

RELATIONSHIP BETWEEN POLARITY-ESTABLISHMENT PROTEINS AND THE ACTIN CYTOSKELETON   A very large number of polarity-establishment and cytoskeletal components have been described that are important for bud formation and growth. The number of interactions between the different proteins that are being reported is increasing at an enormous rate. For example, actin has been found to interact with Bni1p, Bud6p, Bem1p, and a myriad of actin binding proteins, including Sac6p, cofilin, profilin, Tpm1p, Abp1p, and others (5, 79, 90, 186, 214). It is likely that many of these assemble to form complexes that compose or regulate the actin cytoskeleton.

The different classes of proteins involved in polarized growth in yeast usually have one of two localization patterns. Most of the polarity-establishment proteins and components affecting cell polarity localize diffusely along the bud tip (184, 287, 304). Some of these proteins interact with cell signaling proteins as well as cytoskeletal components. For example, two-hybrid studies indicate that Spa2p and a related protein, Sph1p, interact with the MEKs Ste7p, Mkk1p, and Mkk2p (discussed below), and both Spa2p and Bud6p associate with the MEK kinase Ste11p (260, 278); Cdc42p associates with Ste20p (240, 283). Other diffusely localizing proteins, such as Myo2p or Sec4p, might be directly involved in secretion or (for Myo2p) may regulate assembly of the actin cytoskeleton. Perhaps the diffusely staining proteins are involved in assembling and regulating the actin cytoskeleton and polarized secretion. These components are likely then to be particularly important during dynamic remodeling of the cytoskeleton. Consistent with this hypothesis, many of these components (e.g. Spa2p, Pea2p, Bud6p, and Bni1p) have more pronounced defects in polarization of the mating projection than in bud formation and growth (107, 304). In contrast to the diffusely localizing proteins, actin and many of its interacting proteins localize as patches. These presumably represent specialized structures where membrane and cell wall deposition are occurring.

RHO PROTEINS   In mammalian cells, three types of small GTP binding proteins are important for the regulation of the actin cytoskeleton and the formation of polarized structures: Cdc42, Rho, and Rac (reviewed in 117, 256). Rac homologs have not been identified in budding yeast, but there are homologs for both Cdc42p and Rho. A homolog of Cdc42p (YNL180w) is present in the databases, but information about this protein is lacking. Four yeast Rho homologs (Rho1p,

Rho2p, Rho3p, and Rho4p) exist, and at least three (Rho1p, Rho3p, and Rho4p) are important for cell polarity in yeast (134, 198, 208, 209, 322).

RHO1 is an essential gene, and a temperature-sensitive *rho1* strain arrests at the restrictive temperature with very small buds and a cell lysis phenotype (322). Thus, Rho1p is required for maintaining bud growth. This phenotype is similar to *pkc1* temperature-sensitive mutants (see below). Rho1p may also play a role in bud emergence. Strains deleted for Bem2p, a protein that serves as a GAP for Rho1p in vitro, are viable but fail to grow at 37°C (22, 152, 242). At the restrictive temperature, these mutants arrest as large unbudded cells similar to *cdc42* temperature-sensitive cells. Consistent with its role in bud formation and growth, Rho1p localizes as a patch at the incipient bud site and at the bud tips of small budded cells (322). Rho1p interacts physically with, and is required for the activity of, at least two proteins: (β-1-3) glucan synthase, which synthesizes (β-1-3) glucan, a major constituent of yeast cell walls (76, 251); and Pkc1p, yeast protein kinase C, which activates a signaling pathway involved in cell wall biosynthesis (149, 227). In addition, Rho1p interacts with Bni1p, which, as noted above, interacts with actin and profilin and may help organize the actin cytoskeleton (90, 135, 157); Bni1p also interacts with Bem4p, which genetically interacts with Cdc42p and is involved in organization of the actin cytoskeleton (129, 191). Thus, Rho1p appears to control polarized growth in yeast through multiple mechanisms, including the regulation of cell wall synthesis, cell signaling, and organization of the actin cytoskeleton.

In addition to Bem2p, several other regulators of Rho1p have been identified (summarized in Table 1). Of particular interest are two homologous proteins, Rom1p and Rom2p, which contain a domain that is conserved in Rho-GEFs, such as the human dbl protein and yeast Cdc24p; Rom2p serves as a GEF that specifically activates Rho1p in vitro (233). High-copy plasmids of *ROM1* and *ROM2* suppress the growth defects of a *rho1* cold-sensitive mutant, which suggests that they act as GEFs in vivo. Unlike Cdc24p, the GEF for Cdc42p, which localizes around the entire cell periphery, Rom2p has been localized to polarized growth sites, similar to Rho1p (199). Thus, Rom2p localizes specifically at its proposed site of action, and Cdc42p and Rho1p GTPase cycles may have different mechanisms of spatial regulation (199).

Rho2p is highly homologous to Rho1p (53% identity predicted) and may have redundant functions with Rho1p (198). Thus far, the only phenotype that has been identified for *rho2Δ* strains is a slightly increased sensitivity to the microtubule depolymerizing drug benomyl (199). Genetic analysis suggests that Bem2p and Rom2p act as a Rho2p GAP and GEF, respectively. Both *bem2Δ* and *rom2Δ* strains are also hypersensitive to benomyl (199, 308); it is possible that Rho2p has a direct role in microtubule assembly or that microtubule capture sites at the cortex are disrupted in *rho2* cells.

Rho3p and Rho4p are divergent from Rho1p; these proteins share 46% and 41% predicted amino acid sequence identity with Rho1p, respectively, and are 35% identical to one another (208). Deletion of *rho3Δ* and *rho4Δ* reveals that, although *rho3Δ* mutants grow slowly, neither gene is essential for cell viability; however, *rho3Δ rho4Δ* strains are inviable (134, 209). Like *rho1Δ* cells, strains depleted of Rho3p and Rho4p arrest with small buds and lyse, and this phenotype can be suppressed by osmotic stabilizers such as sorbitol (209). *rho3Δ rho4Δ* cells grown in the presence of sorbitol contain delocalized actin, which indicates that these strains exhibit defects in actin polarity.

It is likely that Rho3p and Rho4p are in a different functional family of Rho proteins from that of Rho1p and Rho2p; high-copy plasmids containing *RHO1* and *RHO2* cannot suppress the lethality of *rho3Δ rho4Δ* strains. In contrast, Rho3p and Rho4p may have redundant functions with Cdc42p; high-copy plasmids containing *CDC42* can suppress *rho3Δ* strains (209). Rho3p may mediate polarized secretion, because temperature-sensitive Rho3 alleles are synthetic lethal with *sec4* mutations; *SEC4* encodes a Rab-type GTPase involved in late steps of secretion (134). Unlike the case for Rho1p and Cdc42p, little information is available about targets for Rho3p and Rho4p. Two possible candidates are Bni1p and Bem4p, which interact with these Rho proteins in the two-hybrid system (90, 191).

SEPTINS   One feature that distinguishes yeast from many other eukaryotes is that, in yeast, preparation for cytokinesis occurs early in the cell cycle; the site of bud formation will become the cytokinesis site (193). Many components important for cytokinesis are deposited at the incipient bud site before the bud appears. It is not surprising, then, that many of the components essential for cytokinesis play an important role in bud formation and growth.

An important class of proteins required for cytokinesis are the septins (190, 267). In addition to their role in cytokinesis and bud site selection, septins are important for morphogenesis (96, 121). Septins localize as a ring at the incipient bud site at approximately the same time as actin (97, 151; H Kim, S Ketcham, B Haarer, J Pringle, unpublished information), although it is likely that this localization occurs, at least in part, through independent mechanisms; actin localizes to bud tips in septin mutants, and septins localize normally in cells depolymerized for actin (3, 12). Septins play at least three very important roles during bud growth. First, septins are important for chitin localization, since mutations in these proteins display delocalized chitin (68, 96, 190, 257). Second, septins are important for actin compartmentalization in the isotropic growth phase (Y Barral, M Snyder, unpublished results). When isotropic growth is induced in septin mutants, actin polarity is not maintained, and bud growth aborts. Finally, as described below, septins are important for a cell-cycle checkpoint control

(16). Septin mutants arrest in an apical growth phase, resulting in the formation of highly elongated buds (see below).

MICROTUBULES   Cytological evidence indicates that cytoplasmic microtubule bundles(s) intersect the nascent bud site and that long microtubule bundles extend into the bud (3, 37, 38, 150, 288). This finding led to the long-standing model that either the spindle poles or their associated microtubules are important for bud site selection (38). However, subsequent experiments revealed that components important for bud formation are deposited at the bud site before the microtubules extend to the incipient bud site; thus microtubules cannot direct components to the incipient bud site (288). Moreover, temperature-sensitive tubulin mutants and yeast cells treated with nocodazole, a microtubule depolymerizing drug, still form buds at normal positions both during the first cell cycle (131, 143) and in subsequent cycles (323). The buds that form in strains lacking microtubules are morphologically normal (131, 143). Thus, there is no evidence for the role of microtubules in bud site selection, bud formation, or growth.

It is possible, however, that microtubules play an ancillary role in bud formation and/or growth. Electron microscopic and tubulin immunofluorescence studies indicate that cytoplasmic microtubules emanate from the spindle pole body (the microtubule organizing center), located in the yeast nuclear envelope, to the incipient bud site in unbudded cells and into the bud in budded cells (3, 37, 38, 150, 288). Because the orientation of these microtubules occurs after establishment of yeast polarity, it is likely that these are secondary events; indeed, it has been speculated that these microtubules are important for spindle orientation and nuclear segregation (42, 234, 235, 288). However, it is possible that they also contribute to polarized secretion as well. Consistent with this hypothesis, yeast cells compromised for Myo2p function can be suppressed by multiple copies of the *SMY1* gene, which encodes a yeast kinesin–related protein (183). Thus, although microtubules are not required for bud formation and/or growth, they may assist in either of these processes.

SECRETION   To mediate bud formation and growth, new plasma membrane and cell wall material must be directed to the new growth site, presumably through the secretory pathway. Consistent with this possibility, high concentrations of membrane vesicles are present in the buds of small budded cells (245).

It is likely that distinct classes of secretory vesicles carrying different cargo are involved in protein secretion in yeast and that secretion of at least one class of vesicles requires the actin cytoskeleton. Three lines of evidence support this conclusion. First, *sec4* strains blocked in a late step of the secretory pathway accumulate two types of 100-nm vesicles (120). One class of vesicles contains an endoglucanase and the major plasma membrane ATPase; the other contains invertase and acid phosphatase. Second, *myo2* and *act1* strains often accummulate vesicles without affecting secretion of particular components to the cell surface. *myo2*Δ mutants accumulate secretory vesicles but still deliver invertase with normal kinetics to the cell surface (113, 145). Similarly, actin mutants accumulate secretory vesicles that contain Ypt1p (a small GTPase involved in vesicular transport), but not vesicles containing Sec4p, which is thought to be involved in many types of protein secretion (221). Thus, the actin cytoskeleton may direct a particular class of secretory vesicles to the cell suface.

A third line of evidence for different types of secretory vesicles comes from the analysis of a polarized growth component, Chs3p, an integral membrane protein involved in chitin synthesis (see below). This protein is present as a ring at the incipient bud site and neck of budded cells (54, 270). Chs5p, which is required for many aspects of cell morphogenesis, is necessary for the proper localization of Chs3p but not of other secreted proteins (269, 270). Chs5p is a protein of the *trans*-Golgi network. Chs3p also requires Myo2p for targeting (270). A simplistic model that can account for many of the different observations presented above is that a class of vesicles required for yeast cell polarity is generated from the *trans*-Golgi network, perhaps by Chs5p and other proteins (Figure 7) (270). These vesicles are then translocated by the actin cytoskeleton to polarized growth sites.

Large amounts of data have accumulated on proteins and mechanisms by which components are translocated through the secretory pathway to reach the

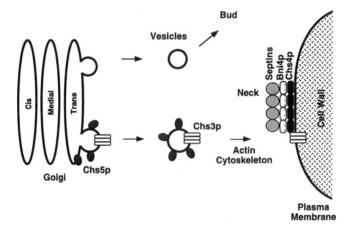

Figure 7   Chs5p containing vesicles target Chs3p to the neck with the help of the actin cytoskeleton. Chs3p is deposited and/or retained at the neck through interaction with Chs4p, which is part of the septin complex.

cell surface. A recent comprehensive review of protein secretion in yeast can be found (147); this process is not discussed here.

CELL WALL SYNTHESIS   The cell wall is the extracellular matrix of the yeast cell, and, as in animal cells, the matrix plays a critical role in cell morphogenesis (55). As noted above, cell wall deposition is thought to occur at actin patches (220), and new cell wall material is presumably added through the dynamic remodeling of the existing matrix and by the integration and synthesis of new material. Detailed reviews describing the many components involved in cell wall synthesis and degradation are available (55, 232).

In addition to serving as one of the end products of polarized secretion, the cell wall contributes to cell polarity in several ways. First, an intact cell wall is required for proper polarization of bud site selection components and actin. As noted above, mutations that affect cell wall synthesis have defects in bud site selection (32, 320), and removal of the cell wall results in actin delocalization. It is possible that the cell wall interacts with the cytoskeleton through transmembrane proteins, such as Axl2p (259) and Chs3p. Because actin polarizes and buds form in an $axl2\Delta$ strain, presumably other membrane proteins exist that anchor underlying cytoskeletal components to the cortical membrane (259). Among such proteins should be an anchor for the septins.

Second, the cell wall contributes to polarity by maintaining the shape of the cell. Diploid cells are more elongated than haploid cells, which are round. Presumably, diploid cells have a longer apical growth phase than do haploid cells. It is clear that the cell wall helps maintain this shape, because enzymatic removal of the yeast cell wall causes the cells to become round.

The cell wall also contributes to cell polarity in yeast through the asymmetric deposition of structural components that are critical to the function of the cells. Most notably, cells contain a polarized distribution of chitin, a cell wall polymer of N-acetylglucosamine that is essential for proper morphogenesis and cell division (277). Chitin staining is pronounced at the incipient bud site in unbudded cells, in the neck of the bud, and in the primary septum (i.e. bud scar) present at the division site in mother cells (55, 124, 277). The chitin ring at the neck is deposited by chitin synthase III (CSIII), a nonessential enzyme whose catalytic activity, Chs3p, is a predicted integral membrane protein that localizes as a ring at the incipient bud site and bud neck during early bud growth and at the neck in cells undergoing cytokinesis (54, 270, 303). Chs3p localizes to the neck through association with Chs4p, a protein required for CSIII activity (68, 299). Chs4p colocalizes with septins to a ring on the mother side of the neck, and septins are required for this localization. Chs4p interacts directly with Chs3p and also, through Bni4p, with the septins (68). As noted above, targeting of Chs3p to the neck requires both Myo2p and Chs5p (270). Thus, an attractive model for Chs3p localization is that membrane vesicles containing Chs3p are generated from the Golgi by Chs5p and directed to, or maintained at, the neck by interaction with Chs4p at the septin ring (Figure 7).

A second chitin synthase, Chs2p, is also polarized at the neck region in yeast (54). At the end of mitosis, a disc of chitin is deposited in the mother-bud neck; this disc forms the primary septum. Chs2p is specifically synthesized and localized to the neck at this time (54). Thus, Chs2p and Chs3p are subjected to different modes of regulation: Chs2p is synthesized and appears at the neck only near the end of cell division; Chs3p localizes to the incipient bud site and neck in small budded cells and then reappears at the neck at the end of cell division.

Chitin and septins are important for the proper formation of the neck. Temperature-sensitive *cdc12* mutants incubated at the permissive temperature are lethal in combination with mutations in *CHS4* or *CHS5* (68; B Santos, M Snyder, unpublished information).

POLARIZED CELL GROWTH DURING THE MATING RESPONSE

In contrast to the predetermined axial and bipolar budding patterns exhibited during vegetative growth, mating cells can initiate projection formation at novel points on the cell surface in response to external signals from potential partners (reviewed in 51, 261). The mating response initiates as cells perceive and respond to peptide pheromones secreted from cells of the opposite mating type; *MAT*a cells produce a-factor, whereas *MAT*α cells secrete α-factor. Each pheromone binds to a specific seven transmembrane–spanning receptor (α-factor binds to Ste2p; a-factor binds to Ste3p) located at the cell surface. Recognition of pheromone elicits a series of intracellular events that includes the dissociation of the receptor-coupled heterotrimeric G protein into both activating (G$\beta\gamma$) and inhibitory (Gα) components, the stimulation of the Ste20p kinase and a MAP kinase signaling cascade (comprised of the MEK kinase Ste11p, the MEK Ste7p, and the MAP kinases Fus3p and Kss1p; see below), and the phosphorylation and activation of the Ste12p transcription factor. These molecular processes lead to arrest of cells in the late G1 stage of the cell cycle, cell agglutination, initiation of polarized growth toward a partner, cell fusion, and nuclear fusion (reviewed in 15, 167, 289, 314).

Selection of Sites of Polarized Cell Growth During the Mating Response

Several lines of evidence indicate that polarization of growth toward a partner depends on detection of pheromone gradients. First, wild-type cells can discriminate between potential partners that do or do not secrete pheromone (139); mating to a pheromoneless partner occurs at an extremely low efficiency, and the ability to discriminate depends on an intact pheromone receptor (139, 141). Second, addition of exogenous pheromone to a mating mixture dramatically decreases mating efficiency, and at high concentrations, exogenous mating factor eliminates mating partner discrimination (74, 201). These defects probably result from projection formation at sites not adjacent to a mating partner. Cells treated with high levels of pheromone initiate projection formation adjacent to the previous site of cytokinesis (axial sites) (74, 195), whereas the sites of projection formation in mating mixtures occur at random positions on the cell surface relative to previous bud sites (195). Third, mating-impaired mutants have been isolated that arrest properly in response to pheromone and form wild-type projections, but these projections form adjacent to previous bud sites, even in mating mixtures (305). Fourth, mutations that constitutively activate or hyperactivate the mating pathway (e.g. *gpa1* and *sst2* strains) cause defects in mating partner discrimination (74, 139). These strains are presumably defective in sensing external pheromone gradients. Finally, direct microscopic analysis has demonstrated that *MAT*a cells are capable of orienting projection formation along a spatial gradient of mating factor (275), and mutants have been identified that fail to track along these gradients (275, 305; LG Vallier, J Segall, M Snyder, unpublished information). All of these observations indicate that the ability to perceive and initiate growth events toward a pheromone gradient is critical for an efficient mating response.

ROLE OF PHEROMONE RECEPTORS IN DIRECTING GROWTH   Pheromone receptors may form a cortical landmark to direct growth along a spatial gradient of pheromone during the mating response. Cells lacking receptors can still mate if downstream signaling components are activated (141, 315); however, mating occurs at a low frequency, and these cells cannot discriminate between mating partners (74, 141). In addition, the Ste2p receptor localizes to the tip of the mating projection and at the conjugation bridge of newly formed zygotes (141). Thus, an attractive hypothesis is that the receptor interacts with components required for the establishment of polarized growth (Figure 8). This hypothesis is supported by the observation that mutations that truncate or completely delete the carboxy-terminal cytoplasmic tail domain of Ste2p cause defects in projection formation (161) and in the orientation of projections in α-factor gradients (LG Vallier, J Segall, M Snyder, unpublished information).

ABSENCE OF PROJECTION FORMATION AT PREVIOUS BUD SITES   During the mating response, it is likely that a cell both constructs a new tag and inhibits the use of previous positional information. Axl2p and Bud4p axial-specific

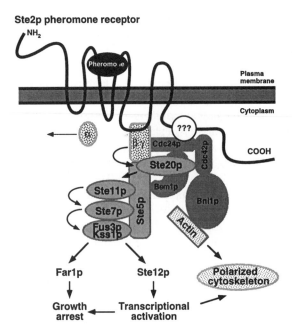

Figure 8 Model for the establishment of pheromone-induced polarized growth. Binding of pheromone to the receptor stimulates downstream responses such as transcriptional activation of pheromone-induced genes, cell-cycle arrest, and polarization of the cytoskeleton and growth components to the site of highest pheromone concentration. Receptor activation not only triggers the mitogen-activated protein kinase pathway (*light modules*) via G proteins (*spotted modules*), it also involves the receptor carboxy-terminal domain. One model is that the carboxy-terminal domain, through direct or indirect interaction (*white circle*) with the polarity-establishment complex (*dark modules*), stimulates the repolarization of the actin cytoskeleton (*speckled rectangle*) to the site of pheromone activation. (Modified with permission from *Trends Cell Biol.*)

components are depleted in cells treated with pheromone (259, 268), which indicates that the axial budding landmark is lost during mating. In addition, the Far1 protein may function to inhibit the use of axial sites. *FAR1* encodes a protein that serves dual roles (305); the amino-terminal portion of Far1p inhibits signaling by the Cln1p-Cdc28p and Cln2p-Cdc28p cyclin-dependent kinase complexes (238, 239, 300), whereas the carboxy-terminal domain plays an additional role in mating (45, 305). Mating-defective strains containing carboxy-terminal mutations in Far1p form projections adjacent to the previous bud site during the mating response; thus, Far1p may function to "erase" the axial landmark or facilitate orientation toward the pheromone source (305).

Polarity Establishment During Projection Formation

After responding to a pheromone gradient and forming a cortical landmark adjacent to a mating partner, polarity-establishment factors (e.g. Cdc24p, Cdc42p, Bem1p) must be able to recognize this site to initiate projection formation. Recent findings suggest that G_β and G_γ subunits of the heterotrimeric G protein (encoded by *STE4* and *STE18*) can form a complex with Ste20p and other proteins required for the propagation of the MAP kinase pathway response as well as with polarity-establishment proteins necessary for projection formation. Genetic analysis and biochemical evidence indicate that Ste4p interacts with both Ste20p and Ste5p, a putative scaffold protein that interacts with the different members of the MAP kinase cascade and is required for signaling (7, 123, 169, 170, 192, 316). Interaction of the $G\beta\gamma$ subunits with Ste20p occurs only in the presence of mating pheromone (176); to date this is the only reported intracellular protein-protein interaction regulated by pheromone signaling. Ste4p also binds the polarity-establishment protein Cdc24p in two-hybrid experiments (327). Biochemical studies also demonstrate that Ste5p and Ste20p can complex with Bem1p and actin, two proteins required for polarized growth in the mating process (175). Thus, a multiprotein complex may link the $G_{\beta\gamma}$ subunits to pathways required for polarized growth and signal transduction.

Ste20p may be the critical component that links the activated $G_{\beta\gamma}$ complex to both mating MAP kinase signaling and cell polarization. Ste20p functions during mating to activate the Fus3p/Kss1p MAP kinase pathway (7, 123, 169, 226, 319) and, as mentioned above, Ste20p interacts with Cdc42p (67, 240, 283, 327). Furthermore, Cdc42p is required for the localization of Ste20p to the tips of mating projections (240). Colocalization at the projection tip and interaction with Ste4p occurs only upon pheromone treatment. However, mutants of Ste20p that do not bind Cdc42p activate expression of the pheromone-inducible *FUS1* gene, display wild-type mating projections, and possess normal Ste20p kinase activity in vitro (170, 240). Thus, Ste20p interaction with Cdc42p is not required for signaling; perhaps this interaction is specifically required for signaling at the proper location in the cell.

Projection Formation and Growth

Many components required for bud formation are also important for projection formation, and the spatial organization of these components reflects that observed in vegetative cells. Membrane vesicles and actin patches accumulate at the projection tip (13, 122), as do several proteins important for cell polarization in vegetative cells, including Bem1p, Spa2p, Pea2p, Bni1p, Bud6p, and Rom2p (10, 90, 199, 287, 288, 304). Mutations in actin and in genes encoding any of the proteins mentioned above cause defects in mating projection morphology (52, 107, 199, 252). Thus, as occurs during budding, once a site is selected through pheromone signaling, the actin cytoskeleton and polarized growth is directed toward that site.

Although mating projections do not have a neck, cells exposed to high levels of mating pheromone contain some constriction at the base of the projection. This region contains the septins and chitin, which presumably are important for formation of the constriction (97, 151). This base of the projection also contains a pheromone-induced protein, Afr1p; Afr1p interacts with septin and is important for proper projection shape (86, 111, 159, 160). Thus, specific components residing at the base of the projection help mediate proper projection shape; some of these proteins also function at the mother-bud neck during vegetative growth.

One important difference between the mating response and budding is that during budding, growth is largely preprogrammed once bud formation begins. In contrast, mating cells appear to continuously determine and direct their growth by sensing pheromone gradients. Yeast cells exposed to pheromone gradients can improve their orientation toward the pheromone source over a period of 12 h (275). In addition, there are yeast mutants, *fig1Δ*, *fig2Δ*, and *fig4Δ*, that exhibit defects in projection formation in mating mixtures but not in the presence of uniform concentrations of pheromone (86). Fig1p, Fig2p, and Fig4p are specifically expressed in the presence of mating pheromone but not during vegetative growth. Thus, these proteins may help mediate proper projection formation during cell-cell communication.

Cells Respond Differently to Varying Levels of Pheromone

Yeast cells exposed to different levels of mating pheromone undergo different morphological responses that may be important for mating (195, 215, 216). Cells exposed to levels of pheromone that do not cause cell-cycle arrest elongate and begin budding at distal sites similarly to pseudohyphal cells (S Erdman, M Snyder, unpublished information). These cells are thought to be spreading to search for potential mates. Cells exposed to minimal levels of mating pheromone that cause cell-cycle arrest will agglutinate but fail to form mating projections (215, 216). This treatment also causes a partial depolarization of F-actin, which may be important for the redirection of polarized growth toward a mating partner (195); it might also facilitate the use of nonaxial, secondary bud sites when cells resume growth (195). Finally, at high levels of pheromone, as long as a gradient is present, cells arrest growth and form mating projections toward the pheromone source (215, 216).

These different morphological responses are likely to be important for the mating process. In nature, when two yeast cells lie at a distance from one another, the cells presumably encounter low levels of pheromone, which will facilitate depolarization of the actin cytoskeleton and use of distal budding sites; this may enable cells to spread across the surface as they divide (195; S Erdman, M Snyder, unpublished information). As they encounter higher levels of pheromone, these yeast cells will arrest, depolarize, and then repolarize toward the pheromone source (195). In addition, low levels of pheromone are sufficient to stimulate production of additional pheromone (1, 140, 216, 293); the resulting increased levels of pheromone will facilitate formation of a mating projection and growth toward the pheromone souce (i.e. the mating partner).

CELL POLARITY DURING CELL FATE DETERMINATION

The formation of a distinct cell type, the bud, allows cells to adopt a cell fate distinct from that of their progenitor cell. As noted above, in nature most yeast strains are believed to be homothallic, i.e. they switch their mating type (reviewed in 125, 223). In this process, mother cells switch but daughter cells do not; *MAT*a cells convert to *MAT*α cells, and *MAT*α cells form *MAT*a cells. This process of generating asymmetric cell fates relies on the polarized structure of dividing yeast cells.

Mating type switching requires the HO endonuclease, which catalyzes the gene conversion of the expressed *MAT* mating locus with stored copies of a- and α-cell type information (223). Switching is restricted to haploid mother cells that express the HO endonuclease at the end of G1 (Figure 9) (223). Expression of the endonuclease is negatively regulated in daughter cells by Ash1p, a zinc finger protein related to GATA transcription factors (27, 282). Deletion of the *ASH1* gene results in both mother and daughter cell switching, and overexpression of certain *ASH1* variants inactivates switching of both mother and daughter cells. Wild-type Ash1p localizes preferentially to the nuclei of daughter cells, where it represses *HO* gene expression (27, 282). The asymmetric localization of Ash1p depends on several proteins, including Myo4p, Bni1p, She1p, She2p, and She3p (27).

Recent analysis of *ASH1* mRNA localization by FISH analysis revealed the molecular basis for the asymmetric localization of Ash1p. *ASH1* mRNA localizes to the tip of large budded cells that have undergone anaphase (189, 295). Proper localization depends, at least in part, on the 3' untranslated region of the RNA and on the integrity of the actin cytoskeleton, as revealed by the fact that mutants in actin, tropomyosin (Tpm1p), profilin, Myo4p, Bni1p, She1p, She2p, and She3p fail to properly localize *ASH1* mRNA (189, 295). Thus, proper localization of the *ASH1* mRNA results in preferential accumulation of the protein in the daughter cell nucleus, which in turn inhibits expression of the

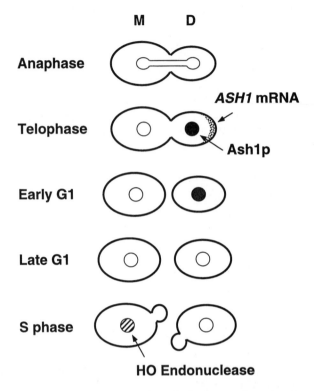

Figure 9 Key events important for cell fate determination in yeast. In postanaphase cells, *ASH1* mRNA, which encodes a repressor of mating type switching, localizes to the bud tip. The protein produced preferentially associates with the daughter nucleus and thereby represses transcription, perhaps through interaction with factors critical for HO expression in yeast (e.g. Swi5p).

HO gene. This is the first example of polarized mRNA localization in yeast, a phenomenon that has been described in a variety of other eukaryotes (see 17 for a recent review). It will be of interest to determine whether other mRNAs are also polarized in yeast, particularly those encoding proteins involved in polarized cell growth, such as actin, Cdc42p, and Spa2p.

CYCLIN-DEPENDENT KINASES COORDINATE POLARIZED GROWTH AND CELL-CYCLE PROGRESSION

Polarized cell growth events are tightly regulated in the yeast cell cycle. During late G1, when cells have sufficient nutrients and are of appropriate size, cells pass a critical stage, termed START (63, 249). When cells traverse START, bud formation begins, and apical growth occurs at the bud tip during S phase (180). Later, during G2 and mitosis, cells switch to isotropic growth, in which growth is restricted to the bud but occurs throughout the bud. Finally, cell growth must be completed prior to cytokinesis. The controls that mediate these events are beginning to be understood.

CDKs Control Cell Polarization in Yeast

Passage through START and entry into apical growth requires the activation of two cyclin-dependent kinases (CDKs): Cdc28p forms a complex with the G1 cyclins Cln1, Cln2, and Cln3p (62, 255; reviewed in 223, 253), and Pho85p associates with the Pcl1p and Pcl2p G1 cyclins (89, 212). These kinases have some overlapping functions, as deletion of certain combinations of Clns and Pcls leads to lethality (89, 212). Activation of Cdc28p-Clnp complexes leads to polarization of the cortical actin cytoskeleton required for bud emergence and apical growth (180), and Cln-depleted cells arrest in G1 as unbudded cells (62, 255). In addition, overexpression of *CLN1* or*CLN2*, but not *CLN3*, or delaying the activation of the Cdc28p/G2 kinase prolongs the apical growth phase and produces cells with elongated buds and a hyperpolarized actin cytoskeleton (8, 180, 230, 273). Thus, the Cln G1 cyclins promote bud formation and apical growth.

Entry into the isotropic growth phase is mediated by the Cdc28p/G2 kinase. This kinase is formed by the association of Cdc28p with the G2 cyclins, Clb1p and Clb2p. Activation of the Cdc28p/G2 kinase reduces the activity of the Cdc28p/G1 kinases (25) and promotes entry into the isotropic growth phase, in which actin is localized throughout the bud (180). Deletion of the G2 cyclins results in the formation of elongated buds with hyperpolarized actin, indicative of a prolonged apical growth phase. In addition, overexpression of Clb2p produces round cells, indicative of a reduction in apical growth (180). Finally, overexpression of stable forms of Clb2 inhibits budding in G1 (180). Thus, Cdc28p-Clns promote apical growth and Cdc28p-Clbs (at least Clb1p and Clb2p) inhibit apical growth and convert cells to an isotropic growth phase. It is unclear whether the role of the Cdc28p/G2 kinase in the regulation of yeast cell polarity is to simply inactivate the Cdc28p/Cln kinase or whether it also has a direct role in promoting isotropic growth.

Equally important to the initiation of bud formation and growth is turning off cell growth. During late mitosis, the distribution of actin changes from throughout the bud to throughout the entire cell; this occurs by a cell-cycle control mechanism that is yet to be defined. Finally, prior to cytokinesis, Cdc28p-Clbp complexes are inactivated through cyclin degradation, and this stimulates the redistribution of actin to the mother-bud neck region and cytokinesis (180). Thus, polarized growth undergoes dynamic changes during the cell cycle.

Roles for G1 Cyclins

Although the yeast Cdc28p-associated G1 cyclins are important for bud emergence and cell polarization, they may not have equivalent roles in promoting entry into S phase and bud formation. The predicted structure of Cln3p is considerably divergent from that of Cln1p and Cln2p (116, 222), and Cdc28p-Cln3p complexes also display weaker kinase activity toward histone H1 (301). In addition, *CLN1* and *CLN2* expression (as well as Cdc28p-Cln1,2p activity) increases dramatically during late G1, whereas the Cdc28p-Cln3p kinase may be activated by posttranslational mechanisms (72, 179, 246, 301, 302, 318). The purpose of Cln3p during early G1 may be to monitor cell size and induce the expression of *CLN1* and *CLN2* and thus drive entry into S phase and bud formation (301). Consistent with this hypothesis, an additional copy of *CLN3* has been shown to decrease the cell size at START (222), and *cln1 cln2* cells do not enter S phase until they reach a much larger size than that of wild-type cells (71). Cln3p regulates the timing and level of *CLN1* and *CLN2* expression (294).

One of the primary functions of the Cdc28p-Cln3p kinase is to activate the SBF and MBF transcription factors (reviewed in 29, 155). SBF is a heterodimer comprised of the DNA binding components Swi4p and Swi6p (11, 247); MBF contains both Swi6p and the Swi4p-related protein Mbp1p (154). These factors promote the expression of *CLN1*, *CLN2*, and other genes required for entry into S, bud formation, and cell-wall synthesis genes (84, 89, 91, 133, 212, 224, 231, 274, 301, 313). In addition, MBF binding sites are found in upstream regulatory sequences of several genes required for polarized cell growth (e.g. *SPA2*, *AXL2*, and *CHS3*) (259, 270). Thus, activation of SBF and MBF is expected to promote synthesis of cell-wall components and other factors important for cell polarization and growth in yeast. It is likely that the Cdc28p-Cln complexes also phosphorylate other important targets directly involved in bud formation and growth; in vivo substrates for these kinases have not been identified.

A possible model for how the different Clns might function is described below. During the early G1 phase of the cell cycle, growth occurs throughout the unbudded cell until it reaches the critical size necessary for passage through START. Cln3p is present during early G1, and the primary role of the Cdc28-Cln3p complex may be to promote expression of SBF-dependent transcripts (71, 294). Since SBF can also be activated by Cln1p and Cln2p, it was proposed that a positive-feedback loop exists in which Cln1p and Cln2p, whose expression is activated by SBF, further stimulate the expression of SBF-dependent transcripts (64, 72). This does not appear to be the case, because SBF- and MBF-dependent transcripts are activated at the proper time and cell size in *cln1 cln2 CLN3* cells (71); these cells display a pronounced delay in bud formation and DNA replication. Thus, the primary function of Cdc28p-Cln3p appears to be the activation of SBF- and MBF-dependent transcription, whereas Cln1p and Cln2p are required for other events necessary for passage through START (71, 294) (see below).

As described above, the Cln proteins (especially Cln1p and Cln2p) promote entry into S phase and budding (71, 272). DNA synthesis is triggered by both the MBF-dependent expression of *CLB5* and *CLB6* (274) and the phosphorylation and subsequent proteolysis of Sic1p, an inhibitor of Cdc28p-Clb signaling (71, 230, 272, 273, 274). The ability of the Cdc28p-Cln kinases to regulate S phase allows cells to couple bud formation and DNA synthesis during the cell cycle.

Following DNA replication and the G2 stage of the cell cycle, cells undergo mitosis. Entry into mitosis is regulated by the Cdc28p/G2 kinase (i.e. Cdc28p complexed with Clb1p and Clb2p). Exit from mitosis is mediated by destruction of the Clbs and an increased expression of *SIC1* (153). The resulting decrease of Cdc28p-Clbp signaling, together with the existing Cdc28p-Cln3p kinase activity, may be sufficient to promote the repolarization of cortical growth components required for cytokinesis.

In summary, different Cdc28p cyclin complexes control distinct aspects of cell polarization and morphogenesis in yeast. Since these kinases also regulate nuclear progression, they coordinate bud formation and growth with other aspects of the cell cycle.

A Budding Checkpoint Controls Cell-Cycle Progression

The influence of nuclear processes such as DNA replication and DNA repair on cell-cycle progression has been an area of active study. In contrast, how cytosolic processes and the cytoskeleton (other than microtubules) influence cell-cycle progression is poorly understood. As noted above, a key cell-cycle control point in the budding process occurs at START, in which nutrients and mating pheromones influence the decision to initiate bud formation. A second important cell-cycle control checkpoint, termed the budding checkpoint, coordinates bud formation and growth with late nuclear division processes (181). This checkpoint regulates the switch from the Cdc28p/Cln G1 kinase to the Cdc28p/Clb G2 kinase. Temperature-sensitive *cdc24* cells fail to form a bud at the restrictive temperature; these cells delay nuclear division by about 2 h (one cell division cycle time) (181). This delay depends on an inhibitory phosphorylation of the tyrosine19 residue of Cdc28p by the Swe1p kinase, which blocks Cdc28p/Clb G2 kinase activity. Cells containing either an alanine to tyrosine substitution of the tyrosine19 or a deletion of the *SWE1* gene are no longer delayed in nuclear division (181, 281). This inhibitory phosphorylation pathway is highly conserved and controls nuclear division in other eukaryotes, including *S. pombe*, where it was first discovered (264, 265, 266). In

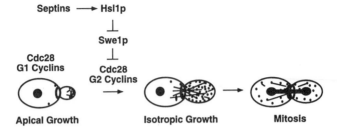

S. cerevisiae Budding Checkpoint

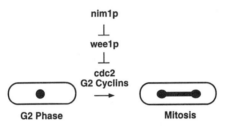

Figure 10 Model for the budding checkpoint of *Saccharomyces cerevisiae*. The switch from apical growth to isotropic growth involves the inactivation of the G1 Cdc28-Cln kinase and activation of the G2 Cdc28p-Clb kinase. Activation of the G2 kinase is controlled by a negative regulatory kinase, Swe1p, which, in turn, is negatively regulated by the homologous kinases Hsl1p, Gin4p, and Kss4p. Organization of septins leads to activation of the Hsl1p kinase and activation of the pathway. (*bottom*) The homologous pathway in *Schizosaccharomyces pombe*. nim1p is a homolog of Hsl1p, wee1p is homologous to Swe1p, and cdc2p is a homolog of Cdc28p. In *S. pombe* this pathway also regulates the transition of G1 kinase to G2 kinase activation.

S. pombe, this pathway regulates a DNA replication checkpoint (Figure 10); its role in monitoring a "polarity checkpoint" is not known (14, 262). In *S. cerevisiae*, the inhibitory phosphorylation has no effect on DNA replication checkpoints, indicating that this pathway may have distinct functions in different organisms.

What might this checkpoint be monitoring? Recent evidence indicates that at least one factor is the organization of septins (Y Barral, M Snyder, unpublished data). Septin mutants arrest in an apical growth phase and exhibit a nuclear division delay, indicative of a delay in activation of the G2 kinase. The apical growth phenotype and nuclear division delay are partially suppressed by either deletion of the *SWE1* gene or overexpression of the *CLB2* gene. In addition, the septins are mislocalized in *cdc24* temperature-sensitive strains. Thus, organization of the septins is monitored by a G2 cell-cycle checkpoint.

How is the organization of the septins monitored? Likely monitors are the Gin4p, Hsl1p, and Kcc4p protein kinases. These proteins are predicted to be protein kinases that function to phosphorylate and inactivate Swe1p. Consistent with this hypothesis, strains containing deletions of one or more of these genes exhibit hyperpolarized cells, suggestive of a G2 delay; this phenotype is suppressed by a deletion of *SWE1* (8; Y Barral, M Porra, S Bilingmaier, M Snyder, unpublished information). Recently, all three proteins—Gin4p, Hsl1p, and Kcc4p—have been localized to a ring at the neck, localization of the septins (Y Barral, M Parra, S Bilingmaier, M Snyder, unpublished information). Thus, a model for how the budding checkpoint is monitored is as follows (Figure 10): Organization of the septins during the apical growth phase in G1 results in localization and activation of the Gin4p, Hsl1p, and Kcc4p kinases at the neck. These kinases, in turn, phosphorylate and inactivate Swe1p, resulting in activation of the Cdc28p/Clb kinase and induction of mitosis and isotropic growth. Thus, proper organization of the septins allows cells to progress to the next phase of the cell cycle. Whether organization of other polarity components, such as actin, is also monitored for cell-cycle control is an important, but unresolved, question.

MAP KINASE SIGNALING ALSO COORDINATES CELL POLARITY AND CELL-CYCLE PROGRESSION IN YEAST

In addition to cyclin-dependent kinases, mitogen-activated protein (MAP) kinase pathways in yeast also control cell polarization and cell-cycle progression (26, 126, 178, 204, 298, 317). MAP kinase signaling modules are used throughout evolution to control cellular responses to external signals (reviewed in 26, 204, 298). A cascade of protein kinases is activated following the perception of an extracellular signal, and this results in a series of phosphorylation events that ultimately activate the MAP kinase. A MAP kinase kinase (also called MEK) phosphorylates and activates the MAP kinase, whereas a MEK kinase (or sometimes Raf) modifies and activates the MEK. In yeast, there are at least five MAP kinase pathways (Figure 11) (reviewed in 126, 178, 298), and many of these regulate cell morphogenesis. The Fus3p and/or Kss1p MAP kinase pathways regulate the mating response and both pseudohyphal growth in diploid cells and invasive growth in haploid cells; the Slt2p pathway regulates cell polarity and cell wall synthesis. In addition, the Hog1p and Smk1p pathways regulate growth in high osmolarity and sporulation, respectively; these

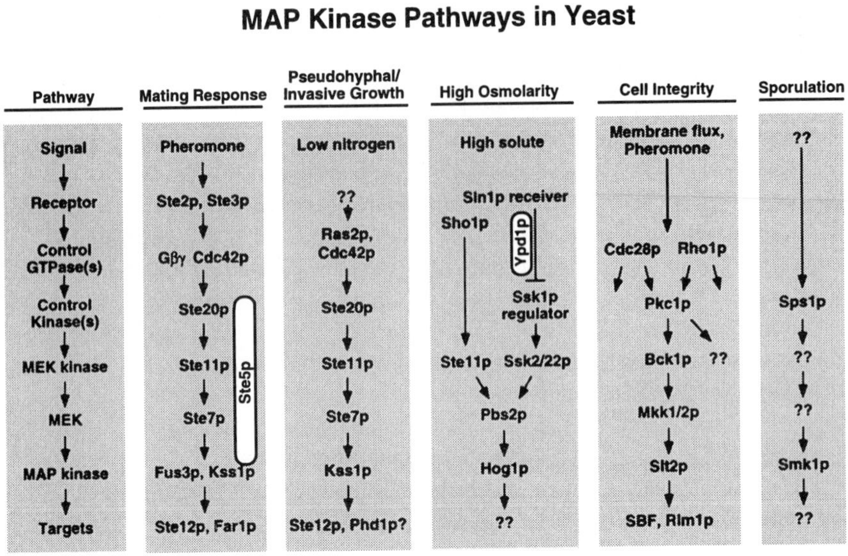

MAP Kinase Pathways in Yeast

Pathway	Mating Response	Pseudohyphal/ Invasive Growth	High Osmolarity	Cell Integrity	Sporulation
Signal	Pheromone	Low nitrogen	High solute	Membrane flux, Pheromone	??
Receptor	Ste2p, Ste3p	??	Sln1p receiver Sho1p		
Control GTPase(s)	Gβγ Cdc42p	Ras2p, Cdc42p	Ypd1p	Cdc28p Rho1p	
Control Kinase(s)	Ste20p	Ste20p	Ssk1p regulator	Pkc1p	Sps1p
MEK kinase	Ste11p	Ste11p	Ste11p Ssk2/22p	Bck1p ??	??
MEK	Ste7p	Ste7p	Pbs2p	Mkk1/2p	??
MAP kinase	Fus3p, Kss1p	Kss1p	Hog1p	Slt2p	Smk1p
Targets	Ste12p, Far1p	Ste12p, Phd1p?	??	SBF, Rlm1p	??

(Ste5p spans the Mating Response MEK kinase–MAP kinase steps.)

Figure 11   Summary of the mitogen-activated protein (MAP) kinase signaling pathways in yeast. MEK, a MAP kinase.

signaling modules may also participate in cell morphogenesis, although their role in this process is less well understood. Recent findings have provided insight into mechanisms that activate MAP kinase signaling in yeast and into the downstream events that are the output of these pathways.

The Mating Response Pathway

As described above, the mating MAP kinase pathway is required for polarized growth events, such as projection formation, that occur during the mating response. Genetic and in vitro evidence has demonstrated that the yeast pheromone–response MAP-kinase cascade consists of Ste11p, Ste7p, and the partially redundant Fus3p/Kss1p protein kinases (MEKK, MEK, and MAP kinases, respectively) (41, 81, 82, 88, 105, 226, 238, 291, 331). Recent evidence indicates that Kss1p can activate the mating pathway only in the absence of Fus3p, and that the presence of a kinase-inactive Fus3p can prevent Kss1p from activating the mating response (197). Thus, Fus3p appears to be the most important MAP kinase during the mating response.

Recent research has provided much insight into the molecular mechanisms for activation of the Fus3p MAP kinase (see above) (reviewed in 167, 289, 314). The interaction between activated $G_{\beta\gamma}$ and Ste20p is critical for stimulation of the mating MAP kinase pathway and polarized growth. In addition, the interaction between Ste4p and Ste5p appears to promote oligomerization of Ste5p, which is required for signaling (136, 137, 321). Ste5p appears to both link heterotrimeric G protein signaling and the MAP kinase pathway and serve as a scaffold protein that interacts with each kinase in the Fus3p MAP kinase module and functions to insulate these components from associations with other kinase modules (53, 123, 136, 169, 202, 250, 291).

Several events that occur downstream of Fus3p and/or Kss1p activation have also been described (e.g. Ste12p and Far1p phosphorylation and regulation; see above). Additional substrates for the Fus3p and Kss1p kinases are likely to exist (292). The result of these downstream signaling events is dramatic cellular responses (e.g. agglutination, projection formation, cell fusion) that culminate in the mating process and in formation of the diploid zygote (reviewed in 15, 167, 289, 314).

The Pseudohyphal Response and Invasive Growth Pathway

Diploid strains grown under nitrogen limiting conditions undergo a dimorphic switch and initiate pseudohyphal growth; changes in cell shape facilitate the production of long filaments that resemble hyphae, and these cells can penetrate the surface of agar plates (110). Haploid invasive growth is a related process in which cells elongate, bud distally, and penetrate a solid medium (258). Initial observations indicated that specific factors that function in the yeast mating response are also necessary for pseudohyphal and haploid invasive growth. Strains lacking either Ste20p, Ste11p, Ste7p, or Ste12p are defective in pseudohyphal and invasive growth, whereas *fus3 kss1* strains and strains lacking pheromone receptors undergo pseudohyphal growth (188, 258). Recent evidence, though, has demonstrated that Kss1p is the MAP kinase that regulates filamentous growth (58, 197). Kss1p displays both inhibitory and activating effects on pseudohyphal growth. In the absence of Ste7p-dependent phosphorylation, Kss1p (and possibly Fus3p) functions to inhibit filamentous growth (58, 197, 258). Phosphorylation of Kss1p relieves this inhibitory effect, and Kss1p can act as an activator of filamentous growth. The previously observed filamentous growth defects of mating pathway components can be suppressed by the absence of Kss1p; a *ste11Δste7Δ kss1Δfus3Δ* strain still undergoes invasive growth (58, 197). These experiments indicate that both activated and inactivated Kss1p play physiologically important roles and that pathways independent of the Kss1p pathway can regulate filamentous growth.

Several components that function downstream of Kss1p in pseudohyphal and invasive growth have been identified. As described above, Kss1p can bind and phosphorylate Ste12p. The inhibitory effect of Kss1p may be mediated by inactivated Kss1p binding to Ste12p (197). During pseudohyphal and invasive growth, Ste12p binds to the Tec1p transcription factor (196); Tec1p is required for filamentous and invasive growth (106). Dig1p and Dig2p are two additional proteins that bind Kss1p, Fus3p, and Ste12p. Dig1p and Dig2p are negative regulators of invasive growth, and phosphophorylation by Kss1p may free Ste12p from a complex containing the Dig proteins and allow for transcriptional activation of genes required for pseudohyphal and invasive growth (57, 243, 296).

The Slt2p Pathway Functions to Maintain Cell Integrity During Polarized Cell Growth

The Slt2p (also called Mpk1p) MAP kinase pathway serves to maintain the integrity of a cell during periods of polarized growth, and this pathway functions downstream of protein kinase C (*PKC1*) signaling in yeast. *SLT2* was originally identified in a screen for mutations that cause cell lysis phenotypes (297), and subsequent genetic screens have identified and ordered the components of this signaling module. *BCK1* (also called *SLK1*) encodes a MEK kinase (59, 172) that is phosphorylated—and presumably activated—by Pkc1p (177). The *MKK1* and *MKK2* gene products are functionally redundant MEKs (138) that are required for the activation of Slt2p (174). Deletion of any component of the Slt2p signaling module (i.e. Bck1p, Mkk1p/Mkk2p, Slt2p) causes a temperature-sensitive growth defect, and growth at the restrictive temperature results in cell lysis (59, 138, 172, 174, 297). In addition, activated alleles of both *PKC1* and *MKK1* can suppress mutations in *RHO1*, which suggests that Rho1p functions upstream of Pkc1p (227); more recent experiments have shown that Pkc1p and Rho1p interact and that Rho1p stimulates Pkc1p kinase activity in a GTP-dependent manner (149). As noted above, Rho1p localizes to sites of polarized growth (322) and thus may target Slt2p pathway components to the cell cortex in order to perform functions necessary to maintain cell integrity.

Substrates for the Slt2p kinase have been identified. Genetic and biochemical experiments as well as phosphorylation-state studies indicate that SBF components are targets of Slt2p (133, 194). Furthermore, proper expression of a subset of SBF-dependent transcripts requires *SLT2* (133, 194). Another likely target of Slt2p is Rlm1p; *RLM1* was isolated in a screen designed to identify components that function downstream of *SLT2* (309). Mutations in *RLM1* cause some phenotypes that are common to Slt2p pathway mutants, and Rlm1p interacts with Slt2p (309). Rlm1p is a member of the MADS box family of transcription factors (309), and the transcriptional activation ability of Rlm1p is dependent on Slt2p (73, 310). The consensus sequence for Rlm1p binding has been identified, and promoters containing these sequences have been shown to be regulated by Rlm1p and Slt2p (73). Rlm1p also interacts with a second Slt2p-like MAP kinase, Mlp1p(310). It is possible that SBF, Rlm1p, or another transcription factor function together with the putative HMGI-like chromatin-associated proteins, Nhp6Ap and Nhp6Bp (158). *NHP6A* and *NHP6B* were isolated as suppressors of Slt2p pathway defects, and strains lacking both of these components display many phenotypes in common with *slt2Δ* strains (60). However, Nhp6Ap and Nhp6Bp do not appear to be direct substrates of Slt2p in vitro, indicating that they function either with or downstream of an Slt2p substrate. In summary, several nuclear factors may function downstream of Slt2p MAP kinase signaling.

The Slt2p pathway functions to maintain the structural integrity of a cell during periods of highly polarized growth as well as during other conditions that elicit significant membrane flux. Mutations in *SLT2* or genes encoding other pathway members cause defects in polarized growth, including an aberrant distribution of the actin cytoskeleton and accumulation of secretory vesicles (59, 210). More recently, Slt2p activity has been shown to increase at bud emergence (326) and other periods of polarized growth, such as during response to pheromone (36, 87, 326). Slt2p kinase activity is also rapidly stimulated by conditions that increase membrane fluidity; increases in temperature, decreases in extracellular osmolarity, or treatment with reagents that induce membrane dynamics all promote Slt2p kinse activity (148). It is likely that these conditions mimic the physiological state of the plasma membrane during the period of increased membrane deposition that is required for polarized growth. An attractive model is that conditions that induce the remodeling of the cell wall or plasma membrane (e.g. bud emergence, mating projection formation) activate the pathway, which in turn, through SBF, triggers new cell wall synthesis.

The Slt2p Pathway and CDK Signaling Coordinate Polarized Growth and Cell-Cycle Progression

The observation that polarized growth occurs at critical stages of the cell cycle (e.g. bud emergence at the G1/S transition) suggested that growth may be coupled to cell-cycle progression (249). Previous genetic evidence indicated that the Slt2p pathway may regulate both polarized growth and cell-cycle events (210). Additional genetic studies together with coimmunoprecipitation experiments, analysis of Slt2p phosphorylation targets in vitro and in vivo, and studies on *SLT2*-dependent transcriptional regulation have provided insight into possible means by which Slt2p signaling may couple these events (133, 194, 203).

Genetic studies suggest that some functions of the Cdc28p CDK can regulate events both upstream of Slt2p signaling as well as downstream or in parallel pathways. Mutations in *SLT2* enhance the phenotype of *cdc28* strains that are defective in progression into S phase (210), and mutations in *PKC1* also cause lethality in combination with *cdc28* (203). In addition, Slt2p is activated during bud formation, a process that requires the Cdc28p kinase, and the proper activation of Slt2p depends on Cdc28p signaling (326). Analysis of dosage suppressors of *slt2* or *pkc1* predicts that Cdc28p CDK signaling can control events either downstream of *SLT2* or in parallel pathways (33, 70, 85, 133, 173, 194, 218). Furthermore, overexpression of *SWI4* suppresses mutations in *PKC1* and *SLT2* (133, 194), and SBF is an important target of Slt2p (see above) (194). *SLT2* is required for the proper expression of a subset of SBF-dependent transcripts, including *PCL1* and *PCL2*, that encode G1 cyclins that bind the Pho85p CDK, and of transcripts of several genes involved in cell wall biosynthesis (133, 194). Consistent with this interpretation, overexpression of *PCL1* or *PCL2* also suppresses the temperature-sensitivity and morphological defects of *slt2* strains (194).

A simplified description for how polarized growth events and cell-cycle progression may be coupled is presented below. Cdc28p activation during late G1 triggers a series of growth events, including increase in membrane flux, and together with Rho1p-mediated events, these processes stimulate the activity of Pkc1p and the downstream Slt2p kinase (87, 148, 149, 203, 227, 326). Activation of Slt2p results in the propagation of the growth response, including an increased expression of genes involved in cell wall biogenesis (133) and the *PCL1* and *PCL2* G1 cyclins that function in polarized growth and cell-cycle progression (89, 194, 212). Thus, the Slt2p pathway is activated by the Cdc28p CDK and activates the Pho85p CDK. This regulatory cascade is expected to rapidly stimulate the growth response and CDK signaling required during the highly polarized growth events that occur at the G1-to-S transition.

The Hog1p *and* Smk1p *Pathways*

Two other MAP kinase pathways, the Hog1p pathway and the Smk1p pathway (named after their respective MAP kinases), also participate in cell morphogenesis in yeast. Both *HOG1* and genes encoding its activators are required for growth in a high-osmolarity medium (31). Hog1p appears to play a role in cell polarity in two ways. First, Hog1p is involved in repressing the pheromone response pathway. Inactivation of *HOG1* results in higher phosphotyrosine content of Fus3p and increased expression of pheromone-responsive genes (118). Second, Hog1p is important for proper polarization after osmotic stress (30). When yeast cells are exposed to osmotic stress, the cells depolarize the actin cytoskeleton from their bud, and actin patches are dispersed throughout the cell. After further incubation, wild-type cells will recover and repolarize their actin cytoskeleton into the existing bud. *hog1* mutants also depolarize their actin cytoskeleton on osmotic stress but fail to repolarize their cytoskeleton to the original bud. Instead, they form a new bud and orient their actin cytoskeleton into this bud. Thus, the Hog1p pathway is needed for proper reorientation of the actin cytoskeleton after osmotic stress.

The Smk1p pathway is important for cell morphogenesis during meiosis. During meiosis, yeast cells form spores that have a highly specialized cell wall composed of several layers (55). *SMK1* and its upstream regulator *SPS1* are specifically expressed during meiosis and are required for the expression of several late meiotic genes (101, 163). *smk1Δ* pathway mutants exhibit heterogeneous spore wall defects (101, 163). Thus, *SMK1*, like *SLT2*, is important for cell wall formation, which indicates that multiple MAP kinase pathways are important for cell morphogenesis at different stages of the yeast life cycle.

CONCLUSION

In yeast, a complex series of events occurs during the selection of sites for growth and the subsequent direction of growth to this site. Both during budding and in the mating response, cortical tags provide positional information that marks sites for growth. The specific components of these landmarks may change according to cell type or growth condition, and these tags, as well as proteins involved in their recognition, must be present during the appropriate stage of the cell cycle. Recently, a large amount of research has been dedicated both to the identification of components that function in polarized growth in yeast as well as to the elucidation of mechanisms for coupling this growth response to cell-cycle progression. Many of the components and signaling pathways that have been identified are conserved throughout evolution (e.g. septins, Cdc42p, Ste20p/PAKs, MAP kinases, CDKs); thus, the identification of molecular mechanisms for regulating polarized growth in yeast are expected to be of general relevance.

> **Visit the *Annual Reviews* home page at**
> **http://www.AnnualReviews.org.**

Literature Cited

1. Achstetter T. 1989. Regulation of α-factor production in *Saccharomyces cerevisiae*: a-factor pheromone-induced expression of the *MFα1* and *STE13* genes. *Mol. Cell. Biol.* 9:4507–14
2. Adames N, Blundell K, Ashby M, Boone C. 1995. Role of yeast insulin-degrading enzyme homologs in propheromone processing and bud site selection. *Science* 270:464–67
3. Adams A, Pringle J. 1984. Relationship of actin and tubulin distribution to bud growth in wild-type and morphogenetic-mutant *Saccharomyces cerevisiae*. *J. Cell Biol.* 98:934–45
4. Adams AE, Botstein D, Drubin DG. 1991. Requirement of yeast fimbrin for actin organization and morphogenesis in vivo. *Nature* 354:404–8
5. Adams AEM, Botstein D, Drubin DG. 1989. A yeast actin-binding protein is encoded by *SAC6*, a gene found by suppression of an actin mutation. *Science* 243:231–33
6. Adams AEM, Johnson DI, Longnecker RM, Sloat BF, Pringle JR. 1990. *CDC42* and *CDC43*, two additional genes involved in budding and the establishment of cell polarity in the yeast *Saccharomyces cerevisiae*. *J. Cell Biol.* 111:131–42
7. Akada R, Kallal L, Johnson DI, Kurjan J. 1996. Genetic relationships between the G protein $\beta\gamma$ complex, Ste5p, Ste20p and

Cdc42p: investigation of effector roles in the yeast pheromone response pathway. *Genetics* 143:103–17

8. Altman R, Kellogg D. 1997. Control of mitotic events by Nap1 and the Gin4 kinase. *J. Cell Biol.* 138:119–30

9. Amberg DC, Busart E, Botstein D. 1995. Defining protein interactions with yeast actin in vivo. *Nat. Struct. Biol.* 2:28–35

10. Amberg DC, Zahner JE, Mulholland JW, Pringle JR, Botstein D. 1997. Aip3p/Bud6p, a yeast actin-interacting protein that is involved in morphogenesis and the selection of bipolar bud sites. *Mol. Biol. Cell* 8:729–53

11. Andrews BJ, Herskowitz I. 1989. Identification of a DNA binding factor involved in cell-cycle control of the yeast *HO* gene. *Cell* 57:21–29

12. Ayscough KR, Stryker J, Pokala N, Sanders M, Crews P, Drubin DG. 1997. High rates of actin filament turnover in budding yeast and roles for actin in the establishment and maintenance of cell polarity revealed using the actin inhibitor latrunculin A. *J. Cell Biol.* 137:399–416

13. Baba M, Baba N, Ohsumi Y, Kanaya K, Osumi M. 1989. Three-dimensional analysis of morphogenesis induced by mating pheromone α factor in *Saccharomyces cerevisiae*. *J. Cell Sci.* 94:207–16

14. Barbet NC, Carr AM. 1993. Fission yeast wee1 protein kinase is not required for DNA damage-dependent mitotic arrest. *Nature* 364:824–27

15. Bardwell L, Cook JG, Inouye CJ, Thorner J. 1994. Signal propagation and regulation in the mating pheromone response pathway of the yeast *Saccharomyces cerevisiae*. *Dev. Biol.* 166:363–79

16. Deleted in proof

17. Bassell G, Singer RH. 1997. mRNA and cytoskeletal filaments. *Curr. Opin. Cell Biol.* 9:109–15

18. Bauer F, Urdaci M, Aigle M, Crouzet M. 1993. Alteration of a yeast SH3 protein leads to conditional viability with defects in cytoskeletal and budding patterns. *Mol. Cell. Biol.* 13:5070–84

19. Bedinger PA, Hardeman KJ, Loukides CA. 1994. Travelling in style: the cell biology of pollen. *Trends Cell Biol.* 4:132–38

20. Bender A. 1993. Genetic evidence for the roles of the bud-site-selection genes *BUD5* and *BUD2* in the control of the Rsr1p(Bud1p) GTPase in yeast. *Proc. Natl. Acad. Sci. USA* 90:9926–29

21. Bender A, Pringle JR. 1989. Multicopy suppression of the *cdc24* budding defect in yeast by *CDC42* and three newly identified genes including the *ras*-related gene *RSR1*. *Proc. Natl. Acad. Sci. USA* 86:9976–80

22. Bender A, Pringle JR. 1991. Use of a screen for synthetic lethal and multicopy suppressee mutants to identify two new genes involved in morphogenesis in *Saccharomyces cerevisiae*. *Mol. Cell. Biol.* 11:1295–305

23. Benton BK, Tinkelenberg AH, Jean D, Plump SD, Cross FR. 1993. Genetic analysis of Cln/Cdc28 regulation of cell morphogenesis in budding yeast. *EMBO J.* 12:5267–75

24. Bi E, Pringle JR. 1996. *ZDS1* and *ZDS2*, genes whose products may regulate Cdc42p in *Saccharomyces cerevisiae*. *Mol. Cell. Biol.* 16:5264–75

25. Blondel M, Mann C. 1996. G2 cyclins are required for the degradation of G1 cyclins in yeast. *Nature* 384:279–82

26. Blumer KJ, Johnson GL. 1994. Diversity in function and regulation of MAP kinase pathways. *Trends Biochem. Sci.* 19:236–40

27. Bobola N, Jansen R-P, Shin TH, Nasmyth K. 1996. Asymmetric accumulation of Ash1p in postanaphase nuclei depends on a myosin and restricts yeast mating-type switching to mother cells. *Cell* 84:699–709

28. Botstein D, Amberg D, Mulholland J, Huffaker T, Adams A, et al. 1997. The yeast cytoskeleton. In *The Molecular and Cellular Biology of the Yeast Saccharomyces*, ed. JR Pringle, JR Broach, EW Jones, 3:1–90. Cold Spring Harbor, NY: Cold Spring Harbor Lab. 1131 pp.

29. Breeden L. 1996. Start-specific transcription in yeast. *Curr. Top. Microbiol. Immunol.* 208:95–127

30. Brewster JL, Gustin MC. 1994. Positioning of cell growth and division after osmotic stress requires a MAP kinase pathway. *Yeast* 10:425–39

31. Brewster JL, Valoir TdV, Dwyer ND, Winter E, Gustin MC. 1993. An osmosensing signal transduction pathway in yeast. *Science* 259:1760–63

32. Brown JL, Bussey H. 1993. The yeast *KRE9* gene encodes an O glycoprotein involved in cell surface β-glucan assembly. *Mol. Cell. Biol.* 13:6346–56

33. Brown JL, Bussey H, Stewart RC. 1994. Yeast Skn7 functions in a eukaryotic two-component regulatory pathway. *EMBO J.* 13:5186–94

34. Brown JL, Jaquenound M, Gulli M-P, Chant J, Peter M. 1997. Novel Cdc42p-binding proteins Gic1 and Gic2 control

cell polarity in yeast. *Genes Dev.* 11: 2972–82

35. Brown SS. 1997. Myosins in yeast. *Cell Motil. Cytoskelet.* 9:44–50

36. Buehrer BM, Errede B. 1997. Coordination of the mating and cell integrity mitogen-activated protein kinase pathways in *Saccharomyces cerevisiae. Mol. Cell. Biol.* 17:6517–25

37. Byers B. 1981. Cytology of the yeast life cycle. In *The Molecular Biology of the Yeast Saccharomyces: Life Cycle and Inheritance*, ed. JN Strathern, E Jones, J Broach, 1:97–142. Cold Spring Harbor, NY: Cold Spring Harbor Lab. 751 pp.

38. Byers B, Goetsch L. 1975. Behavior of the spindle plaques in the cell cycle and conjugation of *Saccharomyces cerevisiae. J. Bacteriol.* 124:511–23

39. Byers B, Goetsch L. 1976. A highly ordered ring of membrane-associated filaments in budding yeast. *J. Cell Biol.* 69: 717–21

40. Byers B, Goetsch L. 1976. Loss of the filamentous ring in cytokinesis-defective mutants of budding yeast. *J. Cell Biol.* 70:35a

41. Cairns BR, Ramer SW, Kornberg RD. 1992. Order of action of components in the yeast pheromone response pathway revealed with a dominant allele of the STE11 kinase and multiple phosphorylation of the STE7 kinase. *Genes Dev.* 6:1305–18

42. Carminati JL, Stearns T. 1997. Microtubules orient the mitotic spindle in yeast through dynein-dependent interactions with the cell cortex. *J. Cell Biol.* 138: 629–41

43. Castrillon DH, Wasserman SA. 1994. Diaphonous is required for cytokinesis in *Drosophila* and shares domains of similarity with the products of the limb deformity gene. *Development* 120:3367–77

44. Chang F, Drubin D, Nurse P. 1997. cdc12p, a protein required for cytokinesis in fission yeast, is a component of the cell division ring and interacts with profilin. *J. Cell Biol.* 137:169–82

45. Chang F, Herskowitz I. 1990. Identification of a gene necessary for cell cycle arrest by a negative growth factor of yeast: *FAR1* is an inhibitor of a G1 cyclin, CLN2. *Cell* 63:999–1011

46. Chant J, Corrado K, Pringle JR, Herskowitz I. 1991. The yeast *BUD5* gene, which encodes a putative GDP-GTP exchange factor, is necessary for bud-site selection and interacts with bud-formation gene *BEM1. Cell* 65:1213–24

47. Chant J, Herskowitz I. 1991. Genetic control of bud-site selection in yeast by a set of gene products that comprise a morphogenetic pathway. *Cell* 65:1203–12

48. Chant J, Mischke M, Mitchell E, Herskowitz I, Pringle JR. 1995. Role of Bud3p in producing the axial budding pattern of yeast. *J. Cell Biol.* 129:767–78

49. Chant J, Pringle JR. 1995. Patterns of bud-site selection in the yeast *Saccharomyces cerevisiae. J. Cell Biol.* 129:751–65

50. Chen G-C, Kim Y-J, Chan CSM. 1997. The Cdc42p GTPase-associated proteins Gic1 and Gic2 are required for polarized cell growth in *Saccharomyces cerevisiae. Genes Dev.* 11:2958–71

51. Chenevert J. 1994. Cell polarization directed by extracellular cues in yeast. *Mol. Biol. Cell* 5:1169–75

52. Chenevert J, Corrado K, Bender A, Pringle J, Herskowitz I. 1992. A yeast gene (*BEM1*) necessary for cell polarization whose product contains two SH3 domains. *Nature* 356:77–79

53. Choi K-Y, Satterberg B, Lyons DM, Elion EA. 1994. Ste5p tethers multiple protein kinases in the MAP kinase cascade required for mating in *S. cerevisiae. Cell* 78:499–512

54. Chuang JS, Schekman RW. 1996. Differential trafficking and timed localization of two chitin synthase proteins, Chs2p and Chs3p. *J. Cell Biol.* 135:597–610

55. Cid V, Duran A, Rey FD, Snyder M, Nombela C, Sanchez M. 1995. Molecular basis of cell integrity and morphogenesis in *Saccharomyces cerevisiae. Microbiol. Rev.* 59:345–86

56. Cleves AE, Novick PJ, Bankaitis VA. 1989. Mutations in the *SAC1* gene suppress defects in yeast Golgi and yeast actin function. *J. Cell Biol.* 109:2939–50

57. Cook JG, Bardwell L, Kron SJ, Thorner J. 1996. Two novel targets of the MAP kinase KSS1 are negative regulators of invasive growth in the yeast *Saccharomyces cerevisiae. Genes Dev.* 10:2831–48

58. Cook JG, Bardwell L, Thorner J. 1997. Inhibitory and activating functions for MAPK Kss1 in the *S. cerevisiae* filamentous-growth signalling pathway. *Nature* 390:85–88

59. Costigan C, Gehrung S, Snyder M. 1992. A synthetic lethal screen identifies SLK1, a novel protein kinase homolog implicated in yeast cell morphogenesis and cell growth. *Mol. Cell. Biol.* 12:1162–78

60. Costigan C, Snyder M. 1994. *NHP6A* and *NHP6B*, which encode HMG1-like proteins, function downstream in the yeast SLT2 MAPK pathway. *Mol. Cell. Biol.* 14:2391–403

61. Cross F, Hartwell LH, Jackson C,

Konopka JB. 1988. Conjugation in *Saccharomyces cerevisiae*. *Annu. Rev. Cell Biol.* 4:429–57

62. Cross FR. 1990. Cell cycle arrest caused by *CLN* gene deficiency in *Saccharomyces cerevisiae* resembles START-I arrest and is independent of the mating-pheromone signalling pathway. *Mol. Cell. Biol.* 10:6482–90

63. Cross FR. 1995. Starting the cell cycle: What's the point? *Curr. Opin. Cell Biol.* 7:790–97

64. Cross FR, Tinkelenberg AH. 1991. A potential feedback loop controlling *CLN1* and *CLN2* gene expression at the Start of the yeast cell cycle. *Cell* 65:875–83

65. Crouzet M, Urdaci M, Dulau L, Aigle M. 1991. Yeast mutant affected for viability upon nutrient starvation: characterization and cloning of the RVS161 gene. *Yeast* 7:727–43

66. Cvrckova F, Nasmyth K. 1993. Yeast G1 cyclins *CLN1* and *CLN2* and a GAP-like protein have a role in bud formation. *EMBO J.* 12:5277–86

67. Cvrckova F, Vergilio CD, Manser E, Pringle JR, Nasmyth K. 1995. Ste20-like protein kinases are required for localization of cell growth and for cytokinesis in budding yeast. *Genes Dev.* 9:1817–30

68. DeMarini DJ, Adams AEM, Faras H, Virgilio CD, Valle G, Chuang JS, Pringle JR. 1997. A septin-based hierarchy of proteins required for localized deposition of chitin in the *Saccharomyces cerevisiae* cell wall. *J. Cell Biol.* 139:75–93

69. Desfarges L, Durrens P, Juguelin H, Cassagne C, Bonneu M, Aigle M. 1993. Yeast mutants affected in viability upon starvation have a modified phospholipid composition. *Yeast* 9:267–77

70. DiComo CJ, Chang H, Arndt KT. 1995. Activation of *CLN1* and *CLN2* G1 cyclin gene expression by BCK2. *Mol. Cell. Biol.* 15:1835–46

71. Dirick L, Bohm T, Nasmyth K. 1995. Roles and regulation of Cln/Cdc28 kinases at the start of the cell cycle of *Saccharomyces cerevisiae*. *EMBO J.* 14:4803–13

72. Dirick L, Nasmyth K. 1991. Positive feedback in the activation of G1 cyclins in yeast. *Nature* 351:754–57

73. Dodou E, Treisman R. 1997. The *Saccharomyces cerevisiae* MADS-box transcription factor RLM1 is a target for the MPK1 mitogen-activated protein kinase pathway. *Mol. Cell. Biol.* 17:1848–59

74. Dorer R, Pryciak PM, Hartwell LH. 1995. *Saccharomyces cerevisiae* cells execute a default pathway to select a mate in the absence of pheromone gradients. *J. Cell Biol.* 131:845–61

75. Doyle T, Botstein D. 1996. Movement of yeast cortical actin cytoskeleton visualized in vivo. *Proc. Natl. Acad. Sci. USA* 93:3886–91

76. Drgonova J, Drgon T, Tanaka K, Kollar R, Chen G-C, et al. 1996. Rho1p, a yeast protein at the interface between cell polarization and morphogenesis. *Science* 272:277–79

77. Drubin DG. 1990. Actin and actin binding proteins in yeast. *Cell Motil. Cytoskelet.* 15:7–11

78. Drubin DG, Jones HD, Wertman KF. 1993. Actin structure and function: roles in mitochondrial organization and morphogenesis in budding yeast and identification of the phalloidin-binding site. *Mol. Biol. Cell* 4:1277–94

79. Drubin DG, Miller KG, Botstein D. 1988. Yeast actin-binding proteins: evidence for a role in morphogenesis. *J. Cell Biol.* 107:2551–61

80. Durrens P, Revardel E, Bonneu M, Aigle M. 1995. Evidence for a branched pathway in the polarized cell division of *Saccharomyces cerevisiae*. *Curr. Genet.* 27:213–16

81. Elion EA, Brill JA, Fink GR. 1991. FUS3 represses CLN1 and CLN2 in concert with KSS1 promotes signal transduction. *Proc. Natl. Acad. Sci. USA* 88:9392–96

82. Elion EA, Satterberg B, Kranz JE. 1993. FUS3 phosphorylates multiple components of the mating signal transduction cascade: evidence for STE12 and FAR1. *Mol. Biol. Cell* 4:495–510

83. Emmons S, Phan H, Calley J, Chen W, James B, Manseau L. 1995. Cappuccino, a Drosophila maternal effect gene required for polarity of the egg and embryo, is related to the vertebrate limb deformity locus. *Genes Dev.* 9:2482–94

84. Epstein CB, Cross FR. 1992. *CLB5*: a novel B cyclin from budding yeast with a role in S phase. *Genes Dev.* 6:1695–706

85. Epstein CB, Cross FR. 1994. Genes that can bypass the *CLN* requirement for *Saccharomyces cerevisiae* START. *Mol. Cell. Biol.* 14:2041–47

86. Erdman S, Lin L, Malczynski M, Snyder M. 1998. Pheromone-regulated genes required for yeast mating differentiation. *J. Cell Biol.* In press

87. Errede B, Cade RM, Yasar BM, Kamada Y, Levin DE, et al. 1995. Dynamics and organization of MAP kinase signal pathways. *Mol. Repro. Dev.* 42:477–85

88. Errede B, Gartner A, Zhou Z, Nasmyth K,

Ammerer G. 1993. MAP kinase-related FUS3 from *S. cerevisiae* is activated by STE7 *in vitro*. *Nature* 362:261–64

89. Espinoza FH, Ogas J, Herskowitz I, Morgan DO. 1994. Cell cycle control by a complex of the cyclin *HCS26 (PCL1)* and the kinase *PHO85*. *Science* 266:1388–91

90. Evangelista M, Blundell K, Longtine MS, Chow CJ, Adames N, et al. 1997. Bni1p, a yeast formin linking Cdc42p and the actin cytoskeleton during polarized morphogenesis. *Science* 276:118–22

91. Fernandez-Sarabia MJ, Sutton A, Zhong T, Arndt KT. 1992. *SIT4* protein phosphatase is required for the normal accumulation of *SWI4*, *CLN1*, *CLN2*, and *HCS26* RNAs during late G1. *Genes Dev.* 6:2417–28

92. Ferro-Novick S, Jahn R. 1994. Vesicle fusion from yeast to man. *Nature* 370:191–93

93. Field C, Schekman R. 1980. Localized secretion of acid phosphatase reflects the pattern of cell surface growth in *Saccharomyces cerevisiae*. *J. Cell Biol.* 86:123–28

94. Field CM, Al-Awar O, Rosenblatt J, Wong ML, Alberts B, Mitchison TJ. 1996. A purified *Drosophila* septin complex forms filaments and exhibits GTPase activity. *J. Cell Biol.* 133:605–16

95. Finegold AA, Johnson DI, Farnsworth CC, Gelb MH, Judd SR, et al. 1991. Protein geranylgeranyltransferase of *Saccharomyces cerevisiae* is specific for Cys-Xaa-Xaa-Leu motif proteins and requires the *CDC43* gene product but not the *DPR1* gene product. *Proc. Natl. Acad. Sci. USA* 88:4448–52

96. Flescher EG, Madden K, Snyder M. 1993. Components required for cytokinesis are important for bud site selection in yeast. *J. Cell Biol.* 122:373–86

97. Ford S, Pringle J. 1991. Cellular morphogenesis in the *Saccharomyces cerevisiae* cell cycle: localization of the *CDC11* gene product and the timing of events at the budding site. *Dev. Genet.* 12:281–92

98. Freeman G, Lundelius JW. 1982. The developmental genetics of dextrality and sinistrality in the gastropod *Lymnaea peregra*. *Wilhelm Roux Arch. Entwicklungsmech. Org.* 191:69–83

99. Freeman NL, Lila T, Mintze RKA, Chen Z, Pahk AJ, et al. 1996. A conserved proline-rich region of the *Saccharomyces cerevisiae* cyclase-associated protein binds SH3 domains and modulates cytoskeletal localization. *Mol. Cell. Biol.* 16:548–56

100. Freifelder D. 1960. Bud position in *Saccharomyces cerevisiae*. *J. Bacteriol.* 124:511–23

101. Friesen H, Lunz R, Doyle S, Segall J. 1994. Mutation of the SPS1-encoded protein kinase of *Saccharomyces cerevisiae* leads to defects in transcription and morphology during spore formation. *Genes Dev.* 8:2162–75

102. Fugita A, Oka C, Arikawa Y, Katagal T, Tonouchi A, et al. 1994. A yeast gene necessary for bud-site selection encodes a protein similar to insulin-degrading enzymes. *Nature* 372:567–69

103. Gabriel M, Kopecka M. 1995. Disruption of the actin cytoskeleton in budding yeast results in formation of an aberrant cell wall. *Microbiology* 141:891–99

104. Gale C, Finkel D, Tao N, Meinke M, McClellan M, et al. 1996. Cloning and expression of a gene encoding an integrin-like protein in *Candida albicans*. *Proc. Natl. Acad. Sci. USA* 93:357–61

105. Gartner A, Nasmyth K, Ammerer G. 1992. Signal transduction in *Saccharomyces cerevisiae* requires tyrosine and threonine phosphorylation of FUS3 and KSS1. *Genes Dev.* 6:1280–92

106. Gavrias V, Andrianopolous A, Gimeno CJ, Timberlake WE. 1996. *Saccharomyces cerevisiae* TEC1 is required for pseudohyphal growth. *Mol. Microbiol.* 19:1255–63

107. Gehrung S, Snyder M. 1990. The SPA2 gene of *Saccharomyces cerevisiae* is important for pheromone-induced morphogenesis and efficient mating. *J. Cell Biol.* 111:1451–64

108. Geli MI, Riezman H. 1996. Role of type I myosins in receptor-mediated endocytosis in yeast. *Science* 272:533–35

109. Gimeno CJ, Fink GR. 1992. The logic of cell division in the life cycle of yeast. *Science* 257:626

110. Gimeno CJ, Ljungdahl PO, Styles CA, Fink GR. 1992. Unipolar cell divisions in the yeast *S. cerevisiae* lead to filamentous growth: regulation by starvation and *RAS*. *Cell* 68:1077–90

111. Giot L, Konopka JB. 1997. Functional analysis of the interaction between Afr1p and the Cdc12p septin, two proteins involved in pheromone-induced morphogenesis. *Mol. Biol. Cell* 8:987–98

112. Goodson HV, Anderson BL, Warrick HM, Pon LA, Spudich JA. 1996. Synthetic lethality screen identifies a novel yeast myosin I gene (MYO5): Myosin I proteins are required for polarization of the actin cytoskeleton. *J. Cell Biol.* 133:1277–91

113. Govindan B, Bowser R, Novick P. 1995. The role of Myo2, a yeast class V myosin in vesicular transport. *J. Cell Biol.* 128:1055–68

114. Govindan B, Novick P. 1995. Development of cell polarity in budding yeast. *J. Exp. Zool.* 273:401–24

115. Haarer B, Pringle JR. 1987. Immunofluorescence localization of the *Saccharomyces cerevisiae* CDC12 gene product to the vicinity of the 10 nm filaments in the mother-bud neck. *Mol. Cell. Biol.* 7:3678–87

116. Hadwiger JA, Wittenberg C, Lopes MD, Richardson HE, Reed SI. 1989. A family of cyclin homologs that control G1 phase in yeast. *Proc. Natl. Acad. Sci. USA* 86:6255–59

117. Hall A. 1994. Small GTP-binding proteins and the regulation of the actin cytoskeleton. *Annu. Rev. Cell Biol.* 10:31–54

118. Hall JP, Cherkasova V, Elion E, Gustin MC, Winter E. 1996. The osmoregulatory pathway represses mating pathway activity in *Saccharomyces cerevisiae*: isolation of a *FUS3* mutant that is insensitive to the repression mechanism. *Mol. Cell. Biol.* 16:6715–23

119. Halme A, Michelitch M, Mitchell EL, Chant J. 1996. Bud10p directs axial cell polarization in budding yeast and resembles a transmembrane receptor. *Curr. Biol.* 6:570–79

120. Harsay E, Bretscher A. 1995. Parallel secretory pathways to the cell surface in yeast. *J. Cell Biol.* 131:297–310

121. Hartwell LH. 1971. Genetic control of the cell division cycle in yeast. IV. Genes controlling bud emergence and cytokinesis. *Exp. Cell Res.* 69:265–76

122. Hasek J, Rupes I, Svobodova J, Streiblova E. 1987. Tubulin and actin topology during zygote formation of *Saccharomyces cerevisiae*. *J. Gen. Microbiol.* 133:3355–63

123. Hasson MS, Blinder D, Thorner J, Jenness DD. 1994. Mutational activation of the *STE5* gene product bypasses the requirement for G protein β and γ subunits in the yeast pheromone response pathway. *Mol. Cell. Biol.* 14:1054–65

124. Hayashibe M, Katohda S. 1973. Initiation of budding and chitin-ring. *J. Gen. Appl. Microbiol.* 19:23–39

125. Herskowitz I. 1988. Life cycle of the budding yeast *Saccharomyces cerevisiae*. *Microbiol. Rev.* 52:536–53

126. Herskowitz I. 1995. MAP kinase pathways in yeast: for mating and more. *Cell* 80:187–97

127. Herskowitz I, Park H-O, Sanders S, Valtz N, Peter M. 1995. Programming of cell polarity in budding yeast by endogenous and exogenous signals. *Cold Spring Harbor Symp. Quant. Biol.* 60:717–27

128. Hicks JB, Strathern JN, Herskowitz I. 1977. Interconversion of yeast mating types. III. Action of the homothallism (HO) gene in cells homozygous for the mating type locus. *Genetics* 85:395–405

129. Hirano H, Tanaka K, Ozaki K, Imamura H, Kohno H, et al. 1997. ROM7/BEM4 encodes a novel protein that interacts with the Rho1p small GTP-binding protein in *Saccharomyces cerevisiae*. *Mol. Cell. Biol.* 16:4396–403

130. Holtzman DA, Yang S, Drubin DG. 1993. Synthetic-lethal interactions identify two genes, *SLA1* and *SLA2*, that control membrane cytoskeleton assembly in *Saccharomyces cerevisiae*. *J. Cell Biol.* 122:635–44

131. Huffaker TC, Thomas JH, Botstein D. 1988. Diverse effects of β-tubulin mutations on microtubule formation and function. *J. Cell Biol.* 106:1997–2010

132. Hyman AA, White JG. 1987. Determination of cell division axes in the early embryogenesis of *Caenorhabditis elegans*. *J. Cell Biol.* 105:2123–35

133. Igual JC, Johnson AL, Johnston LH. 1996. Coordinated regulation of gene expression by the cell cycle transcription factor SWI4 and the protein kinase C MAP kinase pathway for yeast cell integrity. *EMBO J.* 15:5001–13

134. Imai J, Toh-e A, Matsui Y. 1996. Genetic analysis of the *Saccharomyces cerevisiae* RHO3 gene, encoding a Rho-type small GTPase provides evidence for a role in bud formation. *Genetics* 142:359–69

135. Imamura H, Tanaka K, Hihara T, Umikawa M, Kamei T, et al. 1997. Bni1p and Bnr1p: downstream targets of the Rho family small G-proteins which interact with profilin and regulate actin cytoskeleton in *Saccharomyces cerevisiae*. *EMBO J.* 16:2745–55

136. Inouye C, Dhillon N, Durfee T, Zambryski PC, Thorner J. 1997. Mutational analysis of *STE5* in the yeast *Saccharomyces cerevisiae*: application of a differential interaction trap assay for examining protein-protein interactions. *Genetics* 147:479–92

137. Inouye C, Dhillon N, Thorner J. 1997. Ste5 RING-H2 domain: role in Ste4-promoted oligomerization for yeast pheromone signaling. *Science* 278:103–6

138. Irie K, Takase M, Lee K, Levin D, Araki H, et al. 1993. *MKK1* and *MKK2*,

which encode *Saccharomyces cerevisiae* mitogen-activated protein kinase-kinase homologs, function in the pathway mediated by protein kinase C. *Mol. Cell. Biol.* 13:3076–83

139. Jackson CL, Hartwell LH. 1990. Courtship in *S. cerevisiae*: Both cell types choose mating partners by responding to the strongest pheromone signal. *Cell* 63:1039–51

140. Jackson CL, Hartwell LH. 1990. Courtship in *Saccharomyces cerevisiae*: an early cell-cell interaction during mating. *Mol. Cell. Biol.* 10:2202–13

141. Jackson CL, Konopka JB, Hartwell LH. 1991. *S. cerevisiae* α-pheromone receptors activate a novel signal transduction pathway for mating partner discrimination. *Cell* 67:389–402

142. Jackson DA, Cook PR. 1988. Visualization of a filamentous nucleoskeleton with a 23nm axial repeat. *EMBO J.* 7:3667–77

143. Jacobs CW, Adams AEM, Szaniszlo PJ, Pringle JR. 1988. Functions of microtubules in the *Saccharomyces cerevisiae* cell cycle. *J. Cell Biol.* 107:1409–26

144. Johnson DI, Pringle JR. 1990. Molecular characterization of CDC42, a *Saccharomyces cerevisiae* gene involved in the development of cell polarity. *J. Cell Biol.* 111:143–52

145. Johnston GC, Prendergast JA, Singer RA. 1991. The *Saccharomyces cerevisiae* MYO2 gene encodes an essential myosin for vectorial transport of vesicles. *J. Cell Biol.* 113:539–51

146. Kabsch W, Mannherz HG, Suck D, Pai EF, Holmes KC. 1990. Atomic structure of the actin:DNase I complex. *Nature* 347:37–44

147. Kaiser CA, Gimeno RE, Shaywitz DA. 1996. Protein secretion, membrane biogenesis, and endoycytosis. In *The Molecular and Cellular Biology of the Yeast Saccharomyces*, ed. JR Pringle, JR Broach, EW Jones, 3:91–228. Cold Spring Harbor, NY: Cold Spring Harbor Lab. 1131 pp.

148. Kamada Y, Jung US, Piotrowski J, Levin DE. 1995. The protein kinase C-activated MAP kinase pathway of *Saccharomyces cerevisiae* mediates a novel aspect of the heat shock resonse. *Genes Dev.* 9:1559–71

149. Kamada Y, Qadota H, Python CP, Anraku Y, Ohya Y, Levin DE. 1996. Activation of yeast protein kinase C by Rho1 GTPase. *J. Biol. Chem.* 271:9193–96

150. Kilmartin JV, Adams AEM. 1984. Structural rearrangements of tubulin and actin during the cell cycle of the yeast *Saccharomyces*. *J. Cell Biol.* 98:922–33

151. Kim HB, Haarer BK, Pringle JR. 1991. Cellular morphogenesis in the *Saccharomyces cerevisiae* cell cycle: localization of the CDC3 gene product and the timing of events at the budding site. *J. Cell Biol.* 112:535–44

152. Kim YJ, Francisco L, Chen GC, Marcotte E, Chan CSM. 1994. Control of cellular morphogenesis by the Ipt2/Bem2 GTPase-activating protein: possible role of protein phosphorylation. *J. Cell Biol.* 127:1381–94

153. Knapp D, Bhiote L, Stillman DJ, Nasmyth K. 1996. The transcription factor Swi5 regulates expression of the cyclin kinase inhibitor p40-SIC1. *Mol. Cell. Biol.* 16:5701–7

154. Koch C, Moll T, Neuberg M, Ahorn H, Nasmyth K. 1993. A role of the transription factors Mbp1 and Swi4 in progression from G1 to S phase. *Science* 261:1551–57

155. Koch C, Nasmyth K. 1994. Cell cycle regulated transcription in yeast. *Curr. Opin. Cell Biol.* 6:451–59

156. Koch G, Tanaka K, Masuda T, Yamochi W, Nonaka H, Takai Y. 1997. Association of the Rho family small GTP-binding proteins with Rho GDP dissociation inhibitor (Rho GDI) in *Saccharomyces cerevisiae*. *Oncogene* 15:417–22

157. Kohno H, Tanaka K, Mino A, Umikawa M, Imamura H, et al. 1996. Bni1p implicated in cytoskeletal control is a putative target of Rho1p small GTP binding protein in *Saccharomyces cerevisiae*. *EMBO J.* 15:6060–68

158. Kolodrubetz D, Burgum A. 1990. Duplicated NHP6 genes of *Saccharomyces cerevisiae* encode proteins homologous to bovine high mobility group protein 1. *J. Biol. Chem.* 265:3234–39

159. Konopka JB. 1993. AFR1 acts in conjunction with the α-factor receptor to promote morphogenesis and adaptation. *Mol. Cell. Biol.* 13:6876–88

160. Konopka JB, DeMattei C, Davis C. 1995. AFR1 promotes polarized apical morphogenesis in *Saccharomyces cerevisiae*. *Mol. Cell. Biol.* 15:723–30

161. Konopka JB, Jenness DD, Hartwell LH. 1988. The C-terminus of the *S. cerevisiae* α-pheromone receptor mediates an adaptive response to pheromone. *Cell* 54:609–20

162. Kraut R, Chia W, Jan LY, Jan YN, Knoblich JA. 1996. Role of *inscuteable* in orienting asymmetric cell divisions in *Drosophila*. *Nature* 383:50–55

163. Krisak L, Strich R, Winters RS, Hall JP, Mallory MJ, et al. 1994. SMK1, a developmentally regulated MAP kinase, is required for spore wall assembly in *Saccharomyces cerevisiae*. *Genes Dev.* 8:2151–61

164. Kron SJ, Gow NAR. 1995. Budding yeast morphogenesis: signalling, cytoskeleton and cell cycle. *Curr. Opin. Cell Biol.* 7:845–55

165. Kron SJ, Styles CA, Fink GR. 1994. Symmetric cell division in pseudohyphae of the yeast *Saccharomyces cerevisiae*. *Mol. Biol. Cell* 5:1003–22

166. Kupfer A, Swain SL, Janeway CA, Singer CS. 1986. The specific direct interaction of helper T cells and antigen-presenting B cells. *Proc. Natl. Acad. Sci. USA* 83:6080–83

167. Kurjan J. 1993. Pheromone response in yeast. *Annu. Rev. Biochem.* 61:1097–129

168. Lappalainen P, Drubin DG. 1997. Cofilin promotes rapid actin filament turnover in vivo. *Nature* 388:78–82

169. Leberer E, Dignard D, Harcus D, Thomas DY, Whiteway M. 1992. The protein kinase homologue Ste20p is required to link the yeast pheromone response G-protein beta gamma subunits to downstream signalling components. *EMBO J.* 11:4815–24

170. Leberer E, Thomas DY, Whiteway M. 1997. Pheromone signalling and polarized morphogenesis in yeast. *Curr. Opin. Gen. Dev.* 7:59–66

171. Leberer E, Wu C, Leeuw T, Fourest-Lieuvin A, Segall JE, Thomas DY. 1997. Functional characterization of the Cdc42p binding domain of yeast Ste20p protein kinase. *EMBO J.* 16:83–97

172. Lee K, Levin D. 1992. Dominant mutations in a gene encoding a putative protein kinase (*BCK1*) bypass the requirement for a *Saccharomyces cerevisiae* protein kinase C homolog. *Mol. Cell. Biol.* 12:172–82

173. Lee KS, Hines LK, Levin DE. 1993. A pair of functionally redundant yeast genes (*PPZ1* and *PPZ2*) encoding type 1-related protein phosphatases function within the *PKC1*-mediated pathway. *Mol. Cell. Biol.* 13:5843–53

174. Lee KS, Irie K, Gotoh Y, Watanabe Y, Araki H, et al. 1993. A yeast mitogen-activated protein kinase homolog (Mpk1p) mediates signalling by protein kinase C. *Mol. Cell. Biol.* 13:3067–75

175. Leeuw T, Fourest-Lieuvin A, Wu C, Chenevert J, Clark K, et al. 1995. Pheromone response in yeast: association of Bem1p with proteins of the MAP kinase cascade and actin. *Science* 270:1210–13

176. Leeuw T, Wu C, Schrag J, Whiteway M, Thomas DY, Leberer E. 1998. Interaction of a Gβ-subunit with a conserved sequence in Ste20/PAK family protein kinases. *Nature.* 8:191–95

177. Levin DE, Bowers B, Chen C-Y, Kamada Y, Watanabe M. 1994. Dissecting the protein kinase C/MAP kinase signalling pathway of *Saccharomyces cerevisiae*. *Cell Mol. Biol. Res.* 40:229–39

178. Levin DE, Errede B. 1995. The proliferation of MAP kinase signaling pathways in yeast. *Curr. Opin. Cell Biol.* 7:197–202

179. Lew DJ, Marini NJ, Reed SI. 1992. Different G1 cyclins control the timing of cell cycle commitment in mother and daughter cells in the budding yeast *Saccharomyces cerevisiae*. *Cell* 69:317–27

180. Lew DJ, Reed SI. 1993. Morphogenesis in the yeast cell cycle: regulation by Cdc28 and cyclins. *J. Cell Biol.* 120:1305–20

181. Lew DJ, Reed SI. 1995. A cell cycle checkpoint monitors cell morphogenesis in budding yeast. *J. Cell Biol.* 129:739–49

182. Lew DJ, Reed SI. 1995. Cell cycle control of morphogenesis in budding yeast. *Curr. Opin. Genet. Dev.* 5:17–23

183. Lillie SH, Brown SS. 1992. Suppression of a myosin defect by a kinesin-related gene. *Nature* 356:358–61

184. Lillie SH, Brown SS. 1994. Immunofluorescence localization of the unconventional myosin, Myo2p, and the putative kinesin-related protein, Smy1p, to the same regions of polarized growth in *Saccharomyces cerevisiae*. *J. Cell Biol.* 125:825–42

185. Liu H, Bretscher A. 1989. Disruption of the single tropomyosin gene in yeast results in the disappearance of actin cables from the cytoskeleton. *Cell* 57:233–42

186. Liu H, Bretscher A. 1989. Purification of tropomyosin from *Saccharomyces cerevisiae* and identification of related proteins in *Schizosaccharomyces* and *Physarum*. *Proc. Natl. Acad. Sci. USA* 86:90–93

187. Liu H, Bretscher A. 1992. Characterization of *TPM1* disrupted yeast cells indicates an involvement of tropomyosin in directed vesicular transport. *J. Cell Biol.* 118:285–99

188. Liu H, Styles CA, Fink GR. 1993. Elements of the yeast pheromone response pathway required for filamentous growth of diploids. *Science* 262:1741–44

189. Long RM, Singer RH, Meng X, Gonzalez I, Nasmyth K, Jansen R-P. 1997. Mating type switching in yeast controlled by

asymmetric localization of *ASH1* mRNA. *Science* 277:383–87

190. Longtine MS, DeMarini DJ, Valencik ML, Al-Awar OS, Fares H, et al. 1996. The septins: roles in cytokinesis and other processes. *Curr. Opin. Cell Biol.* 8:106–19

191. Mack D, Nishimura K, Dennehey BK, Arbogast T, Parkinson J, et al. 1997. Identification of the bud emergence gene *BEM4* and its interactions with the Rho-type GTPases in *Saccharomyces cerevisiae. Mol. Biol. Cell* 16:4387–95

192. Mackay VL. 1983. Cloning of yeast *STE* genes in 2 micron vectors. *Methods Enzymol.* 101:325–43

193. Madden K, Costigan C, Snyder M. 1992. Cell polarity and morphogenesis in *Saccharomyces cerevisiae. Trends Cell Biol.* 2:22–29

194. Madden K, Sheu Y-J, Baetz K, Andrews B, Snyder M. 1997. SBF cell cycle regulator as a target of the yeast PKC-MAP kinase pathway. *Science* 275:1781–84

195. Madden K, Snyder M. 1992. Specification of sites of polarized growth in *Saccharomyces cerevisiae* and the influence of external factors on site selection. *Mol. Biol. Cell* 3:1025–35

196. Madhani HD, Fink GR. 1997. Combinatorial control required for the specificity of yeast MAPK signaling. *Science* 275:1314–17

197. Madhani HD, Styles CA, Fink GR. 1997. MAP kinases with distinct inhibitory functions impart signaling specificity during yeast differentiation. *Cell* 91:673–84

198. Maduale P, Axel R, Myers AM. 1987. Characterization of two members of the *rho* gene family from the yeast *Saccharomyces cerevisiae. Proc. Natl. Acad. Sci. USA* 84:779–84

199. Manning BD, Padmanabha R, Snyder M. 1997. The Rho-GEF Rom2p localizes to sites of polarized cell growth and participates in cytoskeletal functions in *Saccharomyces cerevisiae. Mol. Biol. Cell* 8:1829–44

200. Manser E, Leung T, Salihuddin H, Zhao ZS, Lim L. 1994. A brain serine/threonine protein kinase activated by Cdc42 and Rac1. *Nature* 367:340–46

201. Marcus S, Caldwell GA, Miller D, Xue CB, Naider F, Becker JM. 1991. Significance of C-terminal cysteine modifications to the biological activity of the *Saccharomyces cerevisiae* a-factor mating pheromone. *Mol. Cell. Biol.* 11:3603–12

202. Marcus S, Polverino A, Barr M, Wigler M. 1994. Complexes between STE5 and components of the pheromone-responsive mitogen-activated protein kinase module. *Proc. Natl. Acad. Sci. USA* 91:7762–66

203. Marini NJ, Meldrum E, Buehrer B, Hubberstey AV, Stone DE, et al. 1996. A pathway in the yeast cell division cycle linking protein kinase C (Pkc1) to activation of Cdc28 at START. *EMBO J.* 15:3040–52

204. Marshall CJ. 1994. MAP kinase kinase kinase, MAP kinase kinase, and MAP kinase. *Curr. Opin. Genet. Dev.* 4:82–89

205. Martin GA, Bollag G, McCormick F, Abo A. 1995. A novel serine kinase activated by rac1/CDC42Hs-dependent autophosphorylation is related to PAK65 and STE20. *EMBO J.* 14:1970–78

206. Martin H, Mendoza A, Rodriquez-Pachon JM, Molina M, Nombela C. 1997. Characterization of *SKM1*, a *Saccharomyces cerevisiae* gene encoding a novel Ste20/PAK-like protein kinase. *Mol. Microbiol.* 23:431–44

207. Mass RL, Zeller R, Woychik RP, Vogt TF, Leder P. 1990. Disruption of formin-encoding transcripts in two mutant limb deformity alleles. *Nature* 346:853–55

208. Matsui Y, Toh-e A. 1992. Isolation and characterization of two novel *ras* superfamily genes in *Saccharomyces cerevisiae. Gene* 114:43–49

209. Matsui Y, Toh-e A. 1992. Yeast *RHO3* and *RHO4 ras* superfamily genes are necessary for bud growth and their defect is suppressed by a high dose of bud formation genes *CDC42* and *BEM1. Mol. Cell. Biol.* 12:5690–99

210. Mazzoni C, Zarzov P, Rambourg A, Mann C. 1993. The *SLT2(MPK1)* MAP kinase homolog is involved in polarized cell growth in *Saccharomyces cerevisiae. J. Cell Biol.* 123:1821–33

211. McCann R, Craig S. 1997. The I/LWEQ module: a conserved sequence that signifies F-actin binding in functionally diverse proteins from yeast to mammals. *Proc. Natl. Acad. Sci. USA* 94:5679–84

212. Measday V, Moore L, Ogas J, Tyers M, Andrews B. 1994. The PCL2 (ORFD)-PHO85 cyclin-dependent kinase complex: a cell cycle regulator in yeast. *Science* 266:1391–95

213. Michelitch M, Chant J. 1996. A mechanism of Bud1p GTPase action suggested by mutational analysis and immunolocalization. *Curr. Biol.* 6:446–54

214. Moon AL, Janmey PA, Louie KA, Drubin DG. 1993. Cofilin is an essential component of the yeast cortical cytoskeleton. *J. Cell Biol.* 120:421–35

215. Moore S. 1987. α-Factor inhibition of the rate of cell passage through the "Start"

step of cell division in *Saccharomyces cerevisiae* yeast: estimation of the division delay per α-factor receptor complex. *Exp. Cell Res.* 171:411–25

216. Moore SA. 1983. Comparison of dose-response curves for a-factor-induced cell division arrest, agglutination, and projection formation of yeast. *J. Biol. Chem.* 258:13848–56

217. Mooseker MS. 1985. Organization, chemistry, and assembly of the cytoskeletal apparatus of the intestinal brush border. *Annu. Rev. Cell Biol.* 1:209–41

218. Morgan BA, Bouquin N, Merrill GF, Johnston LH. 1995. A yeast transcription factor bypassing the requirement for SBF and DSC1/MBF in budding yeast has homology to bacterial signal transduction proteins. *EMBO J.* 14:5679–89

219. Mosch H-U, Fink GR. 1997. Dissection of filamentous growth by transposon mutagenesis in *Saccharomyces cerevisiae*. *Genetics* 145:671–84

220. Mulholland J, Preuss D, Moon A, Wong A, Drubin D, Botstein D. 1994. Ultrastructure of the yeast actin cytoskeleton and its association with the plasma membrane. *J. Cell Biol.* 125:381–91

221. Mulholland J, Wasp A, Riezman H, Botstein D. 1997. Yeast actin cytoskeleton mutants accumulate a new class of Golgi-derived secretory vesicles. *Mol. Biol. Cell* 8:1481–99

222. Nash R, Tokiwa G, Anand S, Erickson K, Futcher AB. 1988. The *WHI1+* gene of *S. cerevisiae* tethers cell division to cell size and is a cyclin homolog. *EMBO J.* 7:4335–46

223. Nasmyth K. 1993. Control of the yeast cell cycle by the Cdc28 protein kinase. *Curr. Opin. Cell Biol.* 5:166–79

224. Nasmyth K, Dirick L. 1991. The role of *SWI4* and *SWI6* in the activity of G1 cyclins in yeast. *Cell* 66:995–1013

225. Nasmyth KA. 1982. Molecular genetics of yeast mating type. *Annu. Rev. Genet.* 16:439–500

226. Neiman AM, Herskowitz I. 1994. Reconstitution of a yeast protein kinase *in vitro*: activation of the yeast MEK homologue STE7 by STE11. *Proc. Natl. Acad. Sci. USA* 91:3398–402

227. Nonaka H, Tanaka K, Hirano H, Fujiwara T, Kohno H, et al. 1995. A downstream target of RHO1 small GTP binding protein is PKC1, a homolog of protein kinase C, which leads to activation of the MPK1 kinase cascade in *Saccharomyces cerevisiae*. *EMBO J.* 14:5931–38

228. Novick P, Botstein D. 1985. Phenotypic analysis of temperature-sensitive yeast

actin mutants. *Cell* 40:405–16

229. Novick P, Osmond BC, Botstein D. 1989. Suppressors of yeast actin mutations. *Genetics* 121:659–74

230. Nugroho TT, Mendenhall MD. 1994. An inhibitor of yeast cyclin-dependent protein kinase plays an important role in ensuring the genomic integrity of daughter cells. *Mol. Cell. Biol.* 14:3320–28

231. Ogas J, Andrews BJ, Herskowitz I. 1991. Transcriptional activation of *CLN1*, *CLN2*, and a new G1 cyclin (*HCS26*) by *SWI4*, a positive regulator of G1-specific transcription. *Cell* 66:1015–26

232. Orlean P. 1996. Biogenesis of yeast wall and surface components. In *The Molecular and Cellular Biology of the Yeast Saccharomyces*, ed. JR Pringle, JR Broach, EW Jones, 3:229–362. Cold Spring Harbor, NY: Cold Spring Harbor Lab. 1131 pp.

233. Ozaki K, Tanaka K, Imamura H, Hihara T, Kameyama T, et al. 1996. Rom1p and Rom2p are GDP/GTP exchange proteins (GEPs) for the Rho1p small GTP binding protein in *Saccharomyces cerevisiae*. *EMBO J.* 15:2196–207

234. Page B, Snyder M. 1993. Chromosome segregation in yeast. *Annu. Rev. Microbiol.* 47:201–31

235. Palmer RE, Sullivan DS, Huffaker T, Koshland D. 1992. Role of astral microtubules and actin in spindle orientation and migration in the budding yeast, *Saccharomyces cerevisiae*. *J. Cell Biol.* 119:583–93

236. Park H-O, Bi E, Pringle JR, Herskowitz I. 1996. Two active states of the Ras-related Bud1/Rsr1 protein bind to different effectors to determine yeast cell polarity. *Proc. Natl. Acad. Sci. USA* 94:4463–68

237. Park HO, Chant J, Herskowitz I. 1993. *BUD2* encodes a GTPase activating protein for Bud1/Rsr1 necessary for proper bud site selection in yeast. *Nature* 365:269–74

238. Peter M, Gartner A, Horecker J, Ammerer G. 1993. FAR1 links the signal transduction pathway to the cell cycle machinery in yeast. *Cell* 73:747–60

239. Peter M, Herskowitz I. 1994. Direct inhibition of the yeast cyclin-dependent kinase Cdc28-Cln by Far1. *Science* 265:1228–31

240. Peter M, Neiman AM, Park H-O, van-Lohuizen M, Herskowitz I. 1996. Functional analysis of the interaction between the small GTP binding protein Cdc42 and the Ste20 protein kinase in yeast. *EMBO J.* 15:7046–59

241. Petersen J, Weilguny D, Egel R, Nielsen

O. 1995. Characterization of fus1 of *Schizosaccharomyces pombe*: a developmentally controlled function needed for conjugation. *Mol. Cell. Biol.* 15:3697–707

242. Peterson J, Zheng Y, Bender L, Myers A, Cerione R, Bender A. 1994. Interactions between the bud emergence proteins Bem1p and Bem2p and rho-type GTPases in yeast. *J. Cell Biol.* 127:1395–406

243. Pi H, Chien CT, Fields S. 1997. Transcriptional activation upon pheromone stimulation mediated by a small domain of *Saccharomyces cerevisiae* Ste12p. *Mol. Cell. Biol.* 17:6410–18

244. Powers S, Gonzales E, Christensen T, Cubert J, Broek D. 1991. Functional cloning of *BUD5*, a *CDC25*-related gene from *S. cerevisiae* that can suppress a dominant-negative *RAS2* mutant. *Cell* 65:1225–31

245. Preuss D, Mulholland J, Kaiser CA, Orlean P, Albright C, et al. 1991. Structure of the yeast endoplasmic reticulum localization of ER proteins using immunofluorescence and immunolocalization microscopy. *Yeast* 7:891–911

246. Price C, Nasmyth K, Schuster T. 1991. A general approach to the isolation of cell cycle regulated genes in the budding yeast *Saccharomyces cerevisiae*. *J. Mol. Biol.* 218:543–56

247. Primig M, Sockanathan S, Auer H, Nasmyth K. 1992. Anatomy of a transcription factor important for the Start of the cell cycle in *Saccharomyces cerevisiae*. *Nature* 358:593–97

248. Pringle JR, Bi E, Harkins HA, Zahner JE, DeVirgilio C, et al. 1995. Establishment of cell polarity in yeast. *Cold Spring Harbor Symp. Quant. Biol.* 60:729–44

249. Pringle JR, Hartwell LH. 1981. The *Saccharomyces cerevisiae* cell cycle. In *The Molecular Biology of the Yeast Saccharomyces: Life Cycle and Inheritance*, ed. JN Strathern, EW Jones, JR Broach, 1:97–162. Cold Spring Harbor, NY: Cold Spring Harbor Lab. 751 pp.

250. Printen JA, Sprague GF Jr. 1994. Protein-protein interactions in the yeast pheromone response pathway: Ste5p interacts with all members of the MAP kinase cascade. *Genetics* 138:609–19

251. Qadota H, Python CP, Inoue SB, Arisawa M, Anraku Y, et al. 1996. Identification of yeast Rho1p GTPase as a regulatory subunit of 1,3-β-glucan synthase. *Science* 272:279–81

252. Read EB, Okamura HH, Drubin DG. 1992. Actin- and tubulin-dependent functions during *Saccharomyces cerevisiae*

mating projection formation. *Mol. Biol. Cell* 3:429–44

253. Reed SI. 1992. The role of p34 kinases in the G1 to S-phase transition. *Annu. Rev. Cell Biol.* 8:529–61

254. Revardel E, Bonneau M, Durrens P, Aigle M. 1995. Characterization of a new gene family developing pleitropic phenotypes upon mutation in *Saccharomyces cerevisiae*. *Biochim. Biophys. Acta* 1263:261–65

255. Richardson HE, Wittenberg C, Cross F, Reed SI. 1989. An essential G1 function for cyclin-like proteins in yeast. *Cell* 59:1127–33

256. Ridley A. 1995. Rho-related proteins: actin cytoskeleton and cell cycle. *Curr. Opin. Genet. Dev.* 5:24–30

257. Roberts RL, Bowers B, Slater ML, Cabib E. 1983. Chitin synthesis and localization in cell division cycle mutants of *Saccharomyces cerevisiae*. *Mol. Cell. Biol.* 3:922–30

258. Roberts RL, Fink GR. 1994. Elements of a single MAP kinase cascade in *Saccharomyces cerevisiae* mediate two developmental programs in the same cell type: mating and invasive growth. *Genes Dev.* 8:2974–85

259. Roemer T, Madden K, Chang J, Snyder M. 1996. Selection of axial growth sites in yeast requires Axl2p, a novel plasma membrane glycoprotein. *Genes Dev.* 10:777–93

260. Roemer T, Vallier L, Sheu Y-J, Snyder M. 1998. The Spa2p-related protein, Sph1p, is important for polarized growth in yeast. *J. Cell Sci.* 111:479–94

261. Roemer T, Vallier LG, Snyder M. 1996. Selection of polarized growth sites in yeast. *Trends Cell Biol.* 6:434–41

262. Rowley R, Hudson J, Young PG. 1992. The wee1 protein kinase is required for radiation-induced mitotic delay. *Nature* 356:353–55

263. Ruggieri R, Bender A, Matsui Y, Powers S, Takai Y, et al. 1992. *RSR1*, a *ras*-like gene homologous to *Krev-1*(*smg21A*/*rap1A*): role in the development of cell polarity and interactions with the Ras pathway in *Saccharomyces cerevisiae*. *Mol. Cell. Biol.* 12:758–66

264. Russell P, Nurse P. 1986. cdc25+ functions as an inducer in the mitotic control of fission yeast. *Cell* 45:145–53

265. Russell P, Nurse P. 1987. Negative regulation of mitosis by wee1+, a gene encoding a protein kinase homolog. *Cell* 49:559–67

266. Russell P, Nurse P. 1987. The mitotic inducer nim1+ functions in a regulatory

network of protein kinase homologs controlling the initiation of mitosis. *Cell* 49: 569–76

267. Sanders SL, Field CM. 1994. Septins in common? *Curr. Biol.* 4:907–10

268. Sanders SL, Herskowitz I. 1996. The Bud4 protein of yeast, required for axial budding, is localized to the mother-bud neck in a cell cycle-dependent manner. *J. Cell Biol.* 134:413–27

269. Santos B, Duran A, Valdivieso MH. 1997. CHS5, a gene involved in chitin synthesis and mating in *Saccharomyces cerevisiae*. *Mol. Cell. Biol.* 17:2485–96

270. Santos B, Snyder M. 1997. Targeting of chitin synthase 3 to polarized growth sites in yeast requires Chs5p and Myo2p. *J. Cell Biol.* 136:95–110

271. Schmidt A, Bickle M, Beck T, Hall MN. 1997. The yeast phosphatidylinositol kinase homolog TOR2 activates RHO1 and RHO2 via the exchange factor ROM2. *Cell* 88:531–42

272. Schneider BL, Yang QH, Futcher AB. 1996. Linkage of replication to start by the Cdk inhibitor Sic1. *Science* 272:560–62

273. Schwob E, Bohm T, Mendenhall MD, Nasmyth K. 1994. The B-type cyclin kinase inhibitor p40SIC1 controls the G1 to S transition in *S. cerevisiae*. *Cell* 79:233–44

274. Schwob E, Nasmyth K. 1993. *CLB5* and *CLB6*, a new pair of B cyclins involved in DNA replication in *Saccharomyces cerevisiae*. *Genes Dev.* 7:1160–75

275. Segall JE. 1993. Polarization of yeast cells in spatial gradients of α-factor. *Proc. Natl. Acad. Sci. USA* 90:8332–36

276. Shapiro L. 1993. Protein localization and asymmetry in the bacterial cell. *Cell* 73:841–55

277. Shaw JA, Mol PC, Bowers B, Silverman SJ, Valdivieso MH, et al. 1991. The function of chitin synthases 2 and 3 in the *Saccharomyces cerevisiae* cell cycle. *J. Cell Biol.* 114:111–23

278. Sheu YJ, Santos B, Fortin N, Costigan C, Snyder M. 1998. SpaZp interacts with cell polarity proteins and signaling comonents involved in yeast cell morphogenesis. *Mol. Cell. Biol.* 18:4053–69

279. Shortle D, Haber JE, Botstein D. 1982. Lethal disruption of the yeast actin gene by integrative DNA transformation. *Science* 217:371–73

280. Shortle D, Novick P, Botstein D. 1984. Construction and genetic characterization of temperature-sensitive mutant alleles of the yeast actin gene. *Proc. Natl. Acad. Sci. USA* 81:4889–93

281. Sia RA, Herald HA, Lew DJ. 1996. Cdc28 tyrosine phosphorylation and the morphogenesis checkpoint in budding yeast. *Mol. Biol. Cell* 7:1657–66

282. Sil A, Herskowitz I. 1996. Identification of an asymmetrically localized determinant, Ash1p, required for lineage-specific transcription of the yeast HO gene. *Cell* 84:711–22

283. Simon M-N, DeVirgilio C, Souza B, Pringle JR, Abo A, Reed SI. 1995. Role for the Rho-family GTPase Cdc42 in yeast mating-pheromone signal pathway. *Nature* 376:702–5

284. Sivadon P, Bauer F, Aigle M, Crouzet M. 1995. Actin cytoskeleton and budding pattern are altered in the yeast *rvs161* mutant: The Rvs161 protein shares common domains with the brain protein amphiphysin. *Mol. Gen. Genet.* 246:485–95

285. Sloat B, Pringle JR. 1978. A mutant of yeast defective in cellular morphogenesis. *Science* 200:1171–73

286. Sloat BF, Adams A, Pringle JR. 1981. Roles of the *CDC24* gene product in cellular morphogenesis during the *Saccharomyces cerevisiae* cell cycle. *J. Cell Biol.* 89:395–405

287. Snyder M. 1989. The SPA2 protein of yeast localizes to sites of cell growth. *J. Cell Biol.* 108:1419–29

288. Snyder M, Gehrung S, Page BD. 1991. Studies concerning the temporal and genetic control of cell polarity in *Saccharomyces cerevisiae*. *J. Cell Biol.* 114:515–32

289. Sprague GF, Thorner J. 1992. Pheromone response and signal transduction during the mating process of *Saccharomyces cerevisiae*. In *The Molecular Biology of the Yeast Saccharomyces*, ed. JR Broach, JR Pringle, EW Jones, 2:657–744. Cold Spring Harbor, NY: Cold Spring Harbor Lab. 810 pp.

290. Stevenson BJ, Ferguson B, DeVirgilio C, Bi E, Pringle JR, et al. 1995. Mutation of *RGA1*, which encodes a putative GAP for the polarity-establishment protein Cdc42p, activates the pheromone response pathway in the yeast *Saccharomyces cerevisiae*. *Genes Dev.* 9:2949–63

291. Stevenson BJ, Rhodes N, Errede B, Sprague GF Jr. 1992. Constitutive mutants of the protein kinase STE11 activate the yeast pheromone response pathway in the absence of the G protein. *Genes Dev.* 6:1293–304

292. Stone EM, Pillus L. 1996. Activation of an MAP kinase cascade leads to Sir3p hyperphosphorylation and strengthens

transcriptional silencing. *J. Cell. Biol.* 135:571–83

293. Strazdis JR, MacKay VL. 1983. Induction of yeast mating pheromone a-factor by α cells. *Nature* 305:543–45

294. Stuart D, Wittenberg C. 1995. *CLN3*, not positive feedback, determine the timing of *CLN2* transcription in cycling cells. *Genes Dev.* 9:2780–94

295. Takizawa PA, Sil A, Swedlow JR, Herskowitz I, Vale RD. 1997. Actin-dependent localization of an RNA encoding a cell fate determinant in yeast. *Nature* 389:90–93

296. Tedford K, Kim S, Sa D, Stevens K, Tyers M. 1997. Regulation of the mating pheromone and invasive growth responses in yeast by two MAP kinase substrates. *Curr. Biol.* 7:228–38

297. Torres L, Martin H, Garcia-Saez MI, Arroyo J, Molina M, et al. 1991. A protein kinase gene complements the lytic phenotype of *Saccharomyces cerevisiae lyt2* mutants. *Mol. Microbiol.* 5:2845–54

298. Treisman R. 1996. Regulation of transcription by MAP kinase cascades. *Curr. Opin. Cell Biol.* 8:205–15

299. Trilla JA, Cos T, Duran A, Roncero C. 1997. Characterization of *CHS4* (*CAL2*), a gene of *Saccharomyces cerevisiae* involved in chitin biosynthesis and allelic to *SKT5* and *CSD4*. *Yeast* 13:795–807

300. Tyers M, Futcher B. 1993. FAR1 and FUS3 link the mating pheromone signal transduction pathway to three G1-phase CDC28 kinase complexes. *Mol. Cell. Biol.* 13:5659–69

301. Tyers M, Tokiwa G, Futcher B. 1993. Comparison of the *Saccharomyces cerevisiae* G1 cyclins: Cln3 may be upstream activator of Cln1, Cln2 and other cyclins. *EMBO J.* 12:1955–68

302. Tyers M, Tokiwa G, Nash R, Futcher B. 1992. The Cln3-Cdc28 kinase complex of *S. cerevisiae* is regulated by proteolysis and phosphorylation. *EMBO J.* 11:1773–84

303. Valdivieso MH, Mol PC, Shaw JA, Cabib E, Duran A. 1991. *CAL1*, a gene required for activity of chitin synthase 3 in *Saccharomyces cerevisiae. J. Cell Biol.* 114:101–9

304. Valtz N, Herskowitz I. 1996. Pea2 protein of yeast is localized to sites of polarized growth and is required for efficient mating and bipolar budding. *J. Cell Biol.* 135:725–39

305. Valtz N, Peter M, Herskowitz I. 1995. *FAR1* is required for oriented polarization of yeast cells in response to mating pheromones. *J. Cell Biol.* 131:863–73

306. Vojtek A, Haarer B, Field J, Gerst J, Pollard TD, et al. 1991. Evidence for a functional link between profilin and CAP in the yeast *S. cerevisiae. Cell* 66:497–505

307. Waddle JA, Karpova TS, Waterston RH, Cooper JA. 1996. Movement of cortical actin patches in yeast. *J. Cell Biol.* 132:861–70

308. Wang T, Bretscher A. 1995. The rho-GAP encoded by BEM2 regulates cytoskeletal structure in budding yeast. *Mol. Biol. Cell* 6:1011–24

309. Watanabe Y, Irie K, Matsumoto K. 1995. Yeast *RLM1* encodes a serum response factor-like protein that may function downstream of the Mpk1 (Slt2) mitogen-activated protein kinase pathway. *Mol. Cell.* 15:5740–49

310. Watanabe Y, Takaesu G, Hagiwara M, Irie K, Matsumoto K. 1997. Characterization of a serum response factor-like protein in *Saccharomyces cerevisiae*, RLM1, which has transcriptional activity regulated by the MPK1(SLT2) mitogen-activated protein kinase pathway. *Mol. Cell. Biol.* 17:2615–23

311. Watts FZ, Shiels G, Orr E. 1987. The yeast MYO1 gene encoding a myosin-like protein required for cell division. *EMBO J.* 6:3499–505

312. Wertman KF, Drubin DG, Botstein D. 1992. Systematic mutational analysis of the yeast *ACT1* gene. *Genetics* 132:337–50

313. White JHM, Green SR, Barker DG, Dumas LB, Johnston LH. 1987. The *CDC8* transcript is cell cycle regulated in yeast and is expressed with *CDC9* and *CDC21* at a point preceding histone transcription. *Exp. Cell Res.* 171:223–31

314. Whiteway M, Errede B. 1994. Signal transduction pathway for pheromone response in *Saccharomyces cerevisiae*. In *Molecular Mechanisms of Signal Transduction in Genetically Tractable Organisms*, ed. RP Dottin, J Kurjan, BL Taylor, pp. 187–236. Orlando, FL: Academic

315. Whiteway M, Hougan L, Thomas DY. 1990. Overexpression of the *STE4* gene leads to mating response in haploid *Saccharomyces cerevisiae. Mol. Cell. Biol.* 10:217–22

316. Whiteway MS, Wu C, Leeuw T, Clark K, Fourest-Lieuvin A, et al. 1995. Association of the yeast pheromone response G protein beta gamma subunits with the MAP kinase scaffold Ste5p. *Science* 269:1572–75

317. Wittenberg C, Reed SI. 1996. Plugging it in: signaling circuits and the yeast cell cycle. *Curr. Opin. Cell Biol.* 8:223–30

318. Wittenberg C, Sugimoto K, Reed SI. 1990. G1-specific cyclins of *S. cerevisiae*: cell cycle periodicity, regulation by mating pheromone, and association with p34CDC28 protein kinase. *Cell* 62:225–37

319. Wu C, Whiteway M, Thomas DY, Leberer E. 1995. Molecular characterization of Ste20p, a potential mitogen-activated protein or extracellular signal-related kinase kinase (MEK) kinase kinase from *Saccharomyces cerevisiae*. *J. Biol. Chem.* 270:15984–92

320. Yabe T, Yamada-Okabe T, Kasahara S, Furuichi Y, Nakajima T, et al. 1996. HKR1 encodes a cell surface protein that regulates both cell wall b-glucan synthesis and budding pattern in the yeast *Saccharomyces cerevisiae*. *J. Bacteriol.* 178:477–83

321. Yablonski D, Marbach I, Levitzki A. 1996. Dimerization of Ste5, a mitogen-activated protein kinase cascade scaffold protein, is required for signal transduction. *Proc. Natl. Acad. Sci. USA* 93:13864–69

322. Yamochi W, Tanaka K, Nonaka H, Maeda A, Musha T, Takai Y. 1994. Growth site localization of Rho1 small GTP-binding protein and its involvement in bud formation in *Saccharomyces cerevisiae*. *J. Cell Biol.* 125:1077–93

323. Yang S, Ayscough KR, Drubin DG. 1997. A role for the actin cytoskeleton of *Saccharomyces cerevisiae* in bipolar bud-site selection. *J. Cell Biol.* 136:111–23

324. Yorihuzi T, Ohsumi Y. 1994. *Saccharomyces cerevisiae MATa* mutant cells defective in pointed projection formation in response to α-factor at high concentrations. *Yeast* 10:579–94

325. Zahner JA, Harkins HA, Pringle JR. 1996. Genetic analysis of the bipolar pattern of bud site selection in the yeast *Saccharomyces cerevisiae*. *Mol. Cell. Biol.* 16:1857–70

326. Zarzoz P, Mazzoni C, Mann C. 1996. The SLT2(MPK1) MAP kinase is activated during periods of polarized growth in yeast. *EMBO J.* 15:83–91

327. Zhao Z-S, Leung T, Manser E, Lim L. 1995. Pheromone signalling in *Saccharomyces cerevisiae* requires the small GTP-binding protein Cdc42p and its activator *CDC24*. *Mol. Cell. Biol.* 15:5246–57

328. Zheng Y, Bender A, Cerione RA. 1995. Interactions among proteins involved in bud-site selection and bud-site assembly in *Saccharomyces cerevisiae*. *J. Biol. Chem.* 270:626–30

329. Zheng Y, Cerione R, Bender A. 1994. Control of the yeast bud-site assembly GTPase Cdc42. Catalysis of guanine nucleotide exchange by Cdc24 and stimulation of GTPase activity by Bem3. *J. Biol. Chem.* 268:24629–34

330. Zheng Y, Hart MJ, Shinjo K, Evans T, Bender A, Cerione RA. 1993. Biochemical comparisons of the *Saccharomyces cerevisiae* Bem2 and Bem3 proteins. Delineation of a limit Cdc42 GTPase-activating protein domain. *J. Biol. Chem.* 268:24629–34

331. Zhou A, Gartner A, Cade R, Ammerer G, Errede B. 1993. Pheromone-induced signal transduction in *Saccharomyces cerevisiae* requires the sequential function of three protein kinases. *Mol. Cell. Biol.* 13:2069–80

332. Ziman M, O'Brien JM, Oullette LA, Church WR, Johnson DI. 1991. Mutational analysis of *CDC42*, a *Saccharomyces cerevisiae* gene that encodes a putative GTP-binding protein involved in the control of cell polarity. *Mol. Cell. Biol.* 11:3537–44

333. Ziman M, Preuss D, Mulholland J, O'Brien JM, Botstein D, Johnson DI. 1993. Subcellular localization of Cdc42p, a *Saccharomyces cerevisiae* GTP-binding protein involved in the control of cell polarity. *Mol. Biol. Cell* 4:1307–16

Annu. Rev. Microbiol. 1998. 52:745–78

SURFACE RECEPTORS AND TRANSPORTERS OF *TRYPANOSOMA BRUCEI*

P. Borst

The Netherlands Cancer Institute, Division of Molecular Biology, Plesmanlaan 121,1066 CX Amsterdam, The Netherlands

A. H. Fairlamb

Department of Biochemistry, University of Dundee, Dundee DD1 5EH, Scotland, United Kingdom

KEY WORDS: adenylate/guanylate cyclase, antigenic variation, flagellar pocket, glucose transporter, transferrin receptors

ABSTRACT

African trypanosomes combine antigenic variation of their surface coat with the ability to take up nutrients from their mammalian hosts. Uptake of small molecules such as glucose or nucleosides is mediated by translocators hidden from host antibodies by the surface coat. The multiple glucose transporters and transporters for nucleobases and nucleosides have been characterized. Receptors for host macromolecules such as transferrin and lipoproteins are visible to antibodies but hidden from the cellular arm of the host immune system in an invagination of the trypanosome surface, the flagellar pocket. The trypanosomal transferrin receptor is a heterodimer that resembles the major component of the surface coat of *Trypanosoma brucei*. The ability to make several versions of this receptor allows *T. brucei* to bind transferrins from a range of mammals with high affinity. The proteins required for uptake of nutrients by trypanosomes provide a target for chemotherapy that remains to be fully exploited.

CONTENTS

0066-4227/98/1001-0745$08.00

INTRODUCTION

Establishment of the basic outlines of antigenic variation and metabolism created a paradox that has haunted trypanosome biology: How do African trypanosomes manage to take up host (macro)molecules by receptor-mediated endocytosis or via transporters without attracting the attention of the host immune system? The basis for antigenic variation is the surface coat, which consists of a single protein species—the variant surface glycoprotein (VSG)—covering the entire trypanosome. The occasional replacement of this coat with a new one allows a subfraction of trypanosomes to escape immune destruction. How can this strategy work if interspersed between VSG molecules are receptors that remain unaltered because they have to specifically bind a host ligand, such as transferrin (Tf) or low-density lipoprotein (LDL)? For African trypanosomes with a wide host range, such as *Trypanosoma brucei*, the problem is compounded by the rapid evolution of host macromolecules. How can a single transferrin receptor (Tf-R) take up transferrins that differ up to 30% in amino acid sequence?

This paradox has been resolved to some extent in recent years. First, all receptor-mediated uptake of macromolecules takes place in an invagination of the trypanosome surface, the flagellar pocket (136). Even if host antibodies bind to these receptors, they cannot be followed up by the cellular arm of the immune response, which recognizes the bound antibody and destroys the parasite.

Second, the trypanosome is able to vary the amino acid sequence of at least some receptors. The best-characterized one is the Tf-R, which consists of a heterodimer anchored into the membrane of the flagellar pocket. *T. brucei* can make some 20 different Tf-Rs that differ slightly in sequence and substantially in their affinity for Tf from different mammals. It seems likely that the Tf-R evolved from a VSG ancestor and that the ability to make multiple Tf-Rs has allowed *T. brucei* to colonize a large range of mammals, as high-affinity Tf binding is required to prevent host antibodies from blocking the receptor. Understanding about other surface receptors—the LDL receptor, the high-density lipoprotein (HDL) receptor, a surface-exposed adenylate cyclase, and the human haptoglobin receptor—is less advanced.

All plasma membrane proteins of *T. brucei* appear to reach the surface by vesicular transport to the flagellar pocket. VSG molecules rapidly spread over the surface, whereas the surface receptors are retained in the pocket; how that happens is a matter of speculation. It is also not known how these receptors are internalized with their cargo. Internalization appears to occur via clathrin-coated vesicles, but whether there are ways to concentrate receptor-ligand complexes in these vesicles remains unknown.

In addition to surface receptors that bind host macromolecules and that have to be surface exposed to meet their party, there are numerous plasma membrane proteins that are hidden from the host immune system by the VSG coat. Some of these are known or suspected to act as transporters of low-molecular-weight substances. The transporter best characterized is the glucose transporter, but several other invariant plasma membrane proteins have been detected.

Trypanosomes are able to constantly sample their surroundings by means of surface receptors and transporters. As a consequence of these environmental signals they are able to modify their surface structures and metabolic pathways to meet their nutritional needs and to evade elimination by their mammalian or insect hosts. The mechanisms by which these environmental signals are transduced is not well understood and is outside the scope of this article. However, homologues of signal transduction proteins in other organisms have been identified in cDNA-expressed sequence tags in *T. brucei rhodesiense* and are likely to be broadly similar to other eukaryotes, as appears to be the case with *Trypanosoma cruzi* (52).

Receptors make potential targets for antitrypanosomal therapy. Surface-exposed receptors could be used as targets for antibody-based attack or to piggyback ligand-bound drugs into the trypanosome. Metabolite transporters might be blocked by substrate analogues. Although there is no dearth of speculation about these knowledge-based routes to parasite destruction, there are only two practical examples: the haptoglobin-mediated killing of serum-sensitive *T. brucei*, and the specific uptake of organic arsenicals by adenosine transporters. In both cases the finding of the plasma membrane proteins involved was serendipitous, but the genes encoding them have not been identified. Obviously, more work needs to be done in this area.

The topic of this review has been discussed critically in two excellent, brief reviews by Overath et al (93, 96). The flagellar pocket of trypanosomes, the place where all endocytosis occurs, has been reviewed (6, 138). Reviews have also been written recently on antigenic variation and expression site associated genes (11, 21, 23, 24, 40, 48, 53, 101, 111, 132).

BACKGROUND, LIFE CYCLE, AND METABOLISM OF *T. BRUCEI*

African trypanosomes require receptors and transporters to extract from their hosts what they need for survival and multiplication. Insect and mammalian hosts provide different environments, each with its specific assortment of foodstuffs and specific anti-parasite defences that must be evaded. When *T. brucei* is taken up by the bloodsucking tsetse fly, it undergoes major genetic reprogramming to adjust its complement of receptors and transporters to insect conditions (107). Even more complex is the adjustment it undergoes in the salivary gland of the fly, where it has to await its chance to get back into a mammal when the fly bites again. On the one hand, the trypanosome has to continue utilizing what the fly offers. On the other hand, it has to anticipate its imminent transfer to the mammalian host by acquiring a VSG surface coat and the ability to switch to a glucose-based metabolism.

It should therefore be obvious that the receptor and transport complement of insect-form and bloodstream-form trypanosomes is very different. Bloodstream trypanosomes also have to cope with an efficient host immune system. If they want to take up host molecules, they have to hide the invariant receptors and transporters from immune attack for their strategy of evading the host immune system by antigenic variation to succeed.

To understand the context in which the trypanosome receptors have to function, we provide a little background knowledge about trypanosome biology, trypanosome metabolism, trypanosome coats, and antigenic variation. These topics have been reviewed in depth elsewhere, and thus we only provide the minimal outline required to understand the sections below on receptors and translocators.

A brief summary of the life cycle of *T. brucei* is given in Figure 1. Pleomorphic forms of the parasite are found in the bloodstream and tissue spaces of mammals; they range from the long slender (LS) to the short stumpy (SS) form. The LS form multiplies and has a repressed mitochondrion without cytochrome-mediated electron transport or associated ATP synthesis. The SS form does not

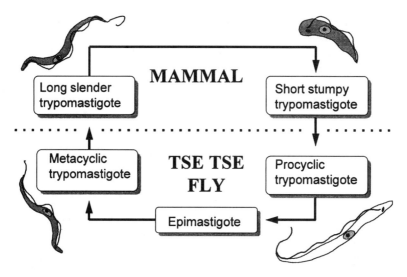

Figure 1   The life cycle of *Trypanosoma brucei*.

multiply and is preadapted to life in the insect, as shown by a partial reactivation of the synthesis of components of the mitochondrial respiratory chain. A differentiation-inducing factor (DIF) that induces LS forms of *T. brucei* to differentiate into SS forms has been detected in trypanosome culture medium. DIF has been purified. It appears to be a hydrophilic, heat-stable, non-peptide compound with a mass of less than 200 Da, but neither its structure nor its receptor is known (M Boshart, personal communication).

The major route for energy generation in bloodstream *T. brucei* is glycolysis. Glucose is oxidized to pyruvate, which is excreted. The reducing equivalents generated in glycolysis are reoxidized via a glycerolphosphate shuttle, involving a SHAM-sensitive, cytochrome-independent mitochondrial glycerol-phosphate oxidase. Under anaerobic conditions, trypanosomes can convert glucose into equimolar amounts of pyruvate and glycerol. Under aerobic conditions, glycerol can also serve as an effective substrate. One may expect that glucose, glycerol, and pyruvate, which can pass the cell membrane at high rates, will require membrane proteins to mediate transport. Glucose transporters have indeed been found and characterized and are discussed in a separate section below. No gene for a glycerol or pyruvate transporter is known. However, biochemical studies favor the presence of energy-independent permeases for glycerol (65) and pyruvate (7).

The total number of plasma-membrane–associated proteins is unknown and can only be guessed at. However, analysis of whole bacterial genomes suggests that somewhere between 9% to 12% of the total genes are known for putative transport and binding proteins (17, 28, 53, 55, 77, 129).

Likewise, in *Saccharomyces cerevisiae*, some 300 genes (5% of the total genome) have a probable transport-related function (88). Based on an estimated 8000 genes in *T. brucei* (of which 1000 are likely to be VSG genes), it is reasonable to predict as an upper limit about 400 genes for all transporter proteins. However, not all these will be located in the plasma membrane; a substantial proportion can be expected to be involved in transport of substances in and out of cellular organelles.

Bloodstream-form trypanosomes are completely covered by a dense coat of VSG. As shown in Figure 2, the VSG dimers are tightly packed and prevent complement from reaching the membrane and immunoglobulin (Ig) from seeing membrane proteins that do not penetrate far into the coat. There is sufficient space between the VSG molecules for low-molecular-weight compounds to reach transporters present in the membrane. Receptors binding host macromolecules, however, must extend far enough through the coat to bind their ligand (Figure 2). These receptors are therefore likely to be exposed to host antibodies.

When trypanosomes are taken up by a tsetse fly, they rapidly transform into insect procyclic–form trypanosomes (107). Reconstruction of this process in vitro initially suggested that it is induced by the combination of a decrease in temperature (from 37° to 25°C) and *cis*-aconitate. More recent work has shown that the *cis*-aconitate is crucial and that the temperature decrease is only required to allow the differentiated insect form trypanosomes to survive (44, 86). It would be interesting to know how the trypanosome recognizes *cis*-aconitate, as the successful induction of differentiation in blood (by a *cis*-aconitate analogue yet to be developed) should rapidly kill the resulting insect-form trypanosomes in their mammalian host. An obvious candidate for the intracellular *cis*-aconitate sensor was aconitase. This possibility has been eliminated by the

-->

Figure 2   (*A*) A schematic representation of the cell surface of bloodstream form *Trypanosoma brucei*, showing a variant surface glycoprotein (VSG) dimer (18), a transferrin (Tf) receptor molecule (113), which is thought to be only present in the flagellar pocket, and a hexose transporter (9). For size comparison, a Tf (*upper left*) and an immunoglobulin (Ig) G (*upper right*) antibody molecule are shown. The dimensions of the VSG dimer are based on the crystal structure of the N-terminal domain of the protein. The VSG is attached to the GPI-anchor via its C-terminal region; the crystal structure of this region is not yet known. The Tf receptor structure is based on its homology with the VSG dimer (see text). The hexose transporter structure is not based on any solid structural information but is derived from a hypothetical model (62) of GLUT1. [Modified from Overath et al (96).] (*B*) Three-dimensional array of VSG dimers on the plasma membrane of bloodstream-form *T. brucei*.

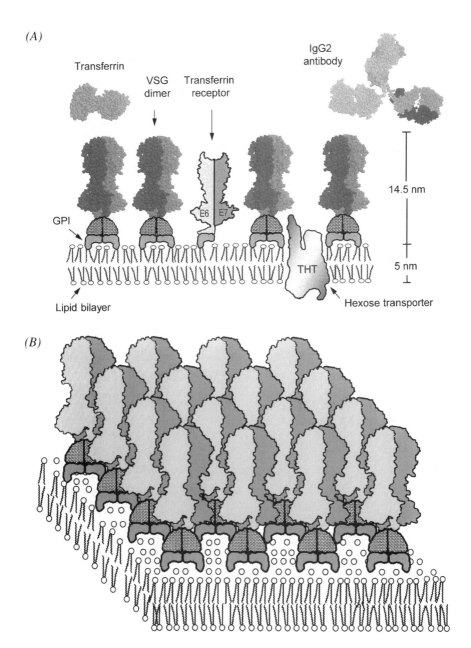

(A)

Transferrin

VSG
dimer

Transferrin
receptor

IgG2
antibody

14.5 nm

5 nm

GPI

E6 E7

THT

Lipid bilayer

Hexose transporter

(B)

surprising finding that the single gene for aconitase of *T. brucei* is dispensable (M Boshart, personal communication). The true *cis*-aconitate sensor remains to be found.

Initial work also suggested that successful differentiation requires (*a*) SS bloodstream *T. brucei* and (*b*) that LS forms will not properly differentiate into viable insect forms. More recent work has shown that this is incorrect. Even highly monomorphic stocks of laboratory strains can differentiate, but differentiation is less synchronous and it takes 24–28 h rather than the 12 h required by SS forms (19, 94).

During differentiation, trypanosomes shed their VSG coat and replace it by a coat of procyclin, also known as PARP (procyclic acidic repeat protein). Presumably the acidic repeats are extended, as indicated in Figure 3. As the PARP molecules are farther apart than the VSG molecules, larger compounds could presumably reach the membrane-embedded transporters more easily in insect-form trypanosomes than in bloodstream-form trypanosomes.

To some extent, the insect-form *T. brucei* looks more like a free-living parasite than does its bloodstream-form counterpart. Its metabolic capacities seem more diverse, and it takes up more of its building blocks in the form of small compounds. Endocytosis is thought to be less important in insect-form than

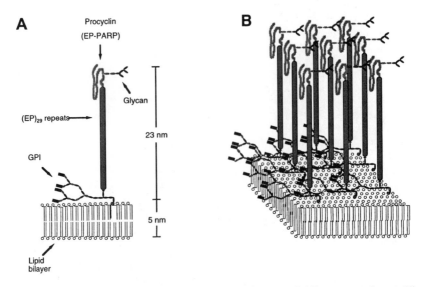

Figure 3   A simple model of the cell surface of insect-form (procyclic) *Trypanosoma brucei*. (*A*) Detailed structure of a single procyclin [EP-PARP (procyclic acidic repeat protein)] molecule, based on the predicted dimensions of the (EP)29 polyanionic rod-like domain. (*B*) Three-dimensional array of procyclin molecules, based on a calculated packing density. (Adapted from 87.)

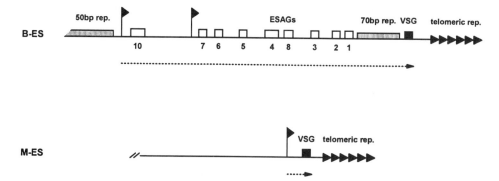

Figure 4 Variant surface glycoprotein (VSG) gene expression sites of bloodstream-form *T. brucei*. The standard bloodstream-form expression site (B-ES) is modeled after that of Revelard et al (105), with additional information on the duplicated promoter and ESAG10 taken from Gottesdiener et al (60, 61). Note that only part of the B-ES contain this promoter duplication and ESAG10. The metacyclic ES (M-ES) is redrawn from that by Rudenko et al (111).

bloodstream-form trypanosomes (see 137), and no indispensable substrates taken up by endocytosis have been identified.

After multiplying in the fly gut to a level of about 300,000 trypanosomes per fly, part of the trypanosomes make their way to the salivary gland, where they ultimately exchange their procyclin coat for a VSG coat and preadapt for entry into the mammal. The VSG coats produced in a single fly are heterogeneous and consist of 20–30 different variant antigenic types (VATs), the metacyclic repertoire (11, 12). It has been argued that this heterogeneity increases the chance that a given trypanosome stock will successfully infect the mammal because, in the wild, potential hosts are already infected and producing antibodies against many different VSGs. Metacyclic VSGs are encoded in specialized telomeric expression sites (Figure 4), each VSG in a separate site. These sites are shut down 4–5 days after mammalian infection, and a standard bloodstream expression site then takes over (see below). In the bloodstream repertoire, usually only a single site expressing a single gene is active at a time in the entire population, resulting in a single VAT predominating each wave of parasitaemia. Recent work indicates that the metacyclic expression sites evolved from bloodstream expression sites (63; JD Barry, personal communication). How these sites are differentially controlled is not known, but control is determined by the promoter sequence (11).

VSG GENE EXPRESSION SITES

VSG genes are expressed in telomeric expression sites together with some 10 expression site–associated genes (ESAGs), as schematically indicated in

Figure 4. A single expression site would suffice to express all VSG genes of the bloodstream repertoire, as each of these genes can in principle move into the expression site by duplicative transposition. It therefore came as a surprise that multiple functional expression sites exist and that trypanosomes can also switch the VSG gene expressed by silencing the active site and activating another one. The current estimate is that some 20 functional expression sites exist, each containing at least 8 ESAGs and the repeats depicted in Figure 4. Although the homology between these expression sites is high, they are far from identical. The divergence between ESAGs in different expression sites varies from <10% (ESAG6 and -7) to approximately 30% (ESAG1) (42).

Control of VSG coats is simple with a single expression site, but with 20 sites it becomes difficult to avoid mixed coats. There must be an elaborate genetic control system to keep 19 sites silent and only 1 active and to allow occasional switching from one site to the next. Such added complexity seemed hard to justify merely for greater flexibility in VSG gene switching. The discovery of ESAGs provided an alternative explanation: As the ESAG set encoded by different expression sites differed significantly, switching expression sites would allow the trypanosome to replace the current ESAGs with a different set. Because some of the proteins encoded by ESAGs had N-terminal signal sequences that might direct them to the surface (see below), it seemed possible that ESAGs encoded surface-exposed proteins involved in substrate uptake and that multiple expression sites might allow antigenic variation of these minor surface proteins. Recent studies on ESAG6 and -7, which encode subunits of the trypanosomal Tf-R, have provided evidence for this concept, as is outlined in the Tf-R section.

THE EXPRESSION SITE–ASSOCIATED GENES OF *T. BRUCEI* AND *TRYPANOSOMA EQUIPERDUM*

The discovery that the gene products of ESAG4, -6, and -7 function at the cell surface has led to speculations about possible surface functions for the other ESAG products. In this section, we review the meager evidence for this speculation. ESAG4 and ESAG6 and -7 are discussed in separate sections below.

ESAG1 was the first ESAG to be discovered (42). It has been studied intensively without yielding a clear function (31, 41), and recent work has shown that the ESAG1 in an active expression site can be disrupted without detectable phenotypic consequences (31). The interpretation of this result is complicated, however, by the presence of transcribed ESAG1 copies outside expression sites (31, 89). Antibodies raised against pESAG1 detect a membrane-associated glycosylated 46-kDa protein in extracts of bloodstream-form (but not insect-form) *T. brucei*. This protein reduces to 36 kDa, the size expected from the gene sequence, in *T. brucei* grown in tunicamycin (41). The properties of pESAG1 suggest that it might be a surface protein, but attempts to determine its cellular location have failed (41).

The presence of ESAG1 copies outside classical bloodstream repertoire expression sites, and the α-amanitin–resistant transcription of these copies, has led Morgan et al (89) to conclude that ESAG1 is a misnomer. This conclusion may be premature, however, as long as the dispersed copies have not been mapped with greater precision. On the basis of their analyses of metacyclic expression sites, JD Barry and coworkers (personal communication) have concluded that these sites were derived in evolution from bloodstream expression sites and that they still contain rudiments of some ESAGs, including ESAG1. In conclusion, the possibility remains open that pESAG1 is an essential surface protein of bloodstream-form *T. brucei* but that its synthesis has become divorced from the bloodstream-form expression sites.

Little is known about ESAG2, -3, and -5. The corresponding proteins have no homology with known sequences and have not been identified in trypanosome extracts. The putative pESAG5 (100) does not even contain clear signal peptides; pESAG2(1) and pESAG3 (1, 76) do, and because the predicted proteins also have a putative hydrophobic segment that spans the membrane, these proteins could be at the surface.

The product of ESAG8 looks like a DNA binding protein and has been located in the nucleus (83). This localization excludes the possibility that pESAG8 acts as a regulatory subunit of the adenylate cyclase encoded by ESAG4, a possibility raised by Ross et al (110). ESAG8bis is nearly identical to ESAG8 (84). Such internal duplications, resulting in extra copies of known genes, have also been found in other expression sites (75).

ESAG9 is part of a divergent gene family with six to eight copies in *T. brucei* and *T. equiperdum* (54). ESAG9 was discovered in a *T. equiperdum* VSG gene expression site, but several members of the ESAG9 gene family appear not to be linked to expression sites in *T. equiperdum* and no copy has been found in any *T. brucei* expression site. ESAG9 can code for a protein with an N-terminal signal sequence and a putative C-terminal GPI-anchor domain that might end up at the trypanosome surface. Whether this protein is actually made, where it is in the cell, and what its function is remain to be determined.

Finally, ESAG10 encodes a putative protein of 666 amino acids with 10 transmembrane segments (59). ESAG10 was recently shown to have significant homology with the BT 1 gene of *Leishmania*, which encodes a biopterin transporter responsible for biopterin uptake of *Leishmania* (43). Whether pESAG10 encodes a biopterin transporter of *T. brucei* remains to be seen.

ESAG10 is located only in *T. brucei* expression sites with a duplicated promoter (Figure 4) and is present in the 13 kb of DNA between the upstream and downstream promoter (59). About half of all promoters of *T. brucei* expression sites contain this duplication, and it may be removed by a 15-kb internal deletion during switches to another expression site (60, 61).

No differences have been detected between trypanosomes using expression sites with or without ESAG10, but this does not mean that the ESAG10 family has no function, as deletion of the duplication does not result in the removal of all ESAG10-related transcripts. Transcription of the ESAG10 family is complex and some transcripts must arise from gene copies outside the active expression site. Why *T. brucei* would find it useful to modulate the expression of ESAG10 in this complex fashion is a mystery worth probing. It is conceivable that supply of folates might differ between mammals and that a meager supply would have to be supplemented by synthesis of folates from biopterin whereas an excess would be deleterious. It is not known, however, whether trypanosomes can convert biopterin into folates or import pterins via a transporter.

A FAMILY OF PUTATIVE SURFACE ADENYLATE/GUANYLATE CYCLASES

A systematic analysis of open reading frames in the AnTat 1.3 A expression site of *T. brucei* by Pays et al (100) uncovered a putative adenylate cyclase gene, called ESAG4. A highly homologous gene was found in a *T. equiperdum* expression site (110), and both genes were shown to complement an adenylate cyclase–deficient yeast mutant (97, 110). ESAG4 is part of a large family of adenylate/guanylate cyclase genes (2, 3) with at least 11 additional members outside expression sites. Several of these outsiders [called GRESAGs (genes related to ESAGs)] are transcribed both in insect- and bloodstream-form trypanosomes.

Hydropathy plots and comparison with cyclases of other organisms suggest that the *T. brucei* cyclases contain a highly conserved intracellular catalytic domain of about 350 amino acids, a single transmembrane domain, and a large, less-conserved N-terminal, presumably extracellular, domain of 900 amino acids. Pays and co-workers have speculated that this extracellular domain has a receptor function and may transduce signals from the host (2). The nature of these signals remains unknown.

Early work by Voorheis & Martin (133, 134) that was extended by Rolin et al (108) showed that basal adenylate cyclase activity in *T. brucei* is low and that it is stimulated by Ca^{2+} in bloodstream-form but not procyclic-form trypanosomes. The expression site–derived pESAG4 is probably responsible for this, as ESAG4 expressed in yeast yields an enzyme stimulated by Ca^{2+} ions and GRESAG 4.1 does not. An antiserum made against a fusion protein containing part of the N-terminal domain of pESAG4 stained only the flagellum, both in bloodstream-form and insect-form trypanosomes (3). Because the latter does not contain pESAG4 itself, the antiserum must have cross-reacted with some of the pGRESAGs.

Interestingly, the antiserum was able to stain live trypanosomes, which suggests that the bulky extracellular domain of pESAG4 (and possibly pGRESAGs) sticks through the VSG coat. This would allow these enzymes to interact not only with bulky ligands that cannot penetrate the coat, but also with host antibodies. It is therefore possible that the presence of ESAG4 encoding the predominant cyclase of bloodstream-form trypanosomes in VSG gene expression sites allows the trypanosome to make a range of cyclases that are optimized either to bind ligands of specific mammalian hosts or to allow antigenic variation of the exposed part to diminish antibody attack. This concept is discussed in more detail below.

THE LOW-DENSITY LIPOPROTEIN RECEPTOR OF *T. BRUCEI*

Bloodstream-form *T. brucei* takes up radiolabeled mammalian LDL at a rate 100- to 1000-fold higher than serum albumin. Uptake is saturable and can be specifically inhibited by cold LDL. Gold-labeled LDL only binds in the flagellar pocket and rapidly appears in endocytic structures (39). These results strongly indicate that uptake of LDL by *T. brucei* is mediated by a specific LDL receptor (LDL-R). From a quantitative analysis of LDL binding, two types of receptor were inferred: a high-affinity receptor with a K_d of 6 nM, present in about 1800 copies, and a low-affinity receptor with a K_d of 250 nM, present in 52,000 copies (38). Overath et al (95) have pointed out, however, that 50,000 copies of a receptor would cover the entire surface of the flagellar pocket, which raises the question of whether the low-affinity binding is related to the LDL-R at all. In subsequent papers, Courtoy and co-workers purified and characterized proteins thought to be the receptor (13, 36–38) but failed to clinch the issue by cloning the gene for the putative receptor.

Purification of the LDL-binding protein initially yielded a 1000-fold purified 86-kDa protein (38), later interpreted to be a fragment of the intact LDL-R of 145 kDa (36, 37). Antibodies against the 86-kDa fragment inhibited binding of LDL at 4°C by about 90% and significantly slowed growth of *T. brucei* in vitro (38). However, immunization of mice with the 145-kDa protein did not affect the course of a *T. brucei* infection (13).

In subsequent work, Bastin et al (13) isolated a 145-kDa membrane protein from insect-form *T. brucei*, from other African trypanosomes, from *Leishmania*, and even from Kinetoplastida that never see a mammal, like the insect trypanosome *Crithidia*, and the plant trypanosome *Phytomonas*. As the isolation was based on immunoaffinity chromatography using antibodies against the *T. brucei* 145-kDa protein, the 145-kDa proteins of all these diverse Kinetoplastida must share conserved sequences. Moreover, all these Kinetoplastida were found to bind human LDL, albeit *Crithidia* fourfold less than *T. brucei*. These results were so unexpected that Bastin et al (13) metabolically labeled the 145-kDa protein of *Crithidia* to show that it was made in the organism and was not a contaminant of the culture medium.

Notwithstanding these controls, we find it hard to believe that a high-affinity LDL-R of *T. brucei* is conserved in *Crithidia* (estimated to have diverged from a common ancestor 300 million years ago) and in *Phytomonas*. The K_d of 6 nM for LDL binding to the high-affinity LDL-R of *T. brucei* implies a precise fit and such a fit is not accidentally conserved in evolution. We therefore think that the 145-kDa protein is not the LDL-R but an abundant, highly conserved protein that happens to copurify with the LDL-R. If there are only 1800 high-affinity LDL-R molecules per trypanosome, a 1000-fold purified preparation is still very impure, and one could argue that the prominent 86-kDa or 145-kDa band cannot be the LDL-R. All analyses on the putative LDL-R have employed polyclonal antibodies rather than monoclonals, and if such antibodies are raised against protein mixtures, they will yield no information about the relevant component in the mixture. Additional experiments are required to identify the true LDL-R in our opinion.

A PUTATIVE HIGH-DENSITY LIPOPROTEIN RECEPTOR

An interesting cDNA, encoding a cysteine-rich cell surface protein located in the flagellar pocket of *T. brucei*, was accidentally picked up by Lee et al (79). The predicted size of this protein is 945 amino acids with a putative N-terminal signal sequence, a large hydrophilic domain containing mainly cysteine-rich 12-amino-acid repeats, a single transmembrane segment, and a putative intracellular domain of 44 amino acids. Expression of a single diploid gene for this cysteine-rich acidic integral membrane protein (CRAM) is about fivefold higher in insect-form than bloodstream-form trypanosomes. A related gene was detectable in other *Trypanosome* species, in *Leishmania*, and in *Crithidia*.

The deduced amino acid sequence of CRAM suggests that it is a conserved receptor for substrates presented to *T. brucei* throughout its life cycle and to its distant relatives as well. Lee et al (79) speculate that it might be the HDL receptor (HDL-R) of *T. brucei*, on the basis of two arguments: First, they find significant homology between the CRAM repeats and the class A repeats of the mammalian LDL-R; second, they find active uptake of HDL by insect-form *T. brucei*, as also found by Black & Vandeweerd (16), whereas uptake by bloodstream-form *T. brucei* had been shown already by Hajduk et al (69).

As CRAM mRNA levels are higher in insect-form than bloodstream-form try-panosomes, Lee et al (79) thought it more likely that the protein is involved in HDL rather than LDL uptake. Experimental evidence for this speculation is incomplete, and it is not clear whether the blood meal in the tsetse fly stomach retains sufficient HDL long enough to make it worthwhile for trypanosomes to have a HDL-R in that stage of the life cycle.

Recent work has shown that the cytoplasmic extension of CRAM is required and sufficient to retain proteins in the flagellar pocket. It seems likely that the extension somehow hooks on to the underlying cytoskeleton, and a search for proteins interacting with the cytoplasmic extension is under way (B Zeng, Y Zhang, MG-S Lee, personal communication).

THE PUTATIVE RECEPTOR FOR TRYPANOSOME LYTIC FACTOR

The serum of humans and related primates contains a factor able to lyse *T. brucei* brucei (see 68). Early work showed this factor, called trypanosome lytic factor (TLF), to be associated with the HDL fraction of human serum (106). Smith et al (121) subsequently found that haptoglobin-related protein and paraoxenase-arylesterase were the components in the HDL subfraction with TLF activity. TLF appears to kill *T. brucei brucei* by generating peroxides. Raper et al (103) reported a second TLF activity in human serum not associated with HDL, but recent work has shown that haptoglobin-related protein is the active ingredient of this TLF2 as well.

Hager et al (67) found that TLF binds to the flagellar pocket wall of *T. brucei brucei*. Binding is saturable and lysis can be inhibited by competition with heat-inactivated TLF, which suggests that TLF binds to a specific receptor. Binding is followed by endocytosis and lysosomal targeting, which appears to result in disruption of lysosomes and death by autodigestion (67). The current hypothesis is that the haptoglobin-related protein binds hemoglobin and that the peroxidase activity of this complex at acidic pH in lysosomes leads to membrane damage. Serum-resistant *T. brucei* variants—i.e. *T. brucei rhodesiense* and *T. brucei gambiense*—bind TLF to the same extent as do sensitive variants but they do not internalize TLF (66). Competition experiments suggest that TLF does not use the putative HDL receptor to gain entry into the trypanosome.

MISCELLANEOUS RECEPTORS

The Putative EGF Receptor

Experiments interpreted to show the presence of a homologue of the mammalian EGF (epidermal growth factor) receptor on the surface of both insect-form and bloodstream-form trypanosomes were reported by Hide et al (71). All evidence was indirect, the site of binding of EGF and the number of binding sites were not presented, the anti-receptor antibodies were not shown to compete with EGF for the receptor, and the putative receptor was not purified. The initial paper by Hide et al (71) has not been followed up, and the gene for the putative receptor has not been cloned. We find it implausible that insect-form *T. brucei* would display a homologue of the mammalian EGF receptor on its surface, and we find the evidence that it does unconvincing.

The Putative Interferon Gamma Receptor

Olsson and coworkers have described a T-lymphocyte triggering factor (TLTF), which stimulates lymphocytes to proliferate and secrete interferon gamma (4, 5, 91), which in turn was thought to act as a growth factor for *T. brucei* multiplication. The gene for TLTF has now been cloned, and TLTF has been shown to localize in small vesicles near the flagellar pocket, in line with the idea that TLTF is secreted and serves to modulate the immune response of the host (131). However, several labs have failed to reproduce the postulated effect of interferon gamma on *T. brucei* multiplication, and no detectable effect of knocking out the murine interferon gamma gene has been found on *T. brucei* infection in mice (JE Donelson, personal communication). Thus, the idea that *T. brucei* should have a receptor for interferon gamma has lost its basis.

SUGAR TRANSPORTERS

T. brucei can metabolize glucose at high rates, especially the bloodstream form, which relies heavily on glucose as its main substrate for metabolic energy generation. Hence, it is not surprising that glucose transporters are present in *T. brucei* to facilitate glucose uptake. In 1992, Bringaud & Baltz cloned a potential hexose transporter gene expressed predominantly in the bloodstream form of *T. brucei* (25). This transporter gene was found to be part of a multi-gene family (26, 27) composed of two groups of tandemly repeated genes, THT1 (six copies) and THT2 (five copies). THT1 and -2 encode proteins that are 80% identical. Bloodstream-form trypanosomes express 40-fold more THT1 than THT2 and the THT1-encoded protein has the property of the glucose transporter present in this life cycle stage. The THT1 genes are not expressed in insect-form *T. brucei*; it expresses only THT2 (8, 26). Similar hexose transporters are present in *Trypanosoma vivax* (135) and *T. cruzi* (127).

Heterologous expression in *Xenopus* oocytes (26) and purification/reconstitution studies (119, 120) of THT1 indicated that transport of D-glucose is by facilitated diffusion rather than Na^+-dependent active transport, as was suggested by Munoz-Antonia et al (90). In addition to D-glucose, THT1 can also transport D-fructose and D-mannose (26, 57, 119), indicating that THT1 is a bona fide low-affinity ($K_m > 0.5$ mM) hexose transporter. The earlier observation that glycerol was also a substrate for THT1 was subsequently shown to be an experimental artefact (65).

Studies by Barrett et al (8) on the THT2 transporter expressed in heterologous systems suggest that this protein corresponds to the high-affinity ($K_m < 0.15$ mM) transporter characterized in the procyclic stage of the life cycle (90, 98). The question of whether THT2 is an active or a facilitated transporter remains to be resolved (8).

The proteins encoded by THT1 and THT2 are thought to be 527 and 529 amino acids long, respectively, and contain the 12 transmembrane segments also present in other hexose transporters. There are no major exoplasmic loops and it is likely that the protein is hidden from antibodies in the VSG coat, however to our knowledge this has not been tested. A more-detailed comparative review of the biochemical and structural properties of glucose transporters in the trypanosomatids has recently been published (126).

The question of how Kinetoplastida are able to display surface proteins only on part of their surface has been clarified to some extent by an elegant study of the developmentally regulated glucose transporter Pro-1 of *Leishmania enriettii* (122). This transporter exists in two forms: iso-1, which localizes to the flagellum and flagellar pocket; and iso-2, which localizes to the remainder of the surface. The two isoforms only differ in their (cytosolic) N terminus. Using chimeric constructs, a flagellar localization signal was localized within the N-terminal segment of iso-1. Conversely, whereas iso-2 attaches to a detergent-extracted cytoskeletal fraction, iso-1 does not. These results show that the exact surface location of plasma membrane proteins in Kinetoplastida is dependent on the interaction of these proteins with the underlying cytoskeleton. Differential targeting to the flagellum or to the parasite body occurs for other kinetoplastid proteins, e.g. the D2 hexose transporter and the myo-inositol/H^+ symporter of *Leishmania* (see 115) and the adenylate cyclase encoded by ESAG4 of *T. brucei* (3) all localize to the flagellum. The functional significance of this peculiar surface location is not known.

NUCLEOBASE AND NUCLEOSIDE TRANSPORTERS

All trypanosomatids are purine auxotrophs and consequently must salvage exogenous purines for nucleic acid synthesis. Although the purine salvage pathways have been extensively studied in *T. brucei*, the transporters responsible for uptake had received scant attention until recently. The stimulus for renewed interest came from the discovery that uptake of melaminophenyl arsenical drugs is mediated by an unusual adenosine transporter (33). Bloodstream

T. brucei contains two high-affinity nucleoside transporters capable of scavenging the low micromolar concentrations of nucleosides present in plasma. The P1-transporter transports adenosine and is inhibited by competitive substrates such as the purine nucleosides inosine and guanosine. P2 specifically transports 6-amino purines (adenosine and adenine) and is potently inhibited in a competitive fashion by the melaminophenyl arsenical, melarsoprol. Taken together with the functional absence of P2 transport activity in a melarsoprol-resistant line, this suggests that uptake of melarsoprol and its analogues is primarily mediated by the P2-transporter (33). The functional absence or alteration in P2-transport in other arsenical-resistant lines has also been found in other isolates of *T. brucei* (117), *T. equiperdum* (10), and *Trypanosoma evansi* (109). The latter study suggested that there may be a third class of purine nucleoside transporters.

Resistance to melaminophenyl arsenicals is frequently associated with cross-resistance to the diamidine class of drugs (14, 50, 56, 102, 139). Although alteration or loss of a common intracellular target could be responsible for such cross-resistance, this seems unlikely given the disparate (but as yet not established) modes of action proposed for the arsenicals and diamidines. Because diamidines, like adenine or adenosine, are able to prevent the lytic action of melarsen oxide in vitro (14), this implicated the P2-transporter in the uptake of diamidines. This appears to be broadly correct because pentamidine inhibits P2-mediated transport of [^{3}H]adenosine and, conversely, [^{3}H]pentamidine transport is strongly inhibited by adenine and melarsoprol. However, inhibition by adenosine is less marked than would be expected if the mechanism involved simple cross-competitive inhibition (32). The melarsoprol-resistant line lacking functional P2-transport showed a 20- to 50-fold decreased rate of transport for pentamidine and accumulated lower amounts of other diamidines in vitro. Since earlier studies (45) demonstrated that other diamidines were competitive substrates for the pentamidine transporter, the principal route of uptake of diamidines appears to be via the P2-adenosine transporter. Whether the residual uptake of diamidines (and presumably melaminophenyl arsenicals) is due to residual P2-transport or a second route of entry has not been established.

Interestingly, in *Leishmania* spp. diamidines appear to be taken up via a putrescine transporter (29, 104). In contrast to *Leishmania* spp. and *T. cruzi* (81), bloodstream *T. brucei* does not possess quantitatively significant putrescine transport activity (51), and therefore, the mechanisms of uptake of diamidines appear to be different in these species.

Evidence has recently been presented for a transporter for S-adenosylmethione (AdoMet) in *T. brucei*, which is distinct from the P1 and P2 transporters (58). Because the K_m for this transporter is >1,000-fold higher than the concentration of AdoMet in plasma, its physiological significance for supplying purines is questionable. Moreover, the only purine analogue capable of signifi-cantly inhibiting uptake (i.e. transport and further metabolism) was sinefungin. Because this compound inhibits transmethylation reactions, the possibility that the authors measured inhibition of incorporation into protein and nucleic acid rather than inhibition of transport requires further investigation.

T. brucei also possesses a number of biochemically distinct purine-selective nucleobase transporters, termed H1–H3. Procyclic forms of *T. brucei* pos-sess a single high-affinity carrier (H1) for hypoxanthine, adenine, guanine, and xanthine (46). Several lines of evidence support the conclusion that the H1 hypoxanthine transporter is an H^+ symporter. Uptake of hypoxanthine by bloodstream forms of *T. brucei* occurs via two additional systems, H2 and H3, with affinities and substrate specificities different from each other and from the procyclic H1 transporter (47). H2, like H1, appears to be a H^+ sym-porter. The plethora of nucleobase and nucleoside transporters with a pre-ference for purines over pyrimidines probably reflects the absolute dependence of trypanosomes on purine salvage. To date, none of these genes has been cloned and sequenced.

Leishmania donovani contains two purine nucleoside transporters, NT1 pref-erentially transporting adenosine and pyrimidine nucleosides and blocked by tubercidin, and NT2 preferentially transporting inosine and guanosine and in-hibited by formycin B. By complementation of *Leishmania* mutants with a de-fective NT1 or NT2 transporter, genes were recently cloned for both transporters (S Landfear, NS Carter, B Ullman, personal communication). The LdNT1 gene encodes a 11-*trans*-membrane protein that is 32% identical to the human adeno-sine transporter. Genomic mapping suggests that LdNT1 is part of a gene family; the other members remain to be cloned. The transporter encoded by LdNT1 ap-pears to be highly adenosine specific and is hardly inhibited by pyrimidine nucleosides, in contrast to NT1-type transport in wild-type *L. donovani*. This suggests that the other member(s) of the NT1 transporter gene family must encode the major transporter that has a high affinity for pyrimidine nucleo-sides. A gene for LdNT2, unlinked to LdNT1, has been cloned but not yet sequenced.

DRUG TRANSPORTERS OF THE ABC TYPE

P-glycoproteins are prominent representatives of a family of adenine nucleotide binding cassette (ABC) transporters that is widely represented in protozoa, fungi, and metazoa (62). Six representatives have been detected in *Leishmania* (22, 80): One is PgpA (92), which belongs to a subclass of ABC transporters known as GS-X pumps (73) because they transport compounds conjugated or complexed with glutathione (GSH) or trypanothione. PgpA is involved in resistance to arsenite and antimonite, probably because it transports the oxyanions complexed to three GSH or trypanothione molecules (64). Recent experiments suggest, however, that PgpA may be located in the vacuolar membrane rather than the cell membrane and that resistance is due to segregation of the oxyanion in the vacuole (M Ouellette, personal communication). Not all arsenite transport in *Leishmania* occurs by GS-X pumps, however. There is evidence for a 37-kDa protein that requires thiols for maximal activity, but it appears to transport free arsenite ions (B Rosen, personal communication).

A second Pgp of *Leishmania*, encoded by the ldmdr1 gene, is involved in a multi-drug–resistant phenotype that is remarkably similar to that caused by overexpression of the human MDR1 Pgp (70, 130). Recent experiments suggest that this protein is not in the cell membrane but may be (largely) mitochondrial (35). There are at least four other Pgp genes in *Leishmania*, but the function and cellular location of the corresponding proteins are not yet known (22).

INVARIANT SURFACE GLYCOPROTEINS IN SEARCH OF A FUNCTION

A systematic search for invariant proteins that can be derivatized by surface labeling with biotin or iodine yielded invariant surface glycoproteins (ISGs) 60, 65, and 75 (140–143) and ISGs 64 and 70 (74). ISGs 70 and 65 are identical (93). ISGs 60, 64, and 65 are present in 50,000–70,000 copies per cell, i.e. at about 0, 5% of the VSG copy number, but at much higher levels than the receptors confined to the flagellar pocket. Indeed, ISGs 65 and 75 are spread over the entire surface of *T. brucei*. Both proteins are predicted to have a large N-terminal glycosylated extracellular domain, a single transmembrane segment, and a small C-terminal intracellular domain. All these ISGs lack homology with known proteins and their function is unknown. They look more like signal transducers than channels or transporters. Since the N-terminal domain of these ISGs is inaccessible to antibody or protease and thought to be embedded in the membrane-proximal part of the VSG coat, similar to the glucose transporter depicted in Figure 2, they might act as receptors for small ligands rather than macromolecules.

THE TRANSFERRIN RECEPTOR OF *T. BRUCEI*

The first evidence for a transferrin receptor (Tf-R) in bloodstream-form trypanosome was obtained by Coppens et al (39), who showed that transferrin (Tf) uptake was rapid, saturable, blocked by cold Tf, sensitive to trypsin, and limited to the flagellar pocket. Schell et al purified a Tf-binding protein from *T. brucei* and made the important observation that a major component of their preparation was encoded by ESAG6 (115). This linked a putative receptor only made in bloodstream-form and not in procyclic-form *T. brucei* to the expression site for the bloodstream repertoire of VSG genes and provided the first well-defined function for an ESAG. pESAG6 was not the only protein required for Tf binding. This became clear when all attempts to reconstruct Tf binding activity in procyclic-form *T. brucei* failed (see 82). The ensuing collaboration between the Overath and Borst labs led to the demonstration that the purified Tf binding protein contains two proteins in equimolar amounts, pESAG6 and pESAG7 (116). There is now ample evidence that a heterodimer consisting of pESAG6 and -7 is necessary and sufficient for binding a single Tf molecule (124, 125), as Tf binding has been reconstructed by transfection of ESAG6 and -7 constructs into procyclic-form *T. brucei* (82), insect cells (34), and *Xenopus* eggs (112). Although pESAG6 can homodimerize, only heterodimers bind Tf (112), despite the fact that pESAG6 and -7 are more than 90% identical in amino acid sequence. pESAG6 has a GPI-anchor and pESAG7 does not and remains in the complex by simply holding on to pESAG6. The position of the anchor in the heterodimer is not critical, as a heterodimer of pESAG6 (with or without anchor) and pESAG7 provided with the pESAG6 C terminus and anchor will bind Tf (112). However, both pESAG6 and -7 sequences are required to bind Tf (34, 82, 112, 124), and the binding affinity for Tf from different mammals is critically dependent on the sequence of both ESAG6 and ESAG7 (21), showing that both subunits contribute binding specificity.

With the discovery of the Tf-R, the uptake of iron by bloodstream-form trypanosomes is known in outline. Tf is an essential growth factor for *T. brucei* (114), and attempts to replace Tf by other sources of iron have failed (W Bitter, P Borst, unpublished data). The Tf-R encoded by ESAG6 and -7 seems essential for Tf uptake, as Fab fragments from anti-Tf-R antibodies can inhibit Tf uptake by 90% (125). Gene disruption experiments have given ambiguous results because of the tendency of *T. brucei* to partially activate "silent" VSG gene expression sites, allowing some pESAG6 and -7 synthesis to occur even though pESAG6 or -7 in the "active" expression site is disrupted (W Bitter, P Borst, unpublished data; Clayton et al, submitted for publication). The number of Tf molecules bound per Tf-R was initially estimated to be two (112); but subsequent papers present conclusive evidence that it is only one (21, 82, 125). The number of receptors per trypanosome was initially thought to be approximately 1000 (115), but a more accurate estimate is approximately 3000 (125). A 10-fold higher value was reported by Salmon et al (112), but this value is a priori implausible (as this high number of receptors could occupy half the flagellar pocket wall), and it was based on inaccurate Tf binding curves obtained with a Tf-R that binds bovine Tf weakly (cf 21, 125).

Tf-R is only attached to the plasma membrane by a GPI-anchor, like VSG, but unlike VSG it does not appear to spread over the entire surface, remaining instead in the flagellar pocket. What retains it there is unclear. It can move out of the pocket, as Tf-R reconstituted in insect-form trypanosomes can be detected all over the surface (82). It is therefore possible that the Tf-R interacts with a transmembrane protein kept in the pocket by cytoskeletal interactions and only made in bloodstream-form trypanosomes (82). It should be noted, however, that the low number of Tf-R molecules per cell makes it difficult to exclude the possibility that some Tf-R spreads over the surface, but remains undetected, because the surface density is very low, as pointed out by Steverding et al (124). It remains therefore possible that no special device exists to trap Tf-R in the pocket beyond ligand interaction and high rates of endocytosis. The spreading of the Tf-R over the surface of insect-form trypanosomes could then be due to the low rate of endocytosis in this stage of the life cycle.

How Tf-R is internalized is also still unclear. Electron micrographs of try-panosomes immunolabeled with anti-Tf-RYIg predominantly show gold par-ticles over the flagellar pocket lumen and only weak labeling of the flagellar pocket membrane and membranes of intracellular endocytotic structures. This has raised the question of whether a free Tf-R-Tf complex is an intermediate in Tf uptake. It is unlikely that this complex would be taken up by fluid en-docytosis, because the rates of fluid endocytosis are too low (39), so a second receptor specifically recognizing the Tf-R-Tf complex would be required for rapid uptake. This would also solve the problem of how a GPI-anchored protein such as the Tf-R can find its way into clathrin-coated vesicles for rapid endo-cytosis. A single transmembrane protein could thus be responsible for keeping the unliganded Tf-R in the pocket as well as for the capture and endocytosis of the liganded receptor. It should be noted, however, that there is a curious discrepancy between the Tf-R seen in the pocket and that found soluble in frac-tionation experiments. Whereas the bulk of the Tf-R is not in the membrane in electronmicrographs, only 26% was found to be soluble (124). As long as this discrepancy has not been explained, it remains uncertain whether a soluble Tf-R-Tf complex is an intermediate in Tf uptake.

After uptake, the Tf-R-Tf complex is routed via endosomes to lysosomes, where Tf is degraded, releasing the iron contained in it. In view of its metabolic stability, the Tf-R probably recycles back to the surface (125). How iron uptake from the endosome and its subsequent storage is regulated remains unclear. Excess iron is toxic, and trypanosomes should have mechanisms to avoid iron overload. Control of cellular iron concentration does not seem to reside at the level of Tf-R concentration, as constant Tf-R levels have been found in cultured *T. brucei* independent of the level of Tf supply (H Gerrits, P Borst, unpublished data). A plausible mechanism is that excess iron is stored in a storage organelle bound to protein. A distinctive Fe-containing organelle with a diameter of about 400 nm has been detected by electron probe microanalysis in *T. cruzi* (118). How the iron is stored in and recruited from these organelles is an interesting question for future research.

WHY MULTIPLE TRANSFERRIN RECEPTORS?

As there are about 20 VSG gene expression sites, each with at least one copy of ESAG6 and ESAG7 (see Figure 4), *T. brucei* can, in theory, make 20 Tf-Rs. That it could make more is unlikely because results of a detailed study suggests that all copies of ESAG6 and -7 are linked to VSG gene expression site sequences (M Ligtenberg, P Borst, unpublished data). The Tf-Rs specified by different expression sites differ in an interesting fashion, as first noted by Zomerdijk et al (144), who found a hypervariable stretch of 21 bp in both subunit genes that looked as if it had been selected for variation at the protein level. Subsequent work has shown that the C-terminal half of both ESAG6 and -7 contain other significant differences between the genes from different expression sites (21, 113), and it is now even doubtful whether the hypervariable stretch noted by Zomerdijk et al (144) represents the most significant functional difference between Tf-Rs encoded by different expression sites (see below). Nevertheless, the hypervariable loop focused attention on the possibility that the Tf-R might undergo antigenic variation (20). Indeed, antibodies against the Tf-R arose during a chronic infection of a rabbit with *T. brucei* (72), and it was surmised that the trypanosome had developed multiple versions of the Tf-R and the ability to switch from one version to the next by switching VSG gene expression sites, only to obtain temporary relief from antibody pressure (20). This hypothesis proved to be an oversimplification, as all attempts to block multiplication of *T. brucei* with antibodies to Tf-R either in vivo or in vitro failed completely (15, 21, 125). This was due to the tight binding of Tf to its receptor, which was apparently able to mask all surface epitopes of the Tf-R. As Tf is present in very high concentrations in mammalian serum, physiological concentrations of Ig were unable to compete with the vast excess of tightly bound ligand (15, 21). These negative results appeared to substantiate the prediction that "we should anticipate that trypanosomes may outsmart us once again" (95).

One problem remained, however: The sequence of Tf evolves rapidly in mammals, human, and pig Tf, differing by as much as 28% in amino acid identity (see 21). How does a parasite, such as *T. brucei*, which infects a very large range of mammals, bind Tf from each of these hosts sufficiently tightly to be able to avoid recognition by host antibodies? It seemed obvious that the diversity of Tf-Rs might provide the trypanosome with the tools to solve this problem (20), and all available evidence now supports this hypothesis (15, 21).

First, there is a large difference between the binding constants for bovine Tf found for Tf-Rs encoded by different VSG gene expression sites (ESs), varying between 108–169 nM (125) or 830 nM (112) for the AnTat 1.3 A expression site (ES) on the one hand and 3–4 nM for the 427 ES 118 (or dominant ES) and ES 221 (15, 125) on the other. Interestingly, the Tf-Rs specified by the 1.3 A ES and the 221 ES differ only in 1 amino acid in their pESAG7 subunit and 4 in their pESAG6 subunit, as pointed out by Borst et al (21). This shows that very minor sequence changes can lead to a more than 20-fold change in K_d.

In addition, Bitter et al (15) found that neither the 118-ES version of the *T. brucei* Tf-R nor the 221-ES version detectably bind canine ES. Growth of *T. brucei* expressing the 221 ES in vitro in medium only containing canine serum results in sluggish growth and eventual selection of trypanosomes that have switched to the V02 ES, which encodes a Tf-R binding canine Tf. Switching is prevented by adding minute amounts of bovine Tf to the growth medium.

These results support the hypothesis that *T. brucei* has evolved multiple alternative versions of its Tf-R to extend its host range. In view of the very high concentrations of Tf present in mammalian blood (25–50 μM), a single receptor might still be adequate to bind sufficient Tf from a range of mammals, but weak Tf binding will result in increased exposure of Tf-R to the immune system with the concomitant risk of complement activation or receptors clogged with Ig and unable to bind Tf. Although this hypothesis is plausible, it needs verification under field conditions, as pointed out by Borst et al (21).

THE TRANSFERRIN RECEPTOR RESEMBLES A VSG DIMER

As the Tf-R is embedded in the VSG coat covering the flagellar pocket, the pESAG6 and -7 subunits of the receptor must be elongated molecules to be able to fit in the coat and bind Tf. Indeed, inspection of the pESAG6/7 sequence suggested the presence of long heptad repeats similar to those that form the backbone of the elongated VSG structure (21, 30). This raised the question of whether the Tf-R heterodimer and the VSG homodimer have a common ancestor. Interestingly, Hobbs & Boothroyd (72) noted already at an early stage that their BS1 (ESAG6) clone specified a protein with significant homology to a VSG. In view of the rapid evolution of VSGs, it seems unlikely that the Tf-R of today would still resemble the VSG gene from which it originated. However, in view of the high rates of segmental gene conversion observed in VSG genes (99, 128) and *T. brucei* housekeeping genes (78), it seems plausible that proteins with similar folding and lateral interaction properties, such as VSG and Tf-R, may have continued to occasionally exchange equivalent segments, keeping the evolutionary origin of the Tf-R visible by database search.

The problem of Tf-R origin and homology to VSG dimers has been analyzed in detail in an elegant recent study by Salmon et al (113). They noted that N-terminal domain of VSG MiTat 1.5 showed an even higher homology to pESAG6 than the MVAT4 studied by Hobbs & Boothroyd (72). Of the 81 amino acids playing a key role in the tertiary structure of the 357-amino acid N-terminal domain of VSG MiTat 1.5 (18, 30), 80% are conserved in pESAG6. In fact, pESAG6 can be modeled to fit exactly the elongated N-terminal domain of VSGs (113).

To test this model experimentally, Salmon et al (113) made use of the *Xenopus* oocyte system. They had previously shown that injection of a mixture of ESAG6 and -7 mRNAs into these oocytes results in the synthesis of a functional Tf-R displayed at the oocyte surface (112). As the mRNA can be synthesized in vitro from plasmid templates, it is relatively simple to produce mutant Tf-R from altered genes. This allowed Salmon et al (113) not only to verify that the pESAG6/7 heterodimer folds like the VSG N-terminal domain, but also to locate the sequences in the molecule involved in binding Tf. Mutations in ESAG6 corresponding to the most-exposed surface loops in VSGs [loops l and m in the VSG structure of Blum et al (18)] indeed had a profound effect on Tf binding, a single V → P substitution at position 233, for instance, increasing the affinity for Tf of the AnTat 1.3A Tf-R 12-fold. Two mutations in the hypervariable region described by Zomerdijk et al (144) had little effect on Tf binding. This region is located underneath the most-exposed surface loops in the structure model of pESAG6/7 (113). Its role in Tf binding is doubtful, and why it varies so much between the pESAG6/7 encoded in different ESs is unclear.

Further evidence for the close functional resemblance of pESAG6/7 and a VSG dimer was obtained by the construction of chimeric proteins. Replacement of the N-terminal 188 amino acids of pESAG6 and -7 by the N-terminal 185 amino acids of VSG MiTat 1.5 resulted in Tf-R with unaltered binding affinity for Tf. This experiment confirms that all residues involved in Tf binding are located in the C-terminal half of pESAG6 and -7, as inferred from the amino acid replacement experiments, and strongly supports the notion that the pESAG6/7 heterodimer folds like a VSG dimer.

Although these results show that the Tf-R is closely related to the VSG coat dimers in which it resides, the evolutionary history of the Tf-R is still a matter of speculation. Whereas pESAG6 is around 400 amino acids, most VSGs are about 50–100 amino acids longer (30). The difference comes from a C-terminal domain of 50–100 amino acids present in VSGs and absent in pESAG6. The structure of this domain is not known, but Carrington & Boothroyd (30) have speculated that it may raise the N-terminal part a bit farther off the cell surface. This does not seem to be an essential feature of coat protein molecules, however, as the VSGs of *Trypanosoma congolense*, a trypanosome species closely related to *T. brucei*, has a much shorter C-terminal domain. It is therefore possible that an ancestral VSG gene with a shorter C-terminal domain gave rise to the Tf-R (113). As it now appears that trypanosomes infecting vertebrates were present early in vertebrate evolution (85), another possibility is that the Tf-R evolved rather late from fully established VSG molecules.

HOW DO AFRICAN TRYPANOSOMES COMBINE ANTIGENIC VARIATION WITH FOOD UPTAKE AND OTHER USEFUL INTERACTIONS WITH THEIR MAMMALIAN HOSTS?

Antigenic variation looks like a simple and effective method for a parasite to elude the immune system of its mammalian host. By regularly changing the surface coat of a small subfraction of the parasite population, the parasite can remain in the mammalian bloodstream to meet its insect vector taking a blood meal. In practice, this strategy is more complex than one would think, as the analysis of antigenic variation in African trypanosomes has shown. There are at least five basic requirements for making antigenic variation work: (*a*) a large repertoire of surface antigens; (*b*) a mechanism for switching the surface antigen expressed in a small subfraction of the trypanosome population, before the host has made antibodies against this antigen; (*c*) the ability to express surface antigens in a defined order to avoid population heterogeneity; (*d*) the ability to combine antigenic variation with substrate uptake (requiring invariant receptors and translocators); and (*e*) the ability to survive in multiple hosts, offering different (macromolecular) substrates. The last two requirements are especially relevant to this review, and there are now rather precise hypotheses explaining how trypanosomes meet these requirements. The translocators described thus far all seem covered by the VSG coat, although this has not been tested by direct experiment for most of them. On the one hand, the VSG molecules are far enough apart to allow rapid diffusion of small molecules toward the translocators in the plasma membrane. On the other hand, the VSG molecules are packed close enough to prevent penetration of antibodies into the coat (see Figure 2).

Receptors for macromolecular ligands have to extend far enough from the membrane to allow ligand binding. We expect each of these receptors to be targeted by host antibody. The studies on the Tf receptor have shown how the trypanosome can mitigate the attack on its receptors. The receptors are hidden from macrophages and T-cells in the flagellar pocket, and high-affinity binding of ligand prevents interference by anti-receptor antibodies. Three other mechanisms can be envisaged: The antigenic variation of the receptor, made possible by multiple copies of the Tf receptor, might in itself help to decrease antibody interference; the high rate of endocytosis of flagellar pocket membrane might continuously deplete the flagellar pocket of antibody; and the trypanosome might be able to space invariant antigens in the VSG coat to prevent bivalent (and hence high-affinity) binding of antibody. More experiments are required to test whether any of these additional mechanisms actually contribute to the defense of *T. brucei* against host antibodies.

OUTLOOK

Parasites depend on their hosts for food. Hence, the receptors and transporters required for uptake of food stuffs are obvious targets for chemo- or immunotherapy. Information on this import machinery in African trypanosomes has been slow in coming, as our review illustrates. One reason is the large evolutionary distance between trypanosomes and the organisms most often used in biochemical studies (123). This has frustrated many attempts to clone transporter genes from trypanosomes by homology with their counterparts isolated from animals, yeast, or bacteria. This frustration will be amply compensated, however, when the genes are cloned by other means or uncovered by the ongoing trypanosome genome project (49). It should be obvious that the higher the divergence between trypanosome genes and their mammalian counterparts, the better the chance will be that inhibitors can be found that inhibit parasite transporters more than host transporters.

There is another reason for optimism. It has become clear in recent years that professional parasites usually cover their metabolic needs by using multiple uptake systems. An example is purine supply, which *T. brucei* can cover by taking up any natural purine base or purine nucleoside. This explains how the organism has managed to evade the development of efficient inhibitors of purine uptake. If there are several doors to get in, blocking one door does not suffice to prevent entry. Once the location and structure of every door is known, however, it should be possible to devise knowledge-based approaches to block all entries or, conversely, to use these entries to selectively deliver toxic compounds, such as purine analogues or novel melaminophenyl- or diamidine-based drugs.

On this basis we expect that the parasite genomes now being sequenced will provide new leads for rational chemo- and immunotherapy that will eventually also benefit the treatment of African trypanosomiasis.

ACKNOWLEDGMENTS

We are grateful to Dr Mike Ferguson and Mr David Hall (Department of Biochemistry, Dundee, UK) for their advice on matters of protein structure and help with Figures 2 and 3. We thank members of the Borst and Fairlamb labs for their comments on the manuscript. The experimental work in the Borst lab is supported by grants from the Netherlands Foundation for Chemical Research (SON), with financial support of the Netherlands Organization for Scientific Research (NWO). Alan Fairlamb is supported by the Wellcome Trust.

Literature Cited

1. Alexandre S, Guyaux M, Murphy NB, Coquelet H, Pays A, et al. 1988. Putative genes of a variant-specific antigen gene transcription unit in *Trypanosoma brucei*. *Mol. Cell. Biol.* 8:2367–78
2. Alexandre S, Paindavoine P, Hanocq-Quertier J, Paturiaux-Hanocq F, Tebabi P, Pays E. 1996. Families of adenylate cyclase genes in *Trypanosoma brucei*. *Mol. Biochem. Parasitol.* 77:173–82
3. Alexandre S, Paindavoine P, Tebabi P, Pays A, Halleux S, et al. 1990. Differential expression of a family of putative adenylate/guanylate cyclase genes in *Trypanosoma brucei*. *Mol. Biochem. Parasitol.* 43:279–88
4. Bakhiet M, Olsson T, Edlund C, Höjeberg B, Holmberg K, et al. 1993. A *Trypanosoma brucei brucei*-derived factor that triggers CD8[+] lymphocytes to interferon-gamma secretion: purification, characterization and protective effects *in vivo* by treatment with a monoclonal antibody against the factor. *Scand. J. Immunol.* 37: 165–78
5. Bakhiet M, Olsson T, Mhlanga J, Büscher P, Lycke N, et al. 1996. Human and rodent interferon-gamma as a growth factor for *Trypanosoma brucei*. *Eur. J. Immunol.* 26:1359–64
6. Balber AE. 1990. The pellicle and the membrane of the flagellum, flagellar adhesion zone, and flagellar pocket: functionally discrete surface domains of the bloodstream form of African trypanosomes. *Crit. Rev. Immunol.* 10:177–201
7. Barnard J, Reynafarje B, Pedersen P. 1993. Glucose catabolism in African trypanosomes. Evidence that the terminal step is catalyzed by a pyruvate transporter capable of facilitating uptake of toxic analogs. *J. Biol. Chem.* 268:3654–61
8. Barrett MP, Tetaud E, Seyfang A, Bringaud F, Baltz T. 1995. Functional expression and characterization of the

Trypanosoma brucei procyclic glucose transporter, THT2. *Biochem. J.* 312:687–91
9. Barrett MP, Tetaud E, Seyfang A, Bringaud F, Baltz T. 1998. Trypanosome glucose transporters. *Mol. Biochem. Parasitol.* 91:195–205
10. Barrett MP, Zhang ZQ, Denise H, Giroud C, Baltz T. 1995. A diamidine-resistant *Trypanosoma equiperdum* clone contains a P2 purine transporter with reduced substrate affinity. *Mol. Biochem. Parasitol.* 73:223–29
11. Barry JD. 1997. The relative significance of mechanisms of antigenic variation in African trypanosomes. *Parasitol. Today* 13:212–17
12. Barry JD, Turner CMR. 1991. The dynamics of antigenic variation and growth of African trypanosomes. *Parasitol. Today* 7:207–11
13. Bastin P, Stephan A, Raper J, Saint-Remy J-M, Opperdoes FR, Courtoy PJ. 1996. An M_r 145000 low-density lipoprotein (LDL)-binding protein is conserved throughout the Kinetoplastida order. *Mol. Biochem. Parasitol.* 76:43–56
14. Berger BJ, Carter NS, Fairlamb AH. 1995. Characterisation of pentamidine-resistant *Trypanosoma brucei brucei*. *Mol. Biochem. Parasitol.* 69:289–98
15. Bitter W, Gerrits H, Kieft R, Borst P. 1998. How *Trypanosoma brucei* may cope with the diversity of transferrins in its mammalian hosts. *Nature* 391:499–502
16. Black S, Vandeweerd V. 1989. Serum lipoproteins are required for multiplication of *Trypanosoma brucei brucei* under axenic culture conditions. *Mol. Biochem. Parasitol.* 37:65–72
17. Blattner FR, Plunkett G, Bloch CA, Perna NT, Burland V, et al. 1997. The complete genome sequence of *Escherichia coli* K–12. *Science* 277:1453
18. Blum ML, Down JA, Gurnett AM, Carrington M, Turner MJ, Wiley DC. 1993.

A structural motif in the variant surface glycoproteins of *Trypanosoma brucei*. *Nature.* 362:603–9

19. Blundell PA, Van Leeuwen F, Brun R, Borst P. 1998. Changes in expression site control and DNA modification in *Trypanosoma brucei* during differentiation of the bloodstream form to the procyclic form. *Mol. Biochem. Parasitol.* 93:115–130

20. Borst P. 1991. Transferrin receptor, antigenic variation and the prospect of a trypanosome vaccine. *Trends Genet.* 7:307–9

21. Borst P, Bitter W, Blundell P, Cross M, McCulloch R, et al. 1997. The expression sites for variant surface glycoproteins of *Trypanosoma brucei*. In *Trypanosomiasis and Leishmaniasis: Biology and Control*, ed. G Hide, JC Mottram, GH Coombs, PH Holmes, 7:109–31. Oxford, UK: Br. Soc. Parasitol./CAB Int.

22. Borst P, Ouellette M. 1995. New mechanisms of drug resistance in parasitic protozoa. *Annu. Rev. Microbiol.* 49:427–60

23. Borst P, Rudenko G, Blundell PA, Van Leeuwen F, Cross MA, et al. 1997. Mechanisms of antigenic variation in African trypanosomes. *Behring Inst. Mitt.* 99:1–15

24. Borst P, Rudenko G, Taylor MC, Blundell PA, Van Leeuwen F, et al. 1996. Antigenic variation in trypanosomes. *Arch. Med. Res.* 27:379–88

25. Bringaud F, Baltz T. 1992. A potential hexose transporter gene expressed predominantly in the bloodstream form of *Trypanosoma brucei*. *Mol. Biochem. Parasitol.* 52:111–22

26. Bringaud F, Baltz T. 1993. Differential regulation of two distinct families of glucose transporter genes in *Trypanosoma brucei*. *Mol. Cell. Biol.* 13:1146–54

27. Bringaud F, Baltz T. 1994. African trypanosome glucose transporter genes: organization and evolution of a multigene family. *Mol. Biol. Evol.* 11:220–30

28. Bult CJ, White O, Olsen GJ, Zhou L, Fleischmann RD, et al. 1996. Complete genome sequence of the methanogenic archaeon, *methanococcus jannaschii*. *Science* 273:1058–73

29. Calonge M, Johnson R, Balana-Fouce R, Ordonez D. 1996. Effects of cationic diamidines on poyamine content and uptake on *Leishmania infantum* in *in vitro* cultures. *Biochem. Pharmacol.* 52:835–41

30. Carrington M, Boothroyd J. 1996. Implications of conserved structural motifs in disparate trypanosome surface proteins. *Mol. Biochem. Parasitol.* 81:119–26

31. Carruthers VB, Navarro M, Cross GAM. 1996. Targeted disruption of expression site-associated gene-1 in bloodstream-form *Trypanosoma brucei*. *Mol. Biochem. Parasitol.* 81:65–79

32. Carter NS, Berger BJ, Fairlamb AH. 1995. Uptake of diamidine drugs by the P2 nucleoside transporter in merarsensitive and -resistant *Trypanosoma brucei*. *J. Biol. Chem.* 270:28153–57

33. Carter NS, Fairlamb AH. 1993. Arsenical-resistant trypanosomes lack an unusual adenosine transporter. *Nature* 361:173–75

34. Chaudhri M, Steverding D, Kittelberger D, Tjia S, Overath P. 1994. Expression of a glycosylphosphatidylinositol-anchored *Trypanosoma brucei* transferrin-binding protein complex in insect cells. *Proc. Natl. Acad. Sci. USA* 91:6443–47

35. Chow LM, Srivastava IK, Vaidya A, Wirth DF. 1997. The role of LeMDR1 in the accumulation of and resistance to mitochondrial drugs in *Leishmania enriettii*. *Mol. Parasitol. Meet. VIII, Sept. 24–28. Woods Hole, MA*, p. 78 (Abstr.)

36. Coppens I, Bastin P, Courtoy PJ, Baudhuin P, Opperdoes FR. 1991. A rapid method purifies a glycoprotein of M_r 145,000 as the LDL receptor of *Trypanosoma brucei*. *Biochem. Biophys. Res. Commun.* 178:185–91

37. Coppens I, Bastin P, Opperdoes FR, Baudhuin P, Courtoy PJ. 1992. *Trypanosoma brucei brucei*: antigenic stability of its LDL receptor and immunological cross-reactivity with the LDL receptor of the mammalian host. *Exp. Parasitol.* 74:77–86

38. Coppens I, Baudhuin P, Opperdoes FR, Courtoy PJ. 1988. Receptors for the host low density lipoproteins on the hemoflagellate *Trypanosoma brucei*: purification and involvement in the growth of the parasite. *Proc. Natl. Acad. Sci. USA* 85:6753–57

39. Coppens I, Opperdoes FR, Courtoy PJ, Baudhuin P. 1987. Receptor-mediated endocytosis in the bloodstream form of *Trypanosoma brucei*. *J. Protozool.* 34:465–73

40. Cross GAM. 1996. Antigenic variation in trypanosomes: secrets surface slowly. *BioEssays* 18:283–91

41. Cully DF, Gibbs CP, Cross GAM. 1986. Identification of proteins encoded by variant surface glycoprotein expression site-associated genes in *Trypanosoma brucei*. *Mol. Biochem. Parasitol.* 21:189–97

42. Cully DF, Ip HS, Cross GAM. 1985.

Coordinate transcription of variant surface glycoprotein genes and an expression site associated gene family in *Trypanosoma brucei*. *Cell* 42:173–82

43. Cunningham ML, Moore J, Seyfang A, Landfear S, Beverley SM. 1997. Pteridine transport in *Leishmania*. *Mol. Parasitol. Meet. VIII, Sept. 24–28, Woods Hole, MA*, p. 236 (Abstr.)

44. Czichos J, Nonnengaesser C, Overath P. 1986. *Trypanosoma brucei*: cis-aconitate and temperature reduction as triggers of synchronous transformation of bloodstream to procyclic trypomastigotes in vitro. *Exp. Parasitol.* 62:283–91

45. Damper D, Patton CL. 1976. Pentamidine transport in *Trypanosoma brucei*-kinetics and specificity. *Biochem. Pharmacol.* 25:271–76

46. De Koning HP, Jarvis SM. 1997. Hypoxanthine uptake through a purine-selective nucleobase transporter in *Trypanosoma brucei brucei* procyclic cells is driven by protonmotive force. *Eur. J. Biochem.* 247:1102–10

47. De Koning HP, Jarvis SM. 1997. Purine nucleobase transport in bloodstream forms of *Trypanosoma brucei brucei* is mediated by two novel transporters. *Mol. Biochem. Parasitol.* 89:245–58

48. Donelson JE. 1995. Mechanisms of antigenic variation in *Borrelia hermsii* and African trypanosomes. *J. Biol. Chem.* 270:7783–86

49. El-Sayed NMA, Donelson JE. 1997. A survey of the *Trypanosoma brucei rhodesiense* genome using shotgun sequencing. *Mol. Biochem. Parasitol.* 84:167–78

50. Fairlamb AH, Carter NS, Cunningham M, Smith K. 1992. Characterisation of melarsen-resistant *Trypanosoma brucei brucei* with respect to cross-resistance to other drugs and trypanothione metabolism. *Mol. Biochem. Parasitol.* 53:213–22

51. Fairlamb AH, Le Quesne SA. 1997. Polyamine metabolism in trypanosomes. In *Trypanosomiasis and Leishmaniasis: Biology and Control*, ed. G Hide, JC Mottram, GH Coombs, PH Holmes, pp. 149–61. Wallingford, UK: CAB Int.

52. Flawia MM, Tellez-Inon MT, Torres HN. 1997. Signal transduction mechanisms in *Trypanosoma cruzi*. *Parasitol. Today* 13:30–33

53. Fleischmann RD, Adams M, White O, Clayton RA, Kirkness EF, et al. 1995. Whole-genome random sequencing and assembly of *Haemophilus influenze Rd*. *Science* 269:496–521

54. Florent IC, Raibaud A, Eisen H. 1991. A

family of genes related to a new expression site-associated gene in *Trypanosoma equiperdum*. *Mol. Cell Biol.* 11:2180–88

55. Fraser CM, Gocayne JD, White O, Adams M, Clayton RA, et al. 1995. The minimal gene complement of *Mycoplasma genitalium*. *Science* 270:397–403

56. Frommel TO, Balber AE. 1987. Flow cytofluorimetric analysis of drug accumulation by multidrug-resistant *Trypanosoma brucei* and *T.b. rhodesiense*. *Mol. Biochem. Parasitol.* 26:183–91

57. Fry AJ, Towner P, Holman GD, Eisenthal R. 1993. Transport of D-fructose and its analogues by *Trypanosoma brucei*. *Mol. Biochem. Parasitol.* 60:9–18

58. Goldberg B, Yarlett N, Sufrin J, Lloyd D, Bacchi CJ. 1997. A unique transporter of S-adenosylmethionine in African trypanosomes. *FASEB J.* 11:256–60

59. Gottesdiener KM. 1994. A new VSG expression site-associated gene (ESAG) in the promoter region of *Trypanosoma brucei* encodes a protein with 10 potential transmembrane domains. *Mol. Biochem. Parasitol.* 63:143–51

60. Gottesdiener KM, Chung H-M, Brown SD, Lee MG-S, Van der Ploeg LHT. 1991. Characterization of VSG gene expression site promoters and promoter-associated DNA rearrangement events. *Mol. Cell Biol.* 11:2467–80

61. Gottesdiener KM, Goriparthi L, Masucci JP, Van der Ploeg LHT. 1992. A proposed mechanism for promoter-associated DNA rearrangement events at a variant surface glycoprotein gene expression site. *Mol. Cell Biol.* 12:4784–95

62. Gould GW, Holman GD. 1993. The glucose transporter family: structure, function and tissue-specific expression. *Biochem. J.* 295:329–41

63. Graham SV, Matthews KR, Barry JD. 1993. *Trypanosoma brucei*: unusual expression-site-associated gene homologies in a metacyclic VSG gene expression site. *Exp. Parasitol.* 76:96–99

64. Grondin K, Haimeur A, Mukhopadhyay R, Rosen BP, Ouellette M. 1997. Co-amplification of the gamma-glutamyl-cysteine synthetase gene gsh1 and of the ABC transporter gene pgpA in arsenite-resistant *Leishmania tarentolae*. *EMBO J.* 16:3057–65

65. Gruenberg J, Sharma PR, Deshusses J. 1978. D-Glucose transport in *Trypanosoma brucei*. D-Glucose transport is the rate-limiting step of its metabolism. *Eur. J. Biochem.* 89:461–69

66. Hager KM, Hajduk SL. 1997. Mechanism

of resistance of African trypanosomes to cytotoxic human HDL. *Nature* 385:823–26

67. Hager KM, Pierce MA, Moore DR, Tytler EM, Esko JD, Hajduk SL. 1994. Endocytosis of a cytotoxic human high density lipoprotein results in disruption of acidic intracellular vesicles and subsequent killing of African trypanosomes. *J. Cell Biol.* 126:155–67

68. Hajduk SL, Hager KM, Esko JD. 1994. Human high density lipoprotein killing of African trypanosomes. *Annu. Rev. Microbiol.* 48:139–62

69. Hajduk SL, Moore DR, Vasudevacharya J, Siqueira H, Torri AF, et al. 1989. Lysis of *Trypanosoma brucei* by a toxic subspecies of human high density lipoprotein. *J. Biol. Chem.* 264:5210–17

70. Henderson DM, Sifri CD, Rodgers M, Wirth DF, Hendrickson N, Ullman B. 1992. Multidrug resistance in *Leishmania donovani* is conferred by amplification of a gene homologous to the mammalian mdr1 gene. *Mol. Cell Biol.* 12:2855–65

71. Hide G, Gray A, Harrison CM, Tait A. 1989. Identification of an epidermal growth factor receptor homologue in trypanosomes. *Mol. Biochem. Parasitol.* 36:51–60

72. Hobbs MR, Boothroyd JC. 1990. An expression-site-associated gene family of trypanosomes is expressed in vivo and shows homology to a variant surface glycoprotein gene. *Mol. Biochem. Parasitol.* 43:1–16

73. Ishikawa T, Li Z-S, Lu Y, Rea PA. 1997. The GS-X pump in plant, yeast, and animal cells: structure, function, and gene expression. *Biosci. Rep.* 17:189–207

74. Jackson DG, Voorheis HP. 1993. The identification, purification, and characterization of two invariant surface glycoproteins located beneath the surface coat barrier of bloodstream forms of *Trypanosoma brucei*. *J. Biol. Chem.* 268:8085–95

75. Kooter JM, van der Spek HJ, Wagter R, d'Oliveira CE, van der Hoeven F, et al. 1987. The anatomy and transcription of a telomeric expression site for variant-specific surface antigens in *T. brucei*. *Cell* 51:261–72

76. Kooter JM, Winter AJ, d'Oliveira C, Wagter R, Borst P. 1988. Boundaries of telomere conversion in *Trypanosoma brucei*. *Gene* 69:1–11

77. Kunst F, Ogasawara N, Moszer I, Albertini AM, Alloni G, et al. 1997. The complete genome sequence of the gram-positive bacterium *Bacillus subtilis*. *Nature* 390:249–56

78. Le Blancq SM, Swinkels BW, Gibson WC, Borst P. 1988. Evidence for gene conversion between the phosphoglycerate kinase genes of *Trypanosoma brucei*. *J. Mol. Biol.* 200:439–47

79. Lee MG-S, Bihain E, Russell DG, Deckelbaum RJ, Van der Ploeg LHT. 1990. Characterization of a cDNA encoding a cysteine-rich cell surface protein located in the flagellar pocket of the protozoan *Trypanosoma brucei*. *Mol. Cell Biol.* 10:4506–17

80. LÈgarÈ D, Hettema E, Ouellette M. 1994. The P-glycoprotein related gene family in *Leishmania*. *Mol. Biochem. Parasitol.* 68:81–91

81. LeQuesne SA, Fairlamb AH. 1996. Regulation of a high affinity diamine transport system in *Trypanosoma cruzi* epimasigotes. *Biochem. J.* 316:481–86

82. Ligtenberg MJ, Bitter W, Kieft R, Steverding D, Janssen H, et al. 1994. Reconstitution of a surface transferrin binding complex in insect form *Trypanosoma brucei*. *EMBO J.* 13:2565–73

83. Lips S, Geuskens M, Paturiaux-Hanocq F, Hanocq-Quertier J, Pays E. 1996. The *esag 8* gene of *Trypanosoma brucei* encodes a nuclear protein. *Mol. Biochem. Parasitol.* 79:113–17

84. Lips S, Revelard P, Pays E. 1993. Identification of a new expression site-associated gene in the complete 30.5 kb sequence from the AnTat 1.3A variant surface protein gene expression site of *Trypanosoma brucei*. *Mol. Biochem. Parasitol.* 62:135–37

85. Maslov DA, Simpson L. 1995. Evolution of parasitism in kinetoplastid protozoa. *Parasitol. Today* 11:30–32

86. Matthews KR, Gull K. 1994. Evidence for an interplay between cell cycle progression and the initiation of differentiation between life cycle forms of African trypanosomes. *J. Cell Biol.* 125:1147–56

87. Mehlert A, Zitzmann N, Richardson JM, Treumann A, Ferguson MAJ. 1998. The glycosylation of the variant surface glycoproteins and procyclic acidic repetitive proteins of *Trypanosoma brucei*. *Mol. Biochem. Parasitol.* 91:145–52

88. Mewes HW, Albermann K, Bahr M, Frishman D, Gleissner A, et al. 1997. Overview of the yeast genome. *Nature* 387:7–8

89. Morgan RW, El-Sayed NMA, Kepa JK, Pedram M, Donelson JE. 1996. Differential expression of the expression site-associated gene I family in African

trypanosomes. *J. Biol. Chem.* 271:9771–77

90. Munoz-Antonia T, Richards FF, Ullu E. 1991. Differences in glucose transport between blood stream and procyclic forms of *Trypanosoma brucei rhodesiense*. *Mol. Biochem. Parasitol.* 47:73–81

91. Olsson T, Bakhiet M, Höjeberg B, Ljungdahl Å, Edlund C, et al. 1993. CD8 is critically involved in lymphocyte activation by a *T. brucei brucei*-released molecule. *Cell* 72:715–27

92. Ouellette M, Fase-Fowler F, Borst P. 1990. The amplified H circle of methotrexate-resistant *Leishmania tarentolae* contains a novel P-glycoprotein gene. *EMBO J.* 9:1027–33

93. Overath P, Chaudhri M, Steverding D, Ziegelbauer K. 1994. Invariant surface proteins in bloodstream forms of *Trypanosoma brucei*. *Parasitol. Today* 10:53–58

94. Overath P, Czichos J, Haas C. 1986. The effect of citrate/cis-aconitate on oxidative metabolism during transformation of *Trypanosoma brucei*. *Eur. J. Biochem.* 160:175–82

95. Overath P, Schell D, Stierhof Y-D, Schwarz H, Preis D. 1992. A transferrin-binding protein in *Trypanosoma brucei*: Does it function in iron uptake? In *Dynamics of Membrane Assembly, NATO ASI Ser.*, ed. JAF Op den Kamp, H63:333–47. Berlin-Springer-Verlag

96. Overath P, Stierhof Y-D, Wiese M. 1997. Endocytosis and secretion in trypanosomatid parasites-tumultuous traffic in a pocket. *Trends Cell Biol.* 7:27–33

97. Paindavoine P, Rolin S, Van Assel S, Geuskens M, Jauniaux JC, et al. 1995. A gene from the VSG expression site encodes one of several transmembrane adenylate cyclases located on the flagellum of *Trypanosoma brucei*. *Mol. Cell Biol.* 12:1218–25

98. Parsons M, Nielsen B. 1990. Active transport of 2-deoxy-D-glucose in *Trypanosoma brucei* procyclic forms. *Mol. Biochem. Parasitol.* 42:197–204

99. Pays E. 1989. Pseudogenes, chimaeric genes and the timing of antigen variation in African trypanosomes. *Trends Genet.* 5:389–91

100. Pays E, Tebabi P, Pays A, Coquelet H, Revelard P, et al. 1989. The genes and transcripts of an antigen gene expression site from *T. brucei. Cell* 57:835–45

101. Pays E, Vanhamme L, Berberof M. 1994. Genetic controls for the expression of surface antigens in African trypanosomes. *Annu. Rev. Microbiol.* 48:25–52

102. Pospichal H, Brun R, Kaminsky R, Jenni L. 1994. Induction of resistance to melarsenoxide cysteamine (Mel Cy) in *Trypanosoma brucei brucei*. *Acta Trop.* 58:187–97

103. Raper J, Nussenzweig V, Tomlinson S. 1996. The main lytic factor of *Trypanosoma brucei brucei* in normal human serum is not high density lipoprotein. *J. Exp. Med.* 183:1023–29

104. Reguera R, Balana-Fouce R, Cubria JC, Alvarez Bujidos ML, Ordonez D. 1994. Putrescine uptake inhibition by arometic diamidines in *Leishmania infantum promastigotes*. *Biochem. Pharmacol.* 47:1859–66

105. Revelard P, Lips S, Pays E. 1990. A gene from the VSG expression site of *Trypanosoma brucei* encodes a protein with both leucine-rich repeats and a putative zinc finger. *Nucl. Acids Res.* 18:7299–303

106. Rifkin MR. 1978. *Trypanosoma brucei*: some properties of the cytotoxic reaction induced by normal human serum. *Exp. Parasitol.* 46:189–206

107. Roditi I. 1996. The VSG-procyclin switch. *Parasitol. Today* 12:47–49

108. Rolin S, Halleux S, Van Sande J, Dumont J, Pays E, Steinert M. 1990. Stage-specific adenylate cyclase activity in *Trypanosoma brucei*. *Exp. Parasitol.* 71:350–52

109. Ross CA, Barns AM. 1996. Alteration to one of 3 adenosine transporters is associated with resistance to cymelarsan in *Trypanosoma evansi*. *Parasitol. Res.* 82:183–88

110. Ross DT, Raibaud A, Florent IC, Sather S, Cross MK, et al. 1991. The trypanosome VSG expression site encodes adenylate cyclase and a leucine-rich putative regulatory gene. *EMBO J.* 10:2047–53

111. Rudenko G, Cross M, Borst P. 1998. Changing the end: antigenic variation at the telomeres of African trypanosomes. *Trends Microbiol.* 3:113–17

112. Salmon D, Geuskens M, Hanocq F, Hanocq-Quertier J, Nolan D, et al. 1994. A novel heterodimeric transferrin receptor encoded by a pair of VSG expression site-associated genes in *T. brucei. Cell* 78:75–86

113. Salmon D, Hanocq-Quertier J, Paturiaux-Hanocq F, Pays A, Tebabi P, et al. 1997. Characterization of the ligand-binding site of the transferrin receptor in *Trypanosoma brucei* demonstrates a structural relationship with the N-terminal domain of the variant surface glycoprotein. *EMBO J.* 16:101–7

114. Schell D, Borowy NK, Overath P. 1991. Transferrin is a growth factor for the

bloodstream form of *Trypanosoma brucei. Parasitol. Res.* 77:558–60

115. Schell D, Evers R, Preiss D, Ziegelbauer K, Kiefer H, et al. 1991. A transferrin-binding protein of *Trypanosoma brucei* is encoded by one of the genes in the variant surface glycoprotein expression site. *EMBO J.* 10:1061–66

116. Schell D, Evers R, Preis D, Ziegelbauer K, Kiefer H, et al. 1993. A transferrin-binding protein of *Trypanosoma brucei* is encoded by one of the genes in the variant surface glycoprotein gene expression site. *EMBO J.* 12:2990

117. Scott AG, Tait A, Turner CMR. 1997. *Trypanosoma brucei*: lack of cross-resistance to melarsoprol *in vitro* by cymelarsan-resistant parasites. *Exp. Parasitol.* 86:181–90

118. Scott DA, Docampo R, Dvorak JA, Shi S, Leapman D. 1997. *In situ* compositional analysis of acidocalcisomes in *Trypanosoma cruzi. J. Biol. Chem.* 272:28020–29

119. Seyfang A, Duszenko M. 1991. Specificity of glucose transport in *Trypanosoma brucei.* Effective inhibition by phloretin and cytochalasin B. *Eur. J. Biochem.* 202:191–96

120. Seyfang A, Duszenko M. 1993. Functional reconstitution of the *Trypanosoma brucei* plasma-membrane D-glucose transporter. *Eur. J. Biochem.* 214:593–97

121. Smith AB, Esko JD, Hajduk SL. 1995. Killing of trypanosomes by the human haptoglobin-related protein. *Science* 268:284–86

122. Snapp EL, Landfear S. 1997. Cytoskeletal association is important for differential targeting of glucose transporter isoforms in *Leishmania. J. Cell Biol.* 139:1775–83

123. Sogin ML, Gunderson JH, Elwood HJ, Alonso RA, Peattie DA. 1989. Phylogenetic meaning of the kingdom concept: an unusual ribosomal RNA from Giardia lamblia. *Science* 243:75–77

124. Steverding D, Stierhof Y, Chaudhri M, Ligtenberg M, Schell D, et al. 1994. ESAG 6 and 7 products of *Trypanosoma brucei* form a transferrin binding protein complex. *Eur. J. Cell Biol.* 64:78–87

125. Steverding D, Stierhof Y, Fuchs H, Tauber R, Overath P. 1995. The TFBP-complex is the receptor for transferrin uptake in *Trypanosoma brucei. J. Cell Biol.* 131:1173–82

126. Tetaud E, Barrett MP, Bringaud F, Baltz T. 1997. Kinetoplastid glucose transporters. *Biochem. J.* 325:569–80

127. Tetaud E, Chabas S, Giroud C, Barrett MP, Baltz T. 1996. Hexose uptake in *Trypanosoma cruzi*: structure-activity relationship between substrate and transporter. *Biochem. J.* 317:353–59

128. Thon G, Baltz T, Giroud C, Eisen H. 1990. Trypanosome variable surface glycoproteins: composite genes and order of expression. *Genes Dev.* 9:1374–83

129. Tomb J, White O, Kerlavage AR, Clayton RA, Sutton GG. 1997. The complete genome sequence of the gastric pathogen *Helicobacter pylori. Nature* 388:539–47

130. Ullman B. 1995. Multidrug resistance and P-glycoproteins in parasitic protozoa. *J. Bioenerg. Biomembr.* 27:77–84

131. Vaidya T, Bakhiet M, Hill KL, Olsson T, Kristensson K, Donelson JE. 1997. The gene for a T lymphocyte triggering factor from African trypanosomes. *J. Exp. Med.* 186:433–38

132. Vanhamme L, Pays E. 1995. Control of gene expression in trypanosomes. *Microbiol. Rev.* 59:223–40

133. Voorheis HP, Martin BR. 1980. 'Swell dialysis' demonstrates that adenylate cyclase in *Trypanosoma brucei* is regulated by calcium ions. *Eur. J. Biochem.* 113:223–27

134. Voorheis HP, Martin BR. 1981. Characteristics of the calcium-mediated mechanism activating adenylate cyclase in *Trypanosoma brucei. Eur. J. Biochem.* 116: 471–77

135. Waitumbi JN, Tetaud E, Baltz T. 1996. Glucose uptake in *Trypanosoma vivax* and molecular characterization of its transporter gene. *Eur. J. Biochem.* 237:234–39

136. Webster P. 1989. Endocytosis by African trypanosomes. I. Three-dimensional structure of the endocytic organelles in *Trypanosoma brucei* and *T. congolense. Eur. J. Cell Biol.* 49:295–302

137. Webster P, Fish WR. 1989. Endocytosis by African trypanosomes. II. Occurrence in different life-cycle stages and intracellular sorting. *Eur. J. Cell Biol.* 49:303–10

138. Webster P, Russell DG. 1993. The flagellar pocket of trypanosomatids. *Parasitol. Today* 9:201–6

139. Zhang ZQ, Giroud C, Baltz T. 1993. *Trypanosoma evansi: in vivo* and *in vitro* determination of trypanocide resistance profiles. *Exp. Parasitol.* 77:387–94

140. Ziegelbauer K, Multhaup G, Overath P. 1992. Molecular characterization of two invariant surface glycoproteins specific for the bloodstream stage of *Trypanosoma brucei. J. Biol. Chem.* 267:10797–803

141. Ziegelbauer K, Overath P. 1992. Identification of invariant surface glycoproteins in the bloodstream stage of *Trypanosoma*

brucei. J. Biol. Chem. 267:10791–96

142. Ziegelbauer K, Overath P. 1993. Organization of two invariant surface glycoproteins in the surface coat of *Trypanosoma brucei. Infect. Immun.* 61:4540–45

143. Ziegelbauer K, Rudenko G, Kieft R, Overath P. 1995. Genomic organization of an invariant surface glycoprotein gene family of *Trypanosoma brucei. Mol. Biochem. Parasitol.* 69:53–63

144. Zomerdijk JCBM, Ouellette M, Ten Asbroek ALMA, Kieft R, Bommer AMM, et al. 1990. The promoter for a variant surface glycoprotein gene expression site in *Trypanosoma brucei. EMBO J.* 9:2791–801

Annu. Rev. Microbiol. 1998. 52:779–806
Copyright © 1998 by Annual Reviews. All rights reserved

COOPERATIVE ORGANIZATION OF BACTERIAL COLONIES: From Genotype to Morphotype

Eshel Ben-Jacob and Inon Cohen
School of Physics and Astronomy, Raymond and Beverly Sackler Faculty of Exact Sciences, Tel-Aviv University, Tel-Aviv 69978, Israel

David L. Gutnick*
Department of Molecular Microbiology and Biotechnology, George S. Wise Faculty of Life Sciences, Tel-Aviv University, Tel-Aviv 69978, Israel

KEY WORDS: pattern formation, colonial development, bacterial motility, chemotaxis, cell-cell communication

ABSTRACT

In nature, bacteria must often cope with difficult environmental conditions. To do so they have developed sophisticated cooperative behavior and intricate communication pathways. Utilizing these elements, motile microbial colonies frequently develop complex patterns in response to adverse growth conditions on hard surfaces under conditions of energy limitation. We employ the term morphotype to refer to specific properties of colonial development. The morphologies we discuss include a tip-splitting (T) morphotype, chiral (C) morphotype, and vortex (V) morphotype. A generic modeling approach was developed by combining a detailed study of the cellular behavior and dynamics during colonial development and invoking concepts derived from the study of pattern formation in nonliving systems. Analysis of patterning behavior of the models suggests bacterial processes whereby communication leads to self-organization by using cooperative cellular interactions. New features emerging from the model include various modes of cell-cell signaling, such as long-range chemorepulsion, short-range chemoattraction, and, in the case of the V morphotype, rotational chemotaxis. In this regard, pattern formation in microorganisms can be viewed as the result of the exchange

*Corresponding author.

0066-4227/98/1001-0779$08.00

of information between the micro-level (the individual cells) and the macro-level (the colony).

CONTENTS

INTRODUCTION

The endless array of patterns and shapes in nature has long been a source of joy and wonder to laymen and scientists alike (76). Discovering how such patterns emerge spontaneously from an orderless and homogeneous environment has been a challenge to researchers in the natural sciences throughout the ages. Many phenomena display the emergence of patterns during diffusive growth, ranging from the growth of snowflakes to solidification of metals, from the formation of a coral reef to cell differentiation during embryonic development.

In the early 1950s, Alan Turing understood that patterns would evolve in systems driven out of equilibrium, where competition and interplay between various tendencies exists (81). We now understand that the diffusion field drives the system towards decorated (on many length scales) irregular fractal shapes. It has become clear that the competition between the drive of the diffusion field and a reverse, stabilizing, drive of microscopic effects (e.g. surface tension and surface kinetics) plays a key role in the determination of the evolved pattern.

Here we describe cooperative patterning during growth of bacterial colonies under hostile conditions of low level of nutrients, a hard surface, or both. Under such conditions, not unlike certain ecosystems in natural environments, complex colonial patterns are observed (10, 11, 14, 18, 40, 58, 59, 61, 63, 74). Drawing on the analogy with diffusive patterning in nonliving systems (5, 8, 49, 54) the above observations can be understood as follows: The cellular reproduction rate, that determines the growth rate of the colony, is limited by the level of nutrients available for the cells. The latter is limited by the diffusion of nutrients towards the colony (for low nutrient substrate). Hence, the colonial growth should be similar to diffusion limited growth in nonliving systems, such as solidification from a supersaturated solution, growth in a Hele-Shaw cell, electrochemical deposition, etc (5, 8). Indeed, for some conditions bacterial colonies can develop patterns reminiscent of those observed during growth in nonliving systems (10, 11, 14, 18, 40, 58, 59, 61, 63).

In general, the bacteria can exhibit richer behavior than abiotic patterning, reflecting the additional levels of complexity involved (10, 12–16, 21, 33). In the former case, the building blocks themselves are living systems; each has its own autonomous self-interest and internal degrees of freedom. At the same time, efficient adaptation of the colony to adverse growth conditions requires cooperative behavior of the bacteria. The bacteria can do so because they possess various modes of communication, such as (*a*) direct cell-cell physical and chemical interactions (34, 64); (*b*) indirect physical and chemical interactions, e.g. production of extracellular "wetting" fluid (44, 62); (*c*) long range chemical signaling, such as quorum sensing (41, 42, 55); and (*d*) chemotactic signaling [chemotactic response to chemical agents that are emitted by the cells (27, 30, 31)].

Studies on pattern formation in abiotic systems demonstrated that different shapes are observed for the same system as the control parameters are varied (e.g. undercooling, supersaturation). Although a number of morphologies are possible under a particular set of conditions, only one is generally observed ("selected"). The commonly accepted morphology selection principle states that the particular morphology selected is the fastest growing one (8, 9). Hence the observed patterns can be organized in a morphology diagram analogous to a phase diagram (of liquid, solid, gas). There is a relatively sharp transition from one shape to the other, as the control parameters are varied and different morphologies are selected.

It has been demonstrated that the concept of morphology diagram can also be applied to the growth of bacterial colonies (6, 10, 12, 14, 59, 61, 63), i.e. the patterns exhibited by a given strain can be organized as a mosaic of regimes (each for a characteristic pattern) on a graph of nutrients and hardness of agar. The sharp transitions between the various regimes imply that at each regime, a characteristic biological feature dominates the growth.

In Figure 1 we refer to an additional concept introduced to describe the patterning of bacterial colonies. Three patterns with different geometrical

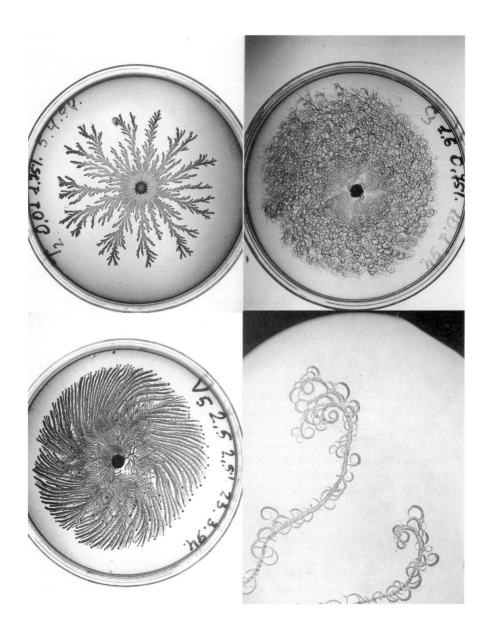

characteristics are shown. The first is best characterized by its branching pattern. In the second, the branches are much thinner and all have a twist with the same handedness, while in the third every branch has a leading droplet consisting of many bacteria at its tip. Microscopic observations (described below) reveal that the dynamics at the cellular level are also very different in each of the patterns. In addition, these geometrical characteristics are inheritable. In order to describe the distinctive characteristic properties of each colonial development, a new concept was introduced: morphotype. The patterns in Figure 1 are representative patterns of the tip-splitting (T), chiral (C), and vortex (V) morphotypes. It should be noted that different bacterial strains (and species) can belong to the same morphotype. Moreover, as we show below, bacterial strains can undergo a transition from one morphotype to another (e.g. branched to chiral). The morphotype transition requires a period of adaptation on the order of days and once the transition has occurred, the new morphotype is stable. It is clear that both pattern formation and morphotype transition require a mode of organized intercellular communication and cooperative multicellular behavior, both of which are characteristic features of a variety of cooperative and cell-density dependent physiological processes currently under study in microbial physiology (42, 48, 56).

THE T AND C MORPHOTYPES

The patterns illustrated in Figures 2–4 were formed by organisms that were initially isolated from cultures of *Bacillus subtilis* (10, 13). 16S RNA sequence analysis in combination with a number of phenotypic and biochemical characterizations was used in the identification of these strains as members of the genus *Paenibacillus* (M Tcherpikov et al, unpublished information) and assigned to a new species, *Paenibacillus dendritiformis*. The strains in these figures belong to two different morphotypes, C and T (7, 17, 19). They showed greater than 99% sequence similarity to each other.

   P. dendritiformis exhibited a profusion of patterns as the growth conditions were varied (Figures 2–4) (12, 14). Observations of similar patterns during

---

Figure 1   Examples of three distinct morphotypes. *Top left*: Branching or tip-splitting pattern of the T morphotype exhibited by *Paenibacillus dendritiformis*. *Top* and *bottom right*: Chiral pattern of the C morphotype exhibited by *Paenibacillus dendritiformis*—a full colony at the *top* and a zoom-in on the *bottom*. Note that all branches have a twist with the same handedness. *Bottom left*: Pattern of the V morphotype. All branches have a leading droplet. The latter is composed of many bacteria moving in a correlated motion around the center. Excluding the bottom right, each pattern is in a single standard petri dish, 8 cm in diameter.

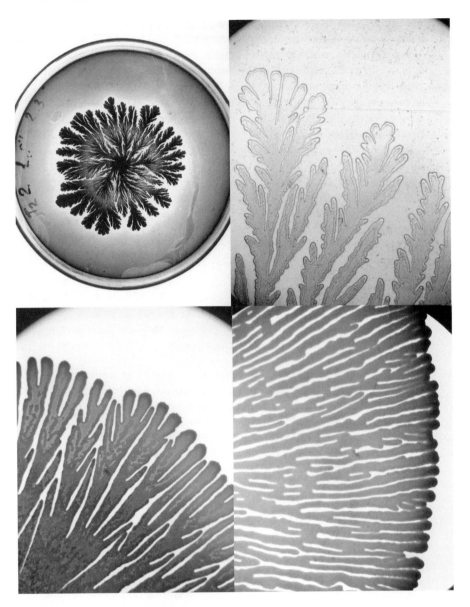

Figure 2   (*a–d*) Examples of branching patterns of the T morphotype; (*e*) closer look at the branches to show the variation in bacterial density within a branch; (*f*) microscopic observations (×50) of a stained colony to show the bacterial distribution; (*g*) growth on harder agar (above 2% concentration) leads to thin branches and weak chirality (global weak twist) (6); (*h*) structures of concentric variations in bacterial density during growth on hard agar (2.25%) and high level of nutrient (15 g/l Bacto-Peptone).

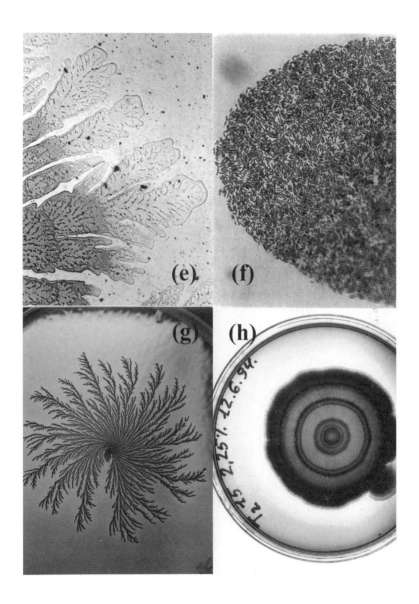

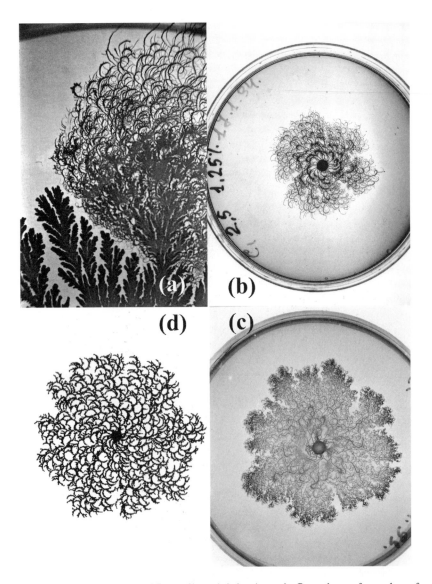

Figure 3   The C morphotype. (*a*) Burst of bacteria belonging to the C morphotype from colony of the T morphotype; (*b* and *c*) examples of chiral patterns exhibited by *Paenibacillus dendritiformis* var. *chiralis*; (*d*) example of results of numerical simulations for the features described in the text.

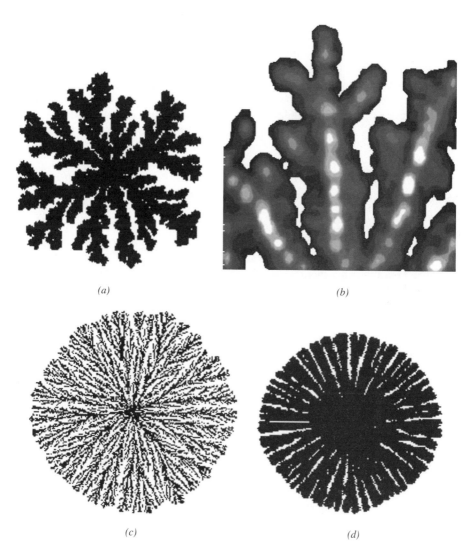

Figure 4   Results of numerical simulations of the communicating walkers model. (*a*) A typical branching pattern (without chemotaxis); (*b*) close look at a branch to show the density variations when short-range attractive chemotactic signaling is included (the *grey* level corresponds to various levels of density); (*c*) the effect of repulsive chemotactic signaling [the same conditions as (*a*) but with repulsive signaling]; (*d*) the effect of chemotaxis on food.

growth of other bacterial strains have been reported (57, 61, 63, 67). Under the microscope, cells were seen to swim in a characteristic random-walk–like fashion within a fluid. This fluid seems to be excreted by the cells, although it could also be drawn from the agar during microbial growth (12, 13). The swimming was confined to the fluid; isolated cells spotted on the agar surface did not move. The boundary of the fluid thus defines a local boundary for the branch of the developing colony. Whenever the cells are active, the boundary propagates slowly as a result of the cellular movement and production of additional wetting fluid. The observations also revealed that the cells were more active at the outer parts of the colony, while closer to the center the cells did not move, and some were observed to sporulate.

The C morphotype also exhibited a morphology diagram with a profusion of most beautiful and complex patterns (Figure 3) (13, 16, 17). In addition to the enormous difference in morphologies between the genetically similar variants, microscopic observations indicated that the cells were longer than those of the T morphotype. Electron microscopic observations indicated that the chiral structure did not stem from twisting of the cell membrane (13). The question then arises whether there is a connection between the increased length of the cells and the formation of chiral patterns.

Chiral asymmetry, first discovered by Louis Pasteur, exists in a whole range of scales, from subatomic particles through human beings to galaxies, and seems to have played an important role in the evolution of living systems (2, 45). Bacteria display various chiral properties. For example, Mendelson et al (64–66, 68) showed that long cells of *B. subtilis* can grow in helices in which the cells form long strings that are twisted around each other.

Since, in the case of *P. dendritiformis* chirality is formed by moving swimming cells, a different mechanism is controlling the handedness. One possibility may have to do with the specific handedness associated with flagellar rotation (37, 73, 77). Ben-Jacob et al have proposed that it is this property of flagellar handedness, coupled with strong cell-cell orientation interactions, that accounts for the observed chirality (16).

Morphotype Transitions

When plated on soft agar (concentrations of about 1% or less) colonies of T cells develop with characteristic tip-splitting morphology. Occasionally, after a period of about 48 h, a morphotype transition occurs, reflected in bursts of growth of C morphotype. Cells of C morphotype can be isolated from such bursts (14, 16). In these growth conditions colonies propagate faster then those of the T cells, reaching the same distance from the spot inoculum with significantly fewer bacteria. Motivated by the "fastest growing morphology" selection principle for nonliving systems, Ben-Jacob et al proposed that the colonial growth velocity (the rate of spreading) is the colonial selective pressure leading to the T → C morphotype transitions (Figure 3). If this hypothesis is correct, one would expect to observe the reverse C → T transitions under those growth conditions for which T spreads fastest. Indeed such reverse transitions are observed during growth on harder agar where T spreads faster than the C morphotype (10, 14).

The T → C transitions are quite frequent. Under proper conditions, about 60% of T colonies show bursts of the C morphotype after about two days. The question arose as to whether this was a true morphotype transition, or whether the C cells were simply present as a small minority within the var. *dendron* population. Recent findings using strains marked with antibiotic resistance markers have demonstrated that under the specific conditions, the cells undergo an actual morphotype transition (M Tcherpakov et al, manuscript in preparation). Moreover, morphotype transition appears to be a more general phenomenon. For example, at least two different isolates of *P. thiaminolyticus* from various collections have also been shown to exhibit characteristics of both the T and C morphotypes, albeit under slightly different conditions of nutrient concentration and agar hardness. Moreover, these strains also undergo the same mode of morphotype transition (M Tcherpakov et al, in press).

MODELING AND SIMULATIONS OF THE COLONIAL DEVELOPMENT

While the modeling of a biological process can be limited to finding a suitable formulation to simulate the process using a computer, ideally one would like to have a generic model that not only leads to a numerical simulation, but also incorporates features that are testable experimentally. In addition, the model should contain elements that allow for a variety of hypothetical situations that might also lead to unanticipated experimental predictions.

The Communicating Walkers Model

To model the colonial development of the T and C morphotypes, Ben-Jacob et al (12, 13) developed "the communicating walkers model." The bacteria are represented in the model by particles dubbed "walkers," which correspond to about 10^2–10^4 bacterial cells.

A walker is specified by its location on the surface and its metabolic state (referred to as an "internal energy"). The latter is increased by consumption of nutrients from the media and subsequently used in order to drive the walker's (bacterial) activities and metabolic processes. For high concentration of nutrients, the food consumption is higher then the spending rate. Hence, the internal energy increases until it reaches a threshold level at which the walker divides. When there is not enough "food," the walker can consume only the available amount, which can be lower than needed for activity. As a result, the internal energy decreases until it drops to zero. The walker then becomes immotile and remains in this state (enters prespore state). In the model, we assume for simplicity a single component of food, which satisfies a simple diffusion equation. As the walkers consume the nutrient, the concentration decreases in front of the colony and additional nutrient diffuses towards the colony. Hence, it is a diffusion-limited growth, as mentioned above.

In laboratory experiments, bacteria swim within the lubrication fluid. In the model, the walkers perform a random walk. At each time step, each of the active (motile) walkers moves a step at a random angle. The walkers are confined within an envelope (defined on a tridiagonal lattice) which represents the boundary of the lubrication fluid. In the event a walker's step would lead to its movement outside the boundary, the step is not performed, and a counter on the appropriate segment of the envelope is increased by one. When a segment counter reaches a threshold N_c, the envelope segment propagates, adding one lattice area to the colony. This requirement of N_c hits represents the colony propagation through wetting of unoccupied areas by the bacterial cells. This feature reflects the local cooperation in the behavior of the bacteria (the analog of a surface tension in nonliving systems). Note that, to a first approximation, N_c represents the agar concentration, since more "collisions" are needed to push the envelope on a harder substrate. Results of numerical simulations of the model are shown in Figure 4.

To test the idea that both flagellar handedness and increased cell size play the major role in the origin of the chiral growth of the C morphotype, Ben-Jacob et al (16) included the additional features of flagellar handedness and cell-cell orientational interaction in the communicating walkers model. To represent the cellular orientation, each walker is assigned an orientation. Every time step, each of the active walkers performs rotation to a new orientation that is derived from the walker's previous orientation. Once oriented, the walker advances a step in either the forward or reverse direction (an experimental observation). As for the T morphotype, the movement is confined within an envelope that is defined on a triangular lattice. Results of the numerical simulations of the model are shown in Figure 4.

The communicating walkers model described here is only one example of a modeling approach. There is another class of models, continuous models, in which the bacteria are represented by the value of their local density (10, 51, 59, 60).

CHEMOTAXIS-BASED ADAPTIVE
SELF-ORGANIZATION

Chemotaxis is the best studied and most prevalent signal transduction system in motile bacteria (38). This process involves changes in the movement of the cell in response to a gradient of certain chemical fields (1, 22, 23, 53). The movement is biased along the gradient either in the forward (in the direction of the gradient) or in the reverse direction. Thus, chemotaxis enables microbial cells in a variety of natural environments to obtain more favorable conditions, such as movement towards nutrients, escape from predators, movement towards specific surfaces, and protection by cellular aggregation. Because space constraints prevent us from including a thorough analysis of the chemotaxis literature, we refer the reader to a number of recent excellent reviews of the field (26, 37, 43, 44, 78).

Usually chemotaxis implies a response to an externally produced field such as attraction towards supplemented nutrients. However, self-generated bacterial chemotactic signaling by the excretion of amino acids and peptides has also been demonstrated (27, 30, 31, 85). In the case of *Escherichia coli* and *Salmonella typhimurium*, this mode of chemoattraction involves membrane receptors such as the Tar receptor for chemotaxis, as well as a new receptor involving chemoattraction on rich medium (27).

At least 50 different gene products are involved in governing the mode by which microbes employ the chemotactic system to modulate their movement. The cell "senses" the concentration of the chemoattractant (or repellent) by measuring the fraction of receptors occupied by the signaling molecules. Thus, at very high concentrations the chemotactic response vanishes because of receptor saturation—the "receptor law." At the lower limit of attractant, the response is also negligible since it is "masked" by noise in the system. Swimming bacteria such as *E. coli* perform chemotaxis by modulating the time gap between tumbling events. Increasing (or decreasing) this time gap when swimming up or down the gradient of attractant (or repellent) bias their movement toward (or away from) favorable (unfavorable) locations. In *E. coli* for example, the tumbling event is controlled by the protein CheY.-Phosphorylated, CheY binds to a switch at the base of the flagellar motor, thereby changing the flagellar rotation from counter-clockwise (the default direction of rotation, which propels the cells in a more or less straight trajectory), to clockwise rotation (which causes the cells to tumble). The signal transduction pathway involving the phosphorylation of CheY involves the action of an autophosphorylating kinase, CheA, whose activity is controlled by chemoreceptors. The product of CheA, CheY $\sim$ P, is then dephosphorylated by the action of CheZ, a phosphatase whose action depends on its interaction with CheY $\sim$ P. Interestingly, an additional mode of regulation has been discovered involving the oligomerization of CheZ, which apparently is mediated by its interaction with CheY $\sim$ P (28). It is the oligomer that catalyzes the dephosphorylation of CheY $\sim$ P. In other swimming species, the details of regulation and signal transduction may vary, but the mechanism of chemotaxis via modulated tumbling time is conserved.

Ben-Jacob et al (19, 32) assumed that for the colonial adaptive self-organization the T morphotype employs three kinds of chemotactic responses. One is the nutritional chemotaxis mentioned above. According to the "receptor law," it is expected to be dominant for a range of nutrient levels. The two other modes of chemotaxis are self-induced, that is there is chemoattraction or chemorepulsion towards or away from signaling molecules produced by the bacterial cells themselves. As we show below, for efficient self-organization it is useful to employ two chemotactic responses operating on different length scales, one regulating the dynamics within the branches (short length-scale) and the other regulating the organization of the branches (long length-scale).

The observations of attractive chemotactic signaling in *E. coli* (15, 27, 30, 31, 80) indicate that it operates during growth at high levels of nutrients. Motivated by the above, we assume that the colony employs an attractive, self-generated, short-range chemotaxis during growth at high levels of nutrients. To test this hypothesis, we add the new feature to the communicating walkers model and compare the resulting patterns with the observed ones. In Figure 4 we show an example of the formation of 3D structures when the attractive chemoresponse is included in the model.

In Figure 4 we also demonstrate the dramatic effect of the repulsive chemotactic signaling emitted by stressed walkers (12, 15, 19–21, 32). The pattern becomes much denser with a smooth circular envelope, while the branches are thinner and radial. This structure enables the colony to spread over the same distance with fewer walkers, thereby providing the developing colony with a distinct biological advantage under certain conditions of nutrient stress.

PATTERN FORMATION IN *ESCHERICHIA COLI* AND *SALMONELLA TYPHIMURIUM*

Physiological Characterization

One of the most dramatic examples of cell-cell communication mediating pattern formation in swimming bacteria was demonstrated by Budrene & Berg (30). They showed that under conditions of energy limitation, cells of *E. coli* swimming in a thin layer of liquid substrate can form various patterns such as concentric rings, sunflower-like structures of spots, and radial arrangements of spots. They showed pattern formation only on minimal media in a Tar-dependent pathway. Spot formation was attributed to cellular aggregation, which in turn required the excretion by the cells of a chemoattractant signaling molecule(s) (aspartate or glutamate). The hypothesis was that cellular aggregation could provide a mechanism for reducing local oxygen concentrations, thereby protecting the cells from damage caused by free radicals and superoxides. Support for the hypothesis included the observation that the addition of H_2O_2 induced cell aggregation and triggered excretion of the chemoattractant.

A similar behavior of pattern formation was also reported in *S. typhimurium* by Budrene & Berg (31) and by Blat & Eisenbach (27). In the latter report, pattern formation was shown to occur not only on minimal media in a Tar-dependent pathway, but also on rich media such as tryptone or Luria broth. To date, this behavior has not been demonstrated in *E. coli*. The patterns formed on rich media were shown to be independent of the Tar receptor, although they were induced by H_2O_2. In this regard, pattern formation in *S. typhimurium* was recently shown to occur in mutants defective in two major regulatory proteins that mediate cellular responses to oxygen stress, $Oxys^R$ and Rpo^S (G Beck, personal communication).

The question remains regarding the significance of the additional pathway for pattern formation on rich media in *S. typhimurium*. One possibility is that since pattern formation in both *E. coli* and *S. typhimurium* is highly sensitive to small variations in agar thickness, temperature, etc, it may be that under more suitable conditions, motile strains of *E. coli* will also be shown to form patterns on rich media via an alternative pathway. It also remains to be determined whether the Tar-independent pathway in *S. typhimurium* is mediated by a new receptor, or whether under specialized conditions, a known receptor is "recruited" for the chemosignaling under this set of conditions.

Modeling the Patterns—Additional Processes

In all of the reports above, the patterns were attributed to aggregation of the cells as a response to chemoattractant signaling. It was not demonstrated that such a mechanism, triggered by oxidative stress, is sufficient to explain all emerging patterns. To test this hypothesis, Ben-Jacob et al composed a model (15, 80) similar to the one for the branching patterns described above. Again, the bacterial cells (*E. coli* or *S. typhimurium* in this case) are represented in the model by walkers that perform random walk, consume food to increase their internal energy, spend this energy for activity, reproduce when food is abundant, and become nonmotile as they approach starvation. In the experiments, the bacteria do not swim in a layer of fluid on top of the agar (as in the branching patterns), but rather swim inside it. Therefore, there is no sharp boundary to the bacterial colony, and in the model, no envelope is present. The addition of a diffusing attractant that is constantly emitted by the bacteria, together with bacterial chemotaxis towards its gradient, leads to creation of spots. Others (29, 80, 85) have reported the same conclusions, and it was verified (85) in an experiment wherein an addition of H_2O_2 induced the constitutive emission of attractant.

This standard approach is insufficient to explain several crucial observations (30). In the experiments, spots appear sequentially in the wake of a spreading broad ring and later "lock" into position as the bacteria turn nonmotile. To capture these effects, one must introduce additional mechanisms. First, Ben-Jacob et al (15, 80) explicitly include in the model a "triggering" field, i.e. a field whose concentration must reach a threshold before attractant emission is activated. Ben-Jacob et al proposed that the value of the threshold for emission may depend on the ambient chemoattractant concentration: This threshold is high if there is no chemoattractant, and the threshold is lower if the chemoattractant concentration is high. This proposal means that both the oxygen metabolism and the attractant pathway affect the emission of the chemoattractant. In the absence of such an autocatalytic effect, the model cannot produce the heretofore observed radial structures. To observe (in the simulations) different patterns, Ben-Jacob et al vary the model's parameters related to the bacterial response to the triggering field and to the precise nature of the chemoattractant signaling. This version of the model is in good agreement with many of the features seen in experiment. The model has one drawback: It can produce radial organization of spots or stripes only when it includes a seemingly nonbiological response to chemoattractant.

One of the features required of a good model is a close relation with the biological knowledge. Therefore Ben-Jacob et al (15) claimed that a model with biological response to chemoattractant can produce radial organization of spots or stripes only with an additional element. They proposed that, as in the branching patterns, starved bacteria emit chemorepellent. Indeed, a model that includes this feature and chemotactic response to this repellent can account for all the salient features of the experiments. All the observed patterns can be obtained by changing the relative strength of the repulsive and attractive effects (Figure 5).

The results of the models led to the proposal that *E. coli* and *S. typhimurium* might employ the following (previously unreported) features as part of the response to energy depletion (15, 80): that (*a*) attractant perception affects its emission, and (*b*) repellant is emitted by starved cells.

Mutants Defective in Pattern Formation

Recently, Eisenbach et al (3) began a genetic study of pattern formation in *S. typhimurium*. Eight mutants were isolated that retained the ability to swarm and perform chemotaxis, but were defective in the Tar-dependent pathway for pattern formation (Figure 6). No mutants were isolated that were defective in

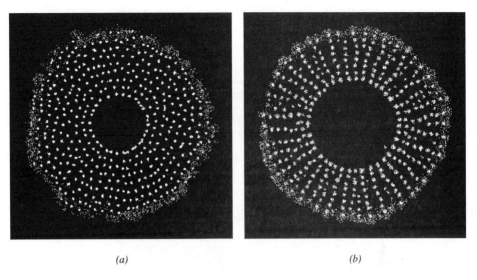

(a)   (b)

Figure 5   Results of numerical simulations of the model for the aggregation patterns of *Escherichia coli* and *Salmonella typhimurium*. (*a*) Typical array of spots; (*b*) radial arrangements of spots when repulsive chemotactic signaling is included.

the Tar-independent pathway. Seven of the eight mutations were tentatively identified. Four of the eight mutants were mutated in a hypothetical reading frame termed *yoj* located just downstream of the gene *ompC*. The *yoj* ORF appears to encode a characteristic signal peptide for a lipoprotein and was found to be homologus to an inner membrane protein, ApbE, which is involved in biosynthesis of thiamin in *S. typhimurium* (4). To test the involvement of Yoj in pattern formation, the gene was cloned behind an inducible *tac* promoter and introduced into the *yoj* mutants defective in pattern formation. Complementation was demonstrated only in the presence of IPTG (3) (Figure 6). In addition to the mutants in *yoj*, two additional mutations were mapped to the *ilv* operon, *ilvG* and *ilvM*, respectively. These two genes encode the two subunits of the enzymes acetohydroxy acid synthase II (AHAS II), which is involved in valine and isoleucine biosynthesis (71). Finally, an additional mutation was mapped to a gene that showed greater than 90% homology to the *pfl* locus in *E. coli* K-12. Pfl, pyruvate-formate lyase, catalyzes the conversion of pyruvate to formate under anaerobic conditions (70, 84).

The roles for these proteins in pattern formation are unknown at this point. One possibility is that mutations in the *yoj* and *pfl* genes are somehow related to oxygen metabolism and stress initiating the chemoattraction. In this regard, it has recently been shown (36) that lowering the flux of metabolites through

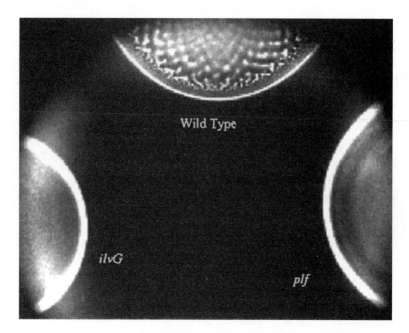

Figure 6 Pattern formation in wild-type *Salmonella typhimurium* and two defective mutants. The figure illustrates the defect in pattern formation in two of the several mutants described in the text. The *ilvG* mutant is defective in one of the subunits of the AHAS II involved in isoleucine-valine biosynthesis. *plf* is the gene encoding pyruvate-formate lyase.

the isoleucine-valine biosynthetic pathway induces a number of stress-response promoters controlled by RpoS. This could account for the selection of pattern forming mutations in the subunits of AHAS II. Further analysis of new mutants will be required in order to decipher the genetic basis of the phenomenon.

THE VORTEX MORPHOTYPE

Bacterial Patterns and Dynamics

More then half a century ago, observations of migration phenomena of *Bacillus circulans* on hard agar surface were reported (39, 75, 83). The observed phenomena include "turbulent like" collective flow, complicated eddy (vortex) dynamics, merging and splitting of vortices, rotating "bagels," and more. This behavior is not unique to *B. circulans*. During studies of complex bacterial patterning, new strains that exhibited behavior similar to *B. circulans* were isolated (14, 18). We refer to the new morphotype produced by these strains as the Vortex (V) morphotype (Figure 7). The strain illustrated in Figure 7 has recently been

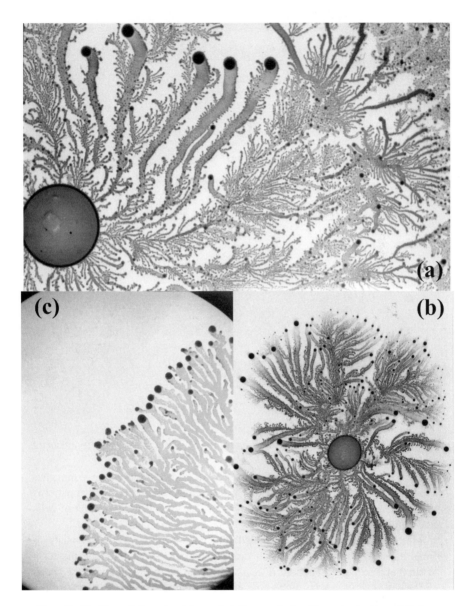

Figure 7   The V morphotype. (*a–c*) demonstrate the complexity of the evolved patterns; (*d*) closer look at the vortices. Note that some rotating "bagels" are also observed. (*e*) Another morphotype, referred to as the spiral vortex (SV) morphotype (21).

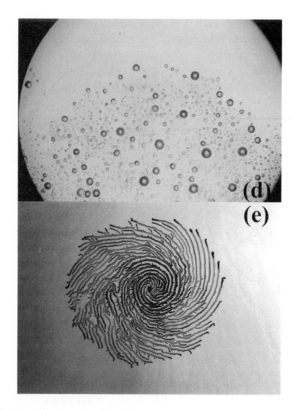

Figure 7   (Continued)

shown to belong to the genus *Paenibacillus* (M Tcherpakov et al, manuscript in preparation).

A wide variety of branching patterns are exhibited by each of the V morphotype strains, as the growth conditions are varied. Some representative patterns are shown in Figure 7. Each branch is produced by a leading droplet of cells and emits side branches, each with its own leading droplet.

Microscopic observations revealed that each leading droplet consists of hundreds to millions of cells that circle a common center (hence the term vortex) at a cell speed of about 10 μm/s. Both the size of a vortex and the speed of the cells can vary according to the growth conditions and the location of the specific vortex in the colony (Figure 7). Within a given colony, both clockwise and anticlockwise rotating vortices are observed. The vortices in a colony can also consist of either a single or multiple layers of cells. We occasionally observed vortices with an empty core, which we refer to as "bagel" shaped. After formation, the number of cells in the vortex increases, the vortex expands, and it translocates as a unit. The speed of the vortices is slower than the speed of the individual cells circulating around its center.

Bacterial cells are also contained in the trails left behind the leading vortices. Some are immobile, while others move, swirling with complex dynamics. The migrating groups of cells are reminiscent of the "worm" motion of slime mold or schools of multicellular organisms. The whole intricate dynamics is confined to the trail of the leading vortex, and neither a single cell nor a group of moving cells can pass out of the boundary of the trail. Only vortices formed in the trails can break out of the trail and create a new branch.

The microscopic observations also revealed that the bacterial motion is performed in a fluid on the agar surface. As is the case of the other morphotypes, this wetting fluid is also assumed to be excreted by the cells and/or extracted by the cells from the agar (12). We did not observe tumbling motion nor movement forward and backward. Rather, the motion was exclusively forward along the long axis of the cell. Moreover, the cells tended to move in the same direction and with the same speed as the surrounding neighboring cells, in what appeared to be a synchronized group movement. Electron microscope observations showed that the bacteria have flagella, which suggests that the motility is likely to be swarming.

Close inspection of these observations enabled the construction of a model for colonial development of swarming bacteria, which is also applicable for gliding bacteria. The model is inspired by the communicating walkers model (12) (see above). Here, the swarming cells are represented by swarmers. Each swarmer has a forward propulsion force. The balance of this force and friction forces tends to set the swarmer's speed to a specific value. In keeping with microscopic investigations, we also include velocity-velocity interactions, which tend to set the swarmer's velocity to the mean velocity of its neighbors (33, 69, 79, 82). In addition, we assume that the swarmers produce an extracellular "wetting" fluid, which they secrete during colonial growth. This extracellular slime also influences the bacterial motion. In the model, the swarmers can move only if the level of the wetting fluid is above a threshold value.

The above features, which are derived directly from the observations, are sufficient to describe the collective migration of bacteria. However, an additional feature has to be considered to explain the emergence of vortices. We propose the new feature to be a rotational chemotaxis, which differs from the chemotaxis normally employed by tumbling bacteria such as *E. coli* (24, 25, 37).

Rotational Chemotaxis and Vortex Formation

Swarming bacteria do not tumble, therefore they must employ a different method to perform chemotaxis. We propose that each individual cell modulates

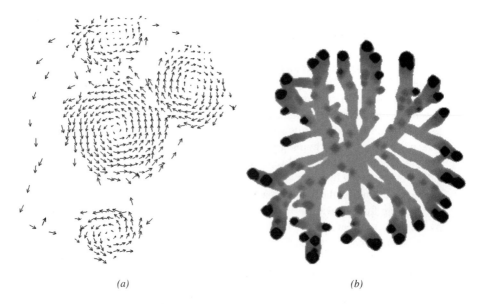

(a)   (b)

Figure 8 Results of numerical simulations of the vortex model described in text. (*a*) Formation of vortices when rotational chemotactic response is included. *Arrows* describe the location and velocity of the swarmers. (*b*) Formation of a colony when food consumption, reproduction, and repulsive chemotactic signaling are included.

its propulsion force according to the local concentration of a chemotactic signaling material. In a group of cells, such a response creates a propulsion force gradient. Together with the velocity-velocity interaction, this imposes a torque or local vorticity on the average motion of the cells. Therefore, a swarmer moving at an angle to the chemical gradient is subjected to a torque that causes the swarmer to twist towards the direction of the local gradient of the chemoattractant.

We have shown that chemomodulation can indeed lead to the formation of stationary vortices (fixed in size and at a fixed location, Figure 8*a*), rotating "bagels", and other elements. All these elements are of a length-scale comparable to or smaller than an individual branch of a colony. During colonial development, these elements are organized to form the observed global pattern.

Modeling the Cooperative Organization of Colonies
When modeling colony formation, we must take into account that bacterial cells in a colony do not move in a predetermined space, and their number and state of activity is not conserved. While the colony expands and changes its shape, cells reproduce, and frequently (in the case of V morphotype) also sporulate. To

provide the means for reproduction, movement, and other metabolic processes, the cells consume nutrients from the environment.

As with the T bacteria, we represent the metabolic state of a swarmer by an "internal energy." When sufficient food is available, the internal energy increases until it reaches a threshold energy, and the swarmer divides in two. When a swarmer is starved for long periods, the internal energy drops to zero and the swarmer "freezes."

As was proposed previously, we assume that starved cells (swarmers for which the internal energy is zero or below a certain level) emit a diffusive compound at a fixed rate, and this chemical acts as a chemorepellent modulator. We assume that the repellent material decays slowly, so that its concentration is almost constant over distances comparable to the typical size of a vortex (long-range chemotaxis). This is in contrast with the attractant concentration, which is assumed to vary considerably within a vortex (short-range chemotaxis). Thus, although the functional form of the repulsion term is similar to the term for attraction, it has a very different effect on the bacterial motion. It affects each vortex as a single unit and provides a mechanism for regulating colonial structure during colonial development. On each vortex, the repulsion acts to push the vortex outward on a curved trajectory. In Figure 8, we show results of numerical simulations where both long-range repulsive and short-range attractive chemotactic signaling terms are included. We also included in the simulations cell differentiation, i.e. a finite probability per unit time (about 1%/div) of the swarmers to become immobile. The immobile swarmers at the trail have a similar probability to regain mobility. This was motivated by microscopic observations revealing the presence of cells left behind the advancing vortices.

MYXOBACTERIA

The myxobacteria provide one of the most complex and fascinating examples of social behavior in the microbial world. This gram-negative motile organism employs coordinated gliding motility and exhibits two distinct life cycles, one involving vegetative growth under conditions of nutrient excess, and the other, sporulation and fruiting-body formation under conditions of nutrient starvation. Both vegetative growth and fruiting-body formation involve a complex spectrum of communal behavior, including cooperative feeding, group motility, cellular aggregation and cohesion, and collective movement. In *Myxococcus xanthus*, two sets of genes, termed A (Adventurous) and S (Social), are involved in vegetative growth and are characterized by mutant phenotypes. Mutants in the A genes are termed S motile, and grow vegetatively only when the cells are present in sufficient number and only within one cell length of each other.

Most of the mutants in this class sporulate. In contrast, S-mutants grow as individual cells, but cannot aggregate, and do not form spores or fruiting bodies (46, 47, 57). Remarkably, two different S mutants were found to form ramified branched tip-splitting patterns almost identical to the patterns associated with the T morphotype shown in Figure 2. Although *M. xanthus* lacks a mechanism for swimming and tumbling, it does respond to both chemoattractants and repellents (57). Moreover, it does contain a full complement of chemotaxis genes (termed *frz*) completely analogous to those present in *E. coli* and *S. typhimurium*. Frz proteins are thought to be involved in a signal transduction pathway early in cellular aggregation that also involves a cell-associated C-factor. This same factor also functions later in fruiting body development (50, 52). Interestingly, one of the Frz proteins was unable to carry out negative chemotaxis away from a variety of repellents and apparently controls the reversal frequency of *M. xanthus* cells. The biological importance of *frz* function was dramatically demonstrated in an experiment in which wild-type *M. xanthus* cells penetrated a prey colony and remained there until all of the food source was digested, while *frz* mutants invariably abandoned the microcolony, leaving the food source behind, thereby demonstrating the importance of the *frz* system of signal transduction for the feeding behavior of *M. xanthus*.

CONCLUDING REMARKS

We have described the mode and dynamics of cooperative organization leading to complex pattern formation in bacteria. Employing concepts from studies of patterning in nonliving systems coupled with morphotype characterization, we introduced a generic modeling approach to study the pattern-formation processes. Emerging directly from the modeling is the introduction of new, experimentally testable concepts such as rotational chemotaxis or the interplay between self-generated short-range, long-range, and nutritional chemotactic forces. Clearly, many questions remain to be unraveled. For example, given the enormous biodiversity in the microbial world, the question of the variety, distribution, and prevalence of morphotypes in various natural environments presents a formidable challenge to the microbial ecologist. Moreover, in our view, major new insights will be forthcoming as we begin to develop an understanding of the genetic basis of cooperative morphotype behavior. Furthermore, it is not unlikely that different strains with similar morphotype characteristics may in fact use alternative physiological and genetic strategies to generate the same type of pattern. Will the generic models developed here be applicable in such circumstances? It is clear that both pattern formation and morphotype transition require a mode of organized cellular communication and cooperative multicellular behavior (72), demonstrating once again that, under specific conditions, certain microbial populations can become transformed from a herd of individual cells to a community of cooperators with mutual interests.

ACKNOWLEDGMENTS

We gratefully acknowledge the contributions of Marianna Tcherpikov and Natalia Goldin as well as the excellent technical assistance of Ina Brainiss and Rina Avigad. Work in our laboratories was sponsored in part by grants from the Israel Science Foundation and US-Israel Binational Science Foundation.

> Visit the *Annual Reviews home page* at
> http://www.AnnualReviews.org.

Literature Cited

1. Adler J. 1969. Chemoreceptors in bacteria. *Science* 166:1588–97
2. Avetisov VA, Goldanskii VI, Kuzmin VV. 1991. Handedness, origin of life and evolution. *Phys. Today* 44(7):33–41
3. Beck G. 1995. *Isolation and characterization of* Salmonella typhinurium *genes involved in pattern formation.* MSc thesis. Weizmann Inst. of Science, Rehovoth, Israel
4. Beck BJ. Downs DM. 1998. The *apbE* gene encodes a lipoprotein involved in thiamine synthesis in *Salmonella typhimurium. J. Bacteriol.* 180:885–91
5. Ben-Jacob E. 1993. From snowflake formation to the growth of bacterial colonies, part I: diffusive patterning in non-living systems. *Contemp. Phys.* 34:247–73
6. Ben-Jacob E. 1997. From snowflake formation to the growth of bacterial colonies, part II: cooperative formation of complex colonial patterns. *Contemp. Phys.* 38: 205–41
7. Ben-Jacob E, Cohen I. 1997. Cooperative formation of bacterial patterns. In *Bacteria As Multicellular Organisms*, ed. JA Shapiro, M Dwarkin, pp. 394–416. New York: Oxford Univ. Press
8. Ben-Jacob E, Garik P. 1990. The formation of patterns in non-equilibrium growth. *Nature* 343:523–30
9. Ben-Jacob E, Garik P, Muller T, Grier D. 1988. Characterization of morphology transitions in diffusion-controlled systems. *Phys. Rev. A* 38:1370–80
10. Ben-Jacob E, Shmueli H, Schochet O, Tenenbaum A. 1992. Adaptive self-organization during growth of bacterial colonies. *Physica A* 187:378–424
11. Ben-Jacob E, Schochet O, Tenenbaum A. 1994. Bakterien schließen sich zu bizarren formationen zusammen. In *Muster des Ledendigen: Faszination inher Entstehung und Simulation*, ed. A Deutsch, pp. 109–26. Cologne: Vieweg Verlag
12. Ben-Jacob E, Schochet O, Tenenbaum A, Cohen I, Czirók A, Vicsek T. 1994. Generic modelling of cooperative growth patterns in bacterial colonies. *Nature* 368:46–49
13. Ben-Jacob E, Schochet O, Tenenbaum A, Cohen I, Czirók A, Vicsek T. 1994. Communication, regulation and control during complex patterning of bacterial colonies. *Fractals* 2(1):15–44
14. Ben-Jacob E, Tenenbaum A, Schochet O, Avidan O. 1994. Holotransformations of bacterial colonies and genome cybernetics. *Physica A* 202:1–47
15. Ben-Jacob E, Cohen I, Schochet O, Aranson I, Levine H, Tsimiring L. 1995. Complex bacterial patterns. *Nature* 373:566–67
16. Ben-Jacob E, Cohen I, Schochet O, Czirók A, Vicsek T. 1995. Cooperative formation of chiral patterns during growth of bacterial colonies. *Phys. Rev. Lett.* 75(15):2899–902
17. Ben-Jacob E, Schochet O, Cohen I, Tenenbaum A, Czirók A, Vicsek T. 1995. Cooperative strategies in formation of complex bacterial patterns. *Fractals* 3:849–68
18. Ben-Jacob E, Schochet O, Tenenbaum A, Avidan O. 1995. Evolution of complexity during growth of bacterial colonies.

In *Spatio-Temporal Patterns in Nonequilibrium Complex Systems*, ed. PE Cladis, P Palffy-Muhoray, pp. 619–34. Reading, MA: Addison-Wesley

19. Ben-Jacob E, Cohen I, Czirók A. 1997. Smart bacterial colonies. In *Physics of Biological Systems: From Molecules to Species*, ed. H Flyvbjerg, J Hertz, OG Mouritsen, K Sneppen, pp. 307–24. Berlin: Springer

20. Ben-Jacob E, Schochet O, Tenenbaum A, Cohen I, Czirók A, Vicsek T. 1996. Response of bacterial colonies to imposed anisotropy. *Phys. Rev. E* 53:1835–45

21. Ben-Jacob E, Cohen I, Czirók A, Vicsek T, Gutnick DL. 1997. Chemomodulation of cellular movement and collective formation of vortices by swarming bacteria and colonial development. *Physica A* 238: 181–97

22. Berg HC. 1993. *Random Walks in Biology*. Princeton, NJ: Princeton Univ. Press

23. Berg HC, Purcell EM. 1977. Physics of chemoreception. *Biophys. J.* 20:193–219

24. Berg HC, Tedesco PM. 1975. Transient response to chemotactic stimuli in *Escherichia coli*. *Proc. Natl. Acad. Sci. USA* 72(8):3235–39

25. Bischoff DS, Ordal GW. 1992. *Bacillus subtilis* chemotaxis: a deviation from the *Escherichia coli* paradigm. *Mol. Microbiol.* 6:23–28

26. Blair DF. 1995. How bacteria sense and swim. *Annu. Rev. Microbiol.* 49:489–522

27. Blat Y, Eisenbach M. 1995. Tar-dependent and -independent pattern formation by *Salmonella typhimurium*. *J. Bacteriol.* 177(7):1683–91

28. Blat Y, Eisenbach M. 1996. Oligomerization of the phosphatase CheZ upon interaction with the phosphorylated form of CheY. The signal protein of bacterial chemotaxis. *J. Biol. Chem.* 271:1226–31

29. Bruno WJ. 1992. *CNLS Newsl.* 82:1–10

30. Budrene EO, Berg HC. 1991. Complex patterns formed by motile cells of *Escherichia coli*. *Nature* 349:630–33

31. Budrene EO, Berg HC. 1995. Dynamics of formation of symmetrical patterns by chemotactic bacteria. *Nature* 376:49–53

32. Cohen I, Czirók A, Ben-Jacob E. 1996. Chemotactic-based adaptive self organization during colonial development. *Physica A* 233:678–98

33. Czirók A, Ben-Jacob E, Cohen I, Vicsek T. 1996. Formation of complex bacterial colonies via self-generated vortices. *Phys. Rev. E* 54:1791–801

34. Devreotes P. 1989. *Dictyostelium discoideum*: a model system for cell-cell

interactions in development. *Science* 245: 1054–58

35. Deleted in proof

36. Deleted in proof

37. Eisenbach M. 1990. Functions of the flagellar modes of rotation in bacterial motility and chemotaxis. *Mol. Microbiol.* 4(2):161–67

38. Eisenbach M. 1996. Control of bacterial choemtaxis. *Mol. Microbiol.* 20:903–10

39. Ford WW. 1916. Studies on aerobic spore-bearing non-phatogenic bacteria, part 2: miscellaneous cultures. *J. Bacteriol.* 1:518–26

40. Fujikawa H, Matsushita M. 1989. Fractal growth of *Bacillus subtilis* on agar plates. *J. Phys. Soc. Jpn.* 58:3875–78

41. Fuqua WC, Winans SC, Greenberg EP. 1994. Quorum sensing in bacteria: the LuxR-LuxI family of cell density-responsive transcriptional regulators. *J. Bacteriol.* 176:269–75

42. Fuqua C, Winans SC, Greenberg EP. 1996. Census and consensus in bacterial ecosystems: the LuxR-LuxI family of quorum-sensing transcriptional regulators. *Annu. Rev. Microbiol.* 50:727–51

43. Garrity LF, Ordal GW. 1995. Chemotaxis in *Bacillus subtilis*: how bacteria monitor environmental signals. *Pharmacol. Ther.* 68:87–104

44. Harshey RM. 1994. Bees aren't the only ones: swarming in gram-negative bacteria. *Mol. Microbiol.* 13:389–94

45. Hegstrom RA, Kondepudi DK. 1990. The handedness of the universe. *Sci. Am.* 262: 108–15

46. Hodgkin J, Kaiser D. 1979. Genetics of gliding motility in *Myxococcus xanthus* (Myxobacterales): genes controlling movement of single cells. *Mol. Gen. Genet.* 171:167–76

47. Hodgkin J, Kaiser D. 1979. Genetics of gliding motility in *Myxococcus xanthus* (Myxobacterales): two gene systems control movement. *Mol. Gen. Genet.* 171: 177–91

48. Kaiser D, Losick R. 1993. How and why bacteria talk to each other. *Cell* 73:873–87

49. Kessler DA, Koplik J, Levine H. 1988. Pattern selection in fingered growth phenomena. *Adv. Phys.* 37:255–87

50. Kim SK, Kaiser D. 1992. Control of cell density and pattern by intercellular signaling in *Myxococcus* development. *Annu. Rev. Microbiol.* 46:117–39

51. Kitsunezaki S. 1997. Interface dynamics for bacterial colony formation. *J. Phys. Soc. Jpn.* 66:1544–50

52. Kuspa A, Kroos L, Kaiser D. 1986.

Intercellular signalling is required for developmental gene expression in *Myxococcus xanthus*. *Dev. Biol.* 117:267–76

53. Lackiie JM, ed. 1986. *Biology of the Chemotactic Response*. Cambridge, UK: Cambridge Univ. Press

54. Langer JS. 1989. Dendrites, viscous fingering, and the theory of pattern formation. *Science* 243:1150–54

55. Latifi A, Winson MK, Foglino M, Bycroft BW, Stewart GS, et al. 1995. Multiple homologues of LuxR and LuxI control expression of virulence determinants and secondary metabolites through quorum sensing in *Pseudomonas aeruginosa* PAO1. *Mol. Microbiol.* 17:333–43

56. Losick R, Kaiser D. 1997. Why and how bacteria communicate. *Sci. Am.* 276:68–73

57. MacNeil SD, Mouzeyan A, Hartzell PL. 1994. Genes required for both gliding motility and development in *Myxococcus xanthus*. *Mol. Microbiol.* 14:785–95

58. Matsushita M, Fujikawa H. 1990. Diffusion-limited growth in bacterial colony formation. *Physica A* 168:498–506

59. Matsushita M, Wakita J-I, Matsuyama T. 1995. Growth and morphological changes of bacteria colonies. In *Spatio-Temporal Patterns in Nonequilibrium Complex Systems*, ed. PE Cladis, P Palffy-Muhoray, pp. 609–18. Reading, MA: Addison-Wesley

60. Matsushita M, Wakita J, Itoh H, Ràfols I, Matsuyama T, et al. 1998. Interface growth and pattern formation in bacterial colonies. *Physica A* 249:517–24

61. Matsuyama T, Matsushita M. 1993. Fractal morphogenesis by a bacterial cell population. *Crit. Rev. Microbiol.* 19:117–35

62. Matsuyama T, Kaneda K, Nakagawa Y, Isa K, Hara-Hotta H, Yano I. 1992. A novel extracellular cyclic lipopeptide which promotes flagellum-dependent and -independent spreading growth of *Serratia marcescens*. *J. Bacteriol.* 174:1769–76

63. Matsuyama T, Harshey RM, Matsushita M. 1993. Self-similar colony morphogenesis by bacteria as the experimental model of fractal growth by a cell population. *Fractals* 1(3):302–11

64. Mendelson NH. 1978. Helical *Bacillus subtilis* macrofibers: morphogenesis of a bacterial multicellular macroorganism. *Proc. Natl. Acad. Sci. USA* 75(5):2478–82

65. Mendelson NH. 1990. Bacterial macrofibres: the morphogenesis of complex multicellular bacterial forms. *Sci. Prog.* 74:425–41

66. Mendelson NH, Keener SL. 1982. Clockwise and counterclockwise pinwheel colony morphologies of *Bacillus subtilis* are correlated with the helix hand of the strain. *J. Bacteriol.* 151(1):455–57

67. Mendelson NH, Salhi B. 1996. Patterns of reporter gene expression in the phase diagram of *Bacillus subtilis* colony forms. *J. Bacteriol.* 178:1980–89

68. Mendelson NH, Thwaites JJ. 1989. Cell wall mechanical properties as measured with bacterial thread made from *Bacillus subtilis*. *J. Bacteriol.* 171(2):1055–62

69. Reynolds CW. 1987. Flocks, herds, and schools: a distributed behavioral model. *Comput. Graphics* 21(4):25–34

70. Rodel W, Plaga W, Frank R, Knappe J. 1988. Primary stractures of *Escherichia coli* pyruvate formate-lyase and pyruvate-formate-lyase-activating enzyme deduced from the DNA nucleotide sequences. *Eur. J. Biochem.* 177:153–58

71. Schloss JV, van Dyk DE, Vasta JF, Kutny RM. 1985. Purification and properties of *Salmonella typhimurium* acetolactate synthase isoenzyme II from *Escherichia coli* HB101/pDUC9. *Biochemistry* 24:4952–59

72. Shapiro JA. 1988. Bacteria as multicellular organisms. *Sci. Am.* 258(6):62–69

73. Shaw CH. 1991. Swimming against the tide: chemotaxis in *Agrobacterium*. *BioEssays* 13(1):25–29

74. Shimkets LJ, Dworkin M. 1997. Myxobacterial multicellularity. In *Bacteria As Multicellular Organisms*, ed. JA Shapiro, M Dwarkin, pp. 220–44. New York: Oxford Univ. Press

75. Smith RN, Clark FE. 1938. Motile colonies of *Bacillus alvei* and other bacteria. *J. Bacteriol.* 35:59–60

76. Stevens FS. 1974. *Pattern in Nature*. Boston: Little, Brown

77. Stock JB, Stock AM, Mottonen M. 1990. Signal transduction in bacteria. *Nature* 344:395–400

78. Stock J, Surette M, Park P. 1994. Chemosensing and signal transduction in bacteria. *Curr. Opin. Neurobiol.* 4:474–80

79. Toner J, Tu Y. 1995. Long-range order in a two-dimensional dynamical XY model: how birds fly together. *Phys. Rev. Lett.* 75(23):4326–29

80. Tsimiring L, Levine H, Aranson I, Ben-Jacob E, Cohen I, et al. 1995. Aggregation patterns in stressed bacteria. *Phys. Rev. Lett.* 75:1859–62

81. Turing AM. 1952. *Philos. Trans. R. Soc. London Ser. B* 237:37–72

81a. Van-Dyk TK, Ayers BL, Morgan RW, La-Rossa RA. 1998. Constricted flux through the branched-chain amino acid biosynthetic enzyme acetolactate synthase triggers elevated expression of genes regulated by *rpoS* and internal acidification. *J. Bacteriol.* 180:785–92

82. Vicsek T, Czirók A, Ben-Jacob E, Cohen I, Schochet O, Tenenbaum A. 1995. Novel type of phase transition in a system of self-driven particles. *Phys. Rev. Lett.* 75:1226–29

83. Wolf G, ed. 1968. *Encyclopaedia Cinematographica*. Göttingen: Institut für Wissenschaftlichen Film

84. Wong KK, Suen KL, Kwan HS. 1989. Transcription of *pfl* is regulated by anaerobiosis, catabolite repression, pyruvate, and *oxrA*: pfl::Mu dA operon fusions of *Salmonella typhimurium*. *J. Bacteriol.* 171:4900–5

85. Woodward DE, Tyson R, Myerscough MR, Murray JD, Budrene EO, Berg HC. 1995. Spatio-temporal patterns generated by *Salmonella typhimurium*. *Biophys. J.* 68:2181–89

SUBJECT INDEX

A

α-toxin
virulence genes of *Clostridium perfringens* and, 333–54

Abcesses
postinjection
nosocomial outbreaks caused by nontuberculous mycobacteria and, 453–85

ABC-type drug transporters
lantibiotics and, 41, 52–53, 55, 59–60
Trypanosoma brucei surface receptors and transporters, 763–64

Accessory functions
lantibiotics and, 41

ace operon
acetate metabolism regulation and, 137, 146–55

Acetate metabolism
regulation of by protein phosphorylation in enteric bacteria
acetate operon, 146–55
alternate pathways, 134–35
comparison of different species, 140–41
conversion of acetate to acetyl CoA, 129–30
differential expression, 148–49
effectors, 153–54
enzymes and regulation, 133–34
FadR, 151
fatty acid catabolism, 130–31
FruR, 151–53
glyoxylate bypass, 131–35
growth of bacteria on acetate, 129–30
IclR, 149–50
inactivation mechanism, 139–40
introduction, 128–29
isocitrate dehydrogenase, 135–41
isocitrate dehydrogenase kinase/phosphatase, 141–46
metabolites, 155
nature and role, 131–32
negative control, 149–51

positive control, 151–54
reversible phosphorylation, 137–39
sensitivity amplification, 145–46
structural genes, 146–48

Acetyl CoA
acetate metabolism regulation and, 129–130, 132–33, 135, 155

Acinetobacter calcoaceticus
acetate metabolism regulation and, 141
hybrid pathways for chloroaromatics and, 302, 304

Acinetobacter spp.
hybrid pathways for chloroaromatics and, 304

Acquired immunodeficiency syndrome (AIDS)
nosocomial outbreaks caused by nontuberculous mycobacteria and, 470–71

Acridines
history of research, 23

Actagardine
lantibiotics and, 45–47, 49–50, 65

Actin
cell polarity and morphogenesis in budding yeast, 697, 700–1, 706–8

Actin-binding proteins
cell polarity and morphogenesis in budding yeast, 697, 700–1

Actinomyces spp.
bacterial multicellularity and, 82, 94
modular organisms and, 110, 118

Actinomycetales spp.
microbial dehalogenation of chlorinated solvents and, 428

Actinomycin D
malarial parasite metabolism and, 575

Actinoplanes spp.
lantibiotics and, 45

Activation
HIV-1 Rev protein and, 510–13
virocrine transformation and, 401–10

Adaptive benefits
bacterial multicellularity and, 81–97
modular organisms and, 119

Adaptive self-organization
chemotaxis-based
cooperative organization of bacterial colonies and, 791–92

Adelberg, EA, 1–38

Adenine
Trypanosoma brucei surface receptors and transporters, 762–63

Adenosine
Trypanosoma brucei surface receptors and transporters, 747, 761–63

Adenosine triphosphate (ATP)
anti-σ factors and, 250–51
Trypanosoma brucei surface receptors and transporters, 748

S-Adenosylmethionine
Trypanosoma brucei surface receptors and transporters, 762

Adenoviruses
nuclear targeting of animal DNA viruses and, 628, 634–35, 657–61

Adenylate cyclase
Trypanosoma brucei surface receptors and transporters, 745–71

Aerobacter aerogenes
acetate metabolism regulation and, 140
thymineless death and, 593

Aerotaxis
anaerobiosis in *Bacillus subtilis* and, 185

African trypanosomiasis
Trypanosoma brucei surface receptors and transporters, 745–71

Aging
Saccharomyces cerevisiae and, 533–55

Agro-food applications
lantibiotics and, 66, 71

Alamethicin
lantibiotics and, 61

Alanine

CUMULATIVE INDEXES

CONTRIBUTING AUTHORS, VOLUMES 48–52

CHAPTER TITLES, VOLUMES 48–52

841